MW00991897

Introduction to Civil Engineering Systems

Introduction to Civil Engineering Systems

A Systems Perspective to the Development of Civil Engineering Facilities

Samuel Labi

WILEY

Cover image: © MichaelRutkowskiPhotography
Cover design: Michael Rutkowski

This book is printed on acid-free paper. ♾

Published by John Wiley & Sons, Inc., Hoboken, New Jersey

Published simultaneously in Canada

Library of Congress Cataloging-in-Publication Data

Labi, Samuel, 1962-
Introduction to civil engineering systems : a systems perspective to the development of civil engineering facilities / Samuel Labi.
pages cm
Includes index.
ISBN 978-0-470-53063-4 (hardback); ISBN 978-1-118-41530-6 (ebk); ISBN 978-1-118-41817-8 (ebk)
1. Civil engineering. 2. Systems engineering. I. Title.
TA145.L38 2014
624–dc23

2013038048

Printed in the United States of America

To my parents, Emmanuel Kwaku Labi and Elizabeth Adjo Labi

CONTENTS

FOREWORD

The book in your hands is a most welcome addition to the range of textbooks in civil and environmental engineering. There is no other up-to-date text that covers the important elements of civil engineering systems. This book provides a significant and needed resource for majors in the field.

Labi's *Introduction to Civil Engineering Systems* fills an important gap. The previous lack of a current text on the subject has been most unfortunate. It is clear that the proper development of our infrastructure requires a holistic, coherent understanding of all the important elements that will make our products successful. Civil engineers have a responsibility for the entire life cycle of what we build: from planning, through design, to the management of the facility over its useful life. A civil engineering curriculum should thus provide its students with the opportunity to learn how to consider carefully the range of issues in civil engineering systems. This text now provides a basis for such a capstone course.

This text has the great merit of being thoughtful, innovative, comprehensive, and forward-looking.

- Beyond procedures and methods, Labi thoughtfully presents issues and discusses their whys and hows.
- He has innovatively structured the material as a coherent cycle of eight phases through the life cycle of a project in a way that makes it much easier to make sense of systems thinking.
- He comprehensively defines the tasks that should be done at each phase in the life cycle and describes the tools for each task.
- Aside from being the most forward-looking work in the field, the text recognizes the great uncertainty about future demands on our systems, and the consequent need for a flexible approach to systems design.

Readers will appreciate that this treatment of engineering systems focuses on civil engineering. Its extensive excellent illustrations display a most interesting array of infrastructure projects. This is a book for us. I hope you will like it!

Richard de Neufville
Professor of Engineering Systems and of Civil and Environmental Engineering
Massachusetts Institute of Technology
Life Member, American Society of Civil Engineers

PREFACE

GENERAL BACKGROUND

The civil engineering discipline involves the development of structural, hydraulic, geotechnical, construction, environmental, transportation, architectural, and other civil systems that address societies' infrastructure needs. The planning and design of these systems are well covered in traditional courses and texts at most universities. In recent years, however, universities have increasingly sought to infuse a "systems" perspective to their traditional civil engineering curricula. This development arose out of the recognition that the developers of civil engineering systems need a solid set of skills in other disciplines. These skills are needed to equip them further for their traditional tasks at the design and construction phases and also to burnish their analytical skills for other less-obvious or emerging tasks at all phases of system development.

The development of civil engineering systems over the centuries and millennia has been characterized by continual improvements that were achieved mostly through series of trial-and-error as systems were constructed and reconstructed by learning from past mistakes. At the current time, the use of trial-and-error methods on real-life systems is infeasible because it may take not only several decades but also involve excessive costs in resources and, possibly, human lives before the best system can be finally realized. Also in the past, systems have been developed in ways that were not always effective or cost-effective. For these and other reasons, the current era, which has inherited the civil engineering systems built decades ago, poses a unique set of challenges for today's civil engineers. A large number of these systems, dams, bridges, roads, ports, and so on are functionally obsolescent or are approaching the end of their design lives and are in need of expansion, rehabilitation, or replacement. The issue of inadequate or aging civil infrastructure has deservedly gained national attention due to a series of publicized engineering system failures in the United States, such as the New Orleans levees, the Minnesota and Seattle interstate highway bridges, and the New York and Dallas sewers, and in other countries. The current problem of aging infrastructure is further exacerbated by increased demand and loading fueled by population growth, rising user expectations of system performance, increased desire for stakeholder participation in decision-making processes, terrorism threats, the looming specter of tort liability, and above all, inadequate funding for sustained preservation and renewal of these systems.

As such, civil engineers of today need not only to develop skills in the traditional design areas but also to continually seek and implement traditional and emerging tools in other related areas such as operations research, economics, law, finance, statistics, and other areas. These efforts can facilitate a more comprehensive yet holistic approach to problem solving at any phase of the civil engineering system development cycle. This way, these systems can be constructed, maintained, and operated in the most cost-effective way with minimal damage to the environment, maximum system longevity, reduced exposure to torts, optimal use of the taxpayers' dollar, and other benefits. Unfortunately, at the current time, graduating engineers enter the workforce with few or no skills in systems engineering and learn these skills informally only after several decades. With limited skill in how to integrate specific knowledge from external disciplines into their work, practicing engineers will be potentially handicapped unless their organizations provide formal training in the concepts of sytems engineering. This text addresses these issues.

THE TEXT

The first part of this text discusses the historical evolution of the various engineering disciplines and general concepts of systems engineering. This includes formal definitions, systems classifications, systems attributes, and general and specific examples of systems in everyday life and in civil engineering. The part also identifies the phases of development of civil systems over their life cycle and discusses the tasks faced by civil systems engineers at each phase. Most working engineers are typically involved in only one or two of these phases, but it is important for all engineers to acquire an overall bird's eye view of all phases so that decisions they make at any phase are holistic and within the context of the entire life cycle of their systems. The next two parts discuss the tasks that civil engineers encounter at each phase and the tools they need to address these tasks. For example, at the needs assessment phase, one possible task is to predict the level of expected usage of the system, and the tool for this task could be statistical modeling or simulation. Certain tools are useful in more than one phase. Given this background, Part IV provides a detailed discussion of each phase of civil systems development and presents specific examples of tasks and tools used to address questions at these phases. Part V presents topics that may seem peripheral but are critical to civil systems development, such as legal issues, ethics, sustainability, and resilience, and discusses their relevance at each phase.

Clearly, this text differs from other texts in the manner in which it presents the material. The systems tasks and tools are presented not in a scattered fashion but rather in the organized context of a phasal framework of system development. Why is it so important to view the entire life cycle of civil engineering systems within a phasal framework? And why do we need to acquire those skills that are needed for the tasks at each phase? One reason is the typically large expense involved in the provision of such facilities. Every year, several trillion dollars are invested worldwide in civil engineering systems, to build new facilities or to operate and maintain existing ones. The beneficial impacts of these investments permeate every sphere of our lives including safety, mobility, security, and the economy and thus need to be identified and measured systematically. Also, adverse impacts such as environmental degradation, community disruption, and inequities are often evident and need to be assessed and mitigated. In summation, given the large expanse and value of civil engineering assets, the massive volume of national and state investments annually to build and operate these systems, and the multiplicity of stakeholders, there is need for a comprehensive yet integrated approach to the planning, design, implementation, operations, and preservation of these systems. A second reason for advocating an organized systems approach is the nature of recent and ongoing trends in the socioeconomic environment: at the current time of tight budgets, increasing loadings and demand, aging infrastructure, global economic changes, and increased need for security and safety, civil engineering systems are facing scrutiny more than ever before and the biggest bang is now sought for every dollar spent on these systems. As such, civil system engineers are increasingly being called upon to render account of their fiduciary stewardship of the public infrastructure and assets. This is best done when the development of such systems is viewed within a phasal framework, when civil engineering system managers acquire the requisite tools needed to address the tasks at each phase, and when these managers provide evidence of organized planning for long-term life-cycle development of their systems.

DIDACTIC STYLE AND RESOURCES

There is a wealth of engineering knowledge that is well documented in textbooks that address specific branches and domain areas in civil engineering and also in other system engineering related disciplines including economics, operations research, and statistics. The author's purpose in writing this text is not to duplicate what already exists but to link the systems concepts from the different disciplines and traditional roles of the civil engineer, and to do this within the context of each system phase, tasks at each phase, and tools for the tasks.

The reader is afforded a clear and understandable text that presents well-explained methodologies and procedures useful for addressing tasks at each phase. Throughout its chapters, the text emphasizes practical applications of the concepts. Theoretical backgrounds are provided only to enable the reader to enhance their understanding of the concepts and to recognize the merits and demerits of alternative theories in solving a particular problem. The chapters and concepts are presented in a sequence and style that are expected to encourage the student to define and solve problems with requisite tools in a manner consistent with engineering and professional excellence. As such, each chapter is an integrated blend of theory and practice, and numerous conceptual and computational illustrations are provided.

As educational experts have acknowledged, students' didactic experience is more fruitful when they are asked to apply the concepts to a real-world problem. As such, a term project, to be carried out by multiperson teams, is recommended as part of any course for which this book is used. A list of possible topics for the term project can be found at the website purposely established for this book. Additional information on each of the 30 chapters, such as updated tools and news items relating to civil systems development at various countries worldwide, Facebook discussions, and YouTube presentations can be found at the book's website.

The subject of civil engineering systems is indeed a broad subject that could fill several texts. As such, there is a limitation to the scope and depth that can be provided in a single text as this. The text therefore provides only a basic fundamental understanding of what civil systems are, the various phases of their development, and the tools needed to address the tasks at each phase. The text serves as a central repository of references for persons interested in further inquiry. Also, recognizing that only a limited number of numerical examples can be included within the covers of this book, the author has provided a set of useful resources at the end of most chapters for the reader who wishes to acquire further knowledge on the subject.

ABET REQUIREMENTS AND AUDIENCE

This text satisfies a significant section of Accreditation Board for Engineering and Technology, Inc. (ABET) requirements for undergraduate civil engineering education such as problem solving, experiments and simulations, data analysis, optimization and financial analysis tools, and use of systems approaches in design of facility components and processes. Also, the text addresses other ABET requirements of socially and environmentally responsible design, engineering practice issues, ethics, licensure requirements, and managerial skills. The text's online resources addresses the requirement of student participation in multidisciplinary project teams.

This book is useful for college instructors and students for courses related to civil engineering systems. Most of the material could be covered in one semester if at least three credit hours per week are used for the course. The book is written primarily for midsenior undergraduate and beginning graduate students. The book should be useful not only in academia but also to practicing civil engineers, civil systems managers and policymakers in general. This includes private and nongovernmental organizations, consultants, international development agencies and lending institutions, public policy makers, government (state, county, provincial, or city) departments, municipal authorities, public works departments, regional planning agencies, metropolitan planning organizations, and other institutions involved in at least one of the phases of civil systems development. These persons will find that the text provides useful fundamentals for understanding and implementing systems perspectives at any of their system development phases of need assessment, planning, design, construction, operations, monitoring, maintenance, or end of life.

ACKNOWLEDGMENTS

The development of a textbook is very much like the development of a civil engineering system—it goes through the initial phases of needs assessment, planning and design of the chapters, feedback from readers, and improvement of subsequent editions. This is particularly true for *Introduction to Civil Engineering Systems*.

First, I recognize the contribution of the pioneers of civil engineering systems: Robert Stark, Robert Nichols, Jeff Wright, Charles Revelle, Earl Whitlach, Lester Hoel, Nicholas Garber, Richard de Neufville, C. Jotin Khisty, Jashmid Mohammadi, Dale Meredith, Kam Wong, Ronald Woodhead, Robert Wortman, Richard Larson, Joseph Stafford, Gerard Voland, and Graeme Dandy. They deserve tremendous credit for blazing the trail, navigating the difficult waters, and thus making it possible for civil engineers to recognize the usefulness of systems concepts to their field.

My appreciation goes to all those who created the materials that served as sources for this text. A large number of books, technical papers, and reports were reviewed during the writing of this text. The work of those who developed these materials served as a valuable knowledge base for developing this book. Without their work, this book would not be the valuable learning resource that it currently is and hopefully will continue to be.

I gratefully acknowledge the support of my colleagues in academia as well as the industry. These include the following professors: Kumares Sinha of Purdue University; Richard de Neufville, Fred Moavenzadeh, and Joe Sussman of MIT; Neville Parker of the City College of New York; Sue McNeil and Nii Attoh-Okine of the University of Delaware; and Adjo Amekudji of Georgia Tech. They also include Dr. Chuanxin Fang of Microsoft Corporation; Mr. Gilbert Kporku of Conterra Limited; and Mr. Arun Shirole, former chief bridge engineer of the New York State Department of Transportation; and Dr. Jung Eun Oh of the World Bank.

I would also like to give credit to my friends, colleagues, and graduate students for enhancing the manuscript in various ways, including Qiang Bai, Zhibo Zhang, Nathee Athigakunagorn and Arash Roshandeh, Charles Atisso, and Rita Adom. No words can describe my appreciation for the support of my family, Grace, Valerie, Rachel, and Chelsea.

The staff at John Wiley & Sons has been extremely supportive, and I am thankful to editors Bob Argentieri and Margaret Cummins for their continual support and understanding.

Finally, for all the good aspects of this book, I duly reserve credit to my helpers, contributors and supporters; for any flaw, the full responsibility is mine.

AUTHOR BIOGRAPHY

Dr. Samuel Labi is an Associate Professor at Purdue University's School of Civil Engineering. He has seven years experience in consulting at Conterra Limited, where his work traversed different civil engineering disciplines and included economic and technical feasibility studies, facilities planning and design, geotechnical investigations, contract administration, and construction and maintentance supervision. Dr. Labi has worked on projects such as the World Bank–sponsored Transportation Rehabilitation Projects in Ghana, in the late 1980s. His national academic awards include the 2002 Milton Pikarski Award for outstanding doctoral dissertation in transportation engineering, the 2007 Bryant Mather Award for best paper in concrete materials awarded by the American Society of Testing and Materials (ASTM), the 2008 K.B. Woods prize awarded by the Transportation Research Board for the best journal paper in design and construction, and the 2014 Frank Masters Award from the American Society of Civil Engineers (ASCE) for innovative or noteworthy contributions to the planning, design and construction of transportation facilities. He is a member of Sigma Xi (Scientific Research Society) and Chi Epsilon (National Civil Engineering Honors Society) and several professional organizations including the American Society of Civil Engineers (ASCE), International Association for Life-Cycle Civil Engineering (IALCCE), Institute for Operations Research and the Management Sciences (INFORMS), and the American Association for the Advancement of Sciences (AAAS). Dr. Labi has taught a number of undergraduate and graduate-level courses at Purdue University and the Massachusetts Institute of Technology, and he has served as a major thesis advisor for several doctoral students. He is an associate editor of the ASCE Journal of Risk and Uncertainty and member of the editorial board of the ASCE Journal of Infrastructure Systems. He is a reviewer for several major international technical journals, conferences, and textbook publishers and a co-author of *Transportation Decision Making—Principles of Project Evaluation and Programming* published by John Wiley & Sons in 2007.

PART I

INTRODUCTION

CHAPTER 1

CIVIL ENGINEERING SYSTEMS AND THEIR EVOLUTION

1.0 INTRODUCTION

This chapter starts by defining civil engineering and describes briefly the current and future practices of the different civil engineering disciplines. The historical evolution of each civil engineering discipline is featured prominently in this chapter because it is also important for today's civil engineers to acknowledge the profession's trailblazers and appreciate their contributions to the growth of the profession. We will show how the evolution of civil engineering, as well as other disciplines related to civil engineering, have been shaped by changes in human value systems, interactions between the profession and socioeconomic forces, advances in science and technology, and innovations in materials, equipment, and the like. Civil systems engineering is not a new practice; on the contrary, over the ages, civil engineers or persons serving in that capacity have always executed their work from a systems perspective, perhaps at times implicitly. In this text, we will argue that the civil engineering discipline could be further enhanced if the development of its systems explicitly incorporates new analytical tools in systems engineering. This is particularly important in the current era, with its high population growth demanding new civil systems and increased need for preserving aging civil infrastructure, at a time when funding constraints are a stark reality, stakeholders are more involved in the development process, and users have higher service expectations.

1.1 CIVIL ENGINEERING SYSTEMS AND HISTORICAL DEVELOPMENTS

1.1.1 The Importance of Studying the History of Engineering Systems

Eminent historians agree that the extent to which the history of a profession is known, preserved, honored, and utilized greatly influences the degree to which the profession knows and comprehends itself; and it also dictates the extent to which the profession is acknowledged and respected by others outside its confines. This applies no less to the engineering profession, where, in spite of its long and rich history, many skilled engineers today tend to be dismissive of their heritage and instead focus solely on state-of-the-art or current trends in their specialized areas of practice. As such, the history of engineering is often relegated to a minor or nonexistent role in professional development conferences and in formal engineering education.

Fortunately, many prominent civil engineers in the current era believe that knowledge of engineering history will lead to a reinforcement of the profession and its stature among other professions, specifically, that civil engineering has a deserved place in the arena of the overall evolution of civilization and the world. Furthermore, knowing the history of engineering systems enables those who practice engineering to better understand the simultaneous relationship between engineering and other sectors of human development, such as health, agriculture, and industry. Also, the history of engineering, from the inspiring narratives of great projects as well as the seemingly small incremental improvements, provides illumination and caveats about what was once thought to be the state of the art and should be recognized as fundamental knowledge rather than irrelevant to the current state of the art (Petroski, 2001).

The purpose of any documentation of history is to interpret the development and activity of humankind (Kirby et al., 1956). As such, the history of engineering systems is but one aspect of the overall narrative of the human experience. However, unlike many other aspects of this experience, the history of engineering records a human activity that is cumulative and progressive because its evolution is characterized by successive building upon previously existing knowledge. The history of engineering systems therefore depicts a dimension of the overall theme of history that mirrors the development of civilization over the millennia. According to engineering historians, the historical evolution of engineering is best understood when it is discussed in the context of other transformative events of history that changed the way humans live: the food production revolution (circa 6000–3000 BC), the emergence of urban communities (circa 3000–2000 BC), the birth of Greek science (600–300 BC), innovations in power generation in Europe (in the Middle Ages), the development of modern science (17th century), the Industrial Revolution (18th century), the invention of electricity and the advent of applied science (19th century), and the current age of automation and information technology (20th and 21st centuries).

1.1.2 Engineering Definitions and General Evolution of Civil Engineering

Civil engineering is best defined in the context of engineering systems in general, and this section will first present general definitions of engineering and then move on to specific definitions of civil engineering. A simple definition of engineering is *the application of science, mathematics, business, and other fields to harness efficiently the resources of nature to develop structures and facilities that benefit the entire society at the current time and in the future*. Other definitions provided by White (2008) and Moncur (2012) include:

- The art of directing the great sources of power in nature for the use and convenience of humans (Thomas Tredgold, 1828)
- A triad of trilogies (first trilogy—pure science, applied science, and engineering; second trilogy—economic theory, finance, and engineering; third trilogy—social relations, industrial relations, and engineering) (Hardy Cross, 1952)
- The art of the organized forcing of technological change ... engineers operate at the interface between science and society (Dean Gordon Brown, year unknown)
- The innovative and methodical application of scientific knowledge and technology to produce a device, system, or process that is intended to satisfy human need(s) (Gerard Voland, 1999)
- The art of organizing and directing men and controlling the forces and materials of nature for the benefit of the human race (Henry Stott, 1907)
- Realization of a figment of imagination that elevates the standard of living and adds to the comforts of life (Herbert Hoover, year unknown)
- Activities that make the resources of nature available in a form beneficial to humans and provide systems that will perform optimally and economically (Llewellen Boelter, 1957)
- Activity other than purely manual and physical work that brings about the utilization of the materials and laws of nature for the good of humanity (Rudolf Hellmund, 1929)
- The professional art of applying science to the optimum conversion of natural resources to the benefit of humans (Ralph Smith, year unknown)

- The practice of safe and economic application of scientific laws governing the forces and materials of nature by organizing, designing, and constructing for the general benefit of humankind (S. Lindsay, 1920)
- The art or science of making practical (Samuel Florman, year unknown)
- Visualization of the needs of society and translating scientific knowledge into tools, resources, energy, and labor to bring them into the service of humans (Sir Eric Ashby, year unknown)
- The professional and systematic application of science to the efficient utilization of natural resources to produce wealth (Theodore Hoover and John Fish, 1941)
- The science of economy, of conserving the energy, kinetic and potential, provided and stored up by nature for the use of humans ... [utilizing] this energy to the best advantage, so that there may be the least possible waste (William Smith, 1908)
- Application, with judgment, of the knowledge of the mathematical and natural sciences, gained by study, experience, and practice, to develop ways to utilize, economically, the materials and forces of nature for the benefit of humankind (The Accreditation Board for Engineering and Technology, Inc., 1993)

For the civil engineering discipline specifically, formal definitions date back to 1828 when the charter of the Institution of Civil Engineers (ICE), in the United Kingdom defined that discipline as: The art of directing the great sources of power in nature for the use and convenience of man, as "the means of production and of traffic in states, both for external and internal trade, as applied in the construction of roads, bridges, aqueducts, canals, river navigation, and docks for internal intercourse and exchange, and in the construction of ports, harbors, moles, breakwaters, and lighthouses, and in the art of navigation by artificial power for the purposes of commerce, and in the construction and application of machinery, and in the drainage of cities and towns" (ICE, 2007).

In 1961, the American Society of Civil Engineers defined civil engineering as: "The profession in which a knowledge of the mathematical and physical sciences gained by study, experience, and practice is applied with judgment to develop ways to utilize, economically, the materials and forces of nature for the progressive well-being of humanity in creating, improving, and protecting the environment, in providing facilities for community living, industry and transportation, and in providing structures for the use of humanity."

In the definitions above, it is possible to discern the recurrence of certain concepts that the reader will recognize later in this text as systems engineering concepts. An example is the ***application of scientific tenets*** (Voland). In fact, classical science, which stipulates that all scientific inquiry should be rooted in hard facts, experimentation and objective analysis, and inferences, is a key aspect of system engineering. The role of science in civil engineering is evidenced in the definitions above through the use of such phrases as *through the aid of science*, *utilization of the laws of nature, art of applying science, application of scientific laws*, and *systematic application of science* are evidential of the role of science in civil engineering. Civil engineering is considered a science because its practice is consistent with the key characteristics of the classical scientific method—hypothesis setting and testing, replicability, refutability, and reductionism (Khisty and Mohammadi, 2001).

Other evidence of systems engineering concepts in the above definitions includes the phrase ***broad range of criteria*** for analyzing and evaluating engineering systems, which includes reference to the engineer as *one who uses the knowledge in all disciplines, including sociology* (Doherty), or *one who operates at the border between science and society* (Brown). This suggests that engineering is not only a science but also goes beyond the tenets of classical science, and thus in the course of their work, engineers typically examine problems from a broad range of criteria, not just those that are science based.

The **optimization of resources**, another systems engineering concept, has long existed in engineering practice as can be observed in the above definitions. For example, in their definitions of engineering, William Smith utilizes words such as *least possible waste* and *best advantage*; Hoover and Fish talk of *efficient utilization*; Ralph Smith makes reference to *optimum conversion*; and Boelter uses words like *perform optimally*.

The **ethical responsibility** of engineers is evident in the description of engineers as *persons who operate at the interface of science and society* (Brown) and the use of phrases such as *benefit of the human race* (Stott), *comforts of life* (Boelter), *good of humanity* (Helmund), *benefit of man* (Smith), *benefit of mankind* (Lindsay), and *needs of society* (Ashby).

Against the background of the definitions, we will now discuss the evolution of civil engineering as a discipline. Historians believe that the discipline took root between 4000 and 2000 BC when humans in ancient civilizations began to abandon their nomadic lifestyles in favor of more permanent shelter, thus generating the need for fixed facilities and structures. The reshaping of caves to protect humans from harsh weather and the use of tree trunks to cross water bodies were early practices related to civil engineering (Straub, 1964). Consequently, a need arose to transport large amounts of goods to and from human settlements for purposes of consumption, trade, and warfare. This also led to the need for roads and water-bearing and water-transporting structures such as aqueducts and canals. The new lifestyle generated other needs such as cultural (tombs for kings), religious (altars and temples), and entertainment facilities (large fighting arenas). Arguably, the first people to develop engineering systems were the Sumerians (located in present-day Iraq, 4500–1700 BC approximately) who constructed an intricate hydraulic system comprised of canals, dams, reservoirs, and weirs that helped transform their arid landscape into a systematic and lush city with beautiful gardens and fertile lands (Kramer, 1963). Other notable large engineering structures that date back several thousand years include the pyramids of Egypt constructed during 2800–2400 BC (Smoothwhirl, 2009) and the Great Wall of China (circa 200 BC).

The BC–AD transition millennium (500 BC to AD 500) was marked by significant advancements worldwide, including ancient civilizations in Persia, Greece, South America, South Asia, China, and Africa. In 3 BC, in what was probably the first scientific approach to the physical sciences applied to civil engineering, Archimedes established the laws of buoyancy and constructed a large screw that raised water from lower levels. Also in that era, impressive civil structures were constructed by a number of ancient civilizations worldwide including qanats (irrigation structures) in present-day Iran, the stupa monasteries in present-day Sri Lanka, and ancient structures in Great Zimbabwe. During the time of the Roman Empire (circa 27 BC to AD 500), extensive civil structures were constructed that included aqueducts, bridges, and dams. Other civilizations that were marked by remarkable achievements in civil engineering included those of Greece, Harrapan (in present-day India and Pakistan), and Maya (in present-day Mexico).

In all these and other civilizations that spanned the course of history, civil engineering systems have been developed in a bid to enhance the quality of life of people, for example, to provide water for irrigation and for drinking; dispose of liquid waste; and transport goods, message-bearing emissaries, and equipment and soldiers for defense purposes. Also, the development of civil engineering systems has proceeded in parallel with the advancements in other devices associated with the use of these systems. For example, the development of horse-drawn chariots provided greater impetus for improvements in road pavement construction.

The development of civil engineering as a profession has been evolutionary and incremental. The etymological root of the word "engineer" is the Latin word "ingenium", which means talent or mental power (Lienhard, 2000), and also was the name given to an ingenious device used by the Roman army to attack fortifications (Dandy et al., 2008). The field of civil engineering is considered the oldest nonmilitary engineering discipline and one of the oldest among all professions worldwide. The earliest engineers that carried out civil works actually were military engineers who possessed expertise in infrastructure of both military and civil purposes. In times of war, these engineers used their expertise to help facilitate conquests or defense by building catapults, observation towers, bridges across rivers, and other military facilities. In times of peace, however, their expertise was used for civilian purposes for the benefit of the populace. At some point in history, a dichotomy was established between military and nonmilitary engineers [Encyclopedia Britannica (EB), 2011]: The term *civil engineer* was used to describe any engineer who did not practice military engineering.

A drawn-out but perceptible watershed in the development of the profession was the formalization of design calculations. Over the centuries, design rules of thumb and empirical formulas used by civil engineers were gradually supplanted or supplemented by standardized design and numerical analyses, and the knowledge acquired through experience was documented and codified. Furthermore, stonemasons and craftsmen, who were mostly self-taught but skilled, acquired specific titles that indicated societal recognition of their skills (EB, 2011). The Renaissance in Europe (1500–1800) was characterized by prosperous urban societies that fueled the demand for infrastructure and technology. This period saw a rapid pace in the development of civil engineering as a profession in France as evidenced by the establishment of state-planned infrastructure by ministers in the Bourbons era (Chrimes and Bhogal, 2001); the first engineering school in modern times, the National School of Bridges and Highways was opened by Perronet in France in 1747; in Paris in 1794 and in Berlin in 1799, the École Polytechnique and the Bauakademie, respectively, were founded. John Smeaton of England was the first person to actually call himself a "civil engineer." In 1818, the Institution of Civil Engineers, the world's first engineering society, was founded in London; and in 1828, it was awarded a royal charter that formally recognized civil engineering as a profession. In the United States, Benjamin Wright, considered the father of American civil engineering, helped design and construct the Erie Canal and several railroads in the 19th century (FitzSimons, 1996). In the 19th and 20th centuries, persons calling themselves civil engineers in the United States and Europe designed and built all types of structures, water supply and sewer systems, railroads, and highways and planned cities. Notable civil engineers in that era included Benjamin Baker, Marc and Isambard Brunel, Gustave Eiffel, John Fowler, John Jervis, Robert Maillart, John Roebling, and Thomas Telford. The American Society of Civil Engineers (ASCE) was founded by 12 engineers in a meeting at the Croton Aqueduct administration offices in New York City on November 5, 1852 (ASCE, 2009). In the 20th century, professional civil engineering organizations with various designations including societies, institutes, and orders were formed in countries worldwide to advance the profession, protect the interests of members, and foster positive interactions with the general public.

Over the last two centuries, the role of civil engineers has been rather explicit and distinguishable from that of other related professions as they have applied their knowledge to plan, design, build, maintain, or/and operate complex civil infrastructure systems that have served humankind in a variety of ways. These systems include buildings for residential, commercial, and industrial purposes; facilities for transporting passengers and freight; and networks for transporting water, storm water, or wastewater. Specifically, civil engineers have responsibilities for constructing and/or managing a wide array of system types, including water and wastewater treatment plants, storm water and wastewater drainage, dams and levees, power plants, highway pavements and bridges, railroads, pedestrian and cyclist facilities, irrigation and shipping canals, river navigation, traffic infrastructure, public transit guideways and terminals, airport runways and terminals, transmission towers and lines, tunnels and industrial plant structures. Depending on the type of system or structure in question, the practice of civil engineering often involves some knowledge of other fields, such as physics, mathematics, geography, geology, soil science, hydrology, and mechanics. Consequently, the development of civil engineering has followed the progress made in these other fields; and in recent decades, the advancement of civil engineering systems management has followed the trends in economics, finance, statistics, and operations research.

The development of civil engineering systems has been a catalyst in the socioeconomic transformations we are experiencing today. Thus, civil engineering is one of the most effective vehicles for quality of life improvements for humankind. Social and economic changes constantly create new demands on civil engineers, who respond by fabricating, maintaining, and operating civil engineering systems to fulfill the needs and desires of society. As such, civil engineers actively and specifically seek engineering decisions that ultimately benefit, or at least minimize conflicts with, the social and economic environments. Dandy et al. (2008) pointed out that the relationship between engineering and society even transcends the physical realm and that part of the cultural character of regions and major cities can be attributed, at least in part, to iconic civil engineering structures at such locations. Examples include the Panama Canal, London's Big Ben Tower, Paris' Eiffel Tower, Rome's Coliseum, Greece's Parthenon, China's Great Wall, Sydney's Harbor Bridge, Egypt's Suez Canal, India's Taj Mahal, and New York's Statue of Liberty.

In the next section, we discuss key historical developments in the different branches of civil engineering in various civilizations over the course of human history.

1.2 CIVIL ENGINEERING SYSTEM—THE BRANCHES

Civil engineering can be classified on the basis of the intended use of the facility [heavy, industrial, commercial, residential, and recreational (Figure 1.1)] and the branches of civil engineering (hydraulic, hydrologic, transportation, architectural, materials, construction, structural, geomatic, and geotechnical engineering). For each facility type, the construction directly involves the civil engineering branches of geomatics, architectural, materials, construction, structural, and geotechnical engineering; and the operations are directly associated with at least one of the following branches of civil engineering: hydraulic, hydrologic, transportation, architectural, and engineering.

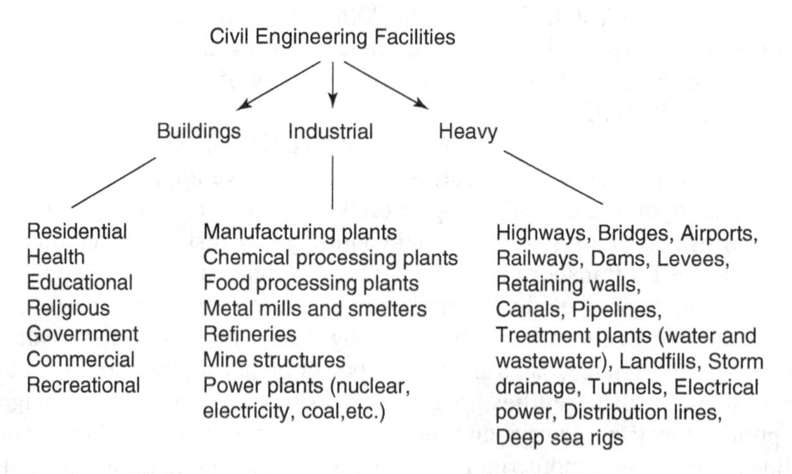

Figure 1.1 Categories of civil engineering facilities.

Phase of Facility Development \ Intended Use of facility[1]	Water (Treatment, Supply, Distribution	Transportation (Road, Rail, Water, Air)	Buildings (Residential, Commercial, Industrial, Recreational)
Initial Stages (Planning and Designing the Facility)	Survey engineers, Structural engineers, Geotechnical engineers		
	Hydraulic engineers Environmental engineers	Transportation engineers	Architectural engineers
Implementation Stage (Constructing the Facility)	Construction engineers, Materials engineers		
	Environmental engineers	Transportation engineers	Building engineers
Usage Stage (Operating and Maintaining the Facility)	Water/Wastewater Treatment plant engineers	Transportation engineers	Building engineers

1. *Of several facility types, only three are shown here: water, transportation, and buildings.*

Figure 1.2 Civil engineering branches categorized by phase of facility development and intended use of the facility.

The interface between the facility type and the civil engineering branch is influenced by the phase of the development of the facility in question. In other words, for any facility associated with a branch, the sequence of development goes through several phases including planning, design, construction, operations, and maintenance (Figure 1.2). Thus, there are engineers who work in phase-based branches, such as geotechnical engineers, who study the feasibility of soil support of a structure and design its foundation; structural engineers, who design the structure to withstand loads; construction engineers, who build the structure; and system operations engineers such as water plant managers, who run the system; and maintenance engineers, who preserve system physical structures. On the other hand, certain civil engineering branches involve a single type of facility (e.g., highways, water treatment plants, etc.); and such engineers are concerned with all phases of these facilities, from planning and design to preservation and operations. In contrast to the phase-based branches, these function-based branches exist on the basis of the intended use of the system and include transportation engineering, hydraulic engineering, and environmental engineering.

Clearly, the expansive breadth of civil system types and the number of systems development phases make it difficult for any individual civil engineer to be skilled in all the different branches and phases. There is necessarily a great deal of specialization, therefore, even within the different branches of civil engineering. Overall, there are at least nine branches of civil engineering, each of which has seen an interesting evolution of development from ancient times to current day. We discuss briefly in the following sections, the nature of work and the pioneers for each branch of civil engineering as well as its historical roots, evolution over time, and future expectations.

1.2.1 Structural Engineering Systems

Any physical object that is intended to support or resist dead or live loads, and to dissipate energy, regardless of its ultimate purpose, is amenable to structural engineering analysis. Thus, structural engineers design and analyze load-bearing architectural or civil engineering structures including buildings, towers, bridges, tunnels, dams, and retaining walls, and noncivil structures including equipment, vehicles (land, sea, or air), and other structures where structural stability of integrity is critical for safety and servicability. One aspect of structural engineering is the decomposition of a structure into its constituent subsystems: columns, beams, plates, arches, shells, and catenaries, even though in some cases, it is more intuitive to analyze the entire multicomponent structure

Figure 1.3 El Alamillo Bridge (Seville, Spain), designed by structural engineer Santiago Calatrava, combines aesthetic performance with structural efficiency (Courtesy of Consorcio Turismo Sevilla).

as a system of systems. Using various materials including steel, concrete, composites, and other materials for their designs, structural engineers investigate the actual or predicted outcomes of their systems in terms of specified mechanical behavior and functionality, for example, performance criteria including safety (e.g., failure of its components), serviceability (e.g., discomfort to system users due to vibration, shaking, or sway), durability (e.g., satisfactory life with minimal maintenance), cost (e.g., optimal use of materials and resources), and in certain cases, aesthetics (Figure 1.3).

Structural analysis helps to ascertain the magnitudes and directions of forces and deformations in a structure due to dead and live loads; structural design determines the dimensions of the structural members to ensure that the structure is capable of supporting the intended loads. Simulation models, which we shall discuss in Chapter 13, are used widely in structural engineering and are intended to replicate, as closely as possible, the actual behavior of the structure as a function of its material properties, structural features, loading, and boundary conditions (Liew and Shanmugam, 2004).

The History of Structural Engineering Systems. The field of structural engineering has existed, albeit as an informal discipline, ever since humans first began to build their own permanent structures. At the height of their civilization (6000–2000 BC), the Sumerians in ancient Mesopotamia (present-day Iraq), designed and constructed large, layered platforms called *ziggurats* (Figure 1.4) for supporting their temples, similar to structures that were built in a later era by the Aztecs of Central America. It has been speculated that one of these Sumerian structures was the Tower of Babel that is described in the Book of Genesis in the Bible. The ziggurat architectural and structural style has inspired a number of modern buildings such as the University of Tennessee's John Hodges Library in Knoxville. The Sumerians also developed key structural elements, such as arcs and domes (which are used in current-day design) and utilized innovative structural techniques

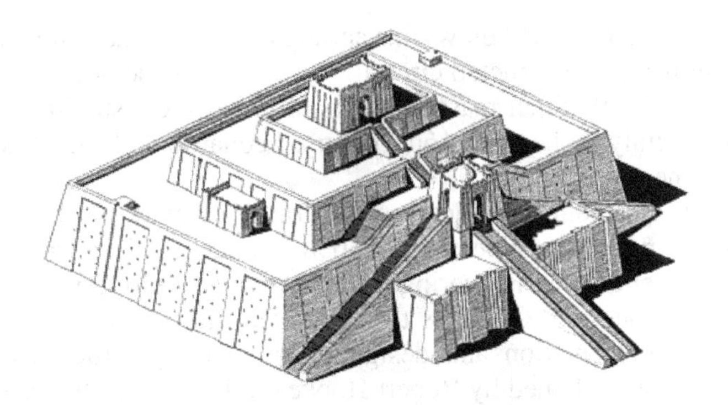

Figure 1.4 Ziggurats, structural systems comprising large, layered platforms, supported worship temples in ancient Mesopotamia and other civilizations several millennia ago (Wikimedia commons/United States Army).

such as buttresses, recesses, and half columns in building their temples and palaces (Shuter, 2008). In the Minoan civilization (circa 2700–1400 BC), column inversion (bottom width smaller than top width) and multiple-storey buildings were significant structural features of that era (Benton and DiYanni, 1998). Some engineering historians believe that the formal discipline of structural engineering began in 2700 BC when Imhotep (considered the first structural engineer in history) built Pharoah Djoser's pyramid. During ancient times and in the medieval era, the design and construction of structures were carried out by artisans similar to Imhotep, particularly carpenters and stonemasons, and officials in royal courts who held titles such as *master builder* (Saouma, 2007) and served as both the architect and structural engineer. According to historians, explicit theories of structures did not exist and there was limited understanding of how structures remained stable; knowledge was accumulated through experience and passed on over time through successive experts.

In ancient Greece (circa 220 BC), Archimedes calculated the areas and determined the centers of gravity of a number of geometric figures and developed calculus and Euclidian geometry, thus providing the mathematical foundations for current structural engineering theory. Also, in ancient Rome, Vitruvius, a famous Roman architect and engineer, in his 15 BC manual of civil and structural engineering, described the techniques used in planning, designing, and building a number of structures (Straub, 1964). During the ancient civilization of Great Zimbabwe in AD 11, a number of formidable civil engineering structures were designed and constructed in a style that "eschewed rectilinearity for flowing curves" (MetArt, 2009). During that era, significant contributors to formal structural engineering included Abu Rayhan al-Biruni and Abd al-Rahman al-Khazini. These Persian scholars helped build the foundations for the theory of structures by pioneering the application of experimental scientific methods to statics and dynamics and by unifying these two areas into the science of mechanics. They introduced algebraic techniques into the field of statics and were first to develop general center-of-gravity theory. In the Tibet region of China in AD 762, structural engineers designed iron bridges that included probably the first suspension bridge in history, which was constructed by engineer Thanstonrgyalpo, the *lcag zam pa* (the builder of iron bridges). Also, in China in the 15th century, bridges that were constructed generally utilized far less material than those of the preceding civilizations worldwide. Also, Chinese bridge builders of that era invented the complete circle structure (the arch of the bridge above being mirrored by a corresponding inverted arch below) that sprung from the same abutments deep under water, such as the Tung-Mei bridge constructed circa 1470 and still in use at least until 1970 and possibly today Needham et al., 2001. Such masonry rings afforded great stability at areas having weak natural foundations. In Italy in the early 16th century, Leonardo da Vinci produced a number of structural and other engineering designs.

The 17th and 18th centuries saw several watermarks in basic sciences that later laid the foundation for structural engineering. In 1658, Galileo published a seminal work that addressed the science of the strength of engineering materials and also pioneered the use of scientific approaches in structural engineering. Galileo's thesis ignited the field of structural analysis (defined as the mathematical representation and design of engineering structures). In 1678, the behavior of materials was first explained by Robert Hooke on the basis of the elasticity of materials, followed by the explanation of the fundamental laws governing structures by Isaac Newton. Several decades later, there were advancements in mathematical methods that facilitated the modeling and analysis of engineering structures. In that era, Leonhard Euler formulated the buckling equation that helped analyze structural elements in compression. Also in that era, Euler and Daniel Bernoulli developed the Euler–Bernoulli beam equation, a basic theory in the design of structures (Bradley, 2007), and the Bernoulli brothers provided analytical tools to analyze structures (Dugas, 1988).

In 1809, the first suspension bridge capable of carrying vehicles was built to cross the Merrimac River (a 250-ft span) in Massachusetts. Advancements such as this, catalyzed by significant discoveries in material science, structural analysis, and the physical sciences, helped structural engineering to evolve into a more formalized profession toward the end of the 19th century, particularly during the Industrial Revolution. In 1873, Carlo Castigliano developed methods for determining displacement as partial derivatives of strain energy through his thesis *Intorno ai sistemi elastici*. Advancements in concrete technology included Joseph Aspdin's 1824 invention of Portland cement, which made concrete construction economically feasible; the 1855 development of modern reinforced concrete by Joseph-Louis Lambot and William Wilkinson; and the 1867 use of steel reinforcement in regions of tensile forces in concrete structures by Joseph Monier (Prentice, 1990; Kirby, 1990; Nedwell et al. 1994). At the end of the 19th century and the early 20th century, advancements in cast iron technology facilitated steel bridge construction in Europe. Also, Vladimir Shukhov established methods for analyzing nontraditional structures such as those with unconventional shapes or thin shells. The new century saw developments in reinforced concrete shear design by Wilhelm Ritter and in the behavior of concrete as a linear-elastic material by Emil Morsch.

The 20th century saw further contributions to reinforced concrete science by innovators that included Swiss engineer Robert Maillart and enhancements in the design and analysis of steel and concrete structural systems through greater understanding of the plastic behavior of concrete. In 1928, the development of prestressed concrete by Eugene Freyssinet helped structural engineers address the weakness of concrete structures in tension. In 1930, Hardy Cross facilitated quick and accurate determination of stresses in complex structures through his moment distribution method, and in the 1950s, John Baker developed the plasticity theory of structures, thus facilitating the design of steel structures (Heyman, 1998). In the late 1960s and early 1970s, Fazlur Rahman Khan introduced innovations such as the "bundled tube" structural design for skyscrapers, which was used for Chicago's John Hancock Center and Sears Tower. Khan also developed the structural concept of X-bracing, which reduced lateral loads on a building by transferring such loads into exterior columns, thus reducing the need for interior columns and making more floor space available.

LEONARDO, THE VISIONARY BRIDGE ENGINEER

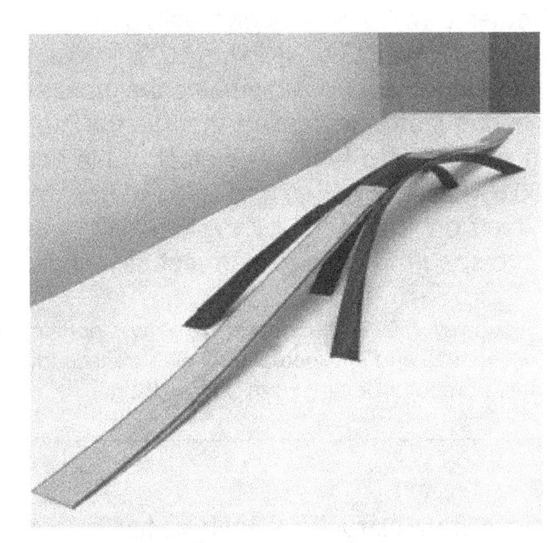

Conceptual model.

Bridge under construction, 2001, Vebjorn Sand, Norway.

In 1502, Leonardo da Vinci wrote a letter to Ottoman Sultan Beyazid II of Istanbul to propose the building of a single-span 720-ft (0.24-km) bridge over the Bosporus at a point known as the Golden Horn. Leonardo's preliminary drawings consisted of the classic keystone arc design; his design was based on the premise that by using a flared foothold and the terrain to anchor each end of the bridge, the arc could be stretched narrow and substantially widened without losing any structural integrity. Believing that such a construction was impossible, the Sultan did not build the bridge.

Since 1952 when the letter was discovered in Turkish National Archives at Istanbul, experts have pondered whether it would have been feasible to construct the bridge. So, by the Sultan's refusal to build the bridge, was Leonardo saved from disaster and professional ignominy? Or was he deprived of the legacy of possibly being the most innovative bridge builder of his time? In any case, engineering historians assert that Leonardo's Golden Horn Bridge design is "an eloquent synthesis of form and function typical of his universal thinking." Leonardo's vision was revisited in 2001 when his design was used to construct a smaller bridge in Norway (see photo). In 2006, a decision was made by the Turkish government to construct a bridge, using Leonardo's design, to span the Golden Horn estuary.

Sources: Atalay and Wamsley (2009). *Image source*: www.leonardobridgeproject.org, an organization that inspires human artistic, spiritual, and intellectual endeavor transcending cultural borders through the construction of Leonardo da Vinci's graceful Golden Horn bridge design.

Since the late 20th century, structural analysis has been enhanced by advancements in computing power. This has fostered the use of computational and numerical methods, including simulation and finite element analysis, to reliably estimate the engineering behavior of structural materials and complex structural configurations. These advancements have made it possible to develop increasingly bold structural systems such as London's Millennium Dome, Greece's Rion-Antirion Bridge, Shanghai's Nanpu Bridge, Japan's Akashi-Kaikyo Bridge, South Korea's Jongro Tower, and Jakarta's Regatta Hotel at the Pantai Mutiara Canal Estate. Others include Toronto's CN Tower, Sweden's Turning Torso Building, Italy's Strait of Messina Bridge, Barcelona's Montjuic Communications Tower, Spain's Alamillo Bridge, Beijing's Bird's Nest Stadium, and Dubai's Burj Khalifa, currently the world's tallest building. The shape of the Burj Khalifa (Figure 1.5) not only takes inspiration from indigenous desert flowers [Landmark Properties (LP), 2009] that also appear as

Figure 1.5 The Burj Khalifa, currently the world's tallest building, is a testament to current advancements in structural engineering (*Source:* Nicolas Lannuzel).

decorative patterns in Islamic architecture but also serves a technical purpose. In order to support the great height of the building, the engineers developed the "buttressed core," a new structural system that consists of a hexagonal core that is reinforced by three Y-shaped buttresses thus facilitating lateral self-support of the building and avoiding twisting [GulfNews (GN), 2010].

The Future of Structural Engineering Systems. The future of structural engineering will be guided in part by innovations in material science, boldness in design, desire for resilience to hazards, and computer technology. Ongoing advancements in structural materials, for example, through research in nanotechnology and materials science, will open up new directions in structural design from the perspectives of sustainability, economy, aesthetics, fire resistance, and durability (Ochsendorf, 2005). The past 50 years have seen strength improvements in structural steel (40%), reinforcing bar (50%), and concrete (at most 100%) (Magnusson, 2007). Further innovations in these materials are expected to continue and could include the development of stainless steel, fiber-reinforced polymers, and other materials for steel construction and concrete reinforcement. Concrete research continues to yield high-performance concrete (HPC), such as translucent concrete with unprecedented compressive and tensile strength. Also, to overcome congestion caused by rebar, stronger rebar alloys with strengths of 75–100 ksi (thus taking up less volume) could be adopted. Thus, the future is expected to be characterized by significant increases in the strength as well as reductions in the sizes of concrete columns and shear walls and steel columns and trusses. Improvements in structural design and analysis will translate into new "geometric freedoms" and will encourage bolder structural and architectural designs involving the complex geometries of exterior and interior elements. Many forms of structural systems that are currently considered impossible or too expensive are expected to become the mainstream. The future of structural engineering will also be shaped by advancements in computers and information technology (Smoothwhirl, 2009) as multidimensional computer simulation and visualization become essential tools for the structural engineer for quickly and efficiently designing and evaluating bridges, tall buildings, and other large or complicated structural systems. In addition to these opportunities, threats loom on the horizon: Future developers of civil engineering structural systems also will need to contend with the impacts of climate change on their structures; namely, altered frequencies and intensities of extreme weather, climate, and sea levels will translate into a myriad of consequences, such as longer droughts, more frequent and severe freeze–thaw cycles, warming of ocean surfaces (resulting in more intense typhoons and hurricanes), larger and more abrupt floods, changing levels of groundwater, and changes in wind speed and profiles (Lenkei, 2007). These, in turn, will accelerate surface deterioration, low cycle fatigue, and accumulated damage, thereby fostering the need to review design codes for planned structures and to adopt adaptation and mitigation measures for existing structural systems (Long and Labi, 2011).

1.2.2 Transportation Systems Engineering

Transportation engineering can be described as the science of providing systems for moving people, goods, and services safely and cost-effectively by sea, land, or air (Fricker and Whitford, 2005). From the modal perspective, transportation engineering therefore has subbranches, such as highway engineering (roads), railroad engineering (freight rail), transit engineering (heavy rail, commuter rail, light rail, monorail, etc.), port engineering (harbors, canals, and other maritime facilities), pipeline engineering, and airport engineering. Other subbranches are typically associated with nonmotorized urban travel and include pedestrian and cyclist management.

For each mode, the functional areas include planning, design, and construction of the system, traffic operations and capacity management, congestion mitigation and safety management, and facility preservation. It can be observed that these functional areas follow a certain sequence or a life-cycle pattern, which we shall discuss further in Chapter 2. Thus, from the phasal perspective, the transportation engineering subbranches could be established also on the basis of the functional area. This explains why at many universities or public agencies, transportation departments are divided not only according to the mode involved (highway division, railway division, etc.) but also on the basis of functional area (planning division, design division, operations division, maintenance division, etc.).

The planning aspects of transportation engineering, for any mode, include facility location, demand assessment, cost estimation, and impact assessment in terms of air quality, mobility, safety, economic development, and other impact types. The traditional technique for forecasting demand is the four-step process: trip generation (how many trips are generated?), trip distribution (what are their destinations?), mode choice (which modes are used by the trip makers?), and traffic assignment (for each mode, what percentage of trip makers use each available route?). More sophisticated demand forecasting techniques consider other aspects of trip makers' backgrounds or the nature of their trips, such as auto ownership, residential or business locations, and trip chaining (linking separate trips together in a tour). Also, at the planning level, the expected system cost is roughly estimated using rules of thumb for other empirical models. Examples include average costs or cost models based on the aggregate characteristics of similar facilities built in the past, expressed per unit dimension or per unit usage such as $/lane-mile or $/passenger-mile, respectively, of the system.

In transportation system design for any mode, engineers determine the appropriate size, materials, orientation, and geometry of transportation facilities. These include the guideway (runway, railway, or highway pavement); terminal; intermediate; or nodal facilities for intermodal overlaps or intramodal directional exchanges or transitory repositories such as highway intersections, rail intersections, terminals, parking garages, and so on. At the design phase, costing is more detailed and yields a relatively more reliable estimate that is based on the cost buildup from the unit costs of the individual pay items of the materials, labor, and equipment used for each specific task. This cost is often used as the basis for bid evaluation.

At the operations phase of any transportation mode, transportation engineers establish optimal operational controls so that the delay or travel time for passengers or freight is minimized. Thus, traffic engineers in any mode develop guidance and information for its users through signs, signals, markings, and more recently newer intelligent transportation systems (ITS) technologies such as advanced traveler information systems (changeable message signs), commercial vehicle facilities [Global Postioning System (GPS)-enabled advisory systems], advanced traffic control systems (arterial signal coordination), and vehicle–infrastructure integration. Engineers strive for safe operations of their systems by including safety elements in their designs or by making continual recommendations for safer facility operations by analyzing crash patterns, frequencies, and severities at various links and nodes of each mode.

History of Transportation Systems. The need for transportation infrastructure arose from the gradual evolution of ancient societies from subsistence lifestyles to communities that produced and exchanged goods and services. The earliest transportation mode was land transport by way of earth tracks through forests and grasslands. First blazed by hunters as game trails, these tracks subsequently evolved into paths for humans and domesticated animals carrying goods to and from trading posts. Increases in trade volume led to widening or strengthening of the tracks to accommodate more frequent and heavier traffic. In this section, we discuss the evolution not only of civil infrastructure but also of the mechanical devices that complemented the use of these civil facilities. In ancient Sumeria, animal-powered wheeled vehicles were developed in 500–400 BC, and this technology spread to other parts of the world. Archeological evidence of this can be seen in areas of the Minoan cities in ancient Crete (2700–1450 BC) that were well connected with stone-paved roads formed using saw-cut blocks (Shuter, 2008). On the Indian subcontinent circa 4000 BC, the critical role of transportation infrastructure in the economy of the Harrapan and Mohenjo-daroor (the Indus Valley civilizations) is evidenced by archeological remnants of paved streets and land transport vehicles such as bullock carts (Carr, 2011). In pre-Columbian South America, several roads and trails, such as the 22,000-km Inca road network system (El Camino Inca) of Peru, were constructed to facilitate commerce, and Inca rope bridges provided access across valleys (Kirby, 1990).

While the origin of highways can be traced to prehistoric tracks and bronze-age ridgeways, it was only after the rise of strong centralized governments that complex road systems emerged (Needham et al., 2001). As empires expanded, the need to control conquered areas generated a large demand for accessibility and mobility through perennial road networks. Circa 300 BC, the Magadha Empire (in present-day India) under ruler Chandragupta Maurya was extended from the Arabian Sea to the Bay of Bengal, and extensive road networks were built to facilitate movement of its military, which was considered to be the largest army in the ancient world. Also during that era, the ancient Romans in the expansion of their empire, had great road engineers whose vocation was one that could ultimately lead them to occupy high (political) offices in the state. The Roman road system was a 50,000-mile network that included almost 30 military highway sections centered in Rome. Even today, their remnants can be seen in areas that were a part of the ancient Roman Empire, from Spain to Syria and from England and the Danube to North Africa. The greatest of the Roman roads was the 360-km-long and 14 Ft-wide Via Appia, or the Appian Way (Figure 1.6) named after

Figure 1.6 The Appian Way, an ancient Roman highway constructed in 312 BC, is still in use today (Courtesy of Paul Vlaar).

ruler Appius Claudius Caecus (Pannell, 1964). This highway, which was constructed with huge lava block paving in a bed of crushed stone cemented with lime, runs from Rome to Brindisi, and parts of it are still in use today. Also in the pre-Christian era, there existed road tunnels in Rome, such as the Petra Pertusa Tunnel on the Via Flaminia and the 2300-ft-long Grotta of Naples, which connected the city with the suburb of Bagnoli. Other civilizations, such as that of ancient Greece, were also known for impressive highway systems: the urban streets and market squares of ancient Greece were mostly paved, and in rocky areas, the roads consisted merely of two wheel ruts carved into the rock, resembling a rail track. Crossing points for vehicles driven in opposite directions were provided at certain intervals. However, the difficulty of letting other travelers pass often led to bitter disputes, the most famous being that between Oedipus and King Laius; and this quarrel led to the patricidal tragedy that was later documented in the journals of Sophocles, the Greek philosopher (Kirby et al., 1956). Another example of excellent highway systems of that era is the Persian Royal Road of the Achaemenid Empire built circa 500 BC by King Darius I (Needham et al., 2001). This highway stretched from the city of Sardis near Izmir (in present-day Turkey), passed through Nineveh, the Assyrian capital that is the site of the present-day city of Mosul in Iraq, and Babylon (near present-day Baghdad, Iraq) and split to join Susa (in present-day Iran) and the Achaemenid capital city of Persepolis (present-day city of Parsa in Iran). At the height of the Ottoman Empire (AD 16–17), many highways, such as the Aleppo (Syria) to Baghdad (Iraq) Road, were constructed, some of which were paved using tar residue derived from distillation of the petroleum obtained in the region's oil fields. In China, in the first two centuries AD, under a succession of emperors, extensive imperial highways were built along the coasts and rivers, using a pavement material and structure similar to what later became used and known in Europe as "water-bound macadam." Several of these roads (notably, including some sections of the link between the Chhin capital in the north and the Szechuan basin in the south) were constructed in straight lines, cut through mountains, and carried on embankments in valleys; erosion control material was provided for embankment slopes (Needham et al., 2001). The Pei-chan Lu (or North Trestle Road) linking Shu to Kuan-chung was aptly named for the massive pillars and beams that supported the road through the ravines it traversed.

In Europe in the 19th century, engineers John McAdam, John Metcalf, Robert Phillips, Thomas Telford, and Pierre Tresaguet made significant contributions to road science, including the use of pavement designs that incorporated self-draining surface slopes and carefully selected sizes of stone aggregate and soil. At the 19th to 20th century transition, advancements in land vehicles, from horse-drawn vehicles to bicycles and motors and electric vehicles, spawned the development of land transportation facilities, such as the provision of impermeable surfaces and systematic drainage facilities, thus reducing the inconveniences of dust and mud bogs. The invention of the stone crusher and steam roller in 1959 increased the speed and economy of road construction (Kirby et al., 1956). The installation of automatic traffic signal systems began in the United States and Europe in the 1920s, following the invention of the traffic light by Garrett Augustus Morgan. The 1920s and 1930s saw a dramatic improvement in highway geometric design, culminating in the construction of high mobility and limited access superhighways or expressways, and interchanges.

In the area of maritime transportation, the Stone Age was characterized by the use of natural harbors that served fishing canoes and boats. Subsequent developments in maritime transportation infrastructure were facilitated by the needs of war and increases in trade volumes. Circa 4000 BC, canals were developed to facilitate inland water transportation in the ancient city-state of Mesopotamia (Shuter, 2008). There is evidence that in Mediterranean ports, galleys (seagoing vessels propelled mainly by oars) were developed circa 3000 BC. Commerce by merchants from the present-day Persian Gulf regions of Bahrain and Failaka was facilitated by an extensive maritime

Figure 1.7 Transportation engineers strive to maximize mobility and accessibility while minimizing travel delay, cost, and environmental degradation (Courtesy of renaissance-downtowns.com).

trade network operating between the Mesopotamian and Harappan (Indus Valley) civilizations. The long-distance sea trade was made possible not only by innovations in sea vehicle technology, such as plank-built watercraft and sail material and design, but also with natural, shallow harbors located at river estuaries. To accommodate larger vessels, natural ports were used and artificial ones were developed through dredging and other earthmoving activities. The first canal system in the world was built circa 2600 BC during the Indus Valley civilization in present-day Pakistan and northern India; this is evidenced by the recent archeological discovery of a massive, dredged canal and a docking facility at the Indian coastal city of Lothal (located in the modern state of Gujarat), dating from 2400 BC (Carr, 2011). In the Mediterranean, where tideless coasts provided natural settings for water-based travel, the Phoenicians, in 1200 BC, developed the port of Sidon for the purpose of maritime trade. In 490 BC, the longest canal of that era, the 1770-km-long Grand Canal of China, was constructed to transport Emperor Yang Guang and his entourage between Beijing and Hangzhou. Also, the ancient Greeks and Romans were adept at harbor building. However, unlike the Greek harbors that were located at places with minimal disruption to the existing currents, land form, and other natural features, the Romans did not shy away from radical disruptions of natural conditions. This difference in design philosophies probably explains why many ancient Roman harbors today are silted up or have succumbed to the sea (Straub, 1964). In the Middle Ages, the start of the 13th century saw the phasing out of galleys and the advent of large ocean-faring ships. These included caravels (a small Spanish, Portuguese, or Arabic sailing vessel rigged on two or three masts), the treasure ship (a large wooden vessel commanded by Chinese Admiral Zheng in the early 15th century), and the man-o-war (an armed naval vessel developed in the late 15th century in the Mediterranean that was propelled primarily by sails). Also, canals were built in the Middle Ages in the Italian city of Venice and in the Netherlands to facilitate inland transportation. Examples include the 240-km-long Canal du Midi in France that was built in 1680. During the Industrial Revolution, the first steamships, and later diesel-powered ships, were developed, and inland canals were built in England and in the United States.

With regard to rail transport, there is archeological evidence that probably the first engineered railway was the Diolkos Wagonway (6 km in length) in Greece circa 600 BC, built to transport boats across the Isthmus of Corinth for several centuries. This railway consisted of grooves that were carved in limestone to serve as the track, and the wheeled wagons were powered by slaves and animals (Lewis, 2001b). After a long break, rail transportation infrastructure reappeared in Europe in 1550 in the form of crude wooden tracks. In the 18th century, the first "modern" railroad on the European continent was a horse-drawn railway established to transport coal between Budweis (in modern-day Czech Republic) to Linz (in Austria). In the 1760s, cast iron plates were used as rails but were replaced decades later by rolled wrought iron rails due to the efforts of British civil engineer William Jessop in Loughborough. In England, mechanized steam-powered rail transportation systems first appeared in the early 19th century. At the start of the 19th century, the first rail-guided steam locomotive was built and operated by engineer Richard Trevithick in Wales, but it proved to be a financially unsustainable venture and led to Trevithick's bankruptcy (Ellis, 1968). With the development of railway systems in Great Britain in the early half of the 19th century, which included contributions by James Watt and George Stephenson, railway transportation gradually spread throughout the world and dominated long distance land transport for nearly a century (Ellis, 1968). This was before other transportation modes (air and highways) became viable or more cost-effective due to inventions in the vehicles used in those modes. In the United States, early railroads differed by their purpose and power sources. These included New York's gravity railroad in (1764), Pennsylvania's Leiper Railroad in (1810), Massachusetts' Granite Railroad in (1826), and the Baltimore and Ohio Railroads (1830). At the close of the 19th century, the development of diesel and electrical energy to replace steam as a rail power source was facilitated by developments in diesel and electrical technology. For example, the development of the pantograph by individuals, such as Granville Woods, and later adopted by engineering companies, including the Baltimore and Ohio Railroads and Siemens & Halske, enabled the conduction of electricity from overhead wires to railcars and led to the operation of the first electric rail system at Coney Island in New York in 1892. Thanks to developments in engine and guideway technology, the 20th century saw yet another generation of rail transportation: Japan's Shinkansen, France's *train à grande vitesse* (TGV), and Western Europe's Eurostar, and magnetically levitated trains (Maglev) in Germany, Japan, and recently China (Osorio and Osorio, 2006).

Air transportation, unlike the other modes, has a history characterized by watersheds that occurred mostly in the last millennium. However, the fascination of transporting people and goods by air dates back several thousand years when catapults were used in warfare and when humans desired to replicate avian flight as in the legends of Daedalus and Icarus in Greek mythology and the Vimanas in Indian mythology. According to engineering historians, the first attempts at flight were probably made in the 6th century in China by Yuan Huangtou who used a kite and Abbas Ibn Firnas in Spain who used a parachute and a controllable glider; in the 17th century, in Turkey, Hezarfen Celebi used a winged glider and Lagari Çelebi used a gunpowder-powered rocket for one-man flights (Darling, 2003). Then in 1783, the Montgolfier brothers in Paris developed hot air balloons for manned flight; and a year afterward, Jean-Pierre Blanchard, seeking to overcome the wind direction limitations of balloons, operated the first human-powered dirigible (NASA, 2002). Some of the notable dirigible developments that subsequently followed were Henri Giffard's machine-powered propulsion in 1852, David Schwarz's rigid dirigible frames in 1896, and Alberto Santos-Dumont's improvements in dirigible speed and maneuverability in 1901. Powered heavier-than-air flight, which was started by the Wright brothers in the United States in 1903, was subsequently enhanced with developments in flight control that made them practical for warfare and ultimately for transporting passengers and goods. Meanwhile, airships were used for a while to transport goods and passengers over great distances but saw sharply diminished use after 1937. The first, second, and third decades of the 20th century saw tremendous advancements in air transportation, and passenger airline service was started during this period as well. World War II was accompanied by several significant innovations in aviation including the first liquid-propelled rockets and the first jet aircraft. The end of the war was marked by a boom in general aviation. At the current time, air transportation is dominated by jet-powered aircraft developed in the mid-20th century.

By the end of World War II, highway engineering had begun to be recognized as a distinct area of engineering (this later evolved into transportation engineering, thereby covering the different modes of travel). Over time, it has been infused with techniques from economics, finance, materials science, and operations research. Important contributions in transportation engineering over the past 100 years include innovations in highway materials by Roy Crum and Prevost Hubbard in the 1920s, development of financial practices for highway engineering systems by Wilfred Owen in 1940, and establishment of relationships between guideway surfaces and operating costs in 1943 by Ralph Moyer. After World War II, there were important contributions as well, such as quantification of the influence of materials on pavement performance in 1946 by a team led by Kenneth Woods; innovations in concrete science by Charles Scholer in 1948; development of techniques for estimating the capacity of multilane highways by O. K. Normann; and pavement design improvements by F. N. Hveem and R. M. Carmany in 1949. Advancements in the 1950s included an analysis of accidents for highway planning purposes in 1950 by Roy Jorgensen and Robert Mitchell; development of the BPR function by Albert Goldbeck for traffic network studies; and technical and financial planning of interstate systems by Herbert Fairbank. Other enhancements in the 1950s included Burton Marsh's work on traffic safety, Ralph Moyer's contributions to urban transportation systems planning, and Tilton Shelburne's research in highway skid analysis. Also in that era, notable contributors to highway engineering systems included Harmer Davis for his research in transportation efficiencies, Guilford St. Clair for transportation finance, Merlin Spangler and Robert Litehiser for highway drainage, and Alan M. Voorhees for identifying patterns in urban travel. The 1960s continued the trend of innovations in materials and a continuation of the innovations in the emerging science of transportation operations. This work included William Goetz in bituminous materials and Bryant Mather and Fred Burggraf in concrete technology and science, as well as Alvin Benkelman who was the inventor of the Benkelman beam device for measuring road surface deflection, and Francis Turner who uncovered patterns in urban transportation system operations.

The Future of Transportation Engineering Systems. As we move deeper into the new millennium, the transportation engineer will be faced with a variety of challenges that will require more explicit adoption of "systems" concepts and approaches for their resolution. Some of these developments include increased population and travel demands that lead to traffic congestion and air pollution in urban areas and increased need for accessibility by rural populations in developing countries; aging transportation facilities, many of which were built several decades ago and have surpassed their design lives; the incorporation of several stakeholders in transportation decision-making processes and higher user expectations; increased threat of terrorist attacks on transportation systems, vulnerability to natural disasters, transportation system resilience to hazards, and postdisaster recovery; and, finally, tightened funding to maintain, rehabilitate, and reconstruct aging transportation infrastructure. Other ongoing and emerging issues to be faced by transportation engineers of the future include sustainability of transportation systems from the perspectives of the environment, safety, sociocultural impacts, land use, energy use, and climate change (Sinha, 2003). At a tactical level, engineers will exploit new technologies for real-time monitoring and optimizing of transportation system operations. Transportation infrastructure engineers will also seek to enable real-time inspection and monitoring of the physical condition and usage patterns of systems to facilitate timely and cost-effective interventions that preserve the system physical and operational integrity. Furthermore, major advances in intelligent transportation systems (e.g., information and communication technologies) and innovations in building materials, nanotechnology, and vehicle technologies (e.g., propulsion and new fuels) are expected to open up new horizons in transportation engineering through increased opportunities for cost reduction, greater mobility, enhanced safety and security, and increased system longevity and economic productivity.

1.2.3 Hydrology and Hydraulic Systems Engineering

Hydrologic systems engineers analyze the occurrence and distribution of water in the air, land, and sea. Their central theme is the cyclical movement of water throughout the Earth through different pathways. These pathways are characterized by the evaporation of water from oceans to form clouds; precipitation in clouds as rain or snow; flow of rainwater (runoff) across land surfaces into rivers, streams, and lakes; in-ground percolation of water into lakes, rivers, or aquifers; return of water to the atmosphere through evaporation from the surfaces of water bodies or through plant transpiration to the atmosphere; and precipitation from the atmosphere and surface water discharge into the ocean. The subbranches of hydrology include hydrogeology (study of the movement of water in subsurface bodies including aquifers), hydrometeorology (study of water and energy transfers between the atmosphere and surfaces of land and water bodies), and surface hydrology (study of hydrologic processes that occur at or near the Earth's surface), and hydroinformatics (the adaptation of computer information technology to hydrology and water resources applications).

Hydraulic system engineers study the mechanical properties of liquids, including energy exchanges due to fluid flow. They also analyze the properties of fluids in motion and the interactions between a flowing fluid and its immediate environment (Lyn, 2004). These engineers plan, design, and manage engineering structures for water supply and distribution and also to control water flow, such as dams, water-crossing bridges, levees, networks for water supply and distribution, urban drainage systems, channels, and transportation canals. Figure 1.8*a* shows the Falkirk wheel, a rotating hydraulic boat lift that connects the Union Canal and the Forth and Clyde Canal in central Scotland. Hydraulic engineers manage irrigation, flood and erosion control, and coastal protection. Hydraulic systems play a critical role in society's need for water conservation, flood control, and drainage. In the recent era that is characterized by wild fluctuations in weather patterns induced by climate change, it has become important to protect facilities located near the coast, lakes, or large rivers from inundation using dikes, sea defense walls, coastal barriers, levees, and other hydraulic systems. Figure 1.8*b* shows a coastal defense barrier at the Isle of Wight in the United Kingdom.

History of Hydrology and Hydraulics. Throughout the ages, proximity to freshwater sources has always served as a main catalyst for human settlement and development. Most major cities in the world are located along the banks of a river. Proximity to water, however, is a double-edged sword: Engineers harness this resource for purposes of irrigation and water supply but also need to protect human settlements from inundation during flood events. The expansion of population and the development of trade over the millennia has led to the increased importance of water for agriculture, water supply, and transportation. As far back as 4000 BC, the Nile River was dammed to enhance the fertility of surrounding land. Also, in the ancient city of Babylon located in the desert empire of Mesopotamia (circa 1760 BC), water resource management was so vital to the empire's socioeconomic fabric that a large system of irrigation canals was constructed and special officials were appointed to supervise the operations of these engineering systems. The officials ensured that the canals were clear of debris, weeds, and silt in order to prevent flooding. The ruler, King Hammurabi, through his provincial governors, personally directed the excavation and dredging of the

(a)

(b)

Figure 1.8 (a) Hydraulic-lift wheel, Falkirk, Scotland and (b) coastal defense structure in operation [Courtesy of (a) AndiW/Wiki Commons and (b) Oikos-team at en.wikipedia].

canals on a regular basis and the construction of high earthen walls near townships to protect them from floods. Also, to prevent neglect of the canals, the king established a set of common laws (probably the world's first), which included clauses that addressed the construction of these and other structures. These clauses struck terror in the hearts of unethical contractors as Hammurabi's strict code of justice dealt a heavy hand to incompetent builders. Those whose structures collapsed and resulted in the deaths of people faced a sentence of death (Prince, 1904). In the Indus Valley civilization era (3000–1500 BC), the hydraulic engineering skills of the Harappans were evident in their documented academic study of tides, waves, and currents and in their dock building (Carr, 2011).

Other hydraulic systems that were designed and constructed before or during these eras include retention basins, canals, irrigation ditches, and dikes in various parts of the world. In ancient Egypt, drinking water was transported to the city of Memphis in 3000 BC, a channel was built to connect the Red Sea and the Nile River in 1950 BC, wells exceeding 300 ft in depth were dug in 1700 BC, and water tunnels in hills were constructed in 1200 BC as part of war preparations (Biswas, 1970). In approximately 600 BC, China's first recognized hydraulic engineer, Sunshu Ao, rose to political prominence in the State of Chu due to his engineering skills, and he ultimately was appointed prime minister. For purposes of irrigation and water supply, Ao constructed the Shao Bei Dam in the Northern Anhui Province and created the Anfeng Tang Reservoir System that is in operation even today (Needham, 1986). Another hydraulic engineer of that era, Ximen Bao, circa 400 BC, diverted the Zhang River from flowing into the Huang He River near Anyang and established a different course that met the Huang He further downstream near the modern-day city of Tianjin. He also created a large canal irrigation system for the agricultural region of Henei.

In the Neo-Babylon Empire under Chaldean rule, circa 700–500 BC, there were significant advances in the engineering of hydraulic structures. One of the earliest aqueducts on record has been attributed to the Assyrian master builder and ruler Sennacherib (circa 700–680 BC) who governed with "a heart of wrath" and exploited the power of water equivocally: During times of peace, he harnessed water bodies to develop the capital city of Ninevah and his Khorsbad palace; and in times of war, he unleashed water as a weapon to flood and destroy enemy strongholds. Sennacherib used 18 freshwater courses from the mountains, 2 dams, and a 3-stage 10-mile-long water canal to develop a sophisticated water supply and distribution system. Water was also transported using an aqueduct reinforced with hardened clay and waterproofed with bitumen. The aqueduct, which ensured a continuous supply of water to the city, crossed valleys on arched bridges.

Ancient Greek philosophers had pondered various aspects of what is now known as the hydrologic cycle. Tartarius (400 BC) suggested that a large underground sea existed that replenished the oceans, and a century later Theophroastus published the first meteorological abstracts (Leonard, 2001). In ancient Rome, Frontinus, a famous engineer who managed the aqueduct systems in that era, authored *De aquaeductu*, an official report to the emperor on the state of Rome's water supply system, including the laws relating to its use and maintenance.

The aqueducts were built by the ancient Greeks and Romans transported water over long distances. Also, in ancient China, remarkable hydraulic systems including canals and irrigation channels were used to harness water for transportation and agriculture purposes for thousands of years (Needham et al., 2001): Dujiangyan, a massive irrigation system involving the Minjiang River in Sichuan, China, was built in 250 BC and is still in use today; also, in the Chin and Han dynasties, great efforts were made to conserve freshwater resources through dike strengthening and other activities under great engineers such as Chia Jang in AD 6. The 1770-km-long Grand Canal of China, which connects Beijing in northern China to Hangzhou in the south (Pannell, 1964), commenced construction in 5 BC. The construction of the Chengkuo Irrigation Canal, the Kuanhsien Irrigation System, the Ling Chhu Transportation Canal, and the Chhien-thang Sea Wall are evidence of the great skills of hydraulic engineers at the time. Interestingly, there existed two rival philosophies that influenced the design of hydraulic systems in China: The Taoist philosophy of greater freedom for natural courses advocated the use of "feminine" activities, such as dredged concavities; and the Confucius philosophy of confining and repressing nature advocated "masculine" activities, such as dike construction. This is similar to the dichotomy between the ancient Greek and Roman philosophies of civil engineering. At the time of the legendary Emperor Yao, his engineer Kun adopted the masculine approach and built several dikes but failed to stem the water flow and ultimately suffered punishment through exile and execution by the emperor. Kun's son, Yu, who adopted the feminine philosophy, was more successful in controlling the floodwaters. The first use of pound locks was by the Chinese several centuries later at the beginning of the Sung dynasty, by Chhaio Wei-Lo, assistant commissioner of transport in Huainan City circa AD 1000.

In the 3rd century BC, Archimedes invented a hydraulic system consisting of a screw, a helical surface surrounding a central cylindrical shaft installed inside a hollow tube (Figure 1.9). When the screw is turned, the bottom end scoops up an amount of water, which slides up in the spiral tube until it finally pours out from the top end of the tube. This innovation was used to draw water from

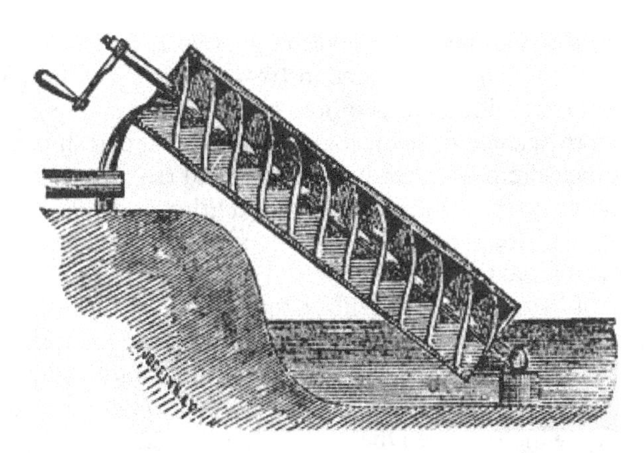

Figure 1.9 Archimedes' screw (Courtesy of Ianmacm at en.wikipedia).

low-lying areas or flooded mines in order to drain those areas or to lift water to higher ground for irrigation purposes.

In ancient India, the era of the Mauryas was characterized by significant hydraulic engineering works including dams and canals. In the 11th century, significant hydraulic systems included the 250-square-mile artificial Bhojpur Lake and the 16-mile Jayamkonda Dam, constructed under the Bhoja and Chola Dynasties, respectively. In ancient Rome in the first century BC, Marcus (Vitruvius) Pollio, an engineer, described a philosophical theory of the hydrologic cycle, in which precipitation fell on the land, infiltrated the earth's surface, and recharged underground streams and water bodies or run-off from the land surface as streams and springs. The Etruscans and Romans were masters of constructing water supply and drainage systems. An example is the draining of Lake Fucino, a drainage project in Italy in AD 45 where a 3.5-mile tunnel and 28-ft drop were constructed through the mountains of Salviano to provide an escape run-off for the trapped lake, a feat that evidenced great skills not only in hydraulic engineering but also in construction engineering and surveying. In the last three centuries BC, the Romans constructed a magnificent system of aqueducts; and under Emperor Tiberious, these systems provided an astonishing 180 million gallons of freshwater for Rome. In AD 97, Julius Sextus Friontius, given responsibility by Emperor Nerva for the water supply of Rome, personally inspected the existing aqueduct system and documented the designs and operational procedures of that system (Landels, 1978).

The Pont du Gard Aqueduct in southern France transported water across the small Gardon River Valley, the aqueduct helped deliver approximately 5 million gallons (20,000 m^3) of water daily from the Uzès Springs to the ancient Roman city of Nemausus (present-day Nimes). At its first level, the aqueduct carries a road. Also, the ancient Sinhalese (circa AD 300–500) utilized concepts of hydrology to build remarkable irrigation systems, large water reservoirs, dams, and water canals, some of which are in use in present-day Sri Lanka. They are also known for having invented a number of hydraulic devices, such as anicut stones and removable pillars, to facilitate water intake, control water flow, or prevent erosion. Also, circa AD 200–600, during the Sassanid era in the Middle East (the last pre-Islamic Persian Empire), the 250-mile Nahwawan Canal was constructed to improve the management of the region's water resources (Needham et al., 2001).

Leonardo da Vinci established the relationships between channel area, water velocity, and flow; and his paper "Treatise on Water" explained the origin of lakes and rivers, water evaporation and condensation, open-channel flow theory, and the relationship between hydraulic head and flow. Subsequent work by da Vinci and Bernard Palissy in the 15th century separately yielded more accurate representations of the hydrologic cycle. In 1598, Giovan Fontana established the link between velocity and discharge; and in 1694, Pierre Perrault established relationships between rainfall intensity and resulting surface flow using his observations of the Seine River. Other pioneers of the modern science of hydrology in the 17th and 18th centuries include Edme Mariotte, who carried out measurements of water velocity and river cross section to determine the relationships between these variables and water flow, and Edmund Halley, who demonstrated that the water evaporation from the Mediterranean Sea surface adequately accounted for the outflow of surface water bodies flowing into that sea.

In the field of hydraulics, Vitruvius, a few decades BC, documented techniques for aqueduct construction and the use of the inverted siphon. During the Renaissance period in Europe, renewed interest in scientific thought spurred advancements in hydraulic science (Biswas, 1970). With regard to water supply systems, probably the greatest breakthrough was the invention of the cast iron pipe in the 17th century, which enabled the conveyance of water under great pressures (Leonard, 2001). The 18th century saw advances in hydraulics that included Daniel Bernoulli's piezometer and the Pitot tube invented in the early 1700s by Henri Pitot, an Italian-born French engineer. Pitot disproved the then commonly accepted notion that at greater depth, the speed of water is greater. In the 1900s, there were further developments in groundwater hydrology and hydraulics, including Darcy's law, which shows how fluids flow through porous media; the Dupuit–Thiem formula, which described underground water flow; and Hagen–Poiseuille's equation, which explained capillary flow patterns. In the 20th century, rational analyses began to replace empiricism, and important contributions were made by Leroy Sherman (the hydrograph), Robert Horton (infiltration theory), and C. V. Theis (aquifer test and equation governing well hydraulics). The 20th and 21st centuries also have been characterized by approaches that are increasingly theoretical in nature, a trend that has been facilitated by increased recent understanding of hydrological processes and also by the advent of computational, mapping, and visualization capabilities such as Geographic Information Systems (GIS).

The timeline for the development of the science of hydrology can be presented into eight-periods (Chow, 1964; Rao, 2002) as shown in Figure 1.10: The *speculation period* when conjectural speculations were rife regarding the different aspects of the hydrologic cycle, and practical knowledge of hydrology was used as a basis for hydraulic structure construction; the *observation period* when hydrological variables received close observation and scrutiny by scientists who had an understanding of the hydrological cycle including Bernard Palissy and Leonardo da Vinci; the *period of measurement* when scientists started measuring hydrologic variables and the science of hydrology was born; the *experimentation period* that laid the building blocks for modern hydrology; the *period of modernization*; the *period of empiricism* when hydrological knowledge was mostly based on empirical observation; the *rationalization period* during which scientists made theoretical contributions relating to the hydrograph, infiltration, and groundwater processes; and the present period called the *theorization period* characterized by the use of information technology to develop and validate complex theories.

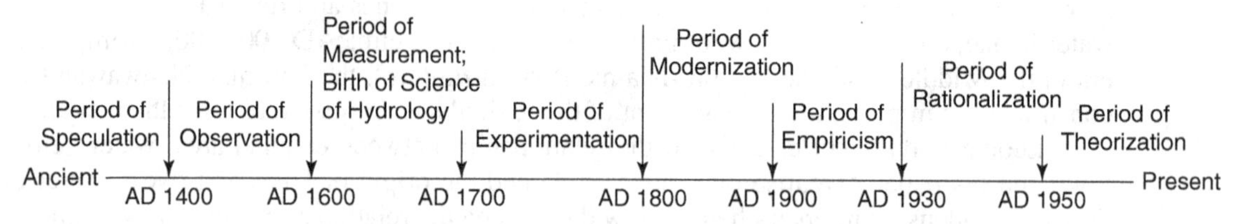

Figure 1.10 Development of hydrology—timeline.

The Future of Hydrologic and Hydraulic Engineering. As we enter the new millennium, the two-way relationships between water resource systems and socioeconomic development will become increasingly visible and important. These impacts will be driven by political and economic uncertainties, increases in world population, and anthropogeny-driven climate changes (Delleur, 2003). Global warming is also expected to lead to rising sea levels and greater volumes of surface runoff, which will necessitate new designs and performance reviews of existing hydraulic systems. In the future, it is also expected that there will be advances in the tools used in analyzing hydrologic and hydraulic systems, including remote sensing, GIS, and hydroinformatics (the application of databases, software, and expert systems). Continuing advancements in the fields of mathematics and computer science will enable future hydraulic engineers to enhance hydroinformatics to encapsulate existing knowledge through genetic programming, data mining, and artificial neural networks for a variety of tasks, including real-time control of urban drainage systems (Abbot et al., 2001). Since the 1993 major flood events in the Mississippi and Missouri basins, the traditional approach of designing the water resource system first and then considering the impacts is gradually giving way to a systems approach in which the hydraulic, environmental, and ecological aspects are all included simultaneously in the planning, design, construction, and operation of such systems (Starosolki, 1991). This trend is expected to continue in the future. Hydraulic engineers will continue to focus on environmental issues, sustainability, and management; and they will seek and utilize more effective ways to engage stakeholders and society in general in their decision-making processes (Chanson, 2007). Furthermore, because of the multiplicity of water sources, the variability of response times in the hydrologic cycle of each source type, and the myriad of current and future uses of water, planners, designers, and operators of hydraulic systems will be expected to deal with a multitude of ongoing and emerging problems associated with physical management of water resources, such as identification of new sources of freshwater, cost-effective water supplies, flows in water bodies, flood protection, hydropower generation, and water transportation.

1.2.4 Environmental Systems Engineering

Environmental systems engineers apply scientific and engineering principles to enhance the quality of the environment (land, air, and water) so that it is healthy for the humans, flora, and fauna that inhabit it, which includes prevention of pollution as well as remediation of polluted areas. Pollutants may be chemical, biological, thermal, radioactive, or even mechanical. The erstwhile terms "sanitary engineering" or "sanitation engineering" are more aligned with public health engineering and thus have a scope that is narrower than environmental engineering. Perhaps the most cross-disciplinary of all civil engineering fields, environmental engineering incorporates physics, mathematics, chemistry, biology, ecology, geology, law, and public health. Environmental engineers are responsible for or involved in recycling, waste disposal, environmental impact assessment and mitigation, water treatment and supply wastewater conveyance and treatment (Figure 1.11), and management of solid or hazardous waste.

History of Environmental Engineering. Historical records suggest that, throughout the ages, people have appreciated the direct relationship between their health and the quality of their environment and thus have sought to apply scientific and engineering concepts to enhance their environments. These applications include the generation and distribution of drinking water and proper disposal of wastewater. Scholars have credited the Sumerians in Mesopotamia, the Minoans in Crete, the Harrapans in the Indus Valley, and the Egyptians for developing early technologies that enhanced environmental quality in those eras. Engineering historians have observed the parallels

Figure 1.11 Digesters at the wastewater treatment plant Deer Island, MA (*Source:* Frank Hebbert/Wiki Commons).

that run across these civilizations: high population density, proximity to major rivers, fluctuations in river levels, and high summer temperatures (Leonard, 2001).

There is archeological evidence that King Menes in ancient Egypt developed water supply and distribution systems that were critical in sustaining that civilization. In ancient Greece, there were elaborate systems for water supply, wastewater discharge, and water transportation. On the Mediterranean island of Crete, the palace of Minos in the Minoan civilization (3400–1200 BC) had latrines outside of residences, stone-constructed systems for drainage, and pressurized systems for water supply (Buescher, 2000). Other environmental engineering feats of the ancient Minoan people of Crete (3000–1500 BC) included detailed systems for storm drainage and disposal of sewage that bear similarities to those of today. The Minoan city streets had water draining facilities, and clay-piped sewer facilities were available to the upper-class citizenry (vestiges of these pipes still carry runoff from heavy rains today). Sewers fitted with ventilating shafts were located beneath certain streets that collected liquid waste from residences. Also, circa 3000–1500 BC, cities in the ancient Harappan civilization, which were located in present-day India, Pakistan, Afghanistan, Turkmenistan, and Iran, filtered their water with charcoal and treated their water with copper containers that helped kill disease-causing germs. They also had a detailed system for draining storm water and residential wastewater as part of their overall city planning, as evidenced by archeological excavations at the city of Mohenjodaro.

Ancient Romans constructed aqueducts to transport water over long distances in a bid to prevent drought and to establish a consistent and healthful supply of water for residents, often using water treatment techniques such as filtration through porous vessels and water treatment by adding chalk and aluminous soil (Leonard, 2001). In ancient Rome, the main sewer, the Cloaca Maxima, not only drained a number of marshes in the valleys but also collected and drained human waste and storm runoff into the Tiber River (Lanciani, 1967). The ancient Greeks sought to protect their water systems from their enemies by laying their aqueducts underground, sometimes to a depth of 60 ft, and the deeper ones were connected to the surface through large wells. For treating their wastewater, the ancient Greeks constructed tri-compartment cisterns to help settlement of the wastewater (Leonard, 2001), a concept similar to present-day septic tanks.

There seems to be little evidence that the engineering knowledge displayed by the Romans was passed on to successive civilizations after the fall of that empire. In medieval Europe, there seems to be very little development of environmental sanitation or water infrastructure in the cities. As such, outbreaks and epidemics of infections and plagues among the dense populations were quite common.

Governmental intervention has been more widespread in environmental engineering than in any other branch of civil engineering. Throughout history, there have been acts of legislation that regulated public actions (and inactions) that could potentially harm the environment, ranging from edicts by the ruler of Babylonia circa 1890 BC, for city inhabitants to desist from actions that tended to clog the canals to laws passed by parliaments in European countries in the 19th century and legislation passed in the United States to restrict water and air pollution in urban areas. Laws were passed in the 15th century in Bavaria, Germany, to reduce the rate of alpine forest degradation, thereby helping to protect the quantity and quality of water supply in the region. In the Middle Ages in London, it took the passage of legislation to help improve the sanitary situation, for example, the Bill of Sewers and other laws were passed in London to regulate the discharge of storm water from gutters and liquid waste from residences; and in 1843, laws were passed in Germany to require the construction of sewers in certain urban areas (Leonard, 2001). In Europe during the Middle Ages, many city inhabitants simply threw their household human waste out the window (Rayburn, 1989) causing aesthetic problems and health hazards. Engineers in London in the mid-19th century proved that provision of good drinking water and effective disposal of wastewater could drastically reduce the incidence of waterborne diseases such as cholera and that additional laws were needed to enhance the city's sanitation.

In 1727, Sir Francis Bacon documented his experiments in water treatment techniques such as boiling, distillation, percolation, and clarification; some of his work is being used in modern-day environmental engineering (Baker and Taras, 1981). A watershed in clean water supply was marked in the United States in 1840 when the 41-mile Croton Aqueduct was constructed in New York City to deliver 95 million gallons of water daily to Manhattan (Leonard, 2001). The Lawrence Experiment Station also was commissioned in Massachusetts to carry out research into wastewater treatment processes. In the 19th and 20th centuries, prominent environmental engineers included Thomas Crapper, who invented the flush toilet; Paul Roberts, who applied the fundamental principles of chemistry and mass transport to water and wastewater treatment and wastewater reclamation; and Abel Wolman, who introduced the field of sanitary engineering and standardized the methods used to chlorinate drinking water supplies. In the 20th century, cesspools were replaced by sewers in most cities, and the processes of water treatment significantly improved, particularly due to disinfection via filtration and chlorination (Marhaba, 2000). For example, chemical coagulation process was first used in 1904 at a municipal water supply system (the Chain of Rocks Purification Plant in Missouri), activated sludge plants were implemented through pilot schemes in New Braunfels and Houston, and full operations began in 1926 at Milwaukee's Sewage Treatment Plant.

The Future of Environmental Engineering. As the world population continues to grow, greater demand is being placed on the quality of the Earth's natural resources and environment. Treatment and disposal methods that were once adequate now require far greater levels of cleanup before discharge into the natural environment (air, land, and surface and ground waters). These substrates are no longer considered free economic goods as has been assumed for decades; that is, their consumption generates to society, a cost that can be measured as the cost of avoiding their contamination or the cost of remediation (Jacko, 2003). In developing countries, the availability of clean drinking water will continue to pose a challenge for governments, and millions will likely die annually from unsanitary water-related conditions unless drastic steps are taken (Leonard, 2001). In developed countries, environmental engineers will continue to wrestle with problems related to anthropogenic air pollution, water resource degradation, ecological damage, and possible contamination from hazardous waste disposal. The role of the environmental engineer will be expanded to explain and mitigate the incidence of pollution-related diseases. Increasing realization of the unsustainable practice of nonrenewable fuel use will spur environmental engineers to play a growing advisory role in the global search for alternative energy sources. As we move into the new millennium, environmental engineering will be increasingly characterized by the application of new sustainable technologies to address these persistent problems. The emerging field of environmental biotechnology is expected to expand to help in pollution detection, remediation, and prevention.

1.2.5 Geotechnical Systems Engineering

Geotechnical engineers study the mechanical behavior of earth materials and, specifically, the state of rest or motion of soil bodies under the action of force systems (Harr, 2004). Geotechnical engineers use the principles of soil mechanics and rock mechanics to carry out at least six general activities: investigate and monitor subsurface conditions and surficial materials at a site, ascertain the relevant geotechnical properties of the site materials, evaluate and monitor the geotechnical integrity of manmade or natural soil/rock slopes and deposits, assess and monitor the risks associated with site conditions, carry out earthwork and structure foundation designs and monitor the geotechnical performance of these designs, and prescribe ground improvements to enhance the geotechnical integrity of a site (Holtz and Kovacs, 1981; Terzaghi et al., 1996).

Geotechnical engineers perform site investigations (surface and subsurface exploration) and laboratory tests in order to acquire information about the mechanical and chemical properties of a site's subsurface characteristics and thus carry out more reliable geotechnical assessments of the site. In such investigations, the site's underlying and surficial soils and bedrock are characterized; then the geotechnical integrity of the site is assessed on the basis of how the underlying soils will behave in response to loading from the proposed structure. Geologic mapping, photogrammetry, and satellite maps are used to obtain additional subsurface data at the sites. This is often supplemented by geophysical techniques such as seismic waves and electromagnetic surveys (including resistivity, magnetometer, and ground-penetrating radar (GPR)). Subsurface exploration usually involves direct soil testing at the site (Figure 1.12*a*) and laboratory tests on the samples retrieved from the site (Figure 1.12*b*). Geotechnical field tests include trial pitting, boring, drilling (small-diameter boring), trenching, cone penetration testing (CPT), and trenching (particularly for locating seismic faults). Large-diameter borings enable direct visual examination of the in situ soil and rock profile, but this technique is used only when it is safe and relatively inexpensive to do so. The engineer examines the soil or rock cuttings expelled from the drill hole during drilling operations, retrieves soil or rock samples at various depths from the drill shafts, and performs tests on the recovered soil or rock. For cone penetration tests, an instrumented probe with a conical tip is used, which is pushed into the soil manually or hydraulically and the rate of penetration is correlated to the soil properties. The dynamic cone penetrometer test (DCPT) is a popular test used to determine the strength of soil intended as subgrades for highway and airport pavement construction.

Geotechnical engineers also assess any risks associated with the site characteristics. Natural hazards include erosion, earthquakes, soil liquefaction, landslides, rock falls, and sinkholes. The risk assessment includes not only the vulnerabilities of the geotechnical system to the natural or built-up environment (including the structure, property, and humans) but also vice versa: the potential geotechnical hazards posed by the proposed structure to the natural or built-up environment.

Another key aspect of geotechnical engineering is the planning, design, and monitoring of earthwork, foundations, and other geotechnical systems for proposed structures or for repair of

(*a*)

(*b*)

Figure 1.12 Site and laboratory geotechnical tests help ascertain the integrity of underlying soils for construction. (*a*) Field sampling (*Source*: www.prlog.org) and (*b*) laboratory tests.

defective or distressed structures and earthworks due to adverse subsurface conditions. Geotechnical engineers develop site-specific foundation design recommendations and criteria for buildings, bridges, and highways. These designs include shallow foundations (footings and slab foundations), deep foundations (piles and drilled piers), lateral structures for earth support (cantilever walls, gravity walls, excavation shoring, and sheet piling), engineered slopes, geosynthetics, and earth structures. Earth structures include embankments, natural channels, dikes, pavements, reservoirs, tunnels, levees, and landfills.

In order to improve a site's geotechnical conditions, engineers often carry out ground improvement by modifying the properties (e.g., permeability, stiffness, and shear strength) of the existing ground. Ground improvement provides support slopes and foundations for several types of civil engineering structures and can reduce construction cost and time (Raju, 2010).

History of Geotechnical Engineering. Throughout the course of human history, soil and rocks have been used as material for various civil engineering and architectural purposes, including building foundations, burial sites, road construction, irrigation, and flood control. Soils of a specific nature were selected for constructing flood control structures in the ancient civilizations of Mesopotamia. Also, during the Indus River civilization (circa 2200 BC), earthen dikes were constructed along the Indus River to prevent flooding. For their temples and other structures, ancient Greeks and Romans used a variety of foundation supports, including strip-and-raft and pad footings.

For many centuries, the field of geotechnical engineering remained more of an art than a science and practitioners used past experience and trial and error. In the 18th century, after a spate of foundation-related engineering problems, scientists begun to seek more scientific approaches to designing foundations and making earthwork recommendations. Classical geotechnical science began in the late 18th century when Charles Coulomb introduced the concepts of engineering mechanics to the analysis and solution of soil problems. Other contributions during this period included Henry Darcy's work on hydraulic conductivity, Joseph Boussinesq's stress distribution theory, Christian Otto Mohr's theory of a two-dimensional stress state, William Rankine's work on the pressure theory, and Albert Atterberg's establishment of metrics to assess soil consistency.

Rapid population growth and increased rural–urban migration in the mid-19th century led to increased demand for taller buildings, extensive transportation systems, and construction of structures at areas hitherto deemed unsuitable due to relatively poor subsoil conditions. At that time, however, building foundation design and construction had advanced very little beyond those of the previous centuries. By 1879, however, critical geotechnical concepts had been developed, such as allowable bearing pressure, concrete and steel spread footings, and the steam pile hammer. As building heights increased due to increased land costs, availability of steel, and the invention of the elevator, the use of deep foundations gained popularity (Parkhill, 1998).

Modern-day geotechnical engineering was born in 1925 with the publication of *Magnum Opus, Erdbaumechanik* by Karl von Terzaghi. Other pioneers in the field included Arthur Casagrande (1902–1981), well known for his ingenious designs of soil testing apparatus and research on seepage and soil liquefaction. Despite the difficult economic conditions of the post-1930s era, federal spending on infrastructure projects helped support research by Terzaghi, Cassagrande, and other engineers. Technological advances that have spurred the development of soil improvement techniques included vibroflotation, vertical sand drains, wick drains, and rubber-tired roller compactors.

The Future of Geotechnical Engineering. As we move into the future, the demands of population growth, the increasing shortage of suitable land, and environmental concerns will mean that civil engineering structures will need to be located at sites previously considered unsuitable due low geotechnical integrity. The increased boldness of structural designs in terms of the heights and sizes of buildings and other structural systems also will pose challenges for engineers involved in designing their foundations. Geotechnical engineers of the future must develop new skills and technologies that would enable such projects to be possible. A case in point is the Chubu Centrair International Airport, which was constructed on a man-made island in Japan's Ise Bay (Figure 1.13). Future enhanced technologies will be required for future civil engineering systems that are slated to be built on artificial islands to prevent excessive settlement and damage during earthquakes. Furthermore, future geotechnical engineers will increasingly include in their analysis, elements of

(a)

(b)

Figure 1.13 The Chubu Centrair International Airport, entirely built on an artificial island, posed a variety of complex challenges in geotechnical engineering [*Source:* (*a*) BehBeh/Wikimedia Commons and (*b*) Gryffindor/Wikimedia Commons].

uncertainty and reliability and will give greater prominence to earthquake science, geosynthetics, the geo-environment, and new, promising techniques for efficient and quick in situ characterization of subsoils (CETS, 1995). Geotechnical engineers will increasingly be called upon to consider resilience in their designs, for example, by developing and adopting cost-effective designs that reduce the vulnerability of civil systems to natural or man-made disasters. Finally, as the Earth seemingly enters a phase of global warming, polar ice caps will melt, leading to increases in sea and land groundwater levels. The resulting change in subsoil pore water pressures is expected to lead to drastic changes in geotechnical conditions, possibly threatening the stability of existing structures. As such, future geotechnical engineers may need to revise their design processes for future geotechnical structures, carry out continual performance reviews of existing geotechnical structures to assess their vulnerability to this threat, and to prescribe and implement remedial actions that may be needed. Future geotechnical engineers will be increasingly called upon to tackle nontraditional problem types as well, such as geo-environmental engineering, geosynthetics design and evaluation, and design of foundations for deep-water offshore structures for which there is relatively little available experience to provide guidance. In such situations, the geotechnical engineer's judgment and experience will be stretched to the limit (ASCE, 2005).

1.2.6 Construction Engineering

Construction engineers plan and manage the construction of architectural or civil engineering structures and systems. They are typically skilled in engineering, management, finance and economics, legal procedure, and human behavior. The tasks undertaken by construction engineers, either directly or through their site representatives, include planning and scheduling, cost monitoring and control, material and equipment procurement, design of mixes (e.g., concrete, asphalt, etc.), quality assurance and quality control of workmanship and materials, site geodetic surveys, and worker safety. In some cases, depending on the type of contract, construction engineers also supervise the design of structures on the site. Experienced construction engineers often assume the role of project manager.

History of Construction Engineering. Construction engineering may be considered the oldest of the civil engineering disciplines. As we learned in Section 1.1.2, the abandonment of nomadic lifestyles generated the need for permanent structures, which required that inhabitants acquire construction skills. For many centuries, the construction engineer was also the architect and the structural engineer and was called the master builder in many early civilizations.

Some of the earliest human feats in construction engineering are evidenced in archeological remains of the city of Babylon, located 50 miles south of Baghdad in modern day Iraq between the Tigris and Euphrates Rivers in 6000–3000 BC. This city is known for its canal networks, organized layouts, and building structures. Many of the houses were two and three stories high, and the city's streets were constructed in a grid fashion relative to the river (at right angles or parallel). Also, there is archaeological evidence that the city had an elaborate sewerage system that consisted of feeder pipes from residences to main sewer pipes located beneath the streets (PMB, 2008). In present-day Malta, there is evidence of the construction engineering feats of the ancient people of the Ghar Dalam phase of the country's history, including the megalithic temples of Malta, circa 5000 BC.

Of the ancient Near East, the ziggurat is the most distinctive infrastructure. Similar to ancient Egyptian pyramids, most of these structures were four-sided, and built to great heights to reach the "realms of the gods." However, unlike the smooth-surfaced pyramids of Egypt, ziggurat exteriors were tiered to facilitate the construction work and supervision and also to accommodate religious rituals essential to the societies at the time. The lower parts of surviving ziggurat remains are indicative of remarkable design and construction engineering techniques. For example, the temple's core (comprising unbaked mud brick) becomes alternatively more or less damp depending on the season, and the constructors provided holes through the temple's baked exterior layer to allow the evaporation of moisture from the core. Also, drains were engineered along the ziggurat's terraces to drain storm water (German, 2012). During the ancient Egyptian civilization circa 3000 BC, several structures to serve various functions were constructed under a succession of kings, notably Menes and Scorpion. These structures included temples, tombs (pyramids), and hydraulic structures (dams, retention basins, canals, irrigation ditches, and dikes). Demonstrating great skill in engineering, the constructors of facilities in that civilization used relatively sophisticated machines, such as the lever, inclined plane, and roller, to transport bulky building materials and ultimately to erect large structures such as the Great Pyramids of Giza and Cheops. Circa 2550 BC, Imhotep, considered by many as the first engineer, used shaped stones, simple construction tools, and mathematics to build the famous stepped pyramid of King Zoser located at Saqqarah in ancient Egypt (Saouma, 2007). Other notable products of ancient construction engineers included the Persepolis in Iran in 500 BC (Figure 1.14*c*), Parthenon by Iktinos in ancient Greece (circa 440 BC), the Great Wall of China (circa 200 BC), and the Coliseum in Rome in AD 72. Also, ancient civilizations such as those of Crete, Greece, and Rome, constructed significant civil engineering structures. In the 7th century

Figure 1.14 A few instances of remaining evidence of the skills of ancient constructors: (*a*) conical tower of Great Zimbabwe, circa AD 11 (*Source:* Vinz at fr.wikipedia), (*b*) ruins of civil structures in Machu Picchu, a pre-Columbian Inca city located in present day Peru (*Source:* Charlesjsharp/WikipediaCommons), (*c*) structures constructed by Darius I in 500 BC, in Persepolis, ceremonial capital of the Achaemenid Empire, present-day Iran (*Source:* Wikimedia Commons).

AD in India, Brahmagupta, an Indian mathematician and astronomer, used arithmetic based on Hindu–Arabic numerals to determine the volume of material associated with excavation projects (Plofker, 2007).

In AD 11, the Kingdom of Great Zimbabwe constructed an impressive huge complex of stone walls that undulated over 1800 acres of terrain in present-day Zimbabwe (Figure 1.14*a*). Constructed from closely fitted granite blocks taken from exposed rocks in the surrounding hills, the walls were given nicely finished surfaces. The rocks were broken into portable sizes and fitted without using mortar by laying them on top of each other. Each rock layer was recessed slightly more than the previous layer to yield a self-stabilizing inward slope (MetArt, 2009).

In the ancient Roman Empire, circa 20 BC to several centuries AD, construction was often guided by the documented manuals written by master builders, the most famous of which was Vitruvius. These manuals included descriptions of the knowledge required of an architect and builder, building materials and their use, rules of building design, design of columns, water collection and supply, and contemporary building equipment such as hoisting gear. The manuals also contained theoretical information such as the basic principles of mechanics. The Pont du Gard Aqueduct, located near Remoulins in southern France, was constructed by engineers from the ancient Roman Empire circa 19 BC to AD 70. The aqueduct was constructed in three years using a workforce of about a thousand people using large stones without any binding mortar. The stones were cut to fit together, thus eliminating the need for cementitious material, and the stones were raised to fit their designed positions using a block-and-tackle technique, the winch for which was powered by a massive human treadmill. The constructors erected a complex protruding scaffolding system comprised of ridges and supports (Straub, 1964).

Yet another example of the remarkable construction skills of past civilizations is Machu Picchu, a pre-Columbian Inca city constructed at the height of the Inca Empire circa 1450. Considered a civil engineering marvel of the ancient world, Machu Picchu is located 2400 m (7880 ft) above sea level in the Urubamba Valley on Peru's side of the Andes Mountains (Figure 1.14*b*). Machu Picchu was constructed by ingenious people who demonstrated their skills in constructing resilient structures for running water supply and distribution, drainage, and food production.

The Future of Construction Engineering Systems. The future landscape of construction engineering is likely to be characterized by applications of advancements in materials science, computer and information technology, automation, project delivery, and supply of materials. Engineers will utilize information technology tools such as computer simulation to enhance the construction process. These tools include computer-aided design (CAD) and computer-aided installation/construction. Emerging technologies in this area include life-sized three-dimensional visualizations that enable the construction engineer to "walk" through the project at any stage of the construction process.

Also, production management principles will be increasingly applied to architectural, engineering, and construction (AEC) systems. These will include new techniques in project management and delivery, including Scalable Enterprise Systems, and new directions in management that are expected to help create and maintain effective teams and auspicious environments. Virtual reality will be used as a tool for seamless integration of processes in the AEC industry. The behavior of construction products and processes will be monitored or optimized using simulation and analytical modeling techniques. Engineers will strive toward adopting concurrency in construction systems. Also, construction engineering systems will be stretched to higher limits due to increasingly bold civil and architectural material processing and designs. For example, three-dimensional "printing" using concrete or other materials is expected to enhance construction efficiency and overall sustainability of civil engineering systems. Construction engineers and managers also will seek to carry out evaluations of alternative contracting approaches or project delivery mechanisms on the basis of a wider range of criteria, such as the impacts on owners, users, and the community, in terms of initial cost, life-cycle cost, environmental sustainability, and economic development. Furthermore, innovations in spatial monitoring, such as global positioning systems and remote sensing, will be increasingly applied in construction supply chain management and project monitoring.

1.2.7 Geomatic Engineering

One of the oldest activities of civil engineering and to this day an indispensable aspect of civil engineering work (Mikhail, 2003), geomatic engineering is the science of accurate establishment of the position of points on the Earth's surface for purposes of establishing reference points and boundaries for natural or man-made objects. Thus, the field includes the design and layout of public infrastructure systems and mapping and control surveys for civil construction projects. The term "geomatics" incorporates the older discipline of surveying with newer spatial data collection and management sciences such as Geographic Information Systems (GIS), GPS, and related forms of Earth mapping. A formal definition of geomatic engineering is "a modern discipline which integrates acquisition, modeling, analysis, and management of spatially referenced data that uses the framework of geodesy to transform spatially-referenced data from different sources into common information systems with well-defined accuracy characteristics." The evolution of surveying to geomatics was spawned by the advancements in digital data processing, and thus the work scope of professional surveyors has transcended beyond those associated with surveying only. Other related and relatively new fields include hydrogeomatics, which evolved from hydrographics and represents the study of surveys of areas on, above, or below the surface of water bodies. Geodetics is the measurement and representation of the Earth in a three-dimensional time-variant space. Geodetic engineers also measure global geodynamical phenomena including the motion of the Earth's crust and movement of the poles.

Geomatic engineers integrate science and technology from both new and traditional disciplines, such as geodesy (or geodetic engineering), photogrammetry, cartography, remote sensing, GPS or Global Navigation Satellite Systems (GNSS), and GIS or geoinformatics, and computer-aided visualization. Geomatic engineers play a critical and continuing role in civil engineering systems development by collecting, archiving, and maintaining diverse spatial data on such systems. Over the past decade, advances in computer science and information technology, remote sensing technologies, and other disciplines have spawned significant advancements in geomatics. In response, a number of university departments and agency divisions that once had names containing the words "surveying," "survey engineering," or "topographic science" now have been renamed to include words such as "geomatics," "geomatic science," or "geomatic engineering." The equipment used for geomatic work has evolved from basic tools such as meniscus straws to compasses and calibrated chains, and then to robotic total stations that are fully computerized, equipped with multiple digital cameras, and capable of long-range laser scanning and intelligent recognition of scanned features (Figure 1.15). Other equipment in current use include LIDAR (light detection and ranging), an optical remote sensor that measures geomatic attributes of targets by illuminating the target with laser beams and interpreting the response.

The History of Geomatic Engineering. Ancient records show that surveying has always been an integral part of civil engineering systems development. In ancient Egypt, a land register existed circa 3000 BC, and the use of that document and simple geometry was critical in reestablishing property boundaries after each seasonal overflow of the Nile River, which washed out the physical markings of these boundaries. Devices used for these surveys included a right-angled triangle with a 3:4:5 side ratio. The skill of the ancient Egyptians in surveying is evidenced in the almost perfect dimensions, shapes, and north–south orientations of their pyramids, such as the Great Pyramid of Giza built circa 2700 BC. Other evidence of human long-standing dependence on surveying skills include Stonehenge (circa 2500 BC) where monuments were "set out" by that era's surveyors using rudimentary tools and techniques. In the BC–AD transition era in ancient Rome, the construction of aqueducts and other large structures was facilitated by the use of three kinds of

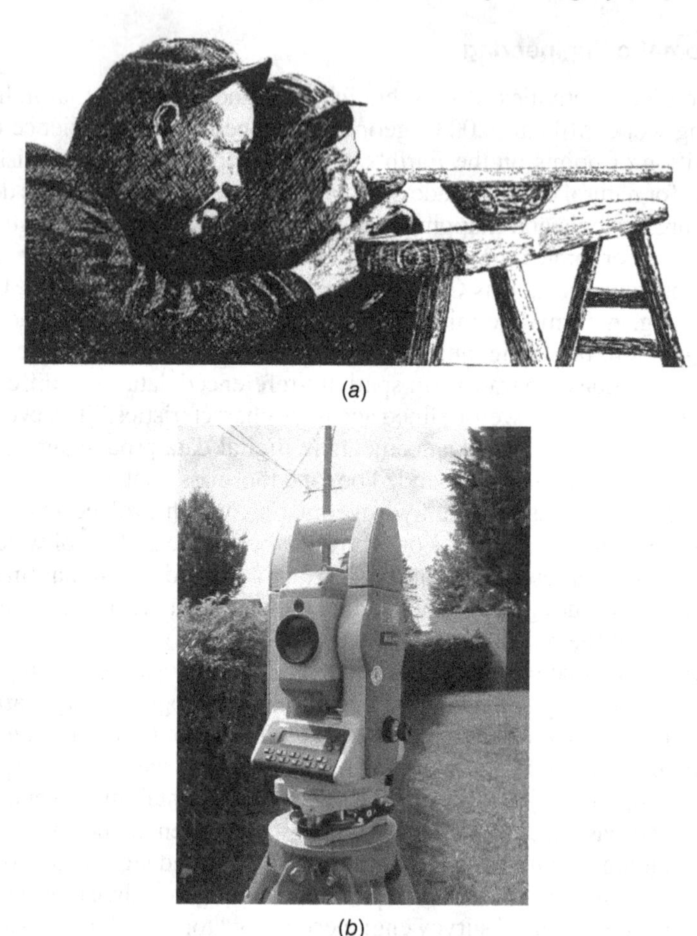

(a)

(b)

Figure 1.15 Basic survey instruments—the ancient and the new. (*a*) Floating sights water-level proto-theodolite, in ancient China. The bamboo tube floats on the convex meniscus of water in a rice bowl (*Source:* Needham et al., 2001). (*b*) Robotic Total Station (PhY/Wikimedia Commons).

surveying instruments (Figure 1.16): *dioptra*, a sighting tube with a sight at both ends and attached to a stand (when fitted with protractors, a dioptra could be used for angular measurements); the *groma*, which consists of "a vertical staff with horizontal cross-pieces mounted at right-angles on a bracket (each cross-piece had a plumb line hanging vertically at each end) and was used to survey straight lines and right-angles"; and the *chorobates*, a leveling tool that comprised a wooden beam fitted with a water level with supports at both ends (Lewis, 2001a).

Ancient Persian scholar Abu Rayhan al-Biruni of Kath (in present-day Uzbekistan) is regarded as the father of geodesy for his important theoretical contributions to the field in that era. In ancient Rome in AD 300, land surveyors, who enjoyed privileged professional status, established the basic dimensional measurements for purposes of administrative division of the empire. In 1086 in England, a surveying document was established by William the Conqueror to contain spatial as well as socioeconomic and physical features data about each land parcel. In continental Europe, the Cadastre, founded by Napoleon Bonaparte in 1808 and considered by that

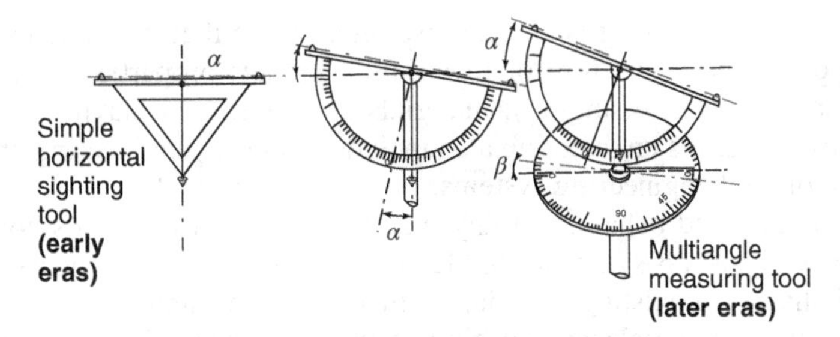

Figure 1.16 Form evolution of the dioptra, an ancient survey instrument (Nerijp/Wikimedia Commons).

ruler to be one of his greatest achievements in civil law, contained the land location, dimensions, value, ownership, and the like and used scales of 1:2500 and 1:1250.

Historically, equipment to measure distances included chains with links of a known length, and a compass was used to measure horizontal angles. Subsequent improvements included the use of carefully scribed disks for enhanced angular resolution, mounting telescopes with reticules precise sighting (such as theodolites), and the use of calibrated circles that allowed surveyors to measure vertical angles. For measuring height, surveyors have traditionally used the altimeter, a barometer that uses changes in air pressure as an indication of changes in vertical distance. Over the years, the need for greater accuracy led to the development of instruments that utilize a calibrated vertical measuring rod to provide a plane for measuring height differences between the instrument and the point in question. Modern instruments include the total station, essentially a theodolite fitted with an electronic device to measure distance. Total stations have evolved from optical-mechanical instruments to computerized and robotic equipment that are linked wirelessly to other offsite systems such as GPS, computers, and printers.

The Future of Geomatic Engineering. In the foreseeable future, the field of geomatic engineering will continue to undergo rapid changes due to technological developments in digital imaging, artificial intelligence, laser sensing, and global positioning systems and other technologies. These ongoing and other emerging technologies are expected not only to revolutionize regular surveying engineering tasks but also to impact a myriad of applications in several other fields of engineering, science, and the humanities where it is valuable to acquire information on near-real-time positioning of systems and phenomena (Mikhail, 2003). Also, the future is expected to see increased use of geomatic techniques for monitoring the stability of large civil structures in terms of their shape deformation arising from internal and external stressors.

1.2.8 Civil Materials Engineering

The choice of material for civil engineering systems construction is influenced by a variety of factors, including the initial and life-cycle maintenance cost, mechanical properties, durability, ease of construction, and aesthetics (Ho, 2003). To make informed decisions on material choices, the materials engineer typically solicits mechanistic or empirical data on the performance of the material in response to environmental factors that include usage and climate/weather. Thus, civil materials engineering involves the investigation of the properties of construction materials such as the raw ingredients for construction (e.g., cement, water, steel, aggregates, subgrade, subbase/base courses, etc.) and mixed products (e.g., asphaltic and Portland cement concrete, etc.) to ascertain their suitability or to recommend ways to enhance their properties for that purpose. The field of civil materials engineering is an interdisciplinary one that investigates the relationships between the composition and structure of materials and their properties. With the current explosion in nanoscience and nanotechnology, materials engineering is playing a more visible role in many institutions. Materials engineering also includes forensic engineering and thus covers the study of the failure of civil engineering systems.

The nature and behavior of any material are governed by its constituent elements and the manner in which it was synthesized. Materials engineers seek to understand the fundamental structure and behavior of existing materials with a view to expanding their uses and the development of new or enhanced materials with specific desired properties. They relate the atomic structure of that material to the properties and performance of the material when it is used in a given application.

Subfields of civil materials science or engineering include nanotechnology, which studies and develops materials at an atomic level; microtechnology, which includes the microfabrication of materials at micrometric level; crystallography, which studies the solid space filling behavior of atoms, the nature of crystal forms or structures, and the characterization of crystal forms as related to their performance and physical properties; materials characterization, which studies the properties of materials using equipment for spectroscopy, thermal analysis, chromatography, and electron microscope analysis. Other subfields include tribology, which is the study of material wear due to external agents such as friction, and surface science, which studies structures and interactions between material phase interface (solid–gas, solid–liquid, or solid–solid).

With regard to material types, subfields in materials engineering include metallurgy (the study of the extraction, processing, and modification of metals and their alloys), biomaterials (materials that are used in or derived from biological systems), and ceramography (which involves ceramic microstructures including transformation-toughened ceramics and polycrystalline silicon carbide).

Recent emerging applications of materials science, particularly of materials characterization, can be found in the area of civil engineering systems monitoring; namely, the deterioration of physical systems (e.g., corrosion of a steel bridge element, age-induced cracking of a pavement, rusting of sewer pipes, etc.) can be quickly detected by available techniques (because they involve a change in the material and hence the material's properties) before these structures suddenly fail and cause possible loss of life or property, or injury.

History of Civil Materials Engineering. Over the ages, the dominance of a new material used in a given era has often defined the progress of that era. For example, we have had the Stone Age, the Bronze Age, and the Steel Age. In ancient Greece and Rome, dry rocky soil yielded building materials that were durable. Harbors were built using large stones that were sunk under their own weight, and quay walls and smaller jetties were constructed of concrete comprised of broken stone, lime, and pozzuolana–a material that has survived the punishing marine environment for over 2000 years and can still be seen along the coasts of Campania, Latium, Pozzuoli, Fornia, and Anzio in Italy. In the Mesopotamia region, where there was little natural stone, engineers in the cities of Assyria and Babylonia depended on brick as the main building material. The ruins of the Tower of Babel, excavated in the early 20th century, revealed a core of unburned bricks surrounded by a shell of burnt brick. As a binding agent, bitumen was sometimes used instead of mortar (Straub, 1964).

Materials science, particularly the engineering of materials to yield new materials of desired physical properties, is one of the oldest forms of engineering and applied science and takes its roots from the manufacture of ceramics and in the last millennium, metallurgy. Indeed, the timeline of materials development include (TMS, 2012) the firing of ceramics (circa 28,000 BC) found at sites in the Pavlov Hills of Moravia; copper metallurgy (by hammering) for decoration by the Old World Neolithic peoples circa 8000 BC; extraction of copper from azurite and malachite and reshaping molten metal in Turkey circa 5000 BC; iron smelting in Egypt in 3500 BC; metal mixing in 3000 BC to produce bronze in Syria and Turkey; invention of glass in northwestern Iran in 2200 BC; production of porcelain in 1500 BC in China; and crucible steel making in India in 300 BC. In AD 400, iron smiths in Delhi, India, forged and erected a 20-ft-high iron pillar that has defied environmental degradation up to today. Other significant watermarks include the publication in the 1540–1600 period of *De La Pirotechnia* by Vannoccio Biringuccio, the first written account of proper foundry practices, *De Re Metallica* by Georgius Agricola, a description of mining and metallurgy practices in the 16th century, and *Della Scienza Mechanica* by Galileo, which scientifically analyzes the strength of materials. In 1755, John Smeaton invented hydraulic cement, thus introducing modern concrete, the dominant construction material of the modern age, and in 1805 Luigi Brugnatelli invented electroplating. In 1827, Wilhelm Albert developed iron wire rope, paving the way for large-scale construction involving steel cables. In 1864 Dmitri Mendeleev developed the Periodic Table of Elements which, to this day, serves as a reference tool for characterizing and identifying basic materials in engineering. This was followed by the invention of dynamite by Alfred Nobel in 1867, which facilitated large-scale civil engineering construction in rock terrain (TMS, 2012).

The field of materials science experienced a major breakthrough in the late 19th century when Willard Gibbs, an American theoretical physicist and chemist, established a relationship between the physical properties of a material and its thermodynamic properties in relation to its atomic structure in various phases. This finding laid the critical basis for understanding material behavior. In the last millennium, advancements in the field were spawned by the need to develop new materials for purposes of space exploration, which included metallic alloys, silica, and carbon materials that are typically used in constructing space vehicles. In the mid-20th century, many materials science departments and divisions in industry and academia were renamed metallurgy departments due to emphasis on metals. However, in recent years, many are reverting to the original name (materials science) because the field has broadened to include a broad array of material classes and types, such as ceramics, polymers, semiconductors, magnetic materials, and other innovative materials that are useful in civil engineering systems design and operations.

The Future of Materials Science and Engineering (MSE). In the near future, the study and application of materials in civil engineering is expected to include geosynthetics (geotextiles, geomembranes, and geogrids). Enhanced versions of these products will be used in embankments on soft foundations or to protect erosion-prone slopes (Holtz, 1991). Due to the adoption of intelligent materials and intelligent designs, there is expected to be an increasing number of energy-efficient buildings and other civil structures (Apelian, 2007). As case in point, Germany's Institute of Solar Energy Systems has developed a technique that uses a thin layer of material containing microencapsulated paraffin to carry out temperature equalization; when temperature inside the building rises above 24°C, the enclosed paraffin in the wall melts, leading to heat reduction in the room. Then at times of low temperature, the paraffin solidifies, releasing the stored heat, leading to energy savings and pollutant reduction. The future seems to be promising for discoveries in intelligent, green, and energy-efficient materials. Another example of future trends in this area is exemplified by roofing system applications such as the Teflon-coated fiberglass membrane roof that was used for the Riyadh International Stadium in Saudi Arabia and self-healing bioconcrete. Future world needs are projected to include recyclable or biodegradable materials. Environmental quality will be enhanced by the use of new biodegradable natural plastics for packaging of goods. Other similar materials including fiber-reinforced polymers (Figure 1.17) will see increased use due to their desirable engineering properties, low life-cycle cost, and contribution to sustainable development. As designers of structural systems demand less weight with greater strength, the focus will be on lightweight structural materials, specifically in the areas of alloys that can be stiffened to the extent needed). The properties of strength, ductility, weight,

Figure 1.17 Use of fiber-reinforced polymers can enhance sustainable design (Courtesy of EVOLO, LLC).

and recyclability, therefore, will continue to guide the development of new materials for future civil engineering systems.

1.2.9 Architectural Engineering Systems

This branch of civil engineering deals with the technological aspects of architectural structures and, therefore, includes study of the behavior and properties of building components and materials, environmental system design and analysis, and building operation. Architectural engineers strive to develop optimal design of buildings and building components and facilities within constraints that include physical space, material strengths, and cost; to achieve this objective, they use domain knowledge in civil and mechanical engineering including structural mechanics, materials science, geomatics, energy science and technology, acoustic science and systems-based tools including economic efficiency, multiple criteria optimization, and computer modeling, simulation, and visualization. Architectural engineers also pay close attention to issues of building resilience to external and internal threats and the sustainability of the buildings physical structure and operations.

The environmental aspect of building systems typically accounts for a dominant fraction of overall building operating cost and includes a wide range of areas, such as heating and air conditioning, lighting and acoustics, building power and energy systems, plumbing and piping, vertical and horizontal transportation, occupant safety, and fire protection. As such, architectural engineers are familiar with a great number of building codes covering these and other related areas. Architectural engineers deal with all phases of building systems development: planning, design, construction, operation, monitoring and inspection, maintenance, and renovation or decommissioning. At each of these phases, the architectural engineer engages in a variety of tasks, including selecting the best option on the basis of cost, occupant safety, environmental impacts, and other considerations.

In countries such as the United Kingdom, Canada, Australia, and some African and Asian countries, architectural engineering is more commonly known as building engineering, building services engineering, or building systems engineering. Architectural engineers work closely with architects and are conversant with architectural features that influence the building performance in terms of cost, energy efficiency, ventilation, and the like.

Many structures in civil engineering are being designed to minimize the use of resources by incorporating features that reduce the needs for energy, water, and lighting. This has been the case for buildings mostly but is also being gradually adopted in other civil structures. The Bahrain World Trade Center towers in Manama, Bahrain, is designed as the world's first skyscraper to incorporate wind turbines in its design; the sail-shaped buildings on either side are designed to funnel wind through the gap to guide wind to three wind turbines installed at three different levels of the building (Figure 1.18). Other notable examples of green buildings are the Centers for Disease Control and Prevention Laboratory Sciences Building in Atlanta, Georgia; Santa Monica's Z6 House (the 6 represents the goal of attaining zero levels of six factors: water, carbon, emissions, waste, energy, and ignorance); Colorado Court Affordable Housing Project, Chicago's Factor-10 House (which is said to consume only a tenth of the environmental resources consumed by an average home); the Lewis Center for Environmental Studies in Oberlin College, Ohio; the Solar Umbrella House in Venice, California; the Resource Center for the Homeless in Austin, Texas; the Wayne L. Morse U.S. Courthouse in Eugene, Oregon; Genzyme's headquarters in Cambridge, Massachusetts; and Toyota's headquarters in Torrance, California (Apelian, 2007; McGrath, 2012). We will acquire a

Figure 1.18 An example of future building systems concepts.

greater appreciation of sustainability in Chapter 28 where we will discuss the principles and benefits of sustainable development.

1.3 FINAL COMMENTS ON THE HISTORICAL EVOLUTION AND FUTURE OF CIVIL ENGINEERING SYSTEMS

A useful conclusion to this chapter would be to discuss the overall context of the evolution of civil engineering systems in terms of their relationships with socioeconomic systems, the different philosophies of design, the cumulative nature of knowledge in civil engineering systems development over the centuries, and future directions in general.

Civil engineering systems that were developed in early civilizations included public buildings, temples, fortifications, roads, irrigation canals, and water supply structures. Over the ages, these systems evolved in their design and construction features to enhance the quality of life of the people through the provision of shelter, water, sanitation, protection, and transportation of goods and services. In the current era, the situation is no different as society continues to depend on good physical infrastructure to enhance the quality of life. In fact, even with the advent of nonphysical infrastructure such as the Internet, there still exists a need, greater than ever before, for physical infrastructure to provide buildings, transportation, clean water, waste disposal and treatment, among others (Dandy et al., 2008).

For the design and construction of civil engineering systems across the civilizations and over the millennia, there existed two philosophies: those that were utilitarian in nature and those that were devotional (Straub, 1964). The lessons from these philosophies are important as civil engineers strive to incorporate sustainability considerations in their system designs. The Romans and Persians developed systems that were consistent with the utilitarian philosophy as they were mainly built for purposes of military strategy and commerce. On the other hand, the civil systems built by the Greeks were primarily of devotional value first and other values second. The Greek animistic conception of nature, which ascribed a living soul to mountains, rivers, and valleys, caused them to shy away from violent interference with natural land forms and obstacles. A parallel to this dichotomy can be found in the two rival moralities that guided the design and construction of hydraulic and other civil engineering systems in ancient China (Needham et al., 2001). The first morality was the Confucius philosophy (confining and repressing nature and thus advocating masculine activities such as dike construction), which was similar to the Roman and Persian approaches. The second was the Taoist philosophy (greater freedom for natural courses and thus advocating the use of feminine activities such as dredged concavities), which was similar to the Greek approach.

Also, the current state of civil engineering systems is the culmination of the collective efforts of several different civilizations at different locations all over the globe and over time. These civilizations, some to a greater extent than others, passed on their knowledge to successive ones through scholarly exchanges, trade, migration, documentation, or oral traditions. The development of civil infrastructure over the centuries began to be characterized by specialization, not only with respect to the type of civil system in question (e.g., transportation engineering, environmental engineering, structural engineering, etc.), but also with respect to the phase of facility development (e.g., construction engineering for the construction phase and traffic engineering for the operations phase).

As we move into the new millennium, the basic needs of society will remain largely unchanged, but the means by which engineers satisfy those needs will change enormously (Dandy et al., 2008). The advancement of civil engineering systems will continue to guide (and be guided by) changes in the socioeconomic and natural environments and will be catalyzed by rapidly expanding frontiers in science and information technology. Social and economic changes constantly create new demands on both engineers and the educational systems that produce them (Labi, 1997; ASCE, 2001). It is therefore important that engineers cultivate the ability to make informed choices, basing their judgments and decisions not only on the analysis of present situations but also on the vision of a preferred future (Berkovski and Gottschalk (1996). Along similar lines of thinking, the National Academy of Engineering, in its 2004 *Vision for the Engineer of 2020* (NAE, 2008), and the American Society of Civil Engineers, in its 2007 *Vision for Civil Engineering in 2025* (ASCE, 2007) and 2008 *Book of Knowledge* (ASCE, 2008), presented their visions of the preferred skill sets of future civil engineers. Other organizations duly recognize that issues that will become critical in the development of future civil engineering systems include sustainable decision making, interoperability between different sectors, climate change impacts, and smart infrastructure.

SUMMARY

Since the dawn of human existence, the economic and cultural prosperity of nations have been very closely linked with the level of technological advancement. In presenting the historical evolution of civil engineering, we have shown that the development of engineering infrastructure has led to, or has been the result of, the socioeconomic advancements of extraordinary civilizations dating as far back as the Sumerian civilization in Mesopotamia in 7000 BC. We also have shown that civil engineers, or persons acting in that capacity, have always executed their work from a systems perspective at least implicitly, and that these subbranches could be enhanced if systems approaches were incorporated in an explicit manner. Many of the basic concepts of systems analysis in civil engineering were discussed in this chapter and were shown to be consistent over time. Civil engineers have applied their knowledge to plan, design, build, maintain, or/and operate complex civil infrastructure systems that have served humankind in a variety of ways, including buildings for residential, commercial, and industrial purposes; facilities for transporting passengers and freight; pipe networks for supplying water; and facilities for waste disposal. For each branch of civil engineering, a brief history, description of typical tasks, and future directions was provided in this chapter, thus setting the stage for Chapter 2 which discusses the general systems perspective, including the phases of and tools for developing systems in any branch of civil engineering.

EXERCISES

1. Discuss any two definitions of engineering and identify the systems engineering concepts that are found in these definitions.
2. Is civil engineering both an art and a science? Explain.
3. N. W. Dougherty stated: "The ideal engineer is a composite.... He is not a scientist, he is not a mathematician, he is not a sociologist or a writer; but he may use the knowledge and techniques of any or all of these disciplines in solving engineering problems." Discuss.
4. According to R. E. Hellmund, engineering "brings about the utilization of the materials and laws of nature for the good of humanity." However, it can be a two-edged sword. Discuss how poor engineering practice could be harmful to society.
5. Discuss the sociological changes in prehistoric times that ultimately led to the need for civil engineering structures.

6. It is desired to extend an existing rail transit line to serve outlying areas of a large city. Identify the various types of civil engineers who likely would be involved over the life of the system.
7. For any one branch of civil engineering, in your own words, discuss the evolution of that branch over the millennia and how developments in other fields fostered advancements in that branch.
8. Discuss the differences between any two rival philosophies of civil engineering systems design, citing examples from past civilizations. Include a discussion of your preferred philosophy in the context of civil engineering systems resilience and sustainability.

REFERENCES

Abbot, M. B., Babovic, V. M., and Cunge, L. A. (2001). Towards the Hydraulics of the Hydroinformatics Era. *J. Hydr. Res.* 39, 339–349.

ABET (1993). *Definition of Engineering*. Accreditation Board for Engineering and Technology, Inc., Baltimore, MD.

Apelian, D. (2007). Looking Beyond the Last 50 Years: The Future of Materials Science and Engineering, *J. Minerals, Metals & Materials Soc.* February 2007 Issue, 65–73.

ASCE (2000). *Judgment and Innovation: The Heritage and Future of the Geotechnical Engineering Profession*. Geotechnical Special Publication No. 111, S. Francisco, and E. Kavazanjian, Jr., Eds., ASCE/Geo Institute, Reston, VA, p. 133.

ASCE (2001). *Engineering the Future of Civil Engineering*. Am. Soc. Civil Eng., Reston, VA.

ASCE (2007). *Vision for Civil Engineering in 2025*. Am. Soc. Civil Eng., Reston, VA.

ASCE (2009). *Brochure on ASCE's History and Heritage Program*. Am. Soc. Civil Eng., Reston, VA.

ASCE (2008). *Civil Engineering Body of Knowledge for the 21st Century*, 2nd ed. Am. Soc. Civil Eng., Reston, VA.

Atalay, B., and Wamsley, K. (2009). *Leonordo's Universe*, National Geographic.

Baker, M. N., and Chow, V. T. (1964). *Handbook of Applied Hydrology*. V. T. Chow, Ed. McGraw-Hill, New York.

Baker, M. N., and Taras, M. J. (1981). *The Quest for Pure Water-The History of the 20th Century*, American Water Works Association, Deriver, CO.

Benton, J. R., and DiYanni, R. (1998). *Arts and Culture: An Introduction to the Humanities*. Prentice-Hall, Upper Saddle River, NJ.

Berkovski, B., and Gottschalk, C. M. (1996). Strengthening Human Resources for the 21st Century: UNESCO Engineering Education and Training Program, *European Journal of Engineering Education* (21)1, 3–11.

Biswas, A. (1970). *History of Hydrology*. North Holland, London.

Buescher, C. (2000). *Water Supply and Wastewater Engineering, Environmental Engineer*. AWWA.

Carr, T. J. (2011). The Harappan Civilization. http://www.archaeologyonline.net.

CETS (1995). *Probabilistic Methods in Geotechnical Engineering*. Commission on Engineering and Technical Systems, The National Academies, Washington, DC.

Chanson, H. (2007). Hydraulic Engineering in the 21st Century: Where to? *J. Hyd. Res.* 45(3), 291–301.

Chrimes, M., and Bhogal, A. (2001). *Civil Engineering – A Brief History of the Profession: The Perspective of the Institution of Civil Engineers*, in *International Engineering History and Heritage – Improving Bridges to ASCE's 150th Anniversary*. ASCE, Reston, VA.

Dandy, G., Walker, D., Daniell, T., and Warner, R. (2008). *Planning and Design of Engineering Systems*, 2nd ed. Taylor and Francis, London, UK.

Darling, D. J. (2003). *The Complete Book of Spaceflight: From Apollo 1 to Zero Gravity*. Wiley, Hoboken, NJ.

Delleur, J. (2003). *Hydraulic Engineering, Handbook of Civil Engineering*. CRC Press, Boca Raton, FL.

Dugas, R. (1988). *A History of Mechanics*. Courier Dover, Mineola, NY.

EB (2011). *Encyclopedia Britannica*. Encyclopedia Britannica Inc., Chicago, IL.

Ellis, H. (1968). *The Pictorial Encyclopedia of Railways*. Hamlyn Publishing, Feltham, UK.

FitzSimons, N. (1996). Benjamin Wright, The Father of American Civil Engineering, in *Civil Engineering History—Engineers Make History*, J. R. Rogers, D. Kennon, R. T. Jaske, F. E. Griggs, Eds. Procs., 1st National Symposium on Civil Engineering History, ASCE, Reston, VA.

Fricker, J. D., and Whitford, R. K. (2005). *Fundamentals of Transportation Engineering*. Prentice Hall, Upper Saddle River, NJ.

German, S. (2012). Ziggurat of Ur, Khanacedemy. smarthistory.khanacademy.org. Accessed Dec. 15, 2012.

GulfNews (GN) (2010). Burj Khalifa: Towering Challenge for Builders. GulfNews.com. 4 January 2010. Retrieved Feb. 10, 2010.

Harr, M. (2004). *Geotechnical Engineering, Civil Engineering Handbook*, 2nd Ed., CRC Press, Boca Raton, FL.

Heyman, J. (1998). *Structural Analysis: A Historical Approach*, Cambridge University Press, Cambridge, UK.

Hjorth, P., Kobus, H., Natchnebel, H. P., Nottage, A., Roberts, R. (1991). Relating hydraulics and ecological processes, in Hydraulics and the Environmental Partnership in Sustainable Development Issue, *J. Hydraulic Res.* 29, 8–19.

Ho, D. W. S. (2003). *Materials Engineering, Handbook of Civil Engineering*. CRC Press, Boca Raton, FL.

Holtz, R. (1991). Geosynthetics in Civil Engineering. In Use of Geosynthetics in Dams, 11th Annual USCOLD Lecture Series, US Commission in Large Dams, Plain Fields NY, pp. 1–19.

Holtz, R., and Kovacs, W. (1981). *An Introduction to Geotechnical Engineering*. Prentice-Hall, Englewood Cliffs, NJ.

ICE (2007). Institution of Civil Engineers' website. Accessed December 26, 2007.

Jacko, R. B. (2003). *Environmental Engineering, Civil Engineering Handbook*. CRC Press, Boca Raton, FL.

Khisty, J. C., and Mohammadi, J. (2001). *Fundamentals of Systems Engineering*. Prentice Hall, Upper Saddle River, NJ.

Kirby, R. S. (1990). *Engineering in History*. Dover, New York.

Kirby, R. S., Withington, S., Darling, A. B., and Kilgour, K. G. (1956). *Engineering in History*. McGraw-Hill, New York.

Kramer, S. N. (1963). *The Sumerians. Their History, Culture, and Character*. University of Chicago Press, Chicago.

Labi, S. (1997). *Vision of a Preferred Future*, in Imprints, College of Engineering Annual Report, Purdue University, West Lafayette, IN.

Lanciani, R., [1967 (first published in 1897)]. *The Ruins of Ancient Rome*. Benjamin Blom, New York.

Landels, J. G. (1978). *Engineering in the Ancient World*. UCLA Press, Los Angeles.

Leinhard, J. (2000). *The Engines of Our Ingenuity*. Oxford University Press, Oxford, U.K.

Lenkei, P. (2007). Climate Change and Structural Engineering. *Periodica Polytechnica, Civil Engineering* 51/2.

Leonard, K. M. (2001). Environmental Engineering, The World's Second Oldest Profession, in *International* Engineering History and Heritage—Improving Bridges to ASCE's 150th Anniversary, Procs., 3rd National Congress on Civil Engineering History and Heritage, Oct. 10–13, Houston, TX.

Lewis, M. J. T. (2001a). *Surveying Instruments of Greece and Rome*. Cambridge University Press, Cambridge, UK.

Lewis, M. J. T. (2001b). *Railways in the Greek and Roman World, Early Railways*. A Selection of Papers from the 1st International Early Railways Conference, A. Guy, J. and Rees , Eds., pp. 8–19.

Long, M., and Labi, S. (2011). Assessing the Vulnerability of Civil Engineering Structures to Long-term Climate Change, Conference on Vulnerability and Risk Analysis, University of Maryland, College Park, MD.

Liew, J. Y. R., and Shanmugan, N. E. (2004). *Theory and Analysis of Structures, Civil Engineering Handbook*, 2nd Ed.CRC Press, Boca Raton, FL.

Lyn, D. A. (2004). Fundamentals of Hydraulics, Civil Engineering Handbook, CRC Press, Boca Raton, FL.

Marhaba, T. (2000). *Crystal Clear: History and Issues in Drinking Water Disinfection, Environmental Protection*, Dollas, TX.

Magnusson, J. (2007). A Beautiful Tomorrow for Structural Engineering? *The Compass* 42(2), Oct. Issue.

McGrath, J. (2012). Sustainable Buildings. http://dsc.discovery.com/tv-shows/curiosity/topics/10-sustainable-buildings.htm. Accessed Dec. 15.

MetArt (2009). *Great Zimbabwe (11th–15th Century), Heilbrunn Timeline of Art History*. Metropolitan, Museum of Art, New York.

Moncur, M. (2012). The Quotations Page. www.quotationspage.com. Accessed Dec. 15.

Mikhail, E. W. (2003). *Surveying Engineering, Handbook of Civil Engineering*. CRC Press, Boca Raton, FL.

Museum of Art (2009). http://www.metmuseum.org. Accessed Nov. 10, 2009.

NAE (2008). *The Engineer of 2020: Visions of Engineering in the New Century*. National Academies Press, Washington, DC.

NASA (2002). *Celebrating a Century of Flight*. NASA Publication SP-2002-09-511-HQ, Washington, DC.

Nedwell, P. J., Swamy, R. N., and Nedwell, P. (1994). Ferrocement: Proceedings of the Fifth International Symposium on Ferrocement UMIST, Manchester, UK, Sept. 6–9.

Needham, J. (1986). *Science and Civilization in China*, Vol. 4, Part 3 Cambridge University Press, Cambridge, UK.

Needham, J., Gwei-Djen, L., and Wang, L. (2001). *Science and Civilization in China*, Vol. 4, *Physics and Physical Technology*, Part 3: *Civil Engineering and Nautics*, Cambridge University Press, Cambridge, UK.

Ochsendorf, J. A. (2005). Sustainable Engineering: The Future of Structural Design, Procs., 2005 Structures Congress & Forensic Engineering Symp.: Metropolis and Beyond, April 20–24, New York.

Osorio, N. L., and Osorio, M. A. (2006). History of Engineering, in *Using the Engineering Literature*, B. A. Osif, Ed. CRC Press, Boca Raton, FL.

Pannell, J. P. M. (1964). *Man the Builder—An Illustrated History of Engineering*. Jarrold and Sons, Norwich, UK.

Parkhill, S. T. (1998). 150 Years of Underground Design and Construction, in Engineering History and Heritage: Procs., 2nd National Congress on Civil Engineering History and Heritage, J. R. Rogers, Ed. Oct. 17–21, Boston, MA.

Petroski, H. (2001). *The Importance of Engineering History*. International Engineering History and Heritage, 1–7.

Plofker, K. (2007). *Mathematics in India. The Mathematics of Egypt, Mesopotamia, China, India, and Islam: A Sourcebook*. Princeton University Press, Princeton, NJ.

PMB (2008). The History of Plumbing. *Plumbing and Mechanical Magazine*, www.theplumber.com/history.html.

Prentice, J. E. (1990). *Geology of Construction Materials (Topics in the Earth Sciences)*. Chapman & Hall, 171, London, UK.

Prince, J. D. (1904). Review: The Code of Hammurabi. *Am. J. Theol.* 8(3), 3601–609.

Raju, V. R. (2010). *Ground Improvement Technologies and Case Histories*. Research Publishing Services, Singapore. Procs., 2005 Structures Congress and Forensic Engineering Symposium, Cambridge, MA.

Rao, A. R. (2002). *Surface Water Hydrology, New Directions in Civil Engineering, The Civil Engineering-Handbook*, 2nd ed., W. F. Chen and J. Y. Richard Liew, Eds. CRC Press, Boca Raton, FL.

Rayburn, W. (1989). *Flushed with Pride*. Pavilion, London.

Saouma, V. E. (2007). Lecture Notes in Structural Engineering, University of Colorado. Retrieved Nov. 2.

Shuter, J. (2008). *The Sumerians*, 2nd ed. Heinemann-Raintree, Chicago, IL.

Sinha, K. C. (2003). *Transportation Engineering, Handbook of Civil Engineering*. CRC Press, Boca Raton, FL.

Smoothwhirl. (2009). History of Civil Engineering. http://www.brighthub.com/engineering/civil/articles.

Straub, H. (1964). *A History of Civil Engineering—An Outline from Ancient to Modern Times*. MIT Press, Cambridge, MA.

Taras, M. J. and Baker, M. N. (1981). *The Quest for Pure Water: The History of the Twentieth Century*, Vol. 1. American Water Works Association, Denver, CO.

Terzaghi, K., Peck, R. B., and Mesri, G. (1996). *Soil Mechanics in Engineering Practice*, 3rd ed. Wiley, New York.

TMS (2012). The Minerals, Metals & Materials Society. www.materialmoments.org. Accessed Dec 15.

Voland, G. (1999). *Engineering by Design*. Prentice Hall, Upper Saddle River, NJ.

White, B. (2008). *Gaither's Dictionary of Scientific Quotations*. Springer, New York, NY.

USEFUL RESOURCES

Levy, M. (2000). *Engineering the City: How Infrastructure Works*, Chicago Review Press, Chicago, IL.

McCuen, R. H., Ezell, E. Z., and Wong, M. K. (2011). *Fundamentals of Civil Engineering*: An *Introduction to the ASCE Body of Knowledge*, CRC Press, Boca Raton, FL.

National Academy of Engineering (2004). *The Engineer of 2020: Visions of Engineering in the New Century*, National Academies Press, Washington, D.C.

Oakes, W. C., Leone, L. L., and Gunn, C. J. (2011). *Engineering Your Future: A Comprehensive Introduction to Engineering*, Oxford University Press, Oxford, UK.

Walesh, S. G. (2000). *Engineering Your Future*, ASCE Press, Reston, VA.

Wood, D. M. (2012). *Civil Engineering: A Very Short Introduction*, Oxford University Press, New York.

CHAPTER 2

FUNDAMENTAL CONCEPTS IN SYSTEMS ENGINEERING

2.0 INTRODUCTION

As we learned in Chapter 1, the civil engineering discipline is founded on a rich and solid heritage that has shaped our current technical knowledge and practices in the profession. The evolution and transformation of the discipline, over the ages, were driven largely by social and economic necessities and developments, as well as advances in other fields and disciplines related to civil engineering. Indications are that ongoing and anticipated social and economic patterns will generate even greater demand for new or improved existing civil engineering systems in the face of diminishing natural resources and funding uncertainty for systems construction, replacement, renewal, and operations. Also, continuing advances in materials science, information technology, operations research techniques, and other fields continue to offer opportunities for efficient resolution of some of these challenges. Thus, in this new millennium, civil engineers are expected to continue drawing upon the knowledge from other fields and to leverage this knowledge to enhance the development of civil engineering systems. The field of systems engineering holds such promise. As researchers and scholars have pointed out in other texts and publications, tremendous opportunities to incorporate "systems" approaches in the development of future civil engineering systems will continue to exist. It is expected that this will be done in a manner that is explicit, compared to the recent past where allusions to (and management of) such facilities as bona fide systems have been only implicit or half-hearted.

In this chapter, we present the fundamental concepts of systems thinking including system definitions, classifications, and attributes. We will tie this discussion to the practice of civil engineering by identifying the eight major development phases of civil engineering facilities, the tasks encountered by engineers at each phase, and the tools needed to carry out those tasks. By the end of this chapter, we will have set the stage for detailed treatment of the phases of civil systems development, and the tasks and tools at each phase, in the remaining chapters of this book.

2.1 WHAT IS A SYSTEM?

We use the word "system" often, perhaps every day, and may often gripe about how some "system" from which we received poor service is not working properly. Systems we encounter in our day-to-day lives include grading systems, online purchasing systems, antilock braking systems, and credit rating systems. In civil engineering, there are numerous examples of systems: environmental remediation systems, arterial traffic control systems, and corrosion control systems, to name a few. Some systems are physical (such as a trickling filter system), others are abstract (such as a traffic signal timing system), and others yet are a combination of physical and abstract subsystems or components. The etymology of the word "system" has its roots in the ancient Latin and Greek languages: from Latin, the words "systemat"- or "systema", and from Greek, the words "systEmat" and "systEma" evolved into the words "synistanai" (to combine) and "syn-" + "histanai" (to cause to stand) (Merriam-Webster, 2011). A system may be defined as "a collection of regularly interacting or interdependent interrelated objects and/or rules, real or abstract, that collectively respond to some external action or serve a certain function." Other definitions have been offered by systems scholars as follows:

- A set of components that work together for the overall objective of the whole (Churchman, 1968).
- A selection of elements, relationships, and procedures to achieve a specific purpose (Wortman and Luthans, 1969).
- A set of objects together with relationships between the objects and their attributes (Weinberg, 1975).
- A number of interconnected components, each of which may serve a different function but all of which are intended for a common purpose (Au et al., 1972).
- An integrated set of operable elements, each with explicitly defined and bounded capabilities, working synergistically to enable a user to satisfy mission-oriented operations needed in a prescribed operating environment with a specified outcome and probability of success (Wasson, 2008).
- A construct of different elements that, when combined, produce results not achievable by the elements acting individually. The value that is intrinsic in the system in its entirety, beyond that contributed independently by its constituent elements, is attributable to the interrelationships between and among the elements (Maier and Rechtin, 2000; INCOSE, 2009).
- A collection of interrelated and interacting components that work together in an organized manner to fulfill a specific purpose or function (Dandy et al., 2008).
- A group of interacting, interrelated, or interdependent elements forming a complex whole (Collins, 2009).

Other definitions of a system have been established by organizations such as the Institute of Electrical and Electronic Engineers (IEEE), the International Institute for Applied Systems Analysis (IIASA), the Research and Development Corporation (RAND), the International Standards Organization (ISO), and the U.S. National Aeronautics and Space Administration (NASA).

As the reader may have observed, there are some common threads that appear in each of these definitions: purpose, role of components, integration of components, working together, boundary conditions, and stochastic nature of inputs and outcomes. In the paragraphs below, we will examine each of these terms more closely, show their relationships to civil engineering facilities, and identify areas of this text where we will encounter further discussion of such relationships.

Purpose. The words "purpose," "output," or "outcome" that recur in the definitions above suggests that every system has a *raison-d'être* (a reason for its existence). The reason for a system's existence often represents at least one benefit to the system owner, the system user, or the community. No system exists for its own sake. As we will demonstrate in Chapter 3, every civil engineering facility has a specific purpose, and the reason for its existence arises from the values, concerns, and wishes of its stakeholders, which in turn translate into more specific statements such as the core mission of the system owner or operator, as well as the goals, objectives, and ultimately performance measures for the system.

Role of Components. Each component of a system should have explicitly specified and bounded capabilities and should be expected to play a role toward the accomplishment of some overall goal, thereby making it possible to analyze, design, fabricate, test, verify, and validate each component, either as an individual, stand-alone entity, or as a part of an integrated system (Wasson, 2008). In any branch of civil engineering, each facility consists of components, some physical, others abstract. For example, a bridge consists of a superstructure, substructure, and deck; the condition and operational performance of each component can be evaluated separately (FHWA, 1995).

Integration of Components. This description refers to the assemblage of the components (physical or abstract, or both) of a system and is characterized by the hierarchies and synergies among these components. Certain descriptions of "integration" go further to state that the components must also be interoperable, which means that they must be compatible with each other in a desired manner, such as their form, fit, and/or function (the 3F's, which we will discuss in Section 2.1.1). As implied in the system definitions in the preceding section, the integration of system components enables the leverage of the capabilities of individual components to accomplish an objective that cannot be achieved by the individual components acting independently. In other words, the overall capabilities of the sum of the components exceed the sum of the capabilities of the components, a basic pillar of systems thinking known as "holism." This is mentioned briefly in the next paragraph.

Working Together, Synergy, and Holism. The physical components or abstract rules of a system alone are not adequate for it to function. For example, consider the sentence: "The chicken crosses the road." Each word in the sentence does not mean much by itself, but there is a meaning when all the words are arranged together in a certain way. Similarly, in an engineering system, each component responds to external stimulus in a manner that is unique due to the inherent characteristics of that component. However, the actual response of a component will depend on the presence of other components and the nature of its relationships with those other components. The physical components may include people, hardware, software, and structural elements of a facility. The abstract components may include policies, laws, rules, and algorithms. An important aspect of working together is that the outcome of the entire system is greater than the sum of the outcomes of the individual components of the system. In Chapter 21, we discuss these issues as part of our discussion on the design phase of systems development.

Boundary Conditions. The mention of system environment (surroundings) in some of the definitions of a system in the preceding section suggests that there is a boundary that separates the system from its environment. In other contexts of systems engineering, that deal with closed systems, the word "boundary" is explicitly stated and defined as the interface between the system and its environment. This is of great interest to civil systems engineers, particularly where the environment can pose a sudden or gradually-evolving threat to the system's operations or survival and also where the system's unintended outcomes (such as air pollution) can pose a threat to the environment. In Chapter 27, we will discuss the threats from the environment that may impair system operations or cause end of life, and in Chapter 28, we will examine the issues of environment sustainability in connection with civil system development.

Stochastic Nature of Inputs and Outcomes. For planning and monitoring purposes, engineers desire that the outcome of a system in terms of its objectives is known with certainty. However, as Wasson's definition points out, every system in reality has a "probability of success," which means that the attainment of the system goal or outcome is always accompanied by some degree of uncertainty. In other words, every specific level of the outcome has a certain likelihood of occurring. In Chapter 5 we will review the concepts of probability that will help us quantify these likelihoods; in other chapters of Part 3, we will examine how probability concepts are used in analytical tools

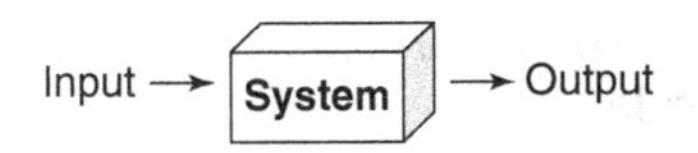

Figure 2.1 Basic representation of a system.

(including modeling, simulation, decision analysis, multiple criteria analysis, reliability and risk analyses, and real options) to address specific tasks at the various phases of civil systems development; and in Part 4, we will see how these probability-related tools could be used to enhance robust decision making at the various phases of development.

From these definitions and our discussion of the key terms they contain, it is possible for us to visualize a system as an integrated collection of components that work together within given boundaries to accept some input and to produce some outputs that are often intended to achieve a certain goal (Figure 2.1).

As we have already discussed in earlier paragraphs of this chapter, there are many kinds of systems that are part of our everyday lives. For example, the human body can be considered a system that consists of various organs (system components) that work together to achieve various overall goals (moving from place to place, learning and working for a living, and staying alive). Other examples include a computer hardware system that consists of a central processing unit, keyboard, and monitor and a computer software system such as a disk operating system that comprises a set of sequential rules and algorithms that govern the way the computer operates. In the field of civil engineering, examples of systems include a structural bracing system for a skyscraper, a water purification system, a foundation support system, a water drainage system, and a formwork system. Obviously, each of these systems can be broken down into smaller component parts or may be themselves a part of larger systems. Clearly, not all systems are physical. Some systems are only abstract, such as the system your school uses to calculate your GPA or the system your bank uses to approve your loan. Furthermore, there are systems that have both physical and abstract components.

Figure 2.2 is a detailed representation of a system that shows the features of its environment. These include the entities that have a stake in the successful operations of the system, intended inputs (resources), unintended inputs (threats and opportunities that may be natural or man-made), default inputs (the goals and objectives of the system), and the outputs of the system. Other features include the constraints to the system's existence or to its operations (including physical space constraints, restrictions on the use of resources, quality of outcomes), which may be intended or

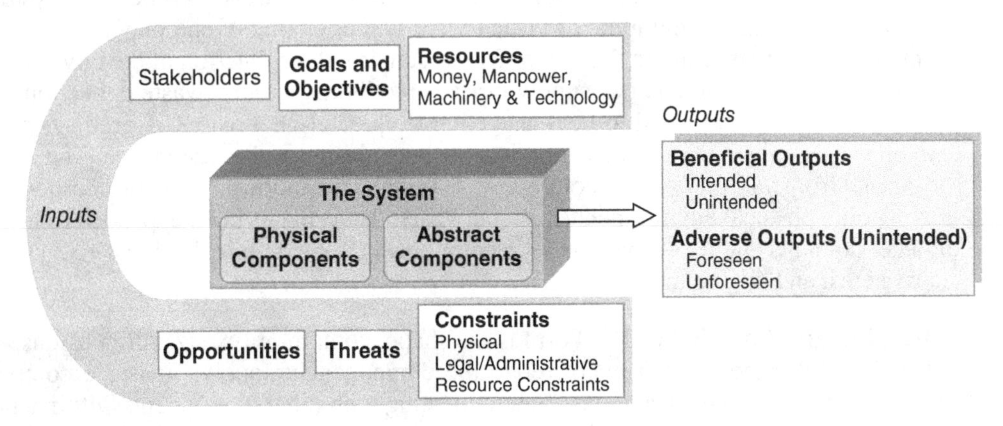

Figure 2.2 A more detailed representation of a system showing features of its environment.

unintended depending on the problem context. Like the inputs, some of the outputs may be intended and others may be unintended.

2.1.1 Some Relevant Terminology

Systems Engineering. Systems engineering is the application of principles from multiple disciplines (such as mathematics, science, and business) to formulate, select, and develop solutions at any phase of a system's life cycle, with the intention of satisfying the given objectives and constraints posed by the system owner/operator, user, community, and other stakeholders in a cost-effective manner.

Process. This term relates closely to the term "system". However, the two terms should not be confused as process necessarily involves the concept of time. A process is defined as "a sequence of interrelated activities that proceed in time" (Dandy et al., 2008). When these activities and sequences are planned, designed, and implemented, the process is described as "man-made," and examples include electoral processes, water treatment processes, and manufacturing processes. On the other hand, when the process is not designed but is a natural outcome of external forces, such as the corrosion of a steel bridge or the deformation of a structural member, it is described as a natural process. Most engineering systems consist of physical abstract components and processes; in such systems, the process is generated through the physical system. For example, for freeway system, the pavements and bridges are physical components, the rules of traffic operation such as the speed limits and lane-use restrictions are abstract components, and the progressive potholing or cracking of the surface asphalt over time or the growth of congestion over time is a process. Often, in studying an existing system or in developing one, there is a need to conceptualize the system by replicating it in order to analyze its current or expected future structure or behavior; this is accomplished with models as we will discuss in Chapters 7 and 8.

Control System. A control system is real or abstract system that has the capability to monitor the operations of a larger system and to alter the operating conditions by making changes to the input factors; for example, a freeway mobility system involves components that detect freeway incidents through monitoring sensors and dispatch tow trucks to the affected location to clear the incident and also provides in-vehicle notification for other freeway users to choose alternative routes to avoid the incident site. Control systems may be run by humans or may be automated.

Ecosystem. An ecosystem describes the relationships and interactions between systems of different types in the same geographic or virtual vicinity. For example, one may refer to the ecosystem of municipal infrastructure in the city of Rio de Janeiro: the collection of the city's streets, bridges, electrical lines and stations, gas supply pipes, streetlights, sewers, waste collection facilities, and other systems that may need to be managed not separately but as one overall system, in order to exploit synergies and to minimize conflicts between design and maintenance. The term ecosystem originated from the context for a community of organisms (biotic and abiotic components) and their surrounding physical environment. In recent years, the term has been expanded to include systems in other domains that may have no biotic components, for example, business ecosystems, research ecosystems, and other similar systems.

Form, Fit, and Function (F3). Form is the shape, configuration, size, dimensions, mass, and/or other visual parameters that characterize the component uniquely; fit is the compatibility of a system component in interfacing or interconnecting with other components within a prescribed set of temporal and spatial limits with ease and without undue adverse effects; function is an operation, activity, process, or action performed by a system component to achieve a specific objective within a prescribed set of performance limits (Wasson, 2008). F3 is an important consideration particularly at the design phase of systems development (Chapter 21).

Cybernetics. Taken from its Greek meaning, "to steer" or "to navigate," cybernetics was originally defined as the scientific study of control and communication between humans and machines (Wiener, 1948); in contemporary literature, the term is more broadly defined as a branch of mathematics dealing with problems of control, recursiveness, and information, that focuses on forms and the patterns that connect (Richards, 2008). Thus, cybernetic considerations in systems analysis is particularly relevant in contexts where the system's outcomes generate a change in its environment and then that change in turn affects some aspect of the system. In Chapters 27 and 28, we discuss the issues of system resilience and sustainability, which are expected to be key cybernetic considerations in studying or managing civil engineering systems.

2.1.2 Hierarchy of Systems

Between the two diametric extremes of the structure of an atom and the structure of the universe, there exists a multitude of intermediate physical systems. Therefore, every system can be considered not only as a **subsystem** (a smaller system of another, larger system) but also as a **supersystem** (a larger system that comprises smaller systems). For example, the reinforced concrete slab on which you might be currently sitting is a subsystem because it is a part of other structural systems that constitute the building in which you are currently located, such as the foundation system, the column support system, and the roofing system. However, it is also a supersystem because it is comprised of smaller systems or subsystems such as the reinforcement system and the aggregate-additives mix system. The biological water treatment process is a subsystem of the overall water treatment system but is itself a supersystem that is comprised of smaller systems of filtration and aeration. Another example is the municipal system of a city, which consists of several subsystems such as the transportation system, the drainage system, and the water supply system. Each of these subsystems, in turn, is comprised of smaller constituent systems; for example, the transportation system consists of the pavement systems, bridge systems, and traffic signal systems. This nature of system hierarchy is also true for abstract systems. In many cases, the term subsystem is a synonym for **component** (Dandy et al., 2008). Interestingly, it is this hierarchical nature of systems that drives one of the key tenets of the systems approach; in order to analyze a system, you must first break it down into its component parts and study not only each component individually but also the interactions between them.

2.1.3 System of Systems (SOS)

The intrinsic hierarchy of any man-made or natural system gives rise to the term **system of systems** (SOS), which can be defined as an integrated collection of individual systems to yield a new and more complex metasystem that has greater performance compared to the sum of performance of the individual systems that comprise the SOS. In other words, the individual systems that constitute an SOS exhibit different behavior when they operate independently; however, working together as a supersystem, their interactions deliver important collective emergent properties. The typical characteristics of system of systems include (Maier, 1998; Boardman et al., 2006; DeLaurentis, 2007): managerial and operational independence of elements, evolutionary development, and emergent behavior. Heterogeneity of component systems, and networking of component systems. Levis et al. (2004) defined an SOS as one that has the following properties:

- The component systems achieve well-substantiated purposes in their own right, even if detached from the overall system.

- The component systems are managed in large part for their own purposes rather than the purposes of the whole.
- The system exhibits behavior, including emergent behavior, not achievable by the component systems acting independently.
- A component's systems, functions, and behaviors may be added or removed during its use.

On the basis of the above properties, most systems in civil engineering (e.g., transportation networks, water treatment plants, water and wastewater distribution networks, construction management schedules, geotechnical stabilization systems, structural frames, and hydroelectric systems) can be described as SOSs.

In general, SOS engineering may be defined as the application of requisite tools in multiple disciplines, such as engineering, social sciences, and business, to analyze or to make decisions at any phase of development. SOS is an emerging discipline and is seeing rapid growth in its theoretical problem setups and applications. At a strategic level, researchers are currently establishing an effective frame of reference for the field and developing a set of unified terminology and SOS architecture. At a tactical or operational level, researchers are studying various techniques for SOS analysis, modeling, and simulation, including probabilistic modeling and design, agent-based modeling, object-oriented simulation and programming, and probabilistic multiple objective optimization. Other researchers are developing tools for numerical and graphical visualization of system behavior. The notion that any system can be viewed within an SOS context was driven primarily by needs in the defense industry. However, the great potential for SOS context application in civil engineering and other engineering disciplines, and indeed other sectors (e.g., health, energy, business, Internet communication, etc.) are evident. SOSs generally exhibit behaviors that are consistent with complex systems; however, not all complex systems are SOSs. Section 2.4.5 provides further discussion of the SOS concept.

2.1.4 Classification of Systems

Systems can be classified in many ways, according to, for example, whether the system is physical or abstract, living or nonliving, natural or human-made, complex or simple. In this section, we discuss some of these classifications.

Classification by Discipline. Systems can be classified by the discipline in which they are applied, for example, engineering systems, political systems, social systems, cultural systems, or biological systems. Within engineering systems, there are electrical systems, civil systems, and mechanical systems, for example; and within civil systems, there are structural systems, transportation systems, and hydraulic systems, to name a few.

Supersystem versus Subsystem (Hierarchical Classification). A civil engineering system may be a subsystem (part of a larger system) or may be comprised of smaller subsystems. In some literature, this dichotomy may be described as high-level systems and low-level systems. High-level systems typically exist at a macrolevel and comprise a combination of subsystems that are physical or abstract or a combination of the two.

Physical versus Abstract. A civil engineering system may be a physical facility (e.g., a bridge, a highway pavement, a sedimentation pond, etc.) or may be abstract (e.g., a set of rules, processes, or algorithms that govern or monitor the physical system condition or performance, such as a system for receiving and cataloging system user complaints.

Table 2.1 Boulding's Classification of Systems

Hierarchical Level	Description	Defining Characteristics	Example
1	Structural systems	Static, spatial frameworks	Atom, crystal
2	Clockwork systems	Predetermined motion	Solar system, clock
3	Control systems	Closed-loop control mechanisms	Thermostat
4	Open systems	Structural self-maintaining	Cell
5	Genetic systems	Collection of cells	Plant
6	Animals	Self-preservation, survival	Mammal, bird
7	Humans	Self-consciousness	Human being
8	Sociocultural systems	Roles, values	Family, clan, tribe, community
9	Transcendental systems	Beyond human knowledge	Religion

Adapted from Khisty et al. (2012).

Boulding's Classification. One of the earliest classifications was by Boulding (1956), who developed a hierarchy on the basis of system complexity (Table 2.1). Each level is upper bound in the sense that it includes, to some extent, the lower levels. Even though the literature contains new classifications schemes for systems, Boulding's classification continues to serve as a classic representation of the lines drawn between what is known theoretically, what is known empirically, and what is not known at all.

Complex (Soft) versus Simple (Hard). Simple systems consist of relatively few elements or subsystems whose interactions are predetermined, organized, and immune from external stimuli; as such, the performance and other attributes of such systems can be reliably predicted using systemic approaches. Complex systems typically consist of a large number of elements or subsystems whose attributes and interactions have a significant random component, are vulnerable to external stimuli, and are influenced by spatial and temporal variations (Khisty et al., 2012). In Section 2.1.5, we will discuss system complexity further in the context of system hardness or softness.

Dynamic versus Static. A civil engineering system may be dynamic (the attributes of the physical or abstract system change with time) or static (little or no temporal variation in the system attributes). An example of the former is urban arterial systems where traffic speeds, flow, and density change constantly by the time of day.

Network versus Nonnetwork. A civil engineering system may have a physical or abstract configuration that is a network (a collection of nodes and links) or is not a network. An example of a physical network is a highway network, and an example of an abstract network is a construction schedule. In Chapter 17, we will discuss network analysis as a valuable tool for making decisions related to systems that have a network configuration.

Solitary versus Conjoint. A solitary system is one with zero interactions with other systems located outside its boundary. Two or more conjoint systems are systems that interact across their boundaries; these systems may be concurrent (operating simultaneously) or concomitant (one operating after, or in response to the other) or both.

Closed versus Open. A closed system is one that does not interact with its environment and thus restricts the transfer of energy or matter across its boundary. Unlike the case for open systems, changes in the environment of a closed system and the adaptability and resilience of these systems to such stimuli are of very little concern. An open system, on the other hand, has at least one interface with its environment and allows transfer of matter, energy, or both, across its boundary to or from the surrounding environment. Examples of open systems include biological and business ecosystems, and civil engineering systems.

Finally, it can be noted that while the development of a civil engineering system comprises multiple phases including design, construction, and operations, the conduction of work at each phase may involve the use of specific abstract or physical systems that carry out specific tasks at that phase. For example, we can have a design review system, a formwork selection system, an automated system for facility inspection, or a system for selecting maintenance treatments at the various phases of system development.

2.1.5 Further Discussion of System Complexity

Simple systems, also known as "hard" systems, consist of relatively few components or subsystems. The attributes of such systems (i.e., their physical structure, condition, operational characteristics, and performance) as well as any interactions between its elements or subsystems are predetermined, organized, and immune from external stimuli such as the system environment. As such, the attributes of such systems are readily observable or predictable. Complex systems (also referred to as "soft" systems) are the antithesis of hard systems as they consist of a relatively large number of components with a large number of interactions (Khisty et al., 2012). Analyzing soft systems typically entails investigation of the behavioral and dynamic interactions between the system components; thus, systemic (holistic) approaches are often useful for such efforts. However, the attributes of a complex system (i.e., its physical structure, condition, operational characteristics, and performance) as well as any interactions between its components can be difficult to predict because they typically have a very influential random component, are very susceptible to influences external to the system, and exhibit marked variation across space and/or time.

In solving problems associated with hard systems, systems engineers first identify the desired end goal and then select the best strategy to achieve that goal. Table 2.2 compares the methodologies for analyzing hard systems and soft systems. In hard systems, goal seeking is considered an adequate model of human behavior associated with civil engineering systems and therefore relies on the language of the problems and the solutions. On the other hand, in soft systems, goal seeking is not considered adequate and thus analysis of these system is associated with deliberations on issues and compromises. Khisty (1995) suggests that both approaches are appropriate for analyzing a majority of civil engineering systems where there exists a broad spectrum of perspectives of system need, a multiplicity of stakeholders and their conflicting interests, and different goals and impacts of the system in terms of technical efficiency, economic efficiency, sustainability, and societal values. The application of both approaches typically is characterized by systematic as well as systemic thinking and a wide array of data items and hypotheses.

It can be readily observed, therefore, that unlike approaches for analyzing purely soft systems, hard system approaches attempt to analyze the human behavior aspects of systems development and not to identify correct or incorrect solutions to a given problem. Instead of viewing the problem from a monochrome lens, hard system approaches consider the varying perspectives and viewpoints of different stakeholders and offer a range of possible solutions. Indeed, where the system has a human activity component, the analysis methods and outcomes are often characterized by uncertainty and fuzziness. The messiness and uncertainty associated with such problems thus can bewilder science-oriented engineers who tend to be more comfortable with analyses that are simple, neat, or tidy (or can be reduced to such). This predicament has been recognized by Checkland and Scholes (1990) and Khisty (1993).

Table 2.2 Differences between Hard and Soft Systems Methodologies

Characteristic	Analysis of Hard Systems	Analysis of Soft Systems
Suitability	For problems that are well structured	For ill-structured problems
Orientation	Systematic goal seeking	Systemic learning
Roots	Simplicity paradigm	Complexity paradigm
Expectations	System can be "engineered"	Systems can be "explored"
	Models are of the world (ontologies)	Models that describe the system (epistemologies)
	Closure is needed	Inquiry never ends
	Seek solution to a problem	Seek accommodation to problems
Questions	How	What and how
Principles	Reductionism	Participants part of research inquiry
	Replicability	Allows reflective learning
	Refutation possible	Process is "recoverable"
	Results homogeneous over time	Results may not be homogeneous over time
Human content	Nonexistent	Significant
Advantages	Involves the use of powerful methods but requires professionals	Available to owners and practitioners
Disadvantages	Not transparent to the public	Fuzziness and uncertainty in results

Adapted from Checkland (1999) and Khisty et al. (2012).

2.1.6 Attributes of a System

The attributes of a system refer to its defining characteristics. In order to analyze a system properly, it is often helpful to have knowledge of its attributes, which include:

1. *Physical structure*: The system shape, size, material, method of construction or installation. For example, the Hoover Dam system has a height of 726 ft (from foundation rock to the roadway on the crest of the dam), a weight of over 6.6 million tons, a arch-gravity design, and portland cement concrete as its dominant material.
2. *Rules or procedures* for operating the system: For example, the Hoover Dam has sluice gates that are made to discharge water at specific times depending on a variety of factors, such as water height, upstream flow, and the expected precipitation and evaporation.
3. *System boundaries*: In order to analyze a system, it is important to establish the system boundaries which are clear and distinct in many cases. In establishing the boundaries of a system, the analyst needs to specify which entities are internal to the system and which are external. The system boundary is what establishes the system environment.
4. *Environment or surroundings* within which the system operates: The collection of all entities that are external to the system but that are either indispensable to or complementary with the system's operation. For closed systems, the environment plays little or no role in the analysis; but for open systems, the environment is often a critical factor in the analysis of the system. For example, for a railway physical system that comprises the tracks, terminals, and the like, the environment could include the weather (ice, snow, wind, sun) and encroaching vegetation, as these entities can affect the performance of that system. In certain types of analysis, a system

could be defined as being closed because the environment is considered an integral part of the system. For example, in analyzing the deterioration of concrete beams, dead load and live load (climate and traffic) may be considered as internal, rather than external, inputs.

5. *Goals or objectives* generally relate to the expected benefits of the system. The goals are desired end states, and objectives are specific statements of goals. The objectives should be realistic (attainable) and measurable. Objectives further give rise to specific performance measures or measures of effectiveness (MOEs). For example, the goals of hydroelectric projects include provision of a surface water reservoir, power, and flood control. Chapter 3 offers a more detailed discussion of the goals of civil engineering systems.
6. *Condition or performance* of the system at any time: In providing some kind of service to society, civil engineering systems may be seen to be performing well or not, depending on the stakeholder in question. From the users' perspectives, system performance is often measured in terms of the direct benefits of the system in terms of delay, convenience, comfort, safety, or out-of-pocket fees, fares, or costs incurred when using the system. From the perspective of the system owner, and sometimes a discerning general public, the system performance is viewed in terms of the physical condition of the system. It is worth noting that while a poor-condition system is not necessarily one with poor service; rather the former, with time, ultimately evolves into the latter.
7. *Performance measures, standards, and criteria* by which attainment of the goals are measured: Feedback and control are essential to the effective performance of a system. From the stated objectives of a system, the civil engineer can develop a list of MOE to ascertain how well (or poorly) a system is performing or the degree to which specified objectives of the system are being achieved. MOEs need not always be associated with utility (benefits) and may also be associated with disutility (such as the costs or adverse impacts of the system on the environment). Also, different MOE's are often generated by different stakeholders. For example, for an water treatment plant, the system owner may be interested in the annual costs of maintenance while the users may be interested in the quality of the water. A performance standard is a threshold (upper or lower limit) beyond which the system performance is deemed undesirable or unacceptable. A performance criterion is a statement involving the performance standard. For example, a performance criterion may be the average annual delay (performance measure) for a proposed urban transportation system should not exceed 15 minutes (performance standard).

2.2 SYSTEMS CONSIDERATIONS IN CIVIL ENGINEERING

2.2.1 The Systems Analysis Approach

Civil engineers utilize the concepts of systems analysis to determine the best course of action at any phase of developing their civil engineering systems. Often, such decisions are made while taking due cognizance of various constraints with the objective of producing an efficient and effective output.

The application of systems concepts occurs not only in engineering but also in other disciplines and indeed in everyday life. As human beings, we are constantly engaged in decision making in all spheres of our lives. An example is the classic apartment choice problem at the start of the academic year, Jane Doe, a college student seeks campus accommodation. Her "problem" is defined as "getting an affordable and nice apartment at a certain maximum distance from campus." After Jane formulates the problem in this manner, she then identifies several alternative apartments by visiting apartment complexes and searching newspaper ads and websites. For each candidate apartment, she makes a mental or written compilation of the costs (rent, utilities, noise) and benefits (proximity to campus, parking availability, etc.). For each cost and benefit criterion, she establishes a threshold value beyond which she would consider the alternative. Also, she has an built-in mental model that weighs the relative importance and trade-offs of each benefit and cost variable. For instance, she may be willing to spend $100 extra to rent an apartment that has off-street parking. On the basis of the relative benefits and costs, she compares and ranks the alternatives. Jane then makes a decision (i.e., selects the best alternative) that gives her the maximum overall benefit within the budget constraints or minimizes her costs given a certain minimum level of benefit she has established. She then implements her decision by moving to that apartment. However, the process does not end there! During the course of the year, she executes the feedback loop: She ascertains the extent to which each cost and benefit item is being realized after she moved in, and whether these are consistent with her initial expectations. She also evaluates her apartment choice by comparing with other apartments of which she has become aware. This new information is fed back into her thought processes and helps her decide on which apartment to rent the following year.

Systems analysis, as shown in the above example, may be defined as a formal inquiry that is carried out explicitly to help a decision maker identify the best course of action among several alternatives. Systems analysis typically involves the following steps (Figure 2.3): (i) establishment of objectives and constraints, and the alternative actions; (ii) analysis including investigation of the likelihood of the impacts of the alternatives in terms of their respective costs and benefits; (iii) statement of the analysis outcomes for each alternative, thus enabling an informed choice of the best alternative; and (iv) ex poste evaluation of the choice after its implementation. At any phase in civil engineering systems development, the task of making a decision to select one of several alternatives may proceed using these steps. For the analysis of the alternatives, the tools described in Chapters 5–18 can be useful. The second step (i.e., establishing the goals and objectives) for the decision problem at hand may not be related to the overall system goals (phase 1 of system development) but rather for the specific task at hand, for example, deciding which formwork material to use for a certain section of a large system under construction.

There are other terms used somewhat loosely to represent systems analysis, depending on the intended use of the analysis and the area of its application. For example, when it is used to make decisions, systems analysis is known as *decision analysis;* when used to ascertain whether an intended action would be consistent with technical or economic objectives, or physical or institutional constraints, it is referred to as *feasibility analysis;* when it is used to rank alternatives, it is called *priority analysis;* when it is used to assess the respective benefits (effectiveness) for a fixed cost or the respective costs for a fixed benefit, it is referred to as *cost-effectiveness analysis;* when it is used to assess the effectiveness of past or proposed policies (often public or governmental) rather than physical interventions, it may be referred to as *policy analysis*.

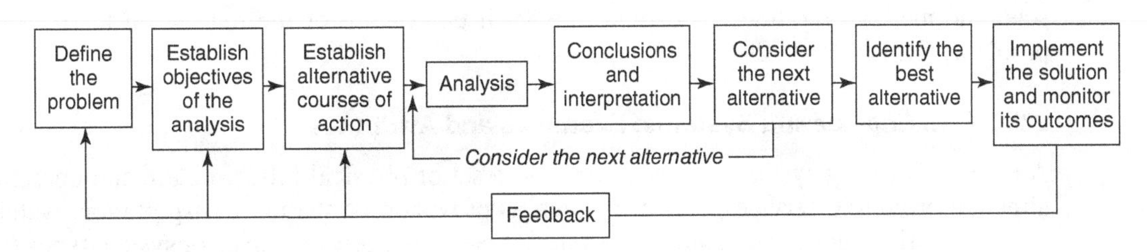

Figure 2.3 Basic steps in systems analysis.

A cost-effectiveness analysis where all benefits and costs are expressed in monetary values, particularly over a time period, is typically termed a *monetary cos–benefit analysis* and yields results such as a benefit–cost ratio and net present value (benefits minus costs). In long-term *risk–benefit analysis*, the discounted monetary costs of the level of each risk factor associated with each alternative action is compared with the discounted sum of the expected benefits from that action, thereby providing a basis for rational comparisons between the alternative actions. The risks considered include those associated with events that have low probability of occurrence but high adverse consequences, as well as those with high probability of occurrence but low adverse consequences. In some cases, risk is considered a cost factor and is considered in cost–benefit analysis rather than carried out as a separate analysis. The formal concept of systems analysis started gaining prominence in the 1960s when it was extensively used by the Research and Development Corporation (RAND) of the United States.

2.2.2 Motivations for Adopting Systems Approaches in Civil Engineering Systems Development

The past few decades have seen dramatic advancements in technology and other fields of endeavor. As a result, engineers have come to realize that robust and long-standing solutions to real-world problems cannot be found using the traditional scientific approach alone (Khisty et al., 2012). This realization hits home particularly in the area of civil engineering systems, where there exists a strong relationship between civil facilities and their surrounding communities through the natural and built-up environment, community safety, sociocultural fabric, and land use. The importance of the human and societal factors in the development of civil engineering systems is one of the key factors that warrant the inclusion of systems approaches in the development of these systems.

Another reason for incorporating systems concepts in civil engineering is the growing complexity of civil engineering infrastructure and facilities. Civil engineering has increasingly seen advances in all phases of project delivery—materials and design, construction processes, and facility operations and preservation. For example, there is increasing incorporation of psychology (human factors), finance and economics (life-cycle costing, financial programming), and other disciplines in civil engineering systems development. As such, engineers, planners, managers, and decision makers involved with the various phases of civil infrastructure development need to make decisions using a systems approach in order to reach a universally acceptable solution. Dandy et al. (2008) recognized that in order to deal with this complexity, a special methodology is needed: breaking down the complex entity into progressively smaller and smaller component parts until each part is simple enough to allow studies of its behavior and how it interacts with other parts. This decomposition approach, which is a basic tenet of systems analysis, facilitates the development of large civil engineering infrastructure. Therefore, in a world of ever-increasing complexity, particularly in the civil engineering field, the systems concept has become more and more relevant. The systems approach is often touted as being broad based and systematic and geared for solving complex problems (Meredith et al., 1985) and involves the application of the scientific method, modified to capture the "holistic" nature of the real world to solve complex problems (Khisty et al., 2012).

2.2.3 Civil Engineering Systems: Examples and Attributes

A civil engineering system may be defined as a set of physical infrastructure and operating rules aimed at providing services to society such as provision of shelter, transportation, water supply and distribution, and waste treatment. Table 2.3 presents a few examples of system types (and their attributes) in the various branches of civil engineering.

Table 2.3 Examples of Civil Engineering Systems and Their Attributes

Branch	Example of System	Physical Structure(s)	Operation	Goals and Objectives (Example)	Performance Measures (Example)
Transportation	Freeway system	Pavement, road signs, guardrails	Vehicular use	Provide mobility around an urban area	Congestion levels Accident frequency
Structures	Steel truss of a bridge	Steel sections and joints	supporting loads	Safe and economical support of live and dead loads	Deflection, shear, corrosion
Hydraulics/ hydrology	Levee system	Walls and pumps	Holding back high water	Prevent flooding	% of time overspill, number of breaches
Environmental	Waste Treatment System	Filtration units, sedimentation ponds, etc.	Waste treatment	Maximize amount of waste treated	Volume of waste treated/ hour, quality of output
Geotechnics	Foundation system for a structure	Footings and rafters	supporting loads	Safe and economical support of structure	Settlement, cracking
Materials	Asphalt mix	Aggregates and bitumen pavement	supporting loads	Safe and economical support of vehicle weights	Rutting, raveling, cracking.
Construction	Critical path scheduling system	—	Scheduling of work	Minimize construction delay Maximize resource utilization	Construction period, resource utilization (%)
Geomatics	Location precision systems	—	Measurements	Minimize error of locating facilities	Percentage of time a point is correctly located within some precision range

2.3 DEVELOPMENT OF CIVIL ENGINEERING SYSTEMS

2.3.1 Prelude

The development of civil engineering systems refers to all the work activities necessary to ensure that a system is provided, runs efficiently, and is preserved in such a manner that it provides cost-effective service. In this context, it is meaningful to identify the various phases of development and the tasks associated with each phase. Figure 2.4 presents a general overview of the phases of civil systems development and the typical tasks that civil engineers undertake at each phase. We discuss this in the next section.

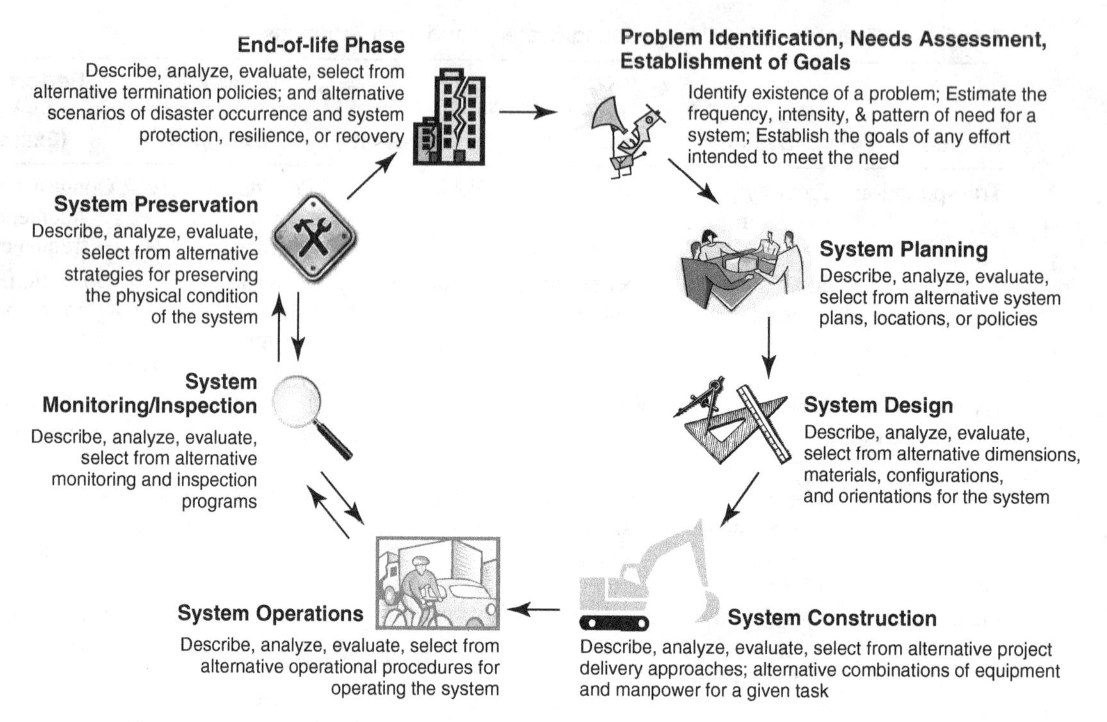

Figure 2.4 Phases of civil systems development and typical tasks at each phase.

2.3.2 Phases of Systems Development

The development of a civil engineering system (Figure 2.4) can be defined in terms of the phases through which it progresses. Typically, development starts with an assessment of the need for the system based on societal or economic interests. After it has been established that a system is needed, the system is planned by selecting the system location, orientation, and configuration such that the impacts from and to the environment are minimized. Then the design of the system is carried out by specifying the material types and the dimensions of the system components. After approval of the design and securing financing, the system is constructed. When construction is completed, the system is commissioned and commences operations (i.e., begins to be used).

It is important that the system ownes knows the level and pattern of demand for the system at the phase of needs assessment. It is also important to address the inevitable physical deterioration of the system, which is often due to climate and loading, in a timely manner. As such, during the operations phase, the system is regularly monitored and inspected by the owner to track the level of usage and the physical condition of its components, respectively. This enables the owner to carry out timely rehabilitation and maintenance of the system. The loop involving the triad tasks of operations, inspection, and preservation continues until the system reaches the end of its physical life and needs to be replaced. At that point, a new needs assessment is carried out to ascertain the continued need for the system, and the cycle continues.

For some systems, such as road signs, this development cycle takes only a few years, while other systems, such as dams, often take a hundred years or more. In Chapters 19–26, we present a more detailed discussion of the phases and their common tasks that are related to the systems approach.

2.3.3 Systems-Analysis Tasks at Each Phase of Systems Development

At each phase of system development, the civil engineer first establishes a plan of work for that phase [this is not the same as the overall system plan (phase 2)]. Also, the tasks carried out by the engineer typically include a description of the way that phase proceeds or will proceed—this is often needed for purposes of communicating some characteristics of the system to the general public or system owner. Often, the engineer will need to analyze some aspect of work at the phase, evaluation between alternatives, and select and implement the best alternative. The trio of processes of analysis, evaluation, and selection of the best course of action could be collectively termed "systems optimization." In Chapter 4, we will discuss in greater detail, the tasks faced by engineers at each phase of civil systems development. A critical task at each phase that is not shown in Figure 2.4 is the feedback task, which can be represented by arrows from any phase to the preceding phases. Lessons learned from an antecedent phase are always needed so that the outputs from the preceding phases can be enhanced. For example, engineers at the maintenance phase can provide feedback to those at the design phase so that the designers can specify materials, dimensions, orientations, and other design features that reduce the frequency or intensity of maintenance. Feedback, a key aspect of systems analysis as we saw in Section 2.2.1, is helpful for developing systems that are intended to be adaptive; also, this feedback is a key aspect of systems dynamics, a useful tool for analyzing system behavior which we will discuss in Chapter 14.

2.3.4 Tools Needed for the Tasks

In order to carry out the systems analysis tasks identified in Section 2.3.3, engineers need to be equipped with a certain set of tools. For example, in order to describe how a system works, the engineer may need to use computer simulation; and in order to arrive at the best option under different objectives and constraints, the engineer may need to use optimization and decision-making tools. In many standard texts that discuss analytical tools, the terms systems analysis, systems decision making, and systems optimization are often used synonymously. One should recognize that systems analysis does not necessarily provide the optimal solution and that the process of optimization typically includes analysis and evaluation of multiple alternatives and ultimately decision making to select the best alternative; thus, it could be argued that system optimization is not necessarily equivalent to systems analysis. Chapters 5–18 are devoted to a discussion of the various tools that civil systems engineers could use in order to execute their traditional tasks and systems-related tasks (Section 2.3.3) more efficiently at the various phases of system development.

2.3.5 Relationship between Civil Engineering Professions and the Phases, Tasks, and Tools

In Section 1.2, we listed the branches of civil engineering, and we identified the types of civil engineering systems associated with each branch. For any type of system in any of the branches, the cycle of development (time from the system's needs assessment to its end of life) shown in Figure 2.4 may be long or short in length but always involves the eight phases shown in the figure. At each phase, professional engineers carry out, various tasks some of which are traditional to the civil engineering branch in question, and others which are related to the systems approach. In carrying out the latter category of tasks, they use the analytical tools mentioned in Section 2.3.4 (Chapters 5–18 provide detailed treatment of these tools).

The traditional and the system approach related tasks are not mutually exclusive. In certain cases, the engineer at that phase may find it necessary to bring in experts that are well versed in a specific subject in order to assist in finding answers to problems beyond the engineer's reach. For example, at the phases indicated, the following specialists may be needed: needs assessment, economists; planning and design, planners and geologists; construction, financial analysts; operations, ergonomic experts; monitoring and inspection, statisticians; maintenance, materials scientists; end of life, environmental scientists. Also, engineers may find themselves working in only one phase for the rest of their career while others may have greater mobility across the phases. In any case, at any phase of the cycle, engineers strive to carry out tasks that contribute to successful execution of the phase in question and, ultimately, for the overall development of the system in general.

2.4 SOME TERMS AND CONCEPTS RELATED TO SYSTEMS THINKING

2.4.1 Systemic Considerations

The word "systemic" means systemwide and is used to describe a certain characteristic that is spread throughout a system or throughout a collection of systems. The term is used in many fields including agriculture, economics, engineering, medicine, and the social sciences. The systemic approach refers to a holistic rather than piecemeal approach and, therefore, is cognizant of the fact that, for a typical system, the effect of the sum of the parts is different (typically superior) to the sum of the effects of the individual parts. Systemic problems are those problems that are best addressed using systemic approaches. In the context of civil engineering systems, derivatives of the term include systemic bias (the inherent tendency of a system or process to yield or to avoid specific outcomes); systemic risk (the uncertainty in performance of an entire engineering system as opposed to the uncertainty associated with any one component or subsystem of the system); and systemic shock, which is a disturbance to a civil engineering system that is strong enough to disrupt the workings between the various components and thus disrupting the system's static or dynamic equilibrium.

2.4.2 Systematic Processes

The word "systematic" is a characteristic that denotes consistency with planning, orderliness, logic, and regularity. In the most basic use of the word, systematic is simply an adjective stating that something is related to a system. In the context of problem solving, the word is generally used when describing an approach that is step by step rather than arbitrary or chaotic.

2.4.3 Systems Engineering

Systems engineering can be considered somewhat similar to systems analysis. Bahill and Gissing (1998) described systems engineering as an interdisciplinary process that ensures that the system users needs are fully met throughout the life of the system, and they identify seven sequential tasks (acronymed SIMILAR): **S**tating the problem (identifying the customers and stakeholders of the system, understanding their needs, identifying and assessing the need for some intervention, establishing requirements, and defining system functions); **I**nvestigating the alternative interventions on the basis of their anticipated performance, cost, and risk; **M**odeling the system to clarify requirements, reveal bottlenecks and fragmented activities, reduce cost, and expose any duplication of effort; **I**ntegration by designing interfaces and assembling the system elements to work together as a whole (aided by coordination and communication); **L**aunching the system by operating it and producing outputs to ensure that the system is performing as expected; **A**ssessing the performance of the system on the basis of evaluation criteria such as technical and economic performance measures; and **R**eevaluating, which is a continuous, iterative process.

2.4.4 Operations Research, Decision Sciences, and Management Sciences

Often considered synonyms of each other, operations research, the decision sciences, and the management sciences are professional interdisciplinary mathematical disciplines that provide rational bases for decision making in many fields, including engineering. Engineers of civil systems typically use concepts in these disciplines which include information technology tools, and analytical tools used to develop and analyze models for making decisions at any phase of their system development. These tools include probability theory, statistics, economics, optimization, modeling and simulation, graph theory, queuing theory, game theory, and decision theory. The objective for their use is often to identify the action that has the highest beneficial impacts in terms of performance, durability, condition, profit, utility, and/or the least adverse consequences as far as cost, loss, and uncertainty in the short term, long term, or both. In the context of civil engineering systems, these techniques can describe the past or current behavior of a system and use the acquired understanding to predict system behavior and, ultimately, to improve system performance or output. In the private sector, companies use these tools to gain competitive advantage; and in the public sector, agencies use these tools to enhance the delivery of public services so that taxpayers are provided the best possible performance of public systems with available resources.

2.4.5 System of Systems (SOS)

As we learned in Section 2.1.3, system of systems is an emerging concept that deals with the large-scale integration of several independent and self-contained systems to form a larger system and the effective running of these so-called supersystems. As mentioned in an earlier section of this chapter, the management of many large scale systems can be enhanced if they are viewed through the lens of an SOS framework. Such large-scale, often complex, systems typically consist of subsystems, each of which is necessary but not sufficient for the entire system to function very efficiently. For example, a wastewater treatment plant is comprised of several different subsystems (e.g., activated sludge, surface aeration, filter beds, and sedimentation systems) and the wastewater cannot be effectively and efficiently treated if the subsystems are separately and independently operated. At the current time, research is in progress to analyze or enhance the integration of existing systems to yield supersystems (Bell and Teh, 2009; Peeta, 2011). SOS is particularly important in the current era of globalization, climate change, and regional security awareness. Examples of civil engineering areas where SOS concepts hold much promise include the monitoring, control, and analysis of (i) transportation traffic inter- and intraactions among modes (land, sea, and air), (ii) land use and environmental quality, (iii) geodetics, (iv) structural and geotechnical interactions among closely spaced structures, (v) construction scheduling systems, and (vi) transportation security and protection of critical infrastructure. The SOS concept and particularly the unique analytical tools associated with it were born out of the realization that the operational efficiency of the sum of the smaller subsystems is often superior to the sum of the operational efficiencies of the individual subsystems (Maier, 1998). Therefore, the operations of the supersystem are associated with different types and scopes of problems and effectiveness than those of their constituents. SOS utilizes tools from a variety of disciplines to examine how a supersystem can be effectively synthesized from smaller systems and operated in a more efficient and cost-effective manner.

2.4.6 Systems Theory

Systems theory, which can be considered the theoretical underpinning for systems engineering, is an interdisciplinary area of study that analyzes the behavior of complex systems in the social sciences, business, engineering, science, and the natural environment. It typically involves a framework for describing a group of entities that work in concert to produce some outcome, while also analyzing,

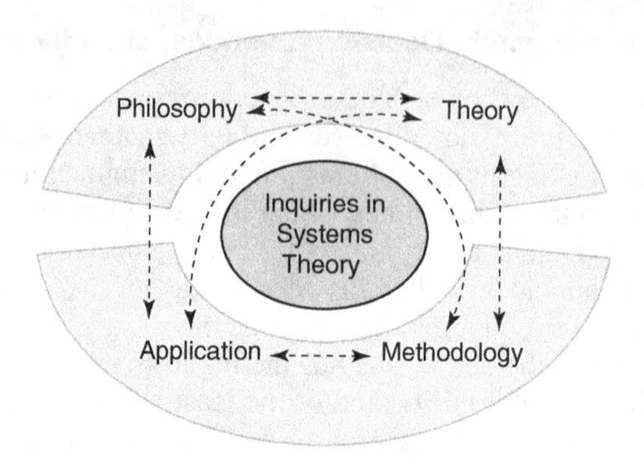

Figure 2.5 Domains of inquiry in systems theory.

evaluating, or optimizing such frameworks through continual feedback and self-correction. Systems theory, which originated in the biological sciences in the 1920s to address a need to describe the interrelated nature of organisms in ecosystems, received a boost from the famous Macy Conferences that took place between 1943 and 1953. Since then, important contributions have been made by researchers including Norbert Weiner, Ludwig von Bertalanffy, Kenneth Boulding, Margaret Mead, Ilya Prigogine, and Jay Forrester, through their studies in the systems-related subjects of complexity, self-organization, connectionism, adaptive systems, chaos theory, cybernetics, systems dynamics, and complex adaptive systems.

Systems inquiry can generally be placed into four integrated and recursive domains (Figure 2.5): philosophy (the epistemology, ontology, and axiology of systems), theory (a set of interrelated principles and concepts that are applicable to all systems), methodology (a set of strategies, techniques, tools, models, and methods that instrumentalize systems philosophy and theory), and application (the use of the domains and their interactions) (Banathy, 1997). Philosophy and theory are associated with knowledge, while method and application are action domains. Thus, systems inquiry can be described generally as action that is based on knowledge.

2.4.7 System Dynamics

System dynamics, a key area in systems theory, was founded by Jay Forrester in the 1950s through his applications of systems in electrical engineering and is the study of the behavior of dynamic complex systems. In system dynamics, it is recognized that the internal relationships in a system, which are often circular, interlocking, and sometimes time delayed, are often just as vital in establishing the behavior of the system as are the system's individual components. Also, the system as a whole behaves in a manner that is very different from the sum of the behaviors of its individual components or subsystems. Thus, chaos theory and social dynamics are key concepts in systems dynamics. Chapter 14 of this text is devoted to more detailed discussion of system dyanamics as a tool for analyzing processes at any phase of civil system development.

2.5 GLOBAL INITIATIVES IN THE STUDY OF SYSTEMS

2.5.1 The Institute for Complex Additive Systems Analysis (ICASA)

ICASA, established by the New Mexico state legislature in 2001, is a cooperative venture among industry, academia, and government that is dedicated to analyzing and understanding the behavior, predictability, and vulnerabilities of the complex interdependent systems often related to large-scale infrastructures. ICASA adopts a strategic and interdisciplinary approach to harnessing research relevant to the information age and applies that research to address real-world problems, develop key relevant technologies, and train and educate the next generation of critical systems thinkers. ICASA research has spawned Complex Additive Systems Analysis, an innovative approach to strategic systems thinking that facilitates the fundamental understanding of current-day large-scale systems on the basis of the premise that these systems develop over time (ICASA, 2011). ICASA researchers seek to comprehend the consequences and the additive effects of design decisions geared toward performance optimization, the evolution of systems design, and the aggregation of individual system components to form a system of systems. The research, among other objectives, is intended to assist in developing strategies for enhancing national security.

2.5.2 The International Institute for Applied Systems Analysis (IIASA)

Founded in 1972 at Laxenburg, Austria, IIASA is a nongovernmental research organization that conducts policy-oriented interdisciplinary scientific research addressing economic, environmental, technological, and social issues in the context of the human dimensions of global change (IIASA, 2011). IIASA researchers use state-of-the-art methodologies and analytical approaches that include systems analysis tools to provide objective and usable information on environmental, social, and economic issues for the use and benefit of the general public, scientists, and national and international agencies and organizations. Research teams comprised of academicians and scientists from various countries carry out systems research in areas such as emissions, transformation, and transportation of pollutants and toxic materials, water resource quality and availability, remediation of biological resources, land use and development, and climate change.

2.5.3 New England Complex Systems Institute (NECSI)

NECSI was founded in 1996 by faculty members of academic institutions in the New England area with the objective of furthering international research and understanding of complex systems and their applications. NECSI seeks to further education, research, dissemination of knowledge, and social and community development worldwide. Consistent with this objective, NECSI promotes the study of complex systems and their application for the enhancement of human welfare. Recognizing that the area of complex systems is a relatively new and growing field of science that seeks to throw light on how system components collectively contribute to a system's collective behaviors and how it interacts with its environment, NECSI researchers utilize and enhance basic concepts and formal approaches for application to real-world problems. NECSI researchers have carried out studies that have used analytical tools that include agent-based modeling, chaos and predictability theories, multiscale analysis and complexity, for applications in ecology, biodiversity, cellular response, evolution, systems biology, health care, infrastructure networks, systems engineering, military conflict, negotiation, ethnic violence, and international development (NECSI, 2011).

2.5.4 The International Council on Systems Engineering (INCOSE)

INCOSE, founded in 1990, seeks to advance the systems engineering discipline and facilitate collaboration to advance scientific and technical knowledge (INCOSE, 2011). Through interdisciplinary research, INCOSE researchers and engineers develop technologically, economically feasible, and socially acceptable solutions to critical problems related to engineering systems. Ultimately, INCOSE aims to advance both the state of the art and the state of the practice of systems engineering in academia, industry, and government.

2.5.5 The Council of Engineering Systems Universities (CESUN)

CESUN was established in 2004 by academic institutions that offer educational and research programs related to engineering systems. With a membership base exceeding 30 institutions in Asia, Australia, Europe, and North America, the Council provides a platform for the member universities to collaborate toward the development of engineering systems as an emerging field of study. CESUN's primary activities include conversations and analysis of critical issues that currently affect (or are expected to affect) the development and management of engineering systems. CESUN also conducts joint projects of shared interest, disseminates educational material, organizes technical meetings, and works with related professional societies (CESUN, 2011).

SUMMARY

Chapter 2 provided an overall understanding of the systems concepts and the case for its application in the development of civil engineering infrastructure. We reviewed terminology and the concepts of systems thinking and we identified and discussed the eight major phases through which a civil engineering system is developed, namely, needs assessment and definition of goals and objectives, planning, design, construction/implementation, operations, inspection/monitoring, preservation, and end of life. We also identified the key systems approach related tasks encountered by the engineer at each phase, that is, analysis, description, evaluation, and selection and decision making. Also, in this chapter, we examined briefly a number of ongoing initiatives, at a global and regional level, toward the advancement of educating systems engineers and conducting systems-related research and applications. Having established in parts of Chapter 2 that establishing the goals of a system is a key initial aspect of the systems development process, we will demonstrate in Chapter 3 how this goal-setting process is not only an interesting undertaking but also rather critical if the system is to perform to the expectations of its users, the owner, and the community at large.

EXERCISES

1. Give an example of a "system" in (a) your preferred branch of engineering and (b) everyday life. For each of these systems, list one of the following system attributes: physical components, if any; abstract components or rules of operation; goals or performance measures.
2. For each of the following civil engineering systems, identify one or more of the following: physical component, abstract component, rules or procedures for operation, the environment, goals or objectives, and measure of condition or performance:
   - **a.** a rail transit system
   - **b.** a hydroelectric power generation system
   - **c.** your university's sport stadium
   - **d.** a pedestrian timber bridge spanning a large creek
3. A nursing home located near a busy freeway suffers from excessive noise from the freeway traffic. Describe fully a solution that you would recommend to resolve the situation using the systems approach.
4. For any two of the following engineering systems, identify whether they are physical or abstract (or both) and provide supporting arguments: a hydroelectric dam, a transportation logistics schedule, an automobile, a disaster evacuation plan, and a cell phone communications mast. Also, for each system, name two inputs and two outputs and identify any two sources of uncertainty that could cause variability in the system outputs.

5. The city of Imaginopolis, which is located near White River, has an elaborate system of levees to protect the city from flooding during high-water events. The levees were built over 50 years ago and many of them are showing signs of structural distress. Using a systems perspective, how would you draw up a program to rehabilitate the levee system under budgetary constraints over a 5-year period?
6. Your campus bus transit corporation is considering switching its fleet from gasoline to hybrid. From the systems perspective, what advice would you give?
7. De Neufville and Stafford (1971) and Khisty et al. (2012) argued that pure science alone is not enough for efficient management of civil engineering systems. Do you agree? Give reasons for your answer.
8. You are the city engineer of a large city. Name and describe any one system of systems (SOS) that falls under your control. List one concern that is likely to be associated with your SOS's physical structure, operation, or physical condition. How would you assess the overall technical performance of (a) any one component system of the SOS and (b) the entire SOS? For the SOS, list and discuss any three costs or "disutilities" associated with the system.
9. Explain Banathy's quadruple-domain presentation of system theory in the context of any one common civil engineering system, and explain, with examples, the nature and role of each domain and its relationship with others.
10. Identify any one busy street on your campus with large volumes of surface traffic comprised of vehicles, cyclists, and pedestrians. Explain how you would carry out problem identification, and if a problem is found to exist, discuss how you would assess the need for a solution.
11. Continuing the previous question, assume that you have established the need for a solution and that you have decided to produce a conceptual design to solve the problem. List and discuss any five conceptual design alternatives.
12. Continuing the previous question, how would you (a) establish the goals and objectives; (b) the criteria to assess the desirability, efficiency, or effectiveness of each design alternative; and (c) choose the best of the conceptual design alternatives.
13. (a) List any five branches of civil engineering. (b) For any one of these branches, list any three systems. (c) For any one of these systems, list the eight phases of development. (d) For any phase, list any three tasks typically faced by the engineer. (e) Discuss how the systems approach could be used to carry out any task at that phase. (f) What role do analytical tools play in the use of the systems approach to carry out that task? (g) What feedback does the engineer typically provide to engineers at preceding phases?

REFERENCES

Au, T., Shane, R. M., and Hoel, L. A. (1972). *Fundamentals of Systems Engineering*. Addison Wesley, Reading, MA.

Bahill, A. T., and Gissing, B. (1998). The Systems Engineering Process, in Re-evaluating Systems Engineering Concepts Using Systems Thinking. *IEEE Transaction on Systems, Man and Cybernetics, Part C: Applications and Reviews* 28(4), 516–527.

Banathy, B. H. (1997). A Taste of Systemics, The Primer Project, http://www.newciv.org/ISSS_Primer/asem04bb.html. Retrieved Dec. 26, 2011.

Bell, S., and Teh, T. H. (2009). *People, Pipes and Places. Networks of Design*. Universal Publishers, Boca Raton, FL.

Boardman, J., DiMario, M., Sauser, B., and Verma, D. (2006). System of Systems Characteristics and Interoperability in Joint Command and Control. Defense Acquisition University, July 25–26.

CESUN (2011). The Council of Engineering Systems Universities (CESUN). Retrieved December 26, 2011, http://cesun.mit.edu/.

Checkland, P. B., and Scholes, J. (1990). *Soft Systems Methodology in Action*. Wiley, New York.

Checkland, P. (1999). *Systems Thinking Systems Practice*. Wiley, New York.

Churchman, C. W. (1968). *The Systems Approach*. Dell Publishing, New York.

Collins (2009). *Collins English Dictionary—Complete & Unabridged*, 10th ed. Harper Collins, New York.

Dandy, G., Walker, D., Daniell, T., and Warner, R. (2008). *Planning and Design of Engineering Systems*, 2nd ed. Taylor & Francis, London.

DeLaurentis, D. (2007). *Research Foundations, School of Aeronautics and Astronautics*, Purdue Univ., West Lafayette, IN.

De Neufville, R., and Stafford, J. H. (1971). *Systems Analysis for Engineers and Scientists*, McGraw Hill, New York.

FHWA (1995). *Recording and Coding Guide for the Structure Inventory and Appraisal of the Nations Bridges*. Federal Highway Admin., Washington, DC.

ICASA (2011). The Institute for Complex Additive Systems Analysis (ICASA). http://www.icasa.nmt.edu/. Retrieved Dec. 26, 2011.

IIASA (2011). The International Institute for Applied Systems Analysis (IIASA). http://www.iiasa.ac.at/. Retrieved Dec. 26, 2011.

INCOSE (2009). International Council on Systems Engineering. http://www.incose.org/practice/fellows consensus.aspx.

INCOSE (2011). The International Council on Systems Engineering (INCOSE). http://www.incose.org/. Retrieved Dec. 26, 2011.

Khisty, C. J. (1993). Citizen Participation Using a Soft Systems Perspective. *Transp. Res. Rec.* 1400, 53–57.

Khisty, C. J. (1995). Soft Systems Methodology as Learning and Management Tool. *ASCE J. Urb. Planning Dev.* 121(3), 91–107.

Khisty, C. J., Mohammadi, J., and Amekidzi, A. A. (2012). *Fundamentals of Systems Engineering*. Prentice Hall, Upper Saddle River, NJ.

Levis, A., Maier, M., Sage, A. (2004). *Academic Audit of AFIT Programs in Systems Engineering, USAF Report SAB TR-05-04*, US Air Force, Washington DC.

Maier, M. W. (1998). Architecting Principles for System of Systems. *Syst. Eng.* 1(4), 267–284.

Maier, M. W., and Rechtin, E. (2000). *The Art of Systems Architecting*, 2nd ed. CRC Press, Boca Raton, FL.

Meredith, D. D. Wong, K., Woodhead, R., and Wortman, R. H. (1985). *Design and Planning of Engineering Systems*. Prentice-Hall, Englewood Cliffs, NJ.

Merriam-Webster (2011). *Merriam-Webster Collegiate Dictionary*, 11th ed., 2003. www.merriam-webster .com. Accessed Dec. 12, 2011.

NECSI (2011). New England Complex Systems Institute (NECSI). http://necsi.edu/. Retrieved December 26, 2011.

Peeta, S. (2011). *A Multi-Period Dynamic Model for Analyzing Infrastructure Dependencies*. Proceedings of the 90th Annual Meeting of the Transp. Res. Board, Washington, DC.

Richards, L. (2008). Definitions of Cybernetics. Reader 1997–2007, 9–11. http://polyproject.wikispaces.com/ file/view/Larry+Richards+Reader+6+08.pdf.

Wasson, C. S. (2008). *Systems Analysis, Design, and Development*. Wiley, Hoboken, NJ.

Weinberg, G. (1975). *An Introduction to General Systems Thinking*. Dorset House, New York.

Wiener, N. (1948). *Cybernetics, or Communication and Control in the Animal and the Machine*. MIT Press, Cambridge, MA.

Wortman, M. S., and Luthans, F. (1969). *Emerging Concepts in Management*. Macmillan, New York.

USEFUL RESOURCES

Blanchard, B. S., and Fabrycky, W. J. (2010). *Systems Engineering and Analysis*, Prentice Hall.

Buede, D. (2009). *The Engineering Design of Systems: Models and Methods*, Wiley.

Defense Systems Management College (DSMC). (2001). *Glossary: Systems Engineering Fundamentals*, 10th ed. Defense Acquisition University Press, Ft. Belvoir, VA.

FAA (2033). *ASD-100 Architecture and System Engineering. National Air Space System—Systems Engineering Manual*. Federal Aviation Administration, Washington, DC.

IEEE (1998). *IEEE Standard for Application and Management of the Systems Engineering Process*. Institute of Electrical and Electronic Engineers (IEEE), New York.

Keating, C. B., and Katina, P. F. (2011). *System of Systems Engineering: Prospects and Challenges for the Emerging Field*, System of Systems Engineering, 234–256, Wiley.

Kossiakoff, A., Sweet, W. N., Seymour, S., Biemer, S. M. (2011). *Systems Engineering Principles and Practice*, Wiley, Hoboken, NJ.

Mar, B., and Palmer, R. (1989). Does Civil Engineering Need System Engineering, *ASCE J. Prof. Issues in Engrg.*, 115(1), 45–52.

Olsson, M., Sjostedt, G. (2004). *Systems Approaches and their Application*, Springer, NY.

Wang, J. X. (2002). *What Every Engineer should know about Decision-making under Uncertainty*, CRC Press, Boca Raton, FL.

Wells, G. D., and Sage, A. P. (2008). *Engineering of a System of Systems, Systems of Systems Engineering*, 44–76, Wiley, Hoboken, NJ.

CHAPTER 3

GOALS AND OBJECTIVES OF CIVIL ENGINEERING SYSTEMS

3.0 INTRODUCTION

Author Robert Heinlein believed that in the absence of clearly defined goals, one becomes enslaved by trivia. This is true in both our personal and professional lives; the absence of clearly defined goals invariably results in poor decisions and poor outcomes. For example, the college student who seeks a place to have lunch may need to consider objectives including the food price, food quality, waiting time, and its distance from the student's office or main lecture room. If the student goes out to seek a lunch location without any goal in mind, she may likely end up having a poor lunch experience. The situation is no different in the case of engineering systems where millions of taxpayer dollars are typically at stake and the engineer has a fiduciary duty to make informed cost-effective and demand-reponsive decisions with limited resources.

The establishment of a system goal is tied directly to the type of system under consideration. As we learned from the various definitions of engineering (Chapter 1), it is clear that civil engineers seek to design and implement systems intended to satisfy societal needs that are consistent with human values. In order to establish an explicit goal for the system, the engineer needs to ascertain the nature of this "need" and an appropriate way to measure the system performance (in other words, the extent to which that need is being fulfilled by the system) at its operations phase.

Outcomes may be categorized as beneficial or adverse and intended or unintended (Figure 3.1). Other perspectives or dimensions include the type of affected stakeholder, whether the outcome is monetary or otherwise, and whether the outcome is a result of the initial phases (planning, design, construction) or the rest-of-life phases (operations, maintenance monitoring, and end of life). Typically, the goals, objectives, or performance of a system are expressed in terms of the intended and beneficial outcomes, and this chapter mainly focuses on this perspective of outcomes. The other dimensions will be discussed only briefly in this chapter but also appear in certain areas in subsequent chapters.

As we shall see in the next section, the intended beneficial outcomes can be represented by a hierarchy that starts from a relatively small set of values, overall goals, and goals, and, finally, a relatively large set of objectives. From the stated objectives of a system, the civil engineer can develop a list of performance measures or measures of effectiveness (MOEs) to ascertain how well a system is performing (or is expected to perform) or the degree to which specified objectives and, ultimately, the goals and values of the system are being achieved. From the MOEs, the engineer establishes performance standards and performance criteria.

It is also important to mention that the establishment and use of objectives do not occur only at the initial phase of development (where the purpose of the system is established) but also at each of the remaining phases of civil systems development: There exist specific goals and objectives associated with that phase. For example, consider the construction of an overhead water reservoir: The overall goal (purpose) of this system is to ensure reliable water supply to the community. Now, let

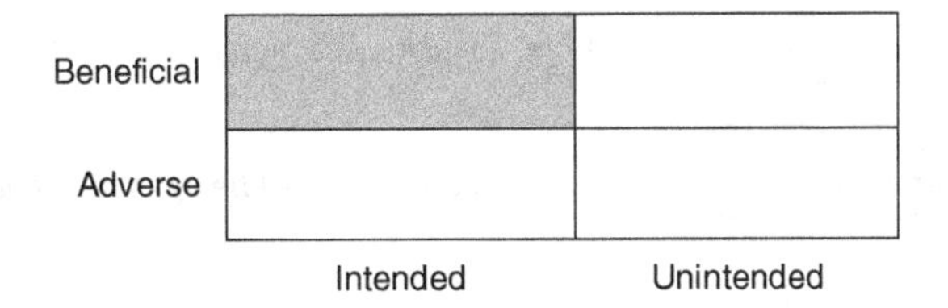

Figure 3.1 Simple classification of system outcomes.

us examine some possible objectives specific to this phase, regarding the water supply system: at the needs assessment phase, the engineer's objective may be to correctly identify that a water supply problem exists; at the planning phase, an objective may be to reliably predict the amount of daily water needs of the community; at the design phase, designers may seek an optimal volume of the tank and economical dimensions of the structural members that meet the stability specifications; at the construction phase, the owner's objective may be to choose a contracting approach that fosters contractor innovation and ensures fastest delivery; at the monitoring phase, the objective may be to minimize the number of inspections of the water tank annually within the given confidence limits; and at the operations phase, the objective may be to ensure that leaks are repaired with minimal delay. The specific objectives at each phase influence, and also are influenced by, the specified tasks and tools at that phase. For example, (a) the tool of statistical analysis is used to carry out the task of describing or modeling system outcomes, in terms of the system objectives, as a function of system inputs (Chapter 7); (b) in the task of assessing the consequences of alternative actions, each phase is carried out on the basis of the system objectives in the form of economic criteria (Chapter 11) or multiple criteria (Chapter 12); and (c) in optimization (Chapter 9), the engineer selects the best action that maximizes some objective associated with that phase or with the system as a whole.

3.1 HIERARCHY OF DESIRED OUTCOMES: VALUES, GOALS, AND OBJECTIVES

As mentioned in the Introduction, the development of system goals and objectives is governed by a hierarchy of desired system outcomes. This hierarchy starts with the basic values of human society followed by the overall goals of *effectiveness, efficiency*, and *equity*, and then goals such as system preservation and longevity, environmental resource protection, economic development, and public safety and security. Under each goal is a set of specific objectives. Then more specific statements, referred to as measures of effectiveness (or performance measures) of the civil engineering system are established for each objective. Figure 3.2 presents the hierarchical interrelationships among the various levels of system outcomes. The number of branches shown at each node is for illustration only; different systems will have different numbers of nodes and branches. In the remainder of this section, we will discuss each level of this hierarchy in some detail.

3.1.1 Values

Values refer to a set of irreducibles that constitute the basic desires and drives that govern human behavior (Khisty et al., 2012): for example, the need to survive, the need to enhance the community's quality of life, the need to belong, the need for order, and the need for security (Figure 3.3). Nations, races, demographic groups, societies, cultures, and other organized groups of humans have, over the years, established values that their members share. The values held by members of these entities typically arise out of the judgment of which conditions, situations, objects, and

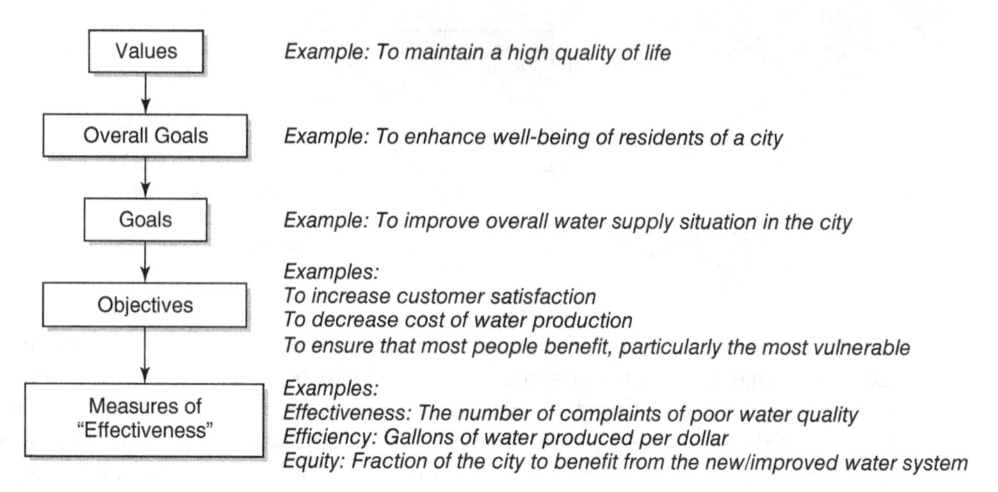

Figure 3.2 Hierarchy of desired outcomes of civil engineering systems (Adapted from Sinha and Labi, 2007).

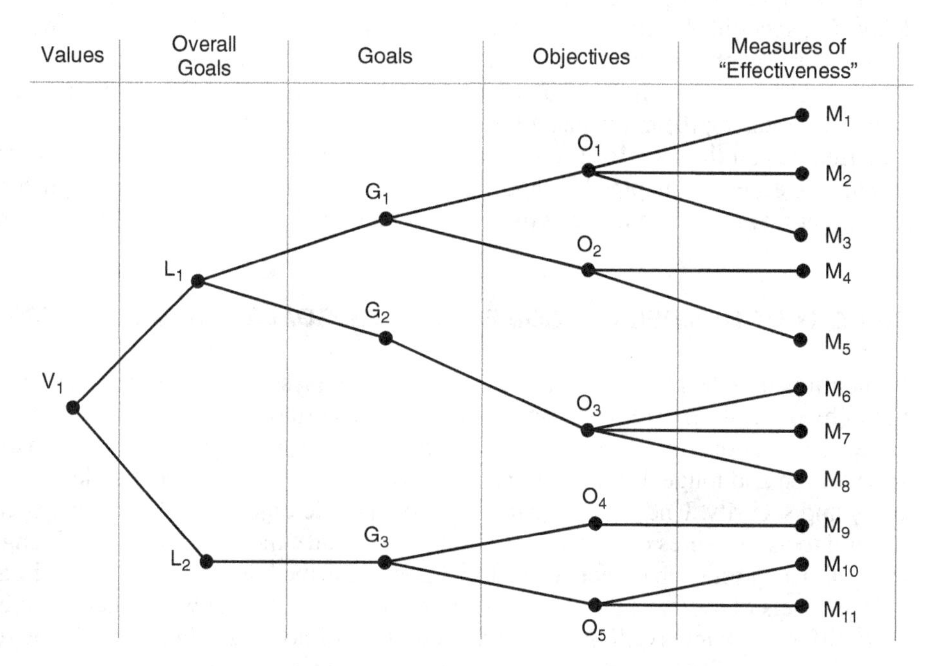

Figure 3.3 Illustration of hierarchical interrelationships among various levels of system outcomes (Adapted from Khisty et al., 2012).

behaviors they consider valuable and those they consider reprehensible. For instance, in some parts of the world, respect and care for the elderly is the highest value that members hold, and thus systems are duly designed to reflect that value; examples include the use of large lettering for public signs that cater to elderly users who often have weak eyesight, the provision of elevators or ramps with hand railings at areas where climbing or descending is needed by the elderly, and publicly displayed instructions to users to allow the elderly the premium use of the system such as bus seats. Also, in certain societies, material wealth, individualism, or communal spirit are highly valued, and civil engineering systems are designed, implicitly or explicitly, to reflect such values. In recent years, several professional engineering societies worldwide, the business community, and governmental agencies have identified sustainability as a key value they hold deeply. In Chapter 28, we will discuss the concept of sustainability and the various performance objectives associated with it. Also in Chapter 29, we discuss ethical issues, which are generally very much related to the values cherished by a society.

3.1.2 Overall Goals

For each value, the engineer establishes *overall goals* that represent broad statements of what the civil engineering system is meant to achieve (Sinha and Labi, 2007). There are three broad overall goals: efficiency (to what extent is the output worth the input?), effectiveness (is the system producing the intended outcomes?), and equity (are diverse or vulnerable segments of the population receiving a fair share of the system benefits?). We further describe each overall goal below.

Effectiveness. From an overall perspective of the system as a whole, effectiveness refers to the degree to which the system is achieving the benefits that were intended. It often deals with those benefits that are difficult to express monetarily, such as accessibility, public health, reduction in pollution, public welfare, and community development. From the perspective of individual phases of system development, effectiveness could refer to the extent to which the specific objectives of that phase are being realized; for example, at the construction phase of a system, the effectiveness of the construction process could be expressed in terms of cost overruns and time delays or absence thereof, adherence to materials specifications, and the workmanship quality of the constructed system.

Efficiency. Efficiency indicates the extent to which the system is providing some output (often monetary) with respect to some input (also often monetary). It is typically expressed as a ratio or difference of costs and monetized benefits (benefit is often considered to be synonymous with effectiveness). When both the costs and benefits are expressed in dollars, this overall goal is termed "economic efficiency," and when costs only are monetized, it is often termed "cost-effectiveness." For revenue-driven civil engineering systems, for example, bridge toll booths, economic efficiency is an important overall goal because the system owner or operator seeks to maximize monetary profits. On the other hand, for other public systems not driven by revenue, such as urban sewerage systems and nontoll highways, cost-effectiveness could be considered a more appropriate overall goal.

Equity. This is related to fairness, an important societal value that is also strongly associated with ethics and environmental justice. When civil engineers consider equity-related goals in their decision making, they help ensure that all segments of the population have a fair share in the benefits of their system or that certain segments do not suffer disproportionately from the adverse impacts of the system, regardless of their gender, age, income level, social status, disability status, residential or working location, etc.). For example, a new highway that connects a high-income suburban area to the downtown area, but that passes through a low-income neighborhood, may offer travel time benefits for road users but may cause socioeconomic disruptions in the communities it traverses.

3.1.3 Goals

A *goal* may be defined as a desired final situation toward which effort is directed. From the perspective of effectiveness, a goal may be expressed in terms of the physical conditions or operational characteristics of a civil engineering system or its effects on its environment. For a system's physical structure, goals may include system condition; for system operations, goals may include safety and convenience; and for system external impacts, the goals may include environmental sustainability and economic development. Goals also may be established from the perspectives of efficiency (to maximize net life-cycle returns from the system) or system equity (to maximize fair use of the system). Tables 3.1a–c present, for each of the three overall goals, examples of goals, objectives and MOEs that are often associated with civil engineering systems.

3.1.4 Objectives

An *objective* is a specific statement that is established from a goal and is designed to achieve that goal. For example, if a goal is to enhance system preservation, then a corresponding objective could be to improve the physical condition of the system over a specified period of time. Table 3.1 presents examples of objectives typically associated with a system *per se* or with any one of its development phases. Identification of objectives is important because it influences the evaluation and decision outcome as to whether to proceed with the tasks at that phase. Civil engineering systems engineers need to establish an array of system objectives that are diverse enough to reflect the different expectations of what the civil engineering system should achieve from the perspective of different *stakeholders*. For civil engineering systems, stakeholders typically include the **system owner** (city engineering offices; public works departments; municipal authorities; county or provincial engineering offices; national, state, or local departments; infrastructure departments of religious bodies; nongovernmental organizations; and private, quasi-private, or public companies that deal in energy, logistics, public transit, water distribution, waste treatment, etc.); **system users** (persons who use the system directly); the **community**, that is, persons who do not necessarily use the system but may be adversely affected by it through its externalities (noise, unsightliness, air pollution, accidents, etc.), persons whose farmland and other property are annexed for public projects, and persons who do not necessarily use the system but are positively affected by the system (through enhanced quality of life, increased accessibility, job opportunities, etc.); and the **taxpayers** and **the general public**.

The development of goals and objectives for civil engineering systems can be achieved by examining the needs, requirements, policies, and mission statements of the system owner or operator, and also by seeking the perspectives of the general public, particularly those likely to benefit or to be adversely affected by the system's implementation. Town hall meetings, Internet sites (including Facebook and Twitter), phone text messaging, and newspaper solicitations have been found to be particularly useful for obtaining stakeholder input.

3.1.5 Measures of Effectiveness

A measure of effectiveness (MOE), often termed a *performance measure*, is simply an objective that is stated in measurable terms. Variations of this term may include *performance criterion, performance attribute, performance indicator*, or *service attribute*. For example, for the goal of system preservation and its associated objective of improving the condition of bridges, the engineer could utilize a Sufficiency Rating (a scale of 0–9, 0 being a failed bridge, and 9 being a bridge in excellent condition) as the performance measure. An example of an MOE for a treatment plant could be the degree to which the output biological oxygen demand is enhanced by a new physical or operational system. In construction engineering, an example of an MOE for a new construction process could be the extent to which project delivery is enhanced through a reduction in the construction period, a reduction in overhead cost, or an increase in worker safety.

A *performance standard* is a fixed value of an MOE that specifies the desired level of some beneficial outcome. Thus, this defines the highest or lowest performance level or cutoff point for

Table 3.1a Typical Goals and Objectives of Civil Engineering Systems for Overall Goal of Effectiveness

Goals	Objectives	MOEs
Enhance service life and reduce probability of failure in the event of disaster	Enhance physical condition Increase system longevity	System condition index (cracking index, corrosion index, etc.) System remaining life System resilience index
Improve **operational effectiveness** so that system provides desired level of service	Decrease congestion, inconvenience, discomfort, and delay to user during operations of the system Decrease frequency and duration of downtime periods Maintain functional performance of the system (e.g., load-bearing, holding back water or earth)	Number of users delayed and delay duration per instance or interval of use Percent times that users face inconvenience Number of times that system shows signs of distress or is put out of service for maintenance Duration of maintenance (downtime) periods
Enhance **safe use** of the civil engineering system for the benefit of its users and bystanders/ pedestrians/ general public Minimize lawsuits (tort liability) associated with the system use	Reduce the frequency and/or rates of fatalities, injuries, and property damage associated with the use of the system Reduce the frequency and payment amounts for settling tort cases regarding system use	Number of fatalities or injuries Rate of fatalities or injuries Work zone incidents during construction Extent of annual property damage Frequency of vandalism Frequency/rate of incidents of crime Annual number of related lawsuits filed or settled Annual amount of related lawsuit payments
Minimize adverse **environmental impacts** (ecology, water quality and quantity, air/noise pollution, etc.) Reduce **energy use,** or enhance **energy efficiency** Minimize damage to **cultural** heritage or resources	Reduce air and noise pollution Reduce environmental (ecological) degradation Improve visual quality of the environment (aesthetics) Improve energy efficiency of the system Avoid damage to sites of cultural interest (such as ancient burial grounds, historic sites, and archeological treasures)	Tons of pollutant emitted per year Percentage of green/open space/ park land Amount of energy consumed per year; energy consumed per user or per instance of use Extent (area) of intrusion of cultural and treasured historical sites
Provide a solid civil engineering **infrastructure base** that will attract new businesses and retain existing ones, thereby enhancing the economic competitiveness of the region	Increase employment Increase business output and productivity Increase the number of businesses	Number of jobs created Number of job losses avoided Increase in gross regional product Increase in business sales

(*continued*)

Table 3.1a *(Continued)*

Goals	Objectives	MOEs
Reduce initial or life-cycle **costs** for owners and users of the civil engineering system	Reduce initial cost Reduce agency expenditure over system life Reduce inconvenience and other costs to the users during system downtime Reduce out-of-pocket costs and other costs to users during normal systems operations	Initial cost of system construction Cost to users during downtime periods Agency cost over life cycle User cost over life cycle Maintenance or operating cost per user or per interval of use Cost incurred by users per instance of use or per unit time period

Table 3.1b For the Overall Goal of Economic Efficiency

Goals	Objectives	MOEs
Enhance **financial performance** and economic attractiveness of the system Enhance **economic viability** and financial feasibility of system Maintain and enhance the **profitability** of the system (for privately-owned or operated civil engineering systems)	Maximize the benefit–cost ratio or net present value associated with actions or investments at any phase (construction, preservation, monitoring/inspection, and operations)	Benefit–cost ratio Net present value Internal rate of return, Payback period Gross or net revenue obtained per user or per unit time period

Table 3.1c For the Overall Goal of Equity

Goals	Objectives	MOEs
Enhance **general quality of life** and community well-being Promote **social equity**	Enhance community cohesion Enhance accessibility to social services Provide opportunities for system use by handicapped and other socially disadvantaged groups Increase recreational opportunities	Number of displaced persons/farms/businesses/homes Benefits per income group Accessibility to the handicapped Ensure proportional benefits to low-income groups

adjudging whether a deficiency exists and if some remedial action is necessary. Synonyms include *performance trigger, threshold*, or *minimum level of service*. For example, for the goal of system preservation, the objective of improving bridge condition, and the performance measure of bridge Sufficiency Rating, the performance standard could be *minimum bridge Sufficiency Rating of 4 units*.

While the term "effectiveness" often connotes a benefit, it could also refer to an adversity. In that case, the term "measure of effectiveness" could be considered a misnomer. To avoid confusion, some texts consider benefits to be synonymous with effectiveness and thus make a distinction between measures of cost (MOC) and measures of effectiveness (MOE) or measures of benefit (MOB). In this text, MOE is used in both contexts.

MOEs are vital in systems development because it is important to ascertain whether the intended goals of the system or of any of specific tasks or processes at each phase, are being achieved, and if so, to what extent. It is preferable to have a quantitative, rather than qualitative, statement of the MOE, as the latter is often fraught with subjectivity, inconsistency, and bias. A quantitative statement, often an index or rating, could have linear or non-linear gradations from the least performance to the highest performance. For economic and technical indicators of system performance, such a scale typically involves a continuous variable. However, sociocultural indicators are often represented by discrete categorical variables that may or may not be ordinal. de Neufville and Stafford (1971) emphasized the importance of selecting appropriate MOEs for engineering systems:

> The choice of measures of effectiveness is crucial because it determines to a great extent the final design. The choice is important because the merits of each particular configuration of a system may appear different from different points of view. What may seem advantageous from one standpoint may not be so from another. Thus, the selection of the preferred design may hinge on the choice of the measures of effectiveness.

The importance of selecting appropriate MOEs was also highlighted by Dandy et al. (2008):

> When a project has more than one objective, a MOE is needed for each specific objective, together with an overall measure that takes account of the separate objectives.... The manner in which MOEs are formulated can be enormously influential in determining the direction that a project will take. In fact, the attempt to define a measure of effectiveness is a valuable starting point in design and planning, even for small projects, because it gives fresh insight into the project.

Also, it is vital to realize that at a given level of an MOE, the value associated with a unit increment in MOE is not always constant across space, time, size of the system, level of system demand (usage), or circumstance of use. To illustrate this nonlinearity in MOE value, consider the following classic hypothetical example: The value (utility) of the initial drinks of water given to a thirsty desert wanderer is not the same as that of the later drinks. In other words, the value of each incremental drink decreases as the number of drinks increase. His second drink may have a value of 0.5 times the value of his first drink, but his fourth drink may have a value less than 0.25 times the value of his first drink. Such nonlinear relationships between resources and outputs add further complexity to the choice of an appropriate measure of system effectiveness, and limits (ranges) or scale adjustments often need to be applied to enhance the applicability of the selected MOEs to specific contexts of system evaluation.

Dimensions of MOEs. There are several ways by which MOEs could be classified. One of these is the level of management: an MOE could established for a project-level or facility-level evaluation (deciding on the best action for a specific system on the basis of system-specific performance) or a network-level evaluation (identifying which systems of a larger network of systems deserve some action on the basis of considerations such as average network performance of the network or minimum performance of any system in the network). Thus, there could be MOEs at the project level or the network level. Also, as pointed out earlier in this chapter, an MOE could also be viewed from the perspective of whether it is perceived as a "benefit" or as a "cost." For example, an increase in system durability is a benefit while construction cost and ecological degradation are costs. Further, a MOE may be monetary (i.e., expressed in dollar values, such as the revenue from system user fees) or nonmonetary, such as increased durability in years. There are others that are intrinsically monetary; often not expressed in dollar values but capable of being expressed as such using established unit rates; for example, an increase in system condition is nonmonetary but can be expressed in monetary value by knowing the relationship between the system condition and the system operating costs. Another MOE dimension is the affected stakeholder—an MOE may cater to a specific stakeholder [e.g., enhanced system longevity (system owner), reduced inconvenience in using system (system user), and avoidance of environmental damage (community)].

Properties of a Desirable MOE. The choice of an MOE can profoundly influence the decision to proceed with the development of a particular system or to make specified improvements to an existing system. Thus, MOE selection can be "tricky business." Fortunately, a number of researchers have established guidelines by which one can select an MOE for a specific systems-related decision problem; these include (Cross and Lynch, 1989; CamSys, 2000; Sinha and Labi, 2007):

- *Appropriateness*: The MOE should reflect one or more goal or objective of the overall system or of the phase in question.
- *Measurability*: The MOE should allow thc systems engineer to assess quantitatively the impact of each alternative in terms of that MOE.
- *Dimensionality*: The MOE should facilitate measurement of the anticipated level of each performance or other attribute that is associated with the proposed system or action. For example, it should be able to measure the effectiveness of actions using appropriate temporal and spatial dimensions that are consistent with the physical structure of the civil engineering system or the reach of its operations or performance. Also, it should also adequately account for the concerns or perspectives of stakeholders. The MOE should be comparable across different geographic regions or time periods.
- *Realistic*: It should be possible to extract or generate useful and reliable data that are related to the MOE, without undue cost, effort, or time.
- *Defensible*: The clarity and conciseness of the MOE should be such that using that MOE, the system can be easily assessed and interpretation of the MOE and its levels can be effectively communicated among decision makers and to an audience of technical and nontechnical personnel as well as the general public. Often, this is possible when the MOE is simple enough to permit a determination of its suitability for a given system under various present and future design and management scenarios.

The next section discusses the various contexts in civil engineering systems development where it is necessary for civil engineers to select MOEs to evaluate existing civil engineering systems or investment decisions for systems.

3.2 COMMON MOEs USED IN DECISION MAKING AT ANY PHASE OF SYSTEM DEVELOPMENT

In the previous section, specifically in Table 3.1, we identified a number of MOEs under various system goals and objectives. Now we proceed to discuss some of these MOEs in greater detail. Needless to say, the reader will be able to recognize that there could be several different ways of categorizing these MOEs, depending on the dimension of interest, for example, monetary versus nonmonetary, initial versus lifecycle, project level versus network level, and so forth. In this section, we discuss the various typical performance measures under the monetary and stakeholder dimensions (Figure 3.4). An additional MOE dimension—the phase of systems development at

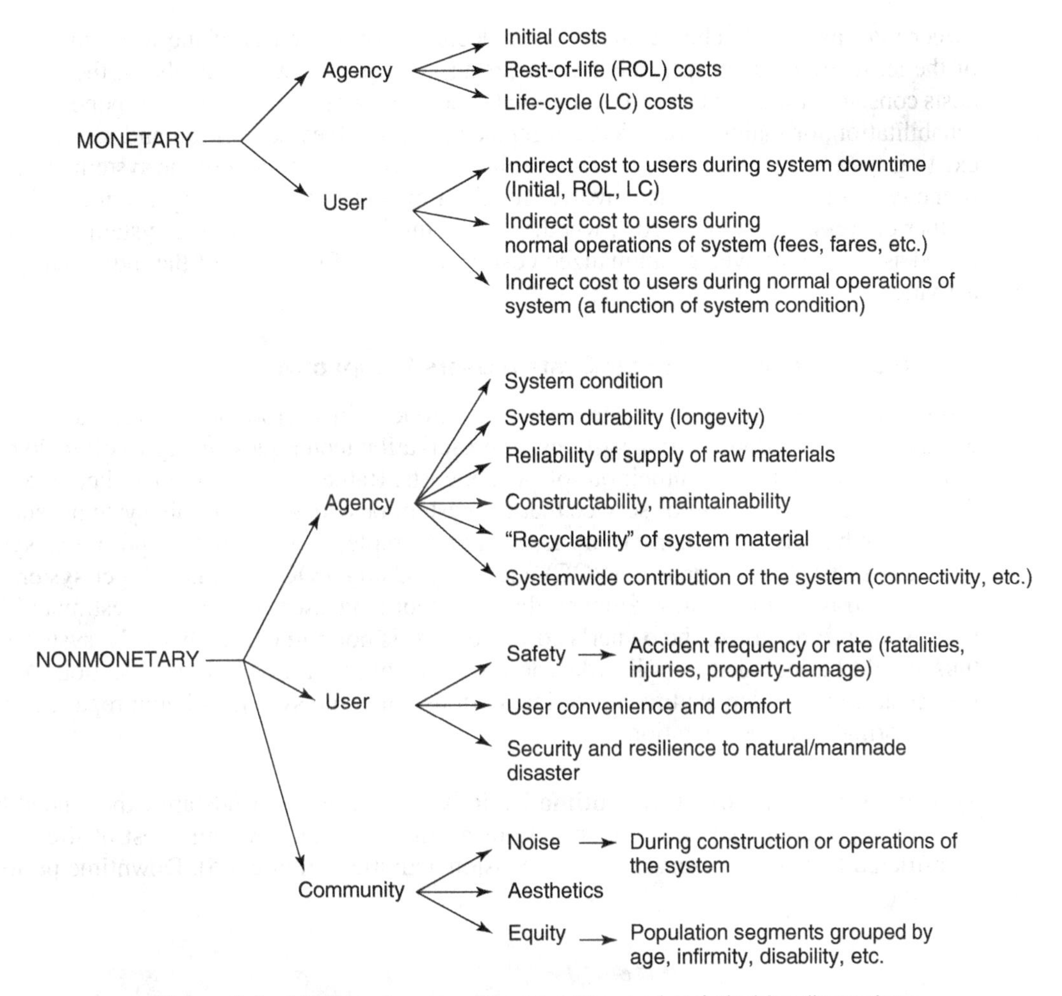

Figure 3.4 MOEs categorized by monetary and stakeholder dimensions.

which the MOE is used—is discussed in Section 3.4, and in that section we provide examples of MOEs that could be used at each of the phases.

3.2.1 Monetary MOEs from the System Owner/Operator Perspective

Monetary performance measures can be categorized into initial costs and rest-of-life costs or benefits. Initial costs pertain to construction costs while rest-of-life costs consist of maintenance and operating costs and, in certain cases, salvage costs. Rest-of-life benefits include reductions in user costs compared to some base case scenario, revenue, and residual worth of the system. The life-cycle cost is the sum of the initial costs and the rest-of-life costs.

Cost to the Systems Owner (Initial or Life Cycle). The use of life-cycle costs are often preferable to initial costs for investment analysis, particularly from a sustainability viewpoint. However, investment decisions that are hinged on life-cycle considerations may call for a relatively large initial outlay that may be out of the reach of a cash-strapped system owner. In such cases, the owner may have little choice but to make decisions on the basis of the least initial cost instead of the least life-cycle cost. In civil engineering systems investment analysis, the system owner's costs consist of the in-house and contractual costs of a system (or system component) replacement, rehabilitation, or maintenance. System replacement involves demolition and reconstruction of an existing system; rehabilitation involves replacement of a component of the system or major retrofit of the system; and maintenance involves repair of localized distresses or prevention of imminent or further distress. The agency cost, whether occurring initially or over the system life, may be estimated as an overall average annualized cost or in terms of the costs of the individual preservation activities.

3.2.2 Monetary MOEs from the System Users Perspective

From the perspective of the system user, differences in performance across alternative plans, designs, materials, construction systems, or preservation techniques/strategies often do not directly translate into differences in their out-of-pocket costs. Rather, such differences become manifest as the differences in the level of service that the system provides. For certain systems, such levels of service can be quantified in monetary terms. For example, in highway transportation, system users incur costs of vehicle operation (VOC) and delay. Using VOC unit rates ($ per system condition level) and travel time values ($/hour), the corresponding user costs can be estimated in dollars. Clearly, therefore, unlike the owner's cost, user cost is not borne directly by the system owner and thus should be included in policy/decision analysis only with due circumspection. Also, system user costs can be calculated for downtime periods when the system is being repaired or modified or for normal systems operations.

System User Costs during Downtime Periods. Downtime periods are experienced by system users during reconstruction of the system or at times during a system's rest of life when its use is restricted to enable preservation or expansion activities (Figure 3.5). Downtime periods, which

Figure 3.5 Civil system downtime, full or partial, can lead to increased user costs of delay, safety, or inconvenience (Floydian/Wikimedia Commons).

are associated with user costs (often convenience, delay, and safety), have durations that vary by activity; for example, repair or expansion actions involving elaborate or complex system designs, materials, and the like generally have longer downtime periods than others. The calculation of system user costs during downtime operations often involves first determining the inputs, namely, system usage volumes and the physical and operational characteristics of the system during normal usage periods as well as during downtime periods. In the context of highway systems, this could include traffic volumes during normal operations and during work zones, the operation hours of the work zone, and the work zone length, as well as the construction duration, capacities, and speeds. The unit monetary values of the user delay, inconvenience, and so forth due to the downtime period are then determined in order to calculate the total monetary user costs over the duration of the downtime period.

System User Costs during Periods of Normal Operations of System. System "normal operations" refer to the regular use of the system for which it was designed. The condition of a system can affect the operating costs of users [i.e., the money that the users (people, mobile components of the system such as vehicles, and so on) incur indirectly in using the system]. The components of these operating costs often include maintenance, repair, and depreciation. In highway systems, for example, the motion of vehicle tires on a rough pavement surface is associated with greater resistance to movement, which leads to higher levels of fuel consumption compared to traveling at a similar speed on a smooth surface, and a bumpy ride, which leads to increased vibration and wear and tear of vehicle parts. Thus, a vehicle that is operated on a rough pavement surface is likely to lose its value faster than one that is operated on a smooth surface pavement. Also, an indirect effect of poor pavement conditions is that road users may be forced to drive at lower speeds, generally leading to higher fuel consumption and air pollution.

3.2.3 Nonmonetary MOEs from the System Owner/Operator Perspective

For the system owner or operator, nonmonetary performance is often synonymous with so-called technical performance, as discussed below.

(a) Physical Condition of the Civil Engineering System. The system's physical condition is probably the most widely used MOE. For each civil engineering system, engineers have established various indices and rating schemes for measuring the system's physical condition. In the case of the wearing surface of bridge deck systems, for instance, common MOEs of physical condition are surface roughness, faulting, cracking, and skid resistance. Some system owners use a customized measure of physical condition that is a weighted combination of two or more indicators. It is worth noting that the "physical condition" MOE measure is related to other MOEs because poor system condition translates into increased cost of safety and other user costs.

(b) Structural Integrity of the System. Often, the physical condition of the system surface alone is inadequate for providing a complete characterization of the overall system quality and integrity. A system with a good surface condition may have significant structural problems beneath that are not immediately obvious. As such, for certain systems, ways have been established to measure the underlying structural integrity of the system (Figure 3.6). For some civil engineering systems, poor recordkeeping of past construction activities has bequeathed a dearth of data and knowledge of the strength-related attributes of system components, thus severely limiting the use of this MOE in analyzing decision alternatives for these systems. For several civil engineering systems, besides the main goal of the system in providing some public service, a main goal (addressed at the design phase) is to ensure that the system is able to support the dead and live loads that are expected. This

Figure 3.6 Measuring strength performance of systems—an example in highway pavements (Courtesy of Quality Engineering Solutions, Inc.).

has been the case since the dawn of time. During King Hammurabi's reign in ancient Mesopotamia (1785–1750 BC), this goal was taken very seriously—as we may recall from Chapter 1, builders were put to death if their structures collapsed and killed the occupants (Dandy et al., 2008).

(c) System Life. Alternatives for design of civil engineering systems typically differ in the lives that they offer. However, it is important to identify what kind of life is to be considered for the evaluation of these system actions. In Section 26.3.1, we discuss the various types of system life including the physical life, functional life, service life, economic life, and technological life.

(d) System Reliability. System reliability is a measure of civil engineering system performance that is closely related to system physical condition, durability, or operational performance (Frangopol et al., 2007). The reliability of a civil engineering system, which is relevant mostly at the system operations phase, is its ability to perform its required functions under a given set of internal and external conditions for a specified period of time. It is often reported in terms of a probability (Kapur and Lamberson, 1977). In Chapter 13, we discuss the concepts of reliability and we provide hints of how they could be incorporated as a measure of system performance.

(e) Reliability of Supply of Resources. For systems design or rehabilitation alternatives that differ by material type, the guaranteed availability of raw materials may be a critical factor that could sway decisions toward a particular material type. For example, in certain countries, the scarcity of cement may render concrete structures an unattractive alternative compared to steel structures. Also, in certain countries and at certain times, the fluctuating prices of oil and its derivatives due to unpredictable supply or demand, leading to lower certainty and reduced reliability of bitumen supply, may render asphaltic concrete pavements unattractive to the system owner compared with

Portland cement concrete pavements. Similarly, at the system construction phase, the availability of certain equipment or labor skills may lead to the choice of one specific contracting approach or construction method over another.

(f) Constructability and Maintainability. A constructability review is a process in construction design whereby the system plans are reviewed by others familiar with construction techniques and materials to assess the constructability of the design. Similarly, a maintainability review is a process whereby system designs are reviewed by others familiar with maintenance processes to ensure that the system would not unduly require excessively frequent and intense maintenance over its life cycle. System designs and material alternatives may differ in their constructability and maintainability levels; and all other factors being equal, design alternatives having the highest possible levels of constructability and maintainability are preferred. Voland (2004) stated that designs that reduce the need for maintenance are critical because many agencies are often unable or reluctant to perform maintenance tasks diligently, leading to failures, worn parts, and disruptions in the system operations.

(g) Reclaimability or Recyclability of Systems Material. A key tenet of sustainability is the reuse of existing material as part of future actions (as we shall see in Chapter 28). For certain system design alternatives, it is possible to salvage old materials from a system that has reached its end of life, for use in the new system. In some cases, the process is carried out in situ, thus eliminating the need to haul away the old material for disposal. Also, such recycling can reduce the movement of construction trucks in the work zone and the need for detours. Further, these processes can reduce the system owner's cost while preserving natural resources because they reuse existing materials and conserve raw materials.

(h) Exposure to Tort Liability. Tort liability is the compensation for damages caused by inaction or negligent actions by the employees of a system owner or operator (as we shall discuss further in Chapter 29). Like all infrastructure, the construction, maintenance, and operations of civil engineering systems can expose its users or the community to the risks of personal injury, fatality, and property damage and can thus give rise to tort liability against the system owner. For example, certain runway or highway pavement surface materials may have relatively little surface friction and promote skidding or may be more prone to rutting and subsequent hydroplaning during wet weather compared to other materials. At the current time, when many public agencies have lost their immunity from liability for damages resulting from exercise of their proprietary functions, system owners and operators are increasingly becoming concerned about their exposure to tort liability.

(i) Resilience to Natural or Man-made Disaster. This is an important MOE because certain design alternatives, by virtue of their material type, structural detail, or orientation, may be more resilient to floods, earthquakes, tsunamis, landslides, or terrorist attacks. Thus, the resilience of a civil engineering system is a function of its design. In Chapter 27, we will discuss in some detail the concepts of overall hazard in terms of external and internal threats, community exposure, and system resilience.

(j) Network Effects (Connectivity, etc.). Any network comprises several interconnected links (see Chapter 17). Typically, a civil engineering system constitutes part or all of a larger network of systems. The system may be a link or a facility located on a link. Each link has a different impact on the network from the perspective of connectivity. All else being equal, for a network of systems, it is preferable to invest limited funds toward preserving a specific system that is most critical from the perspective of network connectivity. The network connectivity impacts of a link may be measured in terms of the consequences of the absence of the link on individual or total network origin–destination travel time and impairment of the possibility of specific routes (Euler, Hamiltonian, traveling salesman, and Chinese postman). Network-related performance criteria are relevant only when the competing investment alternatives are located at different points on a network of systems.

(k) Economic Efficiency. Economic efficiency is often used where it is desired to evaluate system outcomes on the basis of purely monetary impacts. When all costs and benefits (preferably, over the system life cycle) can be translated into dollar values and both can be expressed in their present worth by duly adjusting for the interest rate or inflation or both, analysis of system alternatives can be carried out on the basis of economic efficiency. MOEs for economic efficiency include net present value (NPV), which is the algebraic difference between the monetized benefits and costs; benefit–cost ratio (BCR), which is the ratio between the monetized benefits and the costs; equivalent uniform annual cost (EUAC); equivalent uniform annual return; internal rate of return (IRR); and payback period. In Chapter 11, we provide a more detailed discussion of economic efficiency analysis as a tool for the task of evaluating system economic and financial performance.

(l) Cost-Effectiveness. Cost-effectiveness is used instead of economic efficiency when it is not possible to express some or all the MOEs in dollar values. As such, cost-effectiveness is always expressed as ratios and not as algebraic differences. Examples of cost-effectiveness MOEs are crash reduction per dollar and customer complaints per dollar of operating cost.

3.2.4 Nonmonetary MOEs from the Perspective of the System User and the Community

Prior to the 1970s, civil engineering system decisions primarily were made on the basis of technical and economic considerations, while social and environmental impacts played little or no roles in such decision making. This largely reflected the priorities of society at the time. In the current era, it is widely accepted that engineers have a role to play in the prudent use, conservation, and management of the Earth's resources and that the needs of future generations are duly taken into account in systems evaluation and implementation (Dandy et al., 2008). Therefore in this section, we will discuss a number of nonmonetary MOEs that are related to sustainability or are often of great interest to the system user and the community.

(a) Environmental Effects. It is important to have not only MOEs that measure the impact of a civil engineering system on its environment but also vice versa—MOEs that measure the impact of the environment on the system (Figure 3.7). With regard to the former, the impacts of different system design or operations alternatives may occur through noise, air pollution, aesthetics, and community disruption.

Also, certain materials used to construct or maintain civil engineering systems are associated with significant levels of pollution to soil, groundwater, or proximal surface water courses. For highway pavements, for example, experimental studies have shown that there are different levels of pollutants in runoff from different pavement surface material types (James and Thompson, 1997). It is desired that materials used at any phase should not be harmful to the environment, and engineers are encouraged to anticipate and eliminate any hazards to the environment in specifying materials at the design or other phases (Voland, 2004).

With regard to the impacts of the environment on the civil engineering system, it is important to realize that environmental factors, such as the natural agents of corrosion and oxidation,

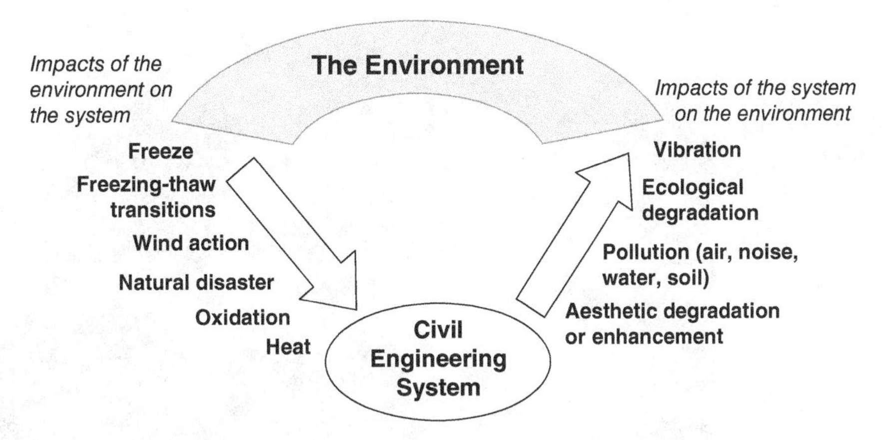

Figure 3.7 Impacts between civil engineering systems and the environment.

moisture, freeze conditions, freeze–thaw transitions, and wind, can have significant impacts on civil engineering systems. Voland (2004) presented three interesting case studies where failure to account for the interactions between a civil engineering design and its environment subsequently led to serious problems: improper waste disposals at Love Canal and Times Beach and the collapse of the Baldwin Dam. For MOEs that are related to the impact of the environment on the system, the issue is how well the system is protected against these agents. So, for example, a design, construction technique or operational policy that builds the system resilience will be favored over one that does not. While resilience is described in the previous section as an agency-related MOE, it is worth noting that system users also could be adversely affected by a system that has low resilience to threats.

(b) Aesthetics. Unlike the case for buildings where architects make efforts to incorporate architectural considerations in design, the design of heavy civil structures has often proceeded without explicit efforts to enhance their visual quality. The choice among different civil engineering system designs has rarely been influenced by aesthetic considerations. Often, this is understandable, as the physical structure for certain civil engineering systems is mostly buried underground and therefore not visible to the eye. In recent years, however, designers of surface civil engineering systems or their surface components are increasingly attentive to aesthetics as one of the MOEs in choosing between alternative system designs. This is particularly relevant in urban settings where the physical system constitutes a major or even dominant component of the visual landscape. Also, by incorporating features (color, texture, shape) of the local environment into the civil engineering system design (e.g., surface material type), the design can be made more context sensitive and compatible with its environment. Detailed descriptions of MOEs and estimation procedures for aesthetic impact assessment of civil engineering systems are available in the literature (Ortolano, 1997; Sinha and Labi, 2007). Figure 3.8 presents the Zhaozhou Bridge over the Jiaohe River (constructed AD600) in the Hebei Province. China's oldest standing bridge and the world's oldest open-spandrel arch bridge, the Zhaozhou Bridge combines functionality and aesthetics.

(c) Community Impacts and Distributive Effects. As will be seen when we discuss ethics in a later chapter, civil engineers have a responsibility to society that transcends their responsibility to clients or colleagues. As such, in evaluating decisions at any phase of systems development, it is critical for the engineer to include the system's social impacts as one of the performance measures

Figure 3.8 The Zhaozhou Bridge, China's oldest standing bridge and the world's oldest open-spandrel arch bridge, combines functionality and aesthetics (Courtesy of Zhao1974/Wikimedia Commons).

so that choices that yield maximum beneficial (or minimal adverse) social impacts are made to have higher priority. In choosing between alternative system locations at the planning phase of civil engineering systems development, the distributive effects of each alternative, especially with respect to sociocultural impacts, can be critical. This is often the case when the civil engineering system (i) requires unusually large areas of urban right of way; (ii) involves the displacement of a large number of households, businesses, community amenities, historic districts, and landmarks; (iii) conflicts with local land-use plans; and (iv) unduly and inequitably reduces the welfare of vulnerable populations (Sinha and Labi, 2007). MOEs for evaluating alternative system locations on the basis of the community and distributive impacts, therefore, could include the number of relocated households, businesses, and public facilities, the area of urban right of way affected, and the level of attainment of environmental justice. The community impacts and distributive effects of a number of civil engineering systems can be assessed using methodologies that are available in the literature (OECD, 2001; The World Bank, 2003).

(d) Sustainability. Civil engineers have come to the realization that through the design and operations of their systems they can influence the conservation and management of the Earth's natural resources. As such, it is becoming common for developers of civil engineering systems to consider goals related to the environment and society (Dandy et al., 2008). In a strict sense, sustainability is not a single goal by itself but rather is typically the encapsulation of numerous goals (Jeon and Amekudzi, 2005) some of which we have discussed in this chapter. Indeed, in systems development, the quest to minimize the impacts on the system environment, ecology, socioculture, and economy and to efficiently utilize natural resources is consistent with sustainability even if the word "sustainability" is not explicitly mentioned in most investment evaluations. Related to this situation, Kates et al. (2005) pointed out that sustainability requires that diverse stakeholders participate in the decision process and their perspectives be duly accommodated; the intent is to facilitate

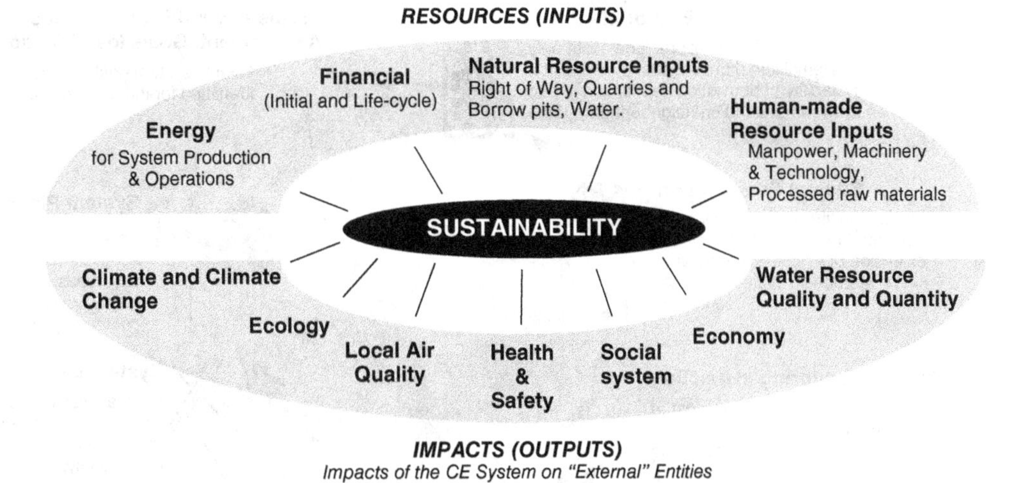

Figure 3.9 Some elements of sustainability in civil engineering systems development.

the reconciliation of different and often conflicting viewpoints so that as much as possible, the best decision is one that achieves multiple values not only simultaneously but also synergistically. Dandy et al. (2008) and IISD (2009) assert that the systems approach provides a simple and consistent basis for investigating sustainability at all level of society, on a global scale as well as at the individual level. Figure 3.9 presents some elements of sustainability in civil engineering systems development. In Chapter 28, we will discuss in greater detail the subject of sustainability.

3.3 GOALS AND MOEs AT EACH PHASE OF CIVIL ENGINEERING SYSTEMS DEVELOPMENT

As we emphasized in the Introduction to this chapter, besides the primary goal or objective for providing a civil engineering system, which is identified at the initial phase of development, each of the remaining phases also has its unique objectives. For example, for a planned freeway system around a city, the overall goal may be to mitigate traffic congestion, and the objectives at the other phases include: At the construction phase, an objective may be to complete the freeway construction project within the original budget and contract period; and at the preservation phase, an objective may be to extend the life of the freeway within available maintenance budget. Figure 3.10 presents examples of MOEs at various phases of civil engineering systems development. These MOEs are discussed below.

3.3.1 The Planning Phase

At the planning phase where the engineer's tasks include selecting an optimal location for the system, MOEs that could be considered include proximity to sensitive ecosystems, the extent of preparatory work required at the site to receive the system, and how well the system would blend in with the aesthetic, community, and social features of that location. Other location-related MOEs include contribution to accessibility or network connectivity, proximity to facilities or routes associated with national security, emergency evacuation, and international borders. In Chapter 20,

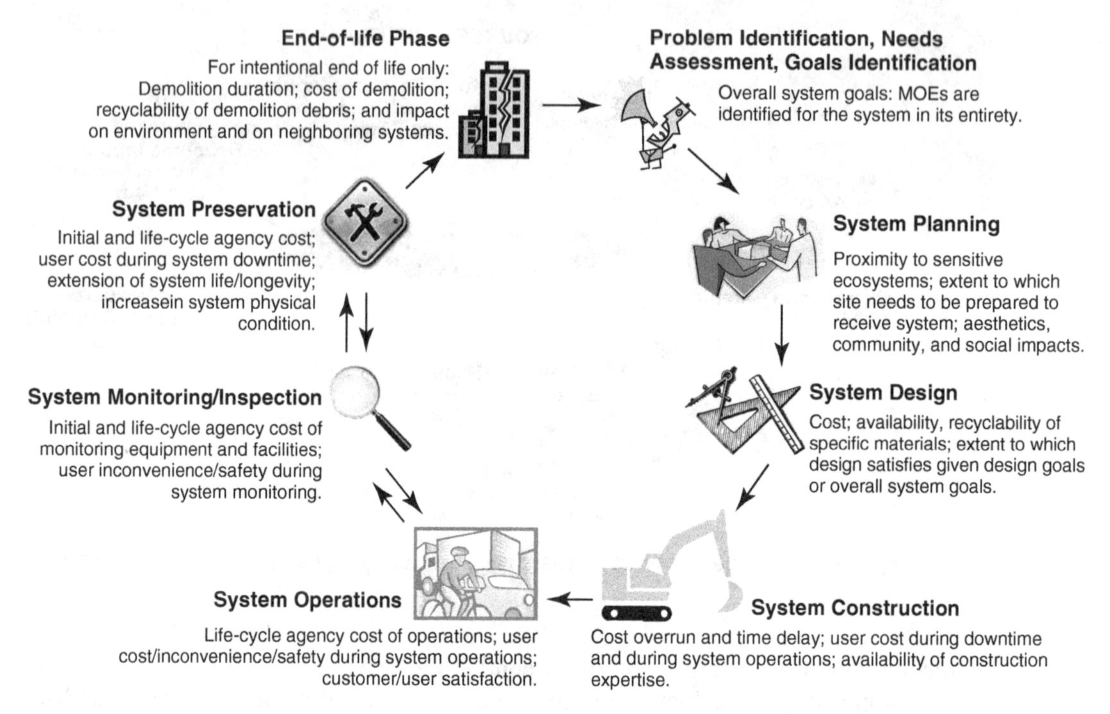

Figure 3.10 Examples of MOEs at various phases of civil engineering systems development.

we provide additional contexts and illustrations where MOEs are needed at the system planning phase.

3.3.2 The Systems Design Phase

Alternative designs often differ from each other in terms of the dimensions of the system components, the constituent materials, and the configuration of the components. As we will see in Chapter 21, the engineer's tasks at this phase include selection of the best design from numerous alternatives. The alternatives can be compared on the basis of MOEs whose levels differ across the design alternatives, and these often include material strength, material durability, resistance to corrosion, cost, and availability of raw materials. Also, in certain cases, architectural considerations (e.g., aesthetics) and the safety and convenience of facility users are useful MOEs in evaluating different design alternatives.

3.3.3 The Systems Construction/Production/Implementation Phase

In Chapter 22, we will discuss typical decisions faced by the system owner at this phase: choice of contracting approach (in-house work execution, traditional design–bid–build, warranty, etc.) and the choice of the appropriate contractor to deliver the project. MOEs that could sway the choice from one contracting approach to another include availability of funding (lack of funds causes many agencies to resort to private–public partnership contracts and nontraditional delivery approaches including design–build–operate). Also at this phase, the contractor faces decisions that include identification and implementation of optimal combinations of materials, manpower, and equipment for each task. MOEs that could influence the distribution of resources (materials, manpower, and equipment) for a specific task include equipment productivity and availability of skilled manpower.

3.3.4 The Systems Operations Phase

At this phase, the engineer seeks the best policy, from a wide array of alternatives, for operating the civil engineering system. For example, a hydraulic engineer may seek the best opening width for a sluice gate in order to ensure that flow does not exceed some critical level; and a city traffic engineer may seek the best combination of signal timings for arterials in a city in order that traveler delay is minimized. The specific MOEs used to make decisions at the system operations phase depend on the system type under consideration. These generally may include the system owner's costs of operating the system (different operational policies may have different staffing requirements, equipment deployments, etc. and thus different costs); the costs, inconvenience, and safety incurred by the system users as the system is in operation; and the extent of satisfaction of the system users (typically captured using opinion surveys). In certain cases, the likelihood or number of tort suits (or total expected settlement amount) could serve as an MOE (i.e., different operational policies can have different impacts on the exposure of the system users to injury and hence the exposure of the system owner to tort liability). In Chapter 23, we present additional examples of tasks at this phase and how MOEs are established to address those tasks.

3.3.5 The System Monitoring and Inspection Phase

As the civil engineering system is used, it suffers both visible wear and tear as well as damage that is not readily visible. Thus, it is important for the engineer to monitor the levels at which the system is being used simultaneously with the physical condition of the system. As we will see in Chapter 24, monitoring the level of usage is important because such information can help the system owner/operator to (i) ascertain whether there is a need to enhance or reduce the capacity of the system to accommodate increased or decreased demand, and (ii) to predict the future time when such capacity changes will be necessary. Also, regular inspections of the facility are important to determine whether some remedial physical work is needed to address defects or to prevent imminent ones. In civil engineering systems, such defects, which may be at the surface or deep within the system structure, may include cracking, corrosion, spalling, deformation, settlement, erosion, and the like. The nature, scope, and scale of these defects depend on the type of material, environment, and level of usage. Therefore, at this phase of systems development, the engineer seeks the best equipment to monitor/inspect the system as well as the optimal frequency of inspection. In making decisions for selecting inspection equipment, the engineer utilizes MOEs that include the initial cost of the equipment; the life-cycle maintenance cost of the equipment and its accessories; the capability of the equipment to detect the defect as efficiently as possible with minimal false negatives; and for intrusive systems, user inconvenience and safety during the inspection process. In making decisions to select the optimal equipment and labor to monitor the level of system usage, the engineer utilizes MOEs that are similar to those for inspection: cost, capability of the equipment to measure the frequency and intensity of usage with as few errors as possible, and user inconvenience and safety during the monitoring process.

3.3.6 The Systems Preservation Phase

In preserving their physical civil engineering systems, many engineers are moving away from investment decisions based only on the initial consequences of implementation to those based on consequences that occur throughout the system life. As such, for a given system design, an engineer may seek the best of the alternative activity profiles available for preserving the system. An activity profile is defined as a strategy or schedule for applying various preservation treatments, namely, in each year, which specific treatment(s), if any, should be applied? In choosing the best profile, the engineer may consider MOEs that include cost (agency and/or user) and benefits (system condition and durability) that are expected from the individual treatments that constitute the preservation activity profile.

In certain cases, however, instead of seeking the best preservation actions over the entire system life, an engineer may still be more interested in more short-run decisions: What treatment could be applied to a system at a given point in time with no apparent consideration of consequences in the longer term? While short-term contexts are generally inappropriate and, thus, generally imprudent, an engineer may need to make decisions in such contexts for a number of reasons, including political expediency. Often, in making decisions in the short-term context, an important consideration is the additional or incremental benefits or costs in terms of the relevant MOEs. In other words, what additional benefit can be accrued by additional spending? Resurfacing, for example, may be a poor choice for a recently constructed road already in excellent condition because it offers little incremental benefits for its high cost. Chapter 25 presents more discussions on system preservation.

3.3.7 The System End-of-Life Phase

System end of life may be intended or unintended. In case of the former, there are opportunities for choosing the most cost-effective technique to end the system life. Intended termination may be for reasons such as facility obsolescence, structural problems, capacity limitations, and other reasons as we have explained in Chapter 26. In cases where the system end of life is intended and the system is demolished (Figure 3.11), the engineer has a choice of the mode of demolition, which in turn is made on the basis of appropriate MOEs. The MOEs may include the time duration for the demolition process, the cost of the demolition technique, the recyclability of demolition debris, and the impact of the demolition on neighboring systems, either through direct impact or spread of the demolition debris or geotechnical problems caused by ground vibration and movements due to the demolition exercise.

Figure 3.11 MOEs for selecting an appropriate termination method for a system can include the impact of the termination method on surrounding facilities (Image credit: Courtesy of Elite Photos, www.elitephotos.co.uk/media).

3.4 DESIRED PROPERTIES OF A SET OF MOEs FOR A GIVEN ANALYSIS

As we have just learned in Section 3.3, at the inception of the systems development cycle where a need for the system is identified, the engineer develops a set of MOEs to serve as a basis for evaluating the operational performance of the system at its operations phase. Also, at each phase of the development cycle, the engineer establishes specific objectives to evaluate the success of various tasks associated with that phase. Keeney and Gregory (2005) and Patidar et al. (2007) provided guidelines for the desired characteristics of any set of MOEs assembled for a specific systems analysis and evaluation problem.

- **Completeness.** First, the set of MOEs should be complete (i.e., it should adequately indicate the degree to which the objective will be met) and should not exclude any key critical performance criteria. In this respect, MOEs should adequately reflect the full spectrum of perspectives of the stakeholders of the civil engineering system or the different aspects of the problem at hand. For example, for the budget-conscious student who seeks a place to have lunch, food price must not be excluded in her MOE list.
- **Operational.** Second, the set of MOEs should be useful and meaningful so that the overall implications of each alternative can be understood). For example, for the budget-conscious student who seeks a place to have lunch, whether or not the restaurant has waiters is likely to be not a meaningful MOE.
- **Free of Redundancy.** Also, the set of MOEs should be free of redundancy, double counting, or overlaps. In this context, the set should be as small as possible to reduce the likelihood of confounding or redundancies. For example, for the student who seeks a place to have lunch, distance from her office (in meters) and the time taken to get there (in minutes) are redundant MOEs because they both represent the difficulty of accessing the restaurant.

In conclusion, the set of MOEs should be comprehensive but small enough to enable an impartial and meaningful analysis. de Neufville and Stafford (1971) cautioned that the choice of MOEs for evaluation is not a trivial exercise, but rather one that vitally influences the outcome of the analysis.

3.5 OVERALL DISCUSSION OF ENGINEERING SYSTEMS GOALS AND OBJECTIVES

As shown in this chapter, there are significant differences in civil engineering goals and objectives across the different civil engineering branches and also across the different system types in each discipline. Even for a given civil engineering system where a main goal is established (at the initial phase where the existence of a problem is identified and the need for a system is investigated), there also are specific goals geared toward specific tasks at each phase. With regard to the main goal established at the initial phase, a certain dynamism exists: During the system operations phase (the longest of the phases), there could be changes in the main goal as years and decades elapse, because society changes its focus on certain issues or certain stakeholders gain less. For example, many civil engineering systems that are currently in operation were conceived and constructed several decades ago without considerations of climate change; at the current time of global warming fears, the performance of such systems is being assessed on the basis of their contribution to climate change; also their resilience to threats posed by climate change are being assessed continually as part of their present-day management.

For a given civil engineering system, there typically exists more than one desired outcome, irrespective of the hierarchy level in question. For example, the goals of a system may include system durability, environmental sustainability, and economic development. As a result, there typically exists more than one performance measure or MOE for decision making. Thus, it is necessary, or at least useful, for engineers to develop skills in decision sciences that involve multiple performance criteria. In Chapter 12, we will discuss a number of tools and techniques for multiple criteria analysis. Another issue associated with multiple objectives or MOEs for a given analysis is that they often conflict with each other (i.e., the attainment of one objective may be associated with nonattainment or diminished attainment of the other objective). The conflict between objectives often occurs across different stakeholders. For example, the construction of a new landfill system near your city may be encouraged by city officials and local industries but opposed by environmentalists, neighboring residents who are concerned with the noise and air pollution due to the landfill operations, and farmers or landowners who may lose their land through eminent domain. Also, for certain systems, increasing the speed of system service may be associated with reduced system safety. Due to such conflicts, the engineer may need to analyze the trade-offs associated with the two MOEs, in other words, how many units of one objective could be bartered for one unit of the other? In cases where the level of attainment of each opposing objective can be quantified, it is possible to examine their trade-offs analytically.

Also, there may exist dependencies between task-related objectives not only within each phase of development but also across phases: The attainment of goals at a given phase can influence the goals at another phase; and failure to achieve goals at one phase also can influence the attainment of goals at a subsequent phase. For example, failure to achieve the safety objectives at the construction phase could lead to diminished opportunity to achieve the objective of overall positive user perceptions of the system at the operations phase.

Often, the objectives of the system owner or operator are typically adopted, at the initial phase, as the main objective; nevertheless, engineers need to avoid bias in establishing the goals and objectives of their systems. For a proposed system that cannot be justified using conventional means, the engineer should be wary of special-interest or lobby groups who try to promote objectives purportedly to enhance public benefit and welfare but mask their true agenda (de Neufville and Stafford, 1971). Thus, all groups should have their say, and the final set of objectives must be arrived at only after a great deal of consideration. Khisty et al. (2012) caution that while establishing MOEs for civil engineering systems may be relatively easier than doing so for human activity systems, engineers should be careful to discern the objectives stated by the stakeholders and their real preferences.

SUMMARY

Without an initial clear definition of the intended goals of a proposed system or the intended goals associated with a task at each phase, no meaningful analysis can be carried out. There are often several interested parties in civil engineering system projects, and each of them must be given due respect, audience, and consideration. The concerns of the stakeholders translate into goals and objectives and, ultimately, measures of system effectiveness. This chapter first reviews the hierarchy of desired outcomes and then discusses each level of this hierarchy in detail. With illustrations, the chapter explains the values, overall goals, goals, objectives, and measures of effectiveness associated with civil engineering systems. The chapter then identifies a number of common MOEs used in decision making at any phase of system development and categorizes them on the basis of whether they are monetary and the stakeholder perspective from which they derive (system owner/operator system user, and the community). In this chapter, we also discuss the goals and MOEs typically encountered at each phase of civil engineering systems development. The chapter provides guidelines for engineers in selecting not only a specific MOE but also a set of MOEs for making decisions regarding the system as a whole or any task in each phase. Each stakeholder may have their unique objective regarding a proposed system or improvement to an existing system; and because there are often several stakeholders, it is possible that certain objectives may conflict with each other. Thus, the chapter ends with a discussion of conflict between objectives, the need for circumspection, and the need to strive to incorporate all stakeholder perspectives and to avoid bias in establishing the goals, objectives, and MOEs for civil engineering systems. Other specific issues that are discussed include the dependencies between task-related objectives within and across the phases of development.

EXERCISES

1. You are asked to design a pedestrian footbridge over a busy street in the downtown area of a city. What are some of the structural engineering considerations that need to be taken into account during the analysis and design of this structural system?
2. A proposal has been made to design and implement physical modifications to the existing bus transit system infrastructure in a major city by providing new buses and terminals. Discuss some of the efficiency, effectiveness, and equity issues associated with this effort.
3. If you were asked to recommend some changes in the operational setup (no bus or terminal improvements) of your campus bus transit system, which factors would you take into consideration?
4. A parking garage is planned for construction in the downtown area of a fast growing city. Using the classifications shown in Figure 3.1, list at least two outcomes in each of the four categories of outcomes (intended vs. unintended, beneficial vs. adverse).
5. The specific objectives at each phase influences, and also is influenced by, the specified tasks and tools at that phase. For each of these two cases: (i) at the phase of needs assessment for a new subway system, and (ii) at the phase of monitoring the condition of a steel tower, what are some of the objectives, and how does this translate into a task for the engineer at each of these phases; also, list at least two analytical tools used to address this task.
6. Explain, with examples, how the values held by a society could influence the establishment of goals and objectives for the design of a new civil engineering system or the operations of an existing system being planned in that society.
7. Explain, with illustrations in the context of a specified civil engineering system in your city, the differences between the three overall goals (effectiveness, efficiency, equity).
8. Major expansion of an existing airport located near a suburb has been planned. List at least 4 key stakeholders associated with the improved system upon completion, and discuss their perspectives (expectations of benefit and concerns of adverse consequences). Citing examples, discuss how these perspectives translate into specific objectives of the airport expansion decision maker, and how they could influence decisions associated with the expansion.
9. Explain why systems scholars including de Neufville and Stafford (1971) and Dandy et al. (2008) stress the importance of selecting appropriate MOEs for engineering systems.
10. A systems engineer seeks to establish MOEs to evaluate alternative strategies for operating an existing urban drainage network. List and discuss any (i) five desirable properties of each individual MOE and (ii) three desirable properties of the set of MOEs for choosing between the strategies.
11. For any civil engineering system in your community, list and discuss any 10 MOEs that could be used in evaluating the system performance. Indicate which of these are monetary or nonmonetary and which are derived from agency, user, or community perspectives.
12. A new waste treatment plant has been planned to replace an existing small aging plant for your hometown. Discuss the MOEs for (i) the overall system, (ii) each phase of the overall system development, and (iii) the operations phase of any of the plant's processes.

REFERENCES

CamSys (Cambridge Systematics, Inc.) (2000). A Guidebook for Performance-Based Transportation Planning. NCHRP Report 446. Transportation Research Board, Washington DC.

Cross, K. F., and Lynch, R. L. (1989). The SMART Way to Sustain and Define Success, *National Productivity Review*, (8)1, 23–33.

Dandy, G., Walker, D., Daniell, T., Warner, R. (2008). *Planning and Design of Engineering Systems*, 2nd Ed., CRC Press, Boca Raton, FL.

de Neufville, R., and Stafford, J. H. (1971). *Systems Analysis for Engineers and Managers*. McGraw-Hill, New York.

Frangopol, D. M., Kawatani, M., and Kim, C. W. (2007). *Reliability and Optimization of Structural Systems: Assessment, Design, and Life-Cycle Performance*. Taylor & Francis, London, UK.

IISD (2009). What Is Sustainable Development? Environmental, Economic and Social Well-Being for Today and Tomorrow. International Inst. for Sust. Dev., http://www.iisd.org/sd/. Accessed Nov. 15, 2009.

James, W., and Thompson, M. (1997). Contaminants from Four New Pervious and Impervious Pavements in a Parking Lot. *Advances in Modeling the Management of Stormwater Impact*, 1–51.

Jeon, C. M., and Amekudzi, A. (2005). Addressing Sustainability in Transportation Systems: Definitions, Indicators, and Metrics, *ASCE J. Infr. Syst.* (11)1, 31–50.

Kapur, K. C., and Lamberson, L. R., (1977). *Reliability in Engineering Design*. Wiley, New York.

Kates, R. W., Parris, T. M., and Leiserowitz, A. A. (2005). What Is Sustainable Development? Goals, Indicators, Values, and Practice. *Environment: Science and Policy for Sustainable Development* 47(3), 8–21.

Keeney, R. L., and Gregory, R. S. (2005). Selecting Attributes to Measure the Achievement of Objectives. *Op. Res.* 53(1), 1–11.

Khisty, J. C., Mohammadi, J., and Amekudzi, A. A. (2012). *Fundamentals of Systems Engineering*. Prentice Hall, Upper Saddle River, NJ.

NYSDOT (1996–2002). *Vulnerability Manuals, Bridge Safety Assurance Program*. New York State Dept. of Transp., Albany, NY.

OECD (2001). *Performance Indicators for the Road Sector*. Organization for Economic Cooperation and Development, Paris.

Ortolano, L. (1997). *Environmental Regulation and Impact Assessment*. Wiley, New York.

Patidar, V., Labi, S., Sinha, K. C., Thompson, P. D., Hyman, W. A., and Shirole, A. (2007). Performance Measures for Enhanced Bridge Management. *Transp. Res. Rec.*, 1991, 43–53.

Sinha, K. C., and Labi, S. (2007). *Transportation Decision-Making: Principles of Project Evaluation and Programming*. Wiley, Hoboken, NJ.

Voland, G. (2004). *Engineering by Design*, 2nd ed. Prentice Hall, Upper Saddle River, NJ.

The World Bank (2003). *Social Analysis Sourcebook: Incorporating Social Dimensions into Bank-supported Projects*. Social Development Department, Washington, DC.

USEFUL RESOURCES

Bourne, M., Mills, J., Wilcox, M., Neely, A., and Platts, K. (2000). Designing, Implementing, and updating performance measurement systems, *International Journal of Operations & Productivity Measurement*, Emerald Group Publishing Ltd., Bingley, UK.

PART II

THE TASKS AT EACH PHASE OF SYSTEMS DEVELOPMENT

CHAPTER 4

TASKS WITHIN THE PHASES OF SYSTEMS DEVELOPMENT

That which we persist in doing becomes easier, not that the task itself has become easier, but that our ability to perform it has improved.

Ralph Waldo Emerson, 1803–1882

4.0 INTRODUCTION

In the previous chapters, we identified various disciplines in civil engineering and defined what a system is in the context of civil engineering. Those chapters also identified the eight major phases through which a civil engineering system is developed and discussed briefly the analytical tasks encountered by the engineer at each phase and the tools needed to carry out the tasks (we herein define an analytical task as that which is associated with the systems approach to problem solving). In this chapter, we examine the analytical tasks in greater detail; and in subsequent chapters (Chapters 5–18), we will identify and study a number of analytical tools that civil systems engineers can utilize in order to carry out these tasks effectively.

As discussed in Chapter 2, the development of a civil engineering system follows a sequence that begins with an assessment of the need for the system, planning and designing the system, system construction or implementation, operating the system while monitoring its use, inspecting its condition, carrying out maintenance as and when needed, and, finally, and system end of life (Figure 4.1). At each of these phases, engineers carry out a number of tasks: Some are traditional tasks, that is, they related directly to the phase in question and require specialized, domain knowledge outside the scope of this book; others are analytical tasks that are related to the system approach in systems engineering. The latter category of tasks includes describing the system or how it works; describing the intended procedure for carrying out that phase successfully; and analyzing, evaluating, and selecting alternative processes or materials at that phase. The description task is often needed for monitoring the manner of the operations or some specific process at any phase; the tasks of systems analysis, evaluation, and selection are needed mostly for decision making. The trio of analysis, evaluation, and selection of alternatives could be collectively termed *optimization*. The feedback task occurs between each pair of precedent and antecedent phases.

Before we proceed, it may be necessary to establish a few definitions. *Alternative* implies mutual exclusivity; in other words, if two actions are described as "alternative," then the occurrence of one action completely precludes the occurrence of another, that is, either you carry out one action or the other, but not both. An *action*, *intervention*, or *decision* is a stimulus on a system that has consequences in terms of cost and is expected to yield some benefits. *Evaluation* broadly means assessing the impacts (benefits and costs) as a basis of deciding whether or not to undertake the action or as a basis of prioritizing that action relative to others.

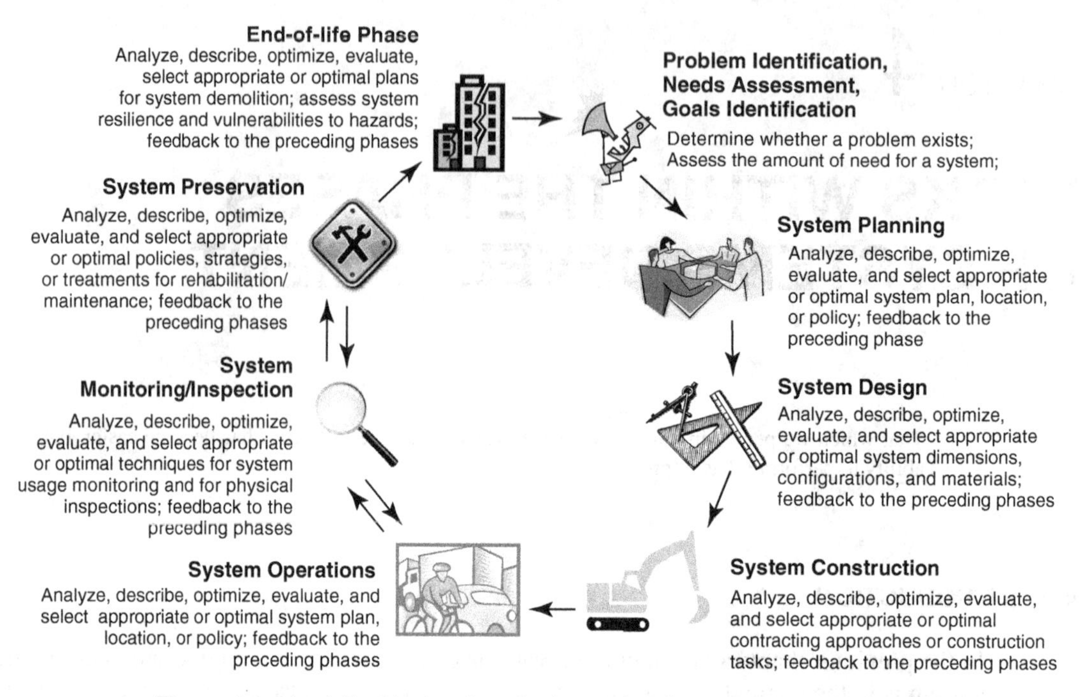

Figure 4.1 Analytical tasks at each phase of civil systems development.

For the different types of systems in each branch of civil engineering, professionals are engaged at each phase of the system development life cycle, working to deliver an effective, efficient, and safe product or service for the benefit of society. In carrying out their tasks at each phase of the cycle, engineers seek to answer several questions associated with the system. Examples of such questions are indicated in the box inset below. As they go about the tasks of planning, designing, constructing, maintaining, inspecting/monitoring, and operating their systems, developers and managers of civil engineering systems address questions such as these on a regular basis, and the burden of their decisions is reflective of the trust placed in them by the taxpayers. In the next section, we provide more specific examples of the typical tasks faced at the various phases of civil engineering systems development.

We continue this chapter with a discussion, with examples, of the task of system description, specifically using models. We will see a number of contexts for model application and how they are classified. Our discussions on the task of systems analysis and system evaluation are presented in the context in each branch of civil engineering. For tasks related to system evaluation, we present the evaluation criteria (impact types) and scopes. We also generally discuss the tasks of optimizing some task at each phase and feedback between phases. Finally, we discuss a number of examples of tasks faced at various phases of systems development.

Some Typical Analytical Tasks Faced by Civil Systems Engineers

- At which year will we need to replace our city's water tank with a larger one to accommodate growing population as has been predicted? How certain are we of the prediction?

- What is the threshold level of freeway congestion beyond which some congestion mitigation action is needed? Is there an existing threshold? Is it too relaxed or too restrictive? Which stakeholders will determine the threshold?
- Of the several possible locations for a planned telecommunications tower, which location can be considered the best?
- Is it always economically worthwhile to use a promising new material to give our civil structure a longer life even though it costs much more than the traditional material?
- What is the best way to describe to a nontechnical audience the structural behavior of our levees in the event of a severe storm?
- For a certain specific project, under which conditions should we choose a flexible pavement instead of a rigid pavement?
- How could we optimize the operations of our waste treatment plant on the basis of numerous different, often conflicting objectives?
- What is the most cost-effective way to measure the current demand for a given civil engineering system?
- At what year will it become economically feasible to replace a certain bridge deck?
- How often should we be inspecting the physical condition of our coastal steel bridges? And which combination of inspections resources (manpower and robots) should we deploy?
- Which optimal sampling techniques and frequencies should be adopted to monitor the condition of an undersea tunnel?
- Which combination of airport pavement layers will provide optimal performance at minimum cost?
- At which traffic level should we pave an unpaved road?
- What is the best way to end the life of a certain existing but functionally obsolete civil engineering structure?
- How could we optimize our infrastructure rehabilitation investments in order to maximize system resilience in the event of a specific natural disaster?

4.1 THE TASK OF DESCRIPTION

Occasionally, a need may arise for the civil engineer to describe some attribute of an overall system, a specific component of the system, or some phase of a system's development to some audience comprising top-level engineers and decision makers, their peers or colleagues, or the general public. A system description is a statement of the physical condition or operational performance of a system at the current time or at a specified future time. It also includes a description of the *process* associated with the system at any phase of its development. In other words a statement made by the engineer regarding an action takes some aspect of the system through an established and typically routine set of procedures that convert it from one form to another. When the task of description is made for a future or anticipated attribute of the system, then it is termed *prediction*.

For purposes of making descriptions or predictions about a system or process, civil engineers typically use *modeling*. Models are tools that represent reality in a similar sense that a portrait represents a human, and the modeler seeks the simplest model that will explain or predict phenomena reliably. Models can be used to describe the attributes (physical structure and/or condition and operational characteristics) of the system or of a process associated with one of the phases of system development. These attributes can be described for past, current, or future situations. But what exactly is a model? A model can be defined as the real or virtual abstraction of the structure, condition, or operation of a system on the basis of the past trends of these system attributes in time and space. Virtual models are abstract, such as mathematical or statistical equations. In certain texts, a model is simply defined as the representation of an aspect of reality. In deciding which model to use to describe a given system or process under a specific situation, it is useful for the modeler to know how models are classified and thus ascertain which conditions make it appropriate to apply a specific type of model. The next few sections, where we discuss the categories of civil engineering models, provide clues for the modeler.

4.1.1 Model Classification

Models used in civil engineering may be classified in a variety of ways, including whether the model describes the system or a process for the system, which system attributes we seek to describe, the model form, the model purpose, the kind of data used to develop the model, and the phase of civil engineering systems development at which the model is being applied. These classification categories are described below.

(a) Model Classification by Subject of the Modeling Process. A model may be describing the system itself at any phase or the process used to develop the system at any phase. For example, we could have a toll booth operating model or a model that predicts the operations cost of a toll booth.

(b) Model Classification by System Attribute. Civil engineers often seek to describe or predict some attribute of the system, such as the system's physical structure, its physical condition, or the way it operates. Examples of physical models include torsional deformation models in structures; examples of condition models include a model that estimates the surface roughness of a airport runway pavement or one that predicts future corrosion levels in structural reinforcement; and an example of operating models is a model that predicts water seepage intensity and directions of water under a foundation footing or the progression of a queue at a manual-payment toll road booth.

(c) Model Classification by Model Form. Real models are actual physical models, typically, three-dimensional (3D) miniatures, exact size, or blown-up replicas of the system under investigation. Real models may be static or dynamic. Unlike dynamic models, static models describe a phenomenon that does not change significantly with time. A simple model classification on the basis of these considerations is presented in Figure 4.2. Figure 4.3*a* presents a 3D virtual model that is computer generated of a hotel in Thailand, and Figure 4.3*b* presents 3D real models of two building frames undergoing earthquake testing.

In reality, no model may be perfectly static or perfectly dynamic: Models that are termed dynamic actually are of a series of static phenomena each having infinitesimal duration while those termed static actually experience some degree of temporal change in the attribute being described.

Examples of virtual mathematical models in civil engineering include static models such as Boussinesq's soil stress model (geotechnical engineering), the moment equation (structures), and equipment cost-effective models (construction engineering), as well as dynamic models such as chemical mixing models (environmental engineering), Darcy's model for percolation of fluids through porous media (hydraulic engineering), queuing models (transportation engineering), and heat energy transfer models (architectural engineering). With regard to numerical mathematical models, the most common numerical models are simulation models that replicate the behavior or operation of engineering systems, which are currently popular due to advances in computing. The theory of probability is important for simulation modeling, examples of which include structural

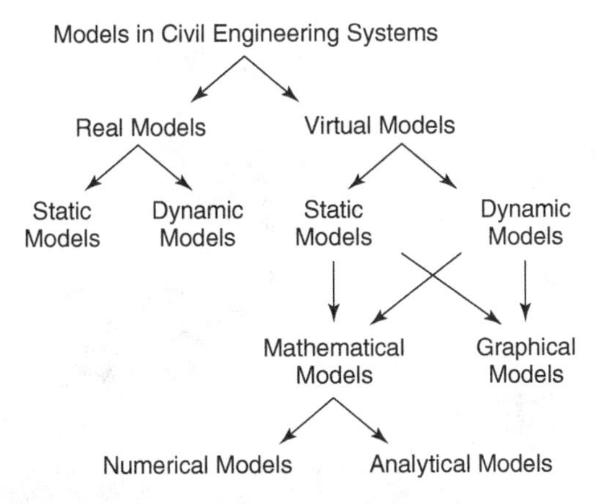

Figure 4.2 Simple classification of models by model form.

(a) (b)

Figure 4.3 Illustration of civil engineering model forms: (*a*) three-dimensional virtual model (Wikimedia Commons), and (*b*) three-dimensional real model (Wikimedia Commons).

element simulation in computational structural mechanics and traffic simulation in civil engineering (Figure 4.4).

(d) The Stark Classification. Stark and Nicholls (1972) presented a three-way classification of civil engineering models:

(i) ***Iconic models*** are scaled versions of the real system (Figure 4.3*b*). Relevant properties of the system are represented by the same properties in the model. The modulus of elasticity of a bridge material, for example, would be represented by the same modulus in the model material. Laboratory models of bridges and buildings for the study of complex structural behavior are examples of iconic models.

(ii) ***Analog models*** use one set of properties to represent another set. For example, electric current can represent heat or fluid flow, fluid systems can represent traffic flow, soap films can represent torsional stress, and contour or equipotential lines can represent ground elevation.

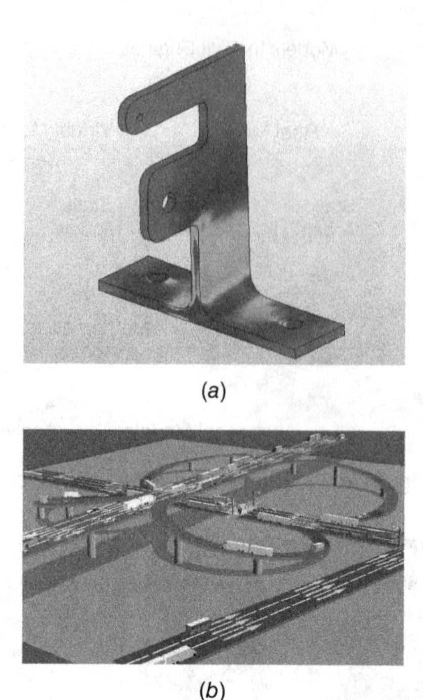

Figure 4.4 Examples of static and dynamic graphical models (*a*) Simulated stresses in a structural element (*b*) highway traffic simulation (*Source:* Ben Masefield, Solveering LLC; Stanracz/Wikimedia Commons).

(iii) ***Symbolic models*** use symbols to represent variables and the relationships between them. The generality and abstractness of symbolic models make them easy to manipulate.

(e) Model Classification by Prescriptive/Descriptive Nature. Models may be classified on the basis of their role in the decision-making process at any phase of system development where the engineer carries out some task that involves description or evaluation of some course of action. In this respect, Revelle et al. (2003) identified two types of models: *prescriptive models* that use the mathematics of decision making to choose a course of action, design, or policy; and *descriptive models* that typically describe the outcome for a given set of inputs. The descriptive model answers the question: *If I follow this course of action, what will happen?* In contrast, the prescriptive model answers the question: *What should I do?* Loosely speaking, another term for the prescriptive model is the *optimization model* in the sense that the policy or design that is found achieves the best value of some objective. Most prescriptive models incorporate some descriptive models. For example, in order to identify (using a prescriptive model), the optimal preservation policy the engineer must first describe or predict (using descriptive models) the expected cost and effectiveness of each of the several alternative preservation treatments or actions that constitute each policy.

(f) Model Classification by Empiricism. Models that are developed on the basis of historical data are termed empirical models, while those based on a theory of the outcome and physical measurements are termed mechanistic models. For example, a statistical corrosion model could be one that uses past data on corrosion rates and steel type, ambient moisture, proximity to the coast, and degree of salt application in winter to develop an empirical model that describes corrosion as a function of the attributes stated. A mechanistic corrosion model is one that uses physics and chemistry principles to describe the precise molecular behavior of the steel in response to attack by corrosive agents. Models that use both mechanistic and empirical data are referred to as hybrid or mechanical-empirical models.

(g) Model Classification by Degree of Certainty Associated with Data. Models may also be categorized by the degree of uncertainty associated with the data they utilize, namely, *deterministic* or *probabilistic*. Deterministic models have input data that do not vary but rather are fixed quantities. For example, in concrete production, a deterministic model would suggest that on the basis of the given quantity and quality of concrete, water, and aggregate, there can be only one outcome in terms of the compressive strength of the concrete. In contrast, *stochastic* models have data elements whose values at any specific time are not known with absolute certainty. Such occurrence values may be studied over a long period of time, and on the basis of such historical information, their occurrence may be characterized by a mean and/or a variance. If adequate historical data were available, it may be possible to ascertain the nature of the distribution followed by the occurrence value of the data elements. For example, in Figure 4.5, a cantilever constructed using material of uncertain strength and material properties experiences a live load of uncertain magnitude at a position that may vary. In real life, this scenario is possible when a cantilever roof is subjected to impact from a falling object from above. In such a case, the model that describes the deflection or other performance of this structural system is stochastic. All three variables are stochastic as they could take any value within a certain range. The deflection will be small if the load is small, the distance from the support is small, and the material strength is high; and deflection will be high if the load is large, the distance from the support is large, and the material strength is low. There are a multitude of possible deflection values within these extremes (Figure 4.6). In the figure, the variation of each input (and also of the output) is represented as a range of values but could also be represented using a probability distribution. In most instances, inputs follow a normal distribution—symmetrical, left skewed, or right skewed—but several other distributions of input variables may be encountered in civil engineering.

According to Rardin (1997), deterministic models are often used by engineers because they are more tractable (easier to analyze) than their stochastic counterparts, and also often produce solutions that are valid enough to be useful for the problem at hand. From the perspectives of the uncertainty surrounding the values of the input data and the prescriptive versus descriptive nature of the data, Revelle et al. (2003) presented a two-dimensional classification of models (Table 4.1).

The classification shown in Table 4.1 covers almost all major common model types that are utilized in systems engineering. Differential or difference equations are models that are deterministic and descriptive. These are models often encountered in calculus or applied mathematics, and they usually involve empirically derived parameters and rate constants that are known or assumed. Furthermore, these models may be linear or nonlinear, "depending on the nature of the system or

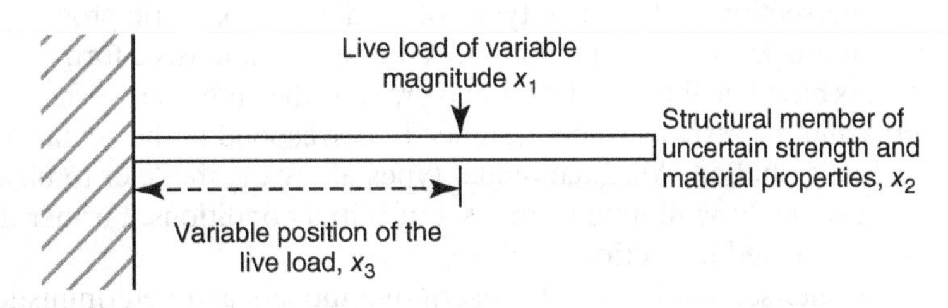

Figure 4.5 Example of civil system with uncertain input variables.

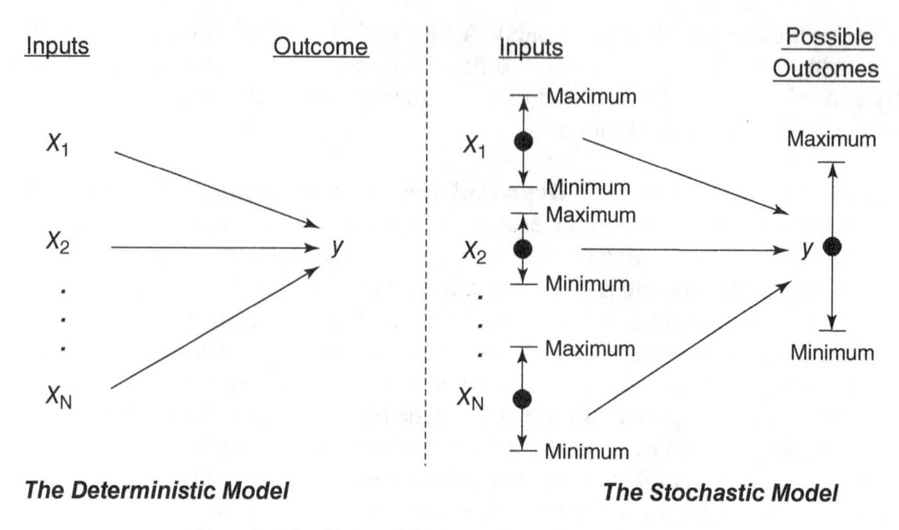

Figure 4.6 Deterministic and stochastic models.

Table 4.1 Model Classification Based on the Level of Certainty and Purpose of the Model

	Deterministic	**Stochastic**
Prescriptive	Linear programming Integer programming Multiobjective programming Dynamic programming	Stochastic programming
Descriptive	Difference equations Differential equations	Stochastic differential equations Monte Carlo simulation

how realistic the model structure or function needs to be" for a specific application of the model (Revelle et al., 2003).

The overlap of descriptive and stochastic models contains several types of models. One is a special kind of differential or difference equation model that has parameters that are random variables. These equations, referred to as stochastic differential equations, become very complex when the equation(s) contain multiple random parameters. The complexity is introduced due to the inclusion of two random parameters because this leads to the need to model the correlation between the parameters in order to fully describe the system or process effectively. Another model form at the intersection of these two types of models is stochastic processes. A third model type at this intersection is known as "simulation," a computer-intensive form of modeling that generates realistic events based on input variables and predicts the outcomes of the system. Here, the statistics of the events and the responses are designed to correspond to the actual statistics of parameters in the system being studied. All three model types allow the modeler to observe a range of possible outputs that evolve through time from a set of initial conditions. Further discussion of these types of models is presented in Section 4.1.2.

Another intersection is that of prescriptive models and deterministic models. Deterministic optimization models are also referred to as mathematical programs because they help engineers and managers in the planning or "programming" of activities across time or space (Rardin, 1997). Specific deterministic optimization models are known by various names depending on whether the mathematical descriptions are linear or nonlinear, static or evolve through time, or have specific shapes or forms. These include linear programming; quadratic programming, which deals with problems having a quadratic objective function; dynamic programming, which considers problems with a number of time or pseudotime stages; multiobjective programming, which operates on more than one objective and derives trade-offs between objectives; and integer programming, which is applied where integer-valued decision are practical. Concepts of optimization and programming are discussed in Chapter 9.

(h) Model Classification by Phase of System Development. The categories of models associated with the phases of civil engineering systems development are (Figure 4.7):

- **Demand models** are utilized at the needs assessment phase, where the engineer seeks to describe or predict the demand for the civil engineering system, either in terms of volume (in the case of highway traffic or water demand) or loading (in the case of structural design). We discuss some deterministic demand models in Chapter 19 where we present techniques to assess the amount of need for a system. Demand models are also used at the system operations phase to verify the level or growth rate of usage or loading.
- **Cost models** describe the amount of money needed to undertake a specific action for the system and are needed at the phases of construction, operations, preservation, inspection, or demolition. Cost models may be an average value or a statistical model that relates cost as a function of system size or improvement scope or type, and other attributes. Also, cost models may be time specific (e.g., average annual maintenance expenditure for a pipeline) or intervention specific (e.g., cost of lining a sedimentation pond with geotextiles). In Chapter 10, we present a number of cost modeling techniques.

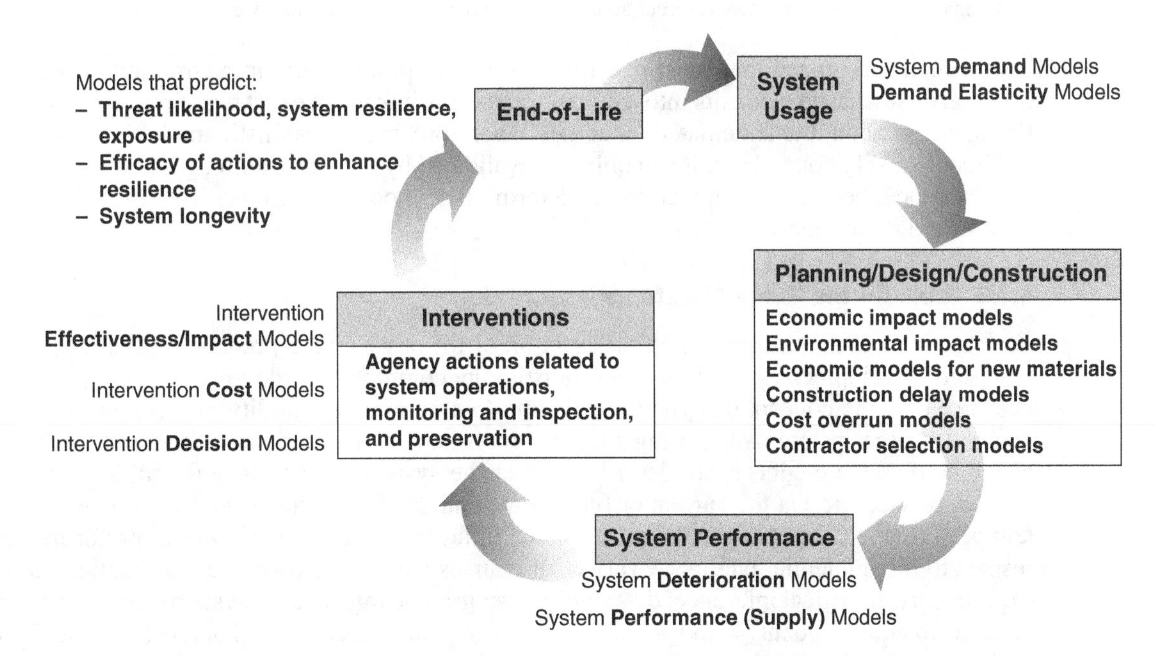

Figure 4.7 Model classification by phase of system development with examples.

- **Deterioration or performance models** describe the pattern or rate at which the system condition or performance erodes with time or with usage. In Chapter 25, we illustrate applications of these models as part of efforts to carry out maintenance of the system in a timely manner.
- **Effectiveness models** describe the effectiveness of a specific action or specific amount of resources in achieving some objective associated with the system, such as the expected increase in system durability after application of a certain maintenance technique or the reduction in construction site fatalities after introducing a new type of scaffolding. We present a number of effectiveness models in Chapter 25.
- **Intervention decision models**, which unlike the four previous models, are prescriptive not descriptive and help determine if, when, and/or how much funding should be used to implement some action or policy at any phase of the system. Decision models are often discrete in nature (provide yes/no outputs) or continuous (optimal year of the action/policy implementation or demand/loading level at which the intervention is warranted).

4.1.2 General Comments about Modeling

As could be discerned thus far in this chapter, modelers of civil engineering system attributes need to acquire useful mathematical skills including analytical methods (including statistical analysis) and numerical methods (including simulation). Therefore, in order to prepare a system description, a good foundation in statistics and probability is indispensable. We will review introductions to these concepts in Chapters 5, 6, and 7. A basic motivation for probabilistic analysis is that the world is never precise or deterministic due to uncertainties and variabilities in natural and man-made conditions. Thus, models that describe (and predict) system attributes must be general enough to cover all possible and practical circumstances of the system. Greene (2011) aptly emphasized this requirement as follows:

> A model can never be truly confirmed unless it is made so broad as to include every possibility. But we may subject it to ever more rigorous scrutiny and, in the face of contradictory evidence, refute it.

Green recognized that most real-life models are probabilistic in nature and stated that the infusion of stochastic elements into a model transforms it from an exact description to a probabilistic statement about the anticipated outcomes. Therefore, the probabilistic model is robust because as Greene (2011) pointed out, it can only be invalidated by an overwhelming amount of contradictory evidence; however, compared to the deterministic model, the probabilistic model is both less precise and more robust.

4.1.3 Tools for the Task of Systems Description

For carrying out the task of system description or for describing a process used at a given phase of system development, probability and statistics are commonly used tools. For example, the engineer may use the concept of hypothesis testing to ascertain the validity of a statement about the demand for the system. Also, using statistics tools such as regression, multivariate analysis, and econometrics, the engineer can describe/predict the system performance (condition, operational level of service, etc.) at the current or future time and determine the factors that influence the system performance, the extent and nature of such influence, and the sensitivity of performance with respect to the influential factors. A variety of courses (stochastic processes, simulation, design of experiments, statistical inference, data mining, geographic information systems, etc.) are offered in colleges to equip students with specific tools to carry out these tasks. Chapters 5, 6, and 7 present typical tools used by engineers to describe their civil engineering systems.

4.2 TRADITIONAL TASKS OF ANALYZING SYSTEMS IN CIVIL ENGINEERING

Ironical as it may be, the "analysis of a system" could have a different meaning compared to "system analysis." The former is associated with or requires the domain (traditional) skills and concepts that are specific to a specific branch of civil engineering (these skills are used to analyze systems in those disciplines; for example, statics in structures, fluid mechanics in hydraulics, soil mechanics in geotechnical, and traffic capacity theory in transportation). The latter is associated with systems engineering and is more consistent with the scope of this text. As stated in the introductory chapters, we do not attempt, within the limited confines of this text, to address all of the concepts that equip an engineer to complete the task of analyzing the system in question. Tools and techniques for carrying out traditional analysis of these systems require domain knowledge and are found in the various texts for each discipline. This text focuses instead on the systems concepts that could complement these analyses. Nevertheless, the sections below identify, for some civil engineering branches, a number of traditional tasks related to domain knowledge areas at that phase and could benefit from the incorporation of system analysis tools.

4.2.1 Construction Engineering

Construction engineers and managers analyze strategies for construction planning and scheduling, equipment and labor utilization, maintenance scheduling for construction equipment, project formwork management, and project monitoring. In analyzing an existing construction plan or scheduling strategy, key considerations include the duration, earliest start time, and the latest end time of each activity. In analyzing plans for equipment utilization, issues include whether the right equipment types/sizes are being used, the project type, the soil/water conditions, the vegetal cover type and extent, the topography, the local regulations, and the project specifications. The analysis often addresses the right mix of labor or equipment to be used; and the considerations may include labor or equipment productivity and cost, job size, and work schedule. In analyzing an existing formwork configuration for its appropriateness, cost, and cost-effectiveness, the key considerations may include the sizes and shapes of the concrete elements, the position of the concrete element, the desired quality of the concrete finish, the weight of the concrete, worker safety, the possibility of formwork recycling, and the strength and cost of form material. Construction engineers also carry out analysis related to project control, construction planning, scheduling and control, and site planning and management. All these traditional tasks could be carried out more efficiently when they are complemented with systems analysis tasks including description, evalution, and so on.

4.2.2 Environmental Engineering

Environmental engineers carry out a wide range of analysis in various areas of environmental engineering and at various phases of environmental system life cycles. In analyzing wastewater and water treatment plants, demand and supply are analyzed to ensure appropriate levels of service and to avoid wastage. Considerations in demand analysis include the initial demand, and the pattern of demand growth (linear, quadratic, exponential, etc.); and the outputs of demand analysis include the expected loads. From the supply perspective, considerations in the analysis may include the size (capacity) of the system (plant) and its constituent subsystems (units) and the costs of expansion and operations. Physical treatment operations that may be analyzed include screening, mixing, sedimentation, filtering, odor control, and aeration. Chemical treatment operations to be analyzed may include those associated with the processes of coagulation, softening, stabilization, demineralization, chemical oxidation, and disinfection. Existing biological treatment operations that could be analyzed include aerobic fixed-film processes, treatment wetland bioremediation, and composting, as well as sludge stabilization subsystems. In the context of solid waste management, the engineer may be required to analyze existing systems for incineration and landfilling. The environmental engineer's tasks include the analysis and modeling of environmental systems and processes including environmental remediation, fate and transport of contaminants in the environment, and physical chemical processes for water quality control. Accompanying these traditional tasks are the tasks of system description, evaluation of alternatives, and so on.

4.2.3 Geotechnical Engineering

In geotechnics, engineers analyze the engineering behavior of earth materials and thus analyze geotechnical structures, materials, and processes. This includes tasks such as investigating existing subsurface conditions and materials in order to assess the suitability of materials or soil formations for ground support and to assess the risk to humans and the environment posed by site conditions and/or natural hazards including earthquakes, soil liquefaction, landslides, sinkholes, and rock avalanches. For example, engineers analyze slope stability to determine whether soil-covered slopes are likely to undergo movement. Also, in designing earthwork systems (tunnels, embankments, dikes, levees, channels, and earth reservoirs), lateral earth support structural systems (cantilever wall, gravity wall, and excavation shoring), and gravity and structure foundations (willow and deep foundations), geotechnical engineers analyze the stresses from current or expected loading and strains in soil materials. Another important aspect of geotechnical systems analysis at the design phase involves the assessment of the impact (in terms of structural stability, cost, and other performance criteria) of different foundation design types and configurations. At the systems operations and preservation phase, geotechnical analysis may involve investigating the sustained ground stability of structures using data obtained from monitoring equipment. In conducting these traditional tasks, geotechnical engineers also carry out the description, evaluation and selection of alternative geotechnical designs and processes.

4.2.4 Hydraulic and Hydrologic Engineering

Hydraulic and hydrologic engineers are tasked with analyzing existing or proposed hydraulic systems such as urban surface drainage systems and sewerage network systems, water distribution networks, and hydraulic systems, including culverts, dams and spillways, levees, hydraulic outlets, and energy-dissipating water structures. They analyze flows in water distribution networks and wastewater collection networks for purposes of storm water drainage management. Hydraulic and hydrologic engineers also carry out other tasks related to catchment flood modeling and management, coastal protection, shoreline management, flood alleviation and estuarine protection planning. The task of analyzing hydraulic systems often involves the control and conveyance of water flow and includes design and evaluation of flow measurement devices, fluid supply, and distribution systems and facilities including pumps and turbines. Clearly, the branch of hydraulic and hydrologic engineering is replete with instances where systems analysis tasks are carried out in conjunction with the traditional tasks discussed above.

4.2.5 Civil Materials Engineering

The tasks of materials engineers include the investigation of the suitability of individual ingredients as well as mixed materials, such as Portland cement or asphaltic concrete mixes, metals, composite materials, and other traditional and nontraditional materials for buildings, highway pavements, and other civil engineering structures. In carrying out this work, analysis considerations include loading, climate and weather, aggregate quality, and desired concrete durability. In recent years, materials engineers analyze the nanoscale properties of steel, concrete, and other construction materials in order to detect changes in these materials as a result of cracking, corrosion, or other modes of deterioration. The traditional tasks of materials engineers also include analysis associated with material properties, including thermodynamics, failure analysis and forensics, mechanical properties, and materials characterization.

4.2.6 Structural Engineering

Structural engineers analyze a wide range of structures and structural configurations. They are tasked with analyzing steel structural systems including trusses and frames and structural concrete elements including beams, columns, and shells. In the course of their tasks, structural engineers consider live loads (traffic, pedestrians, occupants, wind), dead loads, failure mechanisms (bending, shear, torsion, etc.), strength of the materials, and the joint design and vulnerabilities. Most of structural engineering tasks are found in the domain knowledge areas in structural analysis including mechanics of materials, stress analysis, structural mechanics, structural analysis, and steel and concrete design. In recent years, structural engineers are increasingly carrying out systems analysis tasks in addition to their traditional tasks.

4.2.7 Transportation Engineering

The main task faced by transportation engineers involves the analysis of the operations of transportation systems and the physical planning and design of such systems. For example, airport engineers analyze airport location, runway configuration, terminal and passenger flow design; in carrying out these tasks, they consider the type of surrounding land use (for noise impact evaluation), wind direction, types and sizes of expected aircraft, and the passenger demand. In analyzing pavement structures for highways and airports, considerations include traffic loading types and levels, subgrade quality, strength of available base materials, weather effects, desired serviceability (performance), and variability of input parameters. Highway engineers analyze the geometric features of existing highway systems, such as vertical and horizontal curve design, grades, radii, superelevation, lanes, shoulders, curbs, median, climbing lanes, and escape ramps, and intersection layouts. In carrying out this task, they consider stopping and passing sight distances, design speeds, vehicle sizes, safety, human reaction time, and the like. Various initiatives for intelligent transportation systems (ITS) that are typically analyzed include advanced traveler information systems, advanced vehicle control systems, advanced traffic management systems, advanced public transportation systems, commercial vehicle operations, and electronic toll collection systems. Considerations for these analyses include the desired composition of traffic stream, design speeds, vehicle sizes, safety, human reaction time, and productivity of the ITS resources. Operations research related tasks and other systems analysis tasks typically complement the traditional tasks of the transportation engineer.

4.3 THE TASK OF SYSTEM EVALUATION

The evaluation of civil engineering systems essentially involves an assessment of the extent to which the system objectives are being achieved or are expected to be achieved. As such, at any phase of a system's development, the criteria used for evaluating alternatives depend largely on the system goals and objectives (see Chapter 2). Not only must the agency goals and user objectives be considered, but there often are several other stakeholders in the development of civil systems, some of whom are very passionate about their interests. As such, a wide array of evaluation criteria should be considered. Table 4.2 presents the categories and types of possible impacts of civil engineering

Table 4.2 Impact Categories and Types

Category of Impact	Impact Types
"Technical" Impacts	System condition and durability Purpose for which system was built (for rail systems, for example: accessibility, mobility, and congestion mitigation)
Environmental Impacts	Ecology Water quality Air quality Noise Aesthetics
Economic Efficiency Impacts	Life-cycle costs and benefits Initial costs Costs associated with downtime periods Monetary costs and benefits associated with system use
Economic Development Impacts	(Un)employment rate Number of business establishments Gross domestic product Local and regional economies Volume of international trade
Legal Impacts	Tort liability exposure (of system agency due to user injury) Legal conflicts (between contractual participants in system construction, operations, preservation, or demolition)
Sociocultural Impacts	Environmental justice Quality of life

Adapted from Sinha and Labi (2007).

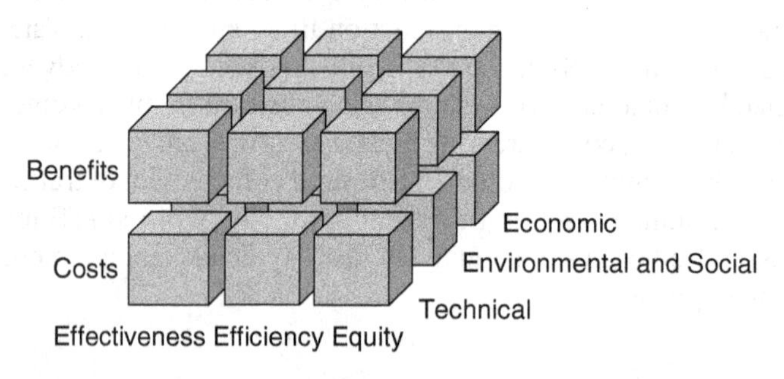

Figure 4.8 Impacts and consequences—a three-dimensional perspective.

systems that could serve as the basis for evaluating the system or evaluating alternative actions at any phase of system development. These impacts include the so-called technical impacts and the environmental, sociocultural, economic efficiency, legal, and economic development impacts.

The criteria that must be considered by engineers as part of the evaluation task can be categorized as follows (Figure 4.8): effectiveness, efficiency, or equity (see Chapter 3); technical, environmental, or economic; and benefits or costs (Figure 4.9).

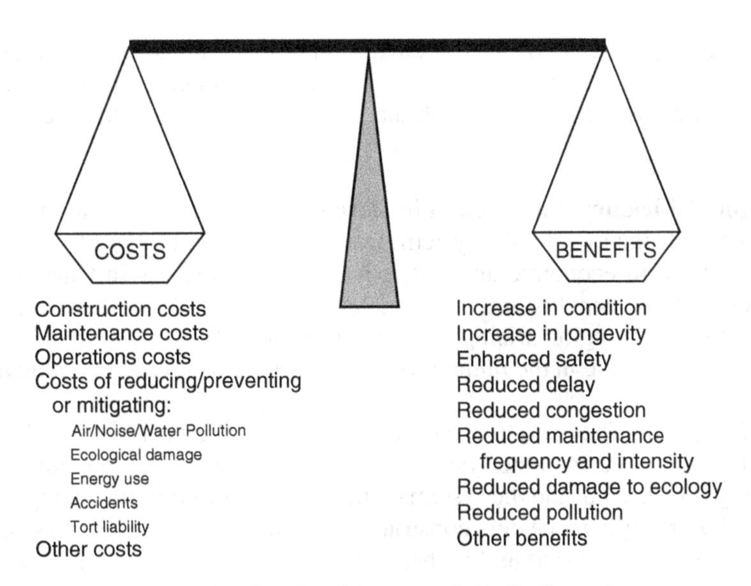

Figure 4.9 Details of the cost–benefit dimension.

4.3.1 The Dimension of Impact Type

As we learned in Chapter 3 and Figures 4.8 and 4.9, the system outcome could be beneficial or adverse, and intended or unintended. The outcome may also be categorized as technical, environmental, economic socio-cultural, and legal as we saw in Table 4.2 and Figure 4.8.

Technical Impacts. Technical impacts typically represent the primary motive for developing a new civil engineering system or enhancing an existing system. For example, most physical intervention actions are carried out either to replace or preserve the civil engineering system. This could lead to an improvement in facility condition, an objective that is not sought for its own sake but that leads to other desired impacts such as increased asset longevity, decreased costs of system usage, and increased resilience of the system to natural or man-made threats.

Environmental Impacts. In the task of evaluation, it is important to identify system improvements that do not lead to undue degradation in environmental quality. The construction and operations of certain civil engineering systems can significantly reduce the quality of surface water, and the noise associated with the construction and operations of some civil engineering systems have been linked to health problems, particularly in urban areas. Disruptions in surface and subsurface hydrological patterns, reductions in permeable land cover, slowing of surface water percolation, and a consequential reduced volume of recharge to underground aquifers are other environmental disturbances from the development of civil engineering systems. Increased surface runoff arising from the construction of such facilities also can lead to greater volumes of surface flow or forced channelization of surface water along unnatural water courses, which can result in greater erosion rates of topsoil and an associated destruction of flora and fauna as well as their habitat. Current studies of ecological impacts now include not only assessment of threats posed by natural features to engineering structures but also focus on the other direction of impact: the effect of the operation of engineering structures on the ecology. Civil engineering projects can impact the visual appeal of their surrounding natural or man-made environment in a positive way, such as a good blend with the surrounding built-up or natural environment, or in an adverse way, such as obscuring an aesthetically pleasant natural or man-made feature. Civil engineers today increasingly seek to ensure that their systems blend in well with the natural, social, and cultural environments where they are situated.

Economic Efficiency and Economic Development Impacts. At any phase of system development, the costs incurred by the system-owner (we discuss this in Chapter 10) are important criteria for evaluation. In economic analysis, which we will discuss in Chapter 11, both the impacts in terms of both the benefits and the costs are considered. Also, there could be economic development impacts in a region arising from the provision of the civil engineering system in the region, for example, increase in the number of businesses and increase in employment.

Legal Impacts. The operations phase of most civil engineering facilities is associated with some amount of risk of harm to the system users and general public. In many countries where sovereign immunity is restricted, and the system owner is liable for property damage, injury, or death resulting from faulty or negligent design, construction, or maintenance of their system. For systems that face increasing system demand and higher user expectations vis-à-vis inadequate funding, this can be a particularly serious issue and may necessitate the inclusion of tort exposure as a criterion for evaluation. We will discuss legal issues in Chapter 29.

4.3.2 Scopes of System Evaluation

The categories and levels of the dimensions of any system evaluation task are presented in Table 4.3. Establishing these parameters has been shown to help in defining the scope of the evaluation and the identification of relevant performance criteria or MOEs for the evaluation (Sinha and Labi, 2007).

Affected Entities. In carrying out the task of system evaluation, the various entities that are affected by the intended or ongoing action must be considered. Impacts on the agency, typically measured in terms of financial impacts (initial life-cycle costs), exposure of the system owner or agency to tort, and public relations are important because they provide an indication of the resources needed or expended and thereby can assist in the internal review process of agency performance. By measuring the actual or anticipated impacts on system users, such as changes in user convenience or enhanced safety/security, an agency is able to gauge how well the primary motives for the system implementation are being attained or are expected to be attained obtained. Also, measuring the impacts on the community, such as air and noise pollution, enables the system owner to ascertain whether any state or federal mandate is being violated.

Geographical Scope. In the task of civil engineering systems evaluation, it is important to establish a study area because the geographical or spatial scope of impacts can influence the analysis results. Such a scope may range from local (point specific), to areawide (city, county, district, state,

Table 4.3 Evaluation Scopes of Impacts

Scope	Levels
Affected entities	Users, community, agency
Geographical scope of impacts	Local, regional, national, international
Temporal scope of impacts	Short, medium, long term

or national). When the geographical scope of the evaluation is widered, the impact of the civil engineering system becomes not only reduced but also harder to measure as one moves farther from the system location. For certain types of impacts and affected entities, specific geographical scopes are recommended.

Temporal Scope. A civil engineering system may have impacts that last only a relatively short time (e.g., dust or noise pollution during the system construction) or may endure for many decades after implementation (e.g., economic development). Thus, the appropriate temporal scope for the task of evaluation will depend on the type of impact under investigation and is also sometimes influenced by, or related to, the affected entity and the spatial scope of the evaluation. Impacts can be short term, medium term, or long term in duration. Besides their classification by duration, the temporal distribution of impacts could also be classified by the time they occur relative to the time of the stimulus: during construction impacts versus postconstruction impacts. For example, topsoil disturbance is a an impact that occurs during construction, while noise pollution is mostly a postconstruction impact because it often occurs during system operation. It is important to distinguish between the temporal scope of the impact and the temporal scope of the evaluation; namely, the impacts may occur only during construction or postconstruction, but in both cases, their evaluation can be carried out *ex ante*, that is, before construction (using simulative or analytical predictive models) or *ex poste*, that is, after construction (using field observations).

4.3.3 Concluding Remarks on Systems Evaluation

The task of evaluating a past existing, or proposed decision, action, or situation is often carried out by engineers at any phase of the system development process. The outcome of the evaluation, which, in some cases leads to the selection of the best alternative, is strongly influenced by the criteria that are being used for the evaluation. These criteria, in turn, reflect the concerns, objectives, and values of the system owner, operator, users, and the community. An issue that often arises is the relative importance or weights between the evaluation criteria. Another issue is the fact that different criteria are often expressed in different units and a uniform scale is therefore needed to bring all of the different units to a common denominator before a true comparison could be made between the alternatives. For doing this, a number of multiple criteria decision-making tools are presented in Chapter 12.

4.4 THE TASK OF FEEDBACK BETWEEN PHASES

The feedback between the phases of system development is a critical task that is often overlooked. It is critical because it presents an opportunity for engineers to refine their work at any phase based on the successes and failures of precedent or antecedent phases. For example, it has been long recognized that in systems design, inadequate consideration is often given to design-related maintenance issues: civil engineers have come to realize that certain aspects of facility design can impair facility maintainability and unduly increase the frequency and intensity of maintenance over a facility's life cycle (Ceran and Newman, 1992). The culprit often, is the institutional inertia that exists in many civil engineering agencies and that precludes the establishment of communication mechanisms. Specifically, it is vital that at any phase, civil engineering systems managers continue to find ways through which they can give increased and explicit consideration to the impacts of the output of their work at any phase on the subsequent phases, such as the impact of design on maintenance. At the planning and design stages of systems development, for example, a scoring procedure could be established to gage the "constructibility" and "maintainability" of the design, and a routine process could be established to improve communication between systems planners/designers and system preservation personnel. Such a routine process could be a part of a simple feedback mechanism geared toward increasing design engineers' awareness of the specific impacts of alternative designs on long-term facility maintenance efforts and costs.

A number of channels exist for carrying out, facilitating, or encouraging the task of feedback. These include meetings of various committees at various levels and jurisdictions of the civil engineering agency; training programs and courses; existing communication mechanisms such as design specifications and manuals, distribution of research, and personnel rotation between design and maintenance offices; and design review panels.

4.5 EXAMPLES OF TASKS AT EACH PHASE OF SYSTEMS DEVELOPMENT

In this section, we present and briefly discuss some of the traditional tasks and the systems analysis related tasks that civil engineers routinely encounter at each phase of system development.

4.5.1 Tasks at the Needs Assessment Phase

As we will see later in Chapter 19, the phase of needs assessment is a critical initial phase of civil engineering systems development. At this phase, the engineer addresses the question of whether there is a need for a new system. Recognizing that such a need is driven by demand for the system, which in turn, is often driven by population growth and other forces, the engineer is tasked with the responsibility of establishing the expected loading on, or demand for, the system either in terms of the raw numbers of users or in terms of some derived quantity. For example, in structural engineering, loading could relate to the force exerted by a live load; in water engineering, loading could refer to the number of gallons demanded for residential, commercial, or industrial use per day, week, or some specified period; in wastewater engineering, loading could refer to the amount of wastewater generated in a given time period; and in transportation engineering, loading could refer to the number of trips by travelers at a given point in time. Also, needs assessment in a smaller, less visible context of that term, may relate not to the entire system *per se*, but to the system phase in question, such as assessing the need for changes in system operational policies, system inspection policies, or system maintenance strategies.

4.5.2 Tasks at the System Planning Phase

At the phase of systems planning, the engineer faces a wide range of tasks, including the evaluation and selection of an appropriate location for the civil engineering system. For a proposed system (or for an enhancement of a part of an existing system), the tasks encountered at this phase include a specification or establishment of an overall system set-up that would ensure harmony with its natural or man-made environment so that (i) the system construction and operations would cause minimal disruption to the environment, and (ii) the environment would cause minimal disruption to the construction, preservation, and operations of the system. In carrying out these tasks, the engineer pays close attention to issues relating to ecology, aesthetics, the context sensitivity of the plans, and the functional relationships that are expected to exist between the proposed system and its environment. The extent to which this task is carried out often depends on whether the proposed change is one related to the system physical structure or operational policies.

Specific tasks at this phase include the development of alternative plans for the system and analysis and evaluation of each plan to ascertain the extent to which the system goals will be met by each plan. Depending on the nature of the goals in question, this task may include evaluation of technical impacts or economic efficiency. Also, the engineer may need to describe each alternative plan to an audience, such as a town hall or committee, using tools such as simulation and to select the optimal plan using optimization techniques. We will discuss more contexts and issues related to planning tasks in Chapter 20.

4.5.3 Tasks at the System Design Phase

According to the Accreditation Board for Engineering and Technology (ABET), design is the process of devising a system, component, or process to meet desired needs. ABET also states that the design process uses science, mathematics, and the like to convert resources optimally to meet a stated objective and involves establishing objectives and criteria, analysis, synthesis, testing, and evaluation. The tasks faced by the engineer at the design phase are clearly evident in this definition, and they are many and varied, as we will see in Chapter 21.

First, the engineer may need to describe the alternative designs for the physical system, which can be done through visual media such as artists' sketches, blueprints, or computer simulation. If the physical structure of the system is expected to change over time, the engineer may need to predict the future physical structure at a specified future time. In cases where there is a need to only modify or expand an existing system, a task may be to describe the proposed physical modifications to the existing structure. Second, the engineer faces the task of analyzing the design: This requires domain knowledge in that branch of civil engineering. For example, in structural engineering, this could mean determining the magnitude of the different stress modes, such as bending moments, shear force, and torsional force in each structural member of the physical system. In the field of environmental engineering, the design task may involve specification of the material types, the dimensions of the configurations of various sub-systems, or the processes to accommodate a certain level of demand: for example, calculating the required area of a sedimentation pond, choosing the disinfectant chemical that should be used, determining how long aeration should be carried out, or selecting the remediation treatment that should be used at a specific polluted site.

At this phase of system development, designs also are continually assessed to ascertain the extent to which a specific design meets the overall system goals or the design goals. Such evaluations are done from the perspective of benefits (to what extent are the objectives achieved) and the monetary and nonmonetary costs of the design. For example, is a design structurally stable to withstand the expected strong winds and hurricanes at a certain location?

Also at this phase, the engineer also brings into play the performance measures established at the needs assessment phase as the basis for the decision-making task. For example, for a physical system, on what basis could a particular physical design alternative be deemed optimal: technical efficiency, initial cost, life-cycle cost, aesthetics, maintainability, environmental compatibility, or sustainability? Also, for a nonphysical system, what rules or procedures should govern the operation of that system, or which changes in operational rules and procedures are needed to the existing system? For a particular operational design alternative, on what basis could it be deemed optimal: availability of skilled operating personnel, operating costs, or harmony of operations with the environment?

Some planning- and design-related tasks in hydraulic engineering include the description, analysis, evaluation, and selection of alternative designs for urban drainage, water distribution, urban sewerage, and hydraulic structures including dams, spillways, hydraulic outlets, energy dissipation structures, culverts, and bridges. In transportation systems design, the tasks include specifying appropriate material types and dimensions for different components of the guideway (runway, pavement, etc.) and geometric designs such as maximum and minimum grades, vertical and horizontal curves, maximum radii, superelevation, lanes, shoulders, median, and curbs. Structural engineers describe, analyze, and optimize different designs for structures on the basis of the material type, dimensions, and configurations for a variety of structural types such as trusses, frames, domes, beams, columns, and shells. In geotechnical engineering, the typical tasks include a description and evaluation of alternative designs for slope stabilization, earth retention, foundations, and geosynthetic applications.

4.5.4 Tasks at the System Construction/Implementation Phase

As we will see in Chapter 22, at the system construction, installation, or implementation phase, the engineer faces tasks that are similar in purpose to those of the preceding phases. As a prelude to the evaluation and selection of the optimal construction process, engineers often face the task of describing the process intended for the construction before that phase is carried out. In cases where the process has already begun or is even completed, the engineer, for purposes of performance audits, reviews, or public inquiries, may be required to describe the process used. Such descriptions involve not only the construction process (phases and timelines for utilization of resources such as equipment, material, and labor) but also may involve initial and more fundamental issues such as a description of the alternative contracting approaches that could be used, as well as their merits and demerits. Systems analysis tasks at this phase include estimation of the consequences of alternative courses of the implementation (i.e., in-house construction versus outside contractor) and the different approaches for contract delivery such as warranty, design–bid, and design–bid–build. Consequences could include the likelihood and/or magnitude of contractual aberrations such as cost overruns, time delays, and quality shortfalls or of contractual benefits such as the cost savings relative to a base case traditional approach. Consideration of these benefits and costs, for each alternative course of implementation, is the task that is encountered at this phase. The engineer addresses questions such as the following: Is the proposed construction process feasible? Are the requisite labor/equipment/funding available? What quality of workmanship can be expected? Are the expected performance levels, delays, and/or cost overruns acceptable? In striving to answer these questions, the construction engineer faces decision-making tasks regarding the construction/implementation process; for example, what is the best planning process for the construction? Should any modifications be made to an existing construction process? Which alternative processes yield the highest benefits or lowest costs? What is the best mix of resources (equipment, materials, and labor) and timing of work to achieve each task of the construction process? What is the optimal contracting approach for a given project of certain characteristics under a given set of attributes? Which downtime (e.g., work zone) management strategy would yield the least cost to the system owner and minimum disruption to the system user?

4.5.5 Tasks at the System Operations Phase

A system is said to be in operation when it is being used for its intended purpose. For example, a physical system, such as a bridge, commences its operations phase when its construction is completed and it is commissioned. Similarly, a nonphysical (or virtual) system, such as a new traffic signal timing plan, enters its operations phase when it is deployed for use at an urban arterial. The operations phase, naturally, is typically the longest of the phases of development. The tasks carried out in this phase for a specific system could therefore span an entire lifetime of an individual's career. Engineers are constantly engaged in carrying out various tasks consistent with the operations of dams, levees, bridges, highways, water treatment plants, tunnels, and other civil engineering systems. Similar to the tasks in other phases, the systems analysis tasks at the operations phase includes description, analysis, evaluation, and selection. In the task of description, the engineer paints a picture of the current or anticipated future operation of the system, often using computer simulation. In analyzing the system's operations, engineers utilize a variety of analytical tools and domain knowledge from traditional tasks to generate information for the tasks of evaluation and decision making. Thus, the tasks of analysis and evaluation include a determination of the extent to which the operational objectives are being achieved, or are expected to be achieved, for each operations strategy alternative and any attendant beneficial or adverse consequences. Specific tasks may include investigation of whether there is a more economical way, compared to current practice, to operate the system; the ratio of the benefits to the costs of operations for the current operational policy or for some policy under consideration; and an assessment of any environmental side effects of the system operation. In the task of decision making at this phase, answers are sought regarding the best operational policy that would yield the maximum benefits within a given budget or the minimum cost needed to maintain a certain specific level of operational performance. We will discuss system operations in further detail in Chapter 23.

4.5.6 Tasks at the System Monitoring Phase

The phase of system monitoring is an irregular phase in the sense that, unlike all other phases, this phase does not involve a physical transformation of the system (we will discuss system monitoring in Chapter 24 in further detail). As such, certain practitioners may consider system monitoring as a "task" faced at the phase of system operations rather than a phase by itself. In this text, we treat it as a phase that runs parallel to the system operations phase because it commences just after the system construction and continues until the system reaches the end of its life. The system monitoring phase comprises two aspects that are continuous throughout the life of the system or are characterized by regular intervals: (i) measuring the usage of the system and (ii) inspecting the physical condition of the system. At this phase, engineers undertake the task of describing the intended (or existing) plan for the system usage monitoring or physical inspection. The tasks involving analysis include determination of the minimum sample size needed to achieve a certain degree of precision in the usage or performance data collected or the reliability associated with the existing (or intended future) data collection scheme. Also, at this phase, the engineer may encounter the task of evaluating monitoring or inspection alternatives and selecting the most cost-effective one for implementation, either for a specific system or for an entire system of systems.

4.5.7 Tasks at the Systems Preservation Phase

In developed countries where most civil infrastructure systems are largely in place, the main challenge is to preserve this infrastructure. Most such systems are publicly owned, and engineers thus face the fiduciary task of making appropriate decisions are consistent with prudent use of taxpayer money. This is often a contentious issue because different stakeholders may have different perspectives of what constitutes a cost-effective decision. For example, road users may not appreciate situations where a road pavement with no visible signs of distress is receiving a treatment (because the engineer is applying a preventive treatment to retard the onset of deterioration) at the same time when some other road pavement in a relatively poor state is not receiving any treatment at that same time (because the engineer may be waiting for the optimal time, such as the following year, for applying the preservation treatment).

At the phase of system preservation, the engineer must be able to describe the existing condition of the system, the past trends of deterioration, and the expected future deterioration trend as a function of usage, climatic condition, age, or other deterioration factors. The engineer is tasked with establishing and documenting the possible preservation options for the system (i.e., the list of standard rehabilitation and maintenance treatments). For each possible preservation treatment, the engineer is responsible for establishing the effectiveness of the treatment in terms of the condition enhancement or added system longevity and the cost of the treatment in terms of the monetary agency cost and the downtime nonmonetary costs suffered by the users.

For system-specific preservation decisions that are needed only at a specific point in time, the engineer uses the information available on the cost and benefits of a treatment to determine its cost-effectiveness in the short or long term and then carries out evaluation to select the optimal category of preservation (i.e., rehabilitation, preventive maintenance, or reactive maintenance) as well as the optimal type of treatment in each category to apply to a given system at a given time. On the other hand, for system-specific preservation decisions that are meant to cover a longer time period such as the entire life of the system, the engineer's task is to first establish the candidate life-cycle activity profiles for preservation (or rehabilitation and maintenance strategies), which is simply a combination of treatment types and timings, and assessing, for each preservation strategy, the implications in terms of facility condition, agency cost, user cost, and other criteria. The analysis and evaluation tasks addressed by the engineer at this phase include quantification of the costs and benefits of each candidate preservation strategy (or an existing strategy) over the system life cycle. Given the plethora of possibilities for preserving a specific system in terms of treatment types and timings over its life cycle, careful decision making is often a necessary task for specifying the best long-term activity profile for the system.

For a system of systems, such as a network of sewers, bridges, or pavements, the task may be to select, at a given year or programming period, a subset of systems that ensure maximum systemwide benefit for a given budget or that yields the minimum cost for a given minimum level of performance. This task involves a number of discrete decision variables and therefore would require binary programming tools for resolution. Also, the task may involve determining the optimal amount of maintenance expenditure over the facility life cycle, which is a continuous decision variable for which the engineer must employ the tool of linear programming tools.

In either case, another task may be to quantify the sensitivity of the choice of optimal preservation strategy or treatment to changes in economic conditions, traffic loading, climatic conditions, agency policy, and other decision factors. This task often calls for analysis involving probabilistic (Monte Carlo) tools. Another task may be to quantify the consequences of deferred or hastened preservation treatment in terms of system condition or the future costs of preservation. In Chapter 25, we will discuss the tasks at the preservation phase in greater detail.

4.5.8 Tasks at the System End-of-Life Phase

System end of life occurs when the system is destroyed through natural or man-made attacks or when the system is deliberately demolished to make way for another similar system. Thus, the engineer at this phase faces the task of assessing the risk of occurrence of natural disasters including earthquakes floods, or landslides and so on, and the consequences of any such occurrence. For example, by using contour lines on maps that show areas of equal earthquake intensity, the risk of earthquake occurrence can be quantified. Also, using maps that show proximity to water bodies and documented history of the patterns of water spread, the risk of flooding could be ascertained numerically. Clearly, the tools of probability will be needed to yield a precise quantitative assessment of such risks. In any case, the engineer faces the task of describing the extent of such risks, analyzing the consequences, and also analyzing and evaluating alternative feasible proactive measures as well as reactive measures in a disaster event. In such cases where the system end of life is not deliberate and is due to external forces, the system engineer is often tasked with establishing a number of scenarios involving the system resilience, the threat likelihood, and the disaster consequences as a basis for analysis and evaluation of alternative types or levels of investments intended to enhance system resilience or to reduce the social consequences of the disaster. We will discuss these issues in further detail in Chapter 26.

In cases where the system end of life is intended by the system owner, the engineer is responsible for describing the alternative processes for the termination (often through computer simulation) as well as the costs and benefits of each alternative. In this way, evaluation could be carried out to identify the best option for termination. In certain cases, the engineer may need to ascertain the sensitivity of the demolition method of choice to the decision factors. Specific options that are often considered in this task may include: demolish the system and abandon the site; demolish the system and reconstruct; sell the entire system and site; lease the system to an interested party; or break the system down into its components and, for each component, either sell, dispose in landfill, or recycle for reuse at same location (for new replacement system or other purpose) or elsewhere.

SUMMARY

At each of the eight phases of civil systems development, engineers carry out not only the traditional tasks using domain knowledge in the civil engineering branch in question but also the systems analysis-related tasks. The systems analysis related tasks include description, analysis, evaluation, and selection of alternatives and feedback. The task of system description involves a statement of the physical condition or operations performance of a system at the current time or at a specified future time. Models used for describing civil engineering systems or processes may be classified in several different ways that are related to the system attributes, model form, model purpose, and data characteristics. The chapter also discussed the task of system evaluation, which is also encountered at any phase of the system development process, involves an assessment of the degree to which some selected system outcomes are being manifest. Therefore, in the task of evaluation, civil engineers determine the potential outcomes of each alternative action and thus decide, for example, which plan, design, contracting or construction approach, operational policy, or maintenance strategy should be adopted. Also discussed in this chapter is the feedback between phases, a critical task that is often overlooked by engineers; Feedback is important because it presents an opportunity for engineers to refine their work at any given phase based on the successes and failures of precedent or antecedent phases. The chapter also discussed examples of the tasks faced at various phases of systems development.

EXERCISES

1. Consider a proposal to construct a pedestrian footbridge over a busy arterial in a college town. At each of the seven phases of the development of this system, describe any two specific tasks that will be faced by the engineer. For any three pairs of these phases, list and discuss feedback tasks that could enhance the process of the system development.
2. For the project in Exercise 1, list, classify, and describe three different models that are used for describing any aspect of the system at any phase.
3. (a) When there is a need to describe a system at the current time or at a certain specified future time, why are probabilistic models considered superior to deterministic models? (b) What is the difference between descriptive models and prescriptive models? Give two examples of each.
4. For the project described in Exercise 1, discuss the scopes that need to be considered in the task of evaluating the alternatives.
5. Describe the feedback task from a general viewpoint. Is this really an important task for the development of civil engineering systems? Argue why or why not.

REFERENCES

Ceran, T., and Newman, R. B. (1992). Maintenance Considerations in Highway Design. NCHRP Report 349, Transportation Research Board, Washington DC.

Greene, W. H. (2011). *Econometric Analysis*, 7th ed. Prentice Hall, Upper Saddle River, NJ.

Rardin, R. L. (1997). *Optimization in Operations Research*, Prentice Hall, Upper Saddle River, NJ.

Revelle, C. S., Whitlatch, E., and Wright, J. (2003). *Civil and Environmental Systems Engineering*, 2nd ed. Prentice Hall, Upper Saddle River, NJ.

Sinha, K. C., and Labi, S. (2007). *Transportation Decision-Making: Principles of Project Evaluation and Programming*. Wiley, Hoboken, NJ.

Stark, R. M., and Nicholls, R. L. (1972). *Mathematical Foundations for Design—Civil Engineering Systems*, McGraw-Hill, New York.

PART III

TOOLS NEEDED TO CARRY OUT THE TASKS

CHAPTER 5

PROBABILITY

When you have eliminated the impossible, whatever remains, however improbable, must be the truth.

Sir Arthur Conan Doyle (1859–1930)

5.0 INTRODUCTION

In a perfect world, all the physical relationships and equations used in civil engineering yield outputs that occur with exactitude. Unfortunately, the real world is far from perfect. The attributes of any civil engineering system (need, usage, structure, operating conditions, etc.) as well as the physical and institutional environments in which they operate, are characterized by some degree of uncertainty. For example, the demand for water, the flow of traffic, the live loading of a structure, annual rainfall, soil conditions, steel strength, and construction costs and periods will hardly ever turn out to be exactly what was expected or predicted at the time of design. To cite another example, the actual strength of five standard steel bars supplied by a manufacturer may not be the exact specified yield strength (e.g., 250 N/mm^2) but may be, in N/mm^2, 250.125, 251.002, 250.995, 249.814, and 250.001. In other words, civil engineering systems data, whether collected in the field or the laboratory, typically will have some degree of randomness and variability. Such inherent uncertainties in civil engineering system characteristics and their environment translate into uncertainty in the expected outcome of the system in terms of the extent to which it meets its performance goals, objectives, or measures of effectiveness and therefore can influence the appropriateness of future system decisions. It is important for engineers involved in the various systems development phases to recognize that these uncertainties exist and are inevitable.

So, how does the engineer treat the issue of uncertainty? Should the engineer simply ignore it? Or should the engineer give it due consideration? The answer will depend on the situation at hand: If the degree of uncertainty is deemed to be small, then the engineer may chose to ignore it by assuming that the variable in question is equal to the **best available estimate** from past records, laboratory simulation and tests, or other means. In this case, the assumption or expectation is that the consequences of any deviation from the actual value of the variable on the decision outcome will be relatively insignificant. This typically has been the case for variables such as material strength properties (elastic constants, tensile or compressive strengths, etc.) and dimensions of physical components of a civil system (such as width and length of a standard structural member). If, on the other hand, there is likely to be significant uncertainty in the civil system attribute in question, the engineer may choose to adopt a **conservative estimate** of the variable in question by using a specified threshold for the variable (minimum strength, maximum stress, some percentile of the attribute value etc.) or a high factor of safety. However, in addressing uncertainties in this manner, some questions may arise (Benjamin and Cornell, 1970):

- *Consistency across situations*: It may be impossible to maintain consistency in the level of conservatism from one situation to another. What is considered conservative in a situation may be deemed liberal in another situation (e.g., different professional bodies specify different minimum concrete compressive strengths). Even for a particular agency, what is conservative at one time may be liberal at another time.

- *Consistency across system components*: It may not be possible to find a value that is conservative for all components of a system, particularly in cases where the characteristics of one subsystem or system component depends on the characteristics of another. For example, a conservative estimate of friction factor of a pipe will produce a conservative (low) estimate of flow in that pipe, but that may produce unconservative (high) estimates of flow on other parallel pipes in the same network.
- *Wasteful resource utilization*: A conservative estimate may result in unduly high costs of the design of the system's physical structure, operations, or life-cycle preservation strategy. For example, a conservatively designed storm drainage system (i.e., large cross sections) may be beneficial in handling flows of all magnitudes but may have such a prohibitive cost that the likelihood of system funding and implementation may be jeopardized.

Only when the situation permits should the engineer treat uncertainty simply by using best estimates or conservative estimates. In many other situations, both approaches do not suffice, and the engineer then must account for inherent uncertainty in a rational and objective manner that involves the principles of probability and statistics.

In addressing the problem of uncertainty, it is important for the engineer to know the likelihood of occurrence of an **event**. In systems engineering jargon, an event may simply represent the instance where a characteristic of the system takes on a certain value. For example, a numerical value of the likelihood of the event that the traffic flow of a proposed road will be 140,000 vehicles per day, that the strength of a certain concrete sample will be at least 25 N/mm^2, that the service life of a certain construction equipment will be 14.5 months. Knowledge of a numerical value of such likelihoods of civil system characteristics can then help the engineer to ascertain the likelihood of obtaining each level of the ultimate outcome or performance of the system. For example, the variabilities in the chloride content of a coastal environment and truck traffic levels are useful in predicting how long a specific bridge deck will last in that environment.

The analysis of random (also called probabilistic, stochastic, or nondeterministic) phenomena in civil engineering systems can be carried out using **probability theory**, an important branch of mathematics. The key element of probability theory, namely, random variables, stochastic processes and events. These are mathematical representations of random events (Harnett, 1975). For example, the toss of a single coin is a random event, but if it is repeated many times, the resulting sequence will exhibit certain statistical patterns that can be described and predicted. In probability theory, probability is represented by a real number between 0 and 1 that indicates how likely a specific event will occur. Within the domain of the available data, an event that has a probability of zero will never occur while an event with a probability of 1 will occur with all certainty, for that domain. For a given event, finding that value, which falls between 0 and 1, constitutes a fundamental aspect of probability and statistical analyses.

Probability theory is a mathematical foundation for statistical analysis and is essential to the description and analysis of engineering systems that on the basis of limited or partial knowledge of their characteristics. As such, the concepts of probability reverberate throughout this text. In Chapters 6, 7 and 8, we will see its usefulness in the tools of statistics, modeling, and simulation. In Chapter 9, we will see briefly that optimization can be carried out in a stochastic context where the input levels are not known with certainty. Also, inputs for cost analysis, engineering economics, and multiple criteria analysis are not always deterministic, and in Chapters 10–12, we shall see how these tools could incorporate probabilistic concepts. Further, the probability concept is a key aspect in the underlying theories of the tools of reliability and risk analysis, systems dynamics, real options, and decision analysis, as we shall learn in Chapters 13–16. Then in Chapters 19–26, we will apply these probability-related tools to answer questions associated with the various phases of civil systems development.

In discussing the fundamental ideas and concepts of probability theory, the use of the mathematical theory of sets provides a convenient foundation. This is because the events of an experiment can be considered as the subsets of a set. Thus, we first discuss set theory as a prelude to the discussion of probability theory. Then the chapter discusses basic concepts in probability such as mutually exclusive and statistically independent events, conditional probability, and random variables. Next, the chapter discusses the issues associated with analyzing probability functions such as the parameters that affect function shapes, mathematical expectation, and variance. Also, various standard discrete and continuous probability distributions are examined.

5.1 SET THEORY

As we have discussed in the previous section, a key aspect of probabilistic analysis is the notion that outcomes of an action, event, or situation can take any one of several possibilities. The collection of all these possibilities is referred to as the **sample space** and each possible occurrence is a **sample point** or **element**. A subset of the sample space, which may include one or more elements, is called an **event**. In the discussion below, we use the term *set* and *event* interchangeably.

A **set** or event is a collection of **elements**, and an element may be an object, situation, or event. For example, $A = \{x, y, z\}$ means the set A contains elements x, y, and z. A **subset** is a set that is contained in a larger set, and a **superset** is a set that contains a subset. A **null** set (or empty set) is a set that contains no elements, and is denoted by symbols $\{\}$ or ϕ. This is also referred to as an *impossible set*. Conversely, the **certain event** is one that will occur with 100% certainty and thus contains all the sample points in the sample space. A **Venn diagram** is a graphical representation of sets and elements.

The **Universal** set (also referred to as universe, population, or sample space) is the set that contains all elements or other sets and also represents the set of all possible outcomes in a given problem or experiment. Uppercase letters are used to denote sets, while lowercase letters are used to denote elements of the set. $\mathrm{A} \subset \mathrm{B}$ means that all members of set A are contained in set B, and $\mathrm{A} = \mathrm{B}$ means that sets A and B contain the same elements.

The **complement** of a set A (denoted as A') is the set of elements outside the set. The union of two sets A and B is the set of elements that belong to either A only or B only or both A and B. This is denoted by $\mathrm{A} \cup \mathrm{B}$. The **intersection** of two sets A and B is the set of elements that belong to both A and B. Two sets are described as being **mutually exclusive** if they have no members in common (Figure 5.1).

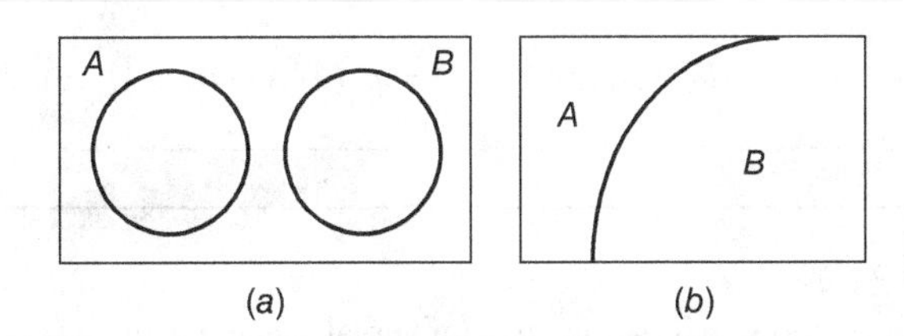

Figure 5.1 Mutually exclusive sets or events.

Two or more mutually exclusive sets that contain all elements of the universal set are described as **collectively exhaustive** (Figure 5.1*b*). In other words, there is no element or event inside the sample space that does not belong to any one of the collectively exhaustive sets in that sample space. The events or sets shown in Figure 5.1*b* are collectively exhaustive. Examples include the dominant material used for constructing interstate highway bridge superstructures: concrete and metal (other material types, such as timber and masonry, are not used for constructing such superstructures).

Example 5.1

Consider the set *A*, *B*, and *C* below.

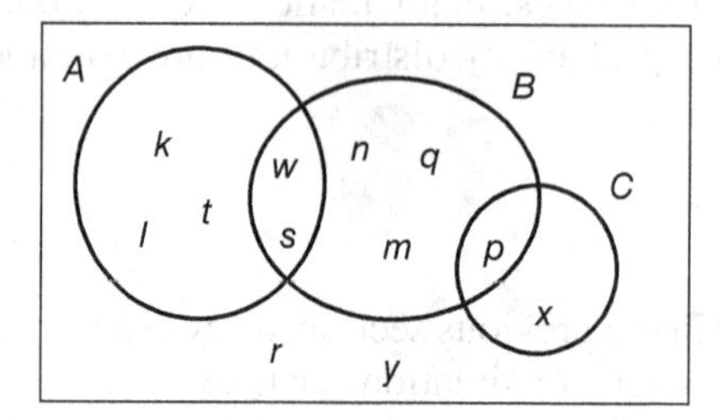

Figure for Example 5.1.

Using the Venn diagram or other means, indicate the elements of the following sets: (a) $A \cap B$, (b) $B \cap C$, (c) $A \cap C$.

Solution

(a) $A \cap B = w, s$. (b) $B \cap C = p$. (c) $A \cap C = \{\ \}$

Example 5.2

Consider the following set *A*.

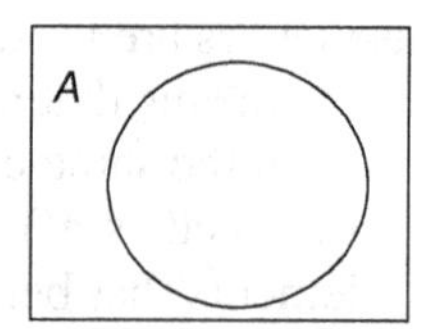

Figure for Example 5.2.

What are the answers to the following?
(a) $A \cap \phi$, (b) $A \cup \phi$, (c) $A \cap A'$, (d) $A \cap A'$, (e) U$'$, (f) ϕ', (g) $(A')'$.

Solution

(a) $A \cap \phi = \phi$. (b) $A \cup \phi = A$. (c) $A \cap A' = \phi$. (d) $A \cup A' = \mathrm{U}$. (e) $U' = \phi$. (f) $\phi' = U$. (g) $(A')' = A$.

Example 5.3

Consider sets *A* and *B* that overlap in a universal space. Using diagrams, shade the following sets: (a) A', (b) $(A \cap B)'$, (c) $(A \cup B)'$, (d) *A* but not *B*, (that is, $A \cap B'$).

Solution

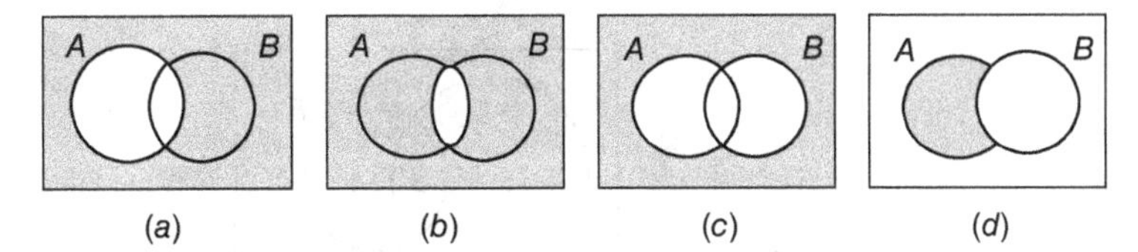

Figure for solution to Example 5.3.

DeMorgans' laws, a fundamental aspect of set theory, are stated as follows:
For two sets: $n(A \cup B) = n(A) + n(B) - n(A \cap B)$.
By dividing each term in the above equation by $n(U)$, we get the general probablity formula

$$\frac{n(A \cup B)}{n(U)} = \frac{n(A)}{n(U)} + \frac{n(B)}{n(U)} - \frac{n(A \cap B)}{n(U)}$$

which yields $p(A \cup B) = p(A) + p(\mathrm{B}) - p(A \cap B)$. This is the general probability formula, a powerful equation whose uses we shall examine in a subsequent section of this chapter.

For three sets: $n(A \cup B \cup C) = n(A) + n(B) + n(C) - n(A \cap B) - n(A \cap C) - n(B \cap C) - n(A \cap B \cap C)$.

Some other laws established by De Morgan are as follows:

$n(A' \cup B') = n(A \cap B)'$ for two sets or events; $n(A' \cup B' \cup C') = n(A \cap B \cap C)'$ for three sets or events.
$n(A' \cap B') = n(A \cup B)'$ for two sets or events; $n(A' \cap B' \cap C') = n(A \cup B \cup C)'$ for three sets or events.

Example 5.4

A construction company's experience has shown that 4 out of a batch of 100 structural units supplied by a certain vendor have fabrication errors and 3 out of 100 have both fabrication errors and the presence of impurities. Past records have shown that on the average, any supply batch contains units of which 8% have impurities. How many units of a randomly selected supply batch have (a) fabrication errors or impurities or both, (b) neither fabrication errors nor impurities, or (c) only impurities?

Solution

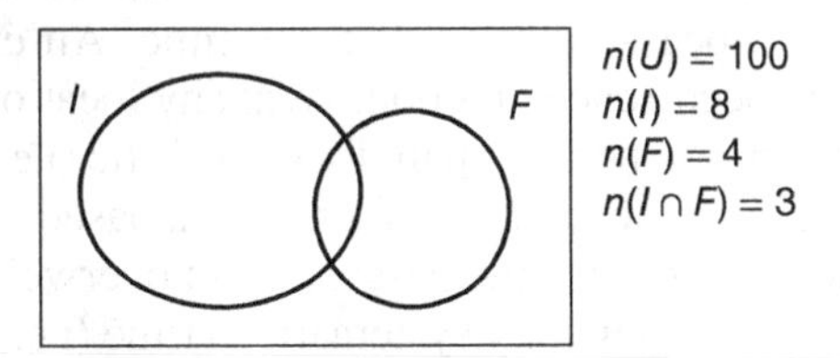

Figure 5.2

Let I represent the event that there are impurities and F be the event that there are fabrication errors.

(a) $n(I \cup F) = n(I) + n(\mathrm{F}) - n(I \cap F) = 8 + 4 - 3 = 9$
(b) $n[(I \cup F)'] = 1 - n(I \cup F) = 100 - 9 = 91$
(c) $n(I \cap F') = n(I) - n(I \cap F) = 8 - 3 = 5$

The completed Venn diagram is presented in Figure 5.3.

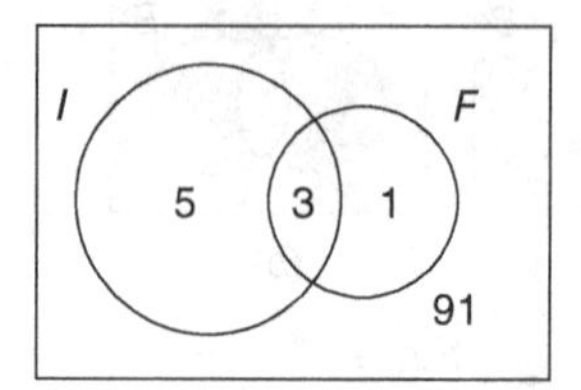

Figure 5.3

5.2 SOME BASIC CONCEPTS IN PROBABILITY

Before we proceed into a discussion of the basic concepts of probability, it is useful to discuss the dichotomy between a systematic event and a random event. A systematic process or event follows a definite, planned, and predictable pattern; an example is the probability that Jane, a 4.0 GPA college student, is awarded a prize if there is a policy to award prizes to all students with GPA above 3.95. A random event or process, on the other hand, follows no definite or predictable pattern and therefore can result in any outcome, for example, selecting students for a market survey by closing your eyes and choosing a few names from the student phone directory. Without a doubt, all engineering tasks would be much easier if all events were perfectly systematic. In reality, however, events and processes in engineering are not perfectly systematic. Nor are they perfectly random. Instead, they lie between the two extremes. Certain events or processes are more systematic than random, while for others, the reverse is true. Examples of the former include the construction period for constructing a tower, the stresses under a foundation footing, the effect of temperature on sewage decay, the bending moments in a given steel structure, and the arrival of trains at an efficient metro station. Examples of the latter include arrival of vehicles at an isolated intersection, terrorist attacks, car crashes on a specific interstate highway location, selecting winners of this week's state lottery, flipping a coin, and accidents at a construction site.

Synonyms of the word "random" include: without bias, without prejudice, and unsystematic. Randomness could be with respect to space, time, or both. Spatial randomness is when at a given time, the event or process can occur at any location—at no location does the event have higher or lower occurrence likelihood than at other locations. Temporal randomness is when at a given location, the event or process can occur at any time. An example of an event that is generally random in space is the occurrence of a pothole at any location within a given highway section. An example of an event that is generally spatially systematic (i.e., at any point in time, we have a fairly good idea of its location) is the geology of a region a based on soils maps. An example of an event that is generally random in time is the passing of an overweight truck over a specific weak bridge. Examples of events that are generally systematic in time (i.e., at any given location, we know when it is going to happen) are the arrival of trains at an efficient metro station and having a specific class lecture at the allocated time and room.

The study of stochastic processes is based on the concept of **random experiments**, where outcomes are random and may therefore result in any of several possible outcomes. The probability of an event is the ratio of the number of successes to the number of all possible outcomes. The set containing all possible outcomes is referred to as the **sample space** or universal set. The probability of a **deterministic event** is equal to 1 (it will definitely happen) or 0 (it will not happen). **Compound sets/events** refer to sets/events that contain two different sets/events. For example, consider the two events: Event A—The rain will fall tomorrow; and event B—the sun will shine tomorrow. An example of a compound event is event C—The rain will fall tomorrow and the sun will shine tomorrow. Another example of a compound event is event D—the rain will fall tomorrow or the sun will shine tomorrow. The **complementary rule** of probability states that the probability that an event does not occur is one minus the probability that it does occur: $p(A') = 1 - p(A)$.

The events that comprise a compound event may be sequential, mutually exclusive, or statistically independent or dependent of each other. Two events are **mutually exclusive (or disjoint)** if they cannot occur at the same time. For mutually exclusive events A and B, p (A happens OR B happens) $= p(A) + p(B)$.

Two events are *non-disjoint* or **sequential** if one follows the other either in time or space. Sequential events may be statistically dependent or independent. Two events are **statistically independent** if the probability of one does not depend on or affect the probability of the other. Two events **are statistically dependent** if the probability of one depends on or affects the probability of the other. We discuss these concepts further in Section 5.2.1.

5.2.1 General Probability Formula for Compound Events

For two compound events A and B, the general probability formula is as follows:

$$p(A \cup B) = p(A) + p(B) - p(A \cap B)$$

This formula is duly modified for mutually exclusive (or disjoint) events and nondisjoint events as shown in Figure 5.4 and explained in the sections below.

5.2.2 Mutually Exclusive (Disjoint) Events

For disjoint events, the intersection term of the general probability equation reduces to zero. That is, $p(A \cap B) = 0$.

Therefore the probability equation becomes

$$p(A \cup B) = p(A) + (B)$$

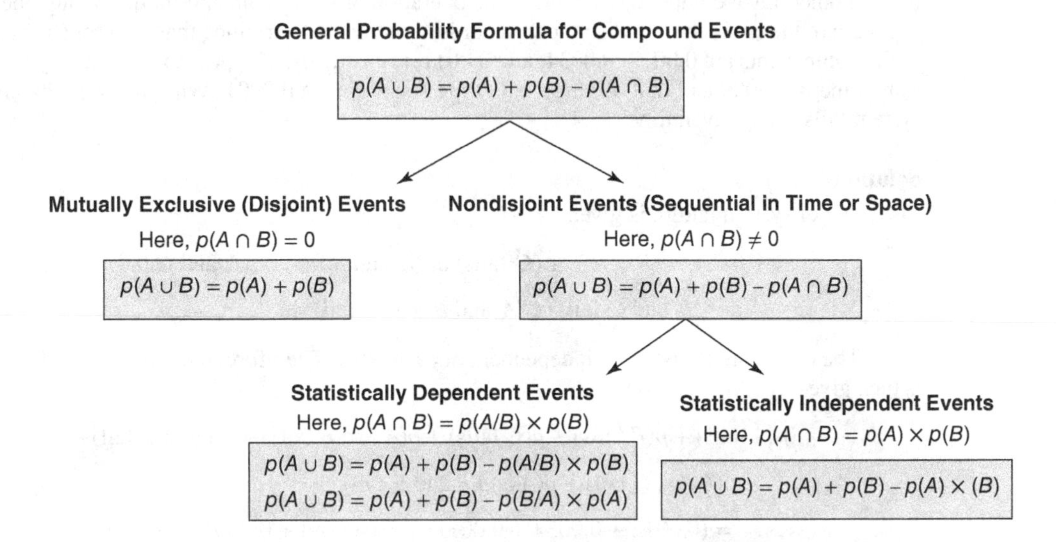

Figure 5.4 The general probability equation and its variations.

Example 5.5

In the year 2010, it is estimated that the probabilities that sewer pipes in a certain city will be, on the average, excellent, good, or poor condition are: excellent (E), 0.22; good (G), 0.38; poor (P), 0.40. (a) Find the probability that in 2010, a randomly selected sewer pipe in the city will be in either excellent or good condition.

Solution

$$p(E \text{ or } G) = p(E \cup G) = p(E) + p(G) - p(E \cap G)$$

But E and G are mutually exclusive, so $p(E \cap G) = 0$.

Therefore

$$p(E \text{ or } G) = p(E) + p(G) = 0.22 + 0.38 = 0.60$$

5.2.3 Nondisjoint Events (Sequential in Time or Space)

For nondisjoint events, $p(A \cap B) \neq 0$. Therefore, the probability equation is

$$p(A \cup B) = p(A) + p(B) - p(A \cap B)$$

The expanded expression for the term $p(A \cap B)$ will depend on whether the events are statistically dependent or otherwise, as seen below.

(a) Statistically Independent Events. For statistically independent events, $p(A \cap B) = p(A) \times p(B)$. Therefore the full probability equation becomes

$$p(A \cup B) = p(A) + p(B) - p(A) \times p(B)$$

Example 5.6

A certain civil engineering system comprises three components: A, B, and C that operate independent of each other. Each component is vital in the operation of the system and failure of any one will result in failure of the entire system. From past records, it has been determined that the probabilities of failure of the components are 0.0012, 0.0034, and 0.0011, respectively. The probability that any two fail at the same time is 0.0009 and that all three fail at the same time is 0.0001. What is the probability that the system fails at any given time?

Solution

The event of system failure is given by:

$$(A \text{ fails}) \text{ or } (B \text{ fails}) \text{ or } (C \text{ fails}) \text{ or } (A \text{ and } B \text{ fail}) \text{ or } (A \text{ and } C \text{ fail})$$
$$\text{or } (B \text{ and } C \text{ fail}) \text{ or } (A \text{ and } B \text{ and } C \text{ fail})$$

The events are statistically independent of each other. Therefore, the probability of system failure is then given by:

$$p(A \text{ fails}) + p(B \text{ fails}) + p(C \text{ fails}) + p(A \text{ and } B \text{ fail}) + p(A \text{ and } C \text{ fail})$$
$$+\, p(B \text{ and } C \text{ fail}) + p(A \text{ and } B \text{ and } C \text{ fail})$$
$$= (0.0012 + 0.0034 + 0.0011) + 3(0.0009) + 0.0001 = 0.0058$$

Example 5.7

An emergency evacuation route for a hurricane-prone city is served by two bridges leading out of the city. In the event of a major hurricane, the probability that bridge A will fail is 0.005, and the probability that bridge B will fail is 0.012. Assuming statistical independence between the two events, find the probability that at least one bridge fails in the event of a major hurricane.

Solution

Let A and B represent the events that bridge A and B, respectively, fail. When either A fails or B fails or both fail, then

$$p(A \text{ or } B) = p(A \cup B) = p(A) + p(B) - p(A \cap B)$$

The failure of A does not influence (and is not influenced by) the failure of B. Thus, the two events can be considered independent of each other. Thus, the probability that when a major hurricane occurs, both events happen is $p(A \cap B) = p(A) \times p(B) = 0.005 \times 0.012$. The probability that either A fails or B fails or both fail $= 0.005 + 0.012 - 0.005 \times 0.012 = 0.017$.

(b) Statistically Dependent Events. For statistically dependent events:

$$p(A \cap B) = p(A/B) \times p(B) \quad \text{or} \quad p(B/A) \times p(A)$$

This is the basis of the basic law of conditional probability, which we shall revisit in the next section. With the last term of the general probability equation expressed as such, the general equation becomes

$$p(A \cup B) = p(A) + p(B) - p(A/B) \times p(B)$$

or

$$p(A \cup B) = p(A) + p(B) - p(B/A) \times p(A)$$

Example 5.8

There are 520 bridges on the freight rail network in the province of Ostenborg. Of these, 200 are classified as "old" rail bridges (i.e., 50 years or more in age). Past records indicate that 50% of all rail bridges in the state exhibit visible signs of structural distress. Of the bridges 70% are old or exhibit structural distress. Find the probability that a randomly selected rail bridge will exhibit structural distress given that it is old.

Solution

Let O be the event that a bridge is "old" and S represent the event that a bridge exhibits signs of structural distress. Obviously, older bridges are generally more likely to show signs of structural distress. Thus, showing structural distress signs is dependent on age.

$$p(O \cup S) = p(\text{O}) + p(S) - p\left(\frac{S}{O}\right) \times p(O)$$

We seek $p(S/O)$.

$$p(O \cup S) = 0.7(520) = 364 \quad p(O) = 200 \quad p(S) = 0.5(520) = 260$$

$$p\left(\frac{S}{O}\right) = \frac{p(O) + p(S) - p(O \cup S)}{p(O)} = \frac{200 + 260 - 364}{200} = 0.48$$

Therefore, if a given bridge is old, there is a 48% chance that it exhibits signs of structural distress.

Conditional Probability. The mathematics of statistically dependent events gives rise to a different area of probability theory called *conditional probability*. Two events are statistically dependent when the occurrence (or nonoccurrence) of one event influences the occurrence of the other. In other words, the occurrence of one event is conditional upon the occurrence of the other. For example, the event that a civil system is overloaded can influence the probability of the event that the system fails. As seen in the above section for statistically dependent events, when two events that comprise a compound event are dependent of each other, the concept of conditional probability arises. $p(B/A)$ is the conditional probability that B occurs given that A has already occurred. As seen in Section 5.2.3(b), $p(A \cap B) = p(A/B) \times p(B)$ or $p(B/A) \times p(A)$. Thus:

$$p\left(\frac{B}{A}\right) = \frac{p(A \cap B)}{P(A)}$$

It can be seen that if $p(A) = 0$, then $p(B/A)$ is undefined. Also, if event A has already occurred, then the probability that B occurs is not $p(B)$ but is $p(B/A)$. Given that an event A has occurred, it becomes equivalent to the sample space with 100% probability. Therefore, as seen in Section 5.2.3(b), for any two events A and B, $p(A \cap B) = p(A/B) \times p(B)$.

Also, $p(B \cap \mathrm{A}) = p(B/A) \times p(A)$. This is known as the **multiplication rule**. Extension of the multiplication rule to multiple sets or events leads to a useful concept known as the total probability rule.

Total Probability Rule. Before we introduce the total probability rule, we first present the phrase "partitioning a given sample space." Figure 5.1*a* illustrates two disjoint sets or events. As we learned earlier in this section (5.1), if there are no other events or sets in the sample space, then the two sets are said to be collectively exhaustive of the sample space. This idea can be extended to N disjoint sets in a sample space, E_1, $E_2, \ldots, E_N$. For example, a construction contract may be completed exactly on schedule (E_1), behind schedule (E_2), or ahead of schedule (E_3) as seen in Figure 5.5; there are no other possibilities. In this example, the events E_1, E_2, and E_3 are said to "partition the given sample space" of the possibilities of contact delivery timeliness.

In the civil engineering branch of construction management, for example, cost overruns and time delays of contract delivery is important and we use this concept to illustrate the total probability rule. It may be of interest to the construction systems engineer to ascertain the relationship between time delay and contract cost overruns. Consider event A (Figure 5.6), which represents the event where a given contract experiences cost overruns.

Obviously, a contract that experiences a cost overrun could also have experienced one of three events: (i) It is finished just on schedule, E_1; (ii) it is finished behind schedule, E_2; or (iii) it is finished ahead of schedule, E_3. Thus, the probability that a contract experiences a cost overrun is the sum of the probabilities that (i) it experiences cost overrun *and* is finished just on schedule, (ii) it experiences cost overrun *and* is finished behind schedule, or (iii) it experiences cost overruns

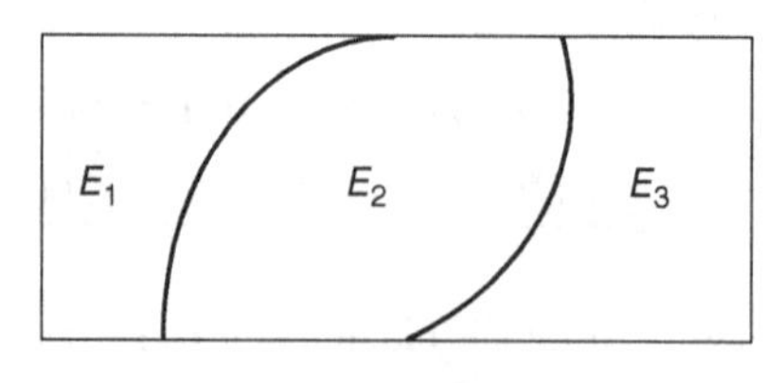

Figure 5.5

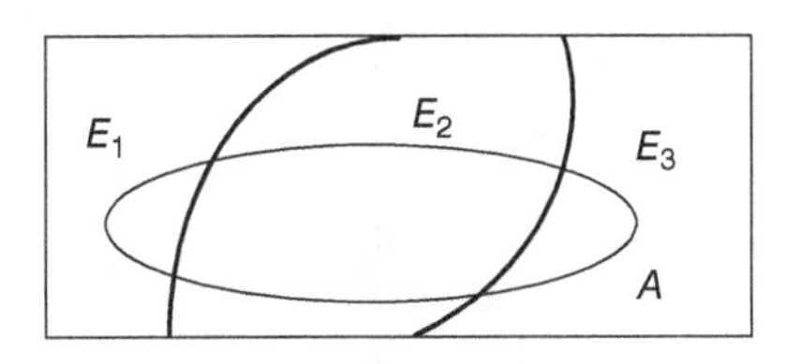

Figure 5.6

and is finished ahead of schedule. This total probability can be written as

$$p(A) = p(A \cap E_1) + p(A \cap E_2) + p(A \cap E_3)$$

From the formula for conditional probability, the probability that the contract experiences cost overrun and is finished ahead of schedule, $p(A \cap E_1)$, is given by $p(A \cap E_1) = p(A/E_1) \times p(E_1)$.

Similarly, the probabilities that the contract experiences cost overrun and is on schedule, and that the contract experiences cost overrun and is finished behind schedule, are given by $p(A \cap E_2) = p(A/E_2) \times p(E_2)$ and $p(A \cap E_3) = p(A/E_3) \times p(E_3)$, respectively. Therefore, the total probability that a contract experiences a cost overrun can be written as $p(A) = p(A/E_1) \times p(E_1) + p(A/E_2) \times p(E_2) + p(A/E_3) \times p(E_3)$.

This is the multiplication rule for the three collectively exhaustive sets. We can now generalize the rule for N sets.

Generally, if sets or events $E_1, E_2, \ldots, E_N$ partition a given sample space, then for any set or event A that intersects each of the events

$$\begin{aligned} p(A) &= p(A \cap E_1) + p(A \cap E_2) + \cdots + p(A \cap E_N) \\ &= p\left(\frac{A}{E_1}\right) p(E_1) + p\left(\frac{A}{E_2}\right) p(E_2) + \cdots + p\left(\frac{A}{E_N}\right) p(E_N) \\ &= \sum_{i=1}^{N} \left[p\left(\frac{A}{E_i}\right) \times p(E_i) \right] \end{aligned}$$

Using this formula, we can find the probability of an event A that intersects each of several collectively exhaustive events. An example problem is provided below. The reader may realize that with a little tweaking of the formula, we can find the probability that any one of the collectively exhaustive events occurs if A has occurred. This is discussed in the next section.

Example 5.9

Students, staff, and visitors typically make up 70, 25, and 5% of spectators during Cumberland College's football game. The probability that a student exhibits unruly behavior is 0.16, while such probability for staff and visitor are 0.01 and 0.005, respectively. What is the probability that a person selected at random from the spectators exhibits unruly behavior during in a football game?

Solution

Let E_1, E_2, and E_3 represent the events that a person selected at random is a staff, student, and visitor, respectively. Let A represent the event that a person exhibits unruly behavior. Figure 5.7*a* illustrates the statistical concept of the problem at hand: the events E_1, E_2, and E_3 partition the sample space (representing all persons in the stadium) and are overlapped by the event A. Figure 5.7*b* reduces the entire problem into a probability tree structure, which makes it easier to solve.

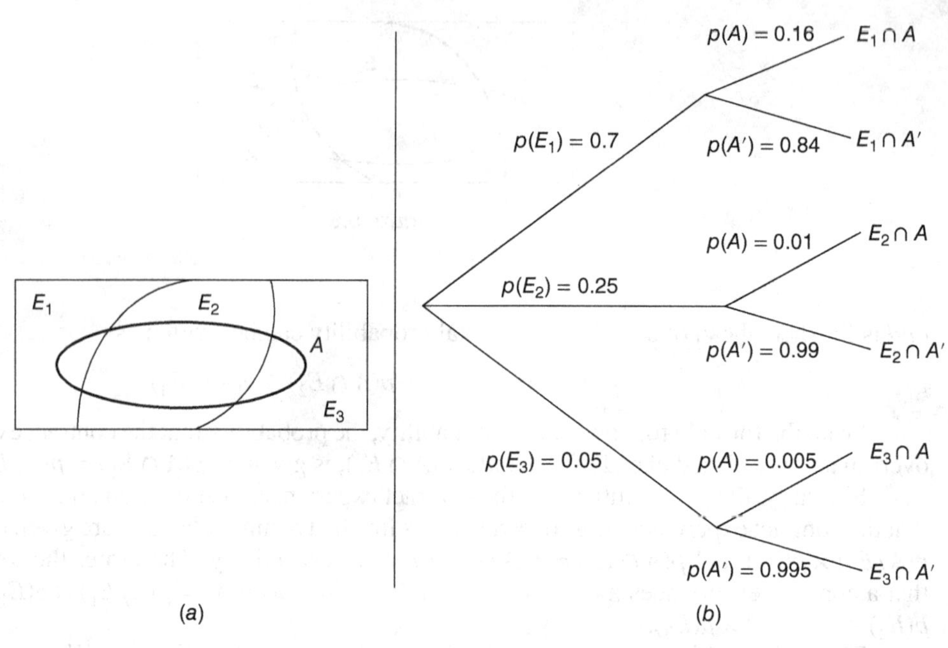

Figure 5.7 Partition diagram and probability tree for Example 5.5: (*a*) partition diagram and (*b*) probability tree.

The events of interest are as follows: A randomly selected person exhibits unruly behavior and is a student; a randomly selected person exhibits unruly behavior and is a staff; or a randomly selected person exhibits unruly behavior and is a visitor. Thus, the probability that a randomly selected person exhibits unruly behavior is calculated using the total probability rule:

$$p(A) = p(A \cap E_1) + p(A \cap E_2) + p(A \cap E_3)$$
$$= p\left(\frac{A}{E_1}\right) \times p(E_1) + p\left(\frac{A}{E_2}\right) \times p(E_2) + p\left(\frac{A}{E_3}\right) \times p(E_3)$$
$$= (0.7)(0.16) + (0.25)(0.01) + (0.05)(0.005) = 0.115 = 11.5\%$$

Bayes Rule. The previous section and example discussed how to calculate the probability of an event A or T that overlaps multiple events that partition a sample space. In certain cases, the problem is rather to find the probability that any of the partitioning events occur given that the overlapping event has already occurred. Consider events $E_1, E_2, \ldots, E_N$ that partition a given sample space; then the probability that E_k occurs given that A has already occurred is given by

$$p\left(\frac{E_k}{A}\right) = \frac{p(E_k \cap A)}{p(A)} \tag{5.1}$$

From the law of commutativity, $p(E_k \cap A) = p(A \cap E_k)$.

Therefore, Equation (5.1) becomes

$$p\left(\frac{E_k}{A}\right) = \frac{p(A \cap E_k)}{p(A)} \tag{5.2}$$

From the formula for conditional probability or the multiplication rule,

$$p(A \cap E_k) = p(A/E_k) \times p(E_k) \tag{5.3}$$

Also, from the total probability equation,

$$p(A) = p(A/E_1) \times p(E_1) + p(A/E_2) \times p(E_2) + \cdots + p(A/E_N) \times p(E_N) \tag{5.4}$$

Substituting (5.3) and (5.4) into (5.2) yields

$$p\left(\frac{E_k}{A}\right) = \frac{p(A/E_k)p(E_k)}{p(A/E_1)p(E_1) + p(A/E_2)p(E_2) + \cdots + p(A/E_N)p(E_N)} \tag{5.5}$$

This is Bayes rule. Using this rule, we can find, for any event E_k ($E_k = E_1, E_2, \ldots, E_N$) that partition a given sample space, the probability that a partitioning set E_k occurs given that an overlapping event A has already occurred, $p(E_k/A)$. For example, we could find the probability that a person is a visitor given that the person exhibits unruly behavior $p(E_3/A)$. If a probability tree is already available for the problem, the values can be simply read off and substituted into the formula. If not, the data given in the problem could be presented in a probability tree format to facilitate the problem solution.

Another variation of the problem is to find the probability that the overlapping event A occurs given that one of the partitioning events, E_k, has already occurred, $p(A/E_k)$. For example, what is the probability that a person at the football game exhibits unruly behavior given that the person is a student, that is, $p(A/E_1)$. In many cases, the answer can be easily read off from a probability tree if one has been constructed for the problem. Otherwise, a formula derived from the general concepts may be used as discussed below.

From the formula for conditional probability, the probability that an overlapping event A occurs given that one of the partitioning events, E_k, has occurred:

$$p\left(\frac{A}{E_k}\right) = \frac{p(A \cap E_k)}{p(E_k)} \tag{5.6}$$

But from a previous equation,

$$p\left(\frac{E_k}{A}\right) = \frac{p(E_k \cap A)}{p(A)}$$

Thus,

$$p(E_k \cap A) = p\left(\frac{E_k}{A}\right) \times p(A)$$

But

$$p(E_k \cap A) = p(A \cap E_k)$$

Therefore, Equation (5.6) becomes

$$\frac{p(A)}{P(E_k)} = \frac{p(A/E_k)}{p(E_k/A)}$$

Described as the **ratio law**, this can be rearranged as follows:

$$p\left(\frac{A}{E_k}\right) = \frac{p(E_k/A) \times p(A)}{p(E_k)} \tag{5.7}$$

Using Equation (5.7), we can determine the probability that an overlapping event A occurs given that one of the partitioning events, E_k, has already occurred.

Example 5.10

Consider the problem in Example 5.9. (a) On an otherwise quiet game one day, a certain gentleman John Doe exhibited unruly behavior. What is the probability that he is a student? (b) Given that a randomly selected person from the stands is a student, what is the probability that the person exhibits unruly behavior?

Solution

Event A represents that an object is thrown. Event E_1 represents that a person selected at random is a student. (a) We seek $p(E_1/A)$:

$$p\left(\frac{E_1}{A}\right) = \frac{p(E_1)p(A/E_1)}{p(E_1)\times p(A/E_1) + p(E_2)\times p(A/E_2) + p(E_3)\times p(A/E_3)}$$

The various values can be read off from the probability tree drawn for this question (see Figure 5.7):

$$p\left(\frac{E_1}{T}\right) = \frac{(0.7)(0.16)}{(0.7)(0.16) + (0.25)(0.01) + (0.05)(0.005)} = 0.976$$

(b) We seek $p(A/E_1)$. This can simply be read off from the initial problem statement or the probability chart as 0.16, or it can be calculated using Equation (5.7):

$$p\left(\frac{A}{E_1}\right) = \frac{p(E_1/A)\times p(A)}{p(E_1)} = \frac{0.976\times 0.115}{0.7} = 0.16$$

5.3 RANDOM VARIABLES

In preceding sections, we studied the rules for finding the probability of a single event or a compound event (combination of two of more events) in an experiment. The compound events we studied comprise only two or three constituent events or sets of the sample space. In this section, we extend this discussion by describing the probability not of just one or two events but of **all** possible events in a given experiment. This is useful when one seeks to find the probabilities associated with not just one or two events (outcomes of an action) but all possible outcomes. This consideration is particularly important in the context of engineering systems where several phenomena are associated with numerical outcomes in terms of measured quantities. For example, as decision makers at given phase of the civil engineering system development cycle, we often seek to evaluate several alternative courses of action and make an appropriate decision. Other more specific examples, as we saw earlier in this chapter include the "time delivery status" of a contract (for this random variable, the possible values are three: on time, behind time, or ahead of time). Another example is the cost overrun amount of a contact (for this random variable, the possible values are infinite).

An event is described as **random** when it cannot be predicted with certainty; that is, it could result in any one of several outcomes. Therefore, a variable whose exact value is not known in advance is termed a random variable. Formally, a **random variable** is defined as a function that assigns a real number to each outcome in the sample space of a random experiment (Montgomery and Runge, 2010). As illustrated in Figure 5.8, there are many types of random variables used in stochastic analysis: continuous versus discrete, categorical discrete versus count discrete, ordinal versus nonordinal, and binary versus multinary. A **continuous variable** is one that has a range that is uncountably infinite. Typically, these are values that are measured, such as distance, weight, volume, and area. A **discrete variable** is one that has a range that is finite or countably infinite. (Table 5.1) These typically are counting numbers (0, 1, 2, 3, etc.) or categorical values.

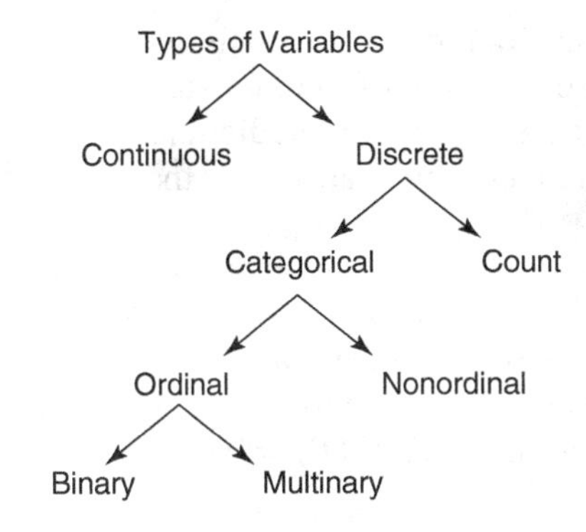

Figure 5.8 Categories of random variables.

Table 5.1 Examples of Discrete and Continuous Random Variables

Continuous	Discrete
• Total time taken to travel a certain transit link	• Level of user satisfaction for an engineering system
• Live load experienced by a certain structural member	• Number of failures of a system in a given time period
• Chemical oxygen demand	• Number of trains arriving at a central transit station in an hour
• Cost overrun of a construction project	• Primary material used for pavement surface construction Portland cement concrete, asphaltic concrete
• Width of a crack in concrete	• Class of a contractor
• Saturation ratio	• Tension status of a beam section Tension, neutral, compression
• Time of concentration for a given watershed	• Metal material composition (pure vs. alloy)
• Thickness of laid asphalt	• Type of structural connection
• Energy efficiency of a building	
• Construction equipment productivity	
• Particle flocculation rate	

Discrete variables may be count or categorical. A **count variable** is a discrete variable that takes on quantitative values, for example, number of times a system breaks down in a year. **Categorical variable** is a discrete variable that takes values that are not quantitative and may be ordinal or nonordinal. An **ordinal (or ordered) variable** is one that is a result of ranking, for example, the condition of a civil system on a tri-level scale: good, fair, and poor. A **nonordinal variable** has values that have no order of relative desirability in a specific context (at least, from the civil engineer's perspective), for example, the socioeconomic status of a system user: high income, middle income, and low income. A **binary variable** has two discrete outcomes, for example, critical flow and noncritical flow of a fluid. A **multinary variable** is one that has several discrete outcomes, for example, the mechanism of failure of a given sewer pipe: corrosion, cracking, or deformation.

Why do we need to know the type of random variable associated with any given stochastic problem? One reason is that in statistical modeling, the selection of appropriate model functional form will depend on the type of response variable. For example, where the response variable is categorical and ordinal, such as the physical condition of a randomly selected bridge deck, an ordered probit model is recommended. A second reason is that knowing the type of explanatory variable can help in identifying the appropriate probability distribution to describe the variation of that variable, a necessary ingredient for stochastically modeling system behavior and processes. Incorrect identification of the type of random variable can lead to the use of an incorrect distribution to describe that process, incorrect predictions regarding system condition or operations, and ultimately, inappropriate recommendations for the required action or intervention at that phase of system development.

General Notation for Random Variables: A capital letter, usually X, is used to denote a random variable. A small letter, usually x, is used to denote any value taken by the random variable. For example, if there is a 80% chance that a certain engineering system will experience over six overloads every year, then if X is a random variable denoting the number of overloads experienced by the system every year, then $p(X > 60) = 0.8$.

5.4 PROBABILITY FUNCTIONS

A probability function is defined as the mathematical relationship between the probability of a random variable and the value of the random variable. An example is an end-of-season table showing the number of scores (points or goals) scored by your school basketball or soccer team in a game (on the x axis) and the number of games or frequency associated with each score. In other words, "how are probability values distributed over various values of the random variable." The probability that X takes on a certain value x is denoted as $p(X = x)$ or $f(x)$. Using the function $p(X = x)$ we can determine the probability that the random variable (X) takes a specific value, x. Probability functions can be expressed in at least one of three ways: a mathematical equation, graph, or table.

Probability, distributions are a key element in the analysis of engineering systems. They help in describing the behavior of a system for purposes of documenting its structure or operational/usage patterns and also for purposes of decision making. For example, in testing to ascertain the veracity (or validity) of a claim about a system, the engineer needs to know the density function of the test statistic for such hypothesis tests. In other analysis, the engineer may seek the density function of the random variable because the theoretical development is based on a specific density function, and the decision-making process and outcome are sensitive to the nature of the assumed density function (McCuen, 1985). Incorrect identification of the density function can therefore translate to inappropriate decision.

There are basically two classes of probability distributions, depending on the nature of the random variable. A probability distribution is described as discrete when its random variable is discrete and is described as continuous when its random variable is continuous. Within each of these two categories, a probability distribution could either have no specific mathematical form (equation) and therefore could take any form or could have a specific functional form (Figure 5.9).

The mathematical relationships between the values of a random variable and their corresponding probabilities are termed **probability mass function** and **probability density function** for discrete and continuous variables, respectively. Also, the mathematical relationship between the cumulative (upper bound) values of the random variable and their corresponding probabilities are termed **cumulative distribution function** and **probability distribution function**, for discrete and continuous variables, respectively (see Figure 5.10).

5.4.1 Probability Mass Functions and Density Functions

The probability mass function, or pmf, which applies to discrete random variables, is typically denoted by $f(X)$ or $p(X = x)$. Because X is a discrete random variable, x can only take positive integer values. In the case of continuous distributions, the random variable cannot take on exact values;

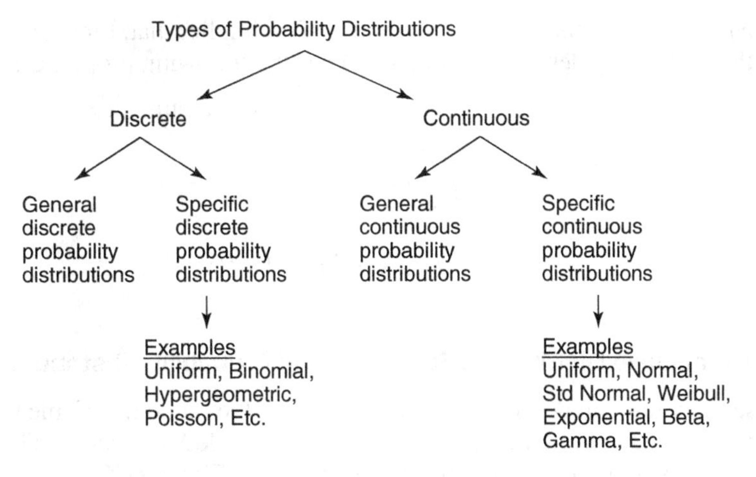

Figure 5.9 Types of probability distributions.

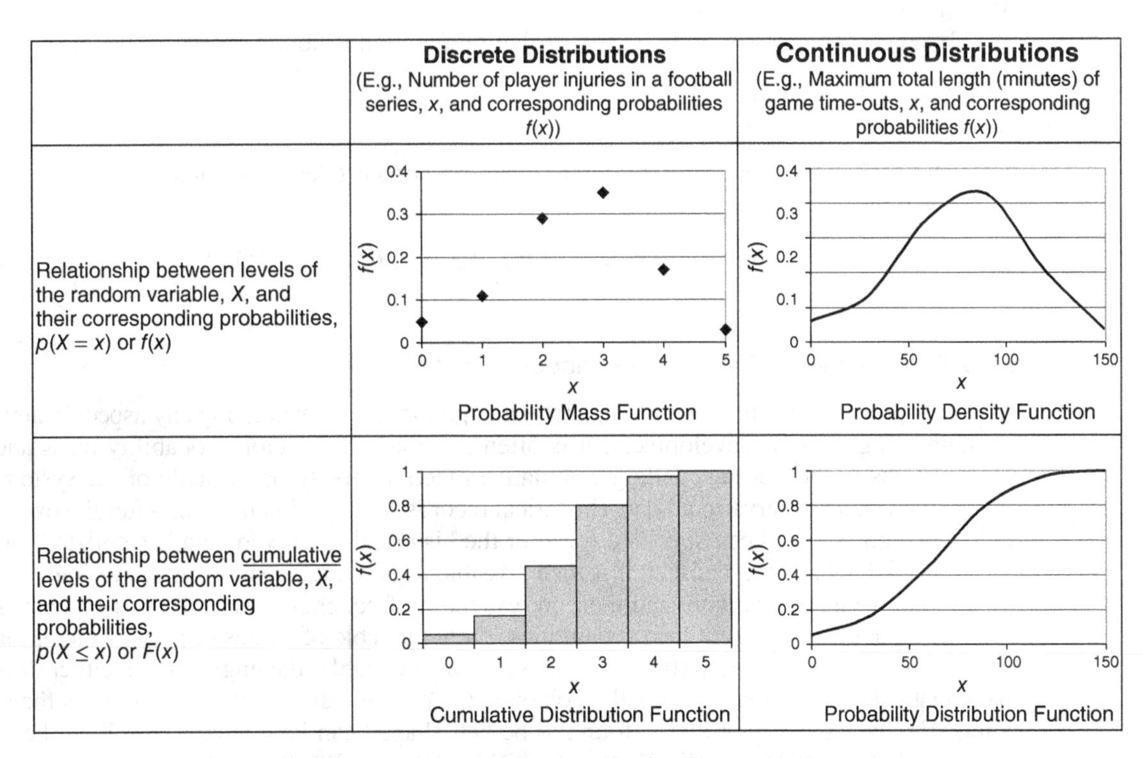

Figure 5.10 Categories of probability functions with illustrations.

as such $p(X = x) = 0$ for every real number x. This implies that, for example, $p(X \geq a) = p(X > a)$. Also, the probability density function (pdf) has the following properties:

$$f(x) \geq 0 \quad \text{for every real number } x$$

$$\int_{-\infty}^{\infty} f(x)\ dx = 1$$

$$p(a \leq X \leq b) = \int_{a}^{b} f(x)\ dx$$

5.4.2 Cumulative Distribution Functions and Probability Distribution Functions

In these functions, we describe the probability that the random variable takes on any value up to a certain upper bound limit. For a discrete random variable X having possible values $x_1, x_2, \ldots, x_n$, the cumulative distribution function (or cdf) is given by $F(x) = p(X \leq x) = \sum x_i$ for all $x_i \leq x$. Also, if $x \leq y$, then $F(x) \leq F(y)$. In other words, every cdf is nondecreasing. For a continuous random variable X, the probability distribution function is given by

$$F(x) = p(X \leq x) = \int_{-\infty}^{x} f(x)\ dx$$

Also, if $x \leq y$, then $F(x) \leq F(y)$.

In other words, every probability distribution function is nondecreasing. This means that

(i) $F(a) = \int_{-\infty}^{a} f(x)\ dx$
(ii) $F(-\infty) = 0$ and $F(\infty) = 1$
(iii) The probability that the continuous random variable takes any value between a and b is given by

$$p(a < X < b) = p(a \leq X \leq b) = \int_{a}^{b} f(x)\ dx$$

5.4.3 Constructing a Probability Function

As part of prerequisites for describing, analyzing, predicting, or evaluating any aspect in any phase of engineering systems development, it is often necessary to develop probability mass and density functions or their corresponding cumulative functions for some attribute of the system. This can be done using observational data (historical records) or experimentation. After the experiment and its outcomes have been specified (or after the historical, observed data has collated), and the random variable has been defined, it is then possible to plot a table showing the various values of the attribute's random variable and their corresponding frequencies and relative frequencies. The relative frequencies represent the probabilities. From the table of values or plot for the probability mass (for discrete) or density (for continuous) random variable, the engineer can either develop a mathematical formula that best fits the points on the plot or can identify what best fits the plotted points. For example, a plot that is found to be bell-shaped can be assumed to follow the normal distribution, and the normal distribution formula can be used as the probability distribution. The construction of a probability distribution (adapted from Harnett, 1975) is illustrated in Figure 5.11.

Before we continue to treat the topic of probability distributions, it is useful to discuss related concepts of mathematical expectation and the factors affecting the shapes of probability distributions.

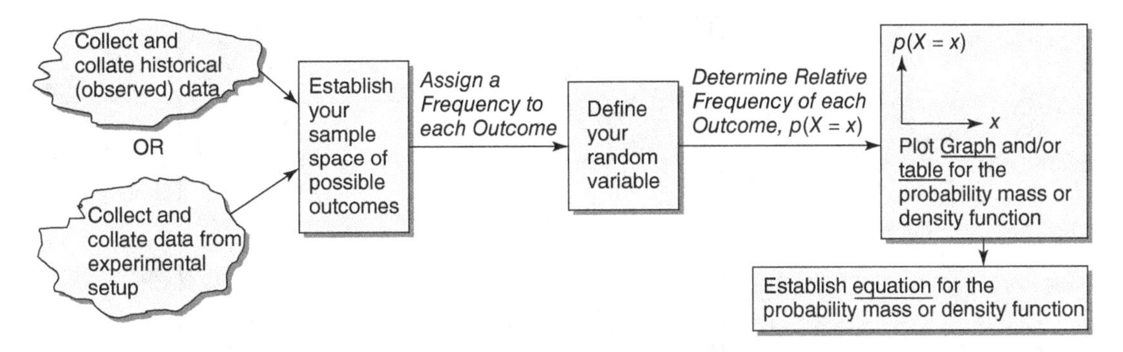

Figure 5.11 How to construct a probability function.

5.4.4 Shapes of Probability Functions

The factors that affect the shape of a probability functions are described as follows.

(a) Continuity (or Discreteness) of the Random Variable. Continuous probability functions are represented by an unbroken line. Discrete probability functions are represented by dots. A cumulative discrete probability function is represented by a step function.

(b) Functional Form of the Equation. The functional form of a probability function may be linear or nonlinear. Nonlinear functional forms include the normal, beta, gamma, weibull, discrete uniform, and so forth (see Appendix 1).

(c) Parameters of the Function. A probability mass or density function can be characterized by one or more parameters. For example, the equation $y = 3 + 2x^{0.5}$ has parameters 3, 2, and 0.5. The parameters of a distribution, which dictate the geometric features of the distribution, are of three kinds: location, scale, and shape parameters (McCuen, 1985).

Location parameter: This determines the abscissa of a location point of the probability distribution. All fractiles of the distribution can be located with reference to this parameter. Any measure of central tendency (see Section 6.2.2), such as the mean, is used frequently as the location parameter of the distribution. However, the lower limit, upper limit, or some other statistic may also serve as a location parameters. Changes in the location parameter do not cause a change in the scale or shape of the distribution. Location parameters are additive.

Scale parameter: This is a parameter that identifies the location relative to some specified point. Typically, measures of dispersion (see Section 6.2.2), such as the range, standard deviation, or variance, are used as scale parameters. Two probability distributions may have the same location parameter and shape but different scale parameters. Thus, scale parameters determine the size of a probability distribution without changing its shape. Scale parameters are multiplicative.

Shape parameter: This is a parameter that controls the geometric configuration of a distribution. Changes in the shape parameter can be reflected in changes in the structure, outline, or balance of the distribution. As such, a shape parameter is typically used to distinguish between individual density functions that belong to a given family of density functions. A distribution may have none, one, or more than one shape parameters. Figure 5.12 illustrates how the shape of a certain distribution function changes in response to changes in the values of its shape parameters α and β.

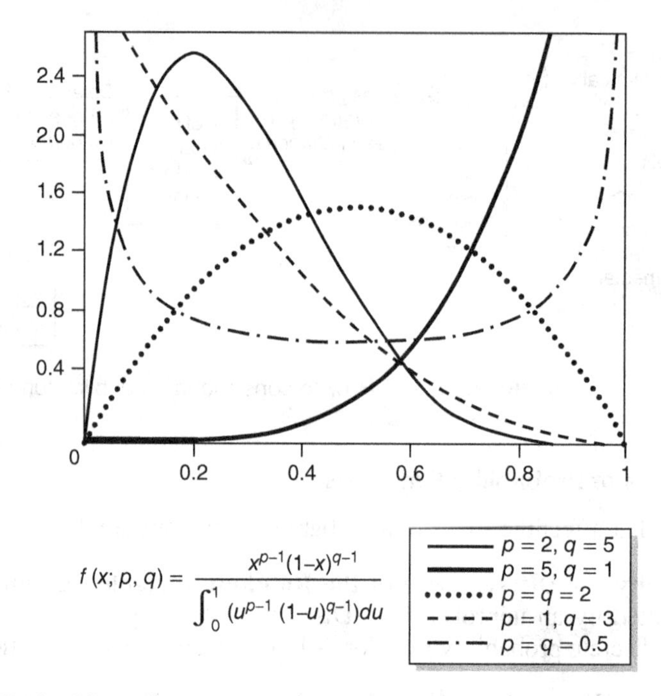

Figure 5.12 Effect of changing parameters on function shape—an illustration.

5.4.5 Expected Value of a Probabilistic Random Variable

The expected value is defined as the mean of the probability distribution of the random variable and often serves as simple basis for making inferences regarding an engineering system. Where the problem involves making a decision on the basis of probabilistic outcomes (performance) of and where there are several possible alternative decisions (or courses of action) of an engineering system, the system engineer chooses the best decision on the basis of which decision yields the lowest expected value of "negative" performance measures or highest expected value of "positive" performance measures. For example, if the outcome or performance of the engineering system is a random variable denoting the number of persons served per day or profit, then the decision maker selects the decision that yields the highest expected value of that random variable. On the other hand, if the outcome or performance is a random variable denoting the area of environment degraded, then the decision maker selects the decision that yields the lowest expected value of that random variable. Clearly, in order to make a decision in this manner, it is useful to know the probability distribution function for the random variable at hand. The determination of the expected value depends on whether the distribution is discrete or continuous, the functional form of the distribution function, and the parameters of the functional form. Where the probability distribution is discrete, the **mean** or **expected value** of the discrete random variable X having possible values $x_1, x_2, \ldots, x_n$ is given by

$$\mu = E(X) = \sum_{i=1}^{n} \left[x_i f\left(x_i\right)\right]$$

How certain are we that we will achieve this mean value of the distribution? The answer can be found in the variance. The **variance** of a discrete random variable X having possible values

$x_1, x_2, \ldots, x_n$ is defined as the expected value of the square of the deviation, that is:

$$V(X) = E\left[(X-\mu)^2\right] = \sum_{i=1}^{n}\left[\left(x_i-\mu\right)^2 f(x_i)\right]$$

Another measure of the certainty or variability of the distribution is the **standard deviation** of the random variable, which is simply given by $\sigma = [V(x)]^{0.5}$. Where the probability distribution is continuous, the **mean** or **expected value** of the continuous random variable X having possible values $x_1, x_2, \ldots, x_n$ is given by

$$\mu = E(X) = \int_{-\infty}^{\infty} \left[x \cdot f(x)\right] dx$$

The **variance** of a continuous random variable x having possible values $x_1, x_2, \ldots, x_n$ is defined as the expected value of the square of the deviation, that is,

$$V(X) = E\left[(X-\mu)^2\right] = \int_{-\infty}^{\infty} (x-\mu)^2 f(x)\, dx$$

Example 5.11

The probability of winning a certain \$5 million lottery is 0.0000003. If Kim partakes in this lottery, what is the expected value of his gross earning?

Solution

Either Kim wins or he does not win. Let X be a random variable denoting his winning status and therefore the amount won. Then we have a discrete random variable with only two outcomes, as follows:

X	**f(x) or p(X = x)**
x_1 = \$0 (Kim wins)	0.9999997
x_2 = \$5,000,000 (He does not win)	0.0000003

Expected value = (\$0 × 0.9999997) + (\$5000000 × 0.0000003) = \$1.5. Therefore, the expected value of Kim's gross earning is \$1.5.

Example 5.12

By investing in a particular stock, Josue can, in one year, make a profit of \$4000 with probability 0.3 or lose \$1000 with probability 0.7. What is Josue's expected gain?

Solution

Assuming that the occurrences of these monetary amounts are discrete:

$$p(X = \$4000) = 0.3, \text{ and } p(X = -\$1000) = 0.7$$

x	**f(x)**
\$4000	0.3
−\$1000	0.7

Expected gain or Expectation = $E(X)$ = (4000 × 0.3) + (−1000 × 0.7) = \$500

Example 5.13

The service life of the deck of a certain bridge type is a random variable X (years), which is described by the following probability density function:

$$f(x) = \begin{cases} x/600 & 0 < x < 30 \\ 0.2 - x/200 & 30 < x < 40 \text{ years} \\ 0 & \text{otherwise} \end{cases}$$

On average, how often would you recommend replacement of the bridge deck?

Solution:

The expected value of the service life of the deck is given by

$$\mu = E(X) = \int_{-\infty}^{\infty} (xf(x))\ dx$$

$$= \int_{-\infty}^{0} xf_1(x)dx + \int_{0}^{30} xf_2(x)dx + \int_{30}^{40} xf_3(x)dx + \int_{40}^{\infty} xf_4(x)dx$$

$$= 0 + \int_{0}^{30} x\frac{x}{600}\ dx + \int_{30}^{40} x\left(0.2 - \frac{x}{200}\right)\ dx + 0 = 23.3\,\text{years}$$

Therefore, as part of your design, you should recommend replacement of the component every 23 years.

Rules of Mathematical Expectation. There are certain rules associated with the computation of expected values of random variables, as seen in Table 5.2.

Table 5.2 Expectation and Variance—Properties

Rule	Description	Math Notation
1	The expected value of a constant is equal to the constant itself	$E[c] = c$
2	The variance of a constant is zero	$V[c] = 0$
3	The expectation of the product of a constant and a variable is equal to the product of the constant and the expectation of the variable	$E[cx] = cE[x]$
4	The variance of the product of a constant and a variable is the product of the square of the constant and the variance of the variable	$V(cx) = c^2V(x)$
5	The expected value of the sum (or difference) of two variables is the sum (or difference) of their expected values	$E[x + y] = E[x] + E[y]$ $E[x - y] = E[x] - E[y]$
6	If x and y are independent variables, then: the expected value of the product of the variables is the product of their expected values	$E[x\,y] = E[x]\,E[y]$
7	Also, if x and y are independent variables, then: the variance of the sum of the variables is the sum of their expected values	$V(x + y) = V(x) + V(y)$

5.5 DISCRETE PROBABILITY DISTRIBUTIONS

5.5.1 General Discrete Distributions

The probability distribution of a random variable X is a description of the probabilities associated with the possible values of X. For a discrete random variable X, the probability distribution or mass function is $f(x)$ or $p(X = x)$.

The cumulative distribution function (cdf) is $F(x) = p(X \leq x)$. For a discrete random variable X having possible values $x_1, x_2, \ldots, x_n$, the cdf is given by $F(x) = p(X \leq x) = \sum f(x_i)$ for all $x_i \leq x$. If $x \leq y$, then $F(x) \leq F(y)$. In other words, every cdf is nondecreasing.

Example 5.14

In a large shipment of precast concrete pipes for an urban drainage project, the probability that any individual precast pipe has a defect is 0.35. The quality control engineer takes a sample of 17 pipes from the shipment for testing. Assuming that the distribution of the precast pipes defects is consistent with the properties of a binomial distribution, find the probability that: (a) *at most* 3 pipes have defects, (b) 5 *or more* pipes have defects, (c) exactly 12 pipes do *not* have defects, and (d) 3, 4, or 5 pipes have defects?

Solution

(a) "At most 3" is the same as "3 or less" and is found in the $p(X \leq x)$ column as $F(3) = 0.10279$.

(b) The probability that 5 or more pipes have defects is the same as 100% minus the probability that 4 or fewer pipes have defects. Therefore, the probability that 5 or more pipes have defects is given by $1 - F(4) = 1 - 0.23484 = 0.76516$.

(c) There are 17 pipes in the sample. Thus, the probability that exactly 12 pipes *do not* have defects is the same as the probability that exactly 5 housings *do* have defects: $p(5) = 0.18486$.

(d) The probability that 3, 4, or 5 housings have defects is given by

$$p(3) + p(4) + p(5) = 0.07006 + 0.13205 + 0.18486 = 0.38697$$

Alternatively, this can be calculated as follows: $F(5) - F(2) = 0.41970 - 0.03273 = 0.38697$.

5.5.2 Special Discrete Distributions

(a) Discrete Uniform Distribution. A random variable X has a discrete uniform distribution if each of the n values in its range, say $x_1, x_2, \ldots, x_n$, has equal probability. Suppose that X has a discrete uniform distribution on the equally spaced values $a, a + c, a + 2c, \ldots, b$ for some constant $c \geq 0$, then the mean of X is $E(X) = (a + b)/2$, and the variance of X is $V(X) = (b - a)^2/12$.

(b) Bernoulli Distribution. A Bernoulli experiment involves a single trial that has the following properties: (i) there are only two outcomes: success or failure, (ii) the probability of success at each trial is constant, (iii) each trial is independent of the other. The Bernoulli distribution is described as the "mother" of all discrete probability distributions.

(c) Binomial Distribution. A binomial experiment consists of a series of Bernoulli trials. The binomial distribution has the following properties: (i) Consists of a number of trials (that are sequential in time or space), (ii) each trial has only two outcomes (success/failure), (iii) the probability of success is constant from trial to trial, and (iv) each trial is independent of the other. The "bi" in binomial means that the random variable for such distributions has a binary outcome as stated in the second property.

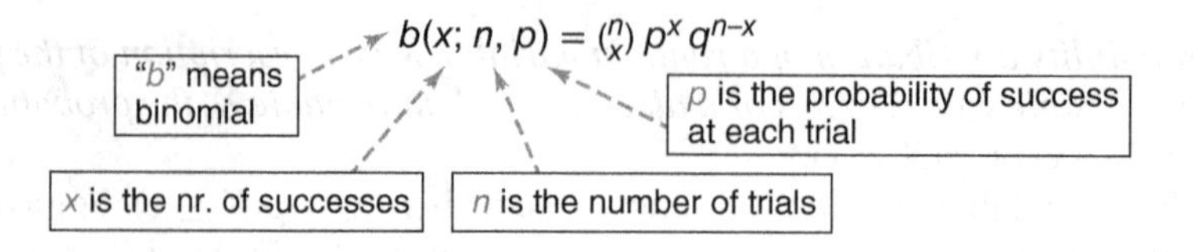

Example 5.15

The probability that a certain engineering system experiences overload in a given year is 0.40. You observe five such systems. What is the probability that three of them experience overload in a given year? Assume that each trial is a Bernoulli process.

Solution

$$\text{Number of trials}, n = 5$$

$$\text{Number of successes}, x = 3$$

$$\text{Probability of each success}, p = 0.40$$

Method 1 (Using Formula)

$$b(x; n, p) = p(X = x) = f(x) = \frac{n!}{x!(n-x)!} p^x q^{n-x} = \binom{n}{x} p^x q^{n-x}$$

$$p(X = 3) = \frac{5!}{3!(5-3)!} (0.4)^3 (1 - 0.4)^{5-3} = 0.23$$

Method 2 (Using Statistical Tables)

$$p(X = 3) = F(3) - F(2) = 0.9130 - 0.6820 = 0.23$$

Shape of the binomial distribution. The parameters of a binomial distribution, n and p, determine the shape of the distribution. In this regard, three different combinations of these parameters can be considered:

1. When n is small and p is large ($p > 0.5$)
2. When n is small and p is also small ($p < 0.5$)
3. When n is large and/or $p = 0.5$

When n is small and p is large, the binomial function is positively (right) skewed as shown in Figure 5.13*a*. When both n and p are small, the binomial function is negatively (right) skewed (Figure 5.13*b*). When n is large and/or $p = 0.5$, the binomial distribution takes the form of a symmetrical distribution (i.e., no skew, with its mean equal to its median). Even when $p \neq 0.5$, the shape of the distribution becomes increasingly symmetrical as n increases.

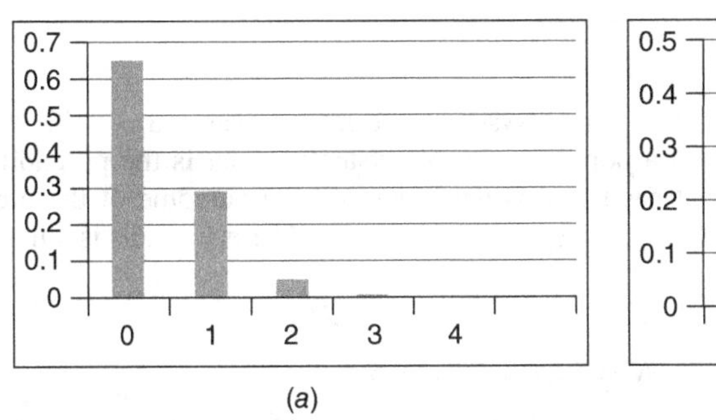

(a)

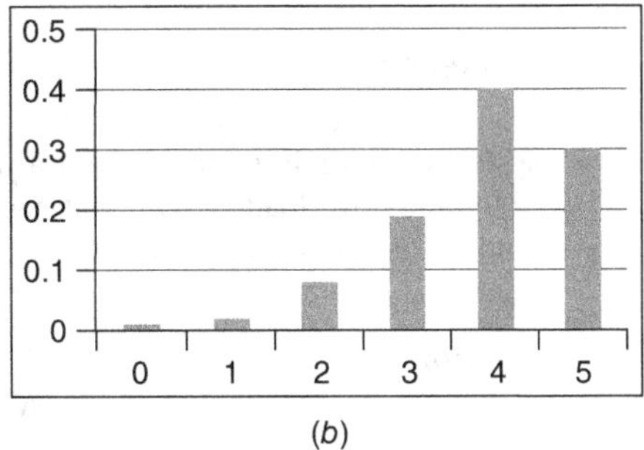

(b)

Figure 5.13 (*a*) Positive and (*b*) negative skews.

(d) Hypergeometric Distribution. In the binomial and multinomial distributions, it is assumed that the sample space is finite or the sampling experiment is carried out with replacement. As such, the probability of "success" or "failure" remains constant from trial to trial. In many cases in civil engineering, however, the sample space is finite and replacement is not possible (such as in destructive testing of materials). As such, the number of objects or events labeled successes and the total number of objects or events change after every trial. A classic example is picking balls from a box without replacement. These cases can be analyzed using the hypergeometric distribution. Besides the changing probabilities, the hypergeometric distribution is similar in all respects to the binomial distribution.

Definition: If we seek the probability of x successes in n trials from a population of N objects (k of which are labeled successes), then such a probability is given by the hypergeometric probability distribution, shown mathematically as follows:

$$h(x; n, p, k) = p(X = x) = \frac{\dfrac{k!}{x!(k-x)!}\,\dfrac{(N-k)!}{(n-x)!(N-k-n+x)!}}{\dfrac{N!}{n!(N-n)!}}$$

Relationship between the Binomial and Hypergeometric Distributions. The binomial distribution has widespread applications in the field of systems engineering, particularly when the application involves sampling from a population. In such applications, to ensure constancy of success probabilities across successive trials, the sampling should be carried out with replacement or should be from an infinitely large population. In real practice, however, the population is finite and sampling is often carried out without replacement (e.g., destructive testing). In such cases, the hypergeometric distribution can be used. Where sampling is without replacement from a finite population, the hypergeometric can be used in place of the binomial when the sample size is 5% or less of the population size.

Conversely, for a problem involving hypergeometric experiments, if the number of trials is very large, the binomial distribution can be used in place of the hypergeometric distribution. If X is hypergeometric with parameters n, N, and k, then the corresponding binomial distribution has parameter $p = k/N$.

Example 5.16

The probability that a certain engineering system experiences overload is 0.40. You observe 5 such systems randomly selected from a population of 50 systems. What is the probability that 3 of the 5 experience overload? (Note that each trial is dependent on the outcome of the previous trial, as the number of systems in each category changes with each trial. Therefore, this is a hypergeometric, not a binomial, problem).

Solution

$$N = 50, k = 20, n = 5, x = 3$$

$$h(3; 50, 0.40, 20) = p(X = 3) = \frac{\dfrac{20!}{3!(20-3)!} \times \dfrac{(50-20)!}{(5-3)!(50-20-5+3)!}}{\dfrac{50!}{5!(50-5)!}} = 0.234$$

(e) Negative Binomial Distribution. The negative binomial distribution which the reverse of the binomial distribution has the same properties as the binomial but with one addition: *Trials are repeated until a certain number of successes is obtained.* So, with the negative binomial distribution, we seek the probability that there will be a certain number of trials before a specified number of successes is achieved. Therefore, the number of trials is not fixed, but the number of successes is fixed. For example, what is the probability that a coin should be flipped five times to get three heads? What is the probability that we get a third head at the fifth flip? If repeated independent trials can result in a success with probability p or failure with probability q, then the probability distribution of the random variable x (the number of trials needed to produce k successes) is given by:

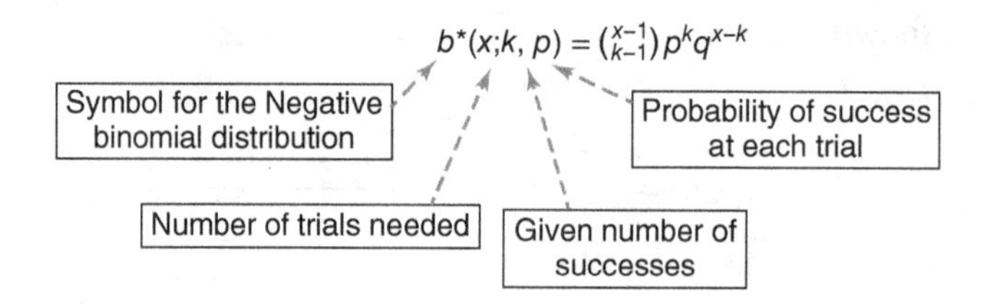

Example 5.17

The probability that a certain engineering system experiences overload is 0.40. What is the probability that six systems must be examined before we encounter the fourth system that experiences an overload? Assume that each trial is a Bernoulli process.

Solution

We seek the probability that x trials are needed to ensure k successes.

Therefore, $x = 6$ and $k = 4$.

$$p(X = x) = \binom{x-1}{k-1} p^k q^{x-k}$$

$$p(X = 6) = \binom{6-1}{4-1} (0.40)^4 (0.60)^{6-4} = 0.092$$

(f) Geometric Distribution. The distribution of the number of Bernoulli trials, say x, needed to elicit only one success is termed a geometric distribution. Examples include: How many times should a trial be carried out in order to achieve one success? What is the probability that a certain number of trials will result in only one success? How many times should I shoot a dart to get the first bull's eye? In the NBA Finals, how many games should Rovers play with the Hoppers to win their first game of that series? How many flips of a coin are needed in order to get your first head? On Friday night, how many drivers should a policeman stop in order to arrest the first drunken driver? What is the probability that an engineering system will suffer one breakdown after 300 hours of operation? What is the likelihood that the Charles River will overflow its banks once in a 20-year period? In a close inspection of steel plates, what is the probability that the inspector finds one blemish in every square yard inspected?

Therefore, the geometric distribution is concerned with (i) the number of times or space intervals until the first occurrence of an event and (ii) the average time or space interval between two consecutive occurrences of an event.

Definition. If repeated independent trials can result in a success with probability p or failure with probability q, then the probability distribution of the random variable x (the number of Bernoulli trials needed to produce one success) is given by:

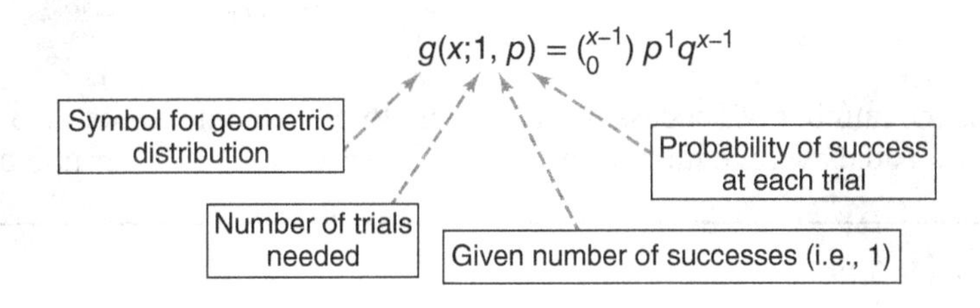

Example 5.18

The probability that a certain engineering system experiences overload is 0.40. What is the probability that six systems need to be examined before we encounter the first overloaded system? Assume that each trial is a Bernoulli process.

Solution

We seek the probability that x trials are needed to ensure one success. Therefore, $x = 6, k = 1$:

$$p(X = 6) = \binom{6-1}{4-1} (0.40)^4 (0.60)^{6-4} = 0.092$$

$$p(X = 6) = \binom{6-1}{1-1} (0.40)^1 (0.60)^{6-1} = 0.031$$

(g) Poisson Distribution. The Poisson distribution is a discrete probability distribution that describes the probability that a certain number of events will occur in a fixed time period. The distribution is based on a given average rate of occurrences, and a key assumption is that the occurrences are independent of the time since the last event. The Poisson distribution can be used not only for occurrences in a time interval but also within an interval of distance, space, or volume.

The Poisson distribution was named after its originator, Siméon-Denis Poisson who, in 1838, published his seminal research work titled "Research on the Probability of Judgments in Criminal and Civil Matters." Poisson's research utilized a random variable that counts the number of discrete occurrences (often referred to as "arrivals") that occur in given a time interval. One of the earliest applications of the Poisson distribution was to describe the number of deaths in the Prussian army due to horse kicks. In recent times, the Poisson distribution has been used widely in all areas of civil engineering including queuing systems, materials quality control, and freeway traffic operations (Good, 1986).

Consider an experiment or situation where some count events occur at random throughout an interval of time or space. This experiment is called Poisson process if the interval can be partitioned into equal-length intervals of lengths that are small enough such that (i) the probability of more than one count in a subinterval is zero, (ii) the probability of one count in a subinterval is the same for all subintervals and is proportional to the length of the subintervals, and (iii) the count in each subinterval is independent of those of other subintervals.

The Poisson process is similar to the binomial process with the exception that "number of trials" is replaced by an interval of time or space.

Definition. Let X be the random variable representing the number of successes in a given time or space interval, t.

Given the three properties of the Poisson distribution function of the random variable X is

$$p(X = x) = \frac{e^{-\lambda t}(\lambda t)^x}{x!}$$

where x is the number of successes, t is the length of time interval or area of space under consideration, and λ is the average number of successes per unit time or per unit area.

Example 5.19

A randomly selected hydraulic engineering system in a certain population of similar systems experiences overload every hour. Find the probability that exactly three of these systems experience overload in any 5-minute period. Assume that each trial is a Bernoulli process.

Solution

The overload or "arrival" rate is $\lambda = 12/60 = 0.2$. Now we seek the probability that exactly three of these systems experience overload in any 5-minute period, that is, we seek x (or r) $= 3$, and we are given that $t = 5$. Calculate $\mu = \lambda \times t = 0.2 \times 5 = 1.0$.

Method a (Using Formula)

$$p(X = x) = \frac{e^{-\lambda t}(\lambda t)^x}{x!} = \frac{e^{-\mu} \times \mu^x}{x!} = \frac{e^{-1.0} \times (1.0)^3}{3!} = 0.0613$$

Method 2

Using statistical tables, (see Appendix 2), $p(X = 3) = F(3) - F(2) = 0.9810 - 0.9197 = 0.0613$. Note that we first had to make sure that both the given mean rate and the question were in the same units (i.e., minutes).

Poisson Approximation to the Binomial. The computation of binomial probabilities may be time consuming and cumbersome. Where n is large and p is not close to 0.5, the Poisson distribution can serve as a good approximation of the binomial distribution. Generally, the approximation of one probability distribution to another can take place reliably when the distributions have similar characteristics. Specifically, it is sought that the two distributions should have similar values of means and of variances. Therefore, if the Poisson distribution (which has a mean of $\mu = \lambda$) is to serve as a good approximation of the binomial (which has a mean of $\mu = np$), then these two values should be equal. That is, mean of Poisson, $\lambda = np$ (mean of binomial).

5.5.3 Summary for Discrete Distributions

Table 5.3, which summarizes the properties of discrete distributions, presents the range of the variable, the probability mass function, and the equations for computing the expected value and the variance. Table 5.4 describes the random variable for each type of discrete distribution and presents the properties or assumptions associated with the distributions. In Figure 5.14, we see the relationships between the common discrete probability distributions.

Table 5.3 Summary for the Discrete Distributions

Random Variable	Distribution Name	Range	Probability Mass Function	Expected Value	Variance
X	General	$x_1, x_2, \ldots, x_n$	$P(X = x) = f(x) = f_X(x)$		
X	Discrete uniform	$x_1, x_2, \ldots, x_n$	$1/n$	$\sum_{i=1}^{n} \frac{x_i}{n}$	$\left[\sum_{i=1}^{n} x_i^2/n\right] - \mu^2$
X	Equal-space uniform	$x = a, a + c, \ldots, b$	$1/n \;\; n = (b - a + c)/c$	$(a + b)/2$	$c^2(n^2 - 1)/12$
# successes in n Bernoulli trials	Binomial	$x = 0, 1, \ldots, n$	$C_x^n p^x (1 - p)^{n-1}$	np	$np(1 - p)$
# Bernoulli trials until 1st success	Geometric	$x = 0, 1, \ldots$	$p(1 - p)^{x-1}$	$1/p$	$(1 - p)p^2$
# Bernoulli trials until rth success	Negative binomial	$x = r, r + 1, \ldots,$	$C_{r-1}^{n-1} p^r (1 - p)^{x-r}$	r/p	$\frac{r(1 - p)}{p^2}$
# successes in a sample of size n from a population of size N containing K successes	Hypergeometric (sampling without replacement)	$x = [n - (N - K)]^+, \ldots, \min\{K, n\}$ and integer	$\frac{C_x^K C_{n-x}^{N-K}}{C_n^N}$	np where $p = K/N$	$np(1 - p)\frac{N - n}{N - 1}$
# of counts in a Poisson process interval	Poisson	$x = 0, 1, \ldots,$	$\frac{e^{-\lambda}\lambda^x}{x!}$	1	1

Adapted from Schmeisser (2010).

Table 5.4 Assumptions Associated with the Various Discrete Distributions

Distribution	Random variable, *X*	Properties
Bernoulli	The number of successes in a single Bernoulli trial	Each trial results in only one of two possible outcomes: success or failure.
Binomial	The number of successes in n Bernoulli trials	Each trial results in only one or two possible outcomes: success or failure. The probability of a success, p, is constant from trial to trial. All trials are statistically independent of each other. The number of trials, n, is a specified constant.
Negative binomial	The number of failures preceding the *rth* success in a sequence of Bernoulli trials	Each trial results in only one of two possible outcomes: a success or failure. The probability of a success, p, is constant from trial to trial. All trials are statistically independent of each other. The sequence of trials ends after the *rth* success.
Geometric	The number of failures preceding the first success in a sequence of Bernoulli trials	Each trial results in only one of two possible outcomes: a success or failure. The probability of a success, p, is constant from trial to trial. All trials are statistically independent of each other. The sequence of trials ends after the first success.
Hypergeometric	The number of successes in a sample of size n (may be approximated by binomial distribution when N is large ($>10n$)	Sampling is performed without replacement from a finite set of size N containing a success. Each member of the sample can result in only one or two possible outcomes—a success or failure. The sample size, n, is a specified constant.
Poisson	The number of event occurrences during a specified period of time	The average rate of occurrences ($\lambda > 0$) is known. Occurrences are equally likely to occur during any time interval. Occurrences are statistically independent.

Adapted from Au et al. (1972) and Schmeisser (2010).

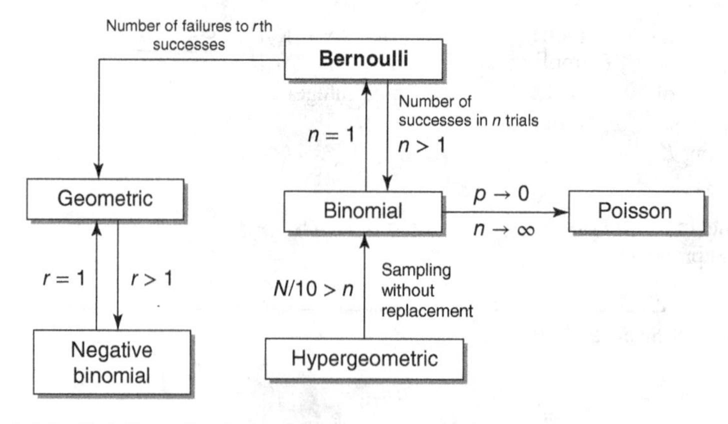

Figure 5.14 Relationships between the common discrete probability distributions.

5.6 CONTINUOUS PROBABILITY DISTRIBUTIONS

5.6.1 General Continuous Distributions

For a continuous random variable X, the probability density function (pdf) has the following properties:

$$\int_{-\infty}^{\infty} f(x)\,dx = 1$$

$f(x) \geq 0$ for every real number x. For continuous random variables, $p(X = x) = 0$ for every real number x. This implies that, for example, $p(X \geq a) = p(X > a)$. The cumulative distribution function, often abbreviated as cdf, of a random variable is $F(x) = p(X \leq x)$:

$$p(a \leq X \leq b) = \int_{a}^{b} f(x)\,dx$$

5.6.2 Special Continuous Distributions

(a) The Continuous Uniform Distribution. If X is a random variable that is uniformly distributed over all values of X in a defined range, then the probability that X takes on a certain value x, is given by

$$f(x; A, B) = \begin{cases} 1/(B - A) & \text{when } A \leq X \leq B \\ 0 & \text{otherwise} \end{cases}$$

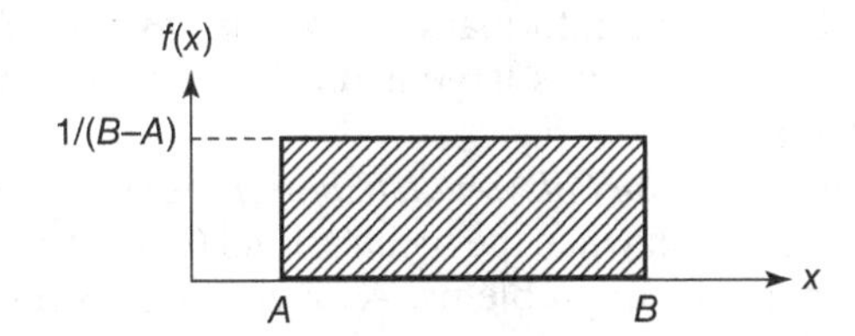

Mean of the uniform distribution = expected value $E(x) = (A + B)/2$, and the variance of the uniform distribution = $(B - A)2/12$.

Example 5.20

An aging filtration unit at a treatment plant can be used for no more than 4 hours at a time. The system used is assumed to follow a uniform distribution with interval [0,4]. What is the probability that any given instance of the system use is at least 3 hours?

Solution

$$p(X \geq 3) = \int_{3}^{4} \left(\frac{1}{4}\right) dx = 0.25$$

(b) Normal Distribution. The normal distribution is probably the most popular distribution in statistics. This distribution, which is bell shaped, represents most phenomena that occur in civil engineering industry and research. If X is a normal random variable with parameters p and q, then the probability density function of x is given by the following equation:

$$f(x) = \frac{1}{\sqrt{2\pi q}} \exp\left[-0.5\left(\frac{x-p}{q}\right)^2\right]$$

The mean μ and variance σ^2 of a normal distribution are given as p and $= q^2$, respectively.

Example 5.21

The time (in hours) that a student spends studying each week follows a normal distribution with mean 20 hours and variance 16 hours. Find the probability that a randomly selected student spends less than 25 hours for study each week.

Solution

$$p(X < 25) = \int_{-\infty}^{25} f(x)\,dx = \int_{-\infty}^{25} \frac{1}{\sqrt{2\pi \times 4}} \exp\left[-0.5\left(\frac{x-20}{4}\right)^2\right] dx = 0.8994$$

(c) Standard Normal Distribution. The standard normal distribution is a special case of the normal distribution where mean = 0 and variance = 1.

If X is the random variable for the normal distribution, then the random variable for the standard normal distribution, z, is given by

$$z = \frac{x-\mu}{\sigma}$$

Normal approximation to the binomial distribution: If X is binomial with parameters n and p, then we can use the standard normal distribution instead of the binomial distribution to solve the problem, particularly when n is large and p is small. The mean is calculated as np, while the variance is calculated as npq.

Normal distribution to the Poisson distribution: A standard normal distribution can be used in place of a Poisson distribution when the mean arrival (occurrence) rate of the random variable is too large and causes computational problems. As λ approaches infinity, the standard random variable Z to be used is $Z = (X - \lambda)/\lambda^{0.5}$.

The exponential and gamma distributions are often used in situations of natural and engineering system attributes and operations where the assumptions of the normal distribution are violated.

Example 5.22

The number of students passing through the Union building main entrance every hour follows a normal distribution with mean 10 and standard deviation 3. Find the probability that at most 8 students pass through the entrance at any randomly selected hour.

Solution

We seek $p(X \leq 8)$. Standardizing this problem yields

$$p\left(\frac{x-\mu}{\sigma} \leq \frac{8-\mu}{\sigma}\right)$$
$$= p\left(z < \frac{8-10}{3}\right)$$
$$= p(z < -0.667)$$

Method 1 (Using Formula)

$$p(Z < -0.667) = \int_{-\infty}^{-0.667} f(z)\,dz = \int_{-\infty}^{-0.667} \frac{e^{-0.5*z^2}}{\sqrt{2\pi}}\,dz = 0.2514$$

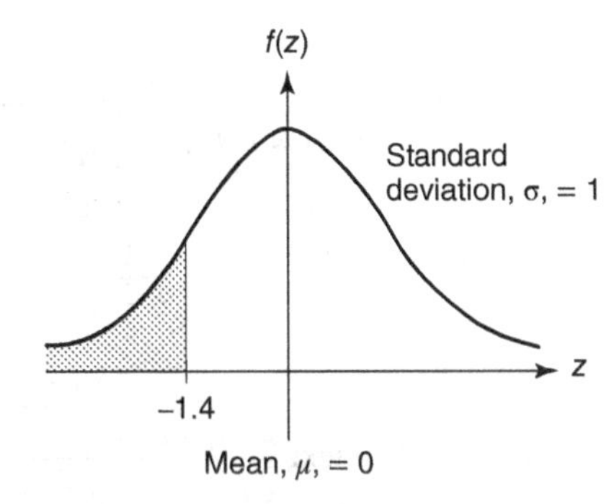

Method 2: (Using Statistical Tables See Appendix 2)

$$p(Z \leq -0.667) = F(-0.667) = 0.2514$$

Other continuous distributions are the student t distribution, the log-normal distribution, the chi-square distribution, the F distribution, the exponential distribution (where we seek the probability that a certain interval will elapse before an event next occurs), the beta distribution, and the gamma distribution (where we seek the probability that a certain interval will elapse before an event occurs a number of times). The exponential distribution is a special case of the gamma distribution.

5.6.3 Summary for Continuous Probability Distributions

Table 5.5 presents a summary for continuous probability distributions and shows the equations for the range, cumulative distribution function, probability density function, expected value, and variance. Figure 5.15 presents the relationships between the common continuous probability distributions.

5.7 COMMON TERMINOLOGY IN PROBABILISTIC ANALYSIS

As we conclude this chapter, it is useful to examine some common terminology used in the literature of probability as it relates to engineering systems.

(The) Bayesians. A school of thought that holds the view that probabilities should be represented by assigned numbers to an event that indicate an individual's degree of belief in a statement on the basis of the evidence irrespective of whether the statement involves a random process.

Certain Event. Event that occurs with all certainty, that is, 100% probability. For example, the event that the Earth will rotate on its axis tomorrow.

Combinations. The total number of ways in which a given number of objects could be arranged or could occur regardless of the order in which they occur. The number of combinations of n objects, taking r at a time, is as follows: $n!/[r!(n-r)!]$.

Complement. The set of elements outside a given set.

Table 5.5 Summary for the Continuous Probability Distributions

Random Variable	Distribution Name	Range	Cumulative Distribution Function	Probability Density Function	Expected Value	Variance
X	General	$(-\infty, \infty)$	$P(X \le x) = F(x) = F_x(x)$	$\left.\frac{dF(y)}{dy}\right\vert_{y=x} = f(x) = f_x(x)$	$\int_{-\infty}^{\infty} xf(x)dx = \mu = \mu_X = E(X)$	$\int_{-\infty}^{\infty} (x-\mu)^2 f(x)dx$ $\sigma^2 = \sigma_X^2 = V(X) = E(X^2) - \mu^2$
X	Continuous uniform	$[a, b]$	$\frac{x-a}{b-a}$	$\frac{1}{b-a}$	$\frac{a+b}{2}$	$\frac{(b-a)^2}{12}$
Sum of random variables	Normal (or Gaussian)	$(-\infty, \infty)$	See text	$\frac{e^{-\frac{1}{2}\left[\frac{x-\mu}{\sigma}\right]^2}}{\sqrt{2\pi}\sigma}$	μ	σ^2
Time to Poisson count	Exponential	$[0, \infty)$	$1 - e^{-\lambda x}$	$\lambda e^{-\lambda x}$	$\frac{1}{\lambda}$	$\frac{1}{\lambda^2}$
Time to rth Poisson count	Erlang	$[0, \infty)$	$\sum_{k=r}^{\infty} \frac{e^{-\lambda x}(\lambda x)^k}{k!}$	$\frac{\lambda^r x^{r-1} e^{-\lambda x}}{(r-1)!}$	$\frac{r}{\lambda}$	$\frac{r}{\lambda^2}$
Lifetime	Gamma	$[0, \infty)$	Numerical	$\frac{\lambda^r x^{r-1} e^{-\lambda x}}{\Gamma(r)}$	$\frac{r}{\lambda}$	$\frac{r}{\lambda^2}$
Lifetime	Weibull	$[0, \infty)$	$1 - e^{-(x/\delta)^\beta}$	$\frac{\beta x^{\beta-1} e^{-(x/\delta)^\beta}}{\delta^\beta}$	$\delta\Gamma\left(1 + \frac{1}{\beta}\right)$	$\delta^2\Gamma\left(1 + \frac{2}{\beta}\right) - \mu^2$

(*Source:* Schmeisser, 2010)

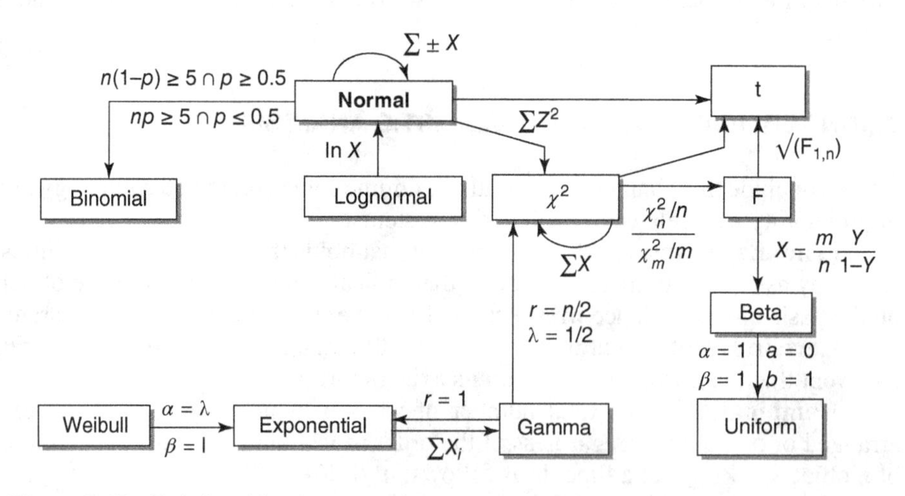

Figure 5.15 Relationships between the common continuous probability distributions.

Complementary Events. Two events are said to be complementary when they are mutually exclusive and when no other event besides them can occur.

Collectively Exhaustive Events. Two or more mutually exclusive sets that contain all elements of the universal set.

Cumulative Distribution Function. The mathematical relationship between the cumulative (upper bound) values of a discrete random variable and their corresponding probabilities.

(The) Frequentists. A school of thought that holds the view that probability deals with random and well defined experiments, and that the probability of a random event represents the relative frequency of the occurrence of the outcome of an experiment, obtained through multiple repetitions of the experiment.

Impossible Event. An event that does not occur with all certainty, that is, 0% probability of occurrence. For example, the event that the sun will cease to exist next week is an impossible event.

Mutually Exclusive. Events that cannot occur at the same instance in time or space. They are all termed disjoint events.

Possible Event. An event that occurs with some probability that is neither 0 nor 100%. Most phenomena associated with engineering systems and their environments are possible events.

Risk Situation. When possible outcomes of an action are known and the probability of each outcome is also known.

Permutations. The number of ways a given number of objects (or events) can be arranged or can occur. The number of permutations of n objects, taking r at a time, is as follows: $n!/(n-r)!$

Probability Mass Function. The relationship between a discrete random variable and the corresponding probabilities of each value of the random variable.

Probability Density Function. The relationship between a continuous random variable and the corresponding probabilities of each value of the random variable.

Probability Distribution Function. The mathematical relationship between the cumulative (upper bound) values of a continuous random variable and their corresponding probabilities.

Sample Space. The set of all possible outcomes in a given problem or experiment.

Sequential or Non Disjoint Events. Events that are not mutually exclusive and thus follow one another either in time or space. Sequential events may be statistically dependent or independent.

Statistically Dependent Events. Events where the probability of an event depends on or affects the probability of the other.

Statistically Independent Events. Events where the probability of an event does not depend on or affect the probability of the other.

Uncertainty Situation. When all its possible outcomes of an action are known but the probability of each outcome is unknown.

SUMMARY

Uncertainties are inevitable at any phase of civil engineering systems development. Thus, the tools to be acquired by engineers should include the methods of incorporating the concepts of probability in their work. Probability is the likelihood that some event has occurred or will occur, and the theory of probability has uncountable applications in civil engineering where it is used to understand system behavior or to draw conclusions about the likelihood of potential events. In this chapter, we discussed the basic fundamentals of probability. Specifically, we discussed the elementary theory of sets and the fundamental theories of probability and showed briefly, using examples, how these concepts can be useful in the tasks of describing, analyzing, and evaluating situations of any phase of civil systems development.

The chapter also introduced the notion of a random variable and showed how it could be described not only by its primary descriptors (such as its mean or variance) but also by specifying its probability distribution. A number of common discrete and continuous probability distributions were discussed and the relationships between the distributions were identified.

Ultimately, the concepts studied in this chapter provide a pedestal for applying probabilistic concepts in the management of civil engineering systems. Such applications are manifest in the description of engineering system behavior in the past, current, or future time, evaluating alternative outcomes that are not deterministic and assessment of outcome risks in conjunction with the outcomes themselves in terms of their costs or benefits. Hopefully, this chapter has prepared us for subsequent chapters where we shall frequently encounter the concepts of probability: the tools of statistics, modeling, and simulation in Chapters 6, 7 and 8; stochastic optimization in Chapter 9; cost analysis, engineering economics, and multiple criteria analysis in Chapters 10–12 and 18; reliability and risk analysis, systems dynamics, real options, and decision analysis in Chapters 13–16, and the various phases of development (Chapters 19–26) where these tools are applied.

EXERCISES

1. Consider a simply supported beam PR as shown below. If a uniformly distributed dead load of 5 N/m is experienced along PQ and a point live load of 50 N can occur at any location of the beam (to the nearest meter), determine the sample space for the reaction force R_R.

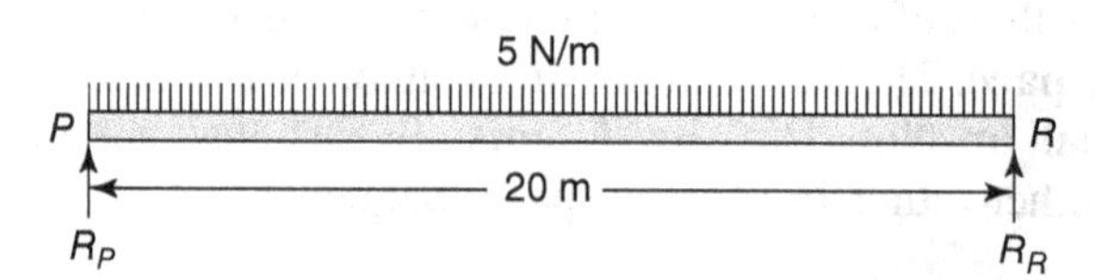

2. Write the formula for the probability that an event belongs to set A, B, or C or belongs to any two or all three, where sets A and C are mutually exclusive to each other, but set B overlaps with both A and C.

3. Consider two events P and Q.

a. Write the general formula used to calculate the probability that either event P occurs or Q occurs or both occur.

b. How does this formula change if:

i. Events P and Q are disjoint (i.e., mutually exclusive of each other).

ii. Events P and Q are nondisjoint events that are statistically independent of each other.

iii. Events P and Q are nondisjoint events that are statistically dependent of each other.

4. A through inspection of 120 airport concrete pavement slabs yielded a number of structural problems. R, C, and F denote the sets of slabs that have steel rebar corrosion, surface cracks, and faulting. The number of slabs in various categories is presented in the figure below.

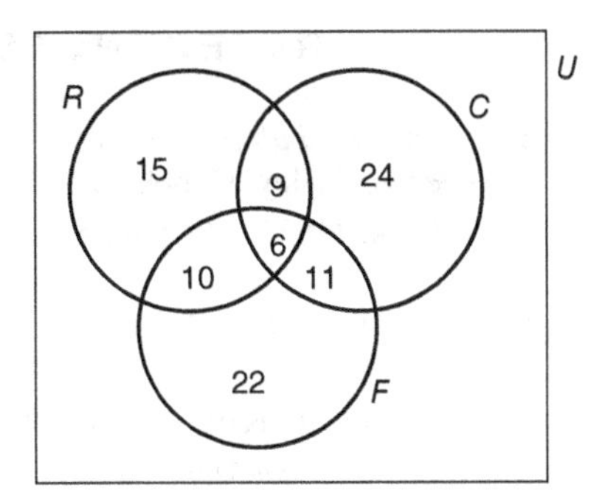

Describe and determine the number of slabs that are in each of the following categories:

(a) $(R \cup C) \cap F$, **(b)** $R \cup (C \cap F)$, **(c)** $R \cap C \cap F$, **(d)** $(R \cup C \cup F)'$, **(e)** $R \cup C' \cup F$,

(f) $R \cap (C \cup F)$, **(g)** $R \cap (C \cup F)$.

5. (a) Give one example each of the following types of discrete variables: (i) Nonordered binary variable, (ii) nonordered multinary variable, (iii) ordered categorical variable. (b) Indicate which of the following random variables are discrete and which are continuous: (i) Time in (minutes) you spent studying last weekend, (ii) number of times you attend classes each week, (iii) the amount of gas you buy in any randomly selected month, and (iv) the performance of a hydraulic system in terms of volume of water pumped per hour.

6. Rewrite the following stochastic events in mathematical notation using a suitably defined random variable:

a. The probability that the deflection at the most critical element of a certain structure exceeds 1.80 inches is 0.85.

b. Of every 100 road construction projects, 28 experience two or more fatalities.

c. Twenty percent of the time, a certain aging engineering system component breaks down at least four times during operation every year.

d. Of all college students 75% read their campus newspaper at least once a week.

7. The number of cars arriving per hour at a toll road booth is assumed to follow a Poisson distribution with mean $\lambda = 7$. What is the probability that three cars will arrive in the next hour.

8. The probability that a certain building system experiences an air-conditioning (AC) defect is 0.14. Four of these building systems are randomly selected from a population of 50 systems for scrutiny. What is the probability that 3 of the 4 buildings experience an AC defect? Note that each trial is dependent on the results of the previous trial, as the number of defective ACs (and hence, the probability of having a defective AC system) changes with each trial until the fourth building is examined.

9. In an NBA championship series, the team that wins four games out of seven will be the winner. The Rovers have a 0.55 probability of winning a game with the Sailors, and the Sailors have a 0.45 probability of winning a game with the Rovers. Find the probability that only six games are needed for Rovers to win the series. How many games should the Rovers play with the Sailors in order to win their first game?

10. The strength of a standard structural steel for a certain construction project follows a normal distribution with mean 20 N/mm^2 and variance 16 N/mm^2. Find the probability that a randomly selected specimen from a supply batch of this steel has strength less than 25 N/mm^2. (a) Use the integral of the probability mass function, (b) transform the normal distribution into a standard normal distribution and use the integral of the probability mass function for that distribution. (c) Use the statistical tables for the cumulative standard normal distribution.

REFERENCES

Ang, A. H-S., and Tang, W. H. (2006). *Probability Concepts in Engineering: Emphasis on Applications to Civil and Environmental Engineering*. Wiley, Hoboken, NJ.

Au, T., Shane, R. M., and Hoel, L. A. (1972). *Fundamentals of Systems Engineering*. Addison-Wesley, Reading, MA.

Benjamin, J. R., and Cornell, C. A. (1970). *Probability, Statistics, and Decision for Civil Engineers*. McGraw-Hill, New York.

Good, I. J. (1986). Some Statistical Applications of Poisson's Work, *Statist. Sci.* 1(2), 157–180.

Harnett, D. L. (1975). *Introduction to Statistical Methods*, 2nd ed. Addison-Wesley, Reading, MA.

McCuen, R. H. (1985). *Statistical Methods for Engineers*, Prentice-Hall, Englewood Cliffs, NJ.

Montgomery, D. C., and Runger, G. C. (2010). *Applied Statistics and Probability for Engineers*, Wiley, Hoboken, NJ.

Schmeisser, B. (2010). *Course Notes for IE 230, Probability and Statistics in Engineering I*. Purdue University, W. Lafayette, IN.

USEFUL RESOURCES

Ayyub, B., and McCuen, R. H. (2011). *Probability, Statistics, and Reliability for Engineers*, CRC Press, Boca Raton, Florida.

Faber, M., Koehler, J., and Nishijima, K. (2011). *Applications of Probability and Statistics in Civil Engineering*, CRC Press, Boca Raton, Florida.

Kottegoda, N., and Russo, R. (1997). *Statistics, Probability and Reliability Methods for Civil and Environmental Engineers*, McGraw-Hill, New York.

Melchers, R. E. (1999). *Structural Reliability Analysis and Prediction*, Wiley, Hoboken, NJ.

Ross, S. (2011). *A First Course in Probability*, Pearson Higher Ed., Upper Saddle River, NJ.

Smith, G. N. (1986). *Probability and Statistics in Civil Engineering: An Introduction*, Nichols Publishing, Dubuque, IA.

Walpole, R. E., Myess, R. H., Myers, S. L., and Ye, K. E. (2011). *Probability and Statistics for Engineers and Scientists*, Pearson, Upper Saddle River, NJ.

CHAPTER 6

STATISTICS

6.0 INTRODUCTION

As discussed in Chapter 4, the description, analysis, and evaluation of past, current, or new designs, concepts, procedures, and materials for engineering systems, as well their processes and environments, are important tasks that engineers face at any phase of the development cycle of civil engineering systems. Statistical analysis, which is often a valuable tool in carrying out such tasks, is generally defined as the *manipulation* (analysis, explanation, interpretation, and presentation) of data in order to *describe* the data in a consistent manner (descriptive statistics) or to *draw conclusions* from the data in order to make informed decisions (inferential statistics). The term *statistics* takes its roots from old languages and arose from the need for government to collect data in order to enhance governance. In Italian and Latin, the words "statista" and "statisticum" mean "statesman" and "state," respectively. In Germany, the science of the state, or analysis of data about the state, *statistik*, was introduced in the 18th century. The mathematical methods of statistics emerged from probability theory. Two major disciplines in statistics are mathematical statistics, which is concerned with the theoretical basis of the subject, and applied statistics, which consists of descriptive and inferential statistics. Descriptive statistics describes data in summarized manner. Inferential statistics investigate data patterns taking due account of any random nature of the data and makes conclusions about the process or population under investigation.

The concepts of statistics are prevalent in the various tasks encountered by developers and managers of civil engineering systems. For example, during the system needs assessment phase, statistical regression tools may be needed to predict the future demand for services provided by a civil engineering system (e.g., water supply, wastewater treatment, air passenger travel between two major cities, and freight logistics between two seaports). At the systems planning phase, statistics is often used to predict the level and variability of the consequences of a system, such as system efficiency and effectiveness, environmental impacts, and estimated construction cost. In current practice, most of these predictions are made using regression models that are developed on the basis of historical data from similar projects. At the operations phase, statistics can be used for a variety of analytical tasks including updating the demand for a system in response to changing socioeconomic conditions or predicting the effectiveness of an operational policy change. The phase of monitoring is often preceded by a statistical determination of the sample size needed to maximize the reliability of the collected data at the least possible cost. At the preservation phase, statistics has been used to develop models to describe or predict the performance or cost of preservation treatments or to investigate the strength of the influence of factors that affect the cost or effectiveness of preservation treatments. At the end-of-life phase, statistical tools may be used to compare the relative cost and benefits of alternative demolition techniques or to predict the vulnerability of a system to structural failure due to earthquake, floods, and other threats.

In this chapter, we present a variety of statistical tools that could be used to accomplish some of the tasks associated with the description or prediction of the nature, behavior, performance, cost, or other attribute of a system at any phase, such as those discussed in the preceding paragraph.

A discussion of the dichotomy between investigations utilizing data that are experimental or observational will be our starting point, followed by the reasons and techniques for sampling. In the area of descriptive statistics, we will show not only a number of simple but effective graphics that the systems engineer can use to convey information to an audience but also numerical measures that could be used to describe a system's characteristics or behavior. Also, we will discuss inferential statistics that will include point estimation, interval estimation, and sampling distributions, thereby serving as a prelude to subsequent chapters including statistical modeling (Chapter 7), risk and reliability (Chapter 13), and the various phase of system development (Chapters 19–26). Hypothesis testing also will be examined by looking at the tools needed by an engineer to test the validity of a priori suppositions, claims, or statements made about an engineering system or process.

6.0.1 Engineering Statistics

Engineering statistics is a collection of statistical concepts that have specific relevance to the field of engineering. These concepts can be broadly categorized as descriptive statistics and inferential statistics (Figure 6.1). Descriptive statistics consists of graphical descriptive statistics (pie charts, bar charts, line graphs, etc.) and numerical descriptive statistics such as measures of central tendency (median, mode, and mean), measures of dispersion (such as standard deviation and range), measures of association (correlation and covariance), and measures of dispersion (percentiles and deciles). Inferential statistics goes further than descriptive statistics to make conclusions about a system based on the analysis of given data on the system. Thus, applications of inferential statistics include (i) investigating the causality between two data items, such as system expenditure and system performance; (ii) testing, constructing, and validating models to explain system characteristics or phase-related processes at any phase of systems development; (iii) designing system physical components and operational rules to take due cognizance of the stochastic nature of design factors; (iv) measuring the ability of an engineering system to perform its intended function (reliability engineering) at the operations phase; and (v) quality control and assurance techniques for system construction or rehabilitation, where statistics is used as a tool to monitor and control conformance to the specifications of civil engineering products and processes.

6.0.2 Experimental versus Observational Studies

In the management of engineering systems, an array of statistics-related duties are often carried out, including investigation of the differences between two alternative materials or processes at any

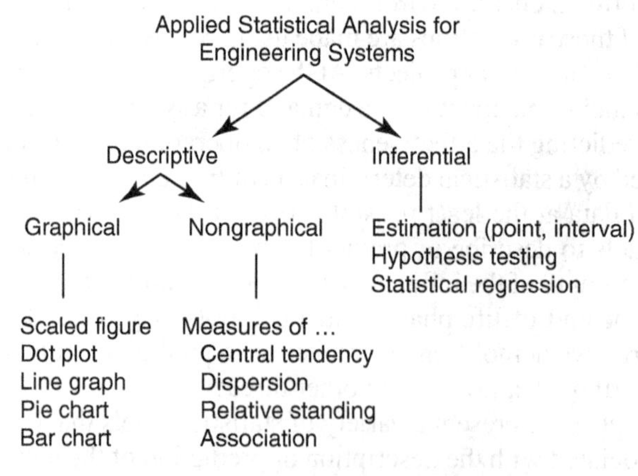

Figure 6.1 Classification of applied statistics.

phase, the causality between two system attributes (e.g., maintenance dollars and performance), or the prediction of some system outcome on the basis of the known or expected levels of other explanatory factors. With regard to causality, the systems engineer may seek to use statistics to make inferences regarding the effect of changes in the values of a factor (typically referred to as explanatory or independent variables, or predictors) on the other factor (typically called the response or dependent variable). For example, in a randomized comparative study, a systems engineer may seek to evaluate the effectiveness of a specific treatment in enhancing the condition or performance of a system or a part thereof. Like most other statistical studies, causal statistical studies can be classified into two broad groups: experimental studies and observational studies. In both types of studies, the effects of the presence (or absence) of an explanatory factor (or changes in its levels) on the response variable are sought. The difference between the two groups lies in the manner in which the study is conducted.

An experimental study involves designing and implementing various alternative configurations (design, material, policy, etc.) of a civil engineering system, taking measurements of each configuration as well as the outcomes (typically, performance) for each configuration, and ascertaining any possible differences in system outcome due to the differences in system configuration or environment. An example of an experimental study is the Long-Term Pavement Performance (LTPP) program, a 20-year study that is comprehensively analyzing over 2400 in-service pavement test sections constructed using asphaltic or Portland cement concrete, at various locations in the United States and Canada. The experiment seeks to ascertain which combinations of pavement thickness, traffic levels, subgrade strengths, rehabilitation and maintenance treatments, and climate conditions are associated with the highest performance of pavements. In experimental studies such as the LTPP, the analyst exercises control over system configuration.

In contrast, an observational study does not involve such control, for reasons that may include the lack of sampling resources or time, ethical reasons, or inability to take a random sample due in part to inaccessibility of the population being studied. Instead, the analyst collects historical data on existing systems, their configurations, and how they have performed in the past and then carries out statistical analysis of that data. An example of an observational study is one that investigates the relationship between transportation air pollution and respiratory illness. The basic steps for an experimental or observational statistical study of an engineering system are illustrated in Figure 6.2. The first step is to plan the investigation and includes clearly establishing the objective of the study. This should be followed by a thorough review of published information on the subject of interest and other closely related subjects, including literature reviews, interviews with experts on the subject, assessing the availability of data, and ethical issues in data collection, among others. The sampling process should be designed with due consideration of the system model and the interactions of independent variables. This step requires skill in the design of experiments. Data is

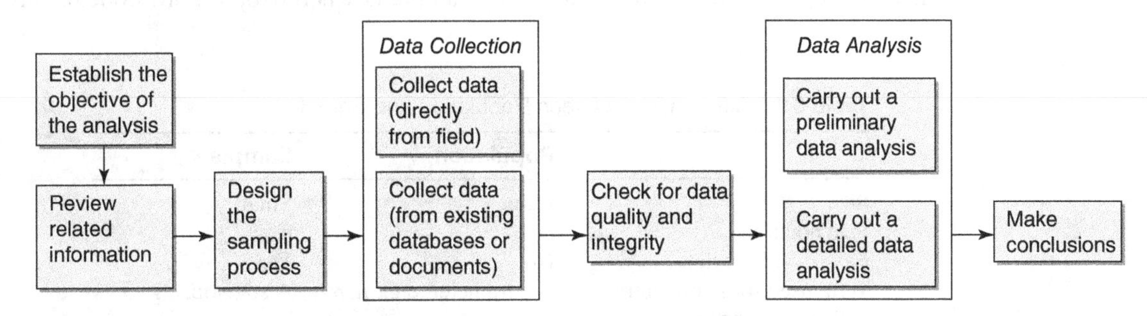

Figure 6.2 Steps for observational or experimental studies.

then collected in the field and/or from existing databases or documents, maps, and the like from which preliminary analysis (descriptive statistics) is carried out, followed by a detailed data analysis. From the analysis, conclusions are made about the sample, and inferences are duly made on the population under investigation.

6.1 POPULATION AND SAMPLING

As a civil engineer, you may be required to answer some questions about a population that is the collection of all elements in a system. For example, a materials engineer may seek to determine the quality of aggregates at a certain quarry; a transportation engineer may need to ascertain the ratio of auto use to transit use; a traffic engineer may seek to investigate whether a new traffic signal at a certain location led to a reduction in crashes; a construction engineer may want to satisfy herself that the strength of concrete being used for a certain project meets the project specification; or an environmental engineer may need to know if the average groundwater pollution in a certain area is violating health standards. The best way to answer these types of questions is to collect the needed data for each individual element in the population of systems or system components. However, it is impractical to do so because of constraining factors of time, resources (personnel, equipment, and funding for the sampling exercise), and the inaccessibility of certain elements of the population, as well as the destructive nature of certain sampling processes. Therefore, a small sample is often taken to represent the population; and, for this reason, it is important to take steps to ensure that all elements of the population have an equal chance of being selected for inclusion in the sample. That way, the sample can be considered a good miniature copy of the population and is **random** and **representative**. This also means that the **sample statistics** (e.g., the sample mean) should mimic closely the **population parameters** (e.g., the population mean); otherwise, any conclusion made on the basis of the sample data may not reflect the entire population. How could the engineer determine whether the sample statistics are sufficiently close to the population parameters so that the sample may be accepted as a true replica of the population? The concepts of bias and efficiency, which will be discussed in Section 6.3, are utilized for this purpose. Table 6.1 shows the differences between a population and a sample.

6.1.1 Types of Samples

Lapin (1990) identified three types of samples: convenience samples, judgment samples, and random samples.

The Convenience Sample. The observations in this kind of sample are those that were made only because it was convenient to do so. A common example is when people draw conclusions based

Table 6.1 Differences between Population and Sample

	Population	**Sample**
Size	Large	Small
Size notation	N	n
Easy to collect data?	No	Yes
Term used for parameters that describe the set	A parameter, e.g., μ, σ	A statistic, e.g., $\bar{x}$, s

on their personal experiences or when infrastructure inspectors collect data on only those parts of a structure that they can see or that are accessible to them. Convenience samples are commonly used in organizations for internal studies but are often not acceptable in research. In some cases, sampling bias is acknowledged, but its consequences are considered unimportant. The cost savings from convenience sampling and its simplicity often outweigh the risk of bias from unrepresentative samples.

The Judgment Sample. In this type of sample, all the observations are made by judgment. This type of sampling is considered appropriate in an attempt to ensure sample diversity and to guarantee representation of all diverse elements of a heterogenous population. Examples include the sampling of the prices of different construction materials in order to develop an updated price index for these materials.

The Random Sample. A random sample is one where every observation in the parent population has an equal chance of being selected for inclusion in the sample. Synonyms include a **probability sample** or **scientific sample**. Probability concepts are applicable only to random samples. Categories of random sampling are discussed in the next section.

6.1.2 Methods of Random Sampling

There are four major ways by which a sample can be carried out to ensure that it is random and yet represents a true miniature copy of the population: simple random sampling, systematic random sampling, stratified (or clustered) random sampling, and any combination of the above.

Simple Random Sampling. This sampling is a simple selection of elements of the population without regard to the nature of the population. The main advantage of simple random sampling is that less effort is required for data collection preparation and for carrying out the data collection. Also, this sampling technique is more appropriate when all elements in the population have similar characteristics besides those under investigation. A disadvantage of simple random sampling is that it may not be truly representative of the population, particularly if the population is heterogeneous.

Systematic Random Sampling. Also referred to as **sequential random sampling**, this sampling method can be either **systematic temporal sampling** (sampling elements from the population within specified time intervals, at the same location) or **systematic spatial sampling** (sampling elements from the population at selected locations at the same time). For example, in taking water samples from a river to test its quality, the analyst could take samples every hour at a specific location (systematic in time) or several analysts could take samples at the same time at different locations (systematic in space), or both.

Stratified Random Sampling. This sampling method first divides the entire population into different subpopulations, or population strata, on the basis of certain ordinal characteristics of the population. Next, a random sample is obtained within each stratum to obtain the desired sample size (Figure 6.3). Subpopulations may be of the same size or of different sizes, depending on the composition of the main population. An advantage of stratified random sampling is that it ensures that each ordered group in the population is represented in the sample and therefore is ideal for populations having diverse but ordered groups. A disadvantage, however, is that relatively more preparation time is needed to calculate the proportions of each group in the population and the determination of their proportions in the sample. Stratified random sampling may be considered a hybrid of the judgment sample and the random sample.

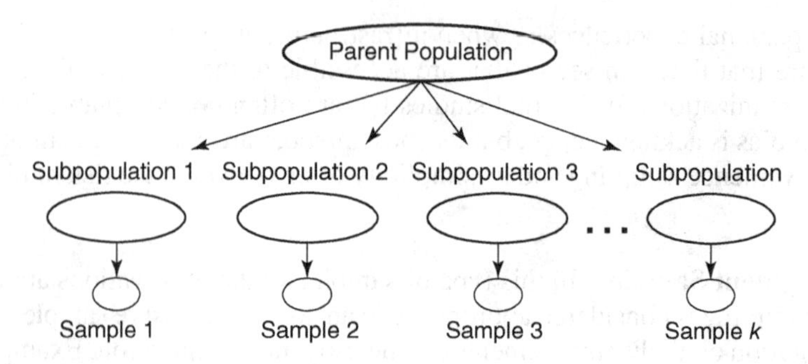

Figure 6.3 Decomposing a population into subpopulations for stratified random sampling.

Cluster Sampling. This sampling scheme, like stratified sampling, divides the population into nonordered groups called clusters. Each cluster is selected randomly, and all observations in the chosen cluster(s) are used for the study. The advantage of this sampling technique is that it is potentially less costly than simple random sampling. For example, in sampling manufactured precast units for quality inspection, it is less wasteful to collect a few supply batches and inspect all of the contents from each batch than to collect all of the supply batches and inspect a few from each. Thus, cluster samples are somewhat similar to convenience samples and therefore may suffer from some sampling bias.

6.1.3 Sample Size versus Sample Cost

As we finish this discussion of sampling, it is important to emphasize two issues related to the cost and benefits of alternative sample sizes. The first issue is incremental cost-effectiveness. As shown in the above discussion, it is expensive to collect data for the entire population. In fact, it may not even be desirable to do so in certain cases because, beyond a certain sample size, there is little incremental gain in reliability from increasing the sample size.

The second issue is the consequences of sample sizes that are too small. This is associated with poor reliability that is also referred to as **error cost** (Lapin, 1990). Larger sample sizes translate into higher costs of data collection but result in smaller error costs. The total cost of sampling can be expressed in terms of these two cost components (error cost and data collection cost). Error costs, which are harder to measure, may be estimated as, for example, the risk of inadequate sampling or the sum of the expected costs of civil infrastructure damage if certain defects are missed due to such inadequacy in sampling. This may be represented as the probability that a defect-related failure occurs and the cost of failure for all assets in the population. An adjustment could be made for the probability that a given sample size will miss the defect. A hyperbolic shape seems to be appropriate for describing the trend of error costs relative to sample size because sampling errors typically decrease rapidly as n increases beyond zero, but the reduction becomes less dramatic for larger sample sizes. Data collection costs are easier to measure and typically comprise a fixed component and variable component. The total cost of sampling can be portrayed as a convex curve achieving minimum cost at the optimal sample size (Figure 6.4).

6.1.4 The Imperative for Data Integrity

The importance of data quality in the management of engineering systems cannot be overemphasized, particularly in the current era when technological advancements have facilitated the

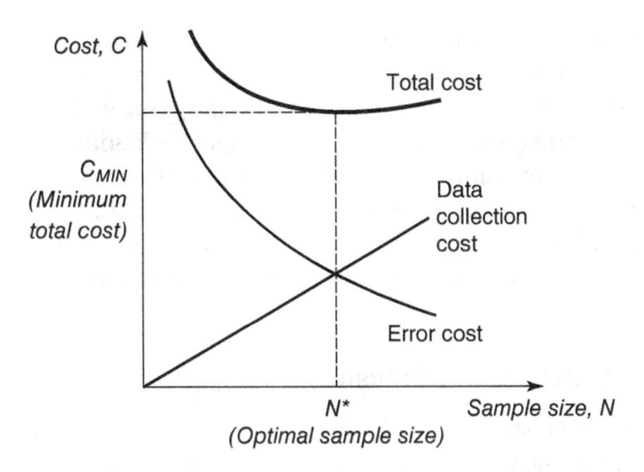

Figure 6.4 Implications of increasing the sample size.

collection and storage of large amounts of data associated with the system performance, levels of usage, environment, and other attributes. Unreliable data will lead to faulty conclusions from statistical analyses and consequently, inappropriate decisions for the management of civil engineering systems. Researchers have indicated that data can be described as high quality when the data (a) help describe reliably the real-world construct to which they refer, and (b) are appropriate for the purposes for which it is intended, often for operations, decision making and planning (Juran, 2010). Owners of civil engineering systems that have a division or department for data management or information technology, typically are able to enforce the integrity of their system attribute data by taking proactive steps to reduce measurement errors and by developing protocols for data entry such as bounds checking of the data, cross tabulation, and by establishing objective criteria for outlier detection. Data quality checks are guided by a need to ascertain their accuracy, correctness, currency, completeness, and relevance (Lee et al., 2006).

6.2 DESCRIPTIVE STATISTICS

Descriptive statistics, used to describe data in a consistent manner, may be graphical (pie chart, bar chart, etc.) or nongraphical (average, most frequent observation, etc.). Such information is important in the development of systems because they help civil engineers communicate or convey technical information to an audience regarding some attribute of a civil engineering system or its processes at any phase, such as the past, current, or future condition, structure, or performance of the system. The term "audience" refers to persons or groups of persons who are may not be civil engineers, such as a board of directors of a public or private organization, congressmen, shareholders in a civil engineering or related organization, journalists and the news media, and the general public. In some cases, the audience may refer to persons or groups of technical persons, such as your supervisor, colleagues (civil or systems engineers) in your organization, or your client, who may be an engineer. Examples of instances where you may need to address such audiences include situations where you are asked to justify your request for increased funding for your system; your board chairman asks you to make a presentation to the board on the attributes (condition, operations, performance, etc.) of your system; journalists seek to ascertain whether tax money is being used wisely on your system; you seek to warn the public about the dangers of certain bad practices on your engineering system, and so forth. In all these instances, there is a vital need for you, as a civil or systems engineer, to convey technical information in a manner that is very easy to understand. If you fail to make the target audience understand, you will likely fail to achieve your intended objective (e.g., improve your organization's image, seek additional funds, describe or justify existing funding for your operations, etc.). Descriptive statistics is related to inferential statistics in the sense that it provides a good starting point upon which detailed inferential statistics can be carried out and also enables technical persons to convey the results of complex inferential statistics to an audience of nontechnical decision makers. There are two types of descriptive statistics: graphical and nongraphical.

6.2.1 Graphical Descriptive Statistics

Graphical descriptive statistics include scaled figures, bar charts, pie charts, dot plots, line graphs, leaf-and-stem plots, box plots, cluster diagrams (dendograms), response surfaces (3D), and scatter plots. In any of these forms of graphical descriptive statistics, there is the *statistical unit* (system event or object under study) and the *data item* (an attribute of the system event or object under study). For example, in Figure 6.5, which describes the traffic characteristics of different highway types, the statistical unit is the highway class and the data items are traffic volume, percentage of truck traffic, and average speed. Figure 6.6 illustrates different forms of graphical descriptive statistics.

6.2.2 Nongraphical (or Numerical) Descriptive Statistics

Unlike graphical descriptive statistics, numerical descriptive statistics use a single number, rather than diagrammatic representations, to describe some attribute of the system. There are four main categories of numerical descriptive statistics: measures of central tendency, dispersion, relative standing, and association (Table 6.2).

Measures of Central Tendency. A given data item typically tends to hover around a certain value, and such propensity can be measured using the mean (the average data item), the mode (the most frequent data item), and the median (the middle-placed data item). These measures of central tendency may be calculated using formula or graphical means.

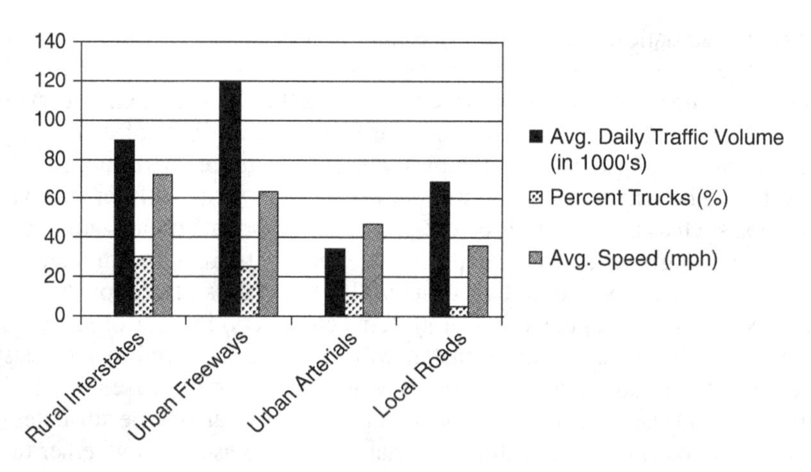

Figure 6.5 Statistical unit and data item—illustration.

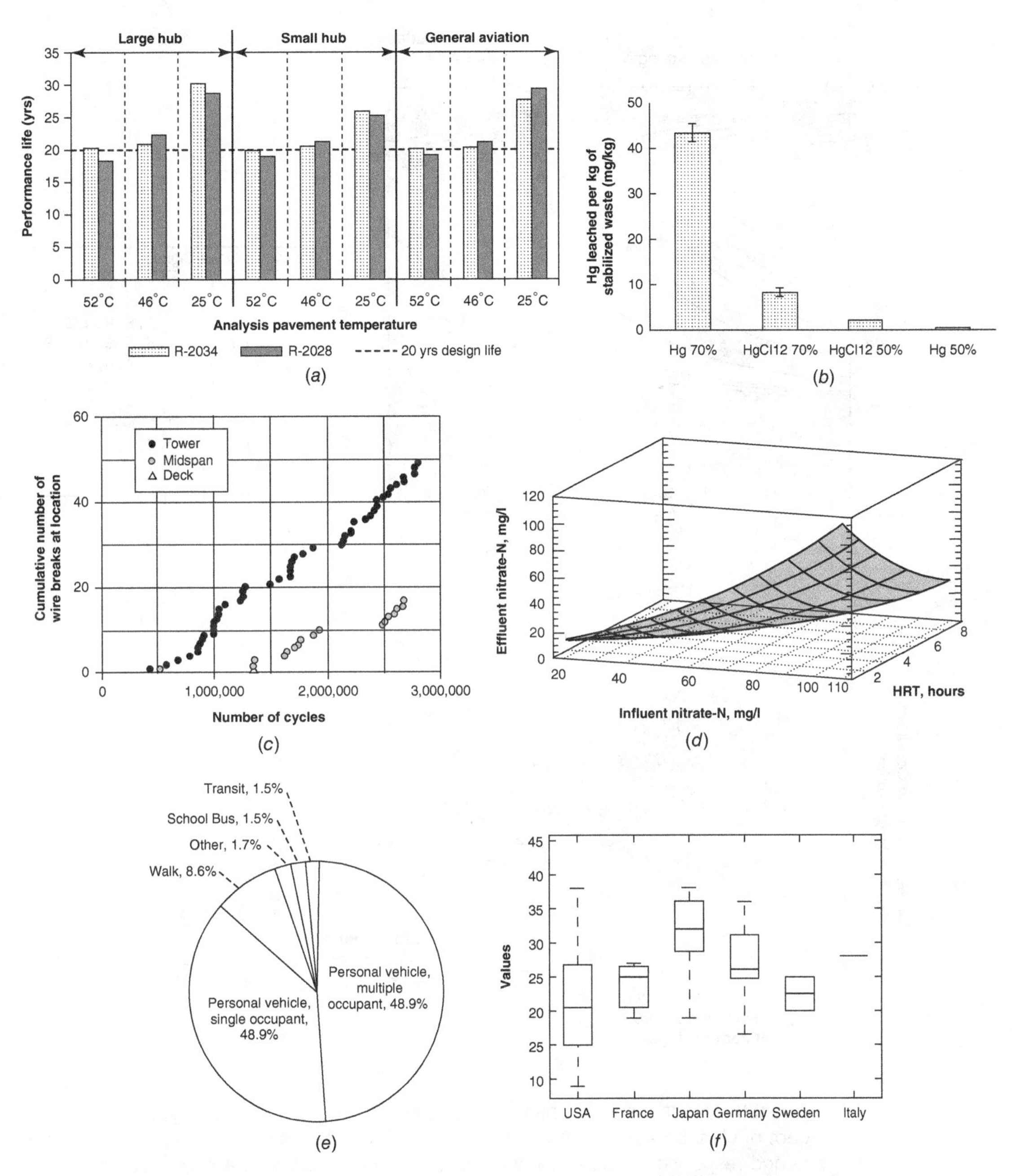

Figure 6.6 Examples of graphical descriptive statistics: (*a*) bar graph—performance of different pavement designs for airports (Hajj et al. (2010); (*b*) bar graph—mercury leach rates of different mercury-based compounds (Randall and Chattopadhyay, 2010); (*c*) scatter plot—susceptibility of bridge grouted stay cables to bending fatigue damage (Wood and Frank, 2010); (*d*) response surface—effluent nitrate concentration as function of influent nitrate concentration and hydraulic retention time of a sulfur-limestone autotrophic denitrification column (Zhang and Zeng, 2010); (*e*) pie chart—personal transportation mode shares (USDOT, 2001); (*f*) box plot—average car mileage per capita at selected countries.

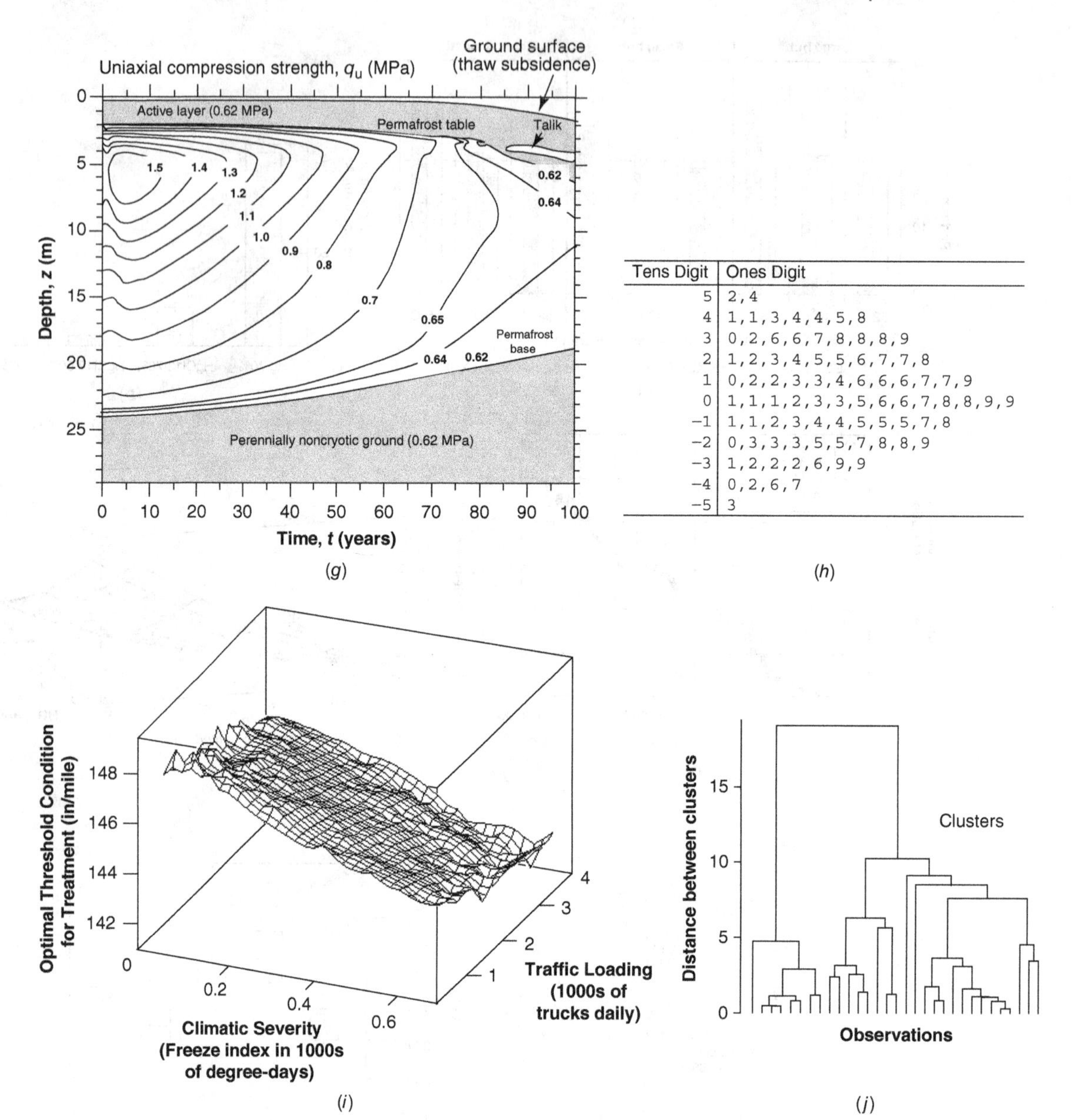

Tens Digit	Ones Digit
5	2,4
4	1,1,3,4,4,5,8
3	0,2,6,6,7,8,8,8,9
2	1,2,3,4,5,5,6,7,7,8
1	0,2,2,3,3,4,6,6,6,7,7,9
0	1,1,1,2,3,3,5,6,6,7,8,8,9,9
−1	1,1,2,3,4,4,5,5,5,7,8
−2	0,3,3,3,5,5,7,8,8,9
−3	1,2,2,2,6,9,9
−4	0,2,6,7
−5	3

Figure 6.6 (*Continued*) (*g*) contour surface—permafrost weakening as a potential impact of climatic warming (Buteau et al. 2010); (*h*) stem and leaf plot; (*i*) contour surface and response surface—variation of total friction factor in a gravel-bed river with bed form (Afzalimehr et al., 2010), and (*j*) cluster dendogram.

Table 6.2 Computation of Numerical Descriptive Statistics

Statistic	Ungrouped[a]	Grouped[b]
Mean	$\bar{x} = \frac{\sum x}{n}$	$\bar{x} = \frac{\sum(f^* x_{MP})}{n}$
Mode	Here, the mode is simply the most frequent observation	Mode $= L + cd_1/(d_1 + d_2)$ Note: This is only an approximation
Median	Median is the data item corresponding to a value of $(n+1)/2$ in the data array.	First find median class as though for ungrouped data. Then calculate median as follows: Median $= L + c\frac{\frac{h}{2} - F}{f_m}$
Mean deviation	$\text{Avg\_Dev} = \frac{\sum \lvert x - \bar{x} \rvert}{n}$	$\text{Avg\_Dev} = \frac{\sum f \lvert x_{MP} - \bar{x} \rvert}{n}$
Variance	$s^2 = \frac{\sum (x - \bar{x})^2}{n - 1}$	$s^2 = \frac{\sum f(x_{MP} - \bar{x})^2}{n - 1}$
Standard deviation	$s = \sqrt{\frac{\sum (x - \bar{x})^2}{n - 1}}$	$s = \sqrt{\frac{\sum f(x_{MP} - \bar{x})^2}{n - 1}}$
Coefficient of variation	Coefficient of variation = (standard deviation)/mean	
Covariance between two data items x and y	$\text{Cov}(X, Y) = \frac{\sum(x - \bar{x})(y - \bar{y})}{n - 1}$	
Correlation (linear)	Spearman's correlation coefficient $\rho = \frac{\text{Cov}(X, Y)}{s_X \, s_Y}$	

[a] Σx is the sum of all observed values; n and N are the number of observations in the sample and population, respectively.
[b] $\Sigma(fx_{MP})$ refers to the sum of the product of the midpoint values and their frequencies, L = lower limit of the median or modal class, N = number of observation in population, n = number of observations in sample (i.e, sample size), F = sum of observations up to but not including the median class, f_m = frequency of the median class, c = width of the class interval, d_1 = frequency of the modal class less frequency of the previous class, and d_2 = frequency of the modal class less frequency of the following class.

Measures of Dispersion. These measures describe the variability or spread of observations and include the mean deviation, variance, standard deviation, coefficient of variation (the ratio of the standard deviation to the mean), range (the simplest measure of dispersion, this is the difference between the lowest and highest observations), interquartile range (difference between the first and third quartile), semi-interquartile range (one-half of the interquartile range), and the quartile deviation. The mean deviation is the average value of the deviations between the individual observed values and the mean observed value. The variance is the average of the square of the deviations between the individual observed values and the mean observed value.

Measures of Relative Standing. These measures, which describe how observations compare to other observations in a given data set, consist of percentiles, quartiles, and deciles. For example, if the 70th percentile of speeds on interstate highways is 60 mph, then 70% of the vehicles on interstate highways are travelling at 60 mph or below. A common way to determine the measures of relative standing is to first construct a frequency table from the observations, construct a cumulative frequency chart, and read off the required decile, quartile, or percentile.

Percentiles: These range from the 1st percentile to the 100th percentile (i.e., there are 100 percentiles). The ith percentile is found first by calculating the $i\%$ of total frequency [i.e., $(i/100)\Sigma f$] and finding the data item (value of x) that corresponds to $(i/100)\Sigma f$ (using a cumulative frequency graph).

Quartiles: There are four quartiles: 1st quartile, 2nd quartile, 3rd quartile, and 4th quartile. The ith quartile is found by first calculating $i/4$ of total frequency (i.e., $i/4\Sigma f$) and finding the data item (value of x) that corresponds to the $i/4^{*}\Sigma f$ (using a cumulative frequency graph).

Deciles: There are 10 deciles: 1st decile, 2nd decile, 3rd decile, ..., 10th decile. The ith decile is found by first calculating $i/10$ of total frequency (i.e., $(i/10)\Sigma f$) and finding the data item (value of x) that corresponds to $(i/10)\Sigma f$ (using cumulative frequency graph).

Example 6.1

For the following data, find (a) the 2nd quartile, (b) the 87th percentile, and (c) the 3rd decile.

Weight (lb)	Frequency
100–109.99	1
110–119.99	3
120–129.99	4
130–139.99	4
140–149.99	8
150–159.99	13
160–169.99	11
170–179.99	12
180–189.99	7
190–199.99	2
200–209.99	1

Solution

We draw a cumulative frequency table and chart (Figure 6.7) as follows:

Weight (lb)	Frequency	Upper Limit	Cumulative Frequency
100–109.99	1	110.5	1
110–119.99	3	120.5	4
120–129.99	4	130.5	8
130–139.99	4	140.5	12
140–149.99	8	150.5	20
150–159.99	13	160.5	33
160–169.99	11	170.5	44
170–179.99	12	180.5	56
180–189.99	7	190.5	63
190–199.99	2	200.5	66
200–209.99	1	210.5	67

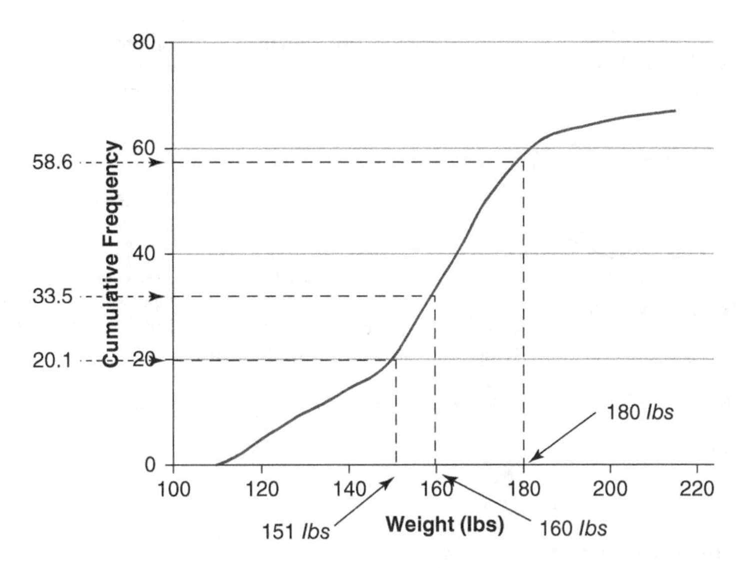

Figure 6.7 Figure for Example 6.1.

2nd quartile = $\frac{1}{2}$ (67) = 33.5. This corresponds to 160 lb; 87th percentile = 87/100 (67) = 58.59; corresponds to 180lb; 3rd decile : 3/10 (67) = 20.1. This corresponds to 151 lb.

Measures of Association. These are numerical descriptive statistics that measure the strength of any relationship between two or more observations. The most common relationship we typically test for is "linearity". For example, is the condition of a system linearly related to its age? The most common measures of association include the covariance and the coefficient of correlation. The covariance of two random variables (e.g., system age and condition) is a measure of how they vary together (i.e., how they "covary"). If the covariance is negative, then a high value of one variable is associated with a low value of the other. If the covariance is positive, then a high value of one variable is associated with a high value of the other. A zero covariance implies no relationship.

6.3 INFERENTIAL STATISTICS

Statistics began several decades ago as a purely descriptive science but has evolved into a useful decision-making tool and is increasingly used to make inferences from data. Civil engineers use inferential statistics to reach conclusions about the demand, condition, performance and other attributes of populations of civil engineering systems. These conclusions are made on the basis of a small sample collected to represent the data.

The mean, variance and other numerical statistics of the attributes of a civil engineering system can be determined from processed data about the system. Because these estimates are for the sample, they are called *sample statistics* and are considered to be not exact but rather approximate estimates only of the true *population parameters*. It is often of interest to determine the extent of the deviation (point estimation) and to establish bounds for the sample statistics (interval estimation). An *estimator* is therefore defined as a sample statistic that estimates the value of the corresponding population parameter (Figure 6.8).

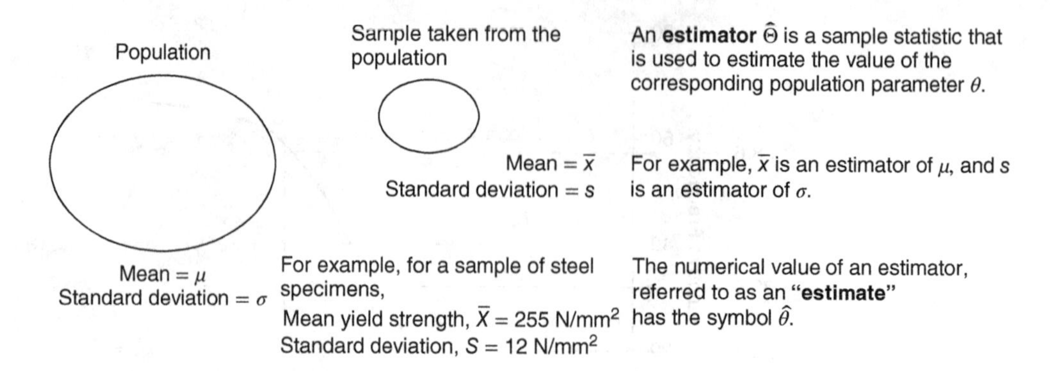

Figure 6.8 What is an estimator?

6.3.1 Point Estimation

Point estimation involves finding the value of a population parameter based on the value of the sample statistic of a sample taken from the population. Two issues associated with point estimation are *bias* (consistent over- or underestimation of estimates) and *efficiency* (closeness of the estimate to the population parameter). It is worth noting, however, that even the most efficient and unbiased estimator is not likely to yield a perfect estimate (i.e., a 100% match to the population parameter). This is illustrated in Figure 6.9.

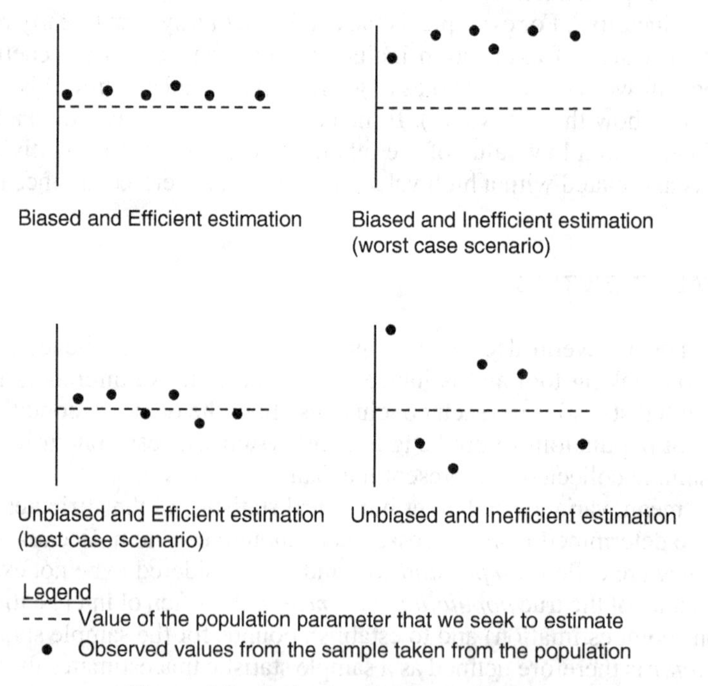

Figure 6.9 Estimator bias and efficiency.

6.3.2 Interval Estimation

Interval estimation involves finding the range of possible values of a population parameter based on the value of the sample statistic of a sample taken from the population and exploring the relationships between that range and other parameters such as the degree of confidence and the sample size.

Confidence Statements. A *confidence statement* describes our confidence (or the probability) that the estimated value of a population parameter lies between a certain lower limit and an upper limit. Thus, every confidence statement has a *confidence interval*, a *significance level*, and a *degree of confidence* (or confidence level). Mathematically, a confidence statement is written as

$$\underbrace{p(\hat{\theta}_L < \hat{\theta} < \hat{\theta}_U)}_{\text{Confidence Interval}} = \underbrace{1 - \alpha}_{\text{Degree of Confidence}}$$

where $\hat{\theta}_L < \hat{\theta} < \hat{\theta}_U$ is the confidence interval; $1 - \alpha$ is the degree of confidence; and α is defined as the significance level (or the probability that the estimated value of the population parameter will NOT fall in the specified confidence interval). The confidence level is given by $100(1 - \alpha)$.

For cases where the parameter of interest is the mean, the confidence statement becomes (Figure 6.10)

$$p(\overline{X}_L < \overline{X} < \overline{X}_U) = 1 - \alpha$$

If we assume a normal distribution and subsequently standardize the above expression, we get

$$p\left(\frac{\overline{X}_L - \mu_{\overline{X}}}{\sigma_{\overline{X}}} < Z < \frac{\overline{X}_U - \mu_{\overline{X}}}{\sigma_{\overline{X}}}\right) = 1 - \alpha$$

If we assume symmetry, and if we "destandardize" the above expression, we get

$$p(\mu_{\overline{X}} - Z_{\alpha/2}\sigma_{\overline{X}} < \overline{X} < \mu_{\overline{X}} + Z_{\alpha/2}\sigma_{\overline{X}}) = 1 - \alpha$$

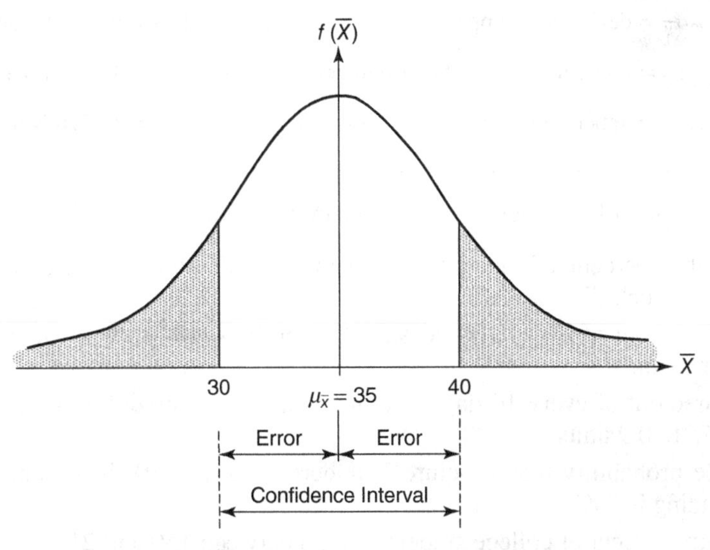

Figure 6.10 Figure for confidence statements.

where $\mu_{\overline{X}}$ is the estimate of the population mean (average values of all batches); $Z_{\alpha/2}$ is the Z value corresponding to one-half of the level of significance; $\sigma_{\overline{X}}$ is the estimate of the population standard deviation; α is the level of significance; $Z_{\alpha/2}\sigma_{\overline{X}}$ is the deviation of upper and lower confidence limits from population mean; $\mu_{\overline{X}} - Z_{\alpha/2}\sigma_{\overline{X}}$ is the lower limit of confidence interval; and $\mu_{\overline{X}} + Z_{\alpha/2}\sigma_{\overline{X}}$ is the upper limit of confidence interval.

Example 6.2

Of the 1000 batches of elements of a structural engineering system, 950 typically experience shear stresses of magnitudes between 30.00 and 40.00 N/mm^2. The mean magnitude of these stress overloads is 35.00 N/mm^2, while the variance is 9 N/mm^2. The sample size (number in each batch) is 25. Assume normal distribution.

Define a suitable random variable for the issue under discussion and identify the degree of confidence, significance level, level of confidence, and confidence interval. State whether or not the confidence interval is symmetrical and sketch the distribution (assume it is bell shaped). Write the mathematical form of the confidence statement, explaining the meaning of all symbols. Determine the values of all symbols and thereby find the lower and upper limits of the confidence interval.

Solution

Let X be a random variable that represents the mean shear stresses of a randomly selected batch. The degree of confidence = 950/1000 = 0.95; degree of confidence is given by $1 - \alpha$, where α is the significance level. Therefore, significance level = 1 − degree of confidence = 1 − 0.95 = 0.05; level of confidence = 100 × degree of confidence = 100 × 0.95 = 95; confidence interval = 40 − 30 = 10 N/mm^2.

The mean stress is exactly in the middle of the confidence interval. This is a feature of symmetrical distributions.

$$\mu_{\overline{X}} = \text{estimate of the population mean(average stresses of all batches)} = 35.00\ \text{N/mm}^2$$

$$Z_{\alpha/2} = Z \text{ value corresponding to one-half of level of significance} = Z_{0.025} = 1.96$$

$$\sigma_{\overline{X}} = \text{estimate of the population standard deviation} = \sigma/\sqrt{n} = 3/\sqrt{25} = 0.6$$

$$\alpha = \text{level of significance} = 0.05$$

$$Z_{\alpha/2}\sigma_{\overline{X}} = \text{deviation of upper and lower confidence limits from population mean} = 1.96 \times 0.6 = 0.576$$

$$\mu_{\overline{X}} - Z_{\alpha/2}\sigma_{\overline{X}} = \text{lower limit of confidence interval} = 35 - 0.576 = 34.424\ \text{N/mm}^2$$

$$\mu_{\overline{X}} + Z_{\alpha/2}\sigma_{\overline{X}} = \text{upper limit of confidence interval} = 35 + 0.576 = 35.576\ \text{N/mm}^2$$

Examples of confidence statements are as follows:

a. Eighty percent of the time, between one-half and two-thirds of all vehicles on I-65 are "semitrucks."

b. For 40 out of every 100 coastal steel bridges, 10–15% of their surface areas suffer from corrosion.

c. Three out of every 10 days in Samara are characterized by smog coefficients ranging from 0.75 to 0.9 units.

d. The probability that structure X undergoes a 6 to 10–inch consolidation settlement under loading is 25%.

e. Sixty percent of college students weigh between 150 and 210 lb.

f. The probability that a college student has a height between 55 and 65 inches is 80%.

g. Ninety percent of college students typically spend 15–25 minutes during each visit to a fast-food restaurant.

The General Confidence Statement for Sampling Distribution of Means. For a given degree of confidence, the confidence interval associated with the estimate of a population parameter is a measure of the error or precision of that estimate. Specifically, for symmetric distributions, error = $\frac{1}{2} \times$ confidence interval. As is possible for any statistic, a general confidence statement can be written for the distribution of sample means. There are two cases for this:

Case 1: A normally distributed population with a known variance and a small sample size; or the type of population distribution is unknown, the variance is known, and the sample size is large. In such cases, we use the normal (Z) distribution to "standardize" the estimator as follows:

a. General (Symmetrical or Unsymmetrical):

$$p\left(\frac{\overline{X}_L - \mu}{\sigma/\sqrt{n}} < Z < \frac{\overline{X}_U - \mu}{\sigma/\sqrt{n}}\right) = 1 - \alpha$$

b. Symmetrical only (Figure 6.11):

$$p\left(-\frac{\text{Error}}{\sigma/\sqrt{n}} < Z < \frac{\text{Error}}{\sigma/\sqrt{n}}\right) = 1 - \alpha$$

$$p\left(\mu - \frac{Z_{\alpha/2} \times \sigma}{\sqrt{n}} < \overline{X} < \mu + \frac{Z_{\alpha/2} \times \sigma}{\sqrt{n}}\right) = 1 - \alpha$$

The range of "acceptable" values of the data item is $\mu_{\overline{X}} \pm Z_{\alpha/2} * \sigma_{\overline{X}}$.

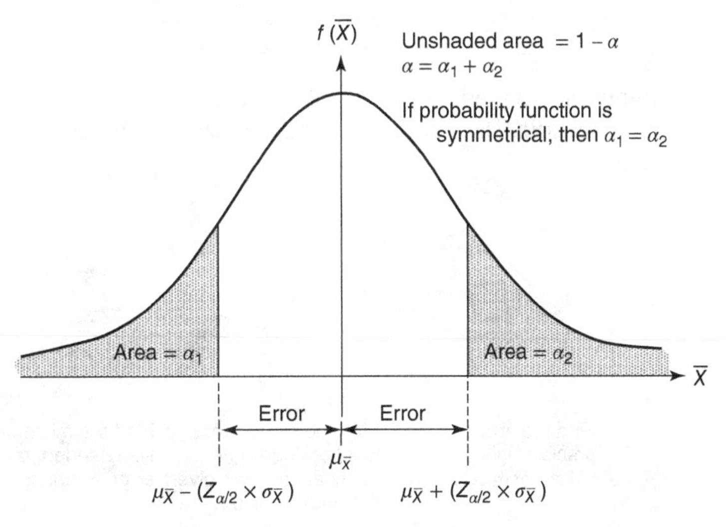

Figure 6.11 Figure for confidence statements.

Case 2: Unknown type of population distribution, unknown variance, and small sample size. In such cases, we use the Student t distribution to "studentize" the estimator as follows:

a. General (Symmetrical or Unsymmetrical):

$$p\left(\frac{\overline{X}_L - \mu}{S/\sqrt{n}} < T < \frac{\overline{X}_U - \mu}{S/\sqrt{n}}\right) = 1 - \alpha$$

b. Symmetrical only:

$$p\left(-\frac{\text{Error}}{S/\sqrt{n}} < T < \frac{\text{Error}}{S/\sqrt{n}}\right) = 1 - \alpha$$

$$p\left(\mu - \frac{t_{\alpha/2} \times S}{\sqrt{n}} < \overline{X} < \mu + \frac{t_{\alpha/2} \times S}{\sqrt{n}}\right) = 1 - \alpha$$

For any one of the above four situations, the three types of questions that may require the application of the confidence statement include *level of significance analysis, adequacy analysis*, and *precision analysis*. These types of analysis could be classified collectively as *reliability analysis*. Reliability analysis help us to investigate the relationships between sample size adequacy, precision, or error inherent with a given sample, as well as the degree of confidence (Figure 6.12). Each type of analysis is discussed in the next sections with examples. Specifically, reliability analysis can be defined as "the determination of the precision, minimum sample size, or level of significance associated with the estimation of a population parameter on the basis of sampling data."

1. Significance Level Analysis (Given All Other Variables, Find α or $1 - \alpha$). Here, we seek to find the significance level or degree of confidence associated with a certain confidence interval of a given estimate and the sample size. For symmetrical distributions, this is given by

$$\alpha = 2Z^{-1}\left(\frac{\sqrt{n}\,\text{Error}}{\sigma}\right)$$

Where Z^{-1} simply means reading the Z table in reverse. For example, Z^{-1} (1.7) is 0.446 (see statistical chart in Appendix 2). A similar expression can be derived where the data is described for the t distribution.

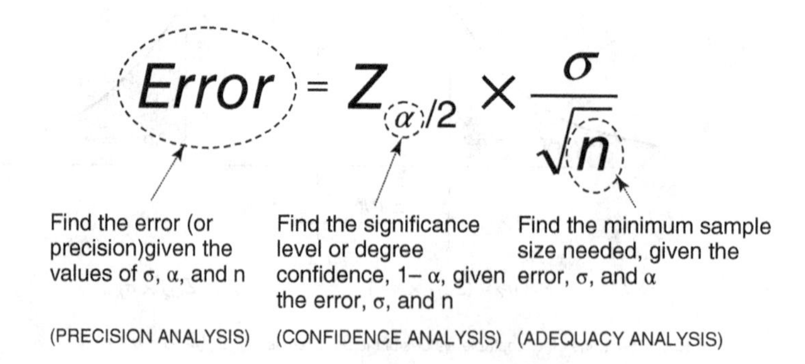

Figure 6.12 Types of reliability analysis.

From Figure 6.6, all other factors remaining the same, a higher error is associated with a lower level of significance and vice versa. As such, we can calculate the maximum level of significance associated with the minimum error and vice versa. A similar analysis could be carried out for the relationship between the level of significance and the sample size, and between the level of significance and the standard deviation:

$$\alpha_{max} = 2 \times Z^{-1}\left(\frac{\sqrt{n}\,\text{Error}_{min}}{\sigma}\right) \qquad \alpha_{min} = 2 \times Z^{-1}\left(\frac{\sqrt{n}\,\text{Error}_{max}}{\sigma}\right)$$

$$\alpha_{max} = 2 \times Z^{-1}\left(\frac{\sqrt{n}_{min}\,\text{Error}}{\sigma}\right) \qquad \alpha_{min} = 2 \times Z^{-1}\left(\frac{\sqrt{n}_{max}\,\text{Error}}{\sigma}\right)$$

Example 6.3

The condition of pipes that constitute a sewer system network has a normally distributed population mean of 34.00 units, with a variance of 50.00 units. We take a random sample of 36 from this population. (a) At what level of significance can we expect the mean of that sample to fall between 32.00 and 36.00 units? (b) What is the proportion of samples for which the mean of that sample can be expected to fall within the interval stated in (a)?

Solution

(a)

$$\text{Error} = Z_{\alpha/2}\frac{\sigma}{\sqrt{n}} \Rightarrow \alpha = 2Z^{-1}\left(\frac{\sqrt{n}\,\text{Error}}{\sigma}\right)$$

where $n = 36$; error = ½ (confidence interval) = ½ (36 − 32) = 2; standard deviation, $\sigma = (50)^{0.5} = 7.07$.

Thus, the level of significance, α, is given by

$$\Rightarrow \alpha = 2 \times Z^{-1}\left(\frac{\sqrt{36} \times 2}{7.07}\right) = 2 \times Z^{-1}(1.7) = 2 \times 0.446 = 0.0892$$

(b) Proportion of samples whose means fall between 32.00 and 36.00 units = probability that the mean of a randomly selected sample falls between 32.00 and 36.00 units = degree of confidence = $1 - \alpha = 1 - 8.92\% = 91.08\%$.

---

2. Adequacy Analysis: Given All Other Variables, Find *n*. Here, we seek to determine the minimum sample size needed to ensure a certain degree of confidence or confidence interval:

$$n = \left(\frac{Z_{\alpha/2}\sigma}{\text{Error}}\right)^2$$

where $Z_{\alpha/2}$ is the Z value corresponding to an area of $\alpha/2$ to the LEFT. Error is the deviation of the estimated parameter for its true value and is the standard deviation of the population.

Example 6.4

How many concrete sample cylinders should a materials engineer at a construction site take in each sample (batch) if the contract specifications require that he should be 90% confident that his estimate of the mean concrete strength does not deviate from the true mean strength of that day's concrete by 3 N/mm^2.

From past experience, he knows that the strengths of concrete produced is normally distributed with a variance of 12.25 N/mm^2.

Solution

Maximum error = 3 N/mm^2. Denote the minimum sample size is n_{min}:

$$n_{min} = \left(\frac{Z_{\alpha/2}\sigma}{\text{Error}_{max}}\right)^2$$

Now, $\sigma = 3.5\ N/mm^2$; $1 - \alpha = 0.90$, hence $\alpha = 0.1$, $\alpha/2 = 0.05$, and $Z_{\alpha/2} = 1.645$.

$$n_{min} = \left(\frac{1.645 \times 3.5}{3}\right)^2 = 3.68$$

Therefore, every test sample should comprise at least four concrete cylinders.

---

3. Given All Other Variables, Find θ_L or θ_U. Find the lower or upper limit of the confidence interval needed to ensure that a certain probability or degree of confidence will be obtained given a certain sample size.

Example 6.5

From a statistical analysis of their past projects, Aussie Construction Corporation determined that the mean and variance of their construction mobilization periods are 25 and 4 days, respectively. During a recent internal audit of the company's operations, 10 past construction projects were sampled at random. Find the 90% confidence interval for the mobilization periods of Aussie's past construction projects. Assume that the population (i.e., mobilization periods of all past projects) is normally distributed.

Solution

$$\text{Error} = Z_{\alpha/2} \times \frac{\sigma}{\sqrt{n}}$$

$1 - \alpha = 0.90$ hence $\alpha = 0.1$, $\alpha/2 = 0.05$, and $Z_{\alpha/2} = 1.645$. Thus,

$$\text{Error} = 1.645 \times \frac{4}{\sqrt{10}} = 2.07$$

The 90% confidence interval is 25 ± 2.07 days, that is, 22.93 – 27.07 days.

---

4. Precision Analysis: Given All Other Variables, Find Error (Precision Analysis for Symmetrical Cases Only). Synonyms for precision include *error* and *deviation*. Precision analysis is the determination of the error (deviation of the estimated parameter value from its true value), given the level of significance (or degree of confidence), standard deviation, and sample size. For the common cases (large samples from a population whose variance is known or small samples from a normally distributed population), the error associated with the estimated parameter is

$$\text{Error} = Z_{\alpha/2}\frac{\sigma}{\sqrt{n}}$$

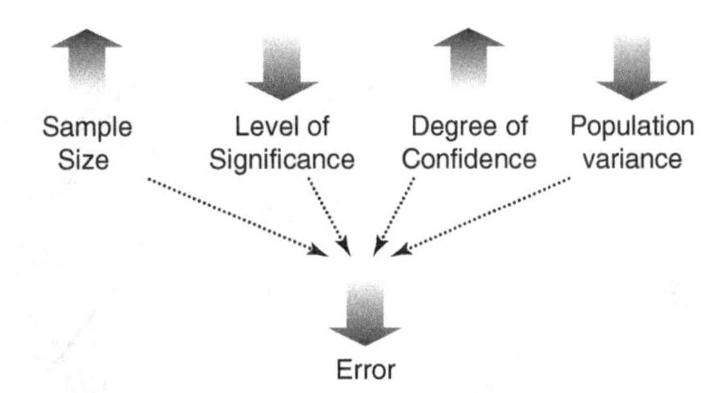

Figure 6.13 Directional relationships between precision and precision factors.

where $Z_{\alpha/2}$ is the Z value corresponding to an area of $\alpha/2$ to the LEFT of the distribution graph; n is the sample size, and σ is the standard deviation of the population.

However, for other cases where:

a. large samples whose population variance is unknown, or
b. for small samples whose parent population is not normally distributed, or
c. for small samples whose parent population has an unknown distribution,

the expression for the statistical error is

$$\text{Error} = t_{\alpha/2}\frac{\sigma}{\sqrt{n}}$$

where $t_{\alpha/2}$ is the t-value with $\upsilon\ (= n - 1)$ degrees of freedom; there is an area of $\alpha/2$ to the right side of the distribution graph.

From the precision equations, it is clear that a higher precision (lower error) is generally associated with greater sample size, lower level of significance, higher degree of confidence, and lower population variance (Figure 6.13).

Example 6.6

In a recent survey of the weights of certain standard structural members, an engineer had a mean of 176 lb and a variance of 2308 lb. Assuming the distribution of the weights of the entire population of that structural member is normal, what is the precision associated with the analyst's experiment at 10% level of significance? Provide a sketch and write the confidence statement for this problem.

Solution

As shown in Figure 6.14, the precision is given by

$$\text{Error} = \mathrm{t}_{\alpha/2} \times \frac{\sigma}{\sqrt{n}}$$

where $\sigma = (\text{variance})^{0.5} = (2308)^{0.5} = 48.04$. Thus, precision $= Z_{\alpha/2}(\sigma/n^{0.5}) = Z_{0.1/2}(48.04/36^{0.5}) =$ 13 lb.

Thus, 90% of the time, we expect our sample means to fall between 163 and 189 lb. Our confidence statement would be as follows: "In 9 out of every 10 samples that we take from this population, the sample means can be expected to fall between 163 and 189 lb."

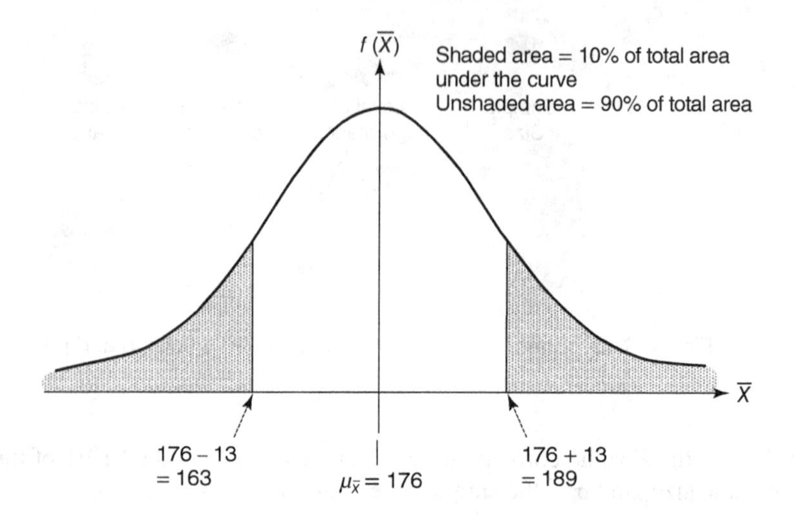

Figure 6.14 Figure for Example 6.6.

6.3.3 Sampling Distributions

A sampling distribution is a description of the frequencies and, therefore, probabilities of various values (or functions) of a given sample statistic. As such, we can have, for all the samples taken from the population, a sampling distribution of the means, of the variances, or of any other statistic. Furthermore, we can have a sampling distribution of the difference between the means of the samples taken from two different populations.

The Central Limit Theorem (CLT). This states that: For a random sample of size n taken from a population with mean μ and variance σ^2, as n is very large, the distribution of the sample means approaches a standard normal distribution with mean 0 and variance 1.

The implication is that "if we sample from a population with an unknown distribution, either finite or infinite, then the distribution of the sample means X will be approximately normal with mean μ and variance σ^2/n, provided that the sample size is large, where μ and σ^2 are the population mean and variance respectively."

Under what conditions is this approximation (i.e., the CLT) valid?

1. If the sample size is large (equal to or greater than 30), in which case the population distribution may or may not be normally distributed.
2. If the sample size is small *and* the population distribution is normally distributed.

Example 6.7

The mean and standard deviation of the weights of identical concrete specimens are 150 and 45 lb, respectively. In an unbiased sample of 50 specimens from this population, what is the probability that the mean weight of the sample exceeds 160 lb?

Solution

We seek the probability that $X > 160$ lb, but in order to compute this probability we need to know the type of probability distribution that describes the sample data (or at least the population data). Let X be the mean of a random sample taken from this population. Then as n is considered large (it exceeds 30), we can apply the central limit theorem. The standardized normal random variable is as follows. We seek

$$p(X > 160) = 1 - p(X < 160) = 1 - p\left(Z < \frac{160.5 - \mu}{\sigma/\sqrt{n}}\right)$$

$$= 1 - p\left(Z < \frac{160.5 - 150}{45/\sqrt{50}}\right) = 1 - p(Z < 1.57) = 5.82\%$$

Therefore, the sample has a 5.82% probability of having a mean exceeding 160 lb.

Example 6.8

The life span of a certain civil engineering system is approximately normally distributed with a mean of 800,000 person-years and a standard deviation of 40,000. For a random sample of 16 such systems, find the probability that the average life will be less than 775,000 person-years.

Solution

The sampling distribution of X will be approximately normal, with the mean of several sample means, $\mu_X = 800$, in thousands and the variance of several sample means $\sigma_X = 40/(16^{0.5}) = 10$, in thousands.

Let X be the mean of the engineer's random sample taken from the entire production population, which has a known mean μ and variance σ^2; n is considered small (as it is less than 30), but we can still apply the central limit theorem because we are told that the population has a normal distribution. The required probability is:

$$p(X < 775.5) = p\left(Z < \frac{775.5 - \mu}{\sigma/\sqrt{n}}\right) = p\left(Z < \frac{775.5 - 800}{40/\sqrt{16}}\right) = p(Z < -2.45) = 0.62\%$$

Therefore, there is a 0.62% chance that the average life of a randomly selected system will be less than 775,000 person-years.

Sampling Distribution of the Difference between Two Means. From the central limit theorem, the variables X_1 and X_2 are both approximately normally distributed with means μ_1 and μ_2 and variances ${\sigma_1}^2/n_1$ and ${\sigma_2}^2/n_2$, respectively. This approximation improves as n_1 and n_2 increase.

The mean and standard deviation of a sampling distribution of the difference between two means are given as follows:

$$\mu_Y = \mu_{\overline{X}_1 - \overline{X}_2} = \mu_{\overline{X}_1} - \mu_{\overline{X}_2} = \mu_1 - \mu_2$$

$$\sigma_Y^2 = \sigma^2_{\overline{X}_1 - \overline{X}_2} = \sigma^2_{\overline{X}_1} + \sigma^2_{\overline{X}_2} = \frac{\sigma_1^2}{n_1} + \frac{\sigma_2^2}{n_2}$$

Example 6.9

Two independent quality tests are conducted to compare steel I-beams supplied by two rival contractors for a large civil engineering project. Eighteen specimens from each supplier were tested for their tensile strength. Assume that the mean strength of both are known to be equal. Also assume that the population standard deviations for both samples = 1. Find the probability that the difference between the strengths of the samples provided by the suppliers differ by more than one unit.

Solution

From the sampling distribution of the difference between the two means, $X_1 - X_2 (= Y)$, it is known that the distribution is approximately normal with the following mean and variance:

The mean of Y,

$$\mu_Y = \mu_{\overline{X}_1 - \overline{X}_2} = \mu_{\overline{X}_1} - \mu_{\overline{X}_2} = \mu_1 - \mu_2 = 0$$

And the variance of Y,

$$\sigma_Y^2 = \sigma^2_{\overline{X}_1 - \overline{X}_2} = \sigma^2_{\overline{X}_1} + \sigma^2_{\overline{X}_2} = \frac{\sigma_1^2}{n_1} + \frac{\sigma_2^2}{n_2} = \frac{1}{18} + \frac{1}{18} = \frac{1}{9}$$

We seek the probability that the difference of the two means exceeds 1:

$$p(Y > 1) = p\left(\frac{Y - \mu_Y}{\sigma_Y} > \frac{1 - \mu_Y}{\sigma_Y}\right)$$

$$= p\left(Z > \frac{1 - \mu_Y}{\sigma_Y}\right) = P\left(Z > \frac{1 - 0}{\sqrt{\frac{1}{9}}}\right)$$

$$= p(Z > 3) = 1 - p(Z < 3) = 1 - 0.9987 = 0.0013$$

Difference between Ordinary Distributions and Sampling Distributions. *Ordinary Distributions*: A typical problem is to find the probability that an element taken from a population will take a value that is within a certain range. For example, find $p(a < x < b)$.

$$p(a < x < b) = p\left(\frac{a - \mu}{\sigma} < Z < \frac{b - \mu}{\sigma}\right)$$

If the population is normally distributed, then the above probability equation can be standardized as follows:

Sampling Distributions: Here, a typical problem is to find the probability that an estimator (sample statistic), such as the sample mean, takes a value that is within a certain range. Depending on the population distribution type and knowledge of the value of the variance, there are two possible cases:

Case 1: Population is normally distributed or sample size is large, and variance is known. For example, find $p(a < X < b)$. The above probability equation can be *standardized* as follows:

$$p(a < \overline{X} < b) = p\left(\frac{a - \mu}{\sigma/\sqrt{n}} < Z < \frac{b - \mu}{\sigma/\sqrt{n}}\right)$$

Case 2: Population type and variance are unknown, and sample size is small.

For example, find $p(a < X < b)$. The above probability equation can be *studentized* as follows:

$$p(a < \overline{X} < b) = p\left(\frac{a - \mu}{S/\sqrt{n}} < t < \frac{b - \mu}{S/\sqrt{n}}\right)$$

6.4 HYPOTHESIS TESTING

A hypothesis can be defined as a tentative assumption established with a view to subsequently test its validity on the basis of empirical data. Hypothesis testing is "a data-based statistical procedure used to make inferences about the attributes of a system" and generally provides an answer to the following question: "Which one of two contrasting claims about a population is correct?" Hypothesis testing always involves two statements:

H_0: The null hypothesis

H_1: The alternative hypothesis (also H_A)

The test involves the comparison of two values: (i) the value calculated from the given sample data (the calculated value of the test statistic) and a threshold value calculated using the given Level of Significance (LOS) LOS (i.e., test value of the test statistic).

In the context of civil engineering applications, there are at least three broad contexts of hypothesis testing applications:

- Is there a significant effect in the condition or operational characteristics of a system in response to a treatment or stimulus?
- Is there a significant difference in the average (mean) or variability (variance) of attributes (such as condition and operational characteristics) of two separate civil engineering systems (populations)?
- Does the quality of a certain civil engineering item (product or service) meet a certain fixed standard or specification?

6.4.1 Types of Hypothesis Tests

Hypothesis tests can be categorized by the number of tails; and if there is one tail, the direction of the tail (Figures 6.15–6.17).

One-tailed (or One-sided) Tests. The test may be upper tailed or lower tailed. If the test is upper tailed, then the hypothesis statement is as follows:

$$H_0:\ \mu \le a$$

$$H_1:\ \mu > a$$

If the test is lower tailed, then the hypothesis statement is as follows:

$$H_0:\ \mu \ge a$$

$$H_1:\ \mu < a$$

Two-tailed (or Two-sided) tests. If the test is two tailed, then the hypothesis statement is as follows:

$$H_0:\ \mu = a$$

$$H_1:\ \mu \ne a$$

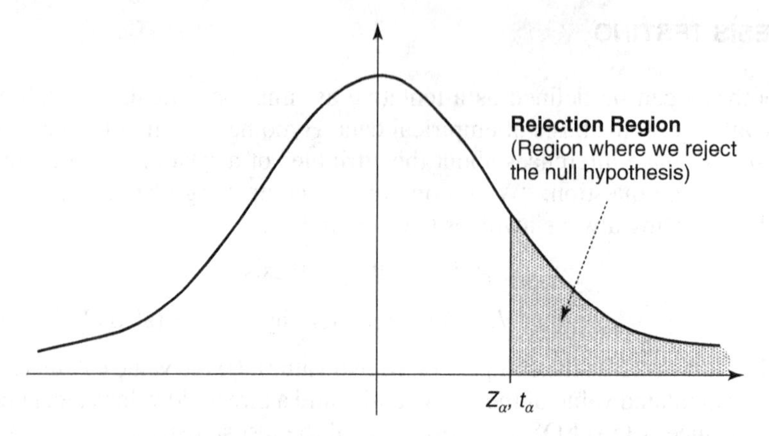

Figure 6.15 Figure for upper-tailed test.

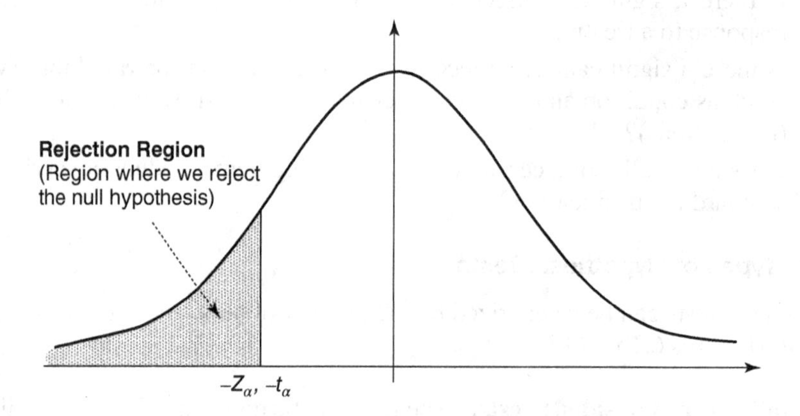

Figure 6.16 Figure for lower-tailed test.

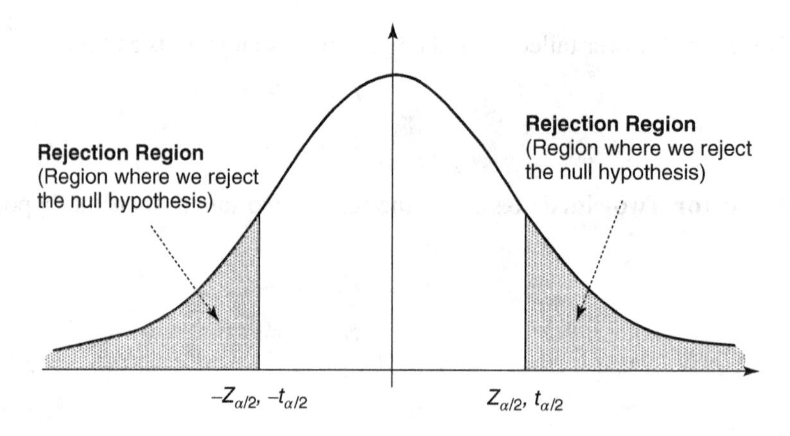

Figure 6.17 Figure for two-tailed test.

6.4.2 Hints in Identifying Appropriate Type of Hypothesis Test

In a given problem, it is important for the analyst to identify the appropriate type of hypothesis test for the problem. How can we tell from the question whether the claim is one sided or two sided? If one sided, how can we tell if the claim is lower tailed or upper tailed?

(a) Two-sided Claims. Some key phrases that are indicative of two-sided claims are: ... is equal to ... ; ... is same as ... ; ... is not equal to ... ; ... is different from .... In such cases, the claim statement is written as follows:

$$\theta = a \quad \theta \neq a$$

(b) Upper-tailed One-sided Claims. Some key phrases that are indicative of upper-tailed one-sided claims are: ... more than ... ; ... superior to ... ; ... in excess of ... ; ... exceeds ...

In such cases, the claim statement is written as follows:

$$\theta = a \quad \theta > a$$

(c) *More Upper-tailed One-sided Claims* (Both the Claim and Counterclaim Contain Inequalities). Some key phrases that are indicative of this situation are: ... more than or equal to ... ; ... is at least superior to ... ; ... equals or exceeds .... In such cases, the claim statement is written as follows:

$$\theta \leq a \quad \theta > a$$

(d) Lower-tailed One-sided Claims. Some key phrases that are indicative of lower-tailed one-sided claims are: ... less than ... ; ... inferior to ... ; ... lower value than .... In such cases, the claim statement is written as follows:

$$\theta = a \quad \theta < a$$

(e) *More Lower-tailed One-sided Claims* (Both the Claim and Counterclaim Contain Inequalities). Some key phrases that are indicative of this situation are: ... less than or equal to ... ; ... equal or inferior to ... ; ... equals or falls short of .... In such cases, the claim statement is written as follows:

$$\theta \geq a \quad \theta < a$$

6.4.3 Possible Errors in Hypothesis Testing

Errors in hypothesis testing may arise because a sample (data from which the hypothesis testing is carried out) may not always be a close copy of its parent population. These errors may occur for the following reasons: (a) We may never achieve perfect randomness, (b) accessibility problems may preclude inclusion of all diverse elements in a sample, (c) sample sizes that would maximize our confidence may be hard to obtain due to time and cost limitations, (d) the distribution type of the population or sample may not perfectly match the assumed distribution, and (e) the collection of sample data may be hampered by equipment error or malfunction and human errors.

Type 1 Error. A type I error is the rejection of the null hypothesis when it is actually true. This is also called an *error of commission*, *seller's risk*, or *false positive*. The size of a hypothesis test is the probability of its type 1 error (i.e., rejecting the null hypothesis when it should not be rejected). The size of the hypothesis test has a value equal to α (the level of significance).

Type 2 Error. A type 2 error is the failure to reject the null hypothesis when it is actually false. This is also called an *error of omission*, *buyer's risk*, or *false negative*. The symbol β represents the probability of a type 2 error (i.e., not rejecting the null hypothesis when it should be rejected). The power of a hypothesis test is equal to $1 - \beta$.

6.4.4 Discussion of Hypothesis Testing Steps

Figure 6.18 presents the steps in hypothesis testing. These steps are discussed below.

Step 1: Establish the Claim and Counterclaim. The claim may be an affirmative statement or, on the contrary, a negating statement. The counterclaim is simply the opposite of the claim. The claim may be a statement involving an equality (=), nonequality ($\neq$), or inequality ($>, <, \geq, \leq$).

Step 2: Formulate Your Hypothesis in Words. This is done by examining the affirmativeness (or otherwise) of the claim, that is, the "direction" of the claim. Your hypothesis should involve two contrasting statements:

H_0: the null hypothesis

H_1: the alternate hypothesis

For example,

H_0: The average system delay is 15 minutes.

H_1: The average system delay is NOT 15 minutes.

Step 3: Identify Your Statistical Parameter. Which statistical parameter is being investigated? Is it the mean, the difference of two means, the standard deviation, the variance? Incorrect identification will result in the choice of an inappropriate distribution for the testing.

Step 4: Rewrite your Hypothesis as Math Notation. For example:

$$H_0: \mu - a = 0 \qquad H_0: \mu - a \geq 0 \qquad H_0: \mu_1 - \mu_2 \leq 0$$

$$H_1: \mu - a \neq 0 \qquad H_1: \mu - a < 0 \qquad H_1: \mu_1 - \mu_2 > 0$$

$$H_0: \sigma - a = 0 \qquad H_0: \sigma_2 - a \geq 0 \qquad H_0: \sigma_1 - \sigma_2 \geq 0$$

$$H_1: \sigma - a \neq 0 \qquad H_1: \sigma_2 - a < 0 \qquad H_1: \sigma_1 - \sigma_2 < 0$$

Step 5: Determine If the Test Is One Tailed or Two Tailed. This is determined using the results from step 4:

If the formulated statements involve = and $\neq$, then the test is two tailed.
If the formulated statements involve $\geq$, or $\leq$, or $>$, or $<$, then the test is one-tailed.
Implication:
If two tailed, use $\alpha/2$ in step 9.
If one tailed, use α in step 9.

Step 6: Select Appropriate Distribution. This step depends on the statistical parameter under investigation (from step 3). See step 6 on the flowchart in Figure 6.18 to select the appropriate distribution.

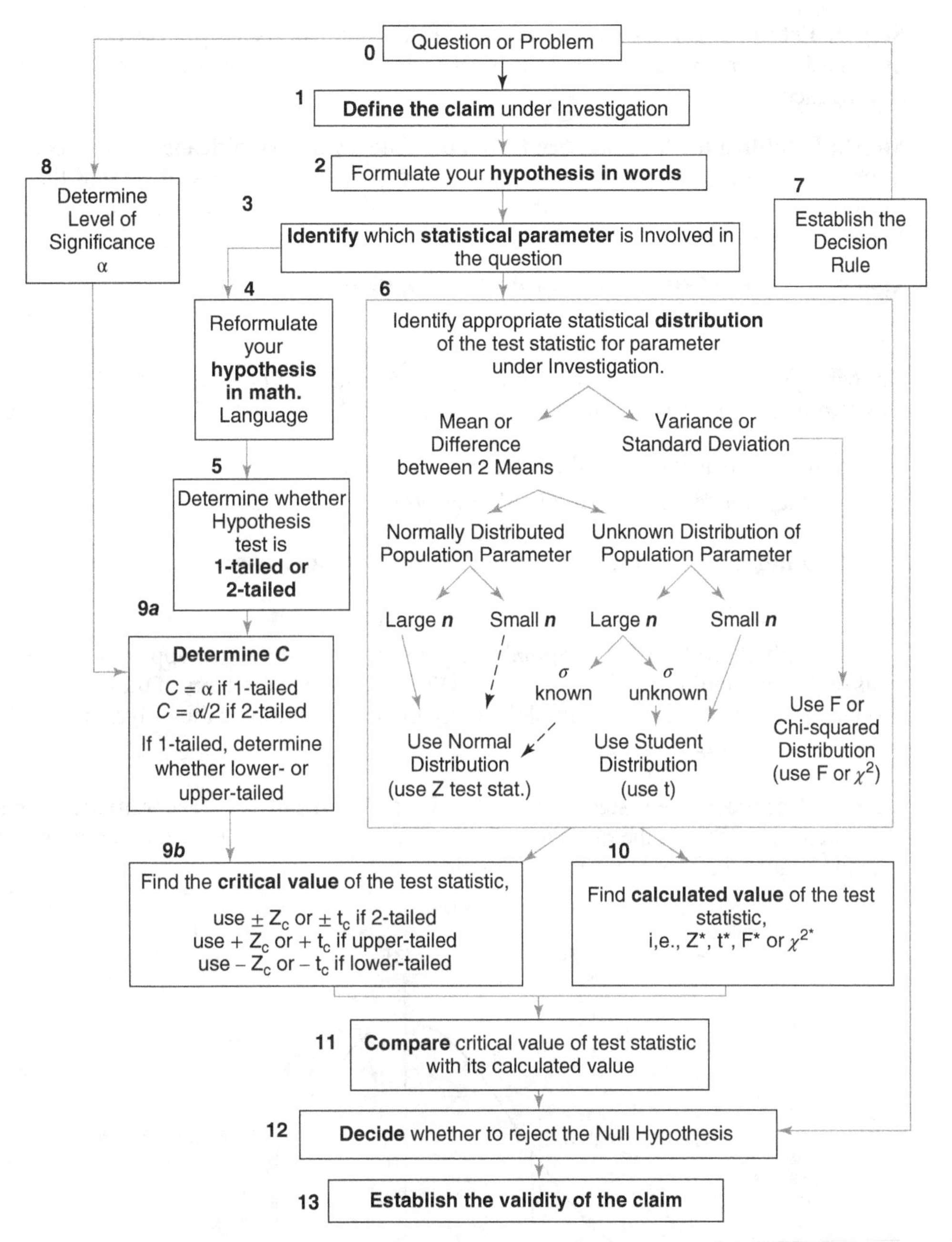

Figure 6.18 Steps for hypothesis testing (broken lines indicate invocation of the central limit theorem).

Step 7: Establish the Decision Rule. The decision rule is simply the declaration that the null hypothesis will be rejected if there is insufficient evidence for us to accept it, at a certain level of significance.

Step 8: Establish the Significance Level, α. The level of significance is provided in the question in one of several possible forms: (i) as a direct value (e.g., you will be told that $\alpha = 0.01$) and (ii) as a value to be derived from the given level of confidence, degree of confidence, probability, or proportion.

Step 9: Find the Critical Value of the Test Statistic.

$$Z_C, t_C, F_C, \chi_C^2$$

Knowing your significance level (from step 8), distribution type (from step 6), and the number of tails (from step 5), you can find the critical value of the test statistic from the statistical tables.

Note that if the test is two tailed, use $\alpha/2$ as your c.

Note that if the test is one tailed, use α as your c.

Step 10: Find the Calculated Value of the Test Statistic.

$$Z^*, t^*, F^*, \chi^{2*}$$

This is simply the value of the original statistic transformed into the appropriate test statistic on the basis of the population distribution type and the estimated parameters of the "population." Note that "population" could mean the sample's true parent population or the universe regarding the claim or assumption being investigated.

Step 11: Compare the Calculated and Critical Values of the Test Statistic. This step is best illustrated by sketching the critical and calculated values of the test statistic on the graph as illustrated in Figure 6.19 for the Z-tailed situation.

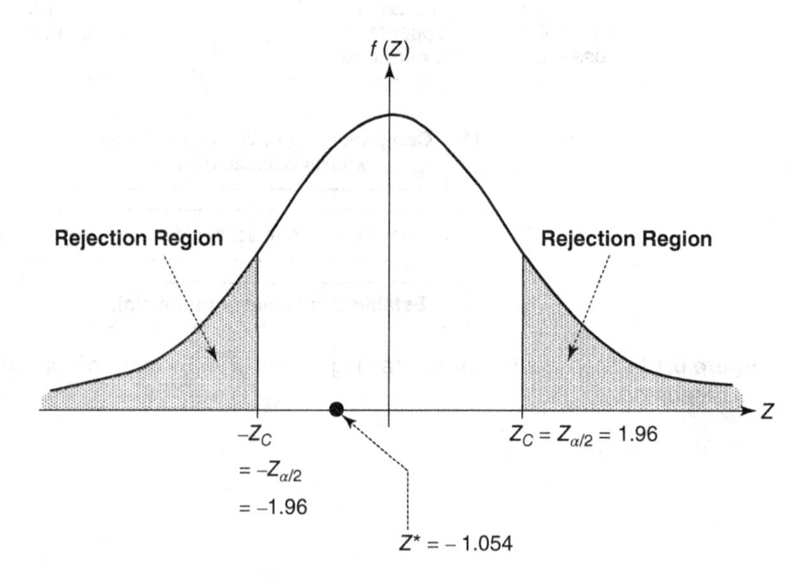

Figure 6.19 Figure for step 11.

Step 12: Decision. Invoking the decision rule (step 7), if the calculated value of the test statistic falls in the rejection region, then we reject the null hypothesis. Otherwise, we fail to reject the null hypothesis.

Example 6.10

It is required that a certain standard structural steel element should be 8 ft in length, with a standard deviation of 6 inches. Elements that are too long or too short are undesirable. A sample of 50 specimens was taken from a large batch of such elements supplied last week by Barrios Company and measured. The mean and standard deviation were calculated as 7.8 and 0.5 ft, respectively. Determine whether the entire batch of supplied elements should be accepted, at 99% confidence.

Solution

The parameter of interest here is the mean. We say the entire batch is OK (should be accepted) if the mean length of the sample is statistically equal to the specified value of 8. If the mean length is too large (statistically), then the production batch is adjudged "not OK." Also, if the mean length is too small (statistically), then also the production batch is adjudged "not OK." First we formulate the hypothesis:

$$H_0:\ \mu - 8 = 0$$

$$H_1:\ \mu - 8 \neq 0$$

Then, we determine the significance level: level of confidence (%) = 99 and degree of confidence = 0.99, $\alpha = 0.01$.

Next, we determine the critical value(s) of the standardized estimate (i.e., the confidence limits beyond which we accept or reject the claim that the calculated value falls within the confidence interval).

$\alpha = 0.01$, therefore, $\alpha/2 = 0.005$. $Z_{\alpha/2} = Z_{0.005} = 2.575$. Therefore, the critical or boundary values of Z are $Z_C = -2.575$ and 2.575 (Figure 6.20).

Next, we determine the calculated value of the test statistic (i.e., the transformed or standardized estimate). Note that μ and σ are from the specified, claimed, or universal, or the "population" values.

$$Z^* = \frac{\overline{X} - \mu}{\sigma/\sqrt{n}} = \frac{7.8 - 8}{0.5/\sqrt{50}} = -2.83$$

Evaluation: At this step, we compare the critical and calculated values of the test statistic (Figure 6.21).

Decision: The calculated value of the test statistic falls in the rejection region, (i.e., $Z^* < |Z_C|$); therefore, we reject the null hypothesis. The mean length of the supplied batch is therefore different (not equal to) the specified length, and the batch should be rejected.

Limitations of Hypothesis Testing. In certain problem contexts in civil engineering, the avoidance of errors within budgetary limitations is a key consideration. In such cases, procedures such as hypothesis testing play a major role in describing system attributes as a basis for decision making because they are associated with efforts to limit the incidence of serious error. This criterion of decision making is not necessarily the best in all problem contexts. In a subsequent chapter

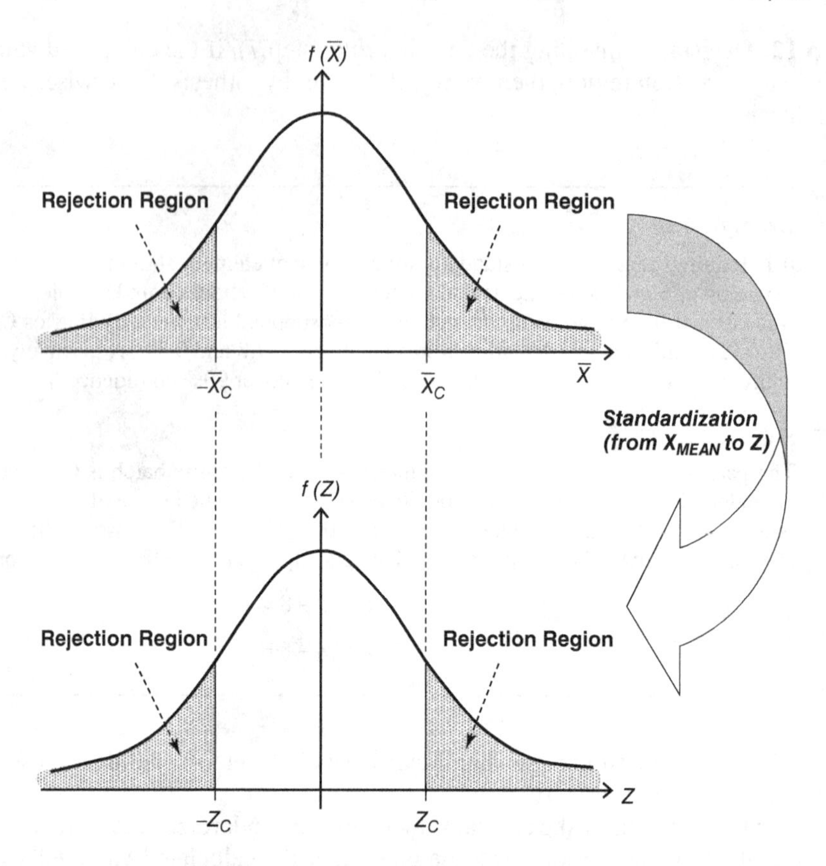

Figure 6.20 Figure 1 for Example 6.10.

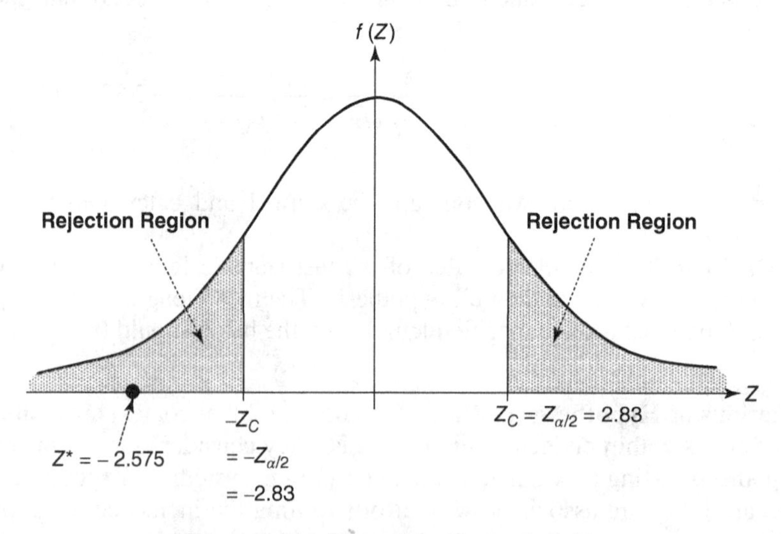

Figure 6.21 Figure 2 for Example 6.10.

of certain texts discuss Bayesian decision making procedures which cover a wider spectrum. In Bayesian procedures, which are based on statistical decision theory (Harnett, 1975), error is just one aspect. The more general procedures in Bayesian approaches allow engineers to establish the probabilities for each level of the population parameter of interest, and it is possible to expand the dimensions of the statistical evaluation to consider the benefits associated with each outcome. Lapin (1990) states that within such an analytical framework, it is possible to consider whether or not to sample at all.

6.5 SOME COMMON TERMINOLOGY AND CONCEPTS IN ENGINEERING STATISTICS

Trimmed Mean/Mode/Median. Trimmed measures of central tendency are calculated such that they are not affected by outliers (Navidi, 2006). This is computed by arranging the sample values in order, trimming a number (or an equal number) of observations from each end, and then computing the mean, mode, or median of those remaining. If $p\%$ of the data is trimmed from each end, then the resulting statistic is referred to as the $p\%$ trimmed mean/mode/median.

Bayesian Inference. An inference method that takes into account the prior probability for an event and thus differs from the classical frequentist approach.

Blocks. Homogeneous groups of experimental units or subjects during design of statistical experiments.

Cohort Effect. The tendency for certain engineering systems constructed in certain years to exhibit relatively higher or lower propensity of some characteristic compared to those constructed in other time periods.

Randomized Block Design. A design for an experiment where the stimuli in each block are randomly assigned to the experimental units.

***Z* Value.** This represents the number of standard errors by which a sample mean is positioned below or above the true population mean.

SUMMARY

In this chapter, we learned a number of basic statistical tools that could be used for a variety of specific tasks associated with the manipulation of data to describe the past, current, or future attributes (condition, operations, structure, or performance) of a civil engineering system. These tasks, which include sampling, descriptive analysis, point and interval estimation, hypothesis testing, and parametric and nonparametric analysis, are encountered at each phase of civil engineering systems development. For example, in order to carry out routine or periodic assessments of the attributes of their systems, engineers need data. However, due to time, cost, and accessibility constraints, they can afford to take only a small sample from a large population of systems or system components in order to investigate the population. This sample must be small to ensure economy but large enough to instill confidence that it adequately reflects the population from which it is drawn. Furthermore, the sample should be random and representative. Systematic sampling and stratified sampling are useful to ensure that a sample is representative of the population. Only a good sample can yield accurate inferences and predictions about the population of systems or system components.

The attributes of a civil engineering system, typically expressed through data obtained from the system, can often be described using a telling graphic such as bar or pie charts and dendograms, or a single number. This number may be a measure of central tendency or dispersion, a measure of relative standing of individual observed values of a data item, or a measure of the association, if any, between different attributes. We also learned about the concept of sampling distribution, which is a mathematical description of the frequencies, and therefore the probabilities, of various values of a given sample statistic. Such a sample statistic could be the sample means or sample variances from a given population or the difference between two means from two different populations. The distribution of the sample means can be approximated to a normal distribution (and consequently to a standard normal distribution) if the distribution of the population is unknown and the sample size is large, or if the distribution of the population is known to be normal, regardless of sample size. Knowing the appropriate distribution for the sample statistic under investigation, we can establish the probability that the statistic takes on any given range of values. Also, we learned about the concept of hypothesis testing, a useful tool for investigating the validity of a claim made about the attributes of a civil engineering system.

In closing, it is important to mention that statistics is an art as much as it is a science. For the processes of collecting, analyzing, or interpreting data, the analyst often uses some judgment to choose one of several alternative statistical techniques, and each alternative may yield a different outcome.

EXERCISES

1. a. What advice (at least two items) would you give to someone who is about to sample a population of civil engineering systems or system components in order to study their behavior, characteristics, operations, and the like?

b. What is meant by the term "randomized comparative treatments"? Give any one example of a study in civil engineering that would involve randomized comparative treatments.

2. a. What is the difference between degree of confidence and confidence interval?

b. Provide a sketch of sample data that are associated with (i) an unbiased and efficient estimate of the mean and (ii) a biased and inefficient estimate of the mean.

c. Write an example of a confidence statement, with supporting sketches, that illustrates each of the following situations: (i) large confidence interval but large degree of confidence and (ii) small confidence interval but small degree of confidence.

3. Interval estimation.

a. For a population having a mean of 10 and a variance of 30, find the probability that a sample size of 9 taken at random from this population will have a mean between 8 and 13.

b. For a normally distributed population, how many samples should we take in order to ensure that there is a 98% chance of obtaining a sample mean falling between 37 and 43? The population mean and variance are 40 and 16, respectively.

c. The mean height of persons at Southern University is 62 inches, with a variance of 960. How confident are you that the mean height of a sample taken from this population will be between 55 and 69 inches?

d. The average time spent by a motorist at a busy I-90 toll booth is 14.5 seconds, with a variance of 12.15 seconds. How many motorists should we sample in order to be 95% sure that the mean of that sample will be between 13 and 16 seconds?

4. Indicate, for each of the confidence statements on the left side of the table below, the degree of confidence, level of confidence, significance level, and confidence interval in the spaces provided on the right side of the table.

	Degree of Confidence	Level of Confidence	Significance Level	Confidence Interval
Eighty percent of the time, between one-half and two-thirds of all vehicles on I- 65 are semitrucks.				
For 40 out of every 100 costal steel bridges, 10–15% of their surface areas suffer from corrosion.				
Of every 10 days in Skyville, 3 have smog coefficients ranging from 0.75 to 0.9 units.				
The probability that structure X suffers a 6- to 10-inch settlement under loading is 25%.				
Sixty percent of college students weigh between 150 to 210 lb.				
The probability that a college student has a height between 55 and 65 inches is 80%.				
Ninety percent of students spend 15–25 minutes during each visit to a fast-food restaurant.				

5. A sample was taken from a population and yielded a mean of 579 lb. Test the following hypothesis, assuming a sample size of 49, a variance of 23,500, and a 10% level of significance.

$$H_0\colon \mu = 570 \text{ lb} \quad H_1\colon \mu \neq 570 \text{ lb}$$

6. As the site engineer at a large construction site, you have been asked to oversee the concrete production process. You are particularly worried about the slump of the concrete. If the slump is too small, it suggests the concrete is too stiff. If there is too much slump, then the concrete is too watery. The contract specifications state that the slump for the particular concrete used should be 1 inch, with a 90% level of confidence. Therefore, during the concrete production process, you instruct the laboratory technician to take 20 random samples of fresh concrete and measure the slump using the appropriate test equipment. The technician obtained the following test results (in inches):

0.92	1.21	1.03	1.10	1.01	0.99	0.89	0.97	1.01	0.99
1.05	1.11	0.95	1.00	1.00	1.04	0.88	1.02	0.97	1.01

Would you accept that day's production of concrete at the given level of confidence? Assume that, from past slump test results, the concrete slumps are known to be normally distributed.

7. Two competing companies provide 36 samples of their standard prestressed concrete beams for consideration in preparation for a construction project. The product strengths from companies A and B are known to have standard deviations of 5 and 7 N/mm^2, respectively. A certain engineer argues that products from company A are stronger than those from B by at least 2 N/mm^2. To check his claim, you randomly take a sample of size 32 beams from each company's production lines and you find that the mean of A exceeds that of B by 0.85 N/mm^2. Because of the sensitivity of the project, you are advised to be sure that even if you repeat the test 100 times, you will get the same answer in 75 of them. Determine whether the engineer's claim is true. Show all detailed calculations and provide a sketch of the graph associated with your final answer.

8. The value of travel time is a concept widely used in measuring the impact of traffic congestion in urban areas. It is believed that the average value of travel time in West Lafayette is less than $16.50 per hour per person. You are asked to verify this and you go out and interview 45 randomly chosen drivers in West Lafayette and obtain a mean travel time value of \$14.15 per hour per person. You formulate the hypothesis below. It is given that the population variance is \$86.00 per hour. We want our conclusion to be consistent in 9 out of every 10 such samples we take. Assume the population is normally distributed. $H_0: \mu \geq \$16.50$ and $H_1: \mu < \$16.50$.

REFERENCES

Afzalimehr, H., Singh, V. P., and Najafabadi, E. F. (2010). Determination of Form Friction Factor. *ASCE J. Hydrol. Eng.* 15(3), 237–243.

Buteau, S., Fortier, R., and Allard, M. (2010). Permafrost Weakening as a Potential Impact of Climatic Warming. *ASCE J. Cold Regions Eng.* 24(1), 1–18.

Hajj, E. Y., Sebaaly, P. E., and Kandiah, P. (2010). Evaluation of the Use of Reclaimed Asphalt Pavement in Airfield HMA Pavements. *ASCE J. Transp. Eng.* 136(3), 181–189.

Harnett, D. L. (1975). *Introduction to Mathematical Methods*, Addison-Wesley, Reading, MA.

Juran, J. M. (2010). *Juran's Quality Handbook*. McGraw Hill, New York.

Lapin, L. L. (1990). *Probability and Statistics for Modern Engineering*. PWI-Kent Publishers.

Lee, Y. W., Pipino, L. L., Funk, J. D., and Wang, R. Y. (2006). *Journey to Data Quality*. MIT Press, Cambridge, MA.

Navidi, W. (2006). *Statistics for Engineers and Scientists*. McGraw-Hill, New York.

Randall, P. M., and Chattopadhyay, S. (2010). Bench-Scale Evaluation of Chemically Bonded Phosphate Ceramic Technology to Stabilize Mercury Waste Mixtures. *J. Environ. Eng.* 136(3), 265–273.

U.S. DOT (2001). *Highlights 2001 NHTS*. USDOT, Bureau of Transportation Statistics, Washington, DC.

Wood, S. L., and Frank, K. H. (2010). Experimental Investigation of Bending Fatigue Response of Grouted Stay Cables. *J. Bridge Eng.* 15(2).

Zhang, T. C., and Zeng, H. (2010). Development of a Response Surface for Prediction of Nitrate Removal in Sulfur–Limestone Autotrophic Denitrification Fixed-Bed Reactors. *J. Environ. Eng.* 132(9), 1068–1072.

USEFUL RESOURCES

Ang, A. H-S., and Tang, W. H. (2007). *Probability Concepts in Engineering: Emphasis on Applications in Civil and Environmental Engineering*, Wiley, Hoboken, NJ.

Kottegoda, N. T., Rosso, R. (2008). *Applied Statistics for Civil and Environmental Engineers*, Wiley-Blackwell, Hoboken, NI.

McCuen, R. H. (1985). *Statistical Methods for Engineers*, Prentice-Hall, Englewood Cliffs, NJ.

Metcalf, A. V. (1997). *Statistics in Civil Engineering*, Hodder Education, Abingdon, UK.

CHAPTER 7

MODELING

7.0 INTRODUCTION

In Chapter 4, we discussed the tasks faced by managers of civil engineering systems, and we recognized that one of the key tasks faced by systems engineers is the description of some system attribute such as the system structure, the way it works, or its condition or performance in the past, at the current time, or at some specified future time (Figure 7.1). For these and other tasks related to analysis and evaluation at any phase, engineers typically use tools including statistical (Chapter 6) and modeling tools.

Examples of models in civil engineering include a model to predict the future need (demand) for urban transit, a model that describes the magnitude and direction of factors that enhance remediation of contaminated soil, a model that predicts the cost of reconstructing a levee 20 years from now, a model that describes the rate of corrosion in a structural steel member, and a model that describes the factors influencing the effectiveness of a building insulation system. The various models in civil engineering can be classified in several ways, including the system attribute that is being modeled, the mathematical form, the intended general purpose of the model (prescriptive versus descriptive), the level of empiricism (deterministic versus stochastic), the nature of the model (numerical vs. analytic, and the phase of civil engineering system development that is relevant to the model (such as demand models, cost models, deterioration models, operational policy effectiveness models, maintenance effectiveness models, and intervention decision models). In this chapter, we will focus on a specific type of analytic models: statistical regression models.

7.1 STEPS FOR DEVELOPING STATISTICAL MODELS

The development of statistical models is more of an art than it is a science because there is no exact answer and the best model is often obtained after several trials. However, there are certain general steps to be followed in statistical model development (Figure 7.2). Different analysts, through experience, may be using variations to this overall framework depending on their experience, agency practices and culture, data availability, and the intended use of the model. In the next section, we discuss the various steps to be followed in a statistical modeling exercise.

7.1.1 Steps in Modeling

Step 1 Definition of Objective

Models are tools used to describe some attribute of a civil engineering system. As such, the first step of any model development process is a definitive statement of what the model is meant to describe or predict. This step will help avoid any communication problems between the data collectors and the modeler and will also help the modeler to establish any a priori expectations of the model outcomes and capabilities.

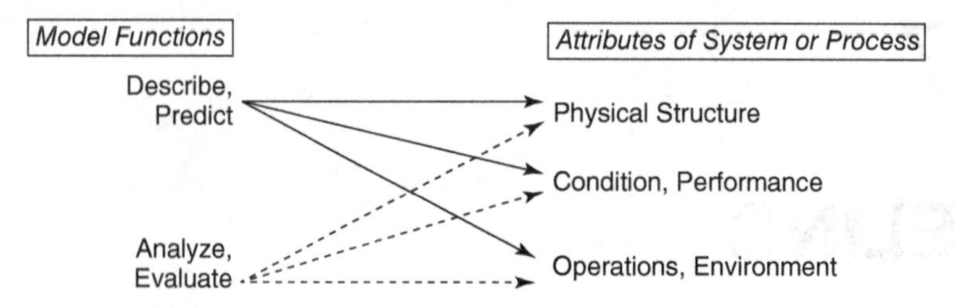

Figure 7.1 Model functions and system attributes.

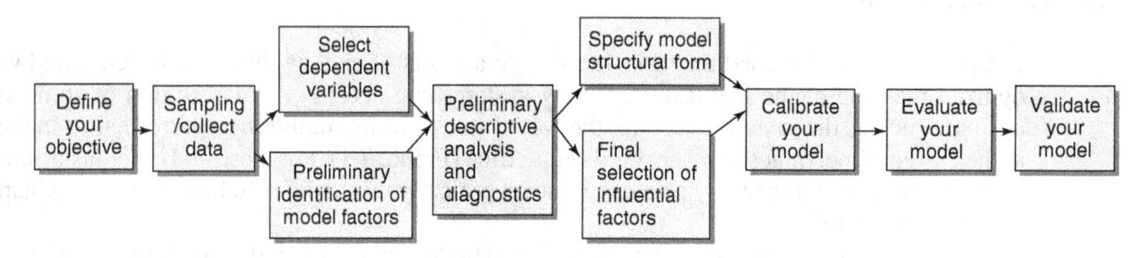

Figure 7.2 General steps for model development.

Step 2 Sampling/Data Collection

As discussed in Chapter 6, it is prudent to use data from a sample of the population instead of the entire population, due to problems that include the lack of time and money for sampling an entire population and the inaccessibility of certain systems or their components. The sample must be not only random (to avoid bias) but also a close replica of the population so that any inferences made from the data are applicable to the population. Most owners or operators of civil engineering systems have assembled databases that are samples that cover part or all of the entire population of their systems.

Step 3 Specify the Response Variable

The response variable should reflect the objective of the modeling process (step 1). This is an important step because it can influence the mathematical form to be used at step 6 of the modeling process. Synonyms for the response variable include regressive, dependent, endogenous, and measured variable. The response variable may be discrete or continuous (Figure 7.3).

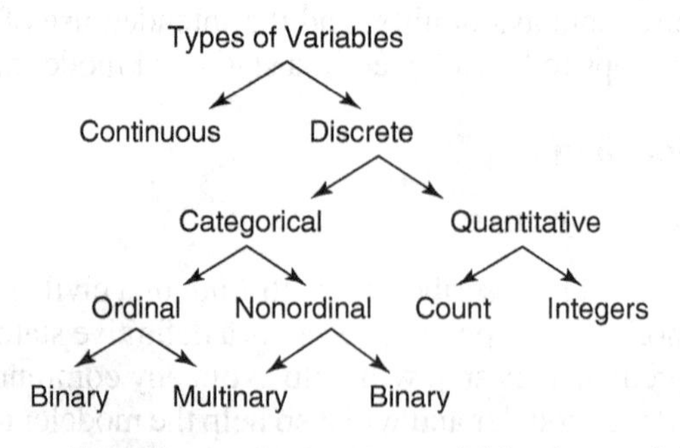

Figure 7.3 Types of response variables for models.

Table 7.1 Categories and Examples of Response Variables Used in Civil Systems Modeling

Nature of Response Variable	Example	Units
Continuous	Surface roughness, corrosion, cracking, or other defect observed on the system	Example, m/km for pavement roughness
	System health index	Example, 0–100 index for bridges
	Annual maintenance expenditure on the system	\$/lane-mile; \$/ft^2; \$/structure, etc.
	System rehabilitation cost	\$/lane-mile; \$/ft^2, etc.
	System service life	Years; accumulated loading
	System vulnerability to threat (natural or artificial)	Example, 0–5 index
Discrete, Count	Number of bridge deck patches per structure	Number
	Number of fatalities associated with the system	Number
	Number of maintenance interventions received by a system	Number
Discrete, Categorical, Ordered	System manager's choice of specific repair option with varying degrees of intensity	Choice
	Level of user satisfaction with system performance	Example, 0–10 index
Discrete, Categorical, Nonordered	System manager's choice of among a set of similar alternative repair options	Choice

As Figure 7.3 and Table 7.1 show, there is a wide range of response variable types that are encountered in civil engineering systems development. The choice of the response variable is dictated by the characteristics of the units of measurement (cost, system physical response, instruments, etc.). Examples of continuous response variables are surface roughness or cost per lane-mile of some preservation action. An example of count variables is the number of crashes at a section or intersection. An example of an ordinal multinary response variable is system repair option (reconstruct/rehabilitate/do nothing). In certain cases of stochastic models, the response variable is the **survival probability** (the likelihood that some system situation remains up to a certain point in time), for example, the likelihood that a system remains in service, which ranges from almost 100% for a new system to almost 0% for an old and dilapidated system; or the **hazard probability** (the likelihood that some system situation does not remain up to a certain point in time), for example, the likelihood that a system suddenly fails, which ranges from almost 0% for a new system to almost 100% for a dilapidated old system.

It is important to note that the type of response variable established in step 3 helps to determine the appropriate model specification (at step 6). Chapter 3 provides greater detail about the performance goals and objectives for each system type and phase, and often these goals ultimately translate into response variables for modeling purposes.

Step 4 Selection of the Explanatory Variables

The explanatory variables are those characteristics of the system or its environment that influence the response we are trying to describe using the statistical model. Synonyms of "explanatory variable" include regressor, exogenous variable, covariate, explanatory factor, input variable, predictor variable, and independent variable. In establishing the independent variables for the model, the systems engineer needs to think about which characteristics of the system or its environment are likely to influence the response variable. If the data collection (step 2) excluded any important variables, then there is a need to go back to collect additional data. A simple model may involve only a few basic independent variables while more complicated models may include several independent variables. For example, models that describe the physical condition of structural systems may include basic variables such as system age, location, and material type; and more complicated versions of these models may include additional independent variables such as design type, level of usage, climatic effects, and maintenance history.

A number of explanatory variables grouped by the development phase at which the model is applied are shown in Figure 7.4. Examples of continuous explanatory variables are age (years), annual or accumulated climatic effects (precipitation in inches), and loading; examples of binary explanatory variables are the climatic freeze zone (1 for freeze, 0 for nonfreeze) and bridge superstructure material (0 for concrete, 1 for steel); and an example of an ordinal independent variable is the quality of the contractor who constructed the system (Class A, B, or C).

Step 5 Carry Out Preliminary Analysis of the Data

The purpose of this step is to identify interesting trends or relationships between the dependent (Y) and independent (X) variables. The tools used in this step, some of which are described in Chapter 6, include scatter diagrams (simple plots of Y versus each X separately), box plots, stem-and-leaf plots, pie charts, analysis of variance (ANOVA), pairwise t tests, etc. If ordinary least squares (OLS) technique is being used for the modeling, there is a key assumption

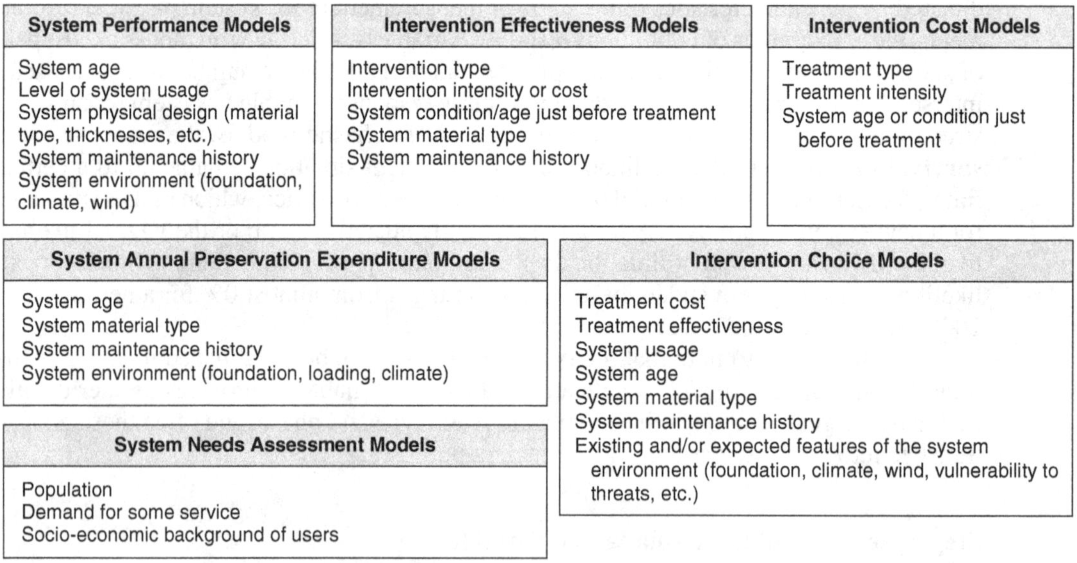

System Performance Models
System age
Level of system usage
System physical design (material type, thicknesses, etc.)
System maintenance history
System environment (foundation, climate, wind)

Intervention Effectiveness Models
Intervention type
Intervention intensity or cost
System condition/age just before treatment
System material type
System maintenance history

Intervention Cost Models
Treatment type
Treatment intensity
System age or condition just before treatment

System Annual Preservation Expenditure Models
System age
System material type
System maintenance history
System environment (foundation, loading, climate)

System Needs Assessment Models
Population
Demand for some service
Socio-economic background of users

Intervention Choice Models
Treatment cost
Treatment effectiveness
System usage
System age
System material type
System maintenance history
Existing and/or expected features of the system environment (foundation, climate, wind, vulnerability to threats, etc.)

Figure 7.4 Examples of explanatory variables used in models at various phases of civil systems development.

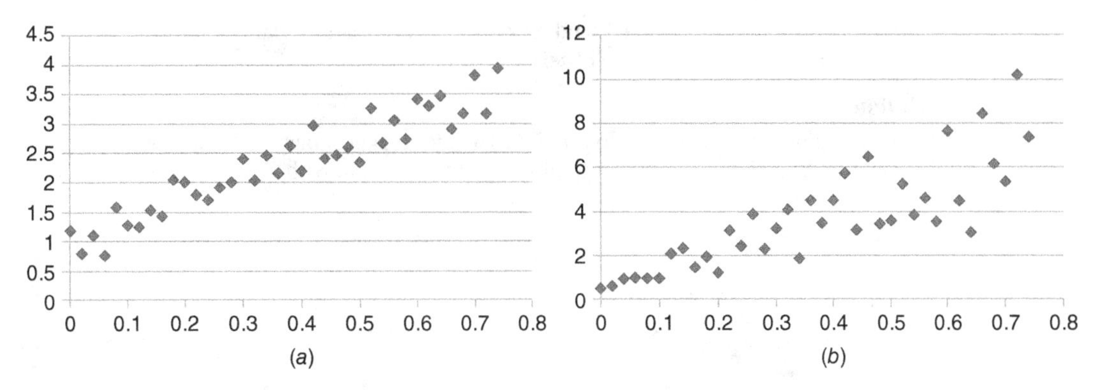

Figure 7.5 Visual detection of heteroscedasticity from data plots: (*a*) homoscedastic and (*b*) heteroscedastic observation.

about the variance of the residuals: If a plot of the variance of the residuals indicates nonconstancy, then the variance of the residuals is termed *heteroscedastic*. A number of graphical and nongraphical tests are available to detect heteroscedasticity. Graphical tests that show an increasing or decreasing bandwidth as the explanatory variable increases are indicative of the presence of heteroscedsticity (Figure 7.5). One nongraphical method is the Durbin Watson Statistic (D value). D values ranging from 0 to 4 indicate homoscedastic variance (a value of 2 indicates perfect homoscedasticity).

A modeler often needs to ascertain whether the data seems to fit any particular distribution. A *normality test*, which can be performed graphically or nongraphically, is used to ascertain whether a sample or any group of data fit a standard normal distribution. In a typical normal or Gaussian distribution, the observations are spread out in such a manner that the most frequently occurring observation is in the middle of the range and other probabilities tail off asymptotically and symmetrically to the left and right directions in a bell-shaped fashion. Also, the closeness of the median to the mean suggests that the data is normally distributed. A simple technique for testing whether your observations are normally distributed is to compare a histogram of the residuals to a normal probability curve: if the former is bell shaped and resembles the latter, then the data is normally distributed. Normality is more readily observed for large data sets than it is for smaller ones. Standard statistics texts offer a variety of formal techniques to detect the presence of normality.

Example 7.1

The table below shows the time (Y, in days) taken to construct several of a certain type of civil engineering system at different locations. The observations show the system size (X_1, in square miles) and the complexity of the construction process, measured in terms of the number of contractors on the project (X_2 = 1 contactor or 2 = 2 contractors).

X_1	7	5.5	2.3	6.2	5	3.2	6.1	0.9	5.4	5.5	5.3	8	6.7	6.9	1.6	3.1	7.2	8.5	6	4	7.1	2.1	7.5	3.6
X_2	1	2	1	2	1	1	1	1	2	2	1	1	2	2	1	1	2	2	2	1	2	1	2	1
Y	39	172	160	212	56	171	131	244	240	260	33	21	98	190	189	131	62	29	152	118	85	221	59	62

(a) Develop a scatter plot for Y and X_1 only. Then develop a scatter plot for Y and both X_1 and X_2. Comment on the trends you observe.

Solution

The scatter plots are provided in Figure 7.6. The plots suggest that combining all observations obscures the data trends, and thus it is better to plot a given X variable with Y while holding all other X variables constant.

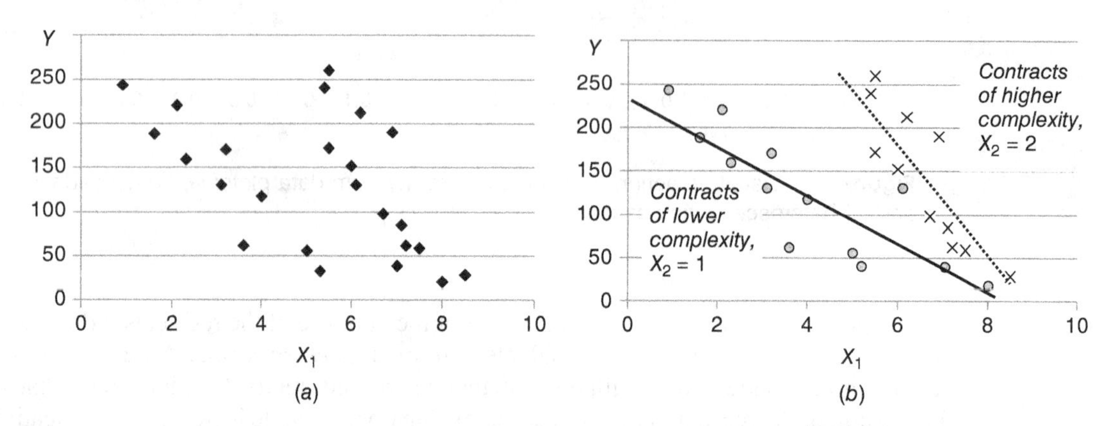

Figure 7.6 Scatter plots for Y versus each X variable, separately, for given ranges of other X variables: (*a*) Y versus X, for all observations combined and (*b*) Y versus X, for each range of X_2 values.

Step 6 Model Specification

This is probably the most important step of the statistical model development process, and a discussion of this step dominates this chapter. In Section 7.2, we will discuss the various categories of model specifications, such as discrete versus continuous, linear versus nonlinear, single equation versus multiple equation, cross-sectional versus time series, and duration versus nonduration.

Often, the choice of *functional form* or *mathematical form* depends on the type of response variable (from step 3), the nature of the response data (e.g., truncation versus no truncation), and other considerations. For discrete response variables, such as a bridge sufficiency rating, a logit or probit functional form could be more appropriate for the model. For continuous response variables, such as system service life or percentage of corrosion, regression or survivor models may be used. The mathematical form for the latter can range from linear to a variety of nonlinear forms: polynomial (including quadratic and cubic), exponential, logarithmic, power, and modified exponential. To ascertain which mathematical form is most appropriate for a given set of data, the raw data must be plotted; and the resulting scatter diagram could be compared with standard curve sketches and any resemblance could be identified for further scrutiny. For models having continuous response variables, Figure 7.7 provides some standard sketches.

Step 7 Final Selection of Independent Variables

In cases where there is an excessive number of independent variables, it may be necessary to drop some of them to simplify the analysis. However, the question arises as to which ones to drop. From the plots in step 6, the analyst may find that certain independent variables have little or no impact on the dependent variable and thus could be dropped without jeopardizing the efficacy of the model.

Step 8 Separate Your Data Set into Two

As recommended in most statistical texts, it is often prudent to break up the original modeling data set into two sets: one for the model calibration (80–90% of the original data set) as

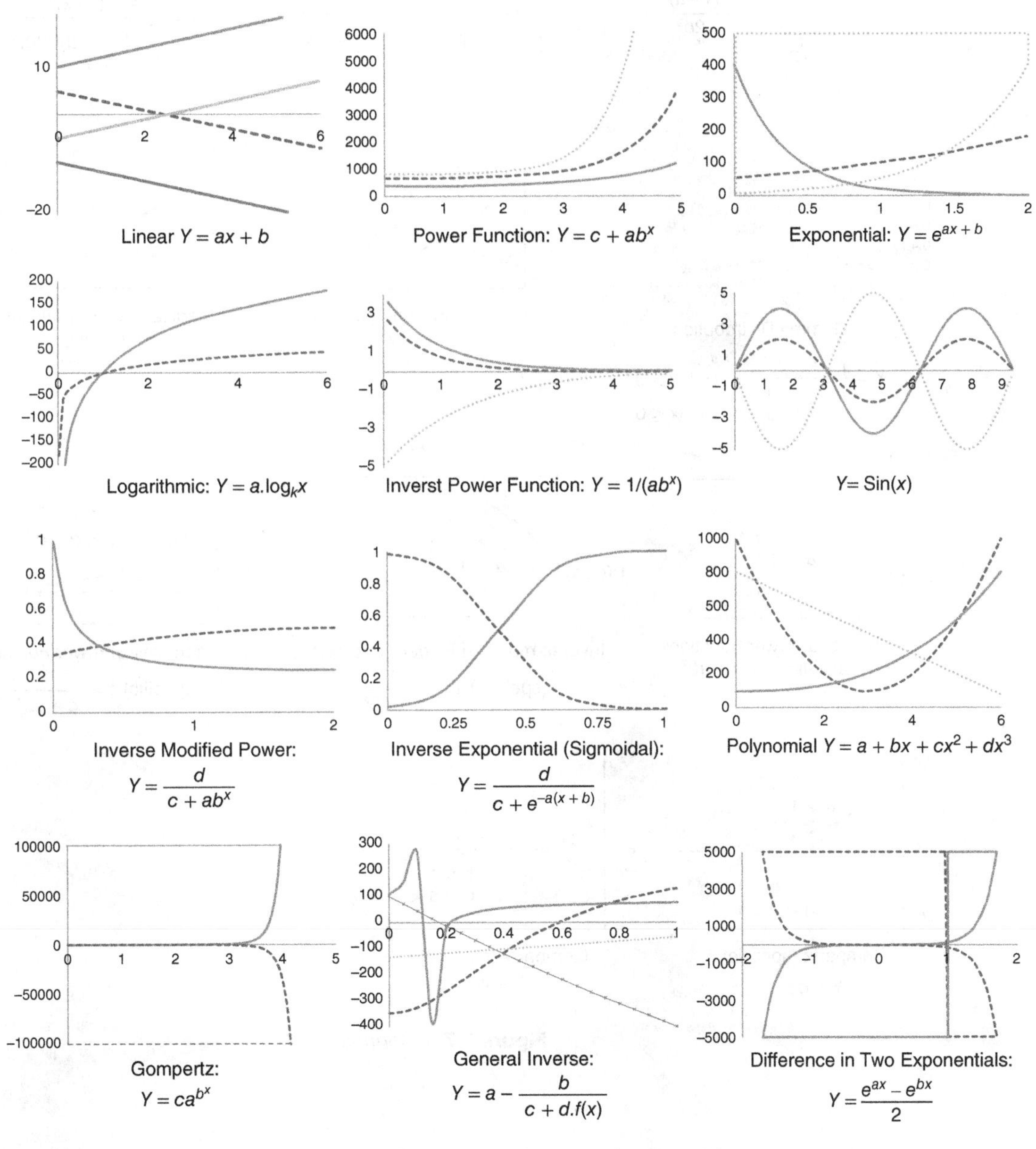

Figure 7.7 Sketches of common functional forms.

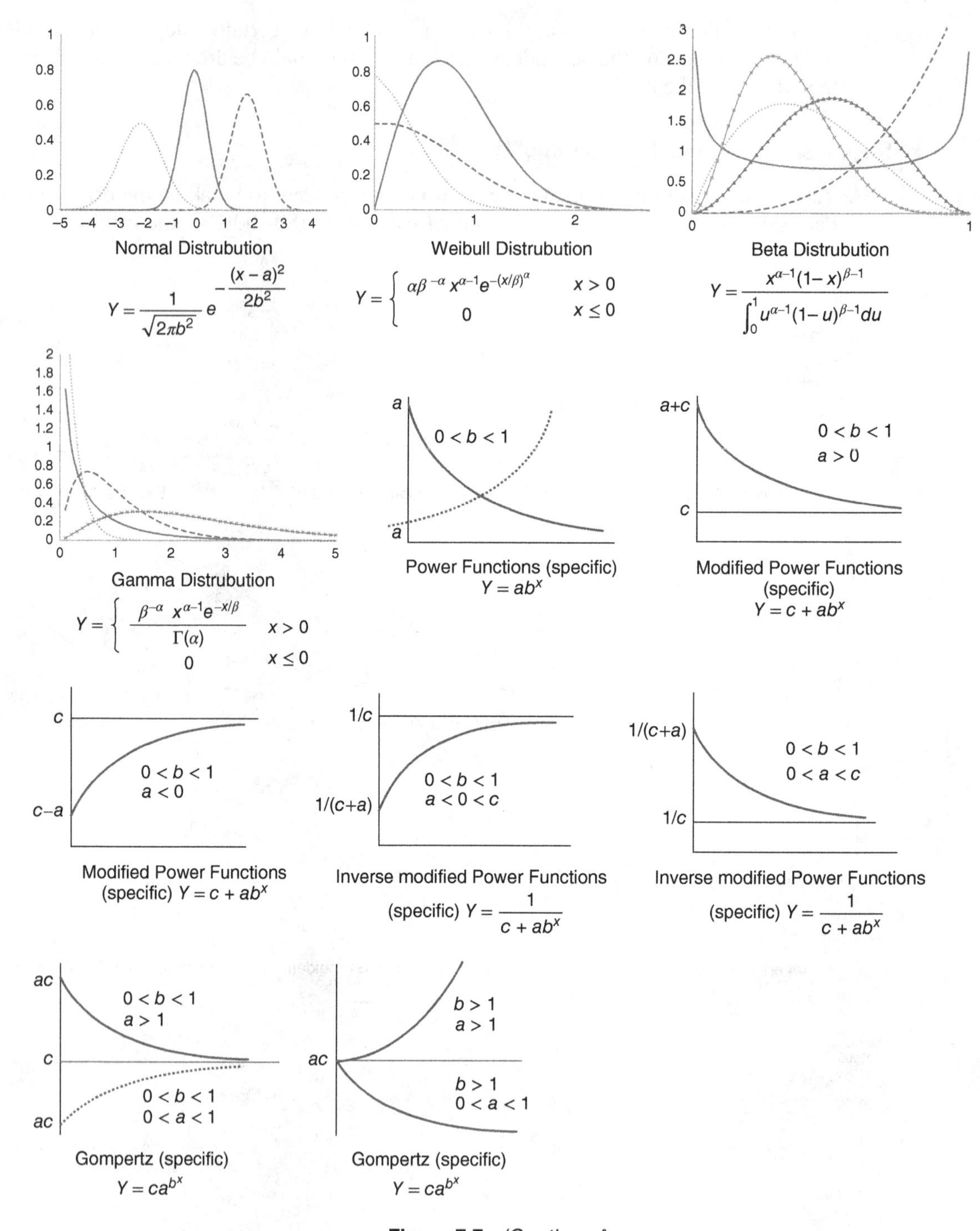

Figure 7.7 *(Continued)*

explained in step 9 and a smaller portion (10–20%) for the model validation as explained in step 10.

Step 9 Model Calibration

"Calibration" simply means determining the best function passing through the points (i.e., for the linear functional form, for example, determining the values of the parameters a and b of the functional form). Mathematically, a "best function" could mean an equation that passes through the points such that the sum of the vertical deviations of various points from the regression line is minimized. An example is given in Figure 7.8 where it is sought to calibrate a linear model by determining the best line that represents the given four observations (statistically, more than 30 observations are recommended for a good model). Note that such a line naturally would also be the best unbiased and efficient line that passes through the points. Regression analysis establishes an empirical relationship between two or more variables. The simplest form is linear regression between one dependent and one independent variable. In more complex forms, the regression model is nonlinear and there are several independent variables (Figure 7.9). The simple linear regression model is given by

$$Y_i = \alpha + \beta X_i + \varepsilon_i$$

where Y is the dependent variable representing the performance measure; X is the vector of explanatory or independent variables; ε is the random error; and α, β are regression parameters.

The model is not expected to be a perfect representation of the underlying phenomenon we are trying to describe. As such, it is often appropriate to include a random error term, ε in the

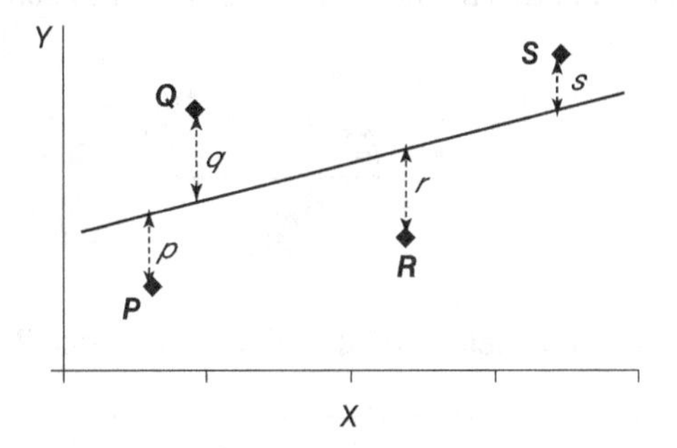

Figure 7.8 Illustration of best-fit line for four observations P, Q, R, and S, in two dimensions.

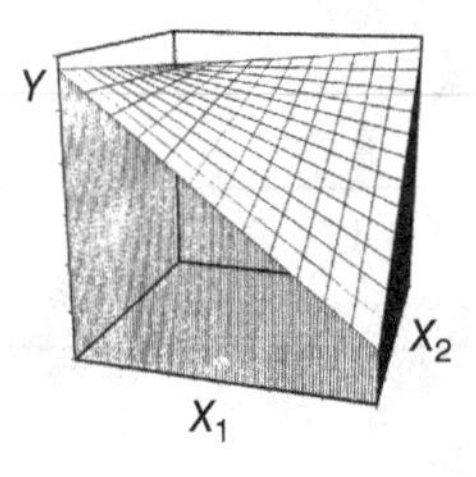

Figure 7.9 Illustration of best-fit plane for observations (points not shown) in three dimensions.

Table 7.2 Calibration Table for a Univariate Model

x_i	y_i	$(x_i - x)$	$(y_i - y)$	$(x_i - x)(y_i - y)$	$(x_i - x)^2$
x_1	y_1				
x_2	y_2				
$\vdots$	$\vdots$				
X_n	Y_n				
Σx_i	Σy_i			$\Sigma[(x_i - x)(y_i - y)]$	$\Sigma (x_i - x)^2$
$\bar{x} = \Sigma x_i/n$	$\bar{y} = \Sigma y_i/n$				

model. Different model forms can be investigated to ascertain which one best fits the data. There are several ways to judge the "goodness of fit" of a regression model the simplest being OLS and maximum likclihood. Scction 7.2.1 provides a more detailed discussion on the tool of linear regression.

For a simple linear model with one explanatory variable, the parameters of the best-fit line can be estimated using the Table 7.2 and the equations below. However, when there is more than one explanatory variable and where the best-fit function is nonlinear, the derivation of the expressions for the parameters can be exceedingly complex and the use of appropriate statistical software is necessary.

As we may realize from Figures 7.7 and 7.8, there is a multitude of lines that could possibly represent the observations in a data set. However, the best line is one for which the sum of the squared deviations is least. It can be shown by calculus that in order to minimize the sum of squared distances, the values of the linear function parameters a (intercept) and b (slope) can be calculated as follows:

$$b = \frac{\sum_{i=1}^{n}[(x_i - \bar{x})(y_i - \bar{y})]}{\sum_{i=1}^{n}(x_i - \bar{x})^2} \qquad a = \frac{\sum_{i=1}^{n} y_i - b \cdot \sum_{i=1}^{n} x_i}{n}$$

For a given set of data, the values of a and b can be determined easily using Table 7.2.

Example 7.2

The table below shows the observed values of X and Y. Calibrate the model to establish a mathematical relationship between X and Y.

Observation	1	2	3	4	5	6	7	8	9	10	11	12	13	14
X	1.5	2.4	2.7	4.1	0.9	3.7	0.5	3.3	2.9	4.1	3.1	4	3.1	4
Y	3.16	3.04	3.36	3.88	2.83	3.76	2.48	3.59	3.34	3.91	3.70	3.68	3.70	3.68

Solution

The calibration table is prepared as follows:

X_i	Y_i	$X_i - X_{AVG}$	$Y_i - Y_{AVG}$	$(X_i - X_{AVG}) * (Y_i - Y_{AVG})$	$(X_i - X_{AVG})^2$
1.5	2.85	−1.16	−0.53	0.62	1.34
2.4	3.47	−0.26	0.09	−0.02	0.07
2.7	3.50	0.04	0.12	0.01	0.00
4.1	3.97	1.44	0.58	0.84	2.08
0.9	2.74	−1.76	−0.65	1.13	3.09
3.7	3.63	1.04	0.25	0.26	1.09
0.5	2.59	−2.16	−0.79	1.71	4.65
3.3	3.76	0.64	0.37	0.24	0.41
2.9	3.51	0.24	0.13	0.03	0.06
4.1	3.68	1.44	0.29	0.42	2.08
3.1	3.44	0.44	0.06	0.03	0.20
4	4.03	1.34	0.65	0.87	1.80
1.9	3.03	−0.76	−0.36	0.27	0.57
2.1	3.16	−0.56	−0.22	0.12	0.31
37.2	**47.35**			**6.53**	**17.75**
2.66	**3.38**				

$$b = \frac{\sum_{i=1}^{n}[(x_i - \bar{x})(y_i - \bar{y})]}{\sum_{i=1}^{n}(x_i - \bar{x})^2} = \frac{6.53}{17.75} = 0.368 \qquad a = \frac{\sum_{i=1}^{14} y_i - 0.368 \cdot \sum_{i=1}^{14} x_i}{14} = 2.371$$

Thus, the model is $Y = 0.368X + 2.371$.

Step 10 Model Evaluation

To ascertain how good the developed model is, the following tests could be used: the coefficient of determination (R^2), the level of significance, the standard error (or t-statistics or p values) of the estimate, the heteroscedasticity of variance, and normality tests. The coefficient of determination, the most common statistic used to evaluate how well the model fits the data, assesses the closeness of the observed data to the model functional form under consideration. The coefficient of determination R^2 is a measure of the fraction of variability in a data set that is explained by the statistical model and shows how well future outcomes (or outcomes for observations outside the modeling data set) are likely to be predicted or estimated by the model. In linear regression, R^2 is the square of the sample correlation coefficient between the actual values of the response variable and their predicted values, which varies from 0 (good fit) to 1 (perfect fit); see Figure 7.10. The adjusted R^2 is more useful than the R^2 if it is calculated for a sample from a population rather than the entire population. For example, if our unit of analysis is an individual bridge system and we have data for all bridges in an overall network, then the adjusted R^2 will not yield any additional information beyond what the R^2 provides.

The standard error of the estimate is a measure of how accurate the predictions are made with the regression model. The regression line or model seeks to minimize the sum of the squared

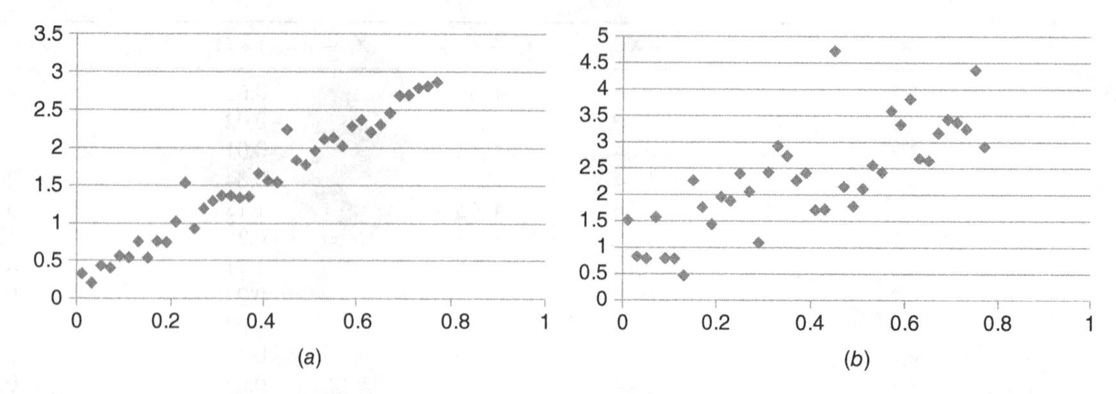

Figure 7.10 Illustration of observation sets with (*a*) low and (*b*) high R^2 values.

Table 7.3 Thresholds for Statistical Significance

If the absolute value of the *t* statistic is ...	Then we say that the variable is significant at ...
1.96 or more	95% confidence
1.64 or more	90% confidence
1.28 or more	80% confidence

errors of the predictions from the true observed values. The best regression model is that with the least value of the standard error of estimate.

Model evaluation also includes a statement of the significance of the explanatory variables in terms of their magnitude and the intuitiveness of their direction. The statistical significance of any variable can be ascertained on the basis of the given level of confidence (Table 7.3). The sign of the *t* statistic indicates the direction of the relationship between the X variable in question and the response variable Y: *a* negative sign suggests that an increase in the value of the X variable is associated with a decrease in the value of the Y variable; and a positive sign suggests that an increase in the value of the X variable is associated with an increase in the value of the Y variable. If these signs are consistent with expectation or engineering judgement, then the model results are considered *intuitive*.

Step 11 Model Validation

The purpose of all models is to increase our understanding of the civil engineering system. As such, the validity of a model is adjudged on how closely it fits to empirical observations and how well it extrapolates to situations or data other than those originally used in the model. A common validation technique is to substitute the values of the independent variables from a validation data set (a set of observations that is external to the calibration data set in time or space) into the calibrated model and determine the corresponding predictions of the response variable—this yields Y_{EST}. Then these values are compared with the actual observed values of the response for those independent variables (denoted by Y_{OBS}). The deviation for each validation observation (each row) is calculated, squared, summed up, divided by n (the number

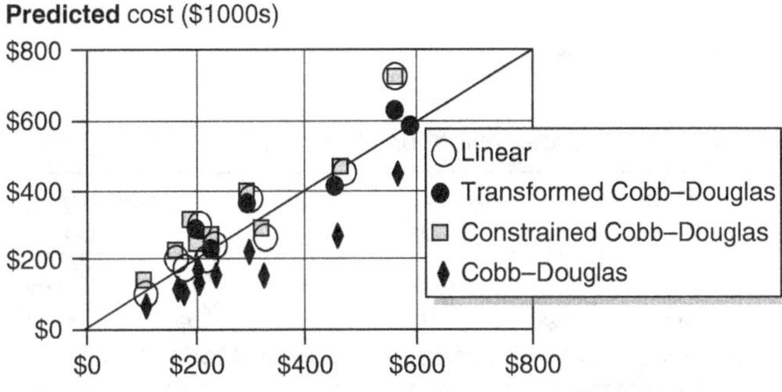

Figure 7.11 Example of validation plot (Adapted from Rodriguez et al., 2006).

of observations), and the square root is calculated to yield the root mean square error (RMSE).

$$\text{RMSE} = \sqrt{\frac{1}{n-1}\sum_{i=1}^{n}(\hat{y}_i - y_i)^2}$$

where n is the number of observations, $\hat{y}$ or y_{EST} is the estimated value of the response variable using the developed model, and y_i or y_{OBS} is the actual value of the response variable, for observation i in the validation data set.

Alternatively, the percentage of deviations (PD) of the estimated responses from the observed (actual) values of the response variable can be calculated as follows:

$$\text{PD} = 100 \times \frac{(\hat{y}_i - y_i)^2}{y_i}$$

where the symbols have the meanings as shown for RMSE.

Validation can also be carried out by preparing validation plots. In the bridge superstructure replacement cost example shown in Figure 7.11, the levels of closeness of four different functional forms are compared with the actual values (represented by the diagonal line, or the "100% validation line"). In Figure 7.11, it is seen that the transformed Cobb–Douglas function (represented by the circles) yields predicted values that are the closest to the observed values and thus appears to be the best model.

Example 7.3

Two engineers separately developed empirical models to predict the user perception of a certain type of civil engineering system using an index ranging from 0.0 (poor) to 5.0 (excellent), as a function of the number of explanatory variables that represent certain system attributes. In order to test their models in a later year, a third engineer sampled 10 similar systems and collected data on the explanatory factors (the vector of X variables) and actual user perceptions (the response variable) of each system. Then, each of the two models was used to estimate the expected user perception for each system. Their results are provided below. Determine which model provides a better description of user perception.

OBS #	A	B	C	D	E	F	G	H	I	J
Y Obs	4.133	2.122	0.499	3.962	3.904	4.812	0.100	6.487	0.699	3.151
Model 1	4.046	0.975	0.163	4.453	3.234	4.630	0.715	6.500	0.512	3.042
Model 2	4.86	0.854	2.541	5.524	4.002	5.214	0.002	4.251	0.468	3.254

Solution

The calculation procedure is provided in Table 7.4.

Table 7.4 Example Showing the Validation Procedure

		Estimated Values of Y		Deviation		Square of Deviation		Percent Deviation	
		Y_{1EST}	Y_{2EST}	$Y_{1EST} - Y_{OBS}$	$Y_{2EST} - Y_{OBS}$	$(Y_{1EST} - Y_{OBS})^2$	$(Y_{2EST} - Y_{OBS})^2$	$100[(Y_{1EST} - Y_{OBS})/Y_{OBS}]$	$100[(Y_{2EST} - Y_{OBS})/Y_{OBS}]$
OBS #	Y_{Obs}	Model 1	Model 2	Model 1	Model 2	Model 1	Model 2	Model 1	Model 2
A	4.133	4.046	4.86	−0.087	0.727	0.008	0.529	0.183	12.788
B	2.122	0.975	0.854	−1.147	−1.268	1.316	1.608	61.999	75.769
C	0.499	0.163	2.541	−0.336	2.042	0.113	4.170	22.624	835.624
D	3.962	4.453	5.524	0.491	1.562	0.241	2.440	6.085	61.581
E	3.904	3.234	4.002	−0.67	0.098	0.449	0.010	11.498	0.246
F	4.812	4.63	5.214	−0.182	0.402	0.033	0.162	0.688	3.358
G	0.100	0.715	0.002	0.615	−0.098	0.378	0.010	378.225	9.604
H	6.487	6.5	4.251	0.013	−2.236	0.000	5.000	0.003	77.073
I	0.699	0.512	0.468	−0.187	−0.231	0.035	0.053	5.003	7.634
J	3.151	3.042	3.254	−0.109	0.103	0.012	0.011	0.377	0.337
					SUM	0.012	0.045	2.587	13.208
					RMSE	0.536	1.247		

From the table, it is seen that:

$$\text{RSME} = [\text{SUM}/N - 1]0.5 = [0.015/(10 - 1)]0.5 = 0.536 \text{ for model 1 and } 1.247 \text{ for model 2.}$$

Also, the sum of percent deviations is calculated as 2.587 and 13.208 for models 1 and 2, respectively. Compared to model 2, model 1 provides outcomes that have lower deviations from the ground truth and thus is superior to model 2.

7.1.2 Sources of Error in Systems Modeling and Suggested Precautions

Errors occur due to uncertainties in the civil engineering systems management environment, such as material imperfections, variability in workmanship (often surrogated by contractor class), climate/weather variations, economic uncertainties, equipment error, and human error or incompetence. Model error could also be caused by misspecifications for example, omitting some key factors. A model may not be truly complete until it adequately incorporates all relevant factors as well as the interactions between/among them. An overzealous modeler may be tempted to include a large number of parameters in a bid to develop a comprehensive model. Even if data were available for such a venture, it is important that requisite care be taken to avoid the duplication of factor effects by choosing between those that have a similar effect; for example, where specific materials is often used for constructing systems of specific sizes, using both the size and material type as explanatory factors in the model may be problematic. In order to simplify the model development process, it may be appropriate to make certain simplifying assumptions, to omit certain duplicate factors, or to consider only the aggregated effects of certain factors.

7.2 MODEL SPECIFICATIONS IN STATISTICAL MODELING

As discussed in step 6 of Section 7.1, the specification of a statistical model is probably the most important step of the statistical model development process. There is a wide variety of specifications for a statistical model. This section provides the tools needed to develop models in each specification category.

7.2.1 Linear Regression

Due to its relatively simple mathematical structure and ease of calibration and interpretation, the linear model is widely used in describing civil engineering systems and their attributes. Washington et al. (2010) advised that, for this reason, linear models serve as a good starting point for modeling but should not be applied when other specifications are more suitable. In this section, we discuss how to estimate linear models, their underlying assumptions, and interpretation of their results. The first assumption is that the response or outcome is a continuous variable. The second assumption, which often is a debilitating one, is implicit from the name of this specification: There should exist a linear-in-parameters relationship between the response variable and each explanatory variable. This assumption can be obviated in cases where it is possible to transform the response variable or the explanatory variables or both. In such cases, even though the underlying relationships are nonlinear, the model is linear. Third, the observations should be independently and randomly sampled. Fourth, there should be an accounting for uncertain aspects of the relationship between the response and explanatory variables: such accounting is done using an *error term* (also called a *disturbance* or stochastic term). Specifically, this term accounts for measurement errors, omitted variables, and inherent random variations in the system attribute under investigation. Fifth, the error term should be independent of the explanatory variables, should have an expected value of zero, and should be independent across observations (i.e., they should not be autocorrelated). Sixth, the explanatory variables should be exogenous (rather than endogenous), that is, their values should be determined by influences that are external to the model; and for this to happen, the explanatory variable should not be correlated with the error term. Finally, the error terms should follow a distribution that is normally distributed (at least approximately). Mathematically, these assumptions can be expressed as shown in Table 7.5. When the engineer is faced with situations where these assumptions are violated, the recourse is to undertake remedial measures and proceed with the linear regression modeling or to adopt alternative techniques that are described in Section 7.2.2.

Table 7.5 Assumptions of Ordinary Least Squares Regression Model—Summary

Assumption or Property	Mathematical Expression
Functional form	$Y_i = \beta_0 + \beta_1 X_{1i} + e_i$
Zero mean of error terms	$E[\varepsilon_i] = 0$
Homoscedasticity of error terms	$\text{VAR}[\varepsilon_i] = \sigma^2$
Nonautocorrelation of error terms	$\text{COV}[\varepsilon_i, \varepsilon_j] = 0 \text{ if } i \neq j$
Exogeneity of explanatory variables	$\text{COV}[X_i, \varepsilon_j] = 0 \text{ for } i,j$
Normal-like distribution of error terms	$[\varepsilon_i] \approx N(0, \sigma^2)$

Source: Washington et al., 2010.

Example 7.4

The table below shows the chloride concentration in soils measured by an environmental engineer at different locations and the corresponding average corrosion index observed for underground water pipes at those locations. Develop a linear model to describe the influence of chloride concentration on the corrosion of underground water pipes. Use your model to predict the expected corrosion of a pipe at a location that has a chloride content of 2.1 units.

Observation	1	2	3	4	5	6	7	8	9	10
Concentration (units), X	0.3	0.5	1.0	1.4	1.7	1.9	2.5	2.8	3.2	3.5
Corrosion Index (0–10 scale), Y	2.1	4.0	3.8	3.9	6.1	7.2	5.9	7.9	9.6	7.8

Solution

First, the normal distribution of the error term is established by plotting the histogram of the standard residuals. Shown as Figure 7.12*a*, this plot indicates some normality behavior even though it is not

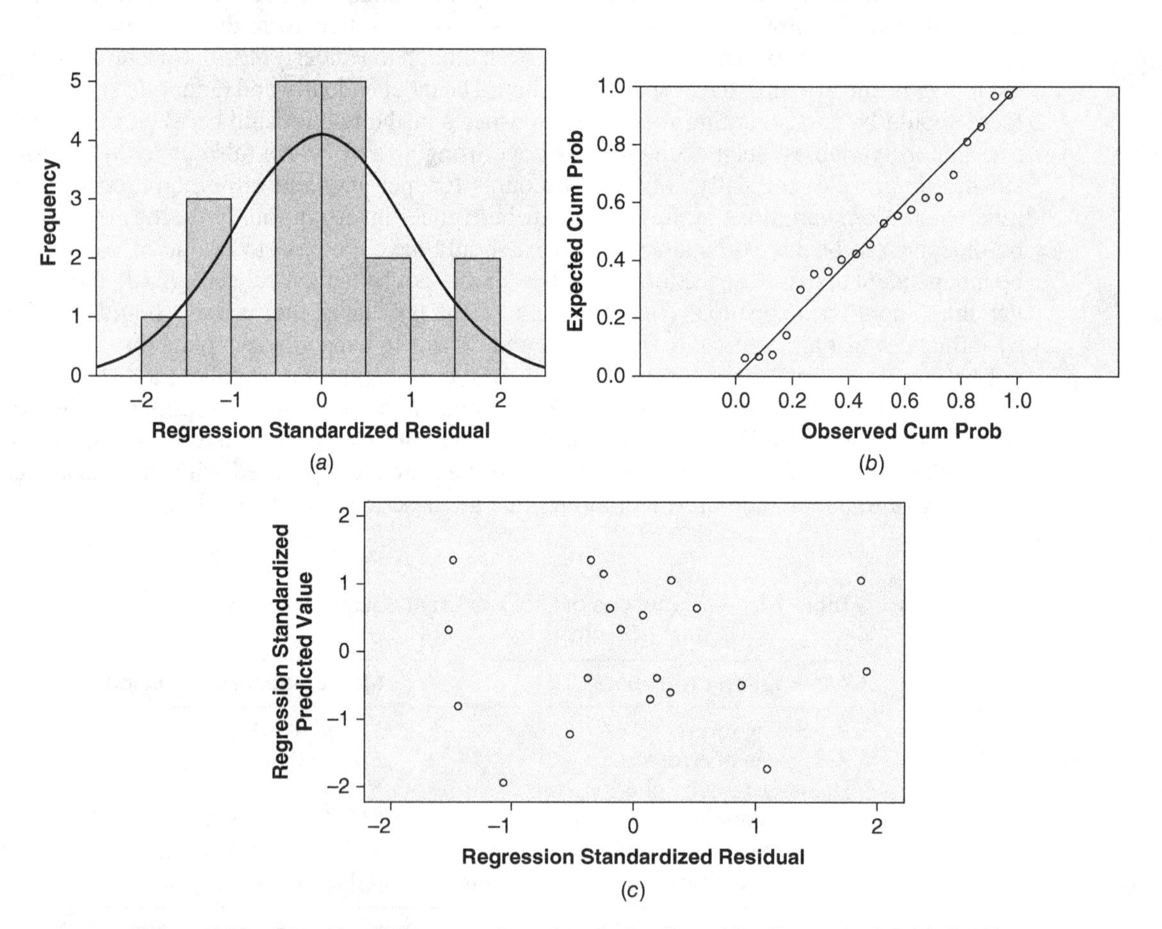

Figure 7.12 Checking the assumptions for application of the linear model: (*a*) distribution of error term, (*b*) normal *P-P* plot of regression standardized residual, and (*c*) scatter plot of predicted value and residual.

Model Summary[b]

Model	R	R Square	Adjusted R Square	Std. Error of the Estimate
1	.935[a]	.875	.868	.7053

a. Predictors: (Constant), X;
b. Dependent Variable: Y

ANOVA[b]

Model		Sum of Squares	df	Mean Square	F	Sig.
1	Regression	62.432	1	62.432	125.519	.000[a]
	Residual	8.953	18	.497		
	Total	71.386	19			

a. Predictors: (Constant), X
b. Dependent Variable: Y

Coefficients[a]

Model		Unstandardized Coefficients		Standardized Coefficients		
		B	Std. Error	Beta	t	Sig.
1	(Constant)	2.294	.396		5.790	.000
	X	1.872	.167	.935	11.204	.000

a. Dependent Variable: Y

Residuals Statistics[a]

	Minimum	Maximum	Mean	Std. Deviation	N
Predicted Value	2.855	8.845	6.365	1.8127	20
Residual	−1.0734	1.3498	.0000	.6865	20
Std. Predicted Value	−1.936	1.368	.000	1.000	20
Std. Residual	−1.522	1.914	.000	.973	20

a. Dependent Variable: Y

Figure 7.13 Model estimation results: SPSS software output.

perfect. Then, in Figure 7.12*b*, the plot of the expected and observed values is shown; and it is seen that all observations are within the bandwidth of 2 units. To check the homoscedasticity of the error terms, a scatter plot of predicted values and residuals is then plotted (Figure 7.12*c*), which shows that all the observations are symmetric. Thus, all the assumptions for the linear regression are valid in this data. The plot of the observations, the best-fit line, and the resulting model are given below. Figure 7.13 presents the results of model estimation as an output of a standard statistical software package (SPSS, 2011). Figure 7.14 shows how the developed model could be presented graphically.

Transforming Nonlinear to Linear Patterns. In most instances in systems engineering, the relationship between the response and explanatory variable follows a nonlinear trend. Fortunately, it is possible to linearize certain nonlinear relationships by manipulating the response or explanatory variables. Nonlinear relationships for which such transformations are possible are referred to as

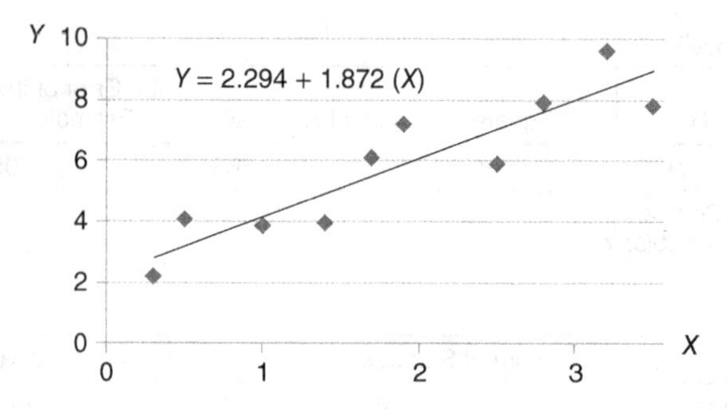

Figure 7.14 Model estimation results: graphical representation of the developed model.

intrinsically linear relationships. Examples of such manipulation could include taking the natural log, squaring, or reciprocal of the variable.

Example 7.5

The table below, which illustrates the task of modeling at the system preservation phase, shows the increase in the condition of a system in response to the maintenance effort (man-hours/ft^2 of surface area). Develop a model that describes the effectiveness of system maintenance efforts in the given units.

Effort (Man-hours/ft^2), X	**2.24**	**1.41**	**1.73**	**2.83**	**2.24**	**2.45**	**2.65**	**2.00**	**3.00**	**2.83**
Effectiveness (increase in system condition), Y	8.36	3.67	5.75	20.45	8.61	10.88	19.93	4.55	20.05	37.63

Solution

Assume that we intend to build a linear model. We first explore the trend by developing a plot of Y versus X. This is shown in Figure 7.15*a*. Then, without finding a suitable linear trend, we transform the Y and X variables in a variety of ways and examine the resulting graphs for any patterns of linearity. The plots suggest that the natural log transformation of the response variable, together with the square transformation of the explanatory variable, provides the best fit to the data and seems to be the best model. This model is

$$\ln \text{(Maintenance effectiveness) or } \ln \text{(Increase in system condition)}$$
$$= 0.3052(\text{Effort}) + 0.6467 \ (R^2 = 0.87).$$

Interactions between Explanatory Variables. In modeling the behavior of a civil engineering system, there is often a combined influence of two or more factors so that their synergistic effect is greater or less than the sum of their individual effects. Figure 7.16 illustrates the different extents of the interaction effects. In one extreme situation, there is no interaction effect; and in the other extreme, the explanatory variables have no effect individually and there is only the interaction effect. In between these two extremes, the interacting explanatory variables maintain their initial effects to a varying degree and the interaction effect may be smaller or greater than the effect of each individual variable.

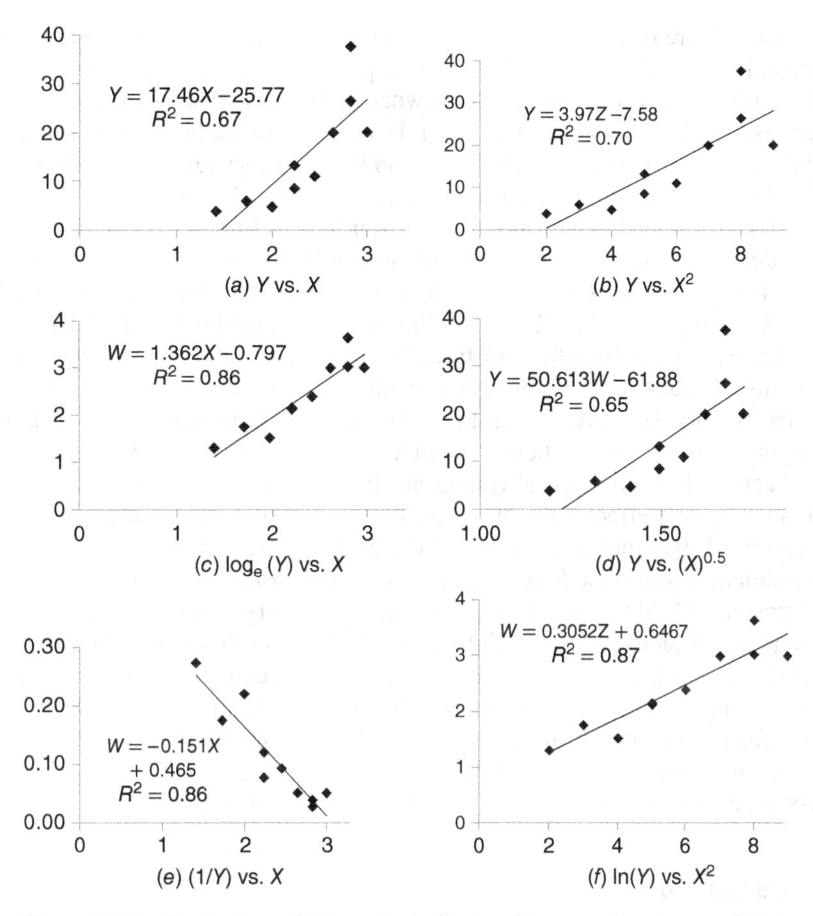

Figure 7.15 Illustration of interaction effects of two explanatory variables.

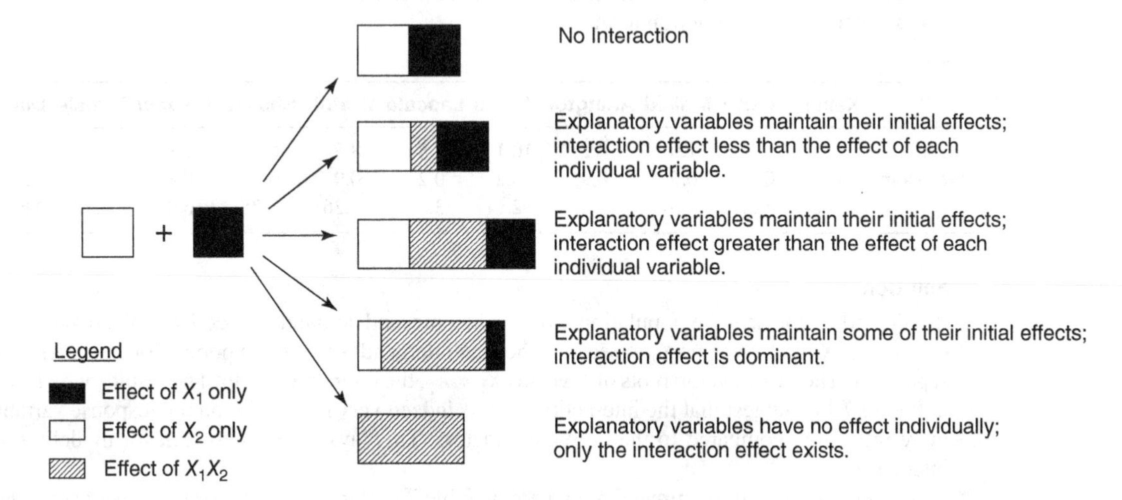

Figure 7.16 Illustration of interaction effects of two explanatory variables.

When there is no interaction effect, then the model is said to consist of only *first-order effects*. A second-order interaction is at least one pair of variables that interact (represented as the term X_1X_2); and a third-order interaction is when at least one trio of variables exist that interact (represented as $X_1X_2X_3$). Generally, the lower the order of interaction in a model, the greater the influence of the constituent variables individually on the response variable (Washington et al., 2010). Also, higher order interactions are included only when their lower order counterparts are also included unless the latter have been found to be insignificant. In other words, the term X_1X_2 is included in the model only when X_1 and X_2 are included, unless X_1 and X_2 are found insignificant individually. For the first four cases where the effects of individual variables exist, these effects may be the same or one may be greater than the other (this is not indicated in Figure 7.16).

An example is the effect of truck loading and rainfall on pavement condition: Where there is no rainfall, truck loads damage pavements; also, where there are no trucks, severe rainfall damages pavements. However, when these two factors occur together, their combined effect is more damaging than the sum of their individual effects (Sinha et al., 1984).

Factors that influence a system attribute often exhibit interactive effects when the factors have a catalyzing effect on each other. In other words, the presence of one factor influences the propensity of the other to have an effect that has more or less on the response variable compared to the situation where the first factor is absent. For example, in a model to describe the progression of corrosion of bridge deck reinforcement, explanatory variables could include the competence of the wearing surface degree of reinforcement protection (coating), the extent of deck cracking, and the extent of deicing salt application. Naturally, in a scenario where only one of these factors exists (poor coating, extensive cracking, or deicing salts), relatively little corrosion would be expected. Their effects individually thus may be minimal. However, for a deck that has any two of these factors, some corrosion can be expected; and for bridge decks where all three factors occur, severe corrosion of the bridge deck reinforcement is almost guaranteed!

Example 7.6

The table below shows the demand for a civil engineering system at the needs assessment phase. The data shows the population (in 100,000s) of different cities in a country and their average household income ($100s per year) and the demand for water. Develop a model separately with and without an interaction term between population and household income and ascertain whether an interaction effect exists. Interpret the interaction term.

City	Xanjin	Puerto	Maleki	Arlington	Duhai	Lascala	Vostok	Shiziko	Kwanju	Tamale	Dia	Xarano
Population	3.3	5.1	11.0	4.2	10.1	13.8	8.8	6.2	7.8	9.1	2.5	12.1
Average income	0.1	0.5	0.6	0.9	0.2	0.2	0.9	0.5	0.8	0.1	0.3	0.6
Demand	0.48	3.20	6.61	3.89	2.11	3.38	7.26	3.75	6.03	1.40	0.61	7.95

Solution

X_1, X_2, and Y represent the population, average income, and demand, respectively. We first develop plots of Y versus X_1 and Y versus X_2 to see how the water demand is related to population and average income separately. Then we develop plots of Y versus $X_1 * X_2$, the interaction term. The results, which are shown in Figure 7.17, suggest that the interaction term is indeed very influential on the response variable, even more influential compared to the individual factors. This lays the groundwork for modeling using an interaction term.

The results of the estimated model are in Table 7.6. For zero autocorrelation between x variables, the Durbin–Watson statistic value should be close to 2.0. If the value is substantially less or greater

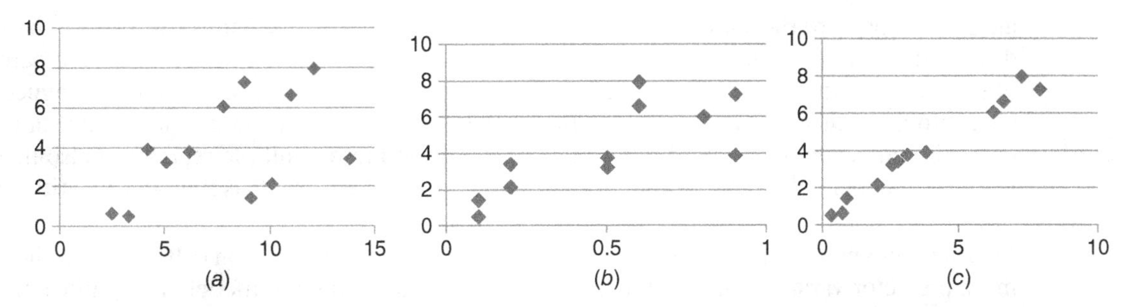

Figure 7.17 Relationship between demand, population and average income, and interaction effects: (*a*) demand (Y) vs. population (X_1), (*b*) demand (Y) vs. average income (X_2), and (*c*) demand (Y) vs. population and average income ($X_1 * X_2$).

Table 7.6 Model Estimation Results

Variable	Coefficient	*t*-stat	R^2	R^2 (adjusted)	Durbin–Watson statistic*
Model I (using X_1 and (X_1X_2) variables)					
Intercept	−0.0575	0.11	0.981	0.974	1.6
X_1	0.07307	1.15			
X_2	0.371	0.32			
X_1X_2	0.8678	6.03			
Model II (using X_1 and (X_1X_2) variables)					
Intercept	0.0862	0.30	0.981	0.976	1.5
X_1	0.05686	1.52			
X_1X_2	0.91097	18.01			
Model III (using X_1 and X_2 variables)					
Intercept	−2.4357	−3.01	0.894	0.87	2.4
X_1	0.39546	5.13			
X_2	6.7927	7.20			

than 2, then the successive error terms are indicative of serial correlation, which could lead to underestimation of the statistical significance of the variables. The issue of autocorrelation is discussed in a subsequent section.

Concluding from the sample size and based on all the statistic tests shown in Table 7.6. Model II appears to be the best result. Thus, the model for describing the demand is

$$\text{Demand} = 0.0862 + 0.05686\ (\text{Population}) + 0.91097\ (\text{Population} \times \text{Income})$$

7.2.2 Multiple Regression

What a systems engineer seeks to predict or describe is often governed by more than just one factor, and as we have seen in the examples discussed in the previous section, simple (single explanatory variable) regression may not be sufficient in building a good model. Therefore, the use of multiple explanatory variables to explain some response is often beneficial. For example, the travel delay along an urban arterial may be influenced not just by the traffic volume but also by the time of day (peak or nonpeak hours), the number of lanes, the signal timing scheme, and the percentage of trucks. Consequently, while multiple regression offers some promise, this tool is plagued by the same problems that potentially plague simple regression, such as nonconstant variance of the error terms, outliers, and nonnormality in error terms. Furthermore, multiple regression also inherently leaves the model vulnerable to a number of statistical problems such as multicollinearity.

Multicollinearity. Also referred to as intercorrelation, multicorrelation (MC) occurs when two or more predictor variables in (or omitted from) a multiple regression model are significantly correlated. While the predictive power or reliability of the model is not often compromised by MC, this problem can weaken the efficacy of the model by misdiagnosing the effect of an individual explanatory variable (e.g., counterintuitive signs). MC is typically encountered in observational data where the analyst lacks the ability to control the variables of interest (Washington et al., 2010). In the case of perfect MC, one of the two correlated variables provides no additional predictive capability over its counterpart. The presence of multicollinearity in a model can be tested by (i) adding or deleting an explanatory variable in/from the model and examining the magnitude of the resulting change in the model coefficient estimates—a large change suggests the presence of MC; and (ii) identifying the explanatory variable deemed insignificant in the regression process and carrying out a F test for their significance as a group—a rejection of the null hypothesis would mean that these variables as a group are significant and thus MC is present. A more formal test is the *variance inflation factor* (VIF):

$$\text{VIF} = 1/(1 - R^2)$$

Multicollinearity is generally considered to be present when the VIF exceeds 6.66.

If the objective of the analysis is merely to predict the response variable Y from a set of explanatory variables, then multicollinearity is not a problem, as the predictions would still be accurate. If, on the other hand, the objective is diagnostic (i.e., to understand how the various X variables impact Y), then multicollinearity poses a formidable problem in that the individual P values can be misleading (the model results could show a high P value for a variable even though the variable is important). Also, the confidence intervals on the model regression coefficients will be very wide (in some cases, the confidence intervals may even include zero, which means one cannot be confident whether a change in the X value is associated with a change in Y). Another consequence of the wide confidence interval is that adding a new variable (or excluding an existing variable from the model) could cause drastic changes in the magnitude, and often even the signs, of the variable coefficients.

Consequences of Multicollinearity.

1. In regression analysis, a regression coefficient is interpreted as a statistic that estimates the impact of a unit change in an independent variable (at given levels of the other variables) on the response variable Y. However, if an independent variable, say X_1 is highly correlated with another independent variable, say X_2, then in the given dataset there exists only observations for which X_1 and X_2 have a particular relationship (i.e., there are no observations for which X_1 changes independently of X_2). In such cases, we lack a precise estimate of the effect of truly ***independent*** changes in X_1. Research has shown that models plagued with MC provide a less reliable estimate of the impact of any one explanatory variable X on the response variable Y while keeping all others constant compared to the situation where explanatory variables are uncorrelated with one another.

2. In testing for the significance of variables, we test whether their coefficients are equal to zero (rejecting the null suggests that the variable is significant). A greater standard error of a variable translates into a greater likelihood of failing to reject the null. When MC is present, the standard errors of the coefficients of the offending variables tend to be large, thus increasing the likelihood that we will fail to reject the null hypothesis (and thus falsely conclude that the variable is not significant, namely, that there is no linear relationship between the independent and dependent variables).
3. Using a different data set would yield very different modeling results. In other words, when MC is present, the estimated regression parameters have large sampling variability and thus the estimated model coefficients will vary widely from one sample to the next (Washington et al., 2010). Furthermore, adding or deleting one of the offending variables could lead to a change the magnitude and even the signs of the model coefficients of the remaining variables.

Addressing Multicollinearity. As clearly seen above, the existence of MC is indeed a serious problem. Unfortunately, most of the data available in civil systems engineering are observational data which tend to be prone to MC. It is therefore important to know there are steps to take, such as the following, to address this problem if it is found to exist.

1. Drop one of the offending explanatory variables. While this is a logical step, doing so may lead to a loss of information. The danger is that omission of an important variable could result in coefficient estimates (for the remaining explanatory variables) that are biased.
2. Collect additional data. Increasing the sample size by collecting additional data, particularly over a wider range of the offending variables, would yield narrower confidence intervals and parameter estimates that are more precise and have lower standard errors.
3. Modify the model specification. The modeler could include an interaction term (the product or ratio of the two offending variables). The interaction term could replace the affected variables entirely or could be used alongside with one or both of them. For example, if bridge length and width are collinear independent variables in a model, then it might be worthwhile to remove one or both of them and use deck area (the product of the length and width) instead. Another example is highway pavement thickness and traffic volume: often observations with higher traffic volumes also have higher pavement thickness. Using both of these variables in seperate terms of a model may yield unintuitive coefficient estimates. A variable representing the ratio (traffic/thickness) may yield more intuitive coefficient signs.
4. "Center" the offending variables. To do this, compute the mean of each affected explanatory variable, and then replace each value with the difference between it and the mean. For example, if the variable is temperature and the mean is 50, then in the data set, replace 45 with −5 and 62 with 12. While this has no mathematical effect on the regression model results, it could help address problems including those associated with rounding.
5. Carry out ridge regression. This is a statistical tool that produces biased but efficient estimators for the model.
6. Leave the model as is. This is a recommended approach where the objective of the analysis is predictive rather than diagnostic. As discussed earlier, MC (while having adverse effects on the interpretation of the explanatory variable coefficients) does not impair the reliability of the forecast provided that the correlated explanatory variables exhibit a similar pattern of multicollinearity as the data that were used to develop the regression model.
7. The best way to deal with MC is at the sampling design phase of the modeling process. Even for an observational study, the data collection strategy could be designed to control

the levels of the suspected offending variables. For example, a suspected high correlation between the variables of pavement thickness and traffic loading could be addressed by ensuring that there are adequate pavement sections in all stratified combinations of thin, moderate, and thick pavements and of low, moderate, and high-traffic volumes are present in the data set.

7.2.3 Cross-Sectional versus Time Series Models

In Chapter 4, we discussed the conceptual differences between time series, cross-sectional, and panel models. In this chapter, we provide specific tools for developing these models.

(a) Cross-Sectional Models. These models estimate a response (dependent variable) at a given point in time on the basis of the values of the explanatory variables that are associated with that same point in time. Thus, the general form of a cross-sectional model is

$$Y_t = f\,(X_{1t}, X_{2t}, \ldots, X_{kt})$$

For example, the extent of usage of a civil engineering system could be modeled as a function of the population, gross domestic product, and minimum wage in the area that is served by the system:

$$D_{2010} = f(\mathrm{POP}_{2010}, \mathrm{GDP}_{2010}, \mathrm{MINWAGE}_{2010})$$

where D_{2010} is the demand for the civil engineering system in year 2010; POP_{2010} is the population in year 2010; GDP is the gross domestic product in year 2010; and $\mathrm{MINWAGE}_{2010}$ is the minimum wage in year 2010.

(b) Time Series Models. A time series is a sequence of data points measured at successive time intervals. Time series models estimate a response (dependent variable) for a given point in time, given past values of the same variable at previous points in time:

$$Y_t = f\,(Y_{t-1}, Y_{t-2}, \ldots, Y_{t-k})$$

For example, the extent of usage of a civil engineering system could be modeled as a function of the past demands at previous years, the population, gross domestic product, and minimum wage in the area that is served by the system:

$$D_{2010} = f(\Delta_{2000}, \Delta_{2001}, \ldots \Delta_{2009})$$

Time series observations are found in every discipline and at every phase of civil engineering systems. Examples include the weekly level of construction progress, the annual flow volume of a surface water body, the surface roughness of a highway pavement, the monthly consumption of water in a growing city, and the number of fatigue cracks on a highly loaded urban bridge. Examples of time series model plots are shown in Figure 7.18. Autocorrelation, an issue often encountered in time series modeling, is discussed below.

Autocorrelation. Autocorrelation, also referred to as serial or lagged correlation, is a statement of the similarity between a given time series and its lagged version, lagged over one or more time periods (Washington et al., 2010) and can be expressed numerically as the correlation between the two data series. A value of +1 means perfect positive correlation, that is, an increase in an observation seen in one time series yields a proportionate increase in the corresponding observation in the other time series. Certain texts consider positive autocorrelation as a specific form of "persistence,"

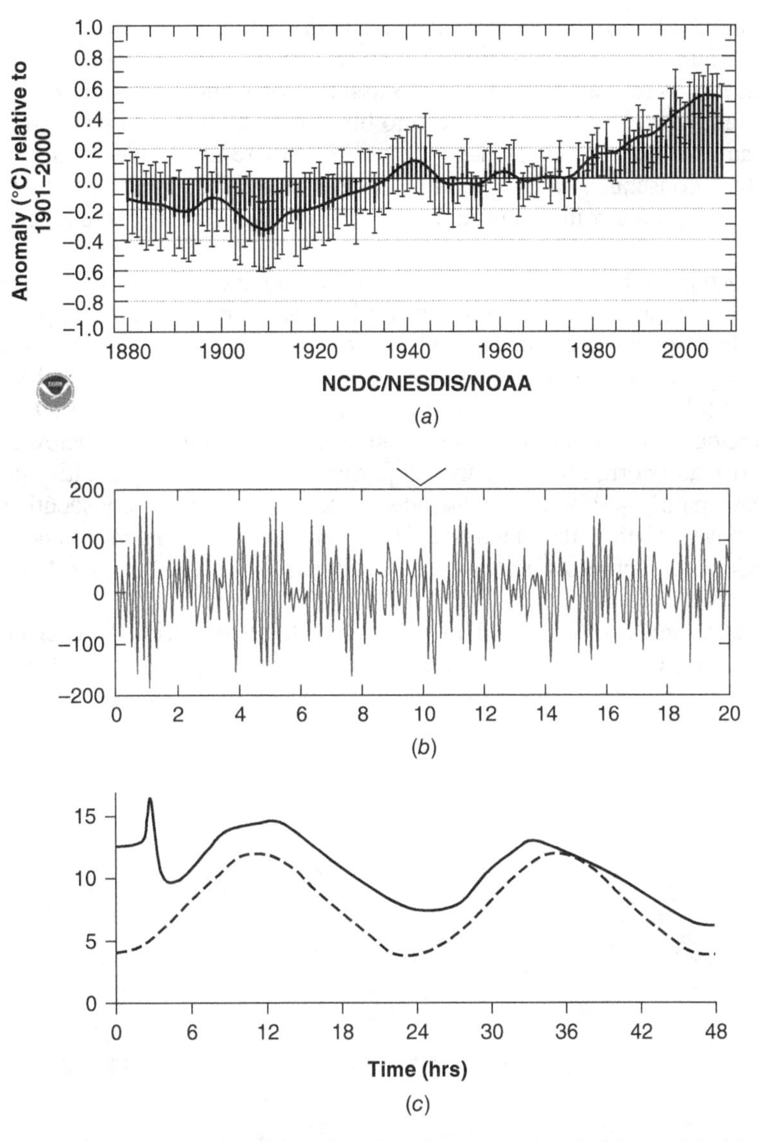

Figure 7.18 Examples of time series plots in civil engineering: (*a*) Mean global (land and ocean) temperature, 1880–2009 (*Source:* NOAA). (*b*) Seismic ground motion excitations under a high-rise building in Vancouver (Koduru and Haukaas, 2010). (*c*) Mortar and water bath temperatures (measuring temperature effects on concrete setting behavior (Wade et al., 2010).

that is, a tendency for a system (from the perspective of the attribute whose data is being modeled) to remain in the same state from across consecutive observations. For example, the likelihood that next year's deterioration of a system will be severe is greater if this year's deterioration is severe. Similarly, a value of −1 means a perfect negative correlation.

Autocorrelation thus manifests the similarity between observations of a systems attribute as a function of the time intervals between them. Also, such cross correlation of a time-varying observation (of a systems attribute) with its preceding or anteceding observations (referred to as autocovariance) can pose a serious limitation to the development of time series models because its existence leads to the violation of OLS assumption of noncorrelation of the error terms. The presence of autocorrelation does not lead to bias in the OLS coefficient estimates; however, when the errors autocorrelations at low lags are positive, there is a tendency to underestimate the standard errors (and consequently, the t scores). To assess the autocorrelation of a time series, at least one of the following tools could be used: the autocorrelation function, the lagged scatter plot, or the time series plot.

The time series plot. To assess autocorrelation, the "departures" or deviations from the "mean" line (a horizontal line on the time series plot that is drawn at the sample mean) are determined. A positively autocorrelated series is where positive deviations from the mean line are generally followed by positive deviations from the mean; or where negative deviations from the mean line are generally followed by negative deviations as illustrated in Figure 7.19 where the trend is seen as wide stretches of consecutive observations that are either above or below the mean line. In contrast, in negative autocorrelation, negative departures are followed by positive departures, and vice versa; these are manifested as few incidences of long stretches of consecutive observations occurring either above or below the mean line. However, because visual assessments of time series autocorrelations from such plots are inherently subjective (LTRR, 2011), other less subjective statistical techniques are recommended.

The lagged scatter plot. This is a plot of the time series data offset in time by one or more time lags (Figure 7.19). Consider a time series of length N with observations x at each period $i(= 1,\ldots,N)$. Generally, the plot of the last $N - k$ observations against the first $N - k$ observations

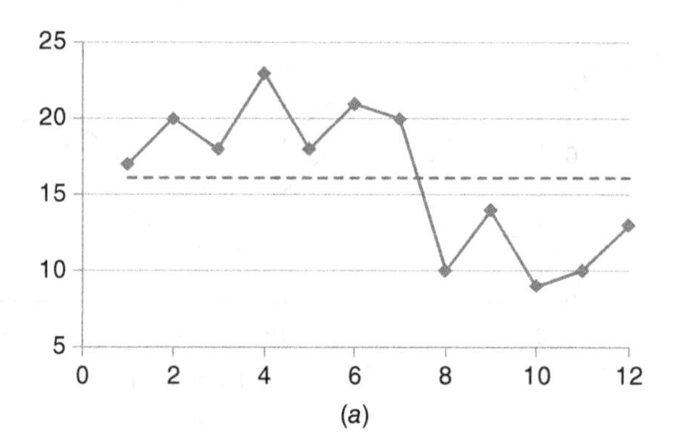

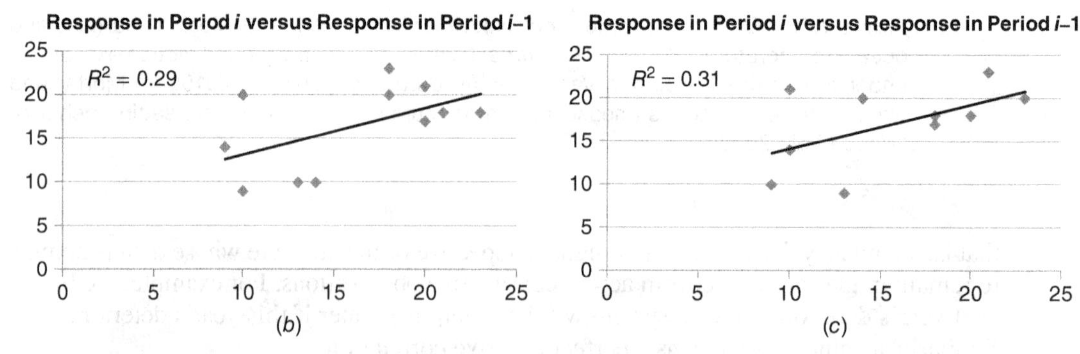

Figure 7.19 Looking for evidence of autocorrelation in time series data: (*a*) time series plot, 1 year lag ($k = 1$), (*b*) lagged scatter plot, 1 period lag ($k = 1$), and (c) lagged scatter plot, 2 period lag ($k = 2$).

is referred to as the scatter plot for lag k. For lag 1, for example, the observations $x_2, x_3, \ldots, x_N$ are plotted against observations $x_1, x_2, \ldots, x_{N-1}$. If the points in the resulting plot are randomly scattered and thus yield a zero correlation coefficient, that suggests a lack of autocorrelation, which would imply that the value of an observation at time t is independent of the value of the observation at other times (LTRR, 2011). On the other hand, a perfectly correlated scatter would yield a correlation coefficient of 1.0. In the lagged scatter plot, if the data cloud is aligned from upper left to lower right, then there is reason to believe that negative autocorrelation exists in the data; and if the data is aligned from lower left to upper right, then positive autocorrelation may exist.

The scatter plot's correlation coefficient is a reflection of the degree to which the two sets of observations are related. For a given lag width, it may be hypothesized that the population from which the sample is taken has zero correlation; to test this hypothesis, the calculated correlation coefficient is compared with the critical level of correlation. For a completely random time series and large sample size, the correlation coefficient between the two observations at the given lag is considered to be approximately normally distributed with mean 0 and variance $1/N$; thus, the threshold level of correlation at 95% significance is $r_{0.95} \cong 0 \pm 2/N^{0.5}$, where N is the sample size (Chatfield, 2004; LTRR, 2011).

The Durbin–Watson statistic. The common test to detect whether a first-order autocorrelation exists is to plot the residuals (from the model) against the original regressors and then against the k lagged values of the residuals. k represents the order of the test. In its simplest form, the test statistic for this regression is given as R^2T (where R^2 is the coefficient of determination and T is the sample size). For a null hypothesis that autocorrelation does not exist, the statistic used for the analysis is asymptotically χ^2 distributed with k degrees of freedom. To illustrate this concept, consider the following table that describes the progress of the construction phase for a pylon network, the work progress is monitored closely by the project manager, and the system owner examines the patterns of progress. In a certain year, the monthly progress (in numbers constructed) of a pylon network is provided in the table below. Determine whether autocorrelation exists in the data. The solution is provided in Table 7.7.

Month	**Jan**	**Feb**	**Mar**	**Apr**	**May**	**Jun**	**Jul**	**Aug**	**Sep**	**Oct**	**Nov**	**Dec**
Progress	17	20	18	23	18	21	20	10	14	9	10	13

Table 7.7 Collation of Autocorrelation Plot Directions

Transition Point	**Direction of Departure with Respect to Mean**	**Direction of Departure with Respect to the Mean in the Following Month**	**Type of Autocorrelation**
January–February	Away	To	Negative
February–March	To	Away	Negative
March–April	Away	To	Negative
April–May	To	Away	Negative
May–June	Away	To	Negative
June–July	To	None	—
July–August	To, away	To	—
August–September	To	Away	Negative
September–October	Away	To	Negative
October–November	To	To	Negative
November–December	To	Not applicable	—

The plots suggest that the data is dominantly negatively autocorrelated.

7.3 SOME IMPORTANT ISSUES IN STATISTICAL MODELING

7.3.1 Addressing the Problem of Statistical Outliers

An outlier is an observation whose position is significantly distant from that of most observations in a given sample. Outlying observations can be indicative of faulty data or erroneous procedures. Where the observations are simulated from a theoretical relationship, they may even suggest that the underlying theory may not be valid under certain circumstances. Outliers may be due to changes in system behavior or its natural or anthropogenic environment, human error, or instrument error. In civil engineering systems, where data for analysis are mostly observational and often cover a wide swath of disciplines (economic, geological, climatic, human, engineering, etc.), it is common to encounter significant percentages of outlying observations in data. Outlier detection is useful in systems management because it can identify and draw attention to unexpected or overlooked system deficiencies or fraud before they develop into a state where they could have consequences that threaten the system, its users, or the environment.

(a) Identifying Outliers. Outlier detection techniques have been used for several decades to detect and to duly remove outlying observations from data sets (Cook, 1977). However, outlier identification is subjective as there does not seem to be any universally accepted mathematical criteria for defining which observations can be considered regular and which are outliers. At the current time, there are at least three outlier identification approaches (Hodge and Austin, 2004):

Approach 1: In this group of techniques, the outliers are determined without any prior knowledge of the trends of the observations. A learning approach that may be characterized as *unsupervised clustering*, techniques that use this approach first process the observations as a static distribution, and then pinpoint the most remote observations, flagging them as potential outliers.

Approach 2: This group of techniques, which identify outliers on the basis of normality as well as abnormality, are characterized as *supervised classification* because they require that each observation in the data is pretagged as "normal" or "abnormal" data.

Approach 3: These techniques model only (or mostly) normality and are similar to semisupervised detection.

Most model-based outlier identification techniques assume that the observations are normally distributed, and they identify observations that the technique considers to be "remote" on the basis of the distribution standard deviation and the distance of the observation from the distribution mean. Common outlier identification techniques include the Peirce, Chauvenet, and Grubbs tests.

Chauvenet's Criterion. In this test, we start by calculating the mean and standard deviation of the observations. Then using the statistical tables for the normal distribution function (see Appendix 2), we find the probability that a given observation will lie at the value of the offending observation. This is done on the basis of how much the offending observation differs from the mean. Then, we find the product of this probability and the number of observations, n. A result of less than 0.5 suggests that the offending data point is an outlier; otherwise, it is accepted as a nonoffending observation. In other words, if for a given observation, the probability that we obtain a certain deviation from the mean is less than $1/(2n)$, then the observation is designated as a true outlier.

Example 7.7 [adapted from Dol and Verhoog (2010)]

The results of a stress experiment yielded the following measurements: 1000, 5000, 1000, 900, 1100, and 1000 N/mm^2. Is 5000 an outlying observation?

Solution

The mean is 1670 N/mm^2 and the standard deviation is 1634 N/mm^2; and 5000 N/mm^2 differs from 1670 N/mm^2 by 3330 N/mm^2, which slightly exceeds two standard deviations from the mean. The probability of having observations that exceed two standard deviations from the mean is roughly 5%. There are six observations; therefore, the value of the statistic (i.e., is, the product of the number of observations and the probability) is $0.05 \times 6 = 0.3$, which is less than 0.5. Thus, consistent with Chauvenet's criterion, the measured value of 5000 N/mm^2 is an outlier.

Grubbs' Criterion. This test detects the outliers one at a time. In the test, we first certify that the data are approximately normally distributed. After an observation is identified as an outlier, it is deleted from the modeling data set. The test is iterated until all observations have been tested. It has been cautioned that the probabilities of detection may change after each iteration; as such, the Grubbs test is not recommended for samples with fewer than six observations as it would likely "identify" most points in a small data set as outliers (Dol and Verhoog, 2010). The detection process is presented below.

A one-sided or two-sided test can be used for the Grubbs test. The test statistic, G^*, which is the maximum absolute deviation from the sample mean, in terms of the number of standard deviations, is written as

$$G = \frac{\max_{i=1,\ldots,N} |Y_i - \overline{Y}|}{s}$$

where $\overline{Y}$ and s denote the sample mean and standard deviation, respectively.

The critical value of G at significance level α, G_C, is

$$G > \frac{N-1}{\sqrt{N}} \sqrt{\frac{t^2_{\alpha/(2N),N-2}}{N-2+t^2_{\alpha/(2N),N-2}}}$$

where $t_{\alpha/(2N),N-2}$ denotes the upper boundary of the t distribution with a significance level of $\alpha/(2N)$ and $N-2$ degrees of freedom. Note that we use $\alpha/(2N)$ instead of α/N if the test is one sided.

At each iteration, the hypothesis that no outliers exist in the remaining data set is rejected at significance level α if G^* exceeds G_C.

(b) Outliers: In-laws or Outlaws?

Outlier Retention. Outliers are seen by some modelers as a nuisance that should be deleted without any fuss. Others, taking a more cautious approach, argue that outlier deletion is only a self-serving attempt by the modeler to shape the data to desired patterns, and thus they consider outliers as important observations that should not be expunged but rather included in the data. Second, the presence of outliers could suggest that there is a problem with the system, particularly in safety-critical environments and where the modeler seeks to detect any natural or man-made threats to the system. It is often cautioned that, even when the data are normally distributed, it can be expected to encounter a fair percentage of outliers, particularly when the sample size is large. In certain cases, it is advised that alternative modeling techniques should be used where there are large numbers of outliers. In situations where we are able to examine the source of the data and identify exactly why some specific observations are outliers, it is possible to account for this effect in the model structure through the use of special modeling specifications (Hodge and Austin, 2004); however, if such observations are few, they may simply be deleted from the modeling data set as discussed below.

Outlier Deletion. There is a school of thought that is adamantly opposed to deletion of outliers. When data is scarce and data sets are small, and also where the preliminary plots are suggestive of nonnormality, outlier deletion could cause more harm than good. Thus, we can identify specific observations as outliers and reject them when the overall data is plentiful, the statistical distribution of the error of the observations is known, and there are mathematical grounds to identify outliers as such. In cases where we find that outliers occurred due to human or instrument error, or a specifically identified irregular situation (where the outlying cause has been identified), they should be deleted. Also, we should record all outlying observations that are excluded from the data analysis and mention their exclusion in the modeling report.

7.3.2 *Cum Hoc Ergo Propter Hoc:* Causation versus Correlation

The Latin phrase *cum hoc ergo propter hoc* is interpreted as "after this, therefore because of this." In other words, because event Y followed event X, event Y must have been caused by event X and thus suggests that any two events that occur together may have a cause-and-effect relationship. However, as we know in systems management and in science and statistics in general, correlation does not necessarily imply causation. The fact that two variables are correlated does not imply that one is a cause of the other. Thus, the conclusion that event Y must have been caused by event X is a logical fallacy because there could be one of at least three counterexplanations why they are highly correlated:

(a) Reverse Causation. It may be the case that correlation exists in the opposite direction.

Example: Statistical data on system failure occurrences: severity of failure (light, moderate, severe, catastrophic) (Y) and the number of times of system inspection in previous 5-year period (X).

Observation: There is a strong correlation between Y and X.

Fallacious conclusion: More system inspections causes a greater severity of system failure.

Counterexplanation: The strong correlation between the extent of system failure and the frequency of inspection does not imply that inspections cause severe failure. System agencies tend to inspect defective systems with greater frequency. As such, if there are a greater number of inspections, that could be because the system has physical condition problems, and thus a greater likelihood of system failure causes more inspections.

(b) Effect of the Lurking Variable. There may be a third unknown factor, Z (called a *lurking* or *common-causal* variable), that may actually be influencing both X and Y.

Example: Annual statistical data on extent of spring-induced pavement cracking (Y) and the extent of water pollution due to deicing salts (X).

Observation: There is a strong correlation between Y and X.

Fallacious conclusion: More saline water pollution leads to more pavement cracking in the spring season.

Counterexplanation: Pavement cracking is induced just after winter weather when the ice lenses beneath a pavement melt and leave a void under the pavement. Also, just after winter, the concentration of deicing salts in surface water bodies is often at their highest levels. Both of these events are caused by greater exposure to a third factor: cold temperatures.

(c) Coincidence. The seeming causation of one event by another may be purely coincidental.

Example: Statistical data on the number of endangered species (X) and Internet usage (Y).
Observation: There is a strong inverse correlation between Y and X.
Fallacious conclusion: People use the Internet more when there are fewer endangered species.

The fallacy in this example is the inappropriate conclusion that a causal relationship exists between Internet usage and the number of endangered species. Events X and Y clearly lack any logical relationship even though they occur simultaneously. For situations such as this, the connection is so remote that it may be considered coincidental.

Thus, if two given data items for a civil engineering system are found to be correlated, further analysis is necessary before a definite conclusion can be made about the existence of a cause-and-effect relationship. While correlation does not require causation, causation requires correlation. Even further, causation requires not only correlation but also a *counterfactual dependence*. For example, consider the failure of a levee, the cause of which was supposed as faulty design. If time travel were possible, then the best way to prove this supposition would be to go back in time and use a different design and then subject the levee to the same forces. Causation could then be ascertained by comparing the performance of the two designs: If the former failed but the latter did not, then the failure was really due to the design. However, we all know that time travel is not possible so such a comparison is not possible. Regardless of how carefully experiments are designed, we can only infer causation but never exactly ascertain that it exists; this is referred to as the *fundamental problem of causal inference* (Holland, 1986). In modeling civil engineering systems, engineers seek, as much as possible, to obtain the best possible representation of two states that differ only in one respect: the aspect under investigation; and the system outcome would be a reflection of the counterfactual dependence (Pearl, 2000). Let us consider another example: At the same highway location, an agency constructs two bridges (one with concrete casting on site and the other with precast concrete) and observed their life expectancies. After several decades of monitoring, their service lives are found to be significantly different, which is strong evidence that the concrete construction type had a causal effect on the bridge longevity. In this case, the correlation between material type and bridge longevity would most likely imply causation between the two variables. This example is for only two observations. In properly designed experimental studies, several observations are used for each class of objects. Within each class, subjects are chosen at random to avoid bias. In reality, conditions will not be perfectly identical. In the given example involving bridges, it is known that bridge structures have different conditions, such as the extent of deicing, traffic volume, quality of steel and concrete, and quality of construction workmanship, but could be placed in groups on the basis of known values of these attributes. In cases such as this, the bridges within a class are likely to exhibit similar behavior in all relevant aspects besides a specific variable under investigation. If the explanatory variable under investigation has a significant effect, then it can be concluded that the variable has a causal effect on the response variable. In an estimated model, this effect can be quantified in statistical terms such as the t statistic or the p value.

7.3.3 Limitations on the Response Variable

Linear regression proceeds with the implicit assumption that the response variable (which is inherent continuous and not discrete) is unrestricted in the domain of continuous values it can take. However, in certain modeling situations, the response variable is naturally limited in that it does not take certain ranges. In other words, the observed values do not cover the entire range of possible values. Such variables are called limited dependent variables (Limdep) or limited outcome variables. Greene (2011) identified at least two types of limited response variables: censored response and truncated response variables.

(a) Censored Response Variables. These variables are observations that are clustered at an upper threshold (termed "right censored"), a lower threshold (termed "left censored"), or both. For these kinds of response variables, the Tobit model is used. The left-censored regression model is given as

$$Y_i = \begin{cases} a + \beta X_i + \varepsilon_i, & \text{if } a + \beta X_i \\ 0 & \text{otherwise} \end{cases}$$

Censoring is common in survival analysis for modeling the longevity of civil systems. Ideally, for a given observation (civil engineering system), both the date of system construction/installation and date of termination/demolition of a system is known, in which case the lifetime can be calculated simply as the time difference between construction and termination. If that particular type of civil engineering system is known to have a longevity of Y^* years, then all observations that exceed Y^* are termed right-censored observations; and all observations whose measurements (in this case, service lives that are less than Y^*) are termed left-censored observations.

(b) Truncated Response Variables. To illustrate the concept of truncated variables, we return to the example used in the previous section for censored variables. In certain cases, there are observations (systems) that are still in existence. So, while the date of construction is known at the time of analysis, the future date of termination is not known. In this case, the observations (i.e., the life that each system has lived to date) are not their full lives but their truncated lives. It should be noted that truncation is different from left censoring: For a left-censored observation, the measurement exists (i.e., the full life has been measured), whereas for a truncated observation, we do not yet know the observation or measurement (i.e., the full life of the system, in this context). It is worth noting that in certain cases of truncation, the measurement (in this example, the expected full life of the system) can be extrapolated using the existing data. In that case, the "simulated" measurements (Figure 7.20) can be used as a normal continuous variable without the need for truncation or censoring.

7.3.4 Exogenous versus Endogenous Variables

Exogeny refers to an attribute (action, object, or effect) coming from outside a system. It is the opposite of endogeny (an attribute generated from within the system). A variable is *exogenous* if it is completely independent of the response variable of the model in which it appears; a variable is *endogenous* if it is explained within the model in which it appears. In regression, the response variable is often referred to as the endogenous variable. For example, in a model that describes or predicts the structural condition of a civil engineering system as a function of the average annual ambient temperature, annual volume of usage, or annual maintenance expenditure, the temperature may be considered an exogenous variable as ambient temperature can be considered completely independent of the system condition. On the other hand, the annual volume of usage may not be truly independent because potential users may choose not to use the system because of its poor condition. Similarly, the maintenance expenditure may not be truly independent of the system condition because high maintenance expenditure in a given year may not be an independent decision by the agency managers but rather due to the poor system condition. Thus, in the context of this example, temperature is exogenous while system usage and maintenance expenditure are endogenous.

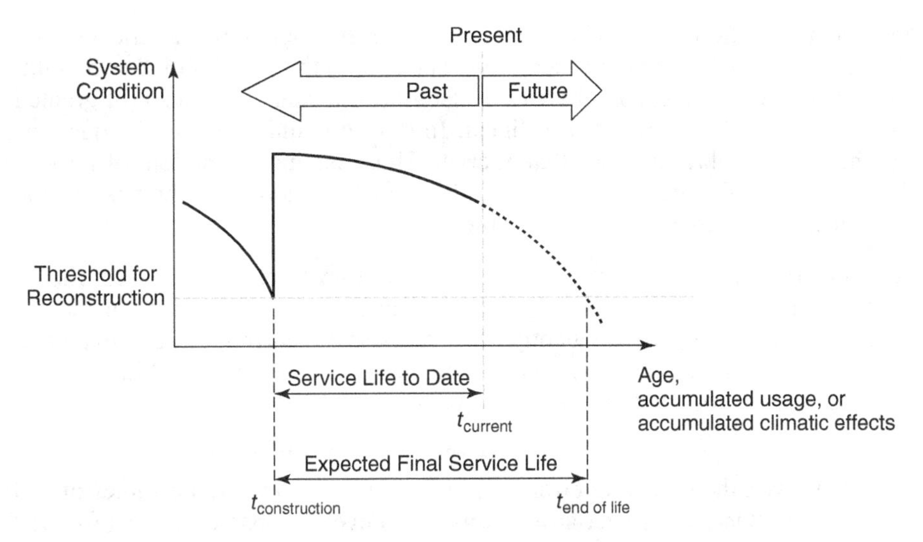

Figure 7.20 Illustration showing the extrapolation of system life to generate simulated longevity measurements that obviate the use of limited response models.

7.3.5 The Dangers of Extrapolation

As stated in the introduction to this chapter, engineers are often tasked with describing some attribute of their systems at any phase of development. Typically, models are used to describe these attributes as they occurred in the past, as they presently are occurring, or as they are expected to occur under current conditions or under a specified set of conditions at the current time or in the future. As seen in this chapter, models are typically developed on the basis of past observed data, but there are a number of hidden dangers in making predictions or estimations of system attributes based on past data.

(a) Gaps in the Data. For a robust model, data must be adequate to cover all possible occurrences of explanatory factors. However, there is often a lacuna in the data, and certain ranges of X were not used in the modeling process. For example, in the simple model shown in Figure 7.21, it is sought to estimate the value of the system response Y when $X = 5$. However, the model data do not

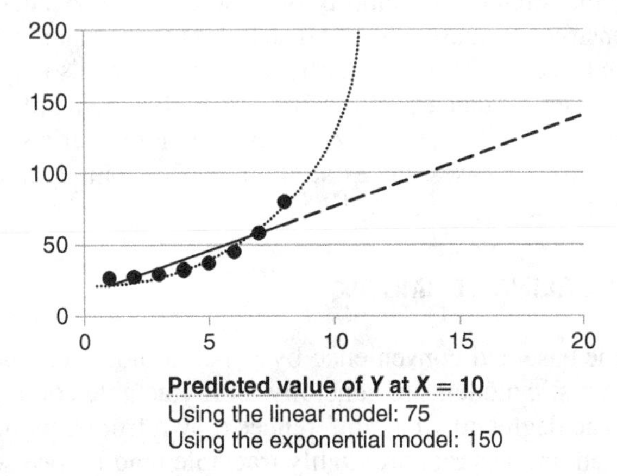

Figure 7.21 Potential pitfalls of inadequate span of data values.

include observations where X was equal to 5 or even where it was a little less or more than 5. Thus, the modeler is not sure how Y behaves when X is in the vicinity of $X = 5$. Although the behavior of Y from 0 to 3 is linear and 8 to 12 is also linear, it cannot be taken for granted that the behavior of Y from $X = 3$ to $X = 8$ is also linear. Indeed, it could very well be quadratic, logarithmic, or another functional form within that interval. Thus, making estimations of Y for observations whose X values fall in that gap on the basis of the model developed for the observations whose X values fall outside that gap could be misleading.

(b) Unavailability of Future Values of Explanatory Factors. In cases where we seek to predict some Y based on a number of explanatory factors X_1, X_2, the values of the X variables may not be known. This happens frequently with cross-sectional models. For example, in the following cross-sectional model, we seek to estimate energy use in year 2050 in a country as a function of population and gross domestic product (GDP).

$$\text{Energy use} = f(\text{GDP, Population})$$

However, the modeling effort is stymied by not knowing the values of GDP and population in year 2050. One way to get around this is to develop separate models for GDP and population, as functions of time or other cross-sectional data, and then plugging the estimated values of GDP and population into the energy use model.

(c) Inadequacy of Data. Observations used in models range between the lowest value and the highest value of the explanatory variables. However, the engineer often may seek to estimate Y for X values that lie outside this range, that is, lower than the lowest X value or higher than the largest X value in the data set. In certain cases, doing so is not considered prudent as analytical models are often reported together with accompanying caveats that the model is only applicable to a certain range of X. Nevertheless, in the absence of any recourse, engineers often may attempt to make such estimations. For the point whose Y variable is being sought, the closer the X variable is to the range of observations of the model, the more confident one will be about the accuracy of the model. In the example in Figure 7.21, the estimated (predicted) value of Y is observed to be very different, depending on which functional form is used to model the given data. If there were adequate data to include values of the X variable at or near the point in question, there could be greater confidence in the predicted value.

(d) Inappropriateness of Data. This problem occurs when engineers attempt to predict future system attributes using models that were developed on the basis of past characteristics. For example, existing models that predict the stability of a levee may have been developed on the basis of data pertaining to construction materials, construction processes, and levee designs several decades ago at the time of the levee design and construction. These models may not be applicable for predicting the stability of new levees that are currently being built or being planned for construction for reasons that include the current and future utilization of superior materials, different construction processes, or better designs. In such cases, the existing models contain data that may be inappropriate for the modeling task at hand.

7.4 GLOSSARY OF MODELING TERMS

Tractability The ease and convenience by which a model can be used to obtain a solution. For example, deterministic models are generally more tractable compared to stochastic models.

Validity The degree to which inferences drawn from a model can be applied to the real-life system. Very often, models that are highly tractable tend to lose some validity and vice versa. As such, in selecting the best model to describe some attribute of a system, the systems engineer often needs to carry out a trade-off between tractability and validity.

Closed Forms A mathematical expression representing a model is said to have closed form if, and only if, it is possible to write the expression analytically in terms of "established" mathematical functions and operators such as constants, single variables, and basic arithmetic operators (+ − ×÷, *n*th root, exponent and logarithm, trigonometric functions, inverse trigonometric functions, etc.). Similarly, an equation or system of equations has a closed-form solution if, and only if, there exists one or more one solutions that can be expressed in closed form. An alternative term for closed-form solution is "analytical solution" (i.e., a solution that can be derived through evaluation of mathematical functions and solution to equations). For complex systems, it is often the case that analytical solutions often do not suffice, and there is often a need to describe these systems using computer simulation.

EXERCISES

1. In civil engineering systems, models may be classified by the criteria shown in the table below. Indicate, by placing A, B, and so forth in the cells below to indicate their classification:

A. A model that helps us to predict the number of transit system users in Pu San in 2025.

B. A model that describes the rate of deterioration of an earth dam.

C. A model that identifies the factors that influence water flow in an old pipe in poor condition.

D. A model that determines the magnitude and direction of the effect of moisture and particle size distribution on the results of a soil shear test.

E. A model that predicts the average delay on an urban arterial highway during peak hours.

F. A model that monitors the physical deformation shape of a structural steel gusset plate over several years upon excess loading.

G. A model that helps track the progress of various tasks at a construction site.

H. A model that helps with the decision to recycle or dispose of material from a terminated civil engineering system.

I. A model that helps with the calculation of the shear forces in a cantilever structure.

J. A model that helps with the assessment of the vulnerability of a civil engineering system to external natural threat.

K. A model that identifies the factors under which a long-term lease of a civil engineering system would be more cost-effective than keeping it in-house.

		Physical Structure		Condition		Performance	
		Deterministic	Stochastic	Deterministic	Stochastic	Deterministic	Stochastic
Needs Assessment	Prescriptive						
	Descriptive						
Planning	Prescriptive						
	Descriptive						
Design	Prescriptive						
	Descriptive						
Construction	Prescriptive						
	Descriptive						
Operations	Prescriptive						
	Descriptive						
Monitoring/Inspection	Prescriptive						
	Descriptive						
Preservation	Prescriptive						
	Descriptive						
End-of-Life	Prescriptive						
	Descriptive						

2. The following data, for a certain type of civil engineering system, shows the variation of the project size (using cost as a size surrogate, in millions of dollars) to the duration of the contracts. Identify three functional forms that provide a close fit to the data.

Project size, $millions	1.9	4.2	5.8	7.8	10.1	12.0	13.8	15.6	17.7	19.5
Contract duration, months	0.6	1	1.4	2.9	4.9	7.5	8.9	9.1	8.9	9.0

3. For the data below, identify the best functional form that provides a close fit to the data.

X	0.25	0.1	0.51	0.15	0.06	0.83	0.27	0.82	0.21	0.25	0.71	**0.05**	0.62	0.32	0.89	0.55	0.95	0.44
Y	**16**	**28**	**10**	**24**	**32**	**12**	**17**	**10**	**22**	**12**	**11**	**36**	**10**	**10**	**14**	**12**	**17**	**12**

4. It is sought to develop a model for the data shown in the table below. Plot the data to ascertain if the data are inherently linear. Using a number of transformations of the X variable, develop an appropriate intrinsically linear model.

X	1.7	10.0	9.1	7.2	6.4	4.2	8.1	1.5	3.1	2.6	5.1	1.8
Y	2.012	1.797	1.813	1.820	1.802	1.827	1.797	2.050	1.869	1.952	1.802	1.994

5. The durability of any civil engineering system is typically determined by factors such as the system environment, constituent material(s), rate of system use, level of maintenance, and so forth. A water supply system is a typical example. It is hypothesized that the duration of any water pipe system depends on pipe attributes such as the pipe material (MATRL), the acidity of the soil (ACIDTY) in which the pipe is buried, the average daily flow in 100s of gallons per hour, (AVFLOW), the average gradient of the pipes (GRADE), and the average annual amount of pipe maintenance done in $100s per mile (MNTCE). From a random sample of water supply pipes in a certain state, data on the above characteristics were collected (see data below). Answers to Exercises (c), (d), and (e) may be provided using the table below.
   a. Provide the two hypothesis statements that could be used to test the significance of any given variable.
   b. Using the hypothesis, upon what basis can we say that any given variable is or is not significant (i.e., it has or does not have any significant influence on the durability of the water pipe). Illustrate your answer with a sketch.
   c. Using the supplied data set, run the model using SPSS or MINTAB. Fill in columns 2 and 3 in the table below. Attach a copy of the output.
   d. From your output, determine which variables are significant at 90% confidence? (Hint: This implicitly involves hypothesis testing where we are simply comparing the absolute value of the calculated t statistic with the critical t statistic of 1.64.) To answer, fill in column 4 of the table below.
   e. Assess the intuitiveness of the result for each variable. Does each sign make sense to you? Explain. To answer, fill in columns 5 and 6 of the table below.
   f. From your output, indicate the goodness of fit of the model: very good/good/fairly good.
   g. A water distribution system is planned for a new suburb on the outskirts of Lafayette. It is intended to use concrete pipes sloping along a gentle terrain with an average gradient of 1.25%, and soil tests indicate that the soil in that area is acidic. From population projections, the expected average hourly flow rate is 200 gallons, and the pipes are expected to receive an annual maintenance of $150 per mile.

Using the model you have developed, predict how long the proposed pipe system is expected to last. Assuming everything else remains the same, what level of maintenance would be needed if the system were to last for 50 years?

Modeling Output Dependent Variable: Durability of Pipe in Years (DURBLTY)

Symbol (Independent Variables)	Coefficient Estimate	*t*-Statistic	Is the Variable Significant[a]?	Intuitiveness of the Result[a] (Yes/No)	Reasons for Intuitiveness or Nonintuitiveness
Constant Term					
MATRL = 0 if concrete = 1 if steel					
ACIDTY = 0 if not acidic = 1 if acidic					
AVFLOW (100s gallons per hr)					
GRADE (%)					
MNTCE ($100s per linear mile)					

Data for Exercise 5

PIPE ID	AVFLOW	MNTCE$	GRADE	MATRL	ACIDITY	DURBLTY
P 15916	11.031	2.25	0.07	0	0	14.55
P 16091	3.48	3	0.06	0	1	52.58
P 16099	3.14	4	0.06	0	1	4.68
P 16563	15.17	2.63	0.05	0	1	11.51
P 10001	10.943	2.5	0.07	0	1	9.35
P 10018	8.3	2.5	0.02	0	1	15.65
P 10019	7.882	3	0.06	0	0	13.7
P 10044	5.9	4.25	0.08	0	1	11.94
P 10045	10.45	2.86	0.05	0	1	8.36
P 10046	11.5	2.6	0.06	0	0	10.94
P 10049	3.86	3.5	0.06	0	1	38.24
P 10117	13.07	2	0	1	0	14.74
P 10119	4.9	2.5	0.07	0	1	10.38
P 10147	3.734	5.75	0.08	0	1	26.05
P 10199	6.68	3	0.06	0	1	14.63
P 10451	7.003	3	0.06	0	0	9.43
P 10455	3.279	4	0.08	0	0	17.57
P 10459	7.578	3	0.06	0	0	13.22
P 10507	4.78	3	0.08	0	1	11.09
P 10512	12.21	1.75	0.06	0	0	12.78
P 10513	0.18	1	0.06	0	0	51.67
P 10514	10.31	3.5	0.06	0	0	19.51

PIPE ID	AVFLOW	MNTCE$	GRADE	MATRL	ACIDITY	DURBLTY
P 10519	10.58	2.19	0.06	0	0	11.44
P 10610	3.4	2.75	0.07	0	0	27.21
P 10660	9.147	5.2	0.06	0	0	116.52
P 10956	12.396	4.5	0.07	0	1	26.23
P 15099	3.74	3	0.06	0	0	38.73
P 15065	3.74	3	0.06	1	0	38.73
P 15160	8.05	2.6	0.05	0	1	11.17
P 15171	9.39	1.75	0.02	0	0	20.95
P 15597	0.6633	2.13	0.07	0	0	64.14
P 15600	9.22	2	0	0	1	19.73
P 15609	7.66	1.75	0.05	0	0	11.29
P 15610	14	2.5	0.02	0	1	26.68
P 15611	8.75	2.41	0.05	0	0	10.52
P 15618	5.57	2.18	0.05	0	1	12.87
P 15619	12.97	2.5	0.02	0	0	23.36
P 15911	4.5	3	0.06	0	1	21.46
P 15919	7.46	1.8	0.06	0	0	11.31

REFERENCES

Chatfield, C. (2004). *The Analysis of Time Series, An Introduction*, 6th Ed. Chapman & Hall/CRC, New York.

Cook, D. R. (1977). Detection of Influential Observations in Linear Regression. *Technometrics (American Statistical Association)*, 19(1).

Dol, W., and Verhoog, D. (2010). Methodology for Data Quality Management, CAPRI Database Extension and Quality Management. Common Agricultural Policy Regionalized Impact, The Hague, Netherlands.

Greene, W. H. (2011). Econometric Analysis, 7th Ed., Prentice-Hall, Upper Saddle River, NJ.

Hodge, V. J., and Austin, J. A. (2004). Survey of Outlier Detection Methodologies. *Artificial Intelligence Review* 22, 85–126, Kluwer Academic Publishers.

Holland, P. W. (1986). Statistics and Causal Inference. *J. Am. Statist. Assoc.*, 81, 396.

Koduru, S. D., and Haukaas, T. (2010). Probabilistic Seismic Loss Assessment of a Vancouver High-Rise Building, *ASCE J. Struct. Eng.*, 136(3), 235–245.

LTRR (2011). GEOS 585A, Applied Time Series Analysis. www.ltrr.arizona.edu/~dmeko/geos585a.html.

Pearl, J. (2000). *Causality: Models, Reasoning, and Inference*. Cambridge University Press, Cambridge, UK.

Rodriguez, M. M., Labi, S., and Li, Z. (2006). Enhanced Bridge Replacement Cost Models for Indiana's Bridge Management System. *Transp. Res. Record*, 1958.

Sinha, K. C., Fwa, T. F., Ting, E. C., Shanteau, R. M., Saito, M., and Michael, H. L. (1984). Indiana Highway Cost Allocation Study, Tech. Report, Purdue University, West Lafayette, IN.

SPSS (2011). *SPSS Statistics*, IBM Corporation, New York.

Wade, S. A., Nixon, J. M., Schindler, A. K., and Barnes, R. W. (2010). Effect of Temperature on the Setting Behavior of Concrete. *ASCE J. Mat. Civil Eng.*, 22(3), 214–222.

Washington, S. P., Karlarfis, M. G., Mannering, F. L. (2010). *Statistical and Econometric Methods for Transportation Data Analysis* Chapman and Hall/CRC, London, Reston, UK.

USEFUL RESOURCES

Ayyub, B. (1997). *Uncertainty Modeling and Analysis in Civil Engineering*, CRC Press, Boca Raton, FL.

Buede, D. M. (2000). *The Engineering Design of Systems: Models and Methods*, Wiley, Hoboken NJ.

Chong, K., Morgan, H. S., Saigal, S., Thynell, S. (2002). *Modeling and Simulation Based Life Cycle Engineering*, CRC Press, Boca Raton, FL.

Close, C. M., Frederick, D. K., Newell, J. C. (2001). *Modeling and Analysis of Dynamic Systems*, Wiley, Hoboken NJ.

Melchers, R. E., and Hough, R. (Eds) (2007). *Modeling Complex Structures*, ASCE Press, Reston, VA.

Nguyen-Dang, H., De Saxce, G., and Moes, N. (2010). *Modeling in Mechanical and Civil Engineering*, LAP Lambert, Saar brucken, Germany.

Washington, S. P. (1999). *Conducting Statistical Tests of Hypothesis*: Five Common Misconceptions Found in Transportation Research, *Transportation Research Record* 1665, 1–6.

CHAPTER 8

SIMULATION

8.0 INTRODUCTION

As we learned in Chapters 2, 5, and 7, one of the key tasks at any phase of civil systems development is to describe a physical system or a process associated with the system. This could be for a past or current situation or for the future, in which case the task is one of prediction. The purpose of this description task is typically to enable the system manager to ascertain the condition or performance of the system and to make any needed recommendations.

At one extreme of systems description, consider a world where everything that happens is perfectly deterministic and can be predicted with absolute certainty, a world where nothing is random or unexpected: The sun rises or sets at the same time each day all year round, the temperature is constant all year round, and the wind always blows in the same direction and with the same intensity. At the other extreme is a world where everything is random: The sun rises and sets at any time, so we could have say, 3 hours of daylight one day and 20 hours of daylight the following day; or one day we could have freezing temperatures and the next day we could have sweltering heat. Even if it were possible, it is not likely that anyone would find a place with either of these two extremes an exciting place to live. In reality (fortunately), real life falls between these two extremes; and most processes and events can be predicted with a fair amount of certainty due to the laws of nature. However, the outcomes are not exactly as predicted by the laws of nature because variability does exist, such as imperfections in the materials used to build systems, unpredictable weather and climate, differences in geotechnical or even geological conditions even over short distances, uncertain economic conditions such as interest rates, and uncertainty in the demand or loading requirements of civil engineering systems.

Therefore, in simulating a civil engineering system, engineers acknowledge that the behavior of the system and its interactions with the environment are dictated by factors that are not constant but rather exhibit a great deal of variability. The deformation of a structural member over time can vary significantly depending on the wind load or live load at any time, the physical condition of the member, and whether other structural members of the system have failed. The physical condition of a system, in turn, is a function of the climate conditions that may accelerate corrosion and other defects. Clearly, therefore, while the overall system behavior may be roughly predictable, there seems to be some variability that cannot be predicted using the classical laws of mechanics or other theoretical laws that govern a system's behavior.

In simulation, the engineer abstracts the system or some part or process thereof so that it is represented by a replica that is more amenable to scrutiny. This is particularly the case where the system is too large to be studied; it is too expensive or impractical to carry out scenarios; the real-life system is inaccessible; it is dangerous, unethical, or unacceptable to use the real-life system; or the real-life system does not yet exist (Sokolowski and Banks, 2009). In such situations, simulation is useful to acquire insights or to describe or predict behavior over time when the real-life system is subjected to different internal or external conditions or different courses of action. The process of developing such a replica is known as *simulation*. The Merriam Webster Collegiate

Dictionary defines simulation as "an imitative representation of the functioning of a real-world system or process over a period of time, for purposes of examining a problem often not subject to direct experimentation."

Simulation may be graphical or nongraphical. In either case, simulation may take any form in terms of complexity. For graphical simulation, one extreme is the simplest form; for example, an artist's sketch of a bridge at different stages of failure; and at the other extreme is a more sophisticated computer-generated graphical simulation of the progressive behavior of a failing bridge structure leading up its collapse. For nongraphical simulation, one extreme is a simple equation, for example, the first law of motion in physics; and the other extreme is a set of equations in systems dynamics (Chapter 14) that characterizes the behavior of each of the multiple components of an engineering or biological ecosystem and the multiple interactions between them. Also, the nature of the simulation object of interest may be a real object or virtual (a process). Regardless of the level of complexity or the nature of the object being simulated, the inputs for any simulation are explicit or implicit statements of the key attributes of the subject being simulated (in the case of an artist's sketch, for example, the statements of the object's attributes are implicit and occur in the mind of the artist and are manifest as the sketch).

Simulation is a useful tool that continues to be applied in several engineering disciplines and other fields for the tasks of testing, training and education, monitoring and optimizing the performance of systems, and predicting future system failure or diagnosing the reasons for failure. For example, driving simulators help researchers study naturalistic driving behavior by providing them with a lifelike experience of the driving environment [road width, alignment (curves), friction, traffic conditions, and weather conditions]. Figure 8.1 shows the outside and inside of a typical driving simulator. This simulator mimics the internal and external conditions of a real vehicle in a virtual driving environment: The user (driver) acquires a realistic impression of sitting in a real vehicle's cab and driving along a real street.

Simulation is used in a wide range of disciplines (Aldrich, 2004); for example, in the training of civilian workers in disaster response and medicine, and also military personnel, particularly where it is too dangerous or costly for trainees to interact with the real world.

(*a*)

(*b*)

Figure 8.1 Simulation of highway system operations: (*a*) Advanced Driving Simulator. (*b*) Inside of the Jentig-50 driving simulator. (*Source:* Peter van Wolffelaar, ST Sotfware B.V.)

Table 8.1 Categorization of Training Simulation Contexts by Status of the Simulation Entities

Category of Training Simulation	Simulation Entity		Example
	Players	Environment	
Live	Real	Real	Actual trainee engineers are placed on the site to carry out tests and other work.
Virtual Type 1	Real	Simulated	Actual trainees use simulated equipment in a simulated environment.
			Actual trainees use real equipment in a simulated environment.
Virtual Type 2	Simulated	Real	Trainees control simulated players that interact with a real environment.
Constructive	Simulated	Simulated	Trainees control simulated players that interact with a simulated environment.

The Modeling and Simulation Coordination Office (M&SCO) of the U.S. Department of Defense categorized training simulation programs by whether the players or the environment are being simulated (Table 8.1).

Simulation tools have also been used in areas outside of engineering. They have been used in business for the so-called management games where decision makers analyze the business consequences of alternative strategic decisions; in finance, where computer simulations are often used to compare the outcomes of scenarios associated with stock markets and portfolios, and in project evaluation, where the analyst mimics the utility, net present value or real options performance of a proposed project or existing facility over a range of discount rates, usage levels, facility deterioration and maintenance, and other factors internal and external of the system. Simulation has also been used in weather forecasting and airplane pilot training. In the social sciences, simulations can be used in a variety of contexts (Hartmann, 1996) including staff training for international development purposes and to deal with fragile regions wracked by conflict; for example, the United Nations Development Program (UNDP) developed the Carana Simulation, which was subsequently adopted by the World Bank (Milante, 2009). Also, meteorological offices predict weather conditions (e.g., spatial variation and intensities of temperature, precipitation, and humidity and the location and intensities of tornadoes and hurricanes) using simulation models built from past weather data. Ironically, there have been applications even in computer science where computer programs have been developed to simulate the behavior and performance of computers under nonnormal conditions.

In civil engineering, the role of simulation continues to become more and more evident for the reasons stated by Sokolowski and Banks (2009) as we learned in the second paragraph of this chapter. In general, Infrastructure Information Modeling (IIM) is the term used to describe the computer simulation of the functional and physical attributes of a civil engineering system [in the case of buildings, Building Information Modeling (BIM)]. By providing data on the system's performance and the environmental consequences in response to different input scenarios associated with the system and the environmental conditions, IIM helps in supporting the decision-making process for the facility throughout the stages of design, construction, operations, monitoring, maintenance, and end of life (NBIMCPC, 2012). Similar to all other analytical and numerical tools, simulation has its merits and demerits as listed in Table 8.2.

Table 8.2 Merits and Demerits of Simulation

Merits	Demerits
Flexibility. Conditions of simulation can be modified easily by simply changing a level of an input parameter.	**Data limitations.** As in any model, the outputs of a simulation are only as good as their input data. Erroneous input data will lead to incorrect outcomes.
Acceptability. Where simulation is accompanied by graphics, it can have visual appeal and easy interpretation, thus enhancing acceptance by nontechnical personnel.	**Input data properties.** May be unable to adequately handle (a) a large number of input variables, (b) complexity of interactions between the variables, and (c) nonquantifiable variables that are encountered.
Real system limitations. Helps overcome the practical realities associated with the real system, including *complexity*, *danger* to the human agents, and *cost*. Also eliminates *system downtime*: in cases where the operations of the real-life system need not be interrupted for the purposes of creating and analyzing system scenarios.	**Nontransferability.** Each outcome of the simulation pertains to a specific system under a specific set of operating and external conditions; thus the simulations may not be perfectly applicable to other systems or conditions. This happens when the input data used to construct the simulation model is very limited. Incorrect model outcomes may thus be obtained when the model is run using input data that are outside the range of the original input data.
Crystal ball. Helps the system owner/operator to quickly visualize the long-term impacts of a current policy or decision.	**Efficiency.** Simulation-based methods for predicting system outcomes are not always as efficient as analytical techniques.
Stability. Can be used to study repeatedly a given situation with little cost for updating its parameters.	
Versatility. Particularly useful where the solution to a problem cannot be found using closed-form analytical techniques.	

Table constructed using information from Smith (1998); Viessman et al. (1989); Ayyub and McCuen (2002); and Cheema (2005).

8.1 SIMULATION TERMINOLOGY

The terms used in simulation often differ from agency to agency. The literature (McCuen, 2002) offers some common meanings for standard terms, which we present below. To illustrate the definitions, we will consider the simple simulation process of flipping a coin and a case where we are simulating the condition of a structural system in terms of its level of deterioration in response to annual loading (number of users per year), climatic severity, structural design type, and material type.

Simulation **trial**: A single instance of simulation, executing a simulation model from input to output, for example, a single flip of a coin or a single instance of calculating the structural condition for given levels of loading, climatic severity, design type, and material type.

Simulation **run** (also, simulation **cycle**): Multiple trials of simulation, for example, a simulation run could comprise 10 flips of a coin or 100 computations of structural condition for multiple different combinations of loading, climatic severity, design type, and material type.

Simulation **factor** (also, **input variable**): A parameter or variable whose value needs to be input in the model before carrying out the simulation.

Simulation **parameter**: A value that remains fixed in the course of a simulation run but could be made to be different for different runs, for example, the structural design type.

Simulation **variable**: A value that changes over a simulation run.

Simulation **output variables**: The variables that represent the end state of the system after each simulation.

Initial conditions: The seed values of the model parameters and variables that represent the starting state of the system or process being simulated.

8.2 CATEGORIES OF SIMULATION

There are numerous ways to classify civil engineering systems simulation: the simulation purpose, target, and nature; the nature of the state variables; the tool used for the simulation; and the phase of system development for which the simulation is carried out. We now discuss these categories.

8.2.1 Classification by Purpose of the Simulation

As stated in the introduction of this chapter, civil engineering system simulation is a tool used for the task of describing some aspect of the system. The purpose of such descriptions may be *predictive*, *didactic*, or *diagnostic* (to ascertain how the system will behave, what is wrong with the system, what will be wrong with the system at a future time, etc.) or *prescriptive* (to identify appropriate remedial measures in order to maintain or restore a favorable situation or to avoid an imminent unfavorable situation).

8.2.2 Classification by Subject of the Simulation

What are we trying to simulate? Like all systems, a civil engineering system can be represented as an object located in (and interacting with) an environment. The environment comprises not only the natural but also the builtup environment and the level and nature of the system's use. The environment includes the fauna and flora, wind, climate (e.g., freeze and freeze–thaw transitions, oxidation, and rain), soil acidity, groundwater levels, wind, frequency and intensity of usage, user characteristics, other structures nearby, and so forth. The system affects the environment through defined processes (see Chapter 26 on sustainability); and the environment affects the system as well, through defined processes such as deterioration modeling or random processes such as disasters (see Chapter 22 on system preservation and Chapter 25 on resilience). The system functions under a given set of internal controls which may be referred to as the system operations phase or process (see Chapter 20). The subject of the simulation may be any one of the processes associated with the system's interactions with the environment or among the system components.

8.2.3 Classification by Nature of the Simulation

Simulation can be either *virtual* (i.e., *mathematical*) which involves the use of an equation to predict the outcome of a process or situation), or *real* (i.e., *physical*), which is an actual three-dimensional representation of the system or process being studied. These terms refer to the manner in which the simulation is done or the outcome of the simulation. In virtual simulation, the output of the mathematical equation may be visualized using computer graphics (Figure 8.2*a*). Real or physical simulation can be carried out on the actual object of simulation or a replica of a size typically

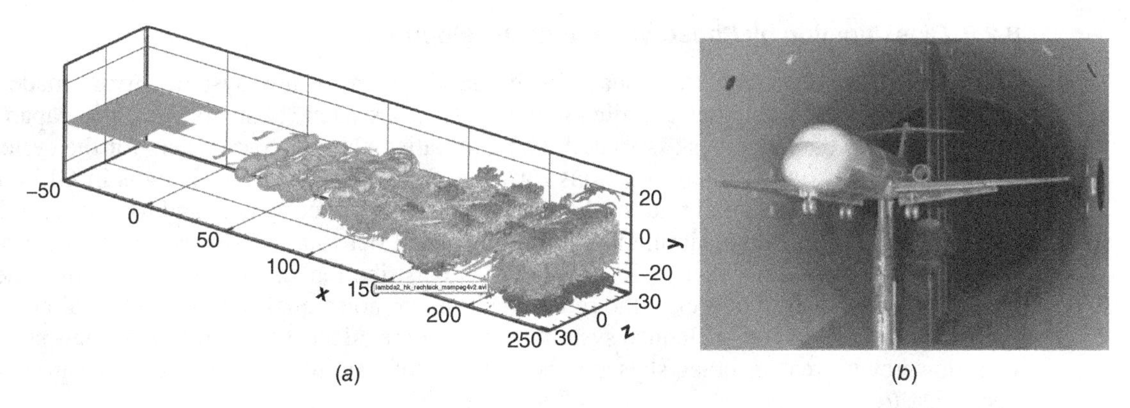

Figure 8.2 Simulation using mathematical and physical means. (*a*) Virtual (numerical) simulation of a vortical mixing process using 3D Navier–Stokes equations (*Source:* Markus Kloker, University of Stuttgart). (*b*) Real (physical) simulation of aircraft aerodynamics using propulsion wind tunnel (*Source:* JeLuF. Creative Commons).

smaller than the actual object (Figure 8.2*b*). In certain situations, both real and virtual models are necessary in order to check the results from each other. Indeed, for most processes in engineering systems, it is often necessary, or at least useful, to carry out both virtual and physical simulations, one of which serving as a validation of the other. For example, the physical aerodynamic patterns actually observed from a wind tunnel (often, using colored smoke) as shown in Figure 8.2(*b*) could be replicated using mathematical models; the image shown in Figure 8.2*a* was developed using pseudospectral methods to solve the Navier–Stokes equation.

In the past decades, research institutions and industries have used expensive equipment such as wind tunnels to carry out physical simulations. However, realizing that such simulations can be carried out using mathematical models implemented and visualized on computers, some of these organizations are gradually dismantling the physical simulation infrastructure and resorting solely to less expensive computer simulations.

8.2.4 Classification by Nature of the State Variables

As we learned in Chapter 5, there are two kinds of variables: discrete and continuous; and in that chapter, we also learned the core difference between these two types of variables. A simulation can be described as discrete or continuous, depending on the manner in which the state variables change. In the simulation of discrete events, the variables that characterize the system state change instantaneously at distinct points in time (Hoeger, 1996). On the other hand, in a continuous simulation, the variables change continuously, typically through a function that uses time as one of its variables. According to Smith (1998), in practice, most simulations incorporate both types of variables; however, one of these is often predominant and therefore drives the classification of the entire simulation.

8.2.5 Classification by Simulation Tool

Digital simulation may be defined loosely as any computer-enabled representation of a system or process. Computer simulation can be used to model all kinds of processes, but it is particularly useful for situations where it is not feasible to use simple closed-form analytic models. *Nondigital simulation* is the replication of a system or process without using a computer.

8.2.6 Classification by Phase of System Development

Planning. This is an important phase in the development of any system. Errors made at this phase reverberate throughout the life of the system. As such, planners seek the capability to analyze reliably the outcomes associated with alternative planning parameters for the system. For example, in analyzing different locations for a system, geospatial analysis tools in GIS could be used to simulate the potential relationships between the proposed system and its natural or builtup environment. For example, simulation could be carried out to ascertain the impact of a planned highway on the wildlife in a region; the traffic redistribution impacts of a new mall near the central business district of a city; the reduction in air or noise quality in a residential area due to emissions associated with a planned system nearby, or the effects of constructing a new skyscraper in a downtown area on other skyscrapers in terms of external air circulation or groundwater pore pressures.

Design. At the design phase of system development, an analyst may develop a mathematical model that predicts the effect of different design parameters on the system outcomes, for example, the extent to which a thicker beam could reduce shearing; how a thicker pavement could enhance service life; and the extent to which water flow is restricted due to siltation in a hydraulic channel. In such cases, the analyst studies different design scenarios by changing the values of the input design parameters and examines the corresponding outputs. In certain cases, the simulation outputs can be visualized as digital images on a computer screen.

Construction. At the construction phase, simulation tools can help reduce cost, boost efficiency, and increase safety. Live training on heavy equipment is often dangerous and expensive, and it is now possible to train people in different virtual environments utilizing virtual heavy equipment as well as simulate a wide range of possible scenarios that the trainee may likely face in their future work. Simulation is also used in training construction project managers to monitor the evolution of the project parameters including cost and schedule, and in certain cases, even safety and work quality. These simulation tools can help trainees fully visualize the consequences of their project control decisions in terms of the project parameters.

Operations. The operations phase of a civil engineering system is by far the longest of the development phases, and significant cost savings can be achieved if appropriate tools, including simulation, are used to enhance efficiency at this phase. For example, in logistics systems that involve the transportation of freight and passengers, simulations are used to study complex routing patterns for large numbers of transportation craft (for land, air, water, or pipeline) traveling over wide expanses of regions in order to maximize profits or services or to minimize costs by identifying the routes that serve a maximum volume of freight/passengers (see Chapter 17) and use the least (or most efficiently use) mobile or fixed transportation assets. In operations of this kind, the performance of the system is simulated on the basis of considerations that include craft capacity, transport time and cost, weather, and the probability of unscheduled downtime.

Monitoring/Inspection. Engineers who monitor and inspect civil engineering systems are concerned with what is actually happening at the site of real-world existing systems. However, that does not preclude a role for simulation tools. Before site inspection is carried out, the tools of simulation can be used to predict what site condition could be expected based on historical trends or the peculiar characteristics of the site and the system. Locations (systems or system components) that are found from the simulation to exhibit any potential problems (e.g., excessive damage, high stresses, or imminent danger to users or the environment) are flagged for closer scrutiny during the visual field inspections. In certain cases, locations flagged for scrutiny may be selected based on the variability of the simulated system conditions. Locations with more stable measurements over an extended period of simulation time or under different simulation input parameter values may not need frequent monitoring in the field; on the other hand, locations that exhibit wildly fluctuating outcomes from the simulation could be flagged for more frequent monitoring. Equipped with such knowledge, the system inspector therefore could consider the flagged locations as sample locations from a larger population for inspection or monitoring purposes. However, where all the members of the population need to be monitored or inspected, the flagged locations could be made to receive greater attention during the inspection or monitoring process.

Maintenance. At the maintenance phase, the engineer typically carries out a simulation of the outcomes of a maintenance activity based on the maintenance resource inputs (expenditures, in the case of a contract, or manpower, materials, and equipment in the case of in-house maintenance). Mathematical models could be developed to ascertain the costs and effectiveness of alternative maintenance strategies, and optimal strategies could be developed from the simulated outcomes of each scenario. Also at the maintenance phase, a finite element model (FEM) and other simulation tools could be used to determine the efficacy of a specific treatment on the basis of the nature of the treatment, exactly what is being treated, and the site characteristics. For example, Fang et al. (2003) used finite element modeling to simulate the effectiveness of crack sealing as a maintenance treatment for cracked pavements in highway systems. Simulation in the maintenance phase goes beyond describing the efficacy of a treatment: Entire life-cycle schedules of maintenance and rehabilitation spanning multiple treatments could be simulated to ascertain their long-term effectiveness in increasing the asset's physical condition or extending its life, as well as the life-cycle costs incurred by the system owner and the direct, indirect, and intangible life-cycle costs incurred by the users during system downtime (e.g., workzones) or during normal operations of the system.

End of Life. In Chapters 26 and 27, we discuss the issues associated with system end of life and the tasks required of the engineer at this phase. Simulation can be used to analyze either of the two system end-of-life categories: intended or unintended. Where the end-of-life is not intended by the system owner, for example, system destruction due to earthquake, flood, internal design errors, or man-made threats, the system analyst can use simulation tools to mimic any of the following: the "attack" processes by the destruction agents, the system failure, the effects on neighboring structures and facilities, evacuation of the affected population, and so on. In such cases of unintended end of life, the simulations also can help identify areas of improvement to avoid or mitigate the disaster or to make adequate preparations to recover if it occurs. Where the end of life of a system is intended, for example, due to system structural or functional obsolescence, and thus the system needs to be demolished, simulation tools that help visualize the demolition process. The simulated impacts of each demolition option can help in choosing the option that minimizes any harmful impacts on neighboring structures and the environment.

8.3 RANDOM NUMBER GENERATION

To characterize randomness, mathematically, engineers use numerical modeling tools, including random number generation (RNG) and Monte Carlo simulation, which can help carry out the "what if" analysis associated with their systems; for example, what if the actual strength of a structural member is not exactly 25 N/mm^2 as assumed in the design but rather is 23.5 N/mm^2; what if the traffic volumes are 50–80% more than expected on a certain urban freeway; or what if a

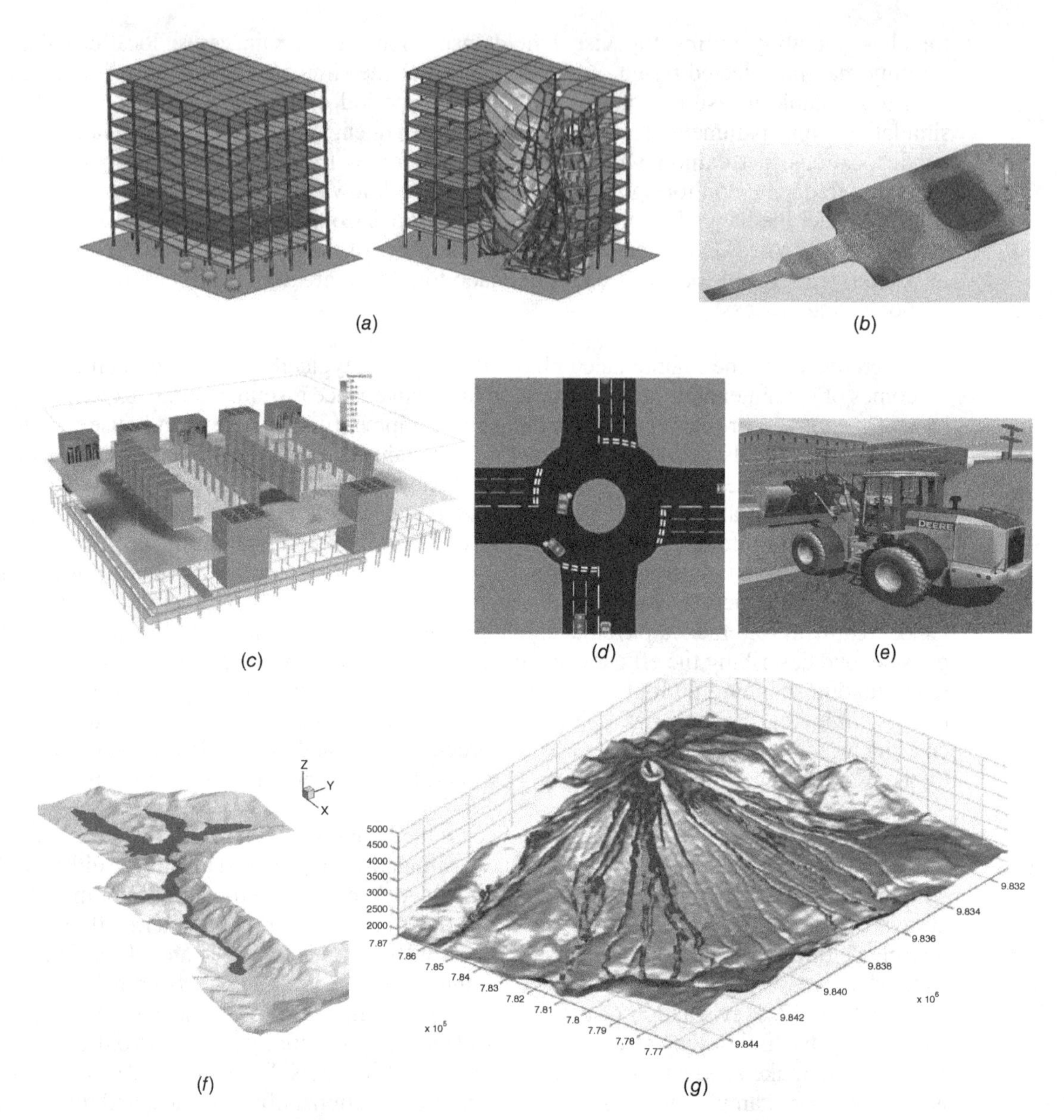

Figure 8.3 Computer simulation illustrations of processes at various phases of civil engineering system development: (*a*) Simulated progressive collapse of a building (*Source:* Sherif El-Tawil, The University of Michigan). (*b*) Simulation of stresses in different parts of a structural element (Courtesy: Ben Masefield, Solveering LLC). (*c*) Simulation of hot and cold currents in a data center (*Source:* sustainability workshop. autodesk.com). (*d*) Urban intersection traffic simulation (Credit/source: Peter Stone/www.fhwa.dot.gov). (*e*) 3D Simulation of earthmoving operations during construction (*Source:* www.simlog.com). (*f*) Flooding conditions simulation, Paute River Basin, Ecuador (Credit/source: Brett F. Sanders, Lorenzo Begnudelli/sanders.eng.uci.edu). (*g*) Simulation of pyroclastic flows and topography, Tungurahua Volcano, Equador (Credit/source: The International Charter: Space and Major Disasters, and IG-EPN).

construction task takes 30% more time than planned in the schedule? In a bid to account for such variability, engineers add, as an input to models to predict system behavior, expressions that represent random occurrences of the factors that influence the outcome. This process is conducted using the principles of random number generation. A formal definition for random number generation is the "process of generating perfectly random data." Traditional applications include areas where an unpredictable result is sought (lottery operators, gambling casino owners), statistical sampling to select a sample from a homogenous population, or where an unpredictable result is expected (due to unpredictable inputs).

When a random number within a certain range is normalized by dividing it by the largest number in the range, the resulting number is a real number that lies in the range [0,1] (Garcia-Aracil et al., 2011; Chung, 2000). When all the random numbers within that range are normalized, the result is a uniform distribution over the range [0,1]. Thus, a random number can be modeled as a random variable that is uniformly distributed over the range [0,1]. An important property of a set of random numbers is that there should be no serial correlation (repeating patterns of the levels of the numbers). The uniform distribution property of random numbers is important in simulation because it serves as a basis for the generation of real numbers that follow any distribution of interest, as we shall see in Section 8.3.4.

8.3.1 Mechanisms for Generating Random Numbers

Long ago, before the concept became commonly used in engineering applications, random numbers were generated as a part of gambling schemes, specifically to select the winner of mechanized gambling operations. In that era, the mechanism of random number generation was **mechanical** in nature and included the drawing of numbered balls from an opaque box, the spinning of roulette wheels, the dealing of cards, throwing dice, or flipping a coin. Many of the present-day lotteries are still operated this way. These mechanical methods were too wieldy and slow for applications in statistics and cryptography, and the next generation therefore was the development of **arithmetic** random number generators. Subsequently, these came to be automated using the computer, and analytical models were used to provide the seed values. Arithmetic number generators were superior to their mechanical counterparts because they were faster, had no requirements for memory for storing numbers, and were capable of repeatability, thus making it more likely for them to achieve the critical properties of uniform nature of the distribution of the generated numbers and absence of any serial correlation in the generated set of numbers (Ayyub and McCuen, 2002).

There are a number of arithmetic random number generators that are often used in systems modeling. These include the linear congruential generator (LCG), midsquare generator, multiplicative generators, general congruence generators, composite generators, Tausworthe generators, and mixed generators (Park and Miller; 1988; Gentle, 2003; Knuth, 1997). The common mechanism for all these generators is similar: It starts with a seed value and uses a mathematical equation to determine the random value in the range [0,1]; the determined value is used as an input in the same mathematical equation to yield another random value, and the process is repeated. The repetition of this process N times yields N random numbers. The difference between the RNG techniques is that their recursive models use different mathematical equations. The **period** of a recursive model is defined as the "number of values that are generated randomly before the stream of values starts to repeat itself" (Gentle, 2003). The efficacy of recursive models in generating random numbers is very sensitive to the period that is used. Experts agree that it is advantageous for random number generators to have large periods; a rule of thumb that is often used is that the period should exceed the number of simulation cycles needed to carry out a given simulation.

In linear congruential generators, a sequence of integers $I_1, I_2, I_3, \ldots$, is defined by the following recursive equation (Ayyub and McCuen, 2002):

$$I_i = (a\,I_{i-1} + b) - \text{Int}[(aI_{i-1} + b)/c]c \quad \text{for } i = 1, 2, 3, \ldots,$$

where $\text{Int}[(aI_{i-1} + b)/c]$ is the integer of the result of the division, a is the multiplier, b is the increment, and c is the modulus; a, b, and c are nonnegative integers. The parameters of this recursive model should also satisfy the following conditions: $0 < c,\ a < c$, and $b < c$.

First, the starting value or seed value, I_0, is provided by the analyst carrying out the simulation, such that $I_0 < c$. Then I_i is determined by dividing the expression $(aI_{i-1} + b)$ by c; and the remainder of this division operation is set as I_i. The I_i value is normalized by dividing by c, as $0 \leq I_i \leq c$.

The random number U_i is defined as $U_i = I_i/c$.

It can be seen that a repetition of this process is bound to yield the same results. As such, this process is not perfectly random. Algorithms such as this are referred to as pseudorandom number generators (PRNGs) because they can create automatically a long sequence of numbers that exhibits significant random properties; however, after a certain point, the sequence repeats. The period of the above random number generator is less than or equal to c. As such, in generators used in practice, the value of the parameter c is typically very large, often exceeding one billion. The advantage of setting a large c value for the simulation is that the resulting number of discrete values becomes very large, thus constituting a closer approximation of a continuous uniform distribution (Ayyub and McCuen, 2002). Multiplicative generators and mixed generators are special cases of linear congruential generators where $b = 0$ and $b > 0$, respectively. Most modern computers have built-in random number generators including the *rand*( ) function that is found in most spreadsheets.

Example 8.1

Using a recursive multiplicative generator defined by the following equation,

$$I_i = (a\,I_{i-1} + b) - \text{Int}[(a\,I_{i-1} + b)/c]c \text{ for } i = 1, 2, 3, \ldots,$$

use the following parameter values: multiplier, $a = 3$, increment, $b = 0$, and modulus, $c = 5$, and a seed value $I_0 = 1$, to generate 10 random numbers. Let the random number U_i be defined as $U_i = I_i/c$.

Solution

The generated random numbers are shown in the table below.

i	0	1	2	3	4	5	6	7	8	9	10
I_i	1	3	4	2	1	3	4	2	1	3	4
U_i	0.1	0.6	0.8	0.4	0.2	0.6	0.8	0.4	0.2	0.6	0.8

8.3.2 Testing of Random Number Generators

In a large number of practical situations, it is desired to generate numbers that have no discernible patterns; thus, subsequent repetition of the number generation process is expected to yield a result that cannot be predicted. These practical situations where a truly random generation of numbers is required include statistical sampling, gambling, cryptography, security coding, and computer simulation. For these and other applications, a good generator of random numbers is desired—one that yields numbers to which no patterns can be recognized or predicted. In order to test the goodness of a random number generator, it is useful to carry out at least two key tests: the *test of uniformity* and the *test of serial correlation* (Ayyub and McCuen, 2002). Each of these tests can be carried out using theoretical or empirical techniques.

In theoretical tests, no random numbers are generated; instead, the recursive model is evaluated in terms of the suitability of its parameters. In empirical tests, random numbers are generated by the generator (say, N random values in the range [0,1]) and are evaluated statistically using the chi-square test or other goodness-of-fit tests to check whether the generated random numbers are consistent with the desirable property of uniformity of the resulting probability distribution. Uniformity tests that are carried out to ascertain whether the generated random numbers are consistent with a uniform continuous probability distribution include the Anderson–Darling test. This test ascertains whether a given sample of data (in this case, a number of random numbers obtained using a generator) is drawn from a given probability distribution (Anderson and Darling, 1954; Rahman et al., 2006). The serial test is carried out to determine whether any serial correlation exists in the generated random numbers; each value in the generated stream of random numbers is considered to originate from a different, albeit identical, uniform distribution (see Washington et al. (2010) for the details of this test).

8.3.3 Varying Degrees of Randomness in Generated Numbers

In the relative frequency charts (probability distributions) shown in Figures 8.4*a*–*c*, the horizontal axis is a specific outcome; for example, the system condition level and the vertical axis are the probability of that outcome given a certain sample, or the percentage of observations that had that condition level. In Figure 8.4*a*, the number generation is perfectly random; and the generated numbers appear to have no pattern. In Figure 8.4*b*, the number generation is distributed random, that is, the generated numbers appear to follow some probability distribution (the normal distribution has been shown in the figure for purposes of illustration). Complete lack of randomness is characterized by a perfectly deterministic situation where the exact outcome is known (Figure 8.5*c*). Figures 8.4*a* and 8.4*c* can be considered as extreme cases of random number generation.

8.3.4 Generating Perfectly Random Numbers

In this section, we show how we could generate random numbers in Microsoft Excel.

(a) Generating Random Numbers between 0 and 10^K.

To generate a random number between 0 and 1, type in: = rand()
To generate a random number between 0 and 10, type in: = 10 rand()

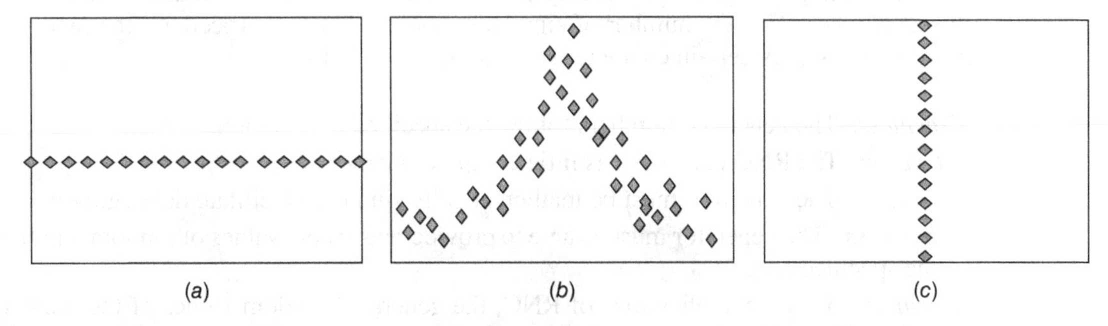

Figure 8.4 Varying degrees of randomness in generated numbers: (*a*) perfectly random, (*b*) according to some probability distribution, and (*c*) perfectly deterministic.

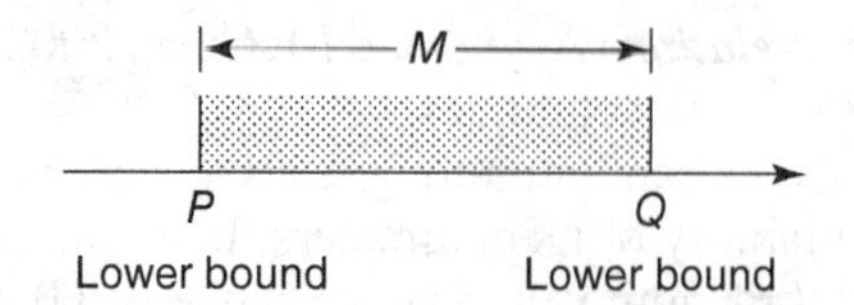

Figure 8.5 Generating random numbers between P and Q.

To generate a random number between 0 and 100, type in: = 100 rand()
Generally, to generate a random number in Excel between 0 and 10^K, type in: = 10^K rand()

(b) Generating Random Numbers between P and Q. To generate a random number between P and Q, type in: $= P + (Q - P)$ rand(), or

$$Z = P + M * \text{rand}()$$

Example 8.2

Generate a random number in Excel between (a) 2 and 5, (b) 16 and 22, and (c) 250 and 605.

Solution

To generate a random number in Excel between 2 and 5, type in: = 2 + (5 − 2) rand()
To generate a random number between 16 and 22, type in: = 16 + (22 − 16) rand()
To generate a random number between 250 and 605, type in: = 250 + (605 − 250) rand()

8.3.5 Generating Random Variables or "Distributed Random" Numbers

As seen in Figure 8.4*b*, it is possible to encounter a set of observations that seem to be random but actually hover around some pattern or some specific probability distribution. In some texts, this is referred to as **random variable generation** instead of random number generation. In this text, we describe such observations as distributed random observations. The generation of a random variable according to a given probability distribution can be considered as a sampling procedure with size N where N is the number of simulation cycles. For this procedure, the random number generally must satisfy certain characteristics as explained below.

Exactness. The generated numbers follow the specified distribution.

Efficiency. The RNG consumes as little computer memory as possible.

Simplicity. The generator must be mathematically simple to facilitate debugging if necessary.

Robustness. The generator must be able to provide reasonable values of random numbers across the specified range.

Synchrony. In some applications of RNG, the generated random values of the input variables need to be synchronized.

In any given problem where developing or selecting a random number generator is being sought, some or all of these properties need to be satisfied to varying degrees; and in achieving the objectives associated with these properties, the analyst will need to deal with their conflicting nature (Ayyub and McCuen, 2002). Thus, after analyzing the trade-offs between the properties for a given problem, the analyst may need to arrive at a compromise. For example, a higher level of exactness may be associated with a lower level of efficiency and vice versa.

There are at least three techniques for generating random numbers that follow some distribution. These are the methods of inverse transformation, composition, function-based, acceptance–rejection, and special properties. In the remainder of this section, we focus on the inverse transportation method only. Readers are encouraged to consult the resources listed at the end of this chapter for details of the other methods.

8.3.6 Generating Random Variables Using the Method of Inverse Transformation

As we observed in Figure 8.4*b*, distributed random number generation yields patterns that fall between the two extremes of perfect randomness and perfect alignment with probability distribution, with the observations seemingly to cloud around some probability distribution. The method of inverse transformation is rather simple and direct (Ayyub and McCuen (2002)): first, a random number u is generated in the range [0,1]; that is, $u \in U[0,1]$, where $U[0,1]$ is a continuous uniform probability distribution, which is followed by the generation of a continuous random variable, X, as follows:

$$x = F_x^{-1}(u)$$

where F_x^{-1} is the inverse of the cumulative distribution function of the random variable X. $F_x(x)$ is in the range [0, 1]; as such, each simulation cycle yields a unique value for x.

For a discrete random variable, a random number is generated, $u \in U[0,1]$; then, the value of the generated random variable X is determined as follows:

$$x_i \text{ such that } i \text{ is the smallest integer with } u \le F_x(x_i)$$

where x_i, $i = 1, 2, 3, \ldots, m$ are m discrete values of the random variable X. The cumulative mass distribution function of X is $F_x(x)$.

Rand Functions. In the sections below, we present the rand() function used in spreadsheets to generate distributed random numbers for a number of continuous probability distributions. (Figure 8.6).

a. **Normal distribution** To generate random numbers that yield a cloud of simulated observations that generally follow a normal distribution with a given mean and standard deviation, type in: = NORMINV(RAND(), mean, standard_dev).
b. **Beta distribution** To generate random numbers that yield a cloud of simulated observations that generally follow a beta distribution with a given mean and standard deviation, type in: = BETAINV(RAND(), alpha, beta, A, B).
c. **Gamma distribution** To generate random numbers that yield a cloud of simulated observations that generally follows a gamma distribution with a given mean and standard deviation, type in: = GAMMAINV(RAND(), alpha, beta).
d. **Student *t* distribution** To generate random numbers that yield a cloud of simulated observations that generally follows a Student *t* distribution with a given mean and standard deviation, type in: = TINV(RAND(), deg_freedom).

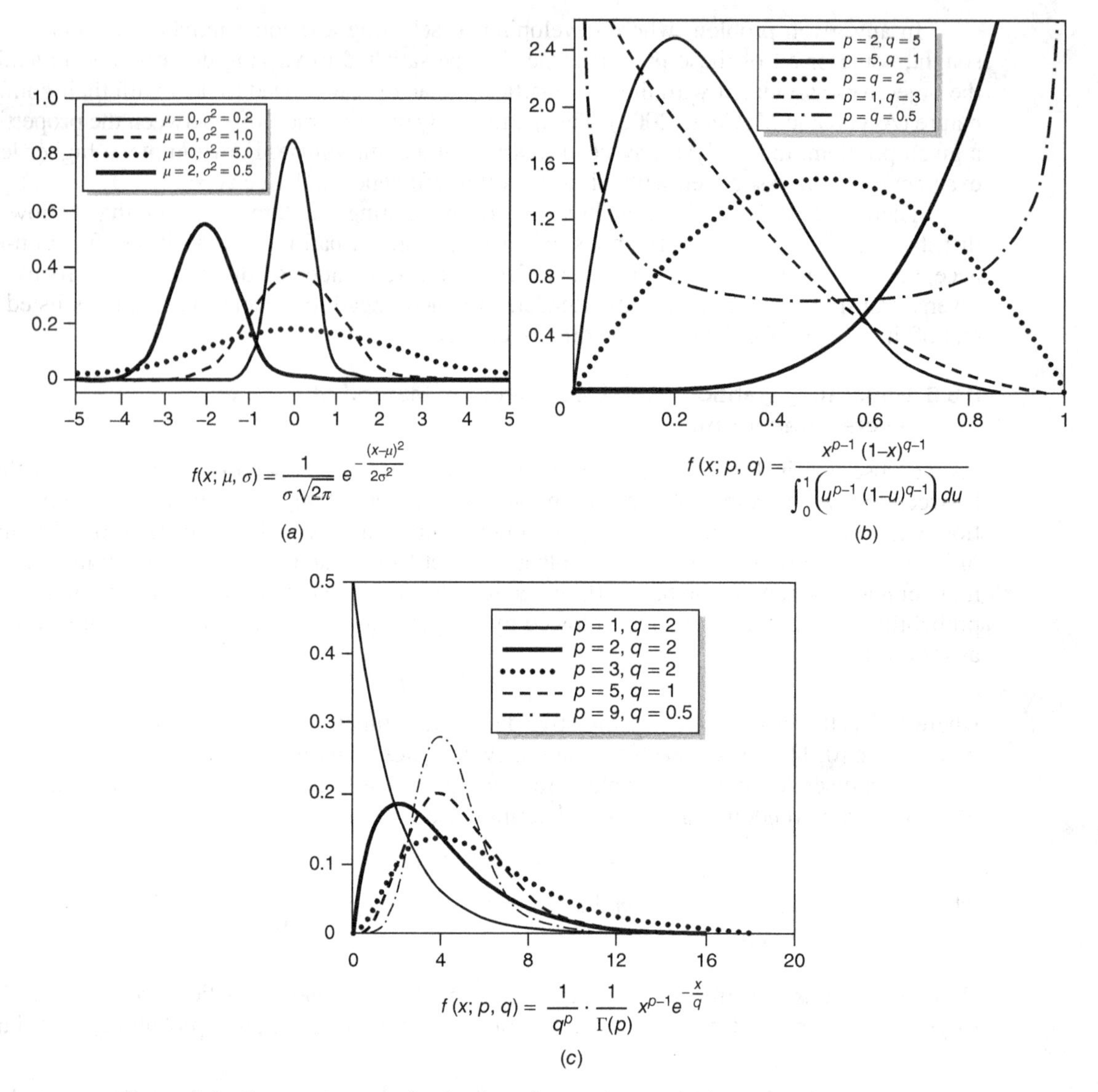

Figure 8.6 Some common distributions for random variable generation: (*a*) normal, (*b*) beta, and (*c*) gamma distribution.

Example 8.3

Using a standard spreadsheet, generate (a) 400 random numbers between 0 and 0.20, (b) 500 normally distributed random variables between 0 and 50, (c) 3000 beta-distributed random variables between 10.15 and 15.50, and (d) 1000 gamma-distributed random variables between 500 and 1500. For each, use generated numbers to plot a relative frequency table and chart.

Solution

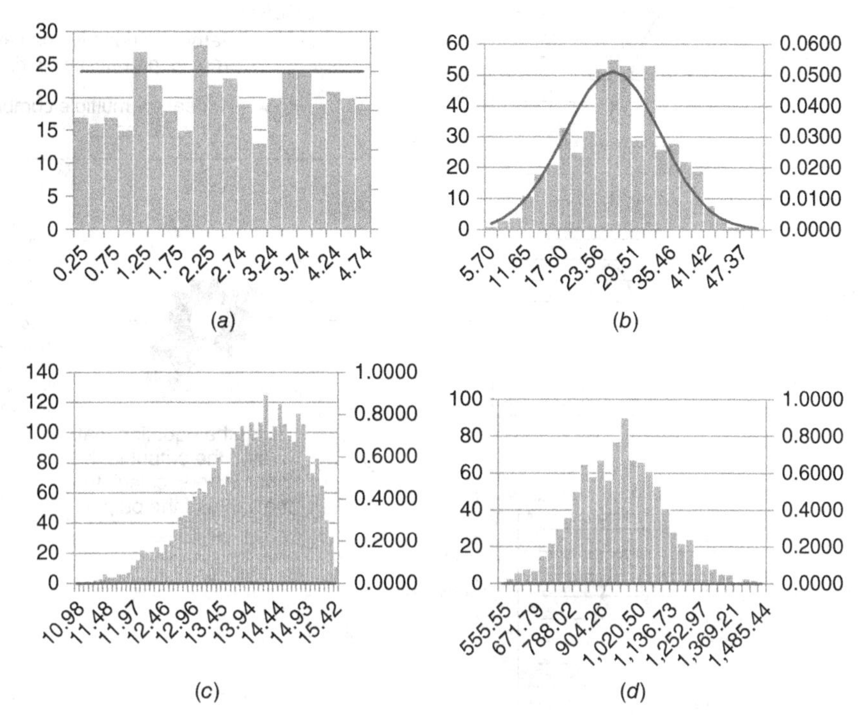

Figure 8.7 Relative frequency charts for Example 8.3: (*a*) uniform, (*b*) normal, (*c*) beta, and (*d*) gamma distribution.

8.4 MONTE CARLO SIMULATION

Monte Carlo simulation is a technique for stochastic simulation of systems. It is useful for predicting the behavior of civil engineering systems where events occur in a manner that is characterized by a great deal of variability or that are very difficult to describe directly with closed-form equations, differential equations, or other mathematical tools. Monte Carlo simulation uses the concept of random number generation to quantify the variability in the explanatory input factors that affect the system attribute under investigation. The variability in the input factors often translates into even greater variability in the output. In a Monte Carlo simulation, the values of each input factor are sampled at random from the probability distribution of that input factor, and the outcome is calculated on the basis of the combination of input factors. The process is repeated continuously, each time using a different combination of random values from the probability distributions of the input factors. Each combination of values from the various input factor samples may be referred to as an *iteration*, and the analyst records the predicted system outcome for each iteration. Depending on the range of values for each input factor, the simulation is carried out over hundreds, thousands, or even millions of iterations, and the result is a cloud of outcomes with their respective relative

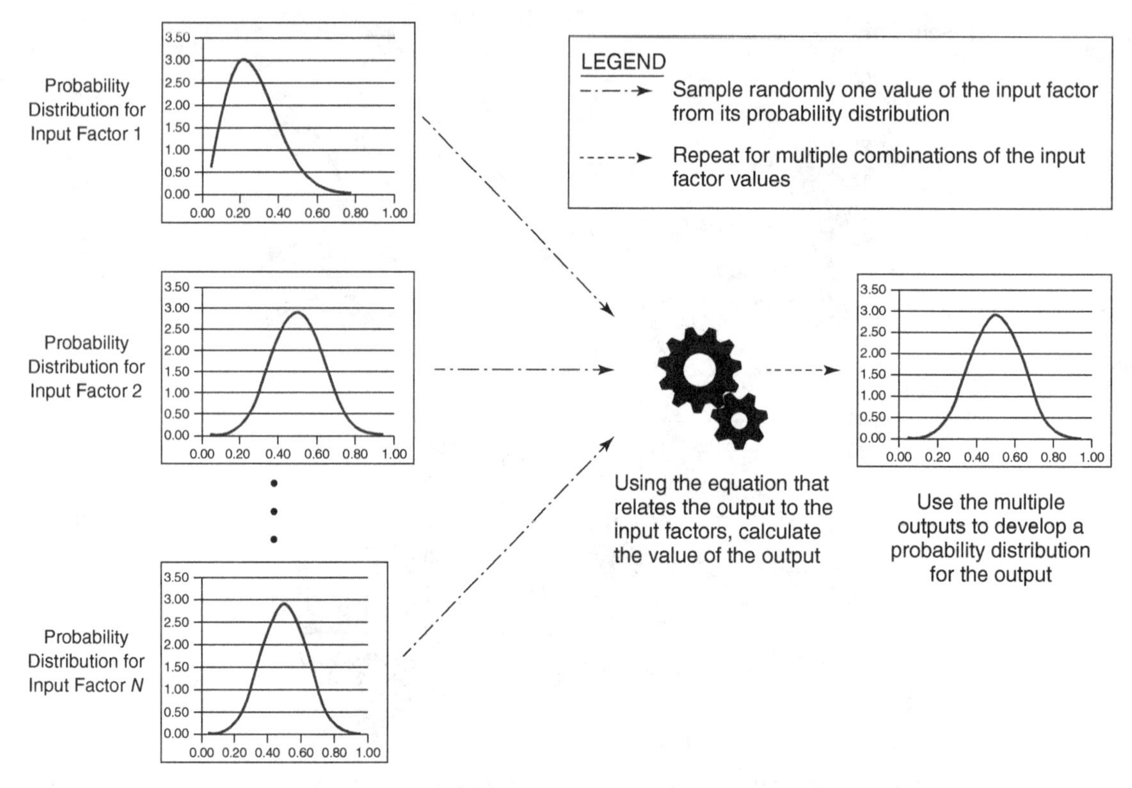

Figure 8.8 Basic concept of Monte Carlo simulation.

frequences or probability distribution (Figure 8.8). Thus, Monte Carlo simulation makes available a window where the system analyst can have a panoramic view of the possible system outcomes in response to the various internal and external conditions (Palisade, 2012). This helps the analyst not only to identify the possible outcomes but also how likely each possible outcome is likely to happen. It is useful as well for analyzing the consequences of such extremes extremely conservative and extremely liberal input factor values.

Obviously, the manner by which a Monte Carlo simulation carries out its sampling from the probability distribution of the input variables plays a large role in the efficacy of the results of a Monte Carlo simulation; this has been the subject of considerable research. Sampling techniques that have been proposed and used in certain applications include Latin Hypercube sampling where combinations of input factor levels are drawn more representatively from the entire range of distribution functions. Another example is the Markov chain Monte Carlo (MCMC) methods, a type of algorithm that is useful for generating samples from specified probability distributions through the construction of a Markov chain. A Markov chain is defined as a set of transitions from one state to another for a finite number of possible states.

8.4.1 Principles of Monte Carlo Simulation

Sawilowsky (2003) stated that in order to ascertain that a Monte Carlo simulation is appropriate and useful, it must be consistent with a number of conditions or good practices. First, the (pseudo-random) number generation must be such that (a) a long "period" elapses before the next iteration and (b) the values of the input factors produced by the random number generator must pass tests for randomness. Second, there should be an adequate number of samples to ensure that reliable results are obtained. Other principles are that the proper sampling technique should be used to obtain the values of each input factor for the iteration, the algorithm used should be valid and relevant for the system attribute that the system analyst seeks to model, and the simulation must adequately mimic the process that is being modeled.

8.4.2 Probability Distributions in Monte Carlo Simulations

The input factors in Monte Carlo simulations typically characterized by different patterns of variability and therefore are characterized by different probability distributions. It is useful for the analyst to ascertain the probability distribution associated with each input factor before carrying out the simulation. This can be done by reviewing the literature associated with the input factor or by collecting data about it and carrying out the relevant statistical tests (see Chapter 6). Common probability distributions that are encountered in civil engineering systems include the continuous (normal, lognormal, uniform, or triangular) and discrete distributions (Palisade, 2012). We discuss some of these distributions in Chapter 5, so we will limit their discussion in this section to a brief statement on each distribution.

Normal Distribution. Also known as the Gaussian function, this is a symmetrical, bell-shaped probability density curve. The analyst specifies the distribution mean, and the variance or standard deviation that describes how the observed value for that input factor will vary with respect to the mean value. The input factor levels most likely to occur are those that fall in the middle of the distribution near the mean value; and the extreme left or right values represent the conservative or liberal occurrences of the input factor. Several natural and man-made phenomena follow a normal distribution.

Lognormal. Certain quantities typically grow exponentially, and thus their probability density curves tend to skew to the right (positive skew). Such quantities are often best described using distributions such as the lognormal. The observations do not take values below zero and can take any positive values including very small or very large values. In the context of civil engineering systems, examples of variables that could be described fairly reliably by a lognormal probability distribution include system demand, population, prices of materials, accumulated precipitation (in infrastructure management), or river discharge volumes (in hydrologic engineering).

Uniform. When the possible values of an input factor are described as uniformly distributed within a specified range, then all the values within that range have an equal chance of occurring. This situation could be described as being more random compared to the normal or lognormal distributions. Uniform distribution is particularly important in simulation because many generators produce random numbers that are uniformly distributed (Ayyub and McCuen, 2002).

Triangular. In this distribution, there is a specified minimum value, a mostly likely value, and a maximum value of the input factor. Values around the most likely value have the greatest probabilities of occurring. A variant of the triangular distribution is the PERT distribution, where the extremes are not as emphasized as they are in the triangular distribution. Examples in civil engineering include the duration of a specific task in a construction project (Palisade, 2012).

Discrete. As we learned in Chapter 5, a discrete distribution is one where there is a probability assigned to each value of a discrete random variable. An example of discrete distribution is the outcome of cost overruns simulation for civil engineering construction projects in a certain state, province, or country; for example, a 60% probability of cost overruns (the difference between the final or as-built cost and the bid contract award amount), a 30% probability of no cost overrun or underrun, and a 10% probability of cost underruns.

8.4.3 Capturing Interactions between the Simulation Inputs: An Advantage of Monte Carlo Simulation

In modeling the outcomes of a system, Monte Carlo simulation can provide certain benefits that deterministic analysis may not always be able to adequately provide (Fishman, 1995; Palisade, 2012). First, deterministic analysis provides only a single point estimate, while the probabilistic results of Monte Carlo simulation provide an indication not only of what could happen but also the likelihood of occurrence of each possible outcome. Second, the use of deterministic models inhibits adequate description or prediction of the impact of each of the several different combinations of input factor values; however, using Monte Carlo simulation, analysts can track and visualize rather easily each combination of input factor levels, the outcome of each combination, and which inputs or combination of inputs have the most profound impacts on the system outcome. More importantly, Monte Carlo simulation facilitates the modeling of interactions, interdependencies, and other relationships between the input variables. This can be done using concepts including *statistical copulas* (which we will discuss shortly in this section) and to incorporate such relationships in the simulation. For example, consider a real-life situation where X alone may cause Z to increase and Y alone may cause Z to decrease; but when both X and Y occur together, Z could remain flat or even decrease. This phenomenon can be captured by Monte Carlo simulation but will likely be missed by deterministic modeling.

Figure 8.9 presents an illustration of the effect of each simulation run for different scenarios of input factor interactions. In (a), there is no accounting for the interaction effects of X_1 and X_2, and thus the overall effect of the two input factors is simply the sum of their individual effects. In (b), there is due accounting for the interaction between X_1 and X_2, and thus the overall effect of the two input factors could be (a) smaller than, (b) the same as, or (c) larger than the sum of their individual effects. Figure 8.10 presents an example in pavement systems engineering, where the overall damaging effect of truck loading and high temperature exceeds the sum of their separate effects.

Thus, in simulation models, the engineer typically seeks to predict or explain some system attribute as a function of a number of explanatory (input) factors that are not fixed in a given situation

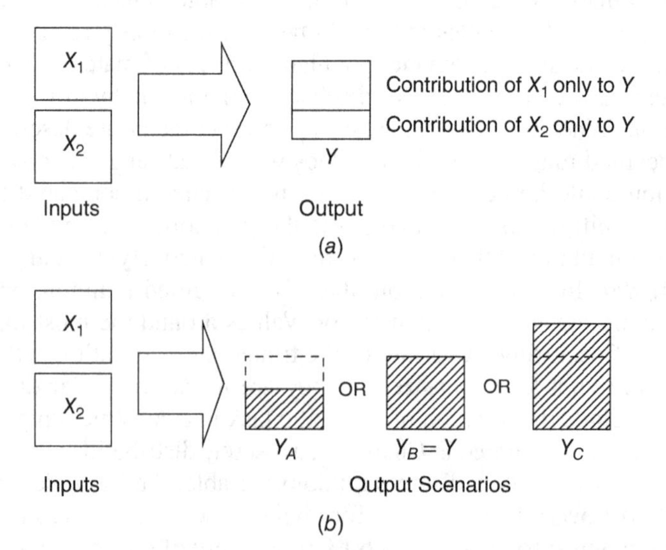

Figure 8.9 Effect of each simulation run for different scenarios of input factor interactions: (*a*) No accounting for the interaction effects of X_1 and X_2 and (*b*) duly accounting for the interacting effects of X_1 and X_2.

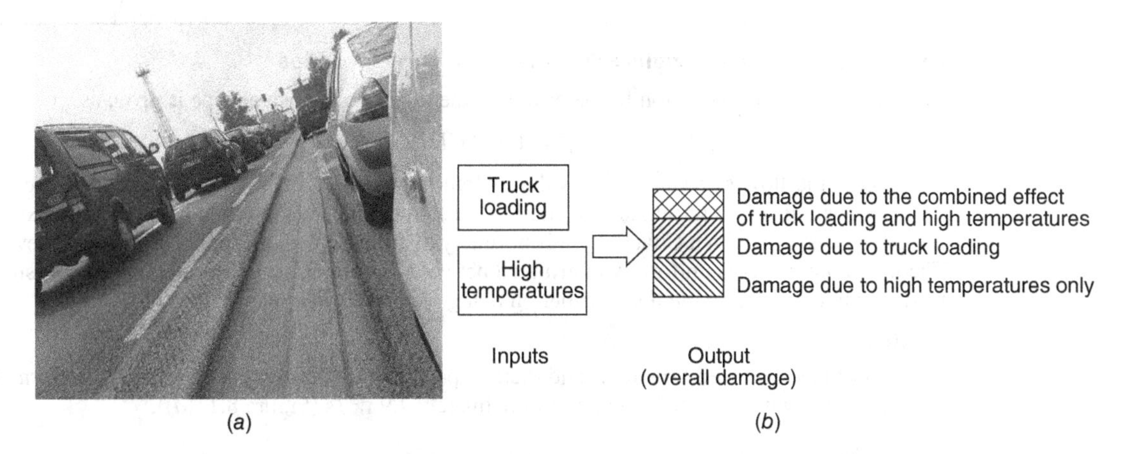

Figure 8.10 The overall damaging effect of truck loading and high temperatures exceeds the sum of their individual effects (*Left*: Courtesy of Burda/Wikimedia Commons).

but could take any value within a certain probability distribution. When two or more of the input factors (covariates) have some interaction effect, the model assumes a significant degree of complexity, and the task of system description becomes more difficult. To develop a reliable model in cases where the interacting covariates are normally distributed, the engineer can carry out simulation using Cholesky's decomposition of the covariance matrix, which is then multiplied to a sample of individually simulated variables (Ford et al., 2011). However, when the covariates each have a distribution that is other than the normal distribution, more advanced techniques are needed to develop the model. In such cases, it may be possible to theoretically derive the joint distribution function (Ang and Tang, 2006); however, the complexity quickly reaches unmanageable proportions when it is sought to derive the joint distribution function for certain distributions such as the bivariate Weibull (Yacoub et al., 2005).

Azam (2011) demonstrated another technique that can be used to simulate the correlation: the rank order correlation technique, where the correlation matrix is developed using the relative ranking of each observation in its respective distribution as the basis. This approach maintains the independent distributions of the random variables, but the resulting correlation matrix is deterministic. Yet another approach is to simulate one random variable and use the value as an independent variable to predict another random variable, such as the envelope method (Kokkaew and Chiara, 2010), or using lookup tables and/or Boolean logic established using expert opinion. Ford et al. (2011) used another technique, **statistical copulas**, which have been applied extensively in insurance and financial risk analysis. A copula is a joint probability distribution that accounts for correlations between the explanatory factors of a model by converting probabilistic dependency structures into uniform random variables while keeping intact the individual marginal distributions. The uniform random values are then utilized in the simulation of the individual marginal distributions, which may be of different functional forms.

8.4.4 A Few Monte Carlo Applications in Different Civil Engineering Branches

In this section, we examine some illustrations of Monte Carlo simulation in various branches of civil engineering systems. These are for purposes of illustration only as there are countless examples in each branch where such simulations are applicable. In Chapters 19–26, we will discuss additional examples of Monte Carlo simulation at the various phases of civil system development.

Example 8.4 Hydraulic Engineering. Flow Rate through a Pipe

The Hazen–Williams equation for determining the flow rate in a water pipe is given as

$$Q = 1.318\ CD^{2.63}S^{0.54}$$

where Q is the flow rate (ft^3/s); D is the hydraulic radius (ft); C is the coefficient of roughness (C decreases with increasing roughness); and S is the slope of the energy grade line (ft/ft). An existing pipe is made of cast iron with a C factor that is normally distributed with mean 97 and standard deviation 12. The hydraulic radius is uniformly distributed between 4.8 and 5.2, and the slope is 0.01. Using Monte Carlo simulation, determine the distribution that governs the flow rate.

Solution

The plot of the simulated flow rates and their respective probabilities is suggestive of a Normal probability Function with expected value of approximately 4.9 ft^3/s (Figure 8.11(*a*)).

Example 8.5 Financing of a Levee System

A new levee system is planned to protect a low-lying city. The input parameters for this analysis are normally distributed with the following parameters: service life; mean 15 years and variance 25; initial cost: mean \$700,000 and variance \$90,000; maintenance cost: mean \$28,000 and variance \$16,000; and interest rate: mean 4% and variance 1%. How much money should the city borrow now in order to finance the entire project including maintenance?

Solution

The output plot suggests that the amount to be borrowed follows a Weibull-like distribution; The most probable value is approximately \$7M (Figure 8.11(*b*)).

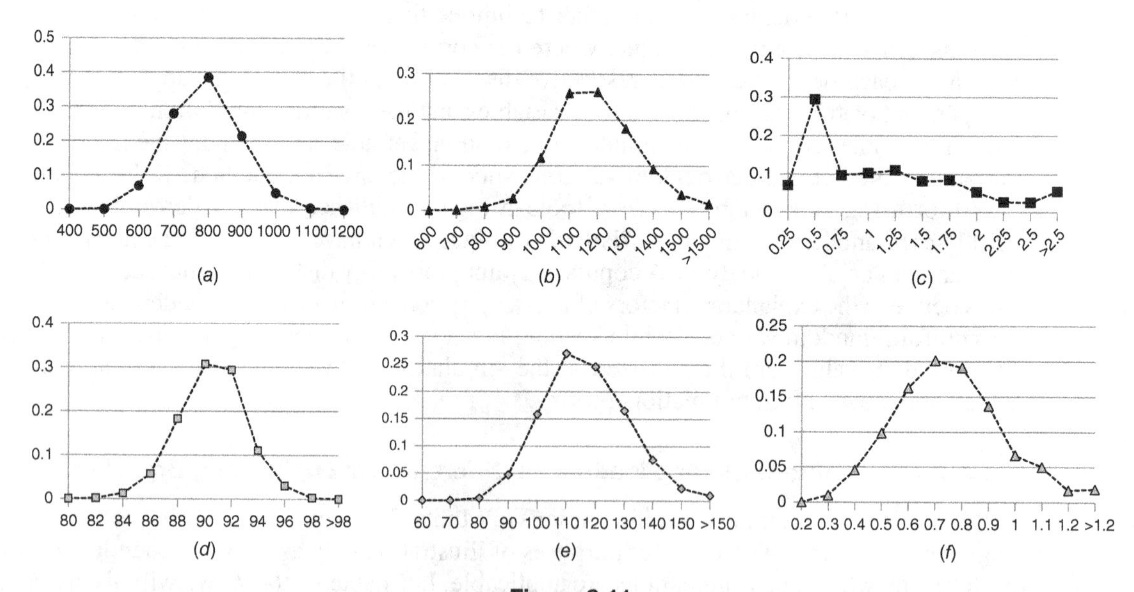

Figure 8.11

Example 8.6 Contaminant Decay in Surface Waters

Many contaminating substances undergo a simple first-order decay process. When released from a point source, their concentrations downstream of the source can be modeled as

$$C(x) = C_0 e^{-(kx/u)} + \frac{W}{KA}(1 - e^{-kx/u})$$

where A = cross-sectional area of the receiving stream (m^2);
C_0 = initial contaminant concentration at the outfall (kg/m^3);
$C(x)$ = contaminant concentration downstream of the outfall (kg/m^3);
u = mean stream velocity (m/s);
K = contaminant decay rate (per second);
x = distance below the outfall (m);
W = (uniformly) distributed load along the stream reach below the outfall (kg/ms).

For a certain stream in Pigo province, these inputs are found to vary spatially or temporally as indicated in the table below. Determine the distribution type and parameters for the resulting contaminant concentration downstream of the outfall.

Variable	Distribution	Parameters
A	Normal	Mean = 20 m^2, Stdev = 5 m^2
C_0	Normal	Mean = 0.05 kg/m^3, Stdev = 0.01 kg/m^3
K	Beta	0.001 s^{-1}, $\alpha = 0.2$, $\beta = 0.3$
u	Normal	Mean = 1 m/s Stdev = 0.25 m/s
W	Uniform	5 kg/ms
x	Constant	5 m

Solution

Figure 8.11(*c*) presents the probabilities of the various levels of contaminant concentration simulated using the model and the input data ranges. It is seen that the plot follows a gamma distribution; the most probable level is 2 kg/m^3.

Example 8.7 Transportation Engineering. Stopping Sight Distance (SSD) for Highway Design

SSD is the distance required by a vehicle traveling at or near the design speed to stop before reaching a stationary object in its path. The SSD (ft) is calculated as follows:

$$\text{SSD} = 1.47Vt + \frac{V^2}{30[(a/32) + G]}$$

where t = brake reaction time (s)
V = design speed (mph)
a = deceleration rate (ft/s^2)
G = grade (%) divided by 100 (for upward grade is positive, and downward grade is negative)

For a given stretch of a certain rural highway, some of these inputs vary as shown in the table below due to differences in the attributes of the drivers and vehicles that use that highway. Determine the distribution type and the most probable value for the resulting stopping sight distance of vehicles using the highway section.

Variable	Distribution	Parameters
Brake reaction time	Normal	Mean = 0.5 s, Stdev = 0.03 s
Design speed	Constant	50 mph
Deceleration rate	Normal	Mean = 2.5 ft/s^2, Stdev = 0.75 ft/s^2
Grade (%)	Constant	1.5

Solution

The plot of the simulated values of the required stopping sight distance suggests a normal distribution with a very slight negative skew. The most probable sight distance that is required is 60 ft (Figure 8.11(*d*)).

Example 8.8 Structural Engineering. Stresses and Deflections in a Pavement Slab

In the design of rigid pavements, computation of stresses for equal stress, the equivalent single wheel load (ESWL), is based on rigid slab analysis such as the well-known Westergaard formulas. These computations provide the maximum bending stress σ_{max} and the maximum deflection δ_{max} as follows (Fwa et al., 1996):

$$\sigma_{max} = \frac{3P(1+\mu)}{2\pi h^2}\left[\ln\left(\frac{2L}{b}\right) + 0.5 - \gamma\right] + \frac{3P(1+\mu)}{64h^2}\left(\frac{b}{L}\right)^2$$

$$\delta_{max} = \frac{P}{8kL^2}\left\{1 + \frac{a^2}{2\pi L^2}\left[\ln\left(\frac{a}{2L}\right) + \gamma - 1.25\right]\right\}$$

where $b = (1.6a^2 + h^2)^{0.5} - 0.675h,\ a < 1.724h$
P = total applied load
h = slab thickness
M = slab Poisson ratio
L = radius of relative stiffness
k = subgrade reaction modulus
a = radius of loaded area
γ = Euler constant = 0.577216

At a certain project site, the values of these inputs exhibit marked variations as indicated in the table below. Determine the distribution type and the most probable value for the resulting maximum bending stress and maximum deflection.

Variable	Distribution	Parameters
Total applied load	Normal	Mean = 5.5 N, Stdev = 0.6 N
Slab Poisson ratio	Normal	Mean = 0.18, Stdev = 0.05
Slab thickness	Normal	Mean = 1 ft, Stdev = 0.01 ft
Radius of relative stiffness	Normal	Mean = 5 m, Stdev = 0.2 m
Radius of loaded area	Normal	Mean = 0.2 ft, Stdev = 0.007 ft
Modulus of subgrade reaction	Normal	Mean = 150 MPa/m, Stdev = 12 MPa/m

Solution

The plot of the concrete slab deflections and their respective probabilities (simulated from the model and data) is suggestive of a normal distribution with a slight positive skew (Figure 8.11(*e*)). The most probable deflection value is 0.58 in.

Example 8.9 Materials Engineering. Evaporation from Fresh Concrete

The drying of fresh concrete leads to the development of plastic shrinkage cracks. This phenomenon is related to the rate of evaporation, which can be estimated using the following equation (Uno, 1998):

$$\text{Evaporation rate} = 5[(T_c + 18)^{2.5} - R(T_a + 18)^{2.5}](V + 4) \times 10^{-6}\ \text{kg/m}^2\ \text{h}$$

where T_c is the concrete temperature in deg °C; R is the relative humidity in %; T_a is the air temperature in degree C; and V is the wind velocity in km/h. At a certain site where concrete has been freshly laid, these factors were found to vary significantly as indicated in the table below. Describe the distribution type for the simulated evaporation rates from the fresh concrete.

Variable	Distribution	Parameters
T_c	Normal	Mean = 25°C, Stdev = 3°C
T_a	Normal	Mean = 18°C, Stdev = 4°C
R	Normal	Mean = 8.2%, Stdev = 2.6%
V	Normal	Mean 25 km/h, Stdev = 5.5 km/h

Solution

Figure 8.11(*f*) presents the probabilities associated with each outcome (that is, each level of water evaporation from fresh concrete). This is approximately a normal distribution with a perceptible positive skew.

SUMMARY

Simulation is a process that mimics the structural behavior, condition, or performance of a system over a period of time. As we learned in Chapter 7, simulation is one of the several modeling tools used to describe these attributes of a system. Early in the chapter, we stated that in carrying out simulation for a civil engineering system, it is implicitly acknowledged that the behavior of the system and its interactions with the environment is dictated by factors that are not constant but rather exhibit a great deal of variability. In other words, while the overall system behavior may be roughly predicted, there seems to be some variability that cannot be predicted using the classical laws of mechanics or other laws that govern a system's behavior. We also discussed the conditions under which simulation is typically carried out: (a) the system is too large to be studied; (b) it is too expensive or impractical to carry out scenarios; (c) the real-life system is inaccessible; (d) it is dangerous, unethical, or unacceptable to use the real-life system; or (e) the real-life system does not yet exist. Simulation may be graphical or nongraphical, and the simulation subject may be a real object or a process. We also reviewed the merits and demerits of simulation as established by past researchers and emphasized the need for due circumspection in the use and interpretation of simulation results. The chapter also reviewed some basic terminology used in simulation and listed the various ways to classify a simulation problem, including the simulation purpose, target, nature, tool, and the phase of systems development where the simulation is being carried out. The chapter also discussed a key aspect of simulation: random number generation (RNG), which we then carried out for perfectly random situations as well as partially random situations (referred to as random variable generation). The method of inverse transformation also was discussed as a useful technique for random variable generation. The chapter concluded with illustrations in the various branches of civil engineering and the use of Monte Carlo simulation in predicting the variability in outcomes of some system attribute in response to the variability in the input factors that influence that attribute. Simulation is a useful tool for systems description and analysis; therefore, we will encounter this tool in interesting applications in subsequent chapters including Chapter 14 (systems dynamics) and the various phases of system development (Chapters 19–26).

EXERCISES

1. Using a spreadsheet, generate (a) 10 random numbers between 0 and 5 (b) 10 random numbers between 0 and 1000 (c) 50 random numbers between 0 and 1000 (d) 10 random numbers between 100 and 1000 (e) 100 random numbers between 5 and 70 (f) 70 random numbers between −15 and −1
2. Use a recursive multiplicative generator defined by the following equation:

$$I_i = (5\, I_{i-1}) - \text{Int}\,(5I_{i-1}/8)10 \text{ for } i = 1, 2, 3, \ldots,$$

   Use the following parameter values: a (i.e., the multiplier) = 5, increment = 0, and modulus = 10, and a seed value I_0 of 3, generate 10 random numbers.
3. **Head loss through a pipe.** The head loss due to friction in a pipe is given by $h_L, = KQ^n$, where K includes the effects of the fluid viscosity, the pipe length, diameter, and roughness. For the Darcy–Weisbach formula (Delleur, 2002), K is given as: $K = 8fL/(\pi^2 gD^5)$ and $n = 2$; L is the pipe length, Q is the discharge, and D is the pipe diameter. The pipe length is 30 ft, the discharge is normally distributed with mean 1.5 ft^3/s and standard deviation 0.7 ft^3/s, the pipe diameter is 1.2 ft and the f factor ranges from 0.1 to 1.2 and could take any value within this interval. Determine the distribution type and parameters for the resulting head loss through the pipe.
4. **Design of subsoil drains to protect drinking water sources.** In certain situations, subsoil drains are required to intercept percolating water so that contamination of the groundwater resource can be minimized and, especially in arid areas, to prevent salinization of the root zone by the upward migration of groundwater. For the former application, the top of the groundwater table should be at least between 1 and 1.5 m deep (Pettygrove and Asano 1984). Two major factors that affect the subsurface drain design are the soil hydraulic conductivity (down to at least 10 ft) and the irrigation rate. The conductivity is measured in situ using prescribed borehole techniques. Typically, the depth of the subsoil drains must exceed the depth of the groundwater table. In the San Joaquin Valley of the United States, the distance between the pipe center lines is estimated from

$$L = \sqrt{\frac{4k(H^2 - d^2)}{q}}$$

   where d is the diameter of the drainpipe (m), k is Darcy's permeability coefficient (m/s), H is the height of the groundwater table, L = m, and q_{perc} is the rate of percolation water flow (m^3/m^2 s).
   If at a certain time, these inputs exhibit marked variations as indicated in the table below, determine the distribution type and parameters for the resulting contaminant concentration downstream of the outfall.

Variable	Distribution	Parameters
d	Constant	0.1 m
H	Normal	Mean = 5 m, Stdev = 0.2 m
K	Normal	Mean = 0.092 m/s, Stdev = 0.005 m/s

5. **Highway geometric design of curve radius.** For vehicles negotiating a horizontal curve, centrifugal forces act to push the vehicle radially outward and such destabilizing forces are counterbalanced by the friction force between the tire and the pavement and the vehicle weight component related to the roadway superelevation. For the laws of mechanics, the following relationship holds:

$$R = \frac{V^2}{15(0.01e + f)}$$

where R is the radius of curve (ft), V is the vehicle speed (mph), f is the side friction (demand) factor, and e is the rate of roadway superelevation. For a certain curve at a highway section in your city, the speed is normally distributed with a mean of 70 mph and a standard deviation of 15 mph; the superelevation is a constant 7%; and the friction factor is normally distributed with a mean of 6 units and a standard deviation of 1.5 units. Plot a distribution of the resulting minimum radii. Determine the distribution type and parameters for the resulting minimum radii that will ensure stability for all vehicles negotiating the curve.

REFERENCES

Aldrich, C. (2004). *Simulations and the Future of Learning: An Innovative (and perhaps Revolutionary) Approach to e-learning*. Pfeifer–Wiley, Hoboken, NJ.

Anderson, T. W., and Darling, D. A. (1954). A Test of Goodness-of-Fit. *J. Am. Statist. Assoc.* 49, 765–769.

Ang, A. H-S., and Tang, W. H. (2006). *Probability Concepts in Engineering: Emphasis on Applications to Civil and Environmental Engineering*. Wiley, Hoboken, NJ.

Ayyub, B. M., and McCuen, R. H. (2002). *Probability, Statistics, and Reliability for Engineers and Scientists*. Chapman & Hall/CRC, London, UK.

Azam, S. (2011). Sample Selection Problems in Injury Data with Linked Hospital Record. Ph.D. Thesis, Purdue University, West Lafayette, IN.

Cheema, D. S. (2005). *Operations Research*, Laxmi Publications, New Delhi, India.

Chung, C-W. (2000). Similarity Search for Multidimensional Data Sequences, Procs., 16th International Conference on Data Engineering, San Diego, CA.

Delleur, J. (2002). *Hydraulic Structures, New Directions in Civil Engineering*. CRC Press, Boca Raton, FL.

Fang, C., Galal, K. A., Ward, D. R., and Haddock, J. E. (2003). Initial Study for Cost-Effectiveness of Joint/Crack Sealing, Tech. Report No. FHWA/IN/JTRP-2003/11. Joint Transportation Research Program, West Lafayette, IN.

Fishman, G. S. (1995). *Monte Carlo: Concepts, Algorithms, and Applications*. Springer, New York.

Ford, K., Arman, M., Labi, S., Thompson, P. D., Shirole, A., and Sinha, K. C. (2011). Methodology for Estimating Life Expectancies of Highway Assets. Final Report, for the National Cooperative Highway Research Program. Washington D.C.

Fwa, T. F., Shi, X. P., and Tan, S. A. (1996). Analysis of Concrete Pavements by Rectangular Thick-Plate Model. *ASCE J. Transp. Eng.*, 146–154.

Garcia-Aracvul, N., Perez-Vidal, C., Sabater, J. M., Morales, R., and Badesa, F. J. (2011). Robust and Cooperative Image Based Visual Serving Systems Using a Redundant Architecture. *Sensors*, 11(2), 11885–11900.

Gentle, J. E. (2003). *Random Number Generation and Monte Carlo Methods*, 2nd Ed. Springer, New York.

Hartmann, S. (1996). The World as a Process: Simulations in the Natural and Social Sciences. In R. Hegselmann, et al. Eds. *Modeling and Simulation in the Social Sciences from the Philosophy of Science Point of View. Theory and Decision Library*. Kluwer, Dordrecht, pp. 77–100. http://philsci-archive.pitt.edu/archive/00002412/.

Hoeger, H. R. (1996). Integrating Concurrent and Conservative Distributed Discrete-Event Simulators. *Simulation*, 67(5), 303–314.

Knuth, D. E. (1997). *The Art of Computer Programming, Section 3.2.1: The Linear Congruential Method*, Vol. 2: *Semi-numerical Algorithms*, 3rd ed. Addison-Wesley, Reading, MA, pp. 10–26.

Kokkaew, N., and Chiara, N. (2010). Modeling Completion Risk Using Stochastic Critical Path-Envelope Method: A BOT Highway Project Application. *Constr. Manage. & Econ.* 28(12), 1239–1254.

McCuen, R. H. (2002). *Modeling Hydrologic Change—Statistical Methods*. Lewis Publishers and CRC Press, Boca Raton, FL.

Milante, G. (2009). Carana, paxsims.wordpress.com/2009/01/27/carana/. Accessed Dec. 20, 2012.

NBIMCPC (2012). National Building Information Model Standard Project Committee. http://www.building smartalliance.org/index.php/nbims/faq/. Accessed March 2, 2012.

Palisade (2012). Monte Carlo Simulation. www.palisade.com.

Park, S. K., and Miller, K. W. (1988). Random Number Generators: Good Ones Are Hard to Find. *Commun. ACM*, 31(10), 1192–1201.

Pettygrove, G. S., and Asano, T. (1984). Irrigation with Reclaimed Municipal Wastewater: A Guidance Manual, Water Resources Control Board, Sacramento, CA.

Rahman, M., Pearson, L. M., and Heien, H. C. (2006). A Modified Anderson-Darling Test for Uniformity. *Bull. Malaysian Math. Sci. Soc.* (2), 29(1), 11–16

Sawilowski, S. (2003). You Think You've Got Trivials? *Journal of Modern Applied Statistical Methods*, 2(1), 218–225.

Smith, R. D. (1998). *Simulation, Encyclopedia of Computer Science*. Nature Publishing, London, UK.

Sokolowski, J. A., and Banks, C. M. (2009). *Principles of Modeling and Simulation*. Wiley, Hoboken, NJ.

Uno, P. J. (1998). Plastic Shrinkage Cracking and Evaporation Formulas, *American Concrete Institute Materials Journal*, 95(4), 365–375.

Viessman, W., Lewis, G. L., and Knapp, J. K. (1989). *Introduction to Hydrology*. Harper and Row, New York.

Washington, S. P., Karlaftis, M. G., and Mannering, F. L. (2010). *Statistical and Econometric Methods for Transportation Data Analysis*, Chapman and Hall/CRC, London, UK.

Yacoub, M. D., Benevides da Costa, D., Dias, U. S., and Fraidenraich, G. (2005). Joint Statistics for Two Correlated Weibull Variates. *IEEE Antennas Wireless Propagation Lett.* 4, 129–132.

USEFUL RESOURCES

GNU Scientific Library (2012). Random Number Generation. http://www.gnu.org/software/gsl/manual/html_node/Random-Number-Generation.html.

Kroese, D. P., Taimre, T., and Botev, Z. I. (2011). Uniform Random Number Generation, in *Handbook of Monte Carlo Methods*. Wiley, Hoboken, NJ.

NIST (2012-2013). Random Number Generation, Computer Security Division, National Institute of Standards and Technology. http://csrc.nist.gov/groups/ST/toolkit/random_number.html.

Vose (2007). Model Risk, Vose Software, Gent, Belgium. ACI Journal (95)4, American Concrete Institute, Detroit, MI.

CHAPTER 9

OPTIMIZATION

9.0 INTRODUCTION

The term *optimization* often conjures up images of complicated and bewildering mathematical equations to a layperson. In reality, however, optimization, far from being an esoteric concept, is a tool that we all use implicitly in the task of making decisions in our daily lives. For example, in seeking a apartment to rent for the coming semester, you first establish your objective (to find a suitable apartment) and set up the factors that you would like to consider in your decision (rent amount, distance from campus, proximity to bus terminals, availability of laundry facilities, etc.). Then, you establish your goals and constraints in terms of these decision factors. Similarly, the stockbroker who selects which stocks to buy or sell, the physician who decides what treatments to apply to a patient, the professor who designs a syllabus for her course, the logistics manager who selects which routes to be traveled by freight vehicles, and even the person seeking to choose someone to marry, carries out optimization, even if implicitly, in order to establish the best solution for the problem within the given constraints.

In the same fashion, in the task of decision making at any phase of systems development, whether it is planning, design, operations, monitoring, maintenance, or end of life, the tool of optimization is used by systems engineers to identify the best solution to a given problem. Such decisions often include what resources are needed, in what quantities, and where and when they are needed, in order to maximize a certain benefit and/or to minimize a certain cost. Resources could include manpower, money, materials, machinery, or any combination of them. The optimal allocation of such resources is typically carried out within the limitations of these resources or others, such as funding level and space (right of way) availability. Thus, in the context of civil engineering systems, optimization may be defined as the identification of the "best" solution that ensures that some optimal system objective (e.g., a maximum overall benefit or a minimum overall cost) is attained under the given constraints at the development phase in question. Figure 9.1 presents examples where optimization tools are useful for the task of making decisions at the various phases of systems development.

At the system planning phase, civil engineers typically may seek the best plan, location, or overall operational policy for the system. For example, what is the optimal route for a new railway such that mobility is maximized, ecosystem fragmentation is minimized, construction cost is minimized (by reducing the need for excavation and filling), user operations cost is minimized (by avoiding steep gradient designs and sharp curves), and so forth. At the system design phase, it may be sought to identify the best dimensions for a physical system or the best combination of resources for a nonphysical system; for example, what dimensions of structural elements provide the least cost while supporting the structure adequately? At the system construction phase, engineers select the optimal project delivery approach for the construction; for example, for a specific project, should the work be carried out in-house by the system owner or should a contract be awarded? If the latter is decided, should it be a warranty contract or a traditional bid–build contract? At the system operations phase, civil engineers encounter frequent applications of optimization; for

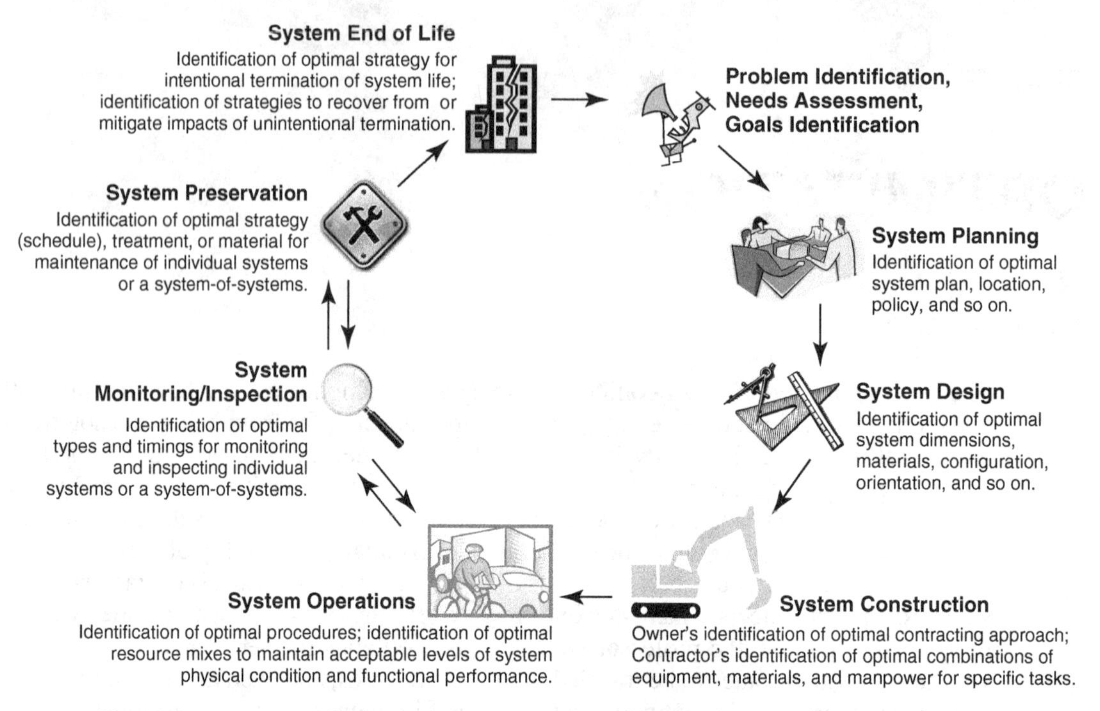

Figure 9.1 Examples of tasks that involve optimization tools at various phases of system development.

example, how much to charge highway users through license fees in order to maximize revenue or how to direct evacuation operations on a highway network to maximize the flow of people out of a disaster area. At the system inspection and monitoring phase, engineers often seek to establish the optimal sample size so that the maximum possible information on system use and condition is obtained at a certain cost. In system preservation, the engineer seeks the optimal annual maintenance spending amounts to yield a certain minimum performance level or to identify the optimal maintenance strategy (which treatments to apply and at which years) that satisfies the constraints of the facility performance and maintenance budget. At the end-of-life or termination phase, particularly where the physical system is deliberately destroyed or decommissioned, it may be required to ascertain the best course of action to take—whether to demolish the system (and if so, whether to sell the physical components, dispose of them, or recycle their constituent materials) or to convert the system to a historical or alternative-use facility.

As we saw in the preceding paragraph, the objective in civil engineering optimization problems typically is to maximize some desired impact or to minimize some undesired impact, or both; these constitute the civil engineer's **objective function**. Where there are multiple objective functions or multiple competing components of a single objective function, the "best" solution is often a compromise between the competing objective functions or objectives. What the civil engineer controls in order to realize the objective are referred to as the **decision variables** (also termed design variables, input variables, or control variables). In seeking the optimal choice of actions, decisions are often made in the face of certain boundaries imposed on the levels of the decision variables; these boundaries are referred to as **constraints**. The common categories (and examples) of constraints in civil engineering include financial (budgets), physical (right-of-way

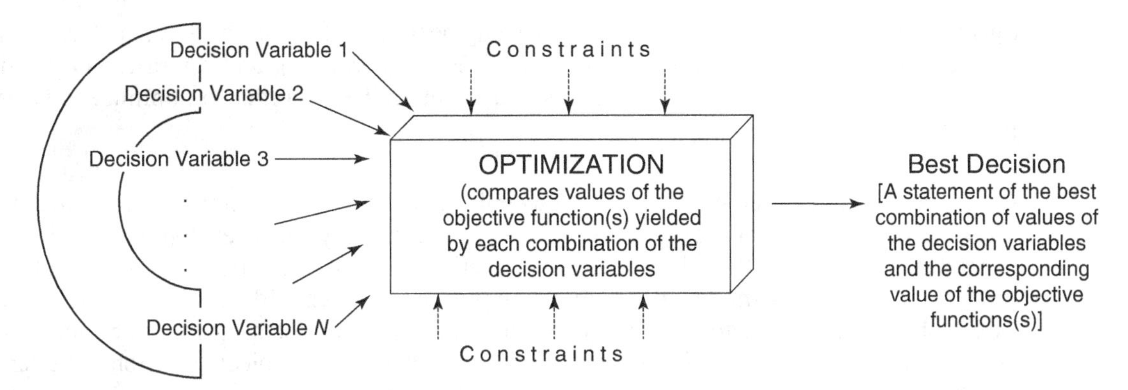

Figure 9.2 Basic concept of optimization.

limitations), institutional (legal and administrative requirements), and political. Thus, in a typical problem structure, both the objective and the constraints are expressed as a function of the decision variables. Figure 9.2 presents the basic concept of optimization.

9.0.1 Relationship between Optimization, Evaluation, and Decision Making

The terms optimization, evaluation, and decision making are often used loosely and even interchangeably, but it is important to establish the nuances in their meanings. Optimization is a tool that the systems engineer uses to quickly assess and compare different alternative candidate solutions and to identify the optimal. In other words, the engineer searches for and finds that specific action, decision, or set of resource inputs and amounts that would yield the maximum overall benefit or a minimum overall cost under the given constraints. Thus, optimization tools greatly facilitate the key systems tasks of evaluation and decision making; simply put, optimization is a tool while evaluation is a task. Evaluation typically utilizes often to a greater extent, several other tools, including costing, simulation, and risk analysis and incorporates several subtasks that include cost analysis, benefit analysis, tradeoff analysis and comparison of costs and benefits of the candidate solutions. Decision making is a task that logically follows the evaluation (i.e., the systems engineer decides which course of action to take on the basis of the evaluation results).

9.0.2 Categorization of Optimization Problems

Optimization problems can be categorized in several ways, including the characteristics of the problem and the solution technique, as well as several other ways that are discussed below.

(a) Phase of the System Development. The primary criterion for categorizing optimization problems is the phase of systems development at which the optimization is carried out. In this respect, as shown in Figure 9.1, the problem may be one of **design optimization** (e.g., where the designer seeks the best dimensions for a physical system or the best control parameters for a nonphysical system, such as determining the best section for a proposed canal that maximizes flow while minimizing cost), **construction optimization** (e.g., where the construction manager identifies the best combination of labor and equipment to complete a task or where the project client seeks the best contracting approach to deliver a specific project), **monitoring/inspection optimization** (e.g., where the engineer identifies the most cost-effective sample size for inspecting the physical condition of system components or for monitoring the system usage), **preservation**

optimization (e.g., where the maintenance engineer identifies the best repair treatments and their timings over the life cycle or remaining life of the system), and **operations optimization** (e.g., where the speed limit for an urban street is determined in order to minimize crashes while maximizing mobility).

(b) Level of Management. There are generally two levels of systems management: **network level** and **facility level** (or project level). At certain phases of systems development, optimization is needed at one or both of these levels. At the network level (also referred to as system level, program level, or system-of-systems level), optimization is used for systemwide management functions such as establishing priorities for actions (i.e., interventions, or projects) at multiple facilities in the network; determining the optimal use of limited funds through project selection; or establishing optimal spending amounts per year for operations, monitoring, or maintenance of the entire network or investigating the trade-offs between alternative projects or groups of projects at any phase. An advantage of such **network-level optimization** is that the system manager acquires a bird's-eye overview of the impacts of decisions on all the systems in a network, including the systemwide performance consequences of different inputs (e.g., spending amounts). A disadvantage is that this level of optimization often utilizes data that are only aggregate in nature and thus the optimal solution may not provide fine details on the recommended decision or action. In **project-level optimization**, however, decisions are sought for a specific individual system, and optimization thus is typically more comprehensive and incorporates more detailed technical data on that specific system. Network-level and facility-level management functions are interdependent and synergistic in that each one feeds off the results of the other.

(c) Attribute of the System under Consideration. Problems in civil engineering systems optimization can also be categorized by the attribute of the system that one seeks to optimize. This may be the **physical component** of a system (e.g., what the best section is for a proposed canal that maximizes flow while minimizing cost), or it may be for the **abstract component** of a system for example, what is the optimal speed limit for a freeway. This categorization scheme should not be confused with the scheme based on the phase of system development: at any phase of development, optimization may be for the physical system or for its use. For example, at the phase of system operation, the civil engineer may seek to select within budgetary constraints a number of critical bridges for retrofit purposes to minimize the risk of failure in the event of a disaster (such optimization addresses a physical component of the highway system); also at that phase, the civil engineer may seek to design a routing system to facilitate the evacuation of persons from a specific area in the event of disaster (such optimization addresses an abstract component of the highway system).

(d) Number of Objectives. The optimization problem may involve only one objective (**single-objective optimization**) or several, often conflicting, objectives (**multiobjective optimization**). This approach should not be confused with multiple-attribute optimization. An objective generally indicates the direction in which we should strive to "do better" and an objective is measured in terms of at least one attribute. For example, in multiple-objective optimization, there could be just one performance attribute that is maximized or minimized in the different objectives (Table 9.1).

(e) Number of Attributes. The optimization problem may involve only one performance attribute (**single-attribute optimization**) or several attributes (**multiattribute optimization**). As seen in Table 9.1, this is different from single- and multiple-objective optimization. In multiple-attribute

Table 9.1 Different Problem Structures Showing Different Objectives and Attributes with Illustrations

	Single-Attribute Optimization	**Multiple-Attribute Optimization**
Single-Objective Optimization	Maximize Z where Z = an objective function involving some specific performance attribute (e.g., average condition of all systems in the network).	Maximize Z where Z is a single-valued function of several performance attribute such as condition, safety, aesthetics, cost, etc. or Maximize Y where Y = an objective function involving some specific performance attribute Subject to: $f(W_1) \geq, \leq, = K_1$ $f(W_2) \geq, \leq, = K_2$ ... $f(W_R) \geq, \leq, = K_R$ where W is some system performance attribute other than that of Y.
Multiple-Objective Optimization	Maximize Z_1 where Z_1 = an objective function involving some specific performance attribute. Maximize Z_2 where Z_2 = an objective function involving same performance attribute but in a different form, for example, range of physical condition of any system in the network. ... Maximize Z_3 where Z_3 = an objective function involving same performance attribute but in a different form, for example, minimum condition of any system in the network.	Maximize Z_1 where Z_1 = an objective function involving some specific performance attribute (e.g., average condition of all systems in the network). Maximize Z_2 where Z_2 = an objective function involving some specific performance attribute different from that in Z_1, for example, average safety performance of the network. ... Maximize Z_3 where Z_3 = an objective function involving some specific performance attribute different from that in Z_1 or Z_2, for example, total cost of maintenance for the network.

optimization, there could be just one objective; and in that case, it is called a multiple-attribute problem because there are constraints associated with each of several different performance criteria, the objective consists of several different performance criteria, or both.

For multiattribute optimization problems in civil engineering systems, the choice of criteria for a specific problem is generally influenced by a gamut of considerations such as agency/owner policy, system type or design, management level in question (network vs. facility), stakeholder concerns, and the like. For example, for highways and streets in older communities, historic system preservation can be an important attribute; for isolated regions, uninterrupted accessibility is often a critical attribute. Also, for systems in urban areas, the agency/owner may prefer interventions and technologies (e.g., surface material) that minimize/eliminate noise or vibration. In the past, the cost (initial or life cycle) has largely served as the sole or dominant attribute in decision making. However, recognizing that the cost attribute alone may not provide robust, sustainable, and universally acceptable decisions, systems engineers are increasingly utilizing more than one performance attribute for optimization in systems management (Pickrell and Neumann, 2000). Compared with single-attribute optimization, multiattribute optimization is considered more realistic and consistent with the functions of systems managers and makes it possible for them to assess the trade-offs inherent with the different criteria; for example, determining what amount of a given attribute could be "bought" or "sold" for a given amount of another.

(f) Discrete/Continuous Nature of the Decision Variable. In every optimization problem, there is at least one control or decision variable whose values influence the output that the systems engineer seeks to optimize. If the decision variable is discrete, then the problem can be described as a **discrete variable optimization** or simply as a **discrete optimization** problem. If the decision variable is continuous, then the problem is a **continuous-variable optimization** problem. Discrete optimization problems may involve count variables (e.g., the number of production units required) or binary variables (e.g., whether or not to select a specific type of resource for a given task or function). In certain cases, if the discrete decision variables are all integers, then the problem is an integer optimization problem. Different programming techniques are used to solve the problem depending on the discrete/continuous nature of the decision variable. Integer programming is used when the decision variables are positive integer numbers; binary programming is used where the decision variables take values of 1 or 0 (e.g., "yes" or "no"; failed" or "not failed") and is also referred to as Zero–one programming. Mixed programming is used when there is a mix of decision variable types: some continuous, others discrete. As seen in Figure 9.3 and Table 9.2, in linear optimization problems, the solution methods include linear programming (LP), where the objective and constraint functions are linear in terms of the decision variables, and nonlinear programming (NLP), where the objective and/or constraint functions are nonlinear in terms of the decision variables. In a later section of this chapter, LP and NLP are discussed in greater detail.

(g) Structure of the Objective Function. The designation of an optimization model as linear or nonlinear depends on the power of the decision variables and the existence of a weighted-sum

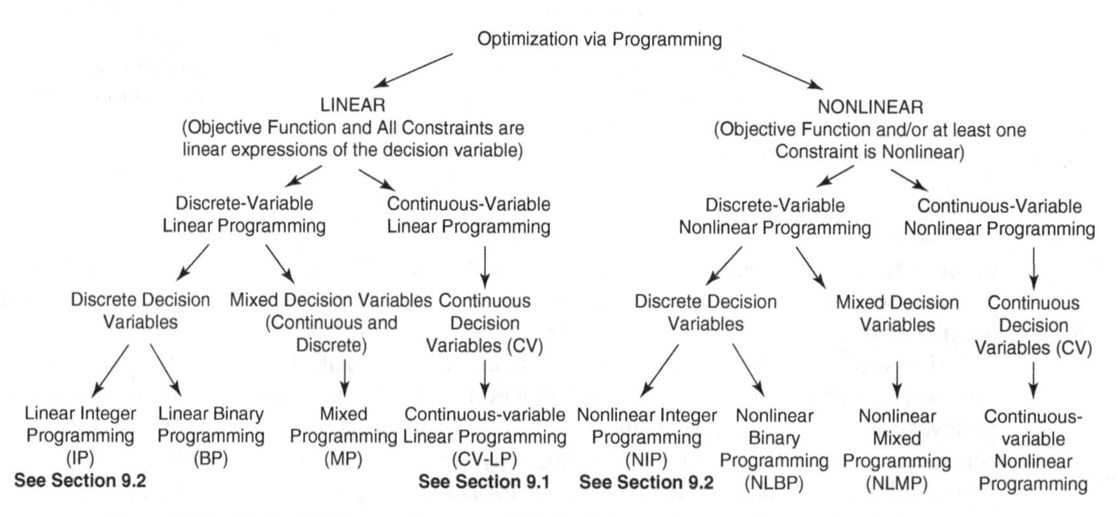

Figure 9.3 Optimization problems categorized by structure of objective function and constraints and nature of decision variable.

Table 9.2 Optimization Problems Categorized by Structure of Objective Function and Constraints and Nature of Decision Variable

Structure of the Objective Function (OF) and/or Constraints	Nature of the Decision Variable (DV)			
		Discrete		
	Continuous	Integer	Binary	Mixed
Linear	Continuous-variable linear programming (CV-LP)	Integer Linear Programming Programming (ILP)	Binary Linear (BLP)	Mixed Linear Programming (MLP)
Nonlinear	Continuous-variable nonlinear programming (CV-NLP)	Integer Nonlinear Programming (INLP)	Binary Nonlinear Programming (BNLP)	Mixed Nonlinear Programming (MNLP)

mathematical form of the function (as we shall see in Section 9.3.2). On the other hand, as seen in the section above, the designation of an optimization problem as discrete or continuous depends on the discrete or continuous nature of the decision variable (Figure 9.3 and Table 9.2). A mathematical program is a discrete optimization model if it includes *at least one* discrete decision variable. On the other hand, only when *all* the decision variables are continuous is the mathematical program described as a continuous optimization model.

(h) Effort of the Search. In simple language, optimization means *searching for* the values of some control variable that would yield some optimum value of some system output. In certain cases, the optimal solution is obtained through engineering judgment and experience. Referred to as **subjective optimization**, this approach is relatively simple and involves minimal analytical effort. At certain agencies, applications of this category of optimization are documented in the form of rules of thumb that have evolved over the years through experiential evidence and continuous refinement and have long served as the basis for their policies and actions. Techniques for subjective optimization include questionnaire surveys of experts; for example, in highway systems, expert opinion has often been used to establish the optimal speed limit at rural freeways that minimizes crashes, pollution, and fuel use, but maximizes productivity. At a higher level of effort, optimization can involve an extensive search before arriving at the optimal solution by examining quantitatively each and every possible scenario, establishing the output of the objective function (i.e., the solution) that corresponds to each scenario, and identifying the best solution. Such **enumeration** or "optimization by repeated simulation" is often a slow and laborious process that is most useful where the relationships between the objective function and the decision variables are not so clear or not explicitly stated in mathematical terms. For example, the engineer may decide to calculate the predicted values of an output performance variable for incremental levels of a decision variable and then identify the value of the decision variable that yields optimum performance. However, for many engineering systems, the typically large number of alternative solutions to any decision problem make enumeration excessively laborious and thus infeasible for application. In such cases, utilizing a higher level of effort is needed, known as **analytical optimization** (Meredith et al., 1985, Dandy et al., 2008), which includes a variety of search techniques (exterior point, interior point, etc.) and accompanying algorithms (Simplex, Tabu, Nelder–Mead, etc.). This optimization category is appropriate when the objective function is very explicit or when calculus or programming techniques can be used to determine the optimal solution with relatively little time and effort. In many cases of engineering system optimization, the results of analytical optimization are duly adjusted using subjective optimization techniques.

(i) Convex versus Nonconvex. A feasible region is the domain of decision variable combinations where all of the optimization constraints are satisfied. In this respect, we could have **convex optimization** or **nonconvex optimization** problems, depending on the configuration of the feasible region, which, in turn, is dictated by the nature of the constraints. A convex optimization problem is one whose feasible region is such that if P_1 and P_2 are any two points in the feasible region, then the segment joining them is also in the feasible region. For example, boundaries such as those shown in Figure 9.4*a* enclose convex feasible regions. However, the regions shown in Figure 9.4*b* are not convex because it is possible to choose at least one pair of points such that not every point on the segment joining them belongs to the region. Many nonlinear optimization problems are convex and thus are guaranteed to yield global optima. The presence of nonconvexity does not mean that there is a need to reformulate the problem as a convex programming problem but means that additional steps are needed to ensure that good solutions are identified. Currently available traditional solvers (e.g., reduced gradient) and newer ones (e.g., heuristic solvers) facilitate the solution of nonconvex optimization problems (Revelle et al., 2004).

(j) Deterministic versus Stochastic Nature of the Problem. In optimization, the desired levels of different inputs to yield some optimal system outcome (objective) are sought. A statement

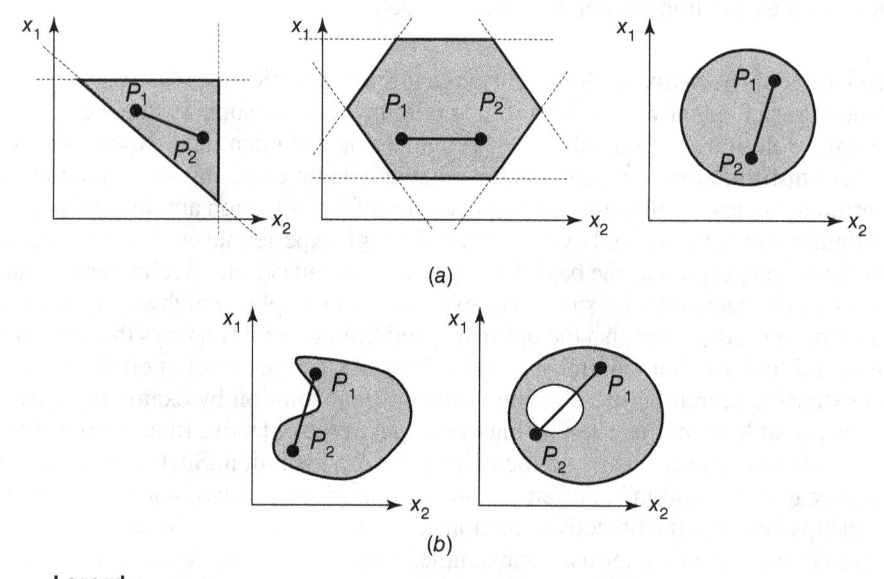

Figure 9.4 (*a*) Convex and (*b*) nonconvex feasible regions.

of whether the problem is deterministic or stochastic depends not only on the level of knowledge about the inputs but also about the relationship by which they translate into the system outcome (i.e., the extent to which the objective is achieved) in response to the inputs. Where each input shows no variation and the outcome for every combination of inputs is known with certainty, the problem is described as a **deterministic optimization** or optimization under certainty; otherwise, it is categorized as stochastic optimization. **Stochastic optimization** problems may be further categorized into risk (where the probability of each possible outcome is known) and uncertainty (where the probability of each possible outcome is not known). In Chapters 5 and 13, we discuss concepts that are related to stochastic optimization.

(k) Direction of the Objective Function. The objective function in a single-objective optimization problem or any one of the objective functions in a multiple-objective problem could be one of maximization or of minimization. Minimization is often used in the context of cost or a cost-driven or disutility-driven expression, such as the cost–benefit ratio; and maximization is often used in the context of benefits or a benefits-driven expression, such as the benefit–cost ratio and the net present value.

9.0.3 Formal Definitions of Common Terms Used in Optimization

We now examine some definitions and mathematical notations that are common in the optimization of engineering systems. These will serve as a precursor to our understanding of complete formulations of optimization problem structures, as we shall see in subsequent sections of this chapter and also in subsequent chapters.

Control or Decision Variable ($x_1, x_2, \ldots, x_n$). A parameter that represents the amount of resources or inputs that characterize each candidate solution. The objective and constraints are expressed in terms of the control variables.

Objective Function $f_O(x_1, x_2, \ldots, x_n)$. A single-valued function of the set of decision variables (x) that represents the expression for the optimal value.

Constraint conditions $f_C(x_1, x_2, \ldots, x_n)$. Mathematical notations of the financial, physical, institutional, or other limitations placed upon the optimal solution.

Feasible Solution Space. The set of all combinations of the control variables that satisfy the constraint conditions.

Optimal Solution. A feasible solution that satisfies the objective function. In the context of civil engineering systems, this is often a statement that specifies which, how many, or how much of each input or resource should be applied to achieve some objective.

9.0.4 Influence of the Nature of Decision Variables on Optimization Technique

As we saw in Section 9.0.2(f), there are two kinds of decision variables: continuous and discrete. Continuous variables are real numbers while discrete variables include integers or binary numbers. Several of the optimization problems encountered in real life deal with continuous variables. Often, in these cases, too little of the decision variable is as undesirable as too much of it. For example, how much salt should we add to a recipe to get the right taste or how much sleep will give us enough rest for the following day? The situation is no different in civil engineering, where, for example, we may seek the right amount of cement to add in a concrete mix, the right year to rehabilitate or expand a water treatment plant, the right slope to provide in designing a sewer, the appropriate length of green light to assign to each phase of traffic signals, and so forth. In civil engineering, other examples of continuous decision variables include time, weight, dimension (length, breadth, width, depth, and height), area, volume, angle, and ratio. In this chapter, we shall discuss the various categories of continuous variable optimization problems and techniques for solving each category. These techniques include calculus methods, where the problem has constraints or where it does not have constraints, as well as linear and nonlinear programming. Specifically, Section 9.1 discusses unconstrained optimization using calculus; Section 9.2 deals with constrained optimization using calculus; Section 9.3 discusses constrained optimization using linear and nonlinear mathematical programming techniques; and Section 9.4 provides the tools for constrained optimization involving binary decision variables.

9.1 UNCONSTRAINED OPTIMIZATION USING CALCULUS

The general unconstrained optimization problem can be written as follows:

$$\text{Max or Min } Z = f(x_1, x_2, \ldots, x_n)$$

where $f(x_1, x_2, \ldots, x_n)$ is a nonlinear function of n decision variables, $x_1, x_2, \ldots, x_n$.

The most common kind of unconstrained optimization problem is the single-variable (or, one-dimensional) problem:

$$\text{Max or Min } Z = f(x)$$

where $f(x)$ is a nonlinear function of x, the decision variable.

The optimal solution, or the coordinates of the maximum or minimum points, can be found using classical differential calculus. As the reader will realize, this can be possible only if $f(x)$ is such that first-order and second-order derivatives exist and are continuous over all values of the decision variable, x.

For a maximum value of the objective function, the following conditions must be satisfied if x^* is the solution:

$$\frac{df(x = x^*)}{dx} = 0 \tag{9.1}$$

$$\frac{d^2f(x = x^*)}{d^2x} < 0 \tag{9.2}$$

For a minimum value of the objective function, the following conditions must be satisfied if x^* is the solution:

$$\frac{df(x = x^*)}{dx} = 0 \tag{9.3}$$

$$\frac{d^2f(x = x^*)}{d^2x} > 0 \tag{9.4}$$

Equation (9.1) states that the objective function should have zero slope when the decision variable is optimum. This equation is a necessary condition for the decision variable x^* to be the maximum or minimum solution. However, it is not sufficient because an x^* that satisfies this condition is still ambiguous with respect to whether it is a minimum, maximum, or inflexion point (Figure 9.5). Note that x^* is an inflexion point if the objective function has zero slope at that point, but x^* is neither a minimum or maximum point of the objective function.

For x^* to be the maximum or minimum solution, the above pairs of equations [(9.1) and (9.2) and (9.3) and (9.4), respectively] are sufficient.

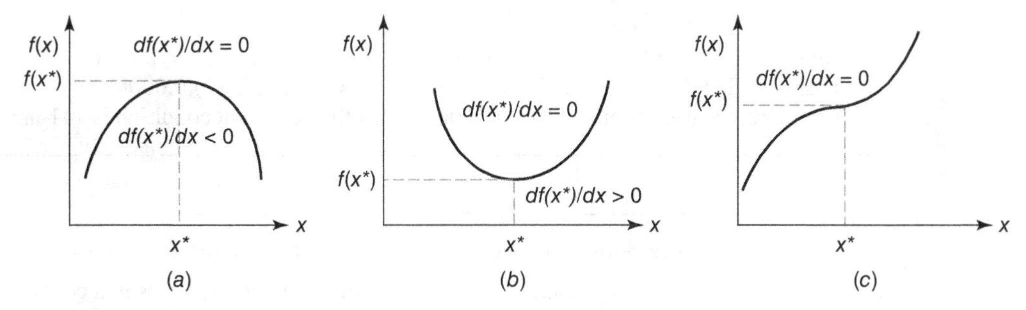

Figure 9.5 Conditions for unconstrained optimization involving one decision variable: (*a*) maximum, (*b*) minimum, and (*c*) inflexion points of the objective function.

Where the problem at hand has no constraints, the optimum (maximum or minimum) value of the decision variable is determined by differentiation. The first derivative identifies the turning point of the decision variable while the second derivative determines whether the identified turning point is a maximum or a minimum. For a turning point, maximum or minimum value of the objective function, $dy/dx = 0$. For the maximum and minimum values of the objective function, d^2y/dx^2 is negative and positive, respectively.

The above discussion is for the relatively simple case involving one decision variable (i.e., univariate problems). The tools for solving optimization problems for unconstrained multivariate optimization problems can be found in Dandy et al. (2008), who present the necessary and sufficient conditions for the optimal solution. For the special case of constrained multivariate optimization problems, the reader may refer to the very interesting and useful tools that are presented in texts including Ossenbrugen (1984), Taha (2003), and Rardin (2002).

Example 9.1 Performance Threshold for Intervention

At the phase of system preservation, a common decision context is "the right time to undertake an action," and this situation is referred to as the optimal timing problem. For example, in preventive maintenance, treatments are applied to a system in fair-to-good condition to arrest the onset of imminent structural distress and to defer the time of rehabilitation. For flexible highway pavements, a popular preventive treatment is thin hot-mix asphalt overlay, and a key determination is the pavement condition at which this treatment must be applied. Figure 9.6 presents the cost-effectiveness corresponding to various candidate trigger levels of pavement condition, x, at which the treatment is applied. What is the optimal timing for applying this treatment?

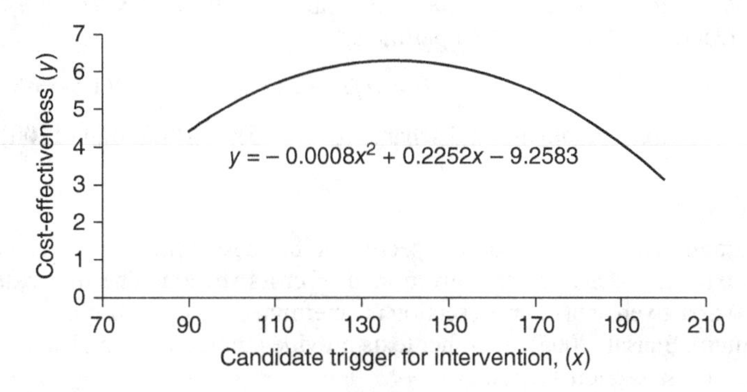

Figure 9.6 Optimal threshold function for highway preventive maintenance.

Solution

Let y represent the cost-effectiveness. At maximum cost-effectiveness, $dy/dx = 0$.

That is, $-(2)(0.0008)x + 0.2252 = 0$. This yields $x = 0.2252/(2)(0.0008) = 141$.

Therefore, the treatment should be applied when the pavement condition is 141 units.

Example 9.2 Maximizing Flow through Urban Drainage Rectangular Channel

It is sought to design a drainage canal to reduce perennial flooding problems in a certain urban area. A rectangular section is preferred to a trapezoidal channel due to right-of-way restrictions. It is also sought to maximize, during operations of the system, the flow through the system. Determine the ratio between the depth of the flow and the breadth that yields the maximum flow (Figure 9.7). Find the maximum discharge if the channel is 3 m wide, $C = 60$, and bed slope i is 1 in 1200.

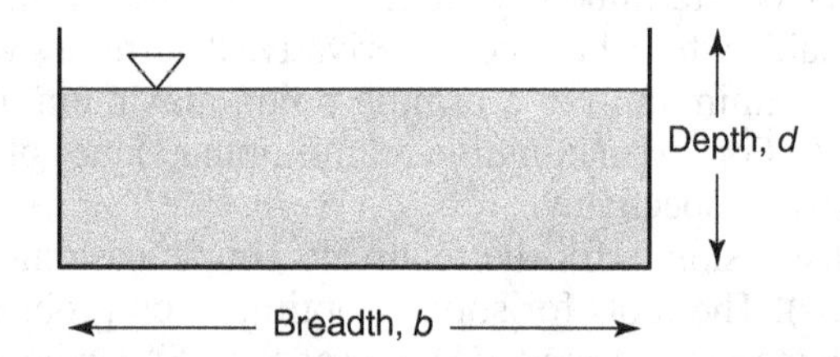

Figure 9.7 Rectangular section of hydraulic system for Example 9.2.

Solution

(a) Consider the rectangular channel of area A. Let b and d be the breadth and depth of flow. Area of flow $= A = bd =$ a given constant. The wetted perimeter $= P = b + 2d$. Thus, $P = A/d + 2d$.

Then the discharge Q, is given as

$$Q = AC\sqrt{\frac{A}{P}i}$$

where A, C, and i are the constants. Therefore, for maximum discharge, P should be minimum.

Perimeter P is minimum when $dp/dd = 0$, that is, $d(A/d + 2d)/dd = 0$. Differentiating this expression yields $d = b/2$.

Thus, for maximum discharge in the channel, the depth of the flow should be kept at one-half the channel's breadth.

(b) $b = 3$ m, $C = 60$, $i = 1/1200$. For maximum flow, $d = b/2 = 3/2 = 1.50$ m. Thus, $A = bd = (3)(1.5) = 4.5$ m^2. Wetted perimeter

$$P = A/d + 2d = 4.5/1.5 + 2(1.5) = 6 \text{ m}$$

Therefore, maximum discharge, $Q = (4.5)(60)([4.5/6][1/1200])^{0.5} = 6.75$ m^3/s.

Comment

In hydraulics, the most economical section or the "best section" of a channel are dimensions that provide the maximum discharge for a given area of cross section. The linear dimensions (width, depth, and radius) of a given area of cross section is determined via optimization so that the discharging capacity is maximum. Bansal (2005) and other texts provide numerous examples of such optimization for different channel cross-section shapes.

Example 9.3 Optimal Tank Design (Adapted from Revelle et al., 2004)

It is sought to design a vertical open-topped cylindrical tank with volume k for storing water to serve a small city. The material thickness of the flat bottom must be twice that of the sides. In order to minimize the material to be used in the construction, what is the optimal ratio between the diameter and the height (Figure 9.8)?

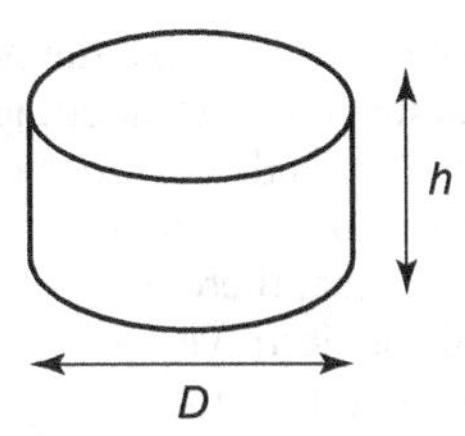

Figure 9.8 Vertical cylindrical steel tank for water storage.

Solution

Let M be the amount of material to be used in the tank construction; M is given by:

$$M = t(\pi Dh) + 2t(\pi D^2/4)$$

where D is the diameter, h is the height, and t is the thickness of the tank walls.

It is sought to minimize M subject to: $(\pi hD^2/4) = k$, where M is the total amount of material used (cubic feet), t is the thickness of the sides (feet), and k is the required volume of a tank (cubic feet). Solving for h in terms of D, $h = (4k)/(\pi D^2)$, and substituting yields the following objective function:

$$\text{Minimize } M = \frac{4tk}{D} + \frac{1}{2}t\pi D^2$$

This objective function is convex in terms of D; thus, any stationary point will be a global minimum. Taking the partial derivative and equating to zero yields

$$\frac{\partial M}{\partial D} = -\frac{4tk}{D^2} + t\pi D = 0$$

Thus

$$D^3 = \frac{4k}{\pi} \quad D = \left(\frac{4k}{\pi}\right)^{1/3}.$$

Substituting for h yields

$$h = \left(\frac{4k}{\pi}\right)^{1/3}$$

Thus, $D/h = 1$. Therefore, for optimal economy in the tank design, the diameter of the base floor must be equal to the tank height.

Example 9.4 Optimal Design of Flood Protection Systems

In designing a flood protection system, the hydraulic engineer seeks a reasonable balance between cost and risk. On one hand, the levee must be high enough to reduce the likelihood or frequency of overtopping (and destruction of property) during flood events. The height of the levee, on the other hand, must not be so high that its costs are prohibitive (Au et al., 1972). Thus, an optimal height is one that balances the benefits (protection of the property) and costs (construction and maintenance) (Figure 9.9).

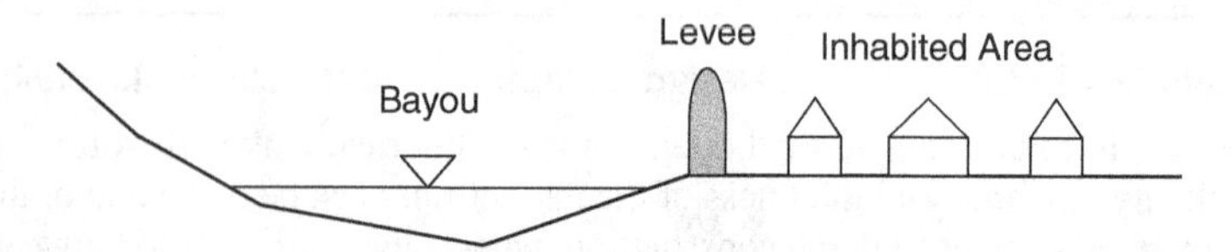

Figure 9.9 Flood protection optimization.

The city authorities seek to minimize the net long-term cost in terms of levee construction and maintenance and property damage due to flooding. For each levee height starting from 0 ft (no levee), the total cost of the levee is C_H, and the expected annual damage cost from flooding is Vp_H, where $V(= 250$ million) is the total value of vulnerable property in the inhabited area; H is the height of the levee; and p_H is the probability that there will be a flood big enough to inundate the inhabited area and that vulnerable property will be destroyed. Clearly, a greater height increases C_H but decreases Vp_H. A lower height has the opposite effect. Obviously, there may be a certain levee height that balances C_H and Vp_H such that the total cost is minimized. Using Table 9.3, determine the optimal height of the levee.

Table 9.3 Example Problem 9.4

Height (ft)	**0**	**2**	**4**	**6**	**8**	**10**	**12**	**14**	**16**	**18**	**20**	**22**	**24**
Wall cost ($M)	0	10	25	40	50	60	70	76	84	91	94	97	99
Overtopping probability	1	0.76	0.58	0.48	0.40	0.32	0.26	0.22	0.20	0.19	0.18	0.17	0.16
Property value ($M)	250	190	145	120	100	80	65	55	50	47	44	43	42

Solution

The expected property damage corresponding to each levee height can be determined as shown below. Also, these points are plotted with the levee height as shown in Figure 9.10.

Height (ft)	**0**	**2**	**4**	**6**	**8**	**10**	**12**	**14**	**16**	**18**	**20**	**22**	**24**
Expected property damage cost ($M)	250	190	145	120	100	80	65	55	50	47	44	43	42

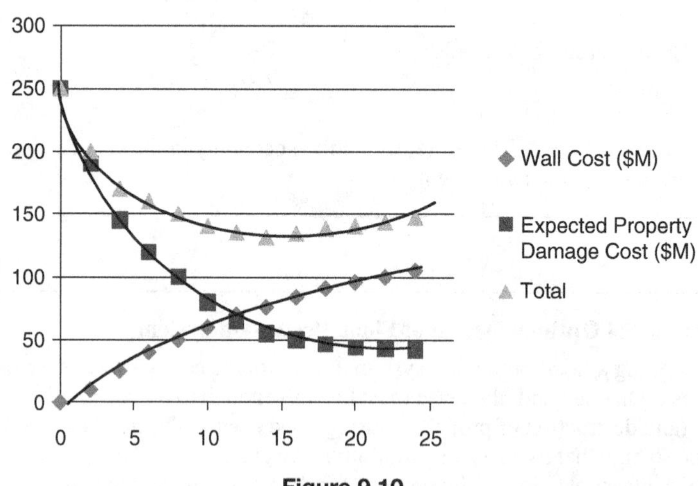

Figure 9.10

An approximate value of the optimal height can be estimated visually from Figure 9.10. However, for a more accurate identification of $h_{optimal}$, a statistical function can be developed to fit the total cost data as follows:

$$TC = 0.2904h^2 - 9.5696h + 209.57$$

At a maximum value of h, $d(TC)/dh = 0$.

Differentiating this function and equating to zero yields $h = 16.5$ ft. Therefore, the optimal height of the levee is 16.5 ft.

9.2 CONSTRAINED OPTIMIZATION USING CALCULUS

9.2.1 Calculus with Substitution

The calculus-with-substitution technique is useful for small nonlinear optimization problems having equality constraints. With an objective function $Z = f(x)$, constraint functions, $g_i(x)$, and a decision variable x, these problems are generally formulated as follows:

$$\text{Optimize } Z = f(x_1, x_2, \ldots x_n) \tag{9.5}$$

Subject to

$$\left.\begin{aligned} g_1 &= f(x_1, x_2, \ldots, x_n) = b_1 \\ g_2 &= f(x_1, x_2, \ldots, x_n) = b_2 \\ &\vdots \\ g_m &= f(x_1, x_2, \ldots, x_n) = b_m \end{aligned}\right\} \tag{9.6}$$

where $m < n$.

This problem has been referred to in the literature as the "classical optimization problem." From a theoretical viewpoint, the constraint equations could be used to solve for m variables in terms of the remaining $(n - m)$ variables; and then these expressions could be substituted into the objective function to yield an expanded but unconstrained optimization problem containing $(n - m)$ variables (Rao, 2009). Then, the solution could be determined using calculus, as shown in the previous section. It is worth mentioning that, in practice, it is typically difficult to solve the constraint equations for the m variables, particularly when there are several constraint equations. In such cases, the method of Lagrange multipliers, described in the next section, is useful in seeking a solution to the optimization problem.

9.2.2 The Lagrangian Technique

For continuous-variable optimization problems with constraints, the Lagrange multiplier method can be used to transform such constrained problems into unconstrained problems. This method involves the following steps:

1. **Express** the objective function and constraints in mathematical notation. For the decision variables $x_1, x_2, \ldots, x_n$,

   $Z = f(x_1, x_2, \ldots, x_n)$, where we seek to minimize or maximize Z subject to the constraint $g(x_1, x_2, \ldots, x_n) = 0$.
2. Introduce a new variable λ such that

$$\Phi(x_1, x_2, \ldots, x_n, \lambda) = f(x_1, x_2, \ldots, x_n) + \lambda g(x_1, x_2, \ldots, x_n)$$

3. Take partial derivatives with respect to each decision variable:

$$\frac{d\phi}{dx_1}, \frac{d\phi}{dx_2}, \ldots, \frac{d\phi}{dx_n}, \frac{d\phi}{d\lambda}$$

In its most general form, the technique absorbs all the constraint functions into the original objective function, thus creating a new problem formulation with an expanded objective function and no constraints. The example below uses, for purposes of illustration, three variables and two constraints; but the formulation can easily be extended to situations with different numbers of variables and constraints.

Example 9.5

Suppose the objective is to maximize or minimize $Z = f(x, y, v)$, and the constraints are

$$g_1 = (x, y, v) = b_1$$

$$g_2 = (x, y, v) = b_2$$

Let λ_1 and λ_2 be the *Lagrange multipliers* for the two constraints, and establish the new and unconstrained objective as

$$L = f(x, y, v) + \lambda_1[b_1 - g_1(x, y, v)] + \lambda_2[b_2 - g_2(x, y, v)] \tag{9.7}$$

where L is referred to as the *Lagrangian function.*

The initial problem has been reduced to an objective function only, and calculus may be used to find the optimal solution. Any number of constraints can be absorbed into the objective function in this manner. However, note that the problem can become cumbersome to solve, particularly if the constraints are nonlinear.

In this example, five partial derivatives need to be determined so that calculus can be applied. These are the partial derivatives with respect to each of the variables, x, y, v, and with respect to each of the multipliers, λ_1 and λ_2. The derivatives take the following general forms:

$$\left.\begin{aligned}
\frac{\partial L}{\partial x} &= \frac{\partial f(x, y, v)}{\partial x} + \lambda_1 \frac{\partial[b_1 - g_1(x, y, v)]}{\partial x} + \lambda_2 \frac{\partial[b_2 - g_2(x, y, v)]}{\partial x} \\
\frac{\partial L}{\partial y} &= \frac{\partial f(x, y, v)}{\partial y} + \lambda_1 \frac{\partial[b_1 - g_1(x, y, v)]}{\partial y} + \lambda_2 \frac{\partial[b_2 - g_2(x, y, v)]}{\partial y} \\
&\vdots \\
\frac{\partial L}{\partial v} &= \frac{\partial f(x, y, v)}{\partial v} + \lambda_1 \frac{\partial[b_1 - g_1(x, y, v)]}{\partial v} + \lambda_2 \frac{\partial[b_2 - g_2(x, y, v)]}{\partial v}
\end{aligned}\right\} \tag{9.8}$$

$$\left.\begin{aligned}
\frac{\partial L}{\partial \lambda_1} &= b_1 - g_1(x, y, z) \\
\frac{\partial L}{\partial \lambda_2} &= b_2 - g_2(x, y, z)
\end{aligned}\right\} \tag{9.9}$$

By equating the five resulting equations to zero, the five unknowns $(x, y, v, \lambda_1, \lambda_2)$ can be determined. There may be more than one solution. In such cases where there is a multiplicity of solutions, the individual solutions could each indicate global minima or maxima, and others indicate a local minimum or local maximum of an irregularly shaped surface. The Lagrange multiplier λ_i

can be interpreted to represent the change in the objective function at optimality caused by a small change in the right-hand-side constant of constraint i; that is,

$$\lambda_i = \frac{\Delta f}{\Delta b_i} \tag{9.10}$$

The Lagrange multiplier thus provides the same interpretation as the dual variable in linear programming. Some example applications of Lagrange multipliers are herein presented.

Example 9.6

We seek to maximize $Z = x_1 + x_2^2 + x_3$, within the following constraints:

$$x_1 + x_2 + x_3 = 8 \quad \text{and} \quad x_1 + 2x_2 + 3x_3 = 12$$

Solution

Let λ_1 and λ_2 be the *Lagrange multipliers* for the two constraints, and establish the new, unconstrained objective as

$$L = x_1 + x_2^2 + x_3 + \lambda_1(8 - x_1 - x_2 - x_3) + \lambda_2(12 - x_1 - 2x_2 - 3x_3)$$

Then, the derivatives are

$$\frac{\partial L}{\partial x_1} = 1 - \lambda_1 - \lambda_2 = 0$$

$$\frac{\partial L}{\partial x_2} = 2x_2 - \lambda_1 - 2\lambda_2 = 0$$

$$\frac{\partial L}{\partial x_3} = 1 - \lambda_1 - 3\lambda_2 = 0$$

$$\frac{\partial L}{\partial \lambda_1} = 8 - x_1 - x_2 - x_3 = 0$$

$$\frac{\partial L}{\partial \lambda_2} = 12 - x_1 - 2x_2 - 3x_3 = 0$$

We first solve the above five derivative equations to yield $x_1 = \frac{23}{4}, x_2 = \frac{1}{2}$, and $x_3 = \frac{7}{4}$, and the corresponding optimal (minimum) value of the objective function Z is $\frac{31}{4}$.

9.3 CONSTRAINED OPTIMIZATION USING MATHEMATICAL PROGRAMMING TECHNIQUES

9.3.1 Introduction

Programming can be described as an optimization technique where one seeks the values of a real (continuous) integer, or binary decision variables from a given domain, that yield a minimum or maximum value of a given real function. The term programming does not refer to computer programming but rather is a relic of the traditional use of the word: the U.S. military referred to the proposed training and logistics schedules prescribed by mathematicians during World War II as "programs" and the term has stuck ever since. There are two broad classes of programming problems: linear and nonlinear.

The general form of a single-objective optimization model or mathematical program is

$$\text{Min or Max } f(x_1, x_2, \ldots, x_n)$$

$$\text{subject to } g_i(x_1, x_2, \ldots, x_n) \begin{Bmatrix} \leq \\ = \\ \geq \end{Bmatrix} b_i \quad i = 1, 2, \ldots, m \tag{9.11}$$

where $f, g_1, \ldots, g_m$ are the given functions of decision variables $x_1, x_2, \ldots, x_n$, and $b_1, b_2, \ldots, b_m$ are the specified constants

---

Example 9.7

Materials X (cost \$10 per unit) and Y (\$15 per unit) are needed for a certain process. The amounts of X and Y used should not exceed 5 and 4 units, respectively. The total cost of X used should not exceed \$50, and the total cost of Y used should not exceed \$55. The total number of units of X and Y used should not exceed 15. It is sought to minimize the total cost of the materials used. Write the full mathematical formulation for this problem.

Solution

Consistent with Equation (9.11), the mathematical formulation for this problem is

$$\text{Min } 10x + 15y$$

Subject to

$$x + y \leq 7$$

$$x \leq 5$$

$$y \leq 4$$

$$10x \leq 50$$

$$15y \leq 55$$

The decision variables are x and y, and there are four constraints, so $m = 4$.

9.3.2 Linear and Nonlinear Programming

Linear Functions. A function is linear if it is a constant-weighted sum of the decision variables. The decision variables must be in the first power; otherwise, the function is nonlinear. Thus, $3x + 9y$ and $0.5x - 32y$, for example, are linear; however, the functions $2x^2 + 5y^3, 3/x - y, (3x)(2y), x - \log_e y$, and $(3x + 5y)/(5x + y)$ are nonlinear (because x or y or both are not to the first power, or the function does not involve a summation).

Example 9.8

Assuming that the x's are decision variables and all other symbols are constants, determine which of the following functions is linear.

(a) $f(x_1, x_2, x_3) = 5x_1 - 2x_2 + 1/x_3$
(b) $f(x_1, x_2, x_3) = 5x_1 + 3x_2 - 6x_3$
(c) $f(x, y) = 3x - 5y^2$
(d) $f(x_1, x_2) = x_1 + \log_e x_2 + x_1 x_2$

(e) $f(x, y, v) = \ln(p)x - e^q y + 3.5v$
(f) $f(x_1, x_2) = (5x_1 + 2x_2)/(x_1 - 3x_2)$

Solution

(a) Nonlinear because it involves negative powers of decision variable x.
(b) Linear because all powers of x are exactly equal to 1.0 and the c_j are constants (i.e., 5, 3, and −6).
(c) Nonlinear because one of the terms has a power that is different from 1.0.
(d) Nonlinear because it involves products, powers not 1, and logarithms of decision variables.
(e) Linear because p and q are constants, so the expression is a weighted sum of the decision variables.
(f) Nonlinear because it involves a quotient, even though both the numerator and the denominator are linear functions.

Linear Programming. Linear Programming (LP) is an optimization technique where the objective function is linear and where each constraint is either a linear equality or linear inequality. An optimization model of the form shown in Equation (9.11) is a linear programming (LP) problem if *both* the objective function and constraint functions $g_1, \ldots, g_m$ are linear in the decision variables. The functions $g_1, \ldots, g_m$ specify a convex polytope (or feasible region bounded by the constraining lines or planes $g_1, g_2, \ldots, g_m$) over which the objective function is to be optimized.

Nonlinear Programming. An optimization model of the form shown in Equation (9.11) is a nonlinear programming (NLP) problem if *either* the objective function or at least one of the constraint functions $g_1, \ldots, g_m$ are nonlinear in the decision variables.

Example 9.9

Identify, with reasons, which of the following optimization problems are linear.

(a) Min Z = $10x + 15y$
Subject to

$$x \leq 5$$

$$x + y^2 \leq 15$$

(b) Min Z = $10x/(1 + 5y)$
Subject to

$$x, y \geq 0$$

$$x \leq 5$$

(c) Min Z = $x + y + v$
Subject to

$$x, y, v \geq 0$$

$$x \leq 15$$

$$xy \leq 15$$

Solution

Problem (a) is nonlinear because its second constraint in nonlinear. Problem (b) is nonlinear because its objective function is nonlinear. Problem (c) is nonlinear because its third constraint is nonlinear.

9.3.3 Linear Programming

Linear Programming is the simplest case of the general mathematical optimization problem and consists of a single **linear** objective function and one or more constraints that are all **linear**. In general, an LP problem has the following three components: objective function, constraint(s), and decision variable(s). An LP problem could be expressed in the form shown in Equation (9.11) or in a similar alternative form as shown below.

$$\text{Max (or Min)}\, Z = \sum_{j=1}^{n} (c_j x_j) \tag{9.12}$$

Subject to

$$\sum_{j=1}^{n} (a_{ij} x_j) \; \{\leq, =, \text{or} \geq\} \; b_i \qquad (i = 1, \ldots, m) \tag{9.13}$$

$$x_j \geq 0 \qquad (j = 1, \ldots, n) \tag{9.14}$$

where Z is the objective function, x_j are the decision variables, c_j are the model coefficients, a_{ij} are the constraint coefficients, and b_i are the right-hand sides, and c_j, a_{ij}, and b_i are assumed to be constant and known. The assumption of nonnegativity conditions (9.14) is implicit in all LP problems.

Linear programming has been used to solve many large-scale problems in engineering, agriculture, economics, and business. In civil engineering, examples include optimizing the mix of ingredients for concrete production, optimizing the assignment of vehicles on transportation system routing operations, optimizing the distribution of water to a region, and the like.

(a) Steps for Graphical Solution of Linear Programming Problems.

Step 1: Identify the objective function and decision variables.

Step 2: Write the constraints with one decision variable as the subject of the equation.

Step 3: Find the boundary functions of the feasible region by transforming each inequality constraint into an equality.

Step 4: Plot the boundary functions and use the inequality constraint functions to indicate the regions that satisfy the constraint.

Step 5: Identify the points at which the boundary functions intersect. These are the **vertices**, or **extreme points**, of the feasible region. These points can be found directly from the plot or using the method of simultaneous equations. Each vertex represents a combination of values of the decision variables and thus is a candidate solution.

Step 6: Substitute the coordinates of each vertex into the objective function. Identify the vertex that yields the desired optimal value of the objective function.

(b) Redundant and Binding Constraints. In some LP problems, not all of the constraints are boundaries of the feasible region. Constraints that are not boundaries are termed **redundant constraints** because they have no effect on the optimal solution. For example, consider the feasible region bounded by the following constraints: $y \geq 10, y \leq 30, x \geq 5$, and $y \leq -5x/4 + 50$. The objective function is Max $Z = 3x + 2y$. The constraints and feasible region for this problem are shown in Figure 9.11. Clearly, the constraints $x \geq 0$ and $y \geq 5$ are redundant as each of them has no effect on the feasible region. In LP problems, redundant constraints are considered a nuisance. Even though they do not inhibit directly the identification of the optimal solution, their presence increases the problem size and therefore reduces the computational speed of the algorithm. As such,

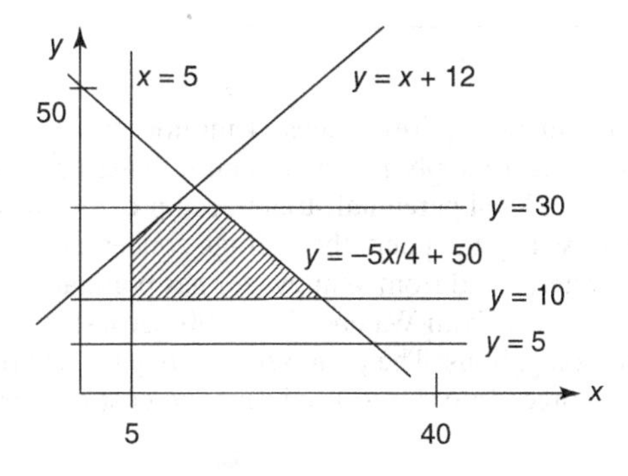

Figure 9.11 Example problem with redundant constraints.

it is advisable to identify and remove all redundant constraints from the formulation before the problem is solved. On the other hand, a **binding constraint** is one that is critical to the identification of the optimal solution. Thus, the optimal solution changes or ceases to exist if that constraint is removed or changed.

Example 9.10

For a certain civil engineering system, it is sought to maximize the benefits, which is given by the expression $Z = 5x + 3y$. The decision variables are x (the amount of resource type 1 to be used) and y (the amount of resource type 2 to be used). The amount of resource type 1 should be at least 3 units, and the amount of resource type 2 should be at least 1 unit. The amount of resource type 2 should not exceed the amount of resource type 1. Also, the total amount of resource types 1 and 2 should not exceed 10 units. The amounts of both resource types should be at least zero. Identify the objective function for this problem and write the constraints. Provide a rough sketch graph for the constraint set and clearly show the feasible region. Also, label all extreme points (or vertices) of the feasible region and indicate their coordinates. Solve the optimization problem.

Solution

The objective function is Max $Z = 5x + 3y$, and the constraints are $x \geq 3; y \geq 1; x + y \leq 10; y \leq x; x \geq 0; y \geq 0$.

A sketch is provided in Figure 9.12, in which the feasible region is the shaded area.

The extreme points are A (5, 5), B (3,3), C (3,1), and D (9,1). The optimal solution is $x_{\text{OPT}} = 9; y_{\text{OPT}} = 1; z_{\text{OPT}} = 48$.

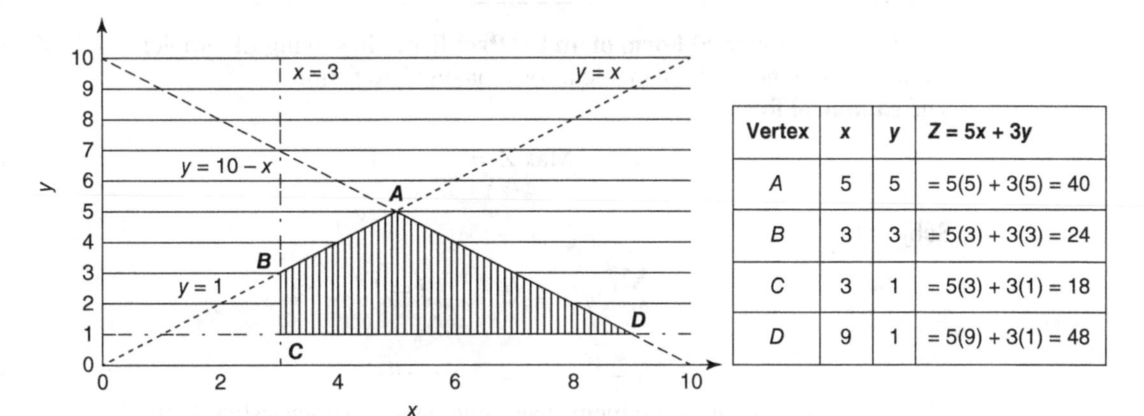

Vertex	x	y	Z = 5x + 3y
A	5	5	= 5(5) + 3(5) = 40
B	3	3	= 5(3) + 3(3) = 24
C	3	1	= 5(3) + 3(1) = 18
D	9	1	= 5(9) + 3(1) = 48

Figure 9.12 Example problem.

Example 9.11

A precast concrete plant requires at least 4 million gallons/day more water than it is currently using. Waterex, a nearby water supply reservoir, can provide up to 10 million gallons per day of such extra supply. Whitewater, a local perennial stream, can provide an additional 2 million gallons a day of extra supply. For water used by the plant, the average concentration of pollution should not exceed 100 units. The water from Waterex and from Whitewater has pollutant concentrations of 50 and 200 units, respectively. The cost of water from Waterex is $1000 per million gallons; and from the Whitewater stream it is $500 per million gallons. The plant seeks to determine how much water should be purchased from each of the two sources in order to minimize the cost of supplying water that, on average, meets the quality standards.

(i) What are the decision variables?
(ii) Write down the constraints.
(iii) Write down the objective function.
(iv) Find the optimal solution using the graphical method and indicate what advice you would give to the plant.

Solution

The decision variables are

Amount of water (millions of gallons) purchased from Waterex, x

Amount of water (millions of gallons) purchased from Whitewater, y

The constraints are

$x + y = 4$ (total water needed is 4 million gallons)

$x \leq 10$ (Waterex can provide up to 10 million gallons)

$y \leq 2$ (Whitewater can provide up to 2 million gallons)

$50x + 200y \leq 100(x + y)$ (total pollutant amount in the mixed water must not exceed 100 units).

$50x + 200y \leq 100(x + y)$, upon simplifying, gives $y \leq 0.5x$.

The optimal solution is $x_{\text{OPT}} = 2.67$; $y_{\text{OPT}} = 1.33$; $z_{\text{OPT}} = 3{,}333$. In other words, the minimum cost is achieved when the amount of water taken from Waterex, x, is 8/3 (i.e., 2.67) million gallons, and that taken from Whitewater, y is 4/3 (i.e., 1.33) million gallons.

(c) Canonical and Standard Form of an LP Problem. In solving LP problems, it is often useful to reformulate them into either canonical form or standard form.

The **canonical form** is

$$\text{Max } Z = \sum_{j=1}^{n} (c_j x_j) \tag{9.15a}$$

Subject to

$$\sum_{j=1}^{n} (a_{ij} x_j) \leq b_i \qquad (i = 1, \ldots, m) \tag{9.15b}$$

$$x_j \geq 0 \qquad (j = 1, \ldots, n) \tag{9.15c}$$

The canonical form of LP problems has a number of characteristics. First, the objective function is of the *maximization* type; second, all constraints are of the "less than or equal to" type; third, the right-hand-side constants, b_i, may be positive or negative; and lastly, all the decision variables are nonnegative.

For transforming an LP into a canonical LP, the following techniques are used:

1. If the objective function is of the minimization type, it can be transformed into the maximization type by multiplying it by -1. For example, Minimize $Z = 3x_1 - 5x_2$ can be rewritten as Maximize $Y = -3x_1 + 5x_2$.
2. "Greater than or equal to" constraints can be transformed into "less than or equal to" constraints by multiplying both sides by -1. For example, the constraint $5x_1 - 4x_2 \geq 12$ can be rewritten as $-5x_1 + 4x_2 \leq -12$.
3. Equality constraints can be converted into two inequalities.
4. Variables that are unrestricted in sign may be replaced by two nonnegative variables.

The **standard form** of an LP problem may be written as follows:

$$\text{Max (or Min)}\, Z = \sum_{j=1}^{n} (c_j x_j) \tag{9.16}$$

Subject to

$$\sum_{j=1}^{n} (a_{ij} x_j) = b_i \qquad (i = 1, \ldots, m) \tag{9.16a}$$

$$x_j \geq 0 \qquad (j = 1, \ldots, n) \tag{9.16b}$$

The standard form has a number of features. First, the objective function may be of the maximization or minimization type; second, the constraints are all **equations**, with the exception of the nonnegativity conditions, which remain as inequalities of the "greater than or equal to" type; third, the right-hand side of each constraint should be nonnegative; and lastly, all variables are nonnegative.

The standard form LP problems are used in a subsequent section of this chapter. For solving LP problems using the Simplex method, the problem should be transformed into standard form using one or more of the following operations (in addition to those listed above):

- A negative right-hand side can be made positive by multiplying through by -1 (and reversing the sign of the inequality).
- Inequality constraints may be converted to equalities by the inclusion of **slack or surplus** variables. For example, the "less than or equal to" constraint,

$$2x_1 + 4x_2 \leq 15 \tag{9.17}$$

May be written as

$$2x_1 + 4x_2 + S_1 = 15$$

where S_1 is a **slack variable**, which must be nonnegative in order to satisfy the original constraint.

- Similarly, the "greater than or equal to" constraint

$$3x_1 + 5x_2 \geq 18 \tag{9.18}$$

may be written as

$$3x_1 + 5x_2 - S_2 = 18 \tag{9.18a}$$

where S_2 is a **surplus variable**, which must be nonnegative.

The term slack variable is used because the right-hand side of Equation (9.13) often represents the availability of a particular resource. If S_1 is positive, then not all of the available resource is being utilized and there is a "slack" in the solution. Constraints of the "greater than or equal to" type often represent minimum requirements; for example, the right-hand side of Equation (9.18) may represent the minimum required output of steel from a particular process. If S_2 is positive, the minimum output is being exceeded and there is a "surplus" output.

It can be seen from Equation (9.16a and b) that the constraint in the LP standard form is a set of m linear equations in n unknowns, including the surplus and slack variables. Regarding the relative number of m and n, there are three possibilities:

1. $n > m$. This is the case in any typical programming problem. The number of unknowns (n) exceeds the number of equations (m) in the constraint set.
2. $n = m$. Here, there exists a unique solution to the constraint set. In this case, there is no need for any effort to find the optimal solution.
3. $n < m$. In this case, the problem is overconstrained to the extent that there is no solution that satisfies all the constraints; in other words, a feasible region does not exist.

A basic solution to an LP problem is one in which $(n - m)$ of the decision variables are equal to zero (Dandy et al., 2008). Also, a basic feasible solution to an LP problem is a basic solution in which all variables have nonnegative values. Therefore, for the LP standard form, the basic feasible solution satisfies the main constraints [Equation (9.15a)] as well as the nonnegativity constraints [Equation (9.15b)].

In Sections 9.3.3(d) to (g), we discuss some general issues in linear programming problems.

(d) Impact of the Nature of the Decision Variable—Integer versus Continuous. As discussed in an earlier section of this chapter, the designation of an optimization model as linear or nonlinear depends on the power of the decision variables and the existence of a weighted-sum mathematical form of the function. On the other hand, the designation of an optimization problem as discrete or continuous depends on the discrete or continuous nature of the decision variable. A mathematical program is a discrete optimization model if it includes *at least one* discrete decision variable. On the other hand, only when *all* of the decision variables are continuous can a mathematical program be described as a continuous optimization model. Table 9.4 presents the various categories of optimization problems on the basis of the linearity of the objective function and constraints and the continuity of the decision variable. The most tractable linear programming problem has continuous variables, linear constraints, and a single linear objective function. If either the constraints

Table 9.4 Optimization Problems: Dimensions of Linearity and Decision Variable Continuity[a]

Linearity of the Objective Function and/or Constraints		Discrete Decision Variable	Continuous Decision Variable	Mixed
Linear constraints	Linear Objective	ILP	CV-LP or simply LP	ILP
	Nonlinear Objective	INLP	CV-LP or simply NLP	INLP
Nonlinear constraints	Linear Objective	INLP	CV-LP or simply NLP	INLP
	Nonlinear Objective	INLP	CV-LP or simply NLP	INLP

[a]NLP, integer linear program; INLP, integer nonlinear program; NLP, nonlinear program; CVLP, continuous variable linear program.

or the objective are nonlinear, it becomes a nonlinear program. Also, the presence of any discrete variables transforms an LP into an integer linear program (ILP) or transform an NLP into an integer nonlinear program (INLP) (Rardin, 2002). Thus, a mixed program (comprised of integer and continuous variables) is an integer program.

Example 9.12

Assuming that all x_j are decision variables, determine which of the following mathematical programs are a linear program (LP), an integer linear program (ILP), a nonlinear program (NLP), or an integer nonlinear program (INLP).

(a) Max $5x_1 + 9x_2$
Subject to

$$x_1 \leq x_2 + 1$$

$$x_1 + x_2 = 8$$

$$x_1, x_2 \geq 0$$

(b) Min $5x_1 - x_2 + 2x_3$
Subject to

$$x_1 x_2 \leq 2$$

$$x_1 + 3x_2 + x_3 = 8$$

$$x_1, x_2, x_3 \geq 0$$

x_1 is integer

(c) Min $x_1 + x_2 + x_3/x_2$
Subject to

$$x_3 \leq x_1$$

$$x_1 + x_2 \leq 8$$

$$x_1 \geq 0$$

(d) Max $13x_1 + 4x_2$
Subject to

$$x_1 + x_2 \leq 10$$

$$x_1, x_2 = 0 \text{ or } 1$$

Solution

(a) The objective function and all of the main constraints of this model are linear and the variable types are continuous. Thus, the model is a linear program (LP).

(b) The objective function is linear, but the overall model is nonlinear due to the product term in the first constraint. The model is an integer nonlinear program (INLP).

(c) The logarithm and quotient terms in its objective function make this model nonlinear. Since all of the variables are continuous, it should be classified as a nonlinear program (NLP).

(d) Except for its discrete variable-type constraints, this model would be a linear program because the objective function and both of the main constraints are linear. Thus, the model is an integer linear program (ILP). Specifically, it is a binary linear program because the decision variable can only take values that have a domain of only two possibilities: 0 or 1.

(e) Feasible Regions. In all LP problems, the feasible solution space is a convex region. As explained in Section 9.0.2(i), a convex region is one in which a straight line joining any two points in the region contains only points in that region. A feasible region that does not satisfy this property is known as a nonconvex region. Three basic theorems in linear programming are (Stark, and Nicholls, 2005):

Theorem 1 *The collection of feasible solutions constitutes a convex set whose extreme points correspond to basic feasible solutions.*

This theorem indicates that we need to be concerned only with convex sets because the only solutions of interest to us must be contained in the class of feasible solutions. Also, the basic feasible solutions correspond to the vertices of the feasible region.

Theorem 2 *If a feasible solution exists, then a basic feasible solution exists.*

Theorem 1 provides the assurance that the convex set contains all of the feasible solutions. Now, if a feasible solution is found to exist (i.e., by trial and error), then there must exist at least one vertex point to the convex set.

Theorem 3 *If the objective function possesses a finite minimum, then at least one optimal solution is a basic feasible solution.*

If the LP problem has been properly formulated, it will satisfy the hypothesis of a finite minimum. Theorem 3 provides the assurance that at least one of the optimal solutions is a basic feasible solution. As such, our search for an optimal solution can focus on the extreme points. This observation is a cornerstone of the simplex method for solving LP problems.

(f) The Convexity Issue in Linear Programming. In all LP problems, the feasible solution space is a convex region. As discussed in Section 9.1.3, in a Euclidian space, a region is defined as convex if "for every pair of points within the region, every point on a straight line segment joining the two points is also within the region." Any region that does not satisfy this property is termed a nonconvex region. Thus, a function is convex if, and only if, the region above its graph is a convex set. Illustrations of convex and nonconvex regions are provided in Figure 9.13. The figure is for two dimensions only, but the concept of convexity applies in any number of dimensions. It is seen that the function Z in (a) is convex because for every pair of points within the region, every point on a straight line segment joining the two points is also within the region. For (b), however, some of the points on the straight-line segment joining the two points do not lie in the region and thus the region is nonconvex.

As illustrated in Figure 9.14, for two decision variables x and y, a constraint function is described as being strictly convex on an interval (x_1, y_1) and (x_2, y_2) if $f[tx_1 + (1-t)x_2] < tf(x_1) + (1-t)f(x_2)$.

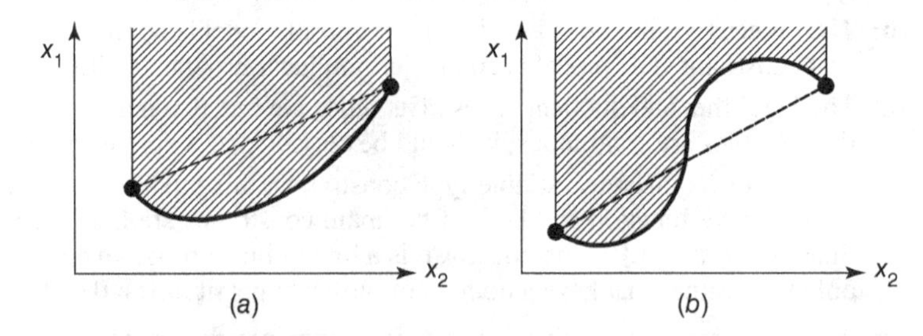

Figure 9.13 (*a*) Convex and (*b*) nonconvex functions in two dimensions.

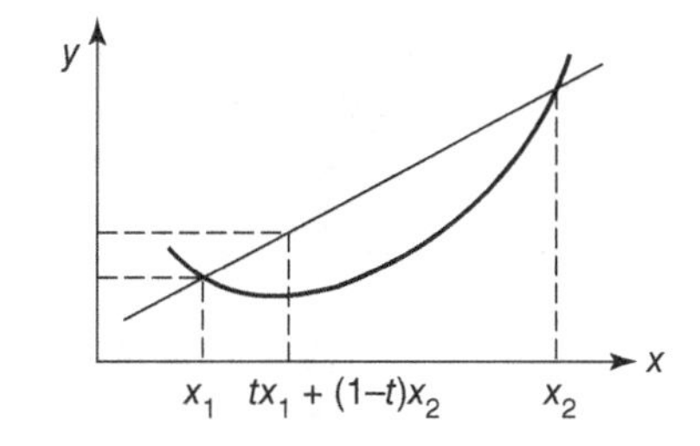

Figure 9.14 Convex constraint function in two dimensions on a given interval.

(g) Peculiarities of Certain LP Problems and Solutions. Solutions to LP problems may be characterized by some peculiarity, such as the multiplicity of optimal solutions (i.e., the existence of alternate solutions), the nonexistence of solutions, or the uniqueness of the solution. Also, the feasible region may be unbounded, or a constraint in the problem may be redundant. In the sections below, we describe, with illustrations, each of these atypical situations.

Problems Having Multiple Optimal Solutions. In some cases, there could be two or more optimal solutions to an LP problem. Consider, for example, the feasible region bounded by the following constraints (Figure 9.15): $y \geq 0, y \leq 30, x \geq 10, x \leq 40, 2y + x \leq 70$, and $y \leq 50 - x$. The vertices of the feasible region are (10, 0), (10, 30), (30, 20), and (40, 10) and (40, 0). If the objective function is $Z = 10x + 10y$, then the optimal solution of 500 is yielded by the points (30, 20) and (40, 10). In other words, there are two alternate optimal solutions. Also, it can be shown that all of the points on the line segment (constraint or boundary) connecting the points (30, 20) and (40, 10) yield the maximum objective function value of 500. Problems such as this one that have two or optimal solutions are said to have *alternate optima*. In LP problems that have several decision variables and constraints, instances of alternate optima are likely to be encountered.

Problems Having No Feasible Solution. Certain LP problems have no feasible region simply because there is no set of decision variables that satisfies all the constraints simultaneously. In such situations, there can be no optimal solution, regardless of the efficiency of the solution methods or algorithm. For example:

$$\text{Max } Z = 5x + 3y$$

Subject to

$$y \geq 4 - x$$
$$y \leq 3 - 3x$$
$$x, y \geq 0$$

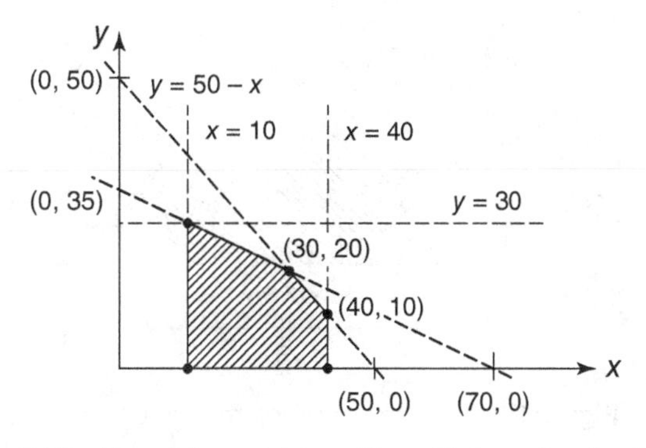

Figure 9.15 Example problem with multiple optimal solutions.

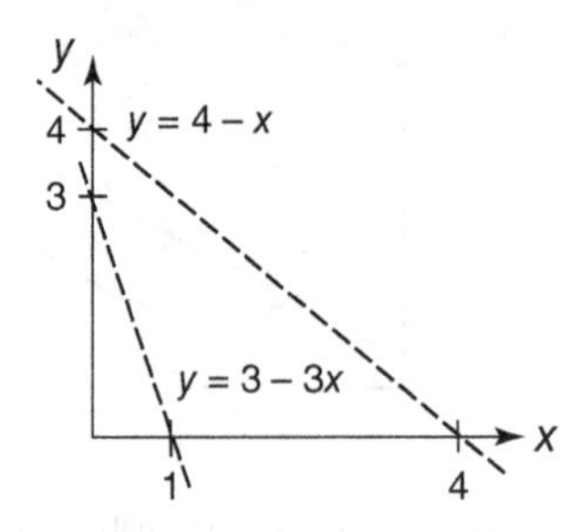

Figure 9.16 Example problem with infeasible solution.

As seen in Figure 9.16, the constraints to this problem do not yield any feasible region. LP problems of this nature are described as *infeasible*. It is relatively easy to detect such situations when the LP problem has only few variables and this can be depicted graphically. When there are several variables, it can be difficult, if not impossible, to manually detect whether the problem would yield an infeasible solution. Care should be taken not to identify a problem as infeasible in haste. Experience in the practice has shown that an LP problem may be identified incorrectly as infeasible because the problem was incorrectly formulated or errors were made in the input data coding or entry. An infeasible solution situation could also result in certain cases where the feasible region exists but is unbounded.

Problems with Unbounded Feasible Regions. In certain cases, the feasible region has no boundary on one or more sides. In many cases, this poses a problem because an optimal solution cannot be found. For example, consider the LP problem:

$$\text{Max } Z = 3x + y$$

Subject to

$$y \leq 0.5x + 2$$

$$y \geq 1$$

$$y \geq 3 - x$$

It can be seen (Figure 9.17) that at all points along the edges that define the unbounded region (particularly the $y = 0.5x + 2$ line), both x and y are increasing. Thus, any point along this line has values of x that is higher than the x values of already identified vertices A and B; similarly, any point

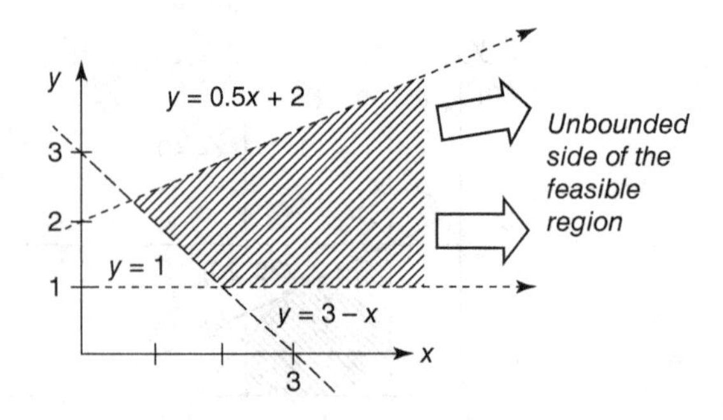

Figure 9.17 Example problem with unbounded feasible region.

along this line has values of y that are higher than the y values of already identified vertices A and B. Thus, as we travel along the line $y = 0.5x + 2$, we can find another solution that yields a superior value of the objective function. So, for this problems of this nature, the objective function could be moved upward and to the right without any limit. Thus, for such LP problems, it is possible to increase the value of the objective function without limit. This situation often arises in cases where the problem is underconstrained.

In a few cases, however, it is possible to find a solution even where the feasible region is unbounded on one or more sides. For example, consider the above case where the objective function is Min $Z = 3x + y$ with the same constraints as given earlier. Because both x and y are increasing at all points along the edges that define the unbounded region ($y = 1$ and $y = 0.5x + 2$), any point along these lines has values of x that are higher than the x values of already identified vertices A and B; similarly, any point along this line has values of y that are higher than the y values of already identified vertices A and B. Thus, as we travel along these lines, we cannot find another solution that gives a better value of the objective function. As such, the minimum value of Z will be given by one of the existing vertices, in this case, vertex A $\left(\frac{2}{3}, \frac{7}{3}\right)$, which yields an objective function value of 4.33.

Problems Having Unique Optimal Solutions. Strictly speaking, the issue of unique optima is not the peculiarity it appears to be because it is the norm rather than the exception. In most optimization problems, there is only one point that satisfies the objective function and all constraint equations simultaneously. In such cases, the optimal solution to the linear program is described as being *unique*. The solution to Example 9.13, for example, is unique. Another example is the feasible region bounded by the following constraints (Figure 9.18): $y \geq 1, y \leq 4, x \geq 0$, and $y \leq -5x/4 + 5$. The vertices of the feasible region are (0, 4), (0, 1), (0.8, 4), and (3.2, 1). If the objective function is $Z = 20x + 10y$, then the optimal solution, $x_{\text{OPT}} = 3.2, y_{\text{OPT}} = 1$, and $Z_{\text{OPT}} = 74$, is unique.

(h) Duality of LP Problems. An LP problem, also termed as a "primal" problem, can be converted into its corresponding "dual" problem. With the latter, an upper bound can be specified for the optimal value of the primal problem. Using the simplex method to solve the original problem automatically yields a solution to the dual problem. The dual variables have an important interpretation that adds considerable information to an LP solution. As such, duality is an important concept in mathematical analysis that often gives valuable insights into the physical problem being solved. In Section 9.3.1, we discussed how an LP problem's canonical form could be expressed as follows:

$$\text{Max } Z = \sum_{j=1}^{n} (c_j x_j) \tag{9.19}$$

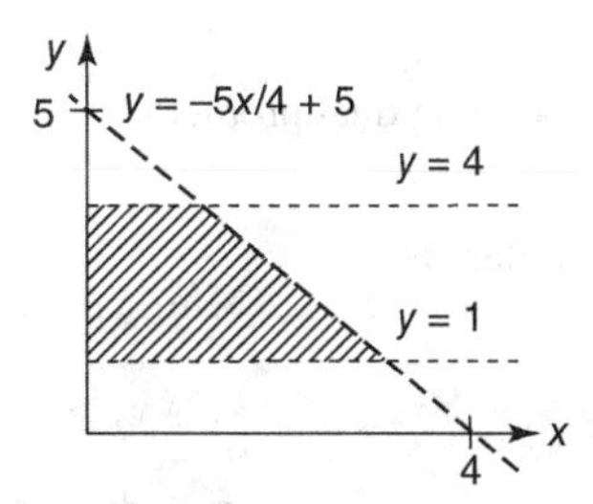

Figure 9.18 Problem with unique optima (a single optimal solution).

Subject to

$$\sum_{j=1}^{n}(a_{ij}x_j) \leq b_i \quad (i = 1,\ldots,m) \tag{9.19a}$$

$$x_j \geq 0 \quad (j = 1,\ldots,n) \tag{9.19b}$$

For an LP problem in canonical form, the dual problem may be written using the following rules (Rardin, 2002):

1. There is a dual variable corresponding to each primal constraint and a primal variable corresponding to each dual constraint.
2. If the primal problem is of the maximization type, the dual problem is then of the minimization type, and vice versa.
3. All of the constraints in the maximization problem are of the "less than or equal to" type. All of the constraints in the minimization problem are of the "greater than or equal to" type.
4. The coefficients of the objective function in the primal problem are the right-hand sides in the dual problem and vice versa.
5. The variables in both problems are nonnegative.

To write the dual to the problem given by Equation (9.19), m dual variables must be defined as follows:

y_i $(i = 1,\ldots,m)$, one corresponding to each primal constraint. Then the dual problem is

$$\text{Min } Z = \sum_{i=1}^{m}(b_i y_i) \tag{9.20}$$

Subject to

$$\sum_{i=1}^{m}(a_{ij}y_i) \geq c_j \quad (i = 1,\ldots,m) \tag{9.20a}$$

$$y_j \geq 0 \quad (j = 1,\ldots,m) \tag{9.20b}$$

Example 9.13

Consider the following primal problem:

$$\text{Max} \quad 5x_1 + 9x_2$$

Subject to

$$x_1 - x_2 \leq 1$$

$$x_1 + x_2 \leq 8$$

$$x_1, x_2 \geq 0$$

What is the corresponding dual problem?

Solution

The dual of the above problem is

$$\text{Min} \quad y_1 + 8y_2$$

Subject to

$$y_1 + y_2 \geq 5$$

$$-y_1 + y_2 \geq 9$$

$$y_1, y_2 \geq 0$$

(i) The Simplex Technique for Solving LP Problems. The simplex technique is a numerical procedure for finding solutions to LP problems. The technique, which is based on the premise that the optimal solution to an LP problem is always a basic feasible solution, involves the following steps:

Step 1 Initialization. Identify, as the initial reference point, an initial basic feasible solution. This is a vertex of the feasible region.

Step 2 Iteration. Consider moving from the reference point to each of the adjacent vertices, one by one, to ascertain whether it is possible to improve the value of the objective function by such a move.

Step 3 Iteration. If such a move (in step 2) is possible, then proceed to that vertex, which now becomes the new reference point. Consider moving from the reference point to each of the adjacent vertices, one by one, to ascertain if the maximum rate of change of the objective function is achieved by such a movement. If yes, move to this point and return to step 2.

Step 4 Continue Steps 2 and 3 until no improvement is possible by moving to any adjacent vertex and an optimum solution is obtained.

Further details on the simplex technique can be found in Rardin (2002). Also, a number of computer packages use the simplex technique to solve large LP problems.

Example 9.14

Use the simplex algorithm to solve the following problem:

$$\text{Max} \quad Z = 3x_1 + x_2$$

Subject to

$$x_1 + 2x_2 \leq 3$$

$$2x_1 + x_2 \leq 8$$

$$x_1, x_2 \geq 0$$

Solution

The problem is rewritten as standard for:

$$\text{Max} \quad Z = 3x_1 + x_2$$

$$x_1 + 2x_2 + s_1 = 3$$

$$2x_1 + x_2 + s_2 = 8$$

Then $z = 0, s_1 = 3, s_2 = 8$ is a feasible solution.

Row		Basic Variable
1	$Z - 3x_1 - x_2 = 0$	$Z = 0$
2	$x_1 + 2x_2 + s_1 = 3$	$s_1 = 3$
3	$2x_1 + x_2 + s_2 = 8$	$s_2 = 8$

Select x_1 as the entering variable, then the largest x_1 could be 3, then:

Row		Basic Variable
1	$-x_2 + s_1 = 3$	$Z = 0$
2	$x_1 + 2x_2 + s_1 = 3$	$x_1 = 3$
3	$-3x_2 - 2s_1 + s_2 = 2$	$s_2 = 2$

Select x_2 as the entering variable, then the largest x_2 could be 0.
Thus, the final solution is when $x_1 = 3$ and $x_2 = 0$, and Z has the maximum value 6.

9.3.4 Nonlinear Programming (NLP)

In Section 9.1.4, we discussed linear programming (LP) where the objective function and all of the constraints are linear. However, the assumption of linearity is not always valid to real systems in civil engineering. In reality, most objective functions and/or constraints tend to be nonlinear. Fortunately, a large number of techniques exist for solving nonlinear optimization problems. On the flip side, the complexity of many practical problems in civil engineering is such that they do not lend themselves readily to the use of formal optimization techniques. Standard texts including Rardin (2000) present the techniques of separable programming and dynamic programming that are useful for solving complex problems. In some cases, it is more prudent to start with the development of a mathematical model of the system (but not to optimize it) followed by computer simulation or enumeration via trial and error to identify the optimal solution.

The general single-objective, nonlinear optimization problem may be written as:

$$\text{Max (or Min) } Z = f(x_1, x_2, \ldots, x_n) \tag{9.21}$$

Subject to

$$h_i(x_1, x_2, \ldots, x_n) \quad \{\leq, =, \text{or} \geq\} \; b_i \qquad (i = 1, \ldots, m) \tag{9.22}$$

This problem differs from the general LP problem in three major respects (Dandy et al., 2008):

1. The feasible solution space for an NLP problem may be either convex or nonconvex. This occurs because the boundaries of the feasible region may, in general, be generated by nonlinear functions. Figure 9.4 illustrates some nonconvex feasible regions.
2. An optimum solution to an NLP problem may occur at any point in the feasible region (not necessarily at an extreme point).
3. It is common for more than one "optimum" to exist. The solution corresponding to the absolute maximum (or minimum) of the objective function is called the **global optimum**; and all other optima are called **local optima**.

Example 9.15

Find the solution to the following NLP problem (Figure 9.19):

$$\text{Max } Z = 2x + 3y$$

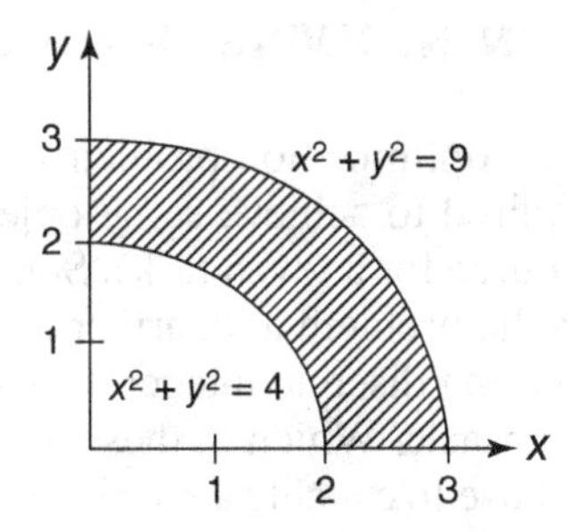

Figure 9.19 Example of an NLP problem.

Subject to

$$x \geq 0$$

$$y \geq 0$$

$$x^2 + y^2 \leq 9$$

$$x^2 + y^2 \geq 4$$

Solution

The feasible region is the collection of points outside a circle of radius 2 units and inside a circle of radius 3 units. The edge of a circle is the edge of a convex set and the objective function is linear; thus, the optimal solution will be found around the edge of the circle of radius 3. Replacing the last two constraints with $x^2 + y^2 = 9$ allows the use of substitution. Substituting y in the objective function gives

$$Z = 2x + 3\sqrt{(9 - x^2)}$$

Setting the derivative with equal to zero gives

$$2 - 3x/\sqrt{(9 - x^2)} = 0$$

$$9x^2/(9 - x^2) = 4$$

$$9x^2 = 36 - 4x^2$$

Thus, $x = 6/\sqrt{13}$; and $y = 9/\sqrt{(13)}$

Often, in order to solve nonlinear problems, one technique is used to convert a nonlinear statement into a linear statement. However, as Revelle et al. (2004) pointed out, the state of off-the-shelf optimization solvers now is such that there simply is no need to replace a nonlinear problem or model with a linearized approximation, except perhaps for programs of truly prodigious size. However, there is a caveat: A convex function, for example, may be more realistically modeled as a piecewise linear function if the vertices of the piecewise linearization correspond, say, to discrete control technologies. This situation does not involve approximation in any sense. As pointed out by Revelle et al. (2004): the maximum number of piecewise linear decision variables taking on values not at vertices is limited by the number of binding constraints in the model. Also, this maximum number, depending on the problem, may or may not produce solutions in which most (continuous) decision variables take on vertex values.

9.4 CONSTRAINED OPTIMIZATION INVOLVING BINARY DECISION VARIABLES

As discussed earlier, discrete optimization problems may involve count variables (e.g., the number of production units required to achieve some objective) or binary variables (e.g., whether to select a specific type of resource for a given task). Sometimes, the decision variables are all binary integers, which then makes the problem a binary or an integer optimization problem. Examples of optimization questions involving discrete integer decision variables are as follows: Which of these items should I put in my knapsack? Which of these items should I put in my shopping cart? Which of these truck types should I use to execute a specific project? Which dish should I have for dinner: Italian, Ethiopian, or Chinese? Which combination of worker types should we use for a specific project? Note that the word "which" is used in discrete optimization problems, while in continuous variable optimization problems, the keywords were "how much" of each decision variable.

To solve optimization problems involving discrete decision variables, different programming techniques are used (i.e., integer programming is used when the decision variables are positive integer numbers, and binary (or zero–one) programming is used when the decision variables take on only values of either 1 or 0. Binary programming problems are often solved using "knapsack" formulations. These are a special category of problems where the decision maker seeks the best possible set of actions to be implemented at a given facility over an extended period of time or for each facility in a network at a point in time or over an extended period of time. The goal is typically to maximize some utility that is comprised of single or multiple performance criteria, subject to one or more constraints.

9.4.1 Knapsack Problem Types

To illustrate the types of knapsack problems, we consider the classic case of a student's knapsack (Figure 9.20). As a student prepares to go to lectures every morning, she faces the task of choosing what SET of items to place in her backpack. The items she could choose from are: calculator, textbook, notebook, Post-it notes, apple, pen, folder, wallet, soft drink, rule, cell phone, and iPod.

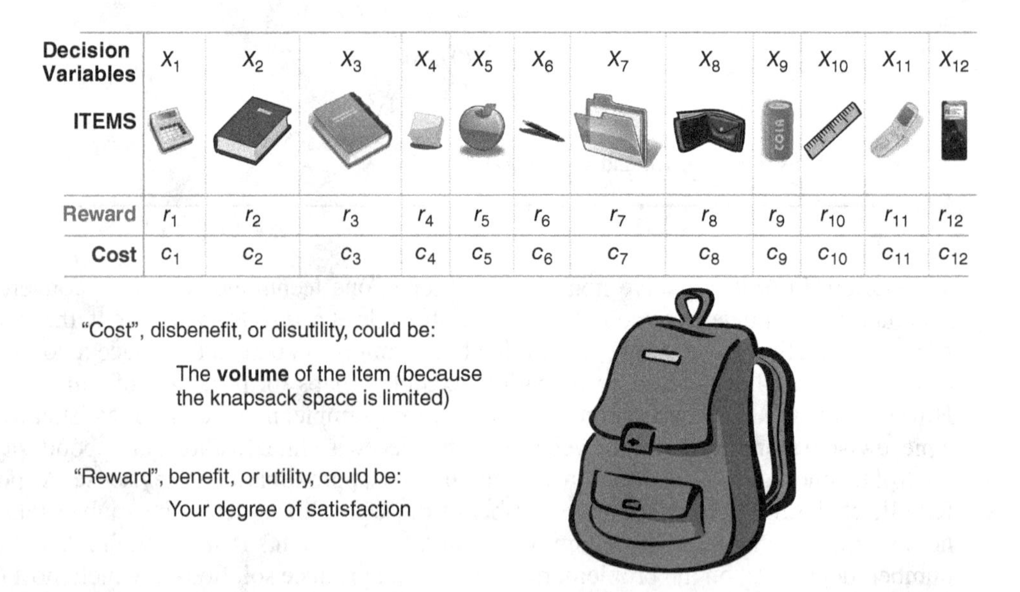

Figure 9.20 Conceptual illustration of the knapsack problem structures.

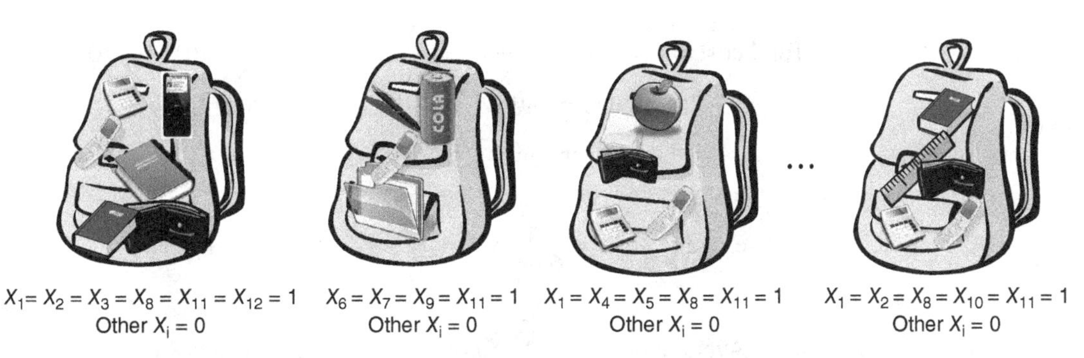

Figure 9.21 A few possible alternatives for the student's knapsack problem.

Each item has a certain "cost" and "benefit." For the student, cost could be the space taken up by an item in her backpack; and benefit could be the amount of satisfaction that an item provides to her. Due to space constraints, she cannot take all 12 items and thus must settle for a subset of the items. Each different subset constitutes an "alternative," and Figure 9.21 presents only four of a large number of possible alternatives. For any alternative set (combination of items), the student's reward is given by

$$r_1X_1 + r_2X_2 + r_3X_3 + r_4X_4 + r_5X_5 + r_6X_6 + r_7X_7 + r_8X_8 + r_9X_9 + r_{10}X_{10} + r_{11}X_{11} + r_{12}X_{12}$$

where X_i and r_i represent the choice of item i and its benefit, respectively.

If the student chooses the first alternative shown in Figure 9.22, then the values of the following reward (or benefit) and the cost functions can be found as follows:

$$\begin{aligned}\text{Total reward} &= r_1(1) + r_2(1) + r_3(1) + r_4(0) + r_5(0) + r_6(0) + r_7(0)\\ &\quad + r_8(1) + r_9(0) + r_{10}(0) + r_{11}(1) + r_{12}(1)\\ &= r_1 + r_2 + r_3 + r_8 + r_{11} + r_{12}\end{aligned}$$

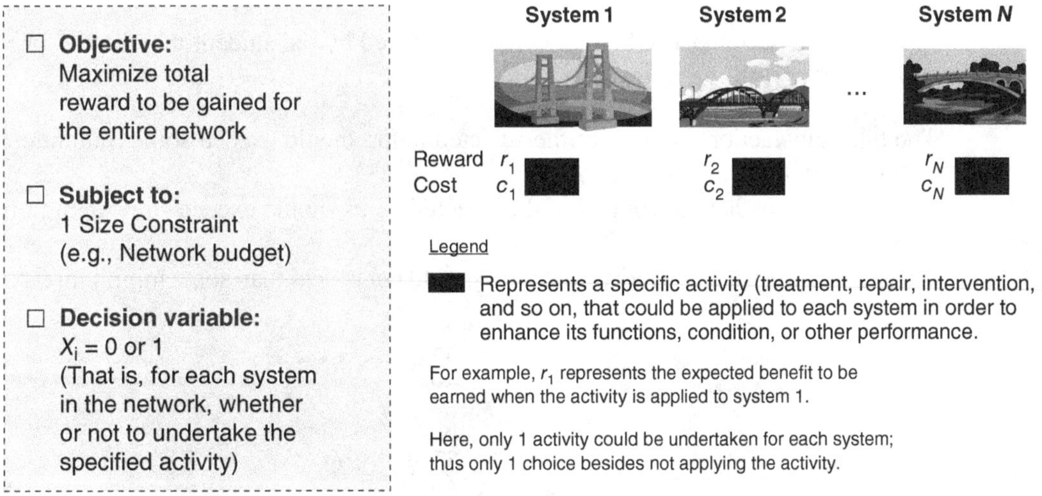

☐ **Objective:**
Maximize total reward to be gained for the entire network

☐ **Subject to:**
1 Size Constraint
(e.g., Network budget)

☐ **Decision variable:**
$X_i = 0$ or 1
(That is, for each system in the network, whether or not to undertake the specified activity)

System 1 | **System 2** | ... | **System *N***

Reward r_1, Cost c_1 | r_2, c_2 | r_N, c_N

Legend

■ Represents a specific activity (treatment, repair, intervention, and so on, that could be applied to each system in order to enhance its functions, condition, or other performance.

For example, r_1 represents the expected benefit to be earned when the activity is applied to system 1.

Here, only 1 activity could be undertaken for each system; thus only 1 choice besides not applying the activity.

Figure 9.22 Simple knapsack problem (KP) for a system of systems.

$$\begin{aligned}\text{Total cost} &= c_1(1) + c_2(1) + c_3(1) + c_4(0) + c_5(0) + c_6(0) + c_7(0)\\ &\quad + c_8(1) + c_9(0) + c_{10}(0) + c_{11}(1) + c_{12}(1)\\ &= c_1 + c_2 + c_3 + c_8 + c_{11} + c_{12}\end{aligned}$$

$$\text{Average reward} = \frac{1}{6}\sum_{i=1}^{12} X_i r_i = r_1 + r_2 + r_3 + r_8 + r_{11} + r_{12}$$

$$\text{Average cost} = \frac{1}{6}\sum_{i=1}^{12} X_i c_i = c_1 + c_2 + c_3 + c_8 + c_{11} + c_{12}$$

Objective Functions. The student may have any one (or more than one) of several objectives that could include the following: maximize the total benefit, minimize the total cost, maximize the sum of the benefit–cost ratios, maximize the average benefit, or minimize the average cost.

Constraints. Cost constraints: The cost constraints faced by the student may be any one or more of the following:

The total space occupied by all items must be less or equal to the knapsack volume, C^*.
The average space occupied by all items must not exceed some maximum threshold, C^{**}.
The space occupied by any individual item must not exceed some maximum threshold, C^{***}.

$$\sum_{i=1}^{N} X_i c_i \leq C^* \tag{9.23}$$

$$\frac{1}{N}\sum_{i=1}^{N} X_i c_i \leq C^{**} \tag{9.24}$$

$$c_i \leq C^{***} \tag{9.25}$$

Benefit constraints: The benefit constraints faced by the student may be any one or more of the following:

The total satisfaction from all of the selected items should exceed some minimum satisfaction threshold, R^*.
The average satisfaction from all of the selected items should exceed some minimum threshold, R^{**}.
The satisfaction from any individual item should not be less than some minimum threshold, R^{***}.

$$\sum_{i=1}^{N} X_i r_i \geq R^* \tag{9.26}$$

$$\frac{1}{N}\sum_{i=1}^{N} X_i r_i \geq R^{**} \tag{9.27}$$

$$r_i \geq R^{***} \tag{9.28}$$

Knapsack problems can be categorized as follows:

(a) Simple Knapsack Problem (KP). Here, the decision maker seeks to maximize the reward gained from the selection of project alternatives from a larger set based on a single "size" constraint, such as budget. In the backpack example, the problem is a simple KP if there is no other choice set besides whether or not to place an item in the knapsack and if the student is constrained in only one dimension: the backpack size or the backpack weight, but not both. Examples in civil engineering systems, at any phase of development, include situations where the decision maker chooses to either implement an action or not to implement it. We now discuss two illustrations: one in a networkwide context and the other in a system-level context.

In the first illustration, let us consider a network of systems (or system of systems) as illustrated in Figure 9.22. Here, there is only one possible activity to undertake for each system in the network: either do the activity or not do it. This could be some repair activity at the preservation phase, installation of real-time sensors at the monitoring phase, changing some policy at the operations phase, and so fourth. In this case, each system corresponds to each item in the knapsack; carrying out the action at that system represents placing the item in the knapsack. The objective could be to maximize the total reward to be gained for the entire network, subject to only one "size" constraint, such as the network budget, or one networkwide performance or benefit. Thus, the optimization problem is to identify which activity to undertake for each system in the network.

In the second illustration, let us consider an individual system that consists of several components, as illustrated in Figure 9.23. There is only one possible activity to undertake for the components of the system: Either implement the activity or not implement it.

Similar to the case discussed for the networkwide problem, the problem at hand may represent for a given individual system that consists of multiple components: there could be some repair activities at the preservation phase, installation of real-time sensors at the monitoring phase, or implementing a new policy at the operations phase. In this case, each component corresponds to each item in the knapsack; and carrying out the action for that component represents placing the item in the knapsack. The objective could be to maximize the total reward to be gained for the entire system (i.e., collection of components), subject to only one "size" constraint, such as the system budget, or one performance or benefit. Thus, the optimization problem is to identify which activity to undertake for each component of the system. Figure 9.24 presents the matrix for the entities (system or system component) vs. the evaluation criteria for the optimization.

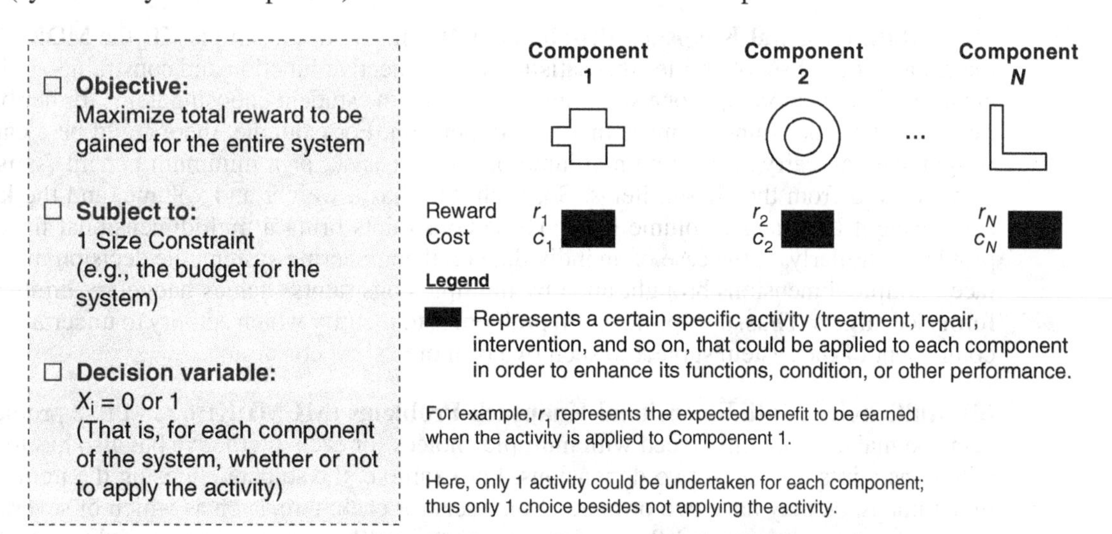

Figure 9.23 Simple knapsack problem (KP) for an individual system.

Decision criteria or performance measures (Only 2 shown here) / System i	System 1	System 2	...	System M
Benefit of the activity for each individual system e.g., increased system durability in years	r_1	r_2	...	r_M
Cost of the activity for each individual system, e.g., cost to the system owner in $	c_1	c_2	...	c_M

Figure 9.24 Matrix of system or system component vs. evaluation criteria.

(b) Multichoice Knapsack Problem (MCKP). In these problems, the decision maker is faced with multiple choices for each alternative. That is, for each item, there could be two or more alternatives for that item. For example, for a calculator, the student may have two brands to choose from. So, for selecting a calculator, she has two options: calculator A or calculator B (assume that she cannot take both). In several instances of civil engineering systems, the problem is not whether to carry out an activity (as in the case of the simple KP), but which of several competing activities to undertake. So, for each system of a network in Figure 9.22 or for each component of a system in Figure 9.23, there exists not just the one black square shown in the figure, but rather two or more squares representing the candidate actions for each entity. For example, at the system preservation phase, the choices for each entity include replacement, rehabilitation, maintenance, and do nothing. Thus, MCKPs have a set of choice constraints in addition to the size (budget or performance) constraints.

The situation is no different for an individual system that is comprised of components. Consider that there is only one possible activity to undertake for components of the system: Either you implement the activity or you do not implement it. Similar to the case discussed for the network-wide problem, this could be some repair activity at the preservation phase, installation of real-time sensors at the monitoring phase, changing some policy at the operations phase, and so forth. In this case, each component corresponds to each item in the knapsack; and carrying out the action for that component represents placing the item in the knapsack. The objective could be to maximize the total reward to be gained for the entire system which is a collection of the components, subject to a "size" constraint, such as the system budget, or one performance or benefit threshold. Thus, the optimization problem is to identify which activity to undertake for each component of the system.

(c) Multidimensional Knapsack Problem (MDKP). As in the simple KP, the MDKP involves the choice of a subset of entities that satisfies some objective function and constraints. In MDKPs, however, there is more than one size constraint in that the student choosing items for her backpack may view her constraints in more than one dimension. For example, there could be a cap on the weight she can carry, a maximum volume of the knapsack, or a minimum benefit (satisfaction) to be derived from the chosen items. So, each item has a weight and volume, and the knapsack has a weight limit and a volume limit. These constraints bring a multidimensional flavor to the problem. Similarly, in the case of an individual civil engineering system, the decision maker could face multiple dimensions brought upon by multiple constraints such as budgetary limits and performance targets. Thus, the optimization problem is to identify which activity to undertake for each component of the system subject to such two or more "size" constraints.

(d) Multichoice Multidimensional Knapsack Problems (MCMDKP). In these problems, the decision maker is not only faced with multiple choices for each alternative but also has to contend with constraints in at least two dimensions. For example, the student choosing the items to place in her knapsack may have two or more choice sets for each item, such as which of several brands of calculator to choose; and the student is constrained by the backpack weight as well as the backpack volume. For civil systems engineers who typically face problems with multiple options for each choice and multiple constraints, the MCMDKP is considered the most realistic of the knapsack problems.

9.4.2 Writing Choice Constraints Mathematically

Consider a collection of M items: $x_1, x_2, \ldots, x_M$. It is sought to select a number of items from this collection. The following constraints can be written mathematically:

a. Either an item is chosen or it is not chosen: $x_i = 1$ if item i is chosen; 0 if it is not chosen.

b. Due to some extenuating reasons beyond the control of the analyst, such as agency policy or political reasons, item k $(\in M)$ must be among the chosen items: $x_k = 1$.

c. The first N items must be chosen: $\sum_{i=1}^{N} x_i = N$.

d. At least p of the first N items must be chosen: $\sum_{i=1}^{N} x_i \geq p$.

e. At least p of the last N items must be chosen: $\sum_{i=M-N+1}^{M} x_i \geq p$.

f. At most p of the first N items must be chosen: $\sum_{i=1}^{N} x_i \leq p$.

g. At least p of the last N items must be chosen: $\sum_{i=M-N+1}^{M} x_i \leq p$.

h. Mutual exclusivity constraint: Either item p or item q is chosen but not both: $x_p + x_q = 1$.

i. Either item p or item q or neither of these two is chosen, but not both: $x_p + x_q \leq 1$.

j. One (and only one) of items in a set of size L, must be chosen: $\sum_{l=1}^{L} x_l = 1$ where x_1 is a member of L.

k. Only one (or none) of items in a set of size L, must be chosen: $\sum_{l=1}^{L} x_l \leq 1$ where x_1 is a member of L.

l. Preclusion constraint: Item p is chosen if and only if item q is chosen: $x_p \leq x_q$.

Example 9.16

A water systems engineer seeks to select a number of water treatment plants for close inspection under a special federal mandate. There are 25 such plants under her jurisdiction, which are numbered sequentially from 1 to 25 according to some criteria. Write the following constraints mathematically:

(a) The first 12, plants must be selected.

(b) At least 3 of the first 10 plants must be chosen.

(c) At most, 5 of the last 12 plants must be selected.

(d) Either plant 13 or plant 21 is selected, but not both.

(e) Plant 6 is selected if and only if plant 19 is selected.

Solutions

(a) $\sum_{i=1}^{12} x_i = 12$, **(b)** $\sum_{i=1}^{10} x_i \geq 3$, **(c)** $\sum_{i=14}^{25} x_i \leq 5$, **(d)** $x_{13} + x_{21} = 1$, **(e)** $x_6 \leq x_{19}$

Example 9.17

As the municipal manager of all public civil engineering systems in Perth County, you have a fiduciary responsibility to optimize the use of taxpayer dollars. For each system in the network, you seek to determine whether or not to undertake repair activity to achieve the maximum benefits for the entire population of systems in your jurisdiction. The overall maintenance budget is \$$C$. Other constraints are that the average benefit for the entire population of systems must not be lower than a certain threshold, R. Also, for only those systems that are selected to receive some repair, the average benefit should not be lower than a certain minimum Q; and the cost of any individual project must not exceed D. Also, the county commissioner has specified that system 2 repair should definitely be carried out. Also, either system 2 or system 3 must be selected for repair, but not both.

Using suitable decision variables, X_i, write a simple but *complete* mathematical formulation for this optimization problem. To help you visualize the problem, we have provided the table and notations below.

System *i* Performance measures	System 1	System 2	···	System *M*
Benefit of the repair activity for each individual system in terms of increased system durability	r_1	r_2	···	r_M
Cost of the repair activity for each individual system, in \$	c_1	c_2	···	c_M

Solution

M is the total number of systems in the population. Of these, let, say, P be the number of only those systems that are selected for repair. The decision variables are $x_i = 0,1$. The objective function is

$$\text{Max } Z = \sum_{i=1}^{M} x_i r_i$$

The constraints are:

1. The total cost of all repaired systems must not exceed the overall maintenance budget, C:

$$\sum_{i=1}^{M} x_i c_i \leq C$$

2. The average benefit for the entire population of systems must exceed a certain threshold, R:

$$\frac{1}{M}\sum_{i=1}^{M} (x_i r_i) \geq R$$

3. For only those systems that are selected to receive some repair, the average benefit should exceed a certain minimum Q:

$$\frac{1}{\sum_{1}^{P} x_i} \sum_{i=1}^{P} (x_i r_i) \geq Q$$

4. The cost of any individual project must not exceed D: $c_i \leq D$, that is, $c_1 \leq D, c_2 \leq D, \ldots, c_M \leq D$.
5. The county commissioner has specified that system 2 repair should definitely be carried out: $x_2 = 1$.
6. Either system 2 or system 3 must be selected for repair but not both: $x_2 + x_3 = 1$.

9.5 SEARCH FOR THE OPTIMAL SOLUTION—QUICKER AND EFFICIENT TECHNIQUES

At this point, we have come to appreciate that optimization involves substituting various combinations of the decision variables and calculating the corresponding values of the objective function. We do this until we obtain the set of decision variables that yields the best value of the objective function. In searching for the optimal solution, there are a number of techniques, ranging from the slow and laborious to the fast and furious. In this section, we discuss techniques that are basic, such as enumeration, as well as those that are quick and efficient.

9.5.1 Enumeration

Enumeration involves the evaluation of all of the alternative solutions one by one according to the criteria and the identification of the optimal solution as the final solution. This technique is slow and thus is often used where the number of combinations is relatively small.

9.5.2 Exterior Point Techniques

For Figure 9.25*a*, the search involves only three points or search stations (A, B, and C), each of which represents a certain combination of the decision variables x and y; but for Figure 9.25*b*, the search is more intensive as it involves eight points (A to H), each of which represents a certain combination of the decision variables x, y, and z. Clearly, as the number of constraints and decision variables increases, the number of search stations also increases. Most solution methods for optimization models use a numerical search, that is, they move from one search station to another; and at each station, they try different combinations of the decision variables until the station that optimizes the objective function is identified. For a few search points, this search could be done manually. For a problem that has thousands or even millions of search stations due to a multitude of

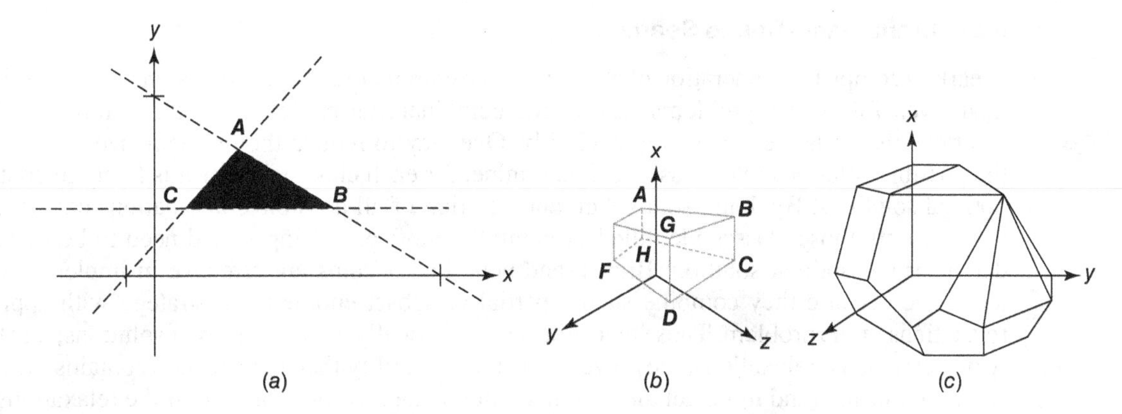

Figure 9.25 Optimization search stations—illustration.

decision variables and constraints, it is desired to carry out this search as quickly as possible. "Improving search" is a term used to describe the process of improving a current solution by checking the neighboring search stations. In certain texts, this is referred to as neighborhood search, local search, local improvement, or hill climbing. If the solution at any search station is superior to that of the previous station, the previous solution is abandoned in favor of the current one, and the process is repeated until all search stations have been checked. As discussed in an earlier section, the effort involved in searching for the optimal solution can be used as an attribute for classifying optimization problems.

In Figure 9.25, the optimal point is often located by traveling along the edges and testing each vertex for optimality. In 9.25*a*, there are 3 vertices to travel; in 9.25*b*, there are 12 vertices to travel; and in 9.25*c*, there are over 24 edges to travel. Algorithms that consider only the edges at the surface of the feasible region are referred to as *exterior point methods*. The simplex algorithm is one such exterior point search method. The simplex search begins at an extreme point of the feasible region, and then moves from one potential solution to the next, all the while retaining feasibility and improving the objective function until the optimum point is reached.

9.5.3 Interior Point Techniques

In recent years, new techniques, collectively called *interior point methods*, while following the improving search paradigm for linear programs, carry out their search by adopting different directions of movement: Instead of staying on the boundaries of the feasible space and passing from one extreme point to another, they proceed directly, by cutting through the feasible space. For interior point methods, there is greater analytical effort associated with each move. However, the number of moves needed to reach the final solution is drastically reduced, and ultimately, particularly for large LP problems, the computation time is much lower. The first commercial interior point method for linear programming was Narendra Karmakar's projective transformation procedure (Karmarkar's algorithm (KA)) which marked a generational improvement over the traditional method for solving LP problems. Given the feasible region as an x-sided solid with y vertices, KA finds the optimal solution by cutting right through the solid in its traversal and generating the solution rather than traversing the region from vertex to vertex along its boundary lines or planes. By providing a solution in polynomial time, KA drastically reduces the solution time for complex optimization problems (in some cases, from several weeks to a few days), particularly in business and engineering applications.

9.5.4 Branch-and-Bound Search

Clearly, a complete enumeration of all the possible combinations of the decision variables is impractical, even for simple problems, due to the combinatorial explosion of the number of discrete solutions that must be considered explicitly. One way to reduce the problem size is to rearrange the possible solutions into classes and determine, for each class, whether it is likely to contain the optimal solutions. By doing so, explicit enumeration of all possible combinations can be avoided because only those classes identified as being the most promising would need to be searched in detail for the optimal solution. Branch-and-bound algorithms are effective in implementing this technique because they combine such a partial or subset enumeration strategy with appropriate relaxations of the problem. These algorithms systematically create classes of solutions; and by analyzing associated relaxations, they investigate the possibility that a given class contains the optimal solution. Further (and more detailed) enumeration is then carried out only if the relaxations fail to yield the optimal solution.

9.5.5 Tabu, Simulated Annealing, and Genetic Algorithm Extensions of Improving Search

For a number of discrete optimization problem types, certain solution search techniques that involve nonimproving moves, have been found useful. One is called **tabu search** because it proceeds by classifying some moves as "taboo" or forbidden. The best single solution encountered at any stage of the process "will always be part of the population, but each **generation** will also include a spectrum of other solutions". Ideally, all will be feasible, and some may be nearly as good in the objective function as the best. Others may have quite poor solution values. Pairs of individuals in the population are combined to yield new solutions. This combining process does not center entirely on the best current solution (Rardin, 2002); as such, local optima are less frequent. Other popular search techniques are the simulated annealing and genetic algorithms.

9.5.6 Optimization Using Results of Repeated Simulation

If the probability distribution of a given system attribute is known, it may be possible to utilize probabilistic approaches to solve optimization problems involving that attribute. For many civil engineering systems, however, the sheer complexity of system processes makes it difficult, and often impossible, to derive the probability distribution, or its moments, for the measure of performance of various alternatives. In such cases, it becomes necessary to resort to the use of numerical methods, such as a *Monte Carlo procedure* (which we discussed in Chapter 8). A Monte Carlo procedure uses an artificial statistical experiment to estimate unknown quantities. Such procedures generally involve the generation of a sequence of numbers which are interpreted as observations of one or more random variables with a particular distribution and then perform operations on such observations so that the law of large numbers can be used to obtain meaningful results. Implementation of the Monte Carlo procedure involves construction of the experiment such that the unknown quantities of interest represent a probability or expected value associated with some random variable(s). If this is carried out and if the number of observations is large, the corresponding frequencies and averages can be used as estimates. The Monte Carlo procedure is a general numerical method, and as such it can be applied to any problem, provided that an appropriate experiment can be designed.

As we learned in Chapter 8, in analyzing a complex system, it may be infeasible to carry out such simulation experiments on the system directly. For example, in order to investigate the impacts of various speed limits on interstate highway crashes, it is not possible to change the speed limit law several times within a week or month as part of the experiment. Doing so will likely expose the system owner to the wrath of the system users. Thus, a mathematical model could rather be developed to mimic the process by duplicating the essential characteristics of the real system and experiments then conducted on the model. The process of using such a model to study the operations and subsequent performance of a real system is referred to as *simulation*. Several different types of models have been used to simulate system performance. If simulation uses a Monte Carlo procedure, then it is referred to as a *Monte Carlo simulation*.

For the purposes of the present section, the nature of Monte Carlo procedures and the application of such procedures in connection with simulation are considered. Certain types of simulation models are more amenable to the application of Monte Carlo procedures than others; therefore, the examples herein discussed (adapted from Au et al., 1972) are confined to the types of models most suitable for Monte Carlo simulation.

Example 9.18 (Discussion)

Exact-change lanes are to be used for the collection of tolls on an expressway. To use an exact-change lane, motorists must have the correct change, which they simply drop into a collection device as they pass through the toll gate. An insufficient number of such lanes will result in the development of excessively long waiting lines, causing correspondingly long delays to motorists. On the other hand, an excessive number of lanes will result in some of these lanes standing idle for a large percentage of the time. The problem is to determine the appropriate number of lanes for the volume of traffic expected on the expressway.

Important measures of performance in this case include the number of motorists waiting to pass through the toll gate, the average or maximum delay or waiting time for a randomly-selected motorist who uses the expressway, and the time that the lanes will spend in an idle or active state. These quantities must be treated as random variables, and their probability must be determined from assumptions made concerning the pattern of arriving vehicles and operation of the system. These probability distributions may or may not be derivable in closed form, depending on the vehicle arrival pattern and the complexity of the behavior of vehicles after arriving. If the probability distribution cannot be derived, then a Monte Carlo procedure can still be used. This would require the development of a model (a simulator), similar to the real system in all important characteristics, which would generate arrivals and pass them through the exact-change lanes in a manner analogous to the operation of the real system. The state of the exact-change lanes, the lost time for each user, and the fluctuations in the number of waiting vehicles could be recorded and used to obtain estimates of the corresponding probability distributions or their moments.

Most applications of Monte Carlo procedures are inherently of a probabilistic nature and can be thought of as a problem of deriving a probability value, an expected value, or an entire distribution. Furthermore, the Monte Carlo approach is normally employed as a last resort, after all attempts to obtain exact results through closed-form solutions have failed.

9.5.7 Heuristic Approaches

The past two decades have seen a considerable amount of research and practice into a growing class of optimization techniques collectively referred to as "heuristics." These techniques, which include genetic algorithms, the evolutionary algorithm, particle swarm optimization, colony optimization, and shuffled complex evolution (Rardin, 2002; Michalewicz and Fogel, 2004) have certain common characteristics (Dandy, 2008). First, these techniques, at any one time, deal with a population of candidate optimal solutions rather than a single solution. They typically do not reach a single optimal solution but progressively improve the population of solutions under consideration and the solutions they reach are near optimal. Further, these techniques use a general technique that can be described as guided-search methods that interact with a model that simulates the system or process under investigation. They typically involve some random processes and thus may be considered as probabilistic optimization techniques. Heuristics are appropriate when an application problem is computationally intensive and exact methods are unable to arrive at a solution within the time available for decision. They involve the use of rule-based algorithms that strive to achieve near-optimal solutions in a reasonable time. Examples of heuristic optimization approaches are discussed below.

Artificial Neural Networks (ANNs). These are networks of cells or processing elements (PEs) that convert the inputs to output using activation and output functions (Figure 9.26). These PEs are connected by weighted connections called **synapses**. Once the network is supplied with data, recursive methods are used to adjust the weights and the network reaches a stable state. To use this method for any problem, a way to represent the problem using the network architecture needs to be established.

Genetic Algorithms (GAs). The concept of genetic algorithms (GAs) has its roots in the sciences of genetics and natural selection. GAs utilize a representation scheme to encode the feasible solutions of the optimization problem and are implemented using strings or **chromosomes**. Each chromosome represents one "member" (or, in the context of optimization, one candidate solution)

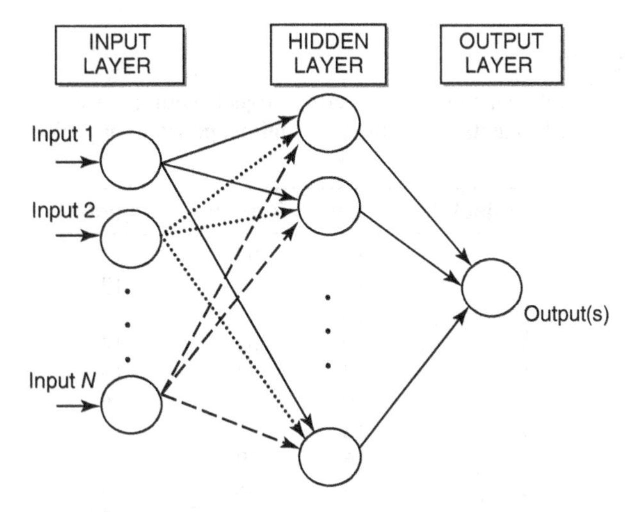

Figure 9.26 Processing element for neural networks.

that is better or worse than other members in the population of candidate solutions. Each chromosome is evaluated for fitness on the basis of some objective function. The chromosomes go through processes of crossover and mutation by exchanging information. The "survival of the fittest" process is applied to progressively enhance the population over successive generations (Hegazy 1999).

The objective function in GA optimization is called the **fitness function**. In GA, optimization is typically carried out to maximizing the fitness function. Thus, if the problem is one of minimization and the objective function is strictly positive, the fitness function is expressed as a product of negative 1 and the objective function or the reciprocal of the objective function. The steps involved in the operation of a simple GA are (Dandy et al., 2008): (1) generation of the initial population; (2) computation of the objective function for each solution in the population; (3) evaluation of the performance of each solution in the population relative to the constraints; (4) computation of the penalty cost for not meeting the constraints; (5) computation of the fitness function for each solution; (6) check for convergence of the population. If convergence occur, the process stops at this step; otherwise, continue with (7) selection of a set of parent strings for the next generation; (8) crossover of pairs of parents; and (9) mutation of selected strings; and then return to the second step to continue.

As an optimization solution technique, GAs have a number of features that are distinct from traditional optimization techniques such as LP (Simpson et al., 1994): First, GAs work directly with a set of solutions rather than a single solution. This set, or population, is spread throughout the solution space so the chance of reaching the global optimum is increased significantly. GAs deal with the actual discrete sizes available so that rounding-off of continuous variables is not required. Because GAs work with a population of solutions, they also identify a number of near-optimal solutions that could correspond to configurations that may have nonquantified objectives, such as environmental or social objectives. Further, GAs only use information about the objective or fitness function and do not require the existence or continuity of the derivatives of the objectives. GAs work by analogy to population genetics and involve operators such as selection, crossover, and mutation. Unlike traditional techniques, GAs do not necessarily converge to a global optimal solution. GAs work in conjunction with a simulation model, and they therefore can handle any nonlinear, discontinuous, or logical set of objective functions or constraints. Any process that can be simulated on a computer can be optimized using GAs (Dandy et al., 2008).

Example 9.19

Consider the following civil engineering projects with the respective costs and benefits indicated. The total budget is 150 units. Determine the optimal portfolio of projects using the GA technique.

Project ID	Cost	Benefit	Project ID	Cost	Benefit
1	10	94	11	18	63
2	18	94	12	19	50
3	10	56	13	14	97
4	20	65	14	12	65
5	18	93	15	10	52
6	10	89	16	16	49
7	13	95	17	13	94
8	16	69	18	12	81
9	13	64	19	18	62
10	19	69	20	11	81

Solution

This is a knapsack problem that could be formulated as follows:

$$\text{Max } Z = \sum_{i=1}^{20} x_i b_i$$

Subject to

$$\sum_{i=1}^{20} x_i c_i \leq 150$$

$$x_i = 0 \text{ or } 1$$

where $x_i = 0$ if project i is not selected and 1 if the project i is selected; c_i and b_i are the cost and benefit, respectively, of project i.

Genetic Algorithm

Step 1: Coding. In this step, the decision variables are translated into the binary coding in GA. In the above exercise, since the decision variables themselves are binary, it is easy to code them as binary bytes as illustrated below.

1	2	3	4	5	6	7	8	9	10	11	12	13	14	15	16	17	18	19	20
0	0	0	0	1	1	1	1	1	0	1	0	1	0	0	0	0	1	1	1

Then, randomly generate 200 initial solutions. Each solution is called a *population.*

Step 2: Crossover. 200 initial populations are randomly matched into 100 pairs. Conduct crossover between the populations in each pair.

Step 3: Mutation. After the crossover, each population mutates according to a prespecified mutation rate (0.05).

Step 4: Fitness evaluation. The total benefit of each population is calculated as the fitness of that population.

Step 5: Selection. Use tournament selection to select 100 populations based on their fitness values.

Step 6: Repeat steps 2 to 5.

Step 7: If the number of iterations reaches the specified maximum number of iteration (in this case, say, 3000) or if all the populations in the previous iteration are the same as the offspring, stop iteration and go to step 8; otherwise, go to step 6.

Step 8: Output the population with the maximum benefit; that population is the final solution.

The result for the genetic algorithm is:

Project ID	1	2	3	4	5	6	7	8	9	10	11	12	13	14	15	16	17	18	19	20
Solution	1	1	1	0	0	1	1	0	1	0	0	0	1	1	1	0	1	1	0	1

where "1" means include the project in the portfolio of projects to be implemented, and "0" means otherwise. It can be shown that this portfolio yields a maximum total benefit of 962 units.

9.6 DISCUSSIONS AND FINAL COMMENTS

9.6.1 Multiplicity of Performance Attributes in Optimization

In any decision-making process, the need exists to construct the preference order, directly or indirectly, so that alternatives can be ranked, and the best alternative can be identified. For decision-making problems that involve a single attribute, the preference order may easily be accomplished. For example, in the case of a decision based on a cost minimization rule (where the lowest-cost alternative is chosen), the preference order is adequately represented simply by the relative costs of the alternatives. Owners of civil engineering systems, however, are gradually finding that evaluation and decision making based on a single attribute typically do not provide acceptable results. For example, if the life-cycle agency cost is used as the sole basis for decision making, the best alternative actions would always be to do little or no action. As such, there is increased impetus for owners of civil engineering systems to make decisions that are based on a wide range of performance attributes. Such complex decision-making problems, however, typically involve conflicting objectives comprised of multiple attributes. It is often true that no dominant alternative may exist that is better than all other alternatives in terms of all of these objectives. For example, it may be difficult to maximize the level of service while minimizing agency costs at the same time.

9.6.2 Scope of Decision-Making Constraints

In civil engineering optimization problems, there is a wide scope of possible constraints. First, constraints may be budgetary or nonbudgetary (e.g., performance constraints). Budgetary constraints are often upper bound while nonbudgetary constraints may be lower or upper bound. Second, for network-level decision making, the constraints in decision making may be defined for individual facilities (e.g., the minimum delay per passenger at a specific transit route should be 3 minutes) or for the entire network of facilities (e.g., the average delay per passenger should not exceed

3 minutes). Also for network-level decision making, the constraints may be defined for individual years (e.g., average physical condition in each year should not exceed x units or maximum budget for each year is $5 million) or for all years within some specified period (e.g., average or total value of some system performance attribute for all years should not exceed some threshold or maximum budget for entire analysis period is $100 million). Third, the leftover values of the constraining variables may be transferable from one year to the next.

9.6.3 Optimization in Various Civil Engineering Disciplines

The development of optimal plans and designs for civil infrastructure has been the goal of engineers, architects, and master builders since the time of the ancient Egyptians and the early civilizations of Greece, as reflected in many of the design practices and codes that have been used over the centuries and millennia. In the design of any civil engineering system, ancient or modern, the common objective has been to build a structure that serves its purpose within resource constraints while satisfying the functionality and user safety requirements. Thus, the design phase, implicitly, is an optimization process. For example, in structural design, the objective is to maximize the efficiency of the structural system by minimizing either the weight or the overall cost of the system, subject to a number of constraints, including maximum shear forces, bending moments, and deflections. From some perspectives, the constraints can be considered as a reflection of the behavioral and safety aspects of the system, while the objective reflects the desires of the designer. Thus, any effort to use formal optimization in an explicit manner, such as formulating the design problem in terms of an optimization model, constitutes a natural extension of the spirit of the traditional design process. In the field of transportation, problems requiring optimization include the scheduling of the repair of fixed facilities or rolling stock over their life cycle or remaining life; selecting which assets to repair in a large network of assets at a given year; which layer types and thicknesses to use for a pavement; allocating shipments over a single-mode or multimodal network; transporting freight from multiple origins to multiple destinations; allocating distribution vehicles to serve multiple destinations; warehousing strategies that schedule planned releases of goods into the distribution system; and facility location in a network. In the fields of hydraulic and ecological engineering, optimization models have been applied to solve problems involving flood control, recreation, habitat preservation, irrigation, water supply and distribution, and hydropower. In construction engineering, the ordering of construction tasks, selection of labor–equipment combinations for a task, and the hauling of construction materials from the source to the destination can be solved using LP. The environmental engineer uses optimization to select or schedule processes for treatment of wastes or for remediation of polluted substrates. The architectural engineer encounters a large number of problems that require the tools of optimization, such as identifying designs that minimize energy use or enhance ventilation in buildings. In all the civil branches, the timing of capacity expansions of an existing system can be addressed using optimization formulations.

9.6.4 Investment Decisions—Contexts and Considerations

Investment decision-making is an area where optimization tools can be useful. In this section, we discuss the different contexts of investment decision-making on the basis of the level of analysis and the nature of the decision variable.

In civil engineering, a large majority of decisions for investments are aimed at deciding whether or not to undertake some activity or selecting one of two or more alternative systems to implement. When these decisions incorporate some element of time, such as when to do some activity, they morph into another kind of decision-making problem known as *work programming*.

Table 9.5 Relationships between Analysis Levels, Nature of the Decision Variable, and Analysis Context[a]

Analysis Level	**Continuity of Decision Variable**	**Question Being Answered**
Network level	Discrete	WWW-What intervention to undertake, at Which facility, and When? (Intervention to is some activity involving operations, monitoring, or repair.)
	Continuous	$—For entire system, how much to spend on O, M, or R?
		%—What percentage of systems in the network to receive O, M, or R?
Project level	Discrete	WW—For a given system, What intervention to undertake, and When?
	Continuous	$—For a given system, how much to spend for operations, monitoring, or maintenance?

[a]Each of these decisions are made for a given year or within some specific multi-year period.

The term programming used here refers to the *scheduling* of work across a time horizon of several years and is not the same as the term used in computer science context (e.g., C++ programming) or in the operations research context of optimization solution techniques (e.g., integer programming). Thus, the use of optimization tools in the task of evaluation could be to specify what actions [pertaining to operations, maintenance, or monitoring (OMR)] to undertake at any specific year or to establish a schedule showing which action to undertake in each of several years over an entire analysis horizon) (Table 9.5).

At the network level (i.e., for an entire system of systems), work programming occurs on a spatial as well as temporal dimension: For example, the engineer may seek to establish (i) which OMR intervention to undertake, at which system, and at which year; (ii) how much to spend on an O, M, or R action at each system in each of year, for all systems in the network individually or combined; or (iii) what percentage of systems in the network should receive some O, M, or R action at each year. The decision variable in context (i) is a discrete variable while those in contexts (ii) and (iii) are continuous variables. At the project level (for an individual system), work programming occurs mostly on a temporal dimension only; the exception is for a multicomponent system where work programming that is both spatial (component related) and temporal, is encountered. At the project level, work programming is typically carried out over the life cycle or remaining life of a given system and helps the system owner to determine, for a specific system, for example, (i) which O, M, or R actions to undertake at each year or (ii) how much to spend on an O, M, or R action in each year. The decision variable in context (i) is a discrete variable while that in context (ii) is continuous.

For those decision contexts above where the decision variable is continuous, the problem could be formulated as a linear programming problem and the simplex procedure can be used to solve the work programming problem. On the other hand, for those decision contexts above where the decision variable is discrete (binary), or in other decision contexts where the decision variable is mixed, the solution is somewhat more complicated and is strongly influenced by the problem size and the nature and number of constraints. According to Revelle et al. (2004), some mixed or discrete formulations can be transformed and solved as linear programs with likely success at achieving discrete solutions. However, for certain problems, it may be difficult to find an efficient method that finds the exact discrete optimal solution, and thus the analyst may need to contend with a reasonably good solution that is close to the optimal. The challenge here is recognizing which technique to use for the problem.

9.6.5 Software for Programming Problems

There is a large number of software packages for solving programming problems. We present a few in this paragraph. The packages differ from each other in terms of their capabilities for specific types of programming problems. It must be noted that these capabilities change from year to year as the vendors of these packages typically carry out continuous research and development to refine the algorithms underlying their packages. Commonly used packages include Lindo and Lingo, CPLEX, FortSP (for linear stochastic optimization), Gurobi, Rogue Wave, MATLAB Optimization Toolbox, Mathematica, MINUIT, and NAG. Other packages for solving programming include KNITRO (particularly suited for large nonlinear optimization problems), Opt++, an object-oriented toolkit for nonlinear optimization algorithms written in C++; the Sparse Non-linear Optimizer (SNOPT) is particularly effective for solving nonlinear problems whose functions and gradients are considered too time consuming to evaluate; the functions must be smooth and may be convex or otherwise. Free or open-source packages include IPOPT, Merlin, OpenOpt, and NLopt. Many of these packages provide interfaces to platforms including Microsoft Excel, GAMS, MATLAB, C, C++, Python, Fortran, Java, AMPL, Mathematica, MPL, and LabVIEW.

SUMMARY

In making decisions at any phase of systems development, civil engineers typically encounter situations where there is a need to employ the tool of optimization to carry out the task of identifying the best solution under given constraints. Depending on the type of civil engineering system in question and the phase of development, these decisions involve the prescription of specific types and quantities of some resource, and where or when they are needed. Often, the objective is to maximize and/or minimize some benefit and/or cost, respectively, to the system owner, user, or community, under constraints often driven by the concerns of these stakeholders. In this chapter, we discussed examples of tasks that require optimization tools at various phases of system development, and we discussed the relationship between optimization, evaluation, and decision making. Also, recognizing that different types of optimization problems require different optimization solution techniques, this chapter identifies several ways by which we could categorize optimization problems, including the number of objectives and number of attributes being considered, the discrete or continuous nature of the decision variable, the structure of the objective function, the convexity of the objective function, and the level of certainty (deterministic vs. stochastic nature of the problem). The chapter then discusses, with examples in a few branches of civil engineering, the concepts of unconstrained and constrained optimization using calculus, constrained optimization using mathematical programming techniques, and constrained optimization involving binary decision variables. For certain types of optimization problems contexts and sizes, the solution search could take several days, weeks, or even months or more. The chapter therefore presents a discussion of quick and efficient techniques that help lead quickly to solutions that are close to optimal. The chapter ends with a discussion of optimization-related issues associated with civil systems management, including the multiplicity of performance attributes, the contexts and considerations in investment decisions and available software packages for solving optimization problems.

EXERCISES

1. A municipality wishes to construct an overhead water tank to serve a complex of newly built student residence halls. Designs under consideration include a cube, a cylinder, a sphere, and a cone (Figure 9.27).

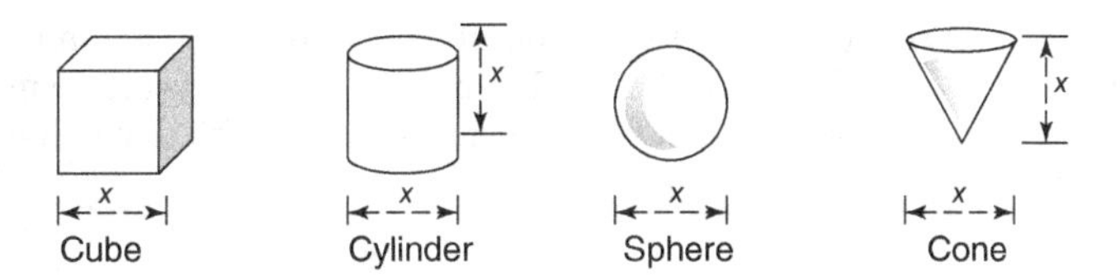

Figure 9.27 Alternative configurations for design.

a. Determine which design (i) has the most capacity, (ii) utilizes the most material, (iii) offers the maximum capacity while utilizing the least building material, and (iv) has the maximum usable (flat) floor area.

b. Which design would you recommend and why? What other design considerations may influence your decision?

c. Does the "useful floor area" attribute play any role in your decision? Why or why not?

2. A trapezoidal channel to carry 135 m^3/min of water is designed to have the best section. Find the bottom width and depth if the bed slope is 1 in 1200, the side slope is at 45°, and $C = 60$.

3. Determine the dimensions of a trapezoidal channel of the best section whose sides have a slope of 3 horizontal to 2 vertical. The proposed lining for the channel has a roughness coefficient $N = 0.012$. The bed slope of the channel is 1 in 5000, and the channel must discharge 10 m^3/s of water.

4. An architectural engineer needs to study the energy efficiencies of at least 1 of 10 large buildings in a certain region. The buildings are numbered sequentially $1, 2, \ldots, 10$. Using decision variables $x_i = 1$, if the study includes building i and $= 0$ otherwise, write the constraint(s) for each of the following requirements:

a. Only five buildings will be selected for the study.

b. At least one of the first four buildings will be selected.

c. At least two of the last three buildings will be selected.

d. A study must definitely be carried out for building 4.

e. Studies must not be carried out for buildings 2 and 9.

f. Building 10 will be studied only if building 3 is studied.

5. The rise in the water tables in a certain coastal city due to global warming is causing geotechnical problems for structures on fat clay foundations. Specifically, the moisture saturation of the otherwise dry clay is causing rapid loss of strength and settlement at five critical civil engineering structures citywide. As the city engineer, you have been given \$10 million to carry out geotechnical retrofits for these structures. Your geotechnical engineer has reported the project costs and benefits of the geotechnical retrofits at these sites (see table below). The benefits are measured in terms of the value of the civil engineering system to the society (on a scale of 1–10).

Civil Structure	**1**	**2**	**3**	**4**	**5**
Retrofit Cost ($M)	2.2	3.9	3.5	6.1	7.2
Benefit Score	3.2	6.9	5.9	9.1	8.4

a. Using Table 9.2 in Section 9.0.2(g), classify this problem.

b. Using appropriate decision variables, indicate the water engineer's objective function and the constraints.

c. Solve the problem using a manual plot on a two-dimensional graph.

d. Solve the problem using GAMS of other appropriate optimization software package.

6. A structural engineer seeks to design a 200-m^3 fuel tank to serve a mining plant. The tank will have a cylindrical body and hemispherical ends. The cost of the curved surface material is twice as much as that of the straight surface material. Find the dimensions of the tank that will minimize its total material cost.
7. Given the following constraint set:

$$x_1 + x_2 \leq 3 \quad (1)$$

$$1 \leq x_2 \leq 2 \quad (2)$$

$$x_1 \geq 0 \quad (3)$$

   a. Clearly show the feasible region on a graph for the constraint set.
   b. Label all extreme points.
   c. If Equation (1) is changed to $x_1 + x_2 = 3$, show the new feasible region on a separate graph and label the extreme points.
8. Consider the following constraint set:

$$x_1 - 2x_2 \leq 2 \quad (1)$$

$$2x_1 + x_2 \leq 9 \quad (2)$$

$$-3x_1 + 2x_2 \leq 3 \quad (3)$$

$$x_1 \text{ is unrestricted in sign} \quad (4)$$

$$x_2 \geq 0 \quad (5)$$

   a. Clearly show the feasible region on a graph for the constraint set.
   b. Label all extreme points.
   c. If a new constraint: $x_1 \geq 0$ is imposed, show the new feasible region and label the extreme points.
9. A transit-mix company markets 2 concrete mixes, *A* and *B*. The company can produce up to 14 truckloads per hour of mix *A* or up to 7 loads per hour of mix *B*. The available trucks can haul up to 7 loads per hour of mix *A* and up to 12 loads per hour of mix *B*, due to the differences in delivery distances. The loading facility can handle not more than 8 truckloads per hour regardless of the mix. The company anticipates a profit of $5 per load on mix *A* and $10 per load on mix *B*. What number of loads per hour of each mix should the company produce?
10. The water engineer of a city's municipal department plans to build a surface water reservoir to address the increasing water demand of the city's fast growing population. The available land for the tank (excluding access land), is 600 ft^2. The planned depth of the tank is 10 ft, and it is desired that its length is at least twice its width. Land limitations dictate that the width cannot exceed 20 ft. The tank will be built with reinforced concrete of 1 ft thickness. The city seeks to reduce the total cost of the concrete construction. The cost of 1 m^3 of concrete construction, which includes not only concrete material costs but also formwork and other related costs, is $150.
    a. Using Table 9.2 in Section 9.0.2(g), classify this problem.
    b. Using appropriate decision variables, indicate the water engineer's objective function and the constraints.
    c. Solve the problem using a manual plot on a two-dimensional graph.
    d. Solve the problem using GAMS or other appropriate optimization software package.
11. A structural engineer seeks to design a tall cylindrical building to house an office complex and communications tower. The total floor space (ft.^{2}) will be 160,000 and each floor will be 12 ft. in height. The engineer seeks to have as many floors as possible. For the purpose of structural stability, the diameter of the building must be at least 25% of the height.
    a. Using Table 9.2 in Section 9.0.2(g), classify this problem.
    b. Using appropriate decision variables, indicate the structural engineer's objective function and the constraints.

c. Solve the problem using a manual plot on a two dimensional graph.

d. Solve the problem using GAMS of other appropriate optimization software package.

12. For construction of a civil engineering facility, a contractor has found natural reserves of sand and gravel at Bloomingdale and Valley Springs where he may purchase such material. The unit cost, including delivery from Bloomingdale and Valley Springs, is \$5 and \$7, respectively. After the material is brought to the site, it is mixed thoroughly and uniformly, and the contract specifications state that the mix should contain a minimum of 30% sand. A total volume of 100,000 m^3 of mixed material is needed for the project. The Bloomingdale Pit contains 25% sand and the Valley Springs Pit contains 50% sand. As the new young construction engineer on the project, you are asked to determine how much material should be taken from each pit in order to minimize the cost of material.
Using Table 9.2 in Section 9.0.2(g), classify this problem.
Define the decision variables.
Write the constraints in mathematical form.
Write the objective function in mathematical form.
Find the optimum solution using the graphical method (how much material should the contractor take from each pit in order to minimize the overall cost of the material?).
If you had not been hired, the contractor would have used 60,000 m^3 from Bloomingdale and 40,000 m^3 from Valley Springs. How much did you save the company by giving them your advice?
Solve the problem using GAMS.

13. The City of Tamale's sewerage system consists of 12 major connecting pipe sections along the streets shown in Figure 9.28. It is planned to install sensor wires along the connector pipes in order to improve the capability of monitoring the operational performance of this system. However, due to budgetary limitations, the sensor wires can be placed along only some, not all, of the 12 connector pipe sections. For each connector pipe section, the anticipated overall benefits (in terms of an index of enhanced confidence in the pipe performance) and costs (in millions of dollars) of the sensor installation if it is carried out along the indicated pipe section. The monitoring budget for that year is \$5 M (i.e., five units of cost) and the city's mayor seeks to maximize the total benefits, under the given budget, to be earned due to the monitoring projects selected in that year. Due to frequent complaints of those served by the Sherman Road Connector, the city mayor requests that monitoring of that pipe section should definitely be among those to be carried out, irrespective of its costs or benefits. As a consultant for the City of Tamale, you are advising the city to identify the optimal solution (i.e., is, which pipes must be monitored and which must not, in order to satisfy the objective and constraints).

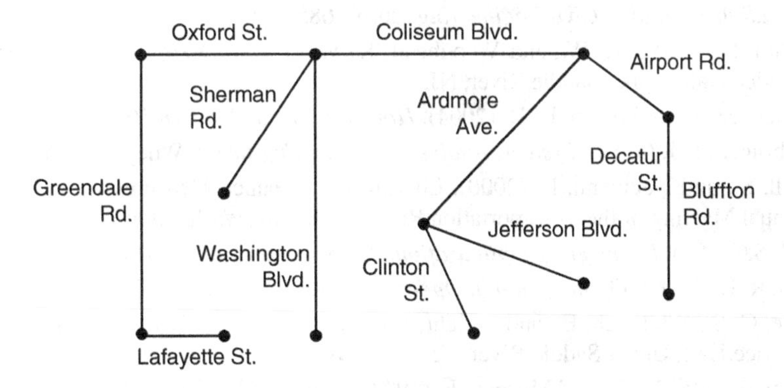

Figure 9.28 An urban street sewer network.

a. Write the stated objective in mathematical notation.

b. Write all the constraints in mathematical notation.

c. Use MS Solver to solve this problem (i.e., find the optimal solution).

d. What is the total systemwide benefit and cost that correspond to the optimal solution?

e. Attach a one-page output printout of the Solver output solution.

f. Which category of knapsack problem is this: KP, MCKP, MDKP, or MCMDKP? Give reasons for your answer.

Sewer Link	Expected Benefit	Expected Cost
Oxford Street Connector	0.8	0.3
Greendale Connector	0.7	0.9
Airport Road Connector	0.2	0.4
Lafayette Street Connector	0.5	0.9
Washington Boulevard Connector	0.1	0.5
Coliseum Boulevard Connector	0.8	0.6
Sherman Road Connector	0.1	1.0
Bluffton Connector	0.6	0.4
Ardmore Avenue Connector	0.9	1.2
Clinton Street Connector	0.3	1.4
Decatur Street Connector	0.4	0.8
Jefferson Boulevard Connector	0.7	0.9

REFERENCES

Au, T., Shane, R. M., and Hoel, L. A. (1972). *Fundamentals of Systems Engineering: Probabilistic Models*, Addison-Wesley, Reading, MA.

Bansal, R. K. (2005). *A Textbook of Fluid Mechanics and Hydraulic Machines*, Laxmi Publications, New Delhi, India.

Dandy, D., Walker, D., Daniell, T., and Warner, R. (2008). *Planning and Design of Engineering Systems*, 2nd ed. Taylor and Francis, New York.

Hassoun, M. (2003). *Fundamentals of Artificial Neural Networks*, Denver, CO.

Hezagy, T. (1999). Optimization of Construction Time-Cost Tradeoff Analysis using Genetic Algorithms, *Canadian Journal of Civil Engineering*, 26(6), 685–697.

Meredith, D., Wong, K. W., and Woodhead, K. W. (1985). *Design and Planning of Engineering Systems*, Prentice Hall, Upper Saddle River, NJ.

Michalewicz, Z., and Fogel, D. B. (2004). *How to Solve It: Modern Heuristics*, 2nd ed., Springer, New York.

Ossenbrugen, P. J. (1984). *Systems Analysis for Civil Engineers*. Wiley, New York.

Pickrell, S., and Neumann, L. (2000). Linking Performance Measures with Decision-making, Procs., 79th Annual Meeting of the Transportation Research Board, Washington, D.C.

Rao, S. S. (2009). *Engineering Optimization: Theory and Practice*, 4th ed. Wiley, Hoboken, NJ.

Rardin, R. L. (2002). *Optimization in Operations Research*. Pearson Education, Upper Saddle River, NJ.

Revelle, C. S., Whitlatch, E., and Wright, J. (2004). *Civil and Environmental Systems Engineering*, 2nd ed. Prentice Hall, Upper Saddle River, NJ.

Simpson, A., Dandy, G., and Murphy, L. (1994). Genetic Algorithms Compared to Other Techniques for Pipe Optimization, *Journal of Water Resources Planning and Management*, 120(4), 423–443.

Stark, R. M., and Nichols, R. L. (2005). *Mathematical Foundations for Design: Civil Engineering Systems*, Dover Publications, Mineola, NY.

Taha, H. A. (2003). *Operations Research. An Introduction*, 7th ed. Pearson Education.

CHAPTER 10

COST ANALYSIS

10.0 INTRODUCTION

In the tasks of analysis, evaluation, and optimization at any phase of civil engineering systems development, a candid assessment of actions is possible only after due consideration of not only the benefits but also the costs of each alternative option. Also, for planned or proposed new civil systems, costing provides a basis for estimating how much capital is needed to construct the system and also for determining how much will be needed to operate, maintain, inspect, and monitor the system over its life cycle; for existing systems, costing helps to estimate the capital required to increase the quantity or quality of the services rendered by the system. At the operations phase, cost analysis can help assess the system benefits (in terms of cost reduction) or the community or user costs compared to a base case scenario, which is often the do-nothing alternative. In our discussion of economic analysis in Chapter 11, we will use these cost and benefit estimates as inputs in economic analysis to ascertain the economic feasibility of a proposed action or to evaluate multiple competing actions at any phase. For these reasons, the practice of civil engineering has always included the estimation or prediction the costs of various actions associated with any phase of the system development process.

This chapter addresses cost engineering, and we present some basic techniques for analyzing these costs. The discussion begins with an identification of the key criteria for classifying the various types of costs associated with civil engineering systems development. Then we will discuss two general types of cost estimation and the approaches they use, namely, the conceptual estimate, for which we will use the aggregate approach to yield rough cost estimates for application at the phases of planning and design, and the detailed estimate disaggregate approach to develop more precise cost estimates using the quantities and unit prices of the factors of production (e.g., labor, materials, equipment use, etc.) to be used at the phases of design and bidding. This chapter also presents techniques for adjusting costs across locations and years. Finally, the issue of cost overruns, the factors that influence agency cost, and other costing issues will be discussed. Recognizing that cost estimates for civil systems generally can be difficult to obtain, Appendix 3 presents, for the benefit of the reader, historical cost values and models established in the literature for systems in different branches of civil engineering.

10.1 SYSTEM COST CLASSIFICATIONS

The various costs encountered in civil systems development can be classified in several ways including the phase of development at which the cost is incurred and the stakeholder who bears the cost (Table 10.1). The table shows that the system owner bears some kind of cost at each of all eight development phases while the user and community costs typically start at the construction phase. Sections 10.1–10.4 discuss the various costs borne by the different stakeholders, while Section 10.5 discusses the costs incurred at each phase. In this section, we present other criteria for classifying the costs of civil engineering systems.

Table 10.1 Primary Classifications of Systems Cost—System Phase and Stakeholder

System Phase / Stakeholder	Needs Assessment	System Planning	System Design	System Construction	System Operations	System Monitoring	System Preservation	System End of Life
System Owner/ Operator	*	*	*	*	*	*	*	*
System User				*	*	*	*	*
Community Affected by System				*	*	*	*	*

*Indicates phases at which the indicated stakeholder typically bears some direct or indirect monetary or non-monetary cost.

10.1.1 Cost Classification by Stakeholder

The system owner's costs mostly refer to the direct expenditure incurred by the system operator or owner in the form of cash payments to its workers or to hired contractors; *user costs* are out-of-pocket cash costs or indirect costs often associated with poor levels of service, borne by the users of the system; and *community costs* are indirect and often intangible costs borne by the groups of persons who live or work in the proximity of the system (Figure 10.1). Community costs may be monetary (e.g., changes in property values at areas proximal to the system) or nonmonetary (e.g., noise, air, or water pollution due to system operations). Further details of the owner, user, and community costs are presented in Sections 10.2, 10.3, and 10.4, respectively.

10.1.2 Cost Classification by Systems Development Phase at Which Cost Is Incurred

As we can see in Figure 10.2, different cost types (monetary or nonmonetary, tangible or intangible, direct or indirect) are incurred by the system owner, the users, or the affected community at many

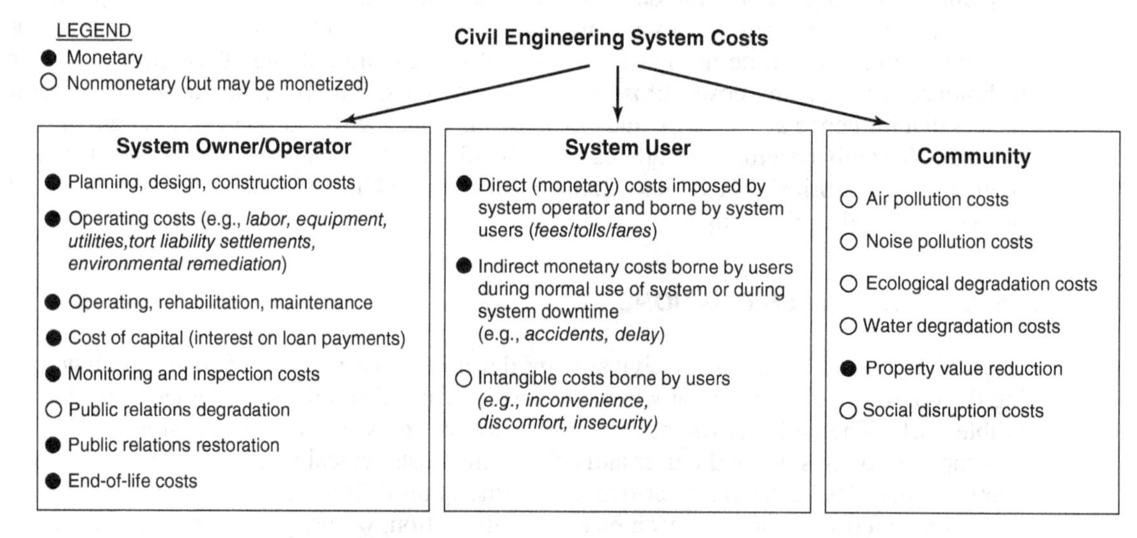

Figure 10.1 System costs categorized by stakeholder.

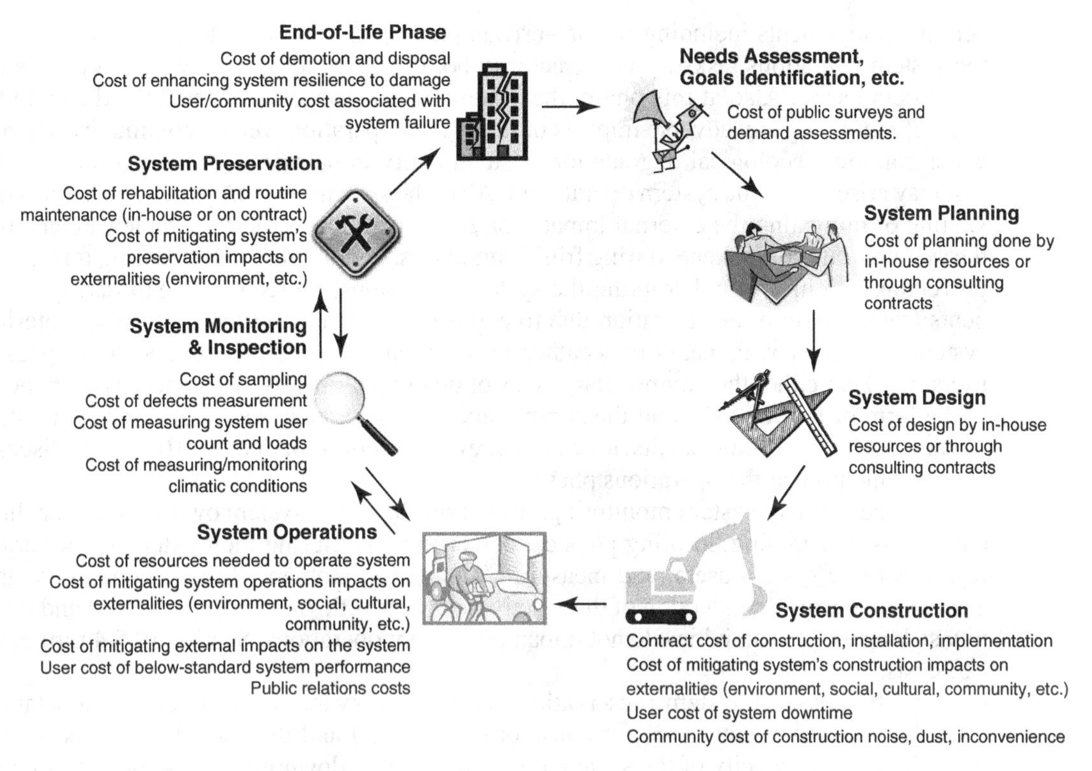

Figure 10.2 Examples of the costs incurred at the systems development phases. [all borne by the system owner/operator unless otherwise indicated.]

phases of the system development cycle. At the phase of planning or designing a new system (or for improvements to an existing system), the system owner carries out planning and design in-house or hires a consultant to do this work. Consulting fees vary according to the type and size of the civil system (as we shall see in Section 10.8.8). At the phases of planning and design, virtually no cost is incurred by the system users or the community. In Section 10.5.1, we provide additional discussion of the costs incurred at this phase.

At the construction phase, the system owner incurs the monetary costs of construction, installation, and/or implementation of the system or its components. This work is often carried out by contract (and not in-house) in the form of direct cash payments to the contractor. At this phase, the community often encounters nonmonetary costs of air pollution (dust) and construction noise. If the project is a reconstruction or rehabilitation activity, then the users suffer the costs of system downtime (in the form of delay, inconvenience, and safety). In response to legislative requirements or as a public relations effort, the system owner or operator typically undertakes activities to mitigate these intangible costs borne by the users and the community due to the system downtime, incurring some monetary costs in the process. We discuss construction costs further in Section 10.5.2.

The system operations phase typically constitutes a dominant fraction, by far, of the system life. At this phase, the system owner incurs the monetary costs associated with operating the system in the form of equipment use, information technology, public outreach, salaries (labor), and in some cases, materials and consumables needed to help the system run effectively. Under certain arrangements including public–private partnerships (PPPs), the system owner outsources the system operations tasks to a contractor (who, in return, may or may not levy a fee or toll on the system users). Also at this phase, the system owner or operator incurs the direct and tangible costs of mitigating any adverse impacts of the system operations on the community (such as noise, water pollution, ecological degradation, visual quality impairment, and sociocultural disruptions that may arise due to the system operations). Also, the system owner/operator bears the cost of preventing or mitigating the external impacts on the system operations; examples include deicing of the system or its components during frigid conditions, cooling the system during torrid conditions, preventing wildlife from disrupting the system operations, protection of exposed system components from long-term deterioration due to exposure to extreme climate and other interference of system operations by fauna, flora, weather, or other natural forces. At the operations phase, system users may bear either the nonmonetary costs of below-par operational standards or monetary costs in the form of fees or tolls, and the community often may bear the costs associated with adverse impacts such as pollution, as discussed in a previous section. In Section 10.5.3, we discuss further the costs incurred at the operations phase.

At the phase of system monitoring/inspection (M/I), the system owner incurs the direct monetary costs of sampling, locating physical defects and measuring their extent and severity, counting/classifying system users, and measuring/monitoring ambient climatic conditions, and so on. The system owner's conduction of these activities poses little or no cost to the user and the community so long as these activities do not impair the system operations. Section 10.5.4 further discusses M/I costs.

At the phase of system preservation, the system owner bears the direct monetary costs of rehabilitation and maintenance (in-house or on contract) and the users bear the costs associated with diminished capacity of the system due to the system downtime. These preservation activities could be preventive (before the onset of deterioration) or corrective (after structural damage has occurred). Also, preservation could be routine (seasonal or annual) or periodic (every 3–6 years). In Section 10.5.5, the costs at the preservation phase are discussed in greater detail.

With regard to intended end of life of the system, the direct cost of such termination is largely borne by the system owner in decommissioning the system (often in preparation for replacing it with one of larger capacity). The indirect costs borne by the users include the inconvenience, delay, or hazards associated with the use of a temporary backup system. The community often also bears indirect and intangible costs in the form of exposure to the dust and toxic materials that may be released during the demolition process (e.g., lead compounds in old buildings) and the vibration and noise associated with the demolition activity. With regard to unintended end of system life due to events including disasters, the costs borne by the system owner may include evacuation costs, litigation costs associated with the disaster response, system recovery costs, and the cost of establishing institutional mechanisms or backup facilities to reduce the user and community exposure or impacts subsequent to the disaster. In Section 10.5.6, the costs at the end-of-life phase are discussed further.

Figure 10.2 presents a number of system costs that are typically incurred at the various phases of development of civil engineering systems.

10.1.3 Preemptive versus After-the-Fact Costs

The cost of an activity or situation associated with a system's condition or operational performance can be categorized as either preemptive or "after the fact." The former is incurred mostly by the system owner or operator to ensure, in a proactive manner, that the system is kept in good physical condition and provides satisfactory performance during its operations. Examples of preemptive costs include the provision of a textured ice-free runway as part of airport operations to prevent airplane skidding and runway runoff incidents during icy weather, the application of a thin hot-mix asphalt overlay to a highway pavement to delay the onset of structural distress, and the spraying of steel water reservoirs to prevent corrosion.

On the other hand, costs that are incurred by the users (and sometimes, the system owner) when the system is already experiencing subpar performance or condition, are described idiomatically as after-the-fact costs. For a system owner, this may include the cost of corrective maintenance (often because the owner failed to carry out earlier preventive maintenance) or the cost of settling tort liability lawsuits filed by system users who suffer injury or inconvenience from a poorly maintained system. For users, after-the-fact costs include indirect consequences such as increased delay, decreased safety (fatality, injury, or property damage), and increased out-of-pocket expenses and other indirect costs due to poorly performing or poorly maintained systems. For the community, after-the-fact costs include the intangible costs borne by residents of areas in close proximity to the system (such as noise and air pollution, vibration), direct damage to abutting property due to system construction, operations, or demolition, and human (nearby residents) casualties attributable to the system use. The after-the-fact approach of costing provides a basis to estimate the benefits (or cost reductions) associated with the indirect or intangible consequences of the civil system. Often, the after-the-fact costs are inversely proportional to the preemptive costs; namely, the more money a system owner spends proactively, the less the frequency or intensity of after-the-fact adversities and their associated costs.

10.1.4 Monetary versus Nonmonetary Costs

A *monetary* cost is one that is expressed in dollar values; otherwise it is considered a nonmonetary cost. The most common examples of monetary costs are those of construction, maintenance, consulting services tort liability, interest payments on loans, and user costs. These occur as cash payments made by the system owner to a contractor, a financial institution, or some other entity, or as out-of-pocket fees and fares paid by the system user. A number of nonmonetary costs associated with system construction or operations are incurred by the agency (public relations problems), the user (discomfort and inconvenience), or the community (social disruption). As we saw in Figure 10.1, certain nonmonetary costs are *intrinsically monetary* because they can be converted into dollar values.

10.1.5 Direct versus Indirect Costs

For the system user, the direct costs include fees and fares, and the indirect costs include delay or some impact that is not incurred directly but may be monetized. Indirect user costs are related to the safety, comfort, and convenience of system users during the system operations or system downtime. From the community perspective, indirect costs may include reduction in property values due to the system's adverse effects on the environment, leading to reduced prices of houses in areas proximal to systems that emit pollution (e.g., noise, air) such as landfills.

10.1.6 Tangible versus Intangible Costs

The tangible costs associated with civil systems are those costs that can be measured; these costs may be monetary or nonmonetary. Intangible costs are those that are difficult to quantify or measure, and they often represent a variety of adverse impacts borne by the system owner/operator, user, or community, such as losses in productivity or customer goodwill and reductions in staff morale.

10.1.7 Internal versus External Costs

Closely related to the concepts of monetization, directness, and tangibility of costs is the externality of costs. An *externality*, or *transaction spillover*, is a cost or benefit (of an action) that is not transmitted through prices (Fang, 2011) and is incurred by a party that did not necessarily agree to the action. A *negative externality* describes an externality that is a cost (such as air pollution) while a *positive externality* refers to an externality that is a benefit (such as economic development). For civil systems, external costs are often borne by the community and are often inevitable. The market-driven approach to addressing negative externalities is to "internalize" them, that is, to quantify and attach a cost value to them. Government agencies make attempts to reduce negative externalities by passing criminal laws (environmental and public health legislation), passing civil tort law (to address any injuries or property damage caused to the community), provision of mitigating infrastructure or services (such as media campaigns), and establishment of *Pigovian taxes* or subsidies. A Pigovian tax is a special tax levied on industries and organizations including owners of civil engineering systems that inadvertently cause negative externalities (e.g., environmental pollution) in the course of their operations.

10.1.8 Recurring versus Nonrecurring Costs

A *recurring* cost is one that is repetitive and thus occurs continuously while a *nonrecurring* cost occurs only once or a few times within a given time horizon. If the time horizon is one life cycle of the system, then the cost of construction is nonrecurring while the cost of operations and maintenance are recurring. If the time horizon is infinity, then construction costs can be considered as recurring because systems do not last forever yet are needed by society continuously; thus they need to be replaced when they reach the end of their service lives.

10.1.9 Fixed versus Variable Costs

Any cost is subject to change. However, for a given system, there is a certain category of costs that tends to remain relatively constant over a specific range of operating conditions. These costs, which generally remain unchanged irrespective of the level of activity or output, are known as *fixed* costs. Examples include worker salaries, insurance and taxes on facilities, license and registration fees, and other user fees (Thuesen and Fabrycky, 1964; DeGarmo et al., 1997). Fixed costs change significantly only when the system is decommissioned or expanded. *Variable* costs, on the other hand, change in proportion with the level of activity or the amount of output. In Section 10.6, we will examine some basic concepts of costing in the context of fixed and variable costs of some activity at any phase.

10.1.10 Initial, Rest-of-Life, and Life-Cycle Costs

Initial costs refer to the costs incurred before the system starts operating. Often, this is meant to consist only of construction costs but may also be extended to include planning and design costs. Rest-of-life costs refer to the costs of maintenance, operation, salvage/disposal, and others costs that may be incurred between the inception of system use and end of the system life. Life-cycle cost is the arithmetic sum of initial and rest-of-life costs (Figure 10.3) and is often estimated on the basis of the nonconstant value of money over time (as we shall discuss in Chapter 11).

10.1.11 Sunk Costs versus Working Costs

In evaluating alternative courses of action on the basis of their respective costs and benefits, we often encounter costs that occur in the initial phases of development or are common to all the alternatives under consideration and, therefore, do not influence the differences in the estimated future costs and benefits of the alternatives. It may be argued that such *sunk costs* play no role in the evaluation

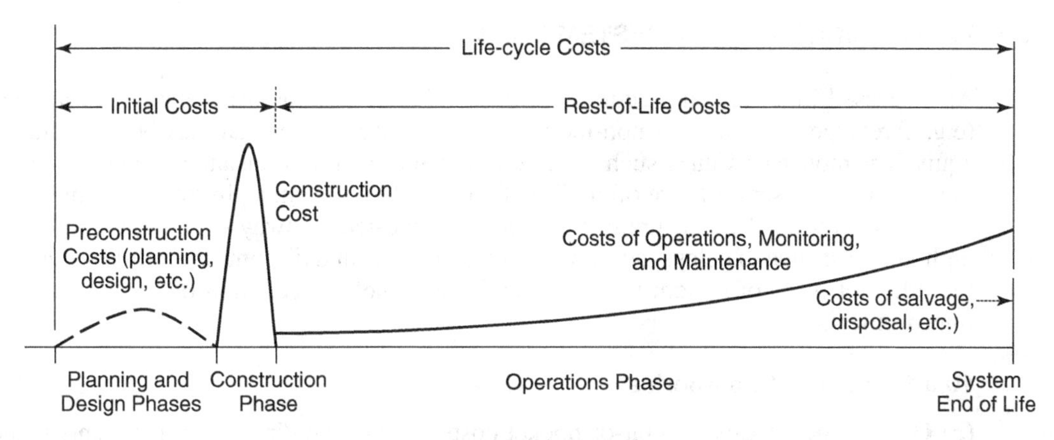

Figure 10.3 Initial, rest-of-life, and life-cycle costs (not to scale).

of alternative actions. For example, in deciding whether to build a steel water tank or a concrete water tank for water storage for a town, the land acquisition cost for the tank is considered a sunk cost as it plays no role in the choice of either material alternative over the other. *Working costs*, on the other hand, play a role in the evaluation of such alternatives.

10.2 COSTS INCURRED BY SYSTEM OWNER/OPERATOR

Owner/operator costs (often termed "agency costs") are the direct cash payments made by the system operator or owner in providing the system and maintaining its condition and level of service and indirect costs associated with user dissatisfaction and poor public relations. Thus, the system owner incurs costs at all the eight phases of systems development. For most civil systems, the owner costs may be categorized into at least eight groups that are related to the development phases: planning costs (for advanced planning and acquiring the right-of-way), design costs, construction costs, operations costs, inspection/monitoring costs, maintenance costs, and end-of-life costs. Many of the cost issues discussed in this chapter pertain to the costs incurred by the system operator or owner at the construction and operations phases.

10.2.1 Initial Costs Incurred by System Owner

The system owner incurs several types of costs at the initial phases of a civil system's life. These costs include the cost of acquiring and preparing the right-of-way, advanced planning costs, preliminary engineering and final design costs, and construction costs. For different types of civil engineering systems, the relative amounts of these costs are different. Also, initial costs are generally lower for reconstruction compared to new construction projects. Additional discussion of the initial costs is provided in Section 10.5.1.

10.2.2 Rest-of-Life Costs Incurred by System Owner

The costs incurred by the system owner or operator after the start of operations of a newly-constructed system include the costs of the system operations, maintenance, monitoring, and end of life. These costs are discussed in greater detail in Sections 10.5.2–10.5.6.

10.3 COSTS INCURRED BY THE SYSTEM USER

As discussed in a previous section, the users of civil systems bear costs that may be monetary (e.g., fares, fees, and tolls) or non-monetary (i.e., consequences that may be translatable into their equivalent monetary values such as time and safety or non-translatable such as discomfort and inconvenience. User costs are often directly related to a system's physical condition; for example, the poor condition of water supply systems and congested highway systems generally translate into high user costs due to delays, accidents, inconvenience, and discomfort. In this section, we discuss the different types of system user costs and user benefits (reduction of user cost) due to system improvements.

10.3.1 User Cost Categories

(a) Out-of-Pocket Costs. Out-of-pocket costs refer to the direct cash payments made by users for the use of the system. In many countries, a majority of the civil systems are provided by the government for the benefit of the general public *pro bono* (i.e., without the collection of any direct user taxes in the form of tolls or fees). On the other hand, there are certain civil systems whose funding is fully or partially borne by the system users through the collection of direct user taxes; These include tolls for bridges and highways, water supply fees, waste collection and treatment fees, and airport taxes. These out-of-pocket costs may be based on the extent to which a user uses the system (e.g., hourly parking rates in a commercial parking garage or weight–distance fees for trucks) or based on blanket fees irrespective of the extent to which the user uses the system. The former, which can be described as pay-as-you-go (PAYG) user costs, are more directly amenable to the price elasticities of supply and demand and therefore can be analyzed using the concepts of classic microeconomics (discussed in Section 10.3.2). As these taxes are based on the extent of system use, certain agencies often find the need to monitor and record, with the aid of technology, the extent to which each user uses the system (see Chapter 24 on system monitoring). Blanket fees, on the other hand generally require less monitoring effort.

Out-of-pocket user fees are often meant to generate funds for the upkeep of the civil system. Recognizing that the different classes of system users cause different levels of damage to (or consumption of) the physical system or to its operational performance, the amount paid by each user class is designed by the system owner to reflect the "damage" inflicted by that class. To establish appropriate PAYG or annual fees (or to update existing fees) for each class of user, the system owner or operator conducts **cost allocation studies**. At any time, the current fairness or equity of the existing fee can be ascertained by comparing the existing fee paid by each user class and the actual current cost responsibility of that class. If there is little or no difference between the actual fee paid and the attributable fee in any given user class, then the user fee structure is described as being **equitable**. However, if users of a certain class pay more than their attributable fee, then the situation is **inequitable** because they are subsidizing other users of the system.

(b) Safety Costs. The safety costs associated with civil systems are often considered nonmonetary (but intrinsically monetary), indirect, and in some cases, even intangible. As we discussed in Section 10.1.3, the costs of user safety (or lack thereof) can be estimated as a preemptive cost or as an after-the-fact cost. Often, the former are incurred only by the system owner/operator when they spend money to ensure that accidents are minimized. On the other hand, the latter are incurred by the other stakeholders (i.e., the users and community when they suffer the consequences of fatality, injury, or property damage in a system accident) or by the system owner when they repair damaged components of the system after an accident or settle tort suits arising from accidents. The after-the-fact approach of safety costing, which provides a basis upon which the safety benefits (accident reduction) of civil systems can be estimated, can be carried out using either the *human-capital cost* method or the *willingness-to-pay* method. The former measures the loss to society due to an accident on the basis of the future earnings potential of the victim and assesses only the market costs, including medical treatment and legal costs; The latter assesses the market and nonmarket costs, including grief, pain and suffering, and reduced quality of life arising from the loss of life or injury to the victim.

(c) Operating Costs Associated with the System User's Equipment. For certain civil systems, using the system takes place with the aid of some equipment (e.g., for highway systems, this is a vehicle). In such cases, the user incurs a cost associated with the operation of their equipment. In cases where such equipment is owned by the system owner (e.g., motorized carts in an airport), the operating costs of such equipment are borne by the owner and not the user. The operating cost of the users' equipment may include of the costs of depreciation, fuel, and consumables (e.g., batteries). These costs are influenced by the system condition, congestion, and design/configuration. This is an indirect but tangible cost. For example, for highway systems, AASHTO's Red Book (AASHTO, 2003) provides methodologies for calculating the user operating costs and for estimating the impact of system improvements on these costs.

(d) Congestion Contribution, Time, Discomfort, and Inconvenience Costs. Each user in the system contributes to congestion of the system and thus causes a cost to all other users of the system. Also, system users typically seek to minimize the time, discomfort, inconvenience, or frustration associated with the use of the system. Generally, these costs are indirect, and in many cases, intangible. In rare cases where they are tangible and have monetary values, such as time or delay, they can be determined as the product of time/discomfort/inconvenience and their respective unit monetary values. For example, time cost = time spent on system (minutes) × value of time ($/minute).

Example 10.1

The vehicle operating cost (VOC) incurred borne by highway users is higher when there is a low pavement condition (often measured in terms of its surface roughness or, specifically, the International Roughness Index (IRI)). A 16-mile interstate project is expected to improve the pavement condition. The initial condition IRI is 120 units. Upon project completion, the condition will be 50 units. The base vehicle operating cost is $143 per 1000 vehicle-miles and traffic volume is 100,000 vehicles/day. Calculate the monetary user costs. Use the Barnes and Langworthy relationship m, the VOC adjustment factor to adjust for the pavement roughness is $= 0.001[(\text{IRI} - 80)/10]^2 + 0.018[(\text{IRI} - 80)/10] + 0.9991$ when $\text{IRI} > 80$ and $m = 1.00$ when $\text{IRI} < 80$.

Solution

The user costs are calculated as follows:

(i) Before improvement: Pavement adjustment factor,

$$m = 0.001[(120 - 80)/10)]^2 + 0.018[(120 - 80)/10] + 0.9991 = 1.087$$

$$\text{VOC} = (1.087)\,(143) = \$155.46 \text{ per 1000 VMT (Vehicle-miles travelled)}$$

(ii) After improvement: Pavement adjustment factor,

$$m = 1.00$$

$$\text{VOC} = (1.00)\,(143) = \$143.00 \text{ per 1000 VMT}$$

$$\text{Change in unit VOC} = 155.46 - 143.00 = \$12.46 \text{ per 1000 VMT}$$

$$\text{Overall change in VOC} = (\$12.46)(100{,}000)(365)(16)/1000 = \$7.27 \text{ million per year}$$

Example 10.2

Safety improvements at a hydroelectric plant construction site are expected to yield a reduction of fatal accidents from 15 to 5 per 100 million instances of potentially hazardous situations. The site is expected to experience 20 million instances of potentially hazardous situations. Assume that the cost of a fatal accident is \$3 million. Calculate the monetary user costs associated with the site safety improvements.

Solution

First, we assume that the frequency of accidents is independent of the unit accident cost. Thus, the monetized reduction in user cost is

$$\text{Number of fatal accidents reduced} \times \text{cost of one fatal accident}$$

$$= (20/100)(15 - 5)(\$3\text{M}) = \$6 \text{ million}$$

Example 10.3

Improvements to a metropolitan transit system are expected to reduce commuter travel time from 20 to 18 minutes per person per day and reduce noncommuter travel time by 3 minutes per person per day (from 25 to 22 minutes). The average traffic volume is 10,000 persons per day. The travel time value of commuters and noncommuters are \$34.50/h and \$17.55/h, respectively. Calculate the reduction in the transit user costs due to the improvement. Assume that the average traffic volume is split into 60% commuters and 40% noncommuters. Discuss any assumptions.

Solution

The reduction in commuter user cost: (2/60)(10,000)(34.5)(365) = \$4.2 million/yr. Reduction in noncommuter user cost = (3/60)(10,000)(17.55)(365) = \$3.2 million/yr. Thus, the total reduction in user cost is \$7.4 million/yr. In actuality, transit demand is not independent of the transit unit costs (in this case, travel time). In other words, the true user benefits will depend on the elasticity of demand in response to some agency-induced service factor (in this illustration, the service factor is travel time. In other cases, it may be trip price or trip comfort). The above example specifies that the number of users is the same before and after the improvement (i.e., the assumption of zero elasticity). If the demand were sensitive to the travel time, then more users will be expected to use the system after the improvement and the user benefits will be even greater. We explore these relations in the next section.

10.3.2 Reduction in Overall User Cost as a Surrogate of System User Benefit

When the system owner or operator makes improvements to a system, the cost incurred by the system users in the form of delay, inconvenience, accidents, or other adverse consequences are reduced. This is often represented as a benefit of the system enhancement. Quantification of this benefit is useful in economic analysis, feasibility studies for proposed system enhancements, and *ex poste* studies for existing systems. A better understanding of these benefits can be grasped on the basis of basic economic theory to which we will now devote the next paragraph (Sinha and Labi, 2007).

In microeconomic theory, **demand** is the quantity that consumers are willing and able to purchase, and **supply** is defined as the quantity of a product (good or service) that a supplier is prepared to make available to the market. A **demand curve** (Figure 10.4*a*) is a graphical representation of the relationship between the amount of a certain good or service that consumers are willing and able to purchase and the price of that product, given that there is no change in all other factors that influence the demand of that product (such as the consumers tastes and preferences, income levels, and

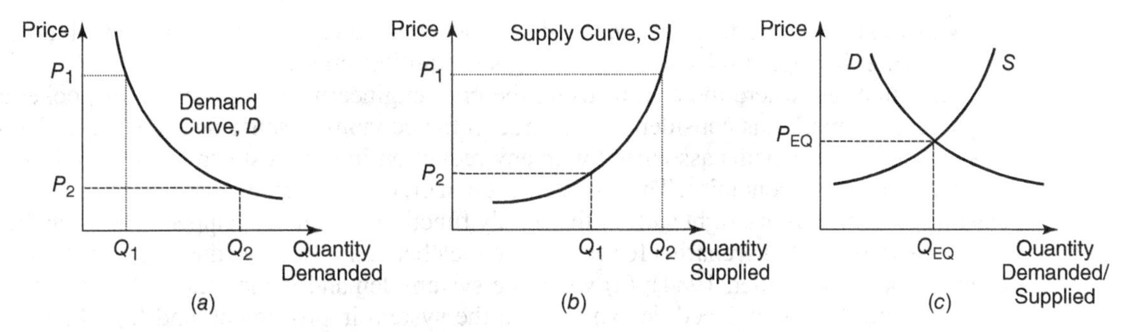

Figure 10.4 (*a*) Demand curve, (*b*) supply curve, and (*c*) demand and supply equilibrium.

the prices of rival or complementary products). The vertical axis is the independent variable (price) and the horizontal axis is the dependent variable (quantity). Price here could be actual cash paid out of pocket by the system user for using the system or some other price of using the system, including delay, convenience, and safety. As we can see in the figure, the demand curve slopes downward: When the price of the product decreases from P_1 to P_2, the quantity demanded increases from Q_1 to Q_2. This is because more people are able to afford the product. Demand is described as **elastic** when a small change in price causes a significant change in the quantity demanded, and **inelastic** when even a large change in price causes no significant change in the quantity demanded. A **supply curve** (Figure 10.4*b*) represents the relationship between the amount of a certain good or service that producers are willing and able to make available on the market and the price of that product, given that all other factors that influence the supply of that product (such as the prices of rival or complementary products) remain the same. As we can see in the figure, the supply curve slopes upward: When the price of the product decreases from P_1 to P_2, the quantity supplied decreases from Q_1 to Q_2, because producers do not find it profitable to produce that good or service.

In a perfectly competitive market, the unit price for the product fluctuates until it finally settles at a point where the quantity demanded by consumers at the current price is equal to the quantity supplied by producers at that price. This point is described as an economic equilibrium for that product with respect to its price and quantity demanded and supplied. As we can see in Figure 10.5*a*, when there is an increase in demand but no change in supply, this results in a shortage, and a new (higher) equilibrium price is reached; and when there is a reduction in demand but no change in supply, this results in a surplus, and a new (lower) equilibrium price is reached. Also, from Figure 10.5*b*, when there is an increase in supply but no change in demand, a surplus results, and

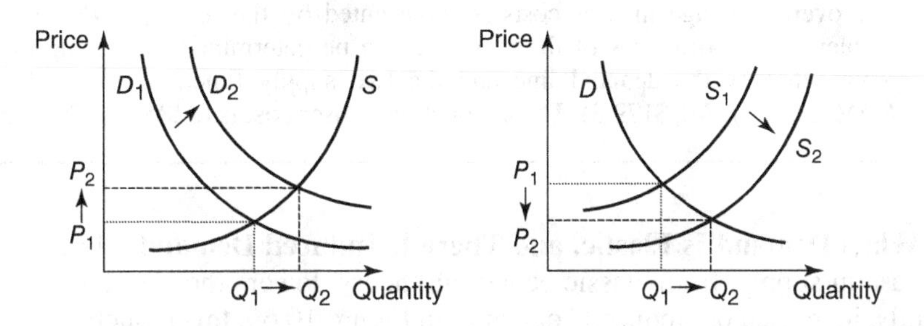

Figure 10.5 Shifts in supply and demand.

a new (lower) equilibrium price is reached; and, when there is a reduction in supply but no change in demand, a shortage results, and a new (higher) equilibrium price is reached.

In situations where the cost of using the civil engineering system is out of pocket and pay as you go, the market is considered to be free in the economics sense and the classic laws apply. In such cases, the benefits associated with any reduction in user cost can be analyzed using basic principles in microeconomics. The replacement of a civil engineering system or improvement of an existing system causes a right shift of the supply function. When this happens, there can be at least three elasticity-related scenarios for estimating the change in user cost due to the system improvement (Dickey and Miller, 1984): (a) when the system demand is inelastic, (b) when the demand is elastic and there is induced demand due to the system improvement, and (c) when demand is elastic and there is generated demand following the system improvement. The discussion below is presented for a composite user cost of the system; however, it can be applied to the individual user cost components as well, such as fares or the cost of safety or delay, and cost of discomfort or inconvenience.

(a) When Demand Is Inelastic. In certain cases, the quantity of system usage demanded is independent of the price of usage or other supply attribute, demand is said to be perfectly inelastic. In such cases, the user benefit arising from an improved civil engineering system is taken as the product of the quantity demanded and the reduction in the unit cost (price) of the system use (Figure 10.6*a*). For illustrative purposes, a "unit" of system demand is any single instance of the use of (or demand for) the system. For example, a technological improvement or system usage policy that decreases the user cost or inconvenience leads to a reduction of the unit cost of each instance of system use. Thus, this stimulus causes a downward shift in the supply function, leading to greater user benefits. On the other hand, adverse stimuli, such as establishing security checks at airport systems that inadvertently increase user convenience, increase the unit cost of each instance of system use, which is reflected by an upward shift in the supply function and equilibrium point, generally leading to a loss of user benefits in the short run. In either case, the quantity demanded remains constant when demand is inelastic.

Example 10.4

Due to migration of some tasks to Internet (online) resources, the supply functions [user costs (UC)] before and after an improvement to a civil engineering system were determined to be

$$UC_{BEFORE} = 5X^2 + 2X + 150 \qquad UC_{AFTER} = 4.8X^2 + 1.9X + 50$$

where X is the level of system demand in millions.

If the level of system demand remains at a constant 5 million, calculate the overall change in user costs due to the improvement.

Solution

The overall change in user costs is represented by the rectangle *PQRS* in Figure 10.6*a*. For this problem, the coordinates of the rectangle can be determined by solving the simultaneous equations represented by the demand line and the two supply functions to yield: $P(0, \$285)$, $Q(5M, \$285)$, $R(5M, \$179.5)$, $S(0, \$179.5)$. Thus, reduction in user cost is $(\$285 - \$179.5)(5M - 0) = \$527.5M$.

(b) When Demand Is Elastic, and There Is Induced Demand. In cases of elastic demand, an increase in supply, from classic economic theory, lowers the user cost of system use and, subsequently, increased or "induced" demand; in Figure 10.6*b*, this reduction in user cost is represented as the trapezoid. For example, for public transportation systems, improved service through the use

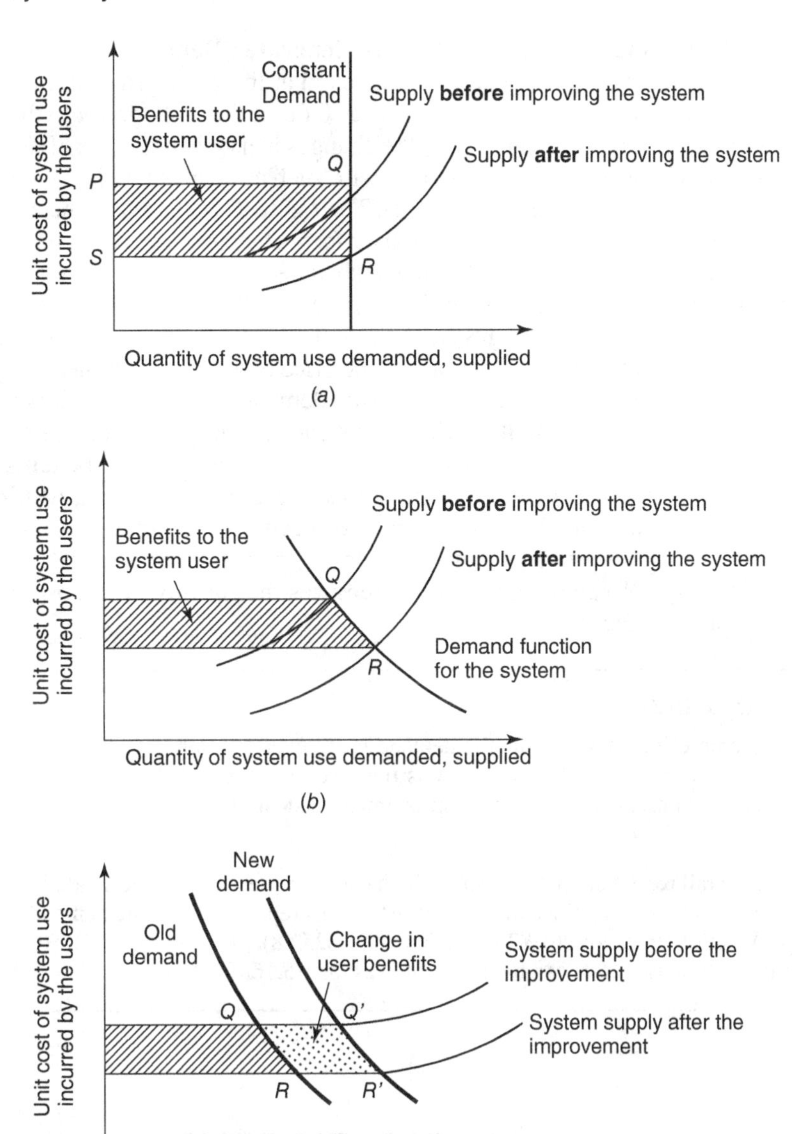

Figure 10.6 User cost changes due to supply improvement under different conditions of demand elasticity: (*a*) Scenario 1: Perfectly inelastic demand, (*b*) Scenario 2: demand is elastic and there is induced demand, and (*c*) Scenario 3: demand is elastic and there is generated demand.

of faster buses and increased reliability would likely reduce the user cost of delay, reflected by a downward right shift of the supply function and equilibrium point and an increase in the trips and decrease in user costs. On the other hand, reduced quality or quantity of transit service, all other factors remaining the same, would be reflected by an upward left shift of the supply curve and equilibrium point (i.e., the number of trips would decrease and cause an increase in the unit user cost).

(c) When Demand Is Elastic, and There Is Generated Demand. In the case of elastic demand, the demand curve shifts due to increased demand at the same price. In that case, the reduction in user costs (shown as the shaded area in Figure 10.6*c*) is due only partly to the system improvement.

For the scenarios discussed above, the changes in user benefits are often in response to endogenous stimuli or changes within the civil system, such as changes in demand (induced or generated) or in supply (quantity or quality of service). The figure also illustrates the effect of *exogenous* stimuli, such as the imposition or removal of user fees, changes in existing user fees or other forms of user cost, establishment or removal of user subsidies, and increased levels of service of competing systems. All of these exogenous stimuli can influence the user demand for the system.

Instances of user cost changes can be found in a variety of civil engineering systems. In water supply systems, for example, an increase in the price of water will immediately cause an increase in the unit cost of system usage. Users with elastic demand will reduce their water consumption while those with inelastic demand will continue the same consumption levels after the change. However, in either case, the end result will likely be a negative gain in user benefits. Another example is user subsidies: When the government subsidizes the use of a civil engineering system, the supply curve is shifted downward because the cost of each unit of consumption is reduced, leading to increased consumption (where demand is elastic) and increased overall benefits (irrespective of demand elasticity). When the government removes the subsidy or imposes new or additional user fees, the opposite effect is experienced.

Example 10.5

In Example 10.4, assume that the demand is not fixed but rather is sensitive to the user cost as follows: $UC_{DEMAND} = 600 - (80X)$ where X is the level of system demand in millions. Calculate the overall reduction in user cost due to the improvement made to the engineering system.

Solution

The overall reduction in user costs, which can be represented as the shaded trapezoid in Figure 10.7, is calculated by solving the simultaneous equations represented by the demand function and the two supply functions to yield: P(0, \$252.8), Q(4.34M, \$252.8), R(5.16M, \$187.4), S(0, \$187.4). Thus, the overall reduction in user costs can be determined as: 0.5 (5.16M + 4.34M)(\$252.8 − \$187.4) = \$310.65M.

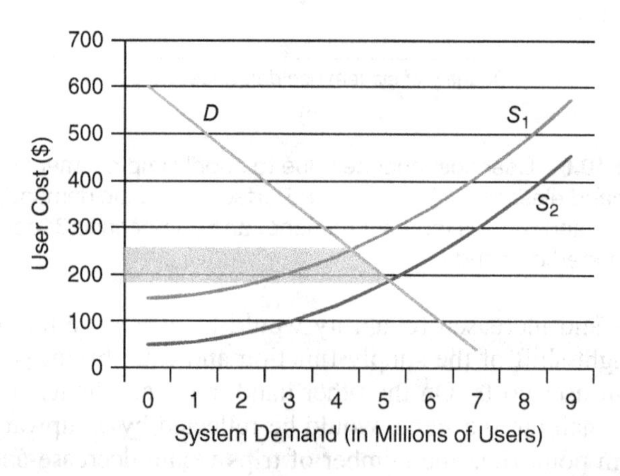

Figure 10.7 Illustration of solution for Example 10.5.

10.4 COSTS INCURRED BY THE COMMUNITY

The costs experienced by the community proximal to a civil engineering system are typically non-monetary, indirect, and in some cases, intangible. As discussed in an earlier section, these external costs or adverse externalities can be *internalized* using any one of three ways depending on whether they are preemptive or after the fact: (i) the cost of curtailing or abating the cost type at the source of generation or incurrence, (ii) the cost of mitigating the social and economic impacts, and (iii) the willingness of the affected community to pay to prevent the externality. We now discuss a number of the types of community costs associated with civil engineering systems, predominantly at the phase of system operations.

10.4.1 Air Pollution and Climate Change Costs

Air pollution (which is mostly short term and local) and climate change (which is long term and global) are among the environmental side effects of the operation of certain civil engineering systems that are now widely recognized as a concern in public health and the long-term sustainability of the built environment. Mobile and stationary sources emit pollutants that can cause or exacerbate health problems upon inhalation or contribute to global warming. Air pollution from civil systems can also degrade vegetation and agriculture.

In the case of highway systems, McCubbin and Delucchi (1998) estimated total annual social costs of air pollution to be $30 billion to $349 billion, and the FHWA reported monetary air pollution costs in cents per vehicle-mile (Table 10.2). Research supported by the European Economic Commission has estimated that, at 1999 conditions, the cost of CO_2 emissions was US$26/ton, a value considered consistent with other estimates of the global abatement costs of meeting the Kyoto Protocol (Friedrich and Bickel, 2001). As discussed in Sinha and Labi (2007), the cost of air pollution can be assessed on the basis of the money spent on cleaning the air at the polluting source or for addressing the social effects of pollution and climate change, or the money that the community is willing to pay to avoid air pollution.

(a) Cost of Cleaning up the Air at the Source of Pollution. In certain cases, it is possible to remove the air pollutant before it disperses into the environment. The cost of doing this, termed "abatement costs," or preemptive costs of air pollution, can be high as it typically would require the installation of high-technology pollutant removers at the source of pollution. With regard to climate change, the causes are varied and include changes in Earth's orbit, solar intensity, and anthropogenic factors. Most of these are natural causes and thus are outside the reach of civil engineers. For those that are anthropogenic, a reduction in emissions from engineering systems, through technology or policy changes, can help address the climate change problem at the source.

Table 10.2 Air Pollution Costs by Vehicles on Highway Transportation Systems

Vehicle Class	Cents/mile
Automobiles	1.1
Pickups/vans	2.6
Gasoline vehicles over 8500 lb	3.0
Diesel vehicles over 8500 lb	3.9

Source: FHWA (2000).

(b) Cost of Mitigating the Impacts of Pollution and Climate Change. The cost of addressing the effects of pollution and climate change is an after-the-fact cost and could be described as the "social damage" effects. For air pollution, these include health-care expenses associated with treatment of respiratory illnesses caused or exacerbated by polluted air, repair of degraded physical infrastructure, compensation for degraded cropland, or remediation of affected forests and polluted groundwater due to acidic depositions formed by chemical reactions and atmospheric gases and pollutants. Compared to air pollution, the effects of climate change are expected to be far more costly to address, as the effects of climate change on agriculture, ecology, infrastructure, and other natural and built-up systems are far-reaching and not yet fully grasped. According to the Intergovernmental Panel on Climate Change, the costs of mitigating climate change impacts is US$0.10 to $20 per ton of carbon at tropical regions and US$20 to $100 at nontropical regions. However, it must be noted that relatively little contribution to climate change may be attributable to civil engineering systems even though the absolute contribution may be significant.

(c) Costs Based on Willingness-to-Pay Approach. On the basis of the assumption that the community is perfectly aware of the adverse impacts of air pollution on their health and property, air pollution costs can be estimated on the basis of the extent to which the community is willing to pay to avoid the problem. These costs can be estimated using past data on the extents to which the community has paid to avoid such community impacts of a system (revealed preferences) or by carrying out questionnaire surveys to ascertain how much individuals in a community would be willing to pay to avoid such impacts (stated preferences).

10.4.2 Noise Costs

Of the systems in the various branches of civil engineering, transportation systems are probably the most culpable for noise pollution (OECD, 1990), particularly at the construction and operations phases of their development cycle. It has been found that transportation noise costs in the United States, Japan, and Europe account for 0.06–0.5% of the total gross domestic product (BTCE and EPA, 1994). As it is for other externalities, the cost of noise can be estimated in terms of the cost of reducing noise emissions at the source or during its transmission, mitigating the effect of excessive noise, or measuring the extent to which affected stakeholders are willing to pay to prevent noise.

For the first approach, the preemptive cost, or the cost of noise barrier construction and maintenance, can serve as a measure of noise pollution cost. In this approach, it is important to realize that the direct costs of barrier construction are often borne by the system owner in order to obviate the indirect health and inconvenience costs to the community. Assuming that the barrier is fully effective, the direct costs of barrier construction that are incurred by the system owner can be considered a proxy for the indirect costs borne by the community. With regard to transportation noise, approximate costs of noise barriers can be estimated on the basis of the barrier material type and dimensions (Sinha and Labi, 2007).

For the second approach, the remediation or medical cost of residents' noise-related ailments can be estimated to represent noise costs. A number of researchers have investigated system noise costs on the basis of medical interventions to mitigate the hearing loss or auditory problems of residents and pedestrians due to such noise.

For the third approach, *hedonic price surveys* can be used (Litman, 2010). This involves estimating the exposure of residential property to noise and measuring the reduction in the values of affected homes. In Scotland, Bateman et al. (2001) determined that each decibel increase in traffic noise leads to a decrease in residential property price by 0.20%. According to Modra (1994), OECD estimated that where noise levels exceed 50 dB(A) Leq (24 hours), property values reduce by 0.5% for each decibel increase. Lee (1982), cited in Litman (2010), estimated that every decibel increase in traffic noise costs each housing unit \$21 per year; and Bein (1997), in research for the British Columbian government, estimated an average annual noise cost of at least \$1000–\$1500 (Canadian dollars) per resident of homes located near busy roads.

Example 10.6

Along a certain 12-mile urban freeway, there are 50 residences (of average value \$200,000) per mile. A sudden increase in traffic volume due to commercial and residential developments near the freeway causes an increase of 5 dB in noise level. Using the OECD rates (Modra, 1994), determine the community cost of the freeway system operations from the perspective of property values.

Solution

Change in decibel level = 5. Thus, the community cost, in terms of the reduction in property value, is = 5(0.005)(50)(12)(\$200,000) = \$3 million.

10.4.3 Social Costs

Social costs of civil engineering systems are represented by the degradation of social values or community cohesion and the disruption of public facilities and services due to the construction or operations of these systems. These costs also include the displacement of persons, farms, and businesses and disruption of desirable growth of communities and regions. Social costs generally, refer to the consequences of any action to human populations that negatively impact the lifestyles of persons in a community in terms of their residential, occupational, and recreational behaviors and social interactions (IOCGP, 2003). The FTA (1993) described social costs as the undesired changes in the demographics, functional and physical layouts, and the erosion of a sense of community neighborhood or belonging. In the developing world, development assistance institutions including the United Nations, the World Bank, the African Development Bank, the Asian Development Bank, and the Inter-American Development Bank require the assessment of social impacts to ensure that funded projects do not lead to undue indirect and intangible costs to people in developing countries in terms of sociocultural and institutional impacts, among others (The World Bank, 2003). As far back as 1968 when the issue of poverty alleviation was highlighted by the then World Bank president, Robert McNamara, the analysis of social impacts has gained an increasingly prominent role in the development agenda of such institutions.

Similar to most other costs of civil systems incurred by the community, social costs are difficult to measure because the nature of social and cultural environments and values differ from place to place and also from time to time. The costs incurred are influenced by the extent of resilience of the affected members of the community and the mechanism of the stimulus.

The World Bank (2003) suggests that assessment of the social costs of civil systems should occur continuously during all the phases of system development, from planning to end of life, and identifies five "entry points" for assessing the social impacts: institutions, stakeholder participation, social diversity and gender, rules and behavior, and social risk. The World Bank further recommends scope selection to be consistent with the context of and circumstances of a specific project. Often, there is a recognizable spatial distribution of civil system externalities, and the affected communities are typically those that are low income, disadvantaged, or marginalized. Thus, the costs borne by the community can be viewed in the context of *environmental justice*, which is discussed in the next paragraph.

Environmental justice, which is related to the balance of the distribution of benefits and costs arising from a civil engineering system, can be defined as the "fair treatment and meaningful involvement of all people regardless of race, color, national origin, or income with respect to the development, implementation, and enforcement of environmental laws in general" (Bass, 1998; Quan, 2002). The U.S. National Academy of Sciences refers to environmental justice as the "equitable distribution of both negative and positive ecological, economic, and social impacts across racial, ethnic, and income groups" (NAS, 2002). In a Utopian situation, all segments of the community, irrespective of their economic, cultural, or social background, would incur similar proportions of cost associated with the civil engineering system. In reality, however, these systems yield very different splits of beneficial and adverse impacts that are often different at the different communities and regions that are affected. For example, the adverse impact of pollution is most intense at the relatively small band of area that is immediately proximal to the facility; unfortunately, these systems are often sited at areas populated by persons with little political representation or power to ensure that the appropriate protective safeguards are in place. Environmental justice, which seeks to address these issues by promoting basic societal values of human rights and fairness, can be viewed through the larger lens of professional ethics (which we will discuss in Chapter 29). Civil system projects can be costly in terms of their violation of the tenets of environmental justice, particularly in regions where a wide gap exists between the haves and have-nots, and civil engineers are in a unique position to speak for the voiceless by designing systems that help to foster environmental justice.

10.4.4 Negative Community Costs

The net change in property values in an area due to the introduction of a civil engineering system is not always negative. While it is true that such systems often cause degradation in resource quality through their construction, operations, and maintenance, and thus a reduction in property values, it is also worth noting that the increased economic development that accompanies a new civil system may be beneficial to the community and may subsequently lead to an overall net increase in living standards and property values. Also, due to advancements in technology, systems are being made more efficient in terms of reduced pollution of the surrounding air, water, and soils. For example, McMillen (2004) suggested that aircraft noise emissions have reduced to such a point that Chicago's O'Hare International Airport can be expanded without reducing local property values, adding that the total residential prices around that airport could potentially increase by as much as $285 million after it receives a new runway. In many developing countries, the increase in property values due to the construction of new civil systems can be very significant, as observed for properties near the Suvarnabhumi Airport in Bangkok, Thailand, and land values at the coast near the offshore oil rigs at Takoradi, Ghana.

10.5 SYSTEM COSTS CATEGORIZED BY PHASE OF SYSTEM DEVELOPMENT

10.5.1 Costs at the Phases of Needs Assessment, System Planning, and Design

System Owner. The needs assessment and planning phase costs include those related to system location feasibility, surveys of prospective users of the system and the community, assessments of environmental impact, and public hearings. If the system engineers seek to compare the costs of alternative actions at a specific phase, then they will not need to include the costs of the preceding phases in the cost analysis (particularly where the costs of the active actions are not influenced differently by the decisions at the preceding phase). Design costs include the costs of engineering analyses, engineering and construction drawing preparations, and developing technical specifications; these costs can constitute a significant percentage of construction costs (see Section 10.8.8).

The costs of right-of-way acquisition and site preparation include the costs of land purchasing, legal requirements, and title acquisition and the administrative costs associated with the site preparatory activities of demolition of structures and relocation of utilities in the right-of-way. In most countries, the relocation of existing underground utility facilities (including communications cable, electricity, water, telephone, and gas) continues to be problematic during the construction of new civil systems, particularly because the exact locations of the existing lines were not properly recorded at the time of construction several decades ago and are therefore unknown at the current time. Where there is a need to relocate existing structures to make way for the proposed system, it is necessary to consider the costs of demolishing these structures, acquiring new land, and reconstructing the affected structures. The demolition costs can be roughly estimated through a field inventory of the project area to determine the dimensions of structures to be demolished and applying the appropriate demolition cost rates. Estimates of the expected system's construction cost at the planning phase are rough approximations and typically are based on the cost of similar projects in the past. The cost associated with this phase may be a simple average or a statistical regression model that predicts the expected cost as a function of the attributes of the construction type, system design and material type, system location, and other variables.

User and Community. At the phase of system planning and design, the system users and community do not incur significant costs. An exception would be when the proposed system is an extension (or staged replacement of) an existing system, in which case the planning and design may involve some geotechnical investigations and other site work that could somewhat disrupt the operations of the existing system.

10.5.2 Costs at the Phase of System Construction

System Owner. The estimation of system construction costs is a complex endeavor that requires highly skilled specialists called *estimators*. Compared to the system cost estimated using aggregate approaches at the planning phase, the cost estimated at the bidding stage of the construction phase is a more accurate representation of the actual cost of the system because it is based on a disaggregate approach that uses specific amounts of the individual resources (labor, equipment use, and materials) and their respective unit prices. In Section 10.7, we present the issues involved in cost estimation at the construction phase. Also, typical cost values incurred at this phase, for different civil engineering systems, are provided in Appendix 3.

User and Community. For existing systems that are under reconstruction, expansion, or upgrade, the system users incur the costs of system downtime at the construction phase. These include the costs associated with user safety, inconvenience, and discomfort. For existing and new systems, construction work can lead to indirect costs associated with noise, dust, and air pollution to the community.

10.5.3 Costs at the Phase of System Operations

System Owner. The operating costs of civil systems arise from efforts necessary to ensure a minimum level of service in terms of system user convenience, security, and user safety. These include the cost associated with sanitation and waste disposal, power, gas, water, communication, and security, vegetation control, and snow/ice management. These costs may be carried out in-house or given out on contract. Operating costs can also include the labor and equipment costs associated with the collection of system users' fares or tolls. Compared to maintenance costs, operating costs have relatively little influence on or from the physical condition of the system. The future operating costs of a system may be estimated from the owner or operator's historical operations cost records in the form of cost functions or simple average values. Cost models could be functions of the supply attributes (type, size, and age of facility) or the demand attributes (level of system usage) or both. Typical cost values incurred at this phase, for a small number of civil engineering systems, are provided in Appendix 3.

User and Community. At the operations phase, the user may incur direct costs through out-of-pocket fees and fares and indirect costs associated with the poor performance or condition of the system. Also at this phase, both the users and the community suffer the indirect or intangible costs associated with any lack of system safety, security, comfort, and convenience. Also, the community may incur the additional costs of air pollution, noise, vibration, and other adverse impacts. As discussed earlier in this chapter, these adverse impacts may lead to an increase in monetary community costs, such as a reduction in property values in the affected areas.

10.5.4 Costs at the Phase of System Monitoring

System Owner. The following monitoring costs are borne by the system owner or operator: (a) costs of inspecting the system periodically to ensure that there are no serious defects that could lead to catastrophic failures and (b) costs of monitoring the levels and patterns of usage of (or demand for) the system to assess any need for capacity enhancements. (c) cost of monitoring the impacts of the system on its environment (d) cost of monitoring the impact of the environment on the system. These costs include the salaries paid to in-house personnel (system inspectors) and the use of in-house equipment. In certain cases, the system owner awards contracts for the inspection and monitoring work. A few typical cost values incurred at this phase are provided in Appendix 3.

User and Community. At the phase of system monitoring, the system users and community do not incur any significant cost. An exception would be when the monitoring activities disrupt the operations of the system and thus lead to delay or inconvenience to the system users.

10.5.5 Costs at the Phase of System Preservation

System Owner. Preservation (or maintenance) costs are incurred by a system owner or operator in order to ensure that the civil engineering system is maintained at a certain minimum level of physical condition. These costs often include rehabilitation, periodic and routine maintenance, and may be classified by the work source (in-house vs. contract), work cycle (periodic vs. routine), or purpose (preventive vs. corrective). Most civil system owners or operators have implemented *management systems* that help track their system performance, maintenance work schedules, and costs and can be used as a decision support for appropriate and timely rehabilitation and maintenance actions. The cost of future maintenance can be predicted from information systems established for the preservation management systems and may range from simple average cost rates (e.g., $\$/ft^2$ of steel surface repair for a water tank or $\$/m^2$ of sedimentation pond liner replacement) to sophisticated statistical functions that model the expected cost of a specific system preservation treatment, or annual preservation expenditures on the basis of supply attributes (e.g., the design type, dimensions, and material type of the physical system), and/or demand attributes (e.g., level of system usage or loading). Examples of such costs are provided in Appendix 3.

User and Community. Similar to the case for the preceding phases, system users incur the costs of system downtime (e.g., preservation work zones) at the phase of system preservation. These include the costs associated with safety, convenience, and comfort. Also, maintenance and rehabilitation activities can lead to indirect costs to the community associated with the noise, dust, safety hazards, and air and water pollution.

10.5.6 Costs at the End-of-Life Phase

System Owner. The costs incurred by the agency depend on whether the system termination was intended or unintended. For intended termination, end-of-life costs include the cost of removing and disposing of the physical system and often include the costs of demolition, and addressing complaints or settling lawsuits filed by owners of nearby systems and the surrounding community affected by the system demolition. Typical cost values incurred at this phase are provided in the Appendix 3. The costs associated with unintended termination include the preemptive cost of retrofitting the physical structure of the system in order to reduce its susceptibility to failure in case of a disaster event and the after-the-fact costs of recovery or repair after any such disaster.

User and Community. Where the end of life is intended and the system is demolished with the intention of replacement, system users incur the intangible costs (inconvenience and discomfort) associated with system downtime, and the community suffers the costs of noise, dust, and safety hazards that are associated with the demolition or disposal. Where the end of life is unintended due, for example, to a disaster event, the users incur the costs associated with system downtime or reduced capacity as a result of the event.

10.6 AVERAGE AND MARGINAL COST CONCEPTS

The cost of a civil engineering system, process, or activity at any phase of system development, consists of a fixed component k (that is not influenced by the output volume) and a variable component $f(V)$ that is influenced by the output volume:

$$\text{Total cost: } \mathrm{TC}(V) = k + f(V)$$

Output is often expressed from the demand perspective or the supply perspective. Examples of demand-side outputs include the number of users served or the consumption in some measurable units in terms of a continuous variable. Examples of supply-side outputs are the number of units or some dimensional attribute of the system (ft^2, miles, line-miles, lane-miles, etc.). For instance, for a new civil system, the fixed component of cost includes the acquisition cost for the right-of-way and the cost of replacing or relocating structures and utilities.

In a cost function, the mathematical form or the ratio of the fixed component to the variable component is a statement of the extent to which the civil engineering system exhibits economies or diseconomies of scale. The ratio of variable cost to fixed cost, VC/FC, is generally low for construction activities and high for maintenance activities. Also, the ratio is generally low for large systems and high for small systems.

The average total cost or, simply, the **average cost**, ATC or AC, is the total cost associated with one unit of output and is calculated as the ratio of the total cost to the output: $\mathrm{ATC} = \mathrm{TC}/V$, where TC is the total cost and V is the volume of output. The average fixed cost, AFC, is defined as the fixed cost associated with one unit of output and is calculated as the ratio of the fixed cost to the output, $\mathrm{AFC} = \mathrm{FC}/V$. Similarly, the average variable cost is the cost of producing one unit of output and is calculated as the ratio of the variable cost to the output, $\mathrm{AVC} = \mathrm{VC}/V$. The average cost concept helps a system owner assess the cost impacts of different output levels at the phases of construction, operations, or maintenance. For example, the cost/person of water supply to a small town may be high compared to that to a large city.

The incremental cost of producing an additional unit of output is known as the **marginal cost**. The analysis of marginal cost is particularly relevant in the management of existing civil engineering systems because a system owner may want to know the incremental cost associated with a planned or hypothetical production of additional units with regard to system expansion, preservation, or operations.

$$\text{Marginal cost: MVC} = \frac{\partial \text{VC}}{\partial V}$$

Average cost and marginal cost are different concepts. For example, suppose an agency spends \$100 million to build 10 levees of a standard dimension and \$105 million to build 11 such levees. The average costs are \$10 million and \$9.54 million, respectively, but the marginal cost of the 11th levee is \$5 million.

Example 10.7

Consider the following general cost function: Total cost, $C = k + f(V)$, where k is the fixed cost component (FC) and $f(V)$ is the variable cost component. V is the output. For each of the following functional forms—linear, cubic, logarithmic, quadratic, exponential, cubic, and power—derive expressions for (a) average fixed cost, (b) average variable cost, (c) average total cost, (d) marginal fixed cost, (e) marginal variable cost, and (f) marginal total cost.

Solution

The cost expressions are provided in Table 10.3. Note that the functions for marginal variable cost or total variable cost may also be derived using the expression $\partial \text{VC}/\partial V$.

This section presented the concepts of average and marginal cost from a monetary cost perspective. In many problem contexts in civil engineering, it may be needed to apply these concepts to nonmonetary criteria such as environmental degradation, safety improvement, and climate change reduction. Also, the selection of the appropriate output variable, V, to be used in the cost analysis must be considered; and the choice of variable is influenced by the type of system in question, the phase of development, and whether the cost is influenced by factors associated with supply or demand. For example, in analyzing the costs of a pipeline installation, the number of pump stations, or a pipeline length could serve as the output variables; in costing bus rapid transit operations, the number of passenger-trips or passenger-miles could be the output variables; in freight operations costing, the output variable could be the ton-miles or ton-trips; and in building systems demolition, the building floor area (ft^2) could be used.

Table 10.3 Expressions for Average and Marginal Cost

	Average Fixed Cost FC/V	Average Variable Cost VC(V)/V	Average Total Cost TC(V) /V	Marginal Variable Cost = Marginal Total Cost. = ∂TC/∂V
Linear $C = k + aV$	k/V	A	$k/V + a$	a
Quadratic $C = k + aV^2$	k/V	$a.V$	$k/V + aV$	$2aV$
Exponential $C = k + ae^V$	k/V	ae^V/V	$k/V + ae^V/V$	ae^V
Cubic $C = k + aV^3$	k/V	aV^2	$k/V + aV^2$	$3aV^2$
Logarithmic $C = k + a \ln V$	k/V	$a \log V/V$	$k/V + a \log V/V$	a/V
Power $C = k + ab^V$	k/V	ab^V/V	$k/V + ab^V/V$	$a \ln(b) b^V$

Source: Khisty et al. (2012).

10.7 COST ESTIMATION AT THE CONSTRUCTION PHASE

An estimate of the cost of carrying out some improvement to an existing system or the cost of a new system varies considerably from the initial stage (planning phase) to the final stage of the construction phase (postconstruction). At any stage between initial and final stages, the estimate of the construction cost is essentially a prognosis by the cost estimator on the basis of the available data (Hendrickson, 2008). The prediction can be carried out on the basis of data and approaches that are either disaggregate or aggregate. Figure 10.8 presents the changes in project costs across the planning–design–construction phases of systems development and indicates which costing approach is appropriate at each phase.

At the planning phase, knowledge about the prospective design is sparse, and thus the cost analysis is coarse. Cost estimators typically refer to two distinct levels of coarseness that reflect the development phases at which they require the cost estimate and consequently indicate the relative confidence placed in the accuracy of the estimate (Wohl and Hendrickson, 1984; Rowings, 2004):

- Conceptual estimates are developed at the planning phase. At the early planning phase (this is often referred to as the *predesign estimate* or *approximate estimate*). At the late planning or early design stages, the conceptual estimate is referred to as a *preliminary estimate*, *budget estimate*, or *definitive estimate*).
- Detailed estimates are developed at the late design stage and the bidding or final stages of the construction phase. At the pre-tender or preaward stages, detailed cost estimation is carried out using the unit costs of key inputs including labor, material, and equipment use. The as-built cost estimate is also a detailed estimate that reflects the actual cost of the work done and hence incorporates any change orders.

Statistical analysis is a useful tool for costing at any level of aggregation. In the disaggregate approach, the cost estimate of any work input can be expressed as a simple average (and standard deviation) or as a statistical function of the attributes of the work input; and in the aggregate approach, the cost estimate of any work output can be expressed as an average value (with variance) or as a statistical function of the system attributes.

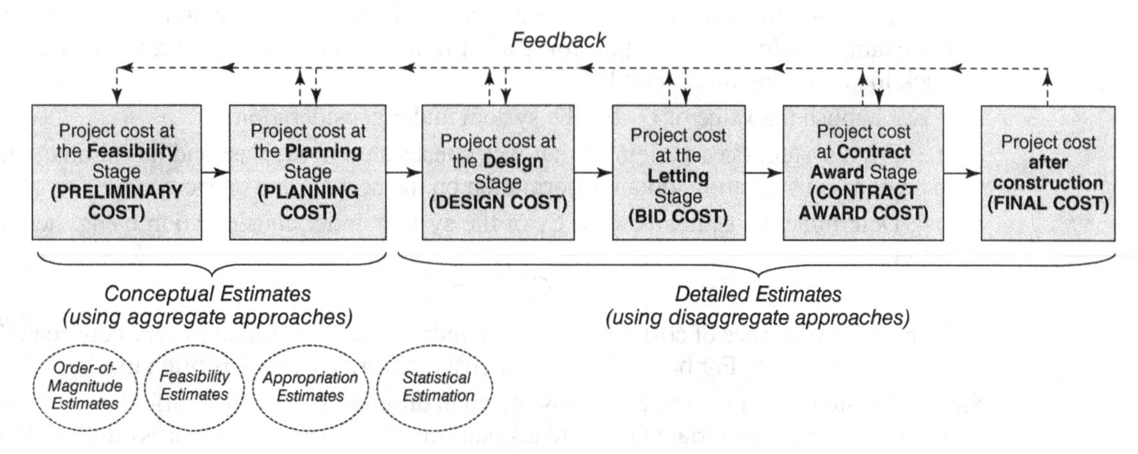

Figure 10.8 Changes in project costs across the planning–design–construction phases.

The estimation of costs is typically carried out by skilled estimators or quantity surveyors. These professionals take pains to ensure that no cost item is omitted or double counted, and that the cost estimate is as close as possible to the true cost, the accuracy depending on whether the estimate is being sought at the planning, design, or construction (contract award) phase. As such, the estimator must not only have an extensive knowledge of the construction process, materials, and practices but also be able to comprehend construction drawings and specifications, and structural details (Rowings, 2004). Also, experienced estimators typically possess skills of communication, business, and economics.

10.7.1 Conceptual Estimation

In the early phases of developing a civil engineering system, the project scope has not been well defined and little information on the design and site conditions are available. In these situations nevertheless, there is often a need for system owner or investors to have some rough idea of what the project will cost. This need is often borne out of the requirement for ascertaining the economic feasibility of the project. Also in such situations, there is little time or opportunity to develop a detailed estimate of the cost. If the aggregate construction costs of past similar systems are available (e.g., $\$/ft^2$ for bridge a deck, $\$/ft^3$ for a water reservoir, $\$/ft^2$ for a levee wall, etc.), these unit costs coupled with the dimensions of the proposed system could be used as a basis for estimating the overall system cost. Often, some adjustment is necessary to account for the situation where the historical projects are at a different location, were constructed several years ago when prices were different, or have a size smaller or larger than the proposed project (see Section 10.8.5). The three main types of conceptual estimates, in order of increasing level of detail, are the order of magnitude, feasibility, and appropriation estimates.

(a) Order of Magnitude (OM) Estimates. This is generally the least accurate of all estimates of a project. In developing this kind of estimate, the estimator seeks to establish a very rough amount for the project cost, often within 25–50% of the true cost. Two techniques for developing OM estimates are the cost capacity factor the comparative cost methods.

Cost Capacity Factor Method. This technique is often used by the process industry (water treatment plants, sewage mechanized quarries, material storage plants, landfills, dams, wind farms, etc.) and involve the following steps (Rowings, 2004):

Step 1. Establish the unit of measurement for the input, output, or throughout capacity, Q. For example, kilowatts of output for a wind farm, barrels per day for a petroleum refinery, truck loads per day for a landfill.

Step 2. Establish the value of Q_1 for the system under consideration.

Step 3. For a system (say, system 1) that was constructed in the past and has similar design to that under consideration, obtain information on the cost C_2 and capacity, Q_2.

Step 4. Determine the estimated cost C_1 of the system under consideration using the equation below.

$$C_1/C_2 = (Q_1/Q_2)^p$$

From past experience of cost estimators, the index p has been found to vary between 0.33 and 1.02, mostly at 0.6. For back-of-envelope cost estimations, $p = 1$ is often used.

Step 5: If system 1 and system 2 are constructed in different years and at different locations, then adjust C_1 using appropriate factors to account for difference in costs across time and location (see Section 10.8.5).

Example 10.8

It is sought to construct a wind turbine that produces 0.5 MW in a given time period. Estimate the cost of this facility if a similar past system produces 0.8 MW and has a cost of \$0.6M. Assume $p = 0.6$.

Solution

$$\text{Cost of proposed turbine } C_1 = C_2(Q_1/Q_2)^p = 0.6(0.5/0.8)^{0.6} = \$0.38\text{M}.$$

Comparative Cost Method. This method, a specific case of the cost capacity Factor method, can be applied to any type of civil engineering system. It requires basic information on the system, such as the intended use of the system (bridge, office, commercial, school, etc.), the intended dominant material type (steel, concrete, asphalt, timber, aluminum, etc.), the design type (e.g., for a bridge, steel deck truss, reinforced concrete T beam, RC I-beam, steel through girder, etc.). The steps used in this technique are as follows (Rowings, 2004):

Step 1. Establish the unit of measurement, M, of the system size. For example, schools (number of students), bridges (deck area), offices and commercial facilities (ft^2), stadiums (number of seats), warehouses (ft^3), health-care facilities (beds).

Step 2. Establish the value of M_1 for the system under consideration.

Step 3. For a system (say, system 1) that was constructed in the past and has similar design to that under consideration, obtain information on the cost C_2 and size M_2.

Step 4. Determine the estimated cost C_1 of the system under consideration using the equation below.

$$C_1/C_2 = (M_1/M_2)$$

Step 5: If system 1 and system 2 are constructed in different years and at different locations, then adjust C_1 using appropriate factors to account for difference in costs across time and location (see Section 10.8.5).

Example 10.9

As part of rehabilitating a large reinforced concrete bridge, it is sought to replace its deck, which has an area 4000 ft^2. The deck of a similar past bridge with area of 2500 ft^2, cost \$0.35M to replace. Estimate the cost of the deck replacement.

Solution

Cost of the proposed deck replacement $C_1 = C_2(M_1/M_2) = 0.35(4000/2500) = \$0.56\text{M}$.

(b) Feasibility Estimates. At the stage of feasibility analysis, the conceptual estimate is more accurate than the order of magnitude estimate but still only expected to be within 20–30% of the true cost. There are three methods for developing feasibility estimates: plant cost ratio, floor area, and total horizontal area methods.

Plant Cost Ratio Method. This method, often used for processing plants and similar systems, is based on the assumption that the proportion of equipment cost relative to the total cost of the system is the same across systems at different locations. Using factor multiplication, the cost of the proposed processing plant, T_1, can be estimated on the basis of the known equipment cost and total cost of similar plants and the intended cost of equipment for the proposed plant.

$$T_1/T_2 = (E_1/E_2)$$

where, T_2 is the total cost of similar past plant, E_2 is the equipment cost of similar past plant, and E_1 is the equipment cost of plant under consideration.

Example 10.10

The equipment portion of a \$5M water treatment plant has a cost of \$3.2M. Calculate the total cost of a similar plant intended for construction whose equipment is expected to cost \$4.7M.

Solution

Cost of proposed plant = $T_2(E_1/E_2) = 5(4.7/3.2) = \$7.34\text{M}$.

Floor Area and Horizontal Area Methods. This method is used for building systems (commercial, residential, and heath-care facilities), where the floor area is a good reflection of the overall cost. Variations of this method include the total horizontal area method (where the cost is assumed to be directly proportional to the amount of horizontal surface) and the finished floor area (where only the finished floor area and not the total floor area, is assumed to be a good reflection of the cost) and the cubic foot of volume method. Using historical data from similar building systems constructed in the past, the cost per square foot is established and uniformly applied to the horizontal area of the proposed structure to derive the total cost of the proposed structure. The computation is similar to that shown for the comparative cost method. In the finished floor area method, it has been cautioned that in order to avoid large errors in the estimate, the proposed system and the historical system should have similar relative proportions of area to height; where this assumption is too restrictive, the cubic foot of volume method, which accounts for floor-to-ceiling height, must be used instead of the floor area.

(c) Appropriation Estimates. Of the different types of conceptual estimates, the most refined is the appropriation estimate, often falling between 10 and 20% of the true cost. This estimate is developed when the system owner is ready to add the cost of the proposed system into a capital building program budget (Potts, 2008). After the feasibility analysis has shown that the benefits exceed the costs and the project is viable, the amount determined as the appropriation estimate is set aside to cover the project implementation. Variations of this type of estimation include the parametric estimation or panel method (which we discuss below), and the plant component ratio method.

Parametric Estimation. Here, records from past similar projects are obtained, and the cost of each parameter or panel is calculated separately and multiplied by the number of panels of each kind. Table 10.4 presents the units of measure for different parameters, in the case of building systems.

(d) Statistical Techniques. Probably the best technique to develop refined conceptual estimates is the statistical method. These model either the total cost or the unit cost of construction, maintenance, or operations as a function of the system dimension (supply based) or output (demand based).

Table 10.4 Units of Measurement for Estimation Parameters for Building Systems

Panel or Parameter Type	Unit of Measurement
Site preparation	Square feet of site area
Foundation and columns	Horizontal square feet of building
Floors	Horizontal square feet of building
Structural components	Horizontal square feet of building
Roofs	Horizontal square feet of roof
Exterior walls	Square feet of wall excluding exterior windows
Interior walls	Square feet of wall excluding interior windows
HVAC	Tons or BTU
Electrical	Horizontal square feet of building
Conveying facilities	Number of floor stops
Plumbing	Number of fixture units
Finishes	Horizontal square feet of building

Source: Adapted from Rowings (2004).

For example, the estimated cost of a proposed building may be estimated on the basis of the floor area, and the cost of transit operations in $/passenger-mile. Statistical tools, such as regression analysis (Chapter 7), can be used to estimate the best parameter values in the mathematical form specified for the cost model.

A statistical cost model consists of a dependent variable, independent variables, and a functional form. The **dependent variable** is the production or output cost expressed in monetary terms, preferably in constant dollars of some base year in order to avoid inflation bias. The dependent variable may be expressed as a unit cost (total cost per unit of output) or total cost. Where the dependent variable is the unit cost, the unit cost for each observation is determined (e.g., $\$/\text{ft}^2$ of bridge deck, $/pavement lane-mile, or $/transit passenger-mile). Then a regression model of the unit cost is developed as a function of the facility design and material type and other physical attributes of the facility. Where the total cost is used as the response variable, the total cost for each observation is determined and a regression model of the total cost is developed as a function of the facility attributes including the facility dimension; the unit cost function can then be derived from the total cost function using differentiation with respect to the facility dimension. Also, using techniques in calculus, it is possible to identify and quantify the sensitivity of the total cost to changes in each of the independent variables; this way, scale economies or diseconomies can be identified and quantified.

The **independent variables** are the factors that affect cost levels and can be categorized as follows: (i) those related to the resource input or output, such as the size of a water tank, tons of material shipped, or passenger-miles served, and (ii) those generally independent of the input or output, such as geographical location and climate. Input- or output-related variables typically constitute the variable component of a cost function, and others variables constitute the fixed component.

The possible **functional forms** include the linear, logarithmic, exponential, quadratic, cubic, and power forms. Nonlinear forms are generally considered more "powerful" than linear forms because they can help identify and quantify any scale economies and diseconomies in the expenditures associated with system construction, operations, and other phases. In general, the cost function is often an increasing function of the output volume. Other functional forms include the production function, a key concept in microeconomics, which relates the resource inputs to the output.

In construction engineering, this is often expressed as (i) a simple table of discrete output and input combinations, an example of which is the input–output ratio of factors such as labor hours and the output, and (ii) a mathematical relationship between the construction output (volume) and a specific factor of production, such as labor, material, or equipment. Mathematically, a production function may have the following general expression:

$$Q = f(X_1, X_2, X_3, \ldots, X_n)$$

where Q is the quantity of output, and $X_1, X_2, X_3, \ldots, X_n$ are input levels of the factors of production such as capital, labor, land, or raw materials.

Common mathematical forms include the additive [Equation (10.1)] and the Cobb–Douglas [Equation (10.2)]:

$$Q = K + AX_1^a + BX_2^b + \cdots + ZX_N^z \tag{10.1}$$

$$Q = AX_1^a \times BX_2^b \times \cdots \times ZX_N^z \tag{10.2}$$

where K is a constant representing some nondimensional characteristic and $a, b, \ldots, z$ are parameters affecting the shape of the relationship between the dependent and independent variables. These values represent the elasticities of the cost variable with respect to the input variables $X_1, X_2, \ldots, X_N$.

Other forms of production functions include a generalized form of the Cobb–Douglas function [termed the constant elasticity of substitution (CES)], and the linear, quadratic, and polynomial forms.

Econometric models, such as Tobit, random effects, and random parameters models, may also be used. After cost functions have been developed, they should be validated using a different set of data. Modeling and validation techniques are presented in Chapter 7. At the planning phase, only a quick rough estimate of cost can be provided because detailed information of the system design is unavailable at that phase. As such, the systems planner prefers to derive an estimated aggregate cost of the finished system, for example, \$/ft for a lineal system, $\$/\text{ft}^2$ for an aerial system or $\$/\text{ft}^3$ for a volume system, using historical data from past similar projects. On the other hand, when the system is due for construction, it is important to develop more precise estimates for the planned activity, using the disaggregate approach. We discuss this approach in the next section.

Some estimators have argued that expressing the production cost function as a collection of discrete output and input combinations may be more practical than as a set of statistical equations because the latter is often characterized by the rather restrictive assumption that the model output (cost) changes in a continuous manner in response to each minute increment in input factor levels.

Example. Statistical Aggregate Cost Functions for Light Rail Transit Systems.

Transit systems consist of fixed facilities and rolling stock (trains). In this case illustration, we use information from Kong et al. (2008), BAH (1991), Cambridge Systematics (1992), and Black (1995). Also, we focus on the fixed facilities only; specifically, we will examine aggregate cost models for the guideway, maintenance yard, stations, and the overall cost of fixed facilities. These costs exclude the costs of right-of-way acquisition, design, engineering, and other administrative costs.

Light Rail Guideway Costs

For light-rail transit systems, the construction of the guideway accounts for 15–40% of overall capital costs. Often, light-rail fixed-facility construction takes due advantage of the existing right-of-way due to the already-established street infrastructure, and this can lead to significantly lower costs. Subway guideways are the most costly, followed by retained-cut guideways, guideways of at-grade levels, and

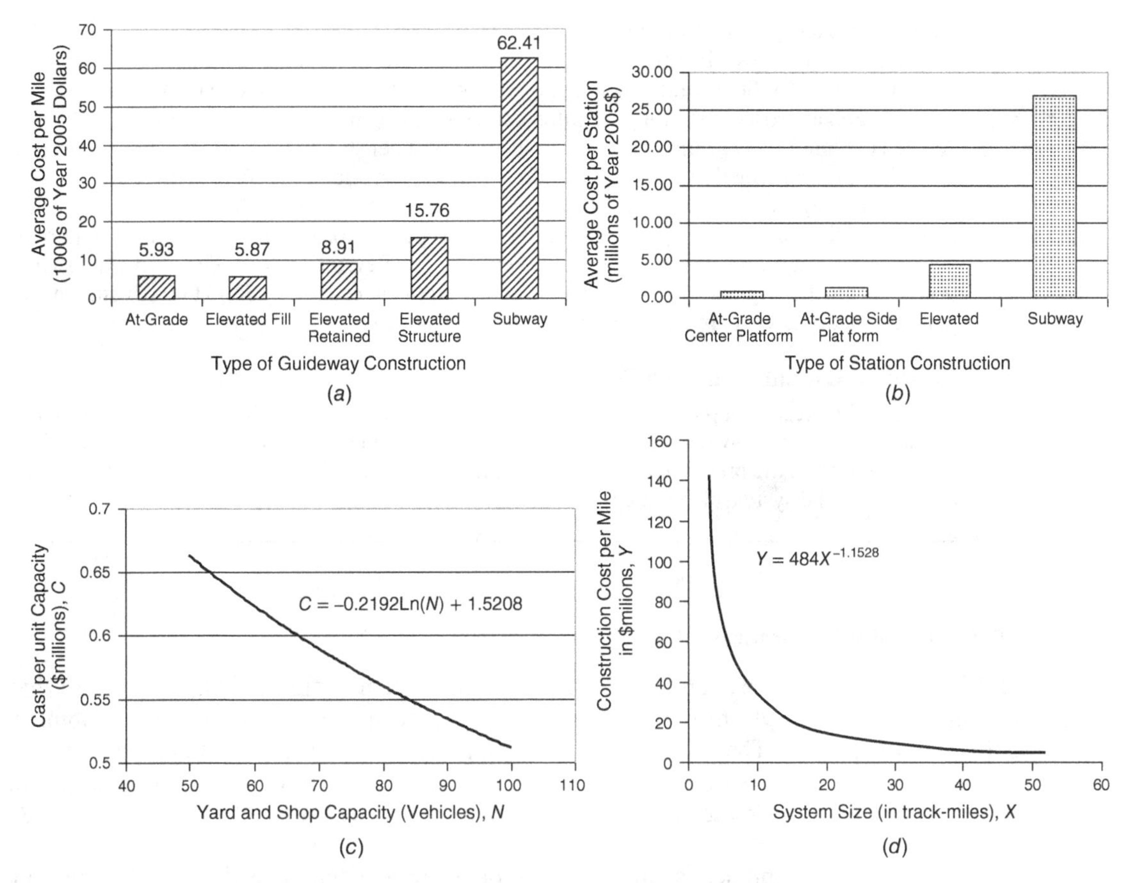

Figure 10.9 Aggregate cost models for light rail transit construction: (*a*) guideway, (*b*) station, and (*c*) maintenance yard and shop, and (*d*) overall cost.

elevated-fill guideways. Right-of-way (ROW) acquisition costs include land purchases for the guideway that vary considerably based on the nature of their construction. The updated average cost of guideway and station construction, is presented in Figures 10.9*a* and *b*. These simple average aggregate values can be used to estimate, at the planning phase, the guideway cost of future light rail transit systems.

Light Rail Maintenance Yard and Shop Costs

Maintenance yard and shop assets include storage and administrative facilities, and their construction costs generally depend on the yard and shop capacities, and the location and dimension of the site. A nonlinear functional relationship between unit construction costs and yard capacity is

$$C = -0.2192\ \ln(N) + 1.5208$$

where C is the unit cost in \$ millions per capacity, and N is the overall capacity of the maintenance yard or shop in terms of the number of vehicles.

As shown in Figure 10.9*c*, which illustrates this model, the unit cost of constructing light-rail transit yard and shop decreases as the yard capacity increases. For a planned new yard or extensions to an existing yard, the use of this model can provide a rough planning-phase estimate of the expected cost.

Light Rail Station Costs

Station costs vary by mode, type, location, and surroundings. Also, the number of levels (floors), platform design (central vs. lateral), and type of equipment [escalators, elevators, Americans with

Disabilities Act (ADA) facilities, etc.] can significantly influence station costs. Also, the cost of at-grade stations is less than those of high-platform stations, as passenger stops can be easily installed by deploying a low-floor light-rail vehicle without major redesign of the existing roadway.

Aggregate costs of stations are often expressed using measures such as \$/mile, \$/ft^2 or \$/station. As seen in Figure 10.9*c*, which presents the average cost per station for light-rail station construction, the type of station construction significantly affects the construction cost. With an average cost of \$2.7 million per station, subway stations are by far the most costly to construct. The next most costly type is the elevated station (averaging approximately \$450,000 per station)—elevated stations require additional construction materials and reinforcement to support the elevated station. The least expensive construction types are the at-grade station where the center platform costs only \$100,000 per station and the side platform station cost (approximately \$130,000 per station).

Light Rail Overall Combined Costs

Figure 10.9*d* presents a plot of the unit costs (per track-mile) versus the system size (in terms of track mileage). The model, which is based on a rather limited amount of data and hence must be used with caution, suggests the presence of scale economies in light-rail construction—the unit construction cost decreases as the system size increases.

10.7.2 Detailed Estimates

Detailed estimates of costs are developed after the system design and scope of work have been established, and the system owner is ready to proceed to invite contractors to submit bids for delivering the project. The process for developing detailed estimates include familiarization with the project characteristics (including assessment of scope, constructability, and risk), examination of the project design, structuring the cost estimate, and determining the cost elements (Rowings, 2004).

Disaggregate approaches are used for developing detailed estimates. In this bottom-up approach, the overall system, component, or process cost is decomposed into constituent *pay items*, and the cost of each pay item is established. Pay items may be priced in dollars per unit dimension including volume, area, and length or per unit weight of resource input or finished item, reported separately for materials only or all-inclusive of labor and supervision and equipment use.

This costing approach, which forms the basis for competitive contract bidding, is more suited when the system project development is past the design phase and thus the estimator possesses reliable knowledge of the quantities of each specific pay item. The disaggregate approach, which is thus more appropriate for estimating the cost of civil systems construction, operations, or maintenance, may seem rather simple but is laborious. Even a small project may have hundreds or thousands of individual pay items, each with its unit price and quantity. These costs are derived after the estimator has decomposed each work activity into its constituent pay items. The pay items are expressed in terms of specific quantities of one or more of the resource inputs (i.e., the material, labor, supervision, and equipment use). The product of the unit price per input quantity and the quantity required or specified yields the total cost of the pay item. The sum of all amounts yields the engineer's estimate for the entire project. Often, there is an adjustment of the profit and contingency line items.

This approach has often been described as a "cost accounting process," a major branch of general accounting that registers the costs of labor, material, and overhead on an item-by-item to determine the total cost of production. In the coarsest form of cost accounting, estimators use cost estimates that are very aggregate in nature and are reported for groups of pay items rather than individual pay items. Table 10.5 shows how the engineer's estimate could be calculated using the list of items and the quantities for the project. In certain contracting mechanisms, this

Table 10.5 Example of Bid Estimate Based on Engineer's List of Quantities

Items	Unit	Quantity	Unit price	Item Cost
Mobilization	Is	1	115,000	115,000
Removal, berm	If	8,020	1.00	8.020
Finish subgrade	sy	1,207,500	0.50	603,750
Surface ditches	If	525	2.00	1,050
Excavation structures	cy	7,000	3.00	21,000
Base course untreated, 3/4″	ton	362,200	4.50	1,629,900
Lean concrete, 10″ thick	sy	820,310	3.10	2,542,961
PCC, pavement, 10″ thick	sy	76,010	10.90	7,695,509
Concrete, ci AA (AE)	Is	1	200,000	200,000
Small structure	cy	50	500	25,000
Barrier, precast	If	7,920	15.00	118,800
Flatwork, 4″ thick	sy	7,410	10.00	74,100
10″ thick	sy	4,241	20.00	84,820
Slope protection	sy	2,104	25.00	52,600
Metal, end section, 15″	ea	39	100	3,900
18″	ea	3	150	450
Post, right-of-way, modification	If	4,700	3.00	14,100
Salvage and relay pipe	If	1,680	5.00	8,400
Loose riprap	cy	32	40.00	1,280
Braced posts	ea	54	100	5,400
Delineators, type I	Ib	1,330	12.00	15,960
type II	ea	140	15.00	2,100
Constructive signs fixed	sf	52,600	0.10	5,260
Barricades, type III	If	29,500	0.20	5,900
Warning lights	day	6,300	0.10	630
Pavement marking, epoxy material				
Black	gal	475	90.00	42,750
Yellow	gal	740	90.00	66,600
White	gal	985	90.00	88,650
Plowable, one-way white	ea	342	50.00	17,100
Topsoil, contractor furnished	cy	260	10.00	2,600
Seeding, method A	acr	103	150	15,450
Excelsior blanket	sy	500	2.00	1,000
Corrugated, metal pipe, 18″	If	580	20.00	11,600
Polyethylene pipe, 12″	If	2,250	15.00	33,750
Catch basin grate and frame	ea	35	350	12,250
Equal opportunity training	hr	18,000	0.80	14,400
Granular backfill borrow	cy	274	10.00	2,740
Drill caisson, 2′ × 6″	If	722	100	72,200
Flagging	hr	20,000	8.25	165,000
Prestressed concrete member				
type IV, 141′ × 4″	ea	7	12,000	84,000
132′ × 4″	ea	6	11,000	66,000
Reinforced steel	Ib	6,300	0.60	3,780
Epoxy coated	Ib	122,241	0.55	67,232.55
Structural steel	Is	1	5,000	5,000

(continued)

Table 10.5 (*Continued*)

Items	Unit	Quantity	Unit price	Item Cost
Sign, covering	sf	16	10.00	160
type C-2 wood post	sf	98	15.00	1,470
24″	ea	3	100	300
30″	ea	2	100	200
48″	ea	11	200	2,200
Auxiliary	sf	61	15.00	915
Steel post, 48″ × 60″	ea	11	500	5,500
type 3, wood post	sf	669	15.00	10,035
24″	ea	23	100	2,300
30″	ea	1	100	100
36″	ea	12	150	1,800
42″ × 60″	ea	8	150	1,200
48″	ea	7	200	1,400
Auxiliary	sf	135	15.00	2,025
Steel post	sf	1,610	40.00	64,400
12″ × 36″	ea	28	100	2,800
Foundation, concrete	ea	60	300	18,000
Barricade, 48″ × 42″	ea	40	100	4,000
Wood post, road closed	ea	100	30.00	3,000
Total				$14,129,797.55

Source: Hendrickson (2008).

list is made available to the bidders; and the bids submitted by the contractors are evaluated to select the best price and also serve as a basis for construction budget control (Hendrickson, 2008). The cost of each pay item may be expressed as a simple average cost (reported with or without the standard deviation) or as a pay item cost model that expresses pay item cost as a function of amount of the pay item, work location, work type, and other factors (see Section 10.7.3).

Examples of publicly-available cost databases for estimating disaggregate costs (e.g., labor, materials, and overhead) include RS Means (2011) and *Rawlinsons Australian Construction Handbook* (2012). The disaggregate costs include those costs for specific project types and locations. Thus, this data helps calculate the costs of construction at any phase of the systems development process and is used for many types of civil engineering systems. The development of detailed disaggregate costs for a project can be expensive as it typically takes several man-hours of skilled estimators; as such it is carried out only when there is reasonable certainty that the project will be implemented (NRC, 2012).

10.7.3 Refining Detailed Cost Estimates Using Statistical Tools

For most system owners, cost estimation simply involves the use of simple averages for each pay item. However, as we discussed in the previous section, more refined estimates of pay item costs can be developed using statistical modeling in order to derive a more accurate estimate of the cost. By duly accounting for the effect of pay item quantity (scale economies), work location, work type, and other attributes that influence pay item cost, such models can reduce the bias associated with the use of simple averages. The **average rate** expresses costs as a dollar amount per unit output or input regarding a pay item, for example, the cost of a certain type of concrete formwork is $\$5/\mathrm{m}^2$. On the other hand, the statistical model expresses costs as a function of the characteristics of a pay item and the work environment. A statistical model can be developed for each pay item, consists of a dependent variable, independent variables, and a functional form.

10.8 ISSUES IN SYSTEMS COSTING

In cost estimation of civil systems, there are a number of costing issues that must be considered as they may significantly affect the magnitude and reliability of cost estimates. These issues include cost progression across the phases; factors affecting the costs of civil systems; cost adjustments across time, location, or scale; cost overruns; the cost of risks due to estimation uncertainties or disasters; and the relative weights among the unit dollar of costs incurred by the different stakeholders.

10.8.1 Progression of Costs across the Phases of Systems Development

Cost-conscious engineers are interested in tracking the escalation of costs across the planning-design-construction phases of civil project development (Figure 10.10) so that more refined predictions of costs can be made for similar systems in the future. For large highway systems, for example, two escalation patterns seem to dominate (Figure 10.11): escalation pattern A (where the cost estimates generally increase as the project goes through the development stages, with the only exception being at the design and letting stage interface, where the estimated cost decreases) and escalation pattern B (where the cost estimates consistently do not decrease as the project proceeds through the successive stages from planning to construction).

10.8.2 Problem of Cost Overruns in Systems Construction

In Section 10.8.1, we discussed the progression of costs across the phases of systems development. In this section, we focus on the construction phase in particular. A cost discrepancy is the deviation of a cost amount between any pair of the development phases of planning, design, and construction. This could be an increase or decrease in project cost at a given stage compared with that at a previous phase (defined as a cost overrun or underrun, respectively). Systems managers are typically interested in the cost discrepancies between the contract award stage (award amount) and the postconstruction stage (final or as-built amount). Generally, across the various phases of project development, cost discrepancies (specifically, cost overruns) have been identified as a pervasive

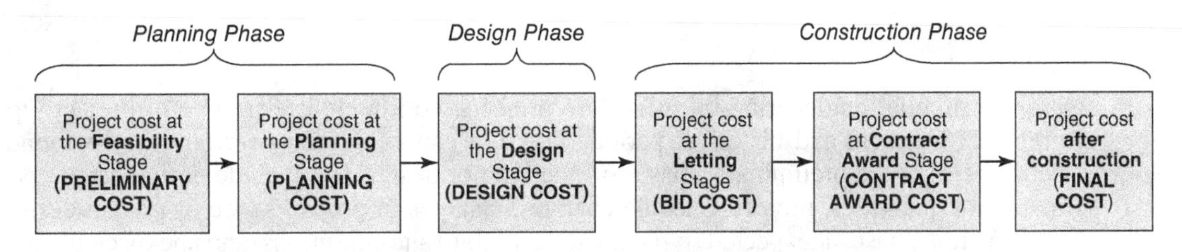

Figure 10.10 Evolution of project costs from planning to construction.

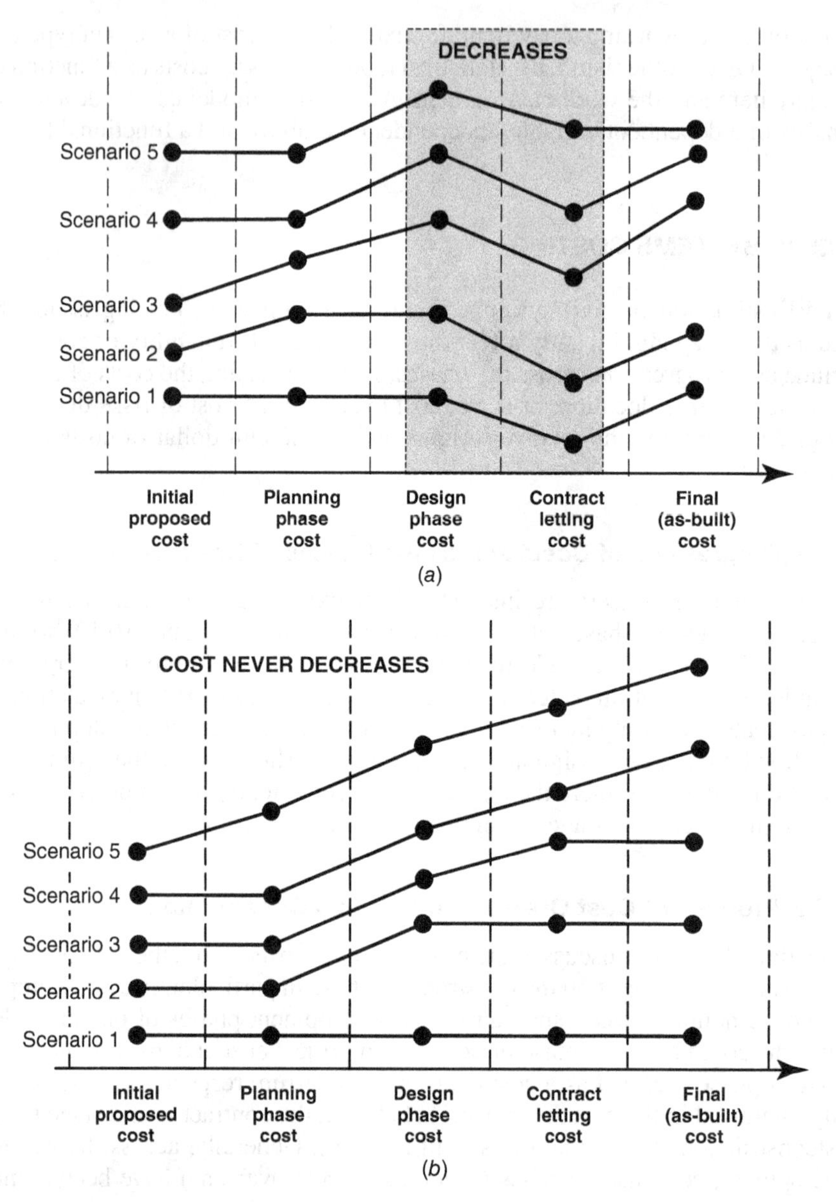

Figure 10.11 Two of several cost escalation patterns for large highway projects (*a*) Pattern A (*b*) Pattern B (not to scale).

problem in civil engineering infrastructure projects worldwide irrespective of project type, geographical location, and historical period. The root causes of cost overruns can be found in the deficiencies at the preliminary phases of planning or design and include design errors, poor estimation of quantities, project schedule changes, changes in project scope or system design at the construction phase, unexpected (typically unfavorable) site conditions, and unexpectedly high costs of materials and labor. The contribution of these root causes to cost overruns has been investigated in past research studies, such as Hufschmidt and Gerin (1970), Jahren and Asha (1990), Semple et al. (1994), Akinci and Fisher (1998), Akpan and Igwe (2001), Chang (2002), Knight and Fayek (2002), and Attala and Hegazy (2003), and Bordat et al. (2003). Most of the problems associated with the root causes are often unknown at the contract award phase as they surface only during or after the project construction has started phase. Thus, between the contract award and final construction stages, it is often difficult to predict cost overrun solely on the basis of the root causes.

10.8.3 Factors Affecting the Agency Costs of a Civil System

The cost estimates presented in Appendix 3 are only average values. In practice, the cost of a specific project may deviate significantly from the expected, average, or historical costs of similar projects for several reasons including:

- *Number of spatial interferences with natural or man-made features*: Systems with a higher frequency or intensity of interference with their natural or man-made environments require a larger number of special system designs to accommodate the features. For example, a railway with a greater number of river crossings would require more crossing structures, leading to a generally higher overall unit cost ($ per mile) of the system.
- *Environmental impacts*: Proposed civil engineering systems that are slated for location at environmentally sensitive areas are generally expected to have higher unit costs because the system owner will need to make extra efforts to comply with legislation that protect the environmental quality or the ecology (e.g., endangered species and their habitats).
- *Existing soil and site conditions*: Where subsoil conditions are poor or highly variable, or where the site is located in areas considered harsh for construction purposes, for example, desert, permafrost, and jungles, the unit costs are generally expected to be higher.
- *Project size*: Larger systems generally have higher overall costs but lower unit costs due to economies of scale. For certain rare kinds of system structures, however, the need for additional stabilizing structures beyond certain sizes may translate to a greater cost increase per unit increase in size, thus reflecting scale diseconomies.
- *Project complexity*: Generally, projects of greater design complexity have more subcontractors, which often translates into higher unit costs.
- *Mechanism used for contracting*: Systems constructed using the traditional mechanism (design-bid-build) for contracting generally have lower unit costs of initial construction (but typically, higher unit costs of operations and maintenance over the system rest-of-life) compared to those constructed using alternative project delivery approaches such as design–build and warranties.
- *Urban/rural location*: Due to right-of-way land values, interference from utilities and human activity, and the delicate nature of work zones, projects that construct, preserve, or monitor civil engineering systems in urban areas generally have higher unit costs compared to their rural counterparts.

Other factors that may affect the costs of civil systems are the extent of bidder competition for contracts, stringency of design standards including ADA requirements, prevailing cost of labor, tightness of specifications for material and workmanship quality, and topographic conditions. In developing empirical cost functions for a specific activity or pay item, the engineer should identify the factors that cause anomalous costs, and data from contracts characterized by such conditions may be excluded from the cost analysis or set aside for special analysis.

10.8.4 Scale Economies and Diseconomies

At any phase of development, most visibly at the construction and operations phases, the cost of producing each additional unit of some output at that phase may decrease or increase as production increases. This may be attributed to efficiencies or inefficiencies associated with the process of production at the phase in question, or may be due to features inherent to the civil engineering system. Scale economy can be defined as the reduction in average cost per unit output for every increment in output. For a given cost function, the economies or diseconomies of scale with respect to any output variable can be investigated by examining the indices of that variable in the cost equation or using simple plots of the output variable in question and the unit cost or total cost (Figure 10.12). Depending on the type of civil engineering system, the development phase, and the "production" type, and level in question, the average cost at a certain output level of production, O_C, may increase or decrease at the same rate, a faster rate, or a slower rate or may remain the same. Some of these possibilities are illustrated as curves *A, B, C,* and *D* in Figure 10.12.

In curve *A*, further increases of output beyond O_C will result in diseconomies of scale (i.e., beyond point O_C each additional output will cost the system owner more to produce). In curve *B*, the unit cost of production remains unchanged from O_C as output increases. In curve *C*, the cost of producing a unit falls after output is increased beyond O_C, reflecting economies of scale. Curve *D* illustrates the case where the unit cost of production falls at a faster rate after a certain point O_C.

In the cost analysis for civil engineering systems, scale economies have been long recognized; however, the development of formal methods for cost adjustments due to scale economies has continued to lag behind the state of the art. In systems evaluation and comparison of alternative materials, designs and processes at any phase, the tendency has been to compare costs using the cost per unit dimension. For example, analysts have compared the empirical costs of concrete and steel bridges on the basis of their aggregate costs (e.g., $\$/\text{ft}^2$ of deck area) or the disaggregate cost of some pay item related to the basic material (e.g., \$/ton of concrete or steel). Unfortunately, this approach implicitly assumes that the reported unit cost of each output parameter or pay item is linearly related to the dimension under consideration—an assumption that is often unrealistic. For example, the unit cost of steel bridges ($\$/\text{ft}^2$) may be lower than that of concrete bridges; this may because the steel bridge unit costs were averaged from a sample that contain very large bridges, hence the lower cost of steel bridges may be due to scale economies. Thus some adjustment of the unit cost is necessary as part of the cost analysis. It is important, in the cost modeling of civil systems to explicitly account for the nonlinear relationship that typically exists between the facility cost and the facility dimensions. As evidenced from the literature, most civil systems often exhibit scale economies, that is, the greater the project dimension, the lower the unit cost (\$/dimension).

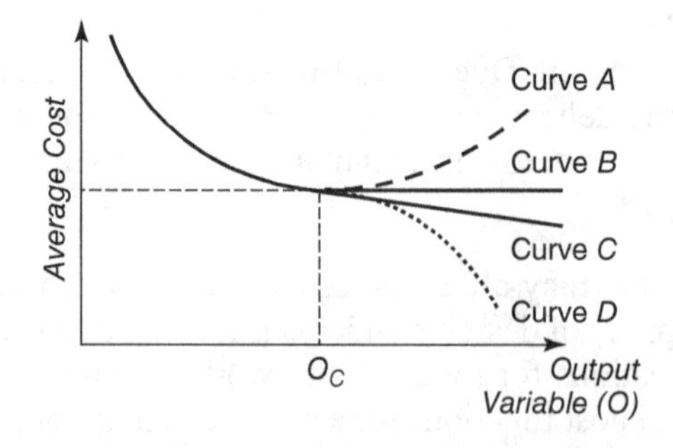

Figure 10.12 Variations of average cost reflecting scale economies and diseconomies (Sinha and Labi, 2007).

In sum, in cost estimation for civil engineering systems, the presence and effects of scale economies have long been realized; however, most past work have traditionally proceeded on the basis of simple average cost in both the early stages of (average aggregate costs, for planning) and the latter stages (average cost of each pay item, for design and bidding). There seems to be little effort to develop universal techniques to adjust cost values to account for scale effects, for purposes of cost estimation and project evaluation. Adjustments of aggregate or disaggregate costs can yield more refined values of cost and can increase confidence in cost estimates. The reader is encouraged to review appropriate sources [including Sinha and Labi (2007)] for a framework and examples for cost adjustments to account for scale economies.

10.8.5 Adjusting Costs due to Temporal and Spatial Variations

(a) Adjusting for Temporal Variation (Current versus Constant Dollars). In cost estimation or project evaluation, it may be necessary to convert the costs of civil engineering systems to their equivalent cost at some specified past, current, or future year. This is generally accomplished using price indices that describe price trends, in terms of inflation or deflation, over time. Inflation refers to the general increase in the price of goods and services. From a conceptual and computational perspective, cost estimates must be prepared in constant (instead of nominal) dollar amounts, so as to remove any bias due to price inflation from the estimated cost. In cases of economic analysis where alternative cost streams over time are being compared (e.g., different life-cycle designs), all costs should be converted to constant dollars before the analysis is carried out; if this is done then the interest rate used in the economic analysis will not need to be adjusted to account for inflation.

The prediction of costs on the basis of historical price trends is useful particularly when long-term economic conditions are stable or at least, predictable. In making cost adjustments over time, it is necessary to recognize that each resource type (e.g., materials, labor, and equipment) have different inflation rates over time. For example, over the past decade or so, the costs of general construction have increased at a faster rate than general inflation; on the other hand, information technology-related costs have reduced significantly. Also, different types of materials (e.g., steel and concrete), labor (skilled and unskilled), and equipment used in system construction or operations have different rates of inflation. Thus, cost adjustments to account for inflation should be carried out with due caution.

A number of cost indices, which account for such relative rates of cost increase across different cost components, exist for adjusting cost data across different years. There are different cost indices for different civil systems (e.g., water retaining structures, highway pavements, wind power systems) and also for different phases of development (construction cost indices and maintenance cost indices). Examples include the Engineering News Record's Construction Cost Index (CCI), the Federal Capital Cost Index (Schneck et al., 1995), the RSMeans City Construction Index, and FHWA's Highway Maintenance and Operating Cost Index. Others include the Building Cost Index (BCI), which accounts for the prices of steel, cement, lumber, and skilled labor, and the Consumer Price Index (CPI), which accounts for personal housing, transportation, medical care, energy, and apparel. Construction cost indices generally account for the prices of cement, steel, lumber, and labor. Indices are developed and used by several governmental agencies such as the U.S. Army Corps of Engineers, the U.S. Bureau of Reclamation, and the U.S. Environmental Protection Agency. Engineering News Record's Construction Cost Index uses 1967 as the base year.

To make a cost adjustment across time, we can use the following standard equation for calculating the cost estimate at the required year:

$$C_{\text{AY}} = C_{\text{BY}} \times \frac{I_{\text{AY}}}{I_{\text{BY}}}$$

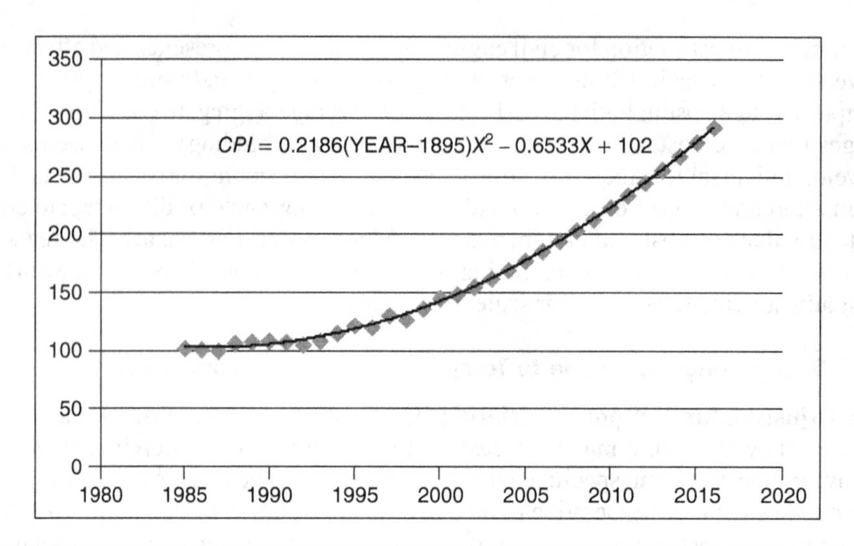

YEAR	1985	1986	1987	1988	1989	1990	1991	1992	1993	1994	1995	1996	1997	1998	1999	2000
CPI	102.0	101.1	100.0	106.6	107.7	108.5	107.5	105.1	108.3	115.1	121.9	120.2	130.6	126.9	136.5	145.6

YEAR	2001	2002	2003	2004	2005	2006	2007	2008	2009	2010	2011	2012	2013	2014	2015	2016
CPI	148.8	155.3	162.3	169.7	177.6	185.9	194.6	203.8	213.3	223.4	233.8	244.67	255.9	267.7	279.9	292.5

Figure 10.13 Example of construction price trends (Federal Aid Highway Construction Price Index).

where C_{AY} is the cost of an activity in the year of analysis, C_{BY} = cost of the activity in the reference year, I_{AY} is the index corresponding to the year of analysis, and I_{BY} is the index corresponding to the reference year.

To determine the values of C_{AY} and C_{BY}, we use the chart in Figure 10.13 or similar charts developed by the costing agency or organization. Note that the y values represent the cost index (in this case, the FHWA CPI) and the x values represent not the raw years but the number of years since 1985. For example, for 2005, $x = 20$.

Using cost indices, the cost of some action at any year can be estimated as follows:

$$\text{COST}_{\text{EAR\_}i} = \text{COST}_{\text{REF\_YEAR}} \frac{\text{COST\_INDEX}_{\text{YEAR\_}i}}{\text{COST\_INDEX}_{\text{REF\_YEAR}}}$$

Given any three of the variables in the above equation, the fourth can be calculated. Note that the reference year, REF_YEAR, may be the base year or any other specified year.

Example 10.11

The Saginaw Bridge on National Highway 55 in Los Amigos Province had a construction cost of $6 million in 1987. What would this project cost if it were built in 1999? Assume that the construction price indices in 1987 and 1999 are 1.2 and 1.6, respectively.

Solution

$$\text{COST}_{1999} = \text{COST}_{1987}(\text{CI}_{1999}/\text{CI}_{1987}) = 6{,}000{,}000(1.6/1.2) = \$8{,}000{,}000$$

There are a number of problems associated with the use of price indices. These include:

1. *Inconsistency with predefined goods at a given time*: In a strict sense, price indices apply to buyers who purchase exactly the types and amount of goods and services upon which the index is calculated. In reality, however, the purchases of most buyers deviate from the price index's predefined package. Therefore, the price index may not be applicable to everyone.
2. *Instabilities in purchasing patterns over time*: Buying patterns, even for an individual buyer, change over time, while the composition of the predefined package (upon which the price index is calculated) remains the same.
3. *Changing quality of goods*: The price index approach of measuring inflation assumes that the quality of a standard product or service stays the same over time. Therefore, changes in the quality of the product (goods) are not considered by this concept. In reality, however, the quality of most goods changes over time because research and development enables superior products to continue to evolve. For example, the quality of a steel bar in 2014 is superior than it was in 2000 (higher tensile strengths, higher corrosion resistance, better bonding to concrete, etc.).

In spite of these problems, the price indices are still considered useful indicators of price inflation and are used to predict future costs.

(b) Adjustment for Spatial Variation. Where requisite cost data at a certain region or location are unavailable, the cost estimator may seek to estimate the cost associated with the construction, operation, or maintenance of a proposed civil engineering system based on the cost of similar phase processes at other locations. However, because the costs of living and production typically vary from one geographic location to another, it is often necessary to adjust cost from one location before it can be used to make cost estimations at other locations. For purposes of illustration, we consider such adjustments in the case of highway systems in the United States. The Federal Highway Administration (FHWA, 2005) has provided state cost factors for highway capital improvements (Table 10.6). Using cost indices, the cost of some action at a different location can be estimated as follows:

$$\text{COST}_{\text{LOCATION\_}i} = \text{COST}_{\text{REFERENCE\_LOCATION}}\left(\frac{\text{COST\_INDEX}_{\text{LOCATION\_}i}}{\text{COST\_INDEX}_{\text{REFERENCE\_LOCATION}}}\right)$$

Example 10.12

The Green River Bridge in Vermont had a construction cost of $3.5 millton. Find the estimated cost of a similar bridge to be built in the state of Georgia.

Solution

$$\text{COST}_{\text{GEORGIA}} = \text{COST}_{\text{VERMONT}}(\text{CI}_{\text{GEORGIA}}/\text{CI}_{\text{VERMONT}}) = 3.5\text{M}(1.27/1.15) = \$3.87\text{M}.$$

Table 10.6 2004 State Cost Factors

State	Cost Factor	State	Cost Factor	State	Cost Factor
Alabama	1.21	Louisiana	1.32	Oklahoma	0.95
Alaska	1.30	Maine	1.10	Oregon	1.25
Arizona	0.95	Maryland	0.83	Pennsylvania	0.95
Arkansas	0.95	Massachusetts	0.78	Puerto Rico	1.23
California	1.56	Michigan	1.24	Rhode Island	0.98
Colorado	1.26	Minnesota	1.11	South Carolina	1.32
Connecticut	0.88	Mississippi	1.51	South Dakota	1.19
Delaware	1.51	Missouri	0.81	Tennessee	0.90
District of Columbia	0.56	Montana	1.19	Texas	1.19
Florida	1.19	Nebraska	1.15	Utah	1.00
Georgia	1.15	Nevada	1.49	Vermont	1.33
Hawaii	0.76	New Hampshire	1.30	Virginia	1.27
Idaho	1.12	New Jersey	0.70	Washington	0.80
Illinois	0.90	New Mexico	0.69	West Virginia	1.39
Indiana	1.28	New York	0.90	Wisconsin	0.70
Iowa	0.94	North Carolina	0.97	Wyoming	1.08
Kansas	0.59	North Dakota	1.42	Wyoming	1.24
Kentucky	1.39	Ohio	0.85		

Source: FHWA (2005).

10.8.6 Cost of Disaster Risks

Risk-based systems costing involves the likelihood and consequences of natural and man-made disasters that can significantly influence the operations and physical structure (and consequently, the costs of the physical preservation and operations of civil systems). Natural disasters include floods, earthquakes, and scouring, while man-made disasters include accidental collisions and terrorist attacks. The probability of system failure can be assessed for each type of vulnerability, and the extent of vulnerability can be assessed. Then the cost of repair in the event of system failure can be used to derive a failure cost or vulnerability cost that could be included in the systems costing (Chang and Shinozuka, 1996; Hawk, 2003). Vulnerability cost can be determined as the product of the probability of disaster occurrence and the cost of damage in the event occurrence. Risks are evident in both the probability of the occurrence and the uncertainties of damage cost in the event of disaster (Sinha and Labi, 2007).

10.8.7 Relative Weight between Stakeholder Costs

The costs of civil engineering systems are borne not only directly by the owner in terms of the contract or in-house costs but also by the users and community who suffer the consequences of facility downtime and various externalities. In cases where all these impacts can be monetized, the total monetary cost of the action to all stakeholders is a sum of the cost to the system owner users, and community. If each dollar of cost incurred by the system owner is valued differently than each dollar of cost incurred by the user or the community, then the question that arises is: What is the appropriate ratio, or *relative weight*, between each dollar incurred by the system owner and that incurred by the user or the community? Some studies have assumed (often implicitly) that $1 of system owner cost is equivalent to $1 of user or community cost and therefore simply add the system owner costs directly to the user and community costs to obtain an overall cost associated with each alternative decision at any phase. However, a trade-off seems to exist between the system owner's expenses and the user/community cost and such tradeoffs may need to be considered in the evaluation; specifically, certain design and preservation options that lead to significantly lower user and community costs often involve larger initial outlays by the system owner (FHWA, 2002). Second, for nonrevenue civil systems, the system owner's costs often play a key role in system planning while little or no consideration is given to user and community costs. A number of researchers have suggested that only a fraction of user or community costs should be considered for addition to the agency costs. As Sinha and Labi (2007) noted, there seems to be little or no consensus on the issue at the current time, and the current practice seems to be that for each alternative under consideration the three costs are simply added together to yield a total unweighted sum.

10.8.8 Preconstruction Costs and Consulting Fees

Many system owners maintain staff that carries out needs assessment, planning (including feasibility studies), or design using in-house resources. Others prefer to engage external consultants to carry out at least one of these activities. Consultants may be engaged to carry out not only these activities but also the preconstruction-related activities (bid document preparation, bid evaluation, preparation of contract award documents, construction supervision) and postconstruction activities (contractor selection or supervision of system monitoring, operations, or maintenance). Goodman and Hastak (2007) presented the following estimates for total consulting fees: preliminary studies including reconnaissance (0.2–0.5%), feasibility studies (1–2%), design (4–6%), and construction supervision (5–10%). The design fee may be much higher for small projects and also for complex projects. The American Society of Civil Engineers identifies six activities where consultants may be engaged: the study and report stage, the preliminary design stage, the final design stage, the bidding stage, the construction phase, and the operation phase. Figures 10.14 and 10.15, from ASCE Manual 45, presents the design consulting fees and total consulting fees, respectively, for new construction of a system and for modifications to the system (ASCE, 2003). The curves, which are intended for budgeting and comparison purposes, reflect the average cost trends from a very large number of civil engineering projects.

Carr and Beyor (2005) noted that consulting fees have reduced by at least 20% over the last several decades due to ASCE's consent, in 1972, to a decree by the U. S. Department of Justice (DOJ) that removed from the society's code of ethics the prohibition of submitting price proposals that "constituted price competition." This settlement stipulated that professional societies refrain from publishing minimum fee curves. The position of the DOJ was that the Sherman Antitrust

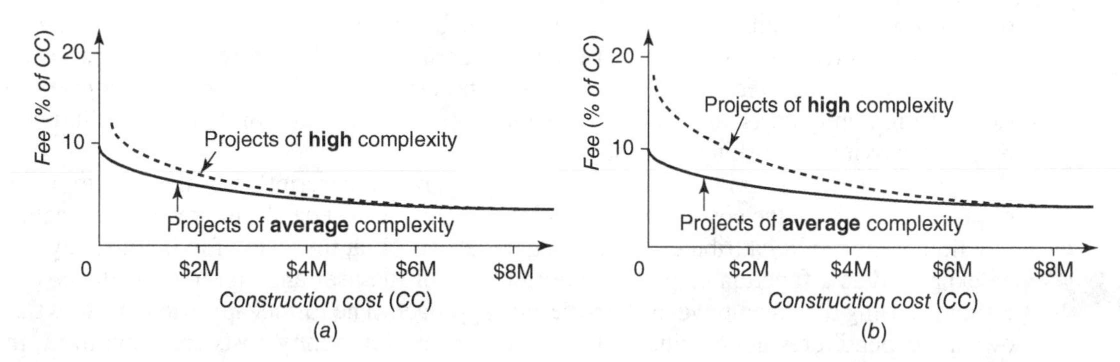

Figure 10.14 Design consulting fees (adapted from ASCE, 2003): (*a*) construction of new system and (*b*) modification to existing system.

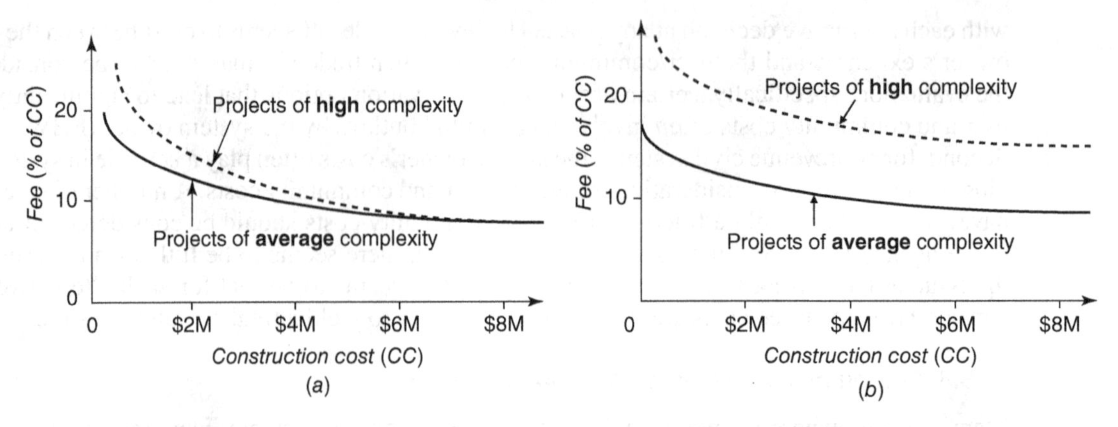

Figure 10.15 Overall consulting fees (adapted from ASCE, 2003): (*a*) construction of new system and (*b*) modification to existing system.

Act is applicable to a number of learned professions including the civil engineering profession. By consenting, the ASCE paved the way for significant fee competition. Also, since that settlement, few government agencies have adopted any meaningful deviation from the old fee curves. Meanwhile, the ASCE, in its Manual 45 periodicals, continues to publish median fee curves based upon the results of various surveys.

SUMMARY

For any civil engineering system, the analysis of the costs of the system, its components, or its associated processes is an inextricable part of the systems management process. The systems engineer must take into account the costs to be incurred not only to the system owner but also to the user and community. Costing is important because it helps determine the amount of capital needed to construct, operate, or maintain the system. Also, in certain cases, costing helps determine the expected benefits of the systems (often benefits can be expressed as the reduction in user costs relative to some base scenario). Costs may be classified by the stakeholder who incurs the cost—the system owner or operator, users, or the community, how the cost varies with the output, the manner of expression of the unit cost, and the phase at which the cost is incurred. The costs incurred by the system owner include the costs of construction, operations, and maintenance. User costs are direct or indirect costs incurred by the system users and are largely due to delay, in convenience, and safety. Community costs, are typically adverse impacts including noise and air pollution suffered by persons living or working near the system.

A preliminary step in systems costing is preparing a description of the system, component, or process for which the cost is being modeled. Aggregate cost can be expressed as a mathematical function of the facility attributes. Also disaggregate costs (at the level of individual pay items) can be expressed as a function of pay item attributes. For the user and community, the costs may be estimated using the preemptive or after-the-fact approach: The former approach involves the system owner's expenditures in ensuring that adverse user or community costs are minimized; the latter approach is typically the costs incurred by all stakeholders, particularly the users due to unfavorable or unsustainable physical or operational conditions of the system.

Estimators of the costs of civil engineering systems are often cognizant of a number of issues including the perils of aggregation of the cost items, the dichotomy between cost estimation at the planning and the construction (bidding) phases, adjustments to account for spatial and temporal variations and scale economies, recognizing and quantifying uncertainties and risk in systems cost estimation, the pervasive issue of cost escalation in public projects, and the relative weights among each dollar of cost incurred by the system owner, users, and the community. The issue of cost escalation and its causes were discussed in some detail. Major culprits for cost escalation include difficult site or subsoil conditions, relocation of existing utilities, sensitivity of the environment, and project size and complexity. To enhance cost estimation for civil engineering systems, historical cost data or models can be found in past as-built project reports, agency websites, and costing manuals published by the public and private sectors. With due circumspection, costs from these sources may be used to develop initial cost estimates for a planned project.

EXERCISES

1. List the eight phases of development of a canal system. Discuss the possible costs incurred by the canal owner, the canal users (shippers), and the surrounding community at each phase of developing this engineering system. In your discussion of each cost, state whether it is monetary or nonmonetary, direct or indirect, tangible or intangible, internal or external, recurring or nonrecurring, initial or rest-of-life, sunk or working cost.
2. Discuss the four user cost categories associated with a tolled urban freeway system. How could these costs change if the freeway operator neglects to carry out maintenance or capacity increases (lane additions) when these interventions are needed.
3. For a water supply system in a perfectly competitive market, discuss, using a graph, the changes in the unit user cost of using the system when demand increases without a change in supply.
4. Discuss the various costs that an agency is expected to incur in the entire life cycle of a proposed expansion to an existing water port.
5. It is proposed to build a pedestrian bridge crossing the busiest part of your campus to separate pedestrian and road traffic at that location. List the various costs that are expected to be incurred at each phase of the development of this system, from assessment of the need for the bridge to the eventual termination and replacement of the bridge several decades thereafter. For each phase, group the costs by the incurring stakeholder. For each cost item listed, write in parentheses whether it is monetary or nonmonetary, direct or indirect, tangible or intangible.
6. Indicate which of the cost items below are preemptive and which are after the fact: (a) cost of installing user safety features at a busy intersection, (b) cost of settling lawsuits filed by complainants who suffered injury due to inadequate safety features at a busy intersection, (c) cost of carrying out the maintenance of a runway pavement in order to prevent the onset of structural failure, (d) cost of carrying out repairs to a pavement that has suffered structural failure, (e) additional cost of using stainless steel used in constructing critical members of a steel bridge, and (f) cost of replacement of corroded critical members of an old steel bridge.
7. Compare and contrast the categories of costs borne by users of a new government-funded untolled highway and those borne by users (customers) of a new water treatment plant that is funded by bills paid by the customers. Your discussion, preferably tabulated, should address the potential out-of-pocket costs, accident costs, safety, time, discomfort, and inconvenience costs, and the operating costs of any user equipment where applicable, at each phase of the system's development.
8. In response to growing energy demands in Benton City, it is proposed to construct a windmill farm to be located near a farming community in the suburbs of Benton. Draw a timewise sequence of the various direct, indirect but tangible, and intangible costs that will be incurred by all of the different stakeholders.

9. For a levee system, the table below shows the observed annual preemptive expenditure (funds spent on preventive maintenance) and the after-the-fact expenditure costs (funds spent on repairing defects in the system). It is seen that the more the system owner spends on preemptive activities, the less the owner spends on after-the-fact activities. Develop and plot a statistical model to relate the two expenditure items. From the developed model, establish and plot the marginal effects model. If the system owner currently spends $55 per square foot of levee surface area on preventive maintenance annually, use your marginal effects model to determine the benefits of increasing the preventive maintenance expenditure by 15%.

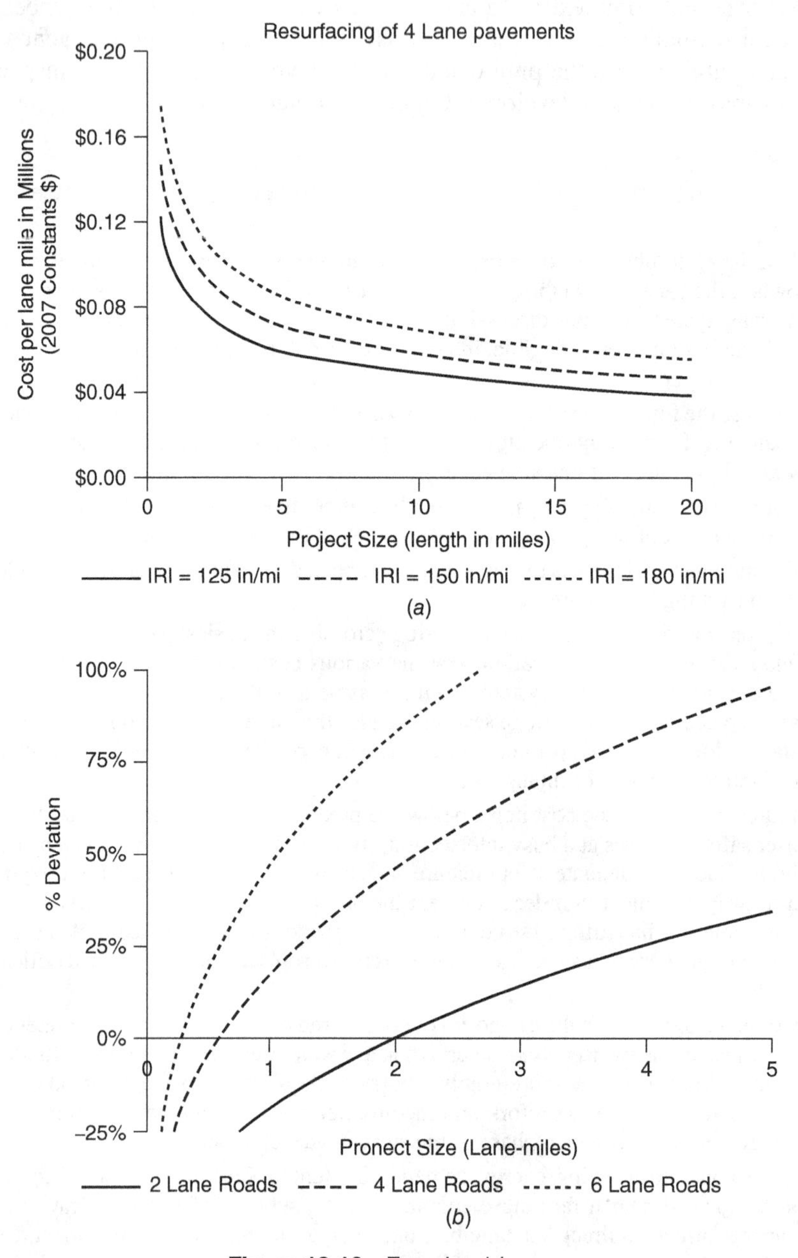

Figure 10.16 Exercise 14

Preventive expenditure ($/ft^2)	8	0	6	2	4	9	12	13	11	12	19	27	30	32	33	35	37	38	42	47	50	52	56	57	57	62	67	75	80	85	93
Repair expenditure ($/ft^2)	149	248	224	201	195	150	181	168	130	92	109	63	27	100	70	75	36	53	48	49	25	46	25	24	57	27	24	25	19	22	24

10. The annual fixed costs of carrying out inspection and monitoring of all concrete dams in a certain province is \$10 million. Also, the dam agency spends \$1.25 for every square foot of dam concrete surface area. Determine the following: (a) the annual variable costs, (b) the total annual costs, (c) the average total costs, and (d) the average marginal costs. Plot a graph of the total, average, and marginal cost functions for the systems' inspection and monitoring program.

11. Discuss the essential differences between the aggregate and disaggregate costing approaches. List the merits and disadvantages of each approach.

12. The fixed operating cost of a water supply treatment plant is \$25,000 per month. From statistical analyses of historical costs, it is shown that the variable costs are represented by the cost function $VC = 0.01W^3 - 2W^2 + 505W$, where W is the monthly usage in cubic meters. If the average water fee is \$0.75/m^3, determine the usage that maximizes the revenues of the system owner transit. Plot a graph of the total costs, total fixed costs, and total variable costs. Provide a plot of the total operating cost function, average operating cost function, and marginal operating cost function. Also, determine and sketch the elasticity function. Comment on the economy of scale implications of the operating costs.

13. The termination (demolition) costs of a certain type of building system is described by the cost function $C = 150F^{2.5}$, where F represents the floor area of the building in millions square feet Plot the average and marginal cost functions for $F = 1$ to 5 in unit increments.

14. Systemville Highway is a 10-mile, four-lane road that is being resurfaced. The current condition of the road pavement is 150 in./mile. Calculate the cost of the project using Figure 10.16*a*. If a cost analyst used the average aggregate cost instead of the aggregate cost model in Figure 10.16*b* to develop the cost estimate, by how much would the analyst be overestimating or underestimating the project cost?

REFERENCES

AASHTO (2003). *User Benefit Analysis for Highways*. American Association of State Highway and Transportation Officials, Washington, DC.

Akinci, B., and Fischer, M. (1998). Factors Affecting Contractors' Risk of Cost Overburden. *ASCE J. Mgmt. Eng.*, 14(1), 67–76.

Akpan, E. O., and Igwe, O. (2001). Methodology for Determining Price Variation in Project Execution. *ASCE J. Constr. Eng. Mgmt.*, 127(5), 367–373.

ASCE (2003). Manuals and Reports on Engineering Practice—No. 45, *Consulting Engineering—A Guide for the Engagement of Engineering Services*. ASCE, Reston, VA.

Attala, M., and Hegazy, T. (2003). Predicting Cost Deviation in Reconstruction Projects: Artificial Neural Networks vs. Regression. *ASCE J. Const. Eng. Mgmt.*, 129(4), 405–411.

BAH (1991). Light Rail Transit Capital Cost Study. Booz-Allen & Hamilton Inc., prepared for Urban Mass Transportation Administration, Office of Technical Assistance and Safety, Fed. Transit Admin., Washington, DC.

Bass, R. (1998). Evaluating Environmental Justice under the National Environmental Policy Act, *Environmental Impact Assessment Review*, 18(1), 83–92.

Bateman, I., Day, B., Lake, I., and Lovett, A. (2001). *The Effect of Road Traffic on Residential Property Values: A Literature Review and Hedonic Pricing Study*. Scottish Executive Development Department, www.scotland.gov.uk/Publications/2001/07/9535/File-1. Accessed April. 20, 2010.

Bein, P. (1997). *Monetization of Environmental Impacts of Roads*. Planning Services Branch, B.C. Min. of Transp. and Hwys, Victoria, Canada. www.th.gov.bc.ca/bchighways.

Black, A. (1995). *Urban Mass Transportation Planning*. McGraw-Hill, New York.

Bordat, C., Labi, S., McCullouch, B., and Sinha, K. (2003). *An Analysis of Cost Overruns, Time Delays, and Change Orders in Indiana*. Tech. Rep. FHWA/JTRP/2004/07, Purdue University, West Lafayette, IN.

Cambridge Systematics, Inc. (1992). The Urban Institute, Sydec, Inc., H., Levinson, Abrams-Cherwony and Assoc., Lea and Elliott. *Characteristics of Urban Transportation Systems, Revised Edition*, DOT-T-93-07. Prepared for Fed. Transit Admin., Washington, DC.

Carr, P. G., and Beyor, P. S. (2005). Design Fees, the State of the Profession, and a Time for Corrective Action. *ASCE J. Manage. Eng.*, 21(3), 110–117.

Chang, A. S-T. (2002). Reasons for Cost and Schedule Increase for Engineering Design Projects. *ASCE J. Mgmt. Eng.*, 18(1), 29–36.

Chang, S., and Shinozuka, M. (1996). Life-Cycle Cost Analysis with Natural Hazard Risk. *ASCE J. Infr. Syst.*, 2(3), 118–126.

DeGarmo, E. P., Sullivan, W. G., Bontadelli, J. A., and Wicks, E. M. (1997). *Engineering Economy*, 10th ed. Prentice Hall, Upper Saddle River, NJ.

Dickey, J. W., and Miller, L. H. (1984). *Road Project Appraisal for Developing Countries*. Wiley, New York.

Fang, H. (2011). Externality vs. Public Goods, Academic Lecture, Duke University.

FHWA (2000). *1997 Federal Highway Cost Allocation Study Final Report Addendum*. USDOT, Washington, DC. www.fhwa.dot.gov, Table 12.

FHWA (2002). *Economic Analysis Primer*. USDOT, Washington, DC.

FHWA (2005). *Price Trends for Federal Aid Highway Construction*. USDOT, Washington, DC.

Friedrich, R., and Bickell, P. (2001). *Environmental External Costs of Transport*, Springer, New York. www.fhwa.dot.gov/policy/hcas/summary/index.htm, Table V-22.

FTA (1993). *Transit Profiles; Agencies in Urbanized Areas Exceeding 200,000 Population, for the 1992 Section 15 Report Year*. Fed. Transit Admin., USDOT, Washington, DC.

Goodman, A., and Hastak, M. (2007). *Infrastructure Planning Handbook*, McGraw-Hill/ASCE, Reston, VA.

Hawk, H. (2003). *NCHRP Report 483: Bridge Life-Cycle Cost Analysis*. Transp. Res. Board, Washington, DC.

Hendrickson, C. (2008). *Cost Estimation—Costs Associated with Constructed Facilities, Project Management for Construction—Fundamental Concepts for Owners, Engineers, Architects and Builders*. http://pmbook.ce.cmu.edu/05_Cost_Estimation.html.

Hufschmidt, M. M., and Gerin, J. (1970). *Systematic Errors in Cost Estimates for Public Investment Projects. The Analysis of Public Output*, J. Margolis, Ed. Columbia University Press, New York.

IOCGP (2003). Principles and Guidelines for Social Impact Assesment in the USA, Impact Assesment and Project Appraisal, 21(3), 231–250.

Jahren, C., and Ashe, A. (1990). Predictors of Cost-Overrun Rates. *ASCE J. Constr. Eng. Mgmt.* 116(3), 548–551.

Knight, K., and Fayek, A. R. (2002). Use of Fuzzy Logic for Predicting Design Cost Overruns on Building Projects. *ASCE J. Const. Eng. Mgmt.*, 128(6), 503–512.

Kong, S-H., Labi, S., Fang, C., and Tsai, I-T. (2008). Estimating the Planning-Level Costs of Rolling Stock and Fixed Facilities for Light Rail Transit Systems. Presented at the 87th Annual Meeting of the Transportation Research Board, Washington, DC.

Lee, D. B. (1982). Net Benefits from Efficient Highway User Charges. *Transp. Res. Record*, 858, 14–20.

Litman, T. (2010). *Transportation Cost and Benefit Analysis—Noise Costs*. Victoria Transp. Policy Inst. www.vtpi.org. Accessed April 21, 2010.

McCubbin, D., and Delucchi, M. (1998). *The Annualized Social Cost of Motor-Vehicle Use in the U.S., 1990-91: Summary of Theory, Data, Methods, and Results*. Inst. of Transp. Studies, Univ. of California, Davis. UCD-ITS-RR-96-3 (1), 55.

McMillen, D. P. (2004). Airport Expansions and Property Values: The Case of Chicago O'Hare Airport. *J. Urban Econ.*, 55, 627–640.

Modra, M. (1994). *Cost-Benefit Analysis of the Application of Traffic Noise Insulation Measures to Existing Houses*. EPA, Melbourne, Australia.

NAS (2002). *Estimating the Benefits and Costs of Public Transit Projects: A Guidebook for Practitioners*, TCRP Report 78 prepared by ECONorthwest and Parsons, Brinckerhoff, Quade & Douglas, Inc., for Transp. Res. Board, Washington, DC.

NRC (2012). *Predicting Outcomes of Investments in Maintenance and Repair of Federal Facilities*. National Academies Press, Washington, DC.

OECD (1990). *Environmental Policies for Cities in the 1990s*. Organization for Economic Cooperation and Development, Paris, France.

Potts, K. (2008). *Construction Cost Management*, Taylor and Francis, London, UK.

Quan, R. (2002). Establishing Chinals Environmental Justice Study, *Georgetown Environmental Law Review* 14, 461–487.

Rawlinsons Australian Construction Handbook (2012). 30th ed., Rawlhouse Publishing, Perth, Australia.

Rowings, J. E. (2004). Construction Estimating, in *The Civil Engineering Handbook*, 2nd ed. W. F., Chen, and J. Y. Liew, CRC Press, Boca Raton, FL.

RS Means (2013). *2013 RS Means Building Construction Cost Data*, 70th edition. R. S. Means Co., Norwell, MA.

Schneck, D. C., Laver, R. S., Threadgill, G., and Mothersole, J. (1995). *The Transit Capital Cost Index Study*. Report prepared for the Federal Transit Admin. by Booz-Allen and Hamilton Inc. and DRI/McGraw-Hill, New York. www.fta.dot.gov/transit_data_info/reports.

Semple, C., Harman, F. T., and Jergeas, G. (1994). Construction Claims and Disputes: Causes and Cost/Time Overruns. *ASCE J. Const. Eng. Mgmt.*, 120(4), 785–795.

Sinha, K. C., and Labi, S. (2007). *Transportation Decision making: Principles of Project Evaluation and Programming*, Wiley, Hoboken, NJ.

Wohl, M., and Hendrickson, C. (1984). *Transportation Investment and Pricing Principles*. Wiley, New York.

World Bank (2003). *Social Analysis Sourcebook: Incorporating Social Dimensions into Bank-Supported Projects*, The World Bank, Washington, DC.

USEFUL RESOURCES

Holm, L., Schaufelberger, J. E., Griffin, D., and Cole, T. (2004). *Construction Cost Estimation*, Prentice Hall, Upper Sadde River, NJ.

Rad, P. F. (2001). *Project Estimation and Cost Management*, Management Concepts Incorporated, Tysons, Corner, VA.

Rawlingson (2013). *Rawlingsons Construction Cost Guide*, Rawlingsons Publications, West Perth, Australia.

CHAPTER 11

ECONOMIC ANALYSIS

11.0 INTRODUCTION

In the previous chapter, we discussed procedures for estimating the system owner's costs at the various phases such as construction, preservation, and operations. We also discussed how to calculate the costs incurred by and the benefits accrued (cost reductions) to users of civil systems and the costs to the community that is affected by the system. At any phase of civil systems development, there are typically several alternative decisions/actions and each alternative has its respective costs and benefits. For example, at the planning phase of railway development, it may be sought to compare the costs and benefits of alternative alignments; at the design phase of bridge design, it may be sought to compare the costs and service lives of traditional steel reinforcement, stainless steel, and fiber-reinforced polymers; at the construction phase, a system owner may seek to compare the long-term costs and benefits of different contracting approaches, such as warranty versus traditional design–bid–build; at the operations phase of urban arterial street systems, a city may seek to compare the reduction in traveler delay time (hence, cost) of alternative signal timing plans. In all these scenarios, the combined value of all the monetary costs and benefits of each alternative is termed the **economic efficiency** of that alternative. Among the alternative actions, the different timings and amounts of incoming and outgoing monies (i.e., costs and benefits) result in different levels of attractiveness of the alternatives even if they have similar levels of initial investment.

As this chapter unfolds, the usefulness of economic analysis as a decision-making tool during most civil system development phases will become more and more apparent. This realization will be based on your day-to-day experiences because the subject of economic analysis for decision making reverberates in all areas of daily living. Should you purchase a monthly parking pass or a semester parking pass or should you pay for parking only when you need it and use a parking meter? Should you lease or buy a car? Should you pay a membership fee to join a discount grocery club or should you simply buy groceries as a nonmember? Should you purchase a newer more expensive car and reduce the likelihood and amounts of annual car maintenance or should you purchase a less expensive older car that is known to incur substantial maintenance over its lifetime?

In this chapter, the basic concepts of economic analysis, such as cash flow diagrams, inflation, opportunity cost, and interest will be first introduced. A discussion will follow of various interest formulae, and we will compare different alternatives on the basis of their cost and benefit streams using different efficiency criteria for economic analysis. The chapter then examines the issues of reduced costs versus added benefit, the difference between life-cycle cost and benefit analysis and remaining service life analysis, perpetuity considerations, and incorporating uncertainty in economic evaluation of civil systems. Finally, the chapter discusses some issues (some of which are controversial) and limitations of economic analysis.

11.1 BASIC CONCEPTS IN ECONOMIC ANALYSIS

11.1.1 Cash Flow Illustrations

It is convenient to illustrate the time stream of incoming or outgoing money within a given time period using a **cash flow diagram** or **cash flow table**. The latter has two columns: one for the time of occurrence and the other for the amount incurred. On a cash flow diagram, the horizontal axis represents time and the vertical arrows represent money inflows or outflows at various time points (Figure 11.1). In cash flow figures and tables, the orientation of the arrows and the sign of the amount, respectively, indicate the direction of movement of the amount. Typically, incoming money (i.e., benefits, revenue, or deposits) are denoted using a positive sign in the cash flow table and an arrow pointing upward of the horizontal time line in the cash flow figure, and vice versa. The **analysis period**, **planning horizon**, or **planning period** is the time interval between a "present" year (presented as $time = 0$), and a "final" year (represented as $time = N$). The analysis period is typically divided into a number of equal time intervals each of which is referred to as a **compounding period** and is often taken as one year. It is conventional practice to consider all transactions occurring within the compounding period as occurring at the end of that period (e.g., for yearly intervals, any amount occurring anytime between January 1 and December 31 is assumed to occur December 31).

11.1.2 Categories and Examples of Cash Flow Diagram Amounts

Table 11.1 presents the categories of amounts typically encountered in cash flow streams at any phase of a civil system. The classifications, which are described in detail in Chapter 10, include the direction of the amount (cost versus benefit), the frequency of incurrence (one-time versus periodic versus recurring), and the party associated with the amount (owners/agency/operator versus system user versus community).

11.1.3 Inflation and Opportunity Cost

The fundamental principle of engineering economic analysis is the **time value of money**. This means that the true worth of a sum of money depends on the time (often, year) at which the value is being considered. Thus, a given amount of money today does not have the same value as the same amount at a future year; for example, \$500 today does not have the same value as \$500 in the year 2020. Why is that so? Typically, the combined forces of inflation and opportunity cost result in changing the value of money over time, and thus the value of money at any time is the net effect of these two forces. **Inflation** is the increase in the prices of goods and services over time and is manifest by a reduction in the purchasing power of a given amount of money as the

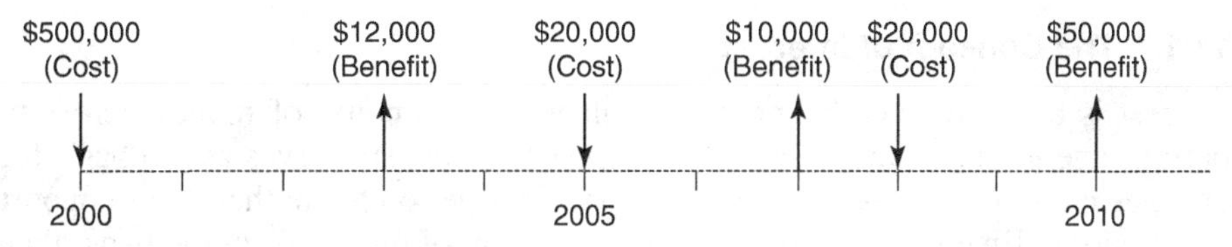

Figure 11.1 Cash flow illustration.

Table 11.1 Categories and Examples of Monetizable Benefits and Costs

		System Owner/ Agency/Operator	System User	Community
Costs	One time	System construction cost	Downtime-related costs during reconstruction	Costs of noise and dust pollution, and ecological damage during construction
	Intermittent/ periodic	System rehabilitation costs	Downtime-related costs during rehabilitation	Costs of noise and dust pollution, and ecological damage during rehabilitation
	Annual/ recurring	Costs of system routine maintenance costs	Downtime-related costs during routine maintenance	Costs of noise and dust pollution, and ecological damage during routine maintenance and system operations.
Benefits	One time	One-time receipt of lease amount from system lessee	—	—
	Intermittent/ periodic	Periodic receipt of lease amounts from system lessee	—	—
	Annual/ recurring	Toll/fare/fee receipts from system users	Reduction in system users' costs of safety, convenience, etc.	Reduction in community costs, monetizeable benefits associated with economic development fostered by the system

years go by. **Opportunity cost** is the income one sacrifices because one decided to spend money on, for example, a civil engineering system, instead of investing it elsewhere and earning returns from that investment. Opportunity cost is strongly related to the interest rate (which is used in financial analysis) and the discount rate (which is used in economic analysis) as we shall discuss shortly in Section 11.1.4. Inflation and opportunity cost each influence the time value of money. Prior to economic analysis, costs are adjusted for inflation so that the amounts used in the analysis are *constant dollars* (as of a certain base year) rather than *current dollars* (dollar value as of the year the amount was received or spent).

11.1.4 The Concept of Interest

Interest is the difference between the values of an amount of money across two different time periods due to opportunity cost. The interest rate concept serves as the basis for determining the changes in the value of an amount from one time period to another due to opportunity cost. Thus, at the current time if one borrows a certain amount of money from the bank, then at a future time, one would owe an amount that is equal to the initial amount plus interest; this is because the value of the amount at the time of payback is not the same as the value of the initial amount. Typically, commercial lending institutions also add a small margin for profit. Interest has also been described as the *price of money*, or the *price of borrowing*.

The **interest rate** is the ratio of the interest paid at the end of the compounding period to the amount incurred at the "present time," that is, the beginning of the compounding period. For example, a simple interest rate of 10% means that, for every dollar borrowed at the initial year, 10 cents must be paid as interest at the end of each year. Due to the profound effect of interest rates on the economy of a nation, interest rates are controlled by central banks in order to remedy current or expected economic problems. For example, in a recession economy, the U.S. Federal Reserve Board typically decreases the interest rate to discourage saving and encourage spending by individuals, businesses, and public agencies; in an overheated economy, however, the central bank increases the interest rate. Also, in stable economic conditions, such as those of developed nations, interest rates are typically lower compared to unstable economies characterized by high inflation and uncertainty of investment returns. This difference is often reflected in the minimum (or marginal) attractive (or acceptable) rate of return (MARR), the least interest rate that a system owner is willing to accept before undertaking an investment. MARR is a function of the perceived risk associated with the investment and the opportunity cost of other options. MARR can range between 5 and 8% in developed nations and 10–15% in developing nations.

(a) Difference between Interest Rate and Discount Rate. There is a subtle difference between the interest rate and the discount rate. The interest rate of money at a certain time is the rate of return that is paid for money in the open money market at that time. Thus, it may be the rate that banks charge their borrowers or the rate that they pay to their depositors. For economic analysis, for example, if the systems engineer is evaluating the feasibility of an action or the relative attractiveness among multiple actions, then the discount rate to be used is the same as the system owner's opportunity cost for capital. On the other hand, for financial analysis, if the systems engineer is determining the annual repayments to amortize a loan, for example, then the interest rate should be used for the analysis.

(b) Types of Interest Rates. Figure 11.2 presents the various types of interest rates. The interest rate could be simple or compound, discrete, continuous, and fixed or variable. In the case of simple interest, the amount of interest at the end of each time interval is the same and is a percentage of the **principal** or initial amount. In the case of compound interest, the amount of interest at the end

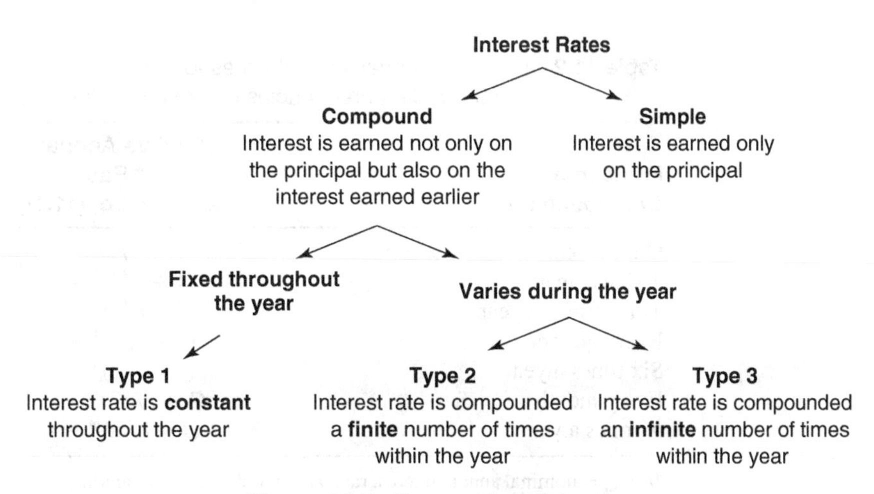

Figure 11.2 Types of interest rates.

of each time interval is a percentage of the total amount owed at the end of the previous period (i.e., the sum of the principal and the previous period's amount of interest) and not the principal. Therefore, amounts borrowed on compound interest involve higher payments for amortization. In engineering economic analysis, the compound interest rate, rather than the simple interest rate is what is used.

A fixed periodic interest rate is used when interest is computed only at the end of each compounding period and with a constant interest rate; the compounding period is one period for such cases. For example, a fixed **annual** rate refers to the period of compounding as one year; in most cases, the compounding period is less than one year, such as weekly, monthly, or quarterly. Interest rates are typically quoted on an annual basis and such a quote is accompanied by the length of the applicable compounding period if it is different from one year. For example, if the interest rate is 6% per interest period and the interest period is 6 months, this rate is often stated as a "12% interest rate compounded semiannually." In such cases, the annual rate of interest of 12% is referred to as the nominal interest rate (denoted as i_n or r); however, the actual annual rate on the principal is not 12% but rather something greater because compounding occurs twice during the year. The actual rate of interest earned on the principal during one year is known as the **effective interest rate**, which is calculated by incorporating the effect due to the number of compounding periods. Effective interest rates are typically expressed on an annual basis. In Table 11.2, we can see how the effective annual interest rate is calculated for different situations regarding how often the compounding is carried out within a year. For interest rate compounding type 2, the interest rate, i, is not fixed but rather is compounded a fixed number of times, m, during each year. In such cases, the effective annual interest rate to be used for the analysis is as shown in Equation (11.1) and Table 11.2. The nominal interest rate is denoted by r or i_n.

$$i_e = \left(1 + \frac{i_n}{m}\right)^m - 1 \tag{11.1}$$

For interest rate compounding type 3 where the interest rate, i, is compounded an infinite number of times during the year, and the effective annual interest rate for the analysis is provided as Equation (11.2).

$$i_e = e^{i_n} - 1 \tag{11.2}$$

Table 11.2 Effective Annual Interest Rates for Various Compounding Frequencies (Interest Rate Type 2)

Frequency of Compounding	**Effective Annual Interest Rate[a] ie [from Eq. (11.1)]**
Once a year	r
Twice a year	$(1 + r/2)^2 - 1$
Three times a year	$(1 + r/3)^3 - 1$
Every quarter	$(1 + r/4)^4 - 1$
Six times a year	$(1 + r/6)^6 - 1$
Every month	$(1 + r/12)^{12} - 1$
m times a year	$(1 + r/m)^m - 1$

[a] r or i_n = nominal annual interest rate; m = number of compounding periods per year.

(c) Choice of Discount Rate. The opportunity cost of a sum of money is the returns lost by diverting the money to another purpose such as a civil systems project. Thus, from a theoretical standpoint, the appropriate discount rate to use for evaluating civil systems investments is the opportunity cost of capital (de Neufville and Stafford, 1971). As most civil systems are public assets, there is lack of competition in their development. For this reason, and also because taxation rates differ for different economic activities, the discount rate is not the same for all situations. The discount rate to be used for analysis can be determined using one of at least two ways: as the risk-free interest rate or the social discount rate (Zhuang et al., 2007).

Discount Rate as the Risk-Free Interest Rate. The basic lower bound value for the discount rate is the risk-free constant dollar interest rates that are available on the market. In other words, an investment is economically feasible when it yields returns that are at least equal to the commonly available investments (e.g., long-term government bonds). The effect of inflation must be considered, and thus the least value of the discount rate (DR) is the risk-free interest rate less the inflation rate:

$$\mathrm{DR_{MIN}} = i_{\mathrm{GB}} - f \tag{11.3}$$

where $\mathrm{DR_{MIN}}$ is the minimum discount rate, i_{GB} is the rate of return from available investments (e.g., government bond) and f is the inflation rate.

Example 11.1

In a certain country, the rate of return on long-term government bonds is 8%. The annual rate of inflation is 4%. If these conditions are expected to remain constant in the long term, what is the minimum discount rate to be used in evaluating the economic feasibility of public infrastructure systems in that country?

Solution

$$\text{Minimum discount rate} = i_{\mathrm{GB}} - f = 8\% - 4\% = 4\%$$

Discount Rate as the Social Discount Rate. Most civil systems are public projects, and thus their evaluation must use a social discount rate. The capital used for such public projects are derived from the private sector where such funds could have been (i) invested or (ii) spent on consumer goods. Thus, to establish the opportunity cost of capital spent on public projects, it is important to determine the appropriate rates of return for private sector investments and also for the money spent on consumer goods. Then, the social discount rate will be the opportunity cost that is some composite function of the capital obtained from these two sources:

$$i_{\mathrm{SD}} = f(O_{\mathrm{PI}}, O_{\mathrm{CS}})$$

Opportunity cost of capital diverted from consumption, O_{CS}. This can be measured as the risk-free long-term interest rate, that is, the current rate of return on government bonds. For individuals who voluntarily purchase bonds instead of consuming the money, the bond's rate of return is perceived to be superior to the time value of the satisfaction they would have derived from the foregone goods or purchases. Conversely, individuals who choose consumption over bond purchases are implicitly stating that the bond rate of return is lower than their time value of the consumption. Thus, the opportunity cost of money taken from consumers, O_{PI}, is equal to or exceeds the bond rate:

$$O_{\mathrm{CS}} \geq i_{\mathrm{GB}}$$

Opportunity cost of capital taken from the private sector, O_{PI}. For a business investment to be attractive to the private sector, its rate of return after corporate taxes must exceed the rates of return that would have been earned from alternative investments such as government bonds. The after-tax rate of return is equal to the product of the real, before-tax rate of return and $(1 - tr)$ where tr is the tax rate. Thus, the opportunity cost of private investment, O_{PI}, is given by

$$O_{PI} \geq i_{GB}/(1 - tr)$$

Along these lines of reasoning, de Neufville and Stafford (1971) offered a practical method for estimating the social discount rate, or the effective opportunity cost of money for public projects, as a simple proportion of the opportunity cost of capital diverted from consumption and the opportunity cost of capital taken from the private sector as follows:

$$i_{SD} = Ai_{GB} + \frac{(1 - A)i_{GB}}{1 - tr} \qquad 0 \geq A \geq 1 \tag{11.4}$$

where A is the proportion of investments being drawn from the consumers; i_{GB} is the interest rate on government bonds; and tr is the tax rate.

Example 11.2

If the current tax rate is 35%, the rate of return of government bonds is 4%, and 30% of investments are drawn from consumers, calculate the social discount rate. Discuss the policy implications of Equation (11.4).

Solution

Using Equation 11.4, the social discount rate is calculated as 5.5%; when 70% of investments are drawn from consumers, the social discount rate is calculated as 4.6%. It can be seen from the equation that a higher tax rate, tr, and a higher percentage of investments drawn from the consumers, A, leads to a lower social discount rate, which favors greater likelihood of economic feasibility for public projects, particularly those with a high initial cost and user benefits that are spread out over long service lives. Thus, planners of civil systems may have a proclivity to support policies that increase the tax rate or draw investments from consumption, or both (such as higher taxes on consumer spending). In Section 11.8 we further discuss the social discount rate in light of ongoing issues such as climate change and sustainability.

11.1.5 Residual Value of Systems

The residual value of a system or its component refers to the value at the end of a specified analysis period, often the design or service life. The two fundamental components of residual value are remaining service life and salvage value. The remaining service life is considered when the analysis period is reached before system's end of life, and the salvage value is considered when the system end of life is reached before analysis period.

(a) Remaining Service Life. In some problem contexts in economic analysis, the analysis period is shorter than the service life. In such cases, the system has some left-over years of service at the end of the analysis period. This residual value is considered as a "benefit" or negative cost in economic evaluation. Remaining service life (RSL) is typically computed as the prorated value of the facility at the end of the analysis period. Failure to account for different RSL values across alternatives can result in bias in the analysis.

(b) Salvage Value. In cases where the system reaches the end of its life within the analysis period, there could be still be some benefit, in terms of the reuse or recycling of the system or its component materials. On the other hand, it may be the case that the system owner incurs a net cost in the system retirement, including salvaging or disposing of the system or its components. Salvage value is the value of the recovered or recycled materials and its consideration often accompanies the assumption that the civil system (or its constituent materials) is removed from service at the end of the analysis period. A difference between the RSL and salvage values is that the RSL value requires continuing operation of the system at the end of the analysis period, whereas the salvage value results from the end of life of the physical component of the system.

11.2 INTEREST FORMULAS (OR EQUIVALENCE EQUATIONS)

11.2.1 Prelude

Interest formulas, also referred to as **equivalence equations**, are the relationships between amounts of money that occur at different points in time and are necessary to transform amounts or series of amounts of money from one time period to another taking due cognizance of the time value of money. A vital aspect of these relationships is the **interest factor**, which is a function of the interest rate and the payment period. Interest factors are expressed as formulas (see Tables 11.2–11.4) or as values derived from formulas. Interest equations typically involve the following five key variables: P is the initial amount; F is the amount at a specified future period; a is the periodic (typically yearly) amount; i is the effective interest rate for the compounding period; and N is the number of compounding periods (or the **analysis period**).

Initial Amount, P: This refers to the money received or incurred at the "initial" period, that is, in year 0 of at the 0th year of a cash flow situation. The initial year may or may not be the current year. Typically, the initial amount is the amount borrowed or invested in a civil system at the initial period and is often a large amount, such as the initial construction cost. The initial amount may be a cost or a benefit and is often associated with the system owner rather than the system user.

Future Amount, F: The future amount refers to a payment or amount incurred (or received) at the end of an analysis period or anytime within the analysis period. A single future amount may be a cost or a benefit and is often associated with the system owner rather than the system user. Examples include the cost of major maintenance carried out periodically.

Periodic Amount, A: This refers to the amount received or incurred every periodic instance within the analysis period. For the owner/operator, examples of periodic amounts are the operating costs such as salaries and annual routine maintenance. Typically, these periodic amounts are a uniform series of amounts, but there are cases where they are either changing (increasing or decreasing) in a systematic manner or fluctuating irregularly. In the latter case, the individual annual amounts are each considered discounted amounts, F, occurring at their respective years. In the former case, the change is often an increase rather than a decrease because with time, systems experience increased demand. Also, for civil systems built using loaned funds or planned to be built using a sinking fund, the periodic amount could represent the amount paid every year either to amortize a loan or to set up the sinking fund to achieve a target amount after a number of years. Periodic amounts are associated with the system owner, user, and the community.

Interest Rate, i: The interest rate, i, is the effective annual interest rate for each period. In Section 11.1, we discussed the interest rate and how it is established.

Analysis Period, N: This refers to the number of compounding periods for or the analysis, that is, the total time over which the economic evaluation is being conducted. Synonyms include planning horizon, planning period, or payment period. Typically, the analysis period is expressed in years. The length of the analysis period depends on the system type and the expected life of the system. Examples of facility lives in civil systems are 35 years for highway rigid pavements, 50 years for reinforced concrete bridges, 70 years for steel bridges, and 100 years for some dams.

These values are rough estimates and actual facility lives may be different from these, depending on the location, system type, environment, and other system attributes. The analysis period may be the overall life of a system or only a part thereof. In the former case, the analysis is referred to as a full life-cycle analysis; and in the latter case, it is referred to as a remaining life analysis. Section 26.4.2 discusses estimation of the life of a civil system.

Appendix 3 presents the formulas, cash flow diagram, and notation for various cases of interest equivalencies and relationships for discrete compounding and continuous compounding of the interest rate.

11.2.2 Illustrations Involving Different Cases of Cash Flow

In this section, different cases of cash flow are illustrated, each one with two economic efficiency variables provided in the question, and we need to determine the value of the third variable.

Case 1: Finding the Future Amount (*F*) to Be Yielded by an Initial Amount (*P*) at the End of a Given Period.

Example 11.3

A city transit agency seeks to replace its ticketing system. The current price is \$20,000. The agency reached an agreement with a vendor to install the new system now (December 2014) and to pay nothing until December 2019 when the agency pays the entire amount in one payment. What price will the agency pay in 2019? Assume 5% interest rate and type 1 compounding of the interest rate.

Solution

This is a single-payment compound amount factor (SPCAF) problem. In other words, we seek F given P. The cash flow diagram (Figure 11.3) and computational formula are as follows:

$$F = P[(1+i)^n] = 20{,}000[(1+0.05)^5] = \$25{,}526$$

Therefore, the firm pays \$25,526 in 2019.

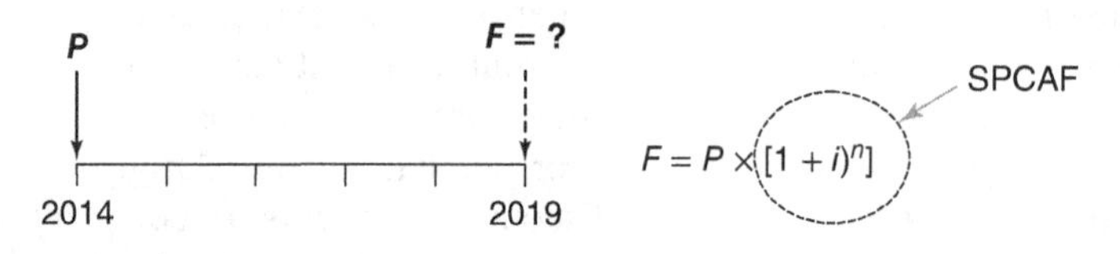

Figure 11.3

Case 2: Finding the Initial Amount (*P*) That Would Yield a Future Amount (*F*) at the End of a Given Period.

Example 11.4

This calculation is common with budgeting and planning tasks in civil systems management and also in managing revenue-generating civil systems. For example, engineers at a water supply plant in Kumasi have determined that 5 years from now, that is, in 2019, an aging aeration unit will reach the end of its service life and will need to be replaced at an estimated cost of \$50,000 at that year. How much should the plant put away now in order to be able to pay for the unit's replacement in 2019? Assume 5% interest rate and type 1 compounding of the interest rate.

Solution

This is a single-payment present worth factor (SPPWF) problem. In other words, we seek P given F. The cash flow diagram (Figure 11.4) and computational formula are as follows:

$$P = \frac{F}{(1+i)^n} = \frac{50{,}000}{(1+0.05)^5} = \$39{,}176$$

Therefore, the plant's owner should set aside \$39,176 now (December 2014) in order to be able to replace the aeration unit 5 years from now in 2019.

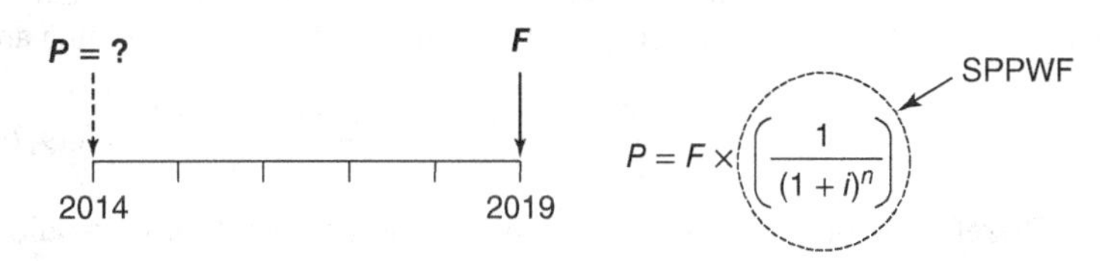

Figure 11.4

Case 3: Finding the Amount of Uniform Annual Payments (A) That Would Yield a Certain Future Amount (F) at the End of a Given Period.

Example 11.5

Assume in Case 2 above that the plant engineers instead plan to pay a uniform amount every year to replace the aeration unit when it is due for replacement in 2019 at a cost of \$50,000 at that year. The plant agrees with the vendor to make five annual payments starting December 2015 until 2019. How much should the plant pay every year? Assume 5% interest rate and type 1 compounding of the interest rate.

Solution

This is a uniform series sinking fund deposit factor (USSFDF) problem. In other words, we seek A given F. The cash flow diagram (Figure 11.5) and computational formula are as follows:

$$A = F \times \frac{i}{(1+i)^n - 1} = 50{,}000 \times \frac{0.05}{(1+0.05)^5 - 1} = \$9048.74$$

Therefore, the plant's owner should pay \$9048.74 every year in order to be able to replace the aeration unit in 2019.

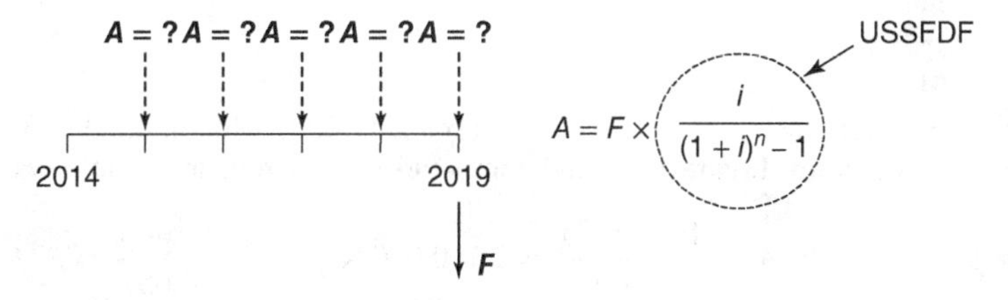

Figure 11.5

Case 4: Finding the Final Compounded Amount (F) at the End of a Given Period due to Uniform Annual Payments (A).

Example 11.6

A city engineer's office has in place a revenue generation scheme that provides $9000 per year for street lighting projects. The engineer's office seeks to ascertain the total amount generated starting 2015 until 2019 when the street lights are due for replacement. By 2019, how much would the engineer's office accumulate for this purpose? Assume 5% interest rate and type 1 compounding of the interest rate.

Solution

This is a uniform series compounded amount factor (USCAF) problem. In other words, we seek F given A. The cash flow diagram (Figure 11.6) and computational formula are as follows:

$$F = A \times \frac{(1+i)^n - 1}{i} = 9000 \times \frac{(1+0.05)^5 - 1}{0.05} = \$49{,}730$$

Therefore, at the end of the analysis period, the engineer's office would have accumulated $49,730.

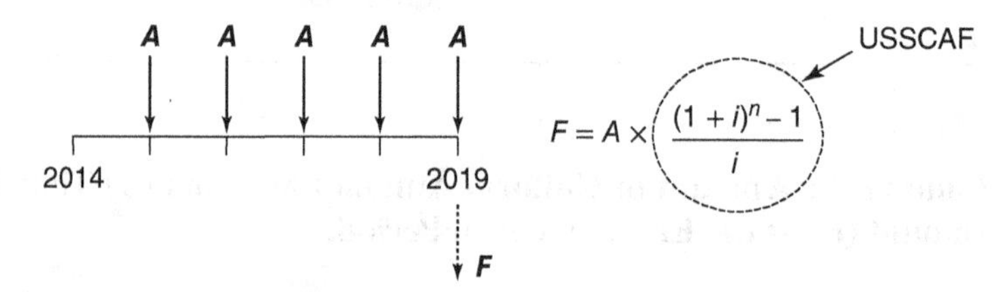

Figure 11.6

Case 5: Finding the Initial Amount (P) That Would Yield Specified Uniform Future Amounts (A) over a Given Period.

Example 11.7

In certain cases of systems management, the owner seeks to know the amount at a certain "present" year that is equivalent to a series of uniform annual amounts over a given analysis period. For example, the managers of the storm drainage system in the City of Espoo seek to estimate how much to set aside at the current time (say, year 2014), that would be equivalent to $2.1 million annual amounts for preventive maintenance of its field facilities over a 5-year period. Assume 5% interest rate and type 1 compounding of the interest rate.

Solution

This is a uniform series present worth factor (USPWF) problem. In other words, we seek P given A. The cash flow diagram (Figure 11.7) and computational formula are as follows:

$$P = A \times \frac{(1+i)^n - 1}{i(1+i)^n} = 2{,}100{,}000 \times \frac{(1+0.05)^5 - 1}{0.05(1+0.05)^5} = \$9{,}091{,}901$$

Therefore, $9,091,901 needs to be procured at the current time to yield the required annual maintenance amount over the remaining service life of the facilities.

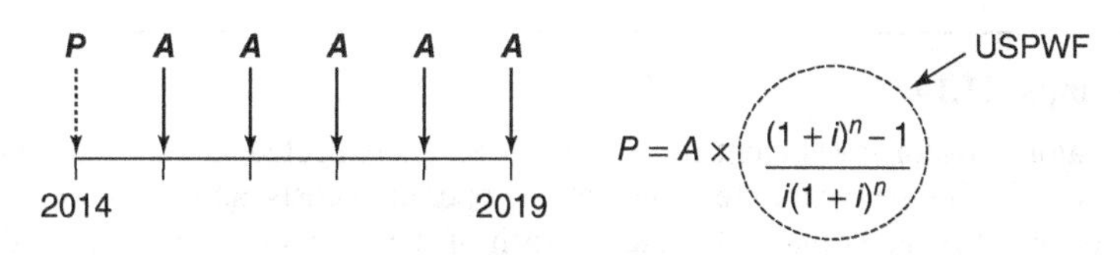

Figure 11.7

Case 6: Finding the Amount of Uniform Annual Payments (*A*) over a Given Period That Would Completely Recover an Initial Amount (*P*), such as a Loan.

Example 11.8

A building engineer in Shanghai seeks to replace aging air-conditioning units that were installed in the early 1970s. In December 2014, the building owner receives a loan of $200,000 to carry out the project, which is to be repaid over a 5-year period. How much will the owner need to pay back to the bank every year, starting December 2015 until December 2019? Assume 5% interest rate and type 1 compounding of the interest rate.

Solution

This is a uniform series capital recovery factor (USCRF) problem. In other words, we seek *A* given *P*. The cash flow diagram (Figure 11.8) and computational formula are as follows:

$$A = P \times \frac{i(1+i)^n}{(1+i)^n - 1} = 200{,}000 \times \frac{0.05(1+0.05)^n}{(1+0.05)^5 - 1} = \$46{,}195$$

Therefore, the building owner pays back $46,195 every year starting December 2015 until December 2019.

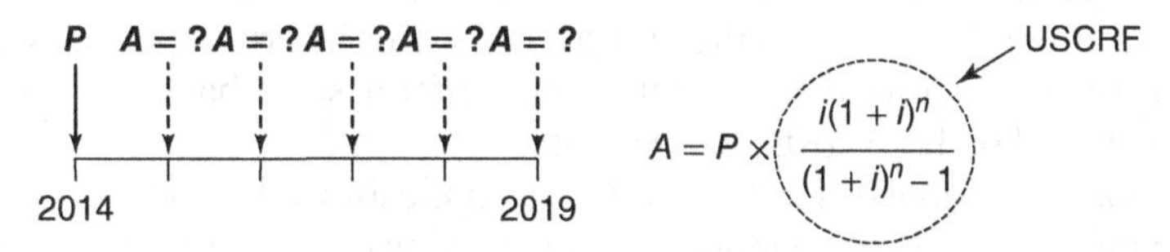

Figure 11.8

11.2.3 Additional Examples

Example 11.9

Five years from now, a water supply department in the city of Darwin intends to rehabilitate its treatment units at a cost of $1.5 million. Ten years from now, the units will be replaced at a cost of $5.8 million. Assuming an interest rate of 8%, find the present worth of each of these costs.

Solution

Note Figure 11.9.

$$\text{PW} = 1.5\text{M} \times \text{SPPWF}\,(8\%, 5) + 5.8\text{M} \times \text{SPPWF}\,(8\%, 10) = \$3.71\text{M}$$

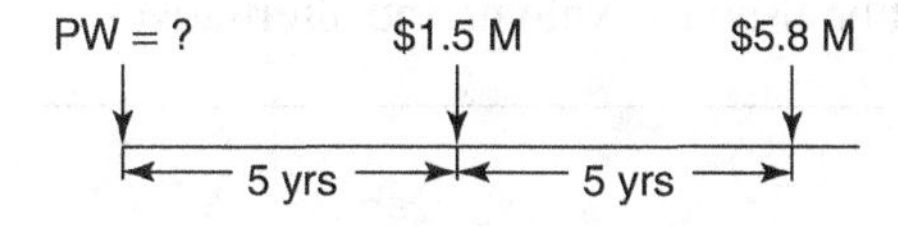

Figure 11.9

Example 11.10

A major corridor investment in the city of Rio is expected to yield \$50,000 per year in reduced crash costs, \$200,000 per year in reduced vehicle operating costs, and \$405,000 per year in reduced travel time costs. What is the combined present worth of these benefits? Assume interest rate is 5% and analysis period is 20 years.

Solution

$$\text{PW} = (50{,}000 + 200{,}000 + 405{,}000) \times \text{USPWF}\ (5\%, 20) = 8.16\text{M}$$

11.2.4 When Annual Amounts Are Not Uniform

When annual amounts are not uniform, but increasing or decreasing in some systematic manner, it is still possible to find their present worth, the equivalent compounded amount, F, at the end of some analysis period, or equivalent uniform annual amounts. Sinha and Labi (2007) provide equations for doing this.

11.3 CRITERIA FOR ECONOMIC ANALYSIS

11.3.1 Introduction

So far in this chapter, we have learned how to bring to present worth or to annualize the present and future amounts in a cash stream. These amounts refer to the benefits and costs of a civil engineering system (in terms of the money spent in construction, maintenance and operations, and benefits in terms of possible revenue and reductions in accidents and inconvenience, and other system performance measures). The question then is how to use this information to assess the economic efficiency of an investment. For this, there are numerous criteria, some based on benefits, some based on costs, and others based on both costs and benefits.

Cost-based Criteria: These criteria are applicable only when all of the alternatives are associated with the same level of benefits, and cost minimization therefore is the sole evaluation criterion: equivalent uniform annual cost (EUAC) and present worth of costs (PWC).

Benefit-based Criteria: These criteria are applicable only when all of the alternatives have the same total cost, and benefit maximization therefore is the sole evaluation criterion: equivalent uniform annual benefit (EUAB) and present worth of benefits (PWB).

Cost and Benefit-based Criteria: These criteria are applicable only when the alternatives have different levels of benefits and costs: equivalent uniform annual return (EUAR), net present value (NPV), internal rate of return (IRR), benefit–cost ratio (BCR), and the payback period.

11.3.2 Present Worth of Costs

In this method, we represent all the costs over the analysis period by a single hypothetical cost that is assumed to occur at the beginning of the analysis period (the 0th year). The method is used in evaluation problems where the benefits of all alternatives are equal (or assumed to be equal) and therefore only their costs are used to identify the best alternative. Another condition is that the same analysis period must be used to evaluate the alternatives.

Example 11.11

A plane purchase is proposed by an airline. For plane type A, the initial purchase cost is \$50M, the average annual cost of maintenance is \$0.25M, and the salvage value is \$8M; for plane type B, the initial cost is \$30M, the average annual maintenance cost is \$0.75M, and the salvage value is \$2M. Both plane types have a useful life of 15 years. At 7% interest rate, identify which alternative should be selected.

Solution

$$\text{PWC}_{\text{A}} \text{ (in millions)} = 50 + 0.25 \times \text{USPWF}(7\%, 15) - 8 \times \text{SPPWF}(7\%, 15) = \$49.38\text{M}$$

$$\text{PWC}_{\text{B}} \text{ (in millions)} = 30 + 0.75 \times \text{USPWF}(7\%, 15) - 2 \times \text{SPPWF}(7\%, 15) = \$36.11\text{M}$$

Therefore, alternative B is more economically efficient.

11.3.3 Equivalent Uniform Annual Cost (EUAC)

This method combines all costs of a civil system project into an equivalent annual cost over the analysis period. This method is useful when the alternatives have same or different service lives but the same level of effectiveness.

Example 11.12

A city is considering the purchase a fleet of buses for its transit services. Bus types 1 and 2, which provide similar levels of performance, are being considered. The city has decided to restrict the analysis period to the "rest period," that is, only that part of the service life where no major maintenance is expected. The initial cost of type 1 has an initial cost of \$100,000, an estimated rest period of 6 years, annual maintenance and operating costs of \$8000, and a \$20,000 salvage value. Type 2 has an initial cost of \$75,000, a 5-year rest period, annual maintenance and operating costs of \$12,000, and salvage value of \$10,000. Find the equivalent annual cost of each alternative and decide which option is more desirable. Assume a 6% interest rate.

Solution

$$\text{EUAC}_{\text{A}} \text{ (in thousands)} = 100 \times \text{CRF}(6\%, 6) - 20 \times \text{SFDF}(6\%, 6) = \$25.47$$

$$\text{EUAC}_{\text{B}} \text{ (in thousands)} = 75 \times \text{CRF}(6\%, 5) - 10 \times \text{SFDF}(6\%, 5) = \$28.03$$

Thus, alternative A is more desirable.

11.3.4 Equivalent Uniform Annual Return (EUAR)

This method converts all of the costs and benefits or returns over the analysis period into a single annual return (benefits less costs) and is particularly useful where the analysis periods are different across the alternatives.

Example 11.13

Ports authority officials are considering two alternative designs for renovating a water port in Takoradi. The first alternative has an initial project cost of \$200M, a design life of 25 years, a salvage value of \$22M, annual operating and maintenance costs of \$80M, and \$175M worth of annual benefits in terms of monetized savings in processing time, safety and security, and vessel operations. The second alternative has an initial project cost of \$175M, an estimated life of 20 years, annual operating and maintenance costs of \$90M, a salvage value of \$15M, and \$155M worth of annual benefits. Which design should be selected? Assume a 4% interest rate.

Solution

$$\text{EUAR}_A \text{ (in millions)} = 175 - 200 \times \text{CRF}(4\%, 25) - 80 + 22 \times \text{SFDF}(4\%, 25) = \$82.73\text{M}$$

$$\text{EUAR}_B \text{ (in millions)} = 155 - 175 \times \text{CRF}(4\%, 20) - 90 + 15 \times \text{SFDF}(4\%, 20) = \$52.63\text{M}$$

Alternative A is more desirable.

11.3.5 Net Present Value (NPV)

The NPV of a cash flow stream over a period of time is the difference between the present worth of all benefits and the present worth of all costs. This reflects the value of the investment at the initial or base year of the analysis. NPV is often considered to be the most revealing of all economic efficiency criteria because it provides a magnitude of net benefits in monetary terms. Of the competing alternatives, that with the highest NPV is considered the most "economically efficient" alternative.

Example 11.14

For the problem in Example 11.3, determine the NPV for each alternative.

Solution

$$\text{NPV}_A \text{ (in millions)} = 75 \times \text{USPWF}(4\%, 25) - 200 - 80 \times \text{USPWF}(4\%, 25) + 22 \times \text{SPPWF}(4\%, 25)$$
$$= \$1{,}292.35\text{M};$$

$$\text{NPV}_B \text{ (in millions)} = 55 \times \text{USPWF}(4\%, 20) - 175 - 90 \times \text{USPWF}(4\%, 20) + 15 \times \text{SPPWF}(4\%, 20)$$
$$= \$715.22\text{M}$$

Alternative A is more desirable.

11.3.6 Internal Rate of Return (IRR)

Prior to investing money in developing a new system or expanding or rehabilitating an existing system, owners of civil engineering facilities (or banks that lend money for the development of such systems) seek to ascertain whether their investment will yield a net rate of return that exceeds some minimum acceptable rate, or whether it will yield a net profit before a specified payback period. Due to the fact that there always exists investment risks or that there is an opportunity to invest elsewhere for possibly greater returns, there is a certain minimum attractive rate of return (MARR) that the investor is willing to accept before he proceeds with the investment. MARR is inversely related to the payback period (generally defined as the time taken for an investment to yield the initial investment). Unlike the payback period, MARR directly takes into account the change in value of the investment over time.

Vestcharge is the economical rate of return or the interest rate at which the net present worth or EUAR is equal to zero. The internal rate of return (IRR) is calculated as the interest rate at which NPV or EUAR is zero; the IRR is then compared to the MARR; if IRR exceeds MARR, then the investment is considered worthwhile.

Example 11.15

An urban public transportation corporation seeks to purchase a new $30,000 ticketing system intended to reduce passenger out-of-vehicle travel time. The system life is 10 years, after which time the value of the system will be $15,000. It is expected that the total savings in travel time will be $5000 annually. The average operating and maintenance costs are $2000 annually. Determine whether the intended investment is economically feasible. The minimum attractive rate of return is 5%.

Solution

Equating the incoming and outgoing cash flows yields:

$$5000 \times \text{USPWF}(i\%, 10) + 15{,}000 \times \text{SPPWF}(i\%, 10)$$
$$= 30{,}000 + 2000 \times \text{USPWF}(i\%, 10)$$

This equation is solved using trial and error or tools such as MS Excel Solver or Matlab to yield $i = 6.25\% > 5\%$. Thus, it is economically more efficient to undertake the project than to do nothing.

11.3.7 Benefit–Cost Ratio Method

The benefit–cost ratio (BCR) is a ratio of the net present benefits to the net present costs incurred over the analysis period. Investments with BCR exceeding 1 are considered to be economically feasible, and the alternative with the highest BCR value is considered the most superior alternative. The U.S. Flood Control Act of 1936 is one of the early instances where the concept of BCR was mentioned as a criterion for evaluating public projects. In fact, up until the 1980s, BCR remained the most popular criterion for project evaluation in the United States.

Example 11.16

For the problem in Example 11.13, use the BCR criterion to identify the superior alternative for each alternative.

Solution

The benefit–cost ratios of each alternative are calculated as follows:

$$\text{BCR}_{\text{A}} = \frac{\text{PWB}_{\text{A}}}{\text{PWC}_{\text{A}}} = \frac{175 \times \text{USPWF}(4\%, 25)}{200 + 80 \times \text{USPWF}(4\%, 25) - 22 \times \text{SPPWF}(4\%, 25)} = 1.90$$

Therefore alternative A is economically more efficient.

$$\text{BCR}_{\text{B}} = \frac{\text{PWB}_{\text{B}}}{\text{PWC}_{\text{B}}} = \frac{155 \times \text{USPWF}(4\%, 20)}{175 + 90 \times \text{USPWF}(4\%, 20) - 15 \times \text{SPPWF}(4\%, 20)} = 1.51$$

The BCR can be considered a useful evaluation criterion, or at least as a preliminary screening measure for small and simple projects. However, as de Neufville and Stafford (1971) and Sinha and Labi (2007) pointed out, it has a number of theoretical flaws, particularly when it is applied to complex systems. In Section 11.4.1, we discuss some of these shortcomings. The incremental benefit–cost ratio (discussed in the next section) overcomes some of the limitations of the BCR.

11.3.8 Incremental Benefit–Cost Ratio Method (IBCR)

This is the second ratio-based method of evaluation and is similar to the BCR method except that incremental attributes are used relative to a base case (often the do-nothing) scenario. The incremental benefit–cost ratio (IBCR) method is applicable if there are at least two alternative projects including the base case. IBCR is also referred to as the **defender–challenger** method of project or investment evaluation. B_i and B_b = total discounted benefits of alternative i and base case, respectively; C_i and C_b = total discounted costs of alternative i and base case, respectively. The steps are as follows:

Step 1. Calculate the benefits and costs of each alternative including the base case.

Step 2. Select the base case as the defender, d.

Step 3. Select the alternative with the least value of total discounted costs, C_i as the challenger, h.

Step 4. Calculate the incremental BCR to compare the defender and challenger:

$$\text{IBCR} = \frac{B_d - B_h}{C_d - C_h} \tag{11.5}$$

Step 5. If the incremental BCR exceeds 1, the challenger becomes the new defender. Otherwise, the defender remains. In either case, select the next alternative with the lowest value of C_i as the new challenger.

Step 6. Repeat steps 2–5 to compare the challenger to the defender until all alternatives have been considered. The remaining defender is chosen as the most preferred alternative.

11.3.9 Payback Period

The payback period is defined as the length of time taken for an investment to generate adequate net benefits to cover the initial cost of the investment, in other words, the time required for cumulative benefits to equal the cumulative costs of an investment. This can be calculated either as the *undiscounted payback period* or the *discounted payback period* (the latter takes the time value of money into account). Investments with short payback periods are generally preferred to those with long payback periods.

Example 11.17

Consider the stream of benefits and costs (in $millions on constant dollar) of an investment as shown in Table 11.3.

Table 11.3

	Year 0	Year 1	Year 2	Year 3	Year 4	Year 5	Year 6
Cash Benefits	0	+20	+20	+20	+20	+20	+20
Cash Costs	−50	−5	−5	−5	−5	−5	−5

Solution

As can be seen from Table 11.4 and the plot in Figure 11.10, the payback period is approximately 3.3 years.

Table 11.4

	Year 0	Year 1	Year 2	Year 3	Year 4	Year 5	Year 6
Net Cash Flow	−50	+15	+15	+15	+15	+15	+15
Cumulative Net Cash Flow	−50	−35	−20	−5	10	25	40

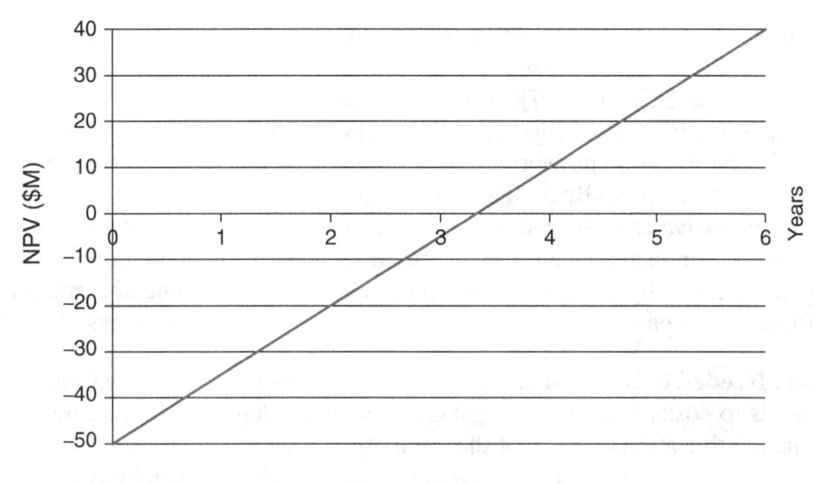

Figure 11.10

11.4 DISCUSSION OF RELATIVE MERITS OF THE ECONOMIC EFFICIENCY CRITERIA

11.4.1 Benefit–Cost Ratio

The BCR is a simple criterion for assessing the economic efficiency of civil systems investments. However, it is subject to a number of limitations (Maas, 1966; de Neufville and Stafford, 1971; Campden, 1986; DeGarmo et al., 1997) as discussed below.

(a) Failure to Account for Inequities. In practice, the BCR fails to account for distributional inequities. In other words, it does not help address situations where one party bears a disproportionate percentage of the project cost while another bears a disproportionate percentage of the benefits. In concept, the benefits aspect of the BCR can be reformulated to accommodate the relative weights of costs borne by the different groups of stakeholders. The same could be applied to the benefits they accrue, as presented in the expression below.

$$\text{BRC} = \frac{f_B(B_{S1}, B_{S2}, \ldots, B_{SM})}{f_C(C_{S1}, C_{S2}, \ldots, C_{SM})} \tag{11.6}$$

where B_{Si} is the benefit of the project to a stakeholder or population demographic segment i; C_{Si} is the cost of the project incurred by stakeholder i; f_B and f_C are the functions of the benefits and costs, respectively.

A case in point is the study by Irfan et al. (2012), which used the following function to assess the economic efficiency of alternative timings for highway preservation:

$$\text{BRC} = \frac{w_1(\text{ISL}) + w_2(\text{ICN})}{v_1(\text{AC}) + V_2(\text{UC})} \tag{11.7}$$

where ISL and ICN are the increase in system life (representing agency preferences or perspectives) and the increase in system condition (representing user preferences or perspectives), respectively; w_1 and w_2 are the relative weights between the agency and the user preferences; UC represents the user cost; AC represents the agency cost; and v_1 and v_2 are the relative weights between the agency and the user costs. Also, even within the system users, there may be inequities that need to be addressed similarly.

(b) Failure to Incorporate Qualitative or Nonmonetary Measures of Investment Performance. DeGarmo et al. (1997) discussed the Bureau of Reclamation's 1967 Nebraska Mid-State Project as an illustration of BCR flaws. The objective of that reclamation project was to divert water from the Platte River for farmland irrigation. The computed BCR was 1.24, suggesting that the project was economically efficient and worth undertaking. However, as the authors pointed out, the favorable ratio was obtained partially on the basis of invalid assumptions such as unrealistically low interest rates, an excessive analysis period, and nonconsideration of the damage to wildlife due to the river diversion. Nonmonetary measures of investment performance can be incorporated in economic efficiency analysis by converting them into their equivalent monetary values using techniques such as willingness to pay (WTP), though these are often difficult to assess.

(c) Care Needed to Distinguish between Added/Reduced Costs or Benefits. The BCR is a ratio of benefits to costs. Thus, even slight changes in the denominator and numerator can significantly alter the relative attractiveness of the competing alternatives. As such, it is important to recognize whether an amount represents an added cost or a reduction in benefits; or whether it is an added benefit or a reduction in cost. In certain contexts, maintenance costs, for example, are considered as negative benefits and thus appear as negative amounts in the numerator, thus reducing the value of the numerator of a BCR function; on the other hand, organizations such as the U.S. Office of Management and Budget (OMB) recommend that maintenance costs should be added to the denominator (that represents the life-cycle costs) in the BCR function (Weisbrod and Weisbrod, 1997). A second example is salvage value: Certain procedures consider salvage value as a negative cost that should appear in the numerator while others argue that it is a benefit that should appear in the numerator.

A third example is related to the user and community benefits, where a question often arises as to which side of the fraction the monetary element should be added or subtracted. A school of thought holds the view that for public projects where the public agency incurs costs to provide benefits to the users and community, the numerator should be reserved only for user and community benefits while the denominator should be for agency benefits. Thus, in the numerator, a positive amount is an increase in user/community benefit and a negative amount is a reduction in user/community cost; in the denominator, a positive amount is an increase in agency cost and a negative amount is a reduction in agency cost, such as salvage.

$$\text{BCR} = \frac{B_U - C_U}{C_A - B_A} \tag{11.8}$$

where B_U, C_U are the monetary benefits and costs to the user/community; and C_A, B_A are the monetary costs and benefits to the agency. Thus, user cost reductions should be subtracted from the numerator [Equation (11.9)], not added to the denominator. Also, costs refer to the expenses borne by the agency that constructs, operates, and maintains the facility, and any reductions thereof should be subtracted from the denominator as shown in Equation (11.10), not added to the numerator.

$$\text{BCR} = \frac{B - \Delta\text{UC}}{C} \tag{11.9}$$

$$\text{BCR} = \frac{B}{C - \Delta\text{AC}} \tag{11.10}$$

(d) Distortion Associated with Relative Amounts of Annual Costs and Benefits. For all intents and purposes, BCR appears to yield a good assessment of the attractiveness of an investment. Researchers have cautioned, however, that this is only true when the annual costs are negligible. For most civil systems, however, there can be significant operations and maintenance costs between the time of construction to the end of system life in order to sustain the level of service at certain minimum standards; for investments involving such systems, BCR tends to underestimate the economic efficiency. We can demonstrate this distortion using a simple example: Consider a commercial building that requires an initial investment of $20 million, projected annual earnings of $200 million (present value), and annual costs of $190 million (present value). For a 20-year period, the BCR is 1.05. Consider another investment, such as a new highway with an initial investment of $200 million, projected benefits of $30 million (present value), and insignificant annual costs of maintenance and operations. For a 20-year period, the BCR is 1.5, which far exceeds that of the commercial building project even though its lifetime net benefit is far less than that of the building project. The failure of the BCR to produce a reasonable evaluation in such situations stems from its inability to distinguish between the capital that was committed for the long term and the annual cash flows which are not committed in such a manner (de Neufville and Stafford, 1971).

To examine such distortions in a formal manner, consider a capital investment with initial cost P, and present worth of annual costs and benefits C and B, respectively. Consider two cases:

Case 1. Annual expenditures are considered as costs and thus are added to the initial cost in the denominator. In this case:

$$\text{BCR} = B/(P + C)$$

Let us refer to this as the BCR.

Case 2. Annual expenditures are considered as reductions in benefits and thus subtracted from the annual benefits in the numerator. In this case:

$$\text{BCR} = (B - C)/P$$

Let us refer to this as the BCR*.

Using a different definition of BCR can produce a different value of this criterion of economic efficiency.

Example 11.18

Consider a project with $10,000 initial cost, $1000 annual costs, and $4000 annual benefits. Assuming a 20-year analysis period and a 5% discount rate, determine the BCR values under Cases 1 and 2. Repeat the calculation for an initial cost of $1000. What do you notice? Repeat for $10,000 initial cost, $3000 annual costs, and $4000 annual benefits. What do you notice?

Solution

It can be seen that different answers are obtained for the two definitions of BCR. When the ratio of annual costs to initial cost increases, BCR* becomes increasingly larger than BCR. Also, when the ratio of annual costs to annual benefits increases, BCR* becomes increasingly smaller than BCR.

To illustrate further the distortion associated with the BCR concept, take the ratio of BCR to BCR*.

$$\frac{\text{BCR}}{\text{BCR}^*} = \frac{1}{(1 + C/P)(1 - C/B)} \tag{11.11}$$

This relationship, for different ratios of annual costs to annual benefits and also for different ratios of annual cost to initial cost, is plotted as Figure 11.11.

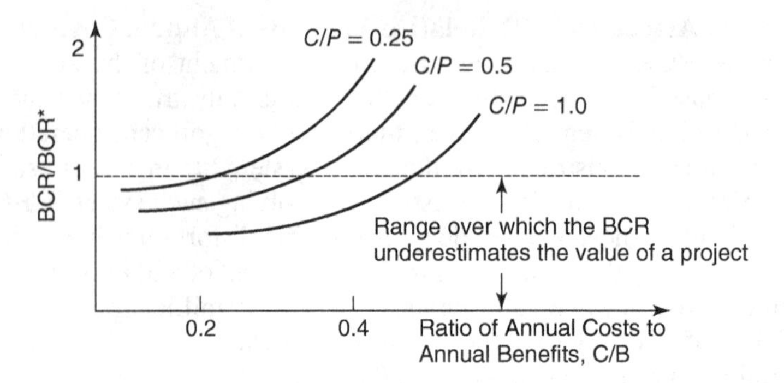

Figure 11.11 Illustration of the distortion associated with BCR concept.

(e) Discussion. A project that has a BCR exceeding 1 will have a positive NPV. Nonetheless, projects that have large benefits and costs have a higher NPV (but may have higher or lower BCR) compared with projects with small benefits and costs. Thus, unless benefit cost ratios are reported along with the corresponding magnitude of costs and benefits associated with each alternative, the ratio itself may not mean much to the system owner or other stakeholders. Due to the numerous limitations discussed above, the use of the BCR, as a criterion for economic evaluation, has waned over the past decades.

11.4.2 Internal Rate of Return (IRR)

The IRR is prone to some of the difficulties associated with BCR, particularly in its inability to account for inequities and its failure to incorporate qualitative or nonmonetary measures of investment performance. However, IRR is a criterion that is considered very appropriate for analyzing the economic efficiency of public projects because it obviates the need to determine the appropriate discount rate for the analysis. Besides having limitations that are similar to those of BCR, IRR has a number of weaknesses (Lorie and Savage, 1955): The evaluation outcome can be rather ambiguous; it may provide a distortion of the project economic efficiency; and for multiple alternatives, it may yield a ranking that is not consistent with other more acceptable evaluation criteria. We herein discuss these limitations briefly.

(a) Ambiguities in Evaluation Outcome. A number of civil systems have high costs at the end of their service lives, for example, systems or their components that have high disposal or demolition costs. Other examples include systems that have low maintenance costs initially but very high maintenance costs as they approach the end of their service lives. Also falling into this category are systems such as natural resource mining where the system owner is required to remediate the environment after completion of the mining operations. In such cases, particularly where the end of the analysis period coincides with the end of the service life, the use of IRR as the criterion for evaluating the economic efficiency of the investment can produce equivocal solutions.

Example 11.19

Consider a system that costs \$200M to construct, annual benefits of \$50M over a 40-year period and \$150M to demolish the system and repair the environment in the 40th year. Plot (i) the cash flow diagram and (ii) a graph of the NPV of the investment for discount rates ranging from 0 to 20%.

Solution

Figure 11.12 highlights the ambiguity of the solutions. From the plots, it can be seen that IRR yields two solutions and thus can be ambiguous in evaluation circumstances such as that described in Example 11.5.

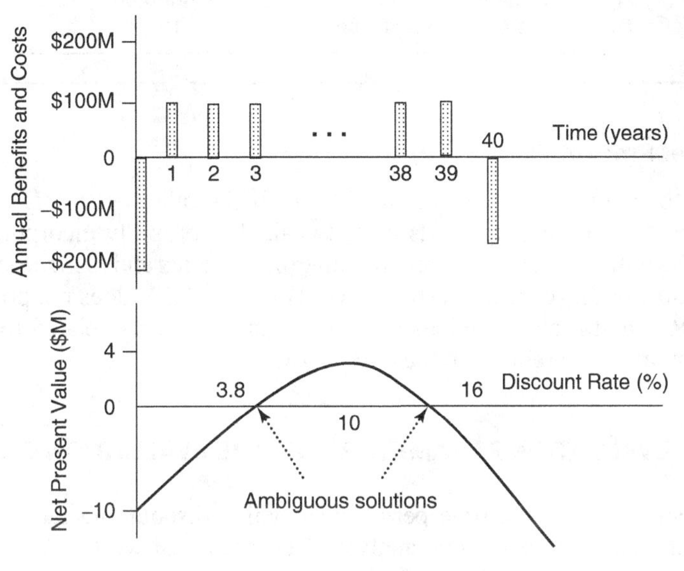

Figure 11.12 Solution to Example 11.5.

(b) Distortion of Economic Efficiency. The use of IRR to evaluate the economic efficiency of certain projects can lead to misestimation of the true efficiency when important costs are not included in the analysis. This is often the case for systems located at public-owned sites. In a typical evaluation, these sites are often considered free resources and thus are excluded from the evaluation. However, if the law allowed, these resources could be invested to yield some returns, and as such, they have opportunity cost.

(c) Inconsistent Rankings with Other Evaluation Criteria. The evaluation outcomes from IRR and NPV can differ, as illustrated in the example below.

Example 11.20

Consider two investments A and B at a 3.5% discount rate (Table 11.5). Identify the superior option and discuss.

Solution

It can be seen that on the basis of NPV, investment B (−723,450 NPV) is inferior to investment A (−336,679 NPV); however, on the basis of IRR, investment A (15.1% IRR) is superior to Investment B (7.8% IRR). Thus, if the discount rate used in the NPV analysis is a reflection of the true opportunity cost of the project, then the use of IRR may lead to an incorrect solution.

Table 11.5

	Investment A	Investment B
Initial Investment	2M	2.5M
Annual Benefits	400,000	250,000
Investment Life (analysis period)	10 years	20 years

11.4.3 Net Present Value (NPV)

The NPV overcomes most of the limitations of the other economic efficiency criteria discussed above. The NPV procedure needs to be tweaked to properly incorporate the inequities of public investments with respect to different population segments and to incorporate qualitative or nonmonetary measures of investment performance. However, NPV does not produce ambiguous outcomes as does IRR; and it also provides a solution that is directly related to the overall outcome of an investment and not a ratio as in the case of BCR.

11.5 EFFECT OF EVALUATION PARAMETERS ON THE EVALUATION OUTCOME

The discount rate and the time period over which discounting takes place can have a profound effect on the output of economic analysis. The base-year worth (or present worth) of future costs or benefits diminishes rapidly for higher discount rates or longer time periods from the base year of the analysis. As a general rule of thumb, the rate can be estimated using the rule of 72, which states, "the quantity $(1 + r)^N$ doubles about every M years, where $M = 72/r$." Thus, for example, the base-year worth is halved in approximately 15 years when the discount rate is 5%. Figure 11.13 shows the reduction in the value of $1000 over a 50-year period at different discount rates. As the figure illustrates, the discount rate greatly influences the present worth. For example, for a 50-year

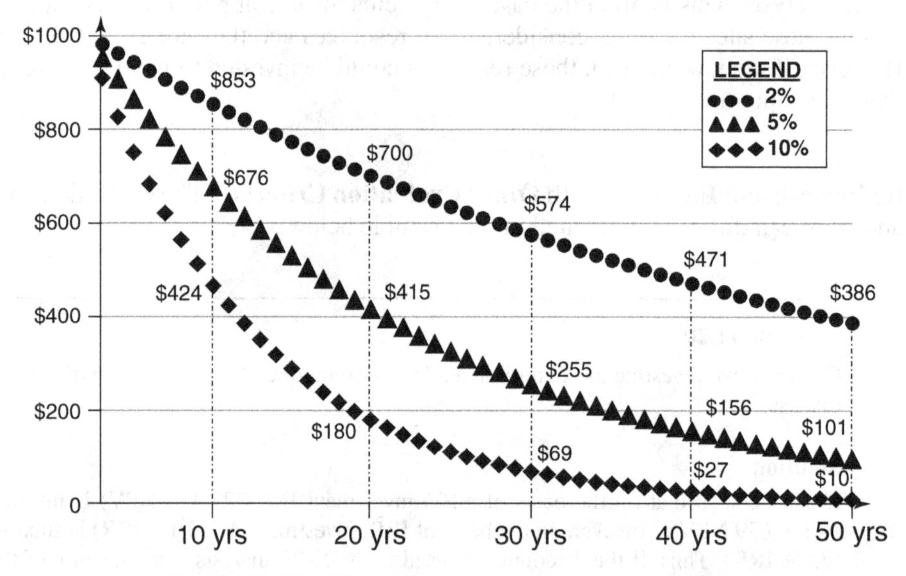

Figure 11.13 Effect of discount rate on present value.

Table 11.6 Effect of Interest Rate on the Relative Attractiveness of Alternatives

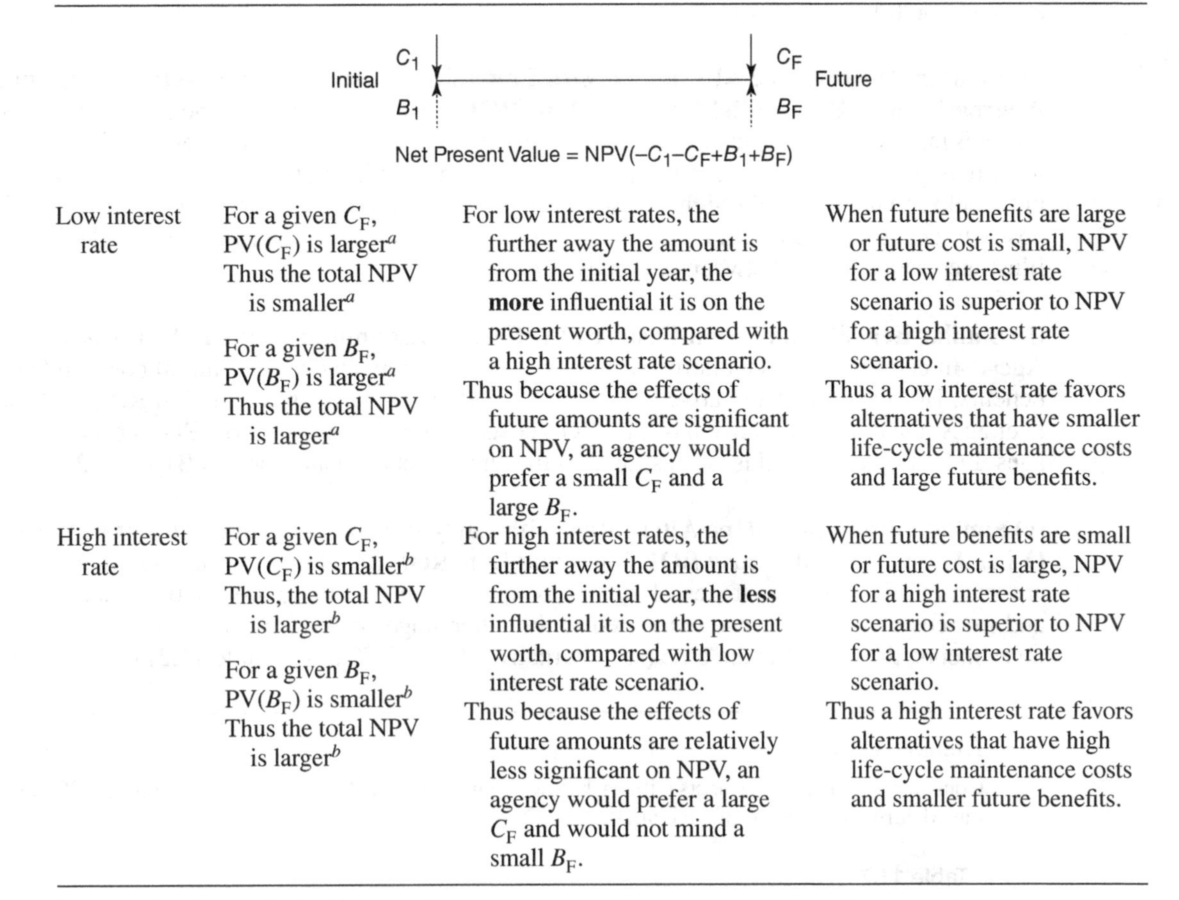

Low interest rate	For a given C_F, PV(C_F) is larger[a] Thus the total NPV is smaller[a] For a given B_F, PV(B_F) is larger[a] Thus the total NPV is larger[a]	For low interest rates, the further away the amount is from the initial year, the **more** influential it is on the present worth, compared with a high interest rate scenario. Thus because the effects of future amounts are significant on NPV, an agency would prefer a small C_F and a large B_F.	When future benefits are large or future cost is small, NPV for a low interest rate scenario is superior to NPV for a high interest rate scenario. Thus a low interest rate favors alternatives that have smaller life-cycle maintenance costs and large future benefits.
High interest rate	For a given C_F, PV(C_F) is smaller[b] Thus, the total NPV is larger[b] For a given B_F, PV(B_F) is smaller[b] Thus the total NPV is larger[b]	For high interest rates, the further away the amount is from the initial year, the **less** influential it is on the present worth, compared with low interest rate scenario. Thus because the effects of future amounts are relatively less significant on NPV, an agency would prefer a large C_F and would not mind a small B_F.	When future benefits are small or future cost is large, NPV for a high interest rate scenario is superior to NPV for a low interest rate scenario. Thus a high interest rate favors alternatives that have high life-cycle maintenance costs and smaller future benefits.

[a]Compared to the case for high interest rate.
[b]Compared to the case for low interest rate.

period, the discounted value is \$386 when the discount rate is 2%; and is only \$10 when the discount rate is 10%.

Table 11.6 discusses and illustrates the effect of the discount rate on the desirability of alternatives. With a lower discount rate, the present values of costs and benefits incurred at subsequent (later) years are relatively high; and with a higher discount rate, the present values of costs and benefits incurred at subsequent (later) years are relatively low. In other words, alternatives that have higher later benefits are preferable when the discount rate is low compared to when the discount rate is high; and alternatives with higher later costs are preferable when the discount rate is high compared to when the discount rate is low. As such, in the evaluation of civil systems, policies that use low interest rates are generally more favorable to alternatives that have (i) relatively lower rest-of-life (ROL) costs and/or (ii) relatively higher ROL benefits.

11.5.1 Effect of Relative Magnitude and Timing of Costs and Benefits on Comparative Evaluation of Alternatives

In a comparative evaluation of multiple alternative actions, the discount rate choice can influence the outcome of the evaluation. The effect of the interest rate depends on the relative magnitude of the initial (construction) cost, the ROL costs and the ROL benefits, and their relative timing. Let us consider the following cases:

(a) Similar Benefits; One Alternative with Low Initial Costs and High ROL Costs, Another Alternative with High Initial Costs and Low ROL Costs. This is often the case when a system owner is faced, during the physical design of the system, with the choice between a design parameter level (e.g., configuration, orientation, or material type) that is relatively inexpensive to construct but would require relatively higher cost to maintain and another parameter level that is relatively costly to construct but would require relatively lower cost to maintain. In Example 11.21, this is illustrated as the choice between alternatives A1 and A2.

(b) Similar Benefits and Initial Costs, But Different Distribution of the ROL Costs across the Ages. In certain cases, the alternatives under consideration have similar initial costs and similar benefits. However, one alternative has low ROL costs at its early ages and high ROL costs at the later ages; the other alternative has high ROL costs at its early ages and low ROL costs at the later ages. In Example 11.21, this is illustrated as the choice between alternatives B1 and B2.

(c) Similar Initial Costs; One Alternative with Early ROL Benefits and Late ROL Costs; the Other Alternative with Early ROL Costs and Late ROL Benefits. Certain system investments are such that in the early stages of system or investment life, there are relatively large benefits and low costs of maintenance compared to the latter stages of system life. For other systems or investments, the reverse is the case (see alternatives C1 and C2 in Example 11.21).

Example 11.21

Consider three pairs of projects with cash flow streams described in Table 11.7. Assuming a 7% discount rate, determine the NPV of each alternative.

Table 11.7

	Alt A1	Alt A2	Alt B1	Alt B2	Alt C1	Alt C2
Initial Cost ($, millions)	30	50	50	50	50	50
Annual Benefit ($, millions): 0–10 yr (11–20 yr)	5	5	5	5	5 (1)	1 (5)
Annual Cost: ($, millions): 0–10 yr (11–20 yr)	3	1	1 (4)	4 (1)	3	3
Analysis Period (years)	20	20	20	20	20	20

Solution

The results of NPV and BCR are shown in Figure 11.14. It can be seen that alternative A1 is desirable over A2, alternative B1 is desirable over B2, and alternative C1 is desirable over C2.

11.5.2 Effect of Discount Rate on the Feasibility of Systems Investments

In evaluating the economic feasibility of providing, replacing, expanding, or renewing a civil system, the discount rate choice can influence the outcome of the evaluation. The effect of the discount rate on project feasibility depends on the relative magnitude of the initial (construction) cost, the

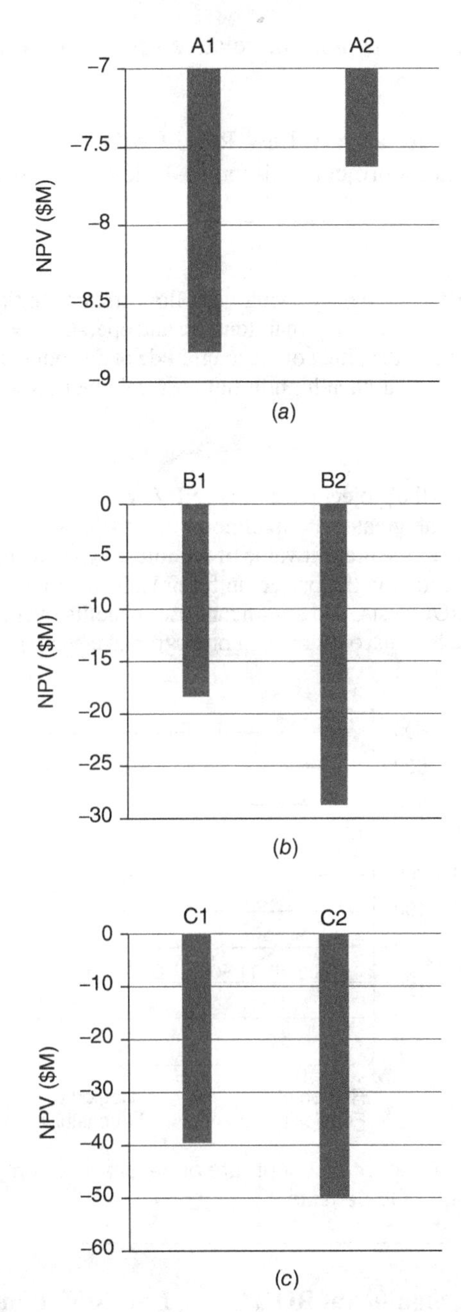

Figure 11.14 Comparison of Alternatives in Example 11.21.

ROL costs and ROL benefits, and the relative timing of the ROL costs and ROL benefits. Let us consider the following cases:

(a) High Initial Cost, Relatively Low ROL Costs, Significant User Benefits over ROL. For this case, let us consider a project with the cash flow streams described in Example 11.22.

Example 11.22

Consider a project with the following cash flow stream: Initial cost of construction = \$250M; annual benefits = \$15M; annual cost of maintenance and operations = \$5M; investment life (analysis period) = 50 years. Comment on the effect of the magnitude of discount rates on the investment evaluation outcome of such investments that have high initial cost, relatively low ROL costs, and significant user benefits over the rest of life.

Solution

Figure 11.15 shows the project feasibility (NPV) versus the discount rate relationship. Clearly, the lower the discount rate, the greater the likelihood that the project will be adjudged to be feasible. A higher discount rate reduces the present value of the future benefits and renders it insignificant compared to the initial cost, thus rendering the project unfeasible. Thus, for civil systems that involve high initial costs, relatively small ROL costs, and significant user benefits over the ROL, some civil system planner may be inclined to use low discount rates in order to make such investments appear feasible.

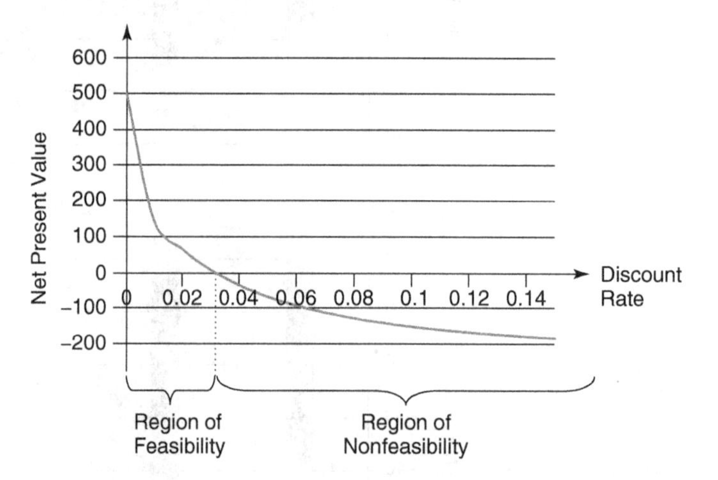

Figure 11.15 Effect of discount rate on feasibility of projects with high initial cost, low ROL costs, high ROL benefits.

(b) Low Initial Cost, Significant ROL Costs, Low ROL Benefits. For civil systems investments where ROL costs are high and ROL benefits are low, a low discount rate will have an opposite effect to (a) above: the costs become significant and the benefits become insignificant, thus impairing project feasibility. A high discount rate will have the opposite effect: the costs become insignificant and the benefits become significant, thus enhancing project feasibility. A high discount rate results in a low present value of the ROL costs and renders it very small compared to the benefits. As such, the project becomes attractive at high discount rates. Therefore, for civil systems that involve a low initial cost, high ROL costs, and low user benefits over the ROL, some civil system planners may prefer to use high discount rates in a bid to enhance the economic feasibility of such investments.

Example 11.23

Consider a project with the following cash flow streams: Initial cost of construction = \$10M; annual benefits = \$2M; annual cost of maintenance and operations = \$6M; investment life (analysis period) = 50 years. Comment on the effect of the discount rate on feasibility of such a project that has low initial cost, high ROL costs, and low ROL benefits.

Solution

As evidenced in Figure 11.16, a lower discount rate increases the present value of the high annual cost to a greater extent compared to what it does to the low annual benefits. As such, the project becomes increasingly unattractive at low interest rates but becomes more attractive at higher discount rates. For civil systems that involve a low initial cost, relatively high ROL costs of operations and maintenance compared to ROL benefits, and low user benefits over the ROL compared to ROL costs, it therefore may be the case that civil system planners may prefer to use high interest rates in a bid to reduce the economic unattractiveness of such investments.

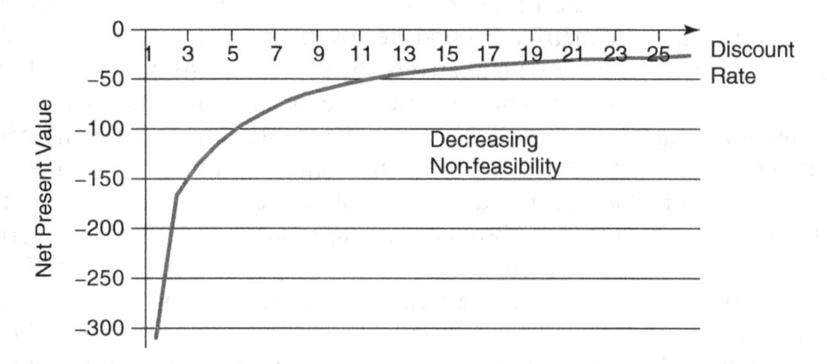

Figure 11.16 Effect of discount rate on feasibility of projects with low initial cost, high ROL costs, and low ROL benefits.

11.6 EFFECT OF ANALYSIS PERIOD ON SYSTEMS EVALUATION

11.6.1 General Effect of the Length of Analysis Period

From the temporal perspective of cost incurrence, the two key components of life-cycle cost analysis (LCCA) are the **initial cost and the rest-of-life (ROL) cost**. The initial cost, which refers to the cost of construction, is taken to occur at the start year (0th year) of the analysis period, and this is not affected by a change in the analysis period. The ROL cost is the stream of expenditures that occur after the start year until the end of the analysis period (which often is the end of facility life). For a given number of maintenance and rehabilitation expenditures, a short analysis period would mean a higher NPV of costs while a long analysis period would mean a lower NPV of costs (Figure 11.17), which is due to the effect of discounting. In Figure 11.17, all of the cost values are the same for both the short analysis period and the long analysis period. Thus, if the interest rate is zero, the total costs per year are equal. However, with the nonzero interest rate, the annualized value of the ROL costs is higher in the case of the short analysis period compared to the long analysis period.

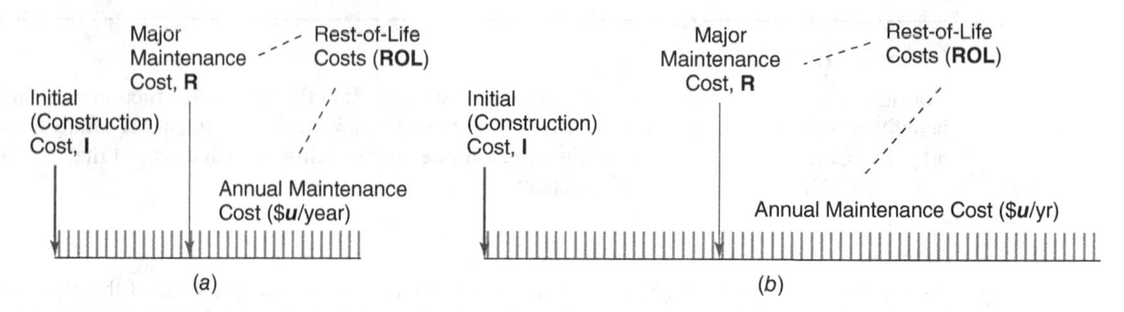

Figure 11.17 Effect of increased analysis period: (*a*) short analysis period (Scenario *S*) and (*b*) long analysis period (Scenario *L*). For 0% interest rate, $EUAC_L = EUAC_S$. For nonzero interest rate, $EUAC_L < EUAC_S$.

11.6.2 ROLI Ratios

In economic analysis, there often arises the need to compare the economic efficiencies of two alternatives that have diametrically opposite cost patterns, one alternative with high initial costs but low ROL costs and the other with relatively lower initial costs but relatively higher ROL costs. The former is generally consistent with durable but expensive materials and the latter with inexpensive materials of average durability (Table 11.8). The table shows, for each alternative, the initial cost (*I*), the rest-of-life cost (ROL), and the ROLI ratio (rest-of-life cost divided by the initial cost). It is useful to examine the effect of different analysis periods on the basis of the ROLI ratios across the alternatives rather than the absolute *I* and ROL amounts across the alternatives.

Table 11.8 shows that alternatives with low initial costs and high ROL costs have a high ROLI ratio. As explained in the preceding paragraph, an increase in the analysis period will generally result in a lower life-cycle cost (i.e., reduced NPV due to discounting effects). When the analysis period is increased, the numerator (ROL) is reduced while the denominator (*I*) remains the same. Thus, for alternatives with high ROLI ratios, an increase in the analysis period will cause a greater reduction in the life-cycle cost compared with alternatives with low ROLI ratios. Thus, all other factors remaining the same, shorter LCCA periods generally favor civil system alternatives that are initially costly to construct but have lower costs of maintenance and other ROL expenditures while longer analysis periods generally favor civil system alternatives that are initially less costly to construct but have higher costs of maintenance and other ROL expenditures. As civil systems increase in age, they incur increased deterioration and the rate of increase may be the same or different across the alternatives.

Table 11.8 ROLI Ratios

	Initial Cost (*I*) (Relative to Other Alternative)	Rest-of-Life (ROL) Cost (Relative to Other Alternative)	ROLI Ratio (ROL/*I*)
Alternative with relatively low initial cost, lower durability, and high rest-of-life cost	Low	High	High
Alternative with relatively high initial cost, high durability, and low rest-of-life cost	High	Low	Low

11.7 PERPETUITY CONSIDERATIONS

Perpetuity is an important consideration in the evaluation of civil systems because such systems are (i) needed perpetually by society and (ii) do not last forever and need to be replaced after a number of decades. For example, Chicago's wastewater treatment plants will always be needed by that city and will need to be reconstructed any time they reach the end of their design life; the Madrid Barajas International Airport will always be needed and its runways and terminals will need to be reconstructed after they reach a certain state. Exceptions include cases where socioeconomic shifts and technology renders such systems unnecessary or obsolete, for example, the future use of aerial personal vehicles for intercity travel may eliminate the need for some intercity highways.

Consider the case of a civil system having a service life of N years (Figure 11.18). A hypothetical single amount, R, occurring at the end of the analysis period can be calculated to represent all postconstruction incoming and outgoing amounts within the analysis period N. As we discussed in the preceding paragraph, civil engineering systems including dams, bridges, airports, roads, and water supply and distribution infrastructure occupy a unique position of permanence in the social and economic environment: Most of these systems will likely be needed by society forever. Therefore, for most civil systems, it is safe to make the assumption that the system will be kept in service in perpetuity, and therefore, the life-cycle investment of R (reconstructions or replacements) will be repeated every N years. In this discussion, we can assume N is constant; however, in reality, N could be decreasing with time due to increasing loading or usage levels and other deterioration factors induced by agents such as climate change; or could be increasing (due to technological advances in materials and maintenance techniques). For civil systems, it is safe to make the assumption that the level of the initial investment (P') is not the same as that of the periodic investments (R), because the latter excludes several one-time costs associated with initial investments. A case in point is water ports construction (where the initial investment includes right-of-way acquisition, geotechnical treatments, deck construction, dredging, etc., while recurring major investments often involve rehabilitation and mainly dredging). Another example is highway construction, where the initial investment includes right-of-way acquisition, embankment construction, utilities relocation, wetlands restoration, and other costs that are typically not incurred in recurring investments such as pavement resurfacing or replacement.

The present worth of all amounts to perpetuity is given as

$$PW_{\infty} = P' + \frac{R}{(1+i)^N} + \frac{R}{(1+i)^{2N}} + \frac{R}{(1+i)^{3N}} + \cdots = P' + \frac{R}{(1+i)^N - 1} \tag{11.12}$$

Example 11.24

A city plans to reconstruct it water supply reservoir. The estimated service life of the structure is 60 years. The reconstruction cost is \$600,000. During its replacement cycle, the reservoir will require two rehabilitations, each with a cost of \$200,000, at the 20th and 40th year. The average annual cost of maintenance is \$5000. At the end of the replacement cycle, the reservoir will be reconstructed and the entire cycle is assumed to repeat perpetually. What is the present worth of all costs to perpetuity? Assume 5% interest rate. Assume P' (the starting nonrecurring cost = 0).

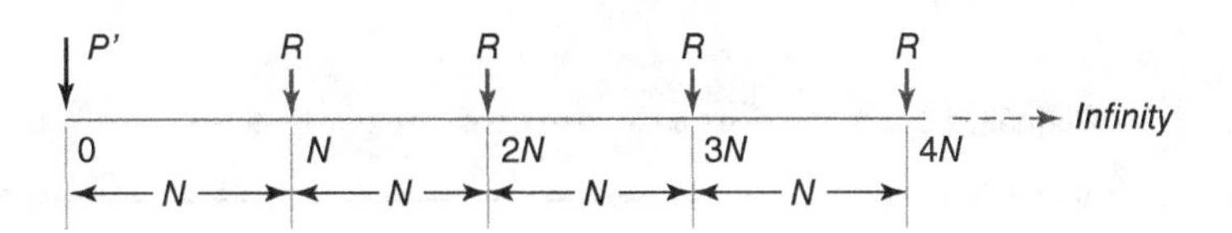

Figure 11.18 System replacement in perpetuity.

Solution

$$\text{Present value of life-cycle cost} = 600{,}000[\text{SPCAF}(5\%, 60)] + 200{,}000[\text{SPCAF}(5\%, 40)] + 200{,}000[\text{SPCAF}(5\%, 20)] + 5000[\text{USCAF}(5\%, 60)] = \$14{,}914{,}087$$

$$\text{PW}_\infty = P' + \frac{R}{(1+i)^N - 1} = 0 + \frac{14{,}914{,}087}{(1+0.05)^{60} - 1} = \$843{,}596$$

11.8 SOME KEY CONSIDERATIONS IN MONETARY LCCA

11.8.1 Comparing Alternatives That Have Different Service Lives

For comparing asset intervention alternatives that have different service lives, there are at least three approaches:

- For each alternative, convert all costs and benefits into EUAR, or
- For each alternative, find net present worth (NPW) over a service life that is a lowest common denominator (LCD), or
- For each alternative, find the present worth of periodic payments to perpetuity.

(a) Conversion of All Costs and Benefits into EUAR, for Each Alternative. This method was discussed in Sections 11.3.3 and 11.3.4.

(b) Finding NPW over a Service Life That Is an LCD, for Each Alternative. For service lives n_1 and n_2, determine the LCD, $n*$, and repeat the entire cost stream over $n*$, for both alternatives (Figure 11.19). Assume total replacement (reconstruction) of the facility after the end of its service life. With the same analysis period of 12 years, we can now proceed to use NPV to evaluate the two alternatives (Figure 11.20).

11.8.2 Incorporating Uncertainty in Economic Evaluation of Systems

Incorporation of uncertainty in economic evaluation of systems involves the use of probability distributions in describing the various factors involved, such as the incoming or outgoing amounts that occur at a present year, a future year, or as annual payments/receipts, and the interest rate.

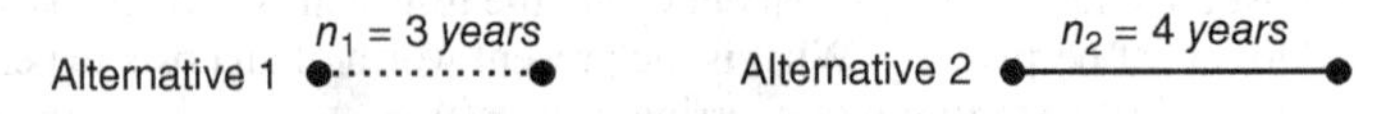

Figure 11.19 Alternatives with different service lives.

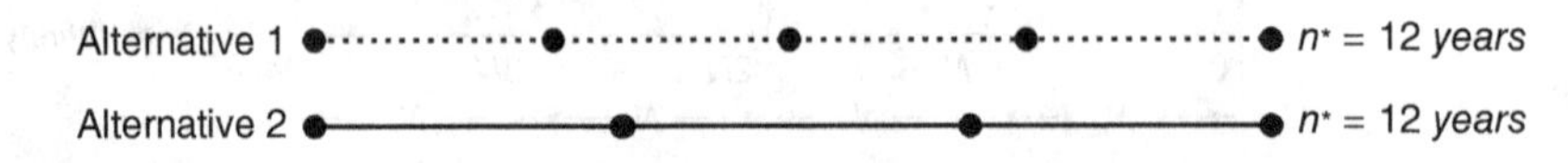

Figure 11.20 Equalizing analysis periods for alternatives with different service lives.

The concept used is the expected value, which is the product of the probability and the cost. The expected value can be calculated as follows:

$$\text{Expected cost in a given year} = (\text{cost}_1 \cdot p_1) + (\text{cost}_2 \cdot p_2) + \cdots + (\text{cost}_n \cdot p_n)$$

$$\text{Expected benefit in a given year} = (\text{benefit}_1 \cdot p_1) + (\text{benefit}_2 \cdot p_2) + \cdots + (\text{benefit}_n \cdot p_n)$$

where cost_i is the level of cost that could be possibly incurred in a given year, benefit_i is the level of benefit that could be possibly incurred in a given year, and p_j is the probability that a cost or benefit level, j, will be incurred in a given year.

11.8.3 Life-Cycle Cost Analysis

Life-cycle cost analysis is a specific application of economic analysis where the streams of benefits and costs extend over one *entire life cycle* or the remaining life of a facility. Used at the project level of systems management, LCCA analysis focuses on a specific facility, such as a bridge or road segment. LCCA evaluates the overall long-term economic efficiency between competing alternative investment options by evaluating the benefits and costs of various alternative preservation and improvement strategies or funding levels over a specified period. The monetized costs and benefits associated with each alternative activity profile are determined and the alternative with the highest NPV is typically selected. In most countries, government legislation mandates the use of LCCA in public infrastructure investment analysis, and national agencies such as the U.S. Federal Highway Administration (FHWA) and the Federal Transit Administration (FTA) encourage the use of LCCA in analyzing all major investment decisions. Fairly recently, greater impetus for application of LCCA in civil systems management in the United States came from the issuance of the Governmental Accounting Standards Board Statement 34 (GASB34), which, in a bid to enhance accountability of the use of public resources, specified guidelines for public agencies to report on the financial and operational performance of their assets (GASB, 1999). Studies and field observations worldwide have shown that cost-effective long-term investment decisions could be made if LCCA principles are used in the decision-making process.

11.8.4 Difference between Life-Cycle Cost Analysis (LCCA) and Remaining Service Life Analysis (RSLA)

Life-cycle cost analysis assesses the impacts of a system over its entire service life. As such, a full-life LCCA is relevant for new facilities. The full service life spans the time of construction to the time of demolition, decommissioning, or salvage, and is often expressed in years.

In cases where the facility already exists and thus has more years left to serve, however, a full-life LCCA may not be applicable. In such cases, RSLA is more appropriate. RSLA helps us determine the most economically efficient set of activities to pursue to preserve and/or to operate the system from the decision year until the time of demolition of decommissioning.

LCCA includes the costs of operations, routine maintenance, periodic maintenance, and the like. LCCA helps us determine the most economically efficient set of design, construction, preservation, or operational activities to carry out over the full or remaining service life of a system.

11.9 ECONOMIC EFFICIENCY ANALYSIS—ISSUES AND LIMITATIONS

11.9.1 Specification of System Life for LCCA Purposes

The life of a civil system often serves as the analysis period for the purposes of economic analysis. This generally can refer to the time until the system must be replaced due to substandard performance, technological obsolescence, regulatory changes, or changes in consumer behavior and values (Lemer, 1996). The primary reasons for which a system owner replaces or retires a system may include the need to accommodate greater demand (capacity or loading); eliminate safety hazards associated with the existing system; avoid the high maintenance costs associated with an existing system; avoid litigation or other problems associated with outdated design practices; respond to changing development patterns that render the system no longer needed; eliminate potential vulnerability inherent in the current design (e.g., fatigue damage); eliminate potential vulnerability to extreme events such as floods, earthquakes, or collision; or to address deterioration that is beyond cost-effective repair/rehabilitation (Thompson et al., 2012). Often, the system owner replaces the system as a result of a combination of some of these factors. In some cases, the system life is longer than usual due to funding limitations in that a cash-strapped owner may choose to apply stopgap preservation actions until funds are available to replace the system. In all of these cases, the system life may be terminated well before or after the initially intended time and thus may not give an indication of the true life of the system.

The actual life of a system can be viewed from a number of perspectives (Ford et al., 2012). The *physical life* is the time period during which the system is physically standing and is capable of performing its intended function even if only partially; the *functional life* is the time period where the system satisfies all of its functional requirements; the *service life* is the time period in which the system is providing the intended type of service, even if at a degraded level of service; the *economic life* is the time period during which it is economically optimal to keep the asset in service rather than retiring or replacing it. The above definitions are structured according to the different criteria for end of system life. Also, a few other "life" definitions are: *Actual life*, the known value of the physical, functional, service, or economic life after the system has actually been retired or replaced; *design life*, a specific target life based on technical and economic considerations established at the design phase of development. In Figure 26.1 in Chapter 26, we will see illustrations of these different definitions of system life, and the different relationships that could exist between the functional life and physical life definitions.

11.9.2 Pareto Efficiency

Traditional economic analysis does not consider equity. Thus an investment may have a high NPV but may benefit only a small segment of the population. In striving to ensure equitable distribution of project benefits, civil engineers increasingly seek to incorporate the concept of Pareto efficiency.

An investment is a Pareto improvement when it makes at least one entity better off without making any other entity worse off and is further described as "Pareto optimal" when no further Pareto improvements can be made. However, as noted by Barr (2004) and Sen (1993), a Pareto-efficient investment makes no explicit statement about the equity of an investment in terms of the social egalitarianism of resource distribution arising from the investment. In discussing the concept of Pareto efficiency and its relationship to economic efficiency analysis, Boardman et al. (2001) stated that the latter utilizes a decision rule with less conceptual appeal but greater feasibility than the actual Pareto efficiency rule. This decision, is based on the well-known Kaldor–Hicks criterion rule (discussed in Section 11.9.3) which states that a policy should be adopted if and only if those who gain could fully compensate those who will lose and still be better off; this criterion provides the basis for the potential Pareto efficiency rule, or more commonly, the net benefits criterion that states that only policies that have positive net benefits should be adopted.

The potential Pareto efficiency rule is often considered justified for at least one of the following reasons: (i) by always choosing policies with positive net benefits, society maximizes the aggregate wealth and therefore helps the poor in society; (ii) different policies will have different gainers and losers, and costs and benefits tend to average out among people so that each person is likely to realize positive net benefits from the full set of policies; and (iii) it appropriately militates against the assignment of excessive weight to organized groups (stakeholders) that are typically associated with representative political systems or the assignment of miniscule weight to unorganized groups in such systems.

11.9.3 The Scitovsky or Kaldor–Hicks (KH) Criterion for Measuring Efficiency

In the Pareto efficiency criterion, an outcome is considered to be more efficient if it results in at least one person being better off and no one being left worse off. The Scitovsky or KH criterion is a more general case of Pareto efficiency: Every Pareto improvement is a KH improvement but not very KH improvement is a Pareto improvement. In the case of the KH efficiency criterion, a stimulus (policy, action, or investment) is considered efficient when the benefiting stakeholders (those that are left better off by the stimulus) compensate adequately the losing stakeholders (those that are left worse off by the stimulus) with the ultimate result that no stakeholder ends up worse off (Sullivan and Sheffrin, 2003). For example, the construction of a civil system that benefits the system owner and the users may result in adverse effects on the community through air or water pollution; such pollution could be cleaned up using funds from user fees and/or owner disbursements. However, by not mandating the payment of such compensation, the KH efficiency criterion leaves open the very practical possibility that an investment outcome that is considered efficient could very well result in at least one stakeholder group being worse off. On the other hand, in the Pareto efficiency criterion, every stakeholder is left either the same or better off.

As we have seen in this chapter, a truly comprehensive economic efficiency analysis of civil systems investments or policies should involve a comparison of benefits and costs to different stakeholders (the system owner, the system user, and the community). For such evaluation problems, the KH criterion provides a solid basis. Thus, the evolution of economic evaluation from a pure cost analysis to one involving both costs and benefits, or to one involving multiple cost and benefit criteria (see Chapter 12), is consistent with the KH criterion. For example, expansion of a toll expressway could be evaluated on the basis of the owner's costs of construction, the user's benefits of reduced delay, the owner's benefits of revenue, and disruption to the socioeconomic fabric of the area such as displaced homes and farmlands. If the overall benefits of the project were found to exceed the costs, the project would, under normal circumstances, be approved for implementation. In most cases, ensuring that the benefits exceed the costs is tantamount to a situation where the stakeholders that are made better off are made to compensate, at least in theory, for those that are made worse off, and thus would be a manifest case of the KH efficiency criterion. In such cases where it is deemed appropriate for all of the stakeholders combined if some are made worse off but others are made better off to a greater degree, the KH criterion can be used.

An issue with KH is that it considers only the absolute levels of investment benefits and not the distribution of benefits. A second issue is the diminishing marginal utility for income: A system improvement that provides one dollar worth of benefit to a high income or wealthy person causes a smaller gain in utility compared to providing one dollar worth of benefit to a low-income or indigent person; similarly, taking one dollar from a rich person causes a smaller loss in utility than taking a dollar from a poor person (Sullivan and Sheffrin, 2003).

11.9.4 Difficulty of Monetizing All Project Costs and Benefits

A serious limitation of economic analysis is the inability to consider benefits and costs that cannot be monetized. It has been argued that intangibles such as the system users' inconvenience, discomfort or the system owners' public relations or reputation really cannot be assigned a monetary value either because it is very difficult or in some cases, even unethical to do so. Other critics contend that economic efficiency evaluations debase democratic principles because such analysis imposes a single goal (namely, economic efficiency) in the evaluation of public projects and policies and that such analysis would be valid only if public policy were determined strictly via benefit–cost analysis results. These critics argue that in a truly democratic environment, public policy is designed solely on the basis of democratic processes that give equal weight to all interests, and therefore economic efficiency analysis is out of place in such environments. Proponents of benefit–cost analysis counter by stating that economic efficiency analyses rarely serve as the only decisive yardstick for policymaking. They add that in using economic efficiency analysis as one of the factors for decision making, it is possible to represent better the interests of less organized and less vocal constituencies with little electoral clout.

11.9.5 Further Discussion on the Social Discount Rate

As discussed in Section 11.2.3, the social discount rate is the rate of return that is used to discount the monetary costs and benefits across time. It is a measure of the opportunity cost associated with the diversion of capital from private hands to large public infrastructure investments including water supply systems, waste disposal systems, highways, or hydroelectric projects. Social discount rates are typically lower for developed nations compared to developing nations. For large civil systems with long design lives that have a large initial investment and benefits (often to the user and the community) that are spread out over a long analysis period, a higher social discount rate decreases the present value of the benefits and thus makes it less likely the investment will be attractive. Because the benefits are spread out far into the long life of the system, a small increase in the social discount rate can greatly influence the attractiveness of the investment.

In analyzing the life-cycle costs and benefits of such civil systems, the social discount rate is used to determine the social marginal cost and the social marginal benefit. However, determination of the social marginal benefit is much more difficult than determination of social marginal cost; while the costs of civil systems are typically monetary and thus are easily measured in universally accepted units, the benefits accrued by the user and community, which include increased user convenience, reduced user delay, and decreased or increased pollution of air, water, or the ecology are much more difficult to monetize (but must be done if we are to proceed with the economic evaluation for the investment). In Chapter 12, we examine ways of providing a dollar value to each of these social consequences of civil system construction or operations.

For large civil systems that have service lives spanning more than one human generation, an important question is the issue of transferring costs or benefits to future generations (Gruber, 2004; Zhuang et al., 2007). Scholars have questioned whether the current generation is paying for or subsidizing the costs so that future generations reap most of the benefits. They also grapple with the issue of whether equal weights should be given to costs incurred in different generations or benefits earned in different generations. One rather extreme position has been that discounting future generations is appropriate because these generations may probably not exist in the future; others hold the view that because the probability of catastrophic events is negligible, equal weight should be given to all generations, implying that the social discount rate should be equal to zero. Due to such considerations, the choice of an appropriate social discount rate has always been plagued with controversy for several decades. The debate took a new turn in 2006 after economists cautioned that global economic output could be jeopardized if steps are not taken to address global warming, a potential catastrophic phenomenon. Ultimately, the choice of the appropriate level of the discount rate can be considered a political decision (Dandy et al., 2008).

SUMMARY

This chapter presented the fundamental concepts of the economic efficiency of investments that provide, replace, expand, or improve civil systems. These include opportunity cost, cash flow diagrams, interest or discount rate, analysis period, and residual values. The five variables in any economic analysis problem are the present amount, the future amount, the annual amount, the interest rate, and the analysis period. A number of equivalence equations, or interest formulas, typically used in economic efficiency evaluation, also were presented. A type 1 equivalence equation involves an interest rate that is compounded and constant throughout the year. Type 2 involves an interest rate that is compounded and changes a finite number of times, and type 3 involves an interest rate that is compounded and changes an infinite number of times. For each of these types of interest rates, there can exist problem contexts that require the determination of the future amount to be yielded by an initial amount at the end of a given period; determining the present amount that would yield a future amount at the end of a given period; determining the amount of uniform annual payments that would yield a certain future amount at the end of a given period; determining the final compounded amount at the end of a given period due to uniform annual payments; determining the initial amount that would yield specified uniform future amounts over a given period; and determining the amount of uniform annual payments over a given period that would completely recover an initial amount. The chapter then presented the various criteria for assessing the economic efficiency of civil systems investments, which included the present worth of all costs, the equivalent uniform annual costs, the net present value and the equivalent uniform annual return, the benefit–cost ratio, the internal rate of return method, and the payback period.

Most civil systems involve periodic payments for rehabilitation, as they do not last forever. Second, civil systems are perpetual, as they are always needed by society. As such, the recurring reconstruction or replacements of civil systems after a number of years or decades, is expected to occur perpetually. Thus, the economic analysis of the replacement cash flows is consistent with perpetuity of periodic payments. The present worth of amounts incurred or received to perpetuity is a function of the interest rate, the interval of the periodic payments, and the magnitude of the periodic amounts. Capitalization means determining the initial capital needed to build and maintain the civil system. This amount is determined by bringing all of the expected costs to their present worth using the present worth factor or using the perpetuity equation. Also, money that is borrowed needs to be paid back. Loan amortization is the payment of a borrowed loan back to the lender (bank) in order for the lender to recover its capital.

In the chapter, we also examined the issue of reduced costs versus added benefit in benefit–cost ratios and the incorporation of uncertainty in the economic evaluation of civil systems. Finally, we discussed a number of controversial issues and limitations of economic analysis.

EXERCISES

Assume that all interest rates indicated are fixed, compounded, and annual, unless otherwise indicated. State any assumptions made.

1. What are some of the economic factors you consider in deciding whether to lease a car, to buy one, or to do neither?
2. A cash flow diagram indicates the amounts paid or received by a stakeholder. For the operations and maintenance of a typical transit system in your community, list the amounts typically encountered in the cash flow stream over the life of the system. Categorize these amounts by their direction (cost versus

benefit), the frequency of incurrence (one-time versus periodic versus recurring), and the party that incurs the cost or receives the benefit (agency versus system user versus community).

3. What is the difference between inflation and opportunity cost? In carrying out economic analysis, why is it useful to only use amounts that have been corrected for inflation?
4. What is interest? Explain the difference between compounding and discounting, and discuss the conditions under which each one is appropriate for analysis.
5. Using a suitable diagram, present and discuss the various types of interest rates. What is the effective annual interest rate for a nominal annual interest rate of 12% that is compounded (a) every quarter and (b) an infinite number of times in a year?
6. Explain at least two ways by which an agency could choose an appropriate interest rate for analyzing its system investments. If the current tax rate is 30%, the rate of return of government bonds is 6%, and 25% of investments are being drawn from consumers, calculate the social discount rate.
7. List the five key factors in interest formula (equivalence equations). Explain how the interest rate and analysis period are decided for a given problem. Also, discuss why each of these factors is, in real life, more probabilistic than deterministic.
8. A company borrows $6000 in December 2013. If the interest rate is 5%, what single payment must be made in December 2019 to repay the entire principal? If the company chooses to make uniform annual payments instead (starting December 2014), what would be the value of such annual payments? Illustrate your answer with a cash flow diagram.
9. A city engineering department borrows money at 1.5% per month. What are the corresponding "nominal" and "effective" rates per annum?
10. You have decided to set up a fund for the education of your future child. What is the amount that will be accumulated in this sinking fund at the end of 20 years if you deposit $1000 in the fund at each of the 20 years? Assume interest rate is 10%.
11. Bingo! Your star shines today, and you win the lottery. You are asked to choose any one of the following payment options: $50,000 now and nothing thereafter, or $6000 each year for 10 years. The interest rate is 2%. Which option would you choose and why? (Note: treat this as a quantitative answer, not an open-ended one). If the interest rate were 20%, would your answer change? Show all calculations.
12. The operator of a metropolitan sewer system wishes to have a sum of $200,000 accumulated 25 years from now for an expected major rehabilitation at that time. The owner plans to invest $2000 per year into this special fund, starting December 2012. What critical annual interest rate must the owner's investment earn?
13. The newly elected city mayor of Gulfsville decides to construct a new levee system near the major river in the town. The estimated life of such a structure will be 40 years. Total initial costs (consulting fees and construction) would be $800,000. Maintenance cost would be $30,000 every 5 years. How much money should the city borrow now in order to carry out the entire project including maintenance? The interest rate is 5%. Illustrate your answer with a cash flow diagram.
14. As a city engineer, you seek the capitalized cost of perpetual service from a water storage tank. Due to the highly corrosive nature of your coastal environment, the tank (which costs $40,000) is maintained at an annual cost of $1000, and is replaced every 10 years. Whenever its service life ends, the tank is intended to be sold to a steel mill as metal scrap for $3000. Using 10% interest rate: (a) Find the capitalized cost of the storage tank investment. (b) Give two reasons why the formula you used in (a) is most appropriate for questions involving civil systems of this type. (c) How much depreciation does this tank undergo in its service life? (d) What is the average annual rate of depreciation? State any assumptions.
15. A city mayor decides to construct a new bridge over the major river in the town. The estimated life of such a structure will be 20 years. There is a 70% probability that the total initial costs (consulting fees and construction) will be $800,000 and a 30% probability that such costs would be $1 million. There is 100% probability that the maintenance costs would be $30,000 every 5 years. How much money should the city borrow now in order to carry out the entire project including maintenance? The interest rate is 5%.
16. The county municipal council owns a toll bridge that costs the county $250,000 every year to operate and $130,000 a year to maintain. The acility brings in revenue of $500,000 per year. There is a proposal to

sign a 10-year lease of the bridge to a private operator who will give the city $1 million upfront with the understanding that the private operator will be responsible for the operation and maintenance of the bridge throughout the lease period and will always keep its level of service above a certain threshold. Assume a 4% interest rate. In your opinion, should the proposal be accepted?

17. Using numerical examples where possible, explain the four shortcomings of the benefit–cost ratio criterion for economic evaluation.

18. Discuss the effect of discount rate on comparative evaluation of alternatives for each of the following situations: (a) Similar benefits; one alternative with low initial costs and high ROL costs, another alternative with high initial costs and low ROL costs. (b) Similar benefits and initial costs, but different distribution of the ROL costs across the ages. (c) Similar initial costs; one alternative with early ROL benefits and late ROL cost; the other alternative with early ROL cost and late ROL benefits. Use numerical examples if necessary to support your discussion.

19. Discuss the effect of discount rate on the feasibility of systems investments. (a) High initial cost, relatively low ROL costs, and significant user benefits over ROL. (b) Low initial cost, significant ROL costs, and low ROL benefits. Use numerical examples if necessary to support your discussion.

20. Discuss the general effect of the length of analysis period on systems evaluation, in the context of the initial cost, the rest-of-life (ROL) costs, and the ratio between these two amounts (ROLI ratio). Compare the ROLI ratios of an alternative with relatively low initial cost, lower durability, and high rest-of-life cost and another with relatively high initial cost, high durability, and low rest-of-life cost. Explain how a change in the analysis period (from short to long) could influence the relative attractiveness of these alternatives, form an economic efficiency viewpoint.

REFERENCES

Barr, N. (2004). *Economics of the Welfare State*. Oxford University Press, New York.

Boardman, A. E., Greenberg, D. H., Vining, A. R., and Weimer, D. L. (2001). *Cost-Benefit Analysis, Concepts and Practice*, Prentice Hall, Upper Saddle River, NJ.

Campden, J. T. (1986). *Benefit, Cost, and Beyond: The Political Economy of Benefit-Cost*, Ballinger Books, Cambridge, MA.

Dandy, D., Walker, D., Daniell, T., and Warner, R. (2008). *Planning and Design of Engineering Systems*, 2nd ed., Taylor & Francis Group, London, U.K.

DeGarmo, E. P., Sullivan, W. G., Bontadelli, J. A., and Wicks, E. M. (1997). *Engineering Economy*, 10th ed., Prentice Hall, Upper Saddle River, NJ.

de Neufville, R., and Stafford, H. (1971). *Systems Analysis for Engineers and Managers*, McGraw Hill, New York.

Ford, K. M., Arman, M. H. R., Labi, S., Thompson, P., Sinha, K. C., and Shirole, A. (2012). Estimating Life Expectancies of Highway Assets, NCHRP Report 713, Transportation Research Board, Washington, DC.

GASB (1999). Statement 34: Basic Financial Statements and Management's Discussion and Analysis for State and Local Governments, Government Accounting Standards Board, Norwalk, CT.

Gruber, J. (2004). *Public Finance and Public Policy*, Macmillan, New York.

Irfan, M., Khurshid, M. B., Bai, Q., Morin, T. L., and Labi, S. (2012). Establishing Optimal Project-Level Strategies for Pavement Maintenance and Rehabilitation—A Framework and Case Study. *J. Eng. Optim.*, 44(5), 565–589.

Lemer, A. (1996). Infrastructure Obsolesence and Design Service Life. *ASCE J. Infratsructure Sys.*, 4(2), 153–161.

Sen, A. (1993). Markets and Freedom: Achievements and Limitations of the Market Mechanism in Promoting Individual Freedoms. *Oxford Economic Papers*, 45(4), 519–541.

Sullivan, A., and Sheffrin, S. M. (2003). *Economics: Principles in Action*. Pearson Prentice Hall, Upper Saddle River, NJ.

Weisbrod, G., and Weisbrod, B. (1997). Assessing the Economic Impact of Transportation Projects, Transportation Research Circular 477, National Research Council, Washington, DC.

Zhuang, J., Liang, Z., Lin, T., and De Guzman, F. (2007). Theory and Practice in the Choice of Social Discount Rate for Cost-Benefit Analysis: A Survey. Asian Development Bank, ERD Working Papers, ADB Publishing, Manila, Philippines.

USEFUL RESOURCES

Dickey, J. W., and Miller, L. H. (1984). *Road Project Appraisal for Developing Countries*, Wiley, New York.

Gramlich, E. M. (1990). *A Guide to Benefit-Cost Analysis*, Prentice Hall, Englewood Cliffs, NJ.

Maas, A. (1966). Benefit-Costs Analysis: Its Relevance to Public Decisions, *Q. J. Econ.*, XXVIII(4), 229–239.

Nas, T. (1996). *Cost-Benefit Analysis: Theory and Application*, Sage Publications, Thousand Oaks, CA.

Stokey, E., and Zeckhauser, R. (1978). *A Primer for Policy Analysis*, W.W. Norton, New York.

USOMB (1992). Guidelines and Discount Rates for Benefit-Cost Analysis of Federal Programs, Circ. A-94, *Fed, Reg.*, 57(218), U.S. Office of Management and Budget, Washington, DC.

Zerbe, R. O., and Dively, D. (1994). *Benefit-Cost Analysis in Theory and Practice*, Harper Collins, New York.

CHAPTER 12

MULTIPLE-CRITERIA ANALYSIS

12.0 INTRODUCTION

Traditionally, decisions at any phase of civil engineering systems development have been governed largely by economic efficiency considerations, which we discussed in the previous chapter. However, governments, organizations, and corporations are increasingly finding that superior and more sustainable long-term solutions can be obtained when evaluation and decision making are based not on only economic returns but on a diverse variety of criteria. This is particularly true in the case of civil engineering systems, where it is typically difficult to adequately quantify all benefits monetarily, the investment impacts have a wide reach spatially and temporally, and numerous stakeholders exist whose concerns often translate into noncommensurate decision criteria.

Figure 12.1 presents some of the criteria that are typically used in decision making at various phases of civil system development. These criteria typically include the civil engineering system's technical efficiency and effectiveness, environmental impacts, compatibility with existing policy and legislation, social and cultural impacts, economic efficiency, and impacts on the local or regional economy. The extent to which each criterion is considered differs for each system type and also from phase to phase.

The diversity often found in a large number of decision criteria is a two-edged sword: It helps to accommodate the concerns of different stakeholders; however, it also gives rise to the analytical problems associated with the differences in measurement units of the different criteria. For example, the system owner's costs are expressed in dollars; the community's air quality impacts are quantified in terms of pollutant emissions or concentrations; and system users' concerns are often expressed in terms of hours of delay or inconvenience. There is thus a need to identify and apply appropriate evaluation techniques that address such dimensional inconsistencies that are associated with multiple-criteria decision problems.

The multiple-criteria nature of evaluation pervades most contexts of decision making anywhere, even in our personal lives; we face situations daily where we must make decisions on the basis of several different considerations. In certain situations, it is possible to convert each consideration into its monetary equivalent. For example, in deciding whether to lease a new car or buy a used car, a student could consider the initial costs of purchasing versus leasing, license and registration costs (which will be higher for a new car), fuel costs (which generally will be lower for a new car), and maintenance costs (which will be lower for a new car). Where a decision must be made solely from economic efficiency considerations, the equivalent annual uniform cost of each alternative, for example, could be used as a basis for the comparison and decision. However, when nonmonetary considerations enter the picture (e.g., the student feels old cars are preferable because they have less risk of being stolen, prefers the new gadgets found only in new cars, or simply likes the "new car" aroma), it becomes difficult to use economic efficiency only. In such decision situations, the student needs to not only intuitively establish the relative importance of each consideration (which we shall refer to as decision criteria or evaluation criteria) but also to make mental projections of each consideration on a commensurate scale that "normalizes" their different units. For this and

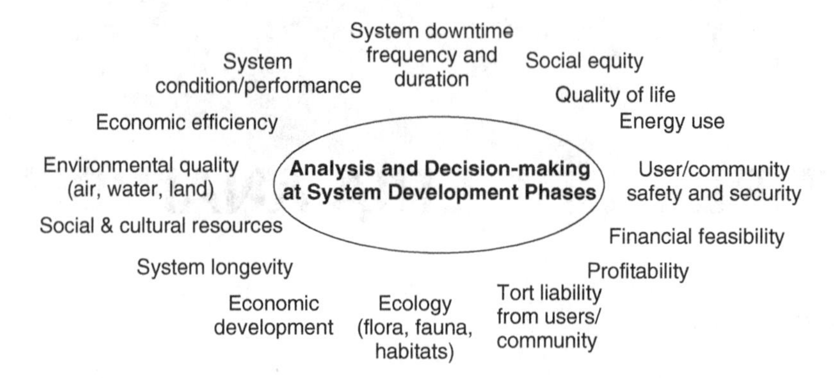

Figure 12.1 Criteria typically considered during decision making at each phase.

other small-scale problems, these processes often occur implicitly in the human brain and yield a final decision in this illustration, a choice by the student to buy a new or used car.

There is a strong rationale for the application of multiple-criteria analysis in civil engineering systems management. At each phase of a system's development, engineers and managers seek to evaluate and select alternative courses of action in a transparent fashion that duly incorporates the perspectives of multiple stakeholders. Analytical tools of multicriteria decision making can help engineers and managers to structure their decision-making processes in a manner that is not only representative of the concerns of the different stakeholders but is also comprehensive, well defined, rational, documentable, and defensible. The use of multicriteria tools can also provide a solid basis upon which systems managers can carry out "what-if" analyses and to examine the trade-offs between competing decision criteria, risk levels, performance thresholds, or funding levels.

In this chapter, we first discuss the Pareto frontier, an important concept in multiple-criteria analysis. Then, recognizing that several different general formulations for multiple-criteria analysis exist, we focus on a specific formulation: the multiattribute utility theory. In presenting this formulation, we discuss the three processes of weighting (specifying the relative importance between the decision criteria), scaling (establishing, for each decision criterion, a scale of desirability associated with the different levels of the criterion), and amalgamation (combining the weighted and/or scaled values of all the decision criteria for each alternative) and then making a decision based on these amalgamated values. We also examine in this chapter, the two categories of approaches by which information is acquired for carrying out any of the three processes of multiple-criteria analysis: approaches that incorporate the decision maker's preferences and those that do not.

12.0.1 Basic Concepts in Multiple-Criteria Decision Making (MCDM)—Illustration

As we saw in the previous section, in our daily lives, we often face situations where we need to make decisions on the basis of a variety of criteria that often have different units. In this section, we use another example to further our understanding of the basic concepts. In making a choice of which of several apartment locations a student should rent next semester, the considerations may include the rent amount (in dollars), the grace period for the rent payment after the end of the month (number of days), and the distance from the nearest campus bus station (feet) (Table 12.1). Most students weigh these choices mentally and make a decision without any formal and mathematical statement of the problem. This is acceptable as long as the problem size is small and we do not need to render account of our final decision to anyone. Otherwise, we must depart from the mental decision-making process in favor of a formal process where we need to carry out at least two of the following tasks: (a) convert the units of each decision criterion into one that is commensurate with other

Table 12.1 Apartment Selection Problem

Apartment Complex	A	B	C	D	E	F	G	H	I	J	K	L	M	N
Rent amount ($)	320	400	700	750	820	500	350	300	800	900	450	300	450	750
Distance to bus stop (ft)	100	250	100	150	120	150	150	300	100	100	150	200	250	150
Rent payment grace period (days)	7	3	1	2	7	3	5	2	2	4	1	4	6	6

criteria; for example, for each alternative in the apartment example, we could normalize the rent amounts by dividing the rent of any apartment location by the lowest rent of any apartment and also carry out similar normalization for the other decision criteria; (b) establish the relative importance among the decision criteria; for example, how much importance does the student attach to the rent amount compared with the distance from the bus stop; (c) establish a way to find the combined overall desirability or undesirability of each apartment alternative; (d) establish a Pareto frontier that shows all the superior alternatives (apartment locations that are equal and most superior) as well as those that are inferior solutions; and (e) identify the best apartment to rent. Certain MCDM frameworks exclude task (a). In carrying out tasks (a), (b), or (c), we may or may not utilize the preferences of the decision maker; if the decision-maker's preferences are utilized, they could be incorporated at the early, intermediate, or later stages of the evaluation process, which we discuss in Section 12.5. Assuming that more than one apartment location turns out to be the best option equally, then these best-choice apartment locations will be seen to lie on what we call the **Pareto frontier** (we discuss this in the next section).

12.0.2 The Pareto Frontier

In multicriteria decision making where several alternative decisions or actions (which we shall often refer to as "candidates" in this chapter) are being examined for purposes of selecting the best altnative (which we shall call the "solution" or "optimal decision" in this chapter), it is typically difficult to identify a single solution that dominates (i.e., is superior to) all others from the perspective of each of all the individual decision criteria. Also, there are many solutions that are not dominated by others; as we learned in Chapter 8, these solutions are described as *Pareto optimal* solutions (Kuhn and Tucker, 1951; Hazen and Morin, 1983). The set of all Pareto solutions is referred to as the *Pareto frontier* (Nakayama et al., 2009). Thus, each solution on the Pareto frontier is not dominated by any other feasible solution. Pareto frontiers are important because they help decision makers compare or trade-off between alternatives and also to examine the real preference structure of the decision makers.

Pareto frontiers may be concave or convex in shape. Also, they could be two dimensional or more than two dimensions. The dimensions represent the decision criteria, and one of dimensions could be the economic efficiency for example, net present value. Examples of Pareto frontiers are presented in Figure 12.2. Figure 12.2*a* illustrates the simplest case of multiple criteria in terms of dimensionality (number of decision criteria). For this bi-objective decision-making problem where it is sought to identify the best alternative that minimizes both decision criteria f_1 and f_2, the points above the curve represent all the feasible solutions. It is seen that the solutions on the curve are not dominated by any other solutions; together, these solutions constitute the Pareto frontier of the problem. All other solutions that are not on the Pareto frontier are dominated by at least one of the Pareto frontier solutions. Where the decision-making problem involves more than two decision criteria, the Pareto frontier is a three-dimensional surface or hyperplane (Figure 12.2*b*). As we learned in Chapter 2 and also in the introduction to this chapter, practicing civil engineers in the real world

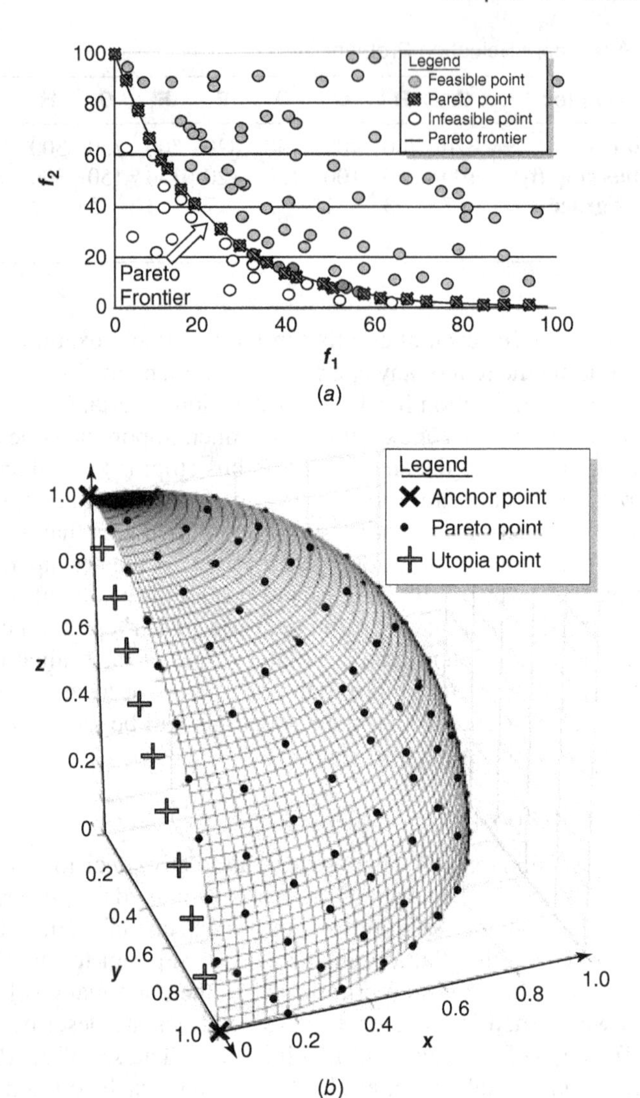

Figure 12.2 Illustration of Pareto solutions and the Pareto frontier (adapted from Bai, 2012; adapted from Utyuzhnikov et al., 2009): (*a*) Convex frontier shown with two decision criteria and (*b*) concave frontier shown with three decision criteria.

encounter more than just one or two decision criteria in their decision-making problems; hence, they typically deal with Pareto frontiers that cannot be visualized in 2D or 3D but are hyperplanes that can be expressed only as mathematical equations.

12.1 FRAMEWORK FOR MCDM ANALYSIS USING MULTIATTRIBUTE UTILITY THEORY

In any decision-making process, there is often a need to directly or indirectly establish the preference order among the competing alternative actions or candidates, thereby making it possible to rank the alternatives and ultimately to identify the best alternative. For decision-making problems that involve a single criterion, the preference order is accomplished rather easily; for example, in the case of decisions to be made on the basis of cost minimization, the preference order is based on "increasing cost" and the lowest-cost alternative is chosen as the best alternative. However, where multiple criteria exist (as shown in the previous apartment selection example), identifying the optimal alternative is more complicated.

One approach is to reduce the multiple-criteria problem to a single-criterion problem that inherently incorporates all the criteria. This section presents a framework for doing this. The framework, which is based on the multiattribute utility theory, represents only one of several ways in which a MCDM problem can be formulated. The framework involves the construction of a *preference order* through a direct elicitation of the preferences of the decision makers and is consistent with *utility theory*. Thus, the underlying assumption is that the decision-makers' preference structure can be represented in the form of a real-value function referred to as a *utility function*. After constructing such a function, the alternatives are evaluated and selected via a simple ranking or optimization process. The key steps of the framework (illustrated as Figure 12.3) are:

Establish decision criteria. The decision makers specify the criteria upon which each alternative will be evaluated.

Weighting. Relative weights are assigned to each of the decision criteria.

Scaling. Since the decision criteria typically have different units, scaling provides a common scale of measurement by converting the unit of each decision criteria to, say, a 0–10 scale. This step involves the development of single-criterion *utility* functions.

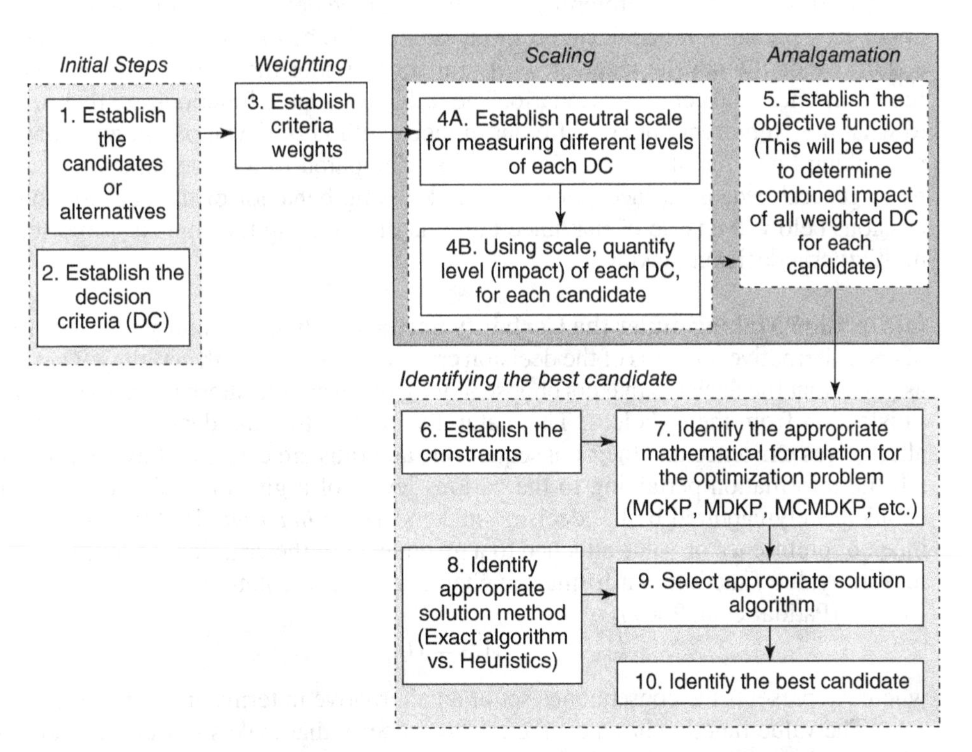

Figure 12.3 Steps for multicriteria decision making using the multiattribute utility theory (MAUT) formulation.

Amalgamation. This step combines the different impacts of an alternative in terms of the various decision criteria, to yield a single representative overall, new decision criterion called *total utility*. Each decision criterion has a relative weight (developed at the weighting step) and a utility function (developed at the scaling step). The exact nature of the amalgamation process is a reflection of the mathematical assumptions about the decision-maker's preference structure.

The very essence of multicriteria decision making is a clear statement of all the relevant decision criteria as they relate to the optimal solution. In order to adequately describe the consequences of the alternative actions and to analyze the trade-offs between the criteria, it is necessary to identify, early in the evaluation process, all the appropriate decision criteria. Chapter 3 provides a comprehensive discussion of civil engineering system goals, objectives, and performance measures (from which the decision criteria originate). Chapter 3 also discusses the desirable properties not only for *individual* performance measures or criteria but also for the *set* of criteria selected for a decision-making task. In Section 12.5, we discuss each of the key steps of the framework shown in Figure 12.3. However, before we do that, it is necessary to take a detour that traverses the terrain of probabilistic considerations. The detour (Section 12.1.1) will be reminiscent of Chapter 5 and Chapter 8, and will show us that like all other problems, multiple-criteria decision-making problems can be analyzed from both certainty and uncertainty perspectives.

12.1.1 Incorporating Certainty or Risk/Uncertainty in the MCDM Framework

Each of the three processes of weighting, scaling, and amalgamation, can be characterized by certainty or risk/uncertainty. Preference structures can be developed for each of these three processes under each scenario of certainty risk or uncertainty. To be mathematically precise, the scaling procedure yields the development of *utility* functions under the risk/uncertainty scenario and *value* functions under the certainty scenario. The utility and value functions are thus similar with regard to what they represent but are different in their method of development and how they capture the risk-taking behavior of the decision makers. Compared to a value function, a utility function is more general because it incorporates the risk-taking behavior of the decision makers. In the discussions below, the issue of risk/uncertainty is discussed in the context of the scaling process for multicriteria decision making.

(a) Decision Making under the Certainty Scenario. In the certainty scenario, the consequences of each alternative, in terms of the decision criteria, are known with certainty. The underlying theory assumes that the decision makers choose among the available alternatives so as to derive maximum satisfaction from their choices. This naturally implies that the decision makers are aware of the alternatives existence and their consequences, and thus are capable of evaluating them effectively. All the information pertaining to the various levels of a given decision criterion are assumed to be adequately captured by the decision-makers' *value function*. This function represents a scalar index of preference or value attached to a given level of the criterion. In other words, the decision-makers' value function is a formal, mathematical representation of their preference structure as follows (Patidar et al. 2007):

$$v(z) = v(z_1, z_2, \ldots, z_j) \tag{12.1}$$

where z represents the consequence set of an alternative in terms of j decision criteria: $z_1, z_2, \ldots, z_i$.

The value function has a characteristic property that makes it useful for addressing the issue of trade-offs among multiple-decision criteria (Keeney and Raiffa, 1976): $v(z') > v(z'')$ if and only if z' is preferred to z''.

An illustration of a multivariate value function is one in a three-dimensional space that assigns a scalar value to each possible combination of system condition and system safety. Such multivariate functions help capture the decision-makers' preferences precisely but may not always be useful from a practical standpoint. Furthermore, the function's complexity increases as the number of dimensions (decision criteria) increases; thus, developing multivariate value functions to capture the decision-makers' preferences could be a daunting exercise. one way to overcome this difficulty is to reduce the dimensionality. To reduce dimensionality, decision theorists have often decompose the multivariate value function into several constituent single-criterion value functions and then develop value functions separately for each criterion. In line with that approach, we present the decomposed functional form and its underlying assumptions.

With regard to the functional form of the value function, a key theorem in value theory states: Given the decision criteria $z_1, z_2, \ldots, z_p$, an additive value function,

$$v(z_1, z_2, \ldots, z_p) = \sum_{i=1}^{p} v_i(z_i) \tag{12.2}$$

exists if and only if the criteria are *mutually preferentially independent* (Keeney and Raiffa, 1976 Patidar et al., 2007); and v_i is a single-criterion value function over the criterion Z_i.

The Concept of Mutual Preferential Independence. Consider two subsets, X and Y, of a larger set of decision criteria $Z \equiv \{Z_1, Z_2, \ldots, Z_p\}$ that are mutually exclusive and collectively exhaustive. The set of decision criteria X is *preferentially independent* of the complementary set Y if and only if the conditional preference structure in the $\mathbf{x}$ space given $\mathbf{y}'$ does not depend on $\mathbf{y}'$ (Keeney and Raiffa, 1976). In other words, the preference structure among the decision criteria in set X does not depend on the levels of the decision criteria in Y. That is, if $(\mathbf{x}_1, \mathbf{y}_0)$ is preferable to $(\mathbf{x}_2, \mathbf{y}_0)$, then $(\mathbf{x}_1, \mathbf{y})$ is preferred to $(\mathbf{x}_2, \mathbf{y})$ for all $\mathbf{y}$. Overall, the set of decision criteria Z is mutually, preferentially independent if every subset X of these decision criteria is preferentially independent of its complementary set of decision criteria (Sinha and Labi, 2007).

For example, in the context of construction engineering systems, consider three decision criteria used to evaluate alternative contracting approaches: time (T), cost (C), and product quality (P). If these three criteria are mutually, preferentially independent, the decomposed value function is

$$v(\text{alternative}) = v(T) + v(C) + v(P)$$

In the value function above, the scaling of each criterion is exemplified by the single-criterion value functions. A common technique to simplify the assessment procedures of the single-criterion value functions as defined above is to represent explicitly the scaling as follows (Patidar et al., 2007):

$$v(\text{alternative}) = w_1 v(T) + w_2 v(C) + w_3 v(P)$$

where w_i are referred to as *relative weights.*

(b) Decision Making under the Risk/Uncertainty Scenario. In this scenario, the problem is characterized by uncertainty. The underlying issues remain the same as those of the certainty case, however the difficulties are more challenging because the exact consequences of each alternative in terms of the decision criteria are not known; each consequence (outcome in terms of a decision criterion) of an alternative is associated with some probability (Patidar et al., 2007).

The usefulness of the concept of utility and multiattribute utility theory are felt particularly in the analysis of decision-making problems that are characterized by risk/uncertainty. The utility of an alternative is a random variable, and the expected utility refers to the first moment or mean of the random variable. Through the assignment of an appropriate utility amount to each possible outcome and calculation of the expected utility of each alternative, the best alternative can be identified as that which yields the maximum expected utility (Keeney and Raiffa, 1976). Thus, a typical application of multiattribute theory involves the following sequence (Goicoechea et al., 1982; Patidar et al., 2007): First, make a number of assumptions regarding the decision-makers' preferences, derive an appropriate functional form on the basis of these assumptions, and then verify the appropriateness of these assumptions through consultation with the decision maker. Next, develop the preference orders, that is, the utility functions, for each decision criterion using the appropriate functional form and the established relative weights of the decision criteria; and finally, establish the preference order for each alternative on the basis of its expected utility.

A multiattribute utility function (Keeney and Raiffa, 1976) captures the decision-makers' preferences regarding the levels of each decision criterion. It is similar to a value function, but it also captures the decision-makers' risk preferences for different levels of each attribute. The expected values of the utility function are then used as a basis to compare the alternatives. The alternative with the maximum expected utility value is then identified as the most preferred alternative. However, as stated in Patidar et al. (2007), establishing such multiattribute functions can be an extremely difficult task due to the multiplicity of dimensions, therefore, is not used to reduce the dimensionality. Instead, several single-criterion (univariate) utility functions are developed separately and then weighted and amalgamated to yield a single-utility function that comprises the multiple decision criteria or attributes. We now discuss the decomposed functional form and its underlying assumptions.

Functional Form of the Utility Function. A key theorem in utility theory states that, given the decision criteria $Z_1, Z_2, \ldots, Z_p$, the multiplicative utility function

$$ku(z_1, z_2, \ldots, z_p) + 1 = \prod_{i=1}^{p} [kk_i u_i(z_i) + 1] \tag{12.3}$$

where u_i is a single-criterion utility function over the criterion Z_i. This function exists if and only if the decision criteria are *mutually utility independent* (Keeney and Raiffa, 1976), where k and k_i are scaling constants.

The Concept of Mutual Utility Independence. Consider two subsets of the set of decision criteria $Z \equiv \{Z_1, Z_2, \ldots, Z_p\}$ that are mutually exclusive and collectively exhaustive: X and Y Sinha and Labi, 2007. The set of criteria X is described as *utility independent* of set Y if and only if the conditional preference order for *lotteries* involving only changes in the levels of attributes in X does not depend on the levels at which the attributes in Y are held fixed.

That is, if $<x_1, y_0>$ is preferred to $<x_2, y_0>$, then $<x_1, y>$ is preferred to $<x_2, y>$ for all y.

The symbol $<>$ represents a lottery that captures the decision-makers' risk preference under conditions of uncertainty. The set of criteria Z are mutually utility independent if every subset X of these criteria is utility independent of its complementary set of criteria. Another key theorem states the existence of an additive utility function: Given the decision criteria $Z_1, Z_2, \ldots, Z_p$, an additive utility function

$$u(z_1, z_2, \ldots, z_p) = \sum_{i=1}^{p} k_i u_i(z_i) \tag{12.4}$$

(where u_i is a single-criterion utility function over Z_i) exists if and only if the additive independence condition is valid among the criteria (Keeney and Raiffa, 1976). This means that the preferences over lotteries on $Z_1, Z_2, \ldots, Z_p$ are influenced only by their marginal probability distributions and not by their joint probability distribution (Sinha and Labi, 2007).

For a given problem, the applicability of the multiplicative or additive functional forms for the multiattribute utility function depends on the nonviolation of the underlying assumptions as stated in the theorems above. The appropriateness of these assumptions can be checked by eliciting information from the decision maker, and an appropriate functional form can be identified using the results of such questionnaire surveys.

Example 12.1

The mayor of a city has asked you to be a consultant to help choose the best design for a structure to support an overhead water reservoir for the city. You have been asked to evaluate three alternative designs on the basis of three performance criteria: cost, durability, and aesthetics. The weights of each performance measure to reflect their relative importance compared to each other are: cost = 0.35; durability = 0.45; aesthetics = 0.20. The utility functions are given in Figure 12.4. The levels of each performance measure for material are given in Table 12.2. Which of the three designs would you recommend, on the basis of their combined performance criteria?

Solution

Table 12.3 presents the scaled, weighed, and amalgamated performance corresponding to each alternative. In Table 12.3, "Scaled" means scaled value (or utility) of the performance measure for that design. "Weighted and scaled" means scaled value (or utility) multiplied by the weight, of the performance measure, for that design material. Clearly, the best choice is Design C.

As the reader may have surmised by now, the specified weights of the performance criteria, the nature of the scaling functions, and the manner of amalgamation greatly influence the choice of optimal alternative. As such, it is important that these three aspects of the multicriteria problem are developed as carefully as possible.

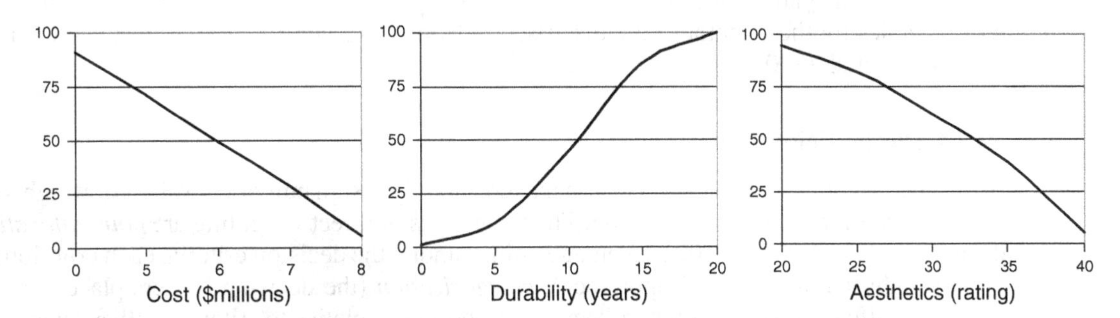

Figure 12.4 Utility functions for scaling the performance criteria.

Table 12.2 Data for Example 12.1

Design	Cost ($Millions)	Durability (Years)	Aesthetics (0–40 Rating)
A	7.7	19	35
B	4.5	8	21
C	5.9	13	28

Table 12.3 Solution to Example 12.1

Alternative \ Performance Criteria	Cost ($Millions)	Durability (Years)	Aesthetics (Rating)	Overall Value or Utility
	Weight = 0.35	Weight = 0.45	Weight = 0.20	
Design A	Scaled = 12 Scaled and weighted = 4.2	Scaled = 97 Scaled and weighted = 43.65	Scaled = 38 Scaled and weighted = 7.6	4.2 + 43.65 + 7.6 = 55.45
Design B	Scaled = 82 Scaled and weighted = 28.7	Scaled = 32 Scaled and weighted = 14.4	Scaled = 92 Scaled and weighted = 18.4	61.5
Design C	Scaled = 52 Scaled and weighted = 18.2	Scaled = 72 Scaled and weighted = 32.4	Scaled = 71 Scaled and weighted = 14.2	64.8

12.2 ESTABLISHING THE WEIGHTS OF DECISION CRITERIA

In the multicriteria decision-making framework illustrated in Figure 12.3, a key task is to assign relative weights to each decision criterion to reflect its relative importance compared to other criteria; for example, to what extent is system user safety more important than system condition? It is important to realize that relative weights may change from time to time and across locations to reflect different circumstances and policies of system owners and operators. As such, it is often useful for the decision maker to carry out sensitivity analysis to investigate the nature of the final solution with respect to different sets of relative weights. The methods often used to establish the weights of decision criteria include: equal weighting, regression-based observer-derived weighting, direct weighting, gamble method, analytical hierarchy process (AHP), and value swinging. details, and examples for these methods are discussed in Hobbs and Meier (2000); Sinha and Labi (2007), and Patidar et al. (2007).

12.2.1 Direct Weighting

In this method of weighting, the decision makers directly assign numerical values of weights to each decision criterion under consideration. The three types of direct weighting are *point allocation* (the decision makers distribute 100 percentage points among the decision criteria, each criterion receiving a weight that reflects its importance); *categorization* (the decision makers place the decision criteria in different categories that reflect their importance relative to criteria in other categories; and *ranking* (the decision makers assign a rank to each decision criterion in the order of its importance, the highest rank of 1 is assigned to the criterion of highest importance).

Ranking provides an ordinal scale of importance while point allocation provides a cardinal scale. Where the decision makers seek to use the weights of the decision criteria in a multivariate value or utility function, it is recommended to use point allocation (which is consistent with cardinality) for direct weighting. In certain cases, the number of decision criteria is too large, and weighting by categorization is recommended. Direct weighting is simple and easy to implement. As Patidar et al. (2007) pointed out, however, direct weighting may not always capture the preferences of the decision makers in an effective manner, compared to other weighting methods.

12.2.2 Weighting Using the Analytic Hierarchy Process

The analytic hierarchy process (AHP), which is based on decomposition, comparative judgments, and the synthesis of priorities, develops weights that reflect the relative importance of multiple decision criteria in a manner that accommodates differences in the decision-makers' opinions. AHP, which is capable of addressing different types of criteria (qualitative, quantitative, tangible, and intangible) establishes a hierarchy at each level of the hierarchy and makes pairwise comparisons of the decision criteria to develop the weights (Saaty, 1980).

Consider, as in Sinha and Labi (2007), $z(i), i = 1, 2, \ldots p$ the set of decision criteria at a given level. Also, consider the quantified judgments on a pair of decision criteria $z(i), z(j)$ that are represented by the following matrix M with entries a_{ij}:

$$\begin{pmatrix} 1 & a_{12} & \ldots & a_{1p} \\ 1/a_{12} & 1 & \ldots & a_{2p} \\ \ldots & \ldots & \ldots & \ldots \\ 1/a_{1p} & 1/a_{2p} & \ldots & 1 \end{pmatrix}$$

Assuming that the measurements are exact, the weight of criterion i relative to j would be given by

$$\frac{w_i}{w_j} = a_{ij} \quad (\text{for } i, j = 1, 2, \ldots, p) \tag{12.5}$$

In reality, however, measurements are not exact, and the matrix is not necessarily consistent. Therefore, to accommodate deviations, the above expression becomes

$$w_i = \frac{1}{p} \sum_{j=1}^{p} a_{ij} w_j \quad (\text{for } i = 1, 2, \ldots, p) \tag{12.6}$$

If a unique solution is to exist, then the above expression can further be reduced (Saaty 1980):

$$M'w' = \lambda_{\max}\, w' \tag{12.7}$$

where M' is the reciprocal matrix, which is a perturbation of M; w' is the eigenvector of M'; and $\lambda_{\max}$ is the largest eigenvalue of the matrix M'.

Thus, the AHP process, using the above theorem, is reduced to one that involves simply developing a matrix of pairwise comparisons between the decision criteria and determining the value of the eigenvector, which reflects the relative weights between the criteria (Patidar et al., 2007). Example 12.2 presents a simple illustration of determining an approximate value of the eigenvector.

Example 12.2

A survey of systems experts in a region indicated that they view system condition as strongly more important than user safety; system condition is moderately more important than community compatibility; and user safety is slightly less important than community compatibility. Construct a matrix based on the results of this preference survey and determine the relative weights of these criteria. Use the ratios indicated in Table 12.4.

Solution

The resulting matrix is as shown in Table 12.5.

Table 12.4 Pairwise Comparison Ratios

IF:	Then the Ratio i/j Is
Criterion i is extremely more important than criterion j	9
Criterion i is strongly more important than criterion j	7
Criterion i is moderately more important than criterion j	5
Criterion i is slightly more important than criterion j	3
Criterion i is equally important to criterion j	1
Criterion i is slightly less important than criterion j	$\frac{1}{3}$
Criterion i is moderately less important than criterion j	$\frac{1}{5}$
Criterion i is strongly less important than criterion j	$\frac{1}{7}$
Criterion i is extremely less important than criterion j	$\frac{1}{9}$

Table 12.5 Matrix for Decision Criteria

	System Condition	User Safety	Community Compatibility
System Condition	1	7	5
User safety	—	1	$\frac{1}{3}$
Community compatibility	—	—	1

The process to derive the relative weights from these numbers is as follows: Enter the corresponding reciprocals of the upper triangle entries in the lower triangle of the matrix; then divide each entry in column j of the matrix M by the sum of the entries in column j. This yields a new matrix denoted as

$$M_{\text{norm}} = \begin{pmatrix} 0.7447 & 0.6364 & 0.7895 \\ 0.1064 & 0.0909 & 0.0526 \\ 0.1489 & 0.2727 & 0.1579 \end{pmatrix}$$

The relative weight w_i is then estimated by calculating the average of the entries in row i of the matrix. Doing this normalizes and averages the scores to yield the relative weights of the decision criteria as follows:

System Condition: 0.724

User Safety: 0.083

Community Compatibility: 0.193

Thus, each element of the eigenvector corresponds to the relative weight of a decision criterion. The eigenvector also can be determined directly using mathematical software. If the software reports the eigenvector in its already normalized form, then it is read directly from the software output and reported as the relative weights.

The AHP is well suited to civil engineering systems decision making because a natural hierarchy of criteria inherently exists within them for decision-making processes. Also, the weights generated using AHP can easily be incorporated in MCDM methods that use additive multivariate value functions or utility functions. A disadvantage of AHP, however, lies in its inability to adequately deal with a large number of decision criteria because the number of pairwise comparisons becomes too large and the decision makers may fail to develop their relative importance in a manner that is realistic and consistent with engineering judgment.

12.2.3 Observer-derived Weighting

Consider the apartment complex selection scenario in the chapter introduction where students are evaluating several apartment location alternatives for the next semester and are making the assessment on the basis of the following decision criteria: rent amount, proximity to nearest bus stop, and rent payment grace period (days). Assume that several students are presented with the respective rent amounts, distance to nearest bus stops, and rent payment grace periods of each alternative apartment complex. On the basis of the values of the three decision criteria for each alternative, the students indicate their overall preference for each apartment complex as a score that ranges from 0 to 10 or 1 to 100. A functional relationship is then established with no explicit weighting of these decision criteria by the students, rather they are mentally (implicitly) assigning weights to the decision criteria as they indicate their preferences for each alternative in the form of an overall preference rating. If the overall rating is denoted by Y (dependent variable) and the values of the decision criteria are denoted by X_1, X_2, and X_3, then a statistical relationship can be developed using the total score, Y, as the response variable and the X's. The resulting coefficients of the regression model will be the implicit weights attached to the decision criteria by the student. This weighting technique, where the decision maker is made to assign weights without realizing that they are making such assignments, is referred to as observer-derived weights (Hobbs and Meier, 2000).

Patidar et al. (2007) stated that this weighting method could be advantageous due to its relative simplicity and that it is a "policy-capturing" technique employed by pollsters and psychologists to develop weights on the basis of the unaided perspectives of the decision makers. A disadvantage is that it merely simulates, but does not improve, the holistic judgments of the decision makers. Second, people tend to consider only a small set of decision criteria when assigning overall scores or making decisions in general; hence, this weighting method may not be very effective for problems with a large number of decision criteria.

Example 12.3

Early in this chapter, we presented Table 12.1, which shows the levels of the following criteria: rent amount ($), distance to nearest bus stop (ft), and rent payment grace period (days) for 14 apartment complexes on a campus. In Table 12.6 below, we have added the total of the average scores assigned by the respondents of a student survey to each alternative. Using these results, derive the relative weights of these decision criteria.

Table 12.6 Total Scores for Each Alternative

Apartment Complex	A	B	C	D	E	F	G	H	I	J	K	L	M	N
Total Score	95	75	40	50	45	65	85	75	40	20	75	85	88	75
Rent Amount ($)	320	400	700	750	820	500	350	300	800	900	450	300	450	750
Distance to Bus Stop (ft)	100	250	100	150	120	150	150	300	100	100	150	200	250	150
Rent Payment Grace Period (days)	7	3	1	2	7	3	5	2	2	4	1	4	6	6

Solution

Using any standard statistical software package to model Y (the total of the average scores the students assigned to each alternative) and X_1 (the rent amount), X_2 (the distance to nearest bus stop), and X_3 (the rent payment grace period), the following regression model is obtained:

$$\text{Score} = 1.077 - 7.9\ (\text{rent amount}) - 4.7\ (\text{distance to nearest bus stop}) + 2.7\ (\text{rent payment grace period})$$

Thus, the relative weights are as follows:
Rent amount: −7.9
Distance to nearest bus stop: −4.7
Rent payment grace period: +2.7

Clearly, the relative weights, as determined using this technique, could be plagued by serious bias due to the different scales of the criteria. As such, if weighting is to be carried out using this method, it is recommended to carry out scaling prior to using this weighting technique.

12.3 SCALING OF THE DECISION CRITERIA

A key step in multicriteria decision making is to establish a common scale of measurement across the decision criteria so that they can be expressed in commensurate units. In that way, for each alternative, the overall desirability of an alternative, in terms of the individual decision criteria, can be reported as a single amount. There is no need for scaling for problem contexts where the units are already commensurate (e.g., problem involving fares or fees incurred by system users and construction and maintenance costs incurred by the system owner, where all the decision criteria are expressed in monetary units). On the other hand, in problem contexts involving decision criteria that have different units of measurement, such as the apartment selection example discussed earlier in this chapter, it is necessary to carry out scaling to reduce all the different units of measurement to the same unit. In the management of civil engineering systems, engineers typically encounter noncommensurate decision criteria, such as the number of jobs (economic development), economic efficiency (NPV, BCR), number of fatalities (safety of system users or the community), facility vulnerability, community cohesion, and environmental quality.

Scaling therefore provides a common scale or unit of measurement for all the decision criteria in a given problem and translates the decision-makers' preferences for each criterion into a dimensionless value. The first step in scaling is to establish the scale for measuring the decision criterion. In the second step, the scale is used to quantify the level of each criterion for each alternative using the value or utility functions. The value and utility function approaches are used when the decision making is carried out under the certainty and risk scenarios, respectively. In this section, we will examine a number of techniques that decision makers could use to carry out scaling.

Before we continue with the discussion on scaling, it would be helpful to discuss some terminological similarities in order to avoid confusion. The various decision criteria encountered in civil engineering systems management typically have different units (e.g., the structure's condition may be expressed in terms of the percentage of corroded area, the total length of cracks, or the structure's remaining life). In some cases, engineers "normalize" the raw values of the decision criteria to account for differences in size or in the level of demand or usage across the different alternatives in order to avoid bias. For example, consider two alternatives that have the same number of surface defects; however, one alternative is larger in size compared to the other. A traditional way to reduce these values for analysis is to express the decision criterion as a value per the size dimension of the structure. Also, consider the apartment complex selection example discussed in Example 12.6. If apartment 10 (with $900 monthly rent) has two bedrooms while all the other apartments have one bedroom, then it is necessary to adjust the rent amount for apartment 10 (e.g., divide by two to obtain a $450 monthly rent for that apartment) before the multicriteria analysis is carried out. Certain texts refer to such adjustments as "scaling." In this text, we refer to them as size adjustments. While such scale adjustments are necessary to avoid comparison bias, they are different from the term scaling that we discuss in this chapter.

As we will discuss in Section 12.5, techniques in multicriteria decision making may or may not be based on the preferences of the decision makers. Scaling techniques are no exception, and Figure 12.5 presents the categories of scaling techniques: the non-preference-based and the preference-based techniques. The outcome of the scaling process is a set of functions, one for each decision criterion, that represents the worth or desirability of each level of the criterion. For example, the most preferred level of the decision criterion is assigned a value of 1.0 (or 100%) and zero (or 0%) is assigned to the least-preferred level. Within the range of the worst to the best, the decision makers can assign a scaled value to represent the impact of each alternative in terms of each decision criterion.

The non-preference-based techniques include linear scaling and monetization. The preference-based techniques are considered by some as subjective because they are developed on the basis of expert opinion through questionnaire surveys. An opposing school counters that it is rather the non-preference-based techniques that are inferior; that linear scaling, which is often consistent with the non-preference methods, is inherently flawed as the costs and benefits of situations and actions are often nonlinear, and monetization is actually a reliable reflection of personal perceptions of value and how much that individual is willing to pay to gain some benefit or to avoid some disbenefits, and thus may even be considered as inherently preference based.

In the preference-based scaling methods, the results of a questionnaire survey of decision makers are used to construct a relationship showing the various levels of a given decision criterion and the desirability of each level of that criterion (on a scale of 0–1, 0–10, or 1–100) from the perspective of the decision makers. The decision-makers group may include the system owner's

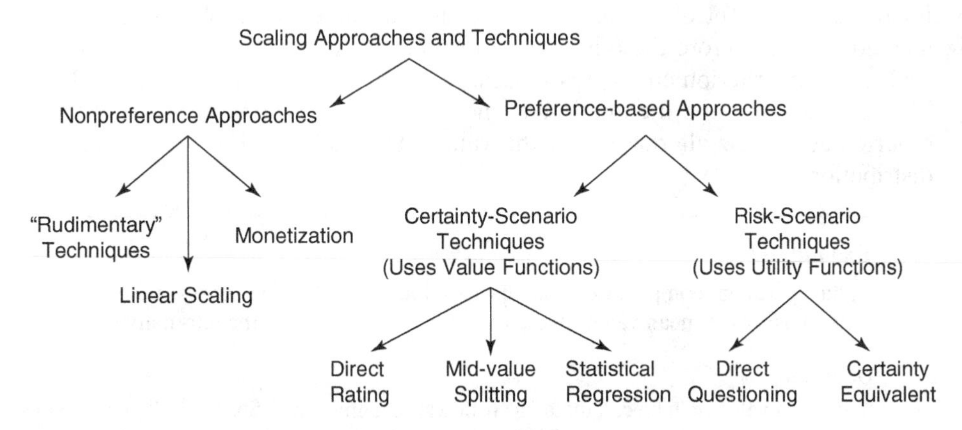

Figure 12.5 Categories of scaling techniques.

managers, subject matter experts, system users, the community, and other stakeholders. If the survey is repeated for each decision criterion, the result is a normalized scale where all the criteria have consistent units—utility or value. This can be used as a basis to compare or combine the different decision criteria that originally had different units.

Consider a decision problem with n decision criteria $(X_1, X_2, \ldots, X_n)$. Assume that $(x_{i1}, x_{i2}, \ldots, x_{in})$ and $(x_{j1}, x_{j2}, \ldots, x_{jn})$ are the outcomes of alternatives i and j in terms of these decision criteria. If it is possible to establish a scalar-valued function $v()$ with the following property:

$$v(x_{i1}, x_{i2}, \ldots, x_{in}) \geq v(x_{j1}, x_{j2}, \ldots, x_{jn}) \Longleftrightarrow (x_{i1}, x_{i2}, \ldots, x_{in}) \succsim (x_{j1}, x_{j2}, \ldots, x_{jn}) \quad (12.8)$$

where the symbol $\succsim$ means "preferred to or indifferent to," then the function $v()$ could be referred to as a value function or utility function (Keeney and Raiffa, 1976).

Scaling functions can be categorized into value functions and utility functions. The difference between the two lies in the level of certainty associated with the outcomes (in terms of the decision criteria) of the alternatives under consideration. For instance, when we rehabilitate a levee, the change in structural performance is not known with certainty. Where there is more certainty than uncertainty regarding the outcome of an action, the resulting scaling function is referred to as a value function; in the uncertainty condition, it is called a utility function. Certainty can be generally considered as a more specific case of uncertainty. So, in a general sense, a value function is a specific case of the utility function where uncertainty is zero. Similar to value functions, utility functions incorporate the innate values that the decision makers attach to the different levels of the decision criterion; however, unlike value functions, utility functions capture the decision-makers' attitudes toward risk (risk averse, risk neutral, or risk taker). We shall discuss this further in Section 12.3.2(b).

In this section, we will first discuss the non-preference-based scaling techniques (rudimentary techniques, linear scaling, and monetization) and at least one preference-based technique (the direct rating method).

12.3.1 Scaling Techniques

(a) Some Rudimentary Scaling Techniques. A number of applications of multicriteria decision making have been carried out in past practice using rudimentary scaling techniques. We use the word "rudimentary" here not to infer that these techniques have served as the fundamental basis for the development of the other techniques, but rather to emphasize that these were the early techniques used before the others were developed. In the rudimentary techniques, the values of each decision criterion are expressed as a ratio of some statistic of all the values of that criterion. That statistic could be the maximum, median, modal, minimum, or mean value of the decision criterion across the alternatives or the value standardized using the normal or other probability distribution.

Example 12.4

In the apartment complex example discussed in Table 12.1, develop scaled values of the three decision criteria using the mean value of that decision criterion across the alternatives.

Solution

The mean values can be calculated as follows: rent amount = \$556.43; distance to bus stop = 162.14 ft; rent payment grace period = 3.79 days. These numbers can be used to find the scaled values as shown in Table 12.7.

Table 12.7 Scaled Values of Each Alternative

	Apartment Complex	A	B	C	D	E	F	G	H	I	J	K	L	M	N
Raw	Rent Amount ($)	320	400	700	750	820	500	350	300	800	900	450	300	450	750
	Distance to Bus Stop (ft.)	100	250	100	150	120	150	150	300	100	100	150	200	250	150
	Rent Payment Grace Period (days)	7	3	1	2	7	3	5	2	2	4	1	4	6	6
Scaled (using the mean)	Rent	0.58	0.72	1.26	1.35	1.47	0.90	0.63	0.54	1.44	1.62	0.81	0.54	0.81	1.35
	Distance	0.62	0.62	0.62	0.62	0.62	0.62	0.62	0.62	0.62	0.62	0.62	0.62	0.62	0.62
	Grace Period	1.85	0.79	0.26	0.53	1.85	0.79	1.32	0.53	0.53	1.06	0.26	1.06	1.58	1.58

Using these scaled values, which indicate the impacts of all the decision criteria in commensurate units, it is now possible to find the overall impact of each alternative by adding up the values for each alternative and determining the best alternative; for example, for apartment 1, the overall impact = −0.58 − 0.62 + 1.85 = 0.65. Such "adding up" is one form of amalgamating the impacts of the different criteria, which we shall discuss in Section 12.5.

It must be noted that these rudimentary techniques are rarely being used in any sophisticated MCDM analysis at the current time. They fell out of favor for good reason: By using such ratio, the decision makers "force" a predefined relationship between the level of the criterion and its desirability, and the forced relationship may not reflect the true perspective of the decision makers. For example, the scaled grace periods shown in the Figure 12.6 force a linear function (see continuous line) on the decision makers (the students), but in reality, they may have an inherent scaling function that is nonlinear, such as the dashed-line S-curve.

(b) Linear Scaling Functions. Linear forms are often a default or last resort when the decision makers lack data for developing a scaling function using other methods. The linear scaling function can range from 0 to 1, 0 to 10, or 0 to 100, depending on the preferences of the decision makers. The length of the initial and final steps of the function, the slope and direction of the function, and the monotonicity or otherwise may also change for different decision criteria. In other words, the initial and/or final steps could be zero, and the slope could be steep or gentle or positive or negative and could be nonmonotonic. Monotonically-increasing linear scaling functions are those where higher

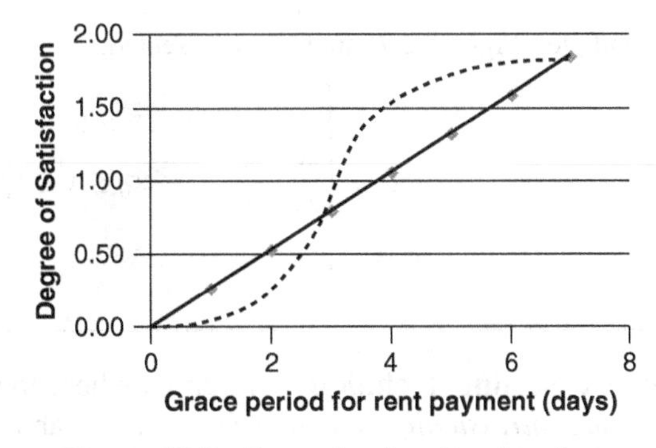

Figure 12.6 Example of scaling function.

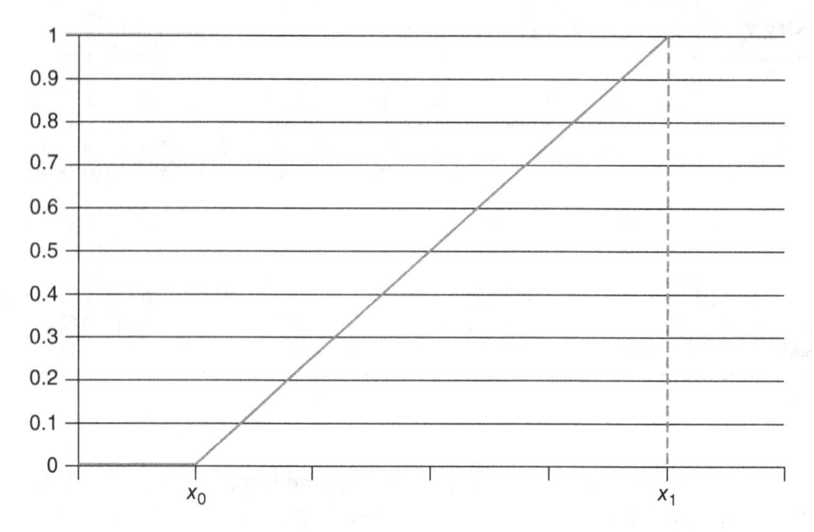

Figure 12.7 Linear scaling function example.

values of the decision criterion are more desirable to the decision makers; for example, these criteria might include system durability and user safety. A scaling function example is presented in Equation (12.8) and Figure 12.7 (Bai et al., 2008). A similar expression for the scaling function can be established for monotonically decreasing linear scaling functions (where higher values of the decision criterion are less desirable to the decision makers), such as accidents and community disruption. In some cases, the linear scaling function monotonically increases up to a point and then monotonically decreases thereafter or monotonically decreases up to a point and then monotonically increases thereafter. This is the case when the decision makers prefer that the decision criterion be too small or too large (a common example that comes to mind is the traffic speed on urban freeway systems) or where the decision makers indicate that a decision criterion is desirable only when it is lower than some threshold or when it exceeds some threshold.

Example 12.5

Write the equations and show the plot for a linear scaling function that indicates 0% value when the decision criterion is between 0 and a point x_0, and increases linearly to 100% when the value is at a maximum of x_1.

Solution

The equation and plot (Figure 12.7) are herein presented.

$$r(x) = \begin{cases} 0 & x \leq x_0 \\ \dfrac{x - x_0}{x_1 - x_0} & x_0 \leq x \leq x_1 \\ 1 & x \geq x_1 \end{cases}$$

(c) Monetization as a Scaling Technique. In cases where the decision criteria associated with a decision problem are *intrinsically monetary* (i.e., translatable to monetary units), the monetary impact of each decision criterion could be established in their respective dollar equivalents. Then, the overall impact for each alternative could be determined as the sum of the monetary consequences of the performance outcomes for that alternative, and the best alternative could be identified rather easily. By using a simple additive function of these monetized amounts, the assumption is that one dollar worth of a decision criterion is the same as one dollar worth of all the other decision criteria. For example, a dollar of agency benefits in terms of system longevity is the same as a dollar in terms of the user's cost of delay or inconvenience. Transforming all the different decision criteria into their monetary equivalents or dollar units is thus a special type of scaling that is appropriately termed "monetization." Decision criteria are described as *intrinsically nonmonetary* when they cannot be expressed in their monetary equivalents for various reasons, including practicality.

In investment analysis, different decision criteria are typically brought to the same dimension or scale using monetization, even though it is often not recognized explicitly as a scaling technique. In project evaluations for certain types of civil engineering systems, decisions are made on the basis of the monetized values of the "relevant" decision criteria while nonmonetized decision criteria, often referred to as "externalities," are often relegated to the background of mere conceptual (and often inconsequential) discussion (Bai, 2012). In problem settings where only a relatively few criteria can be quantified in their monetary equivalents, the use of monetization as a scaling method can severely limit the number of decision criteria that can be considered in decision analysis. The chapters on costing and economic efficiency in this text present the various ways by which certain decision criteria in civil engineering systems management could be monetized and evaluated.

As a scaling technique, monetization has some drawbacks. First, there has not been enough research to quantify all possible civil engineering system decision criteria or performance measures in their monetary equivalents. Second, there could be ethical issues in the attempt to assign monetary values to certain decision criteria, including user safety. Third, the use of monetary values yields a scale that is unbounded, which could cause some computational problems. Fourth, the value of a dollar could be different across the decision criteria and for the different stakeholders.

Example 12.6

For the data presented in Appendix 12.1, assume the following monetized rates of the performance impacts: Environment: \$0.85M/affected acre; economic efficiency: net present value in \$M; user safety hazard: \$20M per fatality; and system durability: \$5M per added year of system life. Determine the scaled values of the first three actions on the basis of their monetized impacts.

Solution

The scaled values of the first three actions on the basis of their monetized impacts, in millions of dollars, are shown in Table 12.8.

Table 12.8 Total Scores for Each Alternative

	Environmental Impact	Economic Efficiency	User Safety Hazard	System Durability
Action 1	38	34	10	25
Action 2	0	61	42	35
Action 3	27	4	8	10

Sample calculation: For action 1, the scaled impact is $\$0.85\text{M/acre} \times (45 \text{ acres}) = \$38\text{M}$. Positive or negative signs could then be attached to each scaled value depending on whether that performance outcome is desired or undesired.

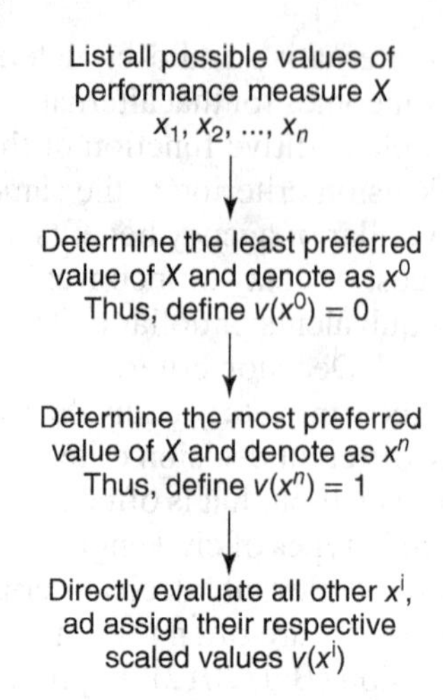

Figure 12.8 Procedure for scaling using direct rating.

(d) Scaling Using Direct Rating. Under the certainty scenario of decision making, the direct rating technique (Keeney and Raiffa, 1976) is probably the simplest preference-based method for establishing a scaling function for a decision criterion. The decision makers indicate directly the value or desirability they attach to each level of the decision criterion on a given scale (0–100%). This method yields reliable results in cases where the decision criterion has only a few outcome levels that are discrete. Thus, it can be used, for example, to measure user satisfaction, which may range from very dissatisfied to very satisfied, but may not be appropriate for the percentage of cracking on a structure. The direct rating process is described in Figure 12.8.

(e) Other Scaling Techniques. Other scaling techniques that involve a significant amount of effort and sophistication include the **midvalue splitting method** (Keeney and Raiffa, 1976), which is associated with the certainty scenario and is based on the identification of the concept of midvalue point and differentially valued equivalent points, and the **direct questioning** and **certainty equivalent** and **gamble** methods, which are used in risk and uncertainty scenarios (Keeney and Raiffa, 1976; Patidar et al., 2007; Bai et al., 2008).

12.3.2 Functional Form of the Developed Scaling Function

In cases where past research has established the probability distributions of the outcome of a given decision criterion, the existing scaling functions can be updated by calibrating the parameters in the distribution function using the new data. A number of utility functions have been developed in past research in civil infrastructure systems for decision making using performance criteria related to system preservation (system condition or longevity), system user safety (in terms of the structural and functional adequacy of the physical system), the environment (air pollution due to the system operations), and system vulnerability to disaster (in terms of earthquake, scour, fatigue, and other hazards).

(a) Shape of the Scaling Function. Irrespective of the scaling method used, there are generally three categories of shapes that a scaling function could belong: monotonically increasing, monotonically decreasing, or nonmonotonic. These functions may be linear or nonlinear. Nonlinear scaling functions may be convex, concave, S-shape, reverse S, and so forth (see Figure 12.9).

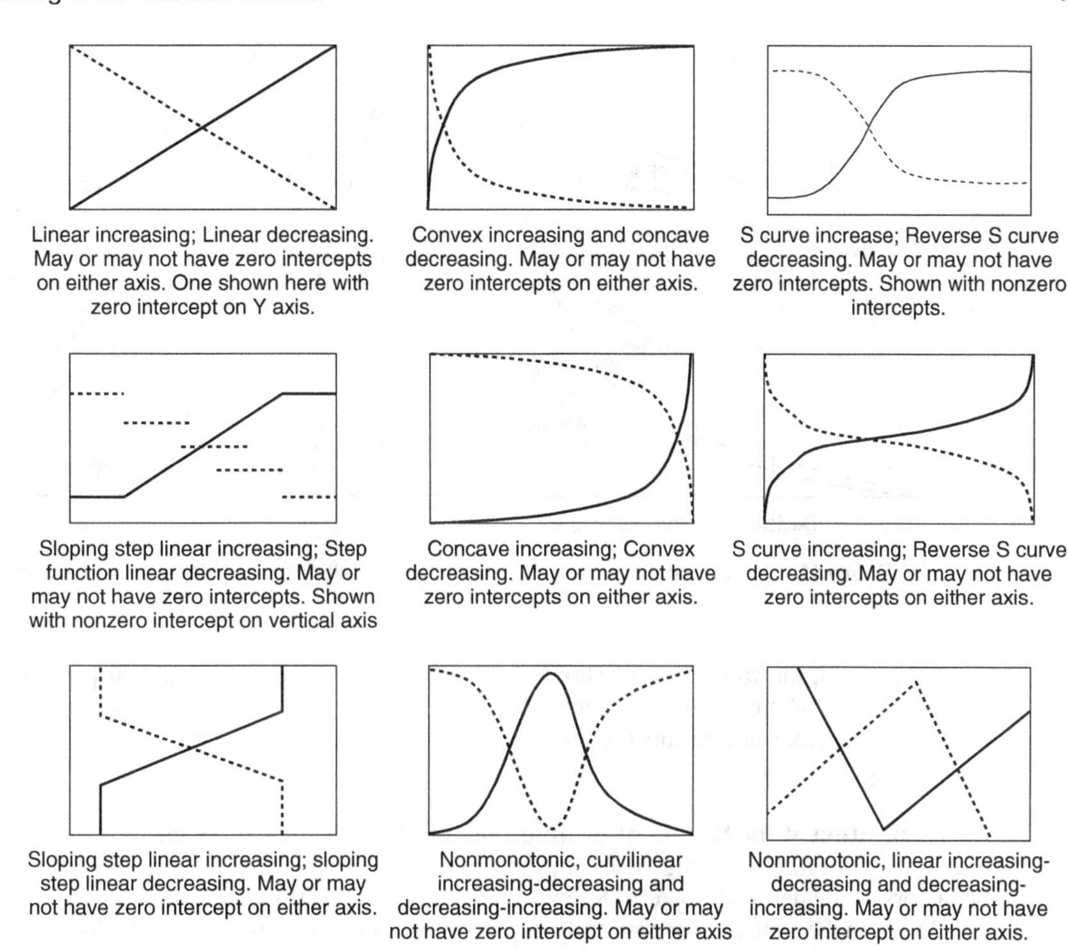

Figure 12.9 Examples of scaling function shapes.

Monotonically Increasing Scaling Functions: These functions represent the decision criterion for which higher values are more desirable to the decision makers. Examples include system condition, system longevity, and reduction in the severity or extent of some type of distress. For example, a higher condition index means a good condition while a lower condition index translates into a poorer condition. Also, a higher change in distress is more desirable while a lower change in distress is less desirable.

Monotonically Decreasing Scaling Functions: These functions typically represent decision criteria whose higher values are less desirable to the decision makers (e.g., corrosion, cracking, number of accidents, user delay, cost of construction operations, or maintenance) or reduction in some positive attribute such as a reduction in facility health. For example, a higher corrosion index is undesirable and is assigned a lower value or utility on the scaling function.

Nonmonotonic Scaling Functions: In certain rare cases, the decision makers do not have a consistent direction of preference for increasing or decreasing the levels of the decision criterion. In such cases, for certain intervals of the decision criterion, the utility or value functions increase while they decrease for other intervals; for example, when the decision makers desire that the decision criterion not be too small or too large or that it should not be lower than some threshold or exceed

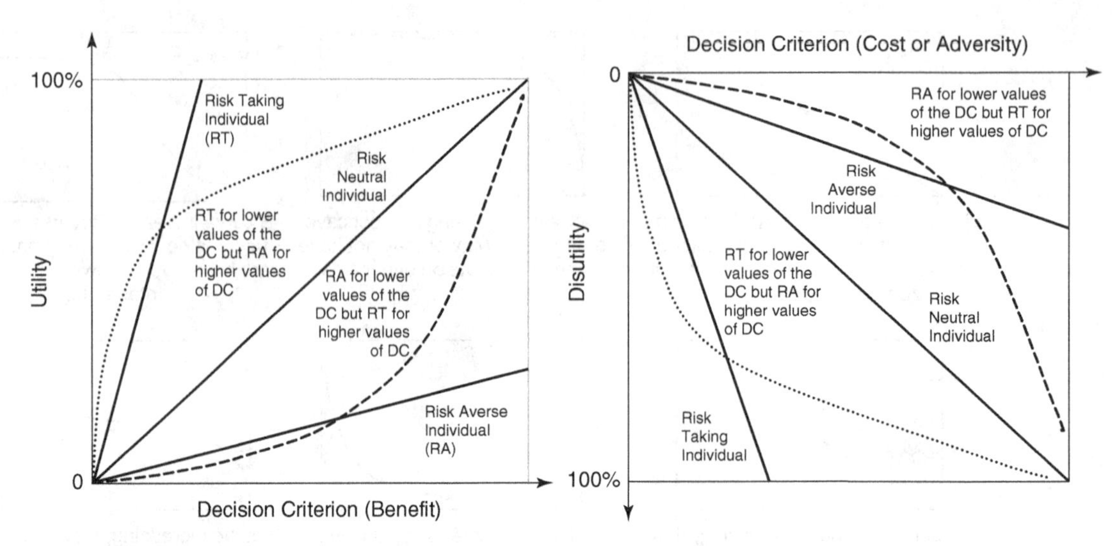

Figure 12.10 How scaling functions reflect the risk-taking nature of the decision maker.

some threshold, the function will show such nonmonotonic trends. A case in point would be a freeway system where a highway agency, from the perspective of energy performance, might desire that the speed limit not be too low or too high as either extreme is associated with higher fuel consumption.

(b) Implication of the Shapes of Scaling Functions. When the preferences of decision makers are used to develop a scaling function, they can provide interesting insights into the risk-taking behaviors of the decision makers (Keeney and Raiffa, 1993). Their collective risk-taking behavior is reflected in the concavity or convexity of the scaling function (Sinha and Labi, 2007). A *risk taker* is one who is willing to gamble with his or her resources to obtain a possibly superior consequence of his or her actions even though that outcome may be less probable than the expected outcome; risk takers have a convex utility function. A *risk-averse* decision maker is one who behaves conservatively and has a concave scaling function. A *risk-neutral* decision maker has a linear scaling function (Figure 12.10). Most owners of civil engineering systems are stewards of public infrastructure, tasked with the prudent management of millions or billions of dollars and therefore generally tend to be risk averse or at best risk neutral.

12.4 AMALGAMATION

In Section 12.3, we discussed, as part of the multicriteria decision-making framework presented in Section 12.2, a number of scaling methods for establishing a consistent measurement unit across the decision criteria. In this way, the analyst can measure all the different performance impacts of each decision alternative in terms of the decision criteria. At this point, the next step is to combine these scaled values for each performance outcome to yield an overall impact for the action in order to identify the best alternative or to rank the alternatives in order of their optimality. The process of combining the different impacts for each alternative is referred to as amalgamation. In the sections below, we discuss how different methods may be used for the amalgamation process.

12.4.1 Weighted Sum or Product Methods

In MCDM problems where multiattribute utility functions are used to obtain the solution, the weighted sum or weighted product methods of amalgamation are used widely in practice. These methods use the relative weights of the decision criteria and scaling functions for each decision criterion. These scaling functions are single-attribute utility functions that are subsequently amalgamated, additively or multiplicatively, to yield a multiattribute utility function. The alternatives are then ranked on the basis of their multiattribute utilities, and the alternative with the highest utility is identified as the optimal solution. In using the multiattribute utility function, two key conditions are assumed to be satisfied: *utility independence* (the utility function of each decision criterion is not influenced by the levels of other decision criteria) and *preference independence* (the choice of one alternative over another on the basis of a specific decision criterion is not influenced by the levels of other decision criteria).

(a) The Weighted Sum Method (WSM). A common method for amalgamation in multiple-criteria decision making, the weighted sum method, uses the additive function form to obtain the final overall desirability (or undesirability) of an alternative. This desirability, often expressed as a utility or value, of any alternative A_i is calculated as (Fishburn, 1967; Triantaphyllou, 2000)

$$U_{A_i} = \sum_{j=1}^{n} w_j a_{ij} \quad i = 1, 2, \ldots, m \tag{12.9}$$

where w_j is the weight of decision criterion j; a_j is the utility or scaled value of decision criterion j for alternative i; n is the number of decision criteria; and m is the number of alternatives. The best alternative is that which yields the maximum U_{Aj}.

When WSM is used, the value of a decision criterion must be dimensionless or have the same units (these are achieved using scaling techniques). As we learned in Section 12.1.1(b), if the scaled values are from preference-based scaling methods, the multiple decision criteria must be *utility independent* and *preference independent.* In addition, in the risk scenario, the expected values of decision criteria are used in the equation above. Example 12.7 illustrates the use of the weighted sum method.

Example 12.7

Consider alternative actions A, B, C, D, and E to rehabilitate a levee system. The decision criteria (P_1, P_2, P_3, and P_4) with weights w_{il} are used to evaluate these alternatives. Table 12.9 presents the scaled impacts of each alternative for each decision criteria. Determine the overall or amalgamated impact for each alternative.

Solution

The amalgamated values of decision criteria for each alternative can be found using the weighted sum method as shown in Table 12.10.

(b) The Multiplicative Utility Function. For each alternative A_i, the multiplicative utility function is defined as follows (Keeney and Raiffa, 1976):

$$U_i = \frac{1}{k}\left(\prod_{j=1}^{n} \left[1 + k w_j u\left(x_{ij}\right)\right] - 1\right)$$

Table 12.9 Data for Example 12.7

Alternatives	P_1 ($w_1 = 0.2$)	P_2 ($w_2 = 0.1$)	P_3 ($w_3 = 0.4$)	P_4 ($w_4 = 0.3$)
A	0.59	0.95	0.06	0.60
B	0.07	0.18	0.81	0.85
C	0.80	0.26	0.06	0.90
D	0.58	0.36	0.13	0.97
E	0.86	0.09	0.15	0.35

Table 12.10 Amalgamated Values for Example 12.7

Alternatives, i	P_1 ($w_1 = 0.2$)	P_2 ($w_2 = 0.1$)	P_3 ($w_3 = 0.4$)	P_4 ($w_4 = 0.3$)	Amalgamated Impact of the Alternative i
A	0.59	0.95	0.06	0.60	0.59 * 0.2 + 0.95 * 0.1 + 0.06 * 0.4 + 0.60 * 0.3 = 0.42
B	0.07	0.18	0.81	0.85	0.07 * 0.2 + 0.18 * 0.1 + 0.81 * 0.4 + 0.85 * 0.3 = 0.61
C	0.80	0.26	0.06	0.90	0.80 * 0.2 + 0.26 * 0.1 + 0.06 * 0.4 + 0.90 * 0.3 = 0.48
D	0.58	0.36	0.13	0.97	0.58 * 0.2 + 0.36 * 0.1 + 0.13 * 0.4 + 0.97 * 0.3 = 0.50
E	0.86	0.09	0.15	0.35	0.86 * 0.2 + 0.09 * 0.1 + 0.15 * 0.4 + 0.35 * 0.3 = 0.35

where $u(x_{ij})$ is the utility of alternative i in terms of the jth decision criterion; w_j is the relative weight of decision criterion j; and k is a scaling constant that is determined from the equation

$$1 + k = \prod_{j=1}^{n} (1 + kw_j)$$

The basic assumption underlying the use of the multiplicative utility function is that all the criteria must be *mutually utility independent.* If $X_1, X_2, \ldots, X_n$ are the n decision criteria, then decision criterion X_i is said to be utility independent if the utility function of X_i is not influenced by the levels of the other decision criteria. Also, $X_1, X_2, \ldots, X_n$ are mutually utility independent if every subset of $\{X_1, X_2, \ldots, X_n\}$ is utility independent of its complement (Keeney and Raiffa, 1976). The alternative with the largest overall utility is the optimal choice.

Example 12.8

Consider the following alternatives A–E that are being evaluated and ranked on the basis of three decision criteria: C_1, C_2, and C_3 with relative weights $w_1 = 0.4$, $w_2 = 0.3$, and $w_3 = 0.3$, respectively (Table 12.11).

Solution

First, the value of the parameter k is obtained by solving the following equation:

$$1 + k = \prod_{j=1}^{n} (1 + kw_j)$$

Table 12.11 Project Impacts in Terms of the Decision Criteria

Project	Project Impacts in Terms of the Respective Decision Criteria		
	C_1	C_2	C_3
A	0.59	0.95	0.2
B	0.07	0.18	0.81
C	0.8	0.26	0.06
D	0.58	0.36	0.13
E	0.86	0.09	0.15

Substituting the w_j values in the equation yields $k = 0$ or $k = -9.1667$.

However, k cannot be 0, thus $k = -9.1667$. Thus, the expression for amalgamating the decision criteria is

$$u_i = \frac{1}{-9.1667}\left(\prod_{j=1}^{n}\left[1 - 9.1667 w_j u\left(x_{ij}\right)\right] - 1\right)$$

Using this equation, the amalgamated value of each alternative can now be calculated as shown in Table 12.12.

Table 12.12 Amalgamated Value of Each Alternative

Alternative	C_1	C_2	C_3	Amalgamated Value
A	0.59	0.95	0.2	0.017002
B	0.07	0.18	0.81	0.159358
C	0.8	0.26	0.06	0.159281
D	0.58	0.36	0.13	0.10988
E	0.86	0.09	0.15	0.212942

From the results, it can be concluded that alternative E is the best choice, followed by B, C, D, and, lastly, A.

(c) The Weighted Product Model (WPM) Method. First developed by Miller and Starr (1969) and subsequently refined and applied by Bridgman (1922) and Triantaphyllou (2000), the WPM method compares two alternatives at a time on the basis of decision criteria to determine the superior alternative. This is accomplished by first taking the ratio of the values of the levels of performance of two alternatives and then using the product formula [see the multiplicative utility function in Section 12.4.1(b)] to obtain the final result for that pair of alternatives. This procedure is repeated for all pairs of alternatives. The WPM amalgamation process therefore yields a set of ratios for each alternative to determine how well it performs overall compared to the other alternatives. Using the results, the decision makers can draw up a list of all alternatives ordered by superiority or can simply identify which alternative is most superior.

$$r_{SL}\left(\frac{A_S}{A_L}\right) = \prod_{j=1}^{n}\left(\frac{x_{Sj}}{x_{Lj}}\right)^{w_j}$$

where x_{Sj} is the level of decision criterion j for alternative S; x_{Lj} is the level of decision criterion j for alternative L; r_{SL} is the ratio between the decision criterion impact of S and L. If $r_{SL} \geq 1$, then alternative S is more desirable than alternative L. If $r_{SL} = 1$, then alternative S is indifferent to alternative L. If $r_{SL} < 1$, then alternative L is less desirable than alternative S; w_i *is* the weight of decision criterion j.

This method is simple and easy to use. As Bai et al. (2008) pointed out, the greatest advantage of the WPM method is that it can use the original raw value and units of the decision criteria, thus eliminating the need for scaling. A disadvantage is that no value of any decision criterion can be zero; otherwise, the expression becomes undefined. Second, the task of pairwise comparison can be burdensome when the number of alternatives is large.

Example 12.9

For the problem posed in the weighted sum method (Example 12.7), use the weighted product method to amalgamate the overall impacts of alternatives C and D and compare the impacts of the two alternatives.

Solution

$$r_{CD}\left(\frac{A_C}{A_D}\right) = \left(\frac{0.8}{0.58}\right)^{0.2} + \left(\frac{0.26}{0.36}\right)^{0.1} + \left(\frac{0.06}{0.13}\right)^{0.4} + \left(\frac{0.9}{0.97}\right)^{0.3} = 3.75$$

which is >1.

The result shows that alternative C is superior to alternative D. It is noted that the solution process actually goes beyond amalgamation as it also carries out prioritization of the alternatives, a step in the MCDM framework that often follows the amalgamation step.

12.4.2 The Analytic Hierarchy Process Method of Amalgamation

In Section 12.2.2, we discussed how the AHP method could be used for weighting different decision criteria. In this section, we discuss how it could also be used for amalgamation (where the overall combined impacts of an alternative in terms of different decision criterion are determined). The AHP method comprises two parts: a pairwise comparison and eigenvector derivation. Consider the matrix X of m alternatives that have impacts in terms of n decision criteria:

$$X = \begin{bmatrix} x_{11} & x_{12} & \cdots & x_{1n} \\ x_{21} & x_{22} & \cdots & x_{2n} \\ \vdots & \vdots & \vdots & \vdots \\ x_{m1} & x_{m2} & \cdots & x_{mn} \end{bmatrix} \tag{12.12}$$

where each cell x_{ij} represents the raw or scaled value of decision criterion j of alternative i. Transforming each cell in the matrix into a ratio of the sum across the decision criteria yields

$$\begin{bmatrix} x_{11}/\sum_{i=1}^{m} x_{i1} & x_{12}/\sum_{i=1}^{m} x_{i2} & \cdots & x_{1n}/\sum_{i=1}^{m} x_{in} \\ x_{21}/\sum_{i=1}^{m} x_{i1} & x_{22}/\sum_{i=1}^{m} x_{i2} & \cdots & x_{2n}/\sum_{i=1}^{m} x_{in} \\ \vdots & \vdots & \vdots & \vdots \\ x_{m1}/\sum_{i=1}^{m} x_{i1} & x_{m2}/\sum_{i=1}^{m} x_{i2} & \cdots & x_{mn}/\sum_{i=1}^{m} x_{in} \end{bmatrix} \tag{12.13}$$

Then the overall desirability of alternative i can be calculated as

$$O_i = \sum_{k=1}^{n} w_k \left(x_{ik} / \sum_{j=1}^{m} x_{jk} \right) \tag{12.14}$$

The AHP method thus can be used to (i) identify the best alternative (that which has the highest value of O_i); (ii) make comparisons between any two alternatives (the alternative with a higher O_i is superior to that with a lower O_i value); and (iii) carry out trade-off analysis between them (e.g., by adopting a certain action instead of another, how much of a certain decision criterion is gained, and how much of another decision criterion is sacrificed).

The AHP method has been used widely for decision making in systems related to several sectors or disciplines, including energy, agriculture, and business. An advantage of the AHP method of amalgamation is that it does not require prior scaling of the decision criterion into a dimensionless unit. A limitation is that the results may be misleading when the decision matrix has missing or zero values.

Example 12.10

For the problem posed in the weighted sum method (Example 12.7), use AHP to determine the amalgamated impacts of the different alternative.

Solution

The calculation process and the final amalgamated values are shown in Table 12.13.

Table 12.13 Amalgamated Value of Each Alternative Using AHP

Alternative	P_1 ($w_1 = 02$)	P_2 ($w_2 = 01$)	P_3 ($w_3 = 04$)	P_4 ($w_4 = 03$)	Amalgamated Value
A	0.59	0.95	0.06	0.60	$0.59/2.9 * 0.2 + 0.95/1.84 * 0.1 + 0.06/1.21 * 0.4 + 0.60/3.67 * 0.3 = 0.16$
B	0.07	0.18	0.81	0.85	$0.07/2.9 * 0.2 + 0.18/1.84 * 0.1 + 0.81/1.21 * 0.4 + 0.85/3.67 * 0.3 = 0.35$
C	0.80	0.26	0.06	0.90	$0.80/2.9 * 0.2 + 0.26/1.84 * 0.1 + 0.06/1.21 * 0.4 + 0.90/3.67 * 0.3 = 0.16$
D	0.58	0.36	0.13	0.97	$0.58/2.9 * 0.2 + 0.36/1.84 * 0.1 + 0.13/1.21 * 0.4 + 0.97/3.67 * 0.3 = 0.18$
E	0.86	0.09	0.15	0.35	$0.86/2.9 * 0.2 + 0.09/1.84 * 0.1 + 0.15/1.21 * 0.4 + 0.35/3.67 * 0.3 = 0.14$
Sum	**2.9**	**1.84**	**1.21**	**3.67**	—

12.4.3 The ELECTRE Method of Amalgamation

There is a category of solution methods for multicriteria decision-making problems that utilizes a full or partial ordinal ranking of the alternatives, which are known as outranking methods, and the most popular of these is the Elimination and Choice Translating Algorithm (ELECTRE) method (Benayoun et al. 1966; Roy and Bertier 1971). ELECTRE carries out pairwise comparisons among the alternatives and thus establishes a set of *outranking relationships*. In the ELECTRE procedure, an alternative is said to outrank another only if (i) the sum of normalized weights (i.e., the *concordance index*) for the alternative exceeds a predetermined threshold value, and (ii) the number of decision criteria for which the latter alternative is superior by an amount greater than a tolerable threshold value (i.e., *discordance index*) is zero. For a decision-making problem with m alternatives, each of which has n decision criteria, the decision matrix can be expressed as

$$X = \begin{bmatrix} x_{11} & x_{12} & \cdots & x_{1n} \\ x_{21} & x_{22} & \cdots & x_{2n} \\ \vdots & \vdots & \vdots & \vdots \\ x_{m1} & x_{m2} & \cdots & x_{mn} \end{bmatrix}$$

Then, the steps to use the ELECTRE procedure are (Triantaphyllou, 2000):

Step 1 (Normalize the Decision Matrix): Transform the value of each criterion to yield dimensionless entries using Equation (12.15):

$$r_{ij} = \frac{x_{ij}}{\sqrt{\sum_{k=1}^{m} x_{kj}^2}} \tag{12.15}$$

Step 2 (Weight the Normalized Decision Matrix):

$$Y = XW = \begin{bmatrix} w_1x_{11} & w_2x_{12} & \cdots & w_nx_{1n} \\ w_1x_{21} & w_2x_{22} & \cdots & w_nx_{2n} \\ \vdots & \vdots & \vdots & \vdots \\ w_1x_{m1} & w_2x_{m2} & \cdots & w_nx_{mn} \end{bmatrix} = \begin{bmatrix} y_{11} & y_{12} & \cdots & y_{1n} \\ y_{21} & y_{22} & \cdots & y_{2n} \\ \vdots & \vdots & \vdots & \vdots \\ y_{m1} & y_{m2} & \cdots & y_{mn} \end{bmatrix} \tag{12.16}$$

Step 3 (Determine the Concordance and Discordance Sets): Concordance Set. The concordance set of two alternatives A_S and A_L, denoted as C_{SL}, is defined as the set of all the criteria for which A_S is preferred to A_L. That is

$$C_{SL} = \{\text{criterion } j, y_{sj} \geq y_{lj}\} \qquad \text{for} \quad j = 1, 2, \ldots, n \tag{12.17}$$

The complementary subset is called the discordance set, denoted as D_{SL},

$$D_{SL} = \{\text{criterion } j, y_{sj} < y_{lj}\} \qquad \text{for} \quad j = 1, 2, \ldots, n \tag{12.18}$$

Step 4 (Construct the Concordance and Discordance Matrices): The following formulas can be used to calculate the entries in concordance and discordance matrices:

$$c_{SL} = \sum_{j \in C_{sl}} w_j \qquad \text{for} \quad j = 1, 2, \ldots, n, \tag{12.19}$$

When $S = L$, c_{SL} is not defined.

$$d_{SL} = \frac{\max_{j \in D_{sl}} |y_{sj} - y_{lj}|}{\max_{j} |y_{sj} - y_{lj}|} \tag{12.20}$$

When $S = L$, d_{SL} is not defined.

Step 5: Determine the Concordance and Discordance Dominance Matrices:

$$c = \frac{1}{m(m-1)} \sum_{\substack{s=1 \\ s \neq l}}^{m} \sum_{\substack{l=1 \\ l \neq s}}^{m} c_{sl} \tag{12.21}$$

Then calculate the concordance dominance matrix F, in which the entries are defined as

$$\begin{aligned} f_{sl} &= 1 \quad \text{if} \quad c_{sl} \geq c \\ f_{sl} &= 0 \quad \text{if} \quad c_{sl} < c \end{aligned} \tag{12.22}$$

Then

$$d = \frac{1}{m(m-1)} \sum_{\substack{k=1 \\ k \neq l}}^{m} \sum_{\substack{l=1 \\ l \neq k}}^{m} d_{kl} \tag{12.23}$$

Calculate the concordance dominance matrix G, in which the entries are defined as:

$$\begin{aligned} g_{sl} &= 1 \quad \text{if} \quad d_{sl} \geq d \\ g_{sl} &= 0 \quad \text{if} \quad d_{sl} < d \end{aligned} \tag{12.24}$$

Step 6 (Calculate the Aggregate Dominance Matrix Q):

$$q_{ij} = f_{ij} \times g_{ij} \tag{12.25}$$

In the matrix Q, if $q_{ij} = 1$, then the alternative A_i dominates (or is superior to) alternative A_j.

Example 12.11

For the problem in Example 12.7, use the ELECTRE method to determine the superior alternative(s) based on their amalgamated impacts.

Solution

First, use the formula in Equation (12.17) to normalize the matrix. Then multiply each element using the appropriate weight. The results are presented in Tables 12.14 and 12.15.

Then the concordance and discordance matrices are shown in Table 12.15.

Using Equations (12.21) and (12.23), the threshold value c for the concordance matrix is obtained as 0.45, and the threshold value d for the concordance matrix is obtained as 0.7348. Thus, the concordance dominance matrix and the discordance dominance matrix become as shown in Table 12.16.

And thus, the aggregate dominance matrix is shown in Table 12.17.

From the aggregate dominance matrix, it can be seen that alternative A is dominated by B, C, and D; the alternative E is dominated by alternative B. Therefore, B, C, and D are the three best choices.

12.4.4 The Goal Programming Method of Amalgamation

For a given alternative, the amalgamation of the different decision criteria simply represents the distance of that alternative to the goal in two, three, or n dimensions depending on the number of

Table 12.14 Scaled, Normalized, and Weighted Matrices

(a) Scaled Decision Criteria Matrix

	C_1	C_2	C_3	C_4
A	0.59	0.95	0.06	0.6
B	0.07	0.18	0.81	0.85
C	0.8	0.26	0.06	0.9
D	0.58	0.36	0.13	0.97
E	0.86	0.09	0.15	0.35

$$r_{ij} = \frac{x_{ij}}{\sqrt{\sum_{k=1}^{m} x_{kj}^2}}$$

------------>

(b) Normalized Decision Criteria Matrix

	C_1	C_2	C_3	C_4
A	0.41	0.89	0.07	0.35
B	0.05	0.17	0.97	0.49
C	0.56	0.24	0.07	0.52
D	0.40	0.34	0.16	0.56
E	0.60	0.08	0.18	0.20

(c) Weighted Matrix

	C_1	C_2	C_3	C_4
A	0.0820	0.0890	0.0286	0.1047
B	0.0097	0.0169	0.3865	0.1483
C	0.1112	0.0243	0.0286	0.1570
D	0.0806	0.0337	0.0620	0.1693
E	0.1196	0.0084	0.0716	0.0611

Table 12.15 Concordance and Discordance Matrices

Concordance Matrix

	A	B	C	D	E
A	—	0.3	0.1	0.3	0.4
B	0.7	—	0.4	0.4	0.8
C	0.5	0.6	—	0.2	0.4
D	0.7	0.6	0.2	—	0.4
E	0.6	0.6	0.4	0.4	—

Discordance Matrix

	A	B	C	D	E
A	—	0.20	1.00	0.52	1.05
B	1.00	—	1.00	1.00	1.00
C	0.81	0.28	—	0.92	1.00
D	1.00	0.22	1.00	—	1.00
E	0.53	0.35	0.45	0.36	—

Table 12.16 Zero–One Concordance and Discordance Dominance Matrices

	A	B	C	D	E
A	—	0	0	0	0
B	1	—	0	0	1
C	1	1	—	0	0
D	1	1	0	—	0
E	1	1	0	0	—

	A	B	C	D	E
A	—	0	1	0	1
B	1	—	1	1	1
C	1	0	—	1	1
D	1	0	1	—	1
E	0	0	0	0	—

Table 12.17 Aggregate Dominance Matrix

	A	B	C	D	E
A	—	0	0	0	0
B	1	—	0	0	1
C	1	0	—	0	0
D	1	0	0	—	0
E	0	0	0	0	—

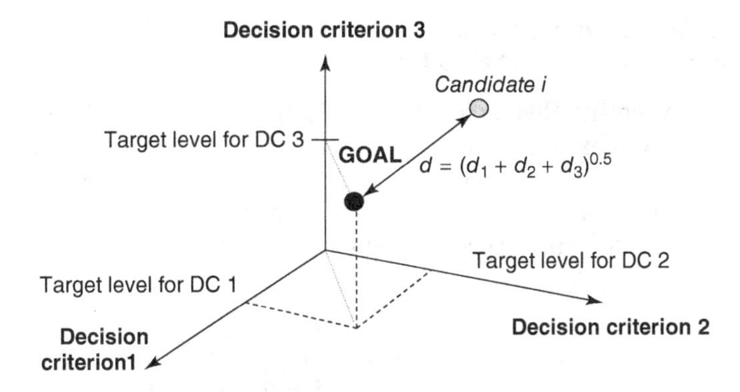

Figure 12.11 Amalgamation of distances from goal (for three decision criteria).

decision criteria. Figure 12.11 depicts, in three dimensions (i.e., for a problem with three decision criteria), the amalgamated performance, d, of an alternative on the basis of the alternative's impact in terms of the decision criteria. In compromise programming (a variation of goal programming) (Zeleny 1973), the solutions closest to the goal are determined by some measure of distance; these solutions are referred to as compromise solutions and are said to constitute a compromise set. In cases where the compromise set is small enough such that the decision makers can identify a satisfactory solution, then the process is terminated. Otherwise, the ideal solution is redefined and the whole process is repeated.

Example 12.12

Consider a problem setting where the City of Megapolis is planning a long-distance transit service connecting suburban areas to downtown Megapolis. Four alternative projects (alternatives) are being evaluated. The city's goal is to have a maximum project cost of \$3M, to serve at least 6000 people, and land used should not exceed 150 acres. The extent to which each alternative achieves the decision criteria are shown in Table 12.18. The scaled values of the decision criteria for each alternative are shown in each cell of the second, third, and fourth rows. The amalgamated values of these scaled values are shown in the fifth row.

Table 12.18 Goal Distance Amalgamation Results

	Alternative A	Alternative B	Alternative C	Alternative D
Cost (\$M)	$4.5 - 3 = 1.5$	$3.1 - 3 = 0.1$	$6.6 - 3 = 3.6$	$5.2 - 3 = 2.2$
Pop. served (1000s)	$2.1 - 6 = -3.9$	$1.9 - 6 = -4.1$	$5.5 - 6 = -0.5$	$4.1 - 6 = -1.9$
Land lost (acres in 100s)	$1.7 - 1.5 = 0.2$	$2.3 - 1.5 = 0.8$	$2.9 - 1.5 = 1.4$	$2.7 - 1.5 = 1.2$
Distance from Goal	$(1.5^2 + (-3.9)^2 + 0.2^2)^{0.5} = 4.18$	$(0.1^2 + (-4.1)^2 + 0.8^2)^{0.5} = 4.178$	$(3.6^2 + (-0.5)^2 + 1.4^2)^{0.5} = 3.89$	$(2.2^2 + (-1.9)^2 + 1.2^2)^{0.5} = 3.14$

Solution

From the results (Table 12.18), D is the least distant from the goal and thus is the most superior.

12.4.5 The Technique for Order Preference by Similarity to Ideal Solution (TOPSIS) Method

Developed by Yoon and Hwang in 1980, TOPSIS seeks the best alternative as that with the shortest distance from the ideal solution and the farthest distance from the worst solution. This method assumes that the preference structure for each criteria is monotonically decreasing or increasing, which means "the more the better" or "the fewer the better respectively." For a decision-making problem with m alternatives, each of which has n decision criteria, the decision matrix can be expressed as

$$X = \begin{bmatrix} x_{11} & x_{12} & \cdots & x_{1n} \\ x_{21} & x_{22} & \cdots & x_{2n} \\ \vdots & \vdots & \vdots & \vdots \\ x_{m1} & x_{m2} & \cdots & x_{mn} \end{bmatrix}$$

Then, the steps to use the TOPSIS procedure are (Triantaphyllou, 2000):

Step 1 (Normalize the Decision Matrix): Similar to the ELECTRE method, each cell of the decision matrix is normalized as follows:

$$r_{ij} = \frac{x_{ij}}{\sqrt{\sum_{k=1}^{m} x_{kj}^2}} \tag{12.26}$$

Step 2 (Weight the Normalized Decision Matrix): The normalized entries in the matrix are multiplied by the relative weights of each decision criterion, yielding

$$U = \begin{bmatrix} w_1 r_{11} & w_2 r_{12} & \cdots & w_n r_{1n} \\ w_1 r_{21} & w_2 r_{22} & \cdots & w_n r_{2n} \\ \vdots & \vdots & \vdots & \vdots \\ w_1 r_{m1} & w_2 r_{m2} & \cdots & w_n r_{mn} \end{bmatrix} \tag{12.27}$$

Step 3 (Identify the Ideal and the Worst Alternatives): Assume there are two alternatives A^{b} and A^{w}, and the cells in the decision matrix are

$$A^{\mathrm{b}} = \{a_{b1}, a_{b2}, \ldots, a_{wn}\} \tag{12.28}$$

where a_{bi} is the most preferred value among $u_{1i}, u_{2i}, \ldots, u_{mi}$ and

$$A^{w} = \{a_{w1}, a_{w2}, \ldots, a_{wn}\} \tag{12.29}$$

where a_{wi} is the least preferred value among $u_{1i}, u_{2i}, \ldots, u_{mi}$.

Step 4 (Calculate the Distance from the Ideal Alternative and the Worst Alternative): The distance from ith alternative to the ideal alternative is defined as

$$D_{i+} = \sqrt{\sum_{k=1}^{n} (u_{ik} - a_{bk})^2} \tag{12.30}$$

The distance from ith alternative to the worst alternative is defined as

$$D_{i-} = \sqrt{\sum_{k=1}^{n} (u_{ik} - a_{wk})^2} \tag{12.31}$$

Step 5 (Calculate the Relative Closeness to the Ideal Alternative): The relative closeness of the ith alternative to the ideal alternative is defined as

$$C_i = \frac{D_{i-}}{D_{i+} + D_{i-}} \quad (12.32)$$

Therefore, the alternative with the highest C_i is considered the best choice.

Example 12.13

For the problem in the WSM illustration (Example 12.7), determine, using TOPSIS, the amalgamated value of each alternative in terms of the different decision criteria.

Solution

First, use Equation (12.26) to normalize the matrix. Then, multiply each element using the weight. Table 12.19 presents the results. A, B, C, and D are the alternative projects. C_1 to C_4 are the decision criteria.

Table 12.19 Scaled, Normalized, and Weighted Matrices

(a) Scaled Decision Criterion Matrix

	C_1	C_2	C_3	C_4
A	0.59	0.95	0.06	0.6
B	0.07	0.18	0.81	0.85
C	0.8	0.26	0.06	0.9
D	0.58	0.36	0.13	0.97
E	0.86	0.09	0.15	0.35

$$r_{ij} = \frac{x_{ij}}{\sqrt{\sum_{k=1}^{m} x_{kj}^2}}$$

----------->

(b) Normalized Decision Criterion Matrix

	C_1	C_2	C_3	C_4
A	0.41	0.89	0.07	0.35
B	0.05	0.17	0.97	0.49
C	0.56	0.24	0.07	0.52
D	0.40	0.34	0.16	0.56
E	0.60	0.08	0.18	0.20

(c) Weighted Matrix

	C_1	C_2	C_3	C_4
A	0.0820	0.0890	0.0286	0.1047
B	0.0097	0.0169	0.3865	0.1483
C	0.1112	0.0243	0.0286	0.1570
D	0.0806	0.0337	0.0620	0.1693
E	0.1196	0.0084	0.0716	0.0611

Based on matrix (c), the ideal alternative is determined as: A^b (0.1196, 0.0890, 0.3865, 0.1693) and the worst alternative is A^w (0.0097, 0.0084, 0.0286, 0.0611). The distances of each alternative from the most ideal and the worst alternatives are shown in Table 12.20.

Table 12.20 Relative Distances Using TOPSIS

	Distance from Ideal Alternative	Distance from Worst Ideal Alternative	Relative Closeness to Ideal Alternative
A	0.0179	0.0002	0.0103
B	0.0003	0.0184	0.9833
C	0.0175	0.0004	0.0218
D	0.0121	0.0003	0.0275
E	0.0138	0.0002	0.0139

From the results, it can be seen that alternative B is the best choice.

12.4.6 The Global Criterion Method

In the global criterion or compromise programming method (Yu, 1973; Cochrane and Zeleny, 1973), the optimal solution is the feasible solution or candidate that is least distant from the global reference point (GRP). The GRP consists of the global optimal values of all the decision criteria. For each feasible solution, its distance from the GRP can be calculated. Then, the optimal solution is identified as the alternative that is least distant from the GRP; mathematically, this is formulated as follows (Miettinen, 1999):

$$\min \left(\sum_{j=1}^{m} \left| f_j(\mathbf{x}) - z_j^* \right|^p \right)^{1/p}$$

$$\text{Subject to} \quad \mathbf{x} \in \mathbf{S} \tag{12.33}$$

where $\mathbf{x}$ is the vector of the decision variables; $f_j(\mathbf{x})$ is the decision criterion for the jth objective; m is the number of decision criteria; z_j^* is the ideal value of the jth decision criterion; $\mathbf{S}$ is the region or set of all feasible solutions; and p is a parameter that could take any value from 1 to ∞. Different p values may result in different solutions. Commonly used p values include 1, 2, and ∞ (Miettinen, 1999). This method is easy to use; the difficulty lies in how to establish the ideal value of the decision criteria, particularly where such ideal values are not stated in the existing policies of the system owner or other decision makers. The simplest case is where the problem involves only two decision criteria and a p value of 1.

Example 12.14

A system owner seeks to select a contractor to carry out structural monitoring of levees on its network. Bids were obtained from four contractors and are being evaluated on the basis of the contract bid amount, the proposed duration of the services, and the contractor's relevant experience. The z_j^* values are the system owner engineer's estimates, which are as follows: bid amount, \$5 million; monitoring duration, 10 weeks; required experience, 16 years. The criteria levels for each contractor are shown in Table 12.21.

Table 12.21 Data for Example 12.14

Contractor	Bid Amount ($millions)	Proposed Monitoring Duration (weeks)	Contractor's Experience (years)
I-Spy Inc.	3	14	8
Coverage Corp.	7	17	11
Triple M	8	11	8
Value Ltd.	9	4	12

Solution

In this problem, the engineer's estimate can be taken as the GRP. For a given contractor, each criterion is a certain distance from the GRP. The number of decision criteria, m, is 3. Assume that $p = 2$, that is, the decision maker seeks the contractor that minimizes the square of the distances from the GRP. Table 12.22 presents the computations.

From Table 12.22, the contractor with the minimum GRP is Value Ltd., and therefore this contractor should be selected. Notice that we used the *square* of the deviations and the square root of the sum of the squared deviations only because p was specified as 2. A different value of p will yield different results and could produce a different optimal solution.

Table 12.22 Computation Results for Example 12.14

Contractor	Deviation for Bid Amount ($millions)	Deviation for Proposed Monitoring Duration (weeks)	Deviation for Contractor's Experience (years)	Sum of Squared Deviations	Root of Sum of Squared Deviations
I-Spy Inc.	$3-5=-2$	$14-10=4$	$8-16=-8$	84	9.17
Cov Corp.	$7-5=2$	$17-10=7$	$11-16=-5$	78	8.83
Triple M	$8-5=3$	$11-10=1$	$8-16=-8$	74	8.60
Value Ltd.	$9-5=4$	$4-10=-6$	$12-16=-4$	68	8.25

12.4.7 Neutral Compromise Solution Method

Proposed by Gal et al. (1999), this method is similar to the global criterion method; the main difference is that this method assumes that the performance target or ideal solution lies in the middle of the range of possible values of each performance objective. The optimal solution is then identified as the alternative that is the minimal distance from the ideal solution (Branke et al., 2008):

$$\text{Minimize } \max_{j=1,2,..,m}\left[\frac{f_j(\mathbf{x})-(z_j^{\min}+z_j^{\max})/2}{z_j^{\min}-z_j^{\max}}\right]$$

$$\text{Subject to } \mathbf{x}\in\mathbf{S}$$

where $z_j^{\min}$ and $z_j^{\max}$ are the minimum and the maximum values, respectively, of the jth decision criterion, all other symbols have the same meanings as those shown in Equation (12.33). The attractiveness of this method lies in its simplicity; however, the assumption that the ideal performance level is the midpoint of the range of performance may be unduly restrictive, if not unrealistic, from a practical point of view. As the saying goes, "there can never be too much of a good thing or too little of a bad thing".

Example 12.15

For the problem given in Example 12.14, find the optimal solution using the neutral compromise solution method.

Solution

In this problem, the midpoint values of each decision criterion, across the alternatives (contractors), is calculated and used as the global reference point (GRP). For a given contractor, each criterion is a certain distance from the GRP. Assume that $p=2$ (i.e., the decision maker seeks the contractor that minimizes the square of the distances from the GRP). Tables 12.23 and 12.24 present the preliminary and intermediate computations and the calculation of the distances.

Table 12.23 Preliminary and Intermediate Computations for Example 12.15

Contractor	Bid Amount $f(x_{j=1})$	Duration $f(x_{j=2})$	Experience $f(x_{j=3})$
I-Spy Inc. (i_1)	3	14	8
Coverage Corp. (i_2)	7	17	11
Triple M (i_3)	8	11	8
Value Ltd. (i_4)	9	4	12
z_j min	3	4	8
z_j max	9	17	12
(z_j min $-z_j$ max)	6	13	4
(z_j min $+z_j$ max)/2	6	10.5	10

Table 12.24 Computation of Distances $[f(x) - (z_j \min + z_j \max)/2]/[z_j \min - zj \max)$

Contractor	Bid Amount	Duration	Experience	Max
I-Spy Inc.	−0.50	0.27	−0.50	0.27
Coverage Corp.	0.17	0.50	0.25	0.50
Triple M	0.33	0.04	−0.50	0.33
Value Ltd.	0.50	−0.50	0.50	0.50

From Table 12.24, the minimum of the above maximum values is 0.27. Thus, the optimal choice of contractor for the system monitoring project is I-Spy Inc. The assumption in this calculation is that too much or too little of each of the decision criteria, is viewed unfavorably by the decision maker, which can be unrealistic in certain cases of evaluation.

12.4.8 Lexicographic Ordering Technique

In this technique, the decision makers first rank the decision criteria in order of the importance they attach to each criterion. Then, for each candidate or alternative, the value of the most important criterion is determined. Next, for the most important criterion, the alternatives are compared with each other to identify which are the most optimal in terms of that criterion. If only one alternative is found to have the optimal value of the most important criterion, then it is selected as the optimal solution to the problem; however, if there are multiple solutions that have the same optimal value in terms of the most important criterion, then their values on the second most important criterion are compared to identify the alternative(s) with the optimal value on the second most important criterion. This process is continued until there is only one solution left or all the criteria have been considered (Fishburn, 1974). The alternative(s) that remains at the end of the process is the optimal solution. This method is easy to use, but it has two shortcomings (Branke et al., 2008). First, it is often difficult for decision makers to rank the decision criteria. Second, the process may terminate prematurely without due consideration of decision criteria other than the most important criterion. For example, in cases where there exists only one alternative for which the most important criterion has the best value, the examination of other alternative ceases even though the identified solution may be inferior in terms of most of the other decision criteria.

Example 12.16

A graduate student is evaluating several possible campus restaurants for lunch. Table 12.25 presents the values of his selection criteria [food price ($) and food quality (rating from 1, excellent], to 5, poor)] for each of eight restaurants. He attaches greater weight to quality compared to food price. Rank the choices and identify the choice with the highest rank.

Solution

First, we rank the decision criteria in order of their importance: quality is the first criterion and price is the second. Therefore, we rank the restaurants in order of food quality:

Bub, E-ats, Turq, O-Jay, Raz, Fria, Pots, Stir

Table 12.25 Data for Example 12.16

	Stir	Bub	Pots	E-ats	O-Jay	Raz	Turq	Fria
Price	3	12	3	8	4	5	9	7
Quality	4	1	3	1	2	2	1	2

Then where multiple restaurants have the same rank in terms of the most important criterion, we rank them based on their values on the second most important criterion as follows:

E-ats, Turq, Bub, O-Jay, Raz, Fria, Pots, Stir

Therefore, the optimal choice, on the basis of these criteria, is E-ats restaurant.

12.4.9 Other Methods for Amalgamation

The **Step Method** (STEM) (Benayoun and Tergny 1969) is an interactive method that assumes that (i) the best compromising solution is the alternative that has the least overall deviation from the ideal solution, which is often the goal established by the decision makers; and (ii) the decision makers have a pessimistic view of the worst component of all the deviations of the individual alternatives from the ideal point. The technique essentially consists of two steps: first, a nondominated solution in the minimax sense to the ideal point for each objective function is sought, and a payoff table is constructed to obtain the ideal criterion vector; and second, comparing the solution vector with the ideal vector of a payoff table by modifying the constraint set and the relative weights of the objective functions. The process ends when the decision makers consider the current solution to be satisfactory.

12.5 PREFRENCE VERSUS NONPREFERENCE APPROACHES FOR WEIGHTING, SCALING, OR AMALGAMATION (WSA)

The various techniques for weighting, scaling, or amalgamation that we have discussed in Sections 12.2, 12.3 and 12.4 respectively can be classified in several different ways. One way is based on whether the decision-makers' preferences are utilized in the analysis: preference and nonpreference approaches. The preference approaches can be further categorized, depending on the stage of the analysis at which such preferences are incorporated (Figure 12.12).

12.5.1 Nonpreference Approach

In the nonpreference approach, no explicit (or very little) preference information from the decision makers is included in the decision-making process. Two common multiple-criteria analysis techniques that are consistent with this approach are the global criterion method and the neutral compromise solution method, which we discussed in Sections 12.4.6 and 12.4.7, respectively.

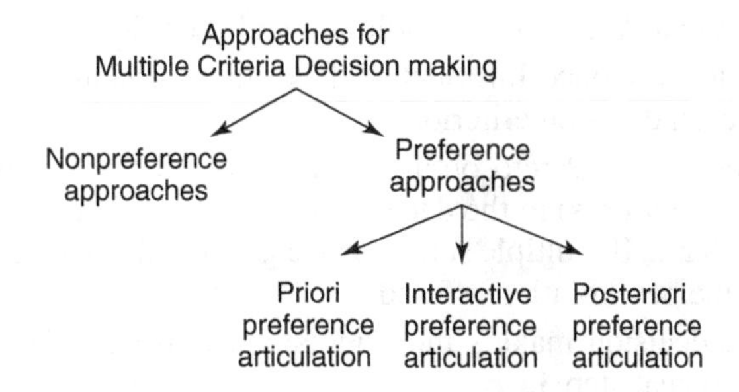

Figure 12.12 Preference and nonpreference approaches for WSA techniques.

12.5.2 Preference Approach

Unlike the nonpreference approach, the preference approach explicitly utilizes the preferences of the decision makers by incorporating these preferences in the process of comparing alternatives to identify the best alternative (Miettinen, 1999). There are three categories of the preference approach, depending on the stage of the process where the decision-makers' preferences are articulated and incorporated (Bai, 2012): the priori, interactive, and posteriori preference sub-approaches. Processes in multicriteria decision making (such as weighting and scaling) are not exclusive to any one sub-approach, that is, they can be used in MCDM solution methods that are consistent with any one of the three sub-approaches.

(a) Priori Preference Articulation. The decision-makers' preferences are first incorporated to carry out any one or more of the MCDM tasks of assigning relative weights to the decision criteria, establishing a scale of utility or satisfaction for each level of a given decision criterion. The decision-making problem is thus transformed from one of multiple criteria to one of a single criterion before the alternatives (candidates) are compared to identify the best alternative or optimal solution. Techniques that can be used in priori preference MCDM include simple weighting, ϵ-constraint, value/utility functions, lexicographic ordering, and goal programming (Miettinen, 1999; Branke et al., 2008; Bai, 2012).

(b) Posteriori Preference Articulation. In the posteriori preference articulation sub-approach for solving multicriteria decision making problems, the Pareto optimal solutions are first generated, and then the decision makers identify one of these as the optimal solution. MCDM techniques that may be used as part of the posteriori articulation approach include weighting, the ϵ-constraint method, and the weighted metrics method. In addition to the above, other MCDM techniques that use posteriori preference articulation include the achievement scalarizing function method (Wierzbicki, 1982; Branke et al., 2008) and the approximation methods (Ruzika and Wiecek, 2005). The posteriori preference articulation techniques have a few limitations. First, identifying all the Pareto solutions for a given problem is often laborious, if not impossible (Miettinen, 1999). Further, even where all the Pareto solutions are identified, the large number of such solutions makes it difficult for decision makers to identify one of the potential solutions as the optimal (Bai, 2012).

(c) Interactive Preference Articulation. In this sub-approach, the timing of the decision-makers' preference and the optimization (alternatives comparison and selection) are intertwined; the preference information is incorporated simultaneously with the optimization process. The decision makers interactively participate in the decision-making process. Branke et al. (2008) described the general process of an interactive method as follows:

Step 1 (Initialization): Determine the range of outcomes of each decision criterion and present them to the decision makers.

Step 2: Generate a Pareto optimal solution and use this as the starting point.

Step 3: Solicit preference information from the decision makers, for example, the aspiration levels for each decision criterion.

Step 4: Generate new Pareto optimal solution(s) that are consistent with the preferences and present the solution(s) to the decision makers as well as any other information regarding the problem at hand. If multiple solutions are generated, solicit the opinion of the decision makers regarding the best solution offered.

Step 5: If the decision makers indicate satisfaction with the solution in step 4, go to step 6; otherwise, repeat step 3.

Step 6: Stop.

The success of MCDM techniques that use the interactive sub-approach hinges on the willingness and ability of decision makers to devote adequate time and patience to providing the analyst with their preferences. This process often may require several sessions involving interactive iterations; in each iteration, there is a need to generate Pareto solutions and to capture decision makers' preferences. Further, this sub-approach requires that the responses of the decision makers ultimately must be consistent (Miettinen, 1999). According to Bai (2012), a large number of MCDM techniques exist that use the interactive preference articulation sub-approach, which include trade-off-based methods, reference point methods, and classification-based methods. Branke et al. (2008) provides a list of commonly used techniques that use this sub-approach.

SUMMARY

This chapter discussed the concept of the Pareto frontier and its importance in conducting an analysis of problems that involve multiple decision criteria. The chapter then presented an MCDM analysis framework that is based on the multiattribute utility functions and whose components use preference or nonpreference approaches at each stage of the framework. The framework comprises the key stages of establishing the decision criteria and providing weights to quantify the relative importance that the decision makers attach to the decision criteria. The chapter recognized that the decision criteria typically encountered in civil engineering systems have different units or metrics, which makes it difficult for decision makers to compare between alternatives on the basis of these decision criteria. As such, the chapter presented and discussed the merits and demerits of a number of alternative techniques that provide, for each decision criterion, a scale of measurement that is common across the different decision criteria. The last part of the framework is the amalgamation step, which combines the different impacts of a given alternative, in terms of the various decision criteria, to yield a single representative combined performance that may be referred to as the overall or total utility.

Two categories of solution approaches for multiple-criteria problems also were discussed: the preference-based and non-preference-based approaches. Unlike the nonpreference approach, the preference approach includes explicit articulation of the decision-makers' preferences and incorporation of these preferences in the optimization process to identify the best decision. Depending on the stage of the process where the decision-makers' preferences are incorporated, the preference approach could be priori preference, interactive preference, or posteriori preference.

In the current era, there is a strong rationale for applying tools for multiple-criteria analysis in civil engineering systems management. At each phase of a system's development, these tools can help system managers evaluate and select alternative courses of action in a transparent fashion that duly incorporates the perspectives of multiple stakeholders, thereby providing an organized structure to the decision-makers' decision-making processes in a clear, well-defined, rational, comprehensive, documentable, and defensible manner. It also can provide a solid basis on which system managers can carry out what-if analyses and to investigate the performance trade-offs between competing alternatives.

EXERCISES

1. Explain why decision making tasks in civil engineering are requiring the tool of multiple-criteria evaluation more and more.
2. Cite an example in your daily personal life where you analyze alternative actions on the basis of multiple criteria in order to select the best action. List the alternatives and the criteria.
3. A graduating student wishes to purchase a vehicle and is considering 20 different brands of vehicles. He is making the decision on the basis of two decision criteria: price and fuel consumption rate (gallons

per mile). Sketch a Pareto frontier for this decision problem, placing five alternatives on the frontier, six above the frontier, and four below the frontier. Interpret each of these three positions in terms of (i) the two decision criteria and (ii) the superiority of the position relative to the other two positions.

4. Using Example 12.16, assume that the weights are 70% and 30% for food price and food quality, respectively. What will be the student's new optimal choice of restaurant?
5. A prospective renter is evaluating seven apartment complexes. Table 12.26 presents the values of her selection criteria [monthly rent amount ($), minimum lease period (MLP) in months, and deposit amount ($)]. She considers the rent amount to be 1.2 times as important as the MLP, and the MLP is 1.5 times as important as the deposit amount. Like most typical renters, she seeks minimal rent amount, MLP, and deposit amount.

Table 12.26 Data for Exercise 12.5

	Riverstop	Knobb	Valley View	Hilltop	Lakeside	Jeffers	Camp Inn
Price	500	600	550	900	700	650	800
MLP	7	8	12	12	12	9	10
Deposit	250	300	350	250	200	200	150

   a. Using the lexicographic ordering technique, determine which should be her optimal choice.
   b. Assuming that her goals are as follows: rent amount of $400, MLP of 6 months, and deposit of $100, determine which should be her optimal choice.

6. A contractor seeks to purchase high-quality steel rebar specified for a pier construction project in a harsh saline marine environment. Table 12.27 presents the values of her selection criteria [reinforcement price per pound ($), distance of the source from the project site (miles), and steel quality (rating from 1, excellent, to 5, poor)] for each of five suppliers. Her goals are as follows: rebar price of $7/lb, distance from the project site of 15 miles, and steel quality rating of 3. Which vendor should she choose?

Table 12.27 Data for Exercise 12.6

Quality	Vendor A	Vendor B	Vendor C	Vendor D	Vendor E
Price	3	12	3	8	4
Distance	52	31	11	27	14
Quality	4	1	3	1	2

7. In Example 12.7, if the weights for the four decision criteria are $w_1 = 0.25$, $w_2 = 0.4$, $w_3 = 0.15$, and $w_4 = 0.2$, then, using the WSM method, which alternative should be the best?
8. In Example 12.7, if the weights for the four decision criteria are $w_1 = 0.2$, $w_2 = 0.4$, $w_3 = 0.3$, and $w_4 = 0.1$, then, using the ELECTRE method, identify the best alternative.
9. In Example 12.12, the city's goal is reset to the minimum project cost is $2M, at least 8000 people should be served, and the minimum land to be lost is 120 acres. Use the goal programming method to identify the best alternative.
10. There are four alternative treatments for a one-mile flexible pavement section: Do-nothing, HMA functional overlay, HMA functional overlay, and resurfacing (partial 3R). The decision maker needs to decide which treatment is the best. Three decision criteria are used to evaluate each alternative: cost, pavement international roughness index (IRI), and user cost savings. The performance of each alternative is presented in Table 12.28.

Table 12.28 Data for Exercise 12.10

Alternatives	Cost ($M)	IRI (inches/mile) Before	IRI (inches/mile) After	User Cost Savings ($M/year)
Do nothing	0	160	160	0
HMA Functional Overlay	0.36	160	80	0.7
Resurfacing	0.46	160	71	0.875

After a weighting survey, the weights for the cost, the IRI, and the user cost savings are 0.35, 0.4, and 0.25, respectively. The following utility functions are used in this evaluation:

Cost: $u(x) = 100(1 - x)$
Pavement surface roughness: $u(x) = 107.29(e^{-0.000044x^2})$
Reduction in user cost: $u(x) = 100x$

Assuming that the additive utility function can be used to evaluate each alternative, identify the best treatment for this one-mile pavement.

11. The table provided in Appendix 12.1 presents sample data on the potential impacts of 36 different actions applied to a civil engineering system. The reader is encouraged to use this dataset to carry out weighting, scaling, amalgamation, and decision making, for any selected number of actions and impact criteria.

REFERENCES AND RESOURCES

Bai, Q. (2012). Trade-off Analysis in Multi-criteria Optimization for Transportation Asset Management. Ph.D. Dissertation, Purdue University, West Lafayette, IN.

Bai, Q., Labi, S., and Li, Z. (2008). Trade-off Analysis Methodology for Transportation Asset Management. Technical Report FHWA/IN/JTRP-2008/31, West Lafayette, IN.

Benayoun, R., and Tergny, J. (1969). Critere Multiples en Programmation Mathematique: Une Solution Dans le Cas Lineaire. *RFAIRO*, 3, 31–56.

Benayoun, R., Roy, B., and Sussman, B. (1966). ELECTRE: Une Methode Pour Guider le Choix en Presence de Points de Vue Multiples, SEMA (Metra International). Note de Travail (49). *Direction Scientifique*, Paris, France.

Branke, J., Deb, K., Miettinen, K., and Skowinski, R. (2008). *Multi-objective Optimization: Interactive and Evolutionary Approaches*. Springer, Berlin, Heidelberg.

Bridgman, P. W. (1922). *Dimensional Analysis*. Yale University Press, New Haven, CT.

Cochrane, J. L., and Zeleny, M. (1973). *Multiple Criteria Decision Making*. University of South Carolina Press, Columbia, SC.

Fishburn, P. C. (1967). *Additive Utilities with Incomplete Product Set: Applications to Priorities and Assignments*. Operations Research Society of America (ORSA), Baltimore, MD.

Fishburn, P. C. (1974). Lexicographic Orders, Utilities and Decision Rules: A Survey. *Manage. Sci.*, 20, 1442–1471.

Gal, T., Stewart, T. J., and Hanne, T. (1999). *Multicriteria Decision Making: Advances in MCDM Models, Algorithms, Theory, and Applications*, Springer, Baston.

Goicoechea, A., Hansen, D. R., and Duckstein, L. (1982). *Multi-criteria Decision Analysis with Engineering and Business Applications*. Wiley, New York.

Hazen, G. B., and Morin, T. L. (1983). Optimality Conditions for Non-conical Multiple-objective Programming. *J. Optim. Theory Appl.*, 40, 25–29.

Hobbs, B., and Meier, P. (2000). *Energy Decision and the Environment: A Guide to the User of Multicriteria Methods*. Kluwer Academic, Boston.

Keeney, R. L., and Raiffa, H. (1976). *Decisions with Multiple Objectives: Preferences and Value Trade-offs*. Cambridge University Press, Cambridge, UK.

Keeney, R. L., and Raiffa, H., and Meyer, R. (1993). *Decisions with Multiple Objectives: Preferences and Value Trade-offs*. Cambridge University Press, Cambridge, UK.

Kuhn, H. W., and Tucker, A. W. (1951). Nonlinear Programming. *Proceedings of the Second Berkeley Symposium on Mathematical Statistics and Probability*. University of California Press, Berkeley, CA.

Miettinen, K. M. (1999). *Nonlinear Multiobjective Optimization*. Kluwer Academic, Norwell, MA.

Miller, D. W., and Starr, M. K. (1969). *Executive Decisions and Operations Research*. Prentice-Hall, Englewood Cliffs, NJ.

Nakayama, H., Yoon, M., and Yun, Y. (2009). *Sequential Approximate Multi-objective Optimization using Computational Intelligence*. Springer, Berlin, Heidelberg.

Patidar, V., Labi, S., Sinha, K. C., and Thompson, P. D. (2007). *Multiple Objective Optimization for Bridge Management. NCHRP Report 590*. National Academies Press, Washington, DC.

Roy, B., and Bertier, B. (1971). La Methode ELECTRE II: Une Methode de Classment en Presence de Criteres Multiples. Note de Travail No. 142. *Direction Scientifique*, Groupe Metra, France.

Ruzika, S., and Wiecek, M. M. (2005). Approximation Methods in Multiobjective Programming. *J. Optim. Theory Appl.*, 126(3), 473–501.

Saaty, T. L. (1980). *The Analytic Hierarchy Process: Planning, Priority Setting, Resource Allocation*. McGraw-Hill, New York.

Sinha, K. C., and Labi, S. (2007). *Transportation Decision Making—Principles of Project Evaluation and Programming*. Wiley, Hoboken, NJ.

Triantaphyllou, E. (2000). *Multi-Criteria Decision Making Methods: A Comparative Study*. Kluwer Academic, Dordrecht, The Netherlands.

Utyuzhnikova, S. V., Fantinia, P., and Guenov, M. D. (2009). A Method for Generating a Well-Distributed Pareto Set in Nonlinear Multiobjective Optimization. *J. Comput. Appl. Math.*, 223(15), 820–841.

Wierzbicki, A. P. (1982). A Mathematical Basis for Statisficing Decision Making. *Math. Model.*, 3, 391–405.

Yu, P. L. (1973). A Class of Solutions for Group Decision Problems. *Manage. Sci.*, 19(8), 936–946.

Zeleny, M. (1973). Compromise Programming. In Cochrane. *Multicriteria Decision Making*, pp. 262–301. University of South Carolina Press, Columbia, SC.

USEFUL RESOURCES

Charnes, A., and Cooper, W. W. (1977). Goal Programming and Multiple Objective Optimization; Part 1. *Eur. J. Oper. Res.*, 1(1), 39–54.

Charnes, A., Cooper, W. W., and Ferguson, R. O. (1955). Optimal Estimation of Executive Compensation by Linear Programming. *Management Science*, 1(2), 138–151.

De Neufville, R., and McCord, M. (1984). *Unreliable Measurement of Utility: Significant Problems for Decision Analysis*. Elsevier, Amsterdam, Holland.

Hwang, C. L., and Masud, A. S. M. (1979). *Multiple Objectives Decision Making—Methods and Applications*. Springer, Berlin.

Hwang, C. L., and Yoon, K. (1981). *Multiple Attribute Decision Making: Methods and Applications*. Springer, New York.

Saaty, T. L. (1977). A Scaling Method for Priorities in Hierarchical Structures. *J. Math. Psych.* 15, 234–281.

Soland, R. M. (1979). Multicriteria Optimization: A General Characterization of Efficient Solutions, *Decision Sci.*, 10, 26–38.

Yu, P. L. (1989). Multiple Criteria Decision-Making: Five Basic Concepts. *Handbooks in Operation Research and Management Science*, 1, 633–699.

APPENDIX 12.1

Sample Data for System Impacts in Terms of Multiple Criteria

	Environmental Impact	Economic Efficiency	User Safety Hazard	System Durability
Action 1	45	34	0.5	5
Action 2	0	61	2.1	7
Action 3	32	4	0.4	2
Action 4	54	32	1.7	3
Action 5	7	76	1.8	8
Action 6	87	120	2.1	12
Action 7	43	17	1.5	3
Action 8	22	45	1.1	4
Action 9	1	3	0.4	2
Action 10	34	32	0.5	4
Action 11	8	14	2.5	3
Action 12	7	8	0.9	3
Action 13	45	87	1.5	6
Action 14	34	38	1.3	4
Action 15	27	55	3.3	6
Action 16	5	10	0.2	4
Action 17	67	77	1.7	6
Action 18	54	87	1.5	7
Action 19	7	12	1.1	3
Action 20	56	45	0.7	4
Action 21	17	27	0.2	3
Action 22	12	18	1.1	4
Action 23	6	23	1.8	3
Action 24	79	98	3.2	10
Action 25	65	67	2.6	5
Action 26	4	15	2.2	4
Action 27	34	39	2.5	4
Action 28	21	31	2.1	5
Action 29	28	45	0.8	3
Action 30	54	76	1.7	5
Action 31	54	61	0.4	8
Action 32	9	5	1.4	4
Action 33	43	32	0.3	2
Action 34	35	29	0.3	2
Action 35	9	31	1.5	3
Action 36	86	107	2.1	8
Average; Standard Deviation; Best; Worst	32.15; 25.29; 0.5; 87	43.36; 31.18; 120; 3	1.42; 0.85; 0.2; 3.3	4.69; 2.33; 12; 2

Units of the impacts are as follows: Environment, acres affected; economic efficiency, net present value in $M; user safety hazard, annual number of fatalities; and system durability, added service life of the system.

CHAPTER 13

RISK AND RELIABILITY

13.0 INTRODUCTION

No one knows the future. The only thing that is certain is uncertainty. We have all heard or made these statements perhaps, but do we really understand them? What are uncertainties and risks, and what role do they play in the management of civil engineering systems? As we learned in Chapter 8, due to the inherent variability in natural processes, real-life engineering systems are characterized by significant variability in their inputs and, subsequently, their outputs. As such, there is no guarantee of system reliability in any sense of the word. The inevitability of uncertainty pervades not only systems thinking but all other areas of human endeavor, captivating scholars and thinkers not only in the current era and recent history, but as far back as the ancient world, as we see in the interesting quotations below:

> When we are not sure, we are alive.
> (*Source*: *Graham Greene, 1904–1991*)

> Our best-built certainties are but sand-houses and subject to damage from any wind of doubt that blows.
> (*Source*: *Mark Twain, 1835–1910*)

> To be uncertain is to be uncomfortable, but to be certain is to be ridiculous.
> (*Source*: *Ancient Chinese Proverb*)

> Only one thing is certain—that is, nothing is certain. If this statement is true, it is also false.
> (*Source*: *Ancient Paradox*)

One thing we can be certain about is that systems engineers will continue to grapple with the issue of the uncertainty of system environments and inherent system capabilities, and subsequently they will be anxious about the ability of their systems to produce the intended outputs or performance levels. In striving to provide optimal designs of physical structures and operational policies that are consistent with serviceability, safety, and economic considerations, civil engineers predict how the designed systems will perform. Unfortunately, it is often the case that proposed future systems have different materials, design, and environmental conditions compared to past systems; as such, the experience and data on past system behavior may not be always relevant to the prediction of future system behavior. Also, current procedures for engineering design, which are generally a result of numerous trial-and-error situations over past decades are associated with not only a lack of precedential experience but also that often yield a level of performance that far exceeds or falls short of expectations when tried in different environments or situations (Harr, 1997). When such variabilities are adequately quantified and predicted, the system engineer is able to accommodate them in the design and can be certain that the resulting designs or operations policies are neither unduly overdesigned nor underdesigned. Thus, the incorporation of reliability in design continues to hold promise in helping designers produce more economical designs that satisfy their expected technical functions without compromising safety and longevity (Frangopol and Moses, 1994).

This chapter prefaces the discussion of reliability by discussing the concepts of certainty, risk, and uncertainty. We identify, in the development of civil engineering systems, a number of sources of uncertainties that are internal (related to the system) or external (related to the system's environment). We then present the basic concepts of reliability and show how uncertainties or reliability in system inputs and outputs can be quantified or measured. Next, we provide four common contexts of reliability analysis in civil engineering and present some numerical examples for each context, followed by a discussion of laboratory and field testing of system reliability and reliability-based design of civil engineering systems. Finally, we discuss briefly the hot-button issue of civil engineering system resilience and its relationship to reliability.

13.1 CERTAINTY, RISK, AND UNCERTAINTY

A key aspect of systems thinking is the systems analysis concept of having inputs to a process, having outputs from the process depending on the inputs, and evaluating the outputs and providing feedback continually to enhance the system. For a given set of inputs, all the possible outcomes may be known, and also the probability of the occurrence of each of these outcomes may be known (Table 13.1). As we discussed briefly in Chapter 12 (and will discuss more fully in Chapter 16), any task in civil engineering systems development can be characterized by one of three levels of certainty: certainty, risk, and uncertainty. In **certainty** situations, both the possible outcomes and their respective occurrence probabilities are known; in **risk** situations, the possible outcomes are known, but the probability of each outcome is not known; and in **pure uncertainty** situations, both the possible outcomes and their respective occurrence probabilities are not known (Knight and McClure, 2009). As we shall see in Chapter 16, the tools used by systems engineers for making decisions are different for the three situations. In the risk situation, the probabilities that are used in the analysis are often derived using deduction (from theoretical models of system behavior) or induction (using relative frequency distributions of past observations). In everyday parlance, the term "risk" is suggestive of an outcome that is undesirable; however, a technical definition of the term accommodates both positive and adverse outcomes.

Risk and uncertainty are due to randomness. Unlike risk, however, the possible outcomes in uncertainty situations do not have quantifiable probabilities and arise due to our imperfect knowledge about natural or man-made events and processes. Uncertainty, according to Schmid (2002), relates to "the questions of how to deal with the unprecedented, and whether the world will behave tomorrow in the way it behaved in the past." In this chapter, our discussions are consistent with the risk situation (known outcomes that have unknown probabilities of occurrence); however, we will use the terms risk and uncertainty synonymously throughout this chapter.

Table 13.1 Difference between Certainty, Risk, and Pure Uncertainty

		Uncertainty	
	Certainty	**Risk**	**Pure Uncertainty**
Possible outcomes	Known	Known	Unknown
Probability of each outcome	Known	Unknown	Unknown

13.2 UNCERTAINTIES IN SYSTEM DEVELOPMENT

Uncertainty, a concept closely related to reliability, can be defined as the chance that the outcome of a decision or action will be different from what is desired or expected. Uncertainty is an inevitable part of our everyday lives (our academic performance, health, sports, etc). Figure 13.1 presents some uncertainties faced by the system owner, user, or community at the various phases of civil systems development. Owners and operators of civil engineering systems are often responsible for ensuring that the risks associated with such uncertainties are minimized. Ideally, civil engineering systems should have zero uncertainty in terms of their safety, longevity, condition, community impacts, and the like. However, as Dandy et al. (2008) pointed out, (i) failure-proof designs that may guarantee 100% protection from adverse events will be prohibitively expensive to construct and operate, and thus may be infeasible from the capital financing viewpoint or may not be cost effective; or (ii) physical systems are subject to stresses (from live loads, demand, earthquakes, floods, etc.), whose frequency or intensity are not within the full control of the engineer, so the possibility always exists that the system will experience a stress exceeding its design stress.

In Figure 13.2, we present a number of sources of internal and external uncertainties in systems development. Internal sources are those that are inherent to the system itself while external sources are caused mostly by the system's environment. External sources of uncertainty could be due to changes related to the political and socioeconomic landscape, the environment, and other forces. External sources related to political and socioeconomic conditions include demographic changes, technology, government legislation or policy, changes in the level or pattern of demand for the services offered by the civil engineering system, all of which often can translate into variations in intensities or frequencies of system loading intensity and, in the case of revenue generating systems, revenue projections that are different from what is expected. Others include volatility in the costs of the factors of production (prices of labor, equipment, materials, and land), fluctuations in the market interest rate, and changing reliability of the supply of raw materials for

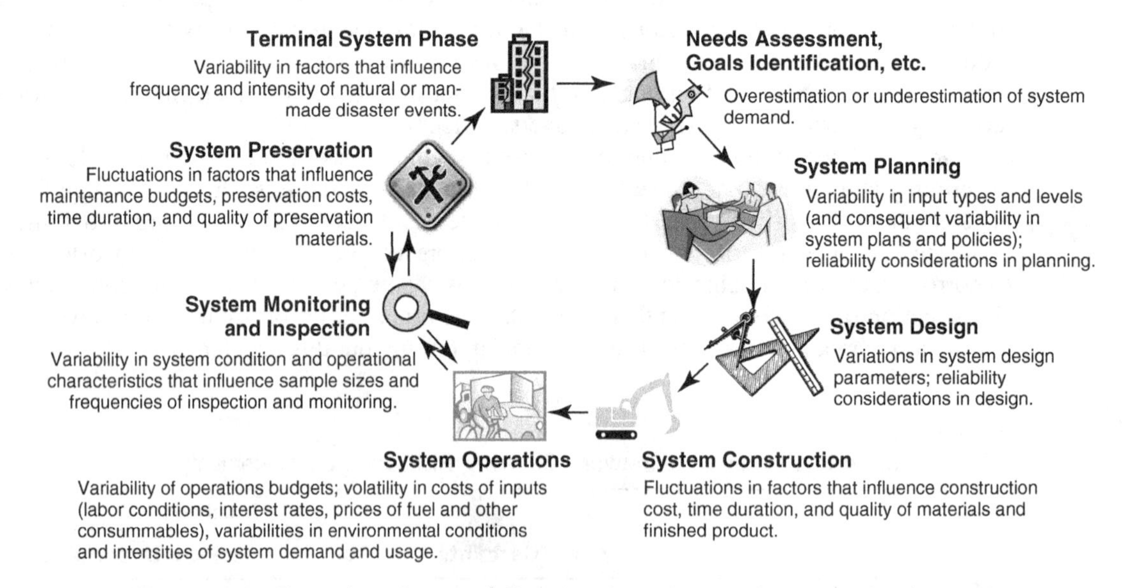

Figure 13.1 Examples of uncertainties at various phases of systems development.

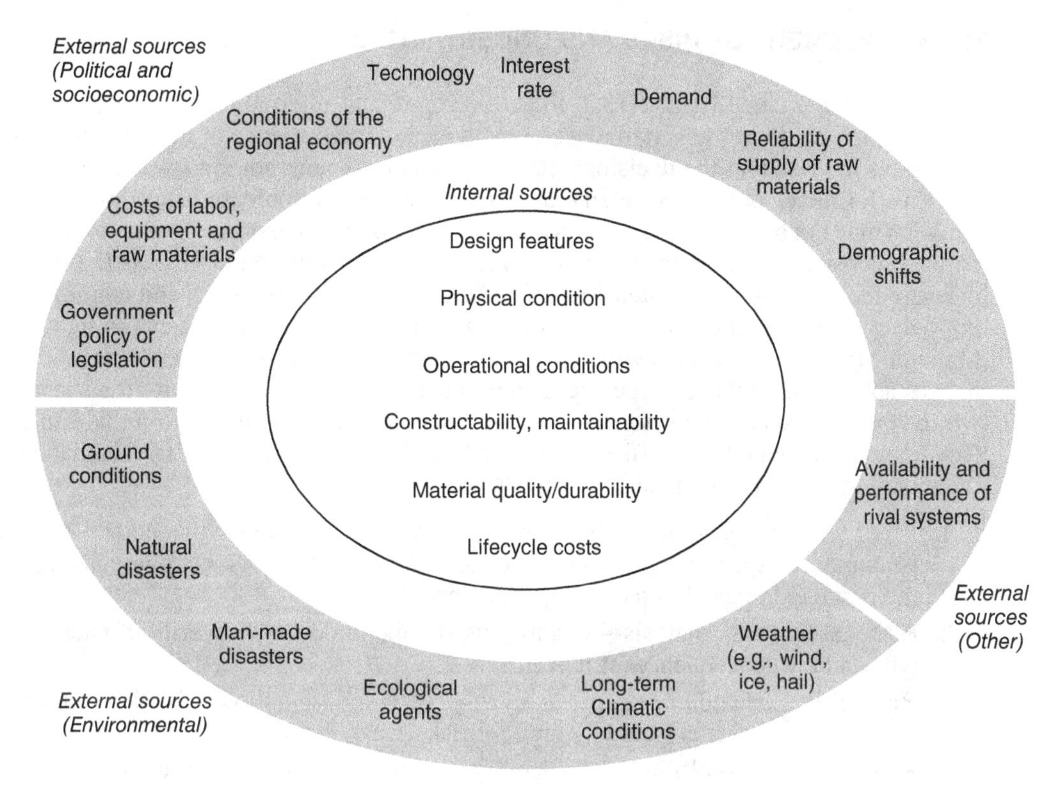

Figure 13.2 Sources of internal and external uncertainties in systems development.

construction, maintenance, or operation of the system. External sources of uncertainty that are related to the environment include changes in the frequency and intensity of climatic conditions (e.g., temperatures, freeze conditions); the levels of water in the ground, lakes, rivers, and coasts; and the intensity and frequency of earthquakes, tsunamis, and other natural disasters as well as man-made disasters including collisions, overloads, vandalism, and other malicious attacks.

Internal sources of uncertainty include the design features; system physical condition and operational performance; system durability (longevity); initial costs of planning, design, and construction; constructability and maintainability; and life-cycle costs. Many of the civil systems that were designed a long time ago, when there was inadequate knowledge about the behaviors of specific design configurations and materials, continue to operate at the current time with some amount of uncertainty regarding the nature of future distress or failure. These uncertainties in the physical condition or operational performance of existing systems continue to pose formidable challenges for engineers responsible for long-term planning and work scheduling for these systems. With regard to system costs, the initial costs of planning, design, and construction always are known for an existing system; however, for future similar systems, a source of uncertainty is the variability of the unit costs, even for identical systems in identical environments, which continues to introduce significant variability in an engineer's cost planning and budgeting. Also, the differences in the constructability and maintainability of similar systems by virtue of their different locations, environments, or operating conditions, introduce uncertainty in the cost and scheduling of the construction and maintenance activities of future similar planned systems.

13.3 THE MANAGEMENT OF RISKS AND UNCERTAINTIES

As shown in Figures 13.1 and 13.2, there are significant internal and external sources of uncertainty at each phase of the system development cycle. The presence of these kinds of uncertainties introduces the risk that the decisions made at each phase may not always lead to the desired levels of performance outcomes; in an era of tight budgets and increased stakeholder oversight, this situation may not be acceptable to the system owner. For this reason, civil systems managers worldwide are constantly attempting to properly identify the possible sources of risk (including hazards), measure the likelihood and intensities of these risks, mitigate the risks and continuously monitor not only the risk likelihoods and intensities but also the effectiveness of any mitigation strategies. Managing the risks associated with the realization of system consequences (costs and benefits) can help establish justifiable contingencies and provide realistic expectations for the consequences that will be experienced by the system owner, the users, and the community. In this context, the U.S. Government Accountability Office (GAO, 2009) and Molenaar et al. (2010) established five largely sequential aspects of risk management:

1. Risk Identification—determining the threat types (risk sources) that potentially influence the attainment of specific objectives associated with each development phase, using tools including past experience, brainstorming, and checklists.
2. Risk Assessment/Analysis—quantifying the likelihood and severity of each specific threat type and the consequences if it occurs.
3. Risk Mitigation and Planning—analyzing risk response options (risk acceptance, avoidance, mitigation, or transfer) and planning for risk management.
4. Risk Allocation—placing responsibility for perceived risks, where applicable, to parties that are best equipped to manage them.
5. Risk Monitoring/Control—continually tracking the likelihoods and intensities of the different threats, and also tracking the costs and effectiveness of any mitigation strategies implemented. Risk control is a set of actions undertaken to remedy situations found through risk monitoring. These may be considered as an overlap with risk assessment and mitigation.

The above steps are merely guidelines. The steps adopted for managing risks for a particular system may differ depending on the system type, location, administrative culture, or complexity. In the area of transportation systems for example, Molenaar et al. (2010) developed a systematic process not only to identify, assess, and analyze risks associated with project costs but also to select mitigation strategies, allocate risks, and carry out monitoring and control of such risks. We now discuss each of the steps in some detail.

13.3.1 Risk Identification

Risk identification simply answers the question: What could go wrong? To do this, the analyst, using tools including past experience, brainstorming, and checklists, determines the threat types (risk sources) that potentially influence the attainment of the specific objectives associated with each development phase. The risks faced by civil systems may derive from sources of uncertainty including (a) individual phases (as we saw in Figure 13.1), (b) man-made versus natural sources, (c) controllable versus uncontrollable sources, and (d) internal versus external sources.

The risks commonly faced in civil systems development, which can be placed in the above categories, include (Ezell et al, 2000; Ayyub, 2003):

Social risks. External uncertainties associated often with the construction and operations phases, social risks emanate from man-made or natural sources and are associated with social values.

For example, protecting the environment or increasing the community's quality of life. In several developed countries, the construction of civil systems are often delayed due to local community resistance.

Political risks. These arise when there is political instability that may influence the implementation of planned systems or the operations or maintenance of existing systems. In some developing countries that have unstable governments, the political risks associated with civil engineering systems can be significant.

Project management risks. These occur when there is poor planning or resource allocations at the phase of development in question. These are typically associated with man-made sources of uncertainty.

Legal risks are those associated with compliance with government regulation (often during the system operations phase) and owner-contractor relationships (often during the construction or maintenance phases).

13.3.2 Risk Assessment/Analysis

In risk assessment, the questions may relate to how likely is it that something will go wrong, and what will happen if it does go wrong. To answer these, the analyst calculates both the likelihood and the severity of each potential threat type and the consequence if it occurs (Table 13.2).

(a) Risk Likelihood Assessment. Tools available to the analyst include statistical models (models that estimate the probability and severity of cost overruns, operational breakdowns, pollutant leaching into groundwater, and differential settlement rates of foundations), maps indicating the historical severities of geohazards including landslides, sinkholes, flooding, and earthquake. Risk can be assessed using qualitative (e.g., fault tree analysis) or quantitative techniques (e.g., probabilistic risk analysis). The likelihood of an event may be expressed as a probability (percentage of times that an event occurs) or the expected number of occurrences within a given time period.

(b) Risk Consequence Assessment. Consequence is typically assessed in terms of the impacts on funding, the environment, the local ecology, and the quality of life in the community. Tools include GIS maps that show buffer zones of population density, demographic patterns, and natural (ecological) resources in relation to potential system degradation due to an event. The consequence of an event may be expressed in terms of technical criteria (inability of the system to perform its required function), cost criteria (amount of funds needed to repair or replace a system or system component), environmental criteria (area of ecology degraded, property damage), and user consequences (delay, inconvenience, injury, or fatality).

(c) Risk Matrices. Statistically, the level of risk has often been calculated as the product of the probability that a threat will occur and the consequences of the threat. From a coarse perspective, there are four possibilities:

- Low threat likelihood, low consequence if threat occurs
- High threat likelihood, low consequence if threat occurs
- High threat likelihood, low consequence if threat occurs
- High threat likelihood, high consequence if threat occurs

This can be represented as shown in Figure 13.3, a risk matrix that defines the various risk levels and is often the final output of the risk assessment step. This can help the system owner visualize the risks associated with any given threat type. Figures 13.3 and 13.4 present a simplified

Table 13.2 Risk Assessment Methods

Method	Scope
Safety Audit/Review	Identifies features and conditions of system components or operating processes that could result in adverse impacts including operational breakdown, user injury, or environmental degradation.
Checklist	Itemized list that covers a wide range of areas regarding system condition/performance to ascertain compliance with standards.
What If/Then Analysis	Identifies hazardous situations that could lead to adverse consequences.
Hazard and Operability (HAZOP) Studies	Identify potential deviations of system condition and operations that could result in adverse consequences; makes recommendations to minimize the frequency and/or consequences of such deviations.
Preliminary Hazard Analysis (PrHA)	An inductive modeling approach that identifies and prioritizes hazards early in the system life; recommends actions that reduce hazard frequencies and/or consequences.
Probabilistic Risk Analysis/Assessment (PRA)	Developed by nuclear systems engineers, this method assesses risks quantitatively; is a comprehensive process that combines several methods of risk assessment.
Failure Modes and Effects Analysis (FMEA)	An inductive modeling approach that identifies the failure modes of systems or system components and the impacts on neighboring or related components.
Fault-Tree Analysis (FTA)	A deductive modeling approach that identifies combinations of human errors and equipment failures that could lead to undesirable consequences.
Event-Tree Analysis (ETA)	An inductive modeling approach that helps identify sequences of events that could lead to undesirable outcomes.
Delphi Technique	Helps decision makers to reach consensus on subjects while ensuring anonymity; solicits judgments regarding risks that are then synthesized and fed back to the experts for further review of earlier judgments. Consensus may be reached after a few iterations.
Interviewing	Identifies risk events on the basis of the knowledge of experienced project managers and subject matter experts. Identify and quantify risk events.
Experience-Based Identification	Identifies events that may pose some risk on the basis of past experience.
Brainstorming	Identifies events that may lead to adverse outcomes by direct interactions with the system stakeholders.

Source: Ayyub (2003).

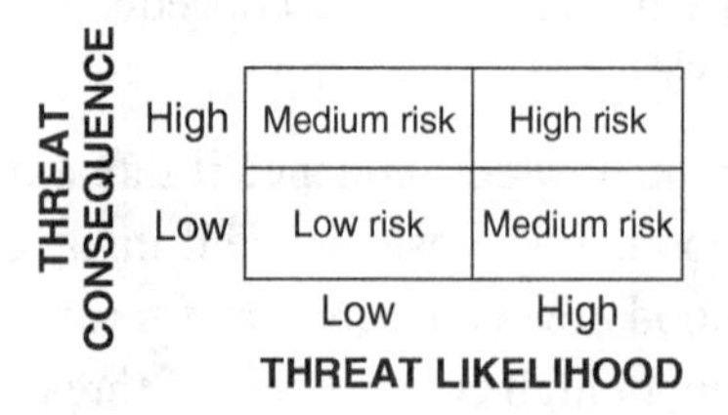

Figure 13.3 Simplified risk matrix (shown with risk levels).

THREAT CONSEQUENCE	Unlikely	Possible	Frequent
Catastrophic	Large asteroid collision	Major earthquake	
Critical			Landslides, Flooding
Significant	Sinkholes	Truck overloading	
Negligible			Earth tremor

THREAT LIKELIHOOD

Figure 13.4 Risk matrix (shown with examples of threats, at a given location).

risk matrix (shown with risk levels) and a risk matrix with examples. The latter is provided with reference to a specific hypothetical location because the threat types and consequences are different for different locations. For any civil engineering system, a risk matrix with the possible threat types, similar to Figure 13.4, should be developed as part of risk assessment.

Risk matrices have been used widely to establish priorities for resource allocations at many national organizations and agencies. It has been used in several application areas of risk management policy and practice including national standards in countries such as Australia and New Zealand, corporate oversight in the United States (due in part to the Sarbanes Oxley Act of the U.S. Congress), and enterprise risk management (ERM) (Cox, 2008). Risk matrices have also been used in risk management practices in civil infrastructure construction projects, safety management of infrastructure operations, and assessment of potential natural and man-made threats to civil infrastructure (Renfroe and Smith, 2007).

Risk matrices have many advantageous features. For example, they are simple, require no special expertise to construct, and are easy to understand and are therefore suitable for communicating issues relating to risk and for personnel training. However, some researchers have cautioned that risk matrices contain mathematical and logical limitations that inhibit effective characterization of risks and evaluation and prioritization of mitigation actions (Cox, 2008). For example, risk matrices assign identical ratings to risks that are very different quantitatively and therefore often make correct comparisons among only very few threat pairs. Second, the categorizations of threat likelihood and also of consequences in risk matrices tend to be too coarse for purposes of decision making; in other words, resources for risk mitigation cannot be effectively allocated on the basis of these coarse categories. Third, risk matrices tend to assign higher qualitative ratings to risks that are quantitatively smaller and thus are inappropriate for assessing risks that have negatively correlated frequencies and severities. The fourth limitation is the subjective nature of both the categorization levels of the likelihood and also for the consequences, and of the ratings; therefore, different users of a risk matrix for the same problem may generate very different risk ratings. Consequently, researchers have proposed a number of good practices in the development of risk matrices. For example, the risk matrix must have consistent likelihood ranges across the entire spectrum of possible scenarios and detailed descriptions of consequence range (Ozog, 2009). Also, it has been recommended that the grid lines (thresholds) in a risk matrix must be established carefully in order to minimize the maximum loss that arises from misclassified risk, and risk matrices must be used with due caution and with clear explanations of any underlying judgments (Cox, 2008).

13.3.3 Risk Mitigation and Planning

In order to lower the risk associated with a threat, the system owner may take steps to reduce the threat likelihood, consequence, or both. As such, risk mitigation efforts may be characterized by any combination of the following strategies (Ford, 2009):

Avoid. This is generally used to address risk situations with high likelihood and high consequence; for example, building a railway in a permafrost region that is beginning to experience significant thawing at many locations. In this case, the railway construction can be abandoned in favor of air transportation. The current dominance of air freight transportation in Alaska is attributed to this specific example.

Reduce. This is generally used to mitigate risk situations with low consequence and high likelihood. In such cases, an alternative action is pursued while taking steps, where possible, to reduce the likelihood or intensity of the threat. For example, in building a pier in a harsh marine environment, the use of traditional steel may be abandoned in favor of stainless steel in order to reduce the likelihood of the threat of corrosion.

Retain. This is often the appropriate strategy to reduce risk in situations with low consequence and low likelihood. In these cases, the risk situation is retained without taking any direct action as (i) the outcome is considered not too harmful if the threat occurs or (ii) the threat is extremely unlikely to occur. For example, in making a decision to store nuclear waste in carefully designed containers deeply-buried at uninhabited areas, a nuclear agency makes the assumption, implicitly or explicitly, that there is a low probability that the containers will leak and that the consequence is low if a leak does occur.

Transfer. This is generally used to address risk situations with high consequence and low likelihood. In this case, the risk is transferred, for example, an insurance policy is acquired so as to reduce the consequence suffered by the system owner in case the threat does occur. Two examples of this strategy are as follows: (a) during system operations, an insurance policy is acquired so as to reduce the consequence suffered by the system owner in case the threat does occur and (b) before system construction, the system owner requires a performance bond from the successful bidder that will guarantee compensation to the owner in case the contractor fails to carry out the contract.

Civil engineering agencies practice risk mitigation constantly, even implicitly at times. Examples include establishing contingency amounts to cover possible cost overruns, priority scheduling of repair for structures that exhibit signs of imminent failure, preparing for extreme events through evacuation response planning, and developing flexible schedules for capacity enhancements or replacing systems or system components to account for the uncertainty in demand or component longevity respectively.

If the risk matrix is intended to be used directly during the subsequent step of risk mitigation, then it must show which risk levels are tolerable and which are not by defining the threshold levels for tolerable and intolerable risk and indicate, for scenarios with intolerable risk levels, how risk could be mitigated to achieve tolerable levels (Ozog, 2009). Figure 13.5 presents a risk matrix with mitigation actions that is often developed to prescribe the appropriate action to prevent or avoid a threat or to reduce the severity of its consequences. This matrix can serve as a decision-support system for palliative actions to mitigate the risk.

13.3.4 Risk Allocation

Risk allocation is the process of delegating responsibility for risks, where applicable, to the parties that are best equipped to manage them. For example, the practice of public–private partnerships is based on the assumption that the risks of defective products are transferred to the private sector as

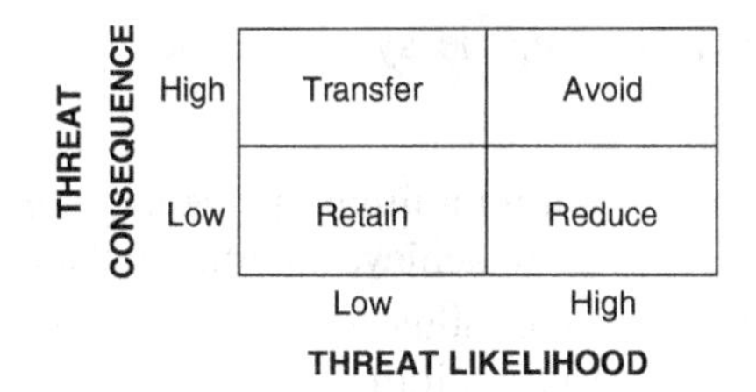

Figure 13.5 Simplified risk matrix shown with mitigation recommendations.

that sector is better equipped, compared to the public sector, to manage these risks. Risk allocation, strictly speaking, may be considered as one of the techniques for risk mitigation.

13.3.5 Risk Monitoring/Control

Risk monitoring refers to the continual tracking of the likelihoods and intensities of the different relevant threats as well as the costs and the effectiveness of any mitigation strategies that have been implemented to counter the threat. Risk control is the set of actions undertaken to remedy situations found through risk monitoring and may be considered as an overlap with risk assessment and mitigation. Often more than one set of actions may exist, and risk control analysts compare these sets in order to identify the optimal set. Examples of good practices in risk monitoring include the evaluation of each phase of system development after that phase is completed (also referred to as *ex poste* evaluation) and providing feedback to engineers and managers working at the preceding phases of the same or similar systems.

13.3.6 Concluding Remarks on Risk

This section provided a basic overview of risk management. In Chapters 22 and 28 where we discuss concepts related to system end of life and system resilience, we will see how some of these risk concepts are applied. In the next section, we will address the subject of reliability, which is a more specific perspective of the concept of risk.

13.4 THE BASICS OF RELIABILITY

Reliability may be defined generally as the ability of a system (or component thereof) to perform its required functions or to achieve its established performance objectives under a given set of conditions and at a given point in time, for example, after a sudden disaster event or during a finite time horizon (Ayyub; 2003; Frangopol and Moses, 1994). The concept of reliability is based on likelihood or probability, which means that the likelihood exists that even new, well-designed civil engineering systems could fail suddenly (an admittedly small likelihood, but still a possibility). Reliability can also be defined as the probability that a system will be in a nonfailure state (Dandy et al., 2008); failure can be defined in simple terms as an event or state of a system in which the system or any of its components does not perform as intended (Wasson, 2006).

Depending on the performance objective and system type in question, a system owner or operator may be interested in specific definitions of reliability. The failure of a system may be viewed from at least one of two perspectives: the system-of-systems (SOS) level or the system-specific level. At each of these levels, performance measures include (individual or average) structural integrity, cost, operational effectiveness, longevity, and the like.

From an SOS perspective, the system owner may view reliability in one or more ways including:

- The likelihood that a certain minimum percentage of systems in the SOS (or of components in a system) will continue to achieve a certain level of performance
- The likelihood that any individual system in the SOS (or of components in a system) will continue to achieve a certain minimum level of performance
- The expected percentage of systems in the SOS (or of components in a system) that will continue to achieve a certain minimum level of performance
- The likelihood that the average performance of all the systems in the SOS (or of components in a system) will exceed a certain minimum

Specific examples are: the likelihood that 95% of all the levees in an area's levee network will not fail during a severe hurricane, the likelihood that no individual intersection in Townsville City's street network experiences an unsatisfactory level of service, or the percentage of individual drainage channels in a city that provide unsatisfactory levels of service.

For a specific individual system, the systems owner may view reliability in many ways including:

- Whether or not the system performs a required function under given conditions for a specific time period
- The extent to which the system performs a required function under given conditions for a specific time period
- The likelihood that the system achieves a certain minimum level of service under given conditions for a specific time period
- The likelihood of each degree to which the system performs a required function under given conditions for a specific time period

For example, the structural reliability of a structural frame may be viewed as whether the frame will be in a nonfailure state or that it supports the expected dead and live loads over its design life; the operational reliability of a freeway system may be viewed as the number of vehicles served at peak periods without undue congestion delays; the safety reliability of a certain construction task is the likelihood that the task will proceed with no loss of life through site accidents; the likelihood that a wastewater network will experience one, two, or three occurrences of leaks in a week; or the probability that a water retaining structure will develop a leak given its age. In the third case, reliability is not a single number but rather a piecewise or continuous function as illustrated in Figures 13.6*a* and 13.6*b*; and in the fourth case, reliability is the probability of failure (leakage, in this case) or absence thereof over a given period of time (Figure 13.6*c*).

The concept of reliability is closely related to the concepts of risk and vulnerability. In certain definitions of risk and vulnerability, each of these terms incorporate both likelihood and consequence. However, reliability generally is concerned with only the likelihood. Reliability can be considered the antithesis of risk and vulnerability: The higher is the likelihood of risk or vulnerability, the lower the reliability, and vice versa. For example, the greater the likelihood of the risk that a levee system is overtopped, the lower the reliability of the system in terms of its functional adequacy. Also, risk generally pertains to both internal (the system) and external (the environment) sources; however reliability generally pertains to internal source only.

As we can see from the examples provided above, establishing the reliability of a system, in terms of any measure of performance or risk factor, draws heavily on the concepts of statistics

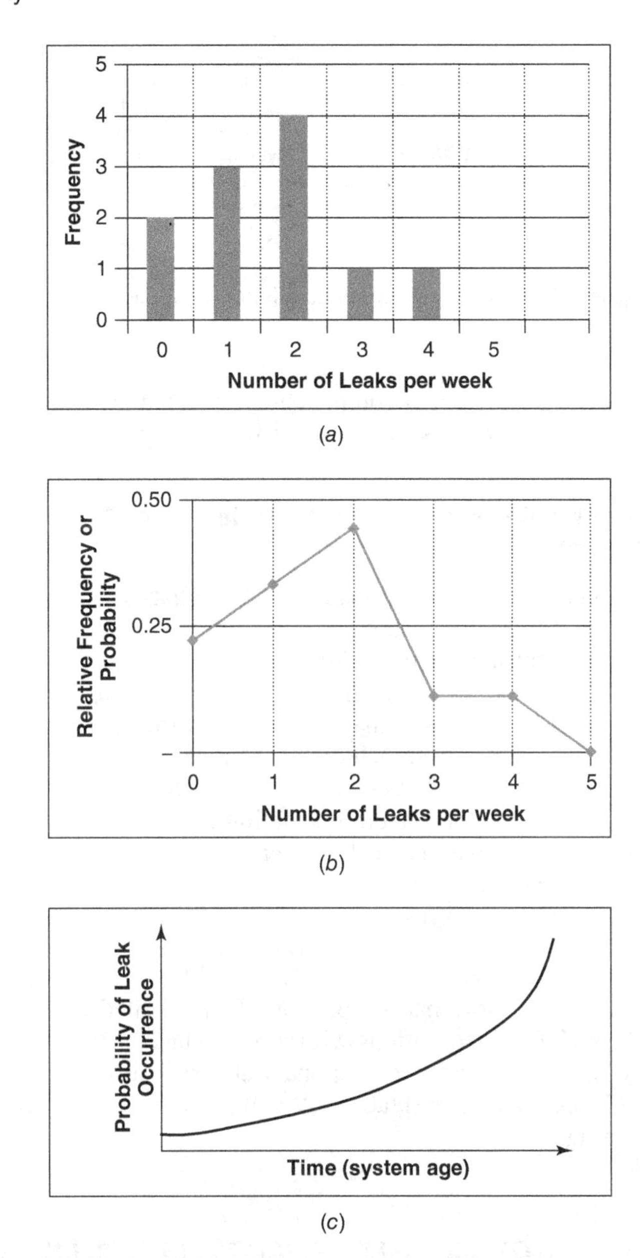

Figure 13.6 Simple reliability functions—Illustration: (*a*) leakage frequency at a given age, (*b*) leakage probability distribution at a given age, and (*c*) trend showing the expected number of leakages (per week) with age.

and probability (see Chapters 5 and 6). Also, a clear definition of the system objectives and performance objectives is a necessary starting point for any analysis of system reliability. At agencies where systems reliability assessment is a core activity required by the system owner or operator, the system engineer is tasked with establishing the reliability requirements for the entire network of systems, each constituent system, or for the individual components of each system; developing a reliability program for all the systems in the agency's inventory; performing analyses to ascertain

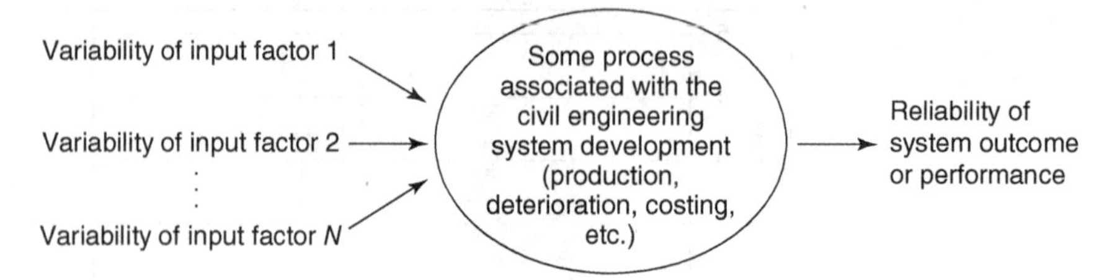

Figure 13.7 Output reliability as a function of the reliability of the input factors.

the extent to which each system or component is expected to meet certain performance targets; and identifying/assessing any threats to system performance.

13.4.1 Roots of the Reliability Problem—Variability of Design Parameters and Operating Conditions

Figure 13.7 presents a schematic illustration of the reliability of system output (e.g., performance, physical condition, longevity, cost, etc.) as the end result of the influence of the variabilities or uncertainties in the contributory input factors $1, 2, \ldots, N$.

The terms reliability and variability are strongly related: the higher the variability, the lower the reliability, generally. As we saw in Chapter 6, the variability, or dispersion, of a random variable can be expressed in several ways, including the range, the mean deviation, the standard deviation, and the coefficient of variation. The *coefficient of variation* (CV) of a probability distribution, which is calculated as the ratio of the standard deviation to the mean of the population, is a reflection of how scattered the data are with respect to the mean and is therefore considered as a "normalized" measure of dispersion of the distribution.

$$V(x) = 100\frac{\sigma(x)}{E(x)}\%$$

For any random variable that is exponentially distributed, the standard deviation is equal to the mean so the C is 100%. Distributions with low variance (CV < 100%) and high variance (CV > 100%) include the Erlang and hyperexponential distributions, respectively. Table 13.3 (adapted from Harr, 1997) presents representative CV values for parameters that are encountered commonly in civil engineering.

13.5 COMMON CONTEXTS OF RELIABILITY ANALYSIS IN CIVIL ENGINEERING

As defined in the introduction to this chapter, reliability is a term that can be used in a great variety of contexts. These contexts vary depending on the type of system in question, the phase of system development in question, and the task at hand. The following are four common contexts in civil engineering related to reliability:

- Of every 100 samples of soil we take at a given site, what is the probability that the moisture content will fall within a specified range? Or, how many fresh concrete cylinders must we sample from a production run in order to be sure that the true compressive strength is within a certain limit?

Table 13.3 Representative CV[a] Values for Common Parameters in Civil Engineering

Parameter Category	Parameter Type		Variability (CV, %)
Soil	Porosity		10
	Specific gravity		2
	Water content, silty clay		20
	Water content, clay		13
	Degree of saturation		10
	Unit weight		3
	Coefficient of permeability at 80% saturation		240
	Coefficient of permeability 100% saturation		90
	Compressibility factor		16
	Preconsolidation pressure		19
	Compression index, sandy clay		26
	Compression index, clay		30
	Standard penetration test		26
	Standard cone test		37
	Friction angle ϕ, gravel		7
	Friction angle ϕ, sand		12
	Cohesion-related strength parameter, c		40
Structural Loads, 50-Year Maximum	Dead load		10
	Live load		25
	Snow load		26
	Wind load		37
	Earthquake load		>100
Structural Resistance	Structural steel	Tension members, limit state, yielding	11
		Tension members, limit state, tensile strength	11
		Compact beam, uniform moment	13
		Beam, column	15
		Plate, girders, flexure	12
	Concrete members	Flexure, reinforced concrete, grade 6	11
		Flexure, reinforced concrete, grade 40	14
		Flexure, cast-in-place beams	8–9.5
		Short columns	12–16
Wood	Moisture		3
	Density		4
	Compressive strength		19
	Flexural strength		19
	Live-load, glue-laminated beams		18
	Snow load, glue-laminated beams		18

[a]CV means Coefficient of Variation.
Source: Harr (1997).

- Of all the asphalt pavements constructed on a state's high-volume traffic interstate highways in 2010, how many of them can be expected to last for, say, 20 years? Or, what is the probability that any randomly selected section of these pavement lasts for 20 years?
- Given that the system consists of potentially vulnerable subsystems that act in concert with each other, how often will the system cease to function suddenly, or what is the probability that the system will do so?
- Given that both the capacity and loading are highly variable, how often will we have an unfavorable situation (i.e., where loading exceeds capacity) or what is the probability that it will happen?

Against this background, we present details, including computational techniques, for quantifying and analyzing reliability in four distinct contexts of civil engineering practice: **sampling** (where the engineer is concerned with the relationship between the sample size and the reliability, precision and the level of confidence in the attribute of the population being sampled), **performance and survival assessment** (where the engineers seeks to be confident that the physical condition or operational performance of a civil engineering system will not exceed some threshold or that the system will continue to survive given that it has survived up to the present time), **progression of failure rates** (includes the prediction of expected failure rates, expected life, and average time between failures), **system resilience** (where the engineer seeks to ascertain that the system will fail suddenly due to a concatenation of failure events associated with subsystems with parallel or series interconnections), and **shortfall avoidance** (where the engineer seeks to make sure that the capacity of the system is as high as possible and exceeds the loading level).

13.5.1 Sampling (Relationship between Reliability, Precision, Level of Confidence, and Sample Size)

In Chapter 6, we discussed some concepts of reliability analysis where we reviewed the relationship between reliability, precision, level of confidence, and sample size. This problem context is encountered frequently by civil engineers involved in quality control and quality assurance of prepared surfaces and manufactured components used in system construction, maintenance, or operations. In problems of this nature, the engineer seeks the minimum sample size needed to ensure a certain degree of confidence or confidence interval:

$$n_{\text{MIN}} = \left(\frac{Z_{\alpha/2} * \sigma}{\text{Error}_{\text{MAX}}} \right)^2$$

where $Z_{\alpha/2}$ is the Z value corresponding to an area of $\alpha/2$ to the left. The precision, or maximum error, is the deviation of the estimated parameter because its true value is the standard deviation of the population.

Example 13.1

How many observations should a quality control engineer take in a test sample if the contract agreement specifies that there should be 90% confidence that the estimate of the mean test parameter does not deviate from the true value by more than 1 unit? From past experience, it has been found that the parameter of produced units is normally distributed with a variance of 16 units.

Solution

Maximum error = 1 unit. Denote the minimum sample size is n_{min}.

The standard deviation, σ, = 4 units. Also, $1 - \alpha = 0.90$, hence $\alpha = 0.1$, $\alpha/2 = 0.05$, and $Z_{\alpha/2} = 1.645$ (see chart in Appendix 2). Thus, $n_{MIN} = [(1.654 \times 4)/1]^{0.5} = 2.57$.

Therefore, the test sample should comprise at least three observations.

13.5.2 Progression of System Condition or Failure/Survival Likelihood over Time

In this context, civil engineers are interested in quantifying the reliability that the system will be in a certain condition state or will survive for a certain period before the initiation of some specific defect or before the system fails. The modeling tools we learned in Chapter 7 can help engineers predict the reliability that a system will be in a certain condition at a given time; that it will not develop some distress if it has not done so as of the time of the data collection; or that it will not fail given that it has not failed up to the time of the data collection. Also, modeling tools are used to acquire better insights into the factors that influence this type of system reliability. As in other areas of systems modeling, reliability, in this context, can be modeled using one of three approaches: mechanistic (which involves the use of laws and theories in the natural sciences), empirical (which is based on an analysis, often statistical, of past observations of the way systems that are similar to the one under investigation), and mechanistic-empirical (which combines both approaches—parts of the reliability model are mechanistic, others are empirical). Work in this area has included the prediction of the growth fatigue crack spot-welded joints in steel structures (Ni and Mahadevan, 2004).

(a) Modeling Approaches

Mechanistic Approaches. Mechanistic approaches, for which example applications are presented in Table 13.4, often utilize the laws of the basic sciences (mainly physics and chemistry) directly and may involve stress–strain relationships, crack propagation, or chemical corrosion processes. Past studies that analyzed various forms of reliability for various contexts of designing different

Table 13.4 Some Mechanistic Methods Used to Assess the Concrete Structure Reliability

Method	Influence or Application	References in (Liang et al., 2002)
Physical-mathematical model	Estimated time to reach some condition threshold	Bazant (1979a,b)
Accelerated test/mathematical model	Service life prediction for concrete	Pommersheim and Clifton (1985)
Model based on survey data of bridge decks exposed to deicing salt, coastal buildings, and offshore structures	Predicted time of deterioration initiation	Guirguis (1987)
Unsteady-state dynamic model	Estimating the life of external vertical walls of reinforced concrete with external thermal insulation	Fukushima (1987)
Corrosion rate model for reinforcing steel	Prediction of the service life of a reinforced concrete building	Morinaga (1990)
Models based on accelerated corrosion tests and field measurements	Predicting the rate of steel corrosion in reinforced concrete structures	Harn et al. (1991)
Gray theory model	Prediction of the remaining service life of harbor structures	Li (1992)

Source: Liang et al. (2002).

civil structures include Lin (1995), Deshmukh and Bernhardt (2000), Lounis (2000), Stewart et al. (2004), Akgul and Frangopol (2004), Biondini et al. (2006), Oh et al. (2007), and Strauss et al. (2008).

If the exact nature of system degradation over time or the sudden failure of a system could be explained exactly in terms of the underlying theories in physics and chemistry, mechanistic approaches would suffice for describing or predicting, in a deterministic manner, the progression of system condition or survival likelihood over time. In practice, however, limited understanding of these theories precludes a precise and theoretical statement of these processes. As such, analysts have little recourse but to use empirical modeling of past observations and data to describe the progression of system condition or survival likelihood over time; these efforts are inherently stochastic due to the seemingly random nature of the data, which in turn is due to the unmeasured "noises" that cloud the patterns followed by the physical and chemical phenomena.

Empirical Approaches. Empirical approaches are suitable for analyzing the performance attributes of civil engineering systems that are related to the system's physical condition; for example, the rate of deterioration, service life, or failure probability of a system can be modeled as a statistical function of system age, accumulated loading or climatic exposure, size, location, design, and material type, using records of deterioration, age at failure, and other attributes. Thus, the reliability of a system may be modeled as a function of time (or some time-related parameter such as accumulated usage), as a function of space (locational), or both. Two key prerequisites to reliability model development are (a) a measurable criterion for failure (including physical failure or operational breakdown) and (b) the unit of "time," which could simply be the age (years since construction or installation or years since some action such as rehabilitation, retrofit, or change in operational policy).

(b) Trade-off between System Performance Level and System Performance Reliability at a given Time. Given a choice, would you prefer to gain $1 million with 1% probability or $10,000 with 99% probability? Or would you prefer to lose $1 million with 1% probability or $10,000 with 99% probability? In each situation, how would your choice change for different probability of each of these two amounts? In making decisions at each phase of system development, system owners and operators constantly engage in such gambles due to inherent uncertainties in system inputs and consequently in system outcomes. Civil engineers and managers that work for system owners and operators are interested not only in the performance of a system (in terms, e.g., in the system's physical integrity, operational functioning, and user safety) but also in the reliability associated with such performance measures. Do they make decisions that reveal their willingness to gain a large benefit that has great uncertainty? Or are they risk averse (prefer to gain small benefits that have greater certainty)? In Chapter 12, we discussed the risk-taking nature of decision makers and how it influences their decision-making process. In the context of the trade-off between performance level and performance variability there could be any one of four scenarios at a given time: (1) a system that provides an excellent level of performance and good reliability of performance, (2) a system that provides an excellent level of performance but poor reliability of performance, (3) a system that provides a poor level of performance but good reliability at that performance level, and (4) a system that provides a poor level of performance and poor reliability of performance (Figure 13.8). Clearly, the first scenario is most desirable to the system owner while the last scenario is least desirable. Whether the system owner prefers the second scenario to the third or vice versa will depend on the importance the system owner attaches to the system performance level compared to the reliability of performance. The computation of the expected value may serve as a normalizing technique to compare across alternatives that differ (performance levels and their uncertainties). In Chapter 16, we will examine how this could be done.

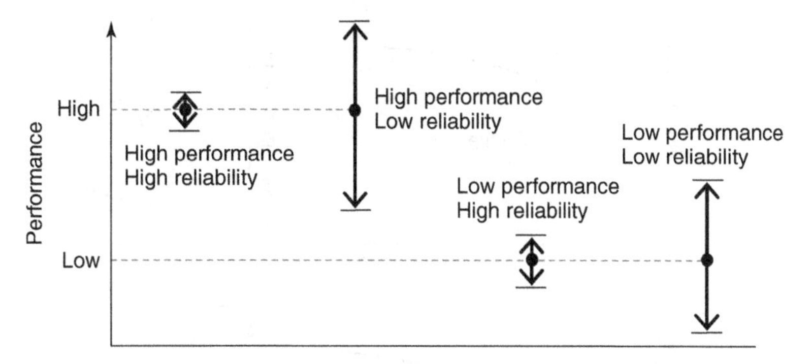

Figure 13.8 Levels and reliability of performance for four different options.

13.5.3 Progression of Failure Rates over Time

Often referred to as classical reliability analysis, this context addresses the statistical characteristics of the longevity of engineering systems and components (Faber, 2007), including the expected failure rate, the expected life, and the average time between failure events.

(a) Failure Rate Plots and Patterns. A failure rate plot is a graph showing the rate of system failure versus some time-related variable such as the time elapsed (or accumulation of loading) from some reference time point such as the time of construction, improvement, or change in operational policy. Depending on the type of system and the definition of failure (i.e., the performance criteria in question), there are several possible rate plot patterns for a civil engineering system (Figure 13.9): *E* represents the time at end of life. As we can see from these plots, the rate of system failure can be significant during the initial years of its life (all patterns with the exception of VI and VII); this may be due to construction-related quality problems. Also, the failure rate may be constant for most of a system's life or decreasing or increasing in a linear or nonlinear manner. Most systems tend to have greater failure rates toward the end of their service lives (Figure 13.10).

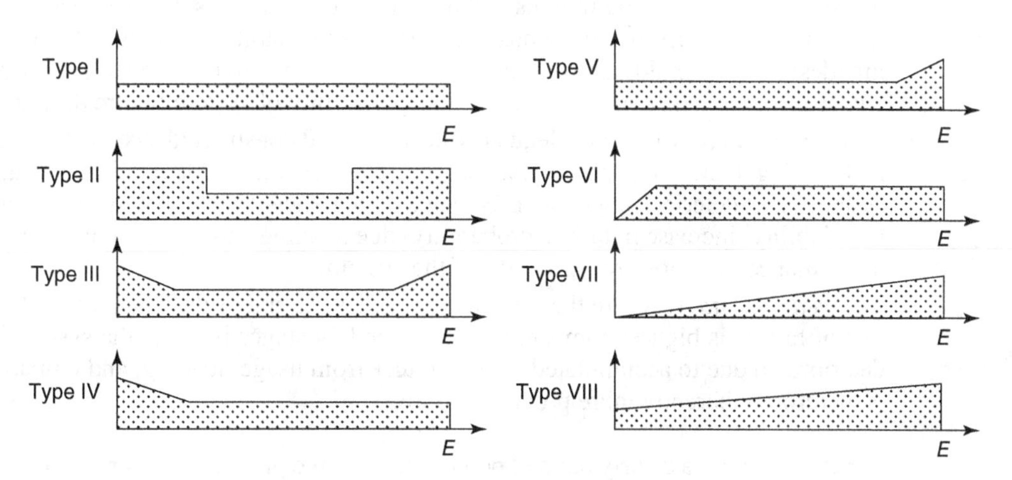

Figure 13.9 A few failure rate plots.

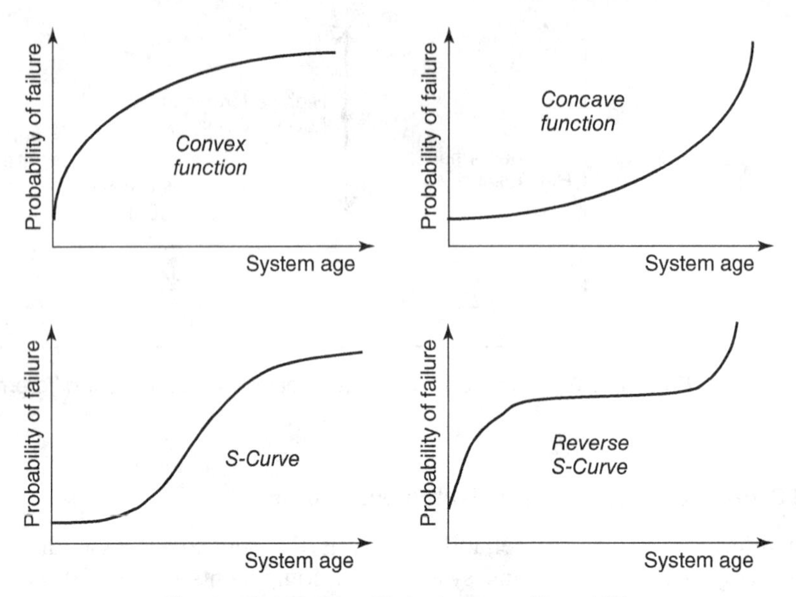

Figure 13.10 Specific instances of type VIII.

The Bathtub Model. The most common type of failure plot in Figure 13.9 is type III, which is also known as the "bathtub" function (Agarwal, 1993; and Dandy et al., 2008). An expanded version for closer scrutiny is provided as Figure 13.11. The three different causes of failure rate [Figure 13.11*a* combines to yield the overall bathtub shape shown in Figure 13.11*b*]. It is noteworthy that the bathtub model is appropriate for modeling the reliability of some, but not all, systems (Nelson, 1990). As indicated in this model, system reliability undergoes the following stages:

1. The *break-in stage*, where the system could fail (not long after being subjected to loading after construction) due to poor design, substandard materials, or faulty construction workmanship. This is a stage of decreasing rate of failure: At this stage, the identification and elimination of defects in system components and subsystems reduce the likelihood of failure. Failures at this stage are termed early failures or "infant mortality" failures. In construction management, the concepts of performance bonds and performance liability periods in traditional contracting (design–bid–build) and warranty-based contracting are intended to safeguard the system owner from the risk, consequences, and responsibility of system failure during this stage.
2. The *mature stage*, where an adequately designed and constructed system operates under normal loading. During this period, the rate of failure remains stable or fairly constant. In reality, systems start deteriorating as soon as they are constructed, such that there is a slight decrease in reliability (increase in failure probability) due to aging and subsequent reductions in initial performance (e.g., lowered strength) of the system.
3. The *wear-out stage*, where the system approaches the end of its design life. At this stage, the rate of failure is highest compared to the preceding stages because the system has aged and deteriorated due to accumulated wear and tear from usage, loading, and climatic effects and is therefore inherently more prone to failure.

For systems with a clearly defined point of failure, the probability distribution function of the failure time can be determined empirically from observed data, using data either from in-service

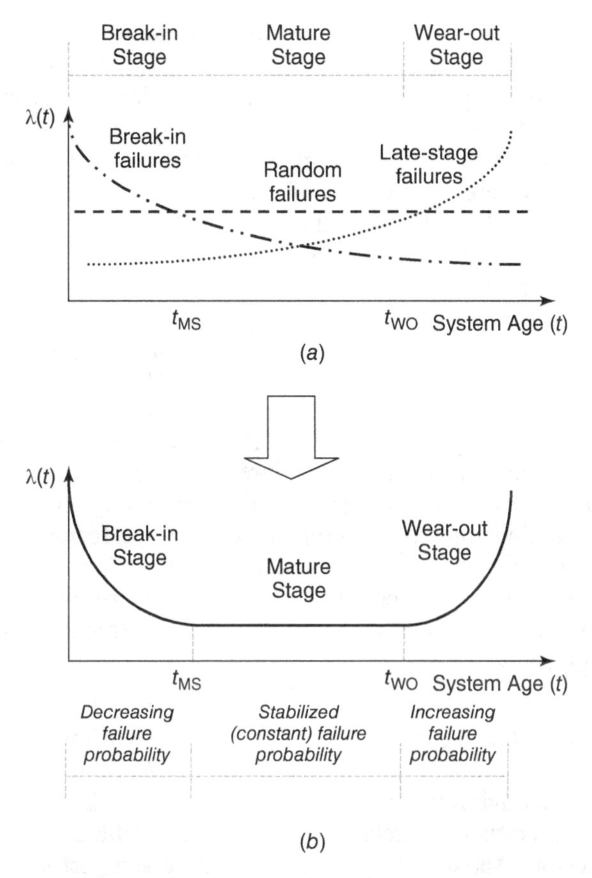

Figure 13.11 Bathtub hazard function indicating the trend of system failure rate: (*a*) components of the bathtub model and (*b*) the bathtub model: trend of failure probability.

systems or from controlled laboratory or field experiments. The break-in stage failure rate is typically modeled using a modified exponential functional form that includes the failure rate parameter (which is constant).

Most work on reliability has been associated with the mature stage (the period of stabilized failure in Figure 13.11) and that this is mainly because of the simplicity associated with using a constant hazard rate (Wasson, 2006). It can be noted that the failure rate at each of the three bathtub stages can be modeled using a decreasing or increasing exponential distribution or the more flexible Weibull distribution, with appropriately-specified parameter values.

(b) General Expressions for Failure and Reliability Functions. Recapitulating what we have learned so far in this section, reliability can be viewed in the context of the progression of system condition or failure/survival likelihood over time. Thus, reliability can be expressed mathematically in several forms, including a failure rate probability function, survival function, and reliability function. An example is

$$R(t) = p(P_t > P^*)$$

where $R(t)$ is the probability density function representing the reliability of the system; P_t is the performance of the system at any time t; P^* is the threshold minimum level of system performance; and t is the length of the period of time (from a base time that is set to zero). This equation means that in this context of reliability, the reliability of a system at a given time is the probability that the system performance exceeds a certain minimum level of performance.

For example, where the measure of performance is the longevity or the service life of a system, reliability would be defined as the probability that the system will operate without failure before reaching design life or a certain specified time or specified number of accumulated cycles of loading, usage, or climatic effects.

$$R(t) = p(T > t) = \int_t^{\infty} f(x)\ dx$$

where $f(x)$ is the density function for failure probability.

It must be noted that the quantified system reliability will be different for different performance indicators that reflect different failure modes, for example, the reliability of a reinforced concrete system in terms of cracking will be different for spalling or corrosion indicators. Thus any statement of reliability must be accompanied by the performance indicator in question. For this reason, we include the subscript C is most of the expressions subsequent to this section.

Reliability in terms of a specific performance criterion C is defined as the probability that a system will perform satisfactorily, in terms of the performance criterion, over a given time period t, is generally given by

$$R_C(t) = P(T \geq t)$$

where T is the time taken for the system to "fail," where failure is defined in terms of the given performance criterion, $C; R_C(t) \geq 0, R_C(\infty) = 0$, and $R_C(0) = 1$.

Due to the fact that reliability is a likelihood or a probability, failure is regarded as a random event. Because failure is the algebraic complement of reliability, the probability that a system will fail (i.e., not perform satisfactorily in terms of the given performance criterion over a given time period t) can be expressed as

$$F_C(t) = 1 - R_C(t) = 1 - P(T \geq t) = P(T < t)$$

where T is the time taken for the system to "fail," where failure is defined in terms of the given performance criterion, $C; F_C(t) \geq 0, F_C(\infty) = 1$, and $F_C(0) = 0$. See Figure 13.12.

It may be realized that this expression represents a cumulative distribution function (see Chapter 5). The corresponding **probability density function** for system failure can be derived using calculus as follows:

$$f_C(t) = -\frac{dR_C(t)}{dt} \quad \text{or} \quad f_C(t) = \frac{dF_C(t)}{dt}$$

Similar to all probability density functions (see Chapter 5), the failure pdf is always positive and the total area bounded by curve (or the total probability) is 1.

$$\int_0^{\infty} f_C(t)\ dt = 1$$

$$f_C(t) \geq 0$$

As it is with any probability distribution, the statistics of the failure function can be determined and interpreted for purposes of system management. These statistics include the average time it takes for the system to fail (i.e., when it starts performing at a substandard level in terms of the given performance criterion), or the probability that the time of failure will be greater than or less than some value or fall between a certain range of values.

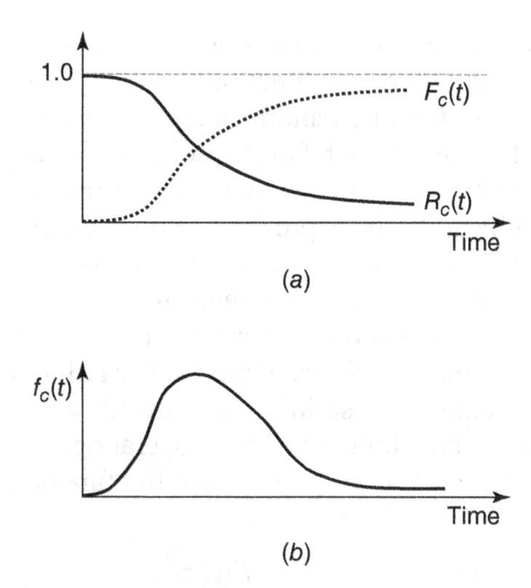

Figure 13.12 Examples of plots showing failure probability and reliability: (*a*) time trend of the probability of system survival and failure and (*b*) probability distribution of failure rate.

(c) Failure Rate Functions. An important parameter in reliability analysis is the mean time to failure (MTTF), which can be calculated as the expected time to failure using the failure pdf.

$$\text{MTTF} = E_C(t) = \int_0^{\infty} tf_C(t)\ dt$$

Also, failure rate is defined as the number of failures per unit of time and is denoted by the symbol $\lambda(t)$.

The average failure rate is the reciprocal of the MTTF: $\lambda = 1/\text{MTTF}$. The exponential function for reliability, which is derived in Wasson (2006) as follows:

$$R(t) = e^{-\lambda t}$$

where $R(t)$ is the probability that the system does not fail in the time interval between 0 and t; λ is the probability of failure per unit time; and t is the system age.

Example 13.2

A certain type of water-retaining structure has a mean failure time of 36 years. Assuming that the tanks have a constant probability of failure, what is the longevity reliability of these systems over a period of (a) 36 years and (b) 12 years?

Solution

$$\lambda = 1/\text{MTTF} = 1/36 = 0.028$$

Thus, R (36 years) $= R(36) = e^{-0.028(36)} = 0.368$; and R (12 years) $= R(24) = e^{-0.028(12)} = 0.717$.

Therefore, approximately 37% of the water tanks are expected to last to the mean time of failure (i.e., 36 years); also, 72% will last 12 years or more.

The Constant Failure Rate (CFR) Function. From Figure 13.9, we can recognize a number of interesting features about failure rate functions for engineering systems: (i) often, there is more than one failure stage because the rate patterns and direction may be different at different ages of the system life, and (ii) the rate at which failure probability changes over time may be constant or nonconstant; also, nonconstant functions may be increasing or decreasing and nonconstant functions may be linear or nonlinear. In the expressions below, we assume that the performance indicator is specified, thus we do not include the subscript C. However, the reader must note that the analysis will yield different results when different indicators are used for the reliability analysis.

Constant failure rate functions are characterized by the exponential probability distribution and describe purely stochastic failures. Unlike other failure distributions, the CFR model is memoryless, thus, it is suitable for describing the reliability of an engineering system whose time to failure is independent of how long it has been operating.

Let $\lambda(t) = \lambda$, where $\lambda > 0$ and $t \geq 0$; then the functions for reliability, failure CDF, and pdf can be written as

$$R(t) = e^{-\lambda t}$$

$$F(t) = 1 - e^{-\lambda t}$$

$$f(t) = \lambda e^{-\lambda t}$$

Also, the values for the average time to failure and the variability thereof are given by

$$\text{MTTF} = \frac{1}{\lambda} \quad \text{and} \quad \sigma^2 = \frac{1}{\lambda^2}$$

Nonconstant Failure Rate Functions. Also referred to as time-dependent failure rate functions, functions of this type take any of several forms. Of these, the Weibull is the most popular because it has a flexible general form with parameters whose specific values influence (i) the final shape of the distribution and (ii) the direction of change of the failure rate: increasing, decreasing, or constant (Table 13.5). The Weibull-distributed failure rate function is given as

$$\lambda(t) = \frac{\beta}{\theta}\left(\frac{t}{\theta}\right)^{\beta-1}$$

where $\theta > 0, \beta > 0$, and $t \geq 0$.

The resulting functions for reliability, failure CDF, and failure pdf are presented in Table 13.6.

Theta (θ) is a **scale parameter** that reflects the mean and the dispersion of the distribution. A higher value of θ generally implies greater reliability and lower probability of failure. Beta (β) is the **shape parameter**.

Table 13.5 Shape Implication of Weibull Parameter Values

Value of Beta Parameter		Description of Probability Density Function of the Failure Rate
$0 < \beta < 1$		Decreasing rate of failure, failure pdf is approximately exponentially distributed
$\beta = 1$		Constant rate of failure
$\beta > 1$	$1 < \beta < 2$	Increasing rate of failure, concave shaped
	$\beta = 2$	Increasing rate of failure pdf is Rayleigh distributed
	$\beta > 2$	Increasing rate of failure, convex shaped
	$3 \leq \beta \leq 4$	Increasing rate of failure, failure pdf is approximately normally distributed

Source: Adapted from Zhang (2012).

Table 13.6 Summary of Key Statistics and Features of Constant and Nonconstant Failure Rate Models

Statistic	**Constant**	**Nonconstant (Note: $\theta > 0, \beta > 0, t \geq 0$)**
Failure rate function	$\lambda(t) = k = 1/\theta$ (constant or flat)	$\lambda(t) = \frac{\beta}{\theta}\left(\frac{t}{\theta}\right)^{\beta-1}$
Reliability function	$R(t) = e^{-\lambda t}$	$R(t) = e^{-(t/\theta)^{\beta}}$
Failure CDF	$F(t) = 1 - e^{-\lambda t}$	$F(t) = 1 - e^{-(t/\theta)^{\beta}}$
Failure pdf	$f(t) = \lambda e^{-\lambda t}$	$f(t) = e^{-(t/\theta)^{\beta}}$

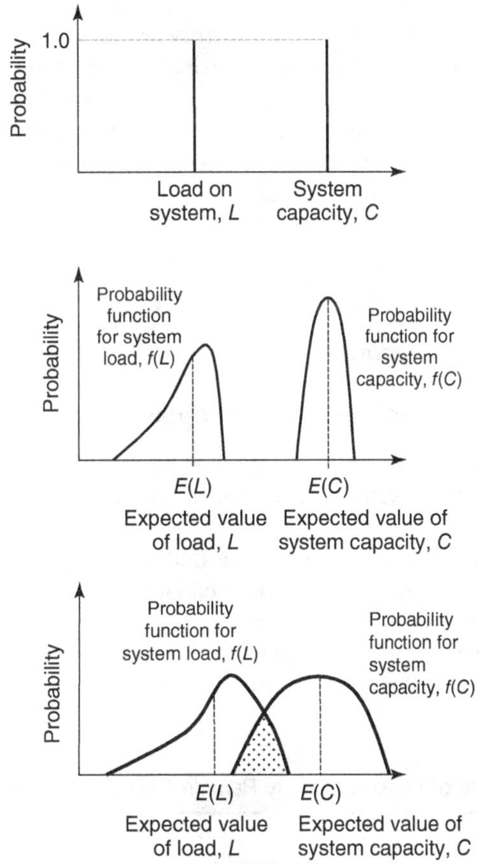

Figure 13.13 Loading and capacity functions.

As shown in Figure 13.13, depending on the values of β, the failure rate $\lambda(t)$ could be increasing or decreasing, and the failure pdf could take shapes including exponential, normal, and Rayleigh. When $\beta < 1$, the failure rate decreases over time; this is indicative of the first stage of the bathtub function or any reliability function where failure rate decreases with time, where the "infant mortality" effect is significant, and defective components or systems fail early and are "weeded out" of the population (Stapelberg, 2009). When $\beta = 1$, the failure rate is constant over time as seen in the second stage of the bathtub function where the system operations has stabilized and any system failures are due to random sources. When $\beta > 1$, the failure rate increases with time; this reflects the "aging" of the system due to wear and tear (see the third stage of the bathtub function). For $\beta > 1$, there are other possibilities of the shape of the failure rate function (Table 13.5): When $1 < \beta < 2$, the function is concave shaped; when $\beta = 2$, the function is Rayleigh distributed, when $\beta > 2$, the function is convex shaped, and when $3 \leq \beta \leq 4$ the function is approximately normally distributed.

Summary Discussion—Modeling Failure Rate over Time. Failure probability plots and patterns indicate the probability of failure or the rate at which a system may fail over the course of time. "Failure" in this context could refer to a physical or operational defect that causes the system to not function or to function at a substandard level. There are several different models that can describe the pattern of progression of failure probability or the rate of progression over time. Most of these patterns are characterized by increasing probability of failure as the system ages. The so-called bathtub function also accounts for the possibility of construction or installation defects early in the life of the system and accelerated probability of failure near the end of the system's life. Failure rate functions, which can help us determine the key failure-related attributes of a system, may indicate a constant or nonconstant rate of failure (i.e., the expected failure rate, the expected life, and the average time between failure events). From the failure rate functions, the reliability function can be derived. Table 13.6 is a summary of the key statistics of constant and nonconstant failure rate (Weibull) models. For the constant rate model, the average time to failure (MTTF) is $1/\lambda$ and the variability of the time-to-failure, σ^2, is $1/\lambda^2$. Other distributions for nonconstant failure rates include the normal and lognormal.

13.5.4 Reliability of Relative Levels of Loading and Capacity

The adequacy of any design in civil engineering essentially represents a comparison between the expected load and the capacity (Table 13.7). The capacity is often referred to as the maximum allowable load.

In all these stress–strength pairs, the underlying common theme is that a load (or stress) is applied, and the system is provided some capacity to accommodate that load, which is the basis for the design of most systems in civil engineering. There is inherent variability in the load (due to variabilities in the factors that produce the load) and also in the system capacity (due to variabilities in the conditions that influence the system capacity). For system loading, sources of variability

Table 13.7 Examples of Load–Capacity Pairs in Civil Engineering

Load	Capacity	Branch of Civil Engineering
Bending or shear stress	Material strength	Structural systems
Flow in a channel	Channel capacity (cross-sectional dimensions)	Hydraulic systems
Highway travel demand	Highway capacity (number of lanes)	Traffic systems
Transit demand	Transit bus/car sizes and headways	Transit systems
Expected quantity of water flow across transportation guideways	Culvert diameter (for pipes)	Hydraulic systems
Column loading	Bearing capacity of soil	Geotechnical systems
Demand for drinking water	Volume of a water tank supplying a town	Water supply systems

include violation of individual user load limits, changes in wind direction and intensity, changes in the magnitude of earthquake acceleration, and changes in environmental conditions (temperature, freeze index, freeze–thaw cycles, and precipitation). For system capacity, sources of variability include the variability of material workmanship quality during the system construction and changes in capacity-related environmental or operations conditions.

In typical design, the designer establishes a **factor of safety**, F, as the ratio of the system capacity C and the system load L, or a **safety margin**, M, as the algebraic difference between C and L:

$$F = C/L$$

$$M = C - L$$

For example, for a system that is expected to experience a load of 80 units and is provided a capacity of 100 units, the factor of safety (FS) and the safety margin are

$$\text{FS} = 100/80 = 1.2 \quad \text{and} \quad M = 100 - 80 = 20 \text{ units}$$

In cases of uncertainty where the loading and capacity are not fixed values but which instead exhibit so much variability that they are best described by probability density function $f_C(c)$ and $f_L(l)$, respectively, then the factor of safety and the safety margin can be calculated using the expected values of their probability functions. In mathematical notation, this can be written as (Harr, 1997):

$$F = \frac{E(C)}{E(L)} = \frac{\int_{-\infty}^{\infty} cf_C(c)\ dc}{\int_{-\infty}^{\infty} lf_L(l)\ dl}$$

$$M = E(C) - E(L) = \left[\int_{-\infty}^{\infty} cf_C(c)\ dc\right] - \left[\int_{-\infty}^{\infty} lf_L(l)\ dl\right]$$

$$= E(C - L) = \int_{-\infty}^{\infty} (c - l)f_{C-L}(c - l)d(c - l) = \int_{-\infty}^{\infty} Mf_M(M)\ dm$$

Figure 13.13 illustrates loading and capacity functions as probability distributions. In the situation shown in Figure 13.13*a*, both the capacity and the demand are fixed values with zero variability, and the capacity exceeds the load. In the situations shown in Figures 13.13*b* and 13.13*c*, both the capacity and the loading are described by probability distributions, with the former exhibiting much lower variability than the former. In Figure 13.13*b*, there is no overlap of the loading and capacity functions; thus, for the situation in Figure 13.13*b*, capacity will always accommodate loading because the worst-case capacity possibility (i.e., when capacity is at its minimum possible level) still exceeds the worst-case loading possibility (i.e., when loading is at its maximum possible level). In Figure 13.13*c*, there is an overlap of the loading and capacity probability functions so a possibility exists that the loading exceeds the capacity (see shaded region). The situation represented by Figure 13.13*c*, where loading exceeds capacity, is the engineer's nightmare: If the system is structural, then it could fail suddenly with possible loss of life; if the system is a hydraulic system, the situation could lead to flooding of inhabited areas and possible loss of life and property; and if the system is a transportation network, it could lead to congestion, user delay, frustration, and possible loss of life.

For the situation represented by Figure 13.13*c*, Figure 13.14 presents the probability distributions for the factor of safety and the safety margin. Because a possibility exists that loading exceeds

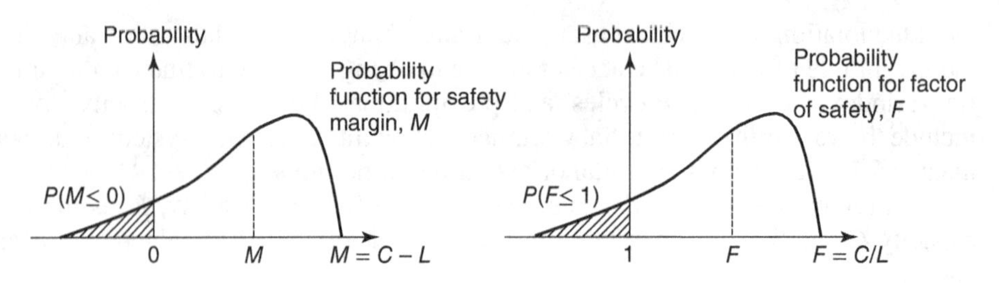

Figure 13.14 Probability distributions for factor of safety and safety margin.

capacity to support the load, the safety margin could be less than 0, and the safety factor could be less than 1.

13.5.5 Likelihood of Abrupt System Failure at a Given Time

In Section 13.5.2, we discussed the reliability of each individual system of a system of systems (SOS) or each individual component of a larger system. In this section, we discuss the reliability of an SOS (that is comprised of multiple systems) or a large system (that is comprised of multiple components or subsystems). The reliability or failure of an SOS or a complex system is dictated by the different failure characteristics of its individual component systems. The reliability of a complex system can be analyzed by first decomposing it into its constituent components. Where the SOS comprises multiple systems (or where the system comprises multiple components) that act in concert with each other through series or parallel connections and are vulnerable to sudden failure, the engineer is interested in the reliability associated with the functioning of the overall SOS or supersystem. In the context of reliability problems in civil engineering, the issue of the reliability of SOSs and multicomponent systems are particularly relevant. The concepts of reliability block diagrams and fault tree analysis diagrams are particularly useful in understanding system reliability in this context.

(a) Reliability Block Diagrams. A reliability block diagram (RBD) is a symbolic analytical logic technique that helps to assess the reliability of large and complex systems (ISI, 2007). RBDs illustrate the physical or functional relationships between system components, subsystems, or events. Specifically, an RBD describes the logical interaction of nonfailures within a system that are needed to ensure that the system remains in operation. The process starts with an input node located at the left side of the block diagram and navigates various block arrangements in series, parallel, or both, and ends with an output node at the extreme right of the block diagram. An RBD contains only one input and one output node. A parallel connection is an indication of redundancy in the system. As illustrated in Figure 13.15, the system configuration could comprise a purely series, a purely parallel, or a combination of series and parallel connections.

For the entire system to operate successfully, at least one path must be maintained between the system input and output nodes. Expressions in Boolean algebra can be used to describe the minimum combination of failures that will lead to a failure of the system. The fewest number of node failures that can cause the entire system configuration to fail is referred to as *minimal cut sets.*

After the RBD is established, the following reliability parameters can be calculated: failure rate, mean time between failures (MTBF), and reliability. The values of these parameters will change if the component or subsystem configuration (and hence the RDB) changes for any reason such as shifts in system owner/operator policy.

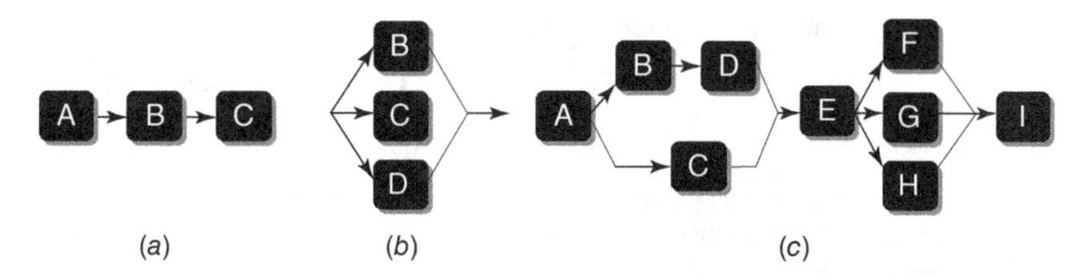

Figure 13.15 Examples of RBD configurations: (*a*) series configuration, (*b*) parallel configuration (components B, C, E), and (*c*) mixed configuration of various components.

Series and Parallel Configurations in RBDs. Figure 13.15*a* presents an example of system components or subsystems arranged in series. This is analogous to a chain that comprises a number of links (each component may be considered as a link). Clearly, the failure of any one link will cause the entire system to fail (or, the failure of any one system will cause the entire system of systems to fail). Figure 13.15*b* presents an example of system components or subsystems arranged in parallel. Parallel systems offer significant redundancy and therefore render the system more resilient to overall failure when the individual components fail. However, parallel systems come with relatively greater cost and thus are typically adopted only when system resilience is critical, for example, pumping stations in a water supply plant. Unlike systems in series configurations, parallel systems do not have constant failure rates per unit of time even if all their components do (Dandy et al., 2008). Figure 13.15*c* presents a mixed configuration.

Example 13.3

For a system comprising N components, derive and interpret the expression for reliability, for (a) a purely series configuration and (b) a purely parallel configuration. Assume that each component has a reliability function governed by the expression $R = e^{-\lambda t}$.

Solution

(a) For a series system, the failure of any one component will lead to the failure of the entire system. From basic probability theory, the probability that the system does not fail is the algebraic product of the probabilities of no failure of the individual components. The reliability of each component i and of the overall system are R_i and R_S, respectively.

$$\begin{aligned} R_S &= R_1 \times R_2 \times \cdots \times R_N \\ &= e^{-\lambda_1 t} \times e^{-\lambda_2 t} \times \cdots \times e^{-\lambda_N t} \\ &= e^{-t(\lambda_1+\lambda_2+\cdots+\lambda_N)} \\ &= e^{-\lambda_S t} \end{aligned}$$

Therefore, for a system that is comprised of components arranged in series, the failure rate of the series system per unit of time, λ_S, is a simple algebraic sum of the failure rates of individual components.

(b) For the purely parallel system, the entire system fails only when all the components fail at the same time. It is assumed that there are no interdependencies between the individual components, that is, the failure of one does not influence (and is not influenced by) the failure or nonfailure of another. From basic probability theory, the probability that the system fails is the algebraic product of the probabilities of failure of the individual components. The reliability of each component i and of the overall system are F_P and F_i, respectively.

$$F_P = F_1 \times F_2 \times \cdots \times F_N$$

But reliability = 1 − failure. Thus:

$$1 - R_P = (1 - R_1) \times (1 - R_2) \times \cdots \times (1 - R_N)$$

$$R_P = 1 - (1 - R_1) \times (1 - R_2) \times \cdots \times (1 - R_N)$$

Therefore, for a system that is comprised of components arranged in series, the failure rate of the series system per unit of time is a simple algebraic sum of the failure rates of the individual components.

Example 13.4

Figure 13.16 represents a mixed system. Show a step-by-step reduction of the configuration of the system to a simple system and derive the final expression for the reliability of this system. Assume that each component has a reliability function governed by the expression: $R = e^{-\lambda t}$.

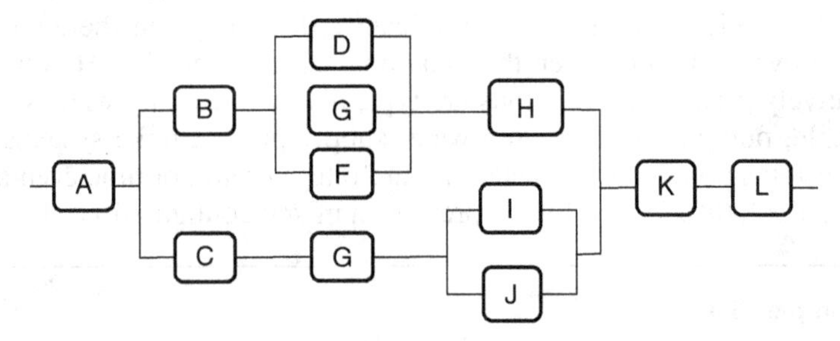

Figure 13.16 Example 13.4.

Solution

Figure 13.17 shows the step-by-step reduction of the configuration of the system to a simple system, and the expression for the system reliability at each step of the reduction process.

Step 1: $R_M = 1 - (1 - R_D)(1 - R_E)(1 - R_F)$ and $R_N = 1 - (1 - R_I)(1 - R_J)$

Step 2:

$$\begin{aligned} R_P &= R_B \times R_M \times R_H \\ &= e^{-\lambda_B t} \times e^{-\lambda_M t} \times e^{-\lambda_H t} \\ &= e^{-t(\lambda_B + \lambda_M + \lambda_H)} \\ &= e^{-\lambda_p t} \end{aligned}$$

Also,

$$\begin{aligned} R_N &= R_I \times R_J \\ &= e^{-\lambda_I t} \times e^{-\lambda_J t} \\ &= e^{-t(\lambda_I + \lambda_J)} \\ &= e^{-\lambda_N t} \end{aligned}$$

Step 3: $R_S = 1 - (1 - R_P)(1 - R_Q)$

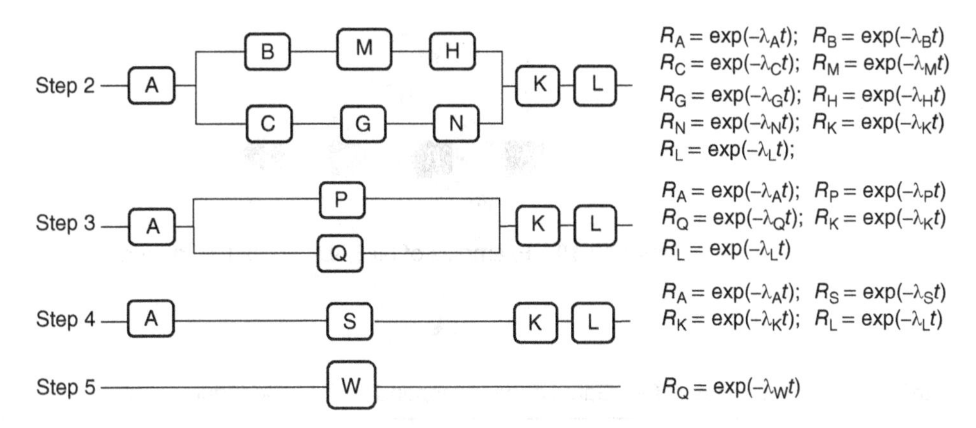

Figure 13.17 Solution for Example 13.4.

Step 4:

$$\begin{aligned} R_W &= R_A \times R_S \times R_K \times R_L \\ &= e^{-\lambda_A t} \times e^{-\lambda_S t} \times e^{-\lambda_K t} \times e^{-\lambda_L t} \\ &= e^{-t(\lambda_A + \lambda_S + \lambda_K + \lambda_L)} \\ &= e^{-\lambda_W t} \end{aligned}$$

(b) Fault Tree Analysis. Fault tree analysis (FTA), also a symbolic analytical logic technique for reliability and safety analysis, was developed in 1962 by Bell Telephone Laboratories for the U.S. Air Force and subsequently adopted by the Boeing Company (ReliaSoft, 2010). Also referred to as a "negative analytical tree," an FTA describes the top event (the state of a system) in terms of basic events (the states of the system components). FTAs are built top-down not in terms of blocks but as events and use a graphic model of various potential pathways through which a system could lead to a failure or undesirable loss event. Using standard symbols for logical relationships (AND, OR, etc.), the pathways link the contributory events and conditions. The basic constructs in FTAs are *gates* and *events* and the two most commonly used gates in FTA are AND and OR. FTA events are identical meaning to RBD blocks, and FTA gates correspond to RBD conditions (Ayyub, 2003; USNRC, 1981).

Consider two events A and B that comprise a top event or system. Assume that the failure (occurrence of either event), causes the entire system (top event) to fail (failure event occurs); in that case, these events are best connected using an OR gate. On the other hand, if both events must occur for the top event to occur, then they are best connected using an AND gate. In Figure 13.18*a*, a system RBD is shown as two blocks in series. In the FTA diagram for this configuration (Figure 13.18*b*) includes two basic events connected to an OR gate, which is the "Top Event." If the Top Event is to occur, then either A or B or both should occur. That is, the system fails when A fails or B fails (ReliaSoft, 2010).

(c) Fault Trees and Reliability Block Diagrams—Differences and Relationships. The basic difference between RBDs and FTAs is that in the former, the focus is on the "success space" (combinations of system successes are examined); in FTA diagrams, on the other hand, the focus is on the "failure space" (combinations of system failures are examined). Traditionally, FTAs have

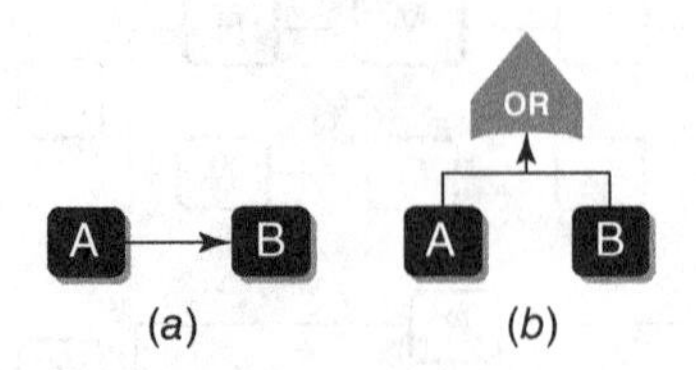

Figure 13.18 Examples of (*a*) RBD and (*b*) FTA diagram.

Table 13.8 Classic Fault Tree Gates and Corresponding RBD Configurations

Term Used for Gate	Symbol in FTA	Description in FTA	Equivalent Description in RBD
OR		The output event occurs **if at least one** of the input events occurs.	Series configuration.
AND		The output event occurs **if all** input events occur.	Simple parallel configuration.
Inhibit		The input event occurs **if all** input events occur **and** an additional conditional event occurs.	Simple parallel configuration of all the events plus the condition.
Voting (i.e., k-out-of-n)	K	The output event occurs if k or more of the input events occur.	k-out-of-n parallel configuration.
Dependency AND	Not used in classic FTA.	The output event occurs **if all** input events occur, however, the events are **dependent** (i.e. the occurrence of each event affects the probability of occurrence of the other events). Not used in classic FTA.	Load-sharing parallel configuration.
Priority AND		The output event occurs **if all** input events occur in a **specific sequence**.	Standby parallel configuration (without a quiescent failure distribution).
XOR		The output event occurs **if exactly one** input event occurs.	Cannot be represented and does not apply in terms of system reliability. In system reliability, this would imply that a two-component system would function even if both components have failed.

Source: ReliaSoft (2010).

been used to identify fixed probabilities. It may be noted that each event in the fault tree has a fixed occurrence probability; RBDs traditionally include reliability models (i.e., distributions that describe the system success over time, see Section 13.5.3) often also include other features such as the effectiveness of system or component repair treatments (ReliaSoft, 2010). It is difficult to convert an RBD, particularly highly complex kinds, into an FTA diagram; however, an FTA diagram is relatively easy to convert into an RBD (with a few specific exceptions). Table 13.8 presents the gate symbols commonly used in FTA and describes their relationship to RBDs (USNRC, 1981).

Example 13.5 (adapted from ReliaSoft, 2010).

Consider the following system configuration shown as Figure 13.19.

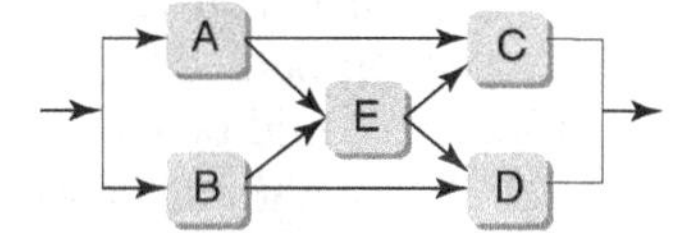

Figure 13.19 Figure for Example 13.5.

To represent this system configuration using an FTA, utilizing duplicate events is required because it is possible to represent gates only for components that are arranged in series or in parallel. Clearly, the system will fail due to any one of the following: failure of components A and B; failure of C and D; failure of A and E and D; or failure of B and E and C. That is, the domain of events for system failure is: (A and B) or (C and D) or (A and E and D) or (B and E and C). These are the *minimal cut sets.* The FTA can be drawn using these events (Figure 13.20). The FTA shown in Figure 13.20 can be converted to an RBD (see Figure 13.21). Components that have the same name are often termed *mirrored blocks*.

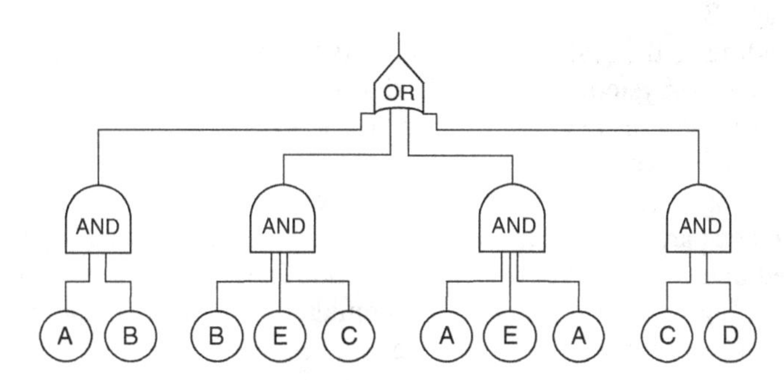

Figure 13.20 FTA for Example 13.5.

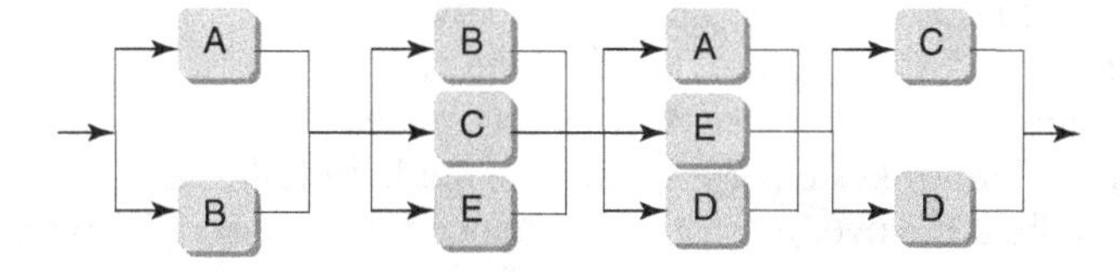

Figure 13.21 RBD for Example 13.8.

13.6 LABORATORY AND FIELD TESTING OF SYSTEM RELIABILITY

Reliability testing aims to provide the probability of system failure (in terms of its structural stability or other performance measure). Such tests also provide an indication of the extent to which reliability will be achieved with a certain degree of confidence and also to reveal, as early as possible, potential performance problems with the design of a civil engineering system. Similar to the sampling of elements in a population, testing the reliability of civil engineering systems or system components is often limited by time, cost, and accessibility. Given the nature and size of typical civil engineering systems, field reliability tests can be very expensive and time consuming. Laboratory tests or computer simulations may be necessary to avoid the high cost and time associated with field testing. The results from a single test are not adequate to generate data for statistical analysis to make inferences about the parent population of SOS or a supersystem. If carried out properly, laboratory or field reliability tests can yield the key parameter of reliability, MTBF. The reporting of this parameter is often accompanied by a level of confidence: For example, the MTBF for leakage failure of a certain pipeline could be stated as 50 months with 95% confidence level.

The reported or required reliability parameter value and confidence level can have a significant influence on the system development cost to the system owner and the risk to the system user: It is generally more costly, at least in the short term, to provide systems that have greater reliability and greater confidence. Consider system A that is designed to cover all risks with the greatest possible confidence. The initial cost of this system will exceed that of another system, say, system B, that is not designed to cover risks or covers risks with much less confidence compared to A. From a purely economic viewpoint, B will be a more attractive option. However, if both systems are vulnerable to risky events, then for B, the cost of mitigating failures will far exceed that for A. Overall, the total costs of B may even exceed that of A. Thus, for situations characterized by significant levels of risk, greater reliability and confidence in design can lead to reduced frequency or intensity of system failure and, subsequently, lower overall cost. Ultimately, the choice of optimal reliability and confidence is a balancing game between the extreme positions of systems A and B.

Depending on the system type, reliability testing may be performed at various levels of system hierarchy: the supersystem (or the SOS), the system, the subsystem, or the component. For example, reliability tests for a bridge system may be carried out for the entire network of bridges, an individual bridge, a bridge or substructure, or smaller structural components such as pin joints, gusset plates, or bearings. Reliability tests for the lowest level of the system hierarchy, in this example, the individual piece parts or small assemblies, are critical so that these local but important problems can be identified and addressed before they cascade into more serious problems and, ultimately, system failure at higher levels. The environmental conditions associated with a given reliability must be reported. These conditions, which depend on the type and orientation of the system, include temperature, humidity, shock, vibration, heat, and groundwater quantity and quality.

Accelerated life testing, a key type of reliability testing, is carried out purposely to mimic, in the laboratory, field conditions that ultimately lead to failure field failure albeit at a faster rate (and thus in much shorter time). This is done by replicating the field stresses and environment that could contribute to failure. Often, this is done at a smaller scale compared to the real-life system. For small system components, such as gusset plates, the real-life scale is used. Accelerated tests, in which the system or system component is made to fail in the laboratory in a manner similar as it would in the field but over a drastically reduced time period, help identify mechanisms of failure and to make predictions of the actual field life of the system or system component on the basis of the laboratory life. Figure 13.22 presents an image of the U.S. Federal Aviation Administration's National Airport Pavement Test Facility at the Hughes Technical Center in New Jersey. This facility

Figure 13.22 Airfield reliability testing at the Hughes Technical Center Pavement Test Facility in New Jersey.

provides high-quality, accelerated test data that can be used to validate the reliability of rigid and flexible pavement designs that are subjected to simulated aircraft traffic loading.

13.7 RELIABILITY-BASED DESIGN OF CIVIL ENGINEERING SYSTEMS

Civil engineering systems are designed and constructed to serve a variety of goals and associated performance measures. These include structural stability, longevity, agency cost, direct or indirect user cost, environmental impacts, and the like. Most studies on reliability and system failure have focused on system structural integrity and longevity. Therefore, most reliability-based designs of civil engineering systems have focused on these two measures of performance. However, the concepts are generally applicable to other performance measures as well. The concept of incorporating reliability into product design is known as design for reliability (DFR). Mostly applicable at the design phase of systems development, DFR typically involves a sequence of deploying a number of practices. The first step in this sequence is the establishment of reliability requirements for the civil engineering system. This is driven by the identified goals and objectives at the first phase of systems development. Then, during the subsequent phases and subphases of system planning, preliminary design, and final design of the system and its subsystems and components, the reliability requirements for the system are duly considered by the system designers working closely with reliability engineers.

Any procedure used for reliability-based design of physical structures and operational polices should have a number of desirable features (Harr, 1997):

Relevant inputs. The procedure should account for the relevant factors of system demand and capacity and any interactions between these factors.

Relevant outputs. It should yield outputs that are related to the anticipated system outcomes or performance.

Practical. It should utilize material characteristics, parameters, and quantities that can be verified with available knowledge.

Inclusive. It should give due consideration to existing indices or ratings related to system performance or performance uncertainty, including the factor of safety or the safety margin.

As we learned in Section 13.5.5, certain types of reliability-based design utilize a reliability model that incorporates block diagrams and fault trees to illustrate and assess the functional relationships between the different components or subsystems of the system being designed. Reliability models utilize predictive models of component reliability (predictive models, in turn, can be developed using historical observations of component failure). Melchers (1999) provides techniques for analyzing the reliability of engineering structures using probability theory by predicting the safety of existing deteriorating structures and the expected safety of proposed structures. Also, Haldar and Mahadevan (2000) provide techniques for carrying out reliability assessment of civil engineering structures using stochastic finite element analysis.

There is increasing awareness of the need for incorporating risk and reliability in civil engineering systems as practitioners increasingly appreciate the fact that many safety incidents could be avoided if greater attention were paid to safety and reliability at the design, construction, and maintenance phases (University of Birmingham, 2012). For civil engineering systems (e.g., bridges, transportation networks, buildings, towers, and levees), the issue of reliability, from a structural or operational perspective, continues to generate much interest in the engineering community and addresses the challenges of these systems, including stability, safety, and performance. The international Civil Engineering Risk and Reliability Association (CERRA) was founded to promote the education, research, and practice of risk and reliability analysis in various disciplines in the civil engineering field. The International Association for Structural Safety and Reliability (IASSAR) is another organization that encourages the research and practice of the scientific principles of safety, risk, and reliability in the design and maintenance of structures and other engineering systems.

13.8 SYSTEM RELIABILITY AND SYSTEM RESILIENCE

In Chapter 27, we discuss the resilience of civil engineering systems. Before we reach that chapter, it will be useful to discuss resilience here briefly because it is important to view the concept of reliability particularly in relation to system resilience. Early in this chapter, we defined reliability as the ability of a system (or component thereof) to perform its required functions or to achieve its established performance objectives under a given set of conditions and at a given point in time. System resilience can be defined as the capability of a failed system to return fully or partially to a nonfailure state, with or without the assistance of agents external to the system. Often, this capability is measured in terms of a probability (fraction of times it has returned to nonfailure state after failure). This probability can be determined using past observations, simulation, or theoretical analysis; and the threshold of system performance at which the system was deemed to have failed depends on the policies of the system owner/operator and the system type. Therefore, the concept of resilience is very much related to the concept of reliability under any of the contexts discussed in this chapter and elsewhere.

Foster (1993) defines resilience as "the ability of a system to accommodate variable and unexpected conditions without catastrophic failure." Therefore, resilience can be viewed through any one of several perspectives at the various phases of system development. For example, a resilient design allows construction of the system to continue even if a particular resource becomes scarce; a resilient operations strategy allows the system to meet the established goals and performance measures (system longevity, stability, economic viability, social and environmental goals, etc.) under unexpected favorable or adverse conditions including new technologies, loading/usage characteristics, the environment, major equipment failures, and disasters. Thus, even though they relate to different directions of the system–environment configuration (see Figure 27.9 in Chapter 27), resilience can be consistent with the concept of sustainability (VTPI, 2010) at all phases of systems development.

System characteristics that introduce or enhance resilience include the diversity, autonomy, and redundancy of the components; the efficiency of its processes, and the structural and operational integrity of critical components. Autonomy means that "the failure of one component or sub-system does not cause other components to fail." Other advanced features that enhance resilience include self-correcting or self-healing capabilities and the ease of repairing the component after it has failed. These characteristics help the system to continue to perform its function even when a link is broken, that is, a component fails or a particular resource (material, equipment, or human expertise) becomes unavailable (VTPI, 2010).

The best opportunities to enhance the resilience of civil engineering systems can be found at the phases of system planning and design. System designers and planners therefore must be encouraged to seek innovative designs that render their systems less vulnerable to damage from sudden or sustained threats. Examples include the use of newly developed long-lasting or stronger materials such as stainless steel in bridge deck reinforcement (Cope et al., 2011). The resilience of civil engineering systems can be increased by enhancing the "adaptive capacity" of the system, either by incorporating greater redundancy in the system design to ensure continuous functioning or by increasing the ability and speed by which the civil engineering system recovers, evolves, or adapts to new often challenging situations (Dalziell and McManus, 2004).

13.8.1 Resilience of a System of Systems

As civil engineering infrastructure and institutions become increasingly networked and consequently, more interdependent, it is becoming obvious that overall supersystems must be equipped with the capability to recover from the shocks induced by natural or anthropogenic stimuli. This need is felt not only in civil engineering systems such as water supply, air transportation, and urban drainage but also in telecommunications, electrical power, and supersystem networks in other engineering and nonengineering disciplines that are networked to civil systems. The owners or operators of these systems seek not only to model the vulnerability of their existing systems to failure, but also to continually evaluate their capability to minimize the consequences of system failure and also restore any failed system to a nonfailure state.

The need for SOS resilience is particularly strongly realized during and after major adverse events when the community can least afford service disruptions because hospitals, security personnel, emergency crews, and response/recovery teams depend on the supply of power, water, communications, and transport access in order to function and to reduce further risks to life and property (Dalziell and McManus, 2004). In New Zealand, a resilience requirement for civil engineering SOSs is mandated by civil defense and emergency management legislation that statutorily requires that all "lifeline services" including road access, water, communications, and electricity must function fully as much as possible during and after an emergency. Dalziell and McManus also stated that encouraging private sector organizations to become more resilient is more difficult compared to doing so for public organizations. This is because planning for greater resilience in private sector systems is not regulated but rather depends on the awareness of individual owners or operators of civil engineering systems of the need for greater resilience and the strategies available to them. It also depends on their willingness and ability to invest in enhancing the resilience of their systems. The development and promotion of persuasive business cases for resilience-enhancing investments, particularly in private sector civil engineering systems, is needed.

SUMMARY

In this chapter, we applied concepts we learned in past chapters, including Chapters 5–8 (probability, statistics, modeling, and simulation). We discussed the uncertainty concepts surrounding the inputs of civil engineering systems development and how these inputs translate into reliability (certainty or lack thereof) that the system will perform its intended function. We discussed risk and reliability, which include risk management, risk identification and quantification, risk modeling, failure analysis, exposure quantification and consequence analysis, evaluation of benefits in terms of risk reduction, risk/cost trade-off analysis, risk-based decision making, and techniques for risk communication.

We then examined four common contexts of reliability analysis in civil engineering: (1) determining the number of samples needed to achieve some precision and level of confidence; (2) describing the probability of system failure or survival over time or accumulated loading in terms of some predefined performance measure; (3) assessing the reliability that the loading will not exceed the system capacity; and (4) assessing the likelihood of the abrupt failure of a complex system at a given time.

We also discussed in this chapter, how systems engineers carry out reliability modeling in order to predict the reliability of their systems or to acquire better insights into the factors that influence reliability by utilizing mechanistic, empirical, or mechanistic-empirical approaches. We examined some empirical approaches, specifically, failure probability plots and patterns and some general expressions for failure and reliability functions. We also discussed reliability models and symbolic analytical logic techniques that include the reliability block diagram and the fault tree diagram. The reliability of each configuration can be analyzed using failure functions. The chapter also discussed how reliability could be influenced by the management level of the decision maker: the system-of-systems and the individual system perspectives.

System resilience, which may be described as the capability of a failed system to return to a nonfailure state, is a key concept that is related to reliability and was discussed in the chapter. A resilient system is able to adapt to changing or unexpected conditions without catastrophic failure. The chapter emphasized that the need for system-of-systems resilience is particularly felt during and after major disasters when the system users and community can least afford service disruptions.

This chapter represents yet another building block in our foundation of systems engineering knowledge as we prepare to discuss other systems concepts in subsequent chapters that address the tasks at the various phases of system development, the system end of life (Chapter 26), and topics relating to the threats to which civil engineering systems are exposed and the system resilience (Chapter 27).

EXERCISES

1. Discuss the possible uncertainties that may be encountered at the various phases of development for (a) a large hydroelectric system planned for construction in a developing country and (b) an oil pipeline, several hundreds of miles in length, planned for construction in a developed country.
2. How many observations should a quality control engineer take in a test sample if the contract agreement specifies that there should be 98% confidence that the estimate of the mean viscosity does not deviate from the true value by more than 2 units? Past experience has shown that the viscosity of produced units is normally distributed with a variance of 8.5 units.
3. A certain type of structural weld has a mean failure time of 16 years. Assuming that it has a constant probability of failure, what is the longevity reliability of this type of weld over a period of (a) 10 years and (b) 16 years?
4. List 12 natural and man-made threats to public safety on your campus. Draw a risk matrix with the following axes: threat likelihood (unlikely, rare, occasional, frequent) and threat consequence (negligible, significant, critical, catastrophic) and place each threat in the appropriate cell of the matrix. For each threat in the matrix, indicate, with reasons, your recommendation of the appropriate risk mitigation strategy.
5. A certain structural joint system has a failure rate of 0.005 failures per year. Determine (a) the probability that a joint of this type will fail in its first year of construction, (b) the number of failures of a structural system that has 50 such joints, and (c) the failure rate that will ensure that no more than 2 of these joints fail in the first year of construction. Use the exponential function.
6. One component of a geotechnical system has a reliability of 0.92%. How many of these systems can be arranged in parallel so that the total system has a reliability of 98%?
7. A landfill system for a hazardous pollutant is lined successively with three different materials: (1) geotextile grade A, (2) geotextile grade B, and (3) compacted clay as illustrated in Figure 13.23*a*. This arrangement could be thought of as three links in a serial arrangement (Figure 13.23*b*). A failure in any one lining disables the entire system. At any given time, the probabilities of failure in the linings are as follows: $p(S_1) = 0.97$; $p(S_2) = 0.92$; and $p(S_3) = 0.99$. If the performance of each lining is independent of the others, find the probability that the lining system fails.

Figure 13.23 Illustration for Exercise 13.7.

8. A sewer system consists of three links in a parallel arrangement (see Figure 13.24). The entire system fails only when all three links fail at the same time. At any given time, the probabilities of failure in the links are as follows: $p(S_1) = 0.78$; $p(S_2) = 0.55$; and $p(S_3) = 0.69$. Assuming that the performance of the links is independent of each other, find the probability that the sewer system fails.

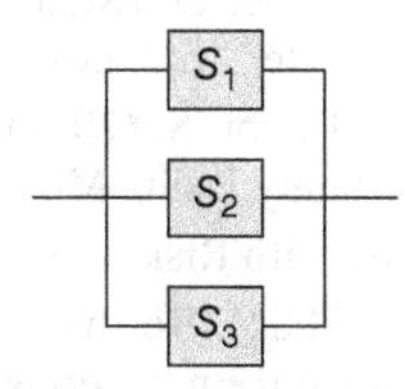

Figure 13.24 RBD for Exercise 13.8.

9. A civil engineering system consists of three subsystems, S_1, S_2, and S_3 in parallel and a fourth subsystem S_4 in series with the parallel subsystems as shown in Figure 13.25. The system can operate properly so long as at least one of the parallel subsystems operates properly and S_4 also operates properly. Historical records show that the probability that each system operates properly is given by 0.85, 0.90, 0.79, and 0.95 for S_1, S_2, S_3, and S_4, respectively. The operation of the subsystems is independent of each other.

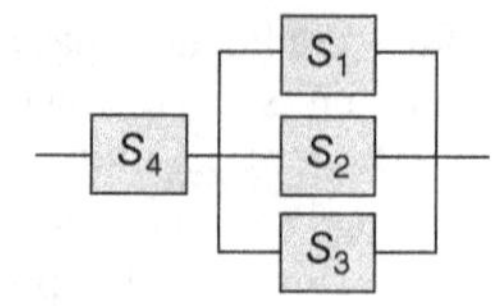

Figure 13.25 RBD for Exercise 13.9.

a. Construct a probability tree to illustrate all the possibilities for the proper operation of the system.
b. Compute the probability that the system operates properly.

10. For the system configuration shown in Figure 13.26, determine the reliability of this system. Assume that each component has a reliability function governed by the expression: $R = e^{-\lambda t}$. The failure rate for each component is provided in Figure 13.26.

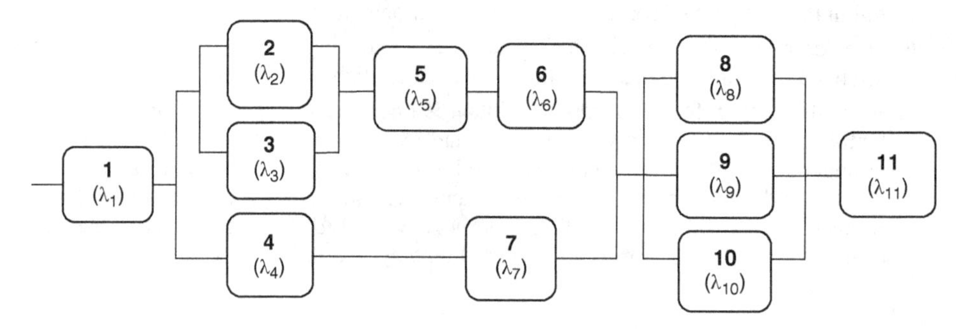

Figure 13.26 RBD for Exercise 13.6.

11. Prove mathematically that the reliability of a parallel system always exceeds the reliability of its most reliable component.

REFERENCES

Agarwal, K. K. (1993). *Reliability Engineering*. Kluwer Academic, Dordrecht, The Netherlands.

Akgul, F., and Frangopol, D. M. (2004). Bridge Rating and Reliability Correlation: Comprehensive Study for Different Bridge Types. *ASCE J. Struct. Eng.*, 130(7), 1063–1074.

Ayyub, B. M. (2003). *Risk Analysis in Engineering and Economics*. Chapman & Hall/CRC Press, Boca Raton, FL.

Biondini, F., Bontempi, F., Frangopol, D. M., and Malerba, P. G. (2006). Probabilistic Service Life Assessment and Maintenance Planning of Concrete Structures. *ASCE J. Struct. Eng.* 132(5), 810–825.

Cope, A., Bai, Q., Samdariya, A., and Labi, S. (2011). Assessing the Efficacy of Stainless Steel for Bridge Deck Reinforcement under Uncertainty Using Monte Carlo Simulation. *Struct. Infrastruct.* 1–14.

Cox, L. A., Jr., (2008). What's Wrong with Risk Matrices? *Risk Analy.* 28(2), 497–512.

Dalziell, E. P., and McManus, S. T. (2004). Resilience, Vulnerability, and Adaptive Capacity: Implications for System Performance. Dept. of Civil Engineering, University of Canterbury, New Zealand. http://www.ifed.ethz.ch/events/Forum04/Erica_paper.pdf.

Dandy, G., Walker, D., Daniell, T., and Warner, R. (2008). *Planning and Design of Engineering Systems*, 2nd Ed., Taylor & Francis, London.

Deshmukh, P., and Sanford Bernhardt, K. L. (2000). Quantifying Uncertainty in Bridge Condition Assessment data. Proceedings of the Mid-continent Transportation Symposium, Ames, IA.

Ezell, B. C., Farr, J. V., and Wiese, I. (2000). Infrastructure Risk Analysis Model. *ASCE J. Infrastruct. Syst.* 6(3), 114–117.

Faber, M. H. (2007). Risk and Safety in Civil Engineering, Lecture Notes, Swiss Federal Institute of Technology, Zurich, Switzerland.

Ford, K. (2009). Uncertainty and Risk Management in Travel Demand Modeling. M.S. Thesis, Purdue University, W. Lafayette, IN.

Foster, H. (1993). *Resilience Theory and System Evaluation in Verification and Validation of Complex Systems: Human Factor Issues*, NATO Advanced Science Institutes, Series F: Computer and Systems Sciences, Vol. 110, pp. 35–60, J. A. Wise, V. D. Hopkin, and P. Stager, Eds., Springer, New York.

Frangopol, D., and Moses, F. (1994). Reliability-Based Structural Optimization, Chapter 13 in *Advances in Design Optimization*, H. Adeli, Ed. Chapman & Hall, New York.

GAO (Governmental Accountability Office) (2009). Transportation Security: Comprehensive Risk Assessments and Stronger Internal Controls Needed to Help Inform TSA Resource Allocation, U.S. Governmental Accountability Office, Washington, D.C.

Haldar, A., and Mahadevan, S. (2000). *Reliability Assessment Using Stochastic Finite Element Analysis*, Wiley, New York.

Harr, M. (1997). *Reliability-based Design in Civil Engineering*. McGraw-Hill, New York.

ISI (2007). Reliability Block Diagram (RBD), ITEM Software, Inc. http://www.reliabilityeducation.com/rbd.pdf.

Knight, F. H., and McClure, J. (2009). *Risk, Uncertainty, and Profit*. Signalman Publishing, Kissimmee, FL.

Liang, M. T., Lin, L. H., and Liang, C. H. (2002). Service Life Prediction of Existing Reinforced Concrete Bridges Exposed to Chloride Environment. *ASCE J. Infrastruct. Syst.*, 8(3), 76–85.

Lin, K-Y. (1995). Reliability-based Minimum Life-Cycle Cost Design of Reinforced Concrete Girder Bridges Exposed to Chloride Environment, Ph.D Thesis, University of Colorado, Boulder, CO.

Lounis, Z. (2000). Reliability-based Life Prediction of Aging Concrete Bridge Decks. Proceedings of the International RILEM Worshop on Life Prediction and Aging Management of Concrete Structures. Cannes, France.

Melchers, R. E. (1999). *Structural Reliability Analysis and Prediction*. Wiley, New York.

Molenaar, K., Anderson, S., and Schexnayder, C. (2010). NCHRP Report 658: Guidebook on Risk Analysis Tools and Management Practices to Control Transportation Project Costs, Transportation Research Board, Washington, D.C.

Nelson, W. (1990). *Accelerated Testing: Statistical Models, Test Plans, and Data Analyses*. Wiley, New York.

Ni, K., and Mahadevan, S. (2004). Probabilistic Fatigue Crack Growth Analysis of Spot-Welded Joints. *Fatigue Fracture Eng. Mat. Struct.* 27(6), 473–480.

NYSDOT (2002). Vulnerability Manuals. Bridge Safety Program, New York State DOT.

Oh, B. H., Lew, Y., and Choi, Y. C. (2007). Realistic Assessment for Safety and Service Life of Reinforced Concrete Decks in Girder Bridges. *ASCE J. Bridge Eng.* 12(4), 410–418.

Ozog, P. (2009). Designing an Effective Risk Matrix. Whitepaper, IO Mosaic Corporation, Sale, NH.

ReliaSoft (2010). Fault Tree Analysis—An Overview of Basic Concepts. weibull.com reliability engineering resource website, ReliaSoft Corporation, http://www.weibull.com/basics/fault-tree/index.htm.

Renfroe, N. A., and Smith J. L. (2007). Whole Building Design Guide: Threat/Vulnerability Assessments and Risk Analysis, National Institute of Building Sciences, Washington, DC.

Schmid, F. A. (2002). The Stock Market: Beyond Risk Lies Uncertainty. *The Regional Economist*, July 2002 issue, http://www.stlouisfed.org/. Accessed Feb. 13, 2013.

Stapelberg, R. F. (2009). Handbook of Rcliability, Availability, Maintainability and Safety in Engineering, Design, Springer, London.

Stewart, M. G., Estes, A. C., and Frangopol, D. M. (2004). Bridge Deck Replacement for Minimum Expected Cost under Multiple Reliability Constraints. *J. Struct. Eng.* 130(9), 1414–1419.

Strauss, A., Bergmeister, K., Hoffman, S., Pukl, R., and Novak, D. (2008). Advanced Life-Cycle Analysis of Existing Concrete Bridges. *ASCE J. Mat. Civil Eng.* 20(1), 9–19.

University of Birmingham (2013). Safety, Risk, and Reliability Management. http://www.birmingham.ac.uk/research/activity/civil-engineering/.

USNRC (1981). *Fault Tree Handbook*, Tech. Rep. NUREG-0492, Systems and Reliability Research, Office of Nuclear Regulatory Research, U.S. Nuclear Regulatory Commission, Washington, DC.

VTPI (2010). Evaluating Transportation Resilience, Evaluating the Transportation System's Ability to Accommodate Diverse, Variable, and Unexpected Demands with Minimal Risk. *TDM Encyclopedia*, Victoria Transport Policy Institute.

Wasson, C. S. (2006). *System Analysis, Design, and Development*. Wiley, Hoboken, NJ.

Zhang, Z. (2012). Risk and Reliability Analysis Lecture, 2012 Infrastructure Management Bootcamp, Georgia Institute of Technology, Atlanta, GA.

USEFUL RESOURCES

Ayyub, B. M., and Klir, G. J. (2006). *Uncertainty Modeling and Analysis in Engineering and the Sciences*, Chapman & Hall/CRC Press, Boca Raton, FL.

Ayyub, B. M., and McCuen, R. H. (2003). *Probability, Statistics, and Reliability for Engineers and Scientists*, 2nd Ed., Chapman & Hall/CRC Press, Boca Raton, FL.

Bernstein, P. L. (1998). *Against the Gods—The Remarkable Story of Risk*. Wiley, New York.

Billington, R., and Allan, R. N. (1987). *Reliability Evaluation of Engineering Systems: Concepts and Techniques*. Plenum Press, New York.

Biondini, F., and Frangopol, D. M. (2009). Lifetime Reliability-based Optimization of Reinforced Concrete Cross-sections under Corrosion, *Str. Safety* 31(6), 483–489.

Blanchard, B. S. (1998). *System Engineering Management*. Wiley, New York.

El-Reedy, M. A. (2000). Assessment of Reliability of Aging Reinforced Concrete Structures. CRC Press (126), December issue.

Ezell, B. C., Farr, J. V., and Wiese, I. (2000). Infrastructure Risk Analysis Model, *ASCE J. Infr. Sys.* 6(3), 114–117.

Haldar, A. (2006). *Recent Developments in Reliability-based Civil Engineering*. World Scientific, Hackensack, NJ.

Haldar, A., and Mahadevan, S. (2000). *Probability, Reliability, and Statistical Methods in Engineering Design*. Wiley, New York.

Harr, M. (2003). Accounting for Variability (Reliability), in *The Civil Engineering Handbook*, 2nd Ed., Eds. Chen, W. F., Liew, J. Y. R., CRC Press, Boca Raton, FL.

Kapur, K. C. and Lamberson, L. R. (1977). *Reliability in Engineering Design*, Wiley, Hoboken, NJ.

Lin, K-Y. (1995). Reliability-based Minimum Life-cycle Cost Design of Reinforced Concrete Girder Bridges, Ph.D. Thesis, University of Colorado, Boulder, CO.

Mahsuli, M., and Haukaas, T. (2012). Computer Program for Multi-model Reliability and Optimization Analysis. *ASCE J. Comput. Civil Eng.* 27(1), 87–98.

Renfroe, N. A., and Smith, J. L. (2011). Threat/Vulnerability Assessments and Risk Analysis, Applied Research Associates, www.wbdg.org, Accessed Jan 1, 2013.

Singh, V., Jain, S., and Tyagi, A. (2007). *Risk and Reliability Analysis: A Handbook for Civil and Environmental Engineers*. ASCE Press, Reston, VA.

CHAPTER 14

SYSTEM DYNAMICS

14.0 INTRODUCTION

Civil engineers often encounter situations where there is a need to explain or predict some system performance as the outcome of multiple factors that not only are themselves outcomes of other factors but that also interact with each other in ways that may be difficult to characterize using the traditional modeling tools we have learned so far in this book. In such problem contexts, any analysis that assumes that the contributory factors are individually endogenous (not derived externally) or independent of each other may miss the critical relationships that are key to the eventual outcomes under investigation.

System dynamics, a simulation-based tool for modeling and analyzing system behavior over time, incorporates the concepts of holism, factor interrelationships, and internal time-delayed feedback loops to yield potentially greater understanding of the timewise behavior of dynamic complex systems. Thus, this tool is useful for explaining the inner workings (and hence predicting the performance) of any system that has circular, interlinked, and time-delayed relationships between or among its constituent subsystems and components. The tool is particularly valuable in situations where these often-overlooked internal relationships can be at least as influential in determining system performance and behavior as the individual subsystems/components of the system or the system environment.

With regard to holism, because there often are properties of the entire system that supersede or complement the properties of the system's individual components, the sum of the behaviors of the components explains only partially the overall picture of the entire system's behavior. In other words, due to such holistic effects, 1 + 1 may be less than or more than 2. For example, consider a structural component that exhibits some outcome ΔY in response to a change in the levels of two influential factors ΔX_1 and ΔX_2, respectively, acting at the same time. The outcome is ΔY_{X1} due to factor X_1 alone without factor X_2, and ΔY_{X2} due to factor X_2 alone without factor X_1. For example, the effect of system loading is often more pronounced when the system experiences severe climate; thus, the effect of simultaneous occurrence of X_1 and X_2 exceeds the sum of their individual effects, that is $\Delta Y > \Delta Y_{X1} + \Delta Y_{X2}$. In Chapter 6, we discussed briefly the concept of holism and why it is useful to consider it, where appropriate, in models that explain system behavior.

With regard to time-delayed effects, it is their incorporation that makes system dynamics particularly different from the modeling tools we learned in Chapter 6 and other analytical tools for the task of system description or prediction. The time-delayed effects can be accommodated in a model using **feedback loops** and **stocks** and **flows**. Feedback refers to the situation where an event or object X affects an outcome Y; also, Y influences X either directly or indirectly through other events; thus, the nature of the X-Y relationship cannot be described or predicted merely on the basis of knowledge of the influence of X on Y alone (SDS, 2009). It is therefore often the case that the system behavior can be reliably described and predicted only through a study of the entire system, particularly, its feedback relationships.

Pioneered in the mid-1950s by Jay Forrester to assist corporate decision makers enhance their comprehension of the processes in their industries, system dynamics has blossomed as a tool for explaining or describing how a system works (and subsequently to predict future system behavior) and thus, for prescribing appropriate actions. This has spread to various applications in both private and public organizations, specifically policy planning and design for developing corporate strategy, analysis of public policy, social dynamics, ecological and biological sciences, medical systems, and energy and environmental systems. As evidenced from recent research and applications, there seems to be great potential for the use of systems dynamics in various tasks at each phase of civil systems development. These include the tasks of description or prediction of outcomes (physical conditions, operational performance, economic or financial performance, and so on) in response to factors or environments characterized by complexity.

In this chapter, we first present some basic concepts in system dynamics, including causal loop diagrams and feedback, stocks and flows, endogeneity, system structure, and the nonlinear nature of input-outcome relationships. This is followed by a presentation of the various patterns by which some variable may vary with time, namely, the linear, exponential, and goal-seeking, and various kinds of the oscillation patterns. We then present a more detailed discussion of two central concepts in system dynamics: *causal loop diagrams* that capture and quantify the feedback between the various entities in a system dynamics situation, and *stocks and flows* that quantify and track the changing levels of some attribute that increases and/or decreases at the same or different times due to external factors. The chapter then presents a framework by which the engineer could analyze a system using the tool of system dynamics. The chapter goes on to show that most phases of civil engineering system development present opportunities where this tool could be used to analyze complex situations that involve interactions between factors and continual feedback. Lastly, the chaos theory, which is closely related to system dynamics, is discussed.

14.1 SOME BASIC CONCEPTS IN SYSTEM DYNAMICS

We can infer from the previous chapter that the analytical tool of system dynamics involves a number of key concepts and assumptions that serve as the basis for this tool. These concepts include visualization of the system processes as a causal loop diagram and subsequently as a set of stocks and their associated inflows and outflows, the recognition of the feedback (via loops) that exists between the outcomes and factors, the endogenous behavior of the factors, the system structure, and the nonlinear nature of the input–outcome relationships.

14.1.1 Causal Loop Diagrams

A causal loop diagram consists of a set of vertices or nodes showing the constituent components or factors. The causal relationships between each pair of nodes, shown as links with arrows, can be labeled as negative or positive. The causal loop diagram is the basis for stock and flow quantification and computations. Section 14.3 of this chapter presents further discussion on causal loop diagrams and feedback loops and some examples.

14.1.2 Feedback

In a causal loop diagram, feedback provides additional information on how the interrelated variables in a system affect one another. Feedback can be positive (compounding, reinforcing, or amplifying) or negative. Section 14.3 discusses further the concept of feedback.

14.1.3 Stocks and Flows

Stocks and flows are one of the basic aspects of a system dynamics model because they facilitate a quantitative description and analysis of such systems. The stock and flows can be implemented and visualized using computer simulation. Unlike flow variables that are measured over a time period, stock variables are measured at a specific point in time. In Section 14.4, we will discuss stocks and flows in further detail.

14.1.4 Endogeneity

A process or physical component of a system is said to experience an **endogenous** stimulus, perturbation, or disturbance when such stimulus originates from within that process or component. On the other hand, when the stimulus is due to factors external to the process or component, then it is described as **exogenous**. Anything that triggers a system behavior or outcome (a classic example is using your hand to swing a pendulum) is considered an exogenous stimulus. Also, time, per se, is not a cause of system behavior. In other words, any stimulus that is applied externally to control the system behavior is condition based and not time based. As such, remedial actions are based on the current state of the system or its process or component and not on the basis of specific time intervals. Considerations of endogeneity significantly influence the efficacy of system dynamics applications in several areas, including policy analysis. In social policy analysis, for example, these considerations help account for the natural compensating and reactive tendencies in social systems that have led to undesired outcomes of several well-intentioned policy initiatives in the past (SDS, 2009). In such cases, feedback and circular causality are often characterized by delay and deceptive patterns. Thus, in striving to facilitate a better comprehension of the dynamic behavior of such systems, it is recommended to identify and incorporate the influence of all endogenous stimuli; that way, it is possible to reveal and account for parts of the system outcomes or behavior that inherently emanate from the structure of the system itself.

14.1.5 System Structure

An extreme consideration of the endogenous viewpoint is the assumption that the entire system has a closed boundary. The term "closed" is indicative not of open versus closed systems in the general system taxonomy (as discussed in Chapter 2) but of the view that the system is **causally closed**. Thus, in developing the system dynamics model to explain a real-life phenomenon, the objective is to assemble a formal structure to mimic the key attributes of the process without incorporating any exogenous stimuli (SDS, 2009). If such a boundary does not exist, all the variables will trace the ultimate sources of their variation to sources that exist outside the system. Therefore, the advantage of setting a causal boundary is that the analyst is encouraged to identify feedback loops within the boundary; the disadvantage is that any significant effects emanating from outside the system (i.e., from exogenous stimuli) might be ignored to the detriment of the model's efficacy. However, if all such exogenous stimuli can be brought into the boundary and made endogenous, then this limitation is overcome as the remaining exogenous sources are either nonexistent or insignificant. Thus, the concept of feedback is a consequence of setting a closed causal boundary and efforts to capture the dynamics only within the boundary (Forrester, 1969). In effect, by specifying that all key causes and effects in a dynamic system must originate from sources that are accommodated in some closed confines (from the perspective of causality), the causal influences are made to form causal loops by feeding back to themselves where appropriate.

14.1.6 Nonlinear Nature of Input–Outcome Relationships

The concepts of *active structure* and *loop dominance* are vital to the ability of feedback loops, and hence to the system dynamics model in general, to adequately explain or predict real-life phenomena. Due to the variability that naturally characterizes dynamic systems, it is useful not only to recognize that the strength of a factor–outcome relationship varies with changing conditions, but also to quantify the magnitude and direction of this variation. In other words, the model must have the capability to shift the dominant or active structure as appropriate.

For a given system of equations, the loop dominance can be shifted thanks to nonlinearities in the system's factor–outcome relationships. For instance, the sigmoidal-shaped pattern of the logistic growth model ($dY/dt = aY - bY^2$) represents the consequence of shifting the loop dominance from a feedback loop that is self–reinforcing (positive) (aY) yielding almost exponential growth to a feedback loop that is balancing (negative) $-bY^2$, (Forrester, 1969; Richardson, 2011). Unlike linear models, nonlinear models have the capability to endogenously change their dominant or active structure and shift their loop dominance. Nonlinear functions are recommended for modeling in system dynamics, particularly because of their capability to shift loop dominance and thus capture the shifting nature of real-life phenomena (SDS, 2009).

14.2 VARIABLE VERSUS TIME (VVT) PATTERNS IN SYSTEM DYNAMICS MODELING

Models in system dynamics address the dynamic behavior of a system over time. A given overall system dynamics model may incorporate multiple types of outcome versus time model patterns that explain the relationships for the different nodes. In Chapter 6, we identified a large number of possible patterns between any pair of dependent and independent variables. In this section, where we examine the behavior of systems over time, our independent variable of interest is time; and we examine a rather limited number of these patterns. In Sections 14.2.1 to 14.2.4, we discuss five distinct VVT patterns that are most common in real-world phenomena. Outside of these five, other more complicated patterns exist; however, they are generally combinations of any two or more of these five patterns. For example, the S-shaped and reverse S-shaped VVT patterns are individual combinations of two patterns (the exponential growth or decline and the goal-seeking patterns).

14.2.1 Linear Pattern

Linear patterns are represented by a straight line with a positive, negative, or zero slope, representing linear growth, linear decline, and equilibrium, respectively (Figure 14.1). Even though the linear patterns are simple and easily understood, the practical reality is that most system behaviors do not follow a linear trend or equilibrium with time. Typically, linear patterns are exhibited when the modeler does not account for feedback in the model.

14.2.2 Exponential Pattern

The exponential pattern indicates a trend where the response increases at an increasing rate or decreases at a decreasing rate with time (Figure 14.2); for example, exponential growth

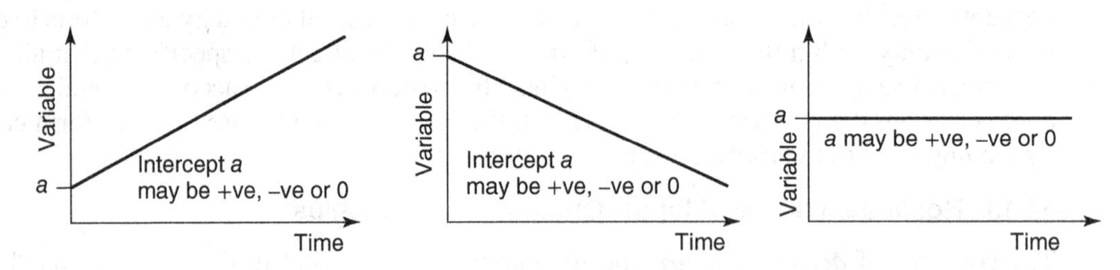

Figure 14.1 Linear family of time paths.

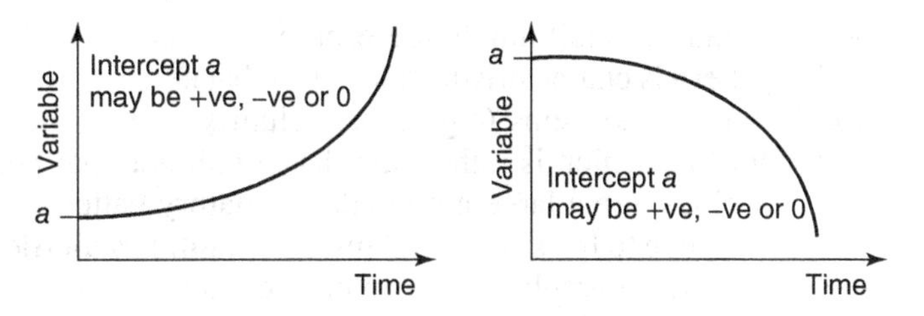

Figure 14.2 Exponential pattern.

(e.g., corrosion in a reinforced concrete beam) and exponential decay (e.g., surface erosion of a gravel road pavement). A majority of the attributes of most engineering systems exhibit this behavior.

14.2.3 Goal-Seeking Pattern

In the goal-seeking pattern (Figure 14.3), the response increases at a decreasing rate or decreases at an increasing rate with time, for example, modified exponential growth (level of usage of a new system). A significant number of attributes of engineering systems exhibit this behavior.

14.2.4 Oscillation Pattern

The oscillation pattern is a common dynamic behavior of certain engineering systems (Figure 14.4). Subtypes of this pattern include the sustained, damped, exploding, and chaotic oscillation patterns. Often, the periodicity (the number of peaks occurring before repetition of the cycle) is of interest. A **sustained oscillation** has a uniform amplitude and a periodicity of one. A **damped oscillation** is where the amplitude of the oscillation becomes progressively smaller with time and ultimately

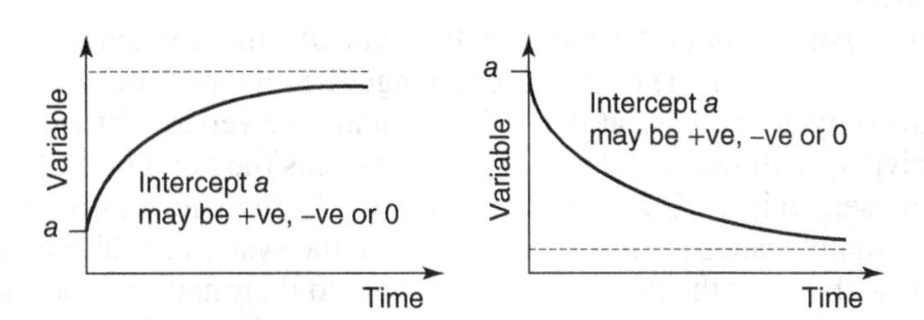

Figure 14.3 Goal-seeking pattern.

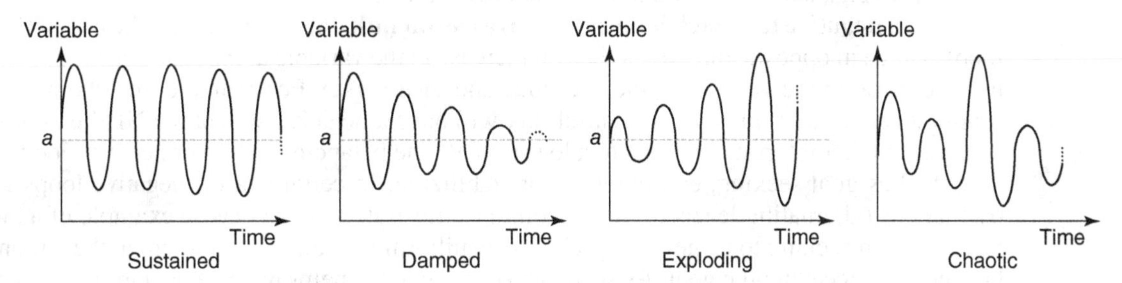

Figure 14.4 Oscillating pattern.

either develops a sustained small-amplitude oscillatory pattern or flattens out as a straight line (equilibrium); this pattern is characteristic of systems that experience dissipative or relaxing forces such as friction in physical systems (e.g., a pendulum swinging in a nonvacuum environment). An **exploding oscillation**, which is rather rare, has amplitude that becomes progressively larger with time until it develops into a large-amplitude oscillatory pattern, flattens out as a straight line (equilibrium), or ceases due to the system collapse. A **chaotic oscillation** has a random pattern that unfolds irregularly (changing amplitudes in an erratic manner) and never repeats and hence has a period equal to infinity.

14.3 CAUSAL LOOP DIAGRAMS (THE CONCEPT OF FEEDBACK)

A causal loop is a relationship expressing how one component (objects, factors, or processes) influences another, and a causal loop diagram (CLD) is a graphical depiction of loops showing the relationships (i.e., feedback and circular causality) between components in a complex system. The diagram comprises nodes that represent the constituent components and the links that represent their interactions. A CLD visualizes the intended or existing structure of a complex system and formalizes and communicates any acquired insights regarding the relationships between the components using causal links and feedback loops. The feedback loop, arguably the most essential and defining characteristic of any system dynamics model, can be defined as "the process through which a signal travels through a chain of causal relations to re-affect itself" (Roberts et al., 1983). In other words, a feedback loop can be identified when it is proven that the outcome of some action, after being dynamically transmitted through various components of the system, returns eventually to its starting point and potentially influences some future action. A feedback loop that tends to amplify the initial action can be termed as a loop with positive, compounding, reinforcing, or amplifying characteristics; one that tends to be contrary to the initial action is termed a balancing loop or negative feedback. Thus, a positive (or reinforcing) loop is the opposite of a negative reinforcing (or balancing) loop. The relationship between each pair of variables is assigned a positive or negative **loop polarity**.

A **positive feedback loop** or **positive causal link** is when the two nodes (at the end of the link or loop) have the same direction of change (i.e., an increase in the variable at the start node is accompanied by an increase at the end node and vice versa). For example, the amount of maintenance a civil system (start node) receives influences the condition of the system (end node): This can be represented in a CLD as an arrow pointing from maintenance to condition. Also, because an increase in maintenance results in an increase in the system condition, this link is positive and thus is denoted with a + at the end of the arrow. Due to their nature to destabilize and disequilibrate in a manner that is either favorable or unfavorable, reinforcing loops are the root of either desirable growth (e.g., hardening of concrete with age) or of progressive deterioration and ultimate failure or collapse (e.g., steel corrosion in a marine environment).

For a **negative feedback loop** or **negative causal link**, the two nodes (at the end of the link or loop) change in opposite directions (i.e., an increase in the variable at the start node is accompanied by a decrease in the variable at the end node and vice versa). For example, for two nodes representing interest rate and NPV, the causal link will have a negative sign at the NPV end of the link because a higher interest rate leads to a lower NPV. The balancing nature of negative loops is often described as goal seeking, equilibrating, or stabilizing. In certain cases, negative loops generate oscillations (alternating levels of the outcome) as illustrated by the classic example of a swinging pendulum: In seeking to achieve its goal of an equilibrium state, the pendulum gathers momentum but ends up exceeding the goal. As such, negative feedback helps balance or stabilize system types that exhibit asymptotic or oscillatory behavior.

To determine if a causal loop is reinforcing or balancing, one could start with an assumption, for example, the variable at node 1 is increased, and the loop is tracked (SSD, 2009). The loop is described as reinforcing if the tracking ends up with the same directional effect as the initial assumption and is described as balancing if the result contradicts the initial assumption. Unlike the situation for balancing loops, the number of negative links in reinforcing loops is an even number (zero is considered an even number) (Casey, 2012; Henten, 2012). For the purposes of identifying the potential dynamic behaviors of the system (also referred to as **reference behavior patterns**), it is important to correctly identify loops in a systems dynamics model that are reinforcing and those that are balancing. Often, reinforcing loops are associated with accelerated increases or decreases of some outcome, while balancing loops are analogous to reaching an upper plateau or lower plain. Together, the multiple causal links and feedback loops, each of which either reinforces or stabilizes, typically generate a multitude of possible outcomes of a complex dynamic system. Where there are significant delays in the individual component relationships, the ultimate system outcome may fluctuate wildly.

Figure 14.5 presents a causal loop diagram for a bridge longevity system dynamics model. In causal loop diagrams, double slashes may be used to indicate relationships that involve some time lapse between the relationship's effects (the end-of-link components or variables) to its causes (the start-of-link variables). These are not shown in Figure 14.5. Negative feedback loops in the system dynamics model are often represented by a C (representing counteracting loops); for feedback loops and links of particular attention or interest, thicker lines are used. Thus, causal loop diagrams visualize the behavior and structure of a complex system, thereby facilitating a qualitative analysis of the system. For additional detailed analysis of a quantitative nature, it is useful to use a stock-and-flow diagram that can be developed from the causal loop diagram.

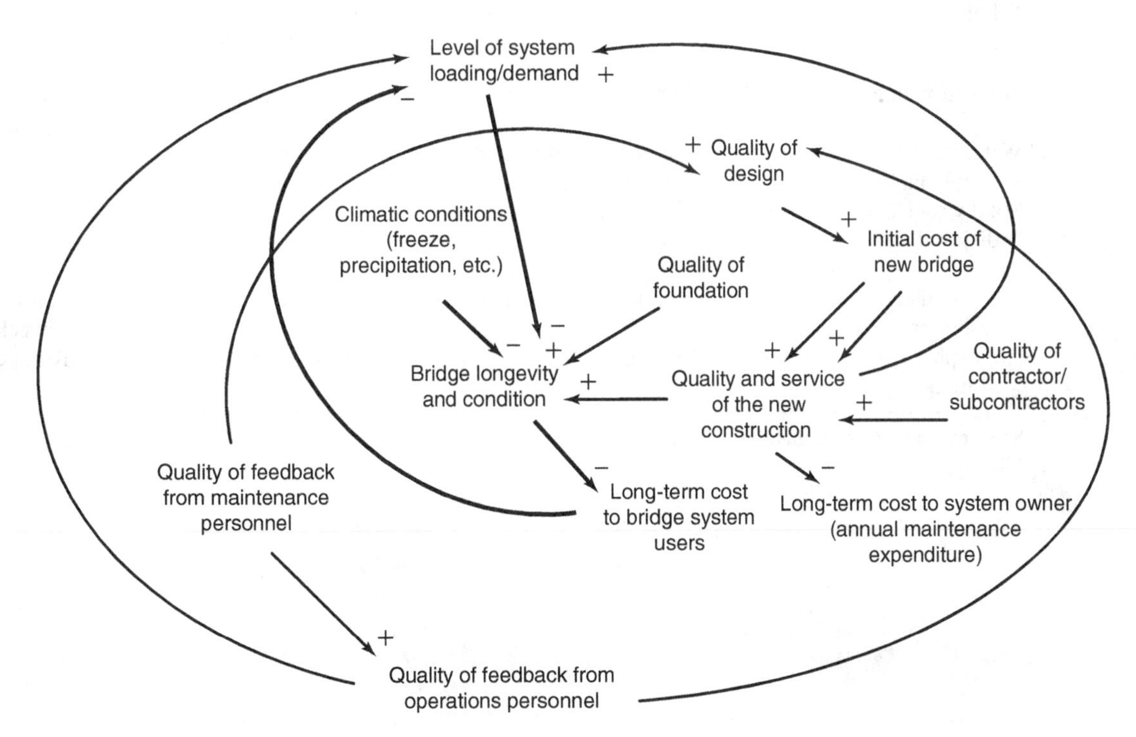

Figure 14.5 Causal loop diagram of a systems dynamics model predicting the longevity of a highway bridge deck.

14.4 STOCKS AND FLOWS

14.4.1 General Concept

Stock-and-flow models, often described as the fundamental "building blocks" of any system dynamics model, are quantitative descriptions of how the system works and can be developed conveniently from causal loop diagrams (Table 14.1). Stocks are an amount, and flows are a rate; therefore, they have different units. For example, stocks could represent how much money you have in your bank account at a given time; flows could represent how quickly money is coming into your account (your income sources or variables) or how quickly money is leaving it (your expenditure items or variables) (Figure 14.6). Stock-and-flow models, which enable detailed quantitative analysis of complex systems, are often implemented using computer simulation. A **stock variable** or **level variable** represents a quantity accumulated or remaining at a specific instance of time, a quantity that may have accumulated or was depleted in the past up to that point. A **flow variable** or **rate variable**, on the other hand, is measured as an increasing or decreasing quantity per unit of time (gallons per second, passengers per hour, vehicles per hour, foundation settlement per year, volume corrosion per year, and so on). A stock can be described

Table 14.1 Examples of Stocks and Flows in Civil Engineering Systems

Stock	Stock Units	Flow: Inflow	Flow: Outflow	Flow Units
Aquifer	Gallons	Recharge via percolation	Human pumping of water; flow into underground streams	Gallons per week
Population in a region	People	Births; immigration	Deaths; emigration	People per year
Water in overhead supply tank	Gallons	Water pumped in from treatment plant	Water piped out to serve the residences and industry	Gallons per second
Solid waste at a disposal site	Tons	Household and industrial waste brought to site	Incineration/decay (engineered or natural) of waste	Tons per week
Water in dam reservoir	Gallons	Precipitation in catchment area	Evaporation; discharge via spillway	Gallons per week
Atmospheric pollutants	Tons	Tons emitted	Tons sequestered	Tons per day

Some examples taken from SSD (2009).

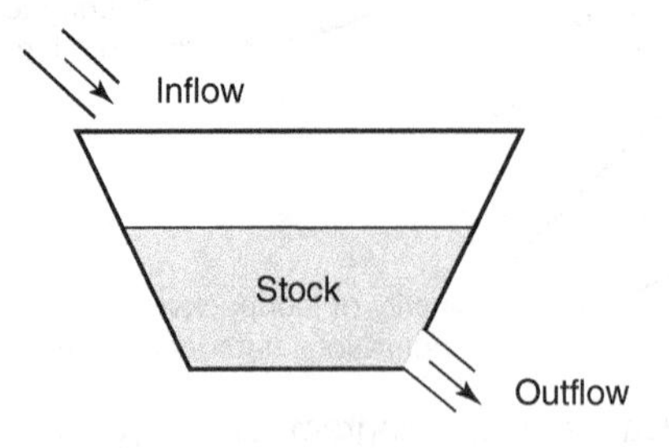

Figure 14.6 Stocks and flows.

mathematically as the net effect of incoming and outgoing flows over time, which changes only due to incoming or outgoing flows (depletion of the stock by outflows and recharge of the stock by inflows).

A \$100 daily payment into your bank account (without any outflows) would yield a linearly increasing account value. Thus, a constant rate of inflow will result in stock that increases linearly. Similarly, a linearly increasing rate of inflow will result in a stock that increases parabolically. The basic formal structure of a system dynamics model is best represented as a set of coupled, first-order, nonlinear equations. The equations may be differential or integral (Richardson, 2011).

$$\frac{d}{dt}x(t) = g(x, k)$$

where g is a nonlinear, vector-valued function, x is a vector of stock variables, and k is a set of parameters.

The simulation of dynamic systems can be facilitated by using time increments of dt and carrying out the simulation for each dt interval. The value of each stock variable is calculated on the basis of its previous value and its net rate of change. For example, let $Q(t)$ represent the amount of some stock variable at a given time t. Then the net flow to the stock derivative can be represented as $dQ(t)/dt$. Similarly, the stock amount can be determined by integrating the expression for the flow within specified limits of time. If, over a time interval, the stock $S(t)$ is increased gradually by inflow $I(t)$ and decreased gradually by outflow $O(t)$, then the change in the stock can be expressed as

$$\frac{dS(t)}{dt} = I(t) - O(t) = I^{\text{NET}}(t)$$

where $I^{\text{NET}}(t)$ is the net flow (i.e., is the difference between inflow and outflow).

When there are J inflow sources and L outflow sources, then the change in the stock can be expressed as

$$\frac{dS(t)}{dt} = I_j(t) - O_l(t) = I^{\text{NET}}(t)$$

where $j = 1, 2, \ldots J$, and $l = 1, 2, \ldots L$.

14.4.2 Examples of Inflow and Outflow Patterns

We now discuss various possibilities of the mathematical form of the stock amount on the basis of the various possible combinations of inflow and outflow patterns.

(1) Case 1 (Linear): Inflow and Outflow Are Constants.

$$I(t) = D \quad \text{and} \quad O(t) = E$$

For example,

$$\frac{dS(t)}{dt} = I(t) - O(t) = D - E$$

$$S(t) = (D - E)t + C$$

where C is a constant number that represents the stock at the beginning.

For example,

$$I(t) = 5 \quad \text{and} \quad O(t) = 3$$

Then

$$\frac{dS(t)}{dt} = I(t) - O(t) = 2$$

$$S(t) = 2t + C$$

(2) Case 2 (Polynomial): Inflow and Outflow Are Linear. Example.

$$I(t) = 3t + 1 \qquad O(t) = 2t$$

Then

$$\frac{dS(t)}{dt} = I(t) - O(t) = t + 1$$

$$S(t) = t^2 + t + C$$

where C is a constant number, representing the stock at the beginning.

(3) Case 3: Inflow and Outflow Are Power Functions.

$$I(t) = a^t \quad O(t) = b^t$$

Then

$$\frac{dS(t)}{dt} = I(t) - O(t) = a^t - b^t$$

$$S(t) = \frac{a^t}{\log_e a} - \frac{b^t}{\log_e b} + C$$

C has its usual meaning.
Example

$$I(t) = 3^t \qquad O(t) = 2^t$$

Then

$$\frac{dS(t)}{dt} = I(t) - O(t) = 3^t - 2^t$$

$$S(t) = \frac{3^t}{\ln 3} - \frac{2^t}{\ln 2} + C$$

where C is a constant number, representing the stock at the beginning.

(4) Case 4: Inflow and Outflow Are Exponential Functions. This is a more specialized case of case 3 where a or $b = e^k$

$$I(t) = e^{kt} \quad O(t) = e^{mt}$$

Then

$$\frac{dS(t)}{dt} = I(t) - O(t) = e^{kt} - e^{mt}$$

$$S(t) = \frac{e^{kt}}{k} - \frac{e^{mt}}{m} + C$$

C has its usual meaning

$$I(t) = e^{3t} \qquad O(t) = e^{2t}$$

Then

$$\frac{dS(t)}{dt} = I(t) - O(t) = e^{3t} - e^{2t}$$

$$S(t) = \frac{e^{3t}}{3} - \frac{e^{2t}}{2} + C$$

(5) Case 5 Multiple Flows and Patterns. In certain situations, there are multiple inflow points and/or multiple outflow points with the same or different patterns. We present herein two examples of such situations.

Example 1.

$$I_1(t) = e^{3t} \qquad I_2(t) = 4t \qquad I_2(t) = \ln t + 6$$

$$O_1(t) = e^{2t} \qquad O_2(t) = 8t + 2$$

Then

$$\frac{dS(t)}{dt} = I_1(t) + I_2(t) + I_3(t) - O_1(t) - O_2(t) = e^{3t} - e^{2t} + \ln t - 4t + 4$$

$$S(t) = \frac{e^{3t}}{3} - \frac{e^{2t}}{2} + \frac{1}{t} - 2t^2 + 4t + C$$

where C has its usual meaning.

Example 2.

$$I_1(t) = \sin(3t), \qquad I_2(t) = 2t$$

$$O_1(t) = \cos(2t), \qquad O_2(t) = 1$$

Then

$$\frac{dS(t)}{dt} = I_1(t) + I_2(t) - O_1(t) - O_2(t) = \sin(3t) - \cos(2t) + 2t - 1$$

$$S(t) = \frac{1}{3}\cos(3t) - \frac{1}{2}\sin(2t) + t^2 + t + C$$

where C has its usual meaning.

14.5 FRAMEWORK FOR SYSTEMS DYNAMICS ANALYSIS

The system dynamics framework follows a series of steps as shown in Figure 14.7; and often, as an analyst proceeds through these steps, the preceding steps are constantly reviewed and refined. For instance, at a later step, the analyst may realize that the initial problem, as defined, is actually only a minor aspect of a much larger problem and thus may need to be redefined. The steps are similar to those we have discussed for general model building in Chapter 6.

The first step is to identify the problem at hand; in other words, the analyst establishes what system outcome is being modeled. This outcome, often related to the system performance or its environment, may be an adversity or a benefit (e.g., failure or deterioration, cost of maintenance or operations, user inconvenience or delay, loss of structural or functional integrity, vulnerability, level of service, durability, structural condition, and so on). The inputs, which may include the characteristics of the system, its users, its environment, and its operating conditions, may themselves be functions of other factors. The next step is to establish a set of hypotheses that concern the system's outcomes in terms of the input factors. Then, the appropriate feedback loops and causal links, and their polarities, as well as the identified stocks and flows, are established. This is followed by a specification of the boundary of causality to ensure that insignificant, exogenous components are excluded. Next, a computer model of the system is developed to simulate the process by which the inputs act and interact to produce the outcome. This is the system dynamics model. To ascertain that the simulations are an adequate representation of what could happen in real life, the developed model is validated by running it for a scenario whose outcomes are already known. If the outcome

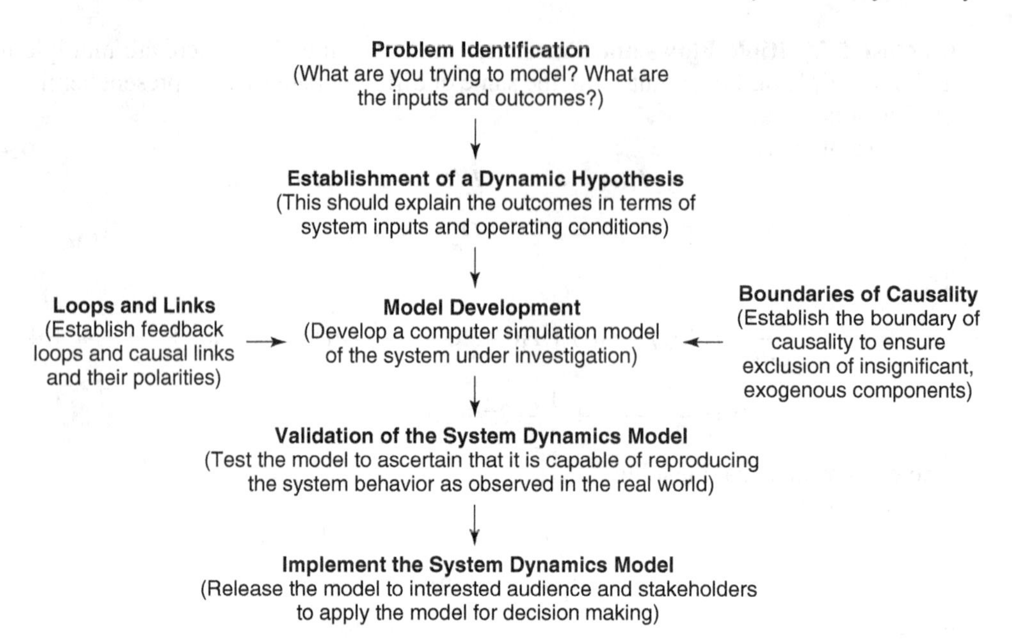

Figure 14.7 General framework for systems dynamics modeling.

predicted by the model is close enough to the actual outcome, then it can be concluded that the model is satisfactory. Further testing of the model may be necessary to ensure that it yields satisfactory predictions of system outcomes for a wide range of input factor levels and combinations. If it passes all the tests, then it can be released for application for modeling complex systems of the type of interest.

14.6 SOME PAST APPLICATIONS OF SYSTEMS DYNAMICS AT CIVIL SYSTEM DEVELOPMENT PHASES

There are numerous processes at various phases of civil engineering systems development that are not only stochastic but also are associated with nonlinear, delayed feedback mechanisms. As such, system dynamics models may be used to describe these processes and their impacts in terms of cost, technical effectiveness, environmental quality, and other performance measures. The rest of this section presents some past applications of system dynamics at the different phases of civil systems development.

Tan et al. (2010) provided insights with particular application to project cost (cash flows) performance at the construction phase of energy systems development and developed a decision tree approach to afford greater flexibility in decision making in the face of future information. In another application in project management at the construction phase of civil engineering system development, Park and Peña-Mora (2004) created a cohesive dynamic project model to assess the disruptive impact of rework (redoing previously completed work) on construction schedules due to the schedule delays and cost overruns often associated with rework.

Another application in systems development has been in the area of assessing the unintended consequences of civil systems development or their preservation practices. Using a systems dynamics model, Friedman (2006) argued that while regional infrastructure development is seen by some as necessary for urban growth, it often has an adverse effect on safety.

The usefulness of systems dynamics during the complex decision-making processes, such as those at the planning phase of civil systems development, is exemplified by the work of Maani and Maharaj (2004) who demonstrated that more effective strategies are developed and superior decisions made when, at the inception of the planning process, the decision maker acquires adequate understanding of the dynamics of the system.

In an application associated with the end-of-life (unintended) phase of system development, Cooke (2003) examined the causal relationships underlying Canada's Westray mine system, including the dynamics that might have helped predict the conditions that ultimately led to a serious explosion that occurred at the mine. Through identification and quantification of the feedback loops and the nonlinear cause–effect relationships that cannot be adequately explained by traditional root-cause evaluations, Cooke developed a dynamical model for the mine system and thus provided revealing explanations for intricate patterns of cause-and-effect relationships that ultimately led to the catastrophe.

14.7 CHAOS THEORY

Chaos theory, a mathematics field that is closely related to system dynamics, analyzes the behavior of dynamic systems that are particularly sensitive to initial conditions. Such sensitivity and the cascading effect of multiple chained events can translate into an explosion of fascinating patterns of possibilities termed the *butterfly effect*. This effect happens where very little, seemingly minor perturbations in the initial conditions mushroom across various aspects of the dynamic system to yield widely divergent outcomes, which makes it difficult to achieve reliable predictions of the system outcomes in the long term (Kellert, 1993) or even in the short term where the outcome is a function of a multitude of influential factors. Thus, even in the case of a system with purely deterministic processes where there are no random elements involved, the system outcome cannot be reliably predicted by the initial inputs and conditions. Inclusion of random processes presents a tremendously expanded range of outcome possibilities in response to a small change in inputs. This behavior is known as *chaos*.

In the contemporary use of the word, chaos is synonymous with a state of disorderliness. In its mathematical definition, however, a dynamic system is described as chaotic if it fulfills the following conditions: It is *sensitive to initial conditions*; it is *topologically transitive*; and it has *dense periodic orbits* (Hasselblatt and Katok, 2003).

Sensitivity to initial conditions, or the *butterfly effect*, refers to the possibility that an arbitrarily small perturbation of a process in a given trajectory may lead to a very different trajectory and thus a large deviation of the ultimate outcome from the otherwise expected future outcome. A 1972 speech by Edward Lorenz was titled "Predictability: Does the Flap of a Butterfly's Wings in Brazil Set Off a Tornado in Texas?" (Lorenz, 1972). The hyperbole in the speech's title is only didactic: The "flap of the butterfly's wing" represents a small perturbation in the initial condition of a given system, and this seemingly minor disturbance triggers a concatenation of events, finally resulting in very consequential, often unexpected outcomes. Thus, a different intensity, pattern, or direction of another "flapping" will yield a large-scale outcome that is very different from the first.

Topological transitivity means that, over a period of time, the system will ultimately evolve to a point where "any given region or open set of its phase space will eventually overlap with any other given region." The interpretation of topological transitivity (also referred to as topological mixing) is not difficult to comprehend: It can be likened to the unpredictable mixing process that takes place when a colored dye or fluid is dropped in a clear fluid.

The implication of the dense periodic orbit is that "every point in space is approached closely by periodic orbits in an arbitrary manner"; topologically transitive processes that do not have this condition may not indicate any significant sensitivity to the initial conditions and therefore are not likely to exhibit chaotic behavior (Devaney, 2003). The irrational rotation of a circle has been cited in the literature as a classic example of a process that is topologically transitive but does not have a dense periodic orbit and hence lacks any sensitive dependence on the initial conditions. In general, chaos theory has been applied in real-time or short-term weather predictions and in ecology population studies.

SUMMARY

A conundrum often faces civil engineers in cases where they struggle to reliably explain or predict some system outcome that is the culmination of a complex medley of multiple, interacting factors. These factors, which are associated with the system, its environment and operating condition, or its users, are often themselves outcomes of other factors, further complicating the prediction process. Any analytical tool that addresses not only the endogenous but also the exogenous nature of these factors by incorporating the key relationships between the factors is more likely to yield a more reliable prediction of system outcomes in such complex environments. System dynamics is a tool that provides such prediction capabilities because it incorporates the concepts of holism, factor interrelationships, and internal time-delayed feedback loops to yield potentially greater understanding of the timewise behavior of dynamical complex systems.

In this chapter, we first learned some basic terms and concepts that are critical to the understanding of system dynamics as an analytical tool for describing how a complex, system works and how some output of such systems could be reliably predicted. We then learned about the various patterns by which some system variable may vary with time, namely, the linear, exponential, and goal-seeking patterns, and four different oscillation patterns.

We also saw how the time-delayed effects could be accommodated in a model using feedback loops and stocks and flows, and thus discussed two central concepts in system dynamics. The first of these is the causal loop diagram, a convenient graphical representation of dynamic systems that captures and quantifies the feedback loops or relationships between the various entities in a dynamic, complex system. The second was stocks-and-flows, which quantify and track the changing levels of some attribute that increases and/or decreases at the same or different times due to external factors; we also reviewed the mathematical relationships that help predict the outcomes of different combinations of inflow and outflow patterns.

We then discussed a framework that could be used to analyze a complex system on the basis of system dynamics. The framework includes a definition of the relationships in the system as dynamic, time-related (but not time-driven) processes, identification of the aspects of the system that inherently (endogenously) influence the system outcomes; consideration of a closed causal boundary that gives rise to loops that represent feedback and circular causality; identification of points representing accumulations (stocks) within the system boundaries and their associated flows (in and out); and establishment of a formal set of models that mimic the behavior of the dynamic system. The stock and flow and causal feedback structures can be visualized using causal loop diagrams, expressed (quantified) as non-linear equations, and implemented using computer simulation.

The chapter went on to provide evidence from past experiences to show that most phases of civil engineering system development present opportunities where system dynamics could be used to analyze complex situations that involve interactions between factors and continual feedback. Lastly, the chaos theory, which is closely related to system dynamics, is discussed.

EXERCISES

1. Give two examples in different branches in civil engineering where the sum of the behaviors of a system components explains only partially the overall picture of the entire system's behavior.

2. Explain the conditions inherent within a complex system that make it necessary to use the tool of system dynamics in analyzing their behavior.
3. Under which conditions is a dynamical system described as chaotic? Provide your answer, with practical illustrations, in the context of its sensitivity to initial conditions, topologically mixing behavior, and possession of dense periodic orbits.
4. Using any example of a dynamic system in civil engineering, explain the concept of stocks and flows. For the system you selected for the illustration, identify the flow variables and the stock variables.
5. For the example you presented in Question 4, state the mathematical expressions for the stock amount for the following scenarios: (a) Case 1 (linear): the inflow and outflow are constants, (b) Case 2 (polynomial): the inflow and outflow are linear, (c) Case 3 (power): the inflow and outflow are linear, and (d) Case 4 (exponential case).

REFERENCES

Casey, T. R. (2012). Mobile Voice Diffusion and Service Connection: A System Dynamic Analysis of Regulatory Policy. *Telecommun. Policy 04*.

Cooke, D. L. (2003). A System Dynamics Analysis of the Westray Mine Disaster. *Syst. Dynamics Rev.* 19(2), 139–166.

Devaney, R. L. (2003). *An Introduction to Chaotic Dynamical Systems*, 2nd ed. Westview Press.

Forrester, J. W. (1969). *Urban Dynamics*. Pegasus Press, Glendale Heights, IL.

Friedman, S. (2006). Is Counter-productive Policy Creating Serious Consequences? The Case of Highway Maintenance. *Syst. Dynamics Rev*. 22(4), 371–394.

Hasselblatt, B., and Katok, A. (2003). *A First Course in Dynamics*, Cambridge University Press, Cambridge, UK.

Henten, A. (2012). Services, Regulation, and the Changing Structure of Mobile Telecommunication Markets. *Telecommun. Policy* 04.

Kellert, S. H. (1993). *In the Wake of Chaos: Unpredictable Order in Dynamical Systems*, University of Chicago Press, Chicago.

Lorenz, E. (1972). Predictability: Does the flap of a butterfly's wings in Brazil set off a Tornado in Texas? 139th meeting of the American Association for the Advancement of Science, Boston, MA.

Maani, K. E., and Maharaj, V. (2004). Systems Thinking and Complex Decision Making. *Syst. Dynamics Rev.* 20(1), 371–394.

Park, M., and Peña-Mora, F. (2004). Dynamic Change Management for Construction: Introducing the Change Cycle into Model-Based Project Management. *Syst. Dynamics Rev.* 19(3), 213–242.

Richardson, G. (2010). The Basic Elements of Systems Dynamics, *Complex Systems in Finance and Econometric*, Springer, New York City.

Roberts, N., Anderson, D., Deal, R., Garet, M., and Shaffer, W. (1983). *Introduction to Computer Simulation*, Addison-Wesley, Reading, MA, p. 16.

SDS (2009). *What Is System Dynamics?* The Systems Dynamic Society. http://systemdynamics.org/.

Tan, B., Anderson, Jr., E. G., Dyer, J. S., and Parker, G. G. (2010). Evaluating System Dynamics Models of Risky Projects Using Decision Trees: Alternative Energy Projects as an Illustrative Example, *Syst. Dynamics Rev.* 26(1), 1–17.

USEFUL RESOURCES

Maani, K. E., and Cavana, R. Y. (2007). *Systems Thinking, System Dynamics: Understanding Change and Complexity*, Prentice Hall, Aukland, NZ.

Meadows, D. (2008). *Thinking in Systems: A Primer*. Earthscan, Oxford, UK.

Morecroft, J. (2007). *Strategic Modeling and Business Dynamics: A Feedback Systems Approach*. Wiley, & Sons, Hoboken, NJ.

Radzicki, M. J., and Taylor, R. A. (2008). *Introduction to System Dynamics*, U.S. Department of Energy, Retrieved 23 June 2013.

Randers, J. (1980). *Elements of the System Dynamics Method*. MIT Press, Cambridge, MA.

Roberts, E. B. (1978). *Managerial Applications of System Dynamics*. MIT Press, Cambridge, MA.

Sterman, J. D. (1984). Appropriate Summary Statistics for Evaluating the Historical Fit of System Dynamics Models. *Dynamica* 10(2), 51–66.

Sterman, J. D. (2001). System Dynamics Modeling: Tools for Learning in a Complex World. *Calif. Manage. Rev.* 43(4), 8–25.

Wolstenholme, E. F. (1990). *System Enquiry: A System Dynamics Approach*. Wiley, Chichester, UK.

CHAPTER 15

REAL OPTIONS ANALYSIS

15.0 INTRODUCTION

At various phases of civil engineering system development, the system owner or operator often encounters strategic decision-making situations like the following: how much capacity to provide initially for a proposed system (often at the needs assessment phase); how much to increase or reduce an existing system's capacity to serve the actual demand level; when to carry out some structural improvements or capacity increase (at the system preservation and operations phases); and the extent or mix of inputs needed for producing some service as part of the system operations.

In a deterministic but utopian world where all economic, political, social, and environmental conditions are known with exactitude, any decision made at the analysis year can be valid for the rest of the life of the system. Unfortunately, as we are all aware and as emphasized in Chapter 13 of this text, uncertainties in real life are the norm rather than the exception; and the existing, expected, or assumed conditions at the time of the analysis, therefore, can change significantly in the years and decades following the analysis. Therefore, an action that was found to be economically unattractive in a given year may become attractive in future years or vice versa.

Clearly, traditional deterministic economic analysis is unable to account adequately and proactively for such uncertainties. If the inputs of traditional economic analysis could be made probabilistic and used in a general framework that incorporates multiple possible pathways for an investment, then a groundwork could be laid for a more realistic and flexible assessment of investments. Such improvements could greatly enhance traditional evaluation of investments.

To appreciate further the shortcomings of the NPV concept, consider the case of your vehicle's spare tire. Assume that you brought the vehicle in 2008 and sold it after 5 years; over the 5-year period, you never used the spare tire. Clearly, for the 5-year analysis period, the purchase of a spare tire and carrying it around in your vehicle trunk (adding significantly to your vehicle weight and hence, fuel usage) means a lot in terms of the total costs of owning and operating the vehicle. In this case, the deterministic NPV for the "keep-a-spare-tire-in-your-car" project will likely be negative; and even for the probabilistic NPV case where after adjusting for probability of tire failure and subtracting the cost of the expected consequences (towing or road service, possible crash, time delay, inconvenience, etc.) the net NPV is likely to be negative. Thus, from an NPV standpoint, having a spare tire will be an infeasible project and should not be pursued. Nevertheless, you kept the spare tire in your trunk! That is because you attach a significant value to the flexibility of being able to exercise the option of changing your tire when necessary, and that *flexibility value* exceeds the NPV loss.

To appreciate further the need for flexibility in decision processes, consider the problem in Example 15.1.

Example 15.1

As the city engineer for Kanocity you are responsible for planning and building a new water treatment plant to serve the city. Due to economic growth, the city population (currently 400,000) is expected to change significantly over the next three decades. The treatment plant size may be small (500,000–1,000,000 population served), medium (1,000,000–2,000,000), or large (over 2,000,000). In order to decide the appropriate size of the plant, what should the planning strategies be? Against the background of uncertainties in population growth, discuss the possible consequences associated with each option.

Solution

One planning strategy would be to build a treatment plant to serve a large population, with the expectation that the city population would grow to over 2,000,000 within the horizon period (the next three decades). Naturally, this is referred to as the **optimistic strategy**. Another strategy would be to build a small plant now; and if the population grows significantly, demolish it and reconstruct a larger one—this is the **conservative strategy**. A third strategy, the **flexible strategy**, is to build a small plant now with options to expand in the future; these options include using a design that is modular or contains extra base strength and is thus capable of incremental expansion without demolishing existing components.

The consequences of each strategy can be profound in terms of the implementation and operations/maintenance costs to the system owner and/or users, and thus can determine the extent of relative superiority across the strategies. The optimistic strategy is appropriate if the population actually grows significantly (in which case the cost of maintaining excess capacity will be outweighed by avoidance of inconvenience posed by inadequate capacity at later years). However, if the population does not grow, then the excess capacity or strength provided at the initial year of implementation is wasted. The conservative strategy is appropriate if the planner lacks adequate data or tools for good prediction of future population, if the system owner lacks financial resources to provide a larger system at the initial year of implementation but may be in a better financial position at future years, or if the system type is such that it is difficult or unduly costly to provide a modular design or excess strength to accommodate future add-ons to the system size. Where these situations do not hold true, the conservative strategy will be very unattractive. Like the conservative strategy, the flexible strategy is appropriate where there is significant uncertainty in the population predictions coupled with a lack of adequate data or predictive capability or where the system owner lacks the financial resources to provide a larger system at the initial year of implementation but may be in a better financial position in future years. Unlike the conservative strategy, however, the flexible strategy is appropriate if the system type is such that it is easy and inexpensive to provide a modular design or excess strength to accommodate future add-ons to the system size.

Discussion

It is clear from Example 15.1 that there is no best strategy for all conditions, but it is true that a large percentage of situations in real life are characterized by conditions that call for flexible strategies when possible. Also, it is worth mentioning that the term *optimistic* as used in this context does not suggest a favored outcome but rather simply means that the expected conditions that justify full implementation are realized, irrespective of whether these conditions are desired or undesired. For example, in preparing for a worst-case scenario of climate change, a planner or designer may adopt an "optimistic" strategy of stronger and deeper foundations for a marine structure. Certainly, the rise in sea levels that accompany climate change cannot be a good thing and is not a favorable outcome; however, designing to accommodate such a condition can be considered "optimistic" because the planner is optimistic in the sense that such conditions even if adverse, will be realized.

15.0.1 What Are "Real Options"?

A real option is the right—but not the obligation—to undertake an action related to a proposed intervention, such as deferring, abandoning, expanding, contracting, or staging the intervention. In the context of civil engineering systems, the intervention may be the construction of the system itself or some action at any phase of any phase of the system's development. As we learned from Example 15.1, the flexibility that accompanies this right can be extremely valuable to the system owner. For example, a water supply agency may decide to scale down or suspend a massive borehole drilling project if the demand for water falls below a certain level and then to proceed with the project at a future time when the demand rises above a certain level. In traditional deterministic discounted cash flow evaluation [which yields a single value of the expected net present value (NPV)], flexibility in decision-making contexts such as this example is not considered adequately but can be addressed by infusing probabilistic considerations such as real options analysis (ROA).

The roots of the real options concept are found in the world of financial engineering, particularly the theory of financial options (Brach, 2003). In the financial world, investment risks are managed by implementing investments that are expected to be profitable and declining those not deemed to be profitable. The profitability of investments is ascertained through a quantification of their economic values (Stulz, 2003), a detailed process that includes an assessment of threats and opportunities (and their inherent uncertainties) in the investment environment and evaluation of different strategies for risk management including *derivatives*. Derivatives are financial instruments that serve as a form of insurance for the underlying financial asset, whereby an appropriate premium is paid to counter any possible threat (loss in value of the financial asset). As the terms imply, *financial options* are related to financial assets (bonds, stocks, commodities, currency, etc.) while *real options* are based on real assets such as civil engineering infrastructure. Therefore, while real options differ from financial options in terms of the type of asset being evaluated, the basic motivation is the same—reducing the risk associated with future threats that render the investment environment uncertain (Arman, 2014).

In the field of civil engineering, there are a multitude of areas where ROA can be applied for more flexible decision making (see Table 15.1). It is useful to recognize in project evaluation

Table 15.1 Examples of Real Options Application at the Various Phases of Civil System Development

Phase	Examples of Application	Sources of Uncertainty
Needs assessment	Whether or not to build the system;	Uncertainty in system demand
	Which year to build the system or project (implement now or defer)	
Planning	Whether to expand a runway and the optimal time to do so	Uncertainty in demand and aviation technology
Design	Whether to use more conservative design parameters dimension, materials, and so on, to accommodate a new function or increased demand.	Uncertainty in demand and cost
Construction	Whether or not to purchase a major piece of equipment	Uncertainties that contractor will, in the future, be awarded contracts that will require the equipment in question
Operations	Whether to lease the operations of a publicly owned system to a private sector concessionaire, and the best way to do this.	Uncertainties in demand, deterioration rate, political conditions, etc.
Monitoring	Whether maintain the water level in a dam or when to release (open spillways) for purposes such as irrigation.	Uncertainty in climate (rainfall)-evaporation rates, and usage levels.
Maintenance	Whether to lease the maintenance of a publicly owned system to a private sector concessionaire, and the best way and time to do this.	Uncertainty in the intensity of deterioration factors (climate, loading, and so on), and costs
End of life	Whether to demolish a system and build replacement, and the best time for doing this.	Uncertainty in demand, rate of deterioration, and cost of replacement.

situations that require the incorporation of the real options concept and to be able to construct NPV profiles for such projects. The attractiveness of the real options concept lies in its ability to explicitly give the system owner or operator the flexibility in making decisions related to the system in light of changing conditions that affect the financial viability of a project related to the system. Specifically, it is a good tool for analyzing investments in uncertain or rapidly-changing conditions where traditional methods do not adequately provide an updated economic value of the investments. In this respect, ROA can help the decision maker to "leverage uncertainty and limit downside risk" (Nembhard and Aktan, 2009; Athigakunagorn, 2014).

The concept of real options does not exist in a vacuum; nor is it a new palliative that addresses uncertainties in all decision-making environments. Rather it can be considered a part of a large portfolio of existing analytical tools and concepts assembled to address certain contexts of decision making. Indeed, in this text, we have learned some of these tools: probability and statistics (Chapters 5 and 6), simulation (Chapter 8), engineering economics (Chapter 11), and risk and reliability (Chapter 13).

Pioneers of ROA applications in civil engineering systems management include Dixit and Pindyck (1994), who identified the adverse effects of uncertainties in economic environments in situations of irreversible types of investments; Trigeorgis (1996b), who advocated for more dynamic and flexible analysis of investments to replace relatively static cash flow approaches; Amram and Kulatilaka (1998), who showed how real options could be applied to analyze strategic investments; and de Neufville (2000), who made the case for dynamic strategic planning for technology policy. Example 15.2 illustrates how to carry out an evaluation involving the option to terminate a project at some point of the project life.

Example 15.2

A metropolitan transit company is evaluating the feasibility of a new initiative for ticketing passengers. The life of the initiative is 2 years. The initiative requires $50K as the initial investment; after year 1, an additional amount of $100K will be invested; and at the end of year 2, there should be revenue streams from the ticket sales. The transit operator is uncertain about the cash inflow amounts because there is no certainty that the commuters will patronize the new system. The probability that commuters will accept the new system in the first year is 70%, and there is an 80% chance that they will accept it in the second year. Assume a 4% rate of return. (a) Draw a decision tree showing the decision paths. (b) Calculate the expected NPV and determine whether the initiative should proceed on the basis of this result. (c) Consider the situation where the system owner has the flexibility to terminate the initiative after the first year and calculate the NPV for that option. Discuss your results.

Solution

The decision tree is presented in Figure 15.1. The traditional expected NPV is

$$E[\text{NPV}] = \text{NPV}_{\text{Year0}} + \text{NPV}_{\text{Year1}} + \text{NPV}_{\text{Year2}}$$

$$E[\text{NPV}] = -50 + \frac{-100}{(1+0.04)^1} + \frac{0.7(0.8 \times 250 + 0.2 \times 200) + 0.3(0.8 \times 90 + (0.2 \times -80))}{(1+0.04)^2}$$

$$= -4.142$$

As the result is negative, using traditional NPV will lead to the conclusion that the initiative should be rejected.

(b) Where the system operator has the option to terminate the initiative after year 1, the operator would proceed with the second stage of the initiative only if the response by the system users

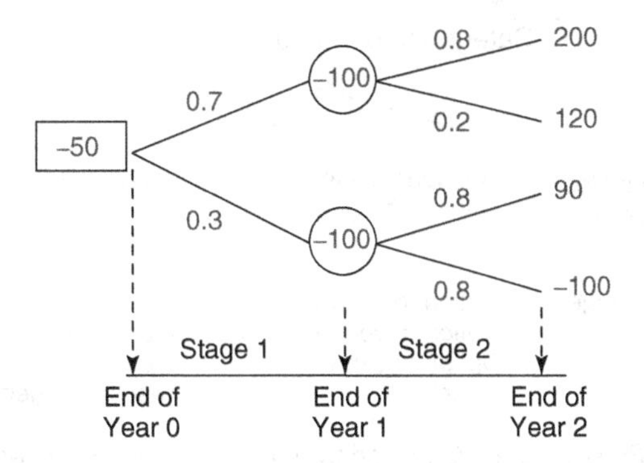

Figure 15.1 Example showing the option-to-terminate decision tree.

was favorable during year 1. If the response was not favorable, the operator would terminate the initiative because proceeding further would lead to incurrence of additional expenditure of $50M, and the expected NPV of the cash inflows after the first stage would be

$$E[\text{NPV}] = \frac{0.08 * -40 + 0.2 * -80}{(1 + 0.04)^1} = -46.15$$

Thus, when the system owner incorporates the option to terminate, the expected NPV is

$$E[\text{NPV}] = -50 + \frac{0.7 * -100}{(1 + 0.04)^1} + \frac{0.7(0.8 * 250 + 0.2 * 200) + 0.3(0)}{(1 + 0.04)^2} + 38$$

As the result is positive, it is clear that incorporating a real option (in this case, the option to terminate) leads to the conclusion that the initiative should proceed as there is an option to terminate in case conditions become unfavorable in the future (in the second phase).

Discussion

Example 15.2 presents a simple example where the system owner has the option to terminate the proposed initiative and how such flexibility could enable a better decision whether or not to proceed with the initiative. In other problem contexts, the options are not to terminate but to defer the implementation of the initiative to some future year when conditions are more favorable.

15.1 REAL OPTIONS TAXONOMY

A **call option** is where the system owner has the right to invest in (purchase) a financial asset or the right to proceed with some action associated with a real asset either at a predetermined future time and price (**European** call option) or at any time over the option life (**American** call option). A **put option** is where the system owner has the right to decline investing in a financial or real asset either at a predetermined future time and price (**European** put option) or at any time over the option life (**American** put option). **Valuation** is the determination of the value of the underlying asset with any flexibility and/or uncertainty that is associated with real option approaches. **Option categories** are those contexts within which an option may exist: new system implementation or continued operations of existing system, project timing for system expansion or strengthering, and so on (see Section 15.2). Option types are specific options that exist under each category (see Figure 15.2). The nonrecoverable upfront expenditure that is incurred to acquire flexibility in a real options problem is referred to as the **option premium** or **option price**, for example, the cost of providing extra

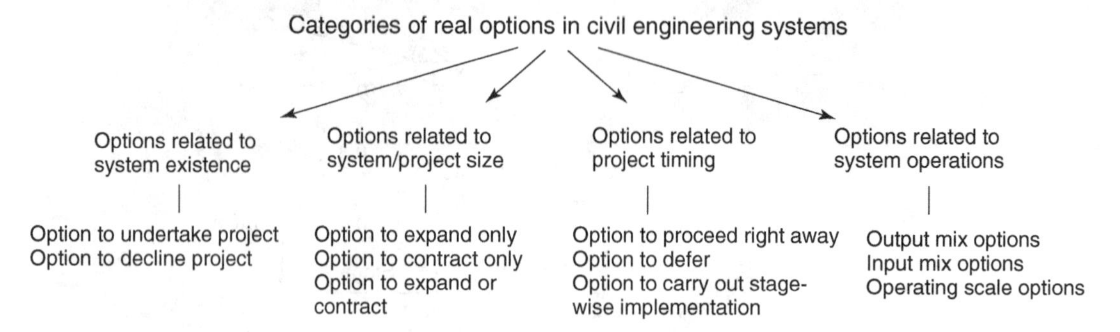

Figure 15.2 Categories of real options in civil engineering systems.

strength in foundations for a building with the expectation that additional floors may be needed in the future to meet growing demand. The **strike price** is the predetermined cost at which an option holder can buy (for call option) or sell (for put option), for example, the construction cost for the additional floors. The **stock price** is the additional revenues or cost avoidance savings from extra demand due to the provision of the flexibility. The **downside risk** is the probability that the project will be in an unfavorable situation, which can be mitigated due to the fact that a project has some flexibility that allows a decision maker to postpone their decision until some uncertainty has been revealed.

15.2 CATEGORIES OF REAL OPTIONS

Figure 15.2 shows a classification of possibilities where flexibility could be embedded in the decision process. This consists of the options related to project existence (option to undertake the project or to abandon it); the options related to project timing (option to start the entire project right away, to defer it, or to implement it in stages); the options related to the project size (option to expand or contract (reduce the scope or scale of) the project); and the options related to project operation (the output and input mix and operating scale options). In Figures 15.3 to 15.7, the letter F mean favorable conditions to undertake the project and U means unfavorable.

15.2.1 Options Relating to Project Implementation

Of the four major categories of real options, the option to implement the project or not to implement it is probably the most consequential: If the former option is exercised, the project is undertaken; otherwise, it does not. If the project refers to the construction of a new system, then the exercise of either option determines whether or not the new system becomes a reality.

(a) Option to Undertake the Project. This option indicates the flexibility of the system owner or operator to implement a project. The option to undertake a project is consistent with the **call option**.

(b) Option to Decline or Option to Terminate the Project. The option to decline or the option to terminate (abandon) include the situation where the system owner or operator has the flexibility to terminate or abandon a project at any point during the life of the project when it is realized that conditions are not favorable for continued implementation of the project. An extreme case is when the equivalent NPV (at the analysis year) of all the revenue and expenditure cash streams over

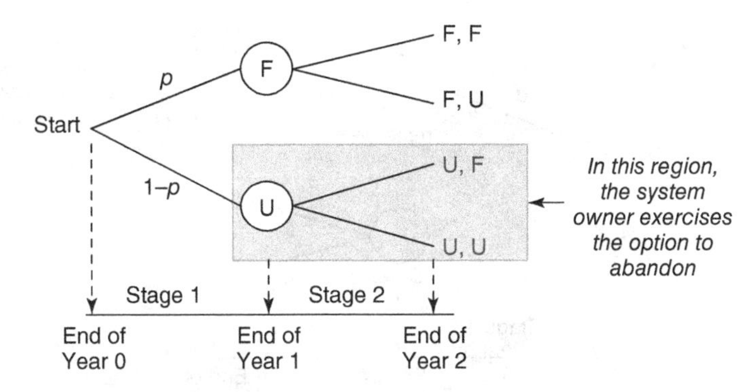

Figure 15.3 Decision tree illustrating the option to abandon.

the remaining life becomes lower than the asset's liquidation value at the analysis year. The act of abandoning a project, selling the cash flows (including the asset's salvage value) over the remaining life of the project is termed the "exercise of a put option." For projects that involve new systems or system components in markets where their acceptance is uncertain, it is important that the system owner has the flexibility to abandon the project. Also, the abandonment option is a necessity for system owners who invest in large capital-intensive projects including railroad systems. Figure 15.3 presents a simple two-stage decision tree illustrating the option to abandon.

15.2.2 Options Relating to Project Timing

In certain cases, the system owner has already decided to undertake the project of interest, and the only question is when to do it. In such cases, there is uncertainty regarding the conditions that influence the eventual decision, for timing, and thus any flexibility in deciding the appropriate timing of the project is valuable.

(a) Initiation or Deferment Options. In this case, the system owner possesses flexibility regarding the implementation date of a project. Consider a plan to develop an underground transit system for a growing city. The city engineer can delay the recommendation to build until market conditions are favorable; for example, the city's population density reaches a certain level, the surface traffic conditions fall below a certain specified minimum level of service, and the average income of commuters rises to the point where they can afford the proposed transit fares. This is consistent with a **call option**. Figure 15.4 presents a simple two-stage decision tree illustrating the call option to initiate the project.

(b) Sequencing Options. The sequencing or stagewise option is a series of initiation/deferment options as discussed in Section 15.2.2(a). This category addresses the flexibility regarding the timing of two or more projects that may be implemented sequentially or in parallel. If sequential implementation is adopted, the system owner acquires the opportunity to observe the outcomes of the initial project (the first stage of the sequence), thus making it possible to identify and address sources of uncertainty relating to the subsequent projects, and ultimately make the option available whether to proceed with the projects in the remaining chain. If the system owner implements all the projects simultaneously (in parallel), they lose the option to avoid further spending in case conditions are found to be unfavorable for the investment.

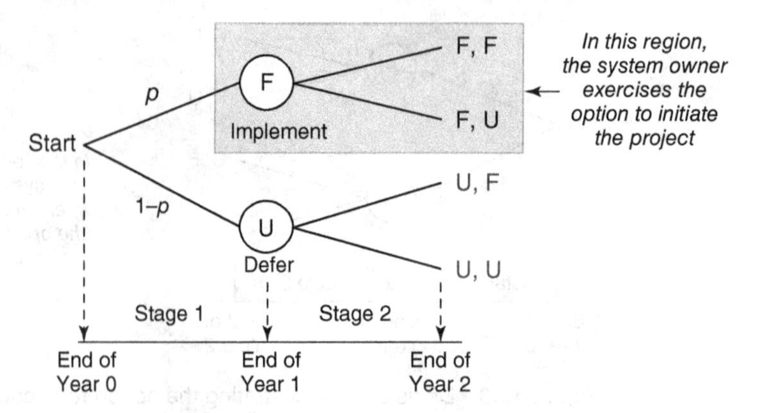

Figure 15.4 Decision tree illustrating the option to initiate.

15.2.3 Options Relating to Project Size

Due to uncertainties in the demand for the system, the system owner may grapple with how much structural or operating capacity to provide. Such uncertainties translate into uncertainty in the project scope, and thus flexibility in the size of the proposed project is extremely valuable to the decision maker. This category of options includes the options to expand or contract and the switching option.

(a) Option to Expand. Often, the system owner finds it useful to design a new system such that it can be easily expanded to accommodate higher demand when needed. In such cases, the system owner exercises the option or right, but not the obligation, to carry out expansion of the system if the future conditions are such that the expansion is merited. The option to expand is equivalent to a **call option**. As we saw in Example 15.1, a system that is built with the option to expand will be more costly that one built without such an option. The difference between the construction or provision costs associated with these two situations is referred to as the **option premium**. Figure 15.5 presents the decision tree illustrating the option to expand.

(b) Option to Contract (Reduce). For certain types of civil engineering systems and certain types of financial or political environments, it is prudent for the system owner to have the flexibility

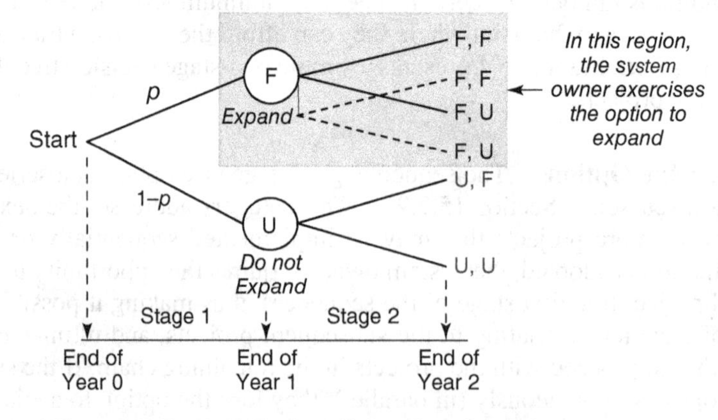

Figure 15.5 Decision tree illustrating the option to expand.

to contract the project if the conditions are favorable to carry out that action. In this case, the system is built in such a way (e.g., using modularization) that the system's output can be reduced at a future time if the conditions become unfavorable for expansion or for same-size operations. Contracting (reducing) a project, thus avoiding future expenditures associated with increase scale of operations, is equivalent to the exercise of a **put option**. A system that is built with the flexibility to contract if needed is worth more than one that is built without such flexibility. The difference in their values is the **value of the option to contract**, and the excess upfront expenditure in providing this option is the **option premium**. Figure 15.6 presents the decision tree illustrating the option to contract.

(c) Option to Expand or Contract. Also known as the *switching option*, this option is consistent with projects that have a flexible design that makes it possible for their entire operation to be scaled up or scaled down as and when required. Specifically, during unfavorable conditions, the system owner has the flexibility to shut down the operation fully or partially (equivalent to a put option) or to resume operations at a future time when conditions become favorable (consistent with a call option). Figure 15.7 presents the decision tree illustrating the option to expand or contract.

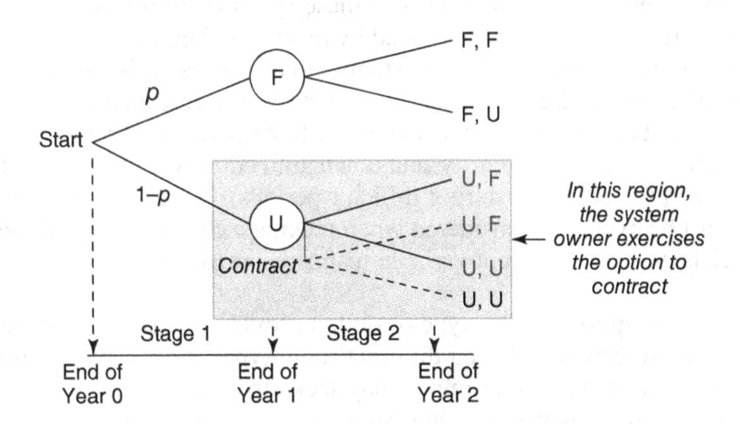

Figure 15.6 Decision tree illustrating the option to contract.

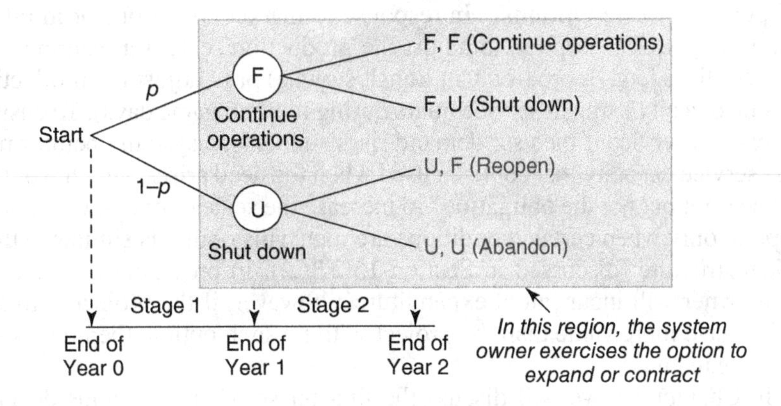

Figure 15.7 Decision tree illustrating the option to expand or contract.

15.2.4 Options Relating to Project Operation

The three categories of options discussed so far (relating to project implementation, project timing, and project size) are typically exercised at any phase of civil system development. There is a certain category of options that are exercised only at the operations phase: output mix options, temporary shutdown options, input mix options, and operating scale options.

(a) Output Mix Options. The output mix option refers to the flexibility to produce different types of outputs from the same facility, often with the intention to improve user safety or convenience or to reduce operating or maintenance costs of the civil engineering system. For example, highway lane assignments by vehicle class can yield different patterns of outputs from the same facility. This type of flexibility is valuable for certain types of civil engineering systems that are characterized by fluctuating demand or where there are generally low levels of demand for a particular type of service, and the system operator seeks the flexibility to offer quickly a different type of service as and when required. For example, the operators of public transportation systems in pursuing such flexibility, ensure that their fleet includes different types and sizes of buses, train cars, or other rolling stock.

(b) Temporary Shutdown Options. Civil engineering systems involve the production of some service to society. Even though most of these systems are for the public good, a few of them are for profit purposes or at least are funded by revenue collected from users. For certain geographical parts or functional aspects of these systems, situations may arise where the costs of labor, material, fuel, and other inputs rise to such an extent that revenues are unable to cover the costs of operations. In such cases, it may not be optimal to operate those parts or aspects of the system for a given period and thus the system is temporarily shut down until costs decrease (unless the system owner decides to subsidize the operations during difficult periods). It has been recommended that this type of option must be explicitly considered when making a choice among different production processes or materials that have different variable-to-fixed cost ratios.

(c) Input Mix Options. This type of option provides the system owner, at the operations phase, the flexibility to choose between different input types, each of which produces the same level of output. A system owner, for example, may seek the flexibility to choose among different fuel types to produce electrical power for the system operation; therefore, a plant that is flexible enough to produce energy from different sources, even though more costly initially, could be valuable in terms of the flexibility it provides.

(d) Operating Scale Options. In response to market conditions or in other situations, a system operator may seek the option to adjust the production scale, for example, in terms of output rate per unit of time (e.g., more frequent transit service) or to adjust the production period length (e.g., extension of a city's transit service hours during football game days). To ensure that there is flexibility to expand service if the user demand rises suddenly, a system operator may ensure that there is excess service capacity that could be used when the need arises, which would give the system operator "the right but not the obligation" to increase the scale of operations; and the operator exercises the option only when certain conditions are met. This option is similar to the option to expand the physical structure (discussed in Section 15.2.3(c)). In providing the excess service capacity, the system owner will incur initial expenditures; however, if the project with the option to expand is found to have more value than the project with no such option, then the extra cost of extra service capacity is justified.

In Chapter 19, we will discuss the first phase of civil systems development by discussing the issues relating to the assessment of the need for a civil engineering system. In that chapter, we will find that the need for a system is largely driven by supply and demand. While demand is largely outside the control of the system owners, supply is often within their control. Supply includes the provision of a new system, replacement (or reconstruction), and increased structural or functional capacity. With regard to system replacement or reconstruction, the rationale is mainly to accommodate higher demand, but it could also be due to other factors, including a desire to keep up with new materials, designs, or technologies; to meet regulatory requirements; or to reduce vulnerability to imminent sudden or incipient disaster. Such uncertainties strengthen the argument for decision-making flexibility not only at the needs assessment phase but also at subsequent phases of civil system development.

15.3 VALUING FLEXIBILITY IN REAL OPTIONS ANALYSIS

15.3.1 Balancing Supply and Demand

As we have seen in the previous sections, there may exist, for civil systems managers, the option to provide upfront excess capacity or strength for a system in anticipation of future demand in terms of capacity or strength, albeit at a higher initial cost (see Figure 15.8a), vis-à-vis providing the amount needed initially but having to replace the entire system or a significant portion thereof at future years as and when necessary (see Figure 15.1b). In between these extremes, a multitude of options exist that differ from each other in terms of which time (year) to provide extra supply and how much supply to provide. There is a need for civil system managers to acquire the requisite analytical tools for examining the trade-offs between certain options in terms of infrastructure performance criteria, including the agency initial cost, the agency maintenance and operating costs, the user costs associated with poor levels of service, the user costs associated with facility downtime during periods of capacity expansion or strengthening, and the costs associated with uncertainty in demand predictions.

15.3.2 The Argument for Flexibility

At the needs assessment, planning, and design phases of civil systems development, engineers make decisions regarding the size of the system to accommodate a certain level of demand. For large public systems, such decisions can be daunting because they often involve the commitment of millions or even billions of taxpayer dollar toward the prescribed, often irreversible course of action. Making an incorrect assessment of the demand can therefore be very costly in terms of financial resources, public relations, and political ramifications. However, the reality is that no demand

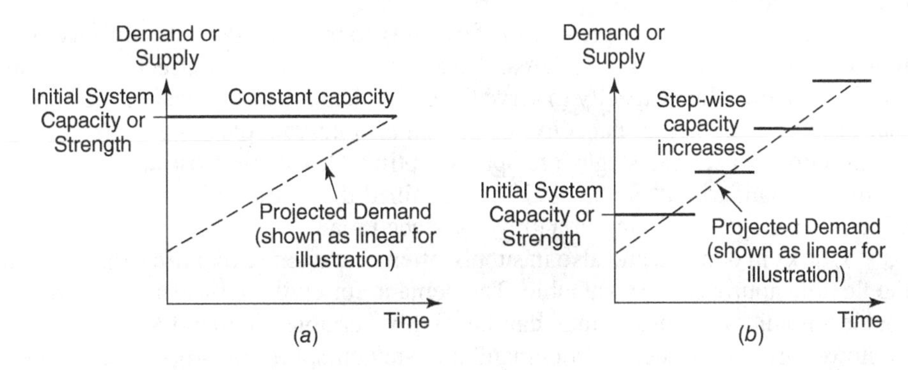

Figure 15.8 Two hypothetical scenarios of demand and supply.

prediction model is perfectly accurate, and future conditions in the socioeconomic environment can invalidate even the best prediction model. Due to such uncertainties, over the life of an infrastructure system, the system functional or structural capacity may need to increase (to accommodate increased user demand) or the functional capacity may need to decrease (to reduce underutilization and wastage). Both situations are associated with the incurrence of agency cost or user cost, or both. For example, providing excess or additional capacity or strength requires agency spending while user cost is reduced; on the other hand, allowing system demand to exceed supply at certain times could cause user costs through user delay, inconvenience, and frustration. Let us consider two scenarios of capacity and supply that are illustrated in Figure 15.8.

In the first scenario, the system owner provides additional capacity or supply only as and when it is needed. Often, due to the time durations associated with facility planning, design, construction, and commissioning, the use of the additional capacity or strength may commence well after the demand exceeded the supply. As such, the demerits of this option are that (i) the system users encounter some inconvenience during the system operations, due to insufficient capacity or use of alternative or temporary systems during the time period when demand exceeds supply; and (ii) system downtime associated with capacity-enhancement work (e.g., work zones in the case of transportation systems) often impairs the operations of the system, causing user delay, inconvenience, or safety hazards during the downtime period. The merits of this approach are (i) supply is provided only when it is needed, thus reducing the magnitude of capital needs at any point in time, which may be more consistent with practical and realistic agency cash flows; and (ii) due to the compounding effect of the interest rate, the initial cost of a one-time provision of large capacity may exceed the sum of discounted future costs.

In the second scenario, a large capacity could be provided at the initial stage. The demerits of this option are that (i) it requires a high upfront cost, and (ii) the future demand may be much lower than anticipated. The merits of this option are that (i) user costs are kept low because there is little user inconvenience such as delay (anytime the usage exceeds or capacity) or detours (when the user load exceeds the system capacity or strength); (ii) the system flexibility is increased to such a level that it is more resilient to handle unexpected upsurges in demand and other shocks; and (iii) providing large capacity at one time takes advantage of scale economies because the constant dollar costs of a one-time provision of large capacity is typically lower than the sum of costs of several capacity increments.

To evaluate these options, it is therefore necessary to quantify all these relative merits and demerits and to take them into account to yield a real decision regarding the preferred option. In order to reach a reasonable balance between these extremes, the provision of additional capacity or strength must be timed in an optimal manner so as to minimize the total agency costs and user costs in the long run.

Investment plans that provide the flexibility to provide adequate capacity or strength to accommodate long-term increases in demand are generally more cost effective than those that provide a large initial, one-time capacity to serve the entire demand at any year of the horizon period or those that provide incremental capacity at very small intervals over the horizon period. Thus, staged construction, where the stages are spaced optimally, can be most appropriate (Wang, 2005); for example, Bhandari and Sinha (1979) determined the traffic volume threshold at which it is cost-effective to stage the paving of a gravel-surfaced road.

Quirks in demand and also in supply often characterize real life situations, and are what make the flexible approach very valuable. The demand for civil engineering systems is derived from the socioeconomic system and thus can be highly variable. Demand scenarios could include faster or slower growth or decline in demand and sudden spikes or drops in demand at certain periods due to natural or man-made causes. For example, an increase in population causes a slow and steady increase in the demand for facility capacity; climate change causes a very slow increase in the demand for structural strength due to higher water tables and a subsequent increase in pore water pressures; and the need to evacuate distressed populations in disaster events causes a sudden increase in the demand for transportation system capacity. Supply scenarios include do-nothing, facility expansion or strengthening, and facility abandonment.

15.3.3 Techniques for Valuing Flexibility

As we learned earlier in this chapter, exercising an option typically comes with a certain cost. For example, when a parking garage is built with extra strength foundations so that at a future time additional floors can be added to accommodate future increases in demand, the cost of extra reinforcement in the foundations and columns is the cost of exercising the option or the "cost of flexibility." If the demand does increase, then the cost of demolishing the existing garage and reconstructing a taller garage is avoided; in this context, the magnitude of the cost avoided is termed "the cost of no flexibility"; the cost of a no-flexibility case is often much higher than that of the flexibility case. If both are deterministic, then the difference represents a fixed option value (Figure 15.9*a*). In reality, both the cost of no flexibility and the cost of flexibility are uncertain and are therefore stochastic distributions (Figures 15.9*b* and 15.9*c*). Thus, the resulting option value itself is a probability distribution that can be derived by simulating the NPVs of the costs associated with the flexibility and the no-flexibility cases (Miller and Clarke, 2004). The decision maker exercises the option only if the expected value of the option is positive. The mean value of the option is equal to the mean of the cost probability distribution for the no-flexibility case less the mean of that or the flexibility case.

To determine the value of real options through the development of cost probability distributions as shown in Figures 15.9 and 15.10 or through some other similar means, a number of techniques have been discussed in the literature or applied in practice. These take root from the techniques used to price financial options and include analytical methods [including the Black–Scholes model (Black and Scholes, 1973), numerical methods including Monte Carlo simulations, system dynamics, and decision tree methods, see Chapter 8, 14, and 16].

The Black–Scholes model determines the value of flexibility associated with a European call option. The use of the model requires a number of very restrictive assumptions that are generally more valid in the context of financial options than they are for real options: the constancy of volatility over the option life, the normal shape of the probability distribution of project returns, and the lognormal shape of the probability distribution of the project's underlying value. Also, most real options are American, rather than European, in nature (Gilbert, 2005).

Simulation techniques, including Monte Carlo analysis, make it possible for the analyst to incorporate multiple sources of uncertainty without any restrictions on their distribution or correlation structures because there is no need to assume that the possible decision pathways are independent. These techniques, which can be applied using standard computing platforms including simple spreadsheets, have been used to value the flexibilities in real option analysis for investments in mining (Cardin et al., 2008), logistics (Billington et al., 2002; Nembhard et al., 2005), parking garage construction (de Neufville et al., 2006), and power plant construction (Kulatilaka and Marcus, 1992). Using Monte Carlo simulation, flexibility can be valued in a few steps (de Neufville et al., 2006). First, a deterministic case is established by projecting the exogenous factors of value over the project lifetime, and a standard discounted cash flow (DCF) analysis is performed for the project for the no-flexibility scenario for this case. Second, a probabilistic case is established by establishing probability distributions for each exogenous factor and simulating their levels over the project lifetime to yield hundreds or thousands of possibilities of factor combinations and outcomes. Third, a DCF analysis is carried out for the no-flexibility scenario for each of the multitude

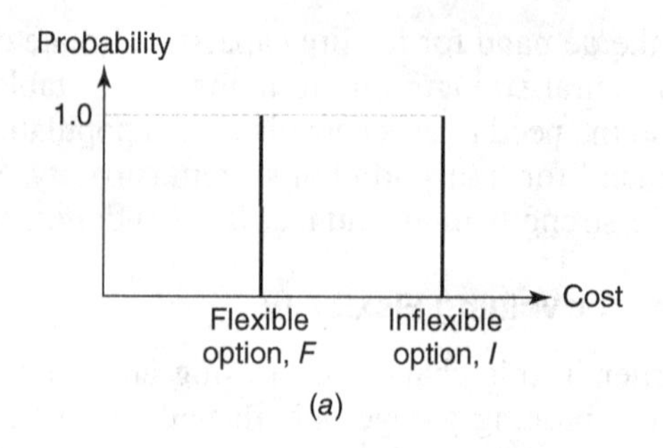

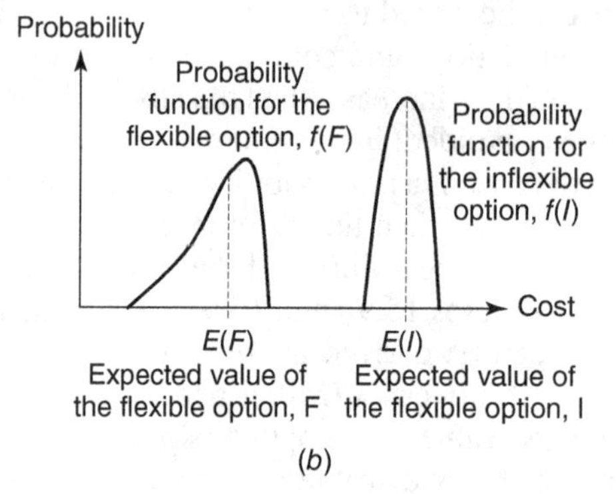

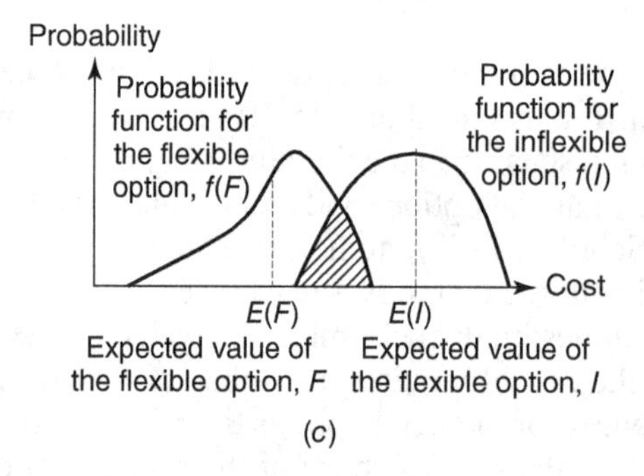

Figure 15.9 Cost probability distributions for flexible and inflexible options (*a*) Deterministic situation, (*b*) probabilitic situation, no overlap, and (*c*) probabilistic situation with over.

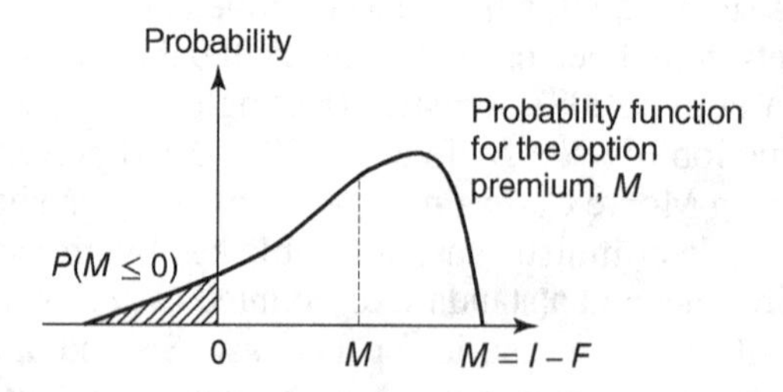

Figure 15.10 Probability distribution for the option premium.

of factor combinations and outcomes. This yields a probability distribution of the possible outcomes in terms of NPV or other criterion of economic analysis. Fourth, a DCF analysis is carried out for the flexibility scenario for each of the multitude of factor combinations and outcomes; and for each combination and outcome, the difference between the NPVs associated with flexibility and no flexibility represents the *value of flexibility* (VOF). This VOF is itself a probability distribution with a mean and serves as a basis for the decision-making process.

In decision tree methods, a tree structure is developed to represent the possible pathways of the decision possibilities and their respective outcomes. Each pathway ends with a value (overall cost or benefit) associated with that pathway; the value of flexibility is determined by comparing the expected value of the decision pathway with flexibility to that with no flexibility. A *binomial lattice* is a decision tree method that is considered very versatile in that it can incorporate multiple sources of uncertainty (albeit less elegantly than the Monte Carlo method) and varying levels of volatility over the option life; and it can be used to determine the value of flexibility in the American- or the European-style options. At each node of the lattice, representing a decision point, or the time at which a decision is to be made, the decision maker exercises (or does not exercise) the flexibility under consideration, and appropriate if-then rules are encoded at each node to guide the decision pathway. The limitations of the binomial lattice technique include computational cumbersomeness because a large number of steps are required to achieve an acceptable level of accuracy. However, this limitation is relatively inconsequential in the decision contexts of real options compared to those for financial options in trading environments (Gilbert, 2005). It is often assumed that the different pathways are independent of each other, which has the desirable effect of greatly reducing the number of candidate pathways and, consequently, the computational complexity and the time associated with the solution search. The solution can be found, and the value of flexibility thereby assessed, through optimization using dynamic programming tools. The disadvantages of the decision tree methods include their unduly restrictive assumptions; for example, the assumptions that there is a market for trading the system or project in question and that the decision pathways are independent—assumptions that may not always hold in the case of civil engineering systems. Also, decision tree methods have generally been found to be inadequate for addressing multiple sources of uncertainty (Cardin and de Neufville, 2008; Buurman et al., 2009).

15.3.4 Potential Benefits of Incorporating Flexibility in Civil Systems Management

Civil engineering systems, unlike assets in the capricious world of finance, are relatively more stable from the perspective of their existence. However, the socioeconomic conditions that influence the economic attractiveness of investments that help provide, improve, or operate civil engineering systems are similar to those that affect the financial world. As such, a need exists to make civil engineering investment decisions that duly accounts for the changing nature of these conditions.

In providing the flexibility for the system owner to reduce any downside risk or to increase the benefits from any upside opportunities in response to changing environmental, financial, and socioeconomic conditions, ROA analysis can be useful for system investments that are long term in nature, involve millions or billions of taxpayer dollars, and are likely to encounter significant levels of risk associated with the technological, market and financial environments. Within the ROA framework, the decision maker is given the flexibility to decide whether to defer, increase, or decrease the resource inputs for the system construction or operations, for example, as future changes in conditions evolve. The advantage of the option of deferring the investment is that an opportunity is created for acquiring, interpreting, and incorporating new information into the decision process. It is useful for the strategic decision maker to acquire such flexibility because it provides a more strategic and long-term vision of the system's management. Also, ROA avoids automatic nonconsideration of investment scenarios that are deemed economically inefficient at the time of the analysis. Further, unlike the discounted cash flow approach, ROA provides a numerical value to flexibility in the decision making. Cardin and de Neufville (2008) presented case studies in various industries, including real estate, mining, aerospace, manufacturing, energy production, and hydroelectricity where flexibility in design can significantly improve value.

15.4 REAL OPTIONS CASE EXAMPLES IN CIVIL ENGINEERING

15.4.1 The Parking Garage

The parking garage problem (Arman, 2014; de Neufville et al., 2006; Zhao and Tseng, 2003) is a classic illustration of real options application. In this example, the benefits of providing strength flexibility in the facility construction are demonstrated. For example, current demand may require a three-level garage; however, it is projected that in, say, 10 years, a four-level garage would be needed. The traditional "conservative" strategy is to construct a three-level garage at the current time, and then at the future year when demand increases, demolish the three-level garage and construct a new four-level garage. The optimistic strategy is to construct a four-level garage at the current time. The flexible strategy is to construct a three-level garage with extra-strength foundations and columns so that an additional level can be added to the existing three-level garage.

The conservative strategy is best only where demand does not increase, thus the cost of providing any extra capacity or strength to hedge against future demand increases is avoided. It is rather expensive to demolish and reconstruct any structure of this kind, and such an action would be indicative of unreliable prediction of demand and poor planning, resulting in a waste of funds belonging to the system owner (who, typically, is the taxpayer), and ultimately, public relations problems for the system operator. Thus, in any scenario where demand does increase, the limitations of the conservative strategy become all too obvious. Also, the conservative strategy is inferior where investments are irreversible, as is the case in this illustration.

The optimistic strategy, consistent with Figure 15.11*a*, results in little or no costs of delay, congestion, or inconvenience to users and easily accommodates any changes in demand without the need for additional investment toward capacity and strength; however, its limitation is that it constitutes an overdesign for most of its life, and the system owner bears the cost of capacity underutilization within that period.

The flexible strategy, in this case, makes it possible for the system operator to carry out stepwise provision of supply and enables accommodation of any increases in user demand. There is a slightly higher cost to the flexibility strategy compared to the conservative strategy because additional strength is provided in the structural members; in this case, the *strike price* is the cost of constructing and maintaining an additional floor, the *stock price* is the cost saving associated with the avoidance of providing an additional floor to serve the extra demand in the future, and the *option price* is the cost of providing additional strength (wider and stronger columns and foundation structural members).

15.4.2 Road Width at Embankment and Cut Sections

In the area of highway planning and design, there are numerous opportunities for the application of real options analysis. From the perspective of right-of-way acquisition, a highway agency may purchase additional right-of-way beyond what is currently required in order to possess the flexibility to expand the highway lanes to serve an increased demand or to sell the extra right-of-way to real estate developers if the demand falls short of expectations (Arman, 2014). Also, at hilltops or low-lying areas on road sections where there is a need to excavate or fill, respectively, the road fill

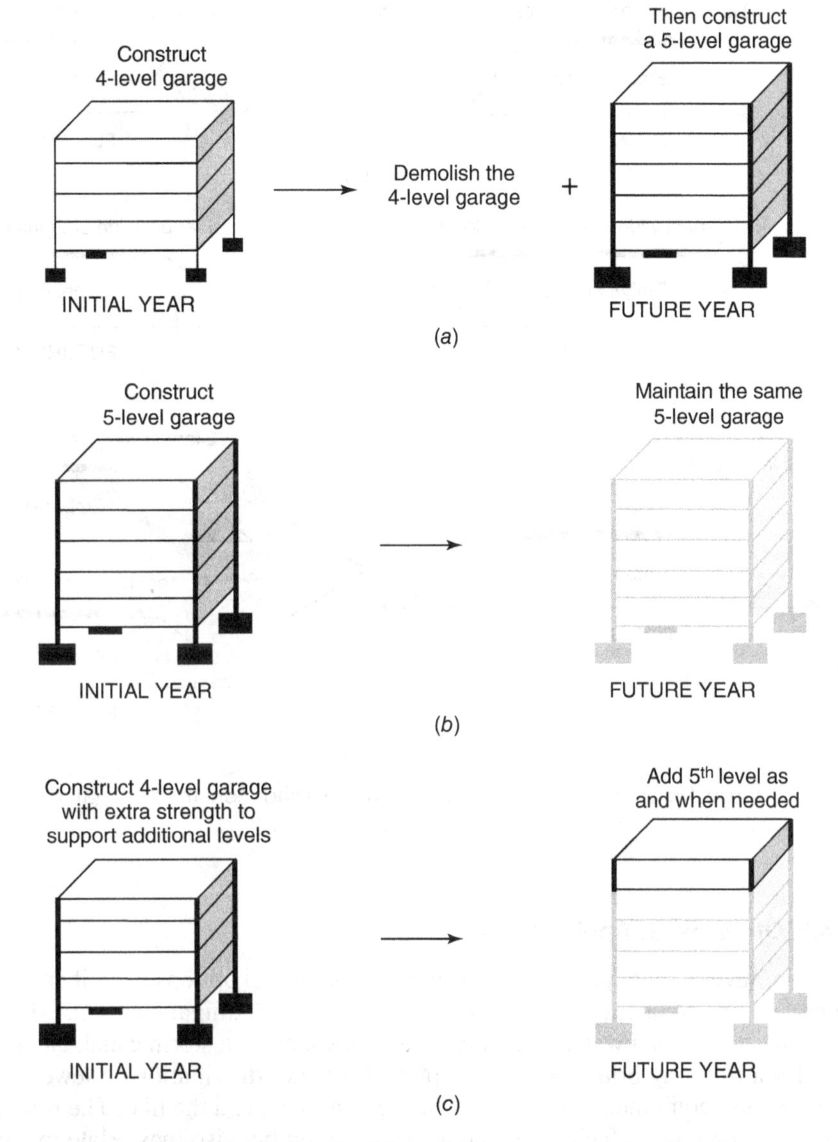

Figure 15.11 Example of real options in parking garage construction: (*a*) conservative strategy, (*b*) optimistic strategy, and (c) flexible strategy.

section or cut section may be constructed wider than initially needed in order to accommodate future additional lanes when demand increases (Figure 15.12). Doing so will remove the need for future roadwork to widen the embankment or cut sections, which avoids not only the future agency costs of doing such work but also the road closures or detours, safety risks, and other user costs incurred during road downtime. In this case, the strike price is the cost of acquiring the additional right of way or the cost of constructing and maintaining the additional embankment width or additional lanes. The stock price is the cost avoided by not having to purchase additional right-of-way or to construct a wider cut section or embankment due to the extra demand. The option price is the cost of constructing a wider embankment.

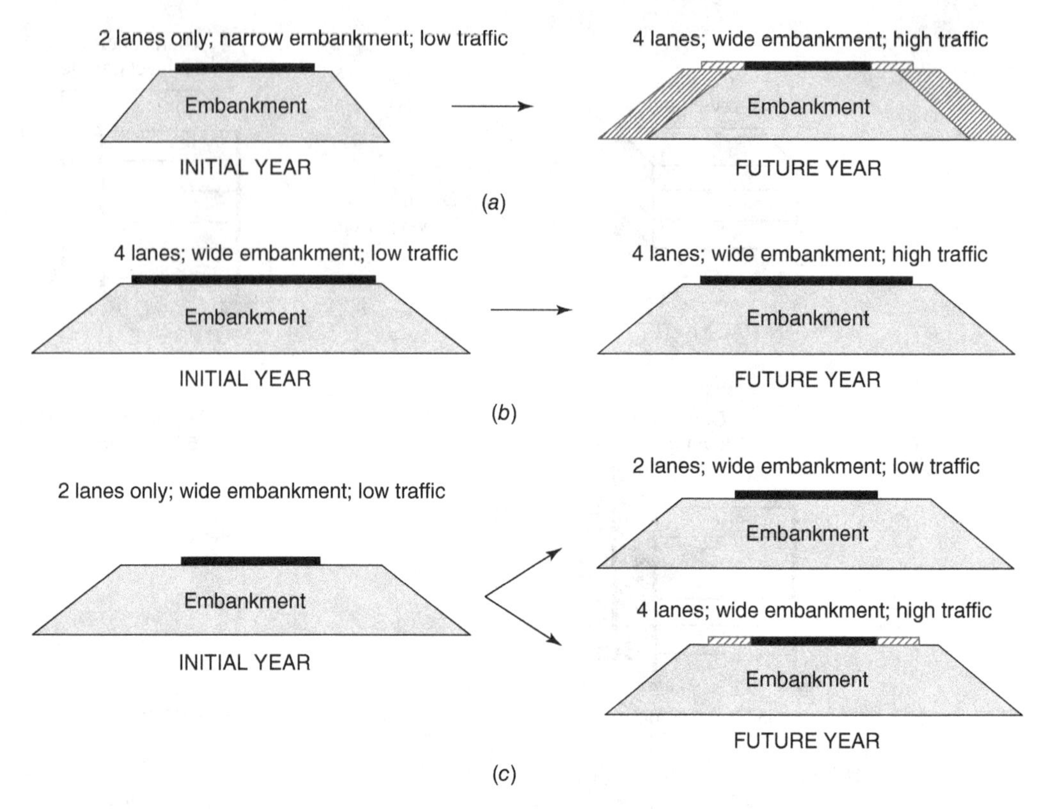

Figure 15.12 Example of real options in road construction: (*a*) traditional option, (*b*) optimistic option, and (*c*) flexible option.

15.4.3 Other Asset Applications

There are several other possible application areas for real options in civil engineering systems from the perspective of facility capacity or strength, or both (Athigakunagorn, 2014). Examples include sizing of an overhead water reservoir for a city, sizing of a storm canal, calculating the number of lanes for a highway bridge, sizing the pipes for water distribution or sewer collection, estimating a contractors' equipment purchase for future contracts, and the like. The problem context may not only be with regard to facility design or construction but also may relate to facility operations and other phases of system development. For example, toll collection booths and toll lanes in a highway road or bridge toll plaza will require flexibility in capacity due to fluctuating traffic demand by the time of day. The flexibility of constructing extra toll booths and toll lanes is indicative of the concept of real options in physical infrastructure while the flexibility to open additional toll booths to serve demand at peak periods is indicative of real options applications in infrastructure operations.

15.5 NUMERICAL EXAMPLE

An agency is considering a new toll road project but is not certain about the level of traffic demand. The construction cost of this project is $125M and the benefit is forecast to be $36M annually for a high-demand scenario, and $24M for a low-demand scenario (assume equal probability of both

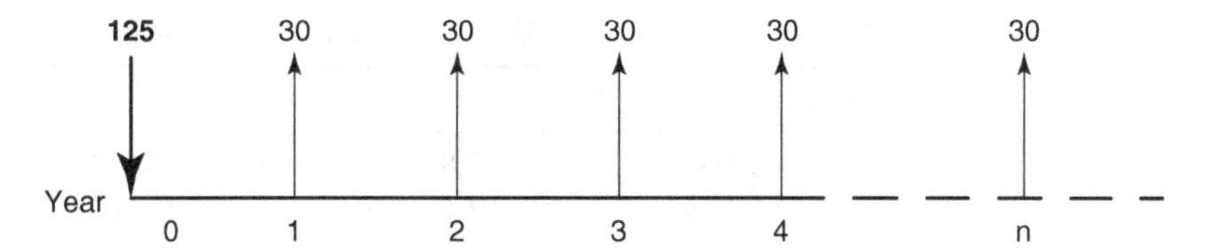

Figure 15.13 Static cash flow of the toll road project.

outcomes). Further assume that the agency's rate of return is 25%. Determine if the project should be implemented using (a) the traditional NPV aproach (b) the real options approach.

(a) Analysis Based on Static NPV. From the static NPV perspective, the average annual benefit ($ 30M) is used to represent both scenarios. The cash flow of this toll road project is illustrated in Figure 15.13.

Hence, the NPV of the project is (in millions)

$$\mathrm{NPV} = -125 + \frac{30}{0.25} = -125 + 120 = -\$5\mathrm{M}$$

Thus, from the perspective of static NPV, the agency should reject this project because the NPV is negative.

(b) Real Options Approach That Considers an "Expanded" NPV. Assume that the agency has the flexibility to postpone the project for 2 years within which more information can be obtained about the traffic demand. This is referred to as the option to defer or the flexibility to defer. This flexibility can be very valuable, particularly if the agency can avoid investing the funds where demand fails to increase to expected levels. To determine the value of such flexibility, the present value of the project ($120M at $t = 0$) is multiplied by an **upstage factor** ($u = 36/30 = 1.2$) and a **downstage factor** (d), which is the reciprocal of $u(d = 1/u = 0.83)$ to determine the project's present value at each stage (Table 15.2).

For example, if the demand increases for 2 years, the present value = 120 × 1.2 × 1.2 = $172.8 million. Then, the agency will adopt the rational position of implementing the project if the value of the project exceeds the cost of the project (i.e., the benefit of the project is positive). Consequently, at the high-demand scenario, the agency will start the project since the net benefit is positive ($172.8M − $125M = $47.8M).

On the other hand, for the low-demand case, the agency will decide not to start the project because the benefit is less than the project's investment cost ($77M < $125M).

To calculate the value of the option to defer, a **backward induction** technique is applied to determine the net value at years 1 and 0. Those two net values will be multiplied by probability

Table 15.2 Net Present Values of Project due to the Traffic Demand Uncertainty (in Millions)

Year	0	1	2
	120	144	172.8
		100	120
			76.92

Table 15.3 Net Value of Option to Defer (in $Millions)

Year	0	1	2	
	11.05	22.98	47.8	← max (0, 172.8 − 125)
		0	0	← max (0, 120 − 125)
			0	← max (0, 76.92 − 125)

(referred to as the **risk-neutral probability**) and discounted back by the risk-free rate. In real options analysis, the risk-neutral probability is used instead of the risk-adjusted probability (as used in decision tree analysis); this is because the risks have been considered in ROA as the demand fluctuations are shown in the various scenarios (as depicted in Table 15.2). Thus, if risk-adjusted probability is used on ROA, that would mean a double counting of the risks and would therefore yield an inaccurate estimate the project value. For the sake of simplicity, assume that (i) the risk-neutral probability in this case is equal to the risk-adjusted discount rate, which is 0.5, and (ii) the risk-free rate is 4%. Therefore, the net value of option when the demand is high at year 1 is

$$\text{Net value of option} = \frac{47.8 \times 0.5 + 0.5}{1.04} = \$22.98\text{M}$$

This process will be iterated back until the net values of each node are obtained (see Table 15.3). At year 0, it can be seen that the value of the option is $11.05M. The net benefit of the project with the option at year 0 is termed the **Expanded NPV** or **NPV+**. This can be calculated as follows:

$$\text{Option value} = \text{Expanded NPV} - \text{Static NPV}$$

$$\text{Expanded NPV} = 11.05 \ - (-5) = \$16.05\text{M}$$

Thus, from the perspective of real options, the project should be implemented because the expanded NPV is positive. It can be observed that by duly considering flexibility and uncertainty, additional value (which was not redized in the traditional NPV approach), can be made visible. It is this additional value that makes the implekentation option a feasible one.

SUMMARY

The real options approach is touted as an improvement over the traditional deterministic discounted cash flow or NPV approach. As Gilbert (2005) pointed out, ROA does come at a cost: It is more detailed and complex compared to DCF, and decision makers who use ROA will expend a greater level of time and effort to reach the problem solution. Nevertheless, the benefits of ROA analysis are not outweighed by the extra effort associated with it (Damodaran, 2005). First, it equips the system owner with the capability to respond to uncertainty as it unfolds, and it fosters a management approach that extends over the entire system life rather than only when the initial decision is made. Also, recognizing that "too much flexibility is too much of a good thing," ROA enables the system owner to identify the optimal level of flexibility to incorporate in the decision process.

EXERCISES

1. Your university campus plans to build a new multistory parking garage to serve the growing needs of students, staff, and visitors. For scenarios characterized by (a) deterministic conditions and (b) uncertain conditions, discuss the strategies for this investment, and explain the conditions under which each strategy would be appropriate. Hint: for the uncertain conditions, the strategies are optimistic, conservative, and flexible.

2. Discuss how considerations in real options could enhance the assessment of system need at the first phase of the system development cycle.
3. "The concept of real options is applicable at only one phase of civil engineering system development – the needs assessment phase." Do you agree with this statement? Argue why or why not.
4. Using an example from everyday life, provide a cogent argument why the real options approach is superior to the probabilistic NPV approach for making decisions in environments characterized by uncertain levels of the decision parameters.
5. Using examples in the financial world, civil engineering, or everyday life, explain the meaning of the following terms: call option, put option, option valuation, option premium or price, stock price, and downside risk.
6. List the four categories of real options applications in civil engineering systems and discuss the specific options that the system owner or operator may face in each category.
7. Explain, with figures, how you would develop a probability distribution for the options premium in a real options problem. How would you interpret this figure?
8. Discuss the benefits and limitations of using real options concepts is the management of civil engineering systems.
9. Consider a project to construct a civil engineering system that has an underlying value of $200M where there system owner has the flexibility to defer the system construction by 0 (now), 1, 2 or 3 years from now. Assume u = 1.17 and d = 1/u. (i) Determine the earliest year to implement the project if the system owner requires that the minimum underlying value of the project should be $300M before the project is carried out (i.e., the system is constructed)? (ii) In what year should the system owner abandon this project if it is specified that the project should be abandoned if its value goes below $150M? (iii) Assume that this decision situation is consistent with the European style options where the system owner can initiate the project only the end of the deferral period (that is, at the end of Year 3), determine the expanded NPV for this project at Year 0, given that the risk-free rate is 4%, the risk-neutral probability (p) is 0.45, and the construction cost of the project is $275M.

REFERENCES

Amram, M., and Kulatilaka, N. (1998). *Real Options: Managing Strategic Investment in an Uncertain World*. Oxford University Press, Oxford, UK.

Arman, M. R. H. (2014). Managing Risks in Highway Winter Operations using Options Theory, Ph. D. Dissertation, Purdue University.

Athigakunagorn, N. (2014). Real Options Applications in Transportation. Ph.D. Dissertation, Purdue University.

Bartolomei, J., Hastings, D., de Neufville, R., and Rhodes, D. (2006). Using a coupled-Design Structure Matrix Framework to Screen for Real Options "in" an Engineering System. Paper presented at INCOSE 2006, 10–14 July, Orlando, FL.

Bhandari, A. S., and Sinha, K. C. (1979). Optimal Timing for Pa*ving Low-Volume Gravel Roads. Transport. Res. Rec.* 702, 83–87.

Billington, C., Johnson, B., and Triantis, A. (2002). A Real Options Perspective on Supply Chain Management in High Technology. *J. Appl. Corporate Finance* 15(2), 32–43.

Black, F., and Scholes, M. (1973). The Pricing of Options and Corporate Liabilities. *J. Polit. Econ.* 81(3), 637–654.

Brach, M. A. (2003). *Real Options in Practice*. Wiley, Hoboken, NJ.

Buurman, J., Zhang, S., and Babovic, V. (2009). Reducing Risk through Real Options in Systems Design: The Case of Architecting a Maritime Domain Protection System. *Risk Analysis* 29(3), 366–379.

Cardin, M., and de Neufville, R. (2008). A Survey of State-of-the-Art Methodologies and a Framework for Identifying and Valuing Flexible Design Opportunities in Engineering Systems. Working paper, http://ardent.mit.edu/real_options/Real_opts_papers/WP-CardindeNeufville2008.pdf. Accessed 02-02-12.

Cardin, M., de Neufville, R., and Kazakidis, V. (2008). A Process to Improve Expected Value in Mining Operations. *Mining Tech.* 117(2), 65–70.

Damodaran, A. (2005). The Promise and Peril of Real Options, Working Paper No. S-DRP-05-02, Stern School of Business, NYU.

de Neufville, R. (2000). Dynamic Strategic Planning for Technology Policy. *Int. J. Tech. Manage.* 19(3/4/5), 225–245.

de Neufville, R., Scholtes, S., and Wang, T. (2006). Real Options by Spread Sheet, Parking Garage Case Example. *ASCE J. Infrastruct. Syst.* 12(2), 107–111.

Dixit, A. K., and Pindyck, R. S. (1994). *Investment under Uncertainty*. Princeton University Press, Princeton NJ.

Gilbert, E. (2005). Investment Basics XLIX. An Introduction to Real options, *Invest. Anal. J.* 60, 49–52.

Kulatilaka, N., and Marcus, A. J. (1992). Project Valuation under Uncertainty: When Does DCF Fail? *J. Appl. Corporate Finance* 5(3), 92–100.

Miller, B., and Clarke, J. (2004). Investment under Uncertainty in Air Transportation: A Real Options Perspective. Presented at the 45th Annual Transportation Research Forum, Evanston, IL.

Nembhard, H. B., and Aktan, M. (2009). *Real Options in Engineering Design*, Operations and management, CRC Press, Boca Raton, FL.

Nembhard, H. B., Shi, L., and Aktan, M. (2005). A Real Options-Based Analysis for Supply Chain Decisions. *IIE Trans.*, 37, 945–956.

Trigeorgis, L. (1996a). *Real Options in Capital Investment: Models, Strategies, and Applications*. Praeger.

Trigeorgis, L. (1996b). *Real Options: Managerial Flexibility and Strategy in Resource Allocation*. MIT Press, Cambridge, MA.

Wang, T. (2005). Real Options 'In' Projects and Systems Design—Identification of Options and Solutions for Path Dependency. Ph.D. Dissertation, Massachusetts Institute of Technology, Cambridge, MA.

Zhao, T., and Tseng, C. (2003). Valuing Flexibility in Infrastructure Expansion. *ASCE J. Infrastructure Systems*, 9(3), 89–97.

Zhao, T., Sundararajan, S., and Tseng, C. (2004). Highway development decision-making under uncertainty: a real options approach. *ASCE Journal of Infrastructure Systems*, 10(1), 23–32.

USEFUL RESOURCES

Amram, M., and Kulatilaka, N. (1999). *Real Options: Managing Strategic Investment in an Uncertain World*. Harvard Business School Press, Boston.

Copeland, T., and Antikarov, V. (2001). *Real Options: A Practitioner's Guide*, Texere, New York.

Damodaran, A. *ROV Spreadsheet Models*. Stern School of Business, New York.

Dixit, A., and Pindyck, R. (1994). *Investment under Uncertainty*. Princeton University Press, Princeton, NJ.

Moore, W. T. (2001). *Real Options and Option—Embedded Securities*. Wiley, New York.

Müller, J. (2000). *Real Option Valuation in Service Industries*. Deutscher Universitäts, Wiesbaden.

Smit, T. J., and Trigeorgis, L. (2004). *Strategic Investment: Real Options and Games*, Princeton University Press, Princeton, NJ.

Trigeorgis, L. (1996). *Real Options: Managerial Flexibility and Strategy in Resource Allocation*, MIT Press, Cambridge, MA.

CHAPTER 16

DECISION ANALYSIS

16.0 INTRODUCTION

Every minute of our daily lives, from awakening until bedtime, is marked with situations where we need to make a decision: whether or not to go out to dinner, how much time to spend studying, whether to lease a car or buy it, how much to donate at a charity event, what size apartment is affordable to rent, whether or not to get married, which cell phone apps to use, or whether to send that urgent message to a friend via text messaging, Twitter, or email. The list is endless. Some decisions are made instinctively, others consciously. Some decisions involve discrete decision variables (because they address the question "which ... "); others involve continuous decision variables (because they address the question "how much ... "). In any case, a decision must be made for life to go on. As Maimonides, the ancient Spanish philosopher, stated: "The risk of a wrong decision is preferable to the terror of indecision." Some decisions yield outcomes that have little impact while others have significant and far reaching consequences. All this is true not only in our everyday lives but also at each phase of the system development process where engineers routinely evaluate alternative courses of action and identify the best alternative for carrying out some task.

In this chapter, we will begin with an examination of some decision-making contexts at each phase of system development and then discuss the basic process of decision making and some relevant tools. The chapter will then present the different ways by which a decision model could be described. One of these classifications, the degree of certainty associated with the decision problem, will serve as a basis for our discussions throughout the chapter; specifically, this chapter will discuss decision making under the two uncertainty conditions (risk and uncertainty, as discussed in Chapter 13) and will show how payoff or regret criteria can be used to determine the best decision. Other ways of modeling discrete decisions will be discussed and the impact of decision-makers' risk preferences on the decision-making process will be illustrated.

16.1 GENERAL CONTEXTS OF DECISION MAKING

There are several contexts of decision making in the development of civil engineering systems, and it is best to discuss these from the perspective of the different phases of systems development. Figure 16.1 presents some decision-making contexts at the various phases, some of which are the same across the phases; for example, at the phases of operations, monitoring, or maintenance, the decision maker may seek to choose the best of multiple alternative policies or specific actions for system operations, monitoring, or maintenance, or to ascertain the feasibility of carrying out some specific action. More specific examples include the following: at the planning phase, the engineer may seek to make decisions regarding the best plan, location, or policy for a system; at the design phase, the engineer makes decisions regarding system orientation, configurations, dimensions, and material types; at the preconstruction phase, the agency decides on the best contracting

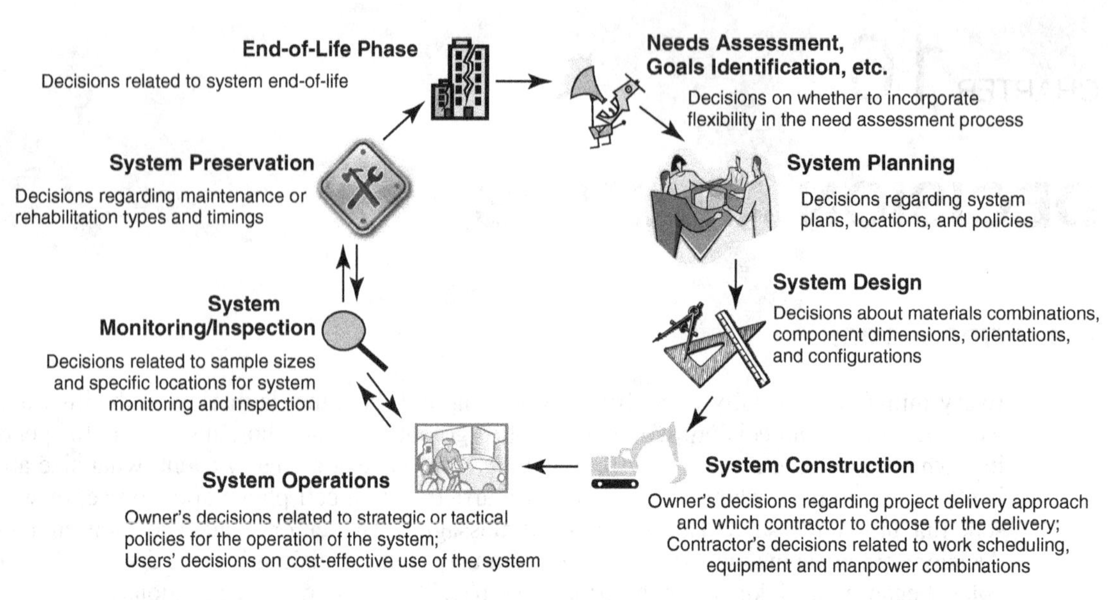

Figure 16.1 A few decision-making contexts at the various phases of system development.

approach, such as traditional design–bid–build or design–build–operate–maintain; at the construction phase, the contractor makes decisions regarding the utilization and scheduling of specific types and quantities of material supplies, equipment, and labor to achieve specific tasks; at the operations phase, system users make decisions on how best to use the system to maximize their comfort, safety, convenience or other benefit, or to minimize any delay, out-of-pocket fees, or some other costs; at the preservation phase, during times of system rehabilitation and maintenance, users typically suffer the inconveniences of system downtime and work zones and make decisions routinely to minimize their inconvenience.

The individual users of a system are responsible for any decisions they personally make regarding how they use the system so, theoretically, they can afford to make decisions that may turn out to be inferior. The owners of a system, on the other hand, cannot afford to make incorrect decisions because it is often the case that they are the caretakers of a public taxpayer-funded system. As such, they shoulder a fiduciary responsibility to weigh all options carefully before making a decision.

16.2 BASIC PROCESS OF DECISION MAKING AND RELEVANT TOOLS

In many cases, the overall framework for decision making in systems analysis follows a certain sequence (Monahan, 2000): Establish and classify the decision criteria in order of importance; establish the alternative actions; evaluate each alternative on the basis of each of the decision criteria; select, as the tentative best alternative, that which maximizes, as much as possible, the decision criteria; implement, using simulation or field tests, the best alternative (and possibly, a few other good alternatives) and evaluate them for other unforeseen consequences; implement the best decision; monitor the ex poste effects of the implemented decision and undertake measures to prevent any adverse consequences. As you can infer, good decision making should be systemic (holistic rather than piecemeal) as well as systematic (step by step rather than random).

Au et al. (1971) described quantitative decision making as a process where the decision maker uses abstraction to transform the problem conditions into a conceptual or mathematical model that encompasses the functional relationship of pertinent variables. This definition is valid even today. Risk analysis, optimization, and statistical analysis are a few of the tools useful for evaluating alternatives and, ultimately, for making decisions. Some of these tools are also used in problem analysis, but there is a clear difference between problem analysis and decision making. Problem analysis is a precursor to decision making: first, the problem is analyzed, and the results of the analyses then serve as inputs to decision making. As Robert H. Schuller advised, "Never bring the problem-solving stage into the decision-making stage," otherwise, you surrender yourself to the problem rather than the solution (Kepner and Tregoe, 1965).

Decision analysis, a tool used at each phase of system development, incorporates all the other. Thus, a broad array of tools are available to the engineer in the task of making a decision at each phase. The engineer uses the concepts of probability and statistics and the related tools of modeling, simulation, an and risk/reliability analysis to describe or predict some attribute or to characterize the variability in the inputs for decision making, and to identify the best decision under given constraints, engineers use optimization tools. The tools of cost analysis, engineering economic analysis, and multiple criteria analysis are used to generate monetary and nonmonetary outcome data for purposes of decision making. To analyze the feedback relationships between a large number of interacting agents, systems dynamics is used; and to introduce flexibility into the decision-making process the tool of real options is used.

16.3 CLASSIFICATION OF DECISION MODELS

A decision model is a relationship, often mathematical, between the conditions associated with a system and its environment and the appropriate course of action on the basis of the conditions; decision models help the decision maker make better-informed decisions. In Section 4.1.2 of Chapter 4, we learned about the several ways a model could be classified. There are many kinds of decision models and any one of them could be described in the ways listed and described in that section. To provide a clearer background for the remainder of our discussion on decision models in this chapter, we will revisit these classifications and discuss them in the specific context of decision analysis.

16.3.1 Phase of Development

As shown in Figure 16.1 and Section 16.1, decisions are made (and therefore decision models are needed) at each phase of system development (e.g., construction-related decision models and operations-related decision models). At each phase, decisions are made by the relevant stakeholder(s) pertaining to that particular phase; however, the decisions made in each phase could also influence the decisions yet to be made at subsequent phases.

16.3.2 Purpose of the Decision Model

Many decision models are applied in a *prescriptive* context because they serve as decision support for the decision maker in choosing a specific course of action. Prescriptive models may be developed using statistical techniques that show, for a past similar situation, what decision was made on the basis of the prevailing conditions at the time; an important caveat, however, is that the past practices may not have been optimal. In other cases, the decision model is *diagnostic*, that is, it provides decision makers with a clear understanding of how the system attributes and operating or and environmental conditions (i.e., the explanatory variables) influenced the past decisions that were made (the dependent variable). In diagnostic modeling of decisions, statistical techniques may be used to explain the sensitivities of the decision with respect to the factors representing the prevailing conditions; also, optimization techniques may be used to develop Pareto frontiers that help quantify the trade-offs between the decision factors.

16.3.3 Stakeholder That Makes the Decision

Models that are used by the system owner to enhance a decision-making task at any phase of systems development are termed *agency decision models*. At the operations phase, users routinely make decisions on how best to use the system to gain the maximum benefit or to incur the lowest possible cost to them individually, and these behaviors are often described or predicted by *user decision models* of a disaggregate nature (here, disaggregate means that the unit of observation is an individual user). For certain kinds of civil systems, user decision models are important to the system owner or agency because they can help the owner to predict or control demand, which in turn influences the level of usage or loading and, ultimately, the deterioration of the physical or operational performance of the system.

16.3.4 Type of Decision Variable

Any decision is related to an occurrence that can be measured in values that could be discrete (a binary number, e.g., do/do not do, or a count) or continuous (see Figure 7.3 in Chapter 7). For example, will you go to the gym tomorrow? How many times will you go to the gym next week? In both these examples, the decision variable is discrete. However, with regard to the question of how much time you will spend at the gym tomorrow, the variable in which the decision is made is continuous. To model a decision whose variable is continuous, regression techniques are often used (see Chapter 7) and to model decisions whose variables are discrete, econometric models including the logit or probit specification (Hensher et al., 2005) is used.

16.3.5 Nature of Decision Attributes

Decisions are made on the basis of the prevailing or expected conditions related to the system, its operations and its environment that are often referred to as decision factor or attributes. In a decision model, these conditions may be *separate*; for example, in the decision to replace an aging levee system, the "conditions" may be the physical state of the levees, the cost involved with their reconstruction, the expected damage to the ecology during reconstruction, the likelihood of major storms in the near future, and the value of the property located in the flood-prone zone. In other cases, the conditions are *combined* as a single composite decision factor, such as their collective utility, cost-effectiveness, risk, and so forth.

16.3.6 Degree of Certainty Associated with Outcome

As we learned in Chapter 4, models may be categorized by the degree of uncertainty associated with the data they utilize (i.e., deterministic or probabilistic). In the specific case of decision models, there could be models characterized by certainty, risk, or uncertainty. In fact, in most systems engineering texts, this is the classification used to categorize decision models and the techniques used to solve decision problems (de Neufville and Stafford, 1971; Au et al., 1971; Dandy et al., 2008; Revelle et al., 2003). Recognizing that this continues to be an elegant way of categorizing decision models, we continue our discussion of decision-making models on the basis of this classification.

16.4 DECISION MAKING ASSOCIATED WITH UNCERTAINTY

In many cases, it is not possible to make a fully informed decision due to the paucity of information. In the current information age, tremendous efforts have been made to address information gaps and to enhance information delivery so that decisions can be made with maximum possible certainty. However, there are and will continue to be many situations where decisions must to be made without full certainty of the inputs and therefore, of the outcomes. On the basis of certainty regarding the decision inputs, decision-making situations can be categorized in the following ways.

16.4.1 Decision Making under Certainty

This is a situation where the decision maker has all the information needed to make a well-informed choice. It is generally much easier to make decisions where the decision outcomes are known with certainty; however, complete possession of the relevant data or information does not guarantee that the decision maker will be able to make the best decision. This category of decision making can be considered only utopian, because in the real world, nothing is really known with certainty: Inputs for decision making, such as materials, weather, climate, economic conditions, and so forth, vary from time to time and from place to place. The variation may be small or large, but it is often enough to translate into variability in the outcomes upon which decisions are based. In traditional engineering calculations, assumptions are typically made that there is negligible variability in material and process characteristics, and outcomes are calculated on the basis of this assumption. Often the true outcome is a close approximation of the outcomes from computation. In certain cases, however, there is so much variability in the inputs (and hence the outputs) that special techniques are necessary to enhance confidence in the final decision, as we will discuss in the next paragraph.

16.4.2 Decision Making under Risk (Uncertainty with Known Probabilities of Outcomes)

In these cases, the decision maker is faced with several choices, and the outcomes of each choice have a certain known probability of being realized. The degree of risk involved with making a decision is related to how much is known about those probabilities. Thus, "risk" is the situation wherein objective data exist and thus are used as a basis for estimating the probability of each outcome of an alternative decision. It is often the case that these "hard" probabilities are based on empirical (observed historical) or experimental tests to describe the outcomes of each decision. Alternatively, "soft" probabilities may be assigned to each possible outcome on the basis of the subjective opinion of experts familiar with the system in question and the outcomes of its associated decisions. Examples of civil engineering conditions that involve risk are: (a) the exact subgrade strength for designing an airport runway pavement is not known due to the spatial variability of the soil at the runway site; (b) the economic superiority of one investment alternative over others is not certain because the interest rate may not remain the same as the rate assumed for the initial economic analysis; and (c) the actual life of a steel water reservoir located near the coast may be lower than expected due to unusually high levels of airborne salt and the uncertainty of maintenance funding. In each of these situations, reasonable probability estimates could be established for each decision factor based on experimental data or past experience on similar projects; that way, the likelihood of the outcomes could be predicted.

16.4.3 Decision Making under Pure Uncertainty (Uncertainty with Unknown Probabilities of Outcomes)

These cases are similar to what is described for the decision making under risk category where there is more than one possible outcome; however, in this category of decision problems, the probability of attaining each outcome is unknown.

16.4.4 Decision Making under Turbulence

From the viewpoint of certainty, this category of decision making is the worst of all: The set of all possible outcomes (possibilities) is unknown, and the probability of each outcome is also unknown.

16.4.5 Discussion

The amount of knowledge or certainty (or lack thereof) surrounding the possible outcomes of a decision is termed "instability." Decision theory is a branch of management science that focuses on the characterization and control of instability in the decision making process. From decision making under certainty to that under turbulence, the degree of instability increases significantly, and the number of variables or decisions that can be accommodated by rational analysis or mathematical solution techniques decreases. A number of probability-based techniques can be used to help manage instability. These include probabilistic risk analysis (a description of the probabilities of possible outcomes associated with random processes that characterize the decision problem); probability trees (descriptions of all the possible pathways or the chain of sequential events or conditions with probabilities assigned to each link in each pathway chain); and decision tree analysis (a description of the outcomes associated with each pathway of sequential decisions over time) (Table 16.1).

Another category related to the above is *decision making under conflict* (Dandy et al., 2008). Such cases, which may be any one of the above situations, may be characterized by the existence of decision makers that may have conflicting objectives. In these decision problems, each decision maker, besides having to contend with possible uncertainty in the problem parameters, also needs to anticipate or to take into account the decision likely to be made by the other decision makers (colleagues or competitors). In the business world, an example would be making the decision that establishes the bid for a certain contract his company seeks; the decision maker in this case would need to consider what his competitors are likely to bid for the same contract. Game theory, a study of strategic, interactive decision making, uses mathematical models to describe decisions made by rational decision makers in an environment characterized by conflict or cooperation. This theory

Table 16.1 Characterization of Decision Problems

Decision making under ...	Outcomes Known?	Level of Outcomes Known?	Probability of Each Level of Outcomes Known?
... certainty	Yes	Yes	Yes
... certainty	Yes	Yes	Yes
... risk	Yes	Yes	No
... uncertainty	Yes	No	No
... turbulence	No	No	No

has helped lead to the development of the concept of expected utility, which has facilitated analysis of decision making under uncertainty.

Strictly speaking, all the categories of decision making could be considered specific instances of the decision-making problem under uncertainty. Problems of certainty are typically easier to model compared to their probabilistic counterparts; thus, in past practice, engineers have tended to represent their uncertainty decision-making problems as ones of certainty (Stark and Nicholls, 1972) and uncertainties are "taken care of" using an appropriate factor of safety.

16.5 INPUTS FOR UNCERTAINTY-BASED DECISION MAKING

For decisions under risk or pure uncertainty, a completely formal statement of the problem requires the following (Au et al., 1971): establishment of all the alternative choices and all the possible outcomes, the likelihood of occurrence of each potential outcome, the unit of value or desirability, a scale that describes the distribution of the value of desirability for different levels of outcomes, and the criteria for the final decision. These are discussed below.

16.5.1 Likelihood of Occurrence of Each Potential Outcome

Likelihood is expressed quantitatively as probability. As explained in Chapter 5, the numerical value of probability can be determined from past observations of the frequency that a particular outcome has occurred under similar conditions (referred to as *statistical probability*). However, where there are inadequate observations to establish statistical probabilities, the decision maker can assign a value of probability to express the degree of (reasonable) belief that the outcome will occur (referred to as *subjective probability*). Then, as actual data from observations become increasingly available, decision makers can update or refine the subjective probabilities.

16.5.2 Scale of Value or Desirability

The relative merits and demerits of each potential outcome of a decision can often be measured on a scale whose units are common across the decision criteria, such as the monetary value or overall utility (satisfaction) of the outcome to the decision maker. Monetary value may be determined on the basis of economic considerations, such as cost and benefit or loss and profit, to one or more stakeholders. Utility, which is also useful in quantifying the relative overall merits of the outcomes of alternatives, helps to describe and incorporate the personal preferences and risk-taking characteristics of the decision maker in the decision-making process. This is the same as the scaling procedure in multiple criteria decision making (Chapter 12).

16.5.3 Criteria for Decision

In many situations of civil systems management, engineers make decisions by selecting the option that yields the least average cost or greatest average benefit over a selected analysis period, such as the facility life cycle (for a new or proposed system) or the remaining life (for an existing system). However, the decision maker may be more interested in other criteria, for example, the variability (uncertainty) of the outcomes. In other words, preference may be given to options that yield moderate benefits with minimal variability compared to those that yield high benefits with a great deal of variability. Ultimately, the selection of a criterion for decision making depends on the extent of involvement of the stakeholders in the decision process and, where there is uncertainty in the outcomes, the extent to which the decision maker is willing to take risks. In extreme cases, there may be situations where it is impossible to quantify some or all of the possible outcomes; and in such cases, the decision maker must make a decision that is not related explicitly to the option outcomes but is considered the most pragmatic decision from a rational viewpoint.

In an ideal situation, every decision problem under conditions of uncertainty should be resolved on the basis of the above considerations. However, the practical reality is that this may not be possible in certain situations; for example, it may be difficult to quantify the likelihood of each outcome due to lack of experiential data form similar conditions; it may be impossible to develop a scale of value of a given outcome cost or benefit; and there may not be an existing policy for determining the criteria to use to compare the options. In many cases such as these, it may be more meaningful to use expert opinion to develop these constructs and then make a decision accordingly. Then, as more objective data become available from the field, these constructs can be refined or updated using the new data.

In the next two sections of this chapter, we will discuss these considerations and how they are used in making decisions under uncertainty and risk situations. To elucidate the use of the relevant analytical tools for such decision situations, we shall first introduce elementary illustrations of decision problems under risk and uncertainty. The basic underlying premise is that the outcome from every "decision" and "state of nature" pair can be described using a matrix (the state of nature determines the outcome of each decision, and the decision maker has no control over the state of nature); and outcomes here are expressed in terms of benefits or costs.

16.6 DECISIONS UNDER RISK

As mentioned in Section 16.4, decision making under risk is where the decision makers are faced with several choices, and they are in full knowledge of the probability of the outcomes of each alternative. This situation is described as decision making under risk because the degree of risk involved with making such a decision is related to how much is known about those probabilities. For civil engineering problems, it is often possible for a decision maker to assign values to the probabilities of the state of nature based on past records and current trends or to perform some type of experimental test to find out more about the actual state of nature. For example, regarding soil strength for a building foundation, information could be collected from soil geology maps, but additional information could be obtained by taking core samples at the site and testing them in a laboratory to supplement information from the maps. However, as Revelle et al. (2003) pointed out, any test has some degree of error associated with it (e.g., samples cannot be taken everywhere and everytime, the number of tests run in the laboratory is typically small, and the tests may be prone to human error). Even after the test is conducted, the state of nature is still not known with absolute certainty. Nevertheless, test results help update the probabilities regarding the true soil strength (state of nature) within certain ranges. The question then becomes: should the decision maker (in this case, the geotechnical engineer) conduct experimental tests to yield potentially better design data, at additional cost, or should she proceed on the design without any additional field tests? To resolve this and other similar uncertainty problems, a useful analytical concept known as a decision tree or table can be helpful.

16.6.1 Decision Trees and Decision Tables

A decision tree is a convenient concept for analyzing staged probabilistic decision problems. The tree begins at an initial time or decision point and each stage proceeds sequentially across the decision points. The various costs or benefits associated with potential decisions along each decision pathway culminates in the payoffs or outcomes (in units that may be dollars, lives saved, utility, or degree of satisfaction or desirability) at the topmost branch at the top of the tree.

The simplest case of decision making under uncertainty is where the decision maker seeks to make a single decision to choose one of two actions knowing the consequences (costs and/or benefits) of each action. We shall illustrate this with a simple example, and then we will go further to examine more complex formulations of the concept that are encountered in real-world decision making.

A certain civil engineer faces the decision to use traditional concrete or self-consolidating concrete (SCC) for constructing a structure in a very harsh climate. Seven out of every 10 times, structures built in mild climates using traditional concrete have a service life of 30 years, and those built in severe climates (which occurs 3 out of every 10 times) have a service life of 20 years. On the other hand, structures built using SCC have a service life of 60 years in mild climates and 40 years in severe climates; this can be represented as Figure 16.2.

As we can see in the figure, the first point in a decision tree is the presentation of the choice (in this illustration, whether to build the facility using traditional concrete or SCC). For each of these two actions, there are consequences (the longevity of the facility), each with a given probability. In the decision tree shown in the figure, the main **stem** is shown as the leftmost box. From this stem, emanate the **branches** (i.e., the options), which end in a circular node. Then, from the nodes emanate the next canopy of branches that show the identified consequences of each option and their respective probabilities. The rightmost part of the decision tree shows the **leaves** of the tree, represented using a rectangle, showing the quantified consequences of each option if it occurs. Then the **flowers** of the decision tree, which occur among the leaves (two, in this illustration), represent the overall consequences of each option.

The overall consequence is often expressed mathematically as the expected value, which is calculated as the sum of the product of the probability and the consequence of each possible path.

For traditional concrete,

$$\text{Overall outcome} = \text{expected value} = 0.7(30 \text{ years}) + 0.3(20 \text{ years}) = 27 \text{ years}$$

For SCC,

$$\text{Overall outcome} = \text{expected value} = 0.6(40 \text{ years}) + 0.4(16 \text{ years}) = 30.4 \text{ years}$$

Thus, it can be considered a better decision to use SCC for the construction.

Let us now consider a number of ways in which the problem illustrated above can become more detailed.

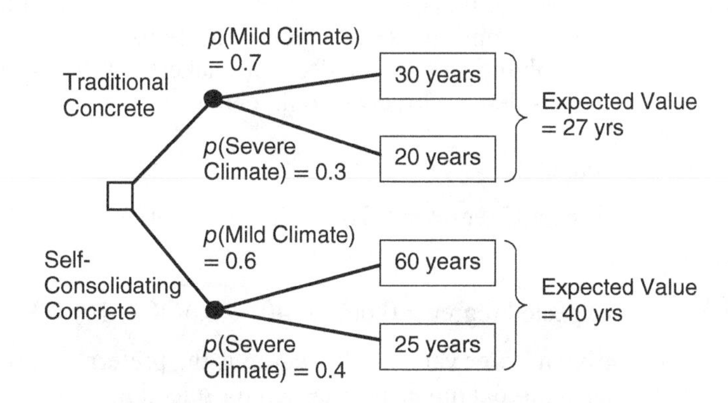

Figure 16.2 Simple decision tree illustration.

Multistage versus Single Stage. In the illustration, the decision problem is single stage because there is only one stage where a decision needs to be made. In more complex situations, there may be multiple branches along the decision tree where decisions need to be made.

Nodal Binary Degree versus Multinary Degree. The degree of a node is the number of branches from the node. In the example given, there are at most two degrees at each node. However, there could be more than two branches, each with its own probability. In an extreme case, there could be an infinite number of branches; in that case, the consequence is a continuous rather than a discrete variable as we discuss below.

Discrete Outcomes versus Continuous Outcomes. In the example provided, all the possible consequences are represented by a binary variable. Thus, the overall consequence, or the expected value (EV) for each action, was given by the summation [Equation (16.1)]:

$$\text{Expected value of action } A_i = \text{EV}(A_i) = \sum_{j=1}^{J} C_j p_j \tag{16.1}$$

where for each action A_i, there are j possible consequences ($j = 1, \ldots, J$) each of magnitude C_j; and probability p_j.

In a more complex situation, the distribution of the consequences is represented by a continuous variable, thus the overall consequence or expected value is given by the integral of the probability function that describes the probability of each potential consequence (tree flower), as shown in Equation (15.2):

$$\text{Expected value of action } i = \text{EV}(A_i) = \int_{x=-\infty}^{+\infty} f(x)\ dx \tag{16.2}$$

where for each action A_i, there are an infinite number of possible consequences represented by the probability density function $f(x)$. In the specific example discussed above, this is actually more realistic than the discrete situation because the observed service life of a system is typically not a set of two or three possible lives but actually is a wide range containing an infinite number of possible values.

Different Criteria for Identifying the Best Option. In the example given, the best option was selected on the basis of the action that gave the maximum expected payoff. In certain cases, the decision maker seeks to choose the option that gives the lowest expected minimum loss and not the maximum benefit as shown in this example. Referred to as the minimum regret criterion), this is the choice of the action that minimizes the regret of the decision makers and is particularly suited for decision contexts involving risk-averse decision makers. In the example given, assuming that a lower service life is considered a loss, we could define "regret" as the shortfall from a target life of, say, 40 years.

For traditional concrete,

$$\text{Expected regret} = 0.7(40 - 30) + 0.3(40 - 20) = 14 \text{ years}$$

For SCC,

$$\text{Expected regret} = 0.6(40 - 40) + 0.4(40 - 16) = 9.6 \text{ years}$$

Thus, SCC, which gives a lower value of the regret, is the preferred option.

As we have seen in this example, the criterion for selecting the best action depends on the risk-taking behavior of the decision makers in terms of their propensity toward a payoff or an aversion to a loss, including maximizing some expected reward (payoff) or minimizing the expected loss. These situations are discussed in the next section.

16.6.2 Criteria for Decision Making under Risk

The criteria for selecting the best option under risk are:

Payoff-related Criteria.

Criterion 1 (Expected Payoff): Choose the alternative that yields the highest expected value of utility. In the example given, the best option was selected on the basis of the action that gave the maximum expected utility (in this illustration, service life).

Criterion 2 (Most Probable Payoff): Choose the alternative that yields the most probable value (i.e., the highest utility in the column with highest probability).

Criterion 3 (Long Shot): Choose the alternative that yields the highest utility in the entire matrix.

Regret-related Criteria.

Criterion 1 (Expected Loss): Choose the alternative that yields the lowest expected loss.

Criterion 2 (Least Probable Loss): Choose the alternative that yields the least likely loss (i.e., the lowest loss in the column with the highest probability).

Criterion 3 (Long Shot Loss): Choose the alternative that yields the lowest loss amount in the entire matrix.

Example 16.1

To satisfy growing economic demand, a port developer is considering three different designs (A, B, or C) for the renovation of a water port of a Pacific island. There is concern, however, that with global warming, the rising seawater levels may pose a threat to the port. Through consultations with climate experts and hydraulic engineers, the project's engineer establishes the following probabilities of the state of nature: Normal (N) conditions is 0.7 and threatening conditions (D) is 0.3, as shown in the payoff matrix in Table 16.2, which lists the net life-cycle benefits over a 100-year period (negative values represent costs mainly due to the rise in seawater level) in millions of dollars. Determine the preferred option for each of the following criteria: expected value, most probable value, and long shot value and expected loss, most probable loss, and long shot loss.

Table 16.2 Data for Example 16.1

	State of Nature (Sea Levels), j		
Alternative i	**Normal $p = 0.4$**	**Threatening $p = 0.5$**	**Very Threatening $p = 0.1$**
Design A	90	50	30
Design B	110	−80	−100
Design C	70	60	40

Solution

(i) For the expected value criterion, we choose the decision that has the highest expected value of utility or lowest expected disutility $E(U_{ij})$. In the example given, the payoffs are a utility and not a disutility because they are net life-cycle benefits. We find the expected value of each criterion as follows:

$$EV_A = (90)(0.4) + (50)(0.5) + (30)(0.1) = 39$$

$$EV_B = (110)(0.4) + (-80)(0.5) + (-100)(0.1) = 34$$

$$EV_C = (70)(0.4) + (60)(0.5) + (40)(0.1) = 32$$

On the basis of this criterion, design A is the best choice.

(ii) For the most probable payoff criterion, we choose the decision that has the most probable value (i.e., the highest utility in the column with the highest probability). The threatening sea levels column has the highest probability ($p = 0.5$), and thus is the most likely state of nature. Of the three benefit values for the designs in that situation, design C has the maximum value and thus is the best choice on the basis of this criterion.

(iii) For the long shot payoff criterion, we choose the decision that has the highest utility in the entire matrix. Of the nine cell entries in the matrix, design B has the maximum benefit ($110 million of net life-cycle benefits) and thus is the best option on the basis of this criterion.

(iv) For the minimum loss criterion, we calculate the expected loss (EL) for each decision outcome. We establish a new table that indicates the decision-maker's opportunity loss. One way of doing this is to add a new row showing the maximum benefits in each state of nature, and subtracting, for each state of nature, the cell entries from the maximum benefit (Table 16.3).

$$EL_A = (20)(0.4) + (10)(0.5) + (10)(0.1) = 9$$

$$EL_B = (0)(0.4) + (140)(0.5) + (140)(0.1) = 14$$

$$EL_C = (40)(0.4) + (0)(0.5) + (0)(0.1) = 16$$

Table 16.3 Solution to Example 16.1

	Loss (Deviation from the Maximum Benefit), at each State of Nature		
Alternative	**Normal Sea Levels, $p = 0.4$**	**Threatening Sea Levels, $p = 0.5$**	**Very Threatening Sea Levels, $p = 0.1$**
Design A	110 – 90 = 20	60 – 50 = 10	40 – 30 = 10
Design B	110 – 110 = 0	60 – –80 = 140	40 – –100 = 140
Design C	110 – 70 = 40	60 – 60 = 0	40 – 40 = 0
Maximum	110	60	40

Using this criterion, the best decision is that which minimizes the likely loss, therefore, the decision maker must choose action A. (In cases where the probabilities of each state of nature are unknown, then what the decision maker seeks to minimize is the regret, rather than the loss, and therefore chooses the alternative that simply gives the least expected regret, least average regret (MINIAVE), or the least maximum regret (MINIMAX) across the alternatives. For these decision situations, we present and discuss appropriate tools in Section 16.7.2 of this chapter.

(v) For the most probable loss criterion, we choose the decision that has the lowest regret amount in the column with the highest probability. The threatening sea levels column has the highest probability ($p = 0.5$), and thus is the most likely state of nature. Of the three loss values for the designs in that situation, design C has the minimum and is therefore the best choice on the basis of this criterion.

(vi) For the long shot loss criterion, we choose the decision that has the lowest loss in the entire matrix. Of the nine cell entries in the matrix, two designs, B and C, have a 0 value of loss and thus are best options on the basis of this criterion. It may be noted that long shot loss often yields ambiguous results as two or more alternatives may have a zero loss value.

16.7 DECISIONS UNDER UNCERTAINTY

As we illustrated using Example 16.1, civil engineers often face decisions characterized by the potential outcomes of the alternative actions. The solutions for these decision problems are aided greatly by the fact that the probabilities of their outcomes are known, often due to analysis of past data on similar actions to similar systems or through expert opinion regarding the likelihood of the outcome of each alternative action. Unfortunately, the engineer does not always have the luxury of knowing these probabilities. There are several instances of decision making in civil systems where a decision needs to be made in the absence of the probabilities of the outcomes of each alternative action.

Notice in Example 16.1 that there was a 70% chance (probability) that the traditional concrete would have a long life. What if this probability was not known? What if the outcome (expected life) for the other action (SCC) also was not known? How then would a decision be made on which concrete type to select? In this section of the chapter, we present, on the basis of two decision criteria (payoff and regret), analytical techniques and criteria that systems engineers can use to make a decision when handicapped by lack of knowledge of the probabilities of the outcomes of alternative actions. These techniques include the optimistic (or maximax payoff) criterion, the pessimistic (or maximin payoff) criterion, the Hurwicz criterion, the LaPlace principle (or equal probabilities), and sensitivity analysis. In Section 16.6.1 where we illustrate these decision-making techniques and criteria, we use the data that are provided in Figure 16.3.

16.7.1 Payoff-based Criteria

(a) Optimistic or Most-Desirable-of-the-Best-Outcomes (MDBO) Criterion. Decision makers who use this criterion first establish the best possible outcome of each action and then identify the best of these and select the corresponding action. Because this criterion is consistent with the maximum of the maximum payout or outcome, it is referred to as the maximax strategy for decision making (Revelle et al., 2003). In making decisions regarding personal matters, humans tend to adopt this strategy for decision making; however, this may not be the best strategy for engineering decision making (Manhart, 2005; Dandy et al., 2008). In Figure 16.3, using this criterion, the decision maker would select design A because it yields the higher of the high longevities, as shown in Table 16.4.

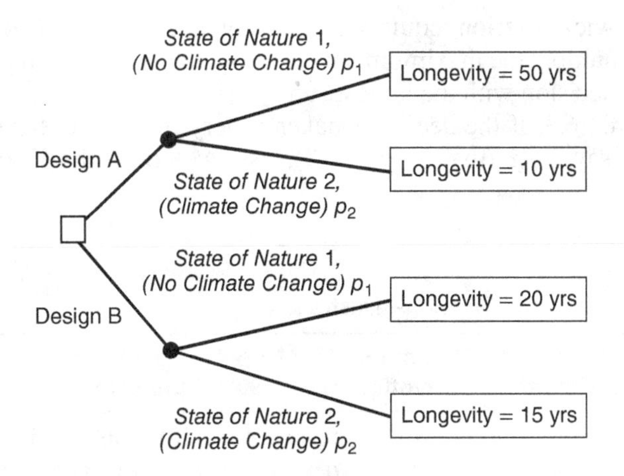

Figure 16.3 Decision tree illustration of decision tools under uncertain conditions.

Table 16.4 The Optimistic Criterion—Illustration

Action Alternatives	Best Value of the Payoffs for each Alternative	Maximum of the Best Values
Design A	50 years	50 years
Design B	20 years	

Table 16.5 The Pessimistic Criterion—Illustration

Action Alternatives	Worst Value of the Payoffs for each Alternative	Least Undesirable of the Worst Payoffs
Design A	10 years	15 years
Design B	15 years	

(b) Pessimistic or Least-Undesirable-of-the-Worst-Outcomes (LUWO) Criterion. Decision makers who use this criterion seek to select the action that minimizes the maximum possible undesired outcome or maximizes the minimum possible desired outcome. Thus, this is often referred to as the least-undesirable-of-worst payoff (LUWO) criterion. In Figure 16.3, the decision maker would select design B because it yields the less undesirable value of the low longevities, as shown in Table 16.5.

(c) The Hurwicz Criterion. It can be seen that the LUWO criterion is extremely conservative and the MDBO criterion is very optimistic. The Hurwicz criterion, which lies is between the two extremes, uses a weighted combination of the best and the worst consequences for each action alternative. The Hurwicz uses a weighted combination of optimism and pessimism and uses a *coefficient of optimism* or *optimism index*, α, and a *coefficient of pessimism* or *pessimism index*, β. The values of α and β depend on the decision-maker's attitude toward the state of nature. The weighted combination for alternative A_i is called the Hurwicz payoff, H_i, for that alternative:

$$H_i = \alpha G_i + \beta L_i$$

where $\beta = 1 - \alpha$; G_i and L_i are the maximum and minimum possible payoffs, respectively, for action alternative A_i.

The Hurwicz criterion requires the decision maker to decide on the value of α. Then, for each decision, we multiply the maximum payoff by α and the minimum payoff by $(1 - \alpha)$ and total these to choose the decision with the largest sum.

In Figure 16.3, if the decision maker is very optimistic, for example, she has an optimism coefficient of say, 90% (that is, $\alpha = 0.9$), then we create the Hurwicz payoff table as shown in Table 16.6.

Table 16.6 The Hurwicz Criterion—Illustration

Action Alternatives	Maximum Payoff	Minimum Payoff	Hurwicz Payoff for 90% Optimism	Hurwicz Payoff for 10% Optimism
Design A	50	10	$0.9(50) + 0.1(10) = 46 \leftarrow$ max	$0.1(50) + 0.9(10) = 14$
Design B	20	15	$0.9(20) + 0.1(15) = 19.5$	$0.1(20) + 0.9(15) = 15.5 \leftarrow$ max

Thus, for an optimistic decision maker, the best action is design A; and for a pessimistic decision maker, the best action is design B.

(d) The LaPlace Principle or Equal Likelihood Criterion. This criterion is similar to the expected value criterion for decisions under risk; it differs in that it assigns equal probabilities to each state of nature. The alternative with the maximum expected value of the payoff is selected as the best action. In Figure 16.3, if we assign equal probabilities to each state of nature, that is, $p_1 = p_2 = p_3$, then the expected values (EV) for each alternative can be found as follows:

$$EV_{\text{DESIGN A}} = (50)(0.5) + (10)(0.5) = 26 \leftarrow \text{max}$$

$$EV_{\text{DESIGN B}} = (20)(0.5) + (15)(0.5) = 17.5$$

The decision maker would choose design A because it has the highest expected value. Notice that the LaPlace criterion is equivalent to the Hurwicz payoff criterion when the coefficient of optimism is 50%.

(e) Sensitivity Analysis. Due to the fact that decisions under uncertainty involve unknown probabilities, the best recourse may be to find the best solution for each of several sets of possible probabilities of the alternative outcomes. For example, the probability that the traditional concrete will have a long life may range anywhere from 0 to 100%. A similar range can be assumed for SCC. A "lazy" way of making the decision is to assume that the probabilities can take any random value within this range, in other words, assuming that the probability follows a uniform distribution. A better way to establish the possible values of this probability would be to assume that the probability that traditional concrete will have a long life follows a certain specific appropriate distribution that has a mean and standard deviation. The concept of using a range of values for this probability is consistent with the theory of fuzzy logic. For example, it may be 60% possible that the probability that traditional concrete will have a long life is 0.7, 15% possible that the probability is 0.6 or 0.8, and 5% possible that the probability is 0.5 or 0.9.

16.7.2 Regret-based Criteria

The decision maker first creates a regret table and chooses the best decision on the basis of the **pessimistic** or **minimax-regret** criterion (also known as the **Savage** criterion), where the action that yields the minimum of the maximum regret is selected. The regret refers to the loss of payoff due to the occurrence of a certain state of nature. It is this regret that the decision maker seeks to minimize. To illustrate this criterion, consider Figure 16.3 and the following scenarios:

Scenario 1: The state of nature 1 occurred (i.e., there was no climate change): In this scenario, if design A had been chosen, the decision maker would have no regrets because she chose the action that yielded the best payoff under this state of nature (i.e., a 50-year life of the system); however, if design B had been chosen, the system longevity would also be 20 years, and the regret would be 50 – 20 = 30 years. Thus, in this scenario, the maximum of the two regrets (30 and 0) is 30.

Scenario 2: The state of nature 2 (climate change) occurred: In this scenario, if design A had been chosen, the decision maker would have regrets because she chose the action that yielded the lower payoff under this state of nature, and the amount of this regret is 5 years (15 – 10) because the highest payoff (15-year service life) would have been earned if the other design had been chosen. Thus, there would have been no regret if design B had been chosen. Thus, in this scenario, the maximum of the two regrets (5 and 0) is 5.

Table 16.7 Scenario Outcomes for Regret-based Criteria

	Action	Possible Payoff	Maximum of the Possible Payoff	Possible Regret	Maximum of the Possible Regrets	Least of the Maximum Regrets
Scenario 1 (state of nature 1) no climate change	Design A	50 years	50 years	0 years	30 years	5 years
	Design B	20 years		30 years		
Scenario 2 (state of nature 2) climate change occurs	Design A	10 years	15 years	5 years	5 years	
	Design B	15 years		0 years		

From these two scenarios, the decision maker identifies which of them have a lower maximum regret and chooses the best action under that scenario. In this problem, scenario 2 is consistent with the lower of the maximum regrets and design A therefore would be chosen as it gives this least amount of the maximum regrets. The results are presented in Table 16.7.

Example 16.2

A developer has $10 million to invest in one of four projects: school buildings (A), water and sewage projects (B), housing (C), or transport projects (D). The amount of profit depends on the market economy and employment situation in the near future. The possible states of nature are poor economy (*P*), normal growth (*N*), and good economic conditions (*G*). Table 16.8 presents the payoff matrix. Determine the optimal action on the basis of the five decision criteria. The cell values indicate the profits.

Table 16.8 Action Consequences under Different States of Nature, Example 16.2

	State of Nature (economy)		
Course of Action	***P***	***N***	***G***
A	−6	10	31
B	−29	21	42
C	0	8	12

Solution

(a) As per the maximin criterion, a decision maker who is pessimistic (or very cautious) about the outlook of a decision, considers the worst-case scenario that might occur and is prepared for it. Such a person chooses C (Table 16.9).

(b) With the maximax criterion, on the other hand, a decision maker who is optimistic about the outcome considers the best of all possible gains and expects the best from nature and therefore chooses B (Table 16.10).

Table 16.9 Computation for the Maximin Criterion

Course of Action	State of Nature (economy) *P*	*N*	*G*	Minimum of Each Row
A	−6	10	31	−6
B	−29	21	42	−29
C	0	8	12	0 ← Choose C

Table 16.10 Computation for the Maximax Criterion

Course of Action	State of Nature (economy) *P*	*N*	*G*	Maximum of Each Row
A	−6	10	31	31
B	−29	21	42	42 ← Choose B
C	0	8	12	12

Table 16.11 Computation for the Hurwicz Criterion

Alternative	Max Gain	Min Gain	Hurwicz Payoff
A	31	−6	$0.6(31) + 0.4(-6) = 16.2$ ← max
B	42	−29	$0.6(42) + 0.4(-29) = 13.6$
C	12	0	$0.6(12) + 0.4(0) = 7.2$

(c) Hurwicz criterion: The following weighted combination for alternative A_i is the Hurwicz payoff, H_i.

$$H_i = \alpha M_i + (1 - \alpha)m_i$$

where M_i and m_i are the maximum and minimum gains for alternative A_i.

If we specify $\alpha = 0.6$, we obtain the results shown in Table 16.11.

The value of α needs to be determined by the decision maker.

If α were specified as 0.90, the payoffs would be

A $(0.9)(31) + (0.1)(-6) = 27.3$

B $(0.9)(42) + (0.1)(-29) = 34.9$ ← max

C $(0.9)(12) + (0.1)(0) = 10.8$

and the choice would be alternative B.

(d) Minimax regret criterion: This criterion proposed by Savage illustrates the use of the concept of opportunity loss resulting from an incorrect decision. The best way to deal with this criterion is to set up a "regret" matrix as shown in Table 16.12.

Notice that if the decision maker had selected B and the state of nature were *G*, she receives the maximum payoff of 42. However, if she had selected C, and the actual state of nature turns out to be *G*, then the *regret*, or the *cost of the mistake*, is $(42 - 12) = 30$. The best way to set up the regret table is to have a row at the bottom of the matrix with the maximum return and then subtract, cell by cell, the cell values from the column maximum return. Then, select the maximum regret in each row and select the minimum regret from this column, as indicated above. Since A is the row with the least regret, we choose alternative A.

Table 16.12 Computation for the Minimax Regret Criterion

	State of Nature (Economy)			Maximum of
Alternative	*P*	*N*	*G*	**Each Row**
A	6	11	11	11 ← Minimax regret
B	29	0	0	29
C	0	13	30	30
Max return	0	21	42	

(e) LaPlace criterion: This criterion is simple to apply. If the three states of nature are equally likely, we assign a probability of $\frac{1}{3}$ to each cell value and find the expected value.

A: $(-6 + 10 + 31)\left(\frac{1}{3}\right) = 11.55 \leftarrow$ Max

B: $(-29 + 21 + 42)\left(\frac{1}{3}\right) = 11.22$

C: $(0 + 8 + 12)\left(\frac{1}{3}\right) = 6.66$

Choose alternative *A* because it has the highest expected value.

Discussion

In navigating through these different criteria in this example problem, we are likely to notice how the specific final decision is strongly influenced by the criterion used for the decision making. The summary of the decision outcomes is as follows:

Maximin – alternative C; Maximax – alternative B;

Hurwicz – alternative A if $\alpha = 0.6$, alternative B if $\alpha = 0.9$;

Minimax regret – alternative A; and LaPlace – alternative A.

16.8 OTHER TECHNIQUES FOR MODELING DECISIONS

16.8.1 Discrete Decisions

The discussions in Sections 16.4–16.6 explained how to make a discrete choice (a decision to select one of several alternatives) based on the payoff and regret outcomes (of the alternatives) that are probabilistic. Unfortunately, the literature refers to these models as payoff or regret models (which reflect the outcome desirability to the decision maker), rather than as one type of discrete choice models (which reflect the ordinal nature of the decision variable or some other mathematical specification of the model). The literature uses the latter term to describe statistical techniques that describe or predict the choices (from a finite set of alternatives) that are made by a decision-making unit consisting of individuals or groups of individuals (e.g., companies, groups, organizations, and agencies). Payoff and regret models are purely prescriptive in their intent and nature, while discrete choice models are essentially descriptive models based on observations and may be used for prescriptive purposes (with caution) and for diagnostic purposes. One of the pioneers in developing discrete choice theory was Daniel McFadden, who was awarded the 2000 Nobel Prize in Economic Sciences for his work.

Discrete choice models provide the mathematical relationship between the choice made by a decision-making unit and the environment within which the choice is made. The decision environment consists of (a) *generic* attributes (i.e., the characteristics of the decision environment that are independent of the alternatives) and (b) *alternative-specific* attributes (i.e., the characteristics of the alternatives from which a choice is to be made). For example, in modeling the choice of whether a user takes the building elevator or the stairs, the generic attributes could be the person's gender, age, and weight building climate, number of floors being travelled; and the alternative-specific attributes could be the reliability of the elevator, the comfort of the elevator, the safety of the staircase, and the height of each step.

Discrete choice models help us carry out diagnostic analysis, that is, to study the factors that influence a choice (or rejection) of any alternative, for example, forecasting how a decision-maker's choices will change due to changes in the levels of the generic attributes or the alternative-specific attributes. Also, discrete choice models help decision makers make future decisions if the attributes can be predicted reliably at a certain future time, and more importantly, if the past observations (that is, decisions) used to develop the model were optimal at the time they were made and also will be optimal at the prediction year.

Sections 16.4–16.7 discussed how to make a discrete choice (a decision to select one of several alternatives) based on payoff and regret outcomes (of the alternatives) that are probabilistic. Literature that include Hensher (2005) and Ben-Akiva and Lehman (1992) discuss how to develop decision models to make a discrete choice based on the utilities of the alternatives.

Other techniques for modeling choices of a discrete nature include discrete optimization (see Section 9.4 of Chapter 9).

16.8.2 Continuous-Variable Decisions

From a general viewpoint, to model decisions involving discrete variables (or "which"), payoff or regret models, discrete-choice models, or discrete optimization can be used; and to model decisions involving continuous variables (or "how much"), regression analysis or continuous-variable optimization are used. For modeling choices of a continuous-variable nature, techniques include regression (see Chapter 7) and continuous-variable optimization (Section 9.3 of Chapter 9).

16.9 IMPACT OF DECISION-MAKER'S RISK PREFERENCES ON DECISION MAKING

When the probabilities of outcomes are unknown, it is useful to express the entire range of payoffs, that is, the outcome for each pair of a decision option and state of nature in the form of a matrix. In the matrix, the payoffs are expressed in terms of utilities (e.g., revenue, profits, or public welfare) or disutilities (e.g., costs borne by the system owner or users or adverse environmental and social effects on the community). The decision maker has no control over the state of nature; however, their perceptions of the different states of nature are what determine the best outcome. In other words, the utilities or disutilites attached by the decision maker to each state of nature is what ultimately "add up" to yield the payoff for each decision option/state of nature pair.

For a given decision option and state of nature, the utility or disutuility, and hence the payoff, assigned by the decision model will differ across different decision makers, depending on their risk characteristics. Decision makers can have any of a broad range of risk attributes including: optimistic, even-handed, risk-taking, and pessimistic. For example, some decision makers are risk averse (would prefer low benefits with high likelihood), and hence would appropriately assign higher payoffs in the decision model. Thus, when faced with a decision characterized by uncertainty, the decision-maker's personal attitudes and biases can affect the final choice of action. As we learned in Chapter 13, the risk behavior of the decision maker can be ascertained from the utility function shape and the parameter values. Possible shapes include linear, concave, convex, and

S-shaped, among others. It can be shown mathematically that a risk-taking decision maker has a strictly convex utility function, a risk-averse decision maker has a strictly concave utility function, and a risk-neutral decision maker has a linear utility function (see Figure 12.10, in Chapter 12). Sinha and Labi (2007) discussed techniques for developing utility functions in such cases of subjective risk characterization. These techniques generally involve a survey of the decision makers, using their responses to establish the utility functions. For objective risk, probability distributions can be used to characterize the levels of such risks.

SUMMARY

At each phase of systems development, engineers make pivotal decisions that influence the outcomes of their systems, in terms of cost and performance. This chapter began by examining the contexts of decision making at each phase and described the basic process of decision making. We then discussed the different ways by which decision models can be classified. One of these ways, the degree of certainty, served as the basis for the ensuing sections of the chapter. On one hand, decisions can be made with the assumption of a deterministic environment, where the decision-related characteristics of the system or its environment, as well as the impacts of any action, are all known with absolute certainty. On the other hand, the practical reality is that most systems operate in an environment of significant uncertainty and decisions therefore often must be made under conditions of risk or of uncertainty.

For decision making under uncertainty, a complete formal statement of the problem requires establishment of all the alternative choices and all the possible outcomes, the likelihood of the occurrence of each potential outcome, the unit of value or desirability, the appropriate scale that describes the distribution of the value of desirability for different levels of outcomes, and the criteria for the final decision. In decision making under risk, the probability of each outcome of an alternative decision is known because it is derived using objective or subjective data. We discussed the decision tree and table and how these concepts could accommodate different problem structures associated with the number of decision stages, the number of decision of action options per node, the nature of the decision outcome variable and whether the expected outcome is a loss or a benefit, and the different final criteria for identifying the best option. In cases of decision making under uncertainty, however, no objective or subjective data exist upon which to base an estimate of the probability of an event; For such decision problems, we use payoff-based criteria.

The chapter ends with a discussion of how to model discrete decisions and the impact of the decision-maker's risk preferences on decision-making processes. In some cases, decision theory simplifies the decision problem, sometimes to a point where the problem becomes "idealized rather than practical." Thus, Dandy et al. (2008) caution that real engineering decision situations are typically quite complex and do not always lend themselves to the simplification needed to achieve one of the idealized situations for which standard analysis is available. Nonetheless, the concepts of decision analysis presented in this chapter provide a useful overall framework for approaching the practical problems of decision making.

EXERCISES

1. Consider a civil engineering system in any branch of civil engineering. Identify and discuss at least one context of decision making at each phase of the development of this system. For any one of these contexts, how would you categorize the decision problem on the basis of the five general ways of classification presented in this chapter?

2. From the perspective of uncertainty, how would you classify decision problems in civil engineering on the basis of whether the following are known about the decision problem: expected outcomes, expected level of each outcome, and the probability of each possible level of each outcome?
3. Discuss the following terms: decision making under conflict, payoff related decision criteria, and regret related decision criteria.
4. Describe, using hypothetical numbers or mathematical symbols, how you would make decisions under risk using (a) payoff-related criteria (expected payoff, most probable payoff), and long shot) or (b) regret-related criteria (expected loss, least probable loss, and long shot loss).
5. Describe, using hypothetical numbers or mathematical symbols, how you would make decisions under uncertainty using (a) payoff-related criteria (optimistic, pessimistic LUWO, Hurwicz, LaPlace) or (b) regret-related criteria (Savage, Hurwicz, minimax, LaPlace).
6. The system owner may be risk averse, risk prone, or risk neutral. Discuss what these terms mean and explain how these risk-taking behaviors could influence civil engineering system decisions.

REFERENCES

Au, T., Shane, R. M., and Hoel, L. A. (1971). *Fundamentals of Systems Engineering*. Addison-Wesley, Reading, MA.

Dandy, G., Walker, D., Daniell, T., and Warner, R. (2008). *Planning and Design of Engineering Systems*, 2nd ed. Taylor & Francis.

de Neufville, R., and Stafford, J. H. (1971). *Systems Analysis for Engineers and Managers*. McGraw-Hill, New York.

Hensher, D. A., Rose, J. M., and Greene, W. H. (2005). *Applied Choice Analysis: A Primer*, Cambridge University Press, Cambridge, UK.

Kepner, C. H., and Tregoe, B. B. (1965). *The Rational Manager: A Systematic Approach to Problem Solving and Decision-Making*. McGraw-Hill, New York.

Monahan, G. (2000). *Management Decision Making*. Cambridge University Press Cambridge, UK.

Revelle, C. S., Whitlatch, E. E., and Wright, J. R. (2003). *Civil Environmental Systems Engineering*. Prentice Hall, Upper Saddle River, NJ.

Sinha, K. C., and Labi, S. (2007). *Transportation Decision-making: Principles of Project Evaluation and Programming*. Wiley, Hoboken NJ.

Stark, R. M. and Nicholls, R. L. (1972). *Mathematical Foundations for Design: Civil Engineering Systems McGraw-Hill*, New York.

CHAPTER 17

NETWORK ANALYSIS TOOLS

17.0 INTRODUCTION

Network theory is the study of graphs that represent either symmetric relations or, more generally, of asymmetric relations between discrete objects. The objects are referred to as nodes and the nodes connected by relations are referred to as links. The nodes or links may be physical or virtual. Network theory and analysis tools are important in the development and management of a certain category of civil engineering systems that either are a network consisting of multiple constituents interacting with each other or are a part of a network of similar or different systems that interact with each other. In fact, several physical or virtual contexts exist in civil engineering that can be represented by a collection of nodes that are connected by links. For example, in construction management activity scheduling, nodes could represent the tasks in a construction process and links could represent the paths between the tasks; in highway transportation, nodes could represent interchanges or intersections and links could represent streets or highways; in logistics and freight management, nodes could represent warehouses; in disaster evacuation, nodes could represent individual persons (points of information dissemination) and links could represent various social communication mechanisms such as Facebook and Twitter. In recent years, network analysis has been used in other applications in civil engineering including data mining (finding patterns in data clouds) and monitoring of risks such as terrorism, construction delays and overruns, or the physical or functional failure of civil systems, by identifying "common threads" or links between irregular situations or seemingly harmless events.

The term *network system* as often used in the literature, may very well be a misnomer. A school of thought holds the view that the network itself is not a system (at least not as defined in Chapter 2) but rather an abstraction of a system, such as a system of water distribution pipes, a social communication system, or a highway system. After such a system has been represented as a set of nodes and links, then it is possible to describe it as a network, for example, water distribution network, pipes, social communication network, or highway network.

Network analysis is particularly important in civil engineering because it helps the system owner or operator to carry out the following general functions for civil engineering systems that have a network-type configuration: (i) monitor the network performance, (ii) control the operations of the network, and (iii) recommend and implement remedial actions to ameliorate the network performance where there are deficiencies. In cases where the system owner seeks to identify optimal ways of carrying out some operation in the network for control purposes, that is, for example, to allocate resources or assign routes to entities that travel in the network, the additional tool of combinatorial optimization is a useful addition for the network analysis. Examples of network-related problem contexts include shortest path problems, transshipment problems, facility location problems, network flow problems, traffic assignment problems, and work scheduling at any phase using tools such as the program evaluation and review technique (PERT) and the critical path method (CPM). In this chapter, we will discuss a number of these useful tools for resolving problems related to the problem contexts.

17.1 FUNDAMENTALS OF GRAPH THEORY

17.1.1 Basic Definitions

A schematic abstraction of connected entities is called a network (or graph), and the mathematical tool used to analyze such systems is commonly termed *graph theory*. The term *graph* simply means that the relationships between nodes and links can be represented graphically (in this chapter, we shall use the terms *graph* and *network* synonymously). A quantitative description of this relationship facilitates mathematical analysis and evaluation of such relationships in terms of link-and-nodal attributes, such as link capacity or the volume of some entity emanating from or consumed at the source nodes and sink nodes, respectively). A graph G can be defined as an ordered triple that comprises a nonempty set of *vertices* denoted by $V(G)$, a set of *edges* $E(G)$, and an *incidence function* ψ_G, that associates each unordered (but not necessarily distinct) pair of vertices of G to each edge of G. If e is an edge of G and v_1, v_2 are the vertices of G such that $\psi_G(e) = v_1v_2$, then e is said to *join* v_1 and v_2; the vertices v_1, v_2 are defined as the *ends* of e. A directed network or linear graph can be represented by $G = [N; A]$, where the elements of N are referred to as nodes, vertices, junction points, or points, and the members of A are called arcs, links, edges, or branches. The *order* and *size* of a graph is the number of vertices and edges, respectively. A *simple graph* is one where there is no loop (multiple edges) between any two vertices.

A network is said to be *planar* (or *emplanable* in a plane) if it can be represented two dimensionally (i.e., on a plane) such that its edges intersect at their ends only. A network is said to be the *subnetwork* of a larger, parent network if all its nodes and links also belong to the parent network. A graph with just one vertex is *trivial*; otherwise, it is *nontrivial*.

In Figure 17.1, edge e_1 is described as being *incident* to vertices v_1 and v_2. Edges e_1 and e_2 are said to be *adjacent* when they are incident to a common vertex. Two edges with identical ends constitute a *loop*, such as e_3 and e_4. The subtle difference between a link and an edge is that an edge that has distinct ends is a *link*. The number of edges incident to a vertex is referred to as the *valency* or *degree* of the vertex.

A graph that contains no cycles is referred to as a *forest* or a *simple acyclic graph*. A simple connected acyclic graph is referred to as a *tree*. A *leaf* is a vertex of degree 1. An *internal vertex* is a nonleaf vertex. An *island* or *isolated vertex* is a vertex of zero degree. A *bridge*, *cut edge*, or *isthmus* is an edge whose removal results in a disconnection of the graph; thus, all the edges in a tree are bridges. The properties of a tree are a graph is a tree if and only if any two of its vertices are joined by exactly one link; a connected graph is a tree if and only if every one of its links is a bridge; and a connected graph is a tree if and only if it has N vertices and N;1 edges (Asmerom, 2012).

A *spanning subgraph* is a subgraph that spans (reaches out to) all the vertices of a graph.

The graph shown in Figure 17.1 can be written mathematically as $G = (V(G), E(G), \psi_G)$ where $V(G) = \{v_1, v_2, v_3, v_4\}, E(G) = \{e_1, e_2, e_3, e_4, e_5, e_6\}$. The term ψ_G is defined completely as: $\psi_G(e_1) = v_1v_2$; $\psi_G(e_2) = v_1v_3$; $\psi_G(e_3) = v_2v_3$; $\psi_G(e_4) = v_2v_3$; $\psi_G(e_5) = v_3v_3$, and $\psi_G(e_6) = v_1v_4$.

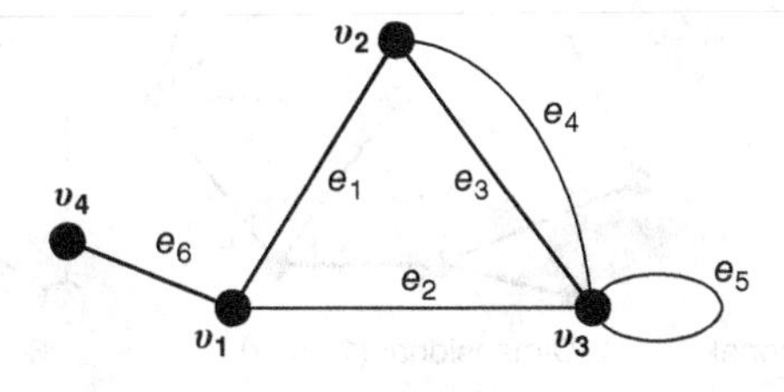

Figure 17.1 Example of a graph showing nodes, links, and loops.

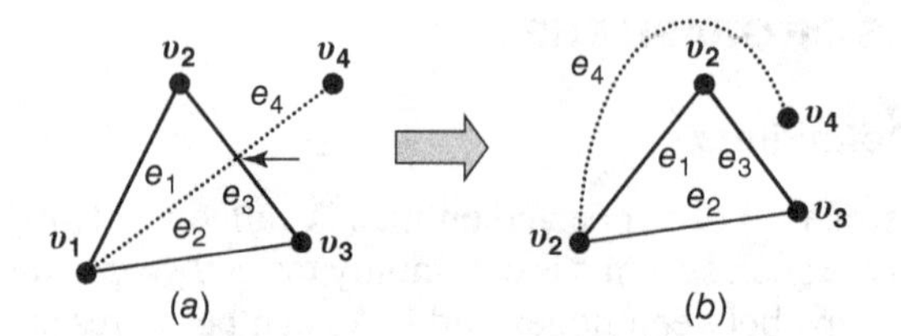

Figure 17.2 Representation of edge intersections.

In some cases, graphs are drawn such that two intersecting edges (such as e_3 and e_4 in Figure 17.2) do not necessarily signify a vertex. A vertex is formed if the two edges are incident at that vertex and is represented by a point. However, where the intersection is drawn only for convenience purposes as an alternative to going around an edge (Figure 17.2*b*), then it is represented by a single cross of the two lines (edges) as indicated by the arrow in Figure 17.2*a*.

17.1.2 Graph Isomorphism

Two graphs G_1 and G_2 are *identical* (denoted as $G_1 = G_2$) if $V(G_1) = V(G_2)$, $E(G_1) = E(G_2)$, and $\psi_{G1} = \psi_{G2}$. Two identical graphs can be represented by diagrams that may or may not be identical. Conversely, it is possible for two nonidentical graphs to be represented by the same diagram.

Two graphs that look exactly the same but have different edge and vertex labels are not identical but rather are *isomorphic* (denoted as $G_1 \cong G_2$). For any two isomorphic graphs, the following *bijections* exist:

θ: $V(G_1) \rightarrow V(G_2)$ and φ: $E(G_1) \rightarrow E(G_2)$ such that $\Psi_G(e) = v_1 v_2$, if and only if $\Psi_G(e) = \theta(v_1)\,\theta(v_2)$. Also, the pair of mappings (θ, φ) is called an isomorphism between the two graphs.

17.1.3 Dimensionality of Networks

A network may have one, two, or more dimensions (Figure 17.3). In a one-dimensional network, which is strictly *linear*, a link can be connected to only one other link in the same direction such as the main urban arterial street in a city. In a two-dimensional network, a link could be connected to one or more other links, such as a city's water distribution network. One- and two-dimensional networks are described as *planar* because they exist at one layer. Multilayer networks with interlinkages between them are described as three dimensional. An example is a multimodal transportation network (each layer represents a different mode—the guideways or pathways for highways, railways, air links, and waterways are the links, and the intramodal and intermodal connections

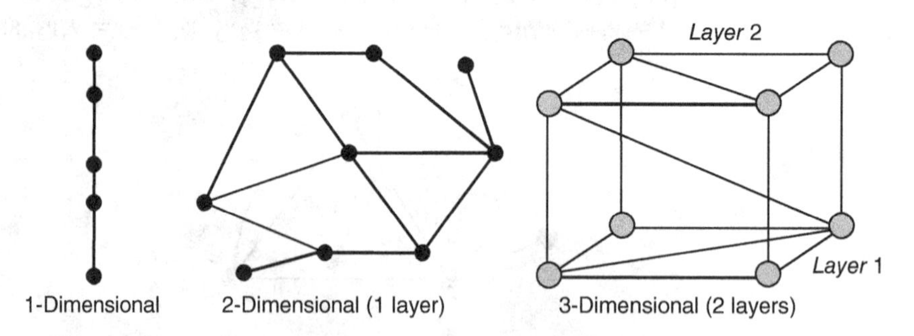

Figure 17.3 Illustration of dimensionality of networks.

between the different mode guideways are the nodes). Networks with four or more dimensions are difficult to represent visually and are best described using mathematical expressions.

17.1.4 Specification of Networks

Network specification is the description of a network (i.e., an indication of which nodes are linked and the edges that link these nodes), and matrices are often used for such specification. Consider a network G that has vertices $v_1, v_2, \ldots, v_V$, and edges $e_1, e_2, \ldots, e_E$: the **incidence matrix** of a network, G, is the (v by e) matrix $\mathbf{L}(G) = [l_{ij}]$ where l_{ij} is the frequency at which that v_j and e_j are incident; and the **adjacency matrix** of a network, G, is the (v by v) matrix $\mathbf{Q}(G) = [q_{ij}]$ where q_{ij} is the number of edges that connect v_i and v_j.

Example 17.1

For the graph shown as Figure 17.4, write the adjacency and incidence matrices.

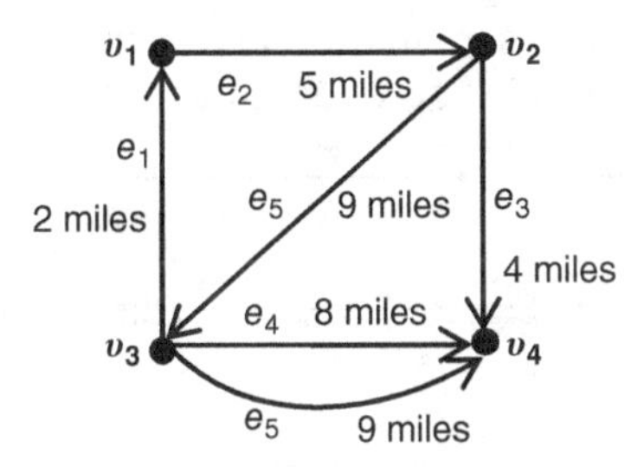

Figure 17.4 Sample graph.

Solution

The adjacency and incidence matrices are shown in Table 17.1.

Table 17.1 Matrices for Example 17.1

(a) Adjacency Matrix (frequency of incidence)

	Node V_1	Node V_2	Node V_3	Node V_4
Node V_1	1	1	1	0
Node V_2	1	1	1	1
Node V_3	1	1	1	2
Node V_4	0	1	2	1

(b) Incidence Matrix ("cost" of the links)

	Link e_1	Link e_2	Link e_3	Link e_4	Link e_5
Node V_1	+2 miles	−5 miles	0	0	−9 miles
Node V_2	0	+5 miles	−4 miles	0	0
Node V_3	−2 miles	0	0	−8 miles	−9 miles
Node V_4	0	1	+4 miles	+8 miles	+9 miles

17.2 TREES, SPANNING TREES, AND MINIMUM SPANNING TREES

As we learned in Section 17.1.1, a connected acyclic simple graph is termed a *tree*. A *leaf* or *pendant vertex* is a vertex of degree 1. A *leaf edge* or *pendant edge* is an edge that is incident to a leaf. A *subtree* of the graph G is a subgraph that is a tree. In certain trees, one of the vertices is designated as the *root* for a specific reason (e.g., a centralized warehouse for distributing some commodity). A *spanning tree* of a graph is a subgraph that contains all the vertices and is a tree. Naturally, only connected graphs can have a spanning tree. The total impedance (distance, or some other cost type) of a spanning tree is the sum of the individual impedances of the links of the spanning tree. The minimum spanning tree concept is useful in the design and evaluation of a certain type of civil engineering networks. These include the infrastructure for installing cables for telephones or electricity, hydraulic structures (irrigation and drainage), water distribution, sewage and waste collection, and highway and street planning. In these and other similar applications, the engineer or planner seeks a set of paths that connects all the nodal points with a minimum total cost, for example, infrastructure for the distribution of electricity or the collection of sewage in a region. The concept of spanning trees is also used in cluster analysis, a statistical tool that places civil engineering systems or users into groups such that there are maximum similarities between groups and minimum similarities between groups.

Example 17.2

For the network shown in Figure 17.5, draw all the spanning trees and identify the minimum spanning tree.

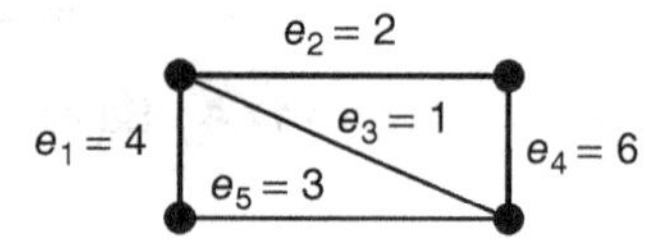

Figure 17.5 Network for Example 17.2.

Solution

The spanning trees shown in Figure 17.6 have the following total impedances (starting from top left, clockwise, of the figure): 14, 11, 7, 10, 6, 12, 9, and 11 units. Clearly, the spanning tree with the least total impedance (cost or distance) is the minimum spanning tree.

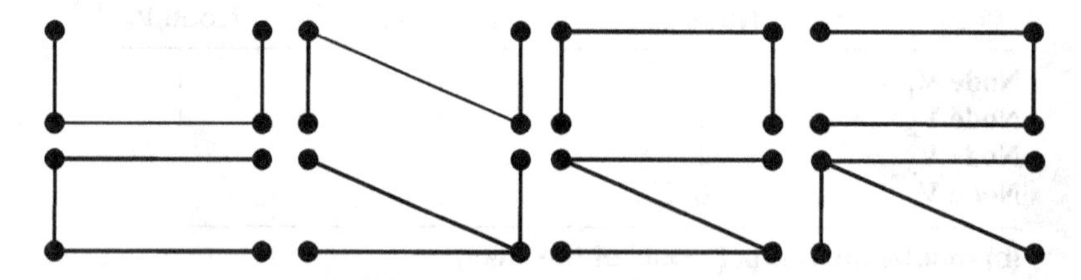

Figure 17.6 Spanning trees for Example 17.2.

17.2.1 Algorithms to Determine the Minimum Spanning Tree

How could one establish the minimum spanning tree, if one exists, for a given network? The "manual" method is to list all the spanning trees and find that which has the least impedance. This is appropriate when the network is small. When the network is large, using this method would take a great deal of time, and the use of an algorithm is helpful. In 1926, Czechoslovakian scientist Otakar Boruvka developed the first algorithm for determining the minimum spanning tree (MST) of a network; this was done to help achieve efficient distribution of electricity in the Moravia region. The most common algorithms currently used to establish MSTs from a given network are Robert Prim's algorithm (Prim, 1957) and Joseph Kruskal's algorithm (Kruskal, 1956). Kruskal's approach selects the minimum-cost link that does not form a cycle.

Boruvka's Algorithm

Step 0. Define a "component" as each node or set of connected nodes.

Step 1. Consider the connected network M with links that have distinct costs.

Step 2. Begin with N as the set of nodes, each regarded as a single component.
As long as N has more than one component,
For each component C of N,
Begin with an empty set of links L.

Step 3. For each node n in N,
Identify the least-cost link from node v to a node outside of C, and add it to L.

Step 4. Add the least-cost link in L to N.

Step 5. Continue until conditions set above are violated.
The resulting tree, W, is the minimum spanning tree of M.

Kruskal's Algorithm

Step 1. Identify the least-cost link in the network (if there is more than one, pick any one).
Mark it with any given color.

Step 2. Identify the least-impedance unmarked (uncolored) link in the network that does not close a colored circuit. Color this link.

Step 3. Repeat Step 2 until you have considered every node in the network (i.e., until you have N; 1 colored links, where N is the number of nodes).
The final set of colored links is the MST.

Prim's Algorithm

Step 1. Pick any node at random to start with; refer to it as, say, A, and color it.

Step 2. Identify the nearest neighboring node of A (call it B).
Color both B and the link A-B.
This is the least-impedance uncolored link in the network that does not close a colored circuit; color this link.

Step 3. Identify the nearest uncolored neighbor to the colored subnetwork.
In other words, identify the closest node to any colored node.
Color this node and the link connecting that node to the colored subnetwork.

Step 4. Repeat Step 3 until all the nodes are colored.
The resulting colored subnetwork is a minimum spanning tree.

Example 17.3

For the network shown in Figure 17.7, using Kruskal's algorithm, determine the minimum spanning tree.

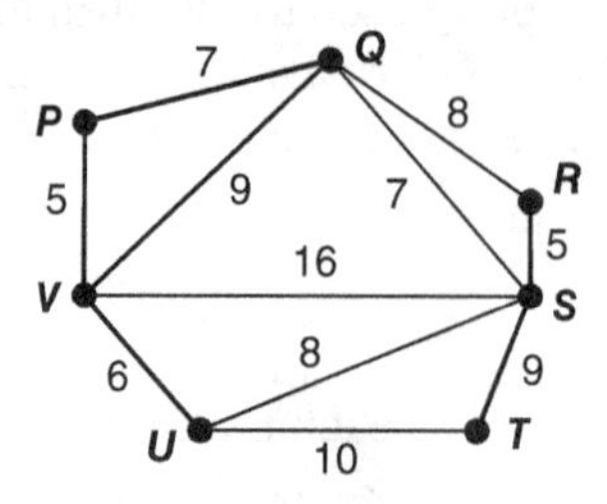

Figure 17.7 Network for Example 17.3.

Solution

See Figure 17.8.

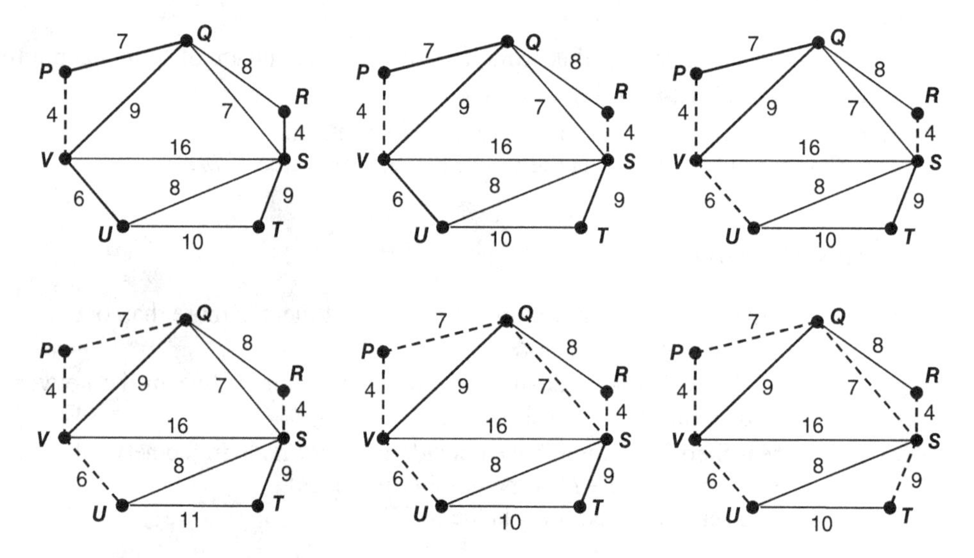

Figure 17.8 Kruskal's algorithm steps for determining the MST for Example 17.3.

Evolution of MST Algorithm Development. The literature from mathematics research indicates a gradual evolution of the efficiency of algorithms that determine the MST for a given network (Chong et al., 2001; Bader and Cong, 2006). Most of the newer algorithms speed up the computation by reducing the number of components that must be connected. $O(\log^2 n)$ time algorithms for MST problems were developed over three decades ago (Hirschberg et al., 1979; Chin et al., 1982), and it was only after several decades that Johnson and Metaxas (1992) developed $O(\log^{1.5} n)$ time algorithms for this problem. Further improvements were made by Chong and Lam (1993) and Chong (1996) whose algorithms had efficiency of $O(\log n \cdot \log\log n)$ time.

Subsequent improvements were made after the realization that randomization of the search could further quicken the search. Karger's algorithm (Karger et al., 1995), which is partly based on

Boruvka's, is a randomized algorithm with $O(\log\ n)$, and Poon and Ramachandran's randomized algorithm (1997) has $O(\log\ n * \text{loglog}\ n * 2^{\log n})$ expected time. Relatively new algorithms include those developed by Karger et al. (1995) include a fast randomized minimum spanning tree, and Chazelle (2000), a deterministic algorithm that has a computational speed of $O(E\ \alpha(E, V))$ time, where α is the inverse of the Ackermann function as described by Pettie and Ramachandran (2002).

Automation of MST Algorithms. An automated tool for determining network MST is available at the Gough website created and maintained by Graham Gough of the University of Manchester. Also, numerous software packages including GRIN and GraphMAGIC, some of which are available on the Internet, use at least one of the above or other algorithms to determine MSTs.

17.3 SHORTEST PATH THROUGH A NETWORK

One of the most common applications of network theory in civil engineering is the identification of the shortest path through a network, in other words, the path that is associated with the least impedance. Impedance is some measure of resistance to the quick and smooth flow of the entity across the links in the network, which could include distance, time, user fees such as tolls, head loss, and so forth. Applications include transportation networks where commuters seek the least-time routes, travelers seek the least-cost routes, and shippers seek paths that have the lowest total distance and/or time. The shortest path through a network can be determined using one of several elegant algorithms that include Tabourier, Dijkstra, Glover, Netsolve, and linear programming. Internet searches of shortest driving distances or travel times, as found in Mapquest, Google Maps, and Yahoo Maps, are based on some of these algorithms.

For a weighted *directed* graph $G = (V, E, w)$ with a *source vertex*, s, and *destination vertex*, t, we can identify the shortest directed path starting at s and ending at t such that we minimize the total distance function:

$$w(m) = \sum_{e \in m} w(e) \tag{17.1}$$

The length and nature of the search for the shortest path in a network is influenced by a number of network characteristics. The search is faster for unweighted networks (where the links implicitly have the same impedance level) compared to a weighted network, directed networks (where each link is one way), and networks with nonnegative weights only.

Applications of the shortest path problems can be found in network tours, where it is sought to visit all of the links or nodes in a network often at a minimal total cost. These tours are discussed in a later section of this chapter.

17.3.1 Shortest Path Problems—Variations and Applications

The variations of shortest path (SP) problems include: the *single-source SP* problem (where we seek the shortest paths starting from a source vertex to all the other vertices in the graph; the *single-destination SP* problem (where we seek the shortest paths starting from all the vertices in the directed graph to a single destination vertex); the *all-pairs SP* problem (where we seek the shortest paths for each pair of vertices)

Again, let us bear in mind that the word "shortest" is expressed in terms of the impedance [i.e., the distance, time, convenience, cost, or head (pressure) loss] or some attribute that impedes the flow of the entity traveling through the network. These variations are implicitly reflected in the different contexts of the shortest path application in civil engineering, as we will see in the next section.

Application Contexts. The shortest path concept is the basic building block of a large number of application contexts in civil engineering network analysis. These contexts include the location of a facility on a network; network connectivity and accessibility, optimal coverage of a network, and origin–destination (O-D) shipment problems; and routing of multiple vehicles in collection/distribution of some link-specific entity. In subsequent sections of this chapter, we discuss specific settings within each of these contexts. Also, we identify a few examples that involve these contexts at the various phases of civil engineering systems development.

Phases. The shortest path concept has wide applications in most phases of civil engineering, some of which are listed in Table 17.2. In separate sections of this chapter, we will discuss some of them in greater detail.

17.3.2 Evolution of Algorithms for Solving Shortest Path Problems

Over the past millennia, civil engineering systems have been constructed during several civilizations to distribute water (using aqueducts), deploy military personnel and equipment (using roads), and other network-related tasks in a manner that involves minimal travel distances to reach some destination. As networks increased in size and complexity, it has been necessary to develop methods that not only make it possible to find the shortest path in a network but also to do so as quickly as possible.

The importance of shortest path algorithms in solving various network-related problems in civil engineering and other disciplines is evidenced by the extensive efforts expended, over the past 60 years, in developing algorithms that solve this problem quickly, particularly for large networks (Schrijver, 2012).

One of the earliest works in this regard was by Shimbel (1953), who worked with information networks, and Ford (1956) who worked at RAND and studied the economics of transportation. Bellman (1958) solved the single-source routing problem where link weights may be negative and also used dynamic programming. Other work was done by Leyzorek et al., (1957) and Dantzig (1960) who successfully used the simplex method for linear programming to identify the shortest path. Also, at the end of the fifties, Moore (1959) helped Bell Laboratories develop a routing system for long-distance telephone calls, and Dijkstra (1959) found solutions to single-source shortest path problems using a simplified yet more efficient version of Ford's algorithm. Also within that period,

Table 17.2 Examples of Shortest Path Application at Various Phases of Civil Systems Development

Phase	Example of Shortest Path Application
Planning	A planner seeks the best location for a proposed facility on an existing network such that the distances to some network reference point are minimized.
Design	A designer seeks to establish a new or improved network for utilities in order to maximize customer access to the utility service.
Construction	A construction manager seeks to perform some multistage construction task within a specified minimum time frame.
Operations	A traffic engineer seeks to advise urban street vehicles via radio and electronic road signs on which routes to take to minimize their travel times.
Monitoring/ Inspection	An inspector seeks to undertake the shortest distance tour of the network by visiting all links (or a sample thereof) in order to measure the demand, defects, or some other link-specific attribute.
Maintenance	A maintenance engineer's crews seek the best route to repair some link-specific networkwide defects while minimizing the cost of travel.

Dantzig, and Pollack and Wiebenson made notable contributions (Schrijver, 2012). Other more recent improvements in the shortest path algorithm are attributed to Fredman and Tarjan (1987), Johnson (1972), Karlsson and Poblete (1983), Gabow and Tarjan (1989), Goldberg and Tarjan (1988) and Ahuja et al. (2003). Innovations in the past few decades include the Floyd–Warshall and Johnson algorithms that solved the all-pairs shortest path problem, perturbation theory, and the A* search algorithm, which uses heuristic methods to identify the shortest path between a specific origin and destination. In all these efforts, the underlying motivation was to speed up the computation time by providing a more efficient algorithm.

17.3.3 The Linear Programming Approach

In using Linear Programming (LP) to determine the shortest path through a network, what we are actually doing is that we are identifying that set of concurrent links from an origin to a destination for which the total impedance is a minimum. Consider a directed network (V, A) that has s is the starting node, t as the terminal node; and c_{ij} as the impedance (in this case, the distance) between i and j where $s, t \in A$, and N is the total number of nodes in the network. The general mathematical formulation is as follows:

$$\text{Minimize } z = \sum_{i=1}^{N}\sum_{j=1}^{N} x_{ij}c_{ij} \tag{17.2}$$

subject to

$$\sum_{j=1}^{N} x_{ij} - \sum_{j=1}^{N} x_{ij} = 0 \quad \text{for all } i \ (i \neq s \text{ and } i \neq t) \tag{17.2a}$$

$$\sum_{j=1}^{N} x_{ij} - \sum_{j=1}^{N} x_{ij} = 1 \quad \text{for } i = s \tag{17.2b}$$

$$\sum_{j=1}^{N} x_{ij} - \sum_{j=1}^{N} x_{ij} = -1 \quad \text{for } i = t \tag{17.2c}$$

$$x_{ij} = 1 \text{ if link } i - j \text{ is on the shortest path 0 otherwise} \tag{17.3}$$

At each node, inward flows are considered as negative and outward flows are positive. To ensure *topological equilibrium*, we enforce the flow conservation constraints at each node: the sum of flows entering a node must equal the sum of flows exiting that node. Thus, Equation 17.2a, b, and c represent the conservation constraints at the intermediate, starting, and terminal nodes respectively.

Example 17.4

For the network shown in Figure 17.9, use LP to determine the shortest path from point S to point T.

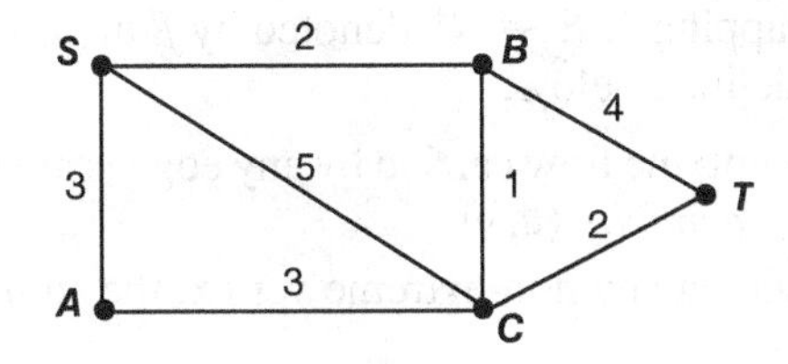

Figure 17.9 Network for shortest path example.

Solution

The LP formulation for the problem is:

$$\text{Minimize} \quad z = \sum_{i=1}^{5}\sum_{j=1}^{5} x_{ij}c_{ij}$$

$$\sum_{j=1}^{5} x_{ij} - \sum_{j=1}^{5} x_{ij} = 0 \quad \text{for all } i \ (i \neq s \text{ and } i \neq t)$$

$$\sum_{j=1}^{5} x_{ij} - \sum_{j=1}^{5} x_{ij} = 1 \quad \text{if } i = s$$

$$\sum_{j=1}^{5} x_{ij} - \sum_{j=1}^{5} x_{ij} = -1 \quad \text{if } i = t$$

where s is the starting point; t is the terminal point; $x_{ij} = 1$ (if link $i - j$ is on the shortest path) or 0 (if link $i - j$ is not on the shortest path), and c_{ij} is the impedance (in this case, the distance) between i and j.

The incidence matrix c_{ij} is:

	S	A	B	C	T
S	0	3	2	5	∞
A	3	0	∞	3	∞
B	2	∞	0	1	4
C	5	3	1	0	2
T	∞	∞	4	2	0

Solving the problem using Excel Solver or appropriate LP optimization software yields the following as the shortest path: $S \to B \to T$ and the corresponding distance is 6 units.

17.4 MAXIMUM FLOW PROBLEM

The maximum network flow problem is one where the engineer seeks to transport some entities from an origin to a destination on the network and seeks the maximum number or amount of some entity that could use the network, given that each link in the network has a certain capacity. Each link of the network may be unidirectional (one way) or bi-directional (two way).

The most generalized case of the maximum flow problem is the circulation problem where it is sought to determine the maximum flow for more than one origin (multi-source) and more than one destination (multi-sink). The origin node (s) and destination node (t) can be termed collectively as the **extreme vertices**. Synonyms for the origin include: source or starting vertex; and synonyms for the destination include: sink or terminal vertex.

For a single origin and destination, the general formation is as follows:

Let $N = (V, E)$ be a network with a set of edges E and vertices V, and has $s, t \in V$ as its origin and destination vertices, respectively.

The **capacity** of an edge can be considered as a mapping c: $E \to R^+$ denoted by *capacity*(u, v). It represents the maximum amount of flow that can be carried by an edge. Similarly, a **flow** can be considered as a mapping f: $E \to R^+$ denoted by *flow*(u, v) subject to the constraints of capacity and conservation as defined below:

1. Capacity constraint: the flow carried by any edge is at least zero but cannot exceed its capacity $0 \leq flow\ (u, v) \leq capacity\ (u, v)$
2. Conservation rule: at any non-extreme vertex, the sum of incoming flows is equal to the sum of outgoing flows.

$$\sum_{u \in in(v)} (flow(u, v)) = \sum_{w \in out(v)} (flow(u, w)) \qquad \forall \text{ vertices } v \neq s, t$$

where $in(v)$ is the set of vertices u such that there exists an edge from u to v; $out(v)$ is the set of vertices w such that there exists an edge from v to w.

The value of flow is defined by:

$$|f| = \sum_{w \in out(s)} (flow(s, w)) = \sum_{u \in in(t)} (flow(u, t))$$

where s is the origin vertex of N. $|f|$ represents the amount of flow passing from the source to the sink.

The **maximum flow problem** is to maximize $|f|$, that is, to route as much flow as possible from origin s to destination t, given the above two constraints.

Example 17.5

For the network presented in Figure 17.10, determine the maximum flow from Larteh to Gladstone. Indicate the flow of each link. As indicated, some of the links are one way and others are two way. The capacity of each link and each direction are shown in Figure 17.9.

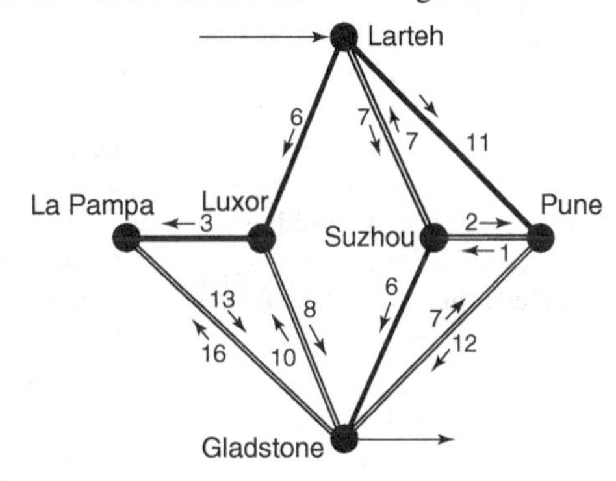

Figure 17.10 Example 17.4.

Solution

The feasible paths and their associated flows are:

Larteh–Luxor–Gladstone
Larteh–Luxor–La Pampa–Gladstone
Larteh–Suzhou–Gladstone
Larteh–Suzhou–Pune–Gladstone
Larteh–Pune–Gladstone
Larteh–Suzhou–Gladstone

The path that yields the maximum flow from Larteh to Gladstone can be found by using enumeration (that is, simply considering the feasible paths one at a time) or by using the mathematical formulation presented above. These are left to the reader as an exercise.

17.5 LOCATING FACILITIES ON CIVIL ENGINEERING NETWORKS

One of the most interesting and widely used applications of network analysis in civil engineering is the selection of locations for civil or other facilities on a civil engineering network. There are several types of facility location problems depending on the attributes of a network system, such as the network type, restrictions on where flows enter or exit the system, restrictions on which parts of the network can be used to site the facilities, the number of facilities to be located, and so forth. Table 17.3 shows some criteria and categories of facility location problems.

Table 17.3 Classification of Facility Location Problems

Classification Criterion	Details	Example
Goal	Set impedance threshold	Locate the facility on the network such that no impedance (distance, time, etc.) of the facility from any specific node or nodes exceeds some specified threshold impedance, I_T. *Specific Example*: Locate a new fire station on your campus such that the distance of any residence to the new station does not exceed 5 miles.
	Minimize average or total impedance	Minimize the average (or total) impedance (distance, cost, time, etc.) per customer (person/vehicle, etc.) (MINI-SUM, MINI-AVE).
	Minimize the worst case	Minimize the maximum impedance (distance, cost, time, etc.) per customer (person/vehicle, etc.) (MINI-MAX). *Specific Example*: Locate a new fire station on your campus such that the maximum distance of the new station (distance to the farthest residence) is as short as possible.
	Multiple criteria (a combination of several criteria that may include the above three)	Minimize the average (or total) impedance (distance, cost, time, etc.) per customer (person/vehicle, etc.) such that no impedance from a specific node or any node exceeds some specified threshold, I_T (MEDI-CENTER). *Specific Example*: Locate a new fire station on your campus such that the average (or total) distance of all residences to the new station is the least possible distance.
Number of Facilities	One facility only	Locate one new state university in your state or province.
	K (a fixed number of facilities, greater than 1)	Locate five new distribution spots for your campus newspaper.
	P (a variable number of facilities)	Determine the number of additional campus bus stops needed and where they should be located.
Possible Facility (Supply) Sites (see Table 17.4)	Nodes only	Locate a new fire station at a suitable intersection in the city or area where your university is located.
	Discrete points along links only	Locate a number of proposed rest stops to be along a major highway.
	Nodes or any point along link	Same as above but potential locations include nodes (freeway interchanges)
Customer (Demand) Sites (See Table 17.4)	Nodes only	Locate a recycling center at all intersections of your city's streets.
	Discrete points along links only	Same as above but along streets and not at intersections
	Nodes or any point along link	Same as above but at intersections as well as along the city streets.

Source: Fricker (1996).

Table 17.4 Types of Minimax Network Problems on Basis of Restrictions on Facility and Customer Locations

MiniMax Problem	Facility Location	Customer Location
Vertex center	Vertex (node) only	Vertex (node) only
General center	Vertex (node) only	Vertex or any point along link
Absolute center	Vertex or any point along link	Vertex (node) only
General absolute center	Vertex or any point along link	Vertex or any point along link

Also, Table 17.4 presents the different types of minimax network problems on the basis of restrictions to the facility and its customer locations. A customer location is a point on a link or at a node at which some service is rendered.

Example 17.6

Figure 17.11 shows a network of roads connecting four cities, G, G, C, K, in a certain province. It is desired to select a location for a water pipe maintenance center to serve the four cities. Assume all roads are bi-directional (no one ways). At what city should the maintenance center be located if the planner seeks to minimize (a) the highest travel time from any city to the maintenance center (this is a *minimax* problem) and (b) the sum of travel times from all other cities to the maintenance center is to be minimized (this is a *minisum* problem).

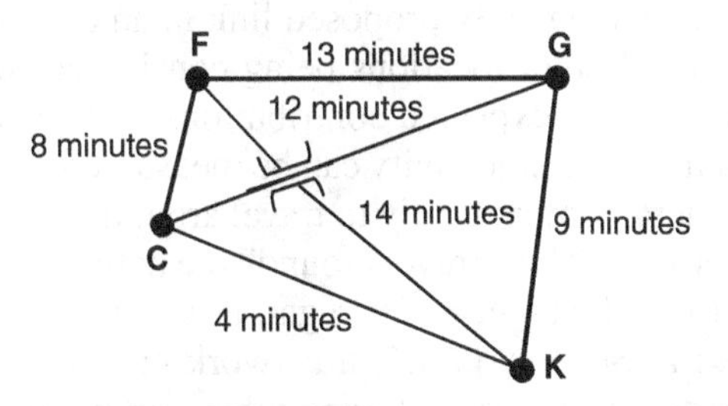

Figure 17.11 Figure for Example 17.6.

Solution

On the basis of minimizing the maximum distance from the facility (maintenance center), locating the maintenance center at C is the best option (min 12). Also, on the basis of minimizing the sum of the distances from the facility (maintenance center), locating the maintenance center at C (min 24) would be the best option (see Table 17.5).

Table 17.5 Solution to Example 17.6

Assumed Location of the Facility	Distances of Facility from Other Cities	Maximum Distance of the Facility to Each City	Sum of Distances of the Facility to All Other Cities
F	0, 13, 14, 8	14	35
G	0, 9, 12, 13	13	34
K	0, 4, 14, 9	14	27
C	0, 8, 12, 4	12	24

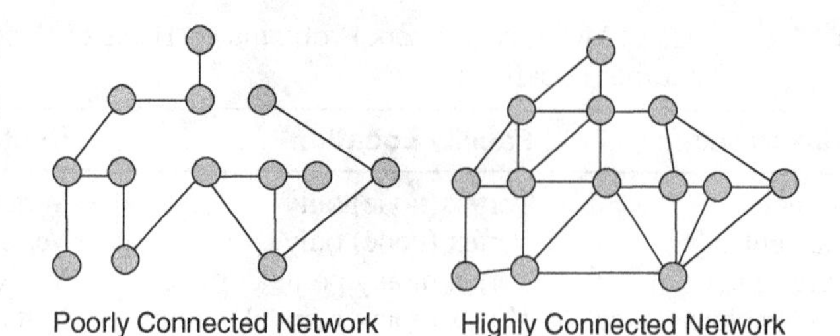

Figure 17.12 General illustration of low and high network connectivities.

17.6 NETWORK CONNECTIVITY

17.6.1 Basic Concepts

For certain kinds of civil engineering networks, the planner's goal is to improve connectivity or accessibility. Generally, for a given number of nodes, the higher the number of links, the greater the connectivity (as illustrated in Figure 17.12).

In the context of connectivity, a civil engineer or planner may be faced with the task of deciding (a) how to design a new network to connect nodes most effectively where no network currently exists, (b) assess the impact of a proposed or newly added link on the connectivity of the overall network, (c) where to add a newly proposed link in an existing network, and/or (d) prioritize several different links at different locations being considered for addition to an existing network by measuring and ranking their expected contributions to the network connectivity.

As a practical matter, connectivity can be measured in terms of the overall cost of traveling on the network. It is generally more costly to travel around a sparsely connected network compared to a well-connected network. This "travel around" the network can be the sum of all the shortest paths from all the nodes to all the other nodes regardless of how many times a node or link is traversed.

As we learned in Section 17.1.1, a network or graph is said to be *connected* if a path can be traced from a vertex of the graph to any other vertex; otherwise, the graph is *disconnected.* In this section, we learn a few more definitions. A graph is *completely disconnected* if there none of its vertex pairs is connected by a link. If the removal of a vertex causes the graph to become disconnected, then that vertex is referred to as a *cut vertex.* A *separating set* or *cut* is a set of vertices which, when removed, results in disconnection of the graph. A graph that contains k internally disjoint paths between any pair of vertices is termed a *k-connected* graph. A graph is described as *k connected* if it is always possible to establish a path from a vertex to every other vertex in the graph even after removing any $k - 1$ vertices (Bondy and Murthy, 2008). Figure 17.13 illustrates a network that is one connected ($k = 1$) but not two connected. The *connectivity* of a graph can be defined as the least number of vertices that are needed to disconnect the graph (Bondy and Murthy,

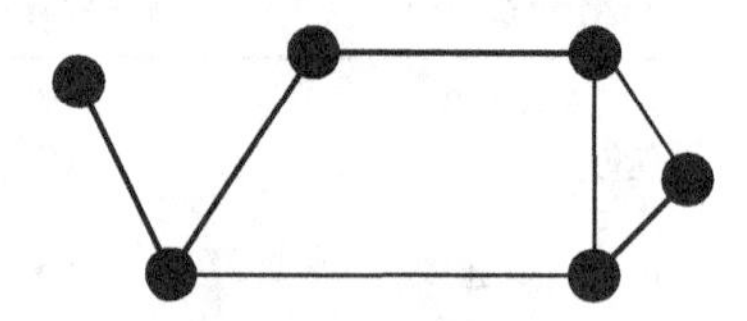

Figure 17.13 Network connectivity terminology illustration.

2008). An edge whose removal results in a disconnected graph is referred to as an *isthmus*, *bridge*, or *cut edge*. A tree can be defined as a network that consists entirely of bridges.

17.6.2 Application Contexts of Network Connectivity Analysis

It is useful to measure the connectivity of a network or the contribution of each link to network connectivity in at least two contexts. First, such analysis is needed when the civil engineer seeks to describe the extent of network vulnerability in terms of the possible effects of network disruption. A network with high connectivity inherently has greater redundancy and hence is more resilient to disaster events that threaten to destroy or impair some of its links. Second, connectivity analysis helps the engineer or planner to assign, for each link, a measure of the topological importance to that link; such a measure could be used as one of the multiple performance measures or evaluation criteria (to complement the traditional evaluation criteria that include costs, benefits, and capacity) in evaluating or prioritizing projects that are located at different links or nodes in the network.

(a) Assigning a Topological Importance Measure to Links for Project Evaluation Purposes. From the preceding discussion, it seems clear that a link in a network could be assigned a network "importance" index as a reflection of the contribution of that link to the network's overall connectivity. Using such an index as well as other evaluation criteria, the proposed candidate investments at different locations can be compared.

(b) Characterizing Network Vulnerability for Each Link for Management Purposes. A network is vulnerable to disruption if any link is threatened; the greater the contribution of the link to network connectivity, the more vulnerable is the network in the event of failure of that link. This context is particularly relevant in disaster situations where it is desired that certain highly weighted links remain intact so that supplies can be adequately delivered to endangered segments of the population or that trapped persons may be evacuated quickly.

17.6.3 Measurement of Network Connectivity and Accessibility

Tables 17.6 and 17.7, synthesized from Woldemariam (2014), present several different ways by which network connectivity and accessibility, respectively, could be measured. To illustrate the computations for each measure, node-to-node cost (distance) data from a simple network shown in Figure 17.14 (Woldemariam, 2014) is used (see fifth column of both tables).

The measures of network connectivity are typically expressed in terms of basic topological features including the number of nodes and links. For the sample network, it can be shown that the **cyclomatic number** is zero; this is suggestive of the absence of circuits in the network. The **diameter** of the sample network, that is, the length of the longest path between any origin and destination pair, is 25 miles.

The **alpha index** of the network is 0%, that is, the network attains 0% of the maximum possible connectivity; the **beta index**, which is an indication of network complexity, is zero because for trees (such as that shown in Figure 17.14) and disconnected graphs, the beta index never exceeds zero; and the **gamma index** is a ratio between the actual number of links in the network and the number of maximum possible links (assuming all node pairs were connected by a link); thus the gamma index indicates the extent of relative connectivity of the network (Xie and Levinson, 2009). For the network shown in the figure, the gamma index is 67%. The **eta index** of the network is 7.5 mile/link. It may be realized that the eta index decreases with increasing number of nodes; Therefore, a lower eta index is indicative of a more developed network. The **pi index** of the network is 1.2 (a higher pi index reflects a more developed network. The sample network has a **theta index** of

Table 17.6 Measures of Network Connectivity

Measure	Description	Equation	Calculated Index (Sample Network)
Degree of node	The number of nodes directly attached to the node	$c_i = \sum_j^n c_{ij}$	$C_1 = 1; C_2 = 3; C_3 = 1; C_4 = 2; C_5 = 1.$
Cyclomatic number (μ)	Maximum number of independent cycles in the network	$\mu = e - v + p$	$\mu = 0$
Diameter	The length of the longest (maximum impedance) path between an origin and destination pair	$\delta(G) = _x\max_y d(x, x)$	$\delta(G) = 25$ miles
Alpha index	The ratio between the actual number of circuits in the network and the maximum number of circuits	$\alpha = \mu/(2u - 5)$ (for planar graphs)	$\alpha = 0$
Beta index	Ratio between number of links and number of nodes in the network	$\beta = e/u$	$\beta = 0.8$
Gamma index	Ratio between the actual number of links and the maximum number of links in the network	$\gamma = e/[3(u - 2)]$ (for planar graphs)	$\gamma = 0.67$
Eta index	Ratio of sum of impedances of all links and all nodes to the number of links in the network	$\eta = M/e$	$\eta = 7.5$ miles per link
pi index	Measures the relation between the entire network and individual links in the network	$\pi = c/d$	$\pi = 1.2$
Theta index	Ratio between the total network impedance and number of nodes	$\theta = M/V$	$\theta = 6$ miles/node
Iota index	Represented by the ratio of the total network impedance and number of weighted nodes	$\iota = M/W$	$\iota = 2.31$ miles per weighted node
Degree of connectivity	The relative position of a network's connectivity in the range of the minimum and maximum connectivity values	$\text{dc} = (1/e)[u(u - 1)/z]$	dc = 2.5

Source: Woldemariam (2014).

6 miles per node, which reflects the average length per node in the network. Also, the **iota index**, which takes into consideration the importance of nodes, is 2.31 miles. In the sample network, the end points and the interior (intersection) nodes were assumed to have two and eight practical functions, respectively. The **degree of connectivity** of the network is 2.5, which reflects the relative position of a network's connectivity relative to the maximum connectivity (which is 1) and the minimum connectivity.

The ***D* matrix** or **Shimbel distance** is a tabular display of the number of links associated with each origin and destination node pair in the network. The accessibility index measures the total length (in units of distance) associated with traversing a node to every other node in a network. For

Table 17.7 Measures of Network Accessibility

Measure	Description	Equation	Calculated Index (Sample Network)
Shimbel Distance (D matrix).	The sum of the number of links in the shortest path between a node and all other nodes in the network	$\sum_{ij} V_{\text{shortest path }(i,j)}$	Node 1 = 8; node 2 = 5; node 3 = 8; node 4 = 6; node 5 = 9
Accessibility index.	A measure of the spatial relation (distance) between node i and all other nodes in the network	$A(i,N) = \sum_{i=1}^{n} d(i,j)$	$A(1,N) = 53; A(2,N) = 38; A(3,N) = 68; A(3,N) = 46; A(5,N) = 67$
Dispersion	A measure of the overall accessibility of a network	$D(N) = \sum_{i=1}^{n}\sum_{j=1}^{n} d(i,j)$	$D(N) = 272$
Degree of Circuity	A measure of the relative location of nodes of a network	$DC = \frac{\sum_{i=j}^{n} (E-D)^2}{v}$	$DC = 0$

Source: Woldemariam (2014).

Notations: C_i = degree of node i; C_{ij} = connectivity between node i and node j (either 1 or 0); n = number of nodes; μ = cyclomatic number; e = number of links; and v = number of nodes; p = number of graphs; $\delta(G)$ = diameter of graph G; α = alpha index for planar graphs; β = beta index; γ = gamma index; η = eta index; M = total network impedance; π = pi index; c = total length or impedance of the entire network; d = total impedance of the network's diameter; θ = theta index; ι = iota index; w = sum of network's nodes weighted by their function; dc = degree of connectivity; $A(i,N)$ = accessibility index; $\sum_{i=1}^{n} d(i,j)$ = summation of impedances between node i and all node j's in the network; $D(N)$ = dispersion of network N; $\sum_{i=1}^{n}\sum_{j=1}^{n} d(i,j)$ = sum on a sum of impedances between node i and all other nodes in the network; E, D = real and straight line impedances, respectively, between nodes; DC = degree of circuity.

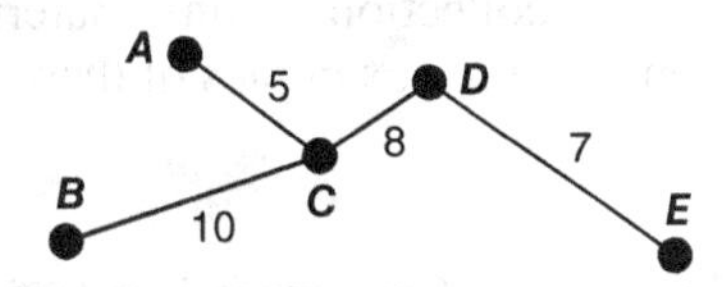

Figure 17.14 Example network for connectivity and accessibility illustration.

the sample network provided, the accessibility of each node can be calculated. The accessibility of the entire network (which is referred to as the **network dispersion**) is 272 miles. The **degree of circuity** of the network is 0, which implies that the real distance between any two nodes in the network is the straight line (the shortest) distance between the nodes. This is the case for the network example because straight-line connections between nodes are assumed; however, in certain application contexts including rolling terrain, the real distances between nodes may be different from the straight-line distances (Woldemariam, 2014).

17.6.4 Discussion

For purposes of planning a new infrastructure network in uncharted territory or for reviewing an existing network for possible upgrades, it is useful to access the network connectivity and accessibility. There are also other applications in disaster management. Damage of a link due to natural or man-made threats (see Chapter 28), including earthquakes, floods, and terrorist attacks, causes a disruption in the network connectivity and accessibility. Often, civil engineers seek to measure the expected degradation in connectivity and accessibility if the threat occurs, and also to identify mitigation actions that partially or fully restore these measures of network performance in a cost-effective manner. There exists methodologies in the literature to measure the network robustness in other words, the degree of criticality of a link for the entire infrastructure network and procedures that optimize allocation of resources among different infrastructure to minimize the probability of failure of critical facilities in the network. For these methodologies and associated case studies, the reader is referred to: Scott et al., (2010); Sullivan et al., (2011); and Murray et al., (2011).

17.7 OPTIMAL COVERAGE OF NETWORKS

This context of application of network analysis is particularly experienced at the system development phases of operations and the monitoring/inspection phase. At these phases, the user or the agency seeks to visit, tour, or "cover" all links or nodes in the network (often this refers to a transportation network) on which the civil engineering system is located or could be accessed. Also, because resources are limited, it is often sought to carry out these tours while minimizing the total tour cost. Depending on the network type in question, the cost associated with the network tour is often reflected as the link distance, link out-of-pocket cost (such as tolls), delays at nodes, or other monetary or nonmonetary costs associated with the network links and/or nodes.

Problems involving optimal coverage of networks can be categorized by the topological characteristics of the tour or the purpose of the tour. From the perspective of topological characteristics, optimal coverage tours include Euler tours and Chinese postman tours that seek to visit each node and Hamiltonian tours and traveling salesman tours that seek to visit each link in the network. From the perspective of the tour purpose, the operations engineer or manager of a shipping agency, for example, may seek to design a routing scheme for single or multiple vehicles to provide some service to entities located at the links or nodes in the network—such as the distribution of some product (mail delivery, logistics, etc.) or collection of some material (solid waste, recyclable products, etc.). In Chapter 23, we will examine some examples of these specific problem contexts.

17.7.1 Euler Tours

The story of the seven bridges of Königsberg is a true occurrence that inspired the creation of network analysis as a branch of mathematics by Leonhard Euler in 1736. In this problem, there were two large islands in the Pregel River in the city of Königsberg, Germany. The islands were linked to the mainland and each other by seven bridge crossings. The problem posed by the king at the time was, starting from any point on the mainland or island, to identify a route that crosses each bridge only once and then returns to the starting point. Figure 17.15 shows a perspective view of the problem, a plan view of the problem, and a schematic graph representation of the problem.

Euler came to the realization that the problem could be addressed in terms of the oddness or evenness of the node degrees (the number of incident edges). In the Königsberg problem, one node has degree 5 and three nodes have degree 3. Euler proved that for a Eulerian circuit to be possible, the graph must be connected, and all the nodes of the graph must be of even degree. Because the Königsberg graph has at least one node of odd degree (actually there are four odd-degree nodes), an Eulerian circuit is not possible.

An *Eulerian path* is one that passes through every edge exactly once. If the starting and ending nodes are the same, then the path is an *Euler cycle* or *Euler circuit*. If the starting and ending nodes are different, it is an *Euler trail, path* or *walk*.

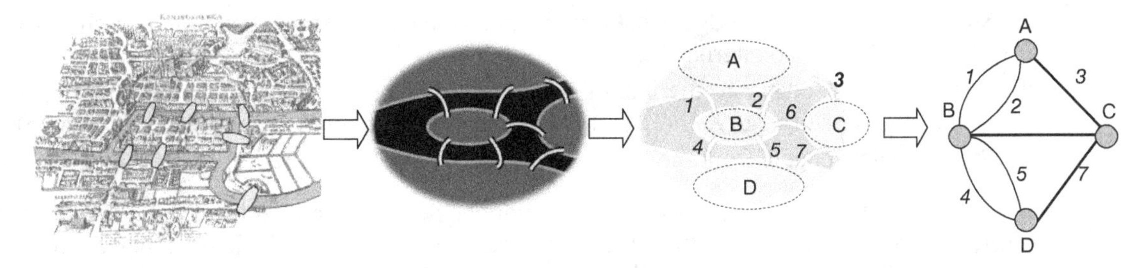

Figure 17.15 The seven bridges of Königsberg (Bogdan Giuşcă/Booyabazooka/ Wikimedia Commons).

In the Königsberg example, if the conditions of the problem were relaxed to permit different starting and ending points, a solution can be found; however, if the starting and ending points differ, then the tour is no longer referred to as a circuit but rather as an Eulerian trail. For an Eulerian trail to be possible, the number of vertices with odd degree must not exceed two.

Example 17.7

Trace an Eulerian circuit for the network in Figure 17.16*a*. Choose any node as your starting point.

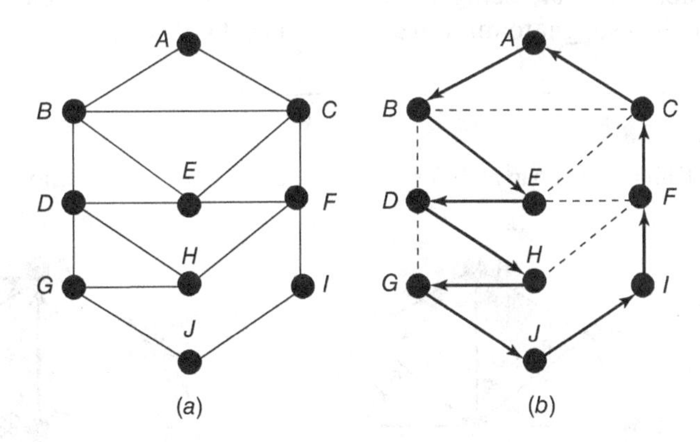

Figure 17.16 Figure for Example 17.7: (*a*) question and (*b*) solution.

Solution

Choosing A as the starting node, the Eulerian circuit is *A-B-E-D-H-G-J-I-F-C-A*.

(a) Algorithms for Euler Tours.

Fleury's Algorithm (Skiena, 1990). Define a bridge as an edge which, if removed, produces a disconnected graph.

Step 1. Check to ascertain that the graph is a connected Eulerian graph (i.e., a graph with at most two vertices having odd degree).

Step 2. Commence at a vertex v that has odd degree (if there is no vertex of odd degree, start with any vertex). Set the incumbent vertex as v. Also, set the incumbent Euler trail as the empty set of edges.

Step 3. Choose the next edge e that is incident on the incumbent vertex; however, choose a bridge only if there is no alternative.

Step 4. Add the edge e to the incumbent Euler trail; set, as the incumbent vertex, the vertex at the other end of edge e. (note that if e is a loop, the incumbent vertex stays the same).

Step 5. Delete edge e from the graph and delete any isolated vertices.

Repeat steps 3 to 5 until there are no more edges to be deleted from the network. The final current path is the Eulerian path. If the network has no vertices of odd degree, then the result of the above sequence is an Eulerian cycle; if the graph had exactly two vertices with odd degree, then the result is an Eulerian trail. Modifications to Fleury's algorithm by Tarjan (1997) and Thorup (2000) have helped improve the efficiency of this algorithm. Other algorithms for finding Euler paths include Hierholzer's algorithm (Fleischner, 1991).

17.7.2 The Chinese Postman Problem

The *Chinese postman problem (CPP)* also referred to as the route inspection problem (RIP) is a classic context in graph theory that designs a path that visits all links or edges in a (connected) undirected network at a minimal travel cost (Kwan, 1962). The problem was studied originally by Chinese mathematician Mei-Ku Kwan in 1962 with the purpose of optimizing a postman's route (DADS, 2010). Note that "cost" is the impedance, such as distance, time, and so forth. Like Euler tours, links, not nodes, are being visited. However, unlike Euler tours, a link may be visited more than once and it is sought to minimize the total cost of travel.

Example 17.8

Trace a Chinese postman path for the network in Figure 17.17*a*. Choose node J as your starting point.

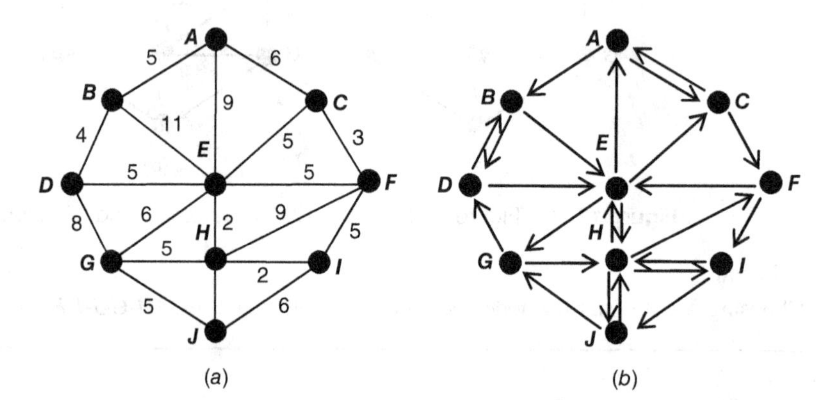

Figure 17.17 Figure for Example 17.8: (*a*) question and (*b*) solution.

Solution

The Chinese postman path starting from J is presented as Figure 17.16*b*:

$$J \rightarrow H \rightarrow I \rightarrow H \rightarrow J \rightarrow G \rightarrow H \rightarrow E \rightarrow H \rightarrow F \rightarrow E \rightarrow G \rightarrow D \rightarrow B \rightarrow D \rightarrow E$$
$$\rightarrow C \rightarrow A \rightarrow B \rightarrow E \rightarrow A \rightarrow C \rightarrow F \rightarrow I \rightarrow J.$$

Mathematical Formulation. As we discussed above, the Chinese postman problem is solved when the traveling entity traverses every link of the network at least once while minimizing the total impedance (in many cases, the distance) of travel. The problem may be formulated as an integer linear program with a variable x_{ij} that represents the number of times the link between nodes i and j is traversed in the direction from i to j, as follows (Revelle et al., 2003):

$$\min \sum_{i=1}^{N} \sum_{j=1}^{N} c_{i,j} x_{i,j} \tag{17.4}$$

subject to

$$\sum_{k=1}^{N} x_{k,i} - \sum_{k=1}^{N} x_{i,k} = 0 \qquad i = 1, 2, \ldots, N \tag{17.5}$$

$$x_{i,j} + x_{j,i} \geq 1 \text{ for all links } (i,j) \tag{17.6}$$

where $x_{ij} \geq 0$ is an integer; N is the number of nodes in the network; $c_{i,j}$ is the distance of the link between node i and node j; if there is no direct link between i and j, then $c_{i,j} = \infty$. The objective function 17.4 is to minimize the total distance traveled. Constraint (17.5) specifies that the number of links entering node i be equal to the number of links exiting node i. Constraint (17.6) specifies that every link must be traversed at least once, in one direction or the other. With regard to computational complexity, the solution to this problem is nondeterministic polynomial (NP) hard.

Example 17.9

Using the mathematical formulations presented in Equations (17.4)–(17.6), determine the Chinese postman path for the network shown in Figure 17.18. Choose node A as your starting point.

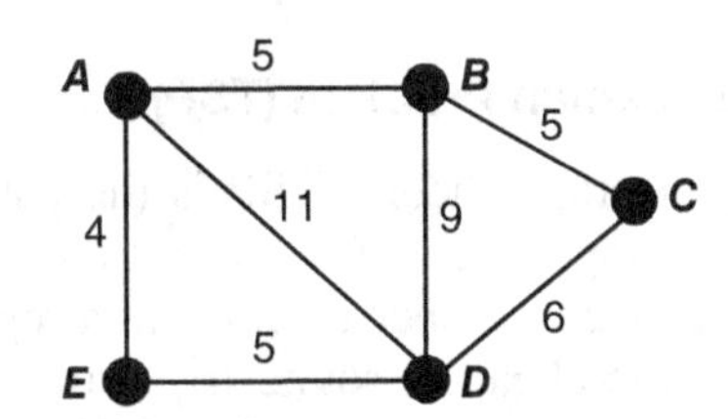

Figure 17.18 Figure for Example 17.9.

Solution

The problem can be formulated as

$$\min \sum_{i=1}^{5} \sum_{j=1}^{5} c_{i,j} x_{i,j}$$

Subject to

$$\sum_{k=1}^{5} x_{k,i} - \sum_{k=1}^{5} x_{i,k} = 0 \qquad i = 1, 2, \ldots, 5$$

$$x_{i,j} + x_{j,i} \geq 1 \text{ for all links } (i,j)$$

where $x_{i,j} \geq 0$ and is an integer; $x_{i,j}$ is the number of times the link between nodes i and j is traversed in the direction from i to j; $c_{i,j}$ is the distance of the link from node i to node j; if there is no direct link from i to j, then $c_{i,j} = \infty$. The distance matrix is

$$\begin{array}{c} \\ A \\ B \\ C \\ D \\ E \end{array} \begin{array}{c} \begin{array}{ccccc} A & B & C & D & E \end{array} \\ \begin{bmatrix} 0 & 5 & \infty & 11 & 4 \\ 5 & 0 & 5 & 9 & \infty \\ \infty & 5 & 0 & 6 & \infty \\ 11 & 9 & 6 & 0 & 5 \\ 4 & \infty & \infty & 5 & 0 \end{bmatrix} \end{array}$$

Solving the problem using Excel Solver or other optimization platform, the $x_{i,j}$ matrix is determined as

$$\begin{array}{c} \\ A \\ B \\ C \\ D \\ E \end{array} \begin{array}{c} \begin{array}{ccccc} A & B & C & D & E \end{array} \\ \begin{bmatrix} 0 & 2 & 0 & 0 & 0 \\ 0 & 0 & 1 & 1 & 0 \\ 0 & 0 & 0 & 1 & 0 \\ 1 & 0 & 0 & 0 & 1 \\ 1 & 0 & 0 & 0 & 0 \end{bmatrix} \end{array}$$

The minimum distance is 50. The Chinese postman route is $A \rightarrow B \rightarrow C \rightarrow D \rightarrow E \rightarrow A \rightarrow B \rightarrow D \rightarrow A$ or $A \rightarrow B \rightarrow D \rightarrow A \rightarrow B \rightarrow C \rightarrow D \rightarrow E \rightarrow A$.

17.7.3 Hamiltonian Tours

A *Hamiltonian path* is a path that passes through each node in a graph or network exactly once. If the starting and ending nodes of the tour are the same, then the path is described as a *Hamiltonian cycle* or *circuit*, otherwise it is a *Hamiltonian trail*. A *Hamiltonian graph* is a graph that contains a Hamiltonian path. For the same graph, there could be multiple Hamiltonian paths or cycles.

17.7.4 The Traveling Salesman Problem (TSP)

Also, a classic problem in graph theory, TSP is the path that visits every node in the network exactly once while minimizing the cost (distance, time, and so on) of travel. In a common application context, consider a given number of cities in a region and the fares of air travel across the city pairs (links); what is the cheapest round-trip route that starts at a city, visits all other cities exactly once, and returns to the starting city? The TSP is one of the most studied problems in network analysis and has applications in several branches of civil engineering including transportation systems.

Example 17.10

For the network in Figure 17.19*a*, trace (i) a Hamiltonian path and (ii) a traveling salesman path. Use node P as your starting point.

Solution

The Hamiltonian path (Figure 17.19*b*) is $P \rightarrow V \rightarrow W \rightarrow Q \rightarrow R \rightarrow X \rightarrow U \rightarrow T \rightarrow S$. The traveling salesman tour (Figure 17.19*c*) is $P \rightarrow V \rightarrow U \rightarrow T \rightarrow S \rightarrow X \rightarrow W \rightarrow R \rightarrow Q \rightarrow P$.

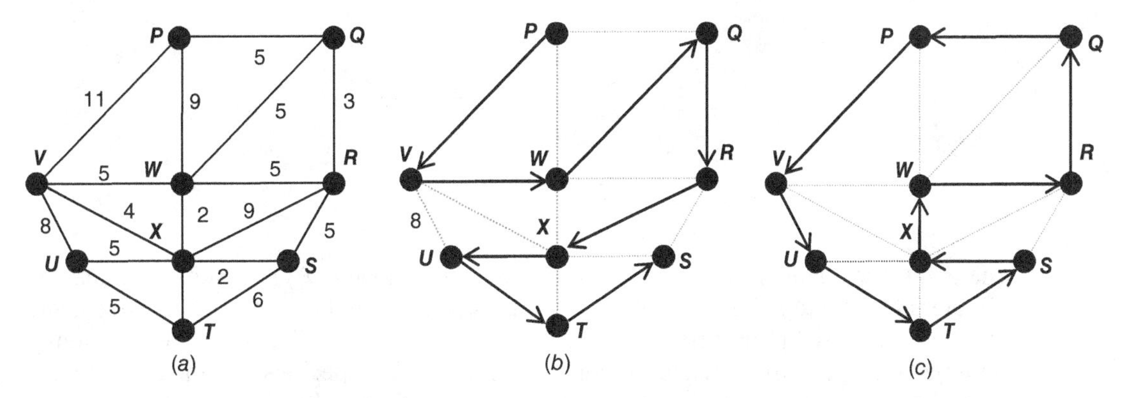

Figure 17.19 Figure for Example 17.10: (*a*) question, (*b*) Hamiltonian path, and (*c*) traveling salesman path.

17.8 OPTIMAL SHIPPING ACROSS ORIGIN–DESTINATION (O–D) PAIRS IN A NETWORK

17.8.1 The Hitchcock Transportation Shipment Model

In the Hitchcock model, there is a number of sources (representing points at which some entities collect some material) and destinations (representing points at which the entities deposit the material for some purpose such as treatment, production, or disposal) (Figure 17.20). At each source node where the material is generated, the sum of the material outflows must be equal to the quantities generated. At each destination node (receiving facility), the total quantity arriving must not exceed capacity of the facility. This application context of network analysis is particularly encountered at the system development phases of construction (where raw materials, precast units, and other materials are transported from and to various site locations) and operations (where shippers, logistics operators, and other system users seek to use the network in some optimal way that suits their purposes).

The unit cost of transportation is known for each link between a source, S, and a destination, D. The problem is to determine the quantities of some material or product to ship from each source to each destination such that the total shipment cost is minimized.

$$\min \sum_{i=1}^{I} \sum_{j=1}^{J} a_{i,j} x_{i,j} \tag{17.7}$$

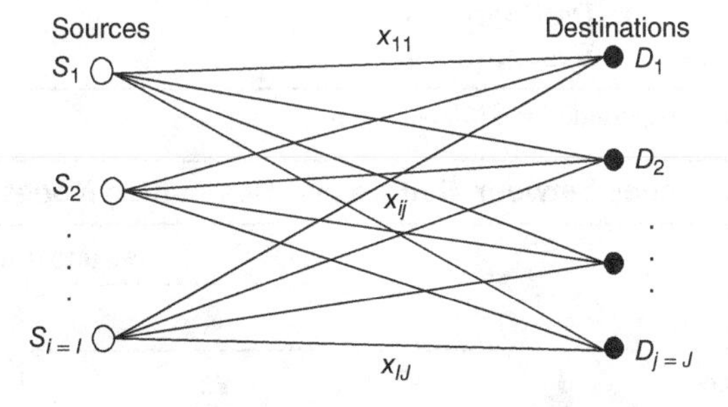

Figure 17.20 Schematic representation of the Hitchcock shipment problem.

subject to

$$\sum_{j=1}^{J} x_{i,j} = g_i \quad i = 1, 2, 3, \ldots, I \tag{17.8}$$

$$\sum_{i=1}^{I} x_{i,j} \leq h_j \quad j = 1, 2, \ldots, J \tag{17.9}$$

where g_i is the quantity of material collected at source node i, $a_{i,j}$ is the cost of transporting a unit of material from i to j; h_j is the capacity of the receiving facility at destination node j; and $x_{i,j}$ is the quantity of material transported from source i to destination j. The objective function (17.7) is to minimize the total cost of transportation. Constraint (17.8) specifies that the quantity leaving node i must be equal to the quantity generated at that node. Constraint (17.9) specifies that the quantity of material that is received by destination node j must be less than or equal to the capacity of that node.

Example 17.11

It is sought to transport a number of types of raw material to a number of factories for production such that the total cost of transportation is minimized. There are four raw material sources at which railcars collect the material and ship to three destinations for production. At each source, there is no excess, that is, every raw material generated is shipped. Also, assume that there is no excess supply at each production point. Table 17.8 presents the quantity of material collected at source node and the capacity of each destination node, and the distances between the nodes (miles) and the transport cost per mile. Determine how much material must be transported from each source to each destination.

Table 17.8 Data for the Hitchcock Shipment Model

(a) Quantity of Material Generated at Nodes and Node Capacities

		Material Generated at Source Node (tons)	Capacity of Destination Node (tons)
Sources	**Source 1**	50	
	Source 2	80	
	Source 3	30	
Destinations	**Destination 1**		100
	Destination 2		25
	Destination 3		15
	Destination 4		20

Cost of transportation = $120 per ton per mile.

(b) Distances between Source and Destination Nodes (miles)

		Destinations			
		1	**2**	**3**	**4**
Sources	**1**	30	22	36	51
	2	14	20	25	48
	3	10	24	22	32

Solution

The problem can be formulated as follows:

$$\min \sum_{i=1}^{3} \sum_{j=1}^{4} a_{i,j} x_{i,j}$$

subject to

$$\sum_{j=1}^{4} x_{i,j} = g_i \qquad i = 1, 2, 3$$

$$\sum_{i=1}^{3} x_{i,j} \le h_j \qquad j = 1, 2, 3, 4$$

$$x_{i,j} \ge 0 \text{ for all } (i,j)$$

where g_i is the quantity of material collected at source node i, $a_{i,j}$ is the cost of transporting a unit of material from i to j; h_j is the capacity of the receiving facility at destination node j; and $x_{i,j}$ is the quantity of material transported from source i to destination j.

The $a_{i,j}$ matrix is

$$\begin{bmatrix} 30 & 22 & 36 & 51 \\ 14 & 20 & 25 & 48 \\ 10 & 24 & 22 & 32 \end{bmatrix} * 120 = \begin{bmatrix} 3600 & 2640 & 4320 & 6120 \\ 1680 & 2400 & 3000 & 5760 \\ 1200 & 2880 & 2640 & 3840 \end{bmatrix}$$

$$g_1 = 50, g_2 = 80, g_3 = 30, h_1 = 100, h_2 = 25, h_3 = 15, \text{and } h_4 = 20.$$

Solving the problem using Excel Solver, the $x_{i,j}$ matrix is determined to be:

$$\begin{array}{c} \\ 1 \\ 2 \\ 3 \end{array} \begin{array}{c} \begin{array}{cccc} 1 & 2 & 3 & 4 \end{array} \\ \begin{bmatrix} 0 & 25 & 15 & 10 \\ 80 & 0 & 0 & 0 \\ 20 & 0 & 0 & 10 \end{bmatrix} \end{array}$$

The amounts that must be shipped are shown in the matrix above, and the corresponding (minimum) cost is = $120[(30 \times 0) + (22 \times 25) + (36 \times 15) + \cdots + (32 \times 10)] = \$388{,}800$.

17.8.2 The Transshipment Model

The transshipment model, a generalized version of the Hitchcock model, includes intermediate nodes located between the sources and the destinations of the network (Figure 17.21). The intermediate nodes represent intermediate facilities such as transfer or processing stations. Mass balance

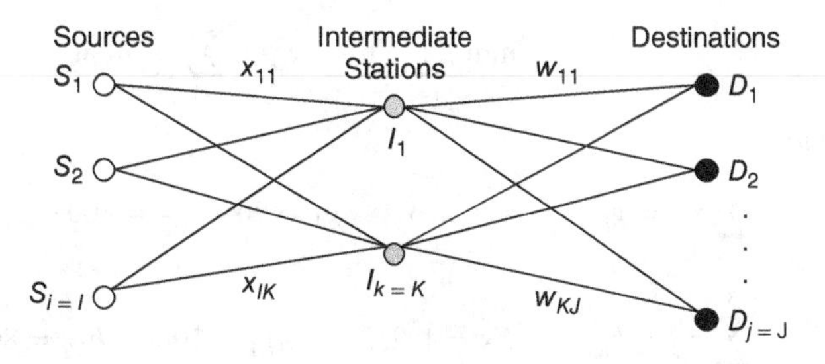

Figure 17.21 Schematic representation of the transshipment problem.

equations are added for each intermediate node, and doing this lends a little more complexity to the problem.

$$\min \sum_{i=1}^{I}\sum_{k=1}^{K} a_{i,k}x_{i,k} + \sum_{k=1}^{K}\sum_{j=1}^{J} b_{k,j}w_{k,j} \tag{17.10}$$

subject to

$$\sum_{k=1}^{K} x_{i,k} = g_i \qquad i = 1, 2, \ldots, I \tag{17.11}$$

$$\sum_{i=1}^{I} x_{i,k} \leq h_k \qquad k = 1, 2, \ldots, K$$

$$\sum_{i=1}^{I} x_{i,k} - \sum_{j=1}^{J} w_{k,j} = 0 \qquad k = 1, 2, \ldots, K \tag{17.12}$$

$$\sum_{j=1}^{J} w_{k,j} = g_k \qquad k = 1, 2, \ldots, K$$

$$\sum_{k=1}^{K} w_{k,j} \leq h_j \qquad j = 1, 2, \ldots, J \tag{17.13}$$

where x_{ik} is the quantity of material transported from source node i to intermediate node k, and w_{kj} is the quantity of material transported from intermediate node k to destination node j. Other symbols remain the same as defined for the Hitchcock formulation.

Example 17.12

It is sought to transport raw materials from three different sources to three factories for production, and then from the production centers to two installation sites, such that the total cost of transportation is minimized. Assume that at each source or intermediate (in this case, production) point, there is no excess, that is, every material generated is shipped. Also, at each intermediate or destination point, there is no excess supply. Table 17.9 presents the quantity of material generated at the source and intermediate nodes, and the capacity of each intermediate and destination node. The distances between the nodes (miles) and the transport costs per mile are given. Determine how much material must be transported from each source to each destination.

Solution

The problem can be formulated as follows:

$$\min \sum_{i=1}^{3}\sum_{j=1}^{3} a_{i,k}x_{i,k} + \sum_{i=1}^{3}\sum_{j=1}^{2} b_{k,i}w_{k,j}$$

subject to

$$\sum_{k=1}^{3} x_{i,k} = g_i \qquad i = 1, 2, 3 \qquad g_1 = 900 \quad g_2 = 400 \quad g_3 = 700$$

$$\sum_{i=1}^{3} x_{i,k} \leq h_{Kk} \qquad Kk = 1, 2, 3 \qquad h_{K1} = 1000 \quad h_{K2} = 800 \quad h_{K3} = 200$$

Table 17.9 Data for the Transshipment Model

(a) Quantity of Material Generated at Nodes and Node Capacities

		Material Generated at Source Node (tons)	Capacity of Intermediate and Destination Nodes (tons)
Sources	**Source 1**	900	
	Source 2	400	
	Source 3	700	
Intermediates	**Intermediate 1**	600	1,000
	Intermediate 2	300	800
	Intermediate 3	300	200
Destinations	**Destination 1**		800
	Destination 2		400

Cost of transportation = $120 per ton per mile.

(b) Distances between Source and Intermediate Nodes (miles)

		Intermediates		
		1	**2**	**3**
Sources	**1**	40	55	46
	2	10	22	25
	3	15	35	50

(c) Distances between Intermediate and Destination Nodes (miles)

		Destinations	
		1	**2**
Intermediates	**1**	7	22
	2	21	25
	3	30	8

$$\sum_{i=1}^{3} x_{i,k} - \sum_{j=1}^{2} w_{k,j} = 0 \qquad k = 1, 2, 3$$

$$\sum_{j=1}^{3} w_{k,j} = g_{Kk} \qquad Kk = 1, 2, 3 \qquad g_{K1} = 600 \quad g_{K2} = 300 \quad g_{K3} = 300$$

$$\sum_{i=1}^{3} w_{k,j} \leq h_j \qquad j = 1, 2 \qquad h_1 = 800 \quad h_2 = 400$$

$$x_{i,k} \geq 0 \quad w_{k,j} \geq 0$$

The $a_{i,j}$ matrix is

$$\begin{array}{c} \\ 1 \\ 2 \\ 3 \end{array}\begin{array}{c} \begin{array}{ccc} 1 & 2 & 3 \end{array} \\ \begin{bmatrix} 40 & 55 & 46 \\ 10 & 22 & 25 \\ 15 & 35 & 50 \end{bmatrix} \end{array} * 120 = \begin{array}{c} \\ 1 \\ 2 \\ 3 \end{array}\begin{array}{c} \begin{array}{ccc} 1 & 2 & 3 \end{array} \\ \begin{bmatrix} 4800 & 6600 & 5520 \\ 1200 & 2640 & 3000 \\ 1800 & 4200 & 6000 \end{bmatrix} \end{array}$$

The b_{ij} matrix is

$$\begin{array}{c} \\ 1 \\ 2 \\ 3 \end{array}\begin{array}{c} \begin{array}{cc} 1 & 2 \end{array} \\ \begin{bmatrix} 7 & 22 \\ 21 & 25 \\ 30 & 8 \end{bmatrix} \end{array} * 120 = \begin{array}{c} \\ 1 \\ 2 \\ 3 \end{array}\begin{array}{c} \begin{array}{cc} 1 & 2 \end{array} \\ \begin{bmatrix} 840 & 2640 \\ 2520 & 3000 \\ 3600 & 960 \end{bmatrix} \end{array}$$

Solving the problem using Excel Solver yields the following $x_{i,j}$ matrix:

$$\begin{array}{c} \\ 1 \\ 2 \\ 3 \end{array}\begin{array}{c} \begin{array}{ccc} 1 & 2 & 3 \end{array} \\ \begin{bmatrix} 300 & 400 & 200 \\ 0 & 400 & 0 \\ 700 & 0 & 0 \end{bmatrix} \end{array}$$

The $w_{i,j}$ matrix is

$$\begin{array}{c} \\ 1 \\ 2 \\ 3 \end{array}\begin{array}{c} \begin{array}{cc} 1 & 2 \end{array} \\ \begin{bmatrix} 600 & 0 \\ 200 & 100 \\ 0 & 300 \end{bmatrix} \end{array}$$

The corresponding amounts that must be shipped are shown in the x_{ij} and w_{ij} matrices above, and the corresponding (minimum) cost is $= 120[(40 \times 300) + (55 \times 400) + \cdots + (50 \times 0)] + [(7 \times 600) + (22 \times 0) + \cdots + (8 \times 300)] = \$9,096,000$.

17.8.3 The Transportation Circulation Minimum-Cost (TCMC) Model

In the Hitchcock model, we learned how to optimize the transportation of material across two sets of nodes: one is the set of source nodes and the other is the set of destination nodes; then in the transshipment model, we learned how to do this for three categories of nodes: the sets of source nodes, intermediate nodes, and destination nodes. Source nodes can only generate material and destination nodes can only receive material. Intermediate nodes can generate as well as receive material. What if we encounter a network where all nodes, with the exception of a single-source and destination node, are intermediate nodes that could generate or receive material? This scenario is not farfetched–it is common in certain cases of real-life logistics operations. To solve network transportation problems of this nature, we use the TCMC model (Revelle et al., 2003). In this model, the network has only one source and only one destination, and also has directed links (Figure 17.22*a*). Also, unlike the Hitchcock and transshipment models, there is no dedicated source or destination: Any node (with the exception of the single-source and single destination) could be connected to any other node and could be a source or a destination. Also, each link has a specified flow ceiling or capacity (upper bound) and flow floor (lower bound).

As Revelle and his co-authors pointed out, the TCMC problem can be rendered simpler and easier to solve by assuming that the single source could be a destination and the single destination could be a source. This is done by establishing a dummy link directed from the destination to the origin. This link, referred to as a *reverse arc*, completes a cycle of flow on the network. It is possible to represent multiple sources and destinations by including a single supersource with arcs leading out of it to all the sources and a single supersink with arcs leading into it from all the sinks. Reverse arcs then flow from the supersink to the supersource.

At each node in the TCMC model, the total incoming flow is equal to the total outgoing flow. These mass balance conditions are established by specifying a set of constraints for each node.

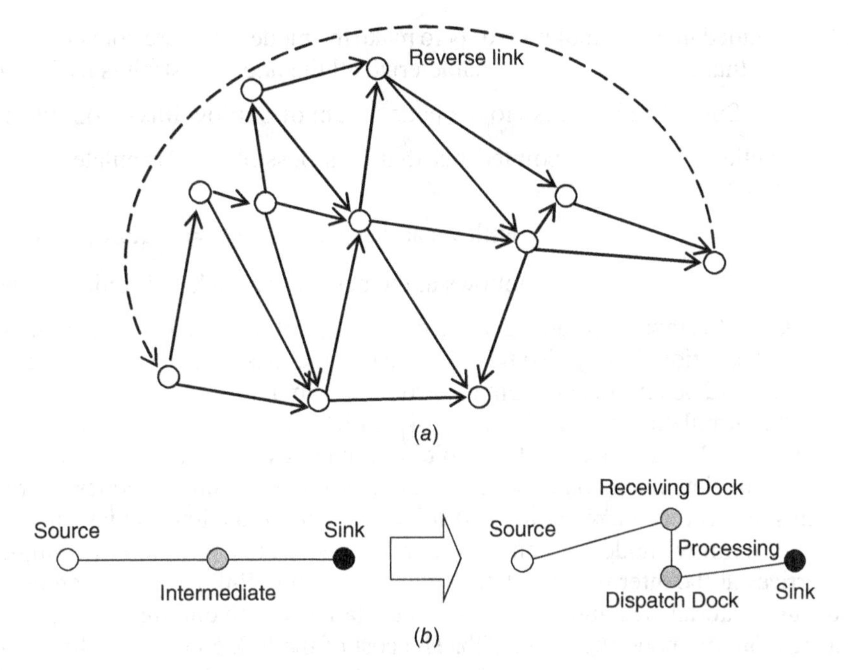

Figure 17.22 Illustrations for (*a*) the transportation circulation minimum-cost (TCMC) model and (*b*) decomposition of intermediate node.

Also, the unit cost of flow is provided for each link in the network. The model below determines the minimum-cost flows within the specified lower and upper bound constraints on flow:

$$\min \sum_{i=1}^{n} \sum_{j=1}^{n} a_{i,j} x_{i,j} \tag{17.14}$$

subject to

$$x_{i,j} \geq s_{i,j} \qquad i = 1, 2, \ldots, n \qquad j = 1, 2, \ldots, n \tag{17.15}$$

$$x_{i,j} \leq c_{i,j} \qquad i = 1, 2, \ldots, n \qquad j = 1, 2, \ldots, n \tag{17.16}$$

$$\sum_{i=1}^{n} x_{i,k} - \sum_{j=1}^{n} x_{k,j} = 0 \qquad k = 1, 2, \ldots n \tag{17.17}$$

where a_{ij} is the unit cost of shipment on the link from node i to node j; c_{ij} and s_{ij} are the upper and lower bounds on flow on the link from node i to node j, respectively; and x_{ij} is the flow in the link. Equation (17.14) is the objective function that minimizes the cost of flow in all links. Equations (17.15) and (17.16) specify that the flow in each link must be at least the lower bound and at most the upper bound. The conservation of mass [Equation (17.17)] specifies that at each node, the inflow must be equal to the outflow.

The linear programming formulation presented above for the TCMC problem is useful for general shipment operations problems where there exist intermediate nodes and material needs to be moved across capacity-bound links between multiple pairs of starting and ending node. The reverse link is included in the formulation only to make the modeling more convenient (Revelle et al., 2003): By adding that link, we can now characterize all the node constraints as follows:

Sum of inflow links to the node – sum of outflow links from the node = 0.

Revelle et al. (2003) pointed out that it is possible to formulate this problem without the reverse link by specifying that:

Sum of the outflows at the starting node = required flow.

Sum of the inflows at the destination node = required flow.

Doing this represents an algebraic elimination of the equality constraint for flow in the reverse link by substitution. It may also be observed that the mass balance constraint at each node is redundant and could be eliminated from the problem formulation.

The formulation in Equations (17.14)–(17.17) includes the link costs and material flow capacities but excludes the nodal costs or holding capacities at the intermediate nodes. To include these, each intermediate node could be represented as a pair of dummy nodes (from a practical perspective, this means that the receiving and shipping docks at the intermediate node is considered as two separate "dummy" nodes). Then these dummy nodes are connected by a single link (thus, material that arrives at the intermediate station via an incoming link is made to enter the receiving dummy node and materials leaving the intermediate station via an outgoing link are made to leave via the dispatch dummy node Figure 17.22b. The cost of the link connecting the two dummy nodes is the cost of processing the material in the intermediate facility (Revelle et al., 2003).

Example 17.13

It is sought to transport raw materials from a source node A to a destination E (Figure 17.23), such that the total cost of transportation is minimized. Table 17.10 presents c_{ij} and s_{ij}, the upper and lower bound constraints on flow on the link from node i to node j, respectively, and x_{ij} is the flow in the link ij. The distances between the nodes (miles) and the transport costs per mile are given in Table 17.10. Determine the minimum-cost flow pattern that satisfies all the upper and lower bound constraints.

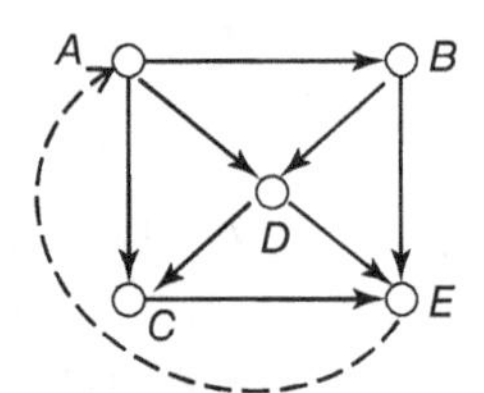

Figure 17.23 Network for Example 17.13.

Table 17.10 Data for the TCMC Model

		j				
		A	**B**	**C**	**D**	**E**
i	**A**	–	2,8,4,70	3,7,5,55	1,5,3,15	–
	B	–	–	–	2,5,7,22	4,6,8,50
	C	–	–	–	–	2,5,5,60
	D	–	–	2,5,7,22	–	4,6,8,10
	E	–	–	–	–	–

In the table, the entries represent lower bound, upper bound, distance, and cost per mile for each link.

Solution

Add a reverse link from E to D. Its lower bound, upper bound, flow, and cost for each link can be set as $(0, \infty, 1, 1)$. The problem can be formulated as follows:

$$\min \sum_{i=1}^{n} \sum_{j=1}^{n} a_{i,j} x_{i,j}$$

subject to

$$x_{i,j} \geq s_{i,j} \qquad i = 1, 2, \ldots, 5 \qquad j = 1, 2, \ldots, 5$$

$$x_{i,j} \leq c_{i,j} \qquad i = 1, 2, \ldots, 5 \qquad j = 1, 2, \ldots, 5$$

$$\sum_{i=1}^{5} x_{i,k} - \sum_{j=1}^{5} x_{k,j} = 0 \qquad k = 1, 2, \ldots, 5$$

where $a_{i,j}$ is the unit cost of shipment on the link from node i to node j, $s_{i,j}$ and $c_{i,j}$ are the lower bound and upper bound on flow on the link from node i to node j, respectively, and $x_{i,j}$ is the flow in the link. The $a_{i,j}$ matrix is

$$\begin{array}{c} \\ A \\ B \\ C \\ D \\ E \end{array} \begin{array}{c} \begin{matrix} A & B & C & D & E \end{matrix} \\ \begin{bmatrix} 0 & 280 & 275 & 45 & \infty \\ \infty & 0 & \infty & 154 & 400 \\ \infty & \infty & 0 & \infty & 300 \\ \infty & \infty & 154 & 0 & 80 \\ 0 & \infty & \infty & \infty & 0 \end{bmatrix} \end{array}$$

Solving the problem using Excel Solver, the $x_{i,j}$ matrix is determined to be

$$\begin{array}{c} \\ A \\ B \\ C \\ D \\ E \end{array} \begin{array}{c} \begin{matrix} A & B & C & D & E \end{matrix} \\ \begin{bmatrix} 0 & 6 & 3 & 4 & 0 \\ 0 & 0 & 0 & 2 & 4 \\ 0 & 0 & 0 & 0 & 5 \\ 0 & 0 & 2 & 0 & 4 \\ 13 & 0 & 0 & 0 & 0 \end{bmatrix} \end{array}$$

The total minimum cost that satisfies all the upper and lower bound constraints is \$6721.

Example 17.14

Formulate the problem without the reverse link by specifying that the sum of the outflows at the starting node must be equal to the required flow and that the sum of the inflows at the destination node also must be equal to the required flow. Assume node O is the starting node; node D is the destination node.

Solution

The formulation is

$$\min \sum_{i=1}^{n} \sum_{j=1}^{n} a_{i,j} x_{i,j}$$

subject to

$$x_{i,j} \geq s_{i,j} \qquad i = 1, 2, \ldots, n \qquad j = 1, 2, \ldots, n$$

$$x_{i,j} \leq c_{i,j} \qquad i = 1, 2, \ldots, n \qquad j = 1, 2, \ldots, n$$

$$\sum_{i=1}^{n} x_{i,k} - \sum_{j=1}^{n} x_{k,j} = 0 \qquad k = 1, 2, \ldots, n-2 \text{ (exclude nodes } O \text{ and } D)$$

$$\sum_{i=1}^{n} x_{i,D} - \sum_{j=1}^{n} x_{O,j} = 0 \qquad \text{(sum of outflows at starting node must equal to sum of inflows at destination node)}$$

where a_{ij} is the unit cost of shipment on the link from node i to node j, s_{ij} and c_{ij} are the lower bound and upper bound on flow on the link from i to j, respectively; and $x_{i,j}$ is the flow on the link from i to j; $x_{i,D}$ is the flow on the link from i to destination node (D); $x_{O,j}$ is the flow on the link from the starting node (A) to j.

17.9 NETWORK APPLICATIONS IN PROJECT MANAGEMENT

Project management is often thought of from the perspective of the construction phase. However, the reality is that engineers at any phase of system development precede the execution of their tasks (which they often call "projects") by carefully crafting and implementing a plan for that project. For example, at the design phase, project managers in the design office schedule and monitor the progress of the various tasks or projects that often include review of the system plan, initial design, detailed design, checks, final design, and design submission. Similar management of phase-related projects is carried out by engineers at all the other phases of development.

In this section, we focus on a specific aspect of project management: project planning, which involves a systematic ordering of different tasks or project activities in order to achieve a goal. For such ordering to be crafted, the project managers need to know the duration of each activity and any precedence relationships between the activities. When this systematic ordering is done effectively, the project manager will be in a better position to ascertain the extent of **activity-specific flexibilities** with respect to time, labor, money, equipment utilization, or some other resource. For example, where the resource in question is time, the following flexibilities could be addressed (Dandy et al., 2008): (a) the maximum and minimum times needed to complete the project if all activities are completed within their allotted time durations, (b) the earliest time to commence an activity or the latest time to finish an activity if the project is to be completed within the specified project period, (c) the activities that are critical for project completion within the specified or minimum time period, and (d) activity-specific flexibilities or the maximum quantity of time by which each activity may be delayed such that the project will be completed within the specified project period.

Three well-known techniques for activity planning in project management are the Gantt chart, critical path method (CPM), developed in the forties by engineers Morgan Walker and James Kelley of Du Pont Remington Rand Univac, respectively; and the Program Evaluation and Review Technique (PERT), developed by the Booz Allen Hamilton Corporation and the U.S. Navy. Of these, CPM and PERT involve the use of network analysis. In CPM, the resources associated with each activity are fixed while PERT considers that the resources vary within a certain range. The Gantt chart is a relatively old technique that shows the sequence of work and time durations (an example is provided in Chapter 20). CPM and PERT were developed to address the key limitation of the Gantt chart, namely its in ability to incorporate precedence relationships in the activity scheduling process.

17.9.1 Network-based Scheduling of Activities

For scheduling using network-based techniques, three key pieces of information are often required (Hendrickson and Tung, 2008): (i) a list of all activities that are needed for the project completion,

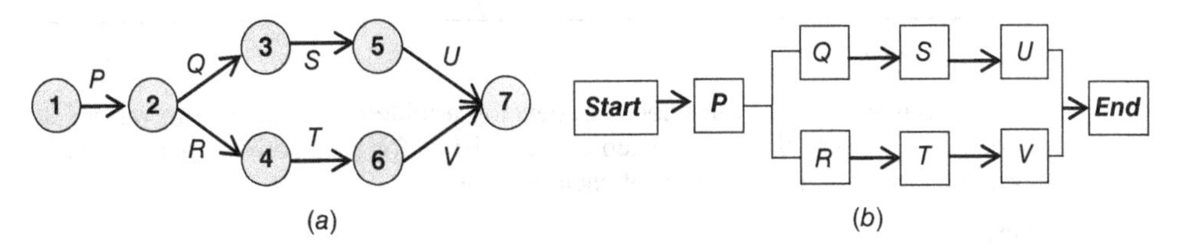

Figure 17.24 Graphical depictions of project activities: (*a*) activity-on-arrow and (*b*) activity-on-node.

(ii) the duration of each activity, and (iii) the dependencies (prerequisites and subsequents) between the activities. The activities and the associated resources that they need for completion may be depicted as a network in one of at least two ways: (a) **activity-on-arrow** (AOA), where the links are the activities and the resources needed for the activity and the nodes are start points or end points of the activities, or (b) **activity-on-node**, where the nodes are the activities and the links to or from each node are the resources needed to execute the activities. Figure 17.24 illustrates these two depictions. In the activity-on-arrow depictions, the links represent the activities and the nodes represent the activity precedences; each activity is represented by its start and end nodes; for example, activity P is 1–2. Activity P has no preceding activity, and Q and R have P as their preceding activity, that is, Q and R cannot commence before P, thus their start nodes are the same as the end nodes of P. For example, for a construction project, the main foundation cannot be constructed before the excavation; for a design project, the detailed design cannot be started before a review of the planning report. Similar precedence relationships can be seen for the activity-on-arrow and the activity-on-node representations.

For AOA network depictions of project tasks, the four basic conditions are:

1. **Activity-link condition.** Each activity must be presented by a single link in the network;
2. **Solitary origin (O)–destination (D) condition.** The network should have one origin node and one destination node to represent the start and end, respectively, of the project.
3. **Precedence condition.** Before any activity can commence, all preceding activities (links) leading to its starting node must be complete.
4. **Maximum number of links condition.** There can only be a maximum of one link between any pair of nodes.

17.9.2 Activity Characteristics

Any project consists of individual tasks, and the characteristics of the tasks are important for developing the overall project schedule. Every task is associated with the expenditure of a certain quantity of resources, for example, money, labor, time, or equipment use. In the context of scheduling, time is the resource to be considered, and will serve as the activity characteristic we consider in this section. The duration of an activity is the time taken from starting to completion. For each activity, the start time may range from the **earliest possible start time** and the **latest possible start time**; similarly, the finish time may have a range between the **earliest possible finish time** and the **latest possible finish time**. The **free float** is the quantity of time that an activity may be delayed without jeopardizing the timing of subsequent activities. The **total float** is the quantity of time that an activity may be delayed without impacting the overall project duration. "Critical" activities, whose delay will lead to delay of the overall project, have zero value for float. The **interfering float** of an activity, which is calculated as the difference between the free and total floats, measures the degree to which the subsequent activities could be delayed without delaying the overall project completion time.

Example 17.15

For any civil system of your choosing, develop a table that presents the various key activities of the project, and estimate the time-related characteristics for each activity: activity description, duration in days, and the preceding activities of each activity.

Solution

Consider a windmill construction project. Table 17.11 presents the activity characteristics for this project.

Figure 17.25 shows the time characteristics for installing the windmill tower and turbine parts, for example, the earliest start time is February 2, 2015. If the activity duration is 50 days, then the earliest finish time is August 19, 2015 and the latest finish time is August 30. Possible actual time of starting and completion are shown as lines *A*, *B*, *C*, and *D*.

Table 17.11 Activity Characteristics for a Windmill Construction Project

Activity Number	Description	Duration (days)	Preceding Activities
1	Acquire permission to enter site.	5	None
2	Order tower and turbine.	7	None
3	Clear topsoil from site.	2	None
4	Excavate for foundation.	1	3
5	Construct concrete foundation base.	30	4
6	Install base steel work.	3	5
7	Trenching for cable installation.	1	None
8	Electrical installations including earth grounding.	1	7
9	Deliver tower and turbine to site.	1	2,6
10	Install the tower and turbine.	3	8,9
11	Test to ensure that entire system is working properly.	2	10

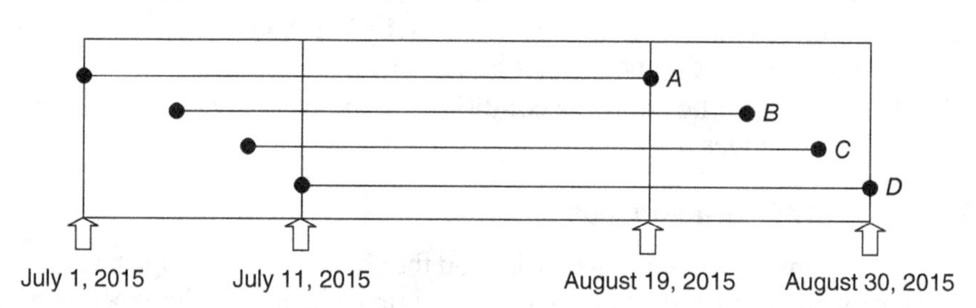

Figure 17.25 Time characteristics of wind turbine installation activity (activity 10 in Table 17.11) showing possible start and end dates.

17.9.3 The Use of Dummy Links and Nodes

In certain cases, dummy nodes and links need to be added to an organizational network so that the rules for AOA depiction can be fully satisfied (Dandy et al. (2008)). In the sections below, we discuss these situations.

(a) Multiple Links between Node Pair. First, consider the issue of multiple links between two nodes. You may recall that for AOA networks, there is a restriction that there can be at most one

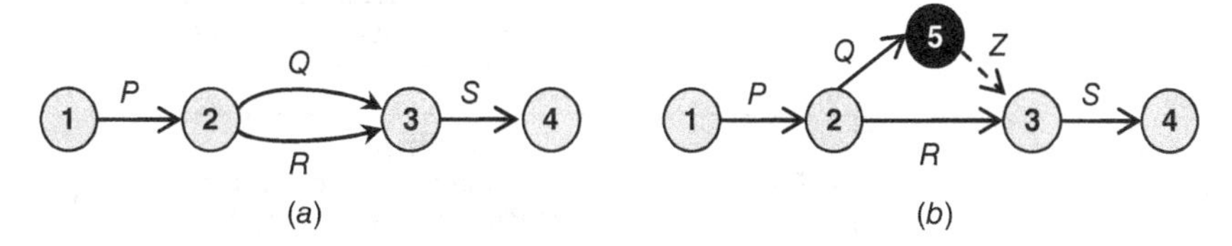

Figure 17.26 Graphical depictions of project activities: (*a*) example network violating MNL rule and (*b*) adjusted network with dummy node and link.

link between a node pair; this ensures that each activity is represented uniquely by its start and end nodes. Consider the small network in Figure 17.26*a* that has two links (two parallel activities) between nodes 2 and 3; activity *P* must be completed before either *Q* or *R* can start; also, both *Q* and *R* must be finished before *D* can start. The project task arrangement is consistent with the precedence rule but inconsistent with the maximum number of links rule. To address this violation, one of the multiple nodes can be represented by a dummy node, say, 5, and a dummy activity, say, *Z*, (see Figure 17.26*b*). The new figure satisfies all the four conditions.

(b) Enforcement of Precedence Condition. In certain organizational networks, a situation may be encountered where there is a quad of project activities, *P*, *Q*, *R*, and *S* (Figure 17.27*a*) such that one of these tasks, say, *P*, must be finished before *Q* is started; however, for *S* to commence, *P* and *R* must be finished. In this situation, a dummy link, *Z*, representing a no-cost activity, can be introduced in the network to satisfy the precedence condition.

17.9.4 The Critical Path Method (CPM)

The critical path is the longest possible continuous pathway from the project starting activity to the finishing activity and therefore helps determine the overall calendar duration needed to complete the project. The word "critical" is apt: A delay of any task in this sequence of activities will cause a delay of the entire project; and this delay will be equal to the activity delay or even more. The critical path method is a process by which the critical path is identified and other related characteristics, such as the floats, are calculated. The CPM is an important tool in the task of planning a project at any phase of the system development process. It is particularly useful for projects involving large and complex civil engineering systems. For a project that is comprised of multiple activities, the CPM is used to determine the following: (i) total time duration needed for the project completion, (ii) the earliest time at which each activity may be started, (iii) the activities that are considered critical to the project, in other words, those activities, if when delayed, will lead to delay of the overall project, (iv) the total float for each activity, (v) total float for all activities, and (vi) the free float for each activity.

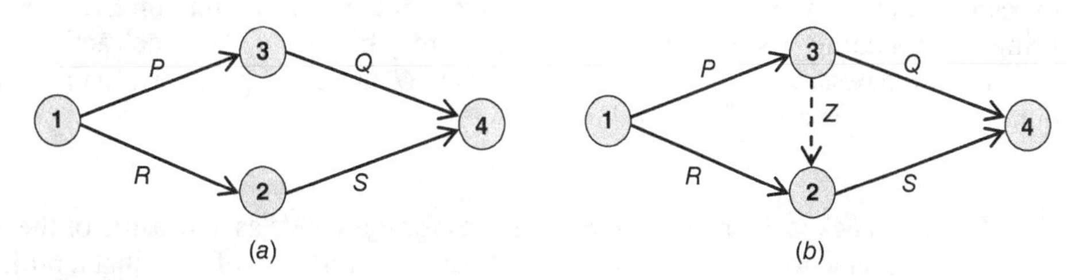

Figure 17.27 Graphical depictions of project activities: (*a*) example network violating precedence condition and (*b*) adjusted network with dummy link.

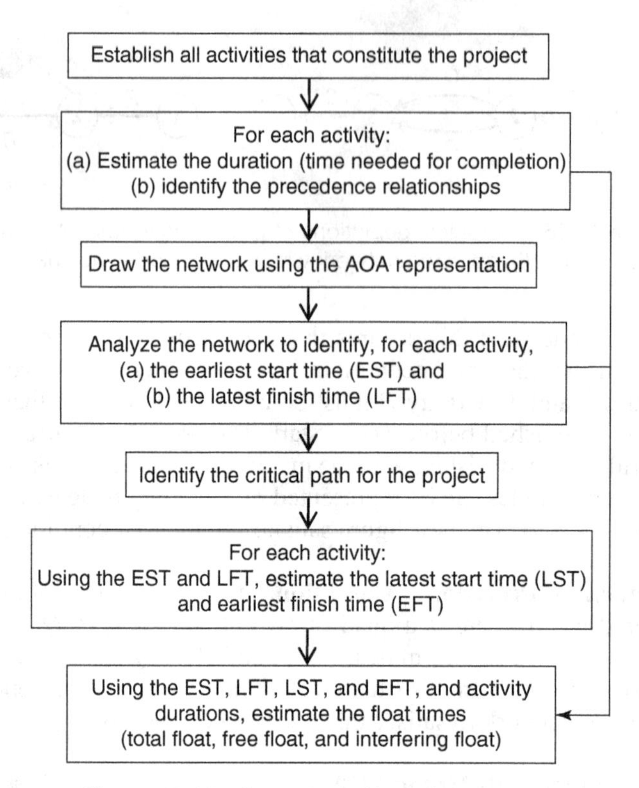

Figure 17.28 Steps for critical path method.

In using the CPM for planning a sequence of activities in project management, the steps to be followed are shown in Figure 17.28.

17.9.5 The PERT Technique

The variations in possible start and end times in CPM are an improvement on the Gantt chart method that incorporates flexibility in schedules that may arise from uncertainties in the project environment, including inclement weather, omissions and errors from preceding phases of system development, labor unrest problems, equipment breakdowns, and other uncertainties. PERT is yet an improvement on the CPM, as it considers the durations of each activity as a random variable characterized by some probability distribution with specific parameter values. The parameter values are typically determined by plotting duration data distributions from past similar activities, and the expected duration is determined as the expected value of the distribution. Alternatively, instead of using the probability distribution, three durations may be provided for each activity: the most likely, optimistic and pessimistic durations, denoted as L, O, and P, respectively. In this case, the expected duration, E, can be estimated as follows:

$$E = (O + 4L + P)/6$$

Float or slack is defined similarly as it was for the CPM: as a measure of the excess time and resources available to complete a task. Free float is the quantity of time that a project activity may be delayed without causing a delay in any subsequent task; total float is the quantity of time that an activity may be delayed without causing a delay in the entire project.

A **predecessor** event is one that immediately precedes another, and a **successor** event is one that immediately follows another. A PERT **event** is the start or end point of an activity; it does not consume any time or resource, and it is not considered to be attained until all of its predecessor activities have been completed. The **lead time** for an event is the time by which a predecessor event must be completed so that there is adequate time for the activities that must elapse before the event reaches completion. The **lag time** of an event is the earliest time by which a successor event can follow the event. **Crashing the critical path** is a term used to describe the reduction of the duration of at least one critical activity. **Fast tracking** refers to the conduction, in parallel, of an increased number of critical activities.

17.9.6 Discussion

The network depictions of project plans, CPM and PERT, are useful tools not only for scheduling activities in a project but also for identifying bottlenecks where special management efforts may be needed, evaluate potential modifications in the schedule, and to make any adjustments to the schedule that are required due to activity reductions, expansions, or modifications (Goodman and Hastak, 2007). If a similar time is scheduled, the project manager also allocates other resources including labor and equipment use. Dandy et al., (2008) and Meredith et al. (1985) show how Gantt charts could be used for such resource allocation.

SUMMARY

This chapter began with a short discussion of the importance of network analysis in civil engineering system development. Knowledge of network analysis tools is vital in several problem contexts and tasks because it helps the system owner or operator to monitor the network performance, control various operations on the network, and recommend and implement remedial actions to improve the performance of a deficient network.

This chapter discussed the fundamentals of graph theory, including basic definitions, and examined the concept of network spanning trees and the minimum flow problem. The shortest path problem in network analysis received some treatment in this chapter due to its wide applicability in several application contexts in civil engineering, including facility location problems, network connectivity assessment, and vehicle routing problems in general. The chapter also discussed the four main categories of optimal coverage of networks: Euler tours, the Chinese postman problem, Hamiltonian tours, and the traveling salesman problem (TSP). Then the chapter discussed three specific problems that involve optimal shipping across O–D pairs in a network and presented linear programming formulations to solve these problems; these included the Hitchcock shipment model, the transshipment model, and the transportation circulation minimum-cost (TCMC) model. Finally, the chapter discussed network applications in project management, namely the Gantt chart, the critical path method (CPM), and the Program Evaluation and Review Technique (PERT).

EXERCISES

1. The nodes or links in a network may be physical or virtual. List an example of networks in civil engineering or other disciplines that have (i) physical nodes and physical links, (ii) physical nodes and virtual links, (iii) virtual nodes and physical links, and (iv)virtual nodes and virtual links.
2. For the network in Figure 17.29, draw all the spanning trees and identify the minimum spanning tree using an appropriate software package.

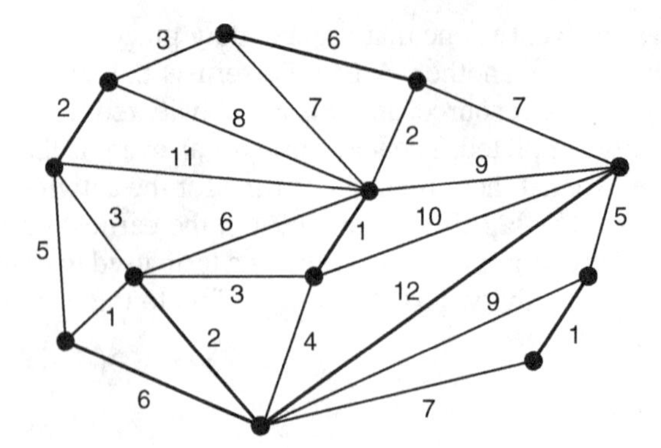

Figure 17.29 Figure for Exercise 2.

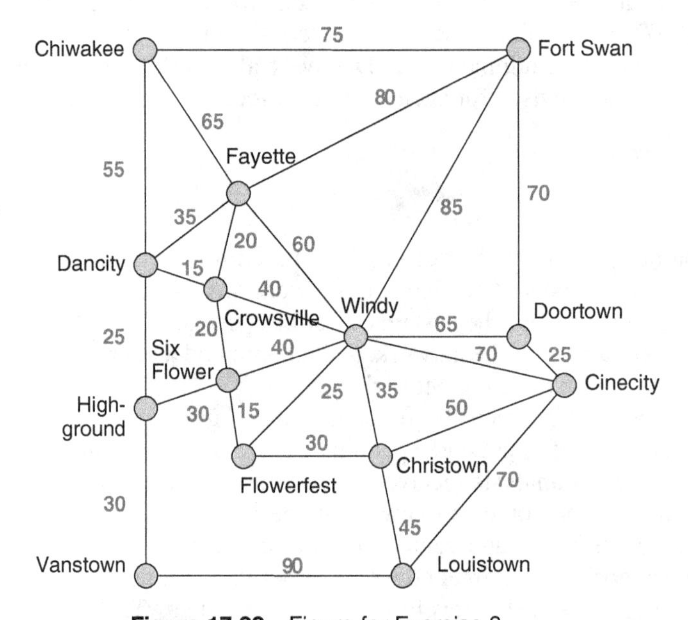

Figure 17.30 Figure for Exercise 3.

3. The network in Figure 17.30 shows the main-route distances between some major cities and towns in and around a certain region. The values shown in the network are approximate distances in miles.
   a. Reproduce the given network in an appropriate network analysis software package such as GraphMagic or GRIN. Use the package to determine the following: (i) the node–node incidence matrix, (ii) the shortest path (route and distance) from Chiwakee to Cinecity, Fort Swan to Christown, Fayette to Louistown, and Chiwakee to Flowerfest, (iii) the minimum spanning tree for the network, and (iv) the traveling salesman cycle over the entire network that could be followed by a pavement inspector.
   b. **i.** For the network, does an Eulerian path from Chiwakee and ending at Louistown exist? If one does not exist, explain why that is so. If one does exist, trace it using manual techniques or a software package.

ii. For the network, does a Eulerian cycle from Chiwakee and ending at Louistown exist? If one does not exist, explain why that is so. If one does exist, trace it using manual techniques or a software package.

iii. (a) Engineer Bridgette Specter, the region's highway bridge inspector, wishes to inspect all the network bridges and must therefore visit all links, but she wishes to minimize her total cost of travel. Trace the path that she needs to take. (b) Using an appropriate software package, determine if a Hamiltonian path (starting from Chiwakee) can be traced in the network. (c) Every week, Elizabeth travels from Chiwakee to Louistown. From the perspective of network connectivity, which link in the network is most important to her? (d) It is proposed to locate a new civil engineering facility at only one of the following cities—Fort Swan, Chiwakee, Windy, Louistown, or Vanstown—such that the sum of distances of the facility from all 14 other nodes is minimized. At which of these 5 cities should the facility be located?

4. Any structural steel truss can be thought of as a three-dimensional network configuration where the structural members are links and the structural joints are nodes. Describe briefly how you would identify the least important structural member (from a structural stability perspective) of such a "network."

5. Using Table 17.12, explain the difference between the four categories of network tours by indicating yes, no, or not applicable.

Table 17.12 Travel Tours

	Euler Tour	Chinese Postman Tour	Hamiltonian Tour	Traveling Salesman Tour
Visits each link only once?				
Visits each node only once?				
Seeks to minimize the cost of the tour?				

6. The network representing the key tasks and durations (weeks) of a construction project is shown in Figure 17.31.

a. For each activity shown in the diagram, determine the expected duration and variability of the duration.

b. Determine the expected duration of the entire project.

c. What is the standard deviation of the entire project duration?

d. Determine the probability that the project duration will be at least 3 weeks earlier than expected.

e. Which project duration had a 95% chance of being realized?

f. If the project is due in 15 weeks, what is the probability that it will be late?

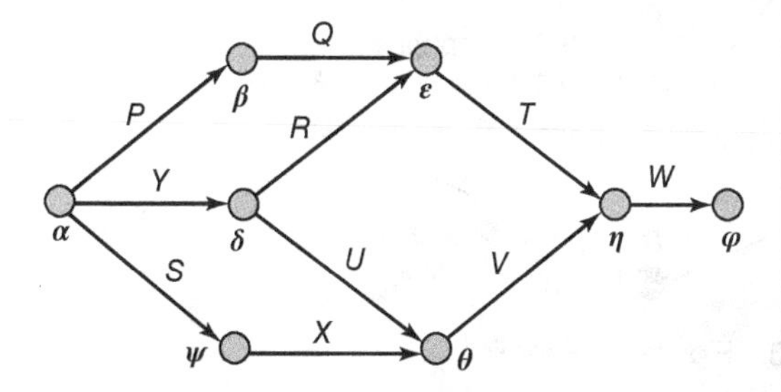

Activity	Link	Optimistic Time, t_{OPT}	Pessimistic Time, t_{PESS}	Most Likely Time, t_M
P	α–β	2	3	4
Q	β–ε	4	6	8
R	δ–ε	3	4	5
S		3	5	7
T	α–ψ	2	4	6
U	δ–θ	4	5	6
V	θ–η	2	3	4
X	ψ–θ	2	4	6
Y		3	4	5
W	η–φ	1	3	5

Figure 17.31 Figure for Exercise 5.

7. The organizational network for a large design project is identical to that shown in Figure 7.30. Using the most likely times as the project activity durations, and ignoring the optimistic and pessimistic times, determine for each activity, the earliest start time, latest finish time, latest start time, and earliest finish time. Also, determine the total float, free float, and interfering float. Identify the critical path and the minimum time needed to complete the project.
8. A mall is under construction in a certain city. The map in Figure 17.32 depicts the position of the mall and the major (arterial) road layout of the city. Determine the total capacity of the arterial network considering traffic entering at the Nimitz Highway Interchange and exiting at the mall intersection. Each link is two way, and the capacities shown are for each of both directions. Determine the flow on each link.

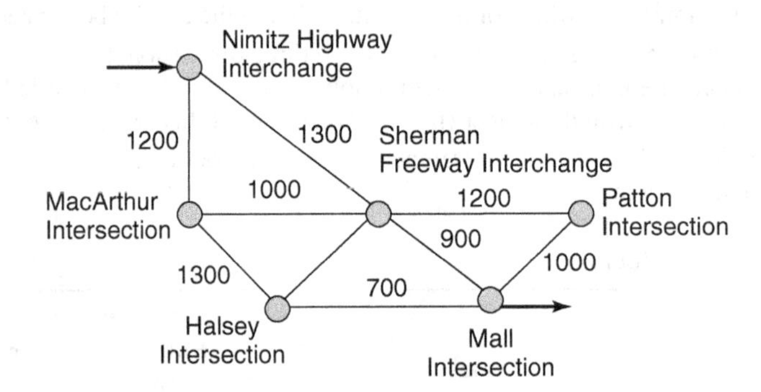

Figure 17.32 Figure for Exercise 8.

9. Figure 17.33 depicts a road network in a developing country. The figure presents the travel times and capacities of each link. A traffic count study showed that during the morning peak, 6000 vehicles move from *A* to *F* and 4500 vehicles move from *B* to *D*. It is sought to find the flow on each link. Formulate this network problem as a linear programming problem for finding the volume of traffic in each direction on each link during peak conditions, and solve the linear programming problem using MS Excel or other software package.

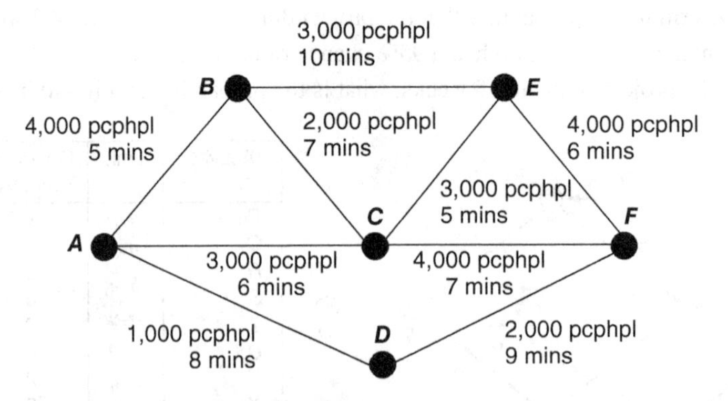

Figure 17.33 Figure for Exercise 9.

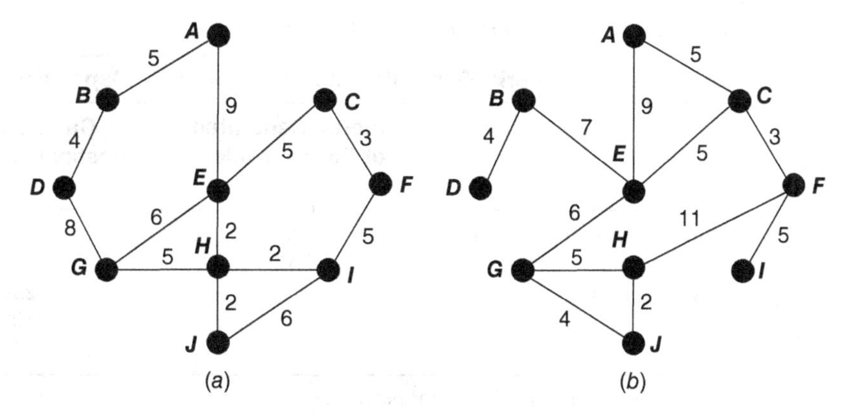

Figure 17.34 Figure for Exercise 17.10: (*a*) question and (*b*) solution.

10. For network shown in Figure 17.34, indicate which one has the greatest topological performance in terms of (i) connectivity and (ii) accessibility. Use any two measures of connectivity and any two for accessibility.

11. For the network given as Figure 17.35, formulate and solve a linear programming optimization problem to find the shortest path from vertex 1 to vertex 6.

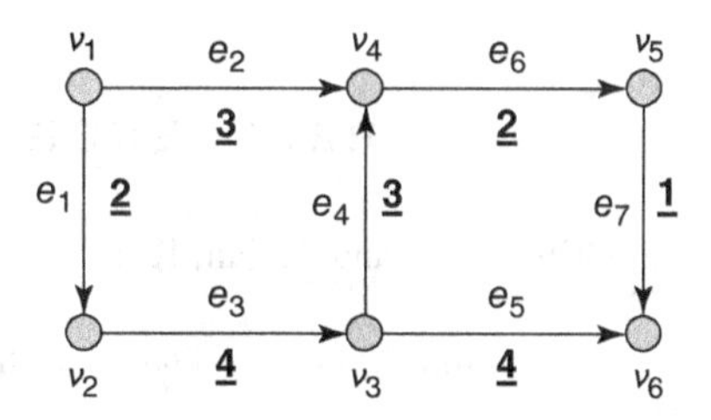

Figure 17.35 Network for Exercise 10.

12. To facilitate operations on a highway system, a logistics company is seeking to minimize the costs of transporting imported products from two seaports to three cities inland. Assume that at each source, there is no excess, that is, every material that needs to be shipped is shipped. The quantity of products generated at the source nodes, the capacities of the destination nodes, the node-to-node link distances (miles), and the transport costs ($ mile) are provided in Table 17.13. How much material must be transported from each source to each destination?

13. For the activity characteristics for a windmill construction project shown in Table 17.11, (a) draw the network for the project, (b) analyze the network to establish the earliest start time (EST) and latest finish time (LFT) for each activity, (c) the critical path(s) for the network, (d) the latest start time (LST) and earliest finish time (EFT) for each activity, and (e) the total float, free float, and interfering float for each activity. Assume that each activity starts at its earliest time. Determine the floating characteristics of this project (the total float, free float, and interfering float).

14. Draw a Gantt chart for the activity characteristics for a windmill construction project shown in Table 17.11. Assume that each activity starts at its earliest time.

Table 17.13 Data for Hitchcock Shipment Model

(a) Quantity of Material Generated at Nodes and Node Capacities

		Material Generated at Source Node (m^3)	Capacity of Destination Node (m^3)
Source nodes	**1**	300	
	2	700	
Destination nodes	**1**		250
	2		500
	3		350

Cost of transportation = \$10 per m^3 per mile.

(b) Distances between Source and Destination Nodes (miles)

		Destination node		
		1	**2**	**3**
Source node	**1**	35	20	14
	2	40	25	30

REFERENCES

Ahuja, R. K., Mehlhorn, K., Orlin, J. B., and Tarjan, R. E. (1990). Faster Algorithms for the Shortest Path Problem, *J. ACM*, 37(2) 213–223.

Ahuja, R. K., Orlin, J. B., and Pallotino, S. (2003). Dynamic Shortest Paths Minimizing Travel Times and Costs. *Networks* 41, 205.

Asmerom, G. A. (2012). Math 131 Home Page. http://www.people.vcu.edu/~gasmerom/MAT131/mst.html.

Bader, D. A., and Cong, G. (2006). Fast Shared-Memory Algorithms for Computing the Minimum Spanning Forest of Sparse Graphs. *J. Parallel Distributed Comput.* 66(11), 1366–1378.

Bellman, R. (1958). On a Routing Problem, *Quarterly of Applied Mathematics* 16, 87–90.

Bondy, A., and Murthy, U. S. R. (2008). *Graph Theory*. Springer, New York.

Chazelle, B. (2000). A Minimum Spanning Tree Algorithm with Inverse-Ackermann Type Complexity. *J. Assoc. Comput. Machinery (ACM)* 47(6), 1028–1047.

Chin, F. Y., Lam, J., and Chen, I. N. (1982). Efficient Parallel Algorithms for Some Graph Problems. *Commun. ACM* 25, 659–665.

Chong, K. W. (1996). Finding Minimum Spanning Trees on the EREW PRAM. In Proceedings of the International Computer Symposium, Taiwan, China, pp. 7–14.

Chong, K. W., and Lam, T. W. (1993). Finding Connected Components in $O(\log n \log\log n)$ Time on the EREW PRAM. In Proceedings of the 4th Annual ACM-SIAM Symposium on Discrete Algorithms (Austin, TX, Jan. 25–27). ACM, New York, pp. 11–20.

Chong, K. W. E., Han, Y., and Lam, T. W. (2001). Concurrent Threads and Optimal Parallel Minimum Spanning Trees Algorithm. *J. ACM* 48(2), 297–323.

DADS (2010). *Dictionary of Algorithms and Data Structures* [online], Paul E. Black, ed., U.S. National Institute of Standards and Technology. Accessed 23 February 2010. Available from http://www.nist.gov/dads/HTML/chinesePostman.html.

Dandy, G., Walker, D., Daniell, T., and Warner, R. (2008). *Planning and Design of Engineering Systems.* Taylor and Francis London, UK.

Dantzig, G. B. (1960). On the Shortest Route through a Network, *Management Sci.* 6, 187–190.

Dijkstra, E. W. (1959). A Note on two Problems in Connection with Graphs, *Num. Math.* 1, 269–271.

Fleischner, H. (1991). X.1 Algorithms for Eulerian Trails, Eulerian Graphs and Related Topics: Part 1, Vol. 2, *Annals of Discrete Mathematics*, 50, Elsevier, New York, X.1–13.

Ford, L. R. (1956). Network Flow Theory, P-923, The RAND Corporation, Santa Monica, CA.

Fredman, M. L., and Tarjan, R. E. (1987). Fibonacci Heaps and their uses in Improved Network Optimization Algorithms, *J. ACM*, 34, 596–615.

Fricker J. D. (1996). Network Analysis Courses Notes, Purdue University, W. Lafayette, IN.

Gabow, H. N., and Tarjan, R. E. (1989). Faster Scaling Algorithms for Network Problems, *SIAM J. Comp.*, 1013–1036.

Goldberg, A. V., and Tarjan, R. E. (1988). Finding Minimum Cost Circulations by Canceling Negative Cycles, *J. ACM.*, 36, 388–397.

Goodman A., and Hastak, M. (2007). *Infrastructure Planning* Handbook, McGraw Hill, New York.

Hendrickson, C., and Tung, A. (2008). *Advanced Scheduling Techniques, Project Management for Construction*, 2nd ed. Prentice Hall, Upper Saddle River, NJ.

Hirschberg, D. S., Chandra, A. K., and Sarwate, D. V. (1979). Computing Connected Components on Parallel Computers. *Communications ACM* 22, 461–464.

Johnson, E. L. (1972). On Shortest Paths and Sorting, Proceedings of the ACM Annual Conference 25, Boston, MA.

Johnson, D. B., and Metaxas, P. (1992). A Parallel Algorithm for Computing Minimum Spanning Trees. In Proceedings of 4th Annual ACM Symposium on Parallel Architectures and Algorithms (San Diego, CA, June 29–July 1), ACM, New York, pp. 363–372.

Karger, D. R, Klein, P. N., and Tarjan, R. E. (1995). A Randomized Linear-Time Algorithm to Find Minimum Spanning Trees. *J. ACM* 42(2), 321–328.

Karlson, R., and Poblete, P. V. (1983). An $O(\mathrm{m} \log \log D)$ Algorithm for Shortest Paths, *Discrete App. Math.* 6, 91–93.

Kruskal, J. B., Jr. (1956). On the Shortest Spanning Subtree of a Graph and the Traveling Salesman Problem. *Proc. Am. Math. Soc.* 7, 48–50.

Kwan, M-K. (1962). Graphic Programming Using Odd or Even Points. *Chinese Math.* 1, 273–277.

Leyzorek, M., Gray, R. S., Johnson, A. A., Ladew, W. C., Meaker, S. R., Petry, R. M., and Seitz, R. N. (1957). A Study of Model Techniques for Communication Systems, Case Institute of Technology, Cleveland, OH.

Meredith, D. D., Wong, K. M., Woodhead, R. W., and Wortman, R. H. (1985). *Design and Planning of Engineering Systems*, 2nd ed. Prentice Hall, Englewood Cliffs, NJ.

Moore, E. F. (1959). The Shortest Path through a Maze, Annals of the Computation Laboratory, Havard University Press, Cambridge, MA.

Murray, A. T., Matisziw, T. C., and Grubesic, T. H. (2006). Critical Network Infrastructure Analysis. *J. Geogr. Syst*, (9), 103–117.

Pettie, S., and Ramachandran, V. (2002). A Randomized Time-Work Optimal Parallel Algorithm for Finding a Minimum Spanning Forest. *SIAM J. Computing* 31(6), 1879–1895.

Poon, C. K., and Ramachandran, V. (1997). A Randomized Linear Work EREW PRAM Algorithm to Find a Minimum Spanning Forest. In *Proceedings of 8th Annual International Symposium on Algorithms and Computation and Lecture Notes in Computer Science* (1350). Springer, New York, pp. 212–222.

Prim, R. C. (1957). Shortest connection networks and some generalizations, *Bell Syst. Tech. J.* 36, 1389–1401.

Revelle, C. S., Whitlatch, E., and Wright, J. (2003). *Civil and Environmental Systems Engineering*, 2nd ed. Prentice Hall, Upper Saddle River, NJ.

Schrijver, A. (2012). On the History of the Shortest Path Problem, *Doc. Math.* 155–167.

Scott, D. M., Novak, D. C., Aultman-Hall, L., and Guo, F. (2006). Network Robustness Index, A New Method for Identifying Critical Links, *J. Transp. Geogr.* (14), 215–226.

Shimbel, A. (1953). Structural Parameters of Communication Networks. *Bull. Math. Biol.* 15, 501–507.

Skiena, S. (1990). *Implementing Discrete Mathematics: Combinatorics and Graph Theory with Mathematical*. Addison-Wesley, Reading, MA.

Sullivan, J. L., Novak, D. C., Aultman-Hall, L., and Scott, D. M. (2010). Identifying Critical Road Segments and Measuring Systemwide Robustness in Transportation Networks, *Trans. Res. A* (44), 323–336.

Tarjan, R. E. (1997). Dynamic Trees as Search Trees via Euler Tours, Applied to the Network Simplex Algorithm. *Math. Program.* 78(2), 169–177.

Thorup, M. (2000). Near-Optimal Fully-Dynamic Graph Connectivity. *STOC*, 343–350.

Xie, F., Levinson, D. (2009). Topological Evolution of Surface Transportation Networks, *Comp. Environ. Urban Sys.*, 211–223.

USEFUL RESOURCES

Ahuja, R. K., Magnanti, T. L., and Orlin, J. B. (1993). *Network Flows: Theory and Algorithms*, Applications, Prentice Hall, Englewood Cliffs, NJ.

Bondy, A., and Murthy, U. S. R. (2008). *Graph Theory*, Springer, NY.

Kerzner, H. (2009). *Project Management: A Systems Approach to Planning, Scheduling, and Controlling*, 10th ed. Wiley, Hoboken, NJ.

O'Brien, J. J., and Plotnick, F. L. (2010). *CPM in Construction Management*, 7th ed. McGraw-Hill, New York.

CHAPTER 18

QUEUING ANALYSIS

18.0 INTRODUCTION

A queue is simply a line waiting to be served. We encounter queues almost every day in our personal and professional lives as civil engineers. In our day-to-day activities, we wait in our vehicles at a drive-through window at pharmacies, fast-food restaurants, and banks; we wait in person in lines for service at supermarkets; and we hold the phone as we wait for the "next available customer service representative" from our credit card provider. Other examples include patients scheduled for the use of hospital surgery rooms, football fans waiting to get into or out of a stadium, candy in a vending machine "waiting" to be bought, students awaiting printing in a computer lab. Examples of queuing systems in civil engineering include vehicles at a stop-controlled intersection or traffic signal; passengers waiting to board a bus, rail, or water or air transit vehicle; airplanes awaiting clearance for takeoff or landing (Figure 18.1), vehicles waiting to pay at highway toll booths, and haul trucks waiting to load or unload.

The entity that arrives and joins the queue is typically a discrete element, such as a car, plane, truck, or person. In certain cases, the queuing entity could be a continuous-flow material such as water (in a reservoir "waiting" to be served daily to houses and apartments) or grain (in a silo "waiting" to be bagged and shipped to food processing plants or for export). In the remainder of this chapter, we will refer to the queuing entity as a "customer."

The arrival of customers to the queue is often random, and the time it takes to serve them is variable as it depends on the unique needs of each customer and the efficiency of the server. It is common therefore to see a queue shrinking at a certain time interval and growing at another interval. Generally, however, queues form when the demand for a service exceeds its supply, even if for a short period of time. Customers find it inconvenient and inefficient to wait in line for service; however, it is also inefficient for the service provider to provide supply that exceeds demand because if that were the case, the servers would be idle most of the time and that would be a waste of resources. As a compromise between these two extreme situations, it is expected that the customers in the queue do a little waiting and the servers are not idle most of the time. Such situations may give rise to questions like the following: How can we predict the time of waiting in a queue or the length of a queue in a given queuing situation? How can we predict the percentage of time that the servers will be idle? In this chapter, we will learn how we can use queuing theory to address questions such as these.

Queuing theory was established in the early 20th century by A.K. Erlang, who developed models to analyze the telephone exchange in Copenhagen. It has since been applied in traffic engineering (highway, air, and transit terminals); planning and design of facilities and systems (schools, post offices, shops, factories, offices, fast-food restaurants, and hospitals) (Gross et al., 2008); and disaster evaluation planning, as well as other disciplines.

Figure 18.1 Air traffic control towers serve plane queues waiting to take off or land. Image shown is Control Tower of Hartsfield–Jackson Atlanta International Airport, currently the world's busiest airport (Courtesy of J. Glover, Atlanta, Georgia).

18.0.1 Components of a Queuing Process

Every *queuing process* (often also referred to as *queuing system*) has physical and operational components. For most queuing processes in civil engineering, the physical components include the queuing entity (customers) and the servers, and the non-physical components include the arrival and service patterns, the number of queues, the queue discipline, the number of servers, the service arrangement, and the queue capacity (Figure 18.2).

18.1 ATTRIBUTES OF A QUEUING PROCESS

On the basis of our discussion in Section 18.0.1, it is clear that a queue can be described in terms of the following six basic attributes: (a) arrival pattern of the customers, (b) service pattern of the

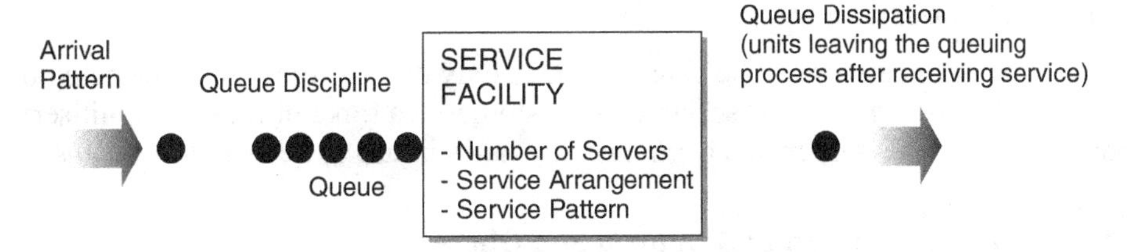

Figure 18.2 General concept of a queuing process.

servers, (c) number of servers, (d) queue discipline, (e) queue process capacity, (f) number of server channels, (g) number of server phases, and (h) number of queues. We will now discuss each attribute and its effects on a queue's characterization and performance.

18.1.1 Arrival Pattern

How can we describe or predict the manner by which customers arrive at a queue? Are the arrivals perfectly deterministic; for example, is there exactly one customer per minute? Or is it a stochastic process where the number of arriving customers per hour falls within a certain range, for example, one to three customers per minute? The arrival pattern is quantified using an **arrival rate** that may be deterministic or probabilistic and may be frequency based or interval based, as explained below.

Deterministic versus Probabilistic. In deterministic arrivals, there is a fixed number of arrivals per unit time or a fixed length of time interval between arrivals; in probabilistic arrivals, the number of arrivals per unit time or the length of the time intervals between arrivals falls within a given range of values, according to some probability distribution.

Frequency versus Interval Based. The arrival rate may be frequency based (average number of arrivals per time, f) or interval based (average time of interval between successive arrivals, λ). It is possible to convert a given interval-based arrival rate to a frequency-based rate, and vice versa, using the relation $\lambda = 1/f$).

Examples of average arrival rates can be found in the following statements: 25 customers arrive at the local fast food drive-through every hour; 2.5 minutes elapse between arrivals of customers at the local bank; the campus bus headway (time intervals between consecutive arrivals) is 10 minutes; the soft drink supplier replenishes the student lounge soft drink vending machine once every week. Where the arrivals are stochastic, the arrival process is fully described by a probability distribution.

The behavior of the customer, in terms of participation in the queue, may be state specific, that is, influenced by the queue length or waiting time (Gross et al., 2008). If the customer arrives and sees a queue that she considers too long, she may decide to *balk* (not join the queue). If she enters the queue and waits for a while, she may lose patience and decide to *renege* (leave the queue).

18.1.2 Service Pattern

The service pattern describes the way the customers are served. In a manner similar to the arrival rate, the service pattern may be deterministic or probabilistic and may be frequency based (number of customers that are served in a given time interval) or interval based (average interval of time that is used to serve each customer). When the service rate depends on the number of customers in the queue, the service pattern is described as being state dependent.

18.1.3 Number of Servers

The number of servers can be single, such as only one customer window being open at a bank or a soft drink vending machine serving one customer at a time; or it can be multiserver in nature, such as several windows open at a bank or many lanes in use at freeway toll booths.

18.1.4 Number of Server Channels and Stages

The servers can be arranged in parallel or in series. Parallel arrangements are referred to as channels, while the series arrangements are referred to as phases. Parallel arrangements (where there are multiple channels of service) include customer windows at a bank or people waiting in a building lobby that has several elevators. Figure 18.3 presents illustrations of queuing processes with different server arrangements: single phase, single channel; two-phase, single channel; single phase, three-channel; and three-phase, two-channel.

18.1.5 Queue Multiplicity

Multiplicity refers to the number of queues being served simultaneously. Examples of single-queue systems include most drive-through windows, banks, narrow toll bridges, and traffic green lights serving only one lane. In civil engineering, examples of multiple-queue systems include multilane toll booths and a traffic green light serving two or more lanes. Typically, the number of queues is less than the number of servers. In rare cases where the number of queues exceeds the number of servers, some extra rules for queue discipline are needed. Common queuing configurations in terms of the number of queues and servers include: 1 queue–1 server, 1 queue–2 servers, 2 queues–2 servers, and 4 queues–1 server. An example of a 4 queue–1 server queuing process is the operation of a four-way stop-controlled street intersection (Figure 18.4).

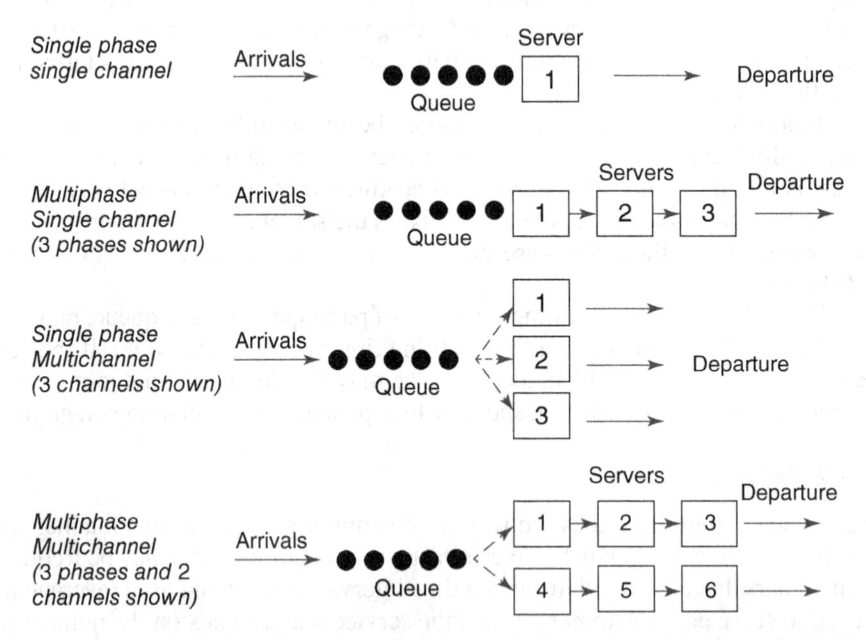

Figure 18.3 Examples of server arrangements (adapted from Khisty et al., 2012).

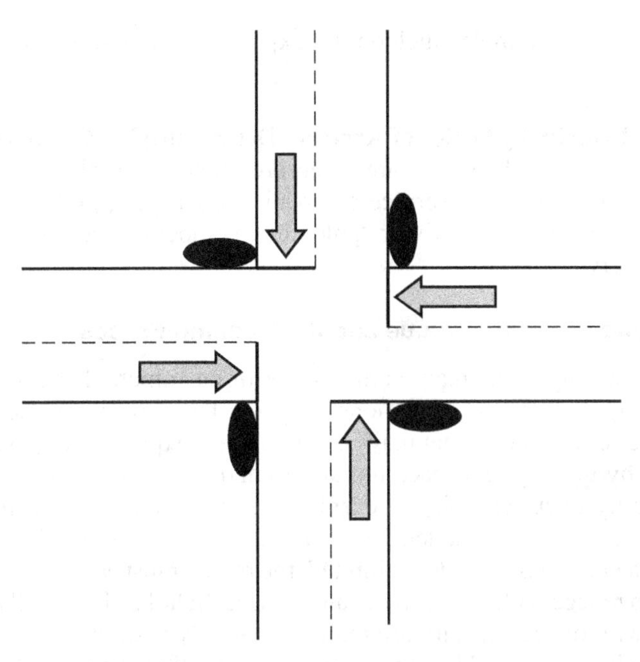

Figure 18.4 Queuing process with four queues–one server.

18.1.6 Queue Discipline

Discipline refers to the rules by which a queue is served; and a serving priority is assigned on the basis of one or more of the following criteria: order of arrival times, order of arrival urgencies, order of expected length of service time, and the order of "desirability".

(a) Serving Priority by Order of Arrival Times. In this respect, a queuing process follows any one of the following common disciplines:

- First in, first out (FIFO); first come, first served (FCFS); or last in, last out. The customer at the front of the line is always served first. FIFO is a nondiscriminatory queue discipline and is considered very fair.
- Last in, first out (LIFO). The customer at the tail end of the queue is served first and the customer in front of the line is served last. Examples include candy dispensers and crowded elevators serving two floors.
- Service in random order (SIRO). Customers are served at random irrespective of the order in which they arrived.

(b) Serving Priority by Order of Arrival Urgencies. In this queue discipline, customers needing attention most urgently are served first, regardless of their arrival time compared to others. Examples include scheduling patients for surgery in order of the severity of their illness or yielding to emergency vehicles (fire trucks, ambulances, and police) at street intersections.

(c) Serving Priority by Order of Expected Service Period. In this discipline, customers (arrivals) whose service will take shorter times are given preference, regardless of the time they joined the queue. Examples include the express lanes at supermarkets (shoppers with less than 10 items).

(d) Serving Priority by Order of Arrival "Desirability". Customers (arrivals) who have a good business history with the server are given preference, regardless of when they joined the queue. Examples include customer service companies identifying preferred callers by their phone numbers and serving them promptly or "gold club members" who always get first priority and special attention in service queues.

18.1.7 Capacities of the Queue and the Queuing Process

The queue capacity is the maximum number of customers that can be in the queue; this is often determined by the physical limitations of space. The capacity of the queuing process is the maximum number of customers that may be waiting in the queue and those that are being served; this is determined by the physical space limitations and the resource (server) limitations. As a matter of practical reality, queue capacity is never infinite; however, some theoretical queuing models include an assumption that the queue length is unlimited. In situations where this assumption is not made and the queue length is considered limited, the model must account for the fact that some customers are forced to renege without receiving any service. Scholars believe that the theoretical assumptions associated with queue capacity are really not unduly restrictive; although capacity limitations, in reality, do exist, they could be ignored in many situations because the real world is not very different from the theoretical world of queuing, and thus it is very unlikely that such extreme situations are encountered in real life (Cooper, 1981).

Example 18.1

Consider the following queuing system taken at a snapshot in time (Figure 18.5). The queue capacity is 7, and the queuing process capacity is 10.

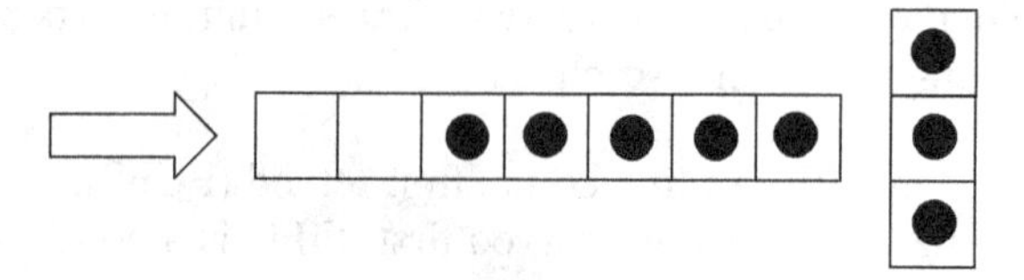

Figure 18.5 Queue capacity illustration.

18.2 PERFORMANCE OF QUEUING PROCESSES

The performance of a queuing process is typically viewed differently by the stakeholders. The customers seek to minimize their *average waiting time*, *maximum waiting time*, *average queue length*, and *maximum queue length*. On the other hand, the service provider primarily seeks to reduce the *percentage of time that each server is idle* and also to minimize the *physical space and operating costs of the queuing process*. However, saving customers time and maintaining their goodwill is also important for a manager conscious of customer satisfaction (Khisty et al., 2012), in which case the service provider also keeps an eye on customer satisfaction. Thus, the overall objective of the provider is to minimize the total cost of the queuing process by minimizing the use

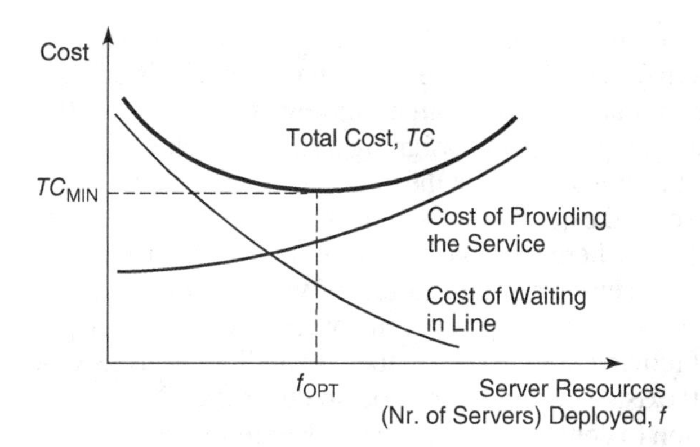

Figure 18.6 Determination of optimal service deployment.

of server resources and maximize user satisfaction by minimizing their waiting times, which are two conflicting objectives.

The number of server resources deployed translates directly to performance. A large number of server facilities and personnel would be beneficial for the customers because that would yield high levels of service and performance; however, doing so may be too costly for the service provider. On the other hand, inadequate server resources while having the benefits of little cost to the service provider often translates into low levels of service and user frustration. In trying to achieve a balance, the service provider continually analyzes the marginal costs and benefits of adding one unit of server resources.

An illustration of the trade-off relationship between the total cost of the service facility and the level of service is shown in Figure 18.6 (Khisty et al., 2012).

Example 18.2

A movie theater manager desires to determine the number of ticket sellers to be hired. It is known that the operating costs for hiring n ticket sellers is $C_0 = 80n + 120$. On the other hand, the cost of time to the customers would decrease as the number of ticket sellers increases, following the function of $C_t = 720/n + 150$. The manager wants to minimize the total costs (i.e., operating costs and time costs). How many ticket sellers should he employ?

Solution

$$\text{Total cost } C = C_0 + C_t = 80n + 120 + 720/n + 150 = 80n + 720/n + 270$$

Let the first-order derivative equal 0 and check that the second derivative is positive. Therefore, $n = 3$ is the minimizer. Total cost $C = 750$. The manager should employ three ticket sellers.

18.3 ROLE OF MARKOV CHAINS IN QUEUING ANALYSIS

A Markov chain is a process with a certain number of possible states, say S, and it continuously undergoes, in a sequence of discrete steps, changes from one state to another in a random fashion. There is a finite amount of time that the process spends in each state. This amount of time, across all states and times, can be described by a random variable that is exponentially distributed with a parameter, say q_i, when it is in state i. When the configuration is in state i and makes a transition, then it has a certain fixed probability, say p_{ij}, of being in state j. A queuing process is consistent with this pattern (Slater, 2000). The mathematical analysis of a queuing system can be significantly simplified when it is analyzed from a Markovian perspective. It is possible to construct a Markov chain for a queuing process by assigning a state to each possible configuration of the queue. Then, we can define the probability that the queue moves from one state to another as the probability that a customer arrives to the queue or is served and departs from the queue. Therefore, state 0 corresponds to a configuration where there are no customers in the queue process, state 1 is the configuration where there is one customer, and so forth. When the queue is in state i, the probability of moving to state $i - 1$ is the probability that a customer is served and departs the queuing process. Also, the probability of moving to state $i + 1$ is the probability that a customer arrives to the system. A special case of state 0 exists where there can be no departure.

The option to construct Markov chains to represent queuing processes leads to opportunities for using standard Markov chain theory techniques to analyze the performance of a queue, such as the probability that the queue has certain queue length, the time spent by a customer in the queue, and the probability that the servers are busy.

18.4 NOTATIONS FOR DESCRIBING QUEUING PROCESSES

We now will discuss some notations for describing a queuing process. We shall first examine the notations for the specific attributes of the queuing process, followed by a look at some general notations for the entire process.

18.4.1 Notations for System Individual Attributes

The notations for describing the arrival or service pattern are as follows:

M is the arrivals or service (departure) rates as described by a Poisson probability density function, and the interarrival times or service durations are described by an exponential probability density function; D is the deterministic (number of units arriving or served per time interval, or intervals between arrivals or durations of service, are constant); E_k is the kth-order Erlang distribution; H_k is the kth-order hyperexponential distribution; and G is any distribution, m is number of server channels, and n is number of queues.

The notations for describing the queue discipline are as follows:

FIFO—First-In–First-Out, or FCFS—Customers are served on the basis of their arrival times; first to arrive is served first; last to arrive is served last): LIFO—Last-In–First-Out, or FCLS—Here too, customers are served on the basis on their arrival times, however, the first to arrive is served last, and the last to arrive is served first; SIRO (service in random order).

18.4.2 Notation for Describing the Overall Queuing Process

The Kendall classification of queuing processes uses six symbols: $A/B/s/q/c/p$, where A is the distribution of the arrival rate or intervals; B is the distribution of the service duration times, s is the number of server channels; q is the queuing discipline (FIFO, LIFO, ...), assumed to be FIFO if not specified; c is the queue process capacity, assumed to be unlimited if not specified; and p is the population size (possible number of customers), assumed to be an open queuing process if not specified.

A notation style typically found in texts consists of only the first four symbols: *A/B/s/q*. Examples of notations for overall system description are as follows:

M/M/1/FIFO—a queuing system with first-in–first-out discipline, where arrivals occur at random, lengths of service times are also random, and there is only one server.

D/M/3/FIFO—a queuing system with first-in–first-out discipline, where arrivals occur at fixed times but lengths of service times are random, and there are three servers.

M/D/1/SIRO—a queuing system where waiting customers are picked at random for service, arrivals occur at random, arrivals lengths of service times are fixed, and there is only one server.

M/D/2/LIFO—a queuing system with last-in–first-out discipline, where arrivals occur at random but lengths of service times are deterministic, and there are two servers.

Other notations are as follows:

λ = average arrival rate ($1/\lambda$ = mean time between arrivals)
μ = average service rate ($1/\mu$ = mean service time)
L = average number of customers in the entire queuing process (comprised of those waiting and those being served)
L_q = average number of customers in the waiting line (average queue length)
W = average time spent by each customer in the entire queuing process (comprises the time spent waiting in the queue and the time taken to be served)
W_q = average waiting time (in the queue)
ρ = service facility utilization factor = percentage of time that the servers are busy
I = percentage of server idle time
P_0 = probability that there is no customer in the queuing process
P_n = probability that there are n customers in the queuing process

18.5 ANALYSIS OF SELECTED QUEUING PROCESS CONFIGURATIONS

To show how one could determine the performance of a queuing process or system in terms of the queue lengths, waiting time, and server utilization, we will use the following four queuing process configurations commonly studied in the literature (Cheema, 2005; Khisty et al., 2012):

1. D/D/1 queuing model, which assumes deterministic arrivals as well as deterministic departures, with one departure channel
2. M/D/1 queuing model, which assumes exponentially distributed arrival times, deterministic departures, and one departure channel
3. M/M/1 queuing model, which assumes both exponentially distributed arrivals and departure times, with one departure channel
4. M/M/N queuing model, which is similar to M/M/1, except that it has multiple departure channels

18.5.1 D/D/1/FIFO Queuing Processes

This is the simplest waiting-line model, and it assumes the following: (a) deterministic arrival, (b) deterministic service time, (c) single-channel single-server, (d) FIFO, and (e) infinite queue length. The simple queuing model's *traffic intensity*, ρ (or *utilization ratio*, r), is an important parameter of the queuing system, and is given as follows:

$$\text{Traffic intensity } (\rho) = \frac{\text{mean rate of arrival } (\lambda)}{\text{mean rate of service } (\mu)}$$

The mean interarrival time is $1/\lambda$; mean service time = $1/\mu$, and the traffic intensity $\rho = 1/\mu/1/\lambda = \lambda/\mu$. When ρ is less than 1.0 (i.e., when $\lambda < \mu$), there will be no queue. Note that if the arrivals and service times where stochastic and $\rho < 1$, a queue may form some of the time.

Example 18.3

On average, 15 planes arrive at an airport every hour and request landing permission. The control tower's average service rate is 3 minutes. Comment on the performance of this queue.

Solution

Mean arrival rate, $\lambda = 15$ planes per hour; mean service time = 3 minutes = $3/60$ hours; therefore, the mean service rate, $\mu = 1/(3/60) = 20$ planes per hour. Traffic intensity = $15/20$ = which is less than one. Thus, there will be no queue in the airspace above the airport.

Analyzing D/D/1/FIFO Queuing Processes Using Graphical Methods. The simplicity of these types of queuing processes facilitates their analysis using simple graphical tools. Figures 18.7*a* and 18.7*b* illustrate linear functions for arrivals and service where the service rate exceeds the arrival rate and where the service rate is less than the arrival rate, respectively. The linear functions of service or arrivals are not fixed at the same value over time but may change with time (as illustrated in Figure 18.7*c*).

In Figure 18.7*a*, the arrival rate is less than the service rate so there is no queue. In Figure 18.7*b*, the arrivals exceed the service rate so a queue gradually builds up over time. At time T^*, the queue length is L (which, in this case, represents the maximum queue length up to that time), and the maximum time spent by any unit in the queue (up to that time) is W. In Figure 18.7*c*, the queue length increases and decreases over time depending on the arrival rate. In this figure, the service rate is constant, but could also be changing over time. The three illustrations show linear deterministic patterns of arrivals. The patterns could also be curvilinear.

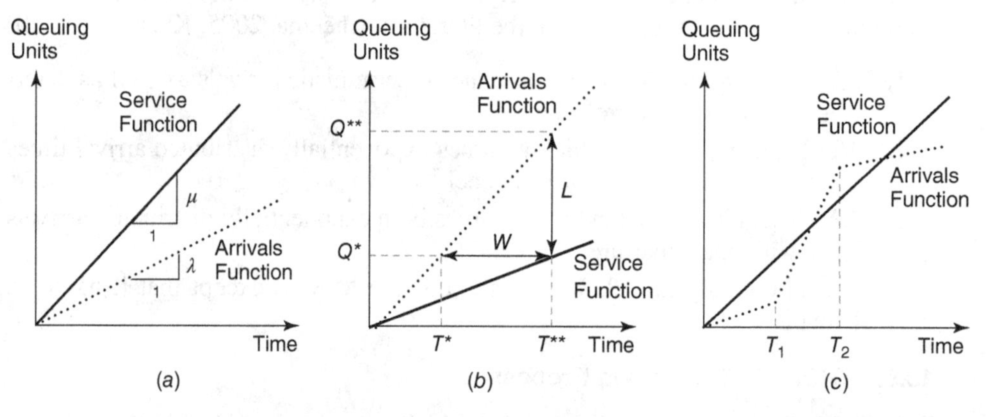

Figure 18.7 Graphical illustration of some D/D/1/FIFO queuing systems: (*a*) Arrivals < Service, (*b*) Arrivals > Service, and (*c*) Changing arrival rates.

Example 18.4

The queuing process by construction trucks loading at the site of a precast concrete vendor is presented as Figure 18.8. Determine (a) when the queue starts to form, (b) the time when the longest queue occurs, (c) the maximum queue length (the highest number of trucks in the queue at any time), (d) the maximum time spent in the queue, (e) the total delay in terms of truck-hours, and (f) the average delay per truck.

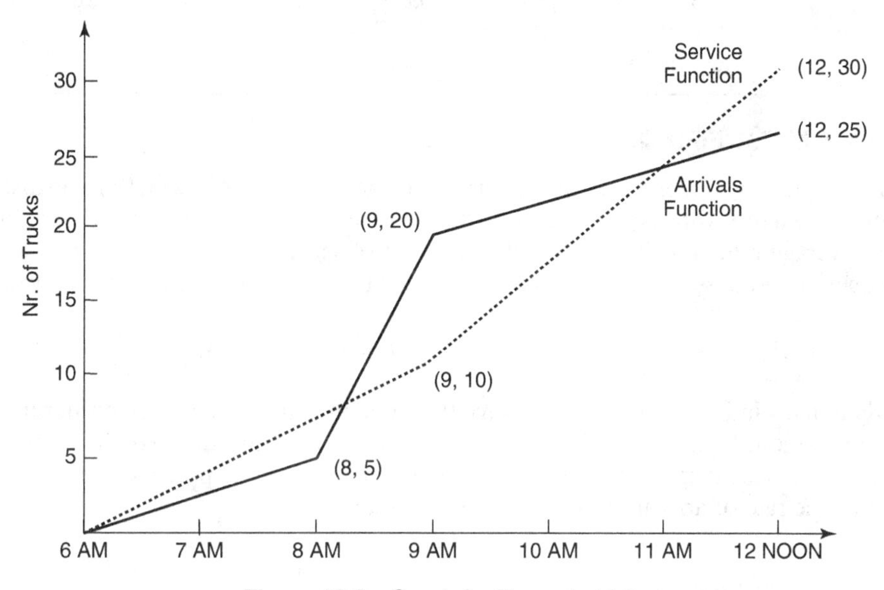

Figure 18.8 Graph for Example 18.3.

Solution

According to the graph, arrival functions and service functions during different periods can be solved.

Arrival function:

$$y = \begin{cases} 2.5t - 15 & (6 \le t \le 8) \\ 15t - 115 & (8 \le t \le 9) \\ \left(\frac{5}{3}\right)t + 5 & (9 \le t \le 20) \end{cases}$$

Service function:

$$z = \begin{cases} \left(\frac{10}{3}\right)t - 20 & (6 \le t \le 9) \\ \left(\frac{20}{3}\right)t - 50 & (9 \le t \le 20) \end{cases}$$

(a) The queue starts when the arrival function starts to exceed the service function:

$$15t - 115 = \left(\frac{10}{3}\right)t - 20$$

$$t = 8.14 \approx 8{:}08 \text{ AM}$$

(b) The longest queue occurs at 9 AM.

(c) The maximum queue length is $20 - 10 = 10$ trucks.

(d) The maximum time spent in the queue happens to the truck arriving at 9 AM. Waiting time = $(20 + 50)/(20/3) - 9 = 1.5$ hours.

(e) The total delay is the area of the trapezoids formed by the arrival function and the service function between $t = 8.14$ and $t = 11$ when the arrival function and the service function cross.

$$\text{Area} = \frac{1}{2} \times (20 - 10) \times (9 - 8.14) + \frac{1}{2} \times (20 - 10) \times (11 - 9) = 14.3 \text{ truck-hours}$$

(f) Average delay = total delay/number of trucks that encounter delay.
Trucks that encounter delay are those arriving between $t = 8.14$ and $t = 11$.
When $t = 8.14, y = z \approx 7$ trucks; when $t = 11, y = z \approx 23$ trucks. Thus, the number of trucks that encounter delay is $23 - 7 = 16$.
Average delay $= 14.3/16 = 0.89$ hour.

18.5.2 M/D/1/FIFO Queuing Processes

In this queuing model, the customers arrive in a manner that follows a Poisson distribution (i.e., their interarrival times are exponentially distributed), and the rate of service is deterministic. Assume that the average rate of arrivals is less than the rate of service. Gross et al. (2008) showed that for such a queuing process, the equations below can be used to analyze the performance.

$$P_0 = 1 - \rho \qquad L_q = \frac{\rho^2}{2(1-\rho)} \qquad L = L_q + \rho \qquad W_q = \frac{L_q}{\lambda} \qquad W = W_q + \frac{1}{\mu}$$

where the symbols and subscripts have the same meanings as defined earlier in Section 18.4.2. As long as the rate of arrivals is less than the service rate; in other words, as long as the utilization ratio, ρ, is less than 1, there will be no queue for this queue process. However, queues may form when the rate of arrivals exceeds the service rate.

Example 18.5

Trucks arrive at a construction batching plant to offload concrete aggregates at a rate of 15 trucks per hour. It takes 3 minutes to offload each truck. Describe the performance of the queue.

Solution

Mean arrival rate $\lambda = 15$ trucks per hour.

The service rate is 1 truck in 3 minutes, $\mu, = 20$ trucks per hour.

Therefore ρ (service facility utilization factor) $= \lambda/\mu = 0.75$.

L_q = average queue length $= \rho^2/[2(1-\rho)] = (0.75)^2/2(1-0.75) = 1.125$ trucks.

L = average number of trucks in the queuing process $= L_q + \rho = 1.875$ trucks.

W_q = average time spent by each truck in the queue $= L_q/\lambda = 1.125/15 = 0.075$ hr $= 4.5$ min.

W = average time spent by each truck in the queuing process $= W_q + 1/\mu = 0.075 + 1/20 = 0.125$ hr $= 7.5$ min.

P_0 = probability that the servers are idle $= 1 - \rho = 0.25 = 25\%$ of the time.

18.5.3 M/M/1/FIFO Queues

For these types of queues, the arrival frequency and service times are both random and are Poisson and exponentially distributed, respectively. The queue discipline is to serve first the customer who arrives first. It is assumed that customers are patient and thus do not leave before they reach their turn and that there is infinite space available for queuing. After the queue is left to operate for a very long time, it reaches a state of equilibrium called the *steady state*. In this state, the behavior of the queuing process is independent of its initial conditions and the elapsed time. There are two possible scenarios for the steady state: When the arrival rate exceeds the service rate, the queue slowly builds up and ultimately ends in gridlock; on the other hand, when the arrival rate is less than the service rate, the queue operates with varying queue lengths, but experiences no gridlock. Unlike M/D/1, M/M/1 may sometimes have a queue even when the arrival rate is less than the service rate; this is due to the randomness of arriving customers because these arrivals follow a probability distribution.

$$W_q = \frac{\lambda}{\mu(\mu - \lambda)}$$

$$W = \frac{1}{\mu - \lambda}$$

$$L_q = \frac{\lambda^2}{\mu(\mu - \lambda)}$$

$$L = \frac{\lambda}{\mu - \lambda}$$

$$I = P_0 = 1 - \frac{\lambda}{\mu}$$

$$P_n = \rho^n(1 - \rho) = \left(\frac{\lambda}{\mu}\right)^n \left(1 - \frac{\lambda}{\mu}\right)$$

where the symbols and subscripts have the same meanings as defined earlier in Section 18.4.2.

Also, let P represent the probability or the percentage of times that a specified event occurs. The equation for P_n can be used to derive other performance characteristics of the M/M/1 queue process as follows:

P(service busy) = P(arriving customer encounters a queue) $= \rho = \lambda/\mu$
P(n or more customers in the entire queuing process) $= \rho^n$
P(less than n customers in the entire queuing process) $= 1 - \rho^n$
P(time spent by a customer in the entire queuing process exceeds t) $= e^{-(t/W_W)}$
P(time spent by a customer waiting in the queue exceeds t) $= \rho \times e^{-(t/W_Q)}$

Example 18.6

Passengers seeking personal transit service at a busy airport terminal arrive in the loading area according to a Poisson distribution with an average arrival rate of four customers per minute. The rate of service is exponentially distributed with an average dispatch rate of 360 customers every hour. The queue discipline is FCFS; and it is assumed that no customer balks or reneges and that the airport terminal has infinite space available for customers who queue for this transit service. Determine the average number of customers in the entire queuing process, the average time spent in the queuing process, and the percentage of time that the servers are idle.

Solution

The arrival rate is λ = four customers per minute and the service rate is six customers per minute.

The utilization ratio is: $\rho = \lambda/\mu = 0.67$.

Average number of customers in the entire queuing process $= L = \lambda/(\mu - \lambda) = 4/(6 - 4)$ = two customers

Average time spent in the queuing process $= W = 1/(\mu - \lambda) = 1/(6 - 4) = 0.5$ minutes

Percentage of time that the servers are idle, $I = 1 - \lambda/\mu = 1 - 0.67 = 33\%$

18.5.4 M/M/N/FIFO Queue

A more general form of the M/M/1/FIFO model considers multiple channels. Also, arrivals are Poisson distributed and service rates are exponentially distributed in this model. Because there are N service channels, the average service rate is determined as $N \cdot \mu$. The following equations can be used to analyze the operational performance characteristics of M/M/N queues.

The probability of having no customer in the entire queuing process is

$$P_0 = \left(\sum_{k=0}^{N-1} \frac{\rho^k}{k!} + \frac{\rho^N}{N!\,(1 - \rho/N)} \right)^{-1}$$

where $k = 1, 2, \ldots N$ server channels.

The probability of having n units in the entire queuing process depends on whether n exceeds N:

$$P_n = \frac{P_0 \rho^n}{n!} \qquad \text{when } n \leq N$$

$$P_n = \frac{P_0 \rho^n}{N^{n-N} N!} \qquad \text{when } n \geq N$$

The average length of queue (number of customers waiting in the queue) is

$$L_q = \frac{P_0 \rho^{N+1}}{N!N} \left[\left(1 - \frac{\rho}{N} \right)^2 \right]^{-1}$$

The average number of customers in the entire queuing process is

$$L = \frac{P_0 \rho^{N+1}}{N!N} \left[\frac{1}{(1 - \rho N)^2} \right]$$

The average waiting time (time spent in the queue) is

$$W_q = \frac{\rho + L}{\lambda} - \frac{1}{\mu}$$

The average time spent in the entire queuing process is

$$W = \frac{\rho + L}{\lambda}$$

The probability that an arriving customer encounters a queue (or, the probability that the number of units in the entire queuing process, n, is greater than the number of departure channels, N) is

$$P_{n>N} = \frac{P_0 \rho^{N+1}}{N!N(1 - \rho/N)}$$

where the symbols and subscripts have the same meanings as defined earlier in Section 18.4.2.

Example 18.7

A queuing process has 5 server channels. Customers arrive at an average rate of 16 customers per minute, and it takes 15 seconds to serve each customer. Assume FCFS queuing discipline and that the interarrival and service times are exponentially distributed. Describe the performance of this queue in terms of the percentage of time that the servers are idle, the average queue length, and the average time spent by a customer in the entire queuing process.

Solution

Average arrival rate, $\lambda = 16$ customers /minute

Average service rate per server channel, $\mu = \frac{60}{15} = 4$ customers per minute

Five server channels are open, therefore, $N = 5$

Average service rate for 5 channels $= N \times \mu = 5 \times 4 = 20$ vehicles per minute

Utilization ratio, $\rho = \lambda/(N^{*}\mu) = \frac{16}{20} = \frac{4}{5} = 0.8$

The percentage of time that the servers are idle is:

$$P_0 = \left(\sum_{k=0}^{N-1} \frac{\rho^k}{k!} + \frac{\rho^N}{N!\,(1-\rho/N)}\right)^{-1} = \left[1 + \frac{4^1}{1!} + \frac{4^2}{2!} + \frac{4^3}{3!} + \frac{4^4}{4!} + \frac{4^5}{5!\left(1-\frac{4}{5}\right)}\right]^{-1} = 0.013$$

The average queue length, or the average number of customers in the queue is:

$$L = \frac{P_0 \rho^{N+1}}{N!N}\left[\frac{1}{(1-\rho/N)^2}\right] = \frac{0.013 \times (4)^6}{5! \times 5}\left[\frac{1}{\left(1-\frac{4}{5}\right)^2}\right] = 2.22$$

The average time spent by customers in the entire queuing process

$$W = \frac{\rho + L}{\lambda} = \frac{4 + 2.22}{16} = 0.39 \text{ minutes}$$

18.6 CONCLUDING REMARKS ON QUEUING ANALYSIS

Classical queuing theory is replete with assumptions that may be too restrictive and thus may not adequately model real-world situations with exactitude. For example, the mathematical models often include the assumption of an infinite number of arriving customers, unconstrained times for service or between arrivals, and infinite queue capacity when, in reality, queues in civil engineering or our personal daily lives rarely reflect these conditions.

In recognizing this dilemma, past researchers have cautioned that the development and implementation of queuing analysis must not be stymied by a lack of closed-form solutions (Gross et al., 2008). In order to put queuing theory to practical use, analysts have been encouraged not to get caught up in the search for exact solutions for queuing problems, but rather to seek robust computational analysis that may yield approximate solutions and to carry out sensitivity analysis as well to examine different performance outcomes under different queuing conditions.

Also, it has been argued that some of the restrictive assumptions associated with the mathematical models may be ignored without sacrificing the integrity of the analysis because there is very little difference between theory and the real world. In other words, it is extremely unlikely that real-world queuing conditions will fall outside those amenable to mathematical analysis. Besides, queuing models have been found to provide satisfactory results even when the queues operate outside the boundary conditions. For example, in daily queues for personal services such as groceries or fast food, there are people who become impatient and leave the queue. However, their numbers are often so small that they do not adversely influence the integrity of the queuing analysis for such situations.

Table 18.1 Summary of Queuing Process Formula

	FIFO M/D/1	FIFO M/M/1	FIFO M/M/N
Average Arrival Rate		λ	
Average Service Rate		μ	
Utilization Ratio		$\rho = \frac{\lambda}{\mu}$	
Average number of queuing units in the queuing process (queue + server), L	$L = L_q + \rho$	$L = \frac{\lambda}{\mu - \lambda}$	
Average number of queuing units in the queue only (i.e., average queue length), L_q	$L_q = \frac{\rho^2}{2(1-\rho)}$	$L_q = \frac{\lambda^2}{\mu(\mu-\lambda)} = \frac{\rho^2}{1-\rho}$	$L_q = \frac{P_0 \rho^{N+1}}{N!N}\left[\frac{1}{(1-\rho/N)^2}\right]$
Average time spent in the queuing process (queue + server), W	$W = W_q + \frac{1}{\mu}$	$W = \frac{1}{\mu - \lambda}$	$W = \frac{\rho + L}{\lambda}$
Average time spent in queue only (mean waiting time), W_q	$W_q = \frac{L_q}{\lambda}$	$W_q = \frac{\lambda}{\mu(\mu-\lambda)}$	$W_q = \frac{\rho + L}{\lambda} - \frac{1}{\mu}$
Probability of waiting more than h units of time in the queue, $T_{t=h}$		$P_{t=h} = \frac{\lambda}{\mu} e^{(\lambda-\mu)15}$	
Probability (or fraction of time) that there are n units in the queuing system, P_n		$P_n = \left(1 - \frac{\lambda}{\mu}\right)\left(\frac{\lambda}{\mu}\right)^n$	$P_n = \frac{P_0 \rho^n}{n!}$ for $n \leq N$ $P_n = \frac{P_0 \rho^n}{N^{n-N} N!}$ for $n > N$
Probability (or fraction of time) that there are more than f units in the queuing system, P_n		$P_{n>f} = \left(\frac{\lambda}{\mu}\right)^{f+1}$	
Percentage of time that the server is idle, $I\,(= P_0)$	$P_0 = 1 - \frac{\lambda}{\mu}$	$P_0 = 1 - \frac{\lambda}{\mu}$	$P_0 = \frac{1}{\left(\sum_{n=0}^{N-1} \frac{\rho^n}{n!} + \frac{\rho^N}{N!(1-\rho/N)}\right)}$

In industrial systems, including those associated with the production of civil engineering products, the complexity of modern production lines due to product-specific specifications and attributes makes it difficult to use classical queuing theory; therefore, for such systems, specialized analytical tools or computer simulations have been developed to analyze and visualize production queues and to undertake control functions to optimize the queue performance (Allen, 1990).

SUMMARY

In this chapter, we discussed the basic attributes that could be used to describe a queuing process, which include the arrival patterns of customers, the service pattern of servers, the number of servers, the queue discipline, the capacity of the entire queuing process, the number of server channels, the number of server stages, and the number of queues. We also determined that the arrival pattern could be deterministic or probabilistic and frequency or interval based. Queue discipline was found to be an important aspect of queuing analysis that involves giving priority to arrivals on the basis of their arrival times, arrival urgencies, order of expected service period, or the order of arrival "desirability."

The primary reason for carrying out queuing analysis is to examine the performance of an existing queue or to predict the performance of a future queue associated with the construction or operations of a civil engineering system. These queues include vehicles at an intersection traffic signal, planes waiting to land or take off, construction trucks waiting to load or unload, and vehicles waiting to pay at toll road booths. In any of these situations, the performance of the queue can be assessed on the basis of a variety of criteria that include the average waiting time of the customers, the average queue length, the percentage of time that a server is idle, the probability that there are a certain number of customers in the queue, and the probability that an arriving customer will encounter a queue.

The number of server resources that are deployed translate directly into performance. A large number of servers is favored by customers; however, doing so may be too costly for the service provider. On the other hand, inadequate server resources, while having the benefits of little cost to the service provider, often translates into low levels of service and user frustration. In striking a balance between these two stakeholders of a queuing process, it can be useful to carry out optimization to ascertain the optimal level of servers to be deployed in a given situation.

This chapter also presented and demonstrated equations to analyze a number of queuing process configurations, including D/D/1, M/D/1, M/M/1, and M/M/N queues. Also, this chapter showed how to analyze D/D/1/FIFO queuing processes using graphical methods.

Table 18.1 presents a summary of queuing process formula.

EXERCISES

1. What is a queuing process or system? Provide examples in three different areas of civil engineering. List and describe each key component of a queuing process.

2. Why is it often necessary to set up multichannel and multiphase queuing systems? Cite examples of such systems in civil engineering in everyday life.

3. Explain why, for deterministic queuing, a queue will never form when the service rate exceeds the arrival rate. Also, explain why, for a probabilistic queue, a queue may form even when the average service rate exceeds the average arrival rate.

4. In the context of airplanes at an airport waiting to take off from the runway, describe the following terms: (a) queue length, (b) waiting time, (c) capacity of the queue, and (d) idle time of the server.

5. Why is it difficult to establish a single measure of effectiveness for a queuing process?

6. Explain how a queuing process can be modeled using the Markov chain concept.

7. On a typical afternoon, three solid waste vehicles arrive at a landfill per hour. It takes 15 minutes to direct each vehicle to the appropriate site and discharge. Assuming that arrival rates and service times are deterministic, there is only one server, and a FIFO queuing discipline is used, ascertain whether a queue can be expected to form at the landfill.

8. During the reconstruction of a major bridge, concrete trucks arrive at the batching plant loading dock at a rate of 24 trucks/hour, starting at 6:00 AM. After 8 hours, the arrival rate declines to 12 trucks/hour and continues at that level for 8 hours. The time required to load each truck is 3 minutes. Assuming that arrival rates and service times are deterministic, there is only one server, and a FIFO queuing discipline is used, describe the performance of the system in terms of (a) the time of queue dissipation, (b) length of longest queue and when it occurs, (c) longest delay and when it occurs, (d) the total delay (person-minutes), (e) the number of trucks that encounter delay, and (f) the average delay time per delayed truck. Plot the graph for the queuing system from 6:00 AM to 10:00 PM, showing the arrival and service functions.

9. Airplanes arrive at a single landing strip at an average rate of 10 planes per hour. On average, a plane requires 4 minutes to land and taxi to its terminal. Assuming that the plane arrival pattern can be described by the Poisson distribution and that the service time is exponentially distributed, calculate the following: (a) the percent of the time that the runway will be idle, (b) the probability that at any given time, there will be 3 planes in the queuing system, (c) the average number of planes in the queuing system, (d) the average queue length, and (e) the average time each plane spends in the queuing system.

10. List all the characteristics (assumptions) of the FIFO M/M/1 queuing model. A queuing process has one server, with a mean arrival rate of four customers every minute and the service rate is five customers per minute. Assuming that arrivals are Poisson distributed and service times are exponentially distributed and that steady-state conditions exist, calculate the following characteristics of this queuing process: (a) utilization ratio, (b) average number of customers in the system, (c) average length of the queue (average number of customers in the queue), (d) average waiting time of each customer in the queuing process, (e) average waiting time of each customer in the queue, and (f) percent of time that the server is idle.

11. The operator of a tolled six-lane (each direction) freeway seeks to identify how many toll booths to keep open during peak periods. If too few toll booths are open, the agency saves money because it pays only for a few toll booth personnel; however, there is excessive user cost due to long delays in the queues that form. On the other hand, if too may toll booths are open, there are short queues and short delay even though the agency spends more for the large number of personnel. From past observations, the daily agency cost and user cost associated with each resource allocation option (number of personnel) have been determined (Table 18.2). Assuming that \$1 of agency cost is equivalent to \$1 of user cost, calculate the total cost associated with each resource allocation option. Plot a graph of total cost, agency cost, and user cost versus the number of resource options. Determine the optimal level of resources for the toll booth operations. During a special day, the city government asks the agency to open all toll booths, and the agency complies. In acceding to the city's request, how much agency cost was traded off, on average, for a unit increase in resource deployment?

Table 18.2 Table for Exercise 11

Number of Personnel	1	2	3	4	5	6
Agency cost (\$1000s)	3.2	4.1	5.8	7.4	10.3	16
User cost (\$1000s)	14.6	11.2	8.1	6.0	5.6	5.1

12. The customer service division of a large civil engineering agency employs 4 people for receiving complaints and other feedback from the system users and the community. During their 8-hour workday, the customer service personnel handle an average of 30 incoming complaints and reports on an FCFS basis. On average, the personnel spend 28 minutes handling each complaint or report. The incoming claim frequency follows a Poisson distribution and the handling follows an exponential distribution. Determine (a) the average number of customers in the queuing process, (b) average number of queuing units in the queue only (i.e., the average queue length), (c) average time spent in the queuing process, (d) average time spent in queue only (mean waiting time), (e) probability (or fraction of time) that there are 2 customers in the queuing system, (f) percentage of time that the server is idle, and (g) the number of hours that the customer service personnel spend with a customer each day.

REFERENCES

Allen, A. O. (1990). *Probability, Statistics, and Queuing Theory with Computer Science Applications*, 2nd ed. Academic, San Diego.

Cheema, D. S. (2005). *Operations Research*, Laxmi Publications, New Delhi.

Cooper, R. B. (1981). *Introduction to Queuing Theory*, 2nd ed. North Holland, Amsterdam.

Gross, D., Shortle, J. F., Thomson, J. M., and Harris, C. M. (2008). *Fundamentals of Queuing Theory*, 4th ed. Wiley, Hoboken, NJ.

Khisty, C. J., Mohammadi, J., and Amekudzi, A. A. (2012). *Systems Engineering with Economics, Probability, and Statistics*, 2nd ed. Ross Publishing, Plantation, FL.

Slater, T. (2000). Queuing Theory. homepages.inf.ed.ac.uk/.

USEFUL RESOURCES

Hall, R. W. (1991). *Queuing Methods for Services and Manufacturing*. Prentice-Hall, Englewood Cliffs, NJ.

Hillier, F. S., and Lieberman, G. T. (1995). *Introduction to Operations Research*, 6th ed. McGraw-Hill, New York.

Narajan Bhat, U. (2008). *An Introduction to Queuing Theory Modeling and Analysis in Applications*. Birkhauser-Springer, New York.

Ng, C. H. (1996). *Queuing Modeling Fundamentals*. Wiley, New York.

PART IV

THE PHASES OF SYSTEMS DEVELOPMENT

CHAPTER 19

THE NEEDS ASSESSMENT PHASE

19.0 INTRODUCTION

From the various definitions of civil engineering presented in Chapter 1, it is clear that civil engineers seek to design and implement different kinds of systems intended to satisfy the physical infrastructure needs of society. Once current or foreseeable problems are identified, the engineer must be able to confidently state that a critical need exists for some intervention (e.g., provide a new system or replace, expand, or improve the strength or condition of an existing system). Often referred to as "gap analysis," the concept of needs assessment has been discussed extensively in other fields, such as personnel management (Steadham, 1980), social sciences (Reviere et al., 1996), and health care (Ratnapalan and Hilliard, 2002). This chapter presents the various stages of the needs assessment phase we will also discuss possible sources of need for systems in different civil engineering disciplines and the various mechanisms for assessing the need for a new system or for improvements to an existing system. This will be followed by a discussion of the mathematical formulations for quantifying system need under different scenarios and patterns of demand and supply.

19.0.1 Stages of the Needs Assessment Phase

Figure 19.1 presents the stages of the needs assessment phase. The first stage determines whether or not a problem exists and the magnitude of the problem. The next stage identifies the relevant stakeholders (i.e., the entities affected by the system). The third stage establishes the goals and objectives of the system. The stages of stakeholder identification and establishment of goals and objectives both were discussed in detail in Chapter 3. The final stage of needs assessment is to quantify the amount of need on the basis of demand and supply trends or functions.

19.0.2 Problem Identification

Problem identification is a key preliminary aspect of needs assessment. In problem identification, the engineer ascertains whether a socioeconomic need or technical deficiency of such scale and scope exists for which some intervention is needed, either a new system (demolishing and replacing an existing system or providing one where none existed previously) or the capacity of an existing system needs to be expanded or its strength and condition improved. For example, is there a need to widen that existing highway to reduce congestion? Are this city's sewer pipes so badly corroded that they need to be replaced? Does that town need a new sewage treatment plant? Does this airport runway need to be resurfaced? Must this city bridge be retrofitted to reduce its vulnerability to structural failure from a possible earthquake? Should barriers be constructed along that freeway to protect the area residents from traffic noise and privacy intrusion? Is there a need to remediate that environment degraded by coal-mining activities? Is there a need to implement geotechnical systems to stabilize this sinking building?

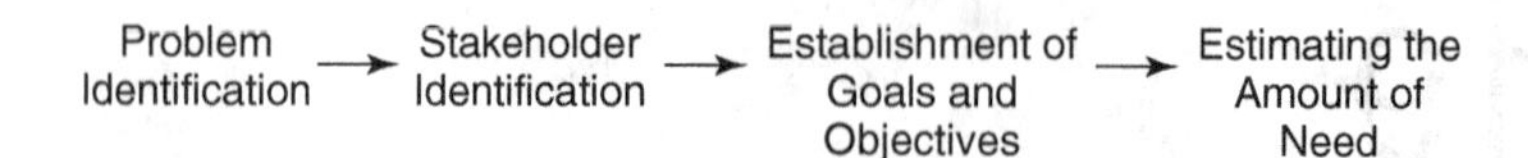

Figure 19.1 Stages of the needs assessment phase.

Problem identification and the other stages of the needs assessment phase is often carried out by people of different disciplines, depending on the developmental sector involved. The need for a new or expanded road system, for example, may be driven by problems that have been identified in the course of transporting or storing agricultural or industrial goods, accessibility to remote areas for health-care delivery, telecommunications, military purposes, or evacuation in times of catastrophic events such as hurricanes and man-made attacks. As such, the "problem" could be identified by at least one of a diverse group of experts such as agricultural economists, health administrators, military personnel, emergency management administrators, local chambers of commerce, or even the general public. Problems also could be identified by civil engineers, particularly when specific physical aspects of an existing deficient system are involved. For example, the need to replace the deck of a particular bridge could be identified by engineering maintenance crews or field inspectors.

In identifying problems or confirming problems identified by others, the civil engineer needs to ascertain whether (and the extent to which) the nature and magnitude of the problem at hand may have changed (or is expected to change) between the time of problem identification and the time of providing the system or some component or process associated with the system. Also, the engineer must consider the source of the original problem statement, as some sources may bias the statement in some way because of a unique perspective (Voland, 1999). In that case, further perspectives to the problem must be sought and the problem reconsidered. It is only when the nature of the problem is formulated correctly and completely that a reliable needs assessment and the subsequent search for solutions can be carried out.

At the needs estimation stage of the needs assessment phase, the characteristics of the need are established, such as the temporal span (how long the system will be needed), the spatial spread (what space will be needed), and the affected entities (which stakeholders are calling for the need to be addressed or are associated with the need). For example, does the need exist at all times or only at certain specific times of the day, week, month, or year? Is it needed by all persons in an area or is it intended for only certain people of specific characteristics within an area (e.g., tourists, low-income persons, children, handicapped, etc.). Such considerations are helpful in the subsequent phases of system planning and design. De Neufville (1990) and Voland (1999) stated that a prerequisite to any engineering design process is to identify as clearly as possible the existing needs that can be addressed using available technology. A new system or system component may originate out of a concern to protect the health and safety of the public or to improve the quality of life for certain demographics such as the elderly and the disabled. An existing system (physical structure or operational process) may need to be redesigned in order to make it more effective in addressing a problem.

19.0.3 Identifying Stakeholders

As discussed in Chapter 3, there often are several interested parties involved (also referred to as "stakeholders") in developments involving civil systems, and each of them must be given due audience and respect. Stakeholders can include an agency (system owner or operator), the system users, the community in which the system is located, environmental protection groups, and the general public. In the task of goals identification at the needs assessment phase of systems development and also in the task of evaluation at the phases subsequent to needs assessment (namely, the planning and design phases), the stakeholders are identified as those who have reported the existence of a problem, those who will potentially benefit from the system, and those who potentially will be adversely affected by the system. However, at the stage of needs estimation during the needs assessment phase, the relevant stakeholders often are only those who may have expressed an explicit need for the system and/or will potentially benefit from the system; such stakeholders typically include individuals, businesses whose operations are directly related to the civil system, and special-interest groups.

19.0.4 Establishment of Goals, Objectives, and Performance Measures

As we found in Chapter 3, an initial and clear definition of the intended goals of a proposed physical or operational system are vital for meaningful needs analysis and working toward meeting the need. As discussed in Section 19.0.3, these goals emanate from the stakeholders. Each stakeholder group has its unique set (often self-serving) of objectives in terms of what the proposed system would offer. Because there is often a plethora of stakeholders (and therefore objectives) associated with a given civil system, it is often the case that certain objectives conflict with each other. As such, developers of civil systems need to exercise caution in specifying their goals and objectives. Often, it is the system owner/operator whose objectives are adopted as the main objective. However, current practice is characterized by increasing attention to all of the stakeholder groups and development of the final set of objectives after much consideration and deliberation. There seem to be significant differences in the ease of establishing goals and objectives for different kinds of systems; for example, establishing the goals for human activity systems (including those related to civil engineering) are considerably more difficult compared with those for mechanical systems (Khisty et al., 2012). Also, the stakeholders' stated objectives may not always be consistent with their real objectives. For a proposed system that cannot be justified using conventional means, special-interest or lobbying groups may try to promote objectives that may be indicative of public benefit and welfare but that mask their true agenda (de Neufville, 1990). The system owner, in some cases, may have murky and fuzzy objectives initially, and civil engineers working for that client will need to streamline any loosely stated goals into clear cut and achievable objectives.

19.1 ASSESSMENT OF SYSTEM NEEDS

19.1.1 What Is Need?

A need may be defined as a deficiency, an unfulfilled requirement, the lack of some product (object, process, or service) that is wanted or deemed necessary, a necessity arising from the circumstances of a situation, or a condition characterized by the lack of something requisite (Random House, 2013). As such, the need for a product reflects the balance between demand and availability (or supply) of that product (Figure 19.2). For example, the need for a new airport to serve a region could be determined as the difference between the demand for the product and the existing supply (in this

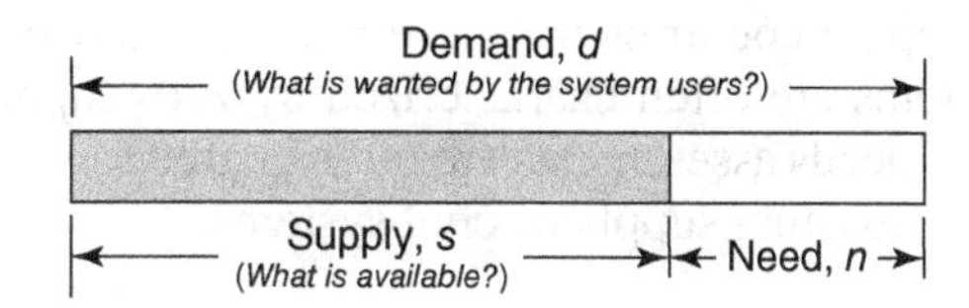

Figure 19.2 System need as the difference between demand and supply.

case, none). Thus, a statement of need, which can be defined as a formal expression of an unfulfilled requirement, helps a system owner to distill a specifically focused requirement from a complex need environment to serve as a first step in fulfilling that requirement (Kaufman and English, 1979; Meredith et al., 1985). The essential feature of a need statement is that it gives an indication of what is wanted and what is available. "What is wanted," or demand, may be described generally as an unconstrained statement of some envisioned ultimate state of the system that is desired by the agency, users, and/or other stakeholders. "What is available," or supply, is an indication of the existing quantity or quality of service provided by the system.

The identification of the system need and its resolution clearly is consistent with the consideration of the goals and objectives of the system and, specifically, the performance criteria upon which such objectives are measured (as discussed in Section 19.0.4 and Chapter 1). It is useful to recognize that the supply may be quantitative, qualitative, or both. As such, the supply may be associated not only with capacity but also with other attributes including structural strength, functionality, and aesthetics. Qualitative supply, even for a given quantity of supply, may also include user safety, security, system reliability, system vulnerability to disruption or disaster, and user comfort. Figure 19.2 presents a conceptual representation of the system need as the difference between the demand and the supply where both are viewed in terms of the same demand–supply attribute, often loading (structural strength) or volume of usage (e.g., flow of passengers, freight, fluids, etc.). The factors that influence demand or supply are marked by great uncertainty. As such, neither demand nor supply is perfectly deterministic. However, to illustrate the concept of needs assessment, much of the discussion in this chapter will be for cases of deterministic supply and demand. In Section 19.6.3, we will discuss the system need as a probabilistic function where supply or demand, or both, are probabilistic.

From the steps of problem identification, stakeholder identification, and definition of goals and objectives, a fairly good approximation of the level of need can be established. At this stage, the engineer determines the various dimensions of the need; the urgency of the need; and the variability of the need by time, location, population segment, and other considerations.

Needs assessment is important in civil systems development because it provides a basis for predicting the need for a proposed civil system in terms of the number of people using (or expected to use) the facility and how long they use it; and it helps establish the appropriate location, size, orientation, configuration, and other features of the proposed system or ascertaining the scope of proposed operational policies. For example, knowing the expected demand or level-of-service expectation at each future year helps in the development of agency cost streams for preserving civil systems whose deterioration or performance are influenced by usage. The needs assessment phase thus provides critical data for quantifying the expected costs or adverse impacts of the proposed system on the environment (e.g., noise, air pollution, etc.), and the benefits or positive impacts (e.g., the total savings in travel time, percentage of population served with potable water, etc.). Ultimately, this information helps the engineer at the subsequent (planning) phase to decide whether to proceed with a proposed system investment or policy change. The type and scope of the need should be established. For example, for physical improvements to an existing structure, is there a need for minor routine work, major repair, or replacement of a part or all of the structure?

In some developing countries that have relatively few civil systems generally, needs assessments for such systems are often characterized by zero supply at the time of the assessment. In developed countries, needs assessments are often geared toward expanding the capacity or increasing the quality of the existing supply of civil systems.

Example 19.1

In year 2012, a certain civil system is capable of serving 850,000 persons per day. The demand in 2012 is 800,000 persons per day and is increasing at the rate of 10% per year. Assuming that the usage is equal to demand and the system capacity does not change significantly in the long term, what is the expected need in year 2020?

Solution

Demand in year 2020 $= 800,000(1 + 0.10)^8 = 866,285$ persons/day
Supply in year 2020 $= 850,000$.Thus, the expected need in year 2020 $= 866,285 - 850,000 = 16,285$ persons/day.

Example 19.2

In the postearthquake recovery period, the demand for a civil system increased by 50%. If the preearthquake demand is 90% of the capacity at that time, what is the amount of need during the postearthquake recovery period as a percentage of the preearthquake demand? Assume that the postearthquake supply is 70% of the preearthquake supply.

Solution

Assume that the preearthquake demand $= x$. Then the preearthquake capacity or supply is $x/0.9 = \left(\frac{10}{9}\right)x$. The postearthquake supply is $0.7x \times 0.9 = \left(\frac{7}{9}\right)x$, and the postearthquake demand is $1.5x$.

Therefore, the need during the postearthquake recovery period is $1.5x - \left(\frac{7}{9}\right)x = \left(\frac{13}{8}\right)x = 72.2\%\ x$.

19.1.2 Initial Need versus Recurring Need

The phase of needs assessment generally refers to the *initial need*, that is, the need for a new system where none existed until now or to replace an existing system because it has reached the end of its life. However, it is often the case that there is a need, during the life of a system, to increase functional capacity to accommodate a growing number of users or to increase structural capacity to accommodate higher loads. Also, in some cases, the requirement to carry out periodic or routine repairs during the system life can be described as a need. Needs related to capacity enhancement and preservation during system life can be described as *recurring needs*. This chapter is mainly concerned with initial need. However, a few sections of this chapter will include recurring needs, namely, the sections on growing need and sudden need and assessment of long-term system needs via demand and supply trends.

19.1.3 Growing Need versus Sudden Need

The need for a civil system could be a growing need or a sudden need as we have just seen in Examples 19.1 and 19.2, respectively. Both needs are illustrated in Figure 19.3 and are discussed below.

1. *Growing needs*. Growing needs account for most cases of civil system needs and typically arise from at least one of two situations: (i) growing demand with fixed supply, where evolving

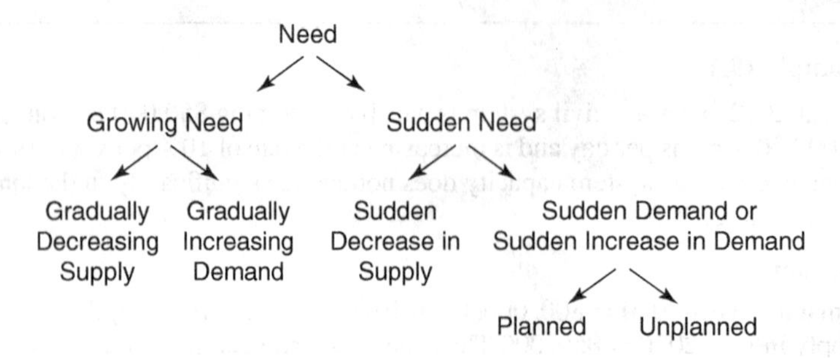

Figure 19.3 Categorizing system needs by suddenness of demand and supply.

social and economic changes gradually translate into the need to provide a new civil system or to replace, retrofit, or expand an existing one; or (ii) gradually decreasing supply with fixed or increasing demand, where the quantity or quality of the service provided by the civil system gradually erodes over time. For example, the deterioration of transit service quality due to neglect and lack of maintenance and aging of the system components. Growing needs typically do not receive much publicity and can creep up slowly until they emerge as critical problems that garner national attention.

2. *Sudden needs.* The sudden needs associated with civil systems are less frequent compared to growing needs, but typically receive much more attention from the press. A sudden need typically arises from at least one of the following two situations: (i) A sudden new demand or sudden surge in existing demand, which may be planned, such as the requirement of providing a structural or geotechnical system to support some load of structure as part of an ongoing design; or unplanned, such as the demand for a new structural component at a bridge that is showing signs of imminent failure. A sudden surge in demand includes situations where natural or man-made events overwhelm the existing system, usually temporarily, requiring an urgent intervention to increase the quantity or quality of supply in order to fulfill that need. Examples include providing transportation infrastructure to evacuate residents from a hurricane area, shelter for earthquake victims, and intelligent communication networks to ease traffic flow during a major sporting event in a city. (ii) A sudden decrease in supply, where the quantity or quality of the service provided by the civil system abruptly reduces at a given point in time, even if demand is constant. Examples include sudden failure of a bridge or water reservoir due to natural and man-made disasters. A sudden decrease could also be due to a terrorist attack, collision, or failure of a system's structural component, such as the I-35 bridge collapse in Minnesota in 2006.

19.1.4 Existing Systems versus New Systems

The need for a civil system also can be classified on the basis of whether it is for a completely new system or an enhancement of an existing system. In Section 19.1.2, we mentioned the case of a need for an entirely new system. In this section, we take a closer look at the need generated by existing systems. For most existing systems, the need to undertake some action is driven to enhance some supply attribute such as strength, capacity, functionality, or aesthetics. Many system owners carry out repair or replacement of their systems or parts thereof because of poor serviceability or functional obsolescence. For replacement decisions for system components or entire systems, the need may be driven by the fact that it is not cost-effective to keep repairing the existing system;

also, the need may be driven by the desire of the system owner to exploit the advantages of new materials, designs, or technologies or to avoid violation of regulations (Thompson et al, 2012). For existing civil engineering systems, most cases of need are related to capacity enhancement, where it is desired to accommodate greater current or anticipated demand. Also, some civil systems may experience an upsurge in need for expansion due to changes in development patterns. Another cause of need arises from unexpected sudden or gradually developing events (Figure 19.3); for example, there could be a need to replace or increase the strength of existing civil systems in response to geotechnical conditions induced by gradual climate change or in response to a sudden earthquake or flood event (Ghosn et al., 2003; Kacin, 2009).

19.1.5 Needs Assessment Examples in Civil Disciplines

The need for all civil systems can be described as being "derived" in the sense that the system is not needed for its own sake but rather to perform a service for the benefit of society. Growing populations, demographic shifts, and expanding economies and development sectors translate constantly into the need for new or improved civil infrastructure including buildings, highways and streets, and water supply systems. Needs assessment in any branch of civil engineering involves acknowledgment of the existence of a need and an explicit statement as to the extent of the gap between what is demanded and what is available. In the paragraphs below, we discuss examples of needs that are typically encountered in the various areas of civil engineering; Table 19.1 presents some examples of civil system supply and demand in these areas.

(a) Structural Engineering Systems. The demand for structural systems is often expressed in terms of the expected loading. This demand is derived from the need for architectural and civil structures that fulfill specific needs in the community. Examples include the need to provide a roof over an open stadium, the need to provide stability to a planned or existing building, and the need to support a proposed overhead tank to serve a growing town. In this context, supply refers to

Table 19.1 Examples of Civil System Supply and Demand

System	Demand	Supply
Public transportation system	Number of transit users	Buses, train cars, terminals
Highway system	Number of vehicles	Highway lanes, ramps
Water supply system	Gallons of water used daily	Overhead of surface water reservoirs
Wastewater treatment system	Gallons of wastewater treated daily	Treatment ponds and tanks
Landfill for solid waste disposal system	Tons of water processed daily	Area of land used for landfill operations
Building system foundations in weak soils	Safety perception via protection from undue settlement	Underpinning
Bridge structural system	Safety and security via protection from sudden failure	Retrofits to reduce vulnerability to failure
Pavement systems	Safety via protection from hydroplaning	Pavement layers that reduce rutting (via resurfacing)
Urban drainage systems	Urbanization, which reduces catchment area and increases runoff	Drainage pipes and manholes
Drainage canals	Sedimentation due to upstream erosion and deposition of silt and other solid waste	Dredging to increase channel capacity

the provision of adequate strength to counter the stresses caused by the expected loads. In recent years, there is a growing call for upgrading and improving physical civil infrastructure, and this has translated into the need for enhancing the individual structural components that constitute larger structural supersystems.

(b) Transportation Engineering Systems. Transportation engineers constantly monitor population growth and shifts and try to determine whether a new transportation system is needed either to replace or augment an existing transportation service between two or more locations, which mode(s) to use, and the potential capacity of the proposed system. Examples of such system needs include a new parallel subway system at a level below that of the existing system (being considered in New York City and Singapore), a new bus rapid transit (BRT) system for a city (as is currently being considered in Accra, Ghana), expansion of an existing airport (the Rafael Hernández Airport in Aguadilla, Puerto Rico), light-rail transit construction (Austin, Texas) or expansion (Boston, Massachusetts), and extension of high-speed rail (in certain parts of China). Transportation needs are typically identified and quantified by analyzing travel demand and capacity, and such studies are often supplemented by interactions with stakeholders (e.g., the general public, commuters, and shippers) using a variety of need assessment mechanisms. In air travel, for example, increased need is often characterized by passenger congestion in the terminals and aircraft delays in landing or takeoff and is met by expanded or new terminals and additional runways (Figure 19.4) In transportation engineering in general, demand is often associated with the volume of some traveling entity (e.g., pedestrians, vehicles, planes, ships) per unit time at a facility, and supply is generally expressed in terms of the maximum flow that can be accommodated per unit time.

(c) Geotechnical Engineering Systems. The demand for geotechnical systems arises from the need to support foundation loads for civil structures or to control soil behavior through activities including earth retention and slope stabilization. The need to develop geotechnical systems to address specific soil conditions is typically realized during geotechnical investigations at construction sites for civil engineering facilities. Such situations are typically encountered during tunneling through unstable soil and constructing foundations for structures at areas of weak soils (Figure 19.5). Any areas needing special geotechnical treatments or systems are typically confirmed via further field and laboratory tests supplemented by field visits. High-profile examples include the

Figure 19.4 Adequate terminal and runways capacities help to satisfy air travel need. Dublin Airport terminals 1 and 2 (Courtesy of ColmDeSpáinn/Wikimedia Commons).

Figure 19.5 Special geotechnical systems are needed in areas of weak soils. Pile driving for a bridge in California (Courtesy of Argyriou/Wikimedia Commons).

need for geotechnical solutions at certain low-lying and marshy locations of France's high-speed railway (Delage et al., 2005) or Thailand's Nakhon Sawan Highway through an area underlain by soft clay (Eide, 1968). Other examples include stabilizing the steep slopes of a roadside to prevent a rock slide as done at the Greenfield Road in Colrain, Massachusetts (Gray and Sotir, 1995), and the Blue Ridge Parkway in North Carolina (Middleton and King, 2003).

(d) Environmental Engineering Systems. Similar to transportation systems, environmental systems are often more geared to address community or regional needs rather than a specific need to address problems at a single facility (as is often the case for structural and geotechnical systems). Estimations of population growth or shifts therefore are a critical part of needs assessment for environmental systems; and the growth in demand is typically monitored and measured against existing supply, over time so that the gap (need) at any time can be identified and quantified. In water and wastewater systems planning, for example, demand is often measured in terms of the expected total flow and loads, and the design of each component in these systems considers several different loading and flow conditions and the most severe condition of the system's design (Sykes, 2002).

(e) Materials Engineering Systems. In a broad sense, the development of a new civil engineering material could be considered the development of a new system as it follows most of the phases of needs assessment, planning, design, implementation, and usage. The need for new materials is driven by the need for certain special properties that are not afforded by existing materials. For example, in concrete production, there is often a need to speed up or retard the hardening of concrete to suit a specific construction time schedule or process, to enhance concrete durability through increased resistance to freeze–thaw effects, to increase the workability or pumpability of fresh concrete, to change the color of concrete, or to minimize corrosion of steel reinforcement. These process changes are accomplished using a variety of new materials, including concrete that is modified using chemical and mineral admixtures.

(f) Architectural Engineering Systems. In every modern building, there is a need to provide systems related to air conditioning, heating, ventilation, fire protection or control, distribution of utilities (water, gas, power, cable), and lighting. These systems are needed for new buildings as well as in existing buildings to accommodate increased usage or physical expansions of the building. The architectural engineering system supply, in terms of the number and size of the systems for heating, air conditioning, ventilation, electrical, plumbing, and other services are driven by the level and pattern of the building's usage.

(g) Hydraulic and Hydrologic Engineering Systems. The need for new or expanded hydraulic systems originates from a wide variety of sources. New irrigation and water distribution systems arise from the need to bring water to agricultural areas and residential areas, respectively. Also, flood control and drainage are the result of the need to address the problem of perennial flooding, particularly in low-lying areas; canal and lock systems come in to play to control the flow of a river through challenging terrain; and coastal and ocean engineering systems are developed to protect shorelines from erosion (Figure 19.6), to harness ocean energy, or to develop ports. In many hydraulic systems, the demand is measured in terms of the expected volume of the water flow (volume per unit time), and the supply is the capacity of the system to channel or contain the water.

Figure 19.6 Persistent coastal erosion causes a need for sea defense systems (*Source:* geograph.org.uk. Sandy Point, Essex, UK. © Copyright Nigel Cox and licensed for reuse under this Creative Commons License).

(h) Construction Engineering Systems. In construction engineering, there is typically a need to develop systems that help reduce the general construction goals of cost and time (completing the project within budget and on time); quality control and assurance (ensuring that all work is completed according to the specified quality of materials and workmanship); safety enhancement (minimizing the number of construction accidents); and conflict minimization (Hancher, 2003). The drive to achieve these goals gives rise to the need for a number of physical and virtual systems for a given project such as an efficient construction planning and scheduling system, an equipment utilization system, an optimal preventive maintenance schedule for construction equipment, and an effective formwork system.

19.1.6 Types of Need

The need for a civil system could be prescribed, perceived, revealed, or stated. We discuss each of these terms below.

Prescribed need. This type of system need is assessed on the basis of the system's deficiencies at the current time or at a specified future time. This type of need is often identified through analysis of empirical demand and supply data for a region (typically for transportation and environmental systems), for a specific location (structural and architectural systems), or through field or laboratory tests (typically for geotechnical and materials systems).

Perceived need. This is what the system stakeholders (typically, the users) *think* they need. Questionnaire surveys and opinion polls can help gauge this kind of need, assuming that the survey respondents are free to state what they think.

Expressed or stated need. This is what the system stakeholders *say* they need, which is often a reflection of the perceived need (what they think they need). The stated need is revealed through a questionnaire survey by direct interaction with the intended users of the proposed system.

Revealed need. This is what the system stakeholders actually *do* to reflect what they actually need. The revealed need may be the same as or different from the stated need. For example, some users may state that they have no need for a certain system, yet they use it when it is made available; or they may express a need for the system but do not use it after it is made available. A revealed need is difficult to assess but can be estimated in an *ex poste* fashion by conducting a field survey of the actual users or stakeholders using the system after the need has been addressed.

19.2 MECHANISMS FOR ASSESSING SYSTEM NEEDS

An assessment of the need for a system (or for a component thereof) depends on the type of system in question. The assessment can be for either the system demand (where the supply is already known), the system supply (where the demand is already known), or both the supply and the demand. Mechanisms for such assessments include laboratory and field tests, questionnaire surveys, interviews, and focus groups (Figure 19.7). Typically, there are inadequate resources to collect supply and demand information from the entire population of relevant civil systems or their components. As such, sampling is often necessary to obtain a random but representative sample. The reader may recall that, in Chapter 6, we discussed sampling techniques for carrying out tasks such as this.

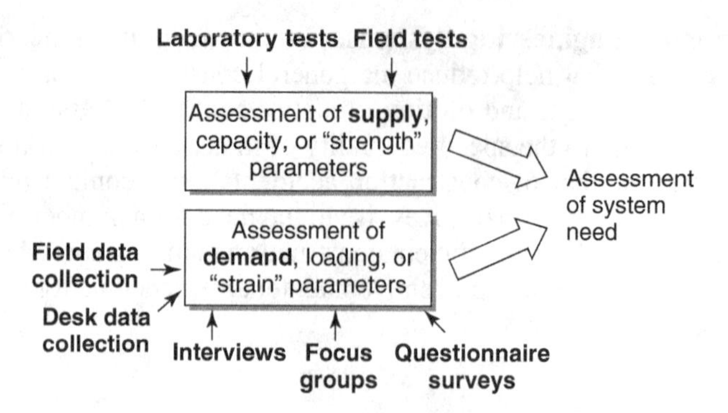

Figure 19.7 Mechanisms for assessing system needs.

19.3 ASSESSING SYSTEM NEEDS USING USER-TARGETED MECHANISMS

19.3.1 Questionnaire Surveys

Questionnaires are a mechanism for administering carefully prepared questions designed to elicit written or oral answers from a target set of responders. Questionnaires can be used to quantitatively measure the extent of need; and the responses can be placed on an ordinal scale from which a frequency distribution of the responses can be established. The nature of the responses is influenced by the design of the questionnaire and the survey questions (Mann, 1998). A well-designed questionnaire can provide useful information to the needs assessor in a standardized format (Ratnapalan and Hilliard, 2002). Questionnaire surveys are a popular form of assessing the needs of a community due to the ease of administering them and their relatively low cost (Steadham, 1980). They show much promise in measuring the needs of a wide array of civil system types and are an attractive mechanism because they are relatively inexpensive to conduct and can sample large numbers of the stakeholder population while providing anonymity to respondents. Needs assessment questionnaires can be designed using resources available in the literature (Mann, 1998; Lockyer, 1998). The drawbacks of questionnaire surveys include poor response rates and the quality of the information generated being unduly influenced by the content and context of the questionnaire (Lockyer, 1998). Ratnapalan and Hilliard (2002) further elaborate that the quality of the information generated by questionnaires is "only as good as the quality of the questions in the questionnaire." Other limitations include bias due to self-assessment and self-reporting by the survey administrator, and the difficulty of generalizing the responses because the responses may be an indication of only the personal views and wants of the responders and may not reflect the needs of the wider population of interest. A possible recourse is to supplement questionnaire surveys with other needs assessment methods to yield a more reliable assessment of the needs of the system stakeholders (Reviere et al., 1996).

19.3.2 Interviews

An interview is a conversation between the needs assessor and a target individual or group of individuals with the objective of measuring the respondents' perspectives on the existence of a problem and the extent of a need for providing a new civil system or expanding/strengthening an existing system. Interviews can be conducted face to face, by e-mail, or by telephone (Ratnapalan and Hilliard, 2002) or using the newest media tools including text message, Facebook, and Twitter. Compared to in-person interviews, telephone or Internet interviews are generally less costly in terms of time and resources. Questionnaire surveys that have set questions with certain questions earmarked for elaboration can be administered easily by telephone. Long telephone interviews should be avoided as they are generally difficult for both the interviewer and the interviewee (Reviere et al., 1996). The merits of the interview method include the personal outreach and the opportunity for faster recognition and deeper understanding of the respondents' perspectives (Steadham, 1980; Crandall, 1998). Its drawbacks include the lack of anonymity for the respondents; the significant amount of resources required in its administration; and the infeasibility of assessing the needs of a large population, which is often the case for civil systems. Also, collating and analyzing the data from a descriptive interview can involve a significant amount of work and may take much more time than would be spent in administering interviews (Ratnapalan and Hilliard, 2002).

19.3.3 Focus Groups

In its first application, the focus group concept was used years ago to help the U.S. Army produce training materials pertaining to military hardware and morale boosting. This concept is used now in a wide variety of areas of modern day life, including supermarket item displays, television programming, and communication between election candidates and the electorate (Tipping, 1998). Focus groups can be a useful method for assessing the needs of the stakeholders of a civil system. The focus group process involves a small number of randomly selected participants who meet the established criteria to be considered system stakeholders and a skilled moderator who conducts the interview while encouraging a sense of synergy while exploring differences in opinion (Ratnapalan and Hilliard, 2002). Unlike the case of individual interviews, focus group members draw strength from each other to express perspectives that may otherwise be viewed as unpopular. Using this method of needs assessment can make adequate qualitative data available in a manner that is cost-effective and timely (Steadham, 1980; Reviere et al., 1996). Experts caution, however, that a sufficient and representative number of focus groups must be interviewed before the data generated therein can adequately serve as a measure of the needs of the population of interest. Similar to questionnaires, the focus group concept is often supplemented with other needs assessment methods (Ratnapalan and Hilliard, 2002).

19.3.4 Discussion

Any one or a combination of the above mechanisms can be used to identify a need and to estimate the extent of the need. In Section 19.4, we will discuss different kinds of needs policies and derive mathematical equations for calculating the need on the basis of system supply and demand.

19.4 ASSESSING LONG-TERM SYSTEM NEEDS VIA DEMAND AND SUPPLY TRENDS

When demand exceeds capacity, the users suffer delay, inconvenience, or in certain cases, safety hazards. These additional costs may or may not be translated into monetary equivalents. Also, the agency also suffers indirectly through worsened public relations that may cost money to alleviate. On the other hand, when supply exceeds demand, users do not pay a penalty; however, the system owner bears the additional cost of capacity underutilization and the opportunity cost because the money used in the capacity provision might have been invested elsewhere to yield some returns. When the demand for a system is always equal to the supply, none of the additional costs that are associated with both extremes exist, but this is an ideal, rarely encountered situation. In this section, we will discuss the patterns of system supply and demand and how a system owner could schedule capacity provisions in an optimal manner to minimize the costs of system capacity underutilization while keeping user delay and inconvenience costs as low as possible. As we learned earlier in this chapter, capacity may refer to volume, strength, or some functional attribute. In this section of the chapter, we will illustrate the concept of capacity and demand relationships using volume as the attribute of capacity or supply. Our discussion is related to capacity-related need; however, the concepts are also applicable to strength-related need and other system performance attributes that are of interest to the system owner, users, or the community.

19.4.1 Continuous Demand and Stepwise Supply

As we saw in Figure 19.1, the need for a typical civil system is the difference between the supply and the demand. If both the supply and demand change over time, so will the need. To simplify our discussion of needs assessment, we use the term "capacity" to represent the supply; however, we should again bear in mind that capacity could be interpreted in terms not only of the quantity but also of the quality of the supply. Often, system demand continually increases due to reasons such as population growth, but supply often cannot be provided continuously to match continuous demand trends and rather is provided in discrete steps (e.g., the floor levels of a parking garage, bus terminals serving a growing city, additional water tanks at a water supply facility, and cooling units for large building complexes). Depending on the supply policy and the demand growth pattern, there will be different levels of needs assessed for the system at any given time. In the discussion below, we present two supply policy scenarios.

(a) Zero-Need Policy (ZNP). The first scenario is the "zero-need" policy, where demand is never allowed to outstrip supply. In this case, additional system capacity is provided as soon as demand approaches supply or capacity. The demand function could be linear or curvilinear. Here are two extreme cases for this scenario: Case 1, where an overabundance of system supply is provided at the initial year to cover current and future demand until the specified horizon year (Figure 19.8*a*), and Case 2, where after some initial system supply, C, additional capacity is provided in very small increments, ΔC, any time the system demand catches up with the system capacity (Figure 19.8*b*). Also, it takes time, t_C, for demand to reach capacity. Obviously, Case 1 would be wasteful because of the excess capacity at all times except the horizon year when demand equals capacity. It is noted, though, that Case 1 has a one-time cost of providing capacity and is therefore economical from that perspective. Case 2 seems to be a parsimonious scenario because it does a great deal to avoid excess capacity. However, this case could be costly because the frequent addition of capacity, often with all its excessive number of contracting and system downtimes, could be just as expensive as operating with excess capacity (Case 1). Case 3, shown as Figure 19.8*c*, seems to be a compromise between the two extremes. In actuality, optimization via life-cycle cost minimization could be applied to determine the optimum intervals of capacity provision and how much capacity should be provided for the ZNP where demand never outstrips supply. In such optimization formulations, it is sought to minimize the present worth of the sum of the costs of operating the system at excess capacity and the cost of providing additional capacity.

Thus, in general, all cases involving the ZNP are inherently wasteful because the installed capacity always exceeds the demand for the system. If the ZNP were applied to a facility that is purely private, such as a toll bridge, then the system owner may set the price of usage (i.e., the toll fare) to cover the excess costs (i.e., the costs of providing, maintaining, and operating the excess capacity); if that happens, then the system users would be paying for the capacity they are not using. Also, if the ZNP were applied to a facility that is purely public, such as a free-use highway or bridge, the system owner (in this case, the government) is often unable to easily transfer the excess costs to the consumer and therefore bears these costs instead.

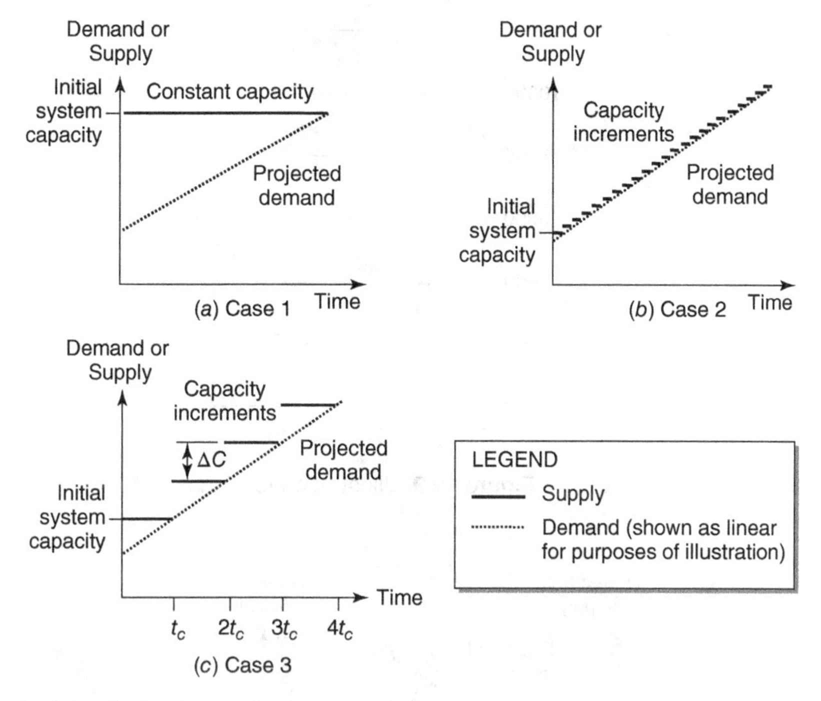

Figure 19.8 Projections of demand and discrete supply for three cases of the zero-need policy (ZNP).

Example 19.3

The demand for a certain civil system provides an initial capacity of 3000 units, which increases according to a pattern governed by the equation: $t^3 + 3t + 3000$. Assuming a zero-need policy, design an appropriate supply scheme for the system over a 20-year analysis period. Initial capacity should not exceed 4000 units with incremental additions of 2000 units thereafter, until the horizon year. The last capacity increment may be less than 2000. Also, indicate the time intervals between the capacity additions.

Solution

The first capacity addition takes place after 10 years.

Demand $= t^3 + 3t + 3000$; when demand $= 6000$, $t = 14.353$. Thus, the second capacity addition takes place after 14 years (Figure 19.9).

When demand $= 8000$, $t = 17.041$. Thus, the third capacity addition takes place after 17 years.

When demand $= 10,000$, $t = 19.077$. Thus, the fourth capacity addition takes place after 19 years.

(b) Finite-Need Policy (FNP). The second scenario is the "finite-need" policy where demand is allowed to outstrip supply by some specified maximum margin or without any margin. The demand function could be linear or nonlinear. In an extreme case for this scenario, the initial system supply is provided at the initial year to cover current demand until the horizon year, regardless of the growth in demand (Figure 19.10*a*). In the second case, after some time, the demand exceeds the supply (thus creating a need); this is allowed to continue up to a certain point when additional capacity, ΔC, is provided (Figure 19.10*b*). It may be the system owner's policy to establish a certain threshold of need beyond which additional supply should be provided.

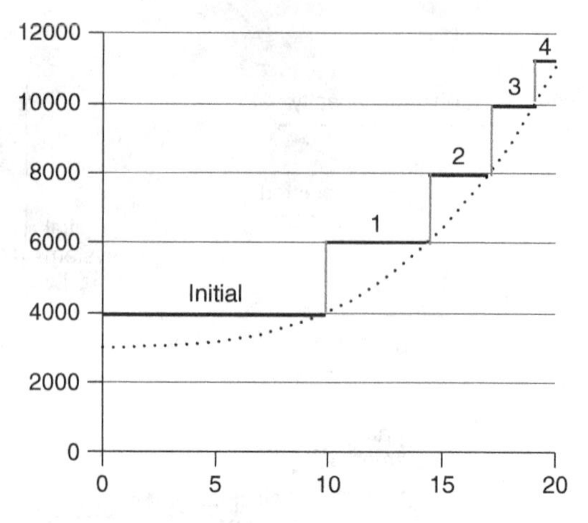

Figure 19.9 Illustration for Example 19.3.

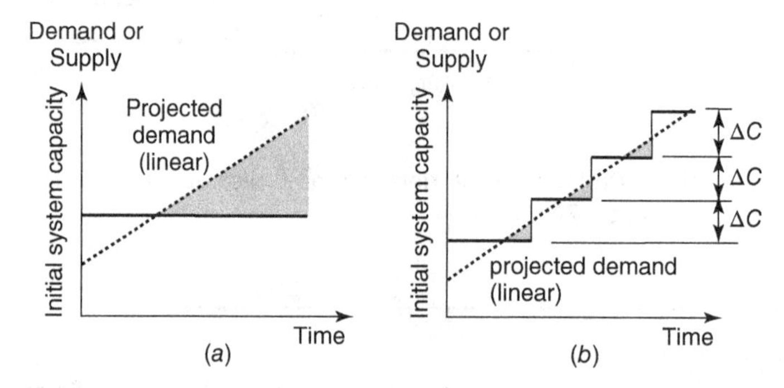

Figure 19.10 Projections of demand and discrete supply, finite-need policy.

19.4.2 Continuous Demand and Supply

Unlike demand, supply is often discrete, as is the case when the system owner or operator provides additional physical capacity, such as through the construction of additional highway lanes, a new parallel bridge to cross a river, construction of new water reservoirs, or installation of new water pipelines. However, in a few cases, supply can be considered continuous; for example, roadway capacity, even without lane additions, increases by increasing the resources used in roadway management strategies such as freeway patrol and assistance, in-vehicle information systems, electronic tolling, and other intelligent transportation systems (ITS) initiatives.

As we saw in Figure 19.2, the need for service provided by a civil system is the difference between the available supply (quantity or quality) provided by the agency and the demand of the users. In the rare case of continuous supply, the agency is able to increase or decrease supply continuously in a similar manner as demand changes. Depending on the demand and supply functions, there may or may not be a need at certain years. In Figure 19.11, we present examples of such possibilities. In these figures, the solid line represents supply and the broken line represents demand. The shapes of the supply and demand functions are presented for purposes of illustration.

In Figure 19.11*a*, both supply and demand are continuous linear functions and, initially, the capacity of the system exceeds the demand. However, the demand for the system is increasing

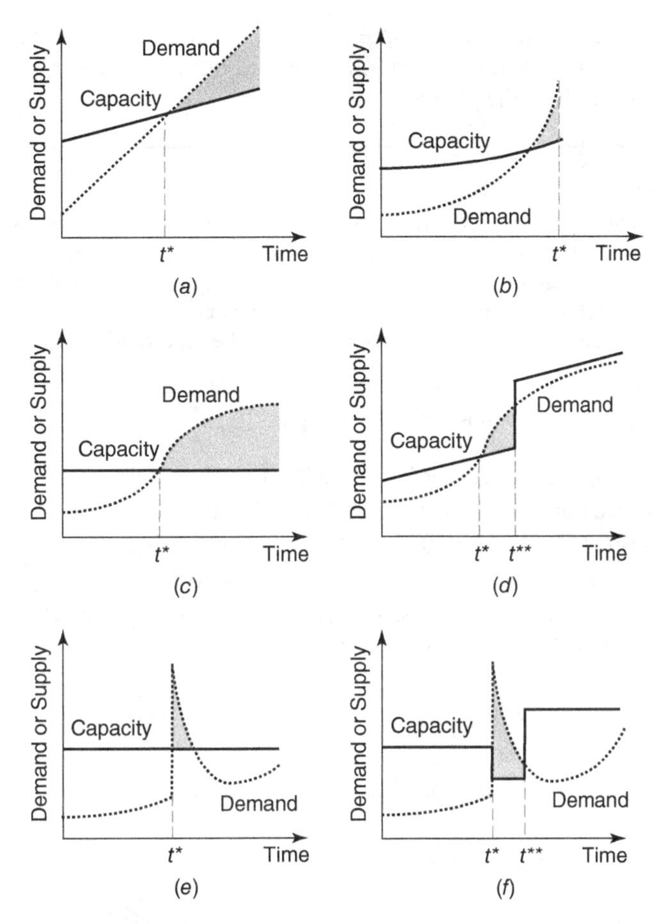

Figure 19.11 Projections of continuous demand and supply—conceptual illustrations.

faster than the rate at which the agency is increasing the system capacity. Consequently, at some year t^*, a system need arises and grows in magnitude. In Figure 19.11*b*, the situation is similar to Figure 19.11*a* with the exception that the system demand and supply both are nonlinear functions of time. In Figure 19.11*c*, the initial capacity exceeds the initial demand but remains constant throughout the analysis period; the demand starts growing slowly initially, but its growth increases with time until it exceeds the supply, thus establishing a need. However, due to congestion, demand growth reduces and remains constant or even decreases after a period of time. In Figure 19.11*d*, the demand and supply both increase, nonlinearly and linearly, respectively, with the supply initially exceeding the demand. After t^* years, the demand outstrips the supply and a need is created. After t^{**} years, the system owner, in response to the increase in demand, provides additional supply by adding physical capacity (thus a vertical jump in the supply function) after which the supply continues to increase thereafter. Situation in Figure 19.11*e* is typical of a disaster or special event (e.g., sporting event days), where the initial capacity exceeds the initial demand and the demand grows gradually. An unexpected disaster event at year t^* causes a spike in the demand that overwhelms the capacity. After the effect of the event subsides, the demand recedes to the preevent level and grows thereafter. Situation in Figure 19.11*f* is similar to Figure 19.11*e*; however, due to the disaster event, the system capacity suffers a reduction (e.g., destruction of a bridge during an earthquake), thus exacerbating the need situation even further. The system operates at a crippled capacity for some time until a physical intervention is carried out to restore the capacity to a level equal to or beyond the level before the event.

Example 19.4

A civil system provides an initial capacity of 15,000 units, which increases at 500 units annually for the next 15 years. The initial, demand is 12,000 units and grows at a rate of 500 units annually for the first 3 years and 1000 units annually for the remaining 12 years. Using a plot of the demand and supply trends, determine (a) the unused capacity at year 2, (b) when a need starts to exist, (c) the level of need at year 10, (d) the demand lead time at year 10, and (e) the total need, in unit-years, over the 15-year period.

Solution

The plot for the problem is provided in Figure 19.12.

(a) Unused capacity at year 2 $= 16,000 - 13,000 = 3000$ units.

(b) A need starts to exist at year 9 as that is when the demand function just crosses the supply function.

(c) Level of need at year 10, $n = 20,500 - 20,000 = 500$ units.

(d) Demand lead time at year 10, $L = 20,500 - 20,000/500 = 1$ year.

(e) Total need, in unit-years, over the 15-year period = area of shaded region $= 0.5(25,500 + 22,500)(15 - 9) = 9000$ unit-years.

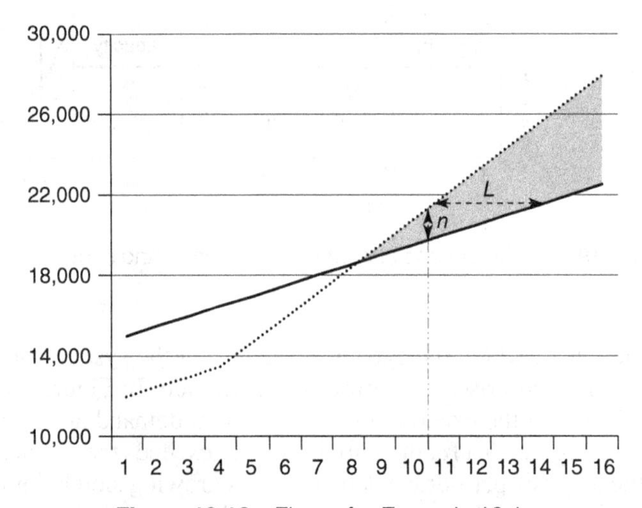

Figure 19.12 Figure for Example 19.1.

Example 19.5

The table below presents the past levels of monthly supply and demand for a certain civil system.

Month	1	2	3	4	5	6	7	8	9	10
Supply	23	25.1	25.9	26.6	26.6	28.4	29.1	29.5	29.8	30.1
Demand	20.8	21.5	22	22.2	23.1	24.2	24.7	26	26.6	27.2

Develop separate regression functions to describe the trends of supply and demand and plot them on the same graph. Using the graph, determine (a) the projected month when a need starts to exist, (b) the projected level of need at year 15, and (c) the total expected need, in unit-years, over the period when a need is expected to exist.

Solution

(a) As shown in Figure 19.13, a need starts to exist when the supply function is equal to the demand function:

$$3.2943 \ln(t) + 21.74 = 0.0544t^2 + 0.1787t + 20.124.$$

This gives $t = 11.85$ months.

(b) The projected level of need at year 15 is

$$\begin{aligned}\text{Demand}_{15} - \text{Supply}_{15} &= [0.0544(15)^2 + 0.1787(15) + 20.124] - [3.2943 \ln(15) + 21.74] \\ &= 4.38 \text{ units}\end{aligned}$$

(c) The total expected need, in unit-years, over the period when a need is expected to exist can be found as the difference between the area under the demand function and that of the supply function, after 11.85 months. This is given by

$$\begin{aligned}\int_{11.85}^{16} f_{\text{DEMAND}} - \int_{11.85}^{16} f_{\text{SUPPLY}} &= \int_{11.85}^{16} \left\{\left(0.0544t^2 + 0.1787t + 20.124\right) - [3.2943 \ln(t) + 21.74]\right\} dt \\ &= \left(\left(0.0544 \cdot \frac{1}{3}t^3 + 0.1787 \cdot \frac{1}{2}t^2 + 20.124t\right)\right. \\ &\quad \left. - \left\{3.2943\,[t \cdot \ln(t) - t] + 21.74t\right\}\right) \Big|_{11.85}^{16} \\ &= -22.1393 - (-33.9048) = 11.7655 \text{ unit} - \text{years}\end{aligned}$$

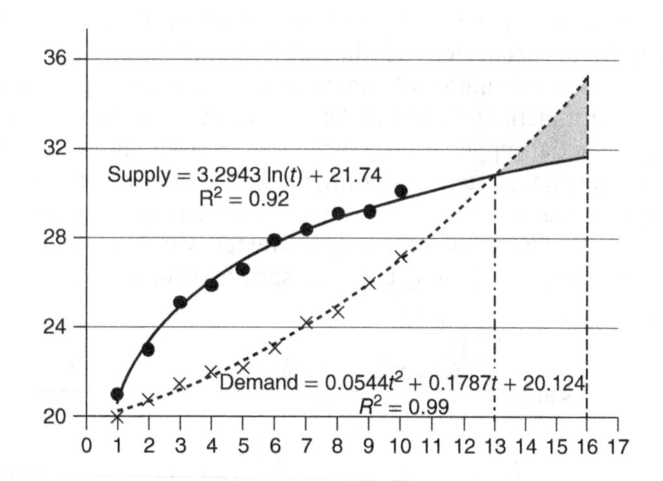

Figure 19.13 Figure for Example 19.2.

19.5 OPTIMAL SCHEDULING OF SUPPLY—MATHEMATICAL DERIVATIONS

As we saw in the last section, in both scenarios where demand exceeds supply or the supply exceeds demand, there are costs incurred by the system owner, user, or both. The perfect scenario would be for the supply to be always exactly equal to demand. However, this is not practical in reality because, as we discussed earlier in this chapter, supply increases, unlike demand, are not in continuous units but occur as large discrete events and are best represented as step functions. As such, it is inevitable that situations will be encountered where the demand exceeds supply, the supply exceeds demand, or both happen over a given horizon period. What we seek, therefore, is a schedule of supply increases, given the pattern of demand projections over time, such that the costs associated with excess supply or excess demand are minimized as much as possible. In this section, we will discuss the basic parameters of this optimization problem and will provide two contexts of the problem, namely, the zero-need and finite-need policies. For each context, we will present a general formulation and a detailed derivation of the relevant expressions. We will do this for linear and nonlinear demand functions and also for different strategies or cases of supply provision.

19.5.1 Prelude

Let NPV_{Agency} and NPV_{User} represent the NPV of the system owner's cost and additional user cost over the analysis period. The system owner's cost includes the initial construction cost and the expansion cost in the analysis. The additional user cost of delay, safety hazard, and inconvenience is incurred by the system users when demand exceeds supply. If the supply is greater than the demand, then it is assumed that there is no additional user cost. The following cost streams and their net present values can be established:

$$NPV_{Agency} = C_0(S_0) + \frac{C(\Delta S_1)}{(1+r)^{T_1}} + \frac{C(\Delta S_2)}{(1+r)^{T_2}} + \cdots + \frac{C(\Delta S_m)}{(1+r)^{T_m}}$$

$$NPV_{User} = \int_{T_1^*}^{T_1} U(S_t, D_t, t)\,dt + \int_{T_2^*}^{T_2} U(S_t, D_t, t)\,dt + \cdots + \int_{T_m^*}^{N} U(S_t, D_t, t)\,dt$$

where t is the time in years; N is the analysis period; r is the effective annual interest rate; T_i is the time (year) of the ith expansion; T_i^* is the time (year) of the ith time that the demand catches up with the supply; m is the total number of expansions in the analysis period; $D(t)$ is the demand function; $S(t)$ is the supply function; $C(\Delta S)$ is the system owner's cost function, a function of additional supply; S_0 is the initial supply; $C_0(S_0)$ is the initial construction cost of the system; and $U(S_t, D_t, t)$ is the user cost function, a function of time t and the demand and the supply at time t.

An important assumption is that after each expansion, the supply exceeds or is equal to the current demand. To choose the best supply strategy, we solve the following optimization problem (that is, find the values of T_i) subject to any specified constraints:

$$\begin{aligned} \text{Min } Z &= NPV_{Agency} + NPV_{User} \\ &= C_0(S_0) + \frac{C(\Delta S_1)}{(1+r)^{T_1}} + \frac{C(\Delta S_2)}{(1+r)^{T_2}} + \cdots + \frac{C(\Delta S_m)}{(1+r)^{T_m}} \\ &\quad + \int_{T_1^*}^{T_1} U(S_t, D_t, t)\,dt + \int_{T_2^*}^{T_2} U(S_t, D_t, t)\,dt + \cdots + \int_{T_m^*}^{N} U(S_t, D_t, t)\,dt \end{aligned}$$

19.5.2 Zero-Need Policy: General Formulation

In the zero-need policy, demand is never allowed to outstrip supply, thus there is no additional user cost. The following supply strategies are considered:

Case 1. At the initial construction, provide adequate supply for the entire analysis period, which means the initial supplied capacity is $S_0 \geq D_0 + qN$.

Case 2. After the first expansion, every period ΔT, carry out an expansion of the system capacity (ΔT could be one year or longer).

Case 3. Carry out system expansion (i.e., add capacity of ΔS) every time the demand catches up with the system capacity.

For each case, each of two different demand function patterns are considered:

$$D(t) \text{ is linear, } D(t) = D_0 + qt$$

$$D(t) \text{ is nonlinear, } D(t) = D_0 + d(t)$$

Figure 19.14 illustrates the supply and demand trends for the zero-need policy.

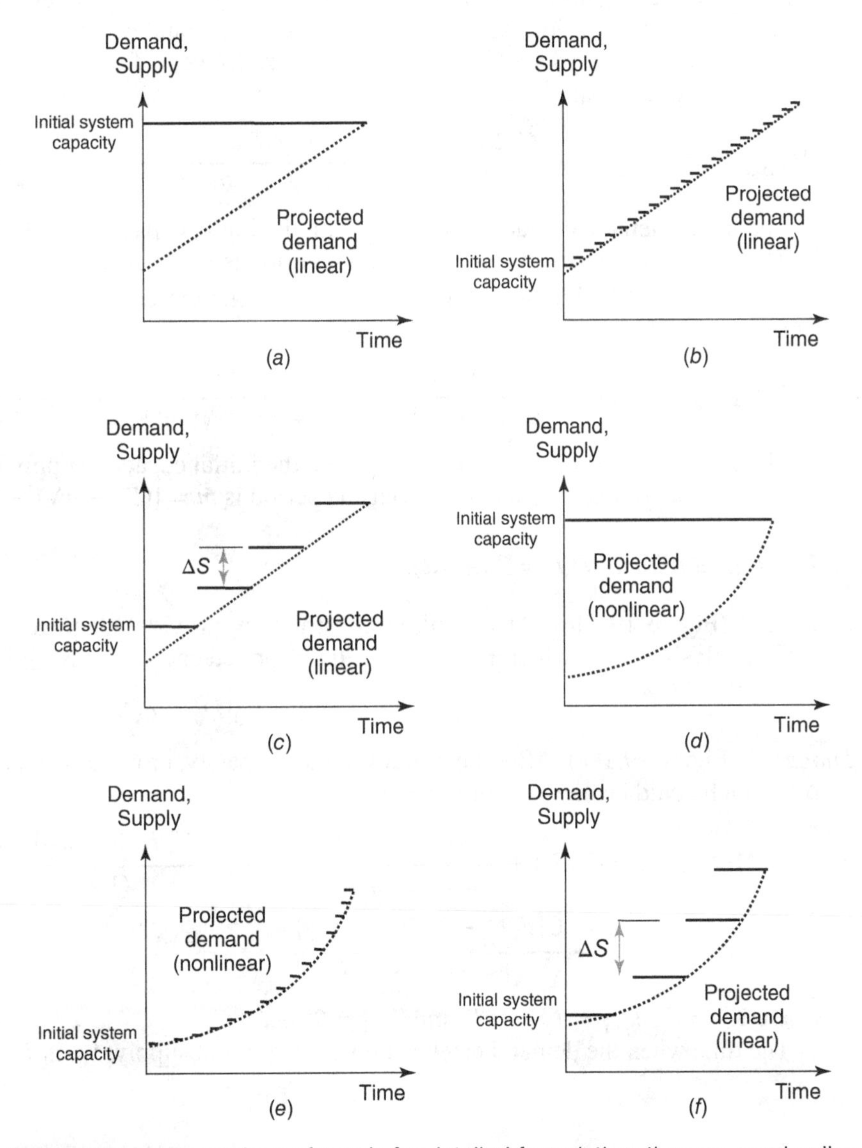

Figure 19.14 Illustrations of trends for detailed formulation, the zero-need policy.

19.5.3 Zero-Need Policy: Detailed Formulation

In the zero-need policy, the demand is never allowed to outstrip supply, thus there is no additional user cost.

Case 1. D(t) is linear, $D(t) = D_0 + qt$.

Subcase 1 (Figure 19.14*a*). At the initial construction, provide adequate supply for the entire analysis period, that is, the initial supplied capacity $S_0 \geq D_0 + qN$:

$$\text{NPV}_{\text{Agency}} = C_0(S_0) \geq C_0\ (D_0 + q_{\text{N}})$$

Subcase 2 (Figure 19.14*b*). At years subsequent to the year of initial capacity supply, provide added capacity (expansion) after every period ΔT (ΔT could be one year or longer); in other words, expand capacity by adding $\Delta T\ q$ of capacity ΔT year after the demand catches up with the initial capacity, S_0:

$$\text{NPV}_{\text{Agency}} = C_0(S_0) + \frac{C(\Delta T\ q)}{(1+r)^{(S_0-D_0)/q}} + \frac{C(\Delta T\ q)}{(1+r)^{(S_0-D_0)/q+\Delta T}} + \cdots + \frac{C(\Delta T\ q)}{(1+r)^{(S_0-D_0)/q+(m-1)\Delta T)}}$$

The time when the demand catches up with the initial capacity supply is $(S_0 - D_0)/q$. The number of expansions in the analysis period is $m = [(N - (S_0 - D_0))/\Delta q]/\Delta T$.

Subcase 3 (Figure 19.14*c*). Add ΔS capacity every time the demand catches up with system capacity:

$$\text{NPV}_{\text{Agency}} = C_0(S_0) + \frac{C(\Delta S)}{(1+r)^{(S_0-D_0)/q}} + \frac{C(\Delta S)}{(1+r)^{(S_0-D_0+\Delta S)/q}} + \cdots + \frac{C(\Delta S)}{(1+r)^{(S_0-D_0+m\Delta S)/q}}$$

The time when the demand catches up with the initial capacity supply is $(S_0 - D_0)/q$. The number of expansions in the analysis period is $m = [(D_0 + qN) - S_0)]/\Delta\text{S}$.

Case 2. D(t) is nonlinear, $D(t) = D_0 + d(t)$.

Subcase 1 (Figure 19.14*d*). At the initial construction, provide adequate supply to cover the entire analysis period, which means the initial supplied capacity is $S_0 \geq D_0 + qN$:

$$\text{NPV}_{\text{Agency}} = C_0(S_0) = C_0[D_0 + d(N)]$$

Subcase 2 (Figure 19.14*e*). After the initial supplied capacity, provide an expansion every period ΔT, which could be one year or longer):

$$\text{NPV}_{\text{Agency}} = C_0(S_0) + \frac{C[d(T_1^* + \Delta T) - d(T_1^*)]}{(1+r)^{T_1}} + \frac{C[d(T_2^* + \Delta T) - d(T_2^*)]}{(1+r)^{T_1+\Delta T}} + \cdots$$
$$+ \frac{C[d(T_m^* + \Delta T) - d(T_m^*)]}{(1+r)^{T_1+(m-1)\Delta T}}$$

where $T_1^* = T_1$, $T_{i+1}^* = T_i^* + \Delta T$, and $T_{i+1} = T_i + \Delta T$.

The time when the demand catches up with the initial supply, S_0, is T_1:

$$D(T_1) = S_0$$

The number of expansions in the analysis period is $m = \{[N - (S_0 - D_0)]/q\}/\Delta T$

Table 19.2 Summary of Expressions Associated with Need Assessment for Various Demand Trends, Zero-Need Policy[a]

	***D*(*t*) Is Linear**			***D*(*t*) Is Nonlinear**		
Variables	**Linear Case 1**	**Linear Case 2**	**Linear Case 3**	**Nonlinear Case 1**	**Nonlinear Case 2**	**Nonlinear Case 3**
T_1^*	—	$(S_0 - D_0)/q$	$(S_0 - D_0)/q$	—	$D(T_1^*) = S_0$	$D(T_1^*) = S_0$
T_i^*	—	$(S_0 - D_0)/q + (i-1)\,\Delta T$	$[S_0 - D_0 + (i-1)\,\Delta S]/q$	—	$T_i^* = T_1^* + (i-1)\,\Delta T$	$D(T_i^*) = S_0 + (i-1)\,\Delta S$
T_m^*		$(S_0 - D_0)/q + (m-1)\,\Delta T$	$(S_0 - D_0 + m\,\Delta S)/q$	—	$T_m^* = T_1^* + (m-1)\Delta T$	$D(T_m^*) = S_0 + m\,\Delta S$
T_1	—	$T_1 = T_1^*$	$T_1 = T_1^*$	—	$T_1 = T_1^*$	$T_1 = T_1^*$
T_i	—	$T_i = T_i^*$	$T_i = T_i^*$	—	$T_i = T_i^*$	$T_i = T_i^*$
T_m		$T_m = T_m^*$	$T_m = T_m^*$	—	$T_m = T_m^*$	$T_m = T_m^*$
m	0	$\left\lfloor \frac{N - (S_0 - D_0)/\Delta q}{\Delta T} \right\rfloor$	$\left\lfloor \frac{(D_0 + qN) - S_0}{\Delta S} \right\rfloor$	0	$\left\lfloor \frac{N - (S_0 - D_0)/q}{\Delta T} \right\rfloor$	$\left\lfloor \frac{(D_0 + d(N)) - S_0}{\Delta S} \right\rfloor$
ΔS_i	—	q	ΔS	—	$d(T_i + \Delta T) - d(T_i)$	ΔS

[a] T_1 is specified by the system owner. Also, for this policy, $\text{NPV}_{\text{user}} = 0$

Subcase 3 (Figure 19.14*f*). Provide additional capacity ΔS (expansion) every time the demand catches up with system capacity:

$$\text{NPV}_{\text{Agency}} = C_0(S_0) + \frac{C(\Delta S)}{(1+r)^{T_1}} + \frac{C(\Delta S)}{(1+r)^{T_2}} + \cdots + \frac{C(\Delta S)}{(1+r)^{T_m}}$$

where $D(T_i) = S_0 + (i-1)/\Delta S$.

The number of expansions in the analysis period is $m = \{[D_0 + d(N)] - S_0\}/\Delta S$.

Table 19.2 presents the expressions associated with need assessment, for different demand patterns (linear or nonlinear) and the different supply strategies (cases), for the zero-need policy.

19.5.4 Finite Need Policy: General Formulation

In the finite-need policy, demand is allowed to outstrip supply some of the time, by some specified maximum margin or without any margin (see Figure 19.13). The following supply strategies are considered:

Case 1. At the initial construction, provide adequate supply for the entire analysis period; this means that the initial supplied capacity is $S_0 \geq D_0 + qN$.

Case 2. After every period ΔT, provide an expansion; the added capacity equals the expected increase in demand in the next ΔT.

Case 3. After every ΔS increase in demand, provide an expansion, that is, add capacity, of magnitude ΔS.

For each case, each of two different demand function patterns is considered:

$$D(t) \text{is linear}, D(t) = D_0 + qt$$

$$D(t) \text{is nonlinear}, D(t) = D_0 + d(t)$$

Figure 19.15 illustrates a few different demand patterns and supply strategies for the finite-need policy.

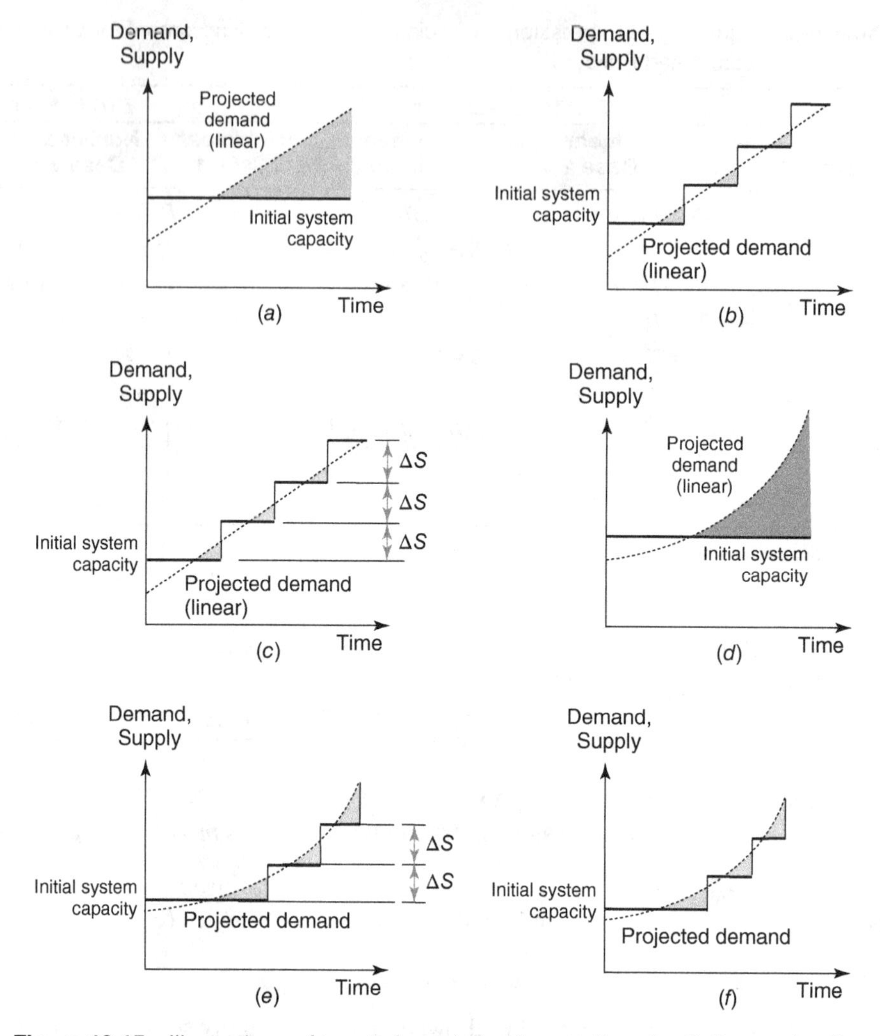

Figure 19.15 Illustrations of trends for detailed formulation, the finite-need policy.

19.5.5 Detailed Formulation for the Finite-Need Policy

In the finite-need policy, the demand is allowed to outstrip supply some of the time by some specified maximum margin or without any margin (see Figure 19.15). First, consider the case where $D(t)$ is linear, that is, $D(t) = D_0 + qt$, for which there are three situations:

Case 1. $D(t)$ is linear, $D(t) + D_0 + qt$.

Subcase 1 (Figure 19.15*a*). At the initial construction, provide enough supply to cover the entire analysis period, which means the initial supplied capacity is $S_0 \geq D_0 + qN$:

$$\text{NPV}_{\text{Agency}} = C_0(S_0)$$

$$\text{NPV}_{\text{User}} = \int_{(S_0 - D_0)/q}^{N} U(S_t, D_t, t)\, dt$$

Subcase 2 (Figure 19.15*b*). Subsequent to the initial capacity supplied, provide additional capacity after every period ΔT; the size of the added capacity should be equal to the expected increase of demand in the next ΔT:

$$\text{NPV}_{\text{Agency}} = C_0(S_0) + \frac{C(\Delta T\, q)}{(1+r)^{T_1}} + \frac{C(\Delta T\, q)}{(1+r)^{T_1+\Delta T}} + \frac{C(\Delta T\, q)}{(1+r)^{T_1+2\Delta T}} + \cdots + \frac{C(\Delta T\, q)}{(1+r)^{T_1+(m-1)\Delta T}}$$

$$\text{NPV}_{\text{User}} = \int_{(S_0-D_0)/q}^{T_1} U(S_t, D_t, t)\, dt + \int_{(S_0+\Delta T q-D_0)/q}^{T_1+\Delta T} U(S_t, D_t, t)\, dt + \cdots$$
$$+ \int_{(S_0+(m-i)\Delta T q-D_0)/q}^{N} U(S_t, D_t, t)\, dt$$

The number of expansions in the analysis period is $m = (N - T_1)/\Delta T$; and T_1 is specified by the system owner.

Subcase 3 (Figure 19.15*c*). Subsequent to the initial capacity supplied, provide an expansion (i.e., additional capacity of ΔS) after every ΔS increase in demand:

$$\text{NPV}_{\text{Agency}} = C_0(S_0) + \frac{C(\Delta S)}{(1+r)^{T_1}} + \frac{C(\Delta S)}{(1+r)^{T_1+\Delta S/q}} + \frac{C(\Delta S)}{(1+r)^{T_1+2\Delta S/q}} + \cdots + \frac{C(\Delta S)}{(1+r)^{T_1+(m-1)\Delta S/q}}$$

$$\text{NPV}_{\text{User}} = \int_{(S_0-D_0)/q}^{T_1} U(S_t, D_t, t)\, dt + \int_{(S_0+\Delta S-D_0)/q}^{T_1+\Delta S/q} U(S_t, D_t, t)\, dt + \cdots$$
$$+ \int_{(S_0+(m-1)\Delta S-D_0)/q}^{N} U(S_t, D_t, t)\, dt$$

where the number of expansions in the analysis period is $m = [(D_0 + qN) - S_0)]/\Delta S$.

Case 2. D(t) is nonlinear: $D(t) = D_0 + f(x)$.

Subcase 1 (Figure 19.15*d*). At the initial point, provide adequate supply for the entire analysis period, which means the initial supplied capacity is $S_0 \geq D_0 + qN$:

$$\text{NPV}_{\text{Agency}} = C_0(S_0)$$

$$\text{NPV}_{\text{User}} = \int_{T_1^*}^{N} U(S_t, D_t, t)\, dt$$

where $D(T_1^*) = S_0$.

Subcase 2 (Figure 19.15*e*). Subsequent to the initial capacity supplied, provide an expansion every period ΔT; the expanded capacity is made to be equal to the expected increase of demand over the next ΔT:

$$\text{NPV}_{\text{Agency}} = C_0(S_0) + \frac{C[d(T_1 + \Delta T) - d(T_1)]}{(1+r)^{T_1}} + \frac{C[d(T_2 + \Delta T) - d(T_2)]}{(1+r)^{T_2}} + \cdots$$
$$+ \frac{C[d(T_m + \Delta T) - d(T_m)]}{(1+r)^{T_m}}$$

The number of expansions in the analysis period is $m = (N - T_1)/\Delta T$; and T_1 is specified by the system owner.

$$\text{NPV}_{\text{User}} = \int_{T_1^*}^{T_1} U(S_t, D_t, t)\, dt + \int_{T_2^*}^{T_2} U(S_t, D_t, t)\, dt + \cdots + \int_{T_m^*}^{N} U(S_t, D_t, t)\, dt$$

Table 19.3 Summary of Expressions Associated with Need Assessment for Various Demand Trends, Finite-Need Policy[a]

	$D(t)$ Is Linear			$D(t)$ Is Nonlinear		
Variables	**Linear Case 1**	**Linear Case 2**	**Linear Case 3**	**Nonlinear Case 1**	**Nonlinear Case 2**	**Nonlinear Case 3**
T_1^*	—	$(S_0 - D_0)/q$	$(S_0 - D_0)/q$	—	$D(T_1^*) = S_0$	$D(T_1^*) = S_0$
T_i^*	—	$(S_0 - D_0)/q + (i-1)\,\Delta T$	$(S_0 - D_0)/q + (i-1)\,\Delta S/q$	—	$D(T_i^*) = S_0 + d(T_i) - d(T_{i-1})$	$D(T_i^*) = S_0 + (i-1)\,\Delta S$
T_m^*		$(S_0 - D_0)/q + (m-1)\,\Delta T$	$(S_0 - D_0)/q + (m-1)\,\Delta S/q$	—	$D(T_m^*) = S_0 + d(T_m) - d(T_{m-1})$	$D(T_i^*) = S_0 + m\,\Delta S$
T_1	—	T_1	T_1	—	T_1	T_1
T_i	—	$T_1 + (i-1)\,\Delta T$	$T_i = T_1 + (i-1)\,\Delta S/q$	—	$T_1 + (i-1)\,\Delta T$	$d(T_i) - d(T_{i-1}) = \Delta S$
T_m		$T_1 + (m-1)\,\Delta T$	$T_i = T_1 + (m-1)\,\Delta S/q$	—	$T_1 + (m-1)\,\Delta T$	$d(T_m) - d(T_{m-1}) = \Delta S$
m	0	$\dfrac{N - T_1}{\Delta T}$	$\dfrac{(D_0 + qN) - S_0}{\Delta S}$	0	$\dfrac{N - T_1}{\Delta T}$	$\dfrac{[D_0 + d(N)] - S_0}{\Delta S}$
ΔS_i	—	$\Delta T q$	ΔS	—	$d(T_i + \Delta T) - d(T_i)$	ΔS

[a] T_1 is specified by the system owner.

where $D(T_1^*) = S_0$ and $D(T_i^*) = S_0 + d(T_i) - d(T_{i-1})$.

Subcase 3 (see Figure 19.15*f*). Subsequent to the initial capacity supplied, provide an expansion with ΔS additional capacity after every ΔS increase in demand:

$$\text{NPV}_{\text{Agency}} = C_0(S_0) + \frac{C(\Delta S)}{(1+r)^{T_1}} + \frac{C(\Delta S)}{(1+r)^{T_2}} + \cdots + \frac{C(\Delta S)}{(1+r)^{T_m}}$$

The number of expansions in the analysis period is $m = \{[D_0 + d(N)] - S_0\}/\Delta S$; and T_1 is specified by the system owner.

We also have

$$d(T_i) - d(T_{i-1}) = \Delta S$$

$$\text{NPV}_{\text{User}} = \int_{T_1^*}^{T_1} U(S_t, D_t, t)\,dt + \int_{T_2^*}^{T_2} U(S_t, D_t, t)\,dt + \cdots + \int_{T_m^*}^{N} U(S_t, D_t, t)\,dt$$

where $D(T_1^*) = S_0$ and $D(T_i^*) = S_0 + (i-1)\Delta S$.

19.5.6 Summary of Derivations of for Optimal Scheduling of Supply

Table 19.3 presents the expressions associated with need assessment, for different demand patterns (linear or nonlinear) and the different supply strategies (cases), for the finite-need policy.

19.6 SOME ISSUES AND CONSIDERATIONS IN NEEDS ASSESSMENT

19.6.1 Need Estimation Adjustments to Account for Supply–Demand Interactions

So far in this chapter, we have addressed need estimation on the basis of supply and demand projections, with the implicit assumption that demand and supply are independent of each other. For an aggregate estimation of need for most civil engineering systems and also for purposes of long-term need assessment, this assumption may be valid. However, for situations where the demand and supply agents are individual entities such as persons (e.g., transit users) or companies (e.g., transit agencies), or where we seek to examine supply and demand in response to short-term stimuli, the laws of classical economics become more visible. For example, where demand and supply are perfectly elastic and demand exceeds supply, the price of the product (in this case, the fare or fee for the service rendered by the system) increases and this causes the agency to increase production (supply) as it seeks to maximize revenue (e.g., increasing the number of trains/buses or increasing the frequency of service); when supply exceeds demand, the price of the product falls and this causes the agency to reduce production or supply (e.g., reducing the number or frequency of the transit trains or buses). Another example is the case of highway systems: When demand exceeds supply (creating a need situation), the price of travel (in terms of user delay due to system congestion) increases, causing the demand to fall as fewer users are willing to pay that price. These examples illustrate the short-term interactions between demand and supply and demonstrate that in certain contexts, the need is not always independent of supply and demand but could actually cause demand or supply to increase or decrease. Thus, some adjustment of the projected need amount may be necessary, particularly in cases where demand is expected to substantially exceed supply.

In cases where a long-term picture is sought, these spikes or reductions in the demand or the supply will only be seen as small kinks in a long-term demand or supply projection. Second, for most civil systems, the quantity of the supply tends to be largely inelastic with respect to price (whereas the quality of service may be relatively elastic); thus, the interactions between demand and supply may not be very consequential in significantly reducing the long-term demand or supply. Therefore, the needs estimation procedure discussed in Section 19.4 can be considered adequate for assessing the long-term need for most civil systems.

19.6.2 Role of Real Options in Needs Assessment

As we have discussed in the previous sections, owners or operators of civil systems make strategic decisions that include (i) how much capacity to provide initially for a proposed system (this is done at the needs assessment phase) and (ii) the extent of the increase or reduction of an existing system's capacity (this done at the operations phase). In Section 19.4, we assumed a number of supply strategies, including the provision of a large capacity upfront to cover demand over an analysis period or adding capacity whenever the demand reaches a specified level or the need (excess demand over supply) reaches a certain trigger. For example, should a downtime parking garage be built far in excess of existing demand in order to accommodate future increases in the demand; or should the floors be added incrementally as the demand increases? Often, the provision of a large upfront capacity may be imprudent, for the demand may fall far short of the expectations, leading to losses associated with maintaining excess capacity and opportunity cost.

Therefore, decisions regarding the initial capacity of civil systems need to consider the practical uncertainties associated with the economic, political, social, and environmental conditions. In order to manage the risks due to those uncertainties, different proactive strategies should be followed, and those strategies may be analyzed using the real options framework we learned in Chapter 15. A real option is not an obligation but rather a right to undertake an investment venture, such as the deferment, abandonment, expansion, contraction, or staging of a capital investment project. For example, a water supply agency may decide to suspend a massive borehole drilling project if the demand for water falls below a certain level or may decide to proceed when the demand rises above a certain level. Referred to as real options, strategic options such as these are typically not considered adequately in traditional discounted cash flow evaluations where deterministic or even probabilistic values of net present worth are used as a basis for the decision. Real options analysis (ROA) is borrowed from the financial options theory practiced in financial engineering.

The concept of real options has been shown to be particularly valuable because it measures the inherent value of flexibility in decision making, specifically in response to unexpected developments with regard to socioeconomic, technical, and environmental conditions.

Thus, ROA is a particularly attractive tool for analyzing investments in rapidly changing or uncertain environments where traditional methods of economic evaluation are unable to accurately measure the economic value of the investments in rapidly changing or uncertain environments. ROA enables decision makers to leverage the uncertainty associated with the system and its environment in a bid to limit the downside risk. The upfront expenditure related to the flexibility in real options is referred to as the option premium. In the field of civil engineering, there are a multitude of areas where ROA can be applied for more flexible decision making. Real options can be incorporated in the needs estimation framework presented in this chapter by developing or giving priority to supply strategies that allow for gradual changes in capacity in response to demand trends.

19.6.3 Probabilistic Demand and Supply

The discussion of demand and supply is very much like the reliability of relative levels of loading and capacity which we discussed in Section 13.4.3 of Chapter 13. What we call the need is what the reliability engineer refers to as the safety margin, the algebraic difference between the supply or capacity S and the demand or load D:

$$\text{Need} = D - S$$

For example, for a system that is expected to experience a demand of 7200 units daily and has a capacity of 5000 units daily, the need is calculated as 7200 − 5000 = 1200 units.

In cases of uncertainty where the loading and capacity are not fixed values but rather exhibit so much variability that they are best described by probability density function $f_S(s)$ and $f_D(d)$, respectively, and the need can be calculated using the expected values of their probability functions. In mathematical notation, this is written as:

$$\text{Need} = E(S) - E(D) = \left(\int_{-\infty}^{\infty} sf_S(s)\, ds\right) - \left(\int_{-\infty}^{\infty} df_D(d)\, dd\right)$$

$$= E(S - D) = \int_{-\infty}^{\infty} (s - d)f_{S-D}(s - d)d(s - d) = \int_{-\infty}^{\infty} nf_N(n)\, dn$$

Figure 19.16 illustrates the demand and supply functions at a given time as probability distributions. In the situation shown in Figure 19.16*a*, both capacity and demand are fixed values with zero variability, and the supply exceeds the demand. In the situations shown in Figures 19.16*b* and 19.16*c*, both supply and demand are described by probability distributions, with the former exhibiting much lower variability than the latter. In Figure 19.16*b*, there is no overlap of the demand and supply functions; thus, for the situation in Figure 19.16*b*, supply will always accommodate demand because the worst-case possibility for supply (i.e., when supply is at its minimum possible level) still exceeds the worst-case possibility for demand (i.e., when demand is at its maximum possible level). In Figure 19.16*c*, there is an overlap of the demand and supply probability functions so a possibility exists that the demand will exceed the capacity (as shown in the shaded region).

The situation represented by Figure 19.16*c*, where demand exceeds supply, represents a situation of great need that may lead to queuing in transportation systems (and the accompanying impacts of delay, safety, and inconvenience), failure in the case of structural systems due to overloading, and flooding in the case of hydraulic systems, leading to inundation of inhabited areas and possible loss of life and property. For the situation represented by Figure 19.16*c*, Figure 19.17 presents the probability distribution for the need: There is a possibility, at any given time, that demand could exceed supply.

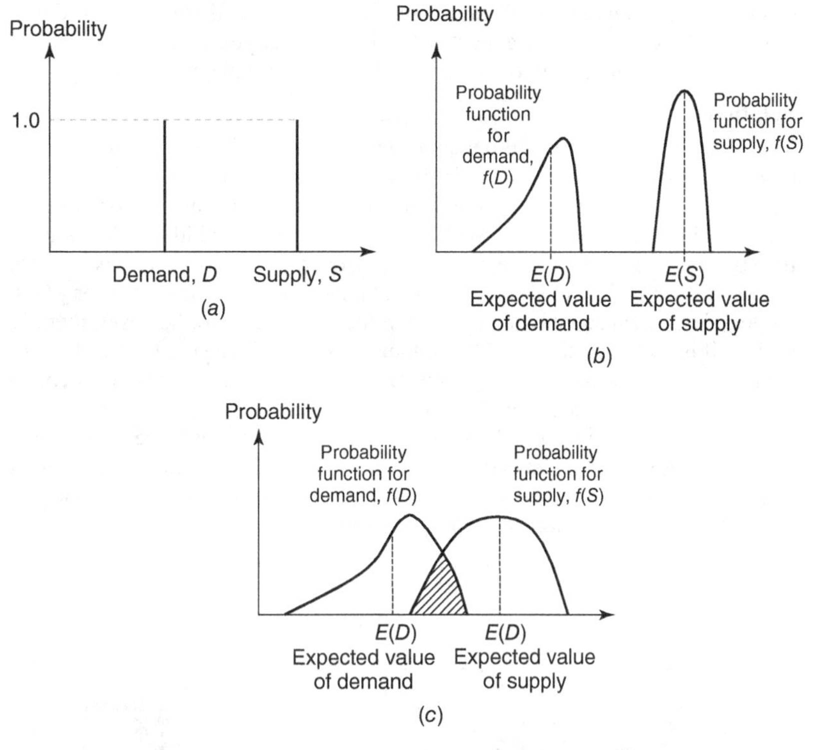

Figure 19.16 Conceptual probabilistic functions for demand and supply at a given point of time: (*a*) deterministic demand and supply, (*b*) probabilistic demand and supply, no overlap, and (*c*) probabilistic demand and supply, with overlap.

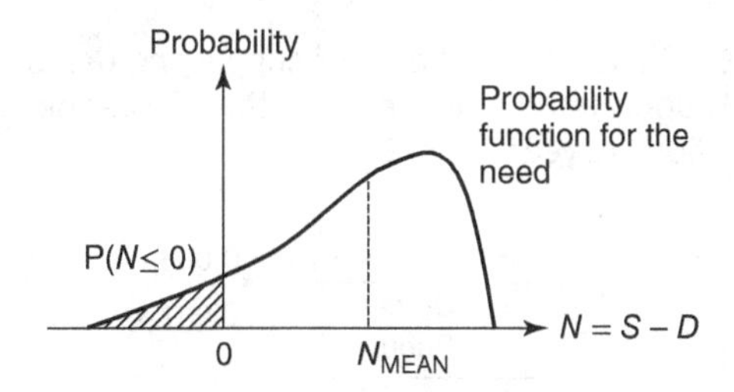

Figure 19.17 Conceptual probability distributions for the need.

Example 19.6

It is sought to analyze the supply and demand trends for a proposed hydraulic engineering system. From past data on similar locations, it is known that the demand at any year is normally distributed with a standard deviation that is 20% of the mean. The mean demand increases at an exponential rate governed by the following function: Demand $= 75,000 \times e^{0.187^{*}\text{Year}}$. In the initial year, a supply ranging from 100,000 to 200,000 gallons will be provided; then in the 8th year, a supply ranging from 550,000 to 700,000 gallons will be provided until the end of the system's life at year 12. Assuming uniform distribution for the supply within the given ranges, (a) provide a sketch of the supply and demand functions, (b) sketch the levels and ranges of supply and demand at years 7 and 9, (c) sketch the distributions of demand and supply and need at year 7, (d) plot the distributions of demand and supply and need at year 9, (e) determine the expected level of need at years 7 and 9, and (f) estimate, over the 12-year period, (i) the total magnitude of need and (ii) the total magnitude of capacity underutilization.

Solution

The sketch of the supply and demand functions is provided as Figure 19.18*a*. The plot shows that the demand across the years falls within the shaded trumpet-shaped band: there are larger variations for later years due to the growing variations with the increasing demand as specified in the question. The supply increases in a discrete fashion: Within a band of constant width and fixed mean for the first 8 years, and then from year 8 to year 12, there is a jump in average supply as well as the variability of supply. Also, at any year, it is possible to sketch or plot the mean and range as shown for years 7 and 9 (Figure 19.18*b*): At year 7, the mean supply is higher than the mean demand; however, there is significant variability in both so it is quite possible that the demand can exceed the supply, thereby creating a need situation. At year 9, however, the mean supply is not only higher than the mean demand, but the lowest supply exceeds the highest demand even though there is significant variability in both demand and supply, and there is thus no chance that there will be a need in year 9. These results are also exhibited in Figures 19.19*a* and 19.19*b*. Figure 19.20 presents the probability distribution of the need at year 7 obtained from the Monte Carlo simulation, which shows that the distribution is normal shaped. For year 9, there is zero need so the probability distribution is simply a straight line along the x axis.

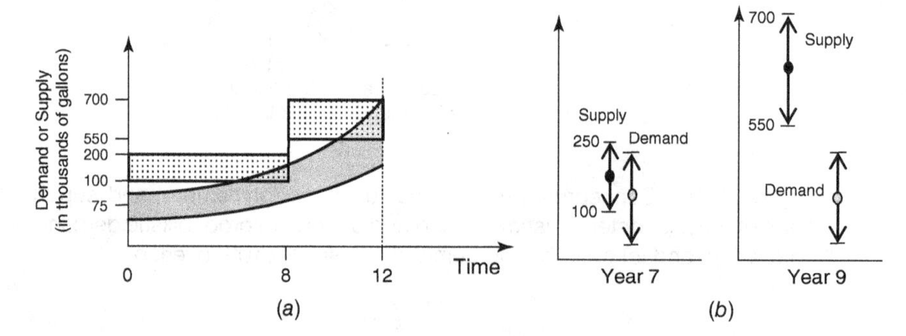

Figure 19.18 Sketch of functions for supply and demand, Exercise 19.3: (*a*) sketch of demand and supply trend functions over the 12-year period and (*b*) detail of supply and demand situations at year 7 and 9.

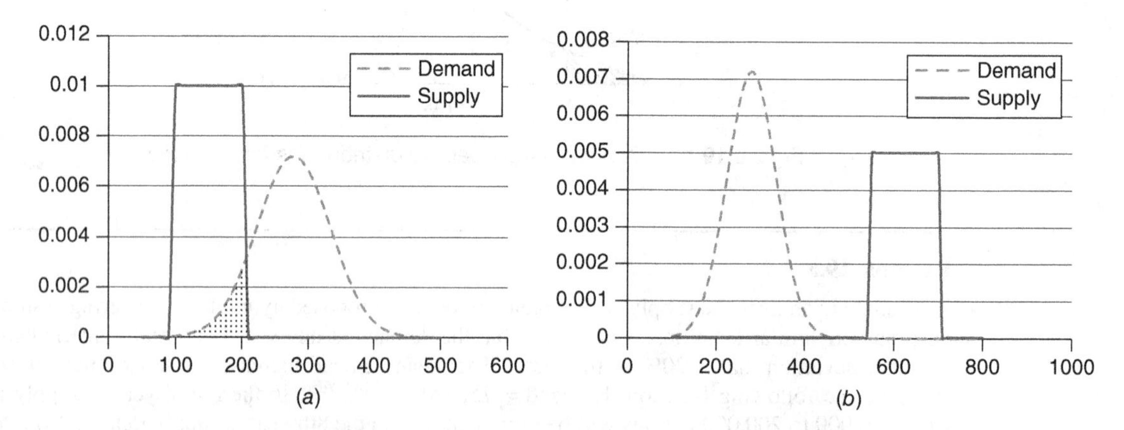

Figure 19.19 Sketch of functions for supply, demand, and need, Exercise 19.3: (*a*) Sketch of demand and supply probability function at year 7 (shaded portion indicates the need) and (*b*) sketch of demand and supply probability function at year 9.

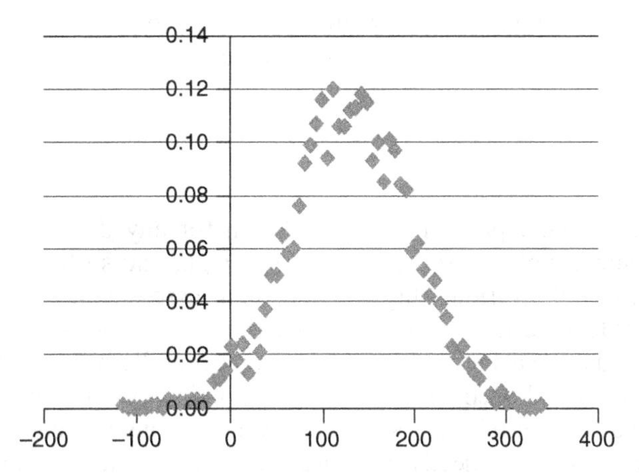

Figure 19.20 Plot of the need probability function at year 7, Exercise 19.3.

19.6.4 Assessment of Need at the Network Level (for a System of Systems)

The discussions thus far in this chapter have been tailored to identifying and quantifying the need for a single system or a component thereof. Most agencies own or operate more than one civil system and may seek to assess the needs not only for a new system but also of system expansion or preservation at each year over an analysis period. Thus, for an entire system-of-systems (SOS), need is often discussed in the context of recurring need instead of initial need (see Section 19.1.2). For an SOS, the recurring need for improvements to functional or structural capacity in terms of the amount of physical work (and the corresponding budget) needed at each year could be determined. As we have seen in the preceding sections, deterioration models and minimum standards of operations/condition are needed to assess the need for a new system or enhancements of an existing system. Then, for the budget needed, the cost functions for each deterioration level are needed. If each individual system in the entire inventory is thus analyzed, the systemwide needs can be determined for the backlogged needs (outstanding work that should have been done in the past), current work needed, and projected future work, which can be assessed both in terms of the physical work needed as well as the corresponding monetary value of such needs. This analysis can be done for each year in a horizon period or for the entire horizon period.

19.6.5 Politics of Needs Assessment

Typically, needs assessments by the owner of a public system are not carried out in a vacuum and independently but rather face a variety of political pressures internal and external of the agency. Given the extensive physical reach of most civil systems, coupled with the very large contract sums involved in their construction, operations, or maintenance, systems owners (agencies) typically interact with political players from both the government and the private sector. As has been noted in the literature (Reviere et al., 1996), the political considerations in needs assessment may range from inputs, restrictions, and modifications of the elements of the needs assessment process, such as the scope or the focus, which may often include the perspective of specific population segments or interest groups. In extreme cases, politicians who earlier proclaimed to the electorate that they had succeeded in addressing a certain need and thus deserve to be reelected may bear pressure on a systems analyst to deny the existence of any further need in that given context. Navigating through these multiple and often powerful political interests requires a great deal of tact and ethics, and a slight misstep could imperil the engineer's career.

SUMMARY

The needs assessment phase is the foundation for any decision making and hence determines whether the subsequent phases of planning, design, and so forth are carried out. In this chapter, we first discussed the various stages of the needs assessment phase: problem identification; identifying stakeholders; establishment of goals, objectives, and performance measures; and assessing the system needs. We also reviewed how needs generally may be categorized (growing needs versus sudden needs and needs associated with existing systems versus those of new systems), and we identified a number of needs situations in the various branches of civil engineering. We examined how to assess system needs using user-targeted mechanisms and quantitative assessments. For the latter, we saw how to derive expressions for optimizing supply schedules on the basis of projected demand and supply, for the zero-need finite-need policies. Recognizing that neither excess demand nor excess supply is desirable, we presented these expressions to prescribe the optimal timing of capacity additions to minimize the costs associated with excess demand or excess supply. We also discussed the issues related to need, including adjustments of need amounts to account for supply–demand interactions, needs assessment at the system-of-systems level, probabilistic needs assessment, the role of real options in needs assessment, and the perilous politics of needs assessment.

EXERCISES

1. For a planned parking garage in a fast growing city, discuss the issues associated with the four stages of the needs assessment phase.
2. Explain, with examples in civil engineering, the difference between a sudden need and a growing need. Your discussion should include causes of these types of needs.
3. Your university plans to move from a semester to a trimester system. Carry out an assessment of the need for additional units of cooling or heating that will be required to serve the expected increase in demand for the campus buildings.
4. The demand for a landfill system at a certain year ranges from 3000 to 6000 tons per day. The capacity of available equipment is expected to be 5700 to 6200 tons per day. Assume a uniform probability distribution of supply and demand within these extremes, and that the means are the midpoints, use Monte Carlo simulation to determine the distribution of the need. State any assumptions.
5. Projections for the demand for water in a certain city in year 2020 follow a normal distribution with average and standard deviations per day of 500,000 gallons and 100,000 gallons, respectively. The existing water supply system can provide 520,000 gallons per day but may face a shortfall up to 10% due to a possible drought. Assume a uniform probability distribution for supply and use Monte Carlo simulation to determine the distribution of the need.
6. A certain urban freeway had a capacity of 1700 passenger car units (pc) per lane per hour in 1990. In 1998, a series of ITS investments and speed limit increases helped to increase the capacity at a linear rate of 40 pc per hour per lane every year for 2 years, after which an additional freeway lane was constructed that added a capacity of 100 pc per hour per lane, and the capacity remained at that level thereafter. The table below presents the demand for 5 years. Develop a statistical regression function for the demand and plot both demand and supply on the same graph. Determine the year after 2010 at which demand will exceed supply. When that happens, what remedial actions would you recommend for the freeway?

Year	1990	1991	1992	1993	1994	1995	1996	1997	1998	1999	2000	2001
Demand	1620	1690	1800	2070	2150	1620	1690	1800	2070	2150	2070	2150

Year	2002	2003	2004	2005	2006	2007	2008	2009	2010
Demand	1620	1690	1800	2070	2150	1620	1690	1800	2070

7. An international sporting event is slated to be held next week in a certain city known for persistent water supply problems. The existing demand is 600,000 gallons per day. Unfortunately, at the fifth day of the games, part of the water system will be shut down in order to complete a longstanding improvement work, reducing the capacity to 500,000 gallons per day; for one day after this improvement, the capacity will increase to 1,100,000 gallons per day. On the ninth day, as the games wind down, this additional capacity will be removed and a capacity of 700,000 gallons per day will remain. The table below presents the expected daily demand in thousands of gallons over the 2-week period during which the games are held. Plot both the demand and supply on a graph and determine the severity of excess capacity and excess demand and the days on which they occur.

Day	1	2	3	4	5	6	7	8	9	10	11	12	13	14
Demand	600	610	625	670	805	1015	1195	1302	1010	790	720	652	625	600

8. The table below presents the past levels of monthly supply and demand for a certain civil system.

Year	2005	2006	2007	2008	2009	2010	2011	2012
Demand	5.50	5.51	5.56	5.6	5.65	5.69	5.83	5.95
Supply	5.6	5.71	5.80	5.87	5.90	6.12	6.11	6.12

Develop separate regression functions to describe the trends of supply and demand and plot them on the same graph. Using the graph, determine (a) the projected month when a need starts to exist; (b) the projected level of need at year 2015; and (c) the total expected need, in unit-years, over the period when a need is expected to exist until year 2015 when the system reaches the end of its service life.

REFERENCES

Crandall, S. J. S. (1998). Using Interviews as a Needs Assessment Tool. *Journal of Continuing Education for Health Professions*, 18, 155–162.

Delage, P., Cui, Y. J., and Antoine, P. (2005). Geotechnical Problems Related with Loess Deposits in Northern France, *Proceedings. International Conf. on Problematic Soils*, Eastern Medit. Univ., Famagusta, N. Cyprus.

De Neufville, R. (1990). *Applied Systems Analysis*. McGraw-Hill, New York.

Eide, O. (1968). Geotechnical Problems with Soft Bangkok Clay on the Nakhon Sawan Highway Project. *Norwegian Geotechnical Institute Publ., Nr*. 78(9), p. 11.

Ghosn, M., Moses, F., and Wang, J. (2003). *NCHRP Report 489:* Design of Highway Bridges for Extreme Events. Transportation Research Board, Washington, DC.

Gray, D. H., and Sotir, R. B. (1995). Biotechnical Stabilization of Steepened Slopes. Presented at the 76th Annual Meeting of the Transportation Research Board, Washington, DC.

Hancher, D. E. (2003). *Construction, The Civil Engineering Handbook*, 2nd Ed., Eds: Chen, W. F., and Liew, J. Y., CRC Press, Boca Raton, FL.

Kacin, J. A. (2009). Fatigue Estimation of a Highway Sign Structure. M.S. Thesis, University of Pittsburgh, Civil and Environmental Engineering, Pittsburgh.

Kaufman, R., and English, F. W. (1979). *Needs Assessment: Concept and Application*. Educational Technology Publications, Englewood Cliffs, NJ.

Khisty, J. C., Mohammadi, J., and Amekudzi, A. (2012). *Fundamentals of Systems Engineering*. Pearson Prentice Hall, Upper Saddle River, NJ.

Lockyer, J. (1998). Getting Started with Needs Assessment: Part 1 The Questionnaire. *J. Continuing Ed. Health Professions* 18, 58–61.

Mann, K. V. (1998). Not Another Survey! Using Questionnaire Effectively in Needs Assessment. *J. Continuing Ed. Health Professions* 18, 142–149.

Meredith, D. D., Wong, K. M., Woodhead, R. W., and Wortman, R. H. (1985). *Design and Planning of Engineering Systems*. Prentice Hall, Englewood Cliffs, NJ.

Middleton, L., and King, M. (2003). A Natural Choice. *Public Roads, March/April Issue*, 66, 5.

Random House (2013). *Random House, Dictionary*, Random House, New York.

Ratnapalan S., and Hilliard, R. I. (2002). Needs Assessment in Postgraduate Medical Education: A Review. *Med Educ Online* [serial online] 2002; 7:8. Available from URL http://www.med-ed-online.org.

Reviere, R., Berkowitz, S., Carter, C., and Ferguson, C. G. (1996). *Needs Assessment: A Creative and Practical Guide for Social Scientists*. Taylor and Francis, Boca Raton, FL.

Steadham, S. V. (1980). Learning to Select a Needs Assessment Strategy. *Training Develop. J.* 30, 56–61.

Sykes, R. (2002). Water and Wastewater Planning. In *Civil Engineering Handbook*, 2nd Ed. CRC Press, Boca Raton, FL, p. 8–1.

Thompson, P. D., Labi, S., Ford, K. M., Arman, M. H., Shirole, A., Sinha, K. C., and Li, Z. (2012). Estimating Life Expectancies of Highway Assets. National Cooperative Highway Research Program Report 713, Vol. 1: Guidebook. Transp. Res. Board, Washington, DC.

Tipping, J. (1998). Focus Groups: A Method of Needs Assessment. *J. Continuing Ed. Health Professions* 18,150–154.

Voland, G. (1999). *Engineering by Design*. Prentice Hall, Upper Saddle River, NJ.

USEFUL RESOURCES

de Neufville, R. (2000). Dynamic Strategic Planning for Technology Policy. *Int. J. Tech. Manage.* 19(3/4/5), 225–245.

de Neufville, R., Scholtes, S., and Wang, T. (2006). Real Options by Spreadsheet, Parking Garage Case Example. *ASCE J. Infrastructure Syst.* 12(2), 107–111.

Eide, A. R., Jenison, R. D., Mashaw, L. H., and Northup, L. L. (2002). *Introduction to Engineering Design and Problem Solving*. McGraw-Hill, New York.

Kalligeros, K. (2006). Platforms and Real Options in Large-Scale Engineering Systems. Ph.D. Dissertation, Engineering Systems Division, Massachusetts Institute of Technology, Cambridge, MA.

Mun, J. (2002). *Real Options Analysis: Tools and Techniques for Valuing Strategic Investments and Decisions*, Wiley, New York.

Zhao, T., and Tseng, C. (2003). Valuing Flexibility in Infrastructure Expansion. *ASCE J. Infrastructure Syst.* 9(3), 89–97.

CHAPTER 20

SYSTEMS PLANNING

20.0 INTRODUCTION

Civil engineering systems can be considered as the thread that brings together the different fabrics of national or regional development including the economy, health delivery, education, agriculture, and energy. In civil systems planning, where alternative plans are developed and evaluated, the planner recognizes that civil systems do not exist in a vacuum but must meet physical, locational, economic, social, and political requirements. As discussed in Chapter 3, a well-designed civil system not only provides benefits to its users and the community in delivering the specific service for which it was intended but also provides government entities with an opportunity to meet broader national or regional objectives such as job creation, energy consumption reduction, and sustainable development. As such, the planning of civil systems must be carried out with adequate knowledge of the demand for the system and also of the system's potential impacts (intended or unintended) on its immediate environment and on development at the regional and national levels.

The responsibility for systems planning is often borne by agencies that have been granted statutory authority for a specific system type. In certain countries, there is a formal and distinct national governmental organization that carries out or supervises planning for public civil systems at a national level that go by appellations such as the National Development Planning Commission or National Infrastructure Planning Agency. In other countries, such as the United States, the responsibility for most civil systems planning is borne by regional, state, or local governments, and the federal government's role is to provide the funds and to ascertain that all the subsequent phases of the system development are consistent with legislation. The agency may carry out the planning task in-house or may outsource part or all of it to consultants instead. Agencies that carry out the planning in-house often have a specific position for persons who carry out civil systems planning, such as a system planner or urban planner in metropolitan planning organizations. In certain organizations, the planning task is carried out by persons in disciplines other than planning. In any case, the system planner not only liaises with other engineering professionals, such as transportation and environmental engineers, but also solicits input from other persons in the community such as city and county personnel and the general public.

The common components of most civil engineering planning processes include the development of a set of alternative plans, documentation of the consequences of each plan, and selection of the best plan. However, there is no universal planning protocol for civil engineering systems because the framework and specific methodologies used by different agencies and individuals for planning vary significantly depending on the system type and size, the agency's mission, the jurisdiction, and other factors. Thus, the systems planning framework and illustrations presented in this chapter are for illustrative purposes only and may be duly modified to reflect the dimensions and practical issues associated with the planning problem at hand.

Besides an explanation of each step of the system planning process, this chapter presents a brief history of civil systems planning and identifies its various dimensions. The rationale and impetus for system planning is then discussed followed by an identification of some evolving and emerging contexts of systems planning. A number of good practices in civil systems planning are presented, and barriers to effective planning are identified. Finally, the chapter presents some numerical exercises in civil systems planning.

20.1 BRIEF HISTORY OF CIVIL SYSTEMS PLANNING

Since the dawn of time, civil engineers have constructed roads, dams, irrigation canals, and other systems that have served the public good. In ancient times, the construction of these systems were preceded by plans and designs developed by master builders with input from the rulers, military commanders, citizenry, and, in some cases, individuals learned in related fields such as scientists and mathematicians. It has been only in recent history, however, mainly during the 20th century, that government entities in most countries began taking steps to create formal processes for planning their civil infrastructure. The impetus for establishing formal planning processes was accelerated by a number of factors: increasing awareness of the need to protect the environment from the adverse effects of the construction and operations of a civil system as well as to protect the system from adverse environmental conditions and to further determine the governmental responsibility for ensuring such protections; greater realization that errors in design and subsequent cost and time overruns during construction could be minimized if the system were planned properly; and the embrace of the trend of using objective, stakeholder-driven performance criteria to assess the full impacts of projects and to compare between alternative plans. Therefore, throughout the 20th century, many countries have passed legislation to facilitate or at least, formalize the planning process. In the United States, for example, these include the 1936 Flood Control Act (which introduced the requirement of benefit–cost analysis for water-related engineering systems), the 1969 National Environmental Policy Act (NEPA) (which expanded the planning process to include due consideration of new agency goals such as minimization of adverse environmental impacts), and 2012 MAP-21 signed by President Obama (which created a streamlined and performance-based program to transform the policy and programmatic framework for surface transportation system investments). Figure 20.1 presents a timeline of some key legislation that has profoundly influenced the planning of civil engineering systems. In Chapter 29, we examine more such legislation.

20.2 DIMENSIONS (PERSPECTIVES) OF CIVIL SYSTEM PLANNING

The nature of the individual planning activities for civil systems can differ from each other in many ways. Thus, planning may be viewed in one or more of several perspective. This section discusses the different considerations associated with each perspective.

20.2.1 The Individual System versus the System of Systems (SOS)

Any individual civil engineering system is often a part of a larger parent network of systems or system of systems. For example, a water pipe section between two stations is only a part of a citywide network of pipes, or an urban highway arterial street is only a part of a city street network. Planning could be for an individual system or for a collection of systems. Even for an individual system, the planner must avoid stand-alone planning: The relationship between the system of interest and the other systems in the network must be duly recognized. Thus, in planning for an individual new system or for improvements to an individual existing system, the planner should study of the effects of the system on other systems in the network and the effects of the existing network on the system. In this text, the scope of interest is mainly the individual system.

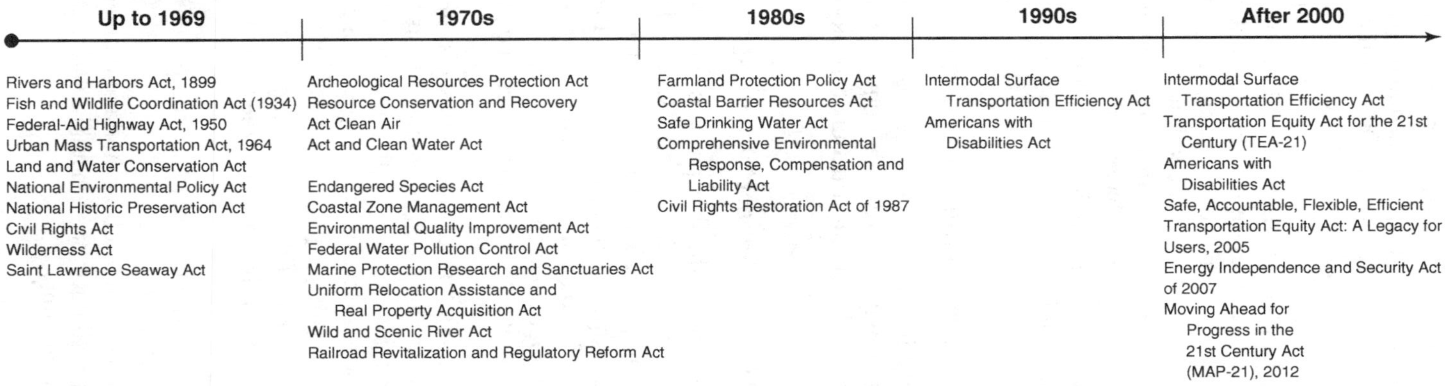

Up to 1969	1970s	1980s	1990s	After 2000
Rivers and Harbors Act, 1899	Archeological Resources Protection Act	Farmland Protection Policy Act	Intermodal Surface Transportation Efficiency Act	Intermodal Surface Transportation Efficiency Act
Fish and Wildlife Coordination Act (1934)	Resource Conservation and Recovery Act	Coastal Barrier Resources Act	Americans with Disabilities Act	Transportation Equity Act for the 21st Century (TEA-21)
Federal-Aid Highway Act, 1950	Clean Air Act and Clean Water Act	Safe Drinking Water Act		Americans with Disabilities Act
Urban Mass Transportation Act, 1964	Endangered Species Act	Comprehensive Environmental Response, Compensation and Liability Act		Safe, Accountable, Flexible, Efficient Transportation Equity Act: A Legacy for Users, 2005
Land and Water Conservation Act	Coastal Zone Management Act	Civil Rights Restoration Act of 1987		Energy Independence and Security Act of 2007
National Environmental Policy Act	Environmental Quality Improvement Act			Moving Ahead for Progress in the 21st Century Act (MAP-21), 2012
National Historic Preservation Act	Federal Water Pollution Control Act			
Civil Rights Act	Marine Protection Research and Sanctuaries Act			
Wilderness Act	Uniform Relocation Assistance and Real Property Acquisition Act			
Saint Lawrence Seaway Act	Wild and Scenic River Act			
	Railroad Revitalization and Regulatory Reform Act			

Figure 20.1 Timeline of selected legislation that has influenced civil systems planning.

20.2.2 Development Phases

On one hand, planning for a civil system may be restricted to cover considerations related strictly to planning, such as the system's location, its functional relationships with its built-up and natural environment, and rough estimations of its cost. On the other hand, planning may traverse a wider swath of the systems development life cycle and may include the preplanning phases of needs assessment and the postplanning of design, construction, operation, and preservation (Meredith et al., 1985); in such cases, the planner will need to develop technical specifications for constructing the system and alternative strategies for maintaining and operating the system. In this textbook, we define planning in the former perspective, namely, system planning, as only a phase of the systems development cycle rather than as a combination of multiple phases. This seems consistent with the school of thought that believes that civil system planners must confine themselves to purely planning tasks while leaving other phases to the professionals of those phases. Nevertheless, systems planners, if they are to develop a good and robust plan, must have a fair idea at least, if not a complete understanding, of how the system they are planning will be designed, constructed operated, and maintained.

20.2.3 Physical Change versus Operational Change

Planning may be for a change in the physical system or merely a change in its operations policies. Changes of a physical nature may involve a new physical facility or an existing facility. For an existing factility, the change may be a reconstruction, capacity expansion, functional improvements, or strengthening.

20.2.4 Planning "Levels" (System Impact Type or Agency Goal)

The system planning phase follows the first phase of systems development where the needs are identified and the goals are established. Thus, the scope of the planning effort is guided by the scope of the needs assessment phase. The planner constantly bears in mind the stated objectives for the system so that the developed plan is consistent with the stated objectives. In this respect, the planning scope may simply refer to the number and types of system impact criteria that are considered in the planning process. Goodman and Hastak (2007) concluded that the number and types of planning criteria have evolved over the decades and can be placed in four "levels" (Figure 20.2):

Level 1. At this level, planners consider the proposed system's location and expected functions, capacities, costs, and in certain cases, revenues. At this level, planning decisions are based on the need for the system (identified through public input or interventions of a legal, political, or institutional nature). This level of planning has long been used by civil systems planners and is still being used, mainly for projects owned by local jurisdictions.

Level 2. The criteria considered at this level comprise those of level 1 and other criteria, such as the economic efficiency of the project in the form of a benefit–cost ratio or its net present worth.

Level 3. At this level, the criteria considered are comprised of those of level 2 and environmental, social, and cultural criteria.

Level 4. This level considers the level 3 criteria and emerging criteria such as environmental sustainability, system resilience or vulnerability to natural or man-made threats including climate change, and uncertainty.

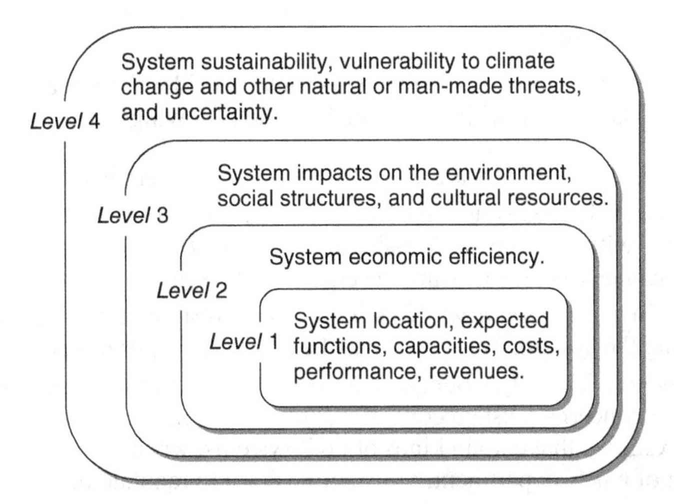

Figure 20.2 Scopes of the planning levels.

20.2.5 Planning Area (Spatial Scope)

The geographical coverage of the system plan may be small or broad, depending on the system type and size, and may transcend natural or artificial boundaries. For most systems, planning is carried out at the local level. For major water supply stems, such as hydroelectric projects, the planning area may be an entire region, state, or province. The geographical coverage for systems planning also could be influenced by political or legislative situations (Goodman and Hastak, 2007). The planning area for the system thus may need to be adjusted so that it is consistent with any existing areawide or regionwide plans, particularly in certain cases where the system is eligible for specific government funds or subsidies. In certain situations, the planning agency uses natural areas (e.g., river basins) or natural borders (e.g., rivers) to establish the planning area.

An appropriately demarcated study area is critical in civil engineering systems planning because the outcome of the planning study can be influenced by the potential spatial reach of each planning alternative. In this context, it is useful to recognize that the spatial scope for the system plan may be local or areawide (e.g., city, county, district, state, province, etc.) Also, for larger spatial scopes, planning effort may become more challenging because the impact of the system becomes not only less visible but also more difficult to assess due to possible interactions with other existing neighboring systems or conditions and the ensuing difficulty of isolating the potential impacts of the proposed system.

20.2.6 Temporal Scope of System Impacts

From a temporal perspective, the system impacts considered in system planning could have short-term, medium-term, or long-term effects. Some impact types are felt for relatively short periods of time (e.g., dust pollution during the system construction) while others may endure for many decades after construction (e.g., noise, air, or water pollution, or economic development due to the operations of the system). Thus, the temporal scope of the planning activity will be influenced by the impact type under consideration.

20.2.7 Stakeholders

In evaluating a system plan, the planner should consider the impact types in conjunction with the concerned stakeholders, which often include the following:

Users: The system users are those for whom the system is provided.

Community: The persons who live, work, or carry out some activity in the proximity of the system without necessarily using it are the system's community, Their concerns are vital if the planner wishes to ensure the equity of the system's costs and benefits across society.

System Owner or Operator: This is the entity responsible for providing, operating and maintaining the system. This entity may be a private corporation or a government (public) agency.

Government: The city, county, state, or federal government is also an important stakeholder whose concerns must be considered during the planning process. The planner must consider, for example, that certain kinds of civil systems, when constructed, may require the establishment of a new department, office, or position to regulate or monitor either some aspect of the system's operations or its impacts.

The planning criteria related to the system owner or operator are typically financial in nature and may include the initial or life-cycle cost of the system, exposure of the owner to tort claims, and public relations. Consideration of these criteria is important in the planning process because they provide an indication of the resources needed or expended and help the system owner conduct internal reviews of the system's performance. Planning criteria related to the system user, which include usage fees or fares, safety, security, and user comfort, are important considerations during planning because they provide a basis for the planner to gauge how well the system goals are being attained. Planning criteria related to the community include air, water and noise pollution, economic development attraction or retention, and possible dislocation or relocation of socio-economic entities (farms, business, residences, and so on).

20.2.8 Civil Engineering Disciplines Associated with the System

Often the planning process for a specific system in a specific branch of civil engineering is carried out with explicit cognizance of the other branches to which the system's construction or operations is related. For example, in planning for a transportation system, the scope of the planning activities may include not only transportation but also the related environmental, structures, and hydraulics engineering systems.

20.3 RATIONALE AND IMPETUS FOR SYSTEM PLANNING

Civil engineers are tasked with the resolution of problems that profoundly influence the way society lives, works, and travels. Thus, in the process of planning, there is a need to evaluate each alternative plan in terms of its broader societal impacts, such as public health, environmental quality, public safety, user convenience, accessibility, and the amount of resources to be expended on the system (Meredith et al., 1985).

The need for careful civil systems planning continues to be accentuated by current and evolving trends on the horizon for civil engineering systems management. These trends, which are discussed in Chapter 1, include the incorporation of the perspectives of more stakeholders in systems decision-making processes. This inclusion is caused by the growing interest and awareness of taxpayers and society in general of the systems that serve them and their sensitivity to the impacts of the construction and operations of such systems on their natural and built environments. This trend has been noticeable in recent years as planners of civil systems have faced increasing scrutiny (regarding the development of their systems) from various stakeholders including advocacy groups and the media. Such scrutiny has often focused on (i) the need for the system, (ii) the consequences of the system in terms of its costs, benefits, or equity across demographic or societal segments, and (iii) the specific aspects of the system's location, orientation, and design. In certain situations, stakeholders have mounted strong opposition to a system's construction or expansion on the basis of at least one of these three considerations and have caused deviations including delay, scope increase or reduction, cost increase, or even cancellation of the planned system. In any of these outcomes, the feasibility of the system is brought into question, and the credibility of the engineer may become shrouded in doubt (Meredith et al., 1985). To prevent these situations or to effectively address them when they occur, the civil engineer needs to follow a well-structured planning process that is cognizant of these issues.

The impetus for thorough planning of civil systems includes legislation, public relations, and a desire by civil systems owners to infuse business-like processes in the management of their systems. As civil engineers continue to plan new systems and manage existing ones, this impetus continues to be reinforced by gradually evolving or sudden events that occur on the landscape of civil systems development. These issues are discussed in the next Section.

20.4 EVOLVING AND EMERGING CONTEXTS OF SYSTEMS PLANNING

Civil systems are developed to meet social and economic needs. It is therefore not unexpected that the planning contexts of civil systems are constantly changing, mostly in response to the constantly evolving needs, objectives, and the dynamic nature of social and economic development. Civil system planning techniques in the current era differ from those of past generations in that they incorporate greater flexibility to accommodate the changing nature of the planning landscape, including uncertainties in future demand and supply conditions, greater realization of funding limitations for system maintenance, and system vulnerability to sudden or gradual natural and man-made threats. We now discuss each of these contexts in some detail.

20.4.1 Increased Awareness of Future Uncertainties

In the not too distant past, the planning of civil systems was carried out generally on the assumption of certainty. Most planners generally made deterministic predictions of the outcomes of civil systems on the basis of past trends at the time. However, in the current environment, which is characterized by greater need to account for uncertainties in the planning considerations, the increasing role of new technologies, and economic fluctuations, the inadequacies of deterministic planning techniques have become painfully obvious.

According to Meyer and Miller (2000), the increased uncertainty associated with this emerging environment has influenced civil systems planning in at least two ways. First, new methods are being developed to accommodate the variabilities in system inputs that ultimately affect the uncertainty of predicted system outcomes in the context of feasibility analysis or comparison of alternatives. These methods include the establishment of multiple different sets of assumed values of system inputs and hypothetical pathways and establishing an associated system outcome for each scenario using analytical tools such as sensitivity analysis and Monte Carlo simulation, which we discussed in Chapter 8. Second, the implementation of civil systems is increasingly being carried out not by a single undertaking but rather using a more flexible approach that involves two or more stages. Thus, it may be more desirable, for example, to first build a conservatively-sized system (that is, of lower capacity or strength) initially on the basis of short-term conditions that have greater certainty. At later stages of the system's use, future additions to its capacity or strength can be implemented as and when they are needed depending on the system demand and other considerations at the time. This type of planning is consistent with incremental needs assessment (discussed in Chapter 19) and the real options concept (discussed in Chapter 15). This piecewise development of civil systems can help the system owner to not only avoid wastage and capacity underutilization (and its concomitant public relations problems) but also reduce excess operations and maintenance costs to the system owner (and the users to whom the excess costs are often ultimately transferred).

20.4.2 Funding Limitations for System Maintenance

Over the past two decades, as system operators increasingly have realized the benefits of maintenance in extending the life-cycle costs of a facility effectively, calls for increased attention to civil system maintenance have become louder. This awareness is taking place in developed countries, rapidly-developing countries, and underdeveloped countries. In developed countries, where a significant portion of public-owned civil systems was developed in the post–World War II period, the physical components of these systems have exceeded or are approaching their design lives and need major rehabilitation or replacement. These systems include hydroelectric dams, highways, tunnels, ports, and bridges. However, given the sheer volume of these systems and the magnitude of the funding necessary for rehabilitation and replacement vis-à-vis tightening funding conditions, it is no surprise that many governments face growing backlogs in the upkeep of such facilities (ASCE, 2013). Occasionally, catastrophic failure of a physical system occurs due to prolonged neglect, causing mayhem and public outrage, and galvanizing the advocacy for increased funding for infrastructure renewal. With regard to planning and design decisions, developing countries that are currently investing billions of dollars in new infrastructure have the advantage of walking the paths blazed by the developed countries; for example, system plans, materials, and designs that involve unduly frequent and intense maintenance and rehabilitation over a system's life cycle are being abandoned for other, more favorable alternatives which, over decades of real life experimentation, have proved to be more cost-effective over the long term.

20.4.3 Developments and Trends That Influence Systems Planning

A number of researchers have pointed out that the past decade has witnessed a gradual evolution, and in some cases sudden adjustment, of societal needs and thus changes in the goals and objectives associated with proposed or existing systems. These changes, which are associated with the first phase of systems development (needs assessment) have profoundly affected the issues to be considered by planners at the second phase (systems planning) as demonstrated by the following examples.

The 2001 Terrorist Attack and 2012 Superstorm Sandy in New York. These catastrophic events served as a rude awakening that civil systems continue to be vulnerable to man-made and natural threats and that each phase of the system development cycle will need to address the issue of system protection, resilience, and recovery from disaster events. In reaction to these events, the planning of civil engineering systems are being modified to enhance overall redundancy, increase resilience, reduce the vulnerability of individual systems to threats, maintain operational integrity of a system or system of systems, and facilitate evacuation of the populace from disaster areas. These events also spurred greater action toward further protection of civil systems against other man-made threats including unintended actions, such as collision and overloading, and natural threats such as tsunamis floods, landslides, and earthquakes. In Chapter 27, we discuss methods that can be used to assess quantitatively, the risks and consequences of such events, thus enabling planners to include system vulnerability and resilience considerations proactively in their system planning processes.

Climate Change, Natural Disaster, and Increasing Frequency of Severe Weather. Climate change may be described as the variations in weather patterns that are manifest as changes in the average weather conditions or changes in the distribution of weather events, such as the increased or decreased intensity and/or frequency of extreme weather events. It has been shown through research that the primary effects of climate change include the incipient melting of polar ice caps causing rises in sea and land groundwater levels, altered frequencies and intensities of extreme weather, increasing frequency of longer droughts, more frequent and severe freeze–thaw cycles, and warming of the ocean surfaces (UN, 2006; IPCC, 2007). These phenomena are resulting in secondary impacts such as more intense typhoons and hurricanes; more frequent, larger, and more abrupt floods; changing levels of groundwater; and changes in wind speed and profiles (Beniston, 2004). Prominent civil engineering organizations have sounded warnings that these changes in climate pose an insidious threat to the stability and operational functions of civil systems worldwide (ASCE, 2007; JSCE, 2009). It is thought that the major floods experienced in the U.S. Gulf region, Australia, and other parts of the world in 2009–2011 are only symptomatic of a much larger looming problem. As such, at the planning phase for new civil systems, engineers increasingly are seeking to adopt planning alternatives that help reduce the anthropogenic contributions to climate change, lessen system exposure to climate change, and facilitate recovery from the effects of climate change (Ali, 2008; Liao, 2008; Lenkei, 2007; Long and Labi, 2011). Time and again, we are reminded of the inadequacy of civil systems to prevent large-scale natural disasters—a failing that is not related to the capabilities of civil engineers but is rather related to the extent to which society is willing and able to fund mammoth investments in civil infrastructure that are needed to counter these natural events. Recent disasters in Haiti and Chile in 2010 and Australia and Japan in 2011 and Superstorm Sandy in New York and New Jersey in 2012, have brought this issue to the forefront of the international conversations.

20.5 PRINCIPLES OF CIVIL SYSTEMS PLANNING

The basic principles of systems planning refer to the **good practices** that promote the establishment of a good plan for civil systems. The system planner should strive to abide by these principles; however, for a specific system plan under consideration, it may be impractical to fulfill all the principles to their fullest extent. We now discuss a number of good practices.

20.5.1 Planning Should Be Farsighted

While operating within the confines of their phase, planners need to duly consider the long-term consequences of their plans, for example, the way the system will be operated and maintained and the end-of-life actions that will be undertaken. Thus, while planners may not necessarily need to carry out detailed designs, develop operational strategies, or identify optimal maintenance schedules, it is their duty to develop plans that will facilitate, at the subsequent phases of the system development, an efficient design, enhanced operations of the system, minimal intensity and frequency of maintenance over the system life, and costeffective and sustainable termination of the system. Past experience has shown that systems planning that considers only the initial consequences is inefficient.

20.5.2 Planning Should Be Geared toward Decisions, Not Merely Plans

It is important for the planner to bear in mind that systems are developed not from plans but from decisions (Goodman and Hastak, 2007). Planning is therefore considered effective only when it succeeds in furnishing the appropriate information to the decision maker for establishing the alternatives and fully comprehending the broad consequences of each alternative.

20.5.3 Planning Should Consider Multiple Considerations

A good plan must be not only technically feasible but also must satisfy a host of other considerations that are related to legal, financial, economic, social, political, institutional, and environmental matters. The planning of civil systems, therefore, is inherently multidisciplinary in nature: The planner must consider the technical feasibility (the extent to which the system is expected to carry out its core mission); the economic feasibility (the extent to which its benefits will outweigh its costs); the financial feasibility (the adequacy of available capital to construct and maintain the system); and the ethical and legal feasibility (the extent to which the system's construction or operations will abide by the existing ethical principles and laws related to the environment or social conditions). It may be relatively easy to establish a system plan that is feasible in all these considerations, but it is more difficult to establish one that is superior in all these respects simultaneously compared to other alternative plans. For example, a plan may be feasible economically because its benefits far outweigh its costs, but it may not be financially feasible because it is not possible to raise the enormous capital that is be needed for its construction. For this reason, civil systems planning is often a delicate balancing act and requires the tools of optimization (Chapter 9) and multiple-criteria analysis (Chapter 12). Figure 20.3 depicts system feasibility on the basis of three considerations only.

20.5.4 Planning Process Should Consider Multiple Alternatives

Consistent with the general principles of systems engineering, a key aspect of planning is the generation of several alternatives and not just one alternative. The development of multiple alternatives is borne out of the realization that civil systems planning is not an exact science, and there may be several feasible solutions for any given planning problem. Each alternative plan should be geared toward the vision of a preferred future and must be assessed on an equal footing with other plans for the system.

20.5.5 Plan and Planning Process Must Be Robust

Changes in government and system owner leadership are common and continue to be an inevitable (even if unsavory) part of life. Changes in government leadership that arise from elections in

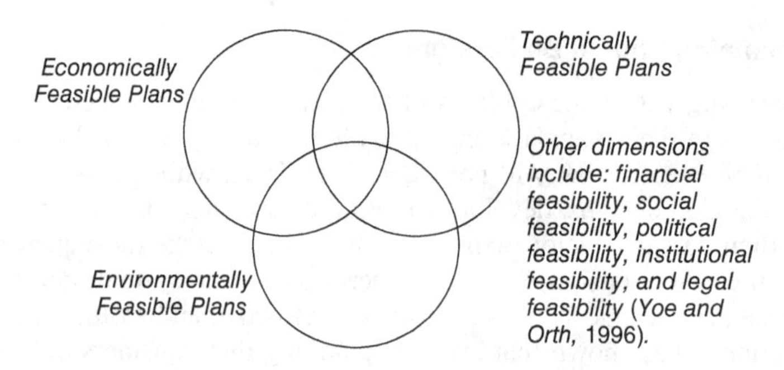

Figure 20.3 System plan feasibility: Illustration showing three considerations.

multiparty political systems or from the overthrow of dictatorships often translate into changes in agency missions and policies and, ultimately, the goals and objectives of a proposed civil system. The impact of such upheavals on the planning process generally depends on the time duration between the planning and construction phases of the system in question: where such time interval is short, the impact is generally minimal as the system construction proceeds before it is disrupted by political changes. Unfortunately, for most civil systems, several years or even decades may elapse between the completion of the plans and the actual construction of the system. Such long "exposure" of finalized plans increases the risk not only of plan obsolescence but also of disruption by political changes at the agency, state/province, or national levels. Therefore, any developed plan, or at the very least, the planning *process* for a system, must be as robust as possible to minimize the disruptive impact of possible future changes in the agency's mission or in the system's overall goals. Robustness is enhanced when the system planner duly incorporates, during the planning process, multiple performance criteria (Chapter 8) reflecting the perspectives of multiple stakeholders and a good dose of risk/uncertainty concepts (Chapter 13) in the planning process.

20.5.6 Plan Must Be Integrated and Holistic

The planning process must be geared toward not only *integration* of the system in question (coordination with other similar or dissimilar systems in the larger network of systems to facilitate harmony in operations) but also *holistic* (in that it must consider the effects between the individual systems on each other so that the sum of their combined outcome is superior to the sum of their individual outcomes). Recognizing that the system being planned will rarely operate in isolation, the system planner must duly incorporate the fact that there are other systems in the overall parent network; and thus the construction, operation, monitoring, and maintenance of systems being planned must be as consistent as possible with those of other individual sister systems in the network. This tie-in implies that the planner should examine not the sum of the consequences of all the systems but the consequences of the sum of all the systems. The planner should remember that the intersystem influences can be significant, and the effect of the sum of the parts therefore is not always equal to the sum of the effects of the parts. A typical example is the planning process for a new road link in an existing road network: The expected traffic flow on the new link will be influenced by the traffic flow and capacity of other links in the network, which are all considered together (not separately) to yield user or system equilibrium. Any inability to incorporate integration and holism in planning could inhibit good planning.

20.5.7 Plan Must Account for Uncertainty

As stated in the previous section, an impetus for system planning is the uncertainty or variability surrounding the inputs into (and therefore output from) the assessment of the behavior or performance of civil systems; and changing population growth, material costs, and technology can profoundly influence the nature of the plans developed. Rapid advances in technology, in particular, could render certain components or features of the system to be obsolete in the future (Dandy et al., 2008); for example, with the widespread availability and use of personal mobile phones, it is now considered redundant to install telephones at regular intervals along some rural highways. It is useful for civil systems planners to recognize and consider uncertainties such as these and also to prepare for worst-case scenarios.

20.5.8 Plan Must Be Sustainable

"Sustainability" in this specific context refers to the survival likelihood of the plan itself rather than the sustainability of the natural or man-made environment after the system is built (which we will discuss in Chapter 28). Many civil system plans do not go on to the design stage as they are "killed off" for some reason. In order to enhance the survival likelihood of the system plan and ultimately, construction of the system, the planner must ensure that the plan is sustainable. According to Goodman and Hastak (2007), a sustainable plan must be one that is based on the lessons of past experiences, duly considers public input, and has government support.

20.5.9 Planing Must Yield Outcomes That Can Be Communicated Effectively

The planning process must culminate in a product that can be explained in sufficient working detail to the personnel responsible for the subsequent (system design) phase and other phases of system development. Also, the developed plan must be in such a form that it can be communicated to critical stakeholders (e.g., the general public legislature, and other government officials) whose approvals are essential if the plan is to go forward to the next phase.

20.6 SYSTEM PLANNING PROCESS

From the needs assessment discussion in Chapter 19, we learned how the civil engineer makes preparations to commence the planning process. The needs assessment phase provides a statement of the problem, establishes the goals and performance measures for the problem, and assesses the need for the system on the basis of projections of system demand and supply. Thus, the needs assessment phase builds a solid foundation for the development of the system plan. The planning phase, in turn, serves as a prelude for subsequent phases including design (Figure 20.4).

For system planning, approaches can range from relatively simple analyses that utilize back of the envelope, rules of thumb, or professional subjective judgment to sophisticated techniques that utilize tools such as simulation, risk analysis, and optimization (Goodman and Hastak, 2007). During the system planning process, the civil systems planner carries out the following activities: (i) establish alternative plans that specify the locations, positions, and orientations to enhance the functional relationships between the proposed system and its environment; (ii) describe each alternative plan for the system using tools such as artistic or computer simulation, often for communicating the plan to the general public or legislators; (iii) evaluate the extent to which each alternative achieves the overall goals of effectiveness, efficiency, and equity from the perspective of technical, social, environmental, financial, or economic considerations; and (iv) select the best plan from the alternatives on the basis of a specified overarching criterion, such as cost-effectiveness or multiple-criteria utility. The so-called technical considerations refer to the primary purposes for which the system is needed; for example, for a proposed transportation system, increased mobility or accessibility; for a proposed hydroelectric system, flood prevention, water supply, or electricity generation.

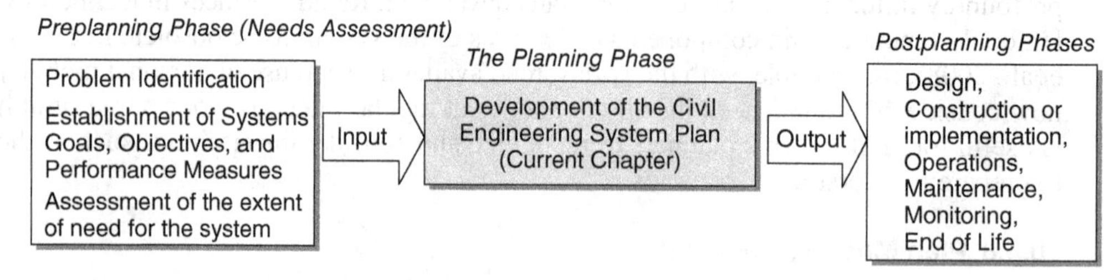

Figure 20.4 Relationships between the planning phase and neighboring phases of systems development.

The steps listed above represent only a general picture of the planning process. The scale, scope, and specific details of the planning process for a given system will depend on the specific civil branch involved (structures, transportation, hydraulics, etc.). The specific designation of the system planner depends on the facility in question; for example, for proposed buildings or development areas, the system planner is the architect or urban planner; and for structures other than buildings, the system planner is the civil engineer. In the course of their work, system planners liaise not only with professionals in other civil engineering disciplines, such as the environmental and transportation engineers, but also with specialists in other external disciplines such as economists, mechanical engineers, landscape architects, ecologists, and natural resource experts, depending on the type of system. Furthermore, the system planner seeks the input of other persons in the community such as city and county personnel and the general public. Again, it is important to recognize that planning for a network of systems or system of systems is different from planning for an individual system. This chapter focuses on the latter.

As most system owners and consulting professionals realize, there is often a thin line between the first phase of system development (problem identification, needs assessment, and goals identification) and the second phase (system planning). In certain cases, the planning phase starts with the traditional tasks of the needs assessment phase (problem identification, quantitative assessment of system need, and/or establishment of the goals and objectives of the proposed system). Also, between the second and third phases of system development, planning and design respectively, the distinction could be even more blurred as planners or engineers, at the advanced or latter stages of system planning, may find themselves carrying out preliminary or conceptual design to establish rough specifications of system dimensions, material specifications, and orientation. Figure 20.5 presents a general framework for planning at each level of an individual civil system.

Step 0: Preplanning Activities

Confirmation of System Goals and Objectives. At this stage, the planner confirms the concerns of all the stakeholders regarding the system goals and objectives by soliciting information about their viewpoints. The goals are established not only to reflect the mission and goals of the civil system owner or operator but also to accommodate the perspectives of the system users, the community, government officials, and the general public. For most civil engineering systems, the community concerns include security, delay mitigation, safety enhancement, and service accessibility. For certain civil systems, the concerns of the community may be more long-term in nature, for example, economic development and environmental improvement. The confirmation of clearly defined concerns and tentative goals early in the planning phase helps not only to reach early compromise between conflicting interests but also to identify specific issues about which consensus has been reached. Often, there are past reports or white papers that could serve as a valuable resource for the planner in the task of developing goals and objectives for a proposed system. After identifying the goals and objectives, the performance measures or measures of effectiveness (MOEs) can be established. MOEs help in the task of comparing and selecting alternative plans in terms of the extent of their technical, social, environmental, and economic feasibility (i.e., how they help to achieve the identified goals and objectives). In the current planning environment, characterized by shifts toward quality of life and sustainable development, nonmonetary impacts are playing an increasingly important role in the plan evaluation process.

Establish Dimensions for the Planning Study. It is important to establish the boundaries of the affected regions for the analysis (e.g., project, corridor, subarea, systemwide, regional, national, or even international). For any given impact category (technical, social, environmental, economic) and the temporal scope of the planning study, different spatial scopes may necessitate the use of

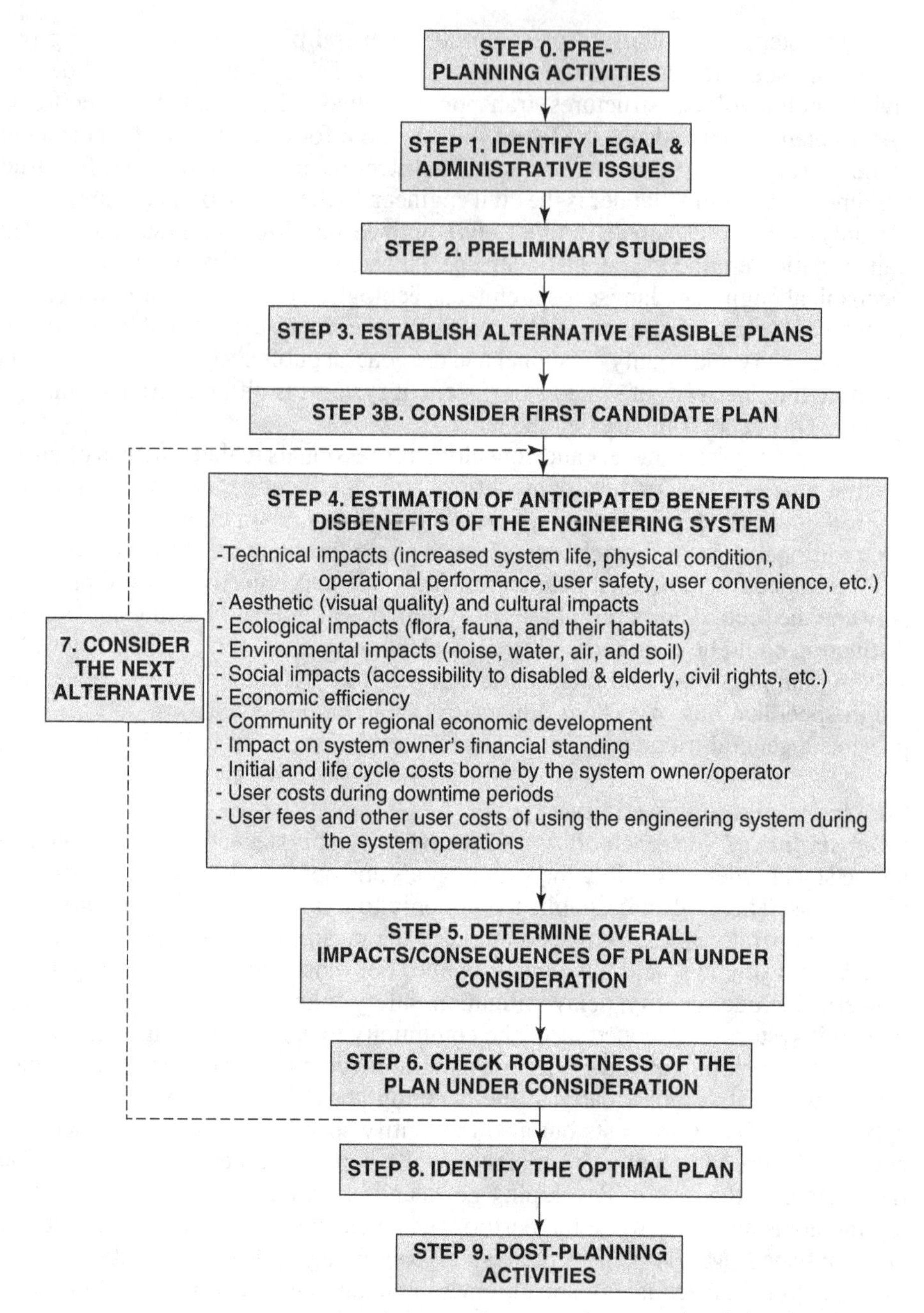

Figure 20.5 Planning of civil engineering systems: The general steps.

different approaches and different MOEs for the planning study. Also, the relative importance of certain categories of system impacts may differ from one spatial or temporal scope to another (Sinha and Labi, 2007).

Concerns of the System Owner and Other Stakeholders. The planner then identifies the potential stakeholders likely to be affected by the system construction, operations, and maintenance, which generally include the system owner who bears responsibility for the facility upkeep, the system users who gain direct benefits (through reductions in delay, safety, and inconvenience), and the community, which typically bears any adverse external effects of the system.

The planner must recognize that local opposition could jeopardize the inception or progress of a proposed civil system and must get the public involved early in the system planning process. The benefits of public involvement in the preplanning phases are numerous, so the following activities are useful in that regard: (i) ensure that no stakeholder, affected party, or interest group related to the public (and their concerns) have been excluded from the planning process; (ii) ensure that no impact of importance to the public has been overlooked by the planners; and (iii) determine the information needed to measure and mitigate the expected impacts that are of interest to the public. For projects involving major civil systems, it is important to involve the public because such projects tend to have large and adverse impacts on the environment and the community. Prior to the solicitation of public input, the planner must establish when and how the public can participate so that the outcome can be favorable. Past experience has shown that when the system planner interacts with the public in a way that demonstrates a sincere appreciation of the value of the public input, public participation produces favorable outcomes. For the planner, public participation can also be used as a mechanism for communicating to the public any favorable impacts of the proposed development that were not obvious (Sinha and Labi, 2007). Also, public participation provides the system planner with knowledge of the public perceptions of the trade-offs among the technical, social, environmental, and economic performance of the system. In the course of soliciting public participation, the system planner must remind the general public that the chosen plan for the system may not satisfy all the interest groups but was chosen as a result of compromises among their conflicting concerns. According to NHI (1995), for effective public involvement, the planner needs to ensure that the solicitation process does the following:

- Offers each interested group an appropriate level of involvement and a type of interaction. These interactions could range from newspaper or website comments to detailed work sessions with the appropriate staff.
- Establishes a proactive (rather than reactive) communication program to provide information to the general public and interest groups through the media, the Internet, and other channels.
- Solicits the advice of representatives from citizen associations, interest groups, and other public bodies.

Step 1: Identify Legal and Administrative Requirements. Legal and administrative issues typically encountered at the planning phase include state statutes, local ordinances, and federal program requirements related to safety, the environment, and equity. Over the decades, numerous regulations and pieces of legislation have been passed to enhance the decision-making processes for civil system construction, operations, and maintenance a bid to further protect the environment, social structures, ecology, and historical and cultural treasures and to ensure equity. It is not surprising, therefore, that the evolution of a civil system throughout the phases of system development, particularly at the phases of planning and construction, involves a medley of administrative hurdles including formal requests and notifications, and submission of reports that detail the studies and outcomes related to engineering, economic, environmental and other aspects of the proposed system (Goodman and Hastak, 2007; Sinha and Labi, 2007).

It is important for the system planner to identify and document the requisite legal and administrative processes because they help the planner to define the various responsibilities of the system owner or operator and oversight or regulatory entities responsible for reviews or approvals of the system owner's actions. It is necessary to identify the requisite legal requirements because the laws that pertain to the system influence the number and type of performance measures established for the planning process and therefore could potentially influence the nature, scope, and outcome of the planning process.

Step 2: Preliminary Studies. At this stage, the planner carries out desk and field reconnaissance studies of the site for the proposed system. The desk study includes acquisition of all past documents pertaining to feasibility studies or design documents for yet-to-be-built or built civil, agricultural, or energy systems at or near the site in question. The documents, which may be related to different disciplines such as engineering, agriculture, and geology, include site reports and maps for soils, climate, and water bodies. Data can be acquired from the archives of consulting firms, contractors, public agencies, libraries, and the media. Often, the desk study needs to be supplemented by site visits and field reconnaissance surveys to identify any unforeseen site problems that could lead to changes in the proposed plan layouts and orientations of the proposed civil system.

Step 3A: Develop Alternative Plans. At this step, the system planner identifies all possible plans and screens them to ascertain that they are appropriate. The resulting set of multiple mutually exclusive plans for the system plans are termed as "alternatives." Is a specific plan relevant to the problem a hand? Is it adequate in addressing the identified need? Are the alternatives too few or too many? In answering these questions, the planner may be guided by the following considerations:

Relevance of each alternative: The planner must ascertain whether the plan is relevant for addressing the stated need at the first phase of the system development process (the needs assessment phase).

Holistic nature of the alternative: Traditionally, the development of planning alternatives has been carried out by considering single physical systems and/or their operating strategies from a stand-alone perspective. However, a multisystem or system-of-systems approach is increasingly being touted for the planning phase, as discussed in Section 20.5.6. This planning approach advocates the consideration of the system not as an independent and isolated entity but as an integral part of a larger system of systems.

Adequacy of the alternative: Does the plan adequately address the intended goals and objectives that have been established by the system owner or operator? At a subsequent step of the planning process, each alternative plan is assessed on the basis of the extent to which it satisfies each planning criterion (technical, environmental, economic, etc.).

Realistic nature of each alternative: As discussed in Section 20.5.3, the planner must check to ensure that each alternative under consideration is not only technically feasible but also realistic. For example, is there adequate right-of-way? Does the country, state, or province possess adequate technological know-how or resources to design, build, operate, and maintain the system?

The number of alternatives: The total number of alternative plans should not be too large or too small. The range of alternatives should be wide enough to permit illustration of the trade-offs associated with the planning criteria or performance measures, so that the selection of the best plan can be made after duly considering the performance trade-offs between competing alternatives.

Transparency of plan development: In the development of planning alternatives, inclusiveness and transparency are vital. The alternative plans should be developed using a clearly defined and transparent process, with input from the stakeholders. At each step, the planner should collaborate with the stakeholders, and the results of each step should be open to public scrutiny, review, and feedback.

Step 3B: Select a Plan for Consideration. The plan under consideration must be from among the set of feasible alternatives plans only.

Step 4: Assessing the Cost and Benefits of Each Alternative Plan. After selecting a specific alternative for consideration (analysis), the benefits and costs of the plan should be estimated. At the current time, a large number of system owners and operators appear to use only the initial costs as the basis for choosing between alternative plans, which may be due to expediency, inertia, lack of data, or simply the fact that the planning phase is considered to be a phase where only approximate numbers may be available and are sufficient to make a decision. Where data are available, the costs to be used in the system plan evaluations should pertain to the system life cycle rather than just the initial costs, thus economic analysis should be used. Costs can be categorized by those that are the same across alternatives and those that are different; only the latter should be considered in the evaluation of different plans. Costs can also be categorized by incurring party: system owner or system user. The system owner's costs include construction costs, maintenance and rehabilitation costs, and operating costs. The system user's costs consist of the downtime-related costs (e.g., delay, inconvenience, and safety costs, which are often intangible or at least indirect) and the costs associated with normal system operation (e.g., the direct costs of any fees or fares and the indirect costs incurred by the user in the course of using the system).

Cost estimation at the planning phase, unlike that of the design phase, are relatively crude. The system planner is often interested in a rough, aggregate cost of system construction, preservation, and operations. These rough costs are often expressed in terms of a unit dimension or output of the system, such as the cost of a new bridge on the basis of $/ft^2 of deck area, the cost of transit operations in $/passenger-mile, or the cost of building a water tank ($/ft^3), or the cost of supplying water ($/ft^3). The rough cost may be a simple average value or a statistical model. Rough or aggregate cost estimates are useful for planning purposes because the detailed costs of individual system components are not known with certainty at the planning phase. Chapter 10 provides rough cost estimates that are useful in the costing stage of the planning phase.

The anticipated benefits of the plan should be assessed, which could be the technical benefits (an increase in the system's remaining life or an enhancement of a condition, user safety, or user convenience; an ecological enhancement or degradation of flora, fauna, or their habitats; environmental impacts (e.g., water, air, noise, vibration, visual quality); social and equity impacts (dislocation of homes and burial grounds, accessibility of the disabled to system use, and civil rights); economic efficiency; and economic development impacts. In certain cases, the benefits of a plan can be estimated as the reduction in costs incurred by the system owner, the users, or the community relative to a base case plan that often is the "do-nothing" or "leave-as-it-is" plan. For transportation systems, for example, typical benefits may include reduced travel times, reduced tort liability, reduced vehicle operating and maintenance costs, increased motorist comfort and safety, reduced rate of pavement deterioration, and reduced or deferred capital expenditures through preservation of capital.

Step 5: Determine Overall Impacts/Consequences of Plan under Consideration. In evaluating the alternative plans, the planner assesses their respective costs, benefits, and overall attractiveness in a bid to select the best plan. An alternative that is found to be optimal at a certain time or under a certain condition may actually be far from optimal at a different time or under different conditions. As such, the process of evaluating the alternative plans must include sensitivity analysis or what-if scenarios for deterministic planning processes and stochastic analysis (including Monte Carlo simulation as discussed in Chapter 13) that incorporates the probability distributions of various input parameters for probabilistic problems. System demand is also an important consideration in evaluation. A large number of time-related physical and operational attributes of civil systems exist that influence (or may be affected by) changes in the demand for the system or the level of system use. Usage forecasts are important because they influence system performance. Therefore, it is important that reliable predictions of system demand be carried out prior to the evaluation of alternative plans. As all good planners realize, the future demand for the system, a key input parameter in system planning, is never known with absolute certainty and thus must be subjected to probabilistic analysis.

During the planning process, the planner encounters streams of monetary and nonmonetary costs and benefits associated with each of the alternative plans, and the challenge is to choose the best plan. First, the evaluation criteria will need to be identified. Then the selected evaluation criteria must be used to identify the best plan. The choice of evaluation criteria will depend on the goals of the planner and, thus, the performance measures under consideration. For each alternative plan, the planner compares that which is sacrificed (cost) to that which is gained (benefits or effectiveness). The outcomes of each alternative include the reduction in subsequent facility preservation or operation costs, economic or financial returns, community benefits, and public satisfaction. Evaluation may be carried out in one of three ways (Sinha and Labi, 2007): the maximum benefits for a given level of investment (the maximum-benefit approach), the least cost for effective treatment of problems (the least-cost approach), and the maximum cost-effectiveness (a function that maximizes the benefit per cost input).

The benefits-only approach for planning is often applicable where the cost of each alternative is the same or for evaluating large capital investment projects that have significant risk and uncertainty. This approach is particularly suitable in planning activities for projects where there is considerable difficulty in identifying the appropriate evaluation criteria (performance measures) for comparing the alternative plans due, for example, to the proposed system's complex nature, relatively long life, and spatial spillover effects.

In situations where the planner expects the benefits to be equal across alternatives or encounters difficulty in measuring the benefits, the evaluation criteria consist of costs only. For several decades, this has been the practice of many agencies, where planners compare alternatives solely on the basis of their initial agency costs or their life-cycle agency and user costs.

To compare the alternatives fairly, the planner may need to consider fully both the costs and the benefits associated with each alternative. In problem contexts where all the planning criteria can be adequately expressed in monetary terms, economic efficiency criteria, such as the net present value or benefit–cost ratio can be used, by using the tools we learned in Chapter 11. If the planning criteria consist of both nonmonetary and monetary performance measures, then the planner can carry out multicriteria decision making to weight and scale and amalgamate the criteria to derive a single objective function using the tools we learned in Chapter 12. A plan is considered superior if it is found to be more cost-effective or yields the optimal value of the objective function compared to the other alternatives.

Step 6: Check Robustness of the Plan. At this step, the planner analyzes the robustness of the plan by carrying out sensitivity analysis of the expected benefits and costs of the plan in response to changes in the input factors that influence these costs and benefits. A plan may have superior benefits and/or low costs compared to other plans, but it could be marked by excessive variability that could make it risky. In certain civil system agencies, the uncertainty of acquiring the expected level of system performance may itself be considered a performance measure and thus may be included in Step 8 of the planning process.

Step 7: Consider the Next Alternative. The system planner then analyzes the next alternative plan using the same process described for steps 4 to 6. This step is repeated until all alternatives are exhausted.

Step 8: Identify the Optimal Plan. The optimal plan is the alternative identified as that which is feasible (Step 6), cost-effective (Step 7), and robust (Step 9). If there are very few alternative plans, simple enumeration may be used to identify the best alternative. Otherwise, other, more sophistical optimization tools we learned in Chapter 9 will be useful in identifying the optimal solution. Where there is significant complexity or uncertainty associated with the planning inputs, the concepts of system dynamics (Chapter 14), risk analysis (Chapter 13), real options (Chapter 15), and decision analysis (Chapter 16), depending on the problem structure, may be used with or without explicit optimization techniques to identify the optimal plan for the system.

Step 9: Postplanning Activities. At the tail end, the activities that are carried out within or external to the planning process, include project appraisal. This activity involves the determination of whether the proposed project to construct the system meets the appropriate criteria for authorization and/or funding. The appraiser may be a public official, a representative of the system owner or operator, or a legislative body. For projects in developing countries that are funded by international lending agencies, the appraisal process is slightly different from those that are funded by the government in their countries or those in developed countries. Even though the outcome of the planning process, not the framework itself, is emphasized in the appraisal process, it is nevertheless important for the appraiser to review the planning process as well as the resultant plan. Also, planners at all levels typically strive to ensure that the appraisal procedures are formulated to conform to their institutional missions and to incorporate specific knowledge and issues that pertain to the system in question and the area it is intended to serve or is located. Goodman and Hastak (2007) outline appraisal guidelines for the planning of civil systems in the United States.

20.7 BARRIERS TO EFFECTIVE PLANNING

20.7.1 Political Influences

Often, certain aspects of an engineering system plan may not be consistent with prevailing political environments; for example, the ecological impact even if undisputed, of a proposed system may be viewed differently by people in different political positions, and the furtherance of the system development may be held hostage as long as that a particular party is in power. To avoid, or at least minimize, such political influences, it is helpful for civil systems planners not only to incorporate the concerns of the general public (particularly where the direct cost bearers or beneficiaries are the general public) but also to solicit the support of all the branches and shades of government potentially involved or interested in the proposed system.

20.7.2 Changing Goals and Objectives

At the back of every civil system plan is a larger goal sought by the system owner, such as an increase in economic development, enhanced quality of life, increased security and safety, greater mobility, and increased accessibility to raw materials for production or social centers. The final plan is often a tightrope balance between these often conflicting goals. As society evolves on an issue, changing perceptions lead to a shift in the relative weights of these goals, and a system plan that was once optimal at the time of construction may not be optimal decades later. This challenge constitutes one of the several uncertainties faced by civil systems planners. For systems whose construction suffers delay, adjustments to the physical structure or its operating policies may be necessary to ensure consistency with the current goals of the key stakeholders. For example, a number of dams in California, Oregon, and Maine that were originally constructed for flood control and water supply several decades ago have been destroyed recently so that the waterways could return to their original environmental purposes, such as serving as migratory channels for aquatic animals (Goodman and Hastak, 2007). Other examples include traffic calming initiatives for urban streets and smart growth planning for urban neighborhoods (Bose and Fricker, 2004; Lewis et al., 2009). More and more cities, such as Boston, are considering relocating their surface streets underground in order to open up surface space for recreation, sociocultural activities, and economic development. However, such modifications to keep up with the times can be expensive. In any case, it is helpful for the system planner, as much as possible, to predict and account for such possible future trends in their plans or to add significant flexibility to accommodate any future unforeseen changes in the socioeconomic environment.

20.7.3 Difficulty of Achieving Holistic Solutions

As discussed in Section 20.5.6, a good plan for a civil system must consider not only the effect of other systems (of the same or different type) in the larger network of systems but also the operational effects between the system in question and other systems. In certain cases, for reasons that include technical (difficulty of quantifying the effects between systems) or institutional (difficulty of relating the different sector practices for privately owned systems and publicly owned systems), it may be difficult to achieve such holism. Nevertheless, the system planner must always seek to identify and where possible, to incorporate any synergies associated with other similar systems or system types with which the main system shares a proximal, institutional, or operational relationship.

20.7.4 Multiplicity of Concerns and Stakeholders

In an earlier section, we discussed how an efficient system is one that duly considers the fact that civil systems are inherently associated with a large number of stakeholders that are concerned with a wide variety of views, often conflicting. When this becomes excessive, it could pose a formidable threat to the implementation of the plan (Sinha and Labi, 2007). The planner must realize that while all views need to be considered, it is often the case that not all these perspectives can be accommodated fully or even partially.

20.7.5 Lack of Government and Public Support

A large number of civil system plans never go on to the design and construction phases as they are abandoned for reasons that include, among others, lack of public or government support. In order to enhance the survival likelihood of the plan and, ultimately, the construction of the system, the planner must ensure that the plan is sustainable. According to Goodman and Hastak (2007), a sustainable plan is one that is based on the lessons of past experiences, considers public input, and has government support. In certain cases, there may be government support but no public support or vice versa, and the project could fail for that reason.

20.8 COMPUTATIONS IN CIVIL SYSTEMS PLANNING

To enhance the task of planning, engineers and planners apply analytical tools such as those we learned in the 14 chapters of Part 3 of this text. In this section of the chapter, we will discuss four examples where the following analytical tools are applied: modeling the expected construction costs or expected benefits of a system at the planning phase, analyzing the economic attractiveness of alternative plans, and determining the best location for the proposed system.

20.8.1 Estimating System Cost at the Planning Phase

As we learned in Chapter 10, for cost estimation in civil engineering, there are four distinct levels of granularity that reflect the phases of system development at which such estimates are often required. At the planning phase, only a conceptual estimate (typically referred to as predesign estimate or approximate estimate) is needed. Average cost values may be used to estimate the aggregate cost at the planning phase; however, a more reliable method to find the planning phase cost is to develop aggregate cost models as a function of the attributes of the system (e.g., the dominant material type, design type, and size) and the environment (e.g., surface or subsurface conditions and climate). Also, there could be other cost variables depending on whether the costs being estimated are the expected initial costs or whether they are the costs expected to be incurred over the system's life cycle. Examples include models that estimate the total cost of system construction, preservation, or operations on the basis of a unit dimension or output of the system, such as the cost of a new bridge in $\$/ft^2$ of deck area or the cost of transit operations in $/passenger-mile. The average costs of various civil systems are provided in Appendix 3. Example 20.1 provides a numerical illustration.

Also, during systems planning, the construction cost is often estimated to occur at a given year. In reality, the construction of civil systems spans several years, and it is often useful to take into account the distribution of construction expenditures over the construction period and the time value of money, as well as the uncertainties in construction costs, time delays, and interest rates. Therefore, in carrying out a rough estimation of the system cost at the planning phase, the engineer often needs to realize that (i) costs are not incurred in one year only, as may be implied in cash flow diagrams, but rather over several years so the need exists to account for the time value of money; and (ii) costs and cost factors are subject to significant variability in the real world due to uncertain economic conditions and the construction environment. These issues are illustrated in Example 20.2.

Further, at the planning phase, engineers may be interested not only in the initial cost (construction cost) but the entire life-cycle cost of a system as well. In that case, the engineer makes projections of the rest-of-life (or the so-called "future") costs that are expected to be incurred by the system owner, including the cost of preservation (rehabilitation and maintenance), operations, and in some cases, interest payments on money borrowed for the system construction. This is illustrated in Example 20.3.

Example 20.1

A rough model for estimating the unit cost of heavy (rapid) rail construction was developed as follows: For heavy rapid rail systems with 40–60% underground (Sinha and Labi, 2007):

$$UC = 3.906\ LM^{-0.702}\ PU^{1.076}\ ST^{-0.358}$$

where *UC* is unit cost (cost per line-mile-station), in millions of 2005$, PU is the percentage of system underground, LM is the number of line-miles, and ST is the number of stations.

In a certain city, 85% of a proposed heavy-rail transit system will be located aboveground. The total length is 12 line-miles; four stations are planned. Determine the estimated project cost.

Solution

$$\text{Cost per line-mile-station} = 3.906(12^{-0.702}\ 15^{1.076}\ 4^{-0.358}) = \$7.66\ \text{million.}$$

Therefore, overall cost of the system = 7.66(12)(4) = $367.68 million.

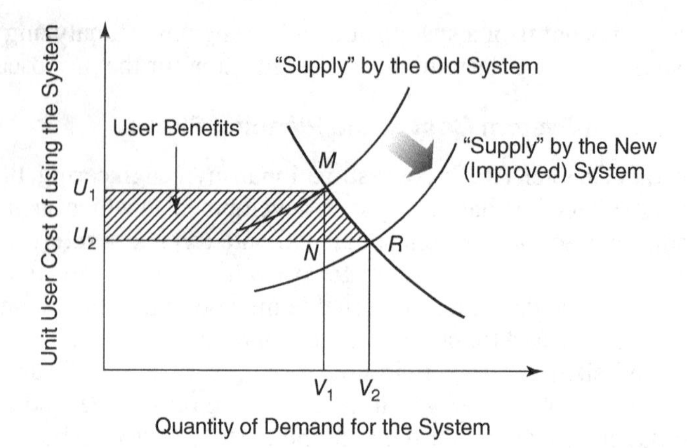

Figure 20.6 User benefits due to new and improved system.

20.8.2 Estimating the Expected Benefits of a System

As we discussed earlier in this chapter, the benefits of a civil engineering system may be the increase in economic production or productivity, enhancement of the quality of life for the residents in a community or region, or for an erstwhile system, a reduction in inconvenience or delay for users or a reduction in adverse community impacts. To measure the benefit of a civil system, the reduction in the costs incurred by the user may be calculated.

An improved quality of service causes a shift of the supply curve. When demand is elastic, and a new system replaces an old system, demand will increase. Due to these two events, from classic economics, the result is a lower cost of providing the service (Figure 20.6). Consequently, the unit cost of using the system, expressed in terms is reduced.

Thus, for the situation where demand is elastic, the user benefits of the new system can be calculated using the following equation: Benefit $= 0.5 \times (U_1 - U_2) \times (V_1 + V_2)$; where U_1 and U_2 are the unit "costs" of providing the service, for the old system and the new system, V_1 and V_2 are the demand values for the old system and the improved system, respectively. In the case where the demand is inelastic, the user benefit incurred by improving the system is calculated as the (quantity) of demand multiplied by the reduction in the unit cost of the system use.

Example 20.2

The actual user cost for an existing system and the anticipated user cost for a proposed system are 3.5 and 2.87 units per million users, respectively. The annual usage of the system is 1.5 million; 1.8 million people are expected to use the system when it is renewed. Determine the benefits due to the reduction in the unit user cost.

Solution

$$\text{User cost for the old system, } U_1 = 3.5 \text{ per million users.}$$

$$\text{User cost for the new system, } U_2 = 2.87 \text{ per million users.}$$

$$\text{Usage of old system, } V_1 = 1.5 \text{ million users}$$

$$\text{Usage of new system, } V_2 = 1.8 \text{ million users}$$

From Figure 20.3, the savings in user cost $= 0.5(U_1 - U_2)(V_1 + V_2) = 0.5(3.5 - 2.87)(1.5 + 1.8) = 1.04$ million units per year.

20.8.3 Analyzing the Economic Attractiveness of Alternative Plans

Planners often need to analyze the economic attractiveness of competing plans. Where all the consequences can be expressed monetarily, this can be done using economic efficiency analysis. Otherwise, tools in multiple-criteria analysis can be used. In Example 20.3, we assume that the impacts are all monetary. Also, the example incorporates elements of probability as the outcomes associated with each alternative are not deterministic.

Example 20.3

Two alternative locations are being planned for a small hydroelectric project. It is estimated that they have equal effectiveness in terms of the amount of power they would harness, so a decision will be made on the basis of their costs. In a simplification of these types of problems, the estimated costs associated with each location are estimated as follows (Figure 20.7):

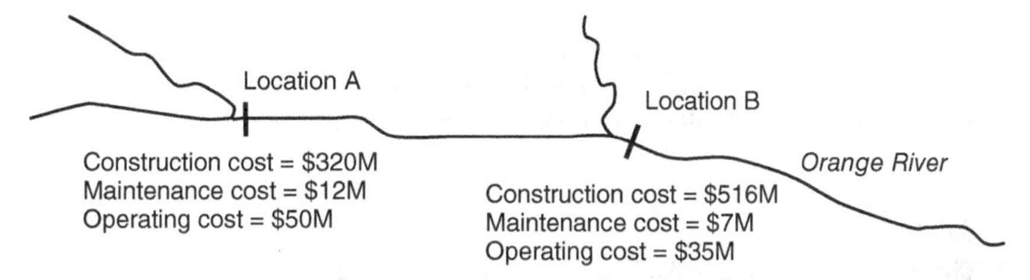

Figure 20.7 Figure for Example 20.5.

The estimated life of the dam, if built at location A, is 70 years with 60% probability and 80 years with 40% probability. The estimated life of the dam, if built at location B, is 60 years with 30% probability, 65 years with 50% probability, and 70 years with 20% probability. Which of the two locations should be selected? State any relevant assumptions. Assume an interest rate of 4%.

Solution

Location plan A: Expected value of the service life $= (70 \times 0.6) + (80 \times 0.4) = 74$ years

$$\text{EUAC (in \$ millions)} = 320 \times \text{CRF}(4\%, 74) + 12 + 50$$

$$= 320 \times \left[\frac{0.04(1+0.04)^{74}}{(1+0.04)^{74} - 1}\right] + 12 + 50 = \$75.7\text{M}.$$

Location plan B: Expected value of service life $= (60 \times 0.3) + (65 \times 0.5) + (70 \times 0.2) = 64$ years

$$\text{EUAC (in \$ millions)} = 516 \times \text{CRF}(4\%, 64) + 7 + 35$$

$$= 516 \times \left[\frac{0.04(1+0.04)^{64}}{(1+0.04)^{64} - 1}\right] + 7 + 35 = \$64.5\text{M}.$$

Location plan B has lower EUAC and should be chosen.

Example 20.4

The highway planner of the Uygur Highway District is considering three alternative plans for maintaining an existing 10-mile two-lane rural road section in the district. The first option is to simply regravel the road every 2 years at the cost of \$0.01M per lane-mile. The second option is to provide a surface

Table 20.1 Data for Example 20.4

	Plan 1 (Regraveling)	Plan 2 (Surface Dressing)	Plan 3 (Asphalting)
Initial cost ($M/lane-mile)	0.01	0.08	0.2
Annual maintenance cost ($M/lane-mile)	0.002	0.004	0.0006
Annual cost of dust control at urbanized centers along highway ($M/lane-mile)	0.001	0	0
Annual cost of erosion repairs ($M/lane-mile)	0.003	0.001	0.0005
Annual cost of vehicle operations ($/vehicle) (Induced demand shown in parenthesis).	0.00004 (45,000 vehs)	0.00002 (60,000 vehs)	0.00001 (70,000 vehs)
Service life (yr)	3	6	10

dressing (chip seal) at the cost of $0.08M per lane-mile. The third option is to provide a 2-inch asphaltic concrete layer at the cost of $0.2M per lane-mile. What is the best plan, from the perspective of economic analysis? Assume that all other benefits are equal. Use a 5% discount rate (Table 20.1).

Solution

The net present worth of each alternative plan is calculated as follows:

NPV for Plan 1 =

$$\left(0.01 \times \frac{0.05 \times 1.05^3}{1.05^3 - 1} + 0.002 + 0.001 + 0.003\right) \times 10 \times 2 + 45,000 \times 0.00004 = \$1.993\ \text{M}$$

NPV for Plan 2 =

$$\left(0.08 \times \frac{0.05 \times 1.05^6}{1.05^6 - 1} + 0.004 + 0.001\right) \times 10 \times 2 + 60,000 \times 0.00002 = \$1.615\ \text{M}$$

NPV for Plan 3 =

$$\left(0.2 \times \frac{0.05 \times 1.05^{10}}{1.05^{10} - 1} + 0.0006 + 0.0005\right) \times 10 \times 2 + 70000 \times 0.00001 = \$1.240\ \text{M}$$

Plan 3 should be chosen.

20.8.4 Determining the Financial Feasibility of a Proposed System

Example 20.5

The City of Resolve has found it necessary to construct a new water supply reservoir to serve its fast growing population. The expected costs and benefits are shown in Table 20.2.

Using Monte Carlo simulation, determine the distribution of the capitalized cost to perpetuity. If the city can raise only $70M from the sale of bonds for the project, determine the probability that it will be feasible to finance the project from bonds only, on the basis of its capitalized cost. Also, determine the distribution of the net present value of the project. The expected life of the system is 20 years. All amounts are in constant dollars.

Table 20.2 Data for Example 20.5

	Mean	Standard Deviation	Probability Distribution
Construction cost	$40 million	$5M	Normal
Annual maintenance/ operating cost	$2 million	$0.25M	Normal
Interest rate	4%	0.1%	Normal

Solution

After performing the Monte Carlo simulation for 500 runs, it can be determined that the average of total cost (costs of construction, maintenance, and operations) is $66.9M, with a minimum and maximum of $88.8M and $46.9M, respectively. Important note: It is expected that different readers will obtain somewhat different average values of the total cost, and the maximum and minimum. If the city raises $70M, then there is approximately 69% probability that this project will be feasible. The probability distribution for the total cost can be developed; this is left as an exercise for the reader. From sensitivity analysis, it can be shown that if the city raises $5M less than expected (in other words, if the amount raised through bond sales is $65M), then the probability that this project will be feasible decreases drastically to 37%; also, if the city raises $5M more than expected (in other words, if the amount raised through bond sales is $75M), then the probability that this project will be feasible increases sharply to 90%.

SUMMARY

A well-planned civil engineering system not only provides benefits to its users and the community in delivering the specific service for which it was intended but also provides governments with an opportunity to meet broader national or regional objectives such as job creation, energy consumption reduction, and sustainable development. As such, planning may be defined as the set of activities that specify how the end product will be achieved while being cognizant of the different elements of development. Civil systems planners responsible for developing alternative plans duly recognize that civil systems do not exist in a vacuum but must meet certain physical, locational, economic, social, and political requirements. The responsibility for system planning is often borne by agencies that have been granted statutory authority for a specific system type. This chapter presented a brief history of civil system planning as well as the different dimensions of civil system planning. The impetus for systems planning, the evolving and emerging contexts of systems planning, and the principles of civil systems planning also were discussed. We then discussed a general framework for civil system planning and discussed the barriers to effective planning and how these barriers could be overcome. Finally, we reviewed a number of computations at the various steps of system planning.

EXERCISES

1. List any five key pieces of legislation related to the planning of civil engineering systems and discuss the way in which each of these have affected the planning process.
2. Discuss any five dimensions for civil systems planning. Which levels of these dimensions are relevant in the case of a new airport being planned in your community?

3. Explain why the planning of civil engineering systems is an important step in the system development process.
4. Civil system planning techniques in the current era differ from those of past generations. Indicate whether you agree with that statement and provide at least two illustrative situations to support your position.
5. List and discuss at least seven practices for a good planning process for civil systems.
6. The university administration seeks to construct a pedestrian bridge over the busiest street on campus. In the context of this proposed facility, list and explain the steps of the civil systems planning process.
7. As the manager of the planning division in a large civil engineering-related organization, it is critical that you monitor any threats to the development of effective plans for the civil systems you design. List and discuss any four obstacles to planning and identify how these threats could be eliminated or at least minimized.
8. Search the print and electronic media for articles that discuss some planning-related aspects of an existing or planned civil system in your locality or elsewhere. These aspects may include the location of the system, the adverse effects of the system, the cost-share of the system construction and/or operations, and so forth.
9. A certain state recently conducted a statewide survey among voters to assess the support for a proposed local options tax for transportation system improvements (LOS-T). The survey results indicated that the LOS-T initiative is supported by 75% of Democrats, 50% of Republicans, and 62% of Independents and other parties. Of the eligible voters in the state, it is known that 35% are Democrats, 25% are Republicans, and 40% are Independents and other parties. (a) What is the probability that a randomly selected voter supports the initiative? (b) Suppose that a randomly selected voter supports the initiative, find the probability that that voter is a Democrat.
10. A new light-rail transit system is proposed for a city between the suburbs and the downtown area. Three criteria have been identified by decision makers upon which the best alternative implementation strategy will be selected: (1) the forecasted transit ridership,(2) the social impact, and (3) the number of jobs created (economic development). The relative weights for ridership, social impact, and jobs created are 0.25, 0.40, and 0.35, respectively. The ridership and the jobs created are measured in thousands; and the socioeconomic impact is measured in terms of the number of residences and businesses displaced (Table 20.3). Assuming the following value functions, use goal programming to determine the best alternative from among four alternatives. Alternative A represents the do-nothing alternative.
Alternatives B–D represent different expansion locations for the light rail system.

Table 20.3 Data for Exercise 10

	Alternative A	Alternative B	Alternative C	Alternative D
Ridership	$V_{rider}(0) = 0$	$V_{rider}(11) = 0.4$	$V_{rider}(15) = 0.65$	$V_{rider}(20) = 1$
Social benefits	$V_{soc}(0) = 1$	$V_{soc}(32) = 0.75$	$V_{soc}(40) = 0$	$V_{soc}(25) = 0.5$
Jobs	$V_{jobs}(0) = 0$	$V_{jobs}(12) = 0.5$	$V_{jobs}(20) = 1$	$V_{jobs}(17) = 0.75$

The following goals have been set by the decision makers: 12,000 for transit ridership, 23 for households and businesses displaced, and 15,000 for jobs created. Determine which of the plans is best.

11. The initial year construction cost of a certain tunnel through the Alps is $20M. Also, for each year of the construction period, the construction cost could be $2M, $3M, and $4M, depending on the difficulty of site conditions. These amounts are in constant dollars so there is no need to adjust for inflation. Depending on the extent of anticipated worker unrest and other factors, the construction may take 1, 2, or 3 years. Further, due to uncertain economic conditions, the interest rate may be 4, 5, or 6%. Draw a cash flow diagram illustrating the best and worst-case scenarios. Determine the cost at the end of the construction period for each combination of annual construction costs. Using your results, develop a nomograph that can help the engineer determine the final construction costs for each combination of conditions.

12. The city of Valparaiso seeks to build a large water reservoir to serve its fast growing population. It has been estimated that the new steel structure will have an initial construction cost of $3M and an operating cost of $0.5M annually. The maintenance cost is expected to be $0.1M for the first 5 years and will grow at an annual rate of 1% until the end of its 30-year service life. The cost of demolition is $1.5M and the demolished structure can be sold as scrap metal for $0.5M. Determine the life-cycle cost of the planned structure.
13. Alternative plans are being considered for an irrigation canal between a natural freshwater reservoir and an agricultural region in Balochistan. One of the plans is for the canal to pass through mountains via tunnels and also to cross a series of ravines and will necessitate the construction of structures to carry the water over those land features. This alternative will involve lining the canal. The other plan involves an unlined but longer canal that will not involve any tunneling or bridges. On the basis of the cost and performance attributes itemized in Table 20.4, determine which plan is more economically feasible. Use a 5% discount rate.

Table 20.4 Data for Exercise 13

	Plan A	Plan B
Distance	3.5	12
Total excavation needed (mils, **ft**^{3})	0.5	0.1
Bridge construction cost ($M)	1	0
Tunnel construction cost ($M)	0.8	0
Annual maintenance of bridge and tunnels ($M)	0.2	0
Lining cost ($M)	2	0
Annual cost of channel maintenance	0.15	1
Annual loss of water due to percolation (mils, **ft**^{3})	0.05	1.5
Service life (yr)	20	5

REFERENCES

Ali, M. M. (2008). Energy Efficient Architecture and Building Systems to Address Global Warming, Engineering Strategies for Global Climate Change. *ASCE Leadership Manage. Eng.* 8, 3.

ASCE (2007). Impact of Global Climate Change, ASCE Policy Statement 360, Energy, Environment, and Water Policy Committee of ASCE, Reston, VA.

ASCE (2013). America's Infrastructure Report Card. http://www.infrastructurereportcard.org/.

Beniston, M. (2004). *Climatic Change and Its Impacts*. Springer, Berlin.

Bose, A., and Fricker, J. D. (2004). Reverse-Engineered Land Use Patterns to Minimize Congestion. *Transport. Res. Rec.* 1831, 141–149.

Dandy, G., Walker, D., Daniell, T., and Warner, R. (2008). *Planning and Design of Engineering Systems*, 2nd Ed., Taylor & Francis, London, UK.

Goodman, A. S., and Hastak, M. (2007). *Infrastructure Planning Handbook*. ASCE Press & McGraw-Hill, New York.

IPCC (2007). *Climate Change 2007: The Physical Science Basis*. Contribution of Working Group I to the 4th Assessment Report of the Intergovernmental Panel on Climate Change, Solomon, S. et al. (eds.). Cambridge University Press, Cambridge, UK.

JSCE (2009). Civil Engineers Confront Global Warming, Mitigation and Adaptation Plans in Japan to Decrease the Risks of Global Warming, Japan Society of Civil Engineers, Tokyo, Japan.

Lenkei, P. (2007). Climate Change and Structural Engineering. *Periodica Polytech.* 47–50.

Lewis, R., Knaap, G., and Sohn, J. (2009). Managing Growth with Priority Funding Areas: A Good Idea Whose Time Has Yet to Come. *J. Am. Planning Assoc.* 75(4), 457–478.

Liao, S. C. (2008). Envisioning and Creating the Future in Response to Global Climate Change, Engineering Strategies for Global Climate Change. *ASCE Leadership Manage. Eng.* 8, 3.

Long, M. K., and Labi, S. (2011). *Assessing the Vulnerability of Civil Engineering Structures to Long-Term Climate Change*. Presented at the *International Conference on Vulnerability and Risk Analysis and Management (ICVRAM)*, University of Maryland, Hyattsville, MD, April 11–13, 2011.

Meredith, D. D., Wong, K. M., Woodhead, R. W., and Wortman, R. H. (1985). *Design and Planning of Engineering Systems*, Prentice Hall, Englewood Cliffs, NJ.

Meyer, M. D., and Miller, E. J. (2000). *Urban Transportation Planning* 2nd Ed., McGraw-Hill, New York.

NHI (1995). Estimating the Impacts of Transportation Alternatives, *FHWA-HI-94-053*, National Highway Institute, Washington DC.

Sinha, K. C., and Labi, S. (2007). *Transportation Decision-Making: Principles of Project Evaluation and Programming*. Wiley, Hoboken, NJ.

UN (2006). *United Nations Framework Convention on Climate Change: Handbook*, United Nations, New York.

Yoe, C. E., and Orth, K. D. (1996). Planning Manual, Institute for Water Resources, U.S. Army Corps of Engineers, IWR Report 96-R-21, Alexandria, VA.

USEFUL RESOURCES

Neil, N. S. (1988). *Infrastructure Engineering and Management*, Wiley, New York.

CHAPTER 21

SYSTEM DESIGN

21.0 INTRODUCTION

Mathematicians, physicists, chemists, and other scientists typically seek unique solutions to the problems that they address. In contrast, engineers are typically involved with problems for which several alternative practical solutions can be designed, and they choose the best solution from several design alternatives. Thus, the engineering design process often prompts the following pertinent questions. How do the outputs of the preceding activities (needs assessment, goals identification, and system planning) produce information for the design process? Which analytical tools and techniques could be used to identify the best of several alternative designs, taking into account the intended goals as well as the various physical, financial, institutional, and other constraints? Can design be made more flexible to accommodate the practical reality of uncertainties in the social, technical, and economic environment? Can multiple-criteria considerations and risk and uncertainty concepts play a role in developing a robust design? For questions such as these, answers often are obtained through an implicit recognition that engineering design is "both an art and a science" as suggested by de Neufville and Scholtes (2011) and Khisty et al. (2012). It is a science because it requires the designer to follow laid-out procedures and make inferences from data in order to establish and test the values of the design parameters; it is an art because multiple solutions may exist for a given design problem, and also because in many cases, engineering design requires originality and creative thinking, which is particularly true in civil engineering where the physical, economic, environmental and institutional conditions of each site are unique.

Engineering design may be defined broadly as a creative problem-solving process in which the engineer works within the budget, time, legal, institutional, and other constraints to convert data, information, and technical know-how to translate ideas into a product or service. In some literature, design has been described as a process that defines the arrangement, orientation, and dimensions of a number of modules, components, modules, and interfaces such that the system effectively and efficiently fulfills its intended function. The Accreditation Board for Engineering and Technology (ABET, 2012) defines design as follows:

> The process of devising a system, component, or process to meet desired needs; it is a decision-making process (often iterative), in which the basic sciences, mathematics, and engineering sciences are applied to convert resources optimally to meet these stated needs.

Civil engineers who carry out system design strive to follow a sequence of activities known as the *engineering design process*, which involves certain general steps and detailed specific steps as you will see in a subsequent section of this chapter. As in all other phases, the system design phase is characterized by a number of tasks (e.g., describing and evaluating alternative designs, choosing the optimal design) and there are tools for carrying out these tasks (e.g., simulation, statistical modeling, and optimization).

Civil engineering institutions and organizations offer detailed courses that teach the student how to carry out design in the different branches of this discipline. As such, this chapter does not dwell on specific design processes but rather provides a broad discussion of the general design process that is applicable to systems design in any branch of civil engineering, along with a few examples of systems design in selected branches. The chapter first presents a classification of design situations in civil engineering. Next, the chapter identifies and discusses the steps of the engineering design process, namely, the tasks of abstraction and synthesis, the configuration design tasks where the interfaces between physical components are considered, the evaluation of alternative designs, and the postdesign steps. The chapter then presents some numerical computations for engineering design in different fields of civil engineering. The chapter goes on to discuss some issues associated with engineering systems design from a traditional viewpoint and then from the viewpoint of emerging challenges in the new millennium. Finally, some numerical design computation examples are presented.

21.1 CLASSIFICATIONS OF ENGINEERING DESIGN

There are several ways of classifying engineering design problems. One way is by the discipline in question: transportation, structural, hydraulic, and so forth. Another way is by whether the design is for a physical system, such as a water pipe, or for a virtual system, such as a traffic signal timing design. A design may also be classified according to whether it is for a new system or for improvements to an existing system. Another classification relates to the extent of human user interaction with the system (none, limited, occasional, or frequent) as this would influence the ergonomic considerations in the design. There is yet another classification that is related to the extent to which a design provides opportunities for creative input. In the sections below, we focus on one classification scheme for engineering design, the Ullman classification.

21.1.1 The Ullman Classification

Ullman's classification (Ullman, 1992) is based on the extent of transformation and creativity associated with a design. In this classification scheme, any design in civil engineering may be described as any one of several design categories in the following ordered range: selection design (lowest level of transformation and creativity), configuration design, parametric design, and original design (highest level of transformation and creativity).

Selection Design. Here, the designer selects standard components supplied by a vendor and then designs an assembly of these components to yield the final product. This category of design is more common in certain mechanical and electrical systems and less common in civil engineering due to the uniqueness of the site conditions in cases of the latter. Nevertheless, a few examples can be found in civil engineering, such as the design of precast structural systems. In selection design, the designer merely does a "mix and match" of the standard components, with very little or no room for creative input or innovation. One of the very few avenues for imaginative input is the selection of unusual combinations of the standard components.

Configuration Design. Similar to selection design, configuration design involves standard components. However, the designer goes beyond mere prespecified selection and assembly of the components and makes a creative decision on the location of each component or arranging them in some innovative manner so that the overall system performance is maximized or the cost minimized (Levin, 2009). For example, in the structural design of airplane frames, the designer encounters the challenge of locating jet engines on an aircraft in order to maximize the structural integrity of the aircraft among other reasons (Hyman, 2002): beneath the wings, cantilevered off the side of

(a)

(b)

Figure 21.1 Different configurations in design: illustration using bridge truss example: (*a*) truss located beneath bridge and (*b*) truss located above bridge (*a*. Courtesy of Leonard G./Wikimedia Commons; *b*. Gregory David Harington/Wikimedia Commons).

the rear fuselage, or mounted in the tail assembly. Other examples include the truss locations for bridges: deck versus through configurations (Figures 21.1*a* and 21.1*b*, respectively); water tanks and pipelines: overhead versus at-ground versus buried configurations. There is relatively little room for innovation and creativity in the location or arrangement of a system component relative to other components.

Parametric Design. In this design category, the designer varies the design parameters for the system and then measures and evaluates the resulting performance. These parameters, which often relate to the system material type, component dimensions (length, thickness, and width), shape, and orientation, are embedded in equations that express the relationships among the performance objectives, design parameters, and constraints. For example, using actual field experiments or computer simulation prior to or as part of the design, a pipeline engineer can vary the diameter of the pipe, the pipe material, the lining material, and the longitudinal slope to elicit maximum flow performance within the constraints of the cost, scouring, and sedimentation. This design category allows for a fairly significant amount of creativity and innovation by the designer.

Original Design. This design category involves the greatest degree of creativity. Original design essentially is groundbreaking or transformational design. There are numerous examples of original design worldwide, and we will examine briefly two original designs in the United Kingdom as classic examples.

1. Unlike typical drawbridges that open up either at one side or in the middle for a water vessel to pass underneath, the Gateshead Millennium Bridge in Great Britian (Figure 21.2*a*) is a V-shaped structure with two arcs hinged at their adjoining ends (one arc is the bridge roadway and the other is the supporting arc). The roadway arc is suspended by cables from the supporting arc that leans back downstream at an angle almost perpendicular to the roadway arc. To allow water vessels to pass underneath, the bridge opens and closes like a giant eyelid by rotating about the horizontal axis that joins its end pivots.
2. The Falkirk Wheel in Scotland (Figure 21.2*b*) is the world's first and only rotating boat lift. By looking at a problem from a new angle, civil engineers designed the wheel to lift water

(a) (b)

Figure 21.2 Examples of groundbreaking designs in civil engineering: (a) Gateshead Millennium Bridge, Great Britain and (b) the Falkirk Wheel, Scotland (a. Courtesy of Mike1024/Wikimedia Commons; b. Lowattboy at en.wikipedia).

vessels from one canal level to the other. In traditional operations, moving a vessel across this vertical distance would require its passage through up to 11 locks over several miles, 8 hours, and significant energy use; the Falkirk Wheel achieves this in a 15-minute movement using very little energy (ICE, 2011).

21.2 ENGINEERING DESIGN PROCESS

The engineering design process is a systematic endeavor that starts from receiving output from the needs assessment and planning phases. The core design process (see shaded area of Figure 21.3) involves the carrying out of the conceptual design and the detailed design. The postdesign steps include design performance feedback and design revisions. Throughout the design process (steps 1–4) where alternative designs are developed, the system designer is tasked with the analysis, description, evaluation, and selection of each alternative and often draws on analytical

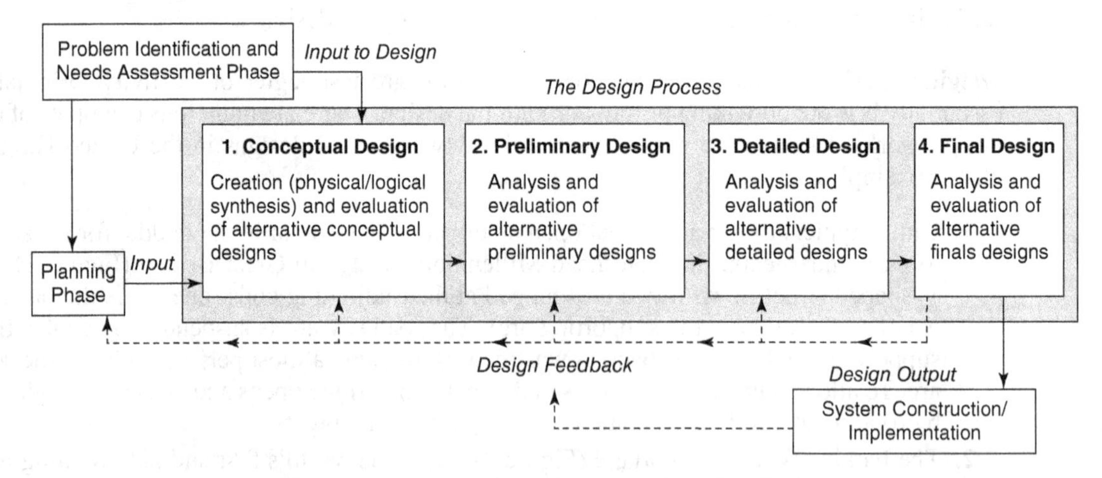

Figure 21.3 General process of engineering design.

tools including simulation, optimization, evaluation, risk and uncertainty analysis, and life-cycle analysis of the costs and the performance benefits, in order to carryout these tasks. We now discuss each step of the process in great detail.

21.2.1 Receiving Output from the Needs Assessment and Planning Phases

The system designer seeks to produce a design that satisfies some specific need that has been identified with respect to the overall goals that were established prior to the system planning and design. As such, the outcomes of the needs assessment and the planning phases for the system serve as a vital input for the system design phase. The system designer is guided by the goals established at the needs assessment phase and used at the system planning phase. These goals may be related to the system user's safety, the system's technical performance, economic efficiency, financial feasibility, and security, and the effects on environmental quality, reliability, durability, public acceptability, and ease of construction and maintenance. These and other goals are discussed in detail in Chapter 3. Also, the planning phase produces a set of broad specifications for the system that serve as a basis for the system design; these specifications include system location, orientation, budgeted cost, and minimum system performance requirements, among many others. Very often, particularly where the same consultant conducts both the planning and design of the system, the line between planning and design is thin, and the distinctions between the two phases may become blurred; in such cases, the last stage of planning (advanced planning), merges seamlessly into the first stage of design (preliminary design), a transition that can accelerate the progress of the planning and design phases.

21.2.2 Conceptual Design Part 1: Logical and Physical Design (Abstraction and Synthesis)

In Part 1 of the conceptual design of the system, the designer carries out abstraction and synthesis of the proposed system. The designer incorporates techniques for cultivating/enhancing creativity to develop the design, accesses domain knowledge in the relevant branch of civil engineering to facilitate abstraction and analyses during the design, and analyzes and uses this information to develop multiple alternatives for the conceptual design. In practice, all these activities are carried out in parallel; but in this section, we shall discuss them sequentially, one at a time.

(a) Incorporating Creative Thinking to Develop the Design. From the output of the system planning phase, the system designer envisions potential design solutions by developing abstract (general) concepts or approaches, and in doing so, relies heavily on creativity and imagination. At this stage of **concept generation**, the designer recalls related past problems or experiences that were either solved or could not be solved (and for what reasons), relevant theories and concepts associated with the past problems, and any fundamental approaches that were used to resolve such problems. Hyman (2002) describes this stage of the design phase as "the most exciting stage" and likens it to a pristine wetland where an adventurer (the designer) can potentially discover "exotic species" (new designs). Further, the designer should expand their realms of thinking to maximize what could be possible and minimize what is probably not possible. The need for system designers to cultivate creative thinking skills is absolutely essential; however, the designer must ultimately develop only solutions that are expected to be acceptable to the stakeholders of the system.

Creative thinking involves the ability to generate ideas and concepts and to synthesize or combine them into forms that are not only useful but also unprecedented. A new idea may come as a flash of insight; however, it is often generated only after a period of careful, even laborious, preparation. Engineering designers use a number of proven techniques to stimulate creative thinking but must be cognizant of creativity blocks that often exist. As system designers proceed through the processes of abstraction and synthesis, they should be able to quickly recognize these blocks, otherwise such barriers could impede the development of innovative design alternatives or identification of the best possible design. The blocks commonly encountered by designers, as identified by Adams (1986), Hyman (2002), and Voland (2004) include: **knowledge blocks** (due to inadequate knowledge about the domain area (branch of civil engineering), the users, or the operating environment); **perceptual blocks** (due to stereotyping elements or failure to recognize alternative interpretations of these elements, improperly delimiting the problem, creating imaginary constraints, exaggerating existing constraints, and information overload that is exacerbated by inability to distinguish what is useful or not useful in the data); **emotional blocks** (due to fear of failure or need for approval, unwillingness or inability to build upon predetermined or prescribed pathways for solutions, and impatience after successive failures, lack of patience to carry out repetitive tasks needed to reach a design solution, or haste to reach a solution); **cultural blocks, taboos, inhibitions, or expectations** (due to cultural predilections within the organizational environment, institutional inertia, and limitations as a result of design expectations or preconceptions held by clients); **environmental blocks** (due to uncomfortable physical surroundings, unsupportive peers, and inflexible or overly critical supervisors); and **expressive blocks** (due to lack of oral or written communication, misdirection due to the use of inappropriate terminology to define the problem or describe possible solutions). If any of these blockages are found to exist, the designer should take steps to eliminate them or to reduce their effects.

Having eliminated or minimized any existing blockages to creativity, the system designer should examine strategies for generating creative designs. First, because each site is different, no single design technique may be the most cost-effective for every design situation or environment. As such, the designer should analyze different designs to ascertain which ones work better for the particular site or situation at hand. Second, it is important to realize that design checking is an important part of the design process; some vital design consideration may be missed by the main designer but could be identified by other designers or assistants who have a neutral, different, or fresh perspective of the issues at hand. As the proverbial saying goes, "two heads are better than one."

At the start of the design process, it is recommended that the designer holds a **brainstorming session** with colleagues, other designers, stakeholders, or members of the general public to identify possible design options. At this brainstorming stage, it is the quantity, not the quality, of design solutions that is sought; and contributors to the brainstorming session should feel free to suggest designs without any encumbrance of practicality constraints. The design ideas resulting from the brainstorming session can be very interesting and revealing. All ideas should be documented. In a subsequent section of this chapter, the brainstorming technique is discussed in greater detail.

Then the designer should investigate if insight may be adapted or gained from a design solution that has been observed already in nature to solve a similar or related engineering problem under consideration. From natural systems (animals, plants, and natural habitats), the designer may observe and learn from the arrangement and configuration of the natural system components. For example, structural engineers designing the frame for a new generation of air borne craft could look for hints in the stability of winged insects under different wind loads (Figure 21.4). This is a creativity technique referred to as **bionics** or **mimicry**, and related disciplines include biomimetics or biomimicry. For example, Mazzoleni and Price (2013) analyzed how organisms adapt to different environments by studying the diversity of animal skin structures) and drew inspiration from their findings to introduce new thinking about the design of building envelopes. Other organs and features of animals and plants have inspired new ways of design. Table 21.1 presents a number of bionic parallels between natural phenomena and engineering system concepts that could help designers in their never-ending search for innovative design solutions.

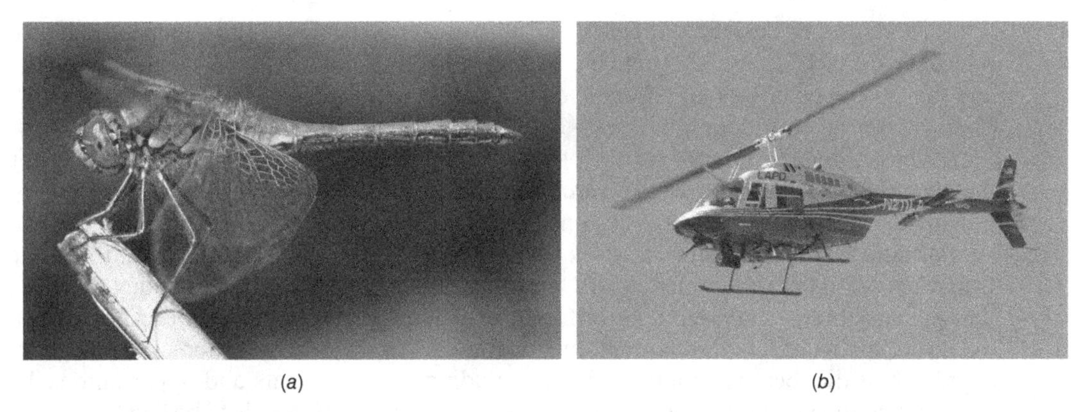

(a) (b)

Figure 21.4 System designer should look for hints in the natural systems (*a*. Courtesy of Darkone/Wikimedia Commons; *b*. Matthew Field (http://www.photography.mattfield.com).

Table 21.1 Some Relationships between Natural Phenomena and Engineering System Concepts

Natural Phenomenon	Corresponding Engineering System Concept or Product
Feather oil of ducks	Antiwetting agent
Bamboo, animal bones	Tubular structures
Muscles attached to bones	Levers and fulcrums
Venus fly-trap	Trigger mechanisms
Evaporation through skin surface	Air conditioning
Squid siphon	Jet propulsion
Animal heart	Pump
Beetle's eye	Retractable landing gear, automatic clasps
Birds legs and claws	Aircraft altitude and ground-speed indicator
Massasuga rattlesnake	Heat sensor
Human brain	Artificial neural networks
Beehives	Storage containers

Source: Voland (2004).

Having used abstraction and creativity techniques to develop a set of partial solutions or components for the proposed system and after modeling to define and clarify these partial solutions, there remains the task of combining or **synthesizing** the partial solutions into a whole in order to generate the overall design or a set of alternative overall designs, all of which must be feasible. This synthesis is necessary because, for a given problem, there are typically many possible solutions and adequate time must be given to ponder the problem so that a large number of alternative solutions can be generated. De Neufville and Stafford (1971) put this in perspective:

> It seems useful to state the desirability of considering more, rather than fewer, alternatives. The development of computers has enormously increased the designer's ability to do this and thereby made it possible for analysis to carry out studies in an effective, orderly fashion. Indeed, a deliberate generation of a wide range of choices is an essential component of systems analysis.

Therefore, as designers identify and investigate possible designs, they should articulate the relevance of each design alternative by tying it back to the identified needs, the intended goals, and the system plan. Design alternatives that do not satisfy critical design goals are subsequently eliminated from consideration.

Throughout the process of the logical and physical design, the system designer must ensure that the designs under consideration are those that fulfill the two basic pillars of systems analysis—holism and parsimony. Holistic designs are those that implicitly recognize that the effect of the sum of different design components is often superior to the sum of the effects of the components. Consistent with the general tenets of Occam's razor (see inset box), parsimony means choosing a design that achieves maximum performance with minimum possible use of resources, for example, the use of high-strength, lightweight metals for steel facility construction. The designer should be cognizant of the local situations at hand, including natural threats and opportunities, the cultural sensitivities and uniqueness of the environment in which the system is to be located, and the system-environment interactions including sustainability and resilience.

OCCAM'S RAZOR VERSUS HICKAM'S DICTUM

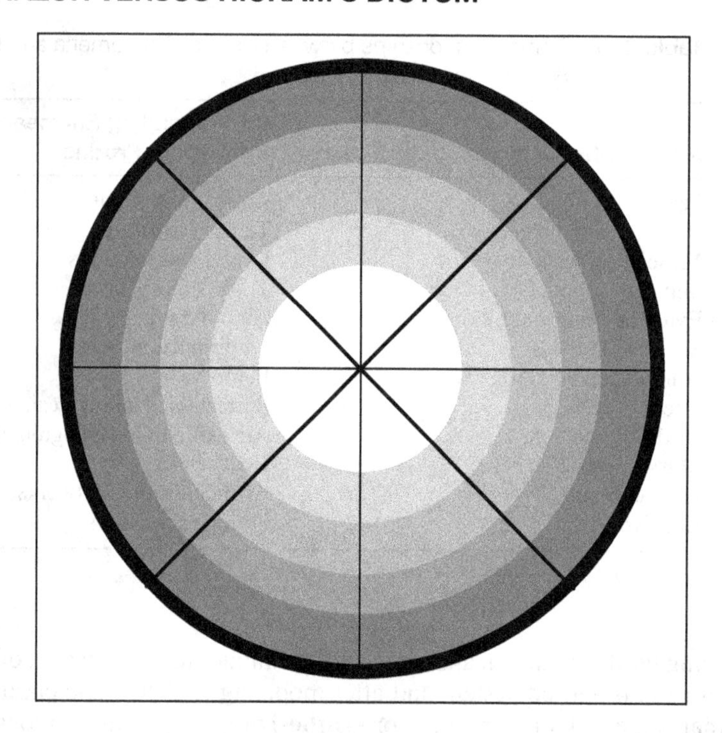

In the 14th century, William of Occam, a Franciscan friar, in supporting his theory that God's existence cannot be deduced by reason alone, stated: "We are to admit no more causes of natural things than such as are both true and sufficient to explain their appearances." Long before that time, there had been a longstanding general rule of thumb *Pluralitas non est ponenda sine necessitate* (translation: "Plurality should not be posited without necessity") that was espoused by Aristotle, Ptolemy, Maimonides, John Scotus, and other early philosophers. Consistent with this rule of thumb, Occam's razor favored parsimony, economy, or succinctness: When you have two competing alternatives that explain the same concept or perform the same function, the simpler one is better; or, when faced with competing hypotheses that are equal in other respects, select the one that makes the fewest assumptions. The term "razor" is used simply to advocate the adoption of the choice that "shaves off" extraneous assumptions. Albert Einstein made a similar remark: "Everything should be made as simple as possible, but not simpler." Contemporary slang puts it succinctly: "Keep it Simple, Stupid (KISS)." In developing a design to solve a problem, one should not clutter the solution with needless detail. As a logical principle, Occam's razor demands that scientists explain existing observations with the simplest possible theory. As new data become available, application of the razor may lead to a different conclusion.

Occam's razor has been applied in several fields including physics, biology, medicine, religion, criminal justice, philosophy, and pattern recognition. In physics, Newton's law of motion was preferred over Kepler's laws of planetary motion (even though they both offered similar descriptions for planetary motion) because Newton's theory made fewer assumptions and was simpler. Also, Albert Einstein, in formulating his special theory of relativity, presented a much simpler theory (compared to that of Hendrik Lorentz) by eliminating consideration of undetectable metaphysical concepts such as ether. In medical science, Osler's rule indicated that one diagnosis was sufficient to explain all the symptoms of a patient; and this rule was considered appropriate in the era it was derived because individuals in that era were, at any time instance, afflicted more by a single type of disease (often infectious) rather than multiple diseases.

In civil engineering systems development, Occam's razor can be applied in any of several phases. At the needs assessment phase, where a socioeconomic problem is identified as a precursor to system planning and design, the razor can be applied to the identification of the existence and nature of the problem. At the design phase, where the engineer needs to describe the design using a model, it is recommended to use the minimum level of detail in the model. A model must never be more complex than it absolutely needs to be to achieve its purpose. At the monitoring and preservation phases of systems development, the engineer may invoke Occam's razor to espouse "diagnostic parsimony," that is, when diagnosing a given system defect, an engineer should look for the fewest possible causes that will account for all the observed defects.

Critics of Occam's razor hold the opposite view. Hickam's dictum (John Hickam, an Indiana University professor of medicine) suggests that a system can have as many defects "as it damn well pleases." In other words, it is often statistically more likely that a defective system has several problems rather than a single problem that explains its multiple defects. Other critics of Occam's razor include Immanuel Kant (who stated that: "the variety of beings should not rashly be diminished.") and Karl Menger (who stated that "entities must not be reduced to the point of inadequacy," and "it is vain to do with fewer what requires more"). Also, Saint's triad, named after a South African radiologist, Charles Saint, states that when the results of a physical examination are not typical of any single condition, multiple underlying diseases should be considered as the diagnosis.

Sources: Michelson (2004), Hilliard et al. (2004), Bradley Fields (2005), Courtney and Courtney (2008), and Maloney (2011).

(b) Incorporating Other Techniques for Cultivating/Enhancing Creativity. As Hyman (2002) points out, creative geniuses are few and far between. Nevertheless, all civil engineering designers can enhance their creativity by adopting techniques that are identified in existing literature (Adams, 1986; Folger and Leblanc, 1995; Hyman, 2002; Voland, 2004). We now discuss these techniques.

Checklisting. Where it is required to make improvements to an existing system or system component, the engineer can use a number of trigger words and questions to spark creative thinking. Trigger questions might include the following. What does the system do? What does it not do currently but could probably do? What other systems are similar to it? Why can other systems perform that function while the system itself cannot? What is wrong with its structure, orientation, position, material, dimension, and so forth? Trigger words may include *lighten, rotate, twist, turn over, bend, make thinner, make thicker, perforate, reverse, fortify, stratify, raise*, and *lower*.

Synectics. There are occasions when the designer faces a system design problem that is so familiar that they face tremendous difficulty in conceiving of any alternative besides the traditional designs. Conversely, a design problem may be so unique that the engineer is unable to relate to it on the basis of their past experiences. In the creativity technique termed synectics, the engineer strives to (i) make the familiar strange, or (ii) make the strange familiar.

Method of Analogies. This technique involves the linkage of the design problem at hand to another problem that resembles it in some way and is easier to solve, is closer to being solved, or is actually solved. In making such comparisons, the engineer should (i) establish a **direct** analogy between the given problem and a solved problem (as in bionics) or an almost solved problem, (ii) make a **fantasy** analogy to imagine the problem in an analogous but more convenient form for the purpose of advancing the design process, such as in synectics, (iii) make a **symbolic** analogy by using a **poetic** metaphor or a literary cliché to view a given problem in a new way, or (iv) make a **personal** analogy by imagining one's self as part of the system, especially under adverse circumstances, in order to gain a new perspective.

Brainstorming. Creativity can be enhanced through brainstorming sessions, where imaginative thinking is encouraged and the focus is placed on the quantity, not the quality, of the ideas; and participants are encouraged to combine or extend ideas to form newer, even outlandish ideas (Dominick et al., 2000). At the start of the session, the group should nominate a moderator (to manage the session) and a secretary (to record the ideas presented during the session). After describing the problem at hand, the moderator asks the participants to provide their contributions. Brainstorming sessions often end naturally when the allotted time is reached or when the flow of ideas slows to an unproductive rate. Then the session secretary presents the collected ideas to the participants either on a chalkboard or through an electronically projected computer screen so that these ideas can be visible as the session proceeds to the subsequent stages. The design team narrows these ideas to the most promising ones for further consideration.

Lateral Thinking. To prelude our discussion in lateral thinking (LT), let us examine the concept of vertical thinking (VT) (Table 21.2). VT refers to the way people typically approach problem solving: moving down a solution path by evaluating information logically and objectively. The process proceeds in sequential steps, with each step justified by logic and fact. Along each step, the designer evaluates new ideas in relation to existing ones and in relation to existing patterns and concepts. Figure 21.1, which presents the sequence of engineering design, is an example of VT and what is necessary as a minimum for design. Dominick et al. (2000) recognized that designers have a natural propensity to VT because as humans, they are creatures of habit and therefore strive to identify patterns and relationships in order to make sense of their environment. However, VT has serious limitations in that it follows a laid-down sequence of thinking, whereby designers often may fail to identify or recognize different ways of viewing a given problem and thus, different, new, or even revolutionary solutions.

Table 21.2 Key Differences between Lateral Thinking and Vertical Thinking

Vertical Thinking	Lateral Thinking
Goal is to select an idea	Goal is to generate ideas
Structured and sequential	Random; jumps around looking at any possibilities
Intended to be analytical	Intended to be proactive
Excludes irrelevant information	Welcomes irrelevant information
Tries to finalize by selecting one final solution	Tries to expand possibilities for solutions
Focuses on whether an idea, or an element thereof, is right or wrong	Does not concern itself with the rightness or wrongness of an idea

Source: Dominick et al., 2000.

Using LT, the designer can break out of the VT straitjacket. LT involves the reorganization of information and reassembling it in different ways that can lead to new and unique ideas (sloane, 2006). In his treatise on lateral thinking, DeBono (1973) identified some of the key factors of LT: recognizing dominant assumptions that polarize the perception of the problem, searching for different ways to view a problem, and relaxing rigid control of thinking. LT techniques include the *reversal method* (taking a design approach and turning it around), *removal of restrictive assumptions* (by bypassing physical or non physical constraints to the design problem), and *random stimulation* (using irrelevant cues or unrelated information as stimuli to fuel free associations and new ideas).

TRIZ. Developed in 1946 by Soviet inventor Genrich Altshuller and a team of collaborators, TRIZ is the acronym for a Russian term ***T****eoriya* ***R****esheniya* ***I****zobretatelskikh* ***Z****adatch*, which translated to English is the "theory of inventive problem solving." TRIZ is a systematic approach for analyzing challenging problems that require creativity and innovation and provides techniques and strategies for developing solutions to such problems (Webb, 2002). More than just another "tired exhortation to think outside the box" (Wallace, 2000), TRIZ is an algorithmic style of generating new ideas or new designs, inventing new systems, and refining existing concepts or designs. TRIZ was developed by studying hundreds of thousands of past inventions and recognizing some common patterns in successful inventions; and TRIZ's algorithm was developed to replicate these patterns. Since its initial publication, the TRIZ algorithm has seen several revisions by Altshuller's disciples. Nevertheless, the algorithm essentially consists of three stages: defining the design problem, identifying technical contradictions, and exploring possible solutions.

Contradiction, a key concept in TRIZ, is defined as a situation where a bid to enhance one performance measure leads to a decrease in the ability of the design to meet another performance measure. Thus, for a problem to be considered inventive, it has to pose at least one technical contradiction. For example, in lightbulbs, the burning filament must be hot enough to produce light but not excessively hot otherwise the filament material is destroyed; this contradiction was resolved by Thomas Edison when he invented the incandescent bulb as he placed the filament in a vacuum (Wallace, 2000). Contradictions in civil systems design often include the strength versus the dead load of civil structures (structural engineering), freeway mobility versus safety (transportation engineering), and concrete workability vs. compressive strength (materials engineering).

The first step in TRIZ analysis is to identify which design parameters contradict each other. Then, the concept of **inventive principle** is invoked (i.e., many of the needed design ideas may have been already used to solve problems in other disciplines in science and engineering). TRIZ's design principles also include **segmentation** (breaking up a problem or situation into multiple components such as a multistage rocket take-off system design) and **multiplication** (deriving solutions by duplicating some component that already is inherent to the problem, such as vaccinations or fighting bushfires by controlled fire-setting). By systematically codifying and tabulating which principles can help solve given combinations of contradicting parameters and tabulating the results, the idea generation process can be accelerated instead of leaving it to trial and error. By reviewing such tables, scientists and engineers are provided with guidance on where to start searching for solutions.

In the current era, several organizations, including Ford Motor Company, Boeing, Daimler Chrysler, Johnson and Johnson, Motorola, Proctor & Gamble, 3M, NASA, and Siemens use TRIZ to develop new designs and strategies (Wallace, 2000; Dominick et al., 2000). Also, the application of TRIZ has become common in risk management, project management, organizational innovation, and Six Sigma processes (Barry et al., 2006). [Six Sigma is a data-driven effort toward six standard deviations between the mean and the nearest specification limit in any process, and thus seeks to identify and remove the causes of defects (errors) and to minimize variability in the outputs at any of the phases of design, manufacture/construction, operations, and maintenance (Pyzdek and Keller, 2009)]. TRIZ does not yield new designs but rather points the designer in the right direction. Dominick et al. (2000) offered the following guidelines for TRIZ application: (i) ideas for design solutions may be realized by examining the broad terrain of scientific disciplines as people in different disciplines and fields may have dealt with problems similar to the problem at hand; (ii) an effective design solution does not necessarily need to add something new, therefore, the designer should be open to reorganization of existing concepts; and (iii) recognition of the design contradictions can be used as a platform for generating new solutions.

Designers of civil systems can use TRIZ concepts to develop innovative solutions to civil engineering design problems or to enhance existing designs much faster compared to brainstorming. At the current time, creativity-enhancing approaches that are modified versions of TRIZ include systematic inventive thinking (SIT), unified structured inventive thinking (USIT), and advanced systematic inventive thinking (ASIT).

Creativity Modeling. From the existing creativity models developed in the professional psychology community, Hyman (2002) synthesized a four-stage creativity model that includes an incubation stage (a formative period during which the mind is relaxed and the individual is engaged in unrelated activity, which frees up the conscious and perhaps unconscious mind to be in a receptive mode) and an illumination stage (conscious recognition of the new idea, the so-called Eureka situation). By following this model explicitly or implicitly, creativity could be enhanced.

Attitudinal Change. Certain human personalities, behavioral attitudes, frames of mind, or moods, by their very nature, are favorable to creativity (Davis, 2004). These include an optimistic and enthusiastic personality, openness to flexibility in thinking, propensity toward adventurism and risk-taking, and tendency to be spontaneous and impulsive. A good sense of humor and even childlike playfulness, while difficult to cultivate where they do not exist in an individual, have been found to be strongly associated with creativity. On the other hand, extreme attributes of cynicism, skepticism, and stiff-necked adherence to procedures are generally enemies of creativity. In this direction, even the more benign attributes of conservatism, analytical mindedness, and risk aversion can seriously impede the flow of creative juices. In any case, a little attitudinal change that reduces the negatives and increases the positives can significantly enhance a designer's imagination and innovative capabilities.

Providing the Engineering Designer with Knowledge in the Humanities. Creative thinking emanates from the human brain. Therefore, efforts to increase creativity may benefit from examining the physiology of the different parts of the brain (Hyman, 2002). It has been shown

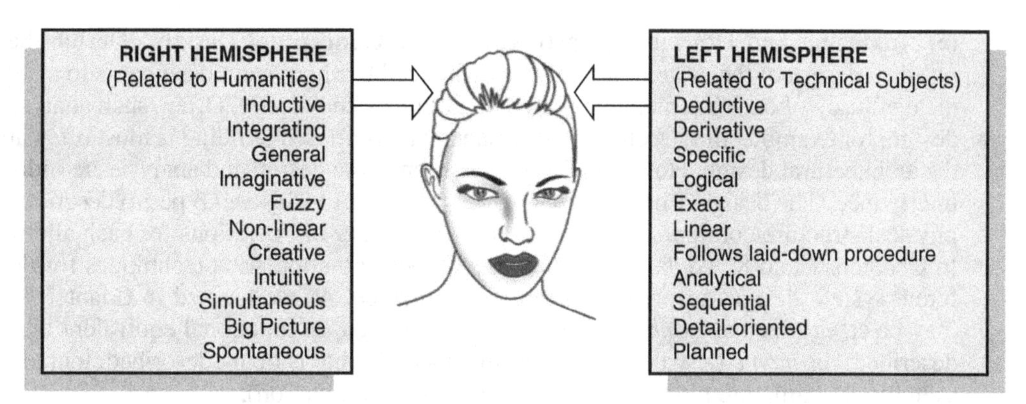

Figure 21.5 Creativity-related functional capabilities of the human brain hemispheres (line drawing courtesy of Wikihow).

through research that the left side of the brain is associated with a certain set of skills that are typically associated with the engineering disciplines while the right side has functions that are more directed to the humanities (Figure 21.5). Thus, the courses typically taken by engineering students tend to build up or reinforce the capabilities of the left side, while leaving the right side undeveloped (or in extreme cases, even suppressing its development). Unfortunately, creativity is often associated with the right side. A plausible explanation is that engineering students, due to the nature of their training, may find it more difficult to exercise the right brain functions that are needed for creativity. It has been argued that the inclusion of humanities in engineering curricula is not only meant to broaden the engineer's knowledge of the world in which we live but, more importantly, to increase their creativity.

Other Creativity Techniques. Other techniques described by Voland (2004) that designers could adopt to enhance their creativity include (i) adaptation of a solution for an unrelated design problem, (ii) consideration of an earlier (often rejected) design for a similar problem at a different time at the same or different location, and (iii) explaining the problem to someone not involved in (or familiar with) the design effort and studying their comments to possibly earn a flash of insight.

(c) Accessing Domain Knowledge to Facilitate Abstraction during Design. In any specific branch of civil engineering, the designer, at the abstraction stage of the design phase, needs to possess the requisite skills or knowledge of the theoretical background, or "domain knowledge," that is associated with the type of system (and thus, the associated branch of civil engineering) under consideration. For example, a structural engineer, no matter how creative he may be, may lack the domain knowledge to carryout a hydraulic design.

(d) Carrying out the Conceptual Design. At this stage, the designer uses the concepts we discussed in Sections 21.2.2(a) to (c). In other words, the designer creatively generates alternative conceptual designs that are acceptable to all stakeholders. The designer should identify and minimize the blockages to creative thinking and generate alternative solutions to the problem. Then the solutions to the different parts of the conceptual design should be synthesized to form one of several alternatives. Each conceptual design should be assessed against the background of the established problems, goals, and system performance measures and must be capable of being described adequately to a technical or nontechnical audience using simulation tools. Conceptual designs must be holistic and parsimonious and make full use of available domain knowledge.

(c) Analyzing and Describing Each Alternative Conceptual Design. During the logical and physical design of the system, a recurring task faced by the system designer is to analyze the technical efficacy of each alternative design using domain knowledge. Often, such analysis follows the design; for example, the structural performance of a proposed building is most often analyzed after the architectural design. However, for many systems, the design and analysis are indistinguishably intertwined. The designer may utilize a variety of model forms and types to describe the proposed physical structure/configuration or the proposed manner of operations for each alternative design. In Chapters 7 and 8, we discussed the model types and simulation techniques for describing different system designs or predicting their performance. As we learned in Chapter 7, these model may be categorized as *symbolic* (an equation), *analogic* (a functional equivalent of what is being described), or *iconic* (a scaled visual resemblance of what is being described, including an artist's sketch, 3-D miniature model, or computer graphical simulation).

21.2.3 Conceptual Design Part 2: Evaluation of Alternative Conceptual Designs

The end product of the first part of the design process is a number of alternative feasible conceptual designs. To choose the best of these, the engineer compares and evaluates the alternative designs primarily on the basis on how well they achieve their intended objectives at reasonable cost to the system operator, end users, or the community. In evaluating the conceptual designs, the designer establishes a set of objective evaluation criteria developed on the basis of the assessed need, the established goals, and the output from the planning phase. As we discussed in Chapter 3, the goals and objectives are primarily "technical" but also are often related to economic, environmental, and sustainability issues including sociocultural considerations, ethics, and aesthetics. Where the evaluation criteria (derived from the goals and objectives) are monetary, life-cycle-based economic efficiency analysis (Chapter 11) can be used; and where they include at least one nonmonetary criterion, the tool of multiple-criteria analysis (Chapter 12) can be used for comparing the alternative conceptual designs. Thus, the design alternatives can be compared on the basis of the extents to which they satisfy either the goals and objectives for the overall system that were established at the needs assessment phase or specific objectives purposely established for design evaluations.

The designer is encouraged to be critical of each alternative or candidate design, make an earnest effort to view each alternative objectively, and actively search for any weaknesses or limitations or strengths. The constructability or maintainability of each design should be a factor in the design evaluation process. The analysis of design alternatives may be refined further by constructing prototypes of the most promising designs (if possible) and testing these designs. The best design, after being identified, should be checked again for any weaknesses and shortcomings and possibly enhanced by incorporating specific promising elements from the rejected design alternatives.

Selection of the best conceptual design requires the art of balancing all the consequences (de Neufville and Stafford, 1971). The selection process may involve the application of value judgments to the performance measures or evaluation criteria. Typically, the selection process boils down to the definition of the utilities of the costs and benefits and the relative weights of the evaluation criteria derived from these costs and benefits. As such, utility theory is vital in the overall design process for civil systems. The design selection process must also consider the distribution of benefits and costs among the various stakeholders of the proposed system, particularly where the alternative designs have different impacts in terms of these costs and benefits. The principle of equity requires that system development should not proceed at the expense of any specific demographic. If such distribution varies across the different designs, then it should become an important consideration during the design process.

21.2.4 Preliminary Design

This is the second stage of design. The output from the conceptual design is the input for the preliminary design. The objective of the preliminary design is to review the conceptual design to ascertain that the design, and indeed the entire system, as conceived, will be capable of fulfilling the stated needs and will realize the established goals and performance objectives with the stated constraints (e.g., the operating environment). Functional block diagrams may be used to identify or clarify the functional relationships between the components of the designed system. At this stage, the designer may reevaluate the adequacy of the selected conceptual design from its technical, financial, environmental, and other perspectives and also to assess the risks associated with its construction or operations from the viewpoints of technical performance, cost, and schedule. In certain organizations, the preliminary stage of engineering design is the same as the conceptual design. At this stage of the design process, the designer also identifies the different materials needed, assesses their quantities, and establishes the dimensions of the system components. Then, the overall cost of the design is roughly estimated. The preliminary design is often archived carefully and used subsequently as a check on the final design.

21.2.5 Detailed Design and Final Design

The detailed and final designs of the system are the last design activities before the construction or implementation phase of the system. Thus, any outstanding design problems must be addressed before the next phase commences. At this stage, the design work is typically carried out in close consultation with the system owner to ensure that the design is compliant with the current design standards and meets with the approval of the system owner, the owner's representative, or other appropriate stakeholders. After ascertaining that the final design has met all the required specifications, codes of practice, and design requirements, the registered engineer certifies the final design.

21.2.6 Postdesign Steps

After successfully testing and evaluating the final design using computer simulation or prototype models where possible, the designer forwards the system design to the next phase of the system development cycle: construction. This does not mean that the designer then ceases all contact with the subsequent phases of system development. The designer should not shift attention from one completed design task to others without stopping to reflect on the ways in which their organizational design process has been enhanced from the just-completed design experience. Also, the designer should solicit, document, and act on feedback (regarding the field performance of the design) from persons involved in the subsequent phases of construction, operations, monitoring, and maintenance in order to learn of success and failure stories regarding various elements of the design; such feedback is critical for designing the next generation of the system at a future time.

Voland (2004) recommends a period of "formal reflection," after the design phase is completed, in order to clarify any aspects of the design experience that could be used to enhance the performance of future tasks. Reflection is particularly valuable if all members of the design team share their final thoughts about a design project once it has been completed, thereby encouraging each person to identify and assess the benefits of the experience. Reflection is important because engineering design is an inherently iterative process, and iteration can occur between any two phases or across all the phases in the system development process; in other words, a designer must not necessarily wait until the system has completed its development cycle (needs assessment through end of life) before returning to the design phase to correct and modify the product of that phase. For example, the engineer may recognize that at the operations phase, that the nature of the socio-economic problem that the system seeks to address may have evolved from that originally described at time of the needs assessment. This may require revision of the statement of need and the system plan; and with this updated needs assessment, the engineer is more likely be successful in recognizing any needed modifications to the system plan and conceptual design. Also, it may be only after a specific design is commissioned and put into operation that the designer becames aware of or acquires a deeper understanding of the behavior of the design. This realization may prompt design revisions for similar future systems.

Within the design phase itself, feedback across the design stages and design revisions is critical. At each design stage, there should be continual critiquing of the design under consideration to identify any possible shortcomings that might prevent the design from being a success and to revise the design to eliminate problem areas (Dominick et al., 2000). According to de Neufville and Stafford (1971), feedback mechanisms are fundamental to any process that is based on the scientific method. The planning and implementation phases of a project can be seen as parts of a continuous design process that should be reexamined at convenient intervals. This is needed to obtain designs that are robust, optimal, and yet context sensitive in today's dynamic environment.

21.3 APPLICATIONS OF SYSTEMS DESIGN IN SELECTED AREAS OF CIVIL ENGINEERING

At the phase of system design, the most visible task is that of the system design itself, where the engineer applies the domain knowledge in that specific branch of civil engineering to create a configuration of system elements to form a system at that location to fulfill a need. As we proceed to discuss applications of selected areas in civil engineering, we need to bear in mind that in order to produce the best design possible, domain knowledge is very necessary but not sufficient and therefore needs to be reinforced with creativity as well as the analytical skills discussed in Chapters 5–18. Table 21.3 presents a few contexts and situations of civil engineering system design.

21.3.1 Systems Design in Environmental Engineering

Designers of environmental engineering systems apply scientific and engineering principles to enhance the quality of the environment (land, air, and water) so that it is healthy for the flora and fauna that inhabit it. For wastewater and water treatment plants, the design considerations include the size of the plant and constituent units (the major determinant is the estimated demand) and the design for the expected flows and loads. The design considerations include the average, maximum, and minimum loads within specified time periods. In designing to meet incremental expansions of plant capacity (see Chapter 19 on needs assessment), design considerations may include how much should be added at each stage and the tolerable excess demand over capacity. The owners or operators of unsubsidized or partially subsidized environmental engineering systems typically set user prices to cover their cost, which implies that consumers generally pay for capacity they do not use; so it is desirable to minimize such excess payments on the grounds of equity and efficiency. In order to do this, design considerations should include the initial demand, the pattern of demand growth (e.g., linear, quadratic, and exponential), the cost of expansion, and the interest rate. The supply schedule can be established using initial demand and reliable growth predictions.

Environmental engineers design subsystems for physical treatment that include screens, mixing devices, sedimentation tanks, filters, and odor control and aeration facilities; the subsystems for chemical treatment include components for coagulation, softening, stabilization, demineralization, chemical oxidation, and disinfection; and the subsystems for biological treatment include the activated sludge, aerobic fixed-film processes, ponds, engineered wetlands, bioremediation and composting, and sludge stabilization. In designing these treatment systems, engineers consider attributes that include the demand, site conditions, costs, the nature of the water or wastewater

Table 21.3 Selected Civil Engineering System Design Contexts

Examples of Civil Engineering Systems	Typical Design Situations
Highway pavement systems	Design of highway pavement layer material types and thicknesses; design of subsoil drains
Transportation network systems	Design of Intelligent Transportation System (ITS) facility locations; design of optimal routes for item distribution/collection; design of routes for transportation facility monitoring; design of emergency routing during disaster events
Construction management systems	Design of formwork system; design of equipment utilization strategy; design of material mixes
Solid waste management systems	Design of landfill systems; design of landfill operations system; design of optimal routes for item distribution/collection
Foundation systems for buildings, bridges, etc.	Design of foundation support systems
Physical systems at airports	Design or airport runway pavement system; design of terminal and parking facilities
Operations systems at airports	Design of airplane landing and take-off systems; design of passenger flow control systems (curbside drop-offs to boarding)
Physical civil structural systems such as bridges	Design of bridge systems, structural frames for residential, commercial, industrial, and recreational facilities
City (municipal) public works systems	Design of city drainage and sewerage system layouts
Hydraulic systems	Design of dimensions and orientations for levees, dams, weirs, channels, etc.
Traffic signal and control systems	Design of signal timings for urban arterials
Physical and operational systems at water and wastewater treatment plants	Design of treatment plant layout; design of plant operations
Highway and railway geometrics	Design of sections (curves and cross sections); design of intersection layout
City transit systems	Design of bus routes; design of bus or train schedules

to be treated, and the standards for the output product. Other systems that are designed by environmental engineers include incineration systems for waste disposal, landfill systems for solid waste disposal, and systems that monitor or to mitigate the pollution related to air, water, or noise.

21.3.2 Systems Design in Geotechnical Engineering

Geotechnical engineers design structures for slope or rock stabilization and earth retention, and foundations for civil structures. They also design geotextile configurations for a variety of geotechnical applications that include strength reinforcement, soil separation, waste containment, filtration, protection from the elements, or drainage. Design considerations often include surcharge loads, mechanical properties of the soil, constructability, maintainability, and cost. Geotechnical structures include pavement subgrades for airports and highways, foundations for buildings and other structures, retaining walls, embankments, soil stabilization systems, and earth levees, dams, and dikes. Geotechnical engineers design these systems to accommodate expected loading from a proposed structure and to repair or prevent damage to structures or the built-up environment due to natural threats (e.g., landslides, sinkholes, subsidence, soil liquefaction, rock falls, and flooding).

21.3.3 Systems Design in Construction Engineering

As we learned in Chapter 1, construction engineering involves the planning and management of the construction of architectural and civil engineering structures. The overall goals of construction project management relate to the following: (1) cost (complete the project within the cost budget), (2) time (complete the project within original contract period), (3) quality (complete all the work within the specified materials and workmanship quality), (4) safety (minimize the number of accidents), and (5) conflict minimization or resolution (Hanscher, 2003). To achieve these goals, construction engineers and managers design construction planning and scheduling systems, equipment utilization systems, optimal preventive maintenance schedules for equipment, and formwork systems, among others.

The design of a construction planning and scheduling system is critical to profit making in any construction company and, from the system owner's perspective, delivering the project in a timely manner. Design considerations include the activities to be carried out; and for each activity, the considerations include the minimum and maximum durations and the earliest and latest starting and ending times. The analytical tools available for design include operations research techniques such as network optimization or specific application platforms such as the critical path method.

The costs of construction equipment can represent as much as 30% of the total cost of a project so the proper use of equipment can lead to significant savings to the contractor. In view of this, designing an equipment utilization system for a project must be carried out with due diligence. This is a two-step process: Determine the equipment type needs and determine the optimal units and sizes of the selected equipment. In the first step, design considerations may include the project type, soil conditions, quantity and type of vegetation to be cleared, nature of the topography, local regulations, and project specifications. The mechanisms and tools for selecting the right mix of equipment include anecdotal sources and expert systems. In the second step, the design considerations may include equipment productivity, job size, equipment purchasing needs, and/or operating cost. Analytical tools that could enhance this design process include dynamic programming and integer programming.

Well-functioning equipment translates into higher productivity and contractor profitability. Therefore, construction engineers seek to design optimal scheduling systems for preventive maintenance of their construction equipment. Too little maintenance reduces spending in the short term but leads to poor equipment condition and performance and, hence, reduced profit in the long run. On the other hand, too much maintenance involves excessive spending even though it leads to excellent condition and performance and, hence, reduced overall profit. Somewhere in between these two extremes is an optimal level of equipment maintenance that guarantees good condition and performance at reasonable maintenance cost. Design considerations include the minimum standards for the equipment condition, the cost of each type of maintenance, and the benefits of each type of maintenance (e.g., improved equipment condition and extended service life) and maintenance budget constraint. The tools for this design task include engineering economics, discrete optimization, and stochastic simulation.

For construction formwork system design, functionality and cost are the primary criteria: the formwork must not be unduly costly but must be of adequate structural integrity to carry the loads imposed by the concrete and any workers, equipment, and materials associated with the work. Design considerations include the size, shape, and position of concrete elements; the desired finish quality of the concrete; the weight of the concrete; worker safety standards; the desirability for formwork recycling; and the strength and cost of formwork material.

21.3.4 Systems Design in Hydraulic Engineering

Hydraulic engineers not only plan and design overall regionwide urban drainage systems, water distribution networks, and urban sewerage network systems but also design specific structures that control hydraulic functions including dams, spillways, and outlet works for dams; hydraulic outlets; canals for drainage and irrigation; levees; energy dissipation structures and culverts for highways and railways; and cooling-water facilities for thermal power plants (Cassidy et al., 1998). Also, hydraulic engineers address the persistent problem of scouring and silt deposition in surface water bodies by designing systems that reduce sediment transport in rivers and minimize the interactions of water with the alluvial boundaries of rivers (Prasuhn, 1987).

21.3.5 Systems Design in Materials Engineering

Designers of civil engineering materials carry out a wide range of functions in materials design including specifying the types and percentages of materials to be mixed in order to satisfy a given set of performance requirements at the construction phase as well as at the operations phase of the system's life cycle. Examples include concrete (where cost, workability at the construction phase, and strength and durability at the operations phase are often key design considerations), stabilized aggregate (where cost, compactability, drainage, or binding ability are typically of interest), and alloys (where ductility, strength, cost, or corrosion resistance are vital).

21.3.6 Systems Design in Structural Engineering

Structural engineers design structural systems that include trusses, frames, domes beams, columns, and shells. In their designs, structural engineers select appropriate materials, dimensions, and configurations, and in doing so, they consider factors including the strength of the materials, live loads (traffic, pedestrians, occupants, wind), dead loads, expected stresses and failure mechanisms (bending, shear, torsion, etc.), durability, and cost.

21.3.7 Systems Design in Transportation Engineering

Transportation engineers design not only physical transportation systems but also the virtual systems associated with the operations of these physical systems. In air transportation, airport terminal and runway capacities are designed on the basis of passenger demand, wind direction, types and sizes of expected aircraft, and type of surrounding land use (to minimize noise impacts). In water transportation, port terminal capacities are designed on the basis of the freight demand and the types and sizes of expected vessels. In highway transportation, pavement thicknesses are designed on the basis of the natural soil strengths; the types and weights of expected vehicles; the highway capacity (number of lanes) is designed on the basis of the traffic volume and desired level of service; and highway geometry (grades, curves, cross-sectional features, and intersection layout) is designed on the basis of safety, vehicle dimensions, and road class. Designers of railway systems specify rail routes, track type and dimensions, track foundation materials and configurations, and vertical and horizontal curve features. Design considerations include traffic type, volume, terrain type, and required travel times.

21.4 CONSIDERATIONS IN CIVIL ENGINEERING SYSTEMS DESIGN

21.4.1 Ongoing Considerations in Civil Engineering Design

(a) Need for Goal-driven Designs. In developing a design in any branch of civil engineering, engineers consider design attributes including materials, dimensions, and configurations that are realistic, economical, environment friendly, and sustainable and that satisfy the functional or structural requirements of the system. Chapter 3 discusses a number of system goals that translate, directly or indirectly, into goals for the system designer. Also, in Chapters 27 and 28, respectively, we will discuss the goals of system resilience and sustainability.

(b) Reality of Design Constraints. Design, essentially, is a juggling act. The designer must navigate a maze of constraints that span an array of attributes or perspectives, such as material appropriateness, physical constraints posed by right-of-way limitations or neighboring natural and man-made systems, user convenience, and public opinion. In doing so, the designer searches for a solution that does not violate different design constraints that may be soft (quantitative boundaries) or hard (often qualitative restrictions). In civil engineering, soft constraints are provided to the designer in the form of design specifications that are numerical minimums or maximums or ranges associated with some physical or operational attribute of the design. Relatively harder constraints may exist in the form of environmental conditions that will affect the system; environmental impacts that are expected to be caused or exacerbated by the system as designed; the ergonomic or human factor requirements of the system; or the administrative, legal, financial, or economic constraints imposed on the system. For example, for a new dam, there could be height restrictions on the dam, a cap on the expected number of displaced persons, a minimum amount of power to be generated, and a range for the expected amount of water to be stored in the dam reservoir (not too little, as that situation could lead to water shortage in the reservoir; and not too much, as that situation could starve downstream populations of water). What makes the design process interesting is that certain constraints are related; and fulfillment of one constraint therefore may lead to the lower likelihood of fulfilling another. This is a challenge that is consistent with the second stage of the TRIZ design creativity enhancement technique that we discussed in Section 21.2.2(b).

(c) Practicality of Ergonomic Considerations in Design. In designing a civil engineering system that properly interfaces with its users, civil engineers consider the attributes of a sample of individual entities that represent the target user population. These attributes, which may be related to the system objectives, may pose constraints on the design. The average and standard deviations of these attributes, or their probability distributions, can be considered in the design. The attributes may include the heights, sizes, and weights of the system users. Also, where the system users are humans, the designer considers the abilities of the system users in terms of their walking speed, strength (to use any user-operated subcomponents of the system), hearing and vision (important for listening to or reading information regarding system use), and strength/endurance. For example, the increase in the geriatric population in many countries may mean that systems designed for public use must be fitted with support railings and bars where necessary and larger print for user operation or guidance signs. Also, some designers consider the aptitude of the prospective users in the acquiring, interpreting, learning, and processing of the information displayed for the benefit of the system user. Some researchers assert that ergonomic considerations in system design can help achieve or sustain the productivity, safety, health, and happiness of the intended users of the system.

(d) Role of Ethics in Design. According to the Accreditation Board for Engineering and Technology (ABET), the ethical, social, economic, and safety considerations in engineering practice are essential for a successful engineering career. In each design task, civil systems designers must realize the weight of the burden that is placed on their shoulders by society in terms of their ethical responsibilities to hold, first and foremost, the welfare of society above all things. Through their output, civil systems designers serve as a voice for the system users and the community, particularly the vulnerable segments of the general population, such as the disabled, the elderly, and the indigent, who often have no voice. At no other phase of civil systems development is the need for such ethical responsibilities so pronounced and obvious.

21.4.2 Emerging Design Considerations in the New Millennium

Engineering system designers help fabricate new and complex systems in response to the needs and desires of society. As such, a vital issue with engineering is adaptability. Social and economic changes constantly create new demands on engineers; it is therefore important that designers of future engineering systems cultivate the ability to make informed choices, basing their designs not only on analyses of the present situation but also on the vision of a preferred future. Engineers change the world and also are changed by the changes they cause. As a result, engineering is an extremely dynamic profession, continuously evolving in response to a changing world. From the systems perspective, Voland (2004) discusses a number of engineering design practices that continue to emerge in this new millennium, which we present below.

(a) The Need for Incorporating Flexibility. De Neufville and Scholtes (2011), in explicit recognition of the uncertainty of future outcomes, provided cogent arguments for the need for flexibility in design. At the phases of planning and preliminary design, the adoption of design flexibility helps prevent waste of resources associated with the "flaw of averages," fixed specifications, narrow forecasts. Also, by using the power of flexibility to intelligently manage risks, the designer can be empowered to accommodate the inevitable changes in the system economic and technical environments and thus to deliver significantly increased value that may otherwise remain untapped. Engineering system designers can thus identify, justify, and implement useful flexibility as part of their work at this phase of system development.

(b) Life-Cycle-Based Design. The lessons of engineering design in the last century have included an increased awareness that designs that are associated with minimal initial cost do not turn out to be the best designs in the long term. In other words, over the life of a system with low initial costs, the high cost of repairs and high user costs often lead to a higher overall cost compared to a system with high initial costs and lower repair and user costs. Thus, system designers increasingly assess their proposed designs from the perspective of both performance and costs over the entire system life cycle. This approach addresses considerations such as the extent of recycling the materials used in the system construction. Life-cycle design in civil engineering is championed by the International Association for Life Cycle Civil Engineering (IALCCE).

(c) Constructability and Maintainability. Poor design often leads to construction difficulties and ultimately, cost overruns and time delay. To avoid these problems and the costly public relations debacles that follow, system designers take great pains to ascertain that their designs can be adequately constructed using technology that is available at the time of construction. Also, they seek to design the civil system such that it can be maintained with minimal resources and difficulty over its life. Certain project delivery mechanisms, such as design–build projects, help reduce these problems because both phases of system design and construction are undertaken by the same contractual entity.

(d) Faster Design Cycles. One of the characteristics of the current engineering environment is that users expect quick resolutions of engineering problems; and civil engineers therefore are continually looking for ways to accelerate the processes of system planning, design, and construction. At the design phase, this need is being addressed using computer technology and concurrent engineering. Continuing advances in information technology reduce the time and effort in communication between design teams and between designers and persons in design-related positions. Also, enhanced computer-aided design (CAD) is allowing engineers to drastically reduce the length of design cycles. Concurrent engineering, where the different phases of engineering design and construction are carried out at the same time, such as design–build projects, has great potential to further reduce the design time.

(e) Reliability-based Design. With the current global emphasis on system sustainability, resilience, and reliability, designers of civil systems are increasingly considering how they could further infuse reliability concepts in their design processes. *Design for reliability* (DFR) is an emerging discipline that refers to the process of incorporating reliability into products at the design phase. First, the system engineer establishes the reliability requirements for the system and then develops a reliability model using block diagrams and fault trees that describe and evaluate graphically the relationships between the system components. In reliability modeling, the failure rates of individual system components are predicted on the basis of empirical observations of past similar systems (Neubeck, 2004). The predictions are not always exactly accurate but are valuable in assessing the relative reliability differences across the design alternatives. A key technique for reliability design is redundancy (i.e., if one part of the system fails, another part takes up its function). A bridge built with sufficient redundancies can stand if one member fails because another member takes up its load-bearing function. Incorporating redundancy in system design is desirable (because it greatly boosts the system reliability) but can be expensive; therefore, it is often recommended to incorporate redundancy in only the critical components of the system. Another design technique that enhances reliability is the *physics of failure* (POF) (Bukowski and Johnson, 1992); this design concept requires an understanding of the physical processes underlying various failure modes including fracture, fatigue, corrosion, and wear. POF involves the basic science concepts of stress and strain of materials and uses analytical tools including simulation and finite element analysis. In Chapter 13, the basic concepts of system reliability are presented, and system resilience is discussed in Chapter 27.

21.4.3 Matching System Performance Goals and Design Specifications

In a bid to satisfy the overarching performance goals of a proposed system, engineers often work within a given set of **design specifications**, or "specs." The requirement to work within these constraints introduces a soft, even artistic, dimension to engineering. Specifications can be expressed as a maximum, minimum, range, or even a probability distribution, and may be categorized as follows (Voland, 2004): **physical specifications** such as the dimensional requirements of the system (e.g., length, width, depth, thickness, area, space, weight, density, space, and time; e.g., the minimum density of a road bed to ensure that it is adequately compacted, the maximum bending moment in a structural member to avoid flexural failure, or the minimum and maximum slopes to prevent sedimentation and scouring for a sewer pipeline); **functional/operational specifications**, such as the sequence of work to carry out a task; **environmental specifications**, such as the maximum allowable noise emitted by the system; **economic/financial specifications**, such as the budget, minimum payback period, and maximum interest rate; and **ergonomic specifications**, such as the minimum strength of the human user to operate some aspect of the system, the level of education or intelligence necessary to read and follow user signs or to operate some aspects of the system, or even the physical dimensions of the system user, such as weight, height, girth, and age.

21.4.4 Engineering Design as a Cognitive Process

Cognitive processes include understanding specific language, logical thinking, and applying knowledge to problem solving, making decisions, and changing preferences in a particular domain of knowledge (Matlin, 2009). Civil systems today are designed by teams whose work is inherently cognitive in nature. These teams, working in a structured environment and following systematic processes, generally follow the design processes outlined in Figure 21.1, and ultimately produce a final design that serves as a solution to the stated need. In the 1950s, Benjamin Bloom developed a scheme for classifying cognitive skills. Known as **Bloom's taxonomy**, this scheme categorizes cognitive thinking and learning into six levels of complexity as shown in Figure 21.6: knowledge,

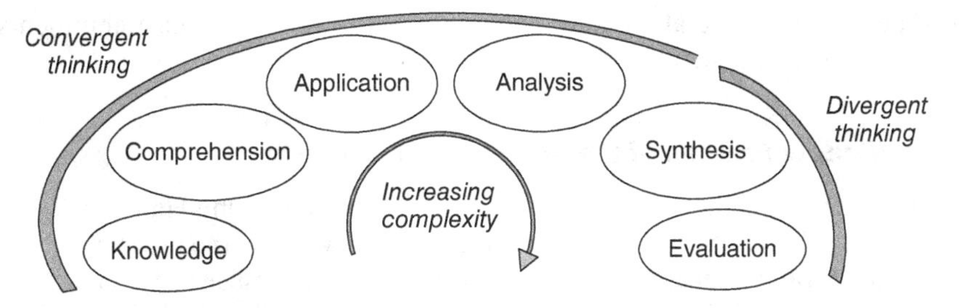

Figure 21.6 Bloom's taxonomy on cognitive learning.

comprehension, application, analysis, synthesis, and evaluation. The least complex cognitive ability is to recall facts, knowledge, or information; for example, what is the compressive strength of concrete? The next level is ability to understand information; for example, if the compressive strength of concrete is 25 N/mm^2, then it would likely fail when subjected to loads that exceed this threshold. The third cognitive level is application (i.e., the ability to use the knowledge acquired at a previous level to address a problem in a new situation); for example, given the strength of concrete, what will be the strength of a proposed concrete member? The fourth level is analytical ability, which consists of breaking down the learned material into individual component parts as a way to understand the subject at hand. This level includes identification of the individual components of the subject matter and the relationship not only between the components but also between each component and the entire sum of components. Analysis is the highest level of convergent thinking; for example, how do the individual sizes and strengths of different concrete members affect the entire strength and stability of the overall structure or can a certain proposed concrete member support the expected loads? The fifth and sixth levels reflect divergent thinking, which is the ability to build on the previous four steps to provide innovative ways of looking at the problem. Synthesis is the cognitive assembly of the components of existing knowledge to form a new design, concept, or idea. Evaluation, the highest level of cognitive thinking, refers to the ability to assess the value of the synthesized concepts and ideas from the perspective of established performance criteria. It is clear that Bloom's taxonomy is consistent with the engineering design process.

21.5 DESIGN FAILURES IN CIVIL ENGINEERING

An important part of civil engineering design is to learn from past failures of similar designs. A number of prominent failures in civil systems were due to errors or problems in planning, design, or the natural or man-made operations environment. We now will discuss a few of these failures that were mostly design related, particularly errors in design due to insufficient knowledge or inadequate testing.

21.5.1 The Mianus River Bridge Collapse

On June 28, 1983, a 100-ft section of Interstate Highway 95 over the Mianus River Bridge in Greenwich, Connecticut, fell into the river, causing the deaths of three people. It was later determined that the following two design flaws, among others, had caused the disaster (Robison, 1994): lack of redundancy in design (only two 9-ft-deep plate girders were used to form the bridge span, thus there was no backup support when one of these failed) and inadequate design to account for skew (the skew of the bridge along a diagonal between its west and east abutments possibly led to the loss of structural integrity).

21.5.2 Window Failures of the John Hancock Tower

In November, 1972, the glass panels of Boston's 60-story John Hancock Tower began to fall from the building's façade. Subsequent investigations showed that the problem was one of both structural and architectural engineering and included the following issues: (i) Wind loads were higher at the lower portion of the building compared to the upper section. The building designers, expecting the opposite to happen, had designed thinner windows at the lower floors; (ii) The epoxy material that connected the reflective coating to the inside of the outer light and to the lead spacer was too rigid, thus preventing the distribution and damping of wind-induced vibrations. Consequently, the outer light in a panel received the brunt of the loads, causing it to crack (Figure 21.7). Investigations of the flawed design of the building's glass façade led to another potentially more serious design error: It was found that the building was too flexible along its longer edge. Along its shorter edge, two tuned dynamic dampers had been installed to offset the building's movement. Each damper consisted of a 300-ton lead mass connected with springs to the structure and riding on a thin layer of oil at a location near the top of the building. As the tower moved in one direction or the other along the shorter dimension, the inertia of these huge masses resisted this motion and effectively damped the motion (Figure 21.7*b*). However, the designers had failed to consider that the weight of the structure would add to the effect of the wind, thereby effectively increasing the bending motion along the longer dimension.

21.5.3 Failures of Suspension Bridges

Between 1820 and 1890, ten suspension bridges in the United States, including the slender Tacoma Narrows Bridge, failed due to wind action. The lesson learned from these failures is that long narrow suspension bridges are vulnerable to aerodynamic instability (self-excitation due to flutter), and thus wind tunnel testing is necessary to produce more stable designs.

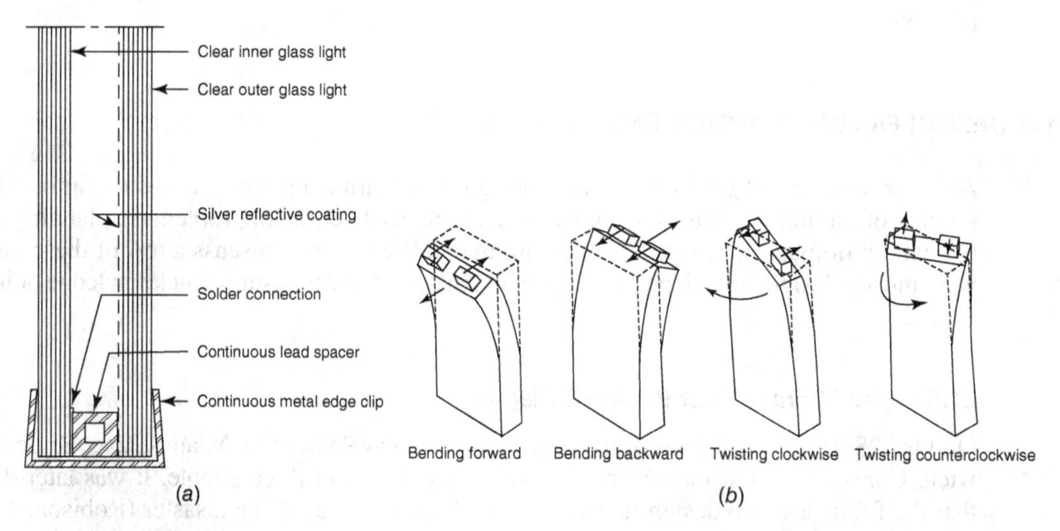

Figure 21.7 The John Hancock Building—design-related issues in 1972: (*a*) window design and (*b*) tuned dynamic dampers.

21.6 SOME DESIGN COMPUTATIONS

Example 21.1

A structural engineer seeks to ascertain that the loads experienced by a structure that she designed recently are the exact wind speed (57 units) as was predicted at the time of the design. Forty-nine readings from an anemometer yielded a mean wind speed of 57.9 units and a variance of 235 units. Using hypothesis testing at 10% level of significance, determine whether the wind speeds experienced are statistically equal to that predicted at the time of the design.

Solution

$$H_0 : \mu = 57 \text{ units}$$

$$H_1 : \mu \neq 57 \text{ units}$$

This is a two-tailed test because of the signs.

Assume a normal distribution, so we can use Z as our test statistic. The decision rule is that we reject the null unless the calculated value of the test statistic (Z*) falls within the rejection region. The level of significance, α is 0.1. Because the test is 2-tailed, $C = \alpha/2 = 0.1/2 = 0.05$. The critical value of the test statistic, $Z_C = Z_{a/2} = Z_{0.05} = 1.645$

$$\text{Standard deviation} = (23{,}500)^{0.5} = 153.3$$

Therefore, the calculated value of the test statistic, $Z^* = \dfrac{(579 - 570)}{(153.3/\sqrt{49})} = 0.410$

From Figure 21.8, it can be seen that Z^* falls outside the rejection region, so we do not reject the null hypothesis. In other words, there is no evidence that alternate hypothesis (that the mean is different from 57 lbs) is true, at the given level of confidence.

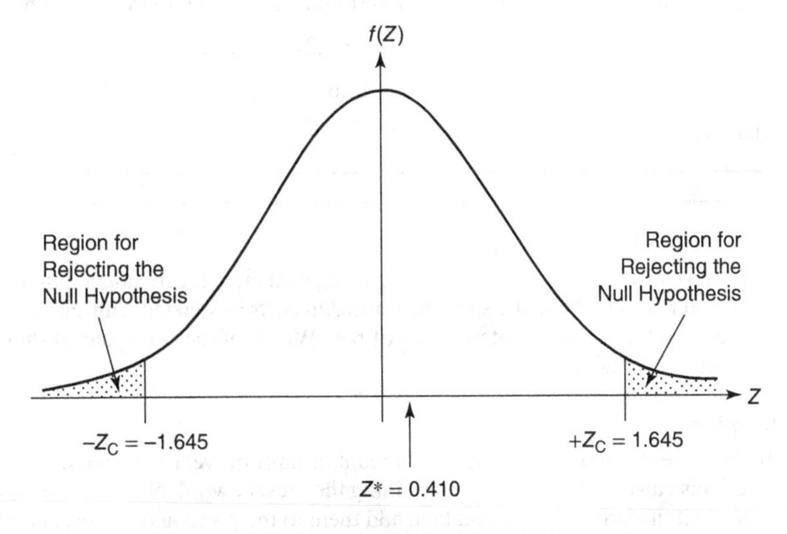

Figure 21.8 Hypothesis test for wind speeds.

Example 21.2

A materials engineer seeks to develop a mix for purposes of constructing a physical structure. The ingredients are materials P and Q and nonzero quantities of both must be used. The quantity of material Q must not exceed three times that of P; the quantity of material P must be at least 1 unit but not more than

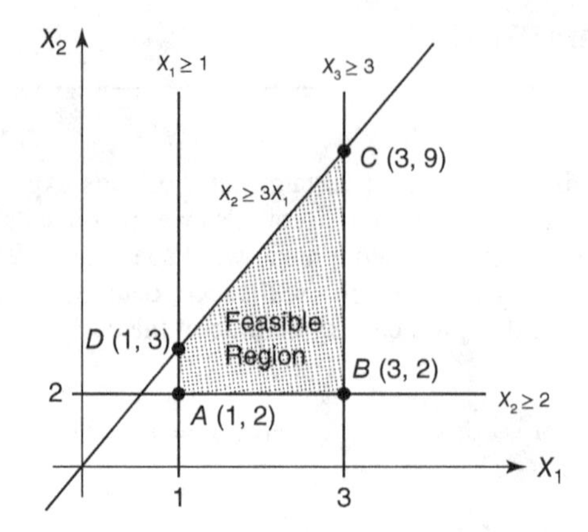

Figure 21.9 Linear programming for optimal mix of materials.

3 units; and the quantity of material Q must be 2 units or more. Let x_1 and x_2 represent the quantities of P and Q used, respectively, (i) Clearly show the feasible region on a rough sketch graph for the constraint set. (ii) Label all extreme points (or vertices) of the feasible region, indicating their coordinates. (iii) Does the inclusion or removal of the fourth and fifth constraints affect the feasible region? (iv) If we seek to maximize $W = 3.2x_1 + 2.5x_2$, solve the optimization problem using the coordinates of each point of the feasible region in your sketch graph. Check your answers using an appropriate software platform.

Solution

The feasible region on a rough sketch graph for the constraint set is provided below, showing all extreme points (or vertices) of the feasible region (Figure 21.9). The fourth and fifth constraints are redundant.

$$W = 3.2x_1 + 2.5x_2$$

$$W_{\mathrm{A}} = 8.2 \qquad W_{\mathrm{B}} = 14.6 \qquad W_{\mathrm{C}} = 32.1 \qquad W_D = 10.7$$

Thus $W_{\max} = 32.1; x_{1,\mathrm{opt}} = 3, x_{2,\mathrm{opt}} = 9$.

Example 21.3

Two alternative geotechnical designs are being evaluated for foundation treatment for a proposed structural system at a problematic site. The estimated costs associated with each alternative system are shown in Table 21.4. Assume an interest rate of 6%. Which of the two systems should be selected? State any relevant assumptions.

Solution

To compare the two alternatives, the amount of uniform yearly payments should be calculated for each case. This can be calculated by determining the present worth of the non-annual costs, determining their equivalent uniform amounts and then add them to the given annual costs (see Chapter 11 for equations and symbols).

For Design A:

$$\text{Initial cost} = 0.2(110{,}000) + 0.4(120{,}000) + 0.4(130{,}000) = \$122{,}000.$$

The present worth of the initial cost (I) and salvage (S) is given by

$$P_A = I - S = I - \left(\frac{1}{(1+i)^N}\right) = 122{,}000 - 20{,}000\left(\frac{1}{(1+0.05)^{15}}\right) = 112{,}380$$

Table 21.4 Data for Evaluating Alternative Geotechnical Designs

	Design A	**Design B**
Initial price ($) (Amounts shown with their corresponding probabilities)	110,000 (20%) 120,000 (40%) 130,000 (40%)	140,000 (20%) 150,000 (50%) 160,000 (30%)
Annual maintenance cost ($)	80,000	60,000
Salvage value ($)	20,000	40,000
Estimated life	15 years	20 years

The equivalent uniform annual cost corresponding to this amount is

$$EUAC_A = A' = P'\left(\frac{i(1+i)^{15}}{(1+i)^{15}-1}\right) = 112{,}380\left(\frac{0.05(1+0.05)^{15}}{(1+0.05)^{15}-1}\right) = 10{,}827$$

Therefore, the total annualized costs of Design A = 10,827 + 80,000 = $90,827.

For Design B:

$$\text{Initial cost} = 0.2(140{,}000) + 0.5(150{,}000) + 0.3(160{,}000) = \$151{,}000.$$

The present worth of the initial cost and salvage is given by

$$151{,}000 - 40{,}000\left(\frac{1}{(1+0.05)^{20})}\right) = 138{,}528$$

The equivalent uniform annual cost corresponding to this amount is

$$EUAC_B = 138{,}528\left(\frac{0.05(1+0.05)^{15}}{(1+0.05)^{15}-1}\right) = 13{,}250$$

Therefore, the total annualized costs of Design B = 13,250 + 60,000 = $73,250.

$EUAC_A > EUAC_B$. Clearly, design B has a lower annualized cost of its life cycle and thus should be selected.

Example 21.4

For purposes of designing of an underwater structure, an engineer seeks to determine the permeability of the soil at the site. She takes a soil sample from the site and places it in the laboratory permeameter (Figure 21.10). The water pressures on the upper flow and downstream sections are 2.4 m and 0.4 m, respectively. Also, the sample cross-sectional area and length are 0.08 m^2 and 0.12 m, respectively.

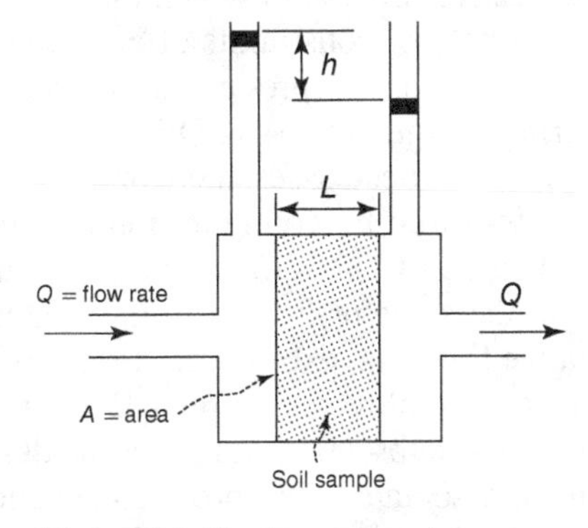

Figure 21.10 Permeameter setup.

The engineer makes several observations of the flow rate; the observations are found to be normally distributed with a mean of 1.8 m^3/day and standard deviation 0.2 m^3/day. Using Monte Carlo simulation, establish a probability distribution for the coefficient of permeability.

Solution

Sample calculation for the simulation:

We use Darcy's equation

$$K = Q/\left(A\frac{dh}{dL}\right)$$

The pressure drop is the difference between the upstream and downstream pressure, or $h = 2.4 - 0.4 = 2.0$ m.

Solving for K, we obtain

$$K = 1.8/\left(0.08\frac{2}{0.1}\right) = 1.35 \text{ m/day} = 1.56 \times 10^{-5} \text{ m/s}$$

From the Monte Carlo simulation, the mean permeability coefficient is 1.56×10^{-5} m/s, with a standard deviation of 0.17×10^{-5} m/s; the median permeability coefficient is 1.56×10^{-5} m/s. The distribution of the permeability coefficient is shown in Figure 21.11.

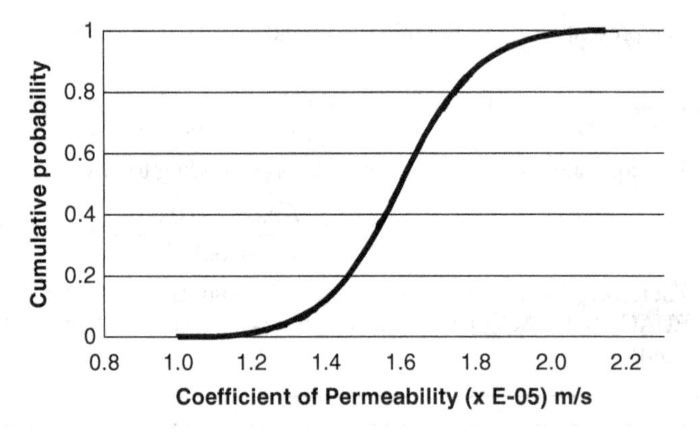

Figure 21.11 Probability distribution for permeability coefficient.

SUMMARY

Engineering design is a creative process in which the engineer, using scientific and creative skills and data, works within a variety of constraints often including material strengths, economy, legal and institutional policies, and space to produce a physical facility or service. In this chapter, we learned that the engineering design process (EDP) is a systematic endeavor that first receives output from the preceding phases of need assessment and planning and carries out physical design via abstraction and synthesis followed by configuration design, preliminary design, detailed design, and final design. The last step in EDP is to solicit and evaluate performance feedback for subsequent design revisions for the same or future similar systems. In this chapter, we learned how to minimize blocks to creative thinking, exploit relationships between natural phenomena and engineering system concepts, and employ design-related techniques including checklisting, synectics, lateral thinking, and TRIZ. The subsequent stages of the design phase, preliminary design, detailed design, and final design are associated with pronounced and visible application of domain knowledge in the branch of civil engineering associated with the system.

The classification of engineering design can help in identifying appropriate techniques to enhance the design process. This chapter discussed the Ullman Classification scheme, where any engineering design endeavor, on the basis of its level of transformation and creativity, can be described as a selection design, configuration design, parametric design, or an original design. The chapter also discussed a few applications of systems design in selected branches of civil engineering and presented a few ongoing and future considerations in civil engineering design. One of the most important of these considerations is the need to incorporate flexibility: in order to accommodate the inevitable uncertainty that characterizes civil engineering systems and their natural and built environments, design engineers must duly consider the nature and extent of such variability so that their designs can be less rigid, more holistic and sustainable, and more cost-effective. The chapter discussed engineering design as a cognitive process and then presented some prominent failures in engineering design. Finally, a few examples of design computations that involve systems-related concepts were presented.

EXERCISES

1. Discuss the similarities and differences between engineering design and basic science.
2. It has been argued that the design of civil engineering systems is as much an art as it is a science. Discuss both sides of this argument.
3. What are the benefits and limitations of having teams, instead of individuals, carry out the tasks associated with the phases of civil engineering design?
4. A new highway route that is currently being designed, between Cityville and another city is generating a great deal of controversy. You have been commissioned by the state to study the problem and provide your recommendations. Identify any five stakeholders in this system. Discuss how you would modify the design to ensure minimal adverse impacts.
5. You are asked to design a pedestrian footbridge over a busy street in a city's downtown area. What are some of the structural engineering considerations that need to be taken into account during the design of this system?
6. Discuss how you would go about the entire sequence of systems design and development for a proposed new waste treatment plant to replace an existing small and old plant for your hometown.
7. Consider the relief supply chain infrastructure for a main island in the event of flood disaster illustrated by Figure 21.12. Suppose that the probabilities that links 1 and 2 are destroyed in the event of a flood are 0.03 and 0.07, respectively. Find the probability that the supply chain is broken in the event of a flood disaster.

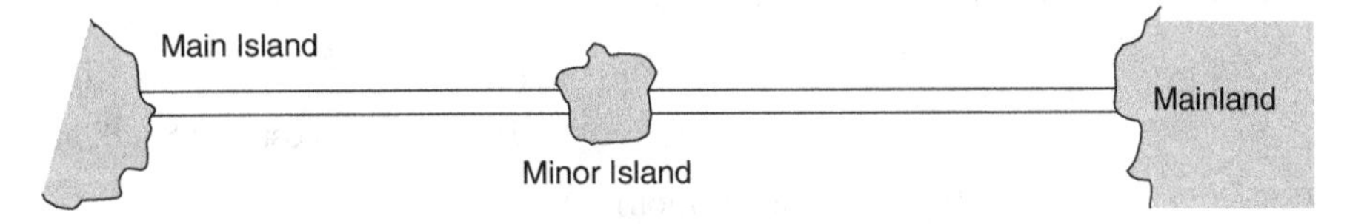

Figure 21.12 Figure for Exercise 21.7.

8. The floodplain in a region is designed for a 10-year flood (i.e., flooding is possible every 10 years). What is the probability that it will take exactly 3 years for the design flood level to be exceeded?
9. A construction engineering schedule system works perfectly with any three of the following component construction activities: A, B, C, D, E, F. (a) In how many ways can the system be configured to work perfectly if its components have to be arranged in a specific order? For example, ABC is not the same as CBA, so these two configurations are "counted" differently. (b) In how many ways can the system be configured to work perfectly if its components may be arranged in any order? For example, ABC is the same as CBA, so these two configurations count only as one.

10. A school wishes to construct an overhead water tank to serve a complex of newly built residence halls. Designs under consideration include a cube, a cylinder, a sphere, and a cone (Figure 21.13). Identify which design (i) has the most capacity, (ii) utilizes directly the most material, (iii) offers the highest benefit-to-cost ratio (maximum capacity while utilizing the least building material), (iv) has the maximum usable (flat) floor area. Which design would you recommend, and why? Besides the capacity and material use, which other design considerations influenced your decision?

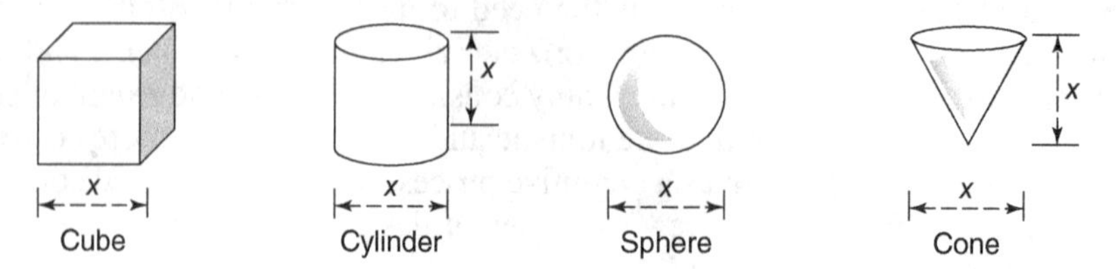

Figure 21.13 Alternative tank designs.

11. For a certain proposed structural system, x_1 and x_2 are nonzero quantities that represent the extents of two design inputs A and B, respectively. It is desired that the sum of 2 units of A and 4 units of B must not exceed 21 units; also, the sum of 5 units of A and 3 units of B must not exceed 18. Maximize the overall structural performance of the system, denoted as Z, where C is the sum of 3 units of A and 2 units of B. (a) On a graph sheet, clearly show the feasible region for the constraint set and label all vertices of the feasible region. Using an appropriate software platform, determine the optimal solution.

12. A small city seeks to develop a new landfill to serve a growing population. The available size of demarcated area is 15 acres, and due to groundwater pollution concerns, the maximum depth is only 40 ft. (a) What is the design life of the landfill? The parameters of the design problem are as follows: target population = 200,000; per capita daily waste generation = 4 lbs; density of waste = 600 lbs/sq. yds. (b) Assuming that the waste density follows a normal probability distribution with mean 600 and standard deviation 20 lb/yd^3; the per capita daily waste generation also follows a normal probability distribution with mean 4.0 and standard deviation 0.5 lb/yd^3, the amount of waste generated per capita is independent of the density of waste, determine the probability distribution of the design life of the landfill.

13. A structural engineer is evaluating three alternative designs on the basis of four performance measures: construction cost, structural integrity, the expected average lifetime condition (resistance to freeze and freeze-thaw cycles), and the effects of the bridge construction on the environment, each having relative importance weights of 0.3, 0.4, 0.2, and 0.1 respectively. For the overall structural integrity (OSI), the index is 0 (worst) to 5 (best); life-cycle cost is in \$millions; lifetime condition is an index ranging from 0 (worst) to 20 (best); and the environmental consequences (EC) is an index ranging from 0 (best) to 50 (worst). The utility functions U_j are provided as follows:

$$U_{\text{OSI}} = 16.33(\text{OSI})$$

$$U_{\text{cost}} = -20.898(\text{cost}) + 175.19 \text{ (where cost is in \$millions)}$$

$$U_{\text{condition}} = 5(\text{condition})$$

$$U_{\text{EC}} = -0.1235(\text{EC})^2 + 3.0745(\text{EC}) + 82.673$$

The levels of each performance criterion (PC) for material are given in the table below.

	OSI	Cost	Condition	Environmental
Design A	3	7.5	18	38
Design B	4	4.0	11	25
Design C	3	5.0	15	30

(a) Using the basic weighted additive method of multiple criteria evaluation, determine which design should be recommended for the new structure. (b) What must be the percentage reduction in Design A's cost for that alternative to be the optimal choice?

14. A number of alternative designs are being considered for an urban drainage system in a certain city. Alternative 1 has a 10-year design life and Alternative 2 has a 15-year design life. The cash flow for each alternative, j, involves an initial construction/installation cost, U_j; an annual stream of operating costs, E_j; an annual stream of maintenance costs, M_j; an annual stream of benefits (revenues), B_j; a rehabilitation cost every fifth year except at the end of design life, H_j; and a salvage cost, S_j, at the end of the design life (the analysis period is taken as the design life).
(a) Assume an interest rate, i. Write out the full expression for the equivalent uniform annual return (EUAR) for any alternative$j$. (b) On the basis of the EUAR criterion of economic evaluation, write the expression that represents the condition for which Alternative 1 is more economically attractive than Alternative 2.

15. Consider two design alternative each with a 10-year design life and the following cash flows: initial construction U_j; annual operating costs, E_j; annual maintenance costs, M_j; and salvage cost, S_j, at the end of the design life (analysis period). The alternatives are the same with only one exception: for Alternative 1, there is an amount (a benefit) in Year 1, while for Alternative 2 that amount (benefit) occurs in Year 9. (a) Which alternative is superior? (b) If the amounts were costs and not benefits, which alternative would be superior? (c) What conclusions can you draw from your answers to (a) and (b) about the influence of the timing of amount occurrence on economic evaluation.

16. A biological reactor (Figure 21.14) (with no solid recycle) must be operated so that an influent BOD of 600 mg/L is reduced to 10 mg/L. The kinetic constants have been found to be $k_s = 500$ mg/L and $\hat{\mu} = 4$ days^{-1}. If the flow is 3 m^3/day, how large should the reactor be? If the flow rate follows a normal probability distribution with mean of 3 m^3/day and standard deviation of 0.2 m^3/day, determine the probability distribution of the required tank size.

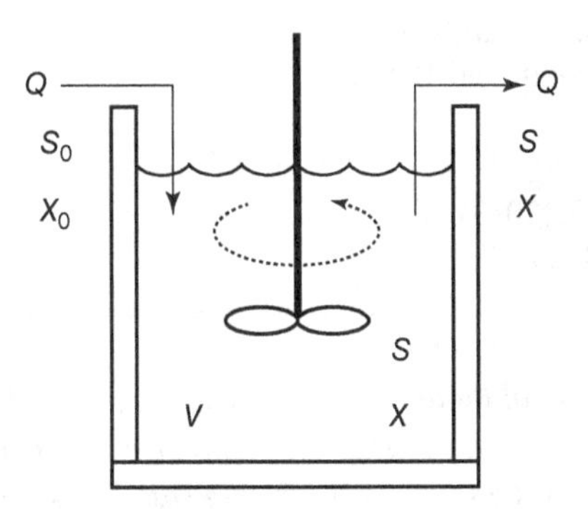

Figure 21.14 Biological reactor for wastewater treatment.

REFERENCES

ABET (2012). Criteria for Accrediting Engineering Programs, ABET, Engineering Accreditation Commission, Baltimore, MD.

Adams, J. L. (1986). *Conceptual Blockbusting: A Guide to Better Ideas*, 3rd ed. Addison-Wesley, Reading, MA.

Barry, K., Domb, E., and Slocum, M. S. (2006). Triz—What is TRIZ. *Triz Journal*. Real Innovation Network. http://www.triz-journal.com/archives/what_is_triz/. Retrieved March 15, 2013.

Bradley Fields, W. (2005). *Hickam's Dictum vs. Occam's Razor: A Case for Occam*. University of Michigan Health System, Ann Arbor, MI.

Bukowski, J. V., and Johnson, D. A. (1992). Software-Reliability Feedback: a Physics-of-Failure Approach. Procs., Reliability and Maintainability Symposium, pp. 285–289.

Cassidy, J. J., Chaudhry, M. H., and Roberson, J. A. (1998). *Hydraulic Engineering*. Wiley, New York.

Courtney, A., and Courtney, M. (2008). Comments on *On the Nature Of Science*. *Physics in Canada*, 64(3), 7–8.

Davis, G. A. (2004). *Creativity is Forever*. Kendall/Hunt Publishing, Dubuque, IA.

De Bono, E. (1973). *Lateral Thinking: Creativity Step by Step*. Harper Colophon, New York.

de Neufville, R., and Scholtes, S. (2011). *Flexibility in Engineering Design*, MIT Press, Cambridge, MA.

de Neufville, R., and Stafford, J. H. (1971). *Systems Analysis for Engineers and Managers*. McGraw-Hill, New York.

Dominick, P. G., Demel, J. T., Lawbaugh, W. M., Freuler, R. J., Kinzel, G. L., and Fromm, E. (2000). *Tools and Tactics of Design*. Wiley, New York.

Folger, H. S., and Leblanc, S. E. (1995). *Strategies for Creative Problem Solving*. PTR Prentice Hall, Upper Saddle River, NJ.

Hancher, D. (2003). Engineering Construction, *The Civil Engineering Handbook*, CRC Press, Boca Raton, FL.

Hilliard, A. A., Weinberger, S. E., Tierney, L. M. Jr., Midthun, D. E., and Saint, S. (2004). Occam's Razor vs. Saint's Triad. *New Eng. J. Med.* 350, 599–603.

Hyman, B. (2002). *Fundamentals of Engineering Design*. Prentice Hall, Upper Saddle River, NJ.

ICE (2011). *Little Book of Civilization*. Institution of Civil Engineers, UK.

Khisty, J. C., Mohammadi, J., and Amekudzi, A. A. (2012). *Fundamentals of Systems Engineering with Economics, Probability, and Statistics*. Prentice Hall, Upper Saddle River, NJ.

Levin, M. S. (2009). Combinatorial Optimization in System Configuration Design. *Autom. Remote Control* 70(3), 519–561.

Maloney, W. J. (2011). Occam's Razor and Hickam's Dictum: The Transformation of a Theoretical Discussion into a Modern and Revolutionary Tool in Oral Diagnostics. *WebmedCentral Dentistry* 2(5): WMC001914.

Matlin, M. (2009). *Cognition*. Wiley, Hoboken, NJ.

Mazzoleni, I., and Price, S. (2013). *Architecture Follows Nature—Biomimetic Principles for Innovative Design (Biomimetics)*. CRC Press, Boca Raton, FL.

Michelson, J. (2004). Critique if (Im)pure Reason: Evidence-based Medicine and Common Sense. *J. Evaluation Clin. Practice* 10(2), 157–161.

Neubeck, K. (2004). *Practical Reliability Analysis*. Prentice Hall, Upper Saddle River, NJ.

Prasuhn, A. (1987). *Fundamentals of Hydraulic Engineering*. Holt, Rinehart, and Winston, New York.

Pyzdek, T., and Keller, P. A. (2009). *The Six Sigma Handbook*, 3rd ed., McGraw-Hill, New York.

Robison, R. (1994). Mianus River Bridge Collapse, in *When Technology Fails*, N. Schlager, Ed. Gale Research, Detroit, MI.

Sloane, P. (2006). *The Leader's Guide to Lateral Thinking Skills: Unlocking the Creativity and Innovation in You and Your Team*, Kogan Page, London, UK.

Ullman, D. G. (1992). *The Mechanical Design Process*. McGraw-Hill, New York.

Vesilind, P. A., Morgan, S. M., and Heine, L. G. (2009). Introduction to Environmental Engineering, Cengage Learning, Stamford, CT.

Voland, G. (2004). *Engineering by Design*, 2nd ed. Prentice Hall, Upper Saddle River, NJ.

Wallace, M. (2000). The Science of Invention, www.salon.com, Accessed January 29, 2013.

Webb, A. (2002). TRIZ: An Inventive Approach to Invention. *Manufacturing Eng. August* 2002, 171–177.

USEFUL RESOURCES

Cross, N. (2008). *Engineering Design Methods: Strategies for Product Design*, Wiley, Chichester, England.

Dym, C., and Little, P. (2008). *Engineering Design: A Product-based Introduction*, Wiley, Hoboken, NJ.

French, M. J. (1985). *Conceptual Design for Engineers*, 2nd ed. Springer, New York.

Lumsdaine, E., and Lumsdaine, M. (1990). *Creative Problem Solving: An Introductory Course for Engineering Students*. McGraw-Hill, New York.

Pahl, G., Bietz, Feldhusen, J., Grote, K. H., Wallace, K., and Blessing, L. T. M. (2006). *Engineering Design: A Systematic Approach*, 3rd ed., Springer, New York.

CHAPTER 22

SYSTEMS CONSTRUCTION

22.0 INTRODUCTION

The fourth phase of civil systems development, construction, can be defined loosely as the process of "translating design into reality." The construction phase implements the system design for a physical or virtual system by providing and assembling its elements at a given location. Synonyms include the terms **realization**, **installation** (which is often used when the system is small compared to the overall parent system of systems, such as a road sign) and **implementation** (which is often used when the designed system is not physical but virtual, such as a new system of signal timings for an existing traffic signal). The construction phase is rather distinct from other phases of systems development in a number of ways. First, the construction phase is almost always given to a contractor to undertake rather than using the system owner's in-house manpower and equipment. Second, it is a complex and often time-consuming process that typically involves a very large workforce characterized by a variety of skills. Third, it is typically the phase where the system owner incurs the highest cost. Like other phases, however, engineers working in this phase have an obligation to provide the best possible product to the system owner while at the same time ensuring safety in the construction work zone (to the construction workers and community) and minimizing any adverse social, community, and economic impacts from the construction process.

Most construction projects require a wide diversity of skilled labor types; and work at this phase is often supervised by a design engineer, construction manager, project architect or construction engineer, depending on the type of civil system being built and the cultural practices of the system owner, and managed overall by a project manager. For this phase to be successful, a carefully designed construction plan is critical. The construction plan can differ significantly depending on the category of civil engineering system in question: building construction (building systems), heavy or civil construction (civil works such as dams, highways, and canals; see Figure 22.1), and industrial construction (civil systems meant to support directly other sectors, such as power transmission infrastructure, mining/petroleum infrastructure including mines and oil rigs, hospitals, and power plants).

In this chapter, we will first identify the key parameters used to measure the performance of a construction project, the parties involved in the activities at the construction phase of system development, and the various stages of the system construction phase. Then, we will review a number of different contracting approaches and how construction costs may be analyzed. The problem contexts at the construction phase that require application of systems concepts also will be discussed, and numerical examples of a few of these contexts that are directly related to the project performance parameters will be presented. Finally, we will identify and discuss a few emerging and evolving issues associated with the construction phase of civil systems development.

22.0.1 Measuring the Performance of a Construction Project

Besides the fact that the construction phase yields a physical realization of the civil system, the construction process itself can be considered a system in its own right because it consists of an

Figure 22.1 Construction work at Brazil's Itaipu Dam (Courtesy of Herr Stahlhoefer/ Wikimedia Commons).

assemblage of different components that come together to produce an outcome (the constructed facility), has performance objectives by which the attainment of the goal is measured, and has an inherent feedback mechanism that involves constant monitoring in a bid to enhance the quality of the finished product. These objectives include **cost** and **time** minimization (the project must be completed within the specified budget and contract period), **quality** maximization (all material quality and dimensional tolerances must be within or exceed the project specifications), **safety** maximization (there should be minimal construction-related injuries or loss of life of the construction workers and the community that are directly associated with or affected by the construction process), and **conflict** minimization (having as few interparty conflicts as possible). These construction performance objectives are important from the perspectives of both the system owner and the contractor. In subsequent sections of this chapter, we will further discuss some of these objectives.

22.0.2 Contractual Parties at the Construction Phase

Any construction process involves a number of distinct parties who must work together harmoniously in order to complete the construction process while maximizing the project performance

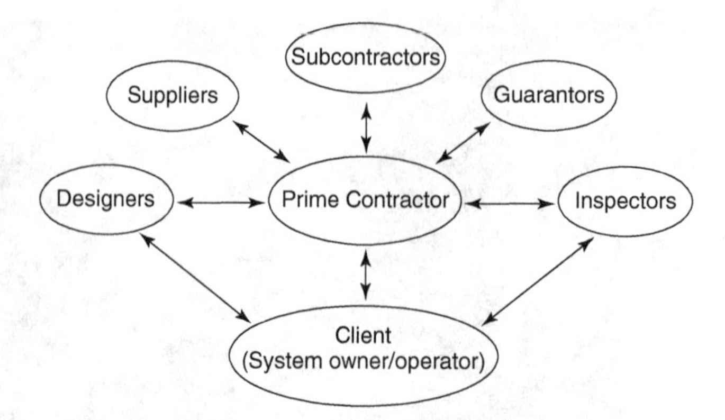

Figure 22.2 Key parties to a construction project.

goals within the constraints of time, money, and specified quality. The **system owner** or **operator** is typically the party that expresses the need for the project, bears the cost, and provides top-level oversight for the project; this party is often referred to as the **client**. The contractor provides and manages the materials, money, manpower, and machinery needed to execute the project successfully. In cases where this contractor is only one of a partnership, joint venture, or larger consortium of contractors (or subcontractors), the main contractor is also referred to as the **general contractor** or **prime contractor**. **Subcontractors** or **specialty contractors** are specialists in certain trades and/or materials who are contracted by the main contractor to execute a specific part of the project. **Suppliers** are vendors who provide the specified materials for the project. **Inspectors** are the owner-appointed parties who are onsite to ensure that the materials, tolerances, and workmanship are within the project's specifications. The **guarantor** is the financial institution that provides the performance bond for the contractor that assures the client that the work will be completed successfully. The **designer** is the party that communicates to the contractor the exact end product desired by the owner, through the technical drawings and specifications. The client is the glue that binds the contractual parties together; and this glue is reinforced by good communication and good relations between the parties, and prompt action by the relevant party where and when required. Without this glue, communication and coordination between the contractual parties is jeopardized, interparty relationships may become combative, resulting in increased risk of project cost and time overruns and substandard quality (Figure 22.2).

The above outlined entities are the typical contractual parties that make up the project team in most project types. The exact constitution of the project team for a given project will depend on the project delivery option and contracting approach used for the project (see Section 2.2). Other stakeholders include the banks that provide the working capital in the form of loans to the contractor, and parties that do not necessarily sign a contract directly related to the construction process such as governmental permitting agencies and the general public.

22.1 STAGES OF THE SYSTEM CONSTRUCTION PHASE

After the design is completed, the system owner's contract department reviews the design, selects the best delivery option and contracting approach to use to deliver the project (we discuss this further in Section 22.2). This office also carries out the contract administration, which is a multistage process that includes the selection of the best contractor to deliver the project. Then, the system

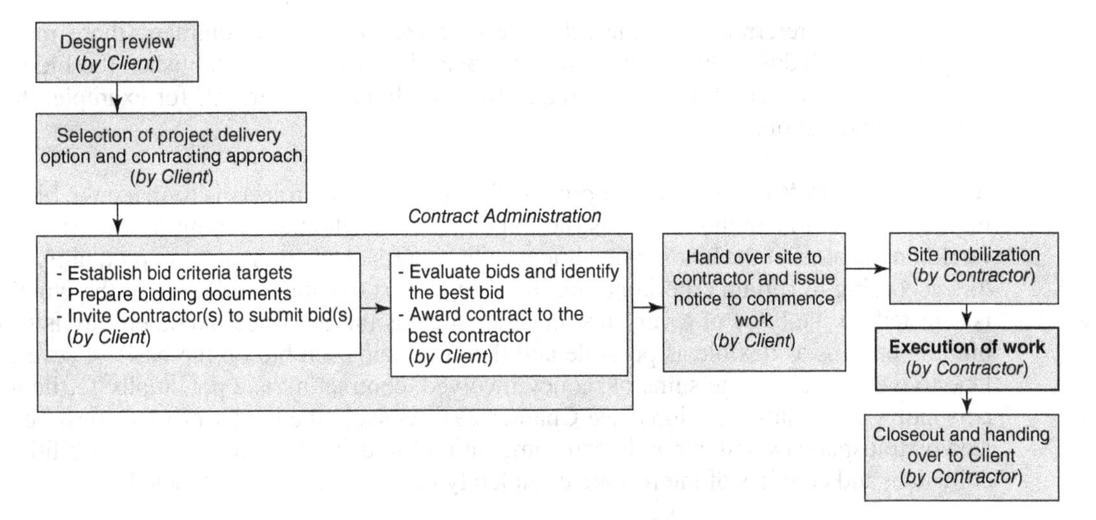

Figure 22.3 Key stages of the construction phase of civil system development.

owner hands the site over to the contractor. During the project execution, materials and testing personnel or consultants hired by the system owner are stationed onsite or visit the site regularly to carry out monitoring or supervision. In the sections below, we discuss the stages of the construction phase in greater detail (Figure 22.3). The described sequence of events is a general pattern of the stages and therefore may be somewhat different for specific projects.

22.1.1 Selection of Project Delivery Option and Contracting Approach

At this stage, the system owner decides on how to deliver the project: whether to use in-house resources or an outside contractor. If it is decided to use the former, then only mobilization and execution are carried out. If the latter is chosen, then another decision to be made is the choice of contracting approach. In Section 22.2, we present a number of contracting approaches.

22.1.2 Design Review

The input for the construction phase is the output of the design phase. The design phase typically produces the technical drawings and specifications. The first stage of the construction phase is to conduct a detailed constructability review of the design to ensure that it can be constructed with the available resources and also to plan a strategy for delivering the project and monitoring the construction process.

22.1.3 Contract Administration Stage

The administration of a contract includes the preparation of the bidding documents, establishing the bidding targets, distributing the invitation to bid, evaluation of the bids, selection of the best qualified bid, and awarding the contract. These steps may vary across the different contracting approaches and even across specific contracts.

(a) Preparation of Bidding Documents and Establishment of Bidding Targets. This step involves the processing of the technical drawings (carried out at the design phase of system development), the technical and general specifications, the contract conditions, and the bill of quantities (also referred to as the schedule of rates). In certain contracts, there may be other supplementary bidding documents. At this stage, the owner also establishes the bidding targets (i.e., the owner's desired number or range for each bidding criterion), for example, the contract amount and duration.

(b) Invitation to Bid. The typical process for construction projects is to make the bidding documents available to a number of construction contractors who then submit tenders (i.e., bid for the work) on the basis of the bidding documents. One approach at this stage is to prequalify all contractors according to certain criteria before they are invited to submit bids so that the winning bidder is selected on the basis of a very few criteria, such as the bid price. Another way is to leave the bidding process as flexible as possible and then evaluate each bid on the basis of several criteria. Due to the typically large sums of money involved, contracting is a potentially fertile ground for corruption and ethics violations (see Chapter 29). As such, the tender process must be characterized by transparency and fair and open competition for all bidders so that the possibilities of fraud, collusion, and conflicts of interest are completely minimized if not eliminated.

(c) Bid Evaluation. In response to the invitation to bid, a number of contractors submit their tenders either as a single lump sum for the entire project or for individual cost items (e.g., labor, material, and so on.). The bidders also submit evidence of their capability to deliver the project, including statements of the quantity and quality of their resources (manpower and equipment), company experience, and a performance bond. Bid evaluation is the process of comparing the submitted bids in order to select the best contractor to deliver the project. The criteria for evaluating the bids include the bid amount and period. For each evaluation criterion, the engineer weighs each submission not only against competing bids, but also against a preestablished value of the criterion known as the engineer's estimate. Due to the number of bid evaluation criteria, multiple-criteria analysis and optimization tools (Chapters 9 and 12) are helpful in identifying the best contractor from the perspective of the criteria. In certain situations, the system owner may eschew competitive bidding in favor of negotiated bidding. For example, in cases of very specialized or urgent work, the system owner may choose to invite a bid from only one contractor. In that case, the single bid is evaluated against only the engineer's estimate.

(d) Contract Award. After the best contractor is identified and notified, the contract award documents are prepared. These often include the bidding documents, bill of quantities general and specific conditions of contract, and the general and technical specifications. The contract award documents also include the required contract forms to be signed by the client and the contractor. The contract is then awarded to the contractor, but the contractor does not gain access to the site or commence work until the **notice to proceed** is received from the owner. In cases of a partnership, joint venture, or larger consortium of contractors, the main contractor also negotiates and signs contracts with subcontractors.

22.1.4 Contractor Mobilization

Once the contractor is granted access to the site, the necessary equipment, personnel, and materials are mobilized for the construction project. Often the contractor is paid advance money (often 5–10% of the contract sum) for site mobilization, which may include the purchase of material storage sites and a central worksite at or near the project site and the development of an area at the worksite for offices, storage areas for equipment, fueling stations for mobile equipment, and materials laboratory.

22.1.5 Execution of Work

Upon notice to proceed from the system owner, the contractor commences work and continues until the end of the agreed construction period. During the execution of the work, the owner monitors the cost, time duration, quality, and safety. If any of these variables falls below the expected levels, the contractor is asked to provide documented evidence of how they plan to promptly bring the situation under control.

22.1.6 Closeout and Handing Over

Upon completion of the project, the contractor turns over the finished project to the system owner. Both parties update their respective databases regarding the costs of pay items, durations of tasks and activities, and productivity rates. For the purposes of internal improvement on future projects, these numbers can be compared with those of similar past contracts and lessons can be learned by both the contractor and the owner.

22.2 PROJECT DELIVERY OPTIONS AND CONTRACTING APPROACHES

For a given project, the delivery option (in-house versus contracting) and the approach to be chosen under the contacting option depend on the extent to which the owner and the contractor will be involved in delivering the project. Most civil system owners refrain from restricting themselves to only one specific project delivery option; rather, any one of several available approaches could be selected on the basis of factors such as the project type (reconstruction projects are typically completed by contracts with outside contractors while maintenance projects are often carried out in-house by the system owner's personnel), size or cost (larger projects tend to have greater private participation), and the funding source (where the system is financed by a source different than the owner, it is often the case that the funding institution prefers to have greater nonowner participation in the project delivery).

As shown in Figure 22.4, the construction project may be delivered by 100% owner participation or by shared owner/nonowner participation (such as public–private partnerships). A 100% owner participation means that the project is delivered entirely in-house by the system owner's manpower and equipment, which is the case for certain types of civil engineering systems. Under the shared option, there could be different **contracting approaches** (the specific contractual arrangements made between the owner and the contractor to execute the work) and **contract types** (how the owner pays the contractor for performing the work) (Schexnayder and Mayo, 2004). The reader should note that different taxonomies are used in different countries and at different agencies; for

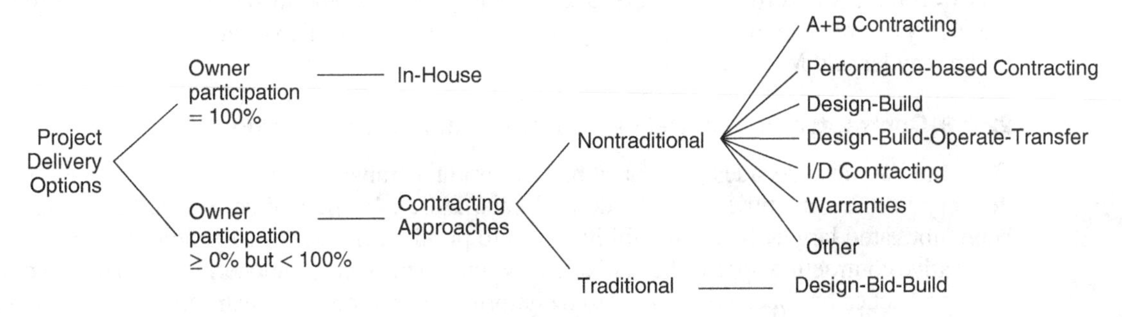

Figure 22.4 Project delivery options and contracting approaches.

example, the definition of project delivery option, contracting approach, and contracting type may differ from agency to agency.

22.2.1 In-house Delivery

In this approach, the work is completed not by private contractors but by the system owner or operator using in-house personnel, equipment, and materials on a force-account basis. In-house delivery may lead to saving money, at least in the short term. The impact on the construction quality is ambivalent: on one hand, in-house work (compared to contact work) may be of inferior quality due, often, to lack of in-house expertise for certain construction tasks, lack of technology, and inefficiency; on the other hand, in-house work may yield a superior product because in-house workers tend to have more familiarity and intimate knowledge of the system.

22.2.2 Contracted Project Delivery—Traditional Contracting Approaches

In most countries, a large number of projects that deliver the construction phase of civil engineering system development use the traditional **design–bid–build** contracting approach where a contractor is selected on the low-bid basis, and the work is done for a fixed price on the basis of the unit prices offered by the contractor. In this approach, the system construction phase is carried out by entities that are independent of those that carried out the design phase. As explained in Section 22.1.3, after the design phase, detailed contract documents (plans, specifications, schedule of rates, conditions of contract) are prepared and sent out for bid. The system owner compares the submitted bids on the basis of established criteria, and the contract is awarded to the best qualified bidder.

The advantages of traditional contracting include the minimal risk posed to the contractor because it specifies all the project details and implicitly absolves the contractor from most responsibility for adverse unforeseen site conditions (Carpenter et al., 2003; Segal et al., 2003). Traditional contracting has been found to be associated with a number of limitations (Hancher, 1999) including greater project delivery time duration and little flexibility. Another issue is that the contractor has little motivation to modify the construction processes in order to accelerate a specific task or to enhance the finished product quality. Also, the contractor has little incentive to adopt new technologies in construction materials, equipment, and processes. Such restrictions on contractor flexibility and the absence of risk or reward to the contractor often diminish the opportunity for project cost reduction. In Florida, it was determined that traditional contracting approaches tend to result in more cost overruns compared to other approaches (FDOT, 2000). Admittedly, certain sources of cost overruns, such as those due to inclement weather and acts of God, are unavoidable; however, cost overruns due to errors in design, planning, and specifications or project management problems are generally avoidable and could be minimized if more flexible contracting approaches are used. In spite of its limitations, traditional contracting remains to be used by many system owners because it is well understood by contractors and the general public and also because low-bid regulations are firmly established in the contracting guidelines of most public system owners and governments (Schexnayder and Mayo, 2004).

22.2.3 Contracted Project Delivery—Nontraditional Contracting Approaches

Over the past few decades, there has been a gradual transformation of the landscape of project delivery in many countries worldwide (Pakkala, 2002). In the United States, these changes have been motivated largely by the flexibility given to public infrastructure agencies to experiment with innovative contracting approaches on federally funded projects (Hancher, 1999). The use of innovative approaches has been determined to be helpful in reducing the construction time, and improved safety and productivity, ultimately reducing the adverse impacts of construction projects on the community and general public (Carpenter et al., 2003). Further, innovative contracting approaches can be more beneficial to the system owner, the community, and society in general by providing greater incentives and flexibility for the contractor to use technologies, materials, and techniques that accelerate the pace of work, yield superior quality products, and reduce noise impacts and other externalities. In the traditional approach, the system owner maintains a large in-house staff in order to closely inspect the construction process and monitor the condition of the finished product periodically in order to address any defects. By shifting some or all of these tasks to the contractor, the system owner can lower its staffing needs. In this section, we discuss a number of alternative contracting approaches typically used by systems owners for delivering their physical infrastructure construction projects. For each **contracting approach**, there can be different **contracting methods** (i.e., the contractual arrangements the owner makes with a contractor to procure a project) and **contract types** (i.e., how the owner pays the contractor for performing the work).

(a) Design–Build. In design–build contracts, unlike the traditional design–bid–build approach, the same contractor carries out both the design and the construction phases, and each entity has the same obligations as in the traditional approach; however, the separation of the design and construction functions are less visible to the system owner (Goodman and Hastak, 2007). Civil systems that are highly specialized in nature, such as nuclear, steel, and chemical plants, tend to be constructed using the design–build approach. This contracting approach arose out of owners desires to avoid being a referee in blame games between the designer and the contractor (Schexnayder and Mayo, 2004). Empirical and anecdotal evidence has shown that design–build contracts, compared to the traditional (design–bid–build) contracts, are less likely to be marked by rancor, cost overruns, time delays, and substandard work quality. Another key advantage of design–build contracts is its "single-point responsibility" (i.e., the contractor is responsible for every design flaw or construction problem) (Murdoch and Hughes, 2008). A disadvantage is that design–build contracts may lack independent checking of the designer's work, and the contractor also may choose to carry out work consistent with short-sighted cost reduction (profit maximization) at the expense of long-sighted sustainable design (Schexnayder and Mayo, 2004).

(b) Variations Involving Design, Build, Operate, and Maintain. In design–build approaches for contracting, the project is delivered by a single contractor or a multicontractor partnership with one prime contractor (McCullouch et al., 2009). Variations include design–build–maintain, design–build–operate–maintain (DBOM). Suitable candidates for the design–build approach include emergency projects where right-of-way, utilities, and environmental regulations issues have been resolved prior to the contracting stage and thus the designer-builder easily takes over the site and carries out the work without undue preconstruction delays. Unlike traditional contracting where the project can commence only when the design phase is fully complete, DBOM allows the project to commence at any stage of design completion between halfway to full completion; this overlap between the two phases often reduces the project duration significantly (Carpenter et al., 2003). Also, the contractor often provides input during the design phase, leading to greater opportunity for using innovative designs, and subsequently, fewer change orders, higher quality, lower costs, faster delivery, and higher safety benefits due to the shorter construction period (Ernzen et al., 2002).

(c) Multiparameter Bidding. In this contracting approach, the system owner considers each contractor's bid amount, proposed contract duration, and other bidding-related criteria, as a basis for evaluating each contractor's bid (Goodman and Hastak, 2007). Where there are only two criteria for the bid, it is referred to as **bi-criteria** or **A + B bidding** (where these criteria are cost and time, it is termed **cost-plus-time bidding**). To identify the most deserving contractor, the system owner applies multicriteria and optimization tools (Chapters 9 and 12). For example, for a two-criteria bid (e.g., cost in dollars and time in days), the evaluator can render the criteria commensurate in their units by converting the expected contract time into dollars by estimating the user cost associated with each day of the contract duration (in dollars/day) and multiply by the expected contract period for each bid. For each contractor, the owner determines the combined weighted amounts associated with the bid amount, construction time, and other parameters where applicable. The motivation for including multiple parameters stems from the desire to minimize the construction period for systems that have very high user traffic or where the service disruption has profound adverse consequences on the users and the community (Herbsman and Glagola, 1998).

(d) Cost-Plus-Fee Contracts. In cases of urgency where there is no time to prepare bidding documents (technical specifications, drawings, etc.), the system owner and the contractor can reach an agreement for the contractor to proceed with the work and the owner to make payments to the contractor as the work progresses. Thus, unlike the typical nature of most contracts, cost-plus-fee contracts are not fixed price but rather involve payment of all the qualifying expenses incurred by the contractor on the work in addition to the agreed fee and a percentage to cover the contractor's profit. The fee is a fixed amount or a fixed or variable percentage of the total cost of the work done (Goodman and Hastak, 2007). Cost-plus-fee contracts are also used in situations where the system owner and contractor have very limited knowledge of the work scope and the cost estimate for the project due to variability, uncertainty, or difficulty in site conditions such as desolate, rugged terrain or war zones. These contracts often contain incentives such as a bonus payment to the contractor if the project meets or exceeds a number of predetermined targets related to schedule, quality, or cost.

(e) Incentives/Disincentives (I/D). Incentives/disincentives contracts are intended to motivate the contractor to complete the project earlier than the time indicated in the contract award agreement by awarding a bonus if the contractor succeeds in doing so (Figure 22.5) and to penalize the contractor if the project duration exceeds the agreed date. The I/D amount is established as the product of the time delay in days and the unit cost of facility usage, in $/day (Arditi et al., 1997). I/D contracts are typically used in the construction of civil engineering systems that have very high public impact, high user volume, or for which the completion time is critical. A major advantage of the cost-plus-time and I/D contracting approaches over traditional approaches is a reduction in the project overall completion time, which is obviously due to the incentives given to the contractor for early completion as contractors strive to avoid payment of penalties in order to increase their profit and to maintain a good public image. I/D provisions in contracts encourage contractors to strive to control the conditions that affect productivity and project cost and duration (Bubshait, 2003). However, as Carpenter et al. (2003) pointed out, efforts to accelerate the project delivery can place an increased burden on the system owner; for example, the contractor's extended daytime work hours or night work may translate into extended hours for the owner's inspection and testing personnel. Also, significant administrative effort is needed for contracts with I/D provisions in order to ascertain when project targets have been reached (Jaraiedi et al., 1995). Also, there may be differences in the interpretations of an I/D contract from the perspectives of the contractor and the owner; and if these are not resolved, they could lead to conflict (Arditi et al., 1997). The question that ultimately arises pertains to what the incentive amount should be. Using a fixed amount or fixed percent of construction cost may lead to either overestimation (which could result in waste of taxpayers' money) or underestimation (which would likely reduce the efficacy of the incentive). The system owner may use the methodology developed by researchers including Shr and Chen (2004) or Sillars and Riedl (2007) to obtain a balanced estimate of the maximum or minimum incentive amount;

Figure 22.5 The 2007 reconstruction of the MacAuthur Maze ramp (Interstate 580), Emeryville, CA. The contractor earned a $5 million bonus for early completion (Courtesy of Wikimedia Commons).

this is a function of the expected construction cost and duration. A contract that contains both I/D provisions and A + B provisions is termed an A + B + I/D contract (Carpenter et al., 2003).

(f) Performance-Based Contracts (PBCs). Performance-based contracts, which are increasingly being used worldwide (World Bank, 2009), define the final product that is expected by the system owner. The contractor is responsible for the realization of the final product and is paid according to the extent to which they comply with the contract-specified performance standards and not specifically on the amount of work done. From the owner's perspective, a drawback of PBCs is that because payments and work done are decoupled, large sums that are independent of the work done during a given time period may need to be paid to the contractor. Also, in PBCs, the system owner retains the burden of having to deploy resources to monitor and measure the contractor's performance. PBCs shift the burden of product quality risk to the contractor (Zietlow, 2005). In cases of subpar performance of the contractor, it may be contractually difficult to "catch up" or to bring in another contractor to address the deficiency. An advantage is that PBCs offer the contractor increased opportunities for higher profit margins due to the flexibility to implement cost-saving processes, technology, or management techniques. Also, the system owner and users can potentially benefit from higher quality product and, subsequently, a lower postconstruction maintenance intensity and frequency; this is because the contractor is incentivized to use innovative practices that are likely to produce a superior product from the onset. Other advantages of PBCs include the transfer of detailed planning, programming, and budgeting functions and the performance attainment risks for the civil engineering system to the contractor (Anderson, 2000; Anastasopoulos et al., 2010).

(g) Warranties. A warranty assures that the finished product will be of such quality that it will provide a specified minimum level of performance and that any deficiency identified within a specified period will cause the product to be replaced or repaired by the contractor at the contractor's own expense (Aschenbrener and DeDios, 2001). Therefore, the contractor is responsible for the quality of the finished product and is liable for any deficiencies resulting from poor quality of workmanship or materials. Warranty practices may yield significant savings to the system owner in the form of reduced or removed responsibility for the product maintenance subsequent to the construction. Also, warranty contracts tend to encourage contractor innovation that typically yields products with lower overall life-cycle costs. Research findings suggest that compared to traditionally-contracted projects, warranty projects can lead to superior product quality and durability and are more cost-effective in the long run (Singh et al., 2007). In considering the use of warranty contracts, system owners need to bear in mind that they may lose in-house expertise as they reduce their participation in the project delivery process. Because most warranties cover premature failure only, there is increasing advocacy for warranty contract clauses to cover a wider range of product quality tests related to the longevity of the finished product. Finally, insurance companies tend to be wary of issuing performance bonds to contractors for warranty contracts due to the higher risk involved (Carpenter et al., 2003). With the increasingly available data from warranty projects, opportunities exist to use systems-related tools to assess the cost-effectiveness of construction warranty projects. Such efforts will need to incorporate project cost, quality, and longevity.

22.3 CONSTRUCTION COST ANALYSIS

The control of costs is probably the most critical aspect of the construction phase of civil engineering system development. Cost estimation, cost accounting, and all other actions geared toward the management and control of project costs are collectively termed **cost engineering**. Cost engineering continues to be a key aspect of construction engineering and management for at least two reasons. First, as we learned in Chapter 1, the construction phase is now dominated by the private sector, which is motivated by profit to seek reliable predictions of project costs, submit realistic bid amounts, and reduce waste during construction. Second, for publicly owned civil systems, the system operators (often, the branch of government referred to as the agency) seeks to demonstrate its prudent and judicious use of taxpayer money and thus take pains to ensure that all project costs are not only predicted reliably but also are kept under control during the construction of the system. Engineers at the construction phase are responsible for identifying cost factors and monitoring and controlling the project cost at each stage of the construction. In this section, we discuss a number of cost factors and the aggregation levels of cost estimation.

22.3.1 Project and Contract-related Factors That Influence Construction Cost

In the sections below, we discuss generally the cost factors that are related to the project (the environment and the system under construction) and the contract. The factors related to the contractor and the design are discussed in Chapter 10.

(a) Interruptions to or from the Natural or Built-up Environment. Projects in environmentally sensitive areas generally have higher costs because the contractor working in such environments would be required to adopt precautionary or mitigation measures, which are often expensive, to protect the surrounding areas from environmental degradation that the system would cause.

(b) Existing Soil and Site Conditions. Greater variability in soil and site conditions translates to higher costs because the design is often carried out for certain target site conditions. A low target may lead to change orders during construction when worse-than-target conditions are encountered during the work. Also, a high target may lead to waste and overdesign particularly in areas where better-than-target conditions are encountered during the work. The project manager or resident engineer changes the design to suit specific site conditions as and where necessary to ensure an effective product or to avoid incurrence of unnecessary costs.

(c) Project Size. Larger projects generally have lower unit costs than smaller projects due to scale economies. For certain kinds of civil structures, however, the contrary is true: Beyond a certain size, the structure would need additional structural or functional components and may translate to a greater cost increase per unit increase in size, thus reflecting scale diseconomies. The degree of scale economies may be different for the linear dimension compared to the breadth or depth dimensions; for example, a 100% increase in length may be accompanied by an 80% increase in cost while a 100% increase in depth may be accompanied by a 60% increase in cost.

(d) Project Complexity. More complex projects typically have higher unit costs because complex projects often have subcontractors, and the communication lapse between the additional contractual parties may contribute to time delays and cost overruns. Construction projects of a complex nature are particularly vulnerable to **tipping points**, defined by Taylor and Ford (2006) as "conditions that, when crossed, cause system behaviors to radically change performance." Tipping point dynamics are often helpful in identifying the failure of certain types of civil engineering construction projects.

(e) Contracting Approach. Projects constructed using traditional contracting approaches generally have lower unit costs of construction compared to those constructed using newer, innovative approaches such as design–build and warranties. However, products delivered using the traditional contracting approach generally have relatively higher unit costs of preservation over the product life cycle (Singh et al., 2007).

(f) Urban/Rural Location. Urban projects generally have higher unit costs compared to their rural counterparts because urban locations are often characterized by relocation of utilities (and the delay encountered in relocation due to poor or nonexistent records of their exact locations). Also, in urban projects, the contractor often needs to carry out measures (install barriers and special formwork or hire safety personnel) to protect workers from urban traffic and also to protect urban residents from construction hazards.

(g) Discussion. Other factors that may affect project costs are the regional or national economic environment (including the prices of basic resources), the degree of competition for contracts, and the design standards, labor costs, specifications for materials, and topographic conditions. Specifying products at a higher standard (e.g., smoothness of finished product surfaces) can cause higher costs for the finished product. For the above reasons, comparing the construction costs across different regions or using data from different regions for cost modeling should be done with extreme caution.

The contractor can influence the construction cost of a project, but this influence diminishes rapidly with the progress of the project [Figure 22.6; adapted from Hendrickson and Au (1989)]: At the beginning of the contract duration, the contractor has the greatest capability to influence costs. Assumptions made by the system designers regarding the site conditions will appear in the contract documents and will influence not only the materials specified in the contract documents but also the contractor's choice of equipment. If the actual site conditions differ from what was assumed, then

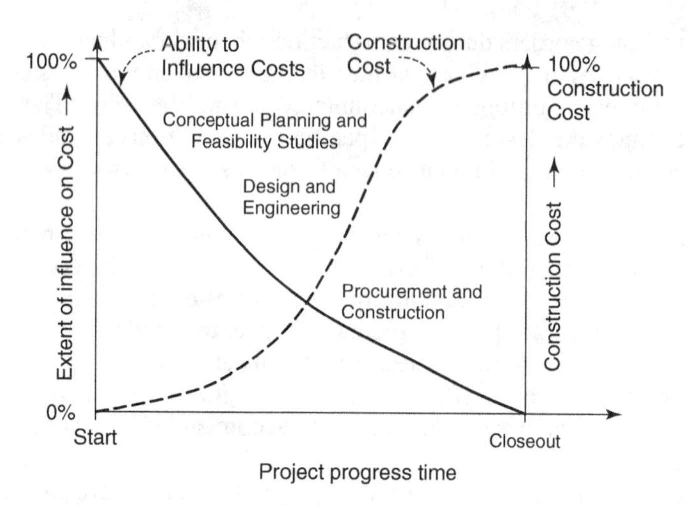

Figure 22.6 Relationship between contract progression and cost–influence capability.

the design will represent a departure from cost-effective practices. Iseley and Gokhale (1997) cited the following example: Sewer projects are often designed on the basis of traditional specifications, equipment and materials; however, advanced materials, construction techniques, and equipment exist that may be safer and more acceptable from environmental and social viewpoints, and more cost-effective, particularly in urban areas where conditions are generally tighter and thus there is great potential benefit for adopting modern construction practices. These include microtunneling and new sewer pipe materials such as glass fiber-reinforced polymers.

22.3.2 Levels of Aggregation

There are two extremes of cost item aggregation for cost estimation: aggregate and disaggregate. Between these extremes, cost estimators typically refer to distinct levels of coarseness that reflect the phase of system development at which such estimates are often required (Table 22.1).

(a) Disaggregate Level (Costing Using Prices of Individual Pay Items). At this level of aggregation, the first step in cost estimation is to decompose each work activity into constituent pay items expressed in terms of a standard dimension (such as linear foot, square foot, or cubic foot.) of the finished product or in terms of specific quantities of materials (such as aggregates, concrete, steel beams, and formwork), equipment, and labor needed to produce the standard dimensions of the finished product. Thus, the pay items are either priced in dollars per length, area, or volume or the weight of the finished product, reported separately for materials or priced per extent of labor and supervision and equipment use. After the various pay items that constitute the work activity have been identified, a unit price is assigned to them (often on the basis of updated historical contract averages or the engineer's estimates), and then the total cost of the work activity is determined by summing up the costs of the constituent pay items. The final output of cost estimation at this level forms the basis for contract bidding. For any project, there typically are thousands of pay items that must be priced separately; thus cost estimation at this level is laborious. This method of costing, which is typically carried out by experienced estimators, is more appropriate for projects that have passed the design phase and for which specific quantities of individuals pay items are known. It is generally not appropriate for cost estimation at the planning phase where the project's design details are not yet known.

Table 22.1 Levels of Aggregation in Cost Estimation

Level of Aggregation	Phase of Development	Description
Very aggregate (very course estimates)	Conceptual estimate in the planning stage	Typically referred to as predesign estimate or approximate estimate, this is a very aggregate estimate of the project cost. At the early planning stages, relatively little is known about the prospective design; therefore the level of identifying the pay items and their costing thereof is very coarse (i.e., aggregate) (Wohl and Hendrickson, 1984).
Semiaggregate	Preliminary estimate in the design stage	Often termed budget estimate or definitive estimate, this is a semiaggregate estimate or hybrid-level estimate that is used when planning level information is supplemented by minimal details about the site conditions or design.
Disaggregate (fine estimate)	Detailed estimate for contract tender or award	This is the estimate determined as the sum of the costs of the individual cost items (materials, labor, equipment use, etc.) and thus is a very disaggregate estimate.
Disaggregate (fine estimate)	As-built cost estimate	This is similar to the detailed estimate for the contract award and is also very disaggregate. It is superior to the contract award estimate because it represents the final, actual cost of the projects, including any cost overruns or underruns.

(b) Aggregate Level (Costing Using Overall Production Output). Cost estimation at this level typically utilize models that estimate the overall cost of constructing a unit dimension of the finished product (e.g., new bridge cost in \$/ft^2 of deck area; residential building in \$/ft^2 of floor area). As we discussed in Chapter 20, this level of cost estimation is typically used at the planning phase of civil systems development (where relatively little is known about the details of the system).

(c) Hybrid of Aggregate and Disaggregate. In certain cases, a hybrid level that is based on output production as well as any known details about the design or site conditions is used to estimate the cost during the planning or design phases. This level has been used to provide approximate estimates of road tunnel construction costs in Greece (Petroutsatou et al., 2012).

It is important to note that at any level of aggregation, the required cost estimate could be expressed as an average number or as a statistical model. For example, from the aggregate perspective, using records of historical contract and cost data, we could determine the simple average cost of a freeway (say, \$1.5 million/lane-mile) or we could develop a cost model that yields the total cost of a freeway pavement for a given contract as shown below:

$$F = f(X_1, X_2, \ldots, X_n)$$

where F is the total cost of the freeway; f is a mathematical function; and $X_1, X_2, \ldots, X_n$ are the values of the factors that influence the cost, for example, contract length, miles of road, number of lanes, urban–rural location, and asphalt type. Note that the differential of this function yields the cost per lane-mile, which can be considered a more reliable cost estimate than just the average cost.

From the disaggregate perspective, using records of historical contract and cost data, we could find the average cost of a ton of asphalt (say, \$130/ton), or we could develop a cost model that yields the total cost of the asphalt material for a given job as shown below:

$$G = g(Y_1, Y_2, \ldots, Y_n)$$

where G is the total cost of a ton of asphalt; g is a mathematical function; and $G_1, G_2, \ldots, G_n$ are the values of the factors that influence the cost of asphalt material, for example, the global price of crude, the quantity (tons) of crude in the contract, and the asphalt type. Note that the differential of this function yields the cost per ton of asphalt material, which can be considered a more reliable cost estimate than just the average cost.

Construction cost models have been developed using a variety of analytical techniques including statistical regression and neural networks (Bode, 1998; Hegazy and Ayed, 1998; Al-Tabtabai et al., 1999).

22.4 GENERAL DECISION CONTEXTS IN CONSTRUCTION THAT MERIT APPLICATION OF SYSTEMS CONCEPTS

The basic responsibilities of engineers at the construction phase include contractor selection, cost engineering, project planning and scheduling, equipment planning and management, design of temporary structures, contract management, human resource management, project safety, and risk analysis (Hancher, 2003). Some of these contexts are the responsibility of the system owner and others are the responsibility of the contractor. In the sections below, we present a few of the contexts associated with these responsibilities, and we discuss how the tools we have learned in Part 2 of this text could help engineers carry out these functions in these contexts. Other contexts include selection of the appropriate contracting approach or project delivery mechanism; prediction of the likelihood that a construction contractor will default, file for bankruptcy, or enter litigation on the basis of the characteristics of the project or the site, the contract language, or the contractor's past experience; and decision support for selecting a subcontractor.

22.4.1 Cost Engineering

As we discussed in Section 22.1(a), cost engineering involves cost estimation, cost accounting, and the development of actions to help control project costs. Engineers at the construction phase typically use systems tools (including model development) to estimate the overall project cost at an aggregate level or to predict the cost of each work item at a disaggregate level such as the asphalt concrete price per ton. Also, engineers monitor regularly the progress of the construction to quickly identify and address unexpected issues that may lead to change orders that significantly increase the construction cost. Further, for the purposes of incorporation into their cost prediction models, engineers duly identify the various factors that may influence the final cost of their project. Using various system-related tools, aggregate or disaggregate costs have been used for estimation of various civil engineering projects worldwide such as a tunnel in Greece (Petroutsatou et al., 2012).

Example 22.1

The historical prices of asphalt paid to contractors for projects in a certain province are presented for different project costs as shown below. For a project planned for next year, it is expected that 800 tons of asphalt will be used. If the agency's engineers wish to base their engineers estimate (EE) for the planned project on the historical prices of the similar project types, should the average be used as the EE for the asphalt price?

Tons (1000s)	1.1	2.5	0.7	1.2	4.3	2	0.6	2.7	1.38	0.9	1.4	3.2	0.5	3.8	1.6
Cost ($/ton)	115	98	120	127	84	95	128	89	108	119	102	89	139	86	97

Solution

The average is \$106.4 per ton. Figure 22.7 presents a plot of the data showing the relationship between the amount of asphalt used (in 1000s of tons) and the unit cost (\$/ton). It can be seen that the unit cost (y) drops as the amount of asphalt (x) increases.

The cost of asphalt for 800 tons is $115.72(0.8)^{-0.229} = \$121.79/\text{ton}$.

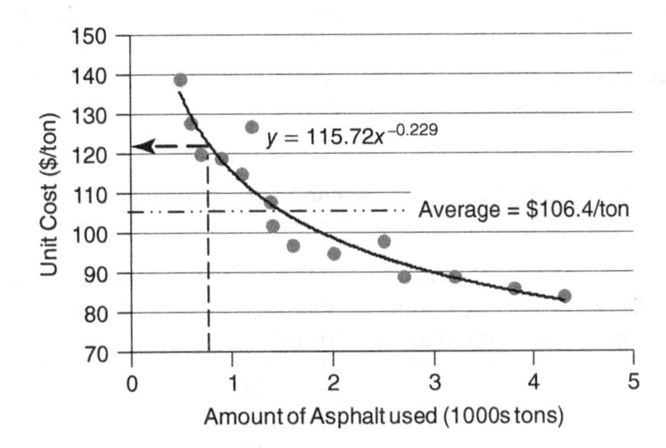

Figure 22.7 Figure for Example 22.1.

Thus, by using the overall historical average (\$106.4/ton) instead of the more appropriate empirical value of \$121.79/ton, the price of each ton would be underestimated by 12.6% (121.76–106.6)/121.76).

22.4.2 Selection of Contracting Approach or Project Delivery Mechanism

As we discussed in Section 22.2, there are a variety of contracting approaches, each with its merits, limitations, and circumstances of appropriate application. For example, the choice of delivering a project in-house with the system owner's resources or by traditional design–bid–build with a private contractor or some innovative contracting approach will depend on the size and complexity of the system, the capability of the system's manpower, the time constraints for project initiation or contract duration, and other factors. Valuable references in the context include Love et al. (1998) who used system-related tools to develop a framework for selecting a suitable procurement method for a building system and Choi et al. (2012) compared the schedule effectiveness of alternative contracting strategies for transportation infrastructure improvement projects. Also, Anastasopolous et al. (2010b) assessed the relative benefits of different contracting approaches and Zhang (2006) presented a number of factors that could be considered in best-value analysis of contracts that involve public–private partnership options.

22.4.3 Contractor Selection

Selecting the right contractor to deliver a construction project is one of the most critical challenges faced by a system owner. Often this choice needs to be made not only on the basis of the lowest cost but also using a variety of criteria (e.g., the proposed construction period, equipment and manpower resources, and past appropriate experience). These criteria have different units and also have different degrees of relative importance for different system owners. The tools of multiple-criteria analysis (see Chapter 12) can be used to address this decision problem. Various types of multiple-criteria decision making and other analytical tools have been applied in this concept for contractor selection worldwide (El-Kashif et al., 2000; Chaovalitwongse et al., 2012).

Example 22.2

Four contractors submitted bids for a portfolio of work to overhaul a city's aging municipal infrastructure. You are evaluating their bids on the basis of four criteria for bid evaluation with the following weights: bid amount: 0.28; proposed construction period: 0.24; contractor's experience: 0.27; and contractor's resource strength: 0.21. From the submitted bids, the levels of each decision criteria for each contractor were determined and tabulated as shown in Table 22.2. Which contractor would you select for the contract? Figure 22.8 presents the engineer's utility functions for the four evaluation criteria, for this project, from the engineer's perspective.

Table 22.2 Evaluation Criteria Values for each Contractor

Criteria/ Contractor	Bid Amount ($millions)	Proposed Construction Period (months)	Contractor's Experience (years)	Contractor's Resource Strength (1–10 scale)
Apex, Inc.	7.1	22	9	5
A-1 Ltd.	6.8	36	15	8
Zenith Co.	5.9	29	5	6
Haute Bros.	6.5	26	11	9

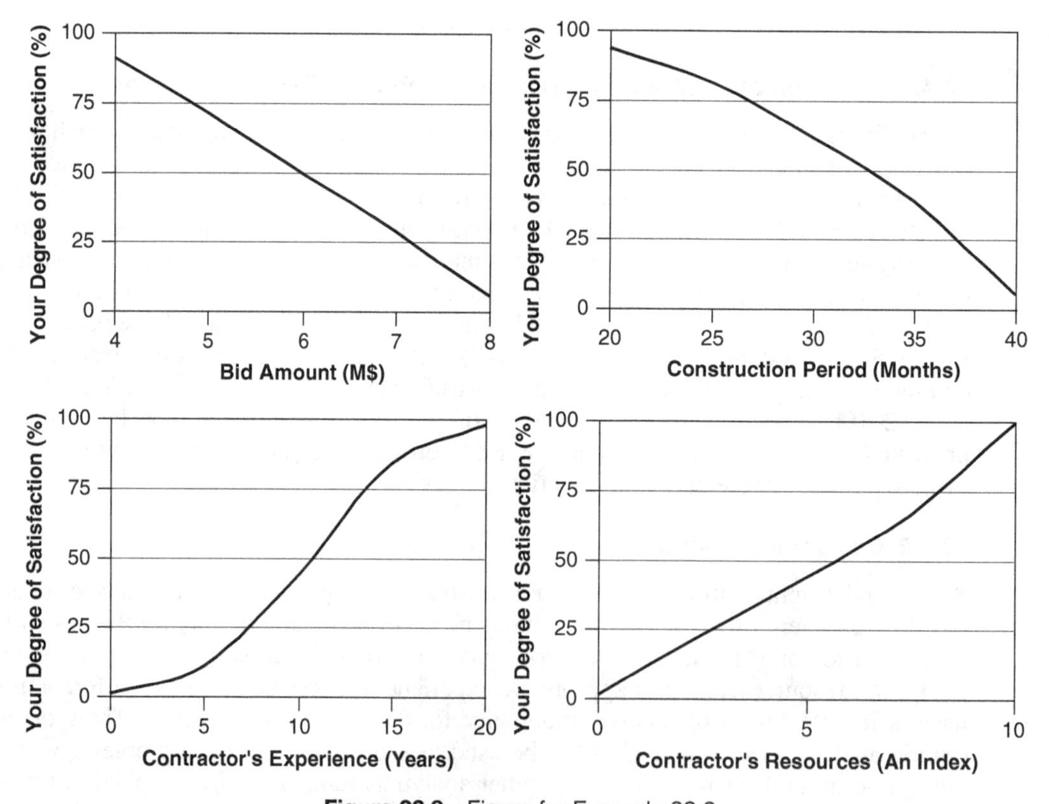

Figure 22.8 Figure for Example 22.2.

Solution

"Scaled" means the scaled value (or utility) of the decision criteria for that contractor. "Weighted and scaled" means the scaled value (or utility) multiplied by the weight of the evaluation criteria for that contractor. As seen in Table 22.3, on the basis of the scaled and weighted utilities of the decision criteria, the Haute Bros. Contractor is the best choice.

Table 22.3 Results for Example 22.2

	Bid Amount	Proposed Construction Period	Contractor's Experience	Contractor's Resource Strength	Overall Value or Utility
Contractor	**Weight = 0.28**	**Weight = 0.24**	**Weight = 0.27**	**Weight = 0.21**	
Apex, Inc.	7.1 25	22 90	9 36	5 42	47.14
A-1 Ltd.	6.8 32	36 30	15 85	8 75	54.86
Zenith Co.	5.9 55	29 62	5 12	6 50	44.02
Haute Bros.	6.5 38	26 78	11 53	9 85	61.52

22.4.4 Selection of Construction Technique

For constructing each part of a facility, there typically are several alternative construction techniques. For example, should the contractor prefabricate a component offsite and install it at the construction site, or should they construct it *in situ*. For analyzing problems of this nature, the engineer may resort to the use of multicriteria techniques (establishing the decision criteria, weighting to ascertain the relative weights across the criteria, and amalgamation and optimization to yield the best choice). In certain multiple criteria decision-making (MCDM) methodologies, each criterion needs to be scaled to render it to a unit that is either dimensionless or common across all the other evaluation criteria. There have been attempts to establish sustainability-related criteria for choosing between construction methods for concrete buildings (Chen et al., 2010). Others have gone further to develop systematic assessments of alternative construction technologies for building systems using weighted value-based decision criteria (Kim et al., 2005; Kadir et al., 2006; Pan et al., 2012). In many cases, the decision criteria included project cost, contract period, product quality, worker health and safety, sustainability of materials, and procurement method. Also, for selecting the appropriate configurations of vertical formwork, a variety of analytical models have been used (Hanna and Sanvido, 1990; Tam et al., 2005). In Chapter 12, we present tools for carrying out multicriteria analysis for purposes of decision making. Also, in Chapter 28, we discuss sustainability issues associated with construction and other phases.

22.4.5 Construction Impact Assessment

As part of the drive toward sustainable construction, it is often desired that the construction processes result in minimal adverse impact to the environment. As such, in recent times, engineers working in this phase are taking a closer look at the impacts of their system construction on the environment. For example, the removal of the topsoil cover prior to or as part of construction not only leads to topsoil erosion but also has the subsequent effects of sedimentation of the surface waters downstream, reduction of channel capacity, and possible flooding. Such impacts can be estimated using the modeling tools of statistical regression (Chapter 7) or simulation (Chapter 8). Using data from projects in Korea, Kim et al. (2012) developed a framework based on artificial neural networks and parametric statistics to estimate the greenhouse gas emissions due to asphalt pavement construction; Son et al. (2012) developed a statistical regression model to predict urban sustainability; and Lee et al. (2012) evaluated the impacts of high-rise glass building systems in terms of economic efficiency, energy use, and CO_2 emissions.

22.4.6 Construction Quality Control and Quality Assurance

Engineers at the construction phase measure and quantify the quality of construction as the work progresses (Figure 22.9). This is often done by testing the materials with specified statistical confidence limits of acceptance (the reader may recall our discussion on statistical tests of hypothesis in Chapter 6). Different materials are tested using specific tests established by the American Society for Testing and Materials (ASTM) and at different time or output intervals. Quality in a construction project could pertain to materials or workmanship or both. At civil system project sites, materials that are tested include ingredients (e.g., cement, water, steel reinforcement, and aggregates) and mixed or finished products such as concrete. In the technical specifications compiled by most civil systems owners, each ingredient or finished product must pass all of several physical, chemical, and biological tests. Workmanship quality can include the tolerances of finished surfaces or some other specific measure, for example, the absence of roughness on the finished surfaces. With the rise in performance-based clauses or contracts, contractors are increasingly judged on the basis of the quality of their finished product, an activity often aided by the use of analytical models (Yasamis-Speroni et al., 2012). Useful and innovative techniques for sampling to ensure product quality have been discussed by Chang and Hsie (1995) and Gharaibeh et al. (2010).

Figure 22.9 Inspectors test the adequacy of cable compaction during the reconstruction of the eastern span of the San Francisco–Oakland Bay Bridge [Courtesy of California Department of Transportation (Caltrans)].

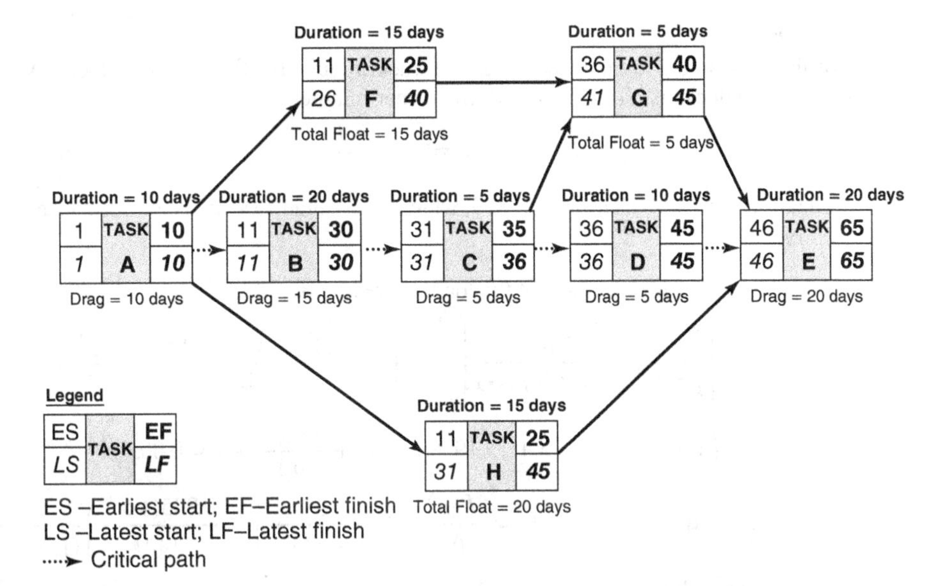

Figure 22.10 Critical path schedule illustration (Courtesy of Nuggetkiwi/Wikimedia Commons).

22.4.7 Project Planning and Scheduling

In this application context of systems concepts, the engineer develops the initial project plans and schedules and monitors the overall progress of the construction project. Engineers working at the construction phase often grapple with the task of optimizing resources so that the intended product quality can be achieved with minimum possible project cost and contract duration. This includes resources to serve construction queues (see Example 22.3) or labor and equipment resources to be used at each point of a schedule of tasks. In a bid to address this time–cost–resource optimization problem, a number of researchers have proposed systems-related analytical solutions; for example, Ashuri and Tavakolan (2012) suggested a fuzzy enabled hybrid genetic algorithm–particle swarm optimization approach. There are several proprietary general public license (GPL) or common public attribution license (CPAL) computer software packages that use the systems-related analytical tools of optimization, network analysis, and probability and can help the construction engineer carry out the functions of planning and scheduling. Figure 22.10 illustrates a construction schedule developed using the critical path method. In Chapter 23, we shall discuss a few numerical examples of the critical path method as applied in system operations.

Example 22.3

At a certain construction site, there five docks for loading material to be carried offsite. The haulage trucks arrive at an average of nine per minute and it takes half a minute to load each truck. Assume FCFS queuing discipline and that the interarrival and service times are exponentially distributed. Describe the performance of this queue in terms of the percentage of time that the loaders are idle, the average time spent by each truck in the entire queuing process, and the average queue length. Determine the percentage of times that there are three trucks waiting to be loaded. As the project manager, would you recommend any resource allocation changes to this queuing process?

Solution

Loading time = 0.5 minute/truck or $\mu = 2$ truck/minute, but there are five docks ($N = 5$). Thus, service rate for five docks is $N\mu = 5(2) = 10$ trucks per minute.

$$\rho = \lambda/\mu = \frac{9}{2} = 4.5$$

Utilization ratio = $\lambda/(N\mu) = \frac{9}{10} = 0.9$, which is less than 1.00 so the performance of the queue is satisfactory.

(a) The percentage of time that the loaders are idle is

$$P_0 = \left[\sum_{k=0}^{N-1} \frac{\rho^k}{k!} + \frac{\rho^N}{N!(1-\rho/N)}\right]^{-1} = \left[1 + \frac{4.5^1}{1!} + \frac{4.5^2}{2!} + \frac{4.5^3}{3!} + \frac{4.5^4}{4!} + \frac{4.5^5}{5! * (1-4.5/5)}\right]^{-1}$$

$$= \left(1 + 4.5 + 10.125 + 15.19 + 17.086 + \frac{15.377}{0.1}\right)^{-1} = 0.005$$

$$\text{Average queue length} = L_q = \frac{P_0\rho^{N+1}}{N!N}\left[\frac{1}{(1-\rho/N)^2}\right] = \frac{0.005(4.5)^6}{5!(5)}\left[\frac{1}{(1-4.5/5)^2}\right] = 6.9 = 7$$

trucks in the queue

$$\text{Average time spent in the system } (W) = (\rho + L)/\lambda = (4.5 + 7)/9 = 1.27 \text{ minutes}$$

(b) Percentage of time that there are three trucks waiting to be loaded = $P_n = \frac{P_0\rho^n}{n!}$ for $n \leq N$. Hint: $n = 3$, P_0 is obtained from answer to (a), and

$$\rho = \lambda/(N^*\mu) = P_3 = \frac{P_0\rho^3}{3!} = \frac{0.005(4.5^3)}{3!} = 0.076$$

22.4.8 Equipment Planning and Management

In this area of construction management, the engineers duties include the selection of appropriate equipment for a specific task, productivity-based planning to accomplish a specific task with a selected set of equipment, determination of the expected life or productivity of the equipment, modeling the depreciation of the equipment, and the general management of the equipment fleet. It has been well established, for example, that the use of systems engineering techniques for decisions not only to select appropriate equipment types (Figure 22.11) and sizes for a given construction task but also whether to lease or purchase, and how to schedule preventive maintenance for the equipment fleet such that overall productivity and profitability are enhanced.

Example 22.4

From her past experience, a contractor has determined that her payloaders have a mean life of 16 years. Assuming that this equipment has a constant probability of failure, determine its longevity reliability over a period of (a) 10 years, (b) 16 years, (c) and 20 years?

Solution

$\lambda = 1/\text{MTTF} = \frac{1}{16} = 0.063$. Thus, R (10 years) = $R(10) = e^{-0.063(10)} = 0.533$; R(16 years) = $R(16) = e^{-0.063(16)} = 0.365$; and R(20 years) = $R(20) = e^{-0.063(20)} = 0.284$. Therefore, approximately 54% of the loaders are expected to last to 10 years or more; 37% will last to the mean time to failure (MTTF) (i.e., 16 years); and 29% will last 20 years or more.

Figure 22.11 Examples of the equipment types typically used for specific work activities: (*a*) crane, (*b*) sheepfoot compactor with dozing functionality, (*c*) excavator with pipe-laying functionality, (*d*) telescopic handler with crane functionality, (*e*) road-header machine for tunnel boring, and (*f*) microtrencher [(*a*) Lokilech/Wikimedia Commons, (*b*) Headwater. Equipment/Wikimedia Commons, (*c*) SA Marais/Wikimedia Commons, (*d*) Norbert Schnitzler/Wikimedia Commons, (*e*) FHWA, and (*f*) SA Marais/Wikimedia Commons].

Example 22.5

The present value of a certain piece of construction machinery is \$3,000,000. After its 30-year service life, the terminal value will be \$200,000 in terms of present dollars. Assuming this equipment depreciates according to the sum-of-the-years (SOY) pattern, find (a) total depreciation at the end of 3 years, (b) the book value at the end of 3 years, (c) plot the book value/depreciation curves, and (d) use your curve to determine the year in which total depreciation exceeds \$1,000,000.

Solution

SOY depreciation in any year t is given by

$$D_t = \frac{N - t + 1}{(N/2)(N + 1)}(P - S)$$

where $N - t + 1$ is the useful remaining life at beginning of year t; N is the planning period or service life; t is the given year; P is the initial amount; and S is the salvage or terminal value.

The accumulated depreciation (ACD) (from the initial year) at the end of any year is $\text{ACD}_t = D_1 + D_2 + \cdots + D_t$.

The book value (BV) at the end of any year is $\text{BV}_t = \text{initial amount} - \text{ACD}_t$;

$P = \$3,000,000$ in today's dollars; $S = \$200,000$ in today's dollars; and $N = 30$ years.

$$\text{Depreciation in year 1} = [(30 - 1 + 1)/\left(\frac{30}{2}\right)(30 + 1)][3M - 2M] = \$180,645$$

$$\text{Depreciation in year 2} = [(30 - 2 + 1)/\left(\frac{30}{2}\right)(30 + 1)][3M - 2M] = \$174,623$$

$$\text{Depreciation in year 3} = [(30 - 3 + 1)/\left(\frac{30}{2}\right)(30 + 1)][3M - 2M] = \$168,602$$

At end of year 3, the total depreciation = \$523, 870, and the book value = \$3M – \$523, 870 = \$2, 476, 130. From the plot (Figure 22.12), it can be seen that the total depreciation will exceed \$1M in year 7.

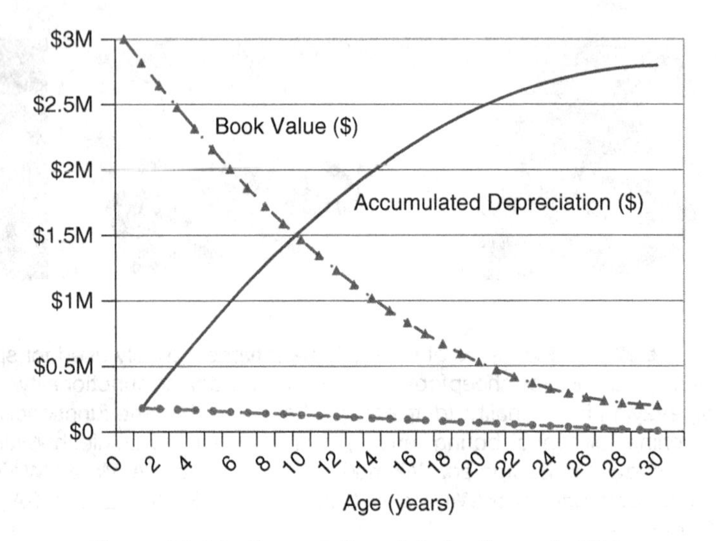

Figure 22.12 Depreciation plots for Example 22.5.

Example 22.6

A piece of equipment has a service life of 10 years. A contractor seeks to determine the best preventive maintenance (PM) schedule for the equipment. At one extreme, the contractor may carry out PM every year. On the other extreme, the contractor opts not to carry out any PM throughout the equipment life. Given the cost of PM (\$10,000 per PM activity) and the function relating cost (x) to benefit (i.e., y), the (increased efficiency of the equipment): $y = -0.0071x^2 + 0.7146x + 7.1968$. Determine the optimal number of PM activities over the service life of the equipment. There can be a maximum of one PM per year.

Solution

The objective is: Maximize the total increased efficiency (y). The decision variable is the number of PM activities, m. Mathematically, this is written as

$$\text{Maximize } y = -0.0071x^2 + 0.7146x + 7.1968$$

$$\text{Subject to } m = x(1000);\ m <= 10, \text{ and } m \text{ is an integer.}$$

Solving this problem formulation in GAMS or any standard optimization package will yield $m = 5$.

22.4.9 Processing of Construction Materials

A common context of decision making at the construction phase involves the determination of the proportions of different materials required in a mix to yield a product of specified characteristics. These products/processes include the production of Portland cement concrete, asphaltic concrete, granular bases and subbases, and soil or subgrade stabilization. Of the several possible combinations of materials fulfilling the specified constraints of size, plasticity, and other properties, the contractor seeks the combination or blend that optimizes some objective function. The objective function may be a simple benefit–cost ratio (cost-effectiveness) or a multiattribute utility function. Analytical tools for solving this problem include linear programming and multicriteria analysis. Applications in past work have been discussed by Neumann (1964), Tubacanon et al. (1980), Lee and Olson (1983), Easa and Can (1985), and Toklu (2005).

Example 22.7

A precast concrete plant requires at least 4 million gallons/day more water than it is currently using. Waterex, a nearby water supply reservoir can provide up to 10 million gallons per day of such extra supply. Whitewater, a local perennial stream, can provide an additional 2 million gallons a day of extra supply. For water used by the plant, the average concentration of pollution should not exceed 100 units. The water from Waterex and from Whitewater has pollutant concentrations of 50 and 200 units, respectively. The cost of water from Waterex is $1000 per million gallons; and from the Whitewater stream, it is $500 per million gallons. The plant seeks to determine how much water should be purchased from each of the two sources in order to minimize the cost of supplying water that, on average, meets the quality standards.

(i) What are the decision variables?
(ii) What are the constraints?
(iii) What is the objective function?
(iv) Find the optimal solution using the graphical method and indicate what advice you would give to the plant.

Solution

The decision variables are

Amount of water (millions of gallons) purchased from Waterex, x

Amount of water (millions of gallons) purchased from Whitewater, y

The constraints are

$x + y = 4$ (total water needed is 4 million)

$x \leq 10$ (Waterex can provide up to 10 million gallons)

$y \leq 2$ (Whitewater can provide up to 2 million gallons)

$50x + 200y \leq 100\ (x + y)$(total pollutant concentration of mixed water must not exceed 100 units).

$50x + 200y \leq 100\ (x + y)$, upon simplifying, gives $y \leq 0.5x$.

As shown in Figure 22.13, the optimal solution is $x_{\text{OPT}} = 2.67$; $y_{\text{OPT}} = 1.33$; $z_{\text{OPT}} = 3333$. In other words, the minimum cost is achieved when the amount of water taken from Waterex, x, is $\frac{8}{3}$ (i.e., 2.67) million gallons, and that taken from Whitewater, y is $\frac{4}{3}$ (i.e., 1.33) million gallons.

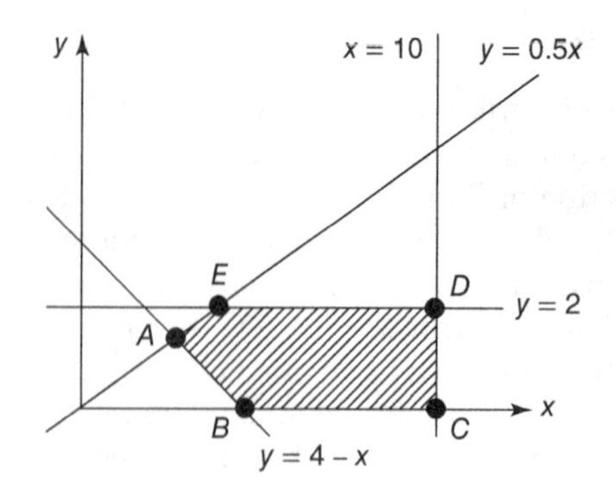

Vertex	x	y	*Total* $ = 1000x + 500y
A	8/3	4/3	= 1000(8/3) + 500(4/3) = 3,333
B	4	0	= 1000(4) + 500(0) = 4,000
C	10	0	= 1000(10) + 500(0) = 10,000
D	10	2	= 1000(10) + 500(2) = 11,000
E	4	2	= 1000(4) + 500(2) = 5,000

Figure 22.13 Illustration for Example 22.7.

22.4.10 Design of Temporary Structures

In designing concrete formwork, scaffolding, shoring, and bracing, the construction engineer often uses the tools of optimization in order to minimize material use within cost constraints and without jeopardizing worker safety (Figure 22.14). The design, erection, and maintenance of these temporary structures can have a profound influence on the pace and quality of construction, and are often undertaken by a specialist subcontractor.

22.4.11 Contract Management

Engineers working for the system owner and the contractor manage their construction contracts throughout the project execution period so that the project goals of quality production, on-time delivery, cost overrun prediction or avoidance, and conflict avoidance or minimization are realized. The management of construction contracts includes negotiation of the contract conditions and ensuring that the contractor does not violate the conditions. Also, any mutually agreed modifications to the contract that may arise from unexpected site conditions, labor unrest, acts of God, or other circumstances are documented and are included as addenda to the contract.

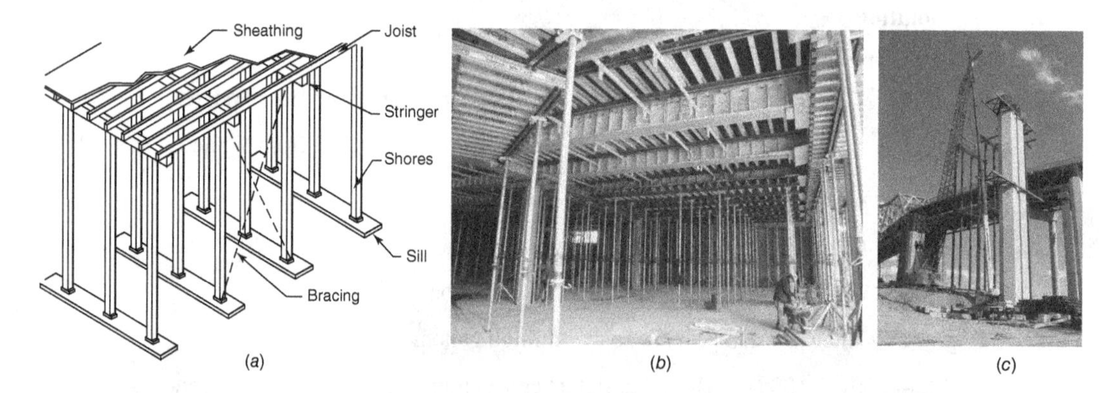

Figure 22.14 Significant benefits in cost and time savings and product quality can be earned when the contractor applies system-related tools in formwork design layout and construction: (*a*) schematic representation of a traditional slab formwork for concrete, (*b*) formwork tables with aluminum and timber joists, and (*c*) emplacement of a falsework section for bridge construction [Courtesy of (*a*) Wikimedia Commons, (*b*) Wikimedia Commons, and (*c*) Leonard G./Wikimedia Commons)].

Example 22.8

A government faced with increasing incidence of cost overruns in public contracts seeks to ascertain the reasons behind this trend and also to predict cost overrun rates for future contracts. The cost overrun rates of 30 construction contracts are indicated below (Table 22.4). Develop statistical models that relate the cost overrun rates (COR) to the contract year, project cost, and project type [water supply (W), highway (H), and energy (E)]. What are your conclusions from the model? Compare the marginal effects of a $1 increase in project cost across the three project types.

Table 22.4 Data for Cost Overrun Modeling Example

Contract #	Year	Cost ($M)	Type	COR (%)
1	2009	20.4	W	2.3%
2	2010	5.2	H	12.3%
3	2008	25.0	H	0.2%
4	2011	16.1	W	5.3%
5	2012	9.8	E	7.7%
6	2012	11.7	H	8.1%
7	2009	17.2	E	6.3%
8	2009	22.1	E	2.1%
9	2010	8.8	H	12.3%
10	2010	12.0	W	9.3%
11	2012	15.5	W	4.3%
12	2011	12.6	W	6.6%
13	2011	11.2	E	9.3%
14	2012	21.5	H	1.9%
15	2012	13.9	H	7.6%
16	2010	5.9	W	12.3%
17	2012	18.7	H	3.1%
18	2009	27.5	H	0.5%
19	2009	14.2	W	5.5%
20	2009	8.1	E	6.7%
21	2012	29.9	H	0.8%
22	2011	7.1	E	7.6%
23	2012	23.6	E	2.2%
24	2011	19.7	H	2.0%
25	2012	26.0	W	0.7%
26	2010	3.2	W	14.1%
27	2011	5.5	W	9.3%
28	2010	17.2	E	2.7%
29	2012	20.9	H	1.3%
30	2012	3.3	H	13.9%

Solution

First, the project costs at each year from 2008, 2009, 2010, and 2011 need to be converted to their corresponding values as some base year. We use 2012 as the base year; however, the reader may choose any year to represent the base year. On the basis of construction cost index trends (see Chapter 10), to obtain the cost at each year in terms of 2012 dollars, we will use a factor of 1.07 for Year 2008 and 2009 costs, 1.05 for Year 2010 costs, and 1.02 is for Year 2011 costs. We then run the model using a standard statistical software package. Assuming a linear regression specification, the coefficients of explanatory variables are determined. The first model run will show that the *Year 2010* and *HIGHWAY* variable are not statistically significant. These are then excluded from the next run which yielded the result shown in the table below.

Variable	Constant	2008	2009	2011	2012	*WATER*	*ENERGY*	*COST*
Coefficient	15.49	−2.073	−1.231	−2.104	−1.755	−0.594	−1.025	−0.494
Standard Error	0.9015	1.8766	1.0401	0.9113	0.8927	0.7347	0.7820	0.0456
t ratio	17.181	−1.105	−1.184	−2.308	−1.966	−0.808	−1.311	−10.837
P[|*T*|>*t*]	0.0000	0.0333	0.2000	0.2000	0.3666	0.3333	0.2666	15.576

The adjusted *R*-square of the model is 0.87, which suggests that the model provides a good fit to the data. It was also found that of the three project types, highway projects generally have the highest average cost overrun rates. From the coefficient of the cost variable (*COST*), it is seen that for each 1M$ increment in the project cost, there is a 0.5% reduction in cost overrun rate on average. This is consistent with past research studies that suggest that higher-cost projects generally have lower cost overrun rates. To determine the marginal effect of project cost across the different project types, one approach is to run the model separately for each project type. After doing so, it can be found that a $1M increment in the project cost is associated with a decrease in the cost overrun rate by 0.515%, 0.524%, and 0.517% for water supply, highway, and energy projects, respectively.

22.4.12 Human Resource Management

This decision context includes the selection of the appropriate types and numbers of construction personnel to complete the project. The tool of optimization has often been used to address this problem at a number of construction sites. Too few workers for a given task will slow the work while too many will lead to worker idling, waste, or possibly, poor quality of work. Ideally, an optimal number can be determined using productivity data from past contracts. The construction sector is one of the most problematic and complex settings at which people are managed. The reader may refer to Dainty and Loosemore (2003) in presenting techniques for strategic and operational management of people within the construction sector. Due to the operational constraints regularly encountered by construction companies, the needs of employees are often superseded by performance concerns, thus often resulting in adverse consequences for the industry's worker morale and welfare and, ultimately, the prosperity and productivity of the entire industry.

22.4.13 Project Safety

As part of construction site safety monitoring and management, construction engineers make predictions about the frequency and severity of safety accidents at a site of given characteristics or estimates of the extent of hazards or potential black spots (Figure 22.15). Carried out as a part of measures to identify and mitigate safety risks, such predictions typically utilize the regression or simulation modeling tools that we discussed in Chapters 7 and 8. The literature contains models that have been developed using these and other similar tools to predict safety performance at construction sites. For example, recognizing that tower cranes are currently dominant equipment in building construction, Shapira et al. (2012) and Lin et al. (2012) developed analytical and simulation models, respectively, for evaluating safety at tower crane construction sites. The reader is also referred to interesting examples discussed by Visscher et al. (2008), Mitropoulos and Namboodiri (2011), Dewlaney et al. (2012), and Ikpe et al. (2012) to review how systems-related analytical tools have been used for quantifying or predicting safety performance at construction sites.

Figure 22.15 Constructing a steel frame. Worker safety continues to be one of the key performance measures of activities at the construction phase.

22.4.14 Risk Analysis

Risk analysis assesses the extent to which the project outcome may deviate from the goals associated with product quality, cost, time, safety, and litigation. In various sections of this chapter, we discuss the prediction and other analytical tasks associated with the enhancement of construction product quality, minimization of cost overruns and time delays, enhancement of construction worker safety, and reduction of the risk of litigation. System-related tools have been used to predict the likelihood of a successful construction phase of civil systems development (Tabish and Jha, 2012) or the general portfolio of risks that a contractor may face (Abdelgawad and Fayek, 2012). In the next section, we will examine the issue of construction management risk with a thicker lens.

22.5 MANAGING RISKS AT THE CONSTRUCTION PHASE

In the previous section, we reviewed a number of problem contexts at the construction phase that require or could be addressed using analytical tools in systems engineering. In this section, we focus on a few of these application contexts for greater scrutiny. These contexts are directly related to project performance criteria, namely, overall risk minimization, avoidance or minimization of cost overruns or time delays, and maximization of product quality.

22.5.1 Construction Risk Management

The outcomes associated with the construction phase are characterized by a large degree of variability due to uncertainties in the factors that influence the engineer's decisions at that phase. These uncertainties are often collectively termed as "risk," and it is the duty of engineers working in this phase to acquire the requisite analytical tools to help them to recognize, quantify, and predict these risks. Generally, these risks may be characterized by their relationship to the natural or built-up environment and those that are inherent (project-related) risks (Figure 22.16).

The importance of risk is underscored by the fact that a significant amount of time and effort in construction management is expended on risk identification and mitigation. Several practices in construction engineering were born out of the need to address risk, such as the requirements for a performance bond from the contractor, insurance for the project, and contractor licensing. Further, the basic aspects of the structure of the contract between the system owner and the contractor structure are driven by risk, such as the general conditions of the contract, the terms of payment, and the owner's selection of the appropriate contracting approach (Section 22.2).

Individuals differ in their comfort with uncertainty based on circumstances and preferences. As such, in making decisions under uncertainty, the construction engineer may exhibit traits that may be consistent with being risk averse (such individuals fear loss and make decisions whose outcomes are almost guaranteed even when the benefits may be small), risk neutral (such individuals are indifferent to uncertainty), and risk lovers (these individuals are prepared to sacrifice everything with the hope of attaining the best possible outcomes).

The minimization of uncertainties may lead to reduction in the likelihood and severity of undesirable performance in terms of the final cost, the construction time, the quality of the finished product in terms of materials, and the tolerances, construction safety, and the frequency and magnitude of interparty conflicts. These various aspects are discussed in the next Section.

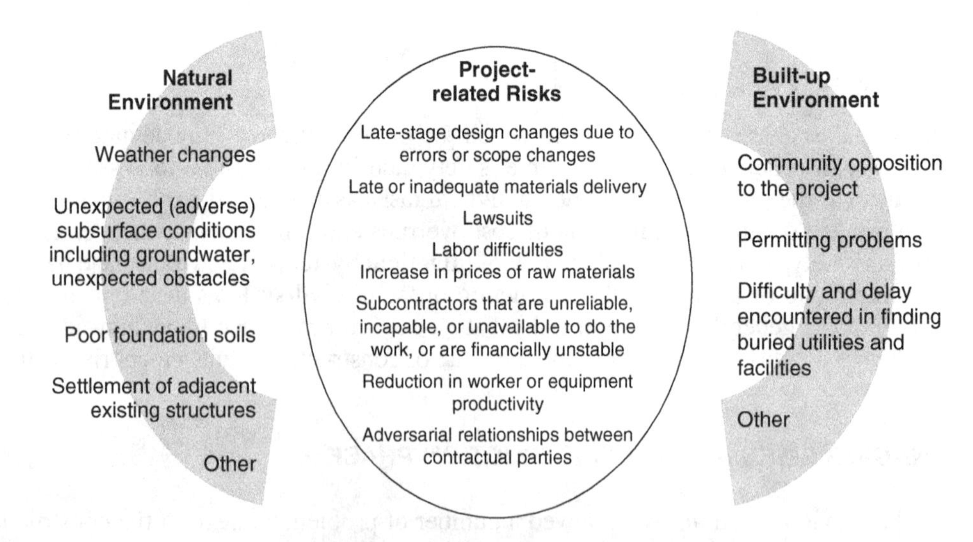

Figure 22.16 Some sources of construction risk.

22.5.2 Construction Cost Overruns

Owners and operators of civil systems worldwide are increasingly concerned about the reliability of the cost estimates of their construction projects as the cost overruns associated with several megaprojects have attracted global attention and criticism. Due to the media coverage of such cost overruns, the public is increasingly questioning the ability of construction engineers to forecast costs accurately. The cost and/or time overruns associated with San Francisco's Bay Bridge replacement, Boston's Central Artery Tunnel project, and Virginia's Springfield Interchange project have prompted widespread public outcry as well as the attention of federal agencies such as the U.S. General Accounting Office. Time and again, the U.S. Congress's Transportation Subcommittee of the House Appropriations Committee has not only called for the improvement the quality of the initial cost estimates of infrastructure construction costs but also iterated the need to track information on the causes of cost overruns (Schexnayder, and Mayo, 2004). Generally, cost discrepancies have been identified as a pervasive phenomenon in civil engineering infrastructure projects worldwide irrespective of the project type, geographical location, and historical period (Flyvbjerg et al., 2003; Anderson et al., 2007). In recognition of the need to address this issue, system agencies including the U.S. National Highway Institute has introduced courses to study uncertainty in cost estimation, introduce the principles of deterministic and probabilistic techniques to measure risks and uncertainties associated with project cost estimates, and to promote the importance of accuracy in construction cost estimation.

Cost Overrun Categories. To acquire a clearer picture of cost overruns, it is useful to consider the early phases of civil systems development (planning–design–construction) as a five-stage process: project feasibility or planning, engineering design, contract letting and bidding, contract award, and final construction (Figure 22.17). Several substages often exist within these stages. A discussion of these stages is pertinent from the cost perspective because the amount and uncertainty of the project cost undergoes significant changes as the project evolves from one stage to the next; and in most cases, from the earlier to the latter phases, the amount of project cost increases while the uncertainty decreases. Construction and facility managers are interested in tracking the cost changes at any stage not only from any precedent stage but to any subsequent stage as well.

A cost discrepancy, which can be defined as the deviation of a cost amount between one phase or stage of project development to another, can be an increase or decrease in project cost at a given phase or stage compared with the cost at a previous phase or stage (defined as a cost overrun or underrun, respectively). Systems owners are typically most interested in the cost discrepancies between the contract award stage (award amount) and the postconstruction stage (final amount).

Types of Cost Overruns (with Respect to Deviation Reference Points)

Cost Overrun I = Final Cost − Planning Cost

Cost Overrun II = Final Cost − Design Cost

Cost Overrun III = Final Cost − Bid Cost

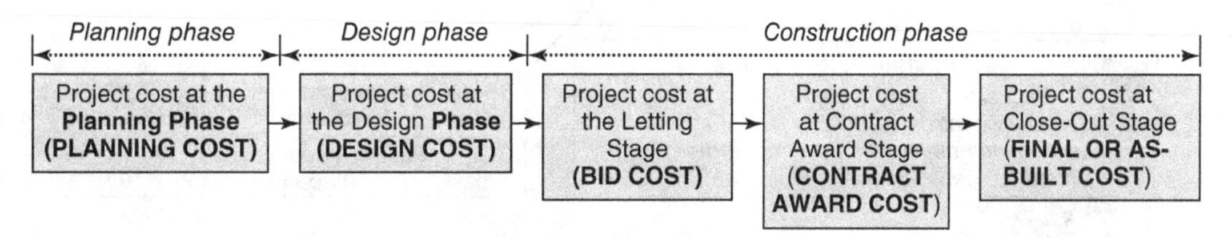

Figure 22.17 Sequence of project cost changes across the various phases of the project development.

Cost Overrun IV = Final Cost – Contract Award Cost
Cost Overrun V = Contract Award Cost – Planning Cost
Cost Overrun VI = Contract Award Cost – Design Cost
Cost Overrun VII = Contract Award – Bid Cost
Cost Overrun VIII = Bid Cost – Planning Cost
Cost Overrun IX = Bid Cost – Design Cost
Cost Overrun X = Design Cost – Planning Cost

Often, cost discrepancy is determined as the difference between the as-built project cost and the contract award amount (Cost Overrun IV). When the cost discrepancy is positive, it is called a cost overrun (sometimes referred to as "cost escalation" in the literature); when it is negative, it is called a cost underrun. In this text, the term cost discrepancy specifically refers to the monetary amount of cost discrepancy rather than the frequency of cost discrepancy occurrence. The former represents the magnitude or severity of a cost overrun. From the cost discrepancy amount, it is possible to calculate the cost discrepancy rate as the amount of cost discrepancy as a percentage of the contract award amount. Compared to cost discrepancy amounts, cost discrepancy rates are a more unbiased measure of the incidence of cost overruns, particularly when projects of different sizes are being compared. Thus, using past data, a system owner can develop models that can predict the likelihood that a contract will incur a cost overrun and the intensity (amount or rate) of the cost overrun (see Figure 22.18).

Cost Overrun Amount or Cost Overrun Magnitude = Final Cost – Contract Award Cost
Cost Overrun Rate (%) = 100(Final Cost – Contract Award Cost)/Contract Award Cost

As one would expect, construction cost estimates become more accurate as a project evolves from the planning stage the as-built. Figure 22.19 [adapted from Meyer and Miller (2001)] illustrates the changes in the uncertainty of the construction cost estimate across the various stages of the construction phase. The quantifiable part of the construction cost increases as the construction phase stages progress and as more detailed information on the construction processes becomes available. At each stage, a contingency amount is set aside to cover unknown construction costs.

Figure 22.20 shows the time trend of cost overrun rates for a number of states in the United States (Bordat et al, 2004); and Figure 22.21 focuses specifically on Wisconsin, showing the

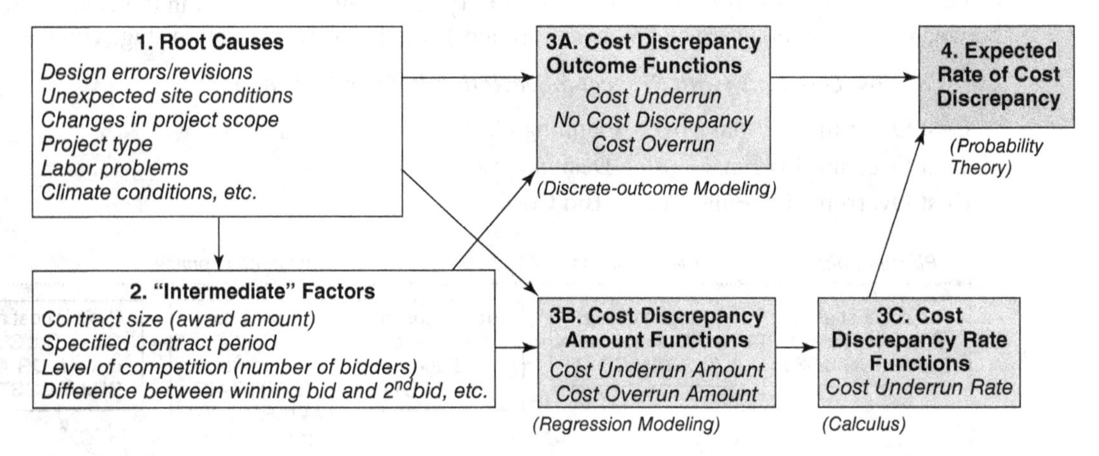

Figure 22.18 Predicting the likelihood and extent to cost overruns (Gritzka and Labi, 2009).

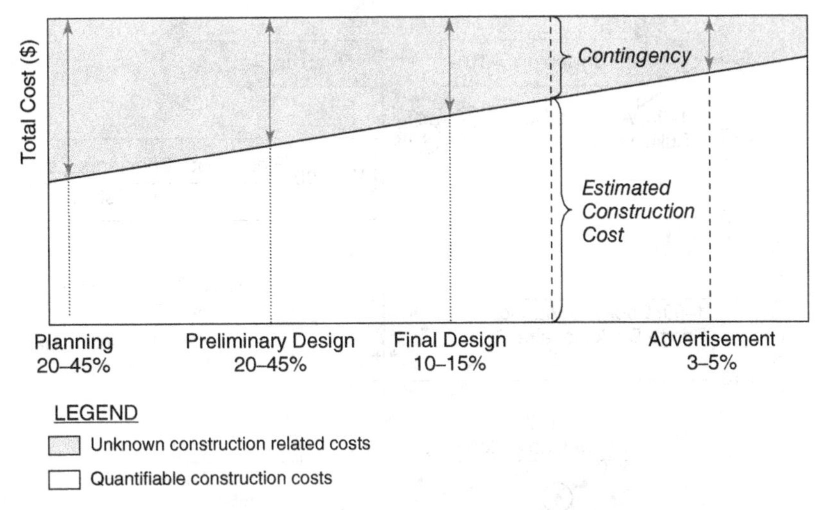

Figure 22.19 Typical progression of cost uncertainty at various stages of project development.

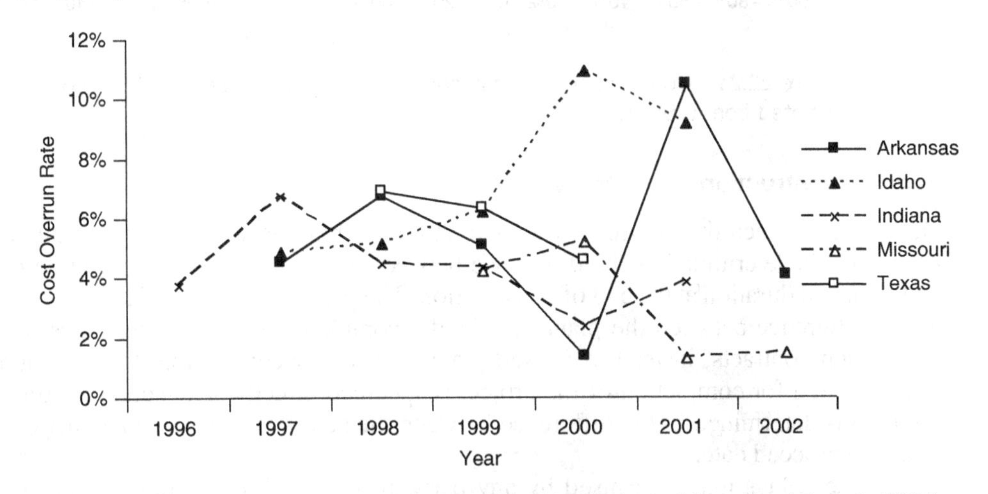

Figure 22.20 Cost overrun rates in various states (adapted from Bordat et al., 2004).

percentage of cost differences plotted against contract sizes that were observed for 131 highway contracts in year 2005; the number of contracts for which the final cost exceeded the engineer's estimate by more than 10% was found to be similar to those for which the engineer's estimate exceeded the final cost by 10% or higher.

Causes of Cost Overruns. Cost overruns can be traced back to issues associated with the preliminary phases of system development: planning and design, such as, design errors, unexpected site conditions, scope changes. The contribution of root causes to cost overruns has been investigated by researchers including Hufschmidt and Gerin (1970), Jahren and Ashe (1990), Akinci and Fisher (1998), Akpan and Igwe (2001), Knight and Fayek (2002), Attala and Hegazy (2003), and Anderson et al. (2007). Most of the problems associated with these root causes are often unknown at the contract award phase—they surface only during the project construction phase. Thus, between the contract award and final construction stages, it is often very difficult to predict cost overruns on the basis of the root causes.

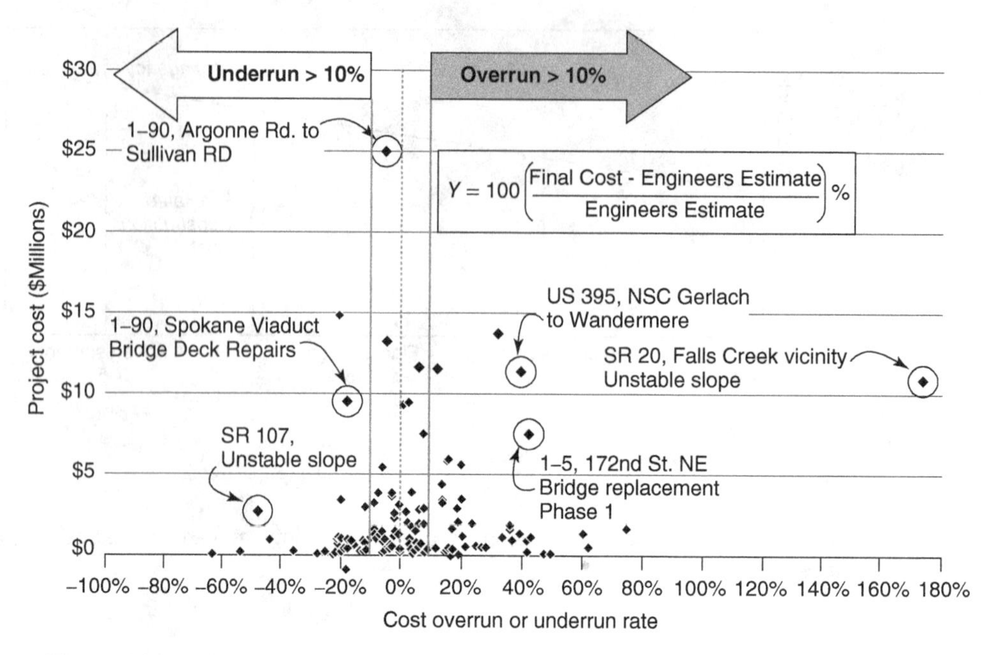

Figure 22.21 Comparison of final cost with engineer's estimate by contract size (adapted from WSDOT, 2008).

22.5.3 Construction Time Delay

Delay occurs when the progress of a contract falls behind its scheduled program. The amount of time delay is determined as the difference between a contract's planned duration of construction and its actual duration at the end of construction. The planned duration of construction is calculated as the difference between the planned calendar completion date and the notice to proceed date. For certain contracts, instead of a fixed planned calendar completion date, the number of work days required for completion of the project is specified directly. The actual contract construction duration is determined as the difference between the actual last day of construction work and the notice to proceed date.

Time delays may be caused by any party affiliated with the contract and may be a direct or indirect result of one or more root causes that are generally related to engineering design and the factors of production. Also, there are certain characteristics of the contract bidding process and project environment that also foster the incidence of time delays. In some cases, contracts are extended with the owner's permission beyond their originally stipulated periods to account for an increased scope of work, unexpected site conditions, and other extenuating circumstances. In other cases, time delays are the responsibility of the contractor, and the owner imposes a penalty (liquidated damages) per day of delay. Figure 22.22 presents the project development sequence showing some of the delay factors associated with the preconstruction and construction stages.

The problem of time delays on civil engineering contracts continues to persist for system owners. Such delays have adverse effects on both the owner and the contractor (either in the form of lost revenues or extra expenses), often raise the contentious issue of delay responsibility, and may result in conflicts that ultimately end up in a court of law. For reconstruction of civil systems located in urban areas, there are also issues of extended downtime of the system and concomitant safety problems, inconvenience to the community, system user dissatisfaction, and possibly political ramifications. At the current time when public agencies seek to burnish their image as

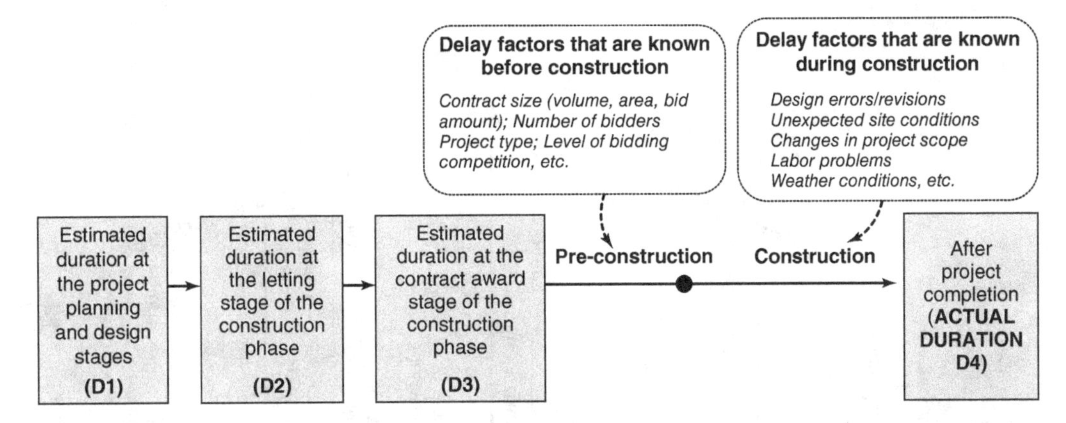

Figure 22.22 Project development sequence showing delay factors at preconstruction and construction stages.

responsible stewards of taxpayer-funded civil engineering infrastructure, they can hardly afford the negative publicity associated with project time delays.

As parts of efforts to address the important issue of time delay, past researchers have developed several categorizations for the problem (Kraeim and Diekmann, 1987; Rowland, 1981): (i) excusable delays with compensation, (ii) excusable delays without compensation, (iii) nonexcusable delays (contractor-responsible delays, see Figure 22.21), (iv) excusable delays, where the contractor is given a time extension but no additional money, (v) concurrent delays, where neither party recovers any damages, and (vi) compensable delays, where the contractor recovers monetary damages.

Majid and McCaffer (1998) found that excusable delays with compensation are due to errors in design, changes in work scope, and failure to provide timely access; excusable delays without compensation are neither the responsibility of the owner/client nor of the contractor and are typically due to acts of God, war, and other extenuating circumstances; and nonexcusable delays are the responsibility of the contractor and often result in payment of liquidated damages as a penalty for the delay.

Other delay causes, which are generally inexcusable, include materials-related delays, labor-related delays, equipment-related delays, financial delays, improper planning, lack of control, subcontractor delays, poor coordination, inadequate supervision, improper construction methods, technical personnel shortages, and poor communication.

In view of the several serious internal and external consequences of project delays, system owners, very early at the contract award stage, constantly seek enhanced tools to predict the occurrence probability and severity of project time delays, often on the basis of known information such as bidding information and project characteristics. Such information are available at agencies that take pains to maintain contract databases containing data including the dates of the notice to proceed, the estimated time of project completion, and the actual last day of work.

Enhanced prediction of time delays and a better understanding of the influential factors are critical to overall management and administration of the construction phase of civil system development. They enable agencies to be better equipped to plan time delay remedies and also to estimate more reliable project durations so that more realistic contract periods can be specified for future civil engineering projects.

The incidence and severity of time delays on civil engineering construction projects have been analyzed extensively. Figure 22.23 presents the contributory factors of nonexcusable delay (Majid and McCaffer, 1998).

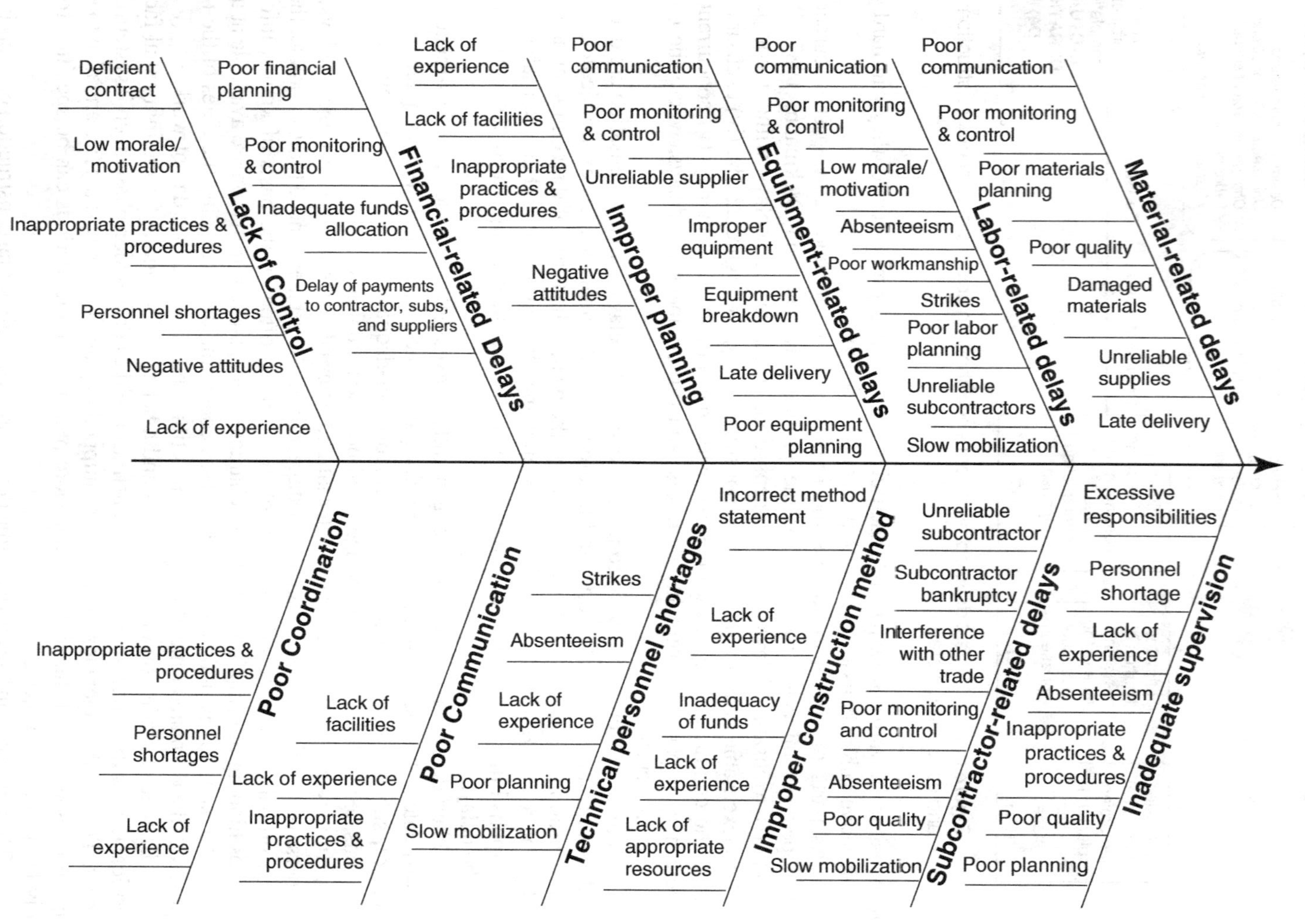

Figure 22.23 Contributory factors of nonexcusable delay (adapted from Majid and McCaffer, 1998).

22.5.4 Construction Quality

With regard to workmanship, the questions that typically arise include the following: Was the correct thickness of road surface laid? Was the steel beam of the required length and breadth? Do the walls of a constructed water tank have the required thickness? With regard to materials, examples of questions include: Was the reinforcement used of the specified quality (tensile strength)? Does the concrete have the required 28-day compressive strength? Do the aggregates pass the required chemical tests? Was the water used for concrete mixing free of harmful chlorides and sulfates? To ascertain that the materials meet the specifications, site inspectors take samples to laboratories for testing or tests may also be conducted in situ, for which a statistical hypothesis is established (H_0 and H_A). For a given degree of confidence, site inspectors determine whether there is evidence to support the hypothesis or otherwise (see steps in hypothesis testing in Section 6.4 of Chapter 6). The material is then rejected or accepted on the basis of the test results.

Example 22.9

As the site engineer at a construction site, you have been asked to oversee the concrete production process. You are particularly worried about the slump of the concrete. If the slump is too small, it suggests the concrete is too stiff. If the slump is too much, then the concrete is too watery. The contract specifications state that the slump for that kind of concrete should be 1 inch, with a 90% level of confidence. So during the concrete production process, you instructed the laboratory technician to take 20 random samples of fresh concrete and measure the slump using slump testing equipment. The technician obtained the following test results (in inches):

0.92	1.21	1.03	1.10	1.01	0.99	0.89	0.97	1.01	0.99
1.05	1.11	0.95	1.00	1.00	1.04	0.88	1.02	0.97	1.01

Would you accept that day's production of concrete at the given level of confidence? Assume that, from the past slump test results, the concrete slumps are known to be normally distributed.

Solution

To solve this problem we may follow the chart presented in Section 6.4. Here, the claim could be attributed to the contractor, who asserts that the concrete produced that day was satisfactory because it met the slump requirements statistically. (Note that the claim could also be the inspector's assertion that the concrete met or did not meet the slump requirements).

The hypothesis then is as follows:

H_0: The concrete produced that day was satisfactory (i.e., the average slump was 1 inch).

H_A: The concrete produced that day was not satisfactory (i.e., the average slump was significantly different from 1 inch).

As you are interested in the average slump, the statistical parameter of interest is the mean.

In math notation, the hypothesis is

$$H_0:\ \mu = 1 \text{ inch}$$

$$H_1:\ \mu \neq 1 \text{ inch}$$

From the formulated hypothesis, it is clear that the test is two tailed. Because (i) we are interested in the mean and (ii) the population parameter is normally distributed, the appropriate statistical distribution to use is the Z distribution.

The decision rule is to reject the null if the calculated value of the test statistic falls in the rejection region.

The level of significance, α, = 1 − Confidence Level = 1 − 0.90 = 0.10. Thus, α is 0.1
Because the test is two tailed, the C value is $\alpha/2 = 0.05/2 = 0.05$.
The critical values of the test statistic are $-Z_C$ and $+Z_C$ that is: $-Z_{0.05}$ and $+Z_{0.05}$.
From the statistical tables, these are determined as follows: −1.645 and +1.645.
The calculated value of the test statistic can be found after calculating the sample mean and standard deviation.

The mean of slumps from the sample is

$$x = (0.92 + 1.05 + 1.21 + 1.11 \ldots)/20 = 1.0075 \text{ inches}$$

The standard deviation of slumps from the sample is

$$\sigma \text{ of}(0.92 + 1.05 + 1.21 + 1.11 \ldots) = 0.0755 \text{ inches}$$

$$Z^* = \frac{x - \mu}{\sigma/\sqrt{n}} = \frac{1.0075 - 1}{0.0755/\sqrt{20}} = 0.4445$$

As can be seen in Figure 22.24, Z^* does not fall in the rejection region, therefore, we fail to reject the null hypothesis. In other words, there is no statistical evidence to conclude that the mean slump differs from 1 inch. Thus, there is no reason to reject the contractor's claim and we conclude that the quality of concrete produced that day, is acceptable.

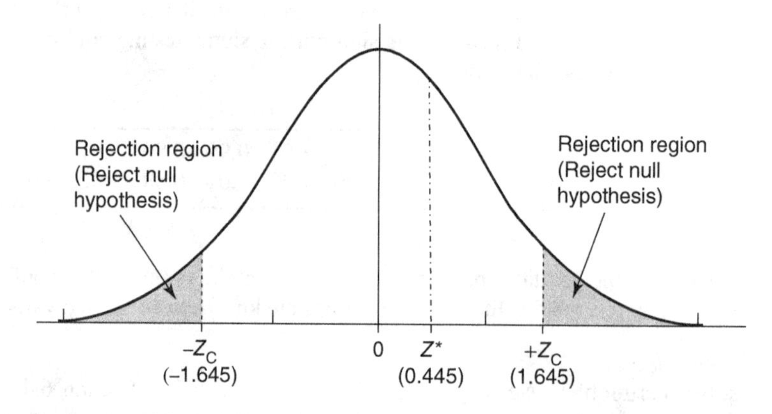

Figure 22.24 Figure for Example 22.9.

22.6 EMERGING AND EVOLVING ISSUES IN CIVIL SYSTEMS CONSTRUCTION

As we conclude our discussion of the construction phase, it is important to discuss a few pertinent issues associated with this phase of civil systems development.

22.6.1 Management of System Operations during Downtimes

In cases where the civil system exists but is being reconstructed, the system is fully or partially closed for the construction work. This period, known as system downtime, can be a source of frustration and inconvenience to system users and is a potential source of lawsuits, particularly if the closure leads to loss of life, injury, or significant loss in business. If the system is one that generates revenue, downtime could also lead to loss of revenue. Therefore, the system owner has a duty to carry out adequate planning to ensure that the system users suffer minimum inconvenience during the replacement of the system. For example, in the highly successful "Hyperfix Project" that reconstructed a heavily trafficked 3-mile section of Interstate Highway 70 near Indianapolis, the Indiana Department of Transportation closed the entire road section to traffic for 85 days; commuter, transit, and commercial traffic were successfully rerouted to other roads on the city network and measures were put in place to mitigate the expected consequences (Sinha et al., 2004).

22.6.2 Exploiting Scale Economies in Construction

Systems owners continue to seek cost reduction in the replacement or construction of their systems. One way to do this is to bundle projects. When the owner possesses a large variety of asset at different locations, there are opportunities for bundling projects such that the volume of similar activities translates into lower unit costs; this concept aims at reducing the overall costs of projects by combining a number of smaller-scale projects into a single contract (Grimm et al., 2006; Estache and Iimi, 2011). The bundling of a project may be carried out on the basis of at least one of three dimensions or criteria: similarity of system or component type, spatial proximity (two or more projects in the same geographical area combined in one contract for purposes of site mobilization), and temporal consistency (projects to be let in the same month, season, or fiscal year may be grouped together as one contract).

22.6.3 Using Information Technology to Facilitate Cost-Effective Delivery of Projects

In the last three decades, there has been a dramatic increase in the application scope and sophistication of information technology to describe and document the work of the many disciplines involved in construction projects. In the current era, virtually all project information is entered into computer software packages, which may be general-purpose tools (spreadsheets and text processing software) or discipline-specific tools (mechanical CAD programs or cost-estimating software) (Fischer and Kunz, 2004).

22.6.4 Increased Awareness of Environmental Stewardship

An opportunity that arises at the construction phase is to use sustainable practices (see Chapter 28). This includes the use of construction materials that are recycled or reused from past projects or are recyclable or reusable for future projects. As we have learned in this chapter, such potential could best be realized through the use of nontraditional contracting approaches where the contractor is granted some flexibility in both design and construction. Also, sustainable construction is promoted when contractors are encouraged to make explicit efforts to ensure that the construction phase does not result in undue harm or inconvenience to the ecology, community, or surrounding businesses.

22.6.5 Increased Exposure to Corrupting Influences

The threat of corruption during the construction phase is all too real because most construction is carried out by contracting to a private-sector contractor. As we have learned in this chapter, this opens up a fertile ground for potential corruption and violations of engineering ethics. Often, the real victims of corruption are the taxpayers or society in general, who are often left with a substandard quality product as the result of a corrupt agreement between the product deliverer and the supervisor of the delivery. Unfortunately, it does not often end there: The lure of illicit monetary gain, if succumbed to, often has a cascading effect on an individual—compromising one ethical principle makes it easier to compromise with other principles. In Chapter 29, we discuss a number of ethical principles that include the engineer's obligations to society. It is therefore important that all aspects of the construction phase are transparent and devoid of conflicts of interest. For example, the tender process must be characterized by fair and open competition for all bidders so that the possibilities of fraud or collusion are completely eliminated. Also, the supervisors of the contractor's work will need to give highest priority to the interest of society and their client (the system owner). Further, the supply of equipment, materials, and other system components or construction services must be solicited from the least cost and highest quality supplier in an open transparent manner.

22.6.6 Increased Potential for Litigation

As civil systems grow in size and complexity, their construction becomes increasingly complex and thus are undertaken by an increased number of contractors working together. Where there is inadequate communication, this can lead to problems on the construction site and great potential for legal action initiated by the system owner, other contractual parties, or stakeholders of the construction process (see Chapter 29). The system owner or contractor can lose significant sums of money, time, and goodwill in lawsuits. This underscores the need for both the system owner and contractor to work together to carry out meticulous precontract and during-contract activities. These include careful investigations of site conditions and, consequently, reliable estimation of project quantities prior to bidding; selection of appropriate contracting approaches that inherently provide flexibility in construction and minimization of conflicts; regular monitoring of the project using construction inspectors; frequent communication with all parties to the contract; and ensuring that the construction process does not pose undue safety hazards to construction workers and the community.

SUMMARY

In this chapter, we first provided an initial general discussion about the construction phase, which included an identification of the key parameters used to measure the performance of a construction project: quality of the completed construction, cost performance (absence of overruns), time performance (absence of time delay), worker safety (minimal site accidents), and absence of conflict among contractual parties. The chapter then discussed the various stages of the system construction phase from the perspective of the system owner: identifying the optimal contracting approach, preparing contract documents, inviting and evaluating bids, selecting and awarding the contract, and handing over the site to the contractor, monitoring the construction, and taking over the site at the end of the construction. Recognizing the current emphasis worldwide on innovative contracting approaches and their demonstrated potential for overcoming the limitations of traditional approaches, the chapter also discussed the different types of contracting approaches. The critical issue of construction cost estimation was discussed as well, focusing on the factors that influence construction cost and the relationship between the phase of development and the level of cost aggregation. Costing is important from the contractor's perspective because efficient cost control translates into a greater profit margin; and from the owner's perspective, efficient cost estimation and tracking helps in the management of cost overruns and facilitates transparency and accountability to the ultimate owner, often the taxpayer. The chapter then examined the application of systems concepts for addressing a variety of problem contexts at the construction phase, including selection of a contractor, a contracting approach, or a specific type of material or work process. As some of these contexts are closely related to the key measures of project performance, the chapter singled out a few of them for closer scrutiny (i.e., cost overruns, time delays, and safety incidents), and statistics-based acceptance testing was touched on briefly. Finally, the chapter discussed a number of emerging or enolving issues associated with the construction phase of civil systems development.

EXERCISES

1. Ultra Construction Company is bidding on three contracts. The probability of winning contracts A, B, and C are 0.35, 0.75, and 0.65, respectively. Find the probability that the company will win (a) all three contracts and (b) any two of the three contracts.
2. Vendors I, II, III, and IV provide all the fresh concrete that ABC Construction purchases. From these vendors, the firm purchase 25, 35, 10, and 30%, respectively, of the concrete it needs on a daily basis. It is known from past experience that vendors I, II, III, and IV provide 80, 95, 70, and 90% perfect concrete. What is the probability that a randomly selected concrete batch is not perfect? Given that a concrete batch is not perfect, what is the probability that it came from vendor III?
3. In a shipment of steel bars, 2 out of every 100 sets are defective. A construction company makes a purchase of 500 bars selected at random from the manufacturer's production. If x is the number of defective bars purchased by the company, (a) identify the probability distribution of x and (b) find the probability that 10 bars are defective.
4. At a certain construction site, the probability of an accident on any given day is 0.005, and the accidents that occur are independent of each other. What is the probability that, in any given period of 40 days, there will be an accident on one day? What is the probability that there are at most 3 days with an accident? Use (a) the binomial distribution and (b) the Poisson distribution to solve this problem. Compare your answers.
5. Quality control of materials is a critical aspect of every civil engineering construction project. During construction of a retaining wall for a large wastewater plant in the City of La Serena, you (the consultant) randomly sampled each of six batches of concrete during a certain day's production. Sampling of each batch consisted of several cylindrical concrete specimens. After 28 days, you tested these concrete specimens for their compressive strengths and calculated the average compressive strength (ACS) of each sample. The following results were obtained:

Sample #	**1**	**2**	**3**	**4**	**5**	**6**	**7**	**8**	**9**	**10**	**11**	**12**	**13**	**14**	**15**
ACS	44.3	42.4	47.8	42.8	45.5	45.8	45.1	43.8	44.4	42.8	45.5	45.8	45.7	41.8	44.3

Sample #	**16**	**17**	**18**	**19**	**20**	**21**	**22**	**23**	**24**	**25**	**26**	**27**	**28**	**29**	**30**
ACS	44.1	47.4	47.2	46.8	41.3	45.9	45.1	48.3	45.7	48.2	45.1	48.5	47.5	48.1	43.4

From the contract specifications, the minimum 28-day strength of that grade of concrete is typically 40 N/mm^2, which should be attained before the concrete is accepted. (a) Provide a simple plot of the mean strength of various samples. (b) From your sketch in (a), describe qualitatively the bias and efficiency of the estimate of the population mean. (c) Compute the estimated mean of the population (calculated as the mean of the sample means). Is your estimated population mean biased from the true population mean? (d) Assess the efficiency of the estimate by computing the variance of the sample means.

6. During construction of a large urban drainage canal, concrete samples were taken to ascertain that the concrete produced is consistent with specifications. According to the contract agreement, the materials engineer should be have 90% confidence that the estimate of the mean slump does not deviate from the specified value by more than 5 mm? From past experience, it has been found that the parameter of produced units is normally distributed with a variance of 12 mm. How many observations should be taken in a test sample?

7. An engineer collected data from a random sample of water tank construction projects to build a statistical linear model that not only estimates the unit cost of water tank construction but also investigates the influence of certain factors on the unit cost of construction ($/ft^3). (i) For each independent variable, use the table below to indicate whether the model result is intuitive and to explain why or why not. (ii) Use the model to estimate the total cost of an elevated steel tank of 50,000 ft^3 capacity.

Independent Variables	Coefficient Estimate	Intuitiveness of the result (Yes/No)	Your Explanation for Intuitiveness or Nonintuitiveness
Constant Term	8.11	–	–
TANK MATERIAL = 0 if concrete = 1 if steel	−2.34		
TANK SIZE (in 10,000s of ft^3)	−0.56		
TANK LOCATION = 0 if underground = 1 if elevated	2.26		

REFERENCES

Abdelgawad, M., and Fayek, A. (2012). Comprehensive Hybrid Framework for Risk Analysis in the Construction Industry Using Combined Failure Mode and Effect Analysis, Fault Trees, Event Trees, and Fuzzy Logic. *ASCE J. Const. Eng. Manage.* 138(5), 642–651.

Al-Tabtabai, H., Alex, P. A., and Tantash, M. (1999). Preliminary Cost Estimation of Highway Construction Using Neural Networks. *Cost Eng. J.* 41(3), 19–24.

Akinci, B., and Fisher, M. (1998). Factors Affecting Contractors Risk of Cost Overburden, *ASCE Jrnl. of. Manage. Eng.* 14(1) 67–76.

Akpan, E. O., and Igwe, O. (2001). Methodology to Determine Price Variation in Projects, *ASCE Jrnl. of. Constt. Eng. & Manage.* 127(5) 367–373.

Anastasopoulos, P., McCullouch, B., Gkritza, K., Mannering, F., and Sinha, K. (2010). Cost Savings Analysis of Performance-Based Contracts for Highway Maintenance Operations. *ASCE J. Infrastruct. Syst.* 16(4), 251–263.

Anderson, S. (2000). Improved Contacting Methods for Highway Construction. National Cooperative Highway Research Program (NCHRP), Project 10–49.

Anderson, S., Molenaar, K., and Schexnayder, C. (2007). Guidance for Cost Estimation and management for Highway Projects, NCHRP Report 574, Transportation Research Board, Washington, DC.

Arditi, D., Khisty, J., and Yasamis, F. (1997). Incentive/Disincentive Provisions in Highway Contracts. *ASCE J. Construct. Eng. Manage.* 123, 302–307.

Aschenbrener, T. B. and DeDios, R. E. (2001). Materials and Workmanship Warranties for Hot Bituminous Pavement. Rep. Nr. CDOT-DTD-2001-18, Cost-Benefit Evaluation Committee, Colorado Dept. of Transportation.

Ashuri, B., and Tavakolan, M. (2012). Fuzzy Enabled Hybrid Genetic Algorithm–Particle Swarm Optimization Approach to Solve TCRO Problems in Construction Project Planning. *ASCE J. Construct. Eng. Manage.* 138(9), 1065–1074.

Attala, M., and Hegazy, T. (2003). Predicting Cost Deviation in Reconstruction Projects: Artificial Neural Networks vs. Regression, *ASCE Jrnl.of Constrt. Eng. & Manage.*, 129(4), 405–411.

Bhargava, A. (2009). A Probabilistic Evaluations of Highway Construction Costs, Ph.D. Thesis, Purdue Univ., W-Lafayette, IN.

Bode, J. (1998). Neural Networks for Cost Estimation. *Cost Eng.* 40(1), 25–30.

Bordat, C., McCollouch, B., Labi, S., and Sinha, K. C. (2004). An Analysis of Cost Overrans and Time Delays in INDOT Projects, FHWA/JTRP/04-07,west Lafayette, IN.

Bubshait, A. (2003). Incentive/Disincentive Contracts and Its Effects on Industrial Projects. *Int. J. Project Manage.* 21, 63–70.

Carpenter, B., Fekpe, E., and Gopalakrishna, D. (2003). Performance-Based Contracting for the Highway Construction Industry. Final Report prepared for Koch Industries Inc., Washington, DC.

Chang, L.-M., and Hsie, M. (1995). Developing Acceptance-sampling Methods for Quality Construction. *ASCE J. Construct. Eng. Manage.* 121(2), 246–253.

Chaovalitwongse, A. W., Wang, W., Williams, T., and Chaovalitwongse, P. (2012). Data Mining Framework to Optimize the Bid Selection Policy for Competitively Bid Highway Construction Projects. *ASCE J. Construct. Eng. Manage.* 138(2), 277–286.

Chen, Y., Okudan, G. E., and Riley, D. R. (2010). Sustainable Performance Criteria for Construction Method Selection in Concrete Buildings. *Automat. Construct.* 19(2), 235–244.

Choi, K., Kwak, Y., Pyeon, J., and Son, K. (2012). Schedule Effectiveness of Alternative Contracting Strategies for Transportation Infrastructure Improvement Projects. *ASCE J. Construct. Eng. Manage.* 138(3), 323–330.

Dainty, A., and Loosemore, M. (2003). *Human Resource Management in Construction: Strategic and Operational Approaches*. Routledge. New York.

Dewlaney, K., Hallowell, M., and Fortunato, B. (2012). Safety Risk Quantification for High Performance Sustainable Building Construction. *ASCE J. Construct. Eng. Manage.* 138(8), 964–971.

Easa, S. M., and Can, E. K. (1985). Optimization Model for Aggregate Blending. *ASCE J. Construct. Eng. Manage.* 111(3), 216–230.

El-Kashif, H., Hosny, O., Ramadan, O., and El-Said, M. (2000). An Integrated Model for Construction Projects Bid Evaluation in Egypt. Proceedings, World Organization of Building Officials, Congress, Calgary, Canada.

Ernzen, J., and Feeney, T. (2002). Contractor Led Quality Control And Quality Assurance Plus Design-Build: Who Is Watching the Quality? *Transport. Res. Rec.* 1813, 253–259.

Estache, A., and Iimi, A., 2011. (Un)bundling Infrastructure Procurement: Evidence from Water Supply and Sewage Projects. *Utilit. Pol.* 19(2), 104–114.

FDOT (2000). Highway Construction and Engineering and Transportation System Maintenance Programs. Office of Program Policy Analysis and Government Accountability, Rep. 9929, Florida Dept. of Transportation.

Fischer, M., and Kunz, J. (2004). The Scope and Role of Information Technology in Construction. CIFE Center for Integrated Facility Engineering, CIFE Tech. Rep. 156, Stanford University.

Flyvbjerg, B., Holm, M. K. S., and Buhl, S. L. (2002). Underestimating Costs in Public Works Projects—Error or Lie. *J. Am. Planning Associ.* 68(3), 279–295.

Gharaibeh, N. G., Garber, S. I., and Liu, L. (2010). Determining Optimum Sample Size for Percent-Within-Limits Specifications. *Transport. Res. Rec.* 2151, 77–83.

Gkritza, K., and Labi, S. (2008). Estimating Cost Discrepancies in Highway Contracts, *ASCE Jrnl of Constr. Eng. & Mangae.*, 134(12) 953–962.

Goodman, A., and Hastak, M. (2007). *Infrastructure Planning Handbook*, ASCE Press & McGraw-Hill, New York.

Grimm, V., Pacini, R., Spagnolo, G., and Zanza, M. (2006). Division into Lots and Competition in Procurement, in *Handbook of Procurement*, N. Dimitri, G., Piga, and G. Spagnolo, Eds., Cambridge University Press, New York.

Hancher, D. E. (1999). Contracting Methods for Highway Construction. *Transport. Res. News* 205, 10–12.

Hancher, D. E. (2003). Construction, in *The Civil Engineering Handbook*, 2nd ed., W. F. Chen, J. Y. R. and Liew , Eds. CRC Press, Boca Raton, FL.

Hanna, A. S., and Sanvido, V. E. (1990). Interactive Vertical Formwork Selection System. *Concrete Int.* 12(4), 26–32.

Hegazy, T., and Ayed, A. (1998). Neural Network Model for Parametric Cost Estimation of Highway Projects. *ASCE J. Construct. Eng. Manage.* 124(3), 210–218.

Hendrickson, C., and Au, T. (1989). *Project Management for Construction*. Prentice-Hall, Englewood Cliffs, NJ.

Herbsman, Z. J., and Glagola, C. R. (1998). Lane Rental: Innovative Way to Reduce Road Construction Time. *ASCE J. Construct. Eng. Manage.* 124(5), 411–417.

Hufschmidt, M. M., and Gerin, J. (1970). Systematic Errors in Cost Estimates for Public Investment Projects, *The Analysis of Public Output*, J. Margolis (Ed.), Columbia University Press, New York.

Ikpe, E., Hammon, F., and Oloke, D. (2012). Cost-Benefit Analysis for Accident Prevention in Construction Projects. *ASCE J. Construct. Eng. Manage.* 138(8), 991–998.

Jahren, C. T., and Asha, A. M. (1990). Predictors of Cost Overrun Rates, *ASCE J. Construct. Eng. & Manage.*, 116(3), 548–552.

Jaraiedi, M., Plummer, R., and Aber, M. (1995). Incentive/Disincentive Guidelines for Highway Construction Contracts, *ASCE J. Construct. Eng. Manage.* 121, 112–120.

Iseley, T., and Gokhale, S. (1997). Trenchless Installation of Conduits, NCHRP Synthesis 242, Transportation Research Board, Washington, DC.

Kadir, M. R. A., Lee, W. P., Jaafar, M. S., Sapuan, S. M., and Ali, A. A. (2006). Construction Performance Comparison Between Conventional and Industrialized Building Systems in Malaysia. *Struct. Surv.* 24(5), 412–424.

Kim, B., Lee, H., Park, H., and Kim, H. (2012). Framework for Estimating Greenhouse Gas Emissions Due to Asphalt Pavement Construction. *ASCE J. Construct. Eng. Manage.* 138(11), 1312–1321.

Kim, S., Yang, I., Yeo, M., and Kim, K. (2005). Development of a Housing Performance Evaluation Model for Multi-Family Residential Buildings in Korea. *Build. Environ.* 40, 1103–1116.

Knight, K., and Fayek, A. R. (2002). Use of Fuzzy Logic for Predicting Design Cost Overruns on Building Projects. *ASCE J. Construct. Eng. Manage.* 128(6), 503–512.

Kraeim, Z., and Diekmann, J. (1987). Concurrent Delays in Construction Projects, *ASCE J. Construct. Eng. Manage.* 113(4), 591–502.

Lee, C., Hong, T., Lee, G., and Jeong, J. (2012). Life-Cycle Cost Analysis on Glass Type of High-Rise Buildings for Increasing Energy Efficiency and Reducing CO_2 Emissions in Korea. *ASCE J. Construct. Eng. Manage.* 138(7), 897–904.

Lee, S. M., and Olson, D. L. (1983). Chance Constrained Aggregate Blending. *ASCE J. Construct. Eng. Manage.* 109(1), 39–47.

Lin, Y., Wu, D., Wang, X., Wang, X., and Gao, S. (2012). Statics-Based Simulation Approach for Two-Crane Lift. *ASCE J. Construct. Eng. Manage.* 138(10), 1139–1149.

Love, P. E. D., Skitmore, R. M., and Earl, G. (1998). Selecting a Suitable Procurement Method for a Building Project. *Construct. Manage. Econom.* 16(2), 221–233.

Majid, M., and McCaffer, R. (1998). Factors of Non-Excusable Delays That Influence Contractors' Performance. *ASCE J. Construct. Eng. Manage.* 14(3), 42–49.

McCullouch, B. G., Sinha, K. C., and Anastasopoulos, P. (2009). Performance-Based Contracting for Roadway Maintenance Operations in Indiana, FHWA/IN/JTRP-2008/12, Joint Transportation Research Program, West Lafayette, IN.

Meyer, M. D., and Miller, E. J. (2001). *Urban Transportation Planning*. McGraw-Hill, Columbus, OH.

Mitropoulos, P., and Namboodiri, M. (2011). New Method for Measuring the Safety Risk of Construction Activities: Task Demand Assessment. *ASCE J. Construct. Eng. Manage*. 137(1), 30–38.

Murdoch, J., and Hughes, W. (2008). *Construction Contracts: Law and Management*, 4Ed., Taylor & Francis, Oxoin, UK.

Neumann, D. L. (1964). A Mathematical Method for Blending Aggregates for a Desired Gradation. Defense Technical Information Center.

Pakkala, P. (2002). *Innovative Project Delivery Methods for Infrastructure, International Perspective*. Finnish Road Enterprise, Helsinki, Finland.

Pan, W., Dainty, A., and Gibb, A. (2012). Establishing and Weighting Decision Criteria for Building System Selection in Housing Construction. *ASCE J. Construct. Eng. Manage.* 138(11), 1239–1250.

Petroutsatou, K., Georgopoulos, E., Lambropoulos, S., and Pantouvakis, J. (2012). Early Cost Estimating of Road Tunnel Construction Using Neural Networks. *ASCE J. Construct. Eng. Manage.* 138(6), 679–687.

Rowland, H. (1981). The Causes and Effects of Change Orders on the Construction Process, Ph.D. Dissertation, Georgia Institute of Technology, Atlanta, GA.

Schexnayder, C. J., Weber, S. L., and Fiori, C. (2003). Project Cost Estimating: A Synthesis of Highway Practice. Report for NCHRP Project 20-07/Task 152.

Segal, G. F., Moore, A. T., and McCarthy, S. (2003). *Contracting for Road and Highway Maintenance*. Reason Public Policy Institute, Los Angeles, CA.

Shapira, A., Simcha, M., and Goldenberg, M. (2012). Integrative Model for Quantitative Evaluation of Safety on Construction Sites with Tower Cranes. *ASCE J. Construct. Eng. Manage.* 138(11), 1281–1293.

Shr, J-F., and Chen, W. T. (2004). Setting Maximum Incentive for Incentive/Disincentive Contracts for Highway Projects. *ASCE J. Construct. Eng. Manage.* 130(1), 84–93.

Sillars, D., and Riedl, J. (2007). Framework Model for Determining Incentive and Disincentive Amounts. *Transport. Res. Rec.* 2040, 11–18.

Singh, P., Oh, J., Labi, S., and Sinha, K. C. (2007). Cost-effectiveness of Warranty Projects in Indiana. *ASCE J. Construct. Eng. Manage.* 133(3), 217–224.

Sinha, K. C., McCullouch, B. G., Bullock, D. M., Konduri, S. Fricker, J. D., and Labi, S. (2004). Evaluation of INDOT Hyperfix Project, FHWA/IN/JTRP-2004/02. Joint Transportation Research Program, West Lafayette, IN,

Son, K., Choi, K., Woods, P., and Park, Y. (2012). Urban Sustainability Predictive Model Using GIS: Appraised Land Value vs. LEED Sustainable Site Credits. *ASCE J. Construct. Eng. Manage.* 138(9), 1107–1112.

Tabish, S., and Jha, K. (2012). Success Traits for a Construction Project. *ASCE J. Construct. Eng. Manage.* 138(10), 1131–1138.

Tam, C. M., Tong, K. L., Lau, C. T., and Chan, K. K. (2005). Selection of Vertical Formwork System by Probabilistic Neural Networks Models. *Construct. Manage. Econo.* 23(3), 245–254.

Taylor, T. R. B., and Ford, D. N. (2006). Tipping Point Failure and Robustness in Single Development Projects. *Sys. Dynam. Rev.* 22(1), 51–71.

Toklu, Y. C. (2005). Aggregate Blending Using Genetic Algorithms. *Computer–Aided Civil Infrastruct. Eng.* (20)6, 450–460.

Tubacanon, M. T., Abuldhan, P., and Chen, S. S. Y. (1980). A Probabilistic Programming Model for Blending Aggregates. *App. Math. Model.* 4, 257–260.

Visscher, H., Suddle, S., and Meijer, F. (2008). Quantitative Risk Analysis as a Supporting Tool for Safety Protocols at Multifunctional Urban Locations. *Construct. Innovat.* 8(4), 269–279.

Wohl, M., and Hendrickson, C. (1984). *Transportation Investment and Pricing Principles*. Wiley, New York.

World Bank (2009). Performance-based Contracting for the Preservation and Improvement of Road Assets, Resource Guide, www-esd.worldbank.org/pbc_resource_guide/.

WSDOT (2008). *Construction Contracts Annual Update*. Washington State Dept. of Transp. http://www.wsdot.wa.gov/biz/construction/performancemeasures.cfm.

Yasamis-Speroni, F., Lee, D., and Arditi, D. (2012). Evaluating the Quality Performance of Pavement Contractors, *ASCE J. Construct. Eng. Manage.* 138(10), 1114–1124.

Zhang, X-Q. (2006). Factor Analysis of Public Clients' Best-Value Objective in Public–Privately Partnered Infrastructure Projects, *ASCE J. Construct. Eng. Manage*. 132 (9).

Zietlow, G. (2005). Cutting Costs and Improving Quality through Performance-Based Road Management and Maintenance Contracts—The Latin American and OECD Experiences, April 24–29, 2005, University of Birmingham (UK).

USEFUL RESOURCES

Flyberg, B., Holm, M. K. S., Buhl, S. L. (2003). How Common and How Large are Cost Overruns in Transport Infrastructure Projects, *Transport Rev.* 23(1), 71–88.

Halpin, D. (2006). *Construction Management*. Wiley, Hoboken, NJ.

Hancher, D. E. (1994). Use of Warranties in Road Construction. National Cooperative Highway Research Program Synthesis of Highway Practice 195, Transportation Research Board, Washington, DC.

Hendrickson, C., and Au, T. (1989). *Project Management for Construction*, Prentice Hall, Upper Saddle River, NJ.

Lopez, R., and Love, P. (2012). Design Error Costs in Construction Projects. *ASCE J. Construct. Eng. Manage.* 138(5), 585–593.

Schexnayder, C. J., and Mayo, R. E. (2004). *Construction Management Fundamentals*. McGraw-Hill, Columbus, OH.

CHAPTER 23

SYSTEM OPERATIONS

23.0 INTRODUCTION

All civil systems are developed in response to a need and therefore are commissioned into use when the construction is completed. Thus, the system operations phase, as naturally expected, is the longest of all the phases of development. Further, this phase is related to the system monitoring/inspection and preservation phases in a parallel manner rather than the sequential manner that is true of all the other phases. In other words, the operation of the system follows after the system is planned, designed, and constructed; however, as the system is being used, it is monitored (either continuously or intermittently for defects and usage patterns) and also maintained (proactively at specified intervals of time to prevent the onset of imminent defects or reactively when needed to address existing defects). Some experts maintain the position that maintenance and monitoring are not phases per se as we have illustrated throughout this text (see cutoff figure in Figure 23.1*a*), and that those activities are simply part of the operations phase (as depicted in Figure 23.1*b*).

In this chapter, we shall define the term operations, discuss some of the general duties of engineers responsible for the operations of civil systems, and a few examples of operations-related tasks in relation to a select number of civil engineering systems. Finally, we will review some numerical problems in civil engineering system operations.

23.1 DEFINITION

A system is considered to be operating when it is being used for the function for which it was constructed. When one mentions the word "operations," it is easy to conjure up mental images of something moving back and forth or cyclically as a motor engine. Thus, it may be difficult to mentally perceive that certain systems in civil engineering are considered to be is operating when they are "merely" playing their role, such as the case of a retaining wall that has the "invisible" function of holding back earth. This is true of many other "static" systems such as towers, pavements, and foundation systems. It is also much easier to conceive the word "operations" in the context of transit systems that involve vehicular movement on a rail track, hydraulic systems that involve the flow of water in a channel or other hydraulic structure, or water treatment systems that involve the sequential treatment of water in multiple treatment phases at different locations within the treatment plant.

A formal definition for system operations is: "The set of continuous activities that ensure the running of a system for the purpose of producing value for the stakeholders."

The "value" could be expressed in terms of various performance measures from the perspective of the agency (ability to support some load, provide some service minimal breakdowns, etc.), the user (minimal disruption, maximum ease of use, etc.), or the community (minimal external impacts such as noise, vibration, or air pollution). In the case of profit-driven systems, another

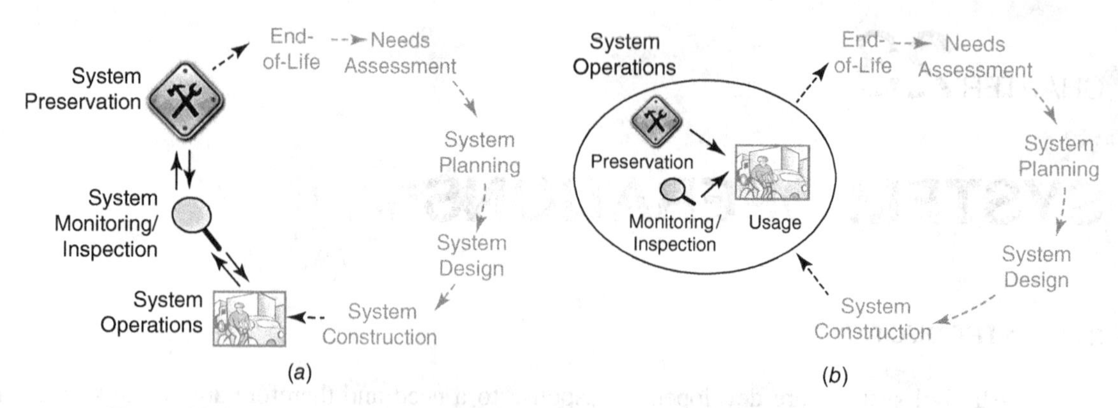

Figure 23.1 Two different perspectives of the operations phase in relation to the preservation and monitoring phases: (*a*) preservation and monitoring/inspection as distinct phases and (*b*) preservation and monitoring/inspection as part of the operations phase.

key performance measure at the operations phase is the generation of revenue. For nonprofit systems, it is often sought to operate the system in a cost-effective and self-sustaining manner so that government subsidies or other external financial interventions are minimized.

At this point, it is appropriate to explain the background of the term "operations research," which denotes a set of specific analytical tools including optimization and simulation, that we studied in Part 2 of this book. During World War II, operations research (OR) was defined as the collection of scientific techniques that provided the military brass with a quantitative basis for decisions regarding war operations. Several persons, working for the British Armed Forces, provided advice on a wide range of war operations including optimal convoy sizes, aircraft color, and bomber formation patterns. In more peaceful applications of operations research, civil engineers apply OR tools extensively, not only at the operations phase of civil systems development, but at all the other phases as well to enhance the engineer's tasks at those phases.

Figure 23.2 provides images of some common civil engineering systems in operation. These examples include the operations of a public transportation system at Curitiba, Brazil, where the system owner (the transit agency) strives to ensure that public satisfaction with travel time, comfort, safety, and cleanliness are maximized within a limited budget; incident clearance on freeway systems where the systems owner seeks to maximize traffic flow and safety by quickly removing all crashes, stalled vehicles, and debris; spilling during the operations of a dam where the operations engineers constantly make decisions on when to open the spillways and for how long; and environmental engineer at a water treatment plant who is engaged in daily operations of running the plant.

For certain types of systems, the engineers responsible for the system operations are also responsible for the system monitoring and inspection and/or system maintenance. This fusion of duties may be due to situations where the operations personnel have intimate day-to-day knowledge about the functioning of the system; thus, they are in the best position to monitor and inspect the system as they go about their duties and also to carry out any needed repairs. In large agencies, however, the tasks of system monitoring, maintenance, and operations are carried out by distinct persons or units. In any case, it is particularly vital that the personnel working in these three phases communicate with each other constantly.

(a)

(b)

(c)

Figure 23.2 Illustrations of civil system operations: the massive investments in taxpayer-funded civil systems underscore the fiduciary responsibility of operations engineers: (*a*) Spilling during dam operations. Knowing when to open the spillways and for how long is a key aspect of dam system operations. (*b*) Bus transit system at Curitiba, Brazil. Maximizing the performance of a system's operations can improve public relations and accountability. (*c*) Water treatment system operations. Environmental engineers at treatment plants run daily tests as part of their routine operations [Courtesy of *(a)*. ZSM/Wikimedia Commons; *(b)*. Mario Roberto Duran Ortiz/Wikimedia Commons; *(c)*. Environmental Protection Agency/Wikimedia Commons].

23.2 GENERAL DUTIES AT THE OPERATIONS PHASE

The duties of the operations engineer are a reflection of the general tasks of system description, analysis, evaluation, and optimization as we discussed in chapter 4. We will present a few of these duties here. Then, in Section 23.3, we will discuss some general contexts where the engineer carries out these duties in the operations of specific types of civil systems.

23.2.1 Resource Allocation

The operations engineer's main duty is to ensure that the system is being used safely and cost-effectively to the satisfaction of the customers or stakeholders. To do this, the operations engineer allocates resources (manpower, facilities, equipment, vehicles, funds, etc.) in order to meet the system's mission and the customers' needs. For resource allocation, the engineer applies mostly the tools of optimization (see Chapter 11).

23.2.2 Provide Base Support Services

The operations engineer is also responsible for overseeing the provision of support services for the system, including system sanitation, pest and vegetation control, grounds maintenance, security, and snow removal and ice control. For many civil engineering systems, these considerations are helpful, if not vital, to the successful operation of the system.

23.2.3 Conduct Constructability Reviews

Where a capacity expansion or major retrofitting of the system is being planned, the operations engineer reviews the construction plans, design, and technical specifications to ensure that the outcome of the project will be consistent with the system owner's construction standards and that the construction process will not pose a serious threat to the ongoing operations of the system.

23.2.4 Operational Performance Measurement

As the civil engineering system typically serves a need posed by at least one stakeholder, the operations engineer is responsible for ascertaining the extent to which such customers are satisfied. Thus, the operations engineer establishes (and updates, where necessary) the quality standards for operational performance and develops feedback mechanisms to assess such performance. That way, relevant information can be acquired and provided to top-level engineers who then assess the extent to which the core mission of the system is being realized.

23.2.5 Conduct Assessment of Natural and Man-made Threats to the Operations of the System

This includes assessment of the likelihood of full or partial failure of the civil engineering system at its operations phase, due to external or internal threats such as inclement weather, earthquakes, landslides, floods, or degraded physical components arising from fatigue, steel corrosion, concrete spalling, or failure modes. Also, any threats of damage due to overloading, vandalism, and collision with moving objects are assessed by the operations engineer. The operations engineer is responsible for developing risk mitigation plans (such as fire hydrant inspections) and response plans in the case of disaster, such as evacuation protocols. We discuss these issues further in Chapter 27.

23.2.6 Updating Estimates of the Amount of Work Needed for Operations

Operations engineers often find a need to adjust the scope or level of their system operations in response to changes in the natural or built-up environment, such as an increase in the levels of system usage or loading or excessively hot or cold weather. These changes may be temporary or long-lasting. In any case, where such changes occur, the effort (and cost) associated with operating the system will be higher or lower than before, and the operations engineer is responsible for quantifying the additional or reduced effort, and in some cases, to estimate the resulting cost or cost reduction of such changes.

23.2.7 Carry Out Duties Associated with Other Phases Related to the Operations Phase

As discussed earlier, in certain system agencies, the operations engineer may be responsible for carrying out the tasks and duties not only for the systems operations but also for the phases of monitoring and inspection (measuring the usage/loading on the system, and tracking the physical condition and defects of the system) and maintenance (applying treatments and schedules to keep the system in good physical condition and for enhancing system durability). At other agencies, there are separate units and personnel for the tasks of monitoring/inspection and maintenance.

23.2.8 Maintain Information Systems for System Operations

A key duty of the operations engineer is to maintain a database of all aspects of the system that are associated with the system operations, including the system inventory (reference points, location, dimensions, materials, for example), system physical condition, past records of operational failures or hiccups, operations-related work done by outside contracting and in-house, and their costs and effectiveness. This responsibility also includes an accounting system that tracks the consumption of in-house resources (materials, man-power, and equipment use) used for the system operations.

23.3 SOME PROBLEM CONTEXTS AT THE OPERATIONS PHASE OF SELECTED TYPES OF CIVIL ENGINEERING (CE) SYSTEMS

23.3.1 Water and Wastewater Treatment Plants

Engineers who operate water/wastewater treatment plants regularly supervise the processes involved in the treatment. These processes include physical processes (screens, chemical reactors, mixing devices, sedimentation tanks, filters, odor control, and aeration systems), chemical processes (coagulation systems, systems for softening, stabilization, demineralization, chemical oxidation system, and disinfection system), and biological processes (activated sludge system, aerobic fixed-film processes and operations, pond design, and treatment wetlands).

Engineers in charge of these operations conduct daily checks and tests to ensure that these systems are operating efficiently, diagnose and supervise repair of malfunctioning units, and optimize the use of resources at the plant (labor, equipment, and materials). In doing so, such engineers have opportunities to use analytical tools including continuous-variable optimization (linear programming formulations), discrete-variable optimization (knapsack formulations), and graphical simulation of the plant operations for educating or informing interested stakeholders and other audiences.

23.3.2 Solid Waste Management

The management of solid waste is inherently consistent with the applications of wide range of analytical tools, including those related to network analysis. For solid waste routing, these include, Chinese Postman tours (traveling along links of a network to collect solid waste from residences, offices, and industries along the links), Hitchcock shipment problems (determining how much material should be routed on each link of a network if the material is collected from a number of possible sources such as waste collection centers and depositing the material at a number of possible destinations for purposes including treatment, disposal, or recycling), and transshipment problems (a generalization of the Hitchcock shipment problem where there are intermediate nodes that represent intermediate facilities such as waste transfer or waste processing stations).

23.3.3 Hydroelectric Systems

Engineers in charge of dam operations are responsible for predicting water levels at dams on the basis of rainfall intensity in catch basins, evaporation rates, upstream water intakes for residential/commercial/industrial use, and so forth. As part of their operations duties, they decide when to open spillways and for how long. Also, there are engineers of other disciplines (such as mechanical and electrical) that are responsible for other aspects of the operations of this system.

23.3.4 Urban Drainage Systems

Hydraulic engineers in charge of urban drainage operations routinely carry out tasks of predicting the volumes of storm water flow (rainfall intensity in catch basin, evaporation rates, and percolation rates). They are also responsible for checking the adequacy of the channel capacities, providing feedback to the hydraulic system designers, and providing recommendations to the hydraulic system maintenance engineers. Thus, they make extensive use of tools such as statistical modeling and continuous-variable optimization.

23.3.5 Airport Runway Systems

Engineers responsible for runway operations at airports regularly check the runway surface to ensure that it is not too rough to cause discomfort and not too smooth to cause slipping when braking. These engineers also identify hazards to runway operations and mitigate risks, provide feedback to runway designers regarding specific designs that enhance or threaten safe and effective operations, and also provide recommendations to runway maintenance engineers.

23.3.6 Building Systems

For residential, commercial, and industrial buildings, the operations engineer often is the building manager. A person in this position is responsible for the duties of ensuring the adequate and uninterrupted supply of heating, ventilation, air-conditioning, electricity, lighting, gas, water, sanitary services, and other specialized utilities. Also, the building manager uses tools including costing, economics, finance, optimization and modeling in order to arrive at defensible recommendations that ensure smooth operations of the building system.

23.3.7 Transportation Logistics Systems

Logistics can be defined as the art and science of effectively and efficiently transporting goods or services from points of origin to points of destination. Similar to all systems, logistics, which is a virtual system, consists of the phases of needs assessment (establishing the need for a logistics system), planning and designing the system, implementing the system, operating it, monitoring its operational performance, and carrying out "maintenance" or tweaking of the system as and when needed. The operational performance of logistics systems may be measured in terms of length (time duration) of delivery, delivery reliability, inventory size, capacity utilization, cost, or cost-effectiveness (which combines multiple performance measures including cost).

23.3.8 Coastal Engineering Systems

In a bid to defend land masses against flooding and erosion or to reclaim land from water bodies, coastal engineers design and construct sea defense and coastal protection systems. However, sea levels are highly variable in both the long term (due to climate change) and the short term; for this reason, the task of coastal engineers does not end with the completion of construction but also includes the operation of these systems to ensure that they continue to provide the intended levels of service. Therefore, the responsibilities of coastal engineers during the operations phase include the monitoring, maintenance, and overall management of these systems. In their work, coastal engineers analyze the mechanics of waves, ocean wave climate, water level fluctuations, and coastal processes. They therefore use extensively descriptive and prescriptive analytical and numerical models including statistical analysis, simulation, and optimization. Also, in evaluating alternative engineering designs and coastal management policies, they use the tools of financial and economic analysis.

23.3.9 Public Transportation Systems

The operation of public transportation systems (buses, trains, ferries, airplanes) is one of the most visible and challenging of all civil engineering systems. Engineers in this branch of engineering constantly juggle resources (such as personnel or equipment) in order to provide acceptable levels of service (including on-schedule arrivals and departures and safe and comfortable rides) to an often-picky clientele. As such, transit operations engineers deal with the management of the fixed physical infrastructure (guideways and terminals), rolling stock (vehicles, trains, planes, and watercraft), scheduling and timetables, safety and security, and financing issues including revenue and subsidies. To carry out these duties, transit engineers use tools including financial and economic analysis, stochastic modeling and simulation, decision analysis, risk and reliability analysis, multiple criteria analysis, and optimization.

23.3.10 Structural Systems

In the context of this text, a structural system is a structure designed by a structural engineer and consists of beams, columns, slabs, domes, trusses, shells, and other structural elements. Structural systems can be found in most other branches in civil engineering where they support loads and enable other systems to carry out their function, for example, building systems, hydraulic systems, aerospace systems and environmental systems. Structural engineers ensure that these systems carry out their functions safely and effectively. Therefore, their work at the operations phase includes monitoring the loading or usage patterns, regular inspection of physical condition, and maintenance recommendation and supervision of the structural system as and when required. In doing so, they use a broad range of analytical tools including economic analysis, reliability analysis, multiple criteria analysis, simulation and stochastic modeling, and optimization.

23.4 NUMERICAL EXAMPLES OF APPLICATION OF TOOLS IN CE SYSTEM OPERATIONS

Example 23.1 Financial Evaluation of Water Supply Operations

A county owns a water tank and is considering leasing it to a private operator for 10 years. The tank brings in revenue of $750,000 annually but has $250,000 annual cost of operation and $130,000 annual maintenance costs. The private operator offers to pay $2 million upfront to the county and to make annual payments of $200,000 to the county for 10 years. The private operator will ensure that the level of service remains consistently above a certain threshold. In your opinion, should the proposal be accepted? Assume a 4% interest rate.

Solution

Assuming a 10-year analysis period for each option, the equivalent uniform annual return (EUAR) can be calculated as follows:

$$\text{EUAR}_{\text{Self-operation}} = \$0.75\text{M} - \$0.25\text{M} - \$0.13\text{M} = \$0.370\text{M} = \$370,000$$
$$\text{EUAR}_{\text{Lease}} = \$2\text{M}[(A/P, 4\%, 10 \text{ years})] + \$0.1\text{M} = 2(0.123) + \$0.2\text{M} = \$0.447\text{M} = \$447,000$$

A/P is the capital recover factor (see Appendix 4). The annualized return from the lease option exceeds that of the self-operation option by 20.8%. Thus, the proposal should be accepted.

Example 23.2 Wave Setup Threat (Coastal Engineering)

Wave setup is defined as the increase in the mean water level between the breaking point and the shore and is due to the presence of waves at the coast (Figure 23.3). After the waves break, the energy flux of the wave is no longer constant but decreases due to the dissipation of energy. This leads to a decrease in the radiation stress (i.e., the stress tensor of excess horizontal-momentum fluxes due to the presence of the waves) after the break point; to balance this, the free surface level increases (Wood and Meadows, 2002; Dean and Walton, 2009). Coastal engineers are interested in wave setup phenomenon particularly during storm events when the wind from the storm creates big waves and thus increase the mean sea level by wave setup, leading to increased risk of damage to coastal structures. The shore protection manual presents the following formula for calculating the wave setup (USACE, 1984):

$$S_w = 0.15d_b - \frac{g^{0.5}(H_0^t)^2 T}{64\pi d_b^{0.66}} \tag{23.1}$$

where d_b is the depth of breaking; H_0 is the unrefracted deep-water wave height. Calculate the worst possible wave setup when the depth of breaking ranges from 1 to 1.5 m and the unrefracted deep-water wave height ranges from 2 to 3 m. Assume wave period (T) is constant = 1 s^{-1}.

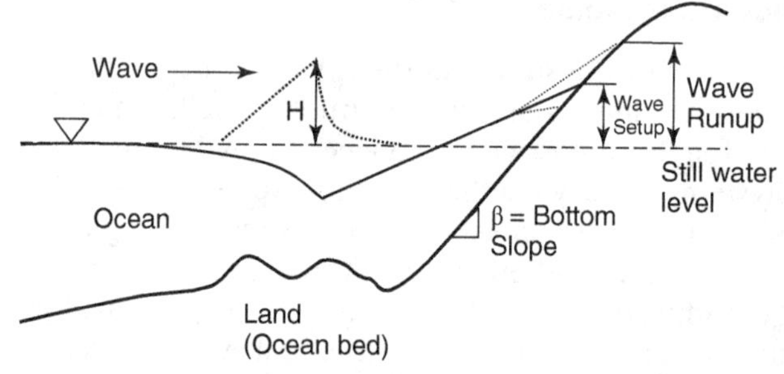

Figure 23.3 Increase in mean sea level by wave setup causes increased risk of damage to coastal structures.

Solution

The worst scenario occurs when the wave setup is at the highest level. This implies that H_0 is at the lowest possible value, which is at 2 m. Also, the worst scenario required d_b at the maximum possible value (1.5 m). Thus, the highest wave setup for this condition can be calculated from Equation 23.1 as follows.

$$S_w = 0.15d_{b,\max} - \frac{g^{0.5}(H_{0,\min}^t)^2 T}{64\pi d_b^{0.66}} = 0.15(1.5) - \frac{9.81^{0.5}(2)^2 1}{64\pi(1.5)^{0.66}} = 0.177 \text{ m}$$

Therefore, the magnitude of the wave setup is 17.7 cm.

Example 23.3 Landfill Operations

On a typical afternoon, three solid waste vehicles arrive at a landfill per hour. It takes 15 minutes to direct each vehicle to the appropriate site and discharge its load. Assuming deterministic arrival rates and service times, only one server, and a first-in–first-out queuing discipline, determine if a queue can be expected at the landfill.

Solution

The arrival rate of the vehicle (λ) is 3 vehicles/hour while the service rate (μ) is 15 minutes per hour (i.e., 4 vehicles/hour). Therefore, for the deterministic case and when λ is less than μ, we can expect there will be no queue at the landfill.

Example 23.4 Airport Runway Operations

Airplanes arrive at a single landing strip at an average rate of 10 planes per hour. On average, an airplane requires 4 minutes to land and taxi to its terminal. Assuming that the plane arrival pattern can be described by the Poisson distribution and that the service time is exponentially distributed, calculate the following: (a) the percent of the time that the runway will be idle; (b) the probability that, at any given time, there will be three planes in the queuing system; (c) the average number of planes in the queuing system; (d) the average queue length; and (e) the average time each plane spends in the queuing process.

Solution

From the given information, the arrival rate (λ) is 10 planes per hour while the service rate (μ) is $\frac{60}{4} = 15$ planes per hour. Then, the utilization rate (ρ) is $\frac{10}{15} = 0.67$.

(a) Percent of the time that the runway will be idle $= p(X = 0) = 0.67^0(1 - 0.67) = 0.33 = 33\%$

(b) Probability that, at any given time, there will be three planes in the queuing process, $p(X = 3) = 0.67^3(1 - 0.67) = 0.01 = 1\%$

(c) Average number of planes in the queuing process, $L = \frac{\lambda}{\mu-\lambda} = \frac{10}{15-10} = 2$

(d) Average queue length $= L_q = \frac{\lambda^2}{\mu(\mu-\lambda)} = \frac{10^2}{15(15-10)} = 1.33$

(e) The average time each plane spends in the queuing system $W = \frac{1}{\mu-\lambda} = \frac{1}{15-10} = 0.2$ hour or 12 minutes

---

Example 23.5 Urban Drainage Demand Assessment

In the rotational method of drainage design, the peak rate of surface flow from a given watershed is assumed to be proportional to the watershed area and the average rainfall intensity over a period of time just sufficient for all parts of the watershed to contribute to the outflow. The rational formula is

$$Q = CiA$$

where Q is the peak discharge (cfs), C is the ratio of peak runoff rate to average rainfall rate over the watershed during the time of concentration (runoff coefficient), i is the rainfall intensity (inches/hour), and A is the contributing area of the watershed (acres). A local agency plans to construct a new drainage system in the city. The area of the city consists of 15 acres of downtown area ($C = 0.8$), 40 acres of residential area ($C = 0.4$), and 60 acres of recreational area ($C = 0.5$). Given that the design rainfall intensity varies randomly between 3.5 and 4.0 inches/hour, determine the distribution of the peak discharge for this area. State any assumptions.

Solution

The rational formula rests on the following assumptions: (a) the rainfall intensity is uniform all over the watershed; (b) the duration of the storm that is associated with the peak discharge is equal to the drainage area's time of concentration; (c) the runoff coefficient is dependent on the rainfall return period but independent of the storm duration; and (d) the runoff coefficient is a reflection of the soil type of the watershed and its antecedent moisture condition, and the rate of infiltration (Rao et al., 2003).

For any watershed area comprised of multiple distinct watersheds, the weighted runoff coefficient can be calculated as follows:

$$C_{\text{effective}} = (C_1A_1 + C_2A_2 + \cdots + C_NA_N)/(A_1 + A_2 + \cdots + A_N)$$

where N is the total number of distinct component areas of the overall watershed. For the watershed in question,

$$C_{\text{effective}} = [0.8(20) + 0.4(50) + 0.5(35)]/(20 + 50 + 35) = 0.504$$

Hence the peak discharge is

$$Q = CiA = 0.504(115)i = 57.96i$$

where i is uniformly distributed between 3.5 and 4.0.

Implementing the expression Q in an appropriate computing platform and carrying out Monte Carlo simulation for the i variable, it may be observed that the peak discharge varies between 203.09 and 231.97 ft^3/s, with a mean and standard deviation of 217.11 and 8.28 ft^3/s, respectively. Note that different readers will obtain different results, but the overall result is expected to be similar and also close to what is reported here.

Example 23.6 Economic Analysis of System Operations

Two alternative ways of operating a highway system are proposed. The first alternative, which is labor intensive (e.g., vehicle patrolling), involves an initial cost of $2 million, annual salaries of $3 million, and fuel costs of $2 million. The second alternative, which is technology intensive (e.g., video camera installations), has initial costs of $35 million, an estimated life of 20 years, annual maintenance costs of $1.5 million, and salvage value of $2 million. Find the equivalent uniform annual return of each alternative and identify the alternative that should be undertaken. Assume a 5% interest rate and a 20-year analysis period. Assume that the alternatives are equally effective.

Solution

The equivalent uniform annual cost (EUAC) is:

$$\text{EUAC}_{\text{Alt 1}} = 2(\text{A/P}, 4\%, 20) + 3 + 2 = 2 \times 0.0802 + 3 + 2 = \$5.160 \text{ million.}$$

$$\begin{aligned}\text{EUAC}_{\text{Alt 2}} &= 35(\text{A/P}, 4\%, 20) + 1.5 - 2(\text{A/F}, 4\%, 20)\\ &= 35 \times 0.0736 + 1.5 - 2 \times 0.0336 = \$4.47 \text{ million.}\end{aligned}$$

A/P and A/F are as defined in Appendix 4. Alternative 2 has lower annual cost than alternative 1. Thus the technology-intensive option should be selected to operate the system.

Example 23.7 Air Pollution Assessment

The following relationship describes the rise of a plume, h (m), above a stack from a momentum source as a function of the wind speed and stack exit conditions (Jacko and Labreche, 2004):

$$\Delta h = D\left(\frac{V_s}{u_s}\right)^{1.4}$$

where D is the stack diameter = 2 m, u_s, is the mean wind speed at the stack height and is normally distributed between 4 and 8 m/s, and V_s, the emission velocity, is uniformly distributed between 5 and 12 m/s. Determine the mean and standard deviation of the plume rise and identify the minimum and maximum heights under these conditions.

Solution

The distribution of Monte Carlo simulation output for 600 iterations is presented in Figure 23.4. The average height and its standard deviation are 3.50 and 1.54 m, respectively. The maximum and minimum simulated flume heights are 9.14 and 1.12 m, respectively. The analytical maximum flume height is when V_s is at maximum and u_s is at minimum. Hence, the analytical maximum flume height is $2\left(\frac{12}{4}\right)1.4 =$ 9.31 m. On the other hand, the analytical minimum flume height is when V_s is at minimum and u_s is at maximum. Therefore, the analytical minimum flume height is $2\left(\frac{5}{8}\right)1.4 = 1.04$ m. Note that these results are generally consistent with the outcomes from the Monte Carlo simulation.

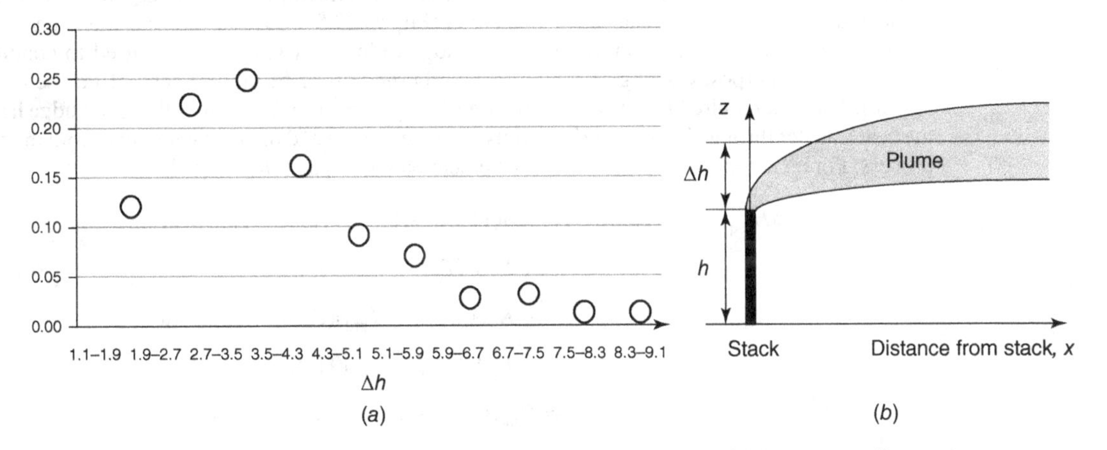

Figure 23.4 (*a*) Plume rise distribution for Example 23.7. (*b*) Plume rise illustration.

Example 23.8 Wastewater Treatment Operations

In wastewater treatment plants, traveling screens are cleaned when their operational performance (measured in terms of head loss) reaches a threshold of approximately 4 inches (10 cm). The head loss through a screen made of vertical, round, parallel wires or rods is (Blevins, 1984)

$$h_L = 0.52\left(\frac{1-\varepsilon^2}{\varepsilon^2}\right)\left(\frac{U^2}{2g}\right) \quad \text{if Re} = \frac{\rho U d}{\varepsilon\mu} > 500 \quad \text{and} \quad 0.10 < \varepsilon < 0.85$$

where d is the diameter of wires or rods in a screen (m); g is the acceleration due to gravity (9.81 m/s^2); s is the distance between wire or rod center in a screen (m), Re is the Reynolds' number; U is the approach velocity (m/s); ε is the screen porosity = $(s-d)/d$; μ is the dynamic viscosity of water ($\text{N}\cdot\text{sec}/\text{m}^2$); and ρ is the water density (kg/m^3). The diameter of wires in the screen is 0.13 m, and the distance between the wire centers in the screen is 0.15 m. Also, the water density is 1000 kg/m^3, the dynamic viscosity of water is 0.8 in 1000 $\text{N}\cdot\text{s/m}^2$ units, and the approach velocity is 2.5 m/s. (a) For a deterministic scenario, determine whether the traveling screen needed to be cleaned during a future 2-year period. (b) For a probabilistic situation, use an appropriate number of Monte Carlo simulations to determine the probability or percentage of the time that the traveling screen needed to be cleaned during a future 2-year that period. Assume that the approach velocity follows a Normal distribution with mean 2.5 m/s and standard deviation 0.85 m/s. Also, due to fluctuations in ambient temperature, the dynamic viscosity of water is not constant but follows a uniform distribution between 0.75 and 0.92 $\text{N}\cdot\text{s/m}^2$.

Solution

(a) For the deterministic situation, the expected head loss at any time, h_L is

$$h_L = 0.52\frac{1-\varepsilon^2}{\varepsilon^2}\frac{U^2}{2g} = 0.52\frac{1-0.15^2}{0.15^2}\frac{1.8^2}{2(9.81)} = 6.83 \text{ cm}$$

For the deterministic situation, the head loss never exceeds 10 cm.

(b) In the probabilistic situation, the head loss is simulated using Monte Carlo simulation. After running the simulation several times (1000 runs were carried out in this solution), it may be observed that 25% of the time, the head loss exceeds 10 cm. Thus it is expected that the screens will need cleaning 25% of the time.

Example 23.9 Aerobic Digestion in Wastewater Treatment Operations

Anaerobic digestion, an important phase of wastewater treatment, uses micro-organisms to break down biodegradable material in the absence of oxygen (Figure 23.5). Used to manage waste and to produce energy, anaerobic digestion converts organic sludge to humus (which may be used to condition soils for agricultural purposes) and gas (which may be burned off or further treated before use as a heating source). The heat required for the anaerobic digestion process is a function of the raw sludge heating and the heat transfer through its boundaries (walls, floor, and roof) and the heat generated from the metabolic process. For an aerobic digester, the complete heat balance is as follows (Sykes, 2003):

$$\begin{aligned}\Delta H_{\text{req}} \text{ (Heal required)} &= c_p\rho Q\,(T_{\text{dig}} - T_{\text{sludge}})\text{(Raw sludge heating)}\\ &+ K_rA_r(T_{\text{dig}} - T_{\text{air}})\text{(Heat through roof)}\\ &+ K_wA_w(T_{\text{dig}} - T_{\text{grd}})\text{(Heat through wall)}\\ &+ K_fA_f(T_{\text{dig}} - T_{\text{grd}})\text{(Heat through floor)}\\ &= H_{\text{met}}Q(X_{\text{ve}} - X_{\text{ve}})\text{(Metabolic heat)}\end{aligned}$$

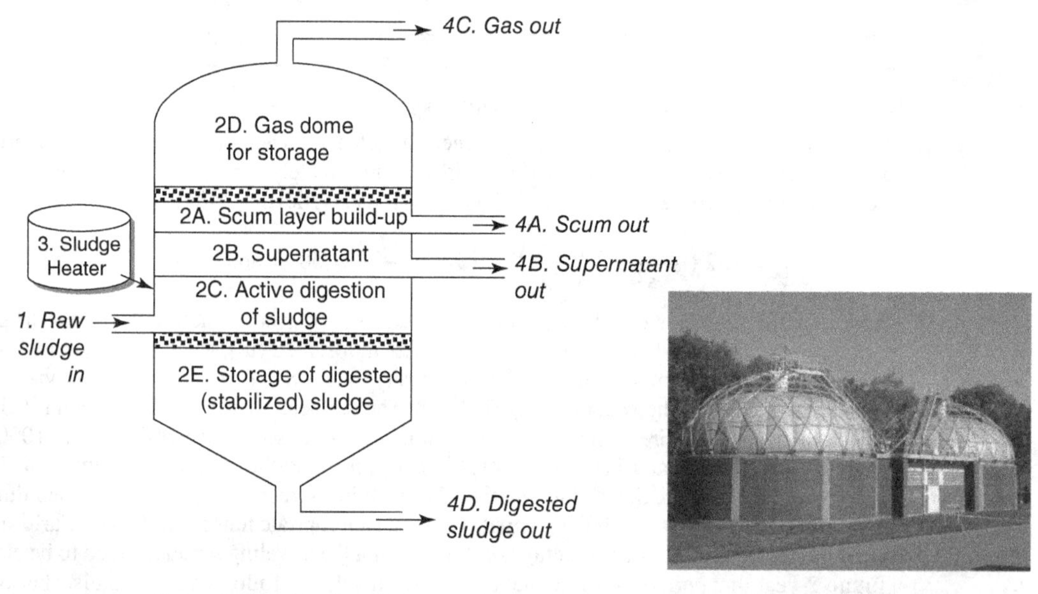

Figure 23.5 Anaerobic digestion operations—schematic and photo of standard single-phase digester.

where ΔH_{req} is the heat requirement (J/s); A is the area normal to heat flux (m^2); C_P is the constant pressure specific heat of water (J/kg); H_{met} is the metabolic heat release (J/kg · Volatile solids); K is the overall heat transfer coefficient ($J/(m^2 \cdot s \cdot K)$; Q is the sludge flow rate (m^3/s); T_{air} is the air temperature (K); T_{dig} is the digester temperature (K); T_{grd} is the ground temperature (K); T_{slu} is the sludge temperature (K); X_{ve} is the effluent concentration of volatile suspended solids (VSS) (kg/m^3); X_{vo} is the influent VSS (kg/m^3); and ρ is the mass density of water (kg/m^3).

In the anaerobic digester of a city's wastewater treatment plant, the area normal to heat flux is 20 m^2 and the sludge flow rate is 0.15 m^3/s. The constant pressure specific heat of water is 4186 J/kg. The metabolic heat release is 1000 J/kg · VS, the overall heat transfer coefficient is 500 $J/(m^2 \cdot s \cdot K)$. The temperatures of the air, digester, ground, and sludge are 50, 100, 25, and 30°C, respectively. The rate of influent and effluent are 500 and 200 VSS (kg/m^3), respectively. The mass density of water is 1000 kg/m^3. Calculate how much additional heat is required from the sludge heater (J/s).

Solution

$$\begin{aligned}\Delta H_{req} &= 4186 \times 1000 \times 0.15 \times (100 - 30) + 500 \times 20 \times (100 - 50)\\ &\quad + 500 \times 20 \times (100 - 25) + 500 \times 20 \times (100 - 25) - 1000 \times 0.15 \times (500 - 200)\\ &= 43{,}953{,}000 + 500{,}000 + 750{,}000 + 750{,}000 - 45{,}000 = 45{,}908{,}000 \text{ J/s}.\end{aligned}$$

Therefore, 45, 908 KJs of additional heat is required from the sludge heater.

Example 23.10 Bridge Tolling Operations

There are three lanes leading to a bridge nonelectronic toll area; each lane has a toll booth. At peak hour, 2200 vehicles seeking nonelectronic toll payment arrive at the toll area. The manual toll collectors take 4 seconds, on average, to serve each vehicle. Assuming that the vehicle arrivals are Poisson distributed and the departure intervals follow an exponential distribution, determine the following queue characteristics at the peak hour: the percentage of time that the servers are idle, the percentage of times that there are 6 vehicles in the queuing process including server, the average number of vehicles in the queuing process (being served or in a queue), and the average time spent by a vehicle in the entire queue process (waiting in line and being served).

Solution

The average arrival rate, $\lambda = 2200/60 = 36.67$ vehicles/minute; the average service rate per server channel, $\mu = 60/4 = 15$ vehicles per minute; the average service rate for 3 channels = $N^* \mu = 3^* 15 = 45$ vehicles per minute

Utilization ratio, $\rho = \lambda/(N \cdot \mu) = 36.67/[(3)(15)] = 0.81$, which is less than 1.0. Also, $\rho N = 2.44$

The percentage of time that the servers are idle is

$$P_0 = \left(\sum_{k=0}^{N-1} \frac{\rho^k}{k!} + \frac{\rho^N}{N!\left(1 - \frac{\rho}{N}\right)} \right)^{-1} = \frac{1}{1 + \frac{2.44}{1!} + \frac{2.44^2}{2!} + \frac{2.44^3}{3!(1-0.81)}} = 0.051 = 5.1\%$$

The percentage of times that there are 6 vehicles in the queuing process including server = P_n.

$$\text{When } n \geq N, P_n = \frac{P_0 \rho^n}{N^{n-N} N!} = \frac{(2.44)^6 (0.051)}{3^{6-3} 3!} \left[\frac{1}{(1 - 0.81)^2} \right] = 0.067 = 6.7\%$$

The average number of vehicles in the queuing process (being served or in a queue) is

$$L = \frac{P_0 \rho^{N+1}}{N! N} \left[\frac{1}{(1 - \rho)^2} \right] = L = \frac{0.051 (2.44)^{3+1}}{3! 3} \left[\frac{1}{(1 - (0.81)^2} \right] = 2.95 \text{ vehicles}$$

The average time spent by a vehicle in the entire queue process is

$$W = \frac{\rho + L}{\lambda} = \frac{2.44 + 2.95}{36.67} = 0.11\text{minutes}$$

Example 23.11 Logistics Operations (The Chinese Postman Tour)

For the city network and nodes (intersection names) shown in Figure 23.6. Determine the path to be followed by a freight company that seeks to travel, starting from Pune on all links to collect goods but minimizes the overall distance of travel, and returns to Pune. Each link should be traveled at least once.

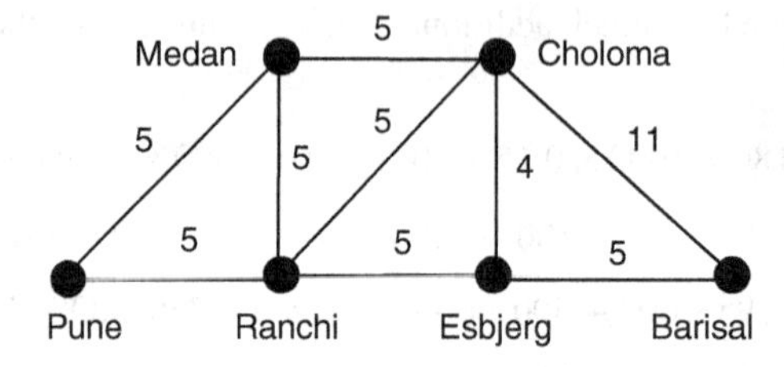

Figure 23.6 Network for Example 23.11.

Solution

The problem can be formulated as follows:

$$\min \sum_{i=1}^{6} \sum_{j=1}^{6} c_{i,j} x_{i,j}$$

Subject to

$$\sum_{k=1}^{6} x_{k,i} - \sum_{k=1}^{6} x_{i,k} = 0 \quad i = 1, 2, \ldots 6$$

$$x_{i,j} + x_{j,i} \geq 1 \text{ for all links } (i,j)$$

where $x_{i,j} \geq 0$ and is an integer; $x_{i,j}$ is the number of times the link between nodes i and j is traversed in the direction from i to j; $c_{i,j}$ is the distance of the link from node i to node j; if there is no direct link from i to j, then $c_{i,j} = \infty$.

The distance matrix is shown in Table 23.1.

Table 23.1 Distance Matrix for Example 23.11

	Pune	Medan	Choloma	Ranchi	Esbjerg	Barisal
Pune	0	5	∞	5	∞	∞
Medan	5	0	3	5	∞	∞
Choloma	∞	3	0	5	4	11
Ranchi	5	5	5	0	5	∞
Esbjerg	∞	∞	4	5	0	5
Barisal	∞	∞	11	∞	5	0

Solving the problem using an appropriate optimization platform or spreadsheet such as MS Solver, the $x_{i,j}$ matrix is determined as shown in Table 23.2.

Table 23.2 x_{ij} Matrix

	Pune	Medan	Choloma	Ranchi	Esbjerg	Barisal
Pune	0	0	0	1	0	0
Medan	1	0	1	0	0	0
Choloma	0	1	0	1	1	0
Ranchi	0	1	0	0	1	0
Esbjerg	0	0	1	0	0	1
Barisal	0	0	1	0	0	0

Thus the minimum total distance is 59 units, and the associated optimal route is

Pune → Ranchi → Esbjerg → Barisal → Choloma → Esbjerg → Choloma → Ranchi → Medan → Choloma → Medan → Pune.

Example 23.12 Logistics Operations (The Classic Transportation Problem)

A food production company seeks to transport corn to a number of factories for corn syrup production such that the total cost of transportation is minimized. There are five farms at which railcars collect the material and ship to two corn processing plants for production. At each farm, there is no excess, that is, all harvested corn is shipped. Also, there is no excess supply at each processing plant. Table 23.3 presents the amount of corn collected at farms and the capacity of each plant, and the distances between the nodes (miles) and the transport costs per mile. Determine how much material must be transported from each farm to each plant.

Table 23.3 Data for Example 23.12

(a) Amount of Material Generated at Nodes and Node Capacities

		Material Generated at Source node (tons)	Capacity of Destination Node (tons)
Sources	Farm 1	110	
	Farm 2	70	
	Farm 3	20	
	Farm 4	250	
	Farm 5	50	
Destinations	Plant 1		300
	Plant 2		200

Cost of transportation = $100 per ton per mile

(b) Distances between Source and Intermediate Nodes (miles)

		Plants	
		1	2
Farms	1	27	11
	2	21	3
	3	15	5
	4	30	8
	5	34	20

Solution

This Hitchcock shipment problem can be formulated as

$$\min \sum_{i=1}^{5}\sum_{j=1}^{2} a_{i,j}x_{i,j}$$

Subject to

$$\sum_{j=1}^{2} x_{i,j} = g_i \qquad i = 1, 2, 3, 4, 5$$

$$\sum_{i=1}^{5} x_{i,j} \le h_j \qquad j = 1, 2$$

$$x_{i,j} \ge 0 \ \text{ for all } (i, j)$$

where g_i is the amount of material collected at source node i, $a_{i,j}$ is the cost of transporting a unit of material form i to j; h_j is the capacity of the receiving facility at destination node j; $x_{i,j}$ is the amount of material transported from source i to destination j. The a_{ij} matrix is presented in Table 23.4.

Table 23.4 Cost Matrix for Example 23.12

	1	2
1	2700	1100
2	2100	300
3	1500	500
4	3000	800
5	3400	2000

$g_1 = 110; g_2 = 70; g_3 = 20; g_4 = 250; g_5 = 50; h_1 = 300; h_2 = 200.$

By solving the problem using Excel Solver, the $x_{i,j}$ matrix is presented in Table 23.5. The corresponding minimum cost is calculated as $954,000.

Table 23.5 x_{ij} Matrix for Example 23.12

	1	2
1	110	0
2	70	0
3	20	0
4	50	200
5	50	0

Example 23.13 Logistics Operations (The Transshipment Problem)

An international courier service seeks to transport mail packages from five different cities to three warehouses for sorting and then from the warehouses to two international airports, such that the total cost of transportation is minimized. Assume that at each city or warehouse every mail package is shipped. Also, at each warehouse or international airport, there is no excess supply. Table 23.6(a) presents the

Table 23.6 Data for Example 23.13

(a) Amount of Material Generated at Nodes and Node Capacities

		Packages Generated at Source Node (tons)	Capacity of Destination Node (tons)
Sources	**Faithville**	100	
	Virtue City	150	
	Peaceburg	250	
	New Harmony	110	
	Fairweather	110	
Intermediates	**Morenz Warehouse**	120	100
	Wabash Warehouse	250	340
	Kievo Warehouse	350	320
Destinations	**Pacific International Airport**		310
	Atlantic International Airport		450

Cost of transportation = $220 per ton per mile

(b) Distances between Source and Intermediate Nodes (miles)

	Intermediates		
Sources	**Morenz**	**Wabash**	**Kievo**
Faithville	19	34	21
Virtue City	43	31	30
Peaceburg	42	11	43
New Harmony	9	43	14
Fairweather	36	33	25

(c) Distances between Intermediate and Destination Nodes (miles)

	Destinations	
Intermediates	**Pacific Int. Airport**	**Atlantic Int. Airport**
Morenz	23	14
Wabash	15	17
Kievo	21	31

amount of mail packages at the cities and warehouses and the capacity of each warehouse and facility at the international airports. The distances between the nodes (miles) and the transport costs per mile are given in Table 23.6(b). For optimal operations, what quantity of mail packages must be transported from each city to each destination? In practical reality, is this optimal level of operations achieved? Explain why or why not. What then is the use of determining the optimal level of operations?

Solution

This transshipment problem can be formulated as

$$\min \sum_{i=1}^{5}\sum_{j=1}^{3} a_{i,k}x_{i,k} + \sum_{i=1}^{3}\sum_{j=1}^{2} b_{k,i}w_{k,j}$$

Subject to

$$\sum_{k=1}^{5} x_{i,k} = g_i \qquad i = 1,2,3,5 \quad g_1 = 100 \qquad g_2 = 150 \qquad g_3 = 250 \qquad g_4 = 110 \qquad g_5 = 110$$

$$\sum_{i=1}^{3} x_{i,k} \leq h_{K_k} \qquad K_k = 1,2,3 \qquad h_{K1} = 120 \qquad h_{K2} = 250 \qquad h_{K3} = 350$$

$$\sum_{i=1}^{3} x_{i,k} - \sum_{j=1}^{2} w_{k,j} = 0 \qquad k = 1,2,3.$$

$$\sum_{j=1}^{3} w_{k,j} = g_{K_k} \qquad K_k = 1,2,3 \qquad g_{K1} = 100 \qquad g_{K2} = 30 \qquad g_{K3} = 320$$

$$\sum_{i=1}^{3} w_{k,j} \leq h_j \qquad j = 1,2 \qquad h_1 = 310 \qquad h_2 = 450$$

$$x_{i,k} \geq 0 \qquad w_{k,j} \geq 0$$

The $a_{i,j}$ matrix is shown in Table 23.7.

Table 23.7 a_{ij} Matrix for Example 23.13

	1	2	3
1	10	0	90
2	0	0	150
3	0	250	0
4	110	0	0
5	0	0	110

The $b_{i,j}$ matrix is shown in Table 23.8.

Table 23.8 b_{ij} Matrix for Example 23.13

	1	2
1	5060	3080
2	3300	3740
3	4620	6820

Solving the problem using Excel Solver yields the $x_{i,j}$ matrix shown in Table 23.9.

Table 23.9 x_{ij} Matrix for Example 23.13

	1	2	3
1	10	0	90
2	0	0	150
3	0	250	0
4	110	0	0
5	0	0	110

The $w_{i,j}$ matrix is shown in Table 23.10, and the corresponding minimum cost is $5,955,400.

Table 23.10 w_{ij} Matrix for Example 23.13

	1	2
1	0	100
2	0	340
3	310	10

In practice, the optimal level of logistics operations is rarely achieved. However, a knowledge of such optimal levels is useful to the operations engineer. Knowing how far the system is operating from the optimal level, the engineer can carry out the required interventions to bring the operations closer to the optimal level.

Example 23.14

The operational performance of a certain structural civil system is influenced by the climatic severity (number of freeze–thaw cycles, or FTC) and the level of demand for the system as shown in the equation below.

$$\text{System operational performance} = 1000e^{-\text{Demand}/10} \times 10(\text{FTC})^{-1}$$

(a) *The deterministic situation.* Table 23.11 gives the expected demand and climatic conditions of three cities in year 2020. Find the average performance of the system.

Table 23.11 Demand and Climate Data for Example 23.14

	City	City B	City C
Demand (system usage, in some unit)	11.6	17.7	21.6
Climatic severity (nr/of freeze–thaw cycles)	182	130	162

(b) *The stochastic situation.* An young city engineer argues that these input factors (population and climate in year 2020) should not be treated as deterministic because they have a wide range of uncertainty. According to the engineer, the population and climate in each city follow some probability distribution with a mean and standard deviation or other parameters. These are shown in Table 23.12. Enter the data in Table 23.12 into a spreadsheet and create an output cell that calculates the operational performance for each city.

Table 23.12 Details of Probability Distributions for Example 23.14

	City A	City B	City C
Demand (system usage)	Uniformly-distributed random number between 10 and 12	Uniformly-distributed random number between 15 and 20	Normally distributed random number with mean 22 and standard deviation 4
Climatic severity	Normally distributed random number with mean 204 and standard deviation 15	Normally distributed random number with mean 134 and standard deviation 13	Uniformly-distributed random number between 150 and 200

(i) Perform a Monte Carlo simulation for each city and calculate the average and standard deviation of the simulated performance. Which city has the largest average performance? Which city has the largest uncertainty (variability) in performance?

(ii) Engineers generally prefer high performance with low uncertainty (variability). So, in your opinion, which of the following "evils" is worse: high performance but small certainty of its attainment or low performance with great certainty of its attainment?

Solution

(a) The operational performance of the systems at the cities are: A, 17.22 units; B, 13.10 units; C, 7.12 units.

(b) (i) The first five outputs of the Monte Carlo simulation shown in Table 23.13.

Table 23.13 Result of Monte Carlo Simulation

	Iteration	1	2	3	4	5	Mean	Variability (Standard Deviation)
City A	Demand	10.97	10.53	10.84	10.74	10.04		
	FTC	176.69	241.47	201.59	177.52	187.45		
	Performance	18.89	14.44	16.77	19.24	19.55	16.93	1.56
City B	Demand	15.51	17.27	17.29	18.85	19.58		
	FTC	145.55	127.99	95.30	126.08	140.91		
	Performance	14.57	13.89	18.62	12.04	10.02	13.05	2.23
City C	Demand	17.72	21.77	23.88	17.94	26.83		
	FTC	189.20	194.42	170.57	150.06	193.28		
	Performance	8.99	5.83	5.39	11.08	3.54	6.76	2.69

City A has the highest average performance (16.93 units) compared to City B (13.05 units) and City C (6.76 units). City C has the highest level of performance uncertainty (standard deviation is 2.69 units) compared to City A (1.56 units) and City B (2.23 units).

(ii) The answer to this part is open ended. Some engineers may prefer low performance as long as they are guaranteed to receive that level of performance, that is, low performance with a great certainty. Persons with such preferences are described as risk averse as they do not like taking risks. On the other hand, there are engineers who prefer high performance even if it comes with relatively low certainty; these are risk-prone individuals, or gamblers.

Example 23.15 Water Resource Planning (Adapted from Reinèr, 2003)

A well is 0.2 m in diameter and pumps from an unconfined aquifer 30 m deep at an equilibrium (steady-state) rate of 1000 m^3/day. Two observation wells are located at distances 50 m and 100 m from the well, and they have been drawn down by 0.3 and 0.2 m, respectively (Figure 23.7). (a) Assuming a deterministic situation, determine the coefficient of permeability and estimated drawdown at the well? (b) Assuming a probabilistic situation where the mean and standard deviation of the pumping rate are 1000 m^3/day and 50 m^3/day, respectively, determine the probability distributions for the coefficient of permeability and estimated drawdown at the well.

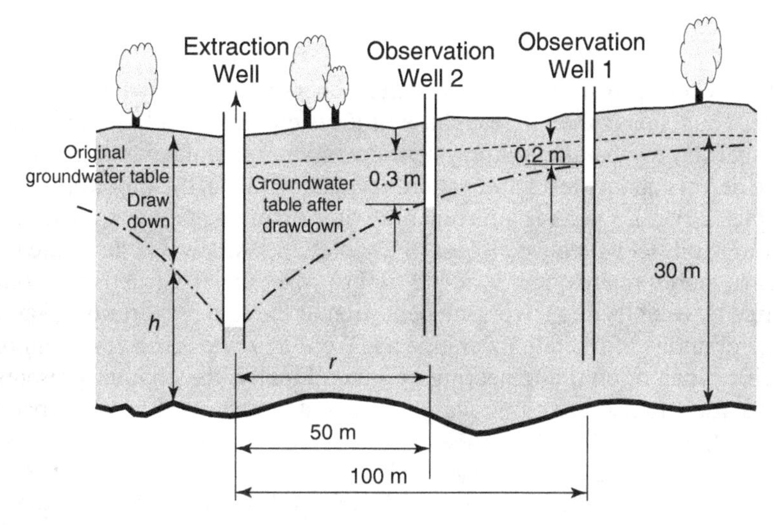

Figure 23.7 Drawdown of water wells in unconfined aquifer.

Solution

(a) The deterministic situation: The coefficient of permeability K is given by

$$K = \frac{Q \ln (r_1/r_2)}{\pi (h_1{}^2 - h_2{}^2)}$$

It is given that $h_1 = 30\text{ m} - 0.2\text{ m} = 29.8\text{ m}$ and $h_2 = 30\text{ m} - 0.3\text{ m} = 29.7\text{ m}$.
Thus

$$K = \frac{1000 \ln (100/50)}{\pi [29.8^2 - 29.72^2]} = 37.1\text{ m/d}.$$

The well radius is 0.5 (0.2 m) = 0.1 m.

$$\text{Thus, } Q = \frac{\pi k (\mathrm{h}_1{}^2 - h_2{}^2)}{\ln (r_1/r_2)} = \frac{\pi (37.1)(29.7^2 - h_2{}^2)}{\ln (50/0.1)} = 1000\text{ m}^3/\text{d}$$

Solving for h_2 yields $h_2 = 28.8$ m. The aquifer is 30 m deep; thus, the drawdown at the well is $30 - 28.8 = 1.2$ m

(b) The probabilistic situation: After carrying out a Monte Carlo simulation, the probability distributions for the coefficient of permeability and the well drawdown can be determined. See Figure 23.8a (cubic meters per day) and 23.8b (meters of drawdown).

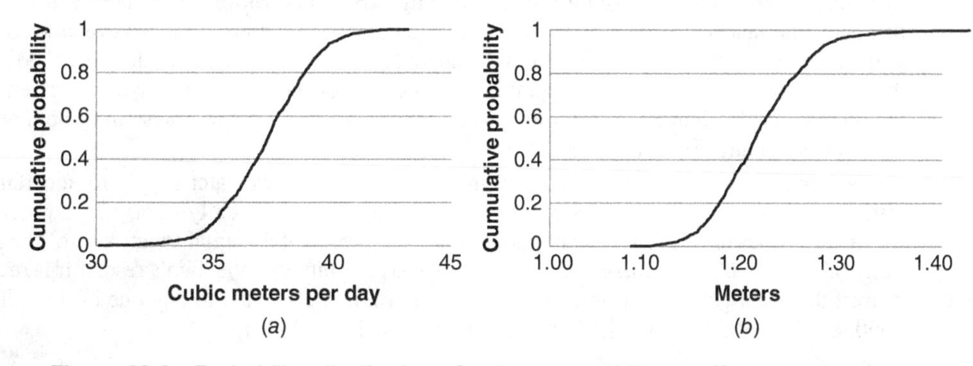

Figure 23.8 Probability distributions for the permeability coefficient and well drawdown.

SUMMARY

A civil engineering system is said to be in its operations phase when it has been commissioned and is in use. Thus, this is the longest phase of a system. At this phase, there are key regular activities that are carried out parallel with system operations: monitoring and inspection of the system for possible defects and its level and patterns of usage (this is the monitoring phase, which we discuss in Chapter 24) and carrying out rehabilitation or maintenance as and when needed (this is the preservation phase, which we discuss in Chapter 25). All civil systems are developed in response to a need and therefore are commissioned for use when the construction is completed. In this chapter, we started by establishing a working definition of the term "operations," and we discussed some of the general duties of operations engineers. We then reviewed a few examples of operations work in specific types of civil engineering systems. Finally, the chapter presents some applications of analytical tools to address numerical problems at the phase of system operations.

EXERCISES

1. The time of service of a certain desgin of hydraulic system equipment ranges between 100,000 and 350,000 hours, with each service life having a uniform probability of occurring. (a) What is the probability density function? (b) What is the probability that a certain equipment of this design lasts at least 200,000 hours? (c) Calculate the mean and variance of this probability distribution.
2. The volume of water pumped daily by an overhead reservoir to a dormitory has mean 2000 gallons and standard deviation 45 gallons. Assuming a normal distribution, find the probability that on any day selected at random, between 1800 and 2200 gallons will be pumped.
3. An engineering system works perfectly with any three of the following components: C_1, C_2, C_2, C_4, C_5, and C_6. In how many ways can the system be configured to work perfectly if the order of the three selected components is (i) important and (ii) not important.
4. The city engineer of the Greater Barranquilla Area (which comprises the cities of San Pedro and Los Amigos, which that are quite far from each other) is interested in knowing percentage of times that sewer systems in each of the two cities are generally running at low, medium, or full capacity so that she can plan on deploying the appropriate resources to manage any possible breakdown. Thirty-five percent of the time, the sewer system at San Pedro operates at full capacity, and 45% of the time it operates at medium capacity; the rest of the time it operates at low capacity. In neighboring Los Amigos, 65% of the time, the sewer system operates at full capacity and 25% of the time, it operates at medium capacity; the rest of the time it operates at low capacity. At any given time, the probability that both systems are operating at full capacity is 15%. (a) Identify any two events in the question whose occurrences you would consider to be disjoint (mutually exclusive). (b) Identify any two events whose occurrences you would consider to be nondisjoint and statistically independent of each other. (c) Identify any two events whose occurrences you would consider to be nondisjoint and statistically dependent of each other. (d) Find the probability that San Pedro's sewer system is operating at either full or medium capacity at any given time. (e) Find the probability that San Pedro's sewer system or the Los Amigos sewer system or both are operating at full capacity at any given time.
5. The inflow and outflow operations of a dam project that provides water supply for a certain city is illustrated schematically in Figure 23.9. From past records, the city's daily demand for water (measured to the nearest thousand gallons), is an amount that is approximately equal to any one of the following values: 10,000, 11,000, 12,000, or 13,000. Also, due to precipitation, the dam's reservoir is recharged (to the nearest thousand gallons) by an amount that is approximately equal to any one of the following values: 1000, 2000, or 3000 gallons. [Adapted from Ang and Tang (2006)].

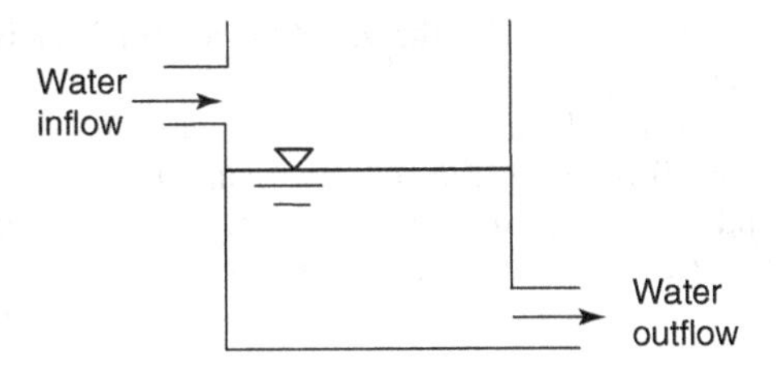

Figure 23.9 Figure for Exercise 5.

a. What are the possible combinations of inflow and outflow of water for the reservoir on a given day?

b. If there are 50,000 gallons in the reservoir at the beginning of a certain day, what are the possible water volumes left in the reservoir at the end of that day?

c. If the amounts of inflow and outflow of water for the reservoir are equally likely, what is the probability that there will be at least 30,000 gallons of water left in the reservoir at the end of the day?

6. Joe asserts that a certain engineering system operates at a user consumption rate that does not exceed 30 units per day, but so far there is no solid evidence to prove it. You take random samples of the system's operational data and obtain a mean of 27 units per day. Test the following hypothesis, assuming a sample size of 25 and a sample variance of 144.45, with a 90% confidence level. The system in question is a new design so there is little experience or knowledge about the distribution type and the variance of the population of its operational characteristics.

$$H_0 : \mu \leq 30 \text{ units per day}$$

$$H_1 : \mu > 30 \text{ units per day}$$

7. For the network shown in Figure 23.10, determine the traveling salesman path for an engineer who is responsible for the operations and inspections of all systems an the network.

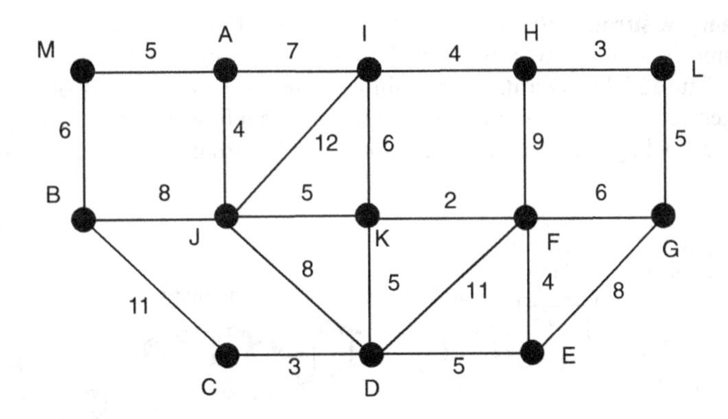

Figure 23.10 Figure for Exercise 7.

8. Safety considerations in highway operations. When a vehicle negotiates a horizontal curve, centrifugal forces act on it to push it radially outward. The centrifugal force is counterbalanced by the friction force between the tire and the pavement and the vehicle weight component related to the roadway superelevation. From the laws of mechanics, the following relationship holds (Easa, 2003):

$$R = \frac{V^2}{15(0.01e + f)}$$

where R is the curve radius (ft); V is the vehide speed (mph); e is the rate of roadway superelevation (%); and f is the side friction factor.

For a certain curve at a highway section in your city, the side friction factor is found to be normally distributed with mean 0.14 and variance 0.05. Using the outputs of a Monte Carlo simulation, determine the distribution and cumulative distribution of the simulated resulting maximum speeds of the vehicles traveling along that stretch. What is the highest speed limit that should be placed before the curve in order to ensure stability for (1) all vehicles negotiating the curve? (ii) 90% of all vehicles negotiating that curve? Comment an your results.

9. Water resource management operations. An environmental engineer seeks to ascertain the concentration of pollutants at a certain point in a certain major river due to industrial discharge of a pollutant at a point source 6 miles upstream. The river has a velocity of 0.5 m/s and a cross-sectional area of 30 m^2. The initial contaminant concentration at the outfall is 0.00056 kg/m^3, and the contaminant decay rate is 2/s. The distance below the outfall is 3 m and the uniformly distributed load along the stream reach below the outfall is 2 kg/ms. (a) Determine the contaminant concentration 6 miles downstream of the outfall (kg/m^3). *Hint:* The contaminant concentration at any distance downstream of the outfall is given by the following first order decay process (Sykes, 2003)

$$C(x) = C_0 e^{-(kx/u)} + \frac{W}{KA}(1 - e^{-hx/u})$$

where the cross-sectional area of the receiving stream, $A = 30\ m^2$; initial contaminant concentration at the outfall, $C_0, = 0.00056\ \text{kg/m}^3$; contaminant decay rate, $K = 2/\text{s}$; mean stream velocity $= 0.5\ \text{m/s}$; uniformly distributed load along the stream reach below the outfall, W, $= 2\ \text{kg/ms}$; distance below the outfall, $x = 6(1609) = 9654$ m.

10. Water treatment system operations. An individual wastewater treatment process uses activated carbon to remove color from the water. The color is reduced as a first-order reaction in a batch adsorption system. How long will it take to remove 90% of the color if (a) the rate constant k is a fixed value of 0.35 day^{-1} (b) the rate constant k is a stochastic normally distributed variable with mean 0.35 day^{-1} and standard deviation 0.05 day^{-1}?

11. Adequacy assurance during system operations. It has been determined that a community requires a maximum flow of 10 mgd of water during 10 hours in a peak day, beginning at 8 A.M. and ending at 6 P.M (Figure 23.11). During the remaining 14 hours, it needs a flow of 2 mgd. During the entire 24 hours, the water treatment plant is able to provide a constant flow of 6 mgd, which is pumped into the distribution system. How large must the elevated storage tank be to meet this peak demand?

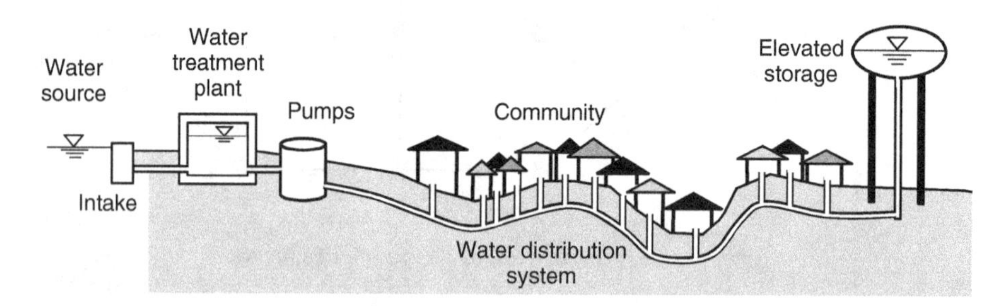

Figure 23.11 Water distribution.

12. Wastewater treatment operations. In the year 2020, a town is expected to have a population of 30,000. It is also expected that the town will send 0.5 m^3 per person per day to the wastewater treatment plant. It is sought to have an average detention time of 2.5 hours and an average overflow rate of 20 m^3/day per square meter. If the circular primary clarifier at the current treatment plant has dimensions 800 m^2 and 2.5m depth, determine whether the current clarifier is adequate to serve the demand in year 2020. If the following problem parameters are stochastic with the following means and standard deviations (per capita demand in m^3/person/day, 0.5 and 0.1; clarifier detention time in m^3/day, 2.5, 0.2).

REFERENCES

Ang, A. H., and Tang, W. H. (2006). *Probability Concepts in Engineering*, John Wiley & Sons, Hoboken, NJ.

Blevins, R. D. (1984). *Applied Dynamics Handbook*, Van Nostrand Reinhold, New York.

Dean, R. G., and Walton (2009). Wave Setup, in Y. C. Kim, Ed., *Handbook of Coastal and Ocean Engineering*. World Scientific, Singapore, pp. 1–23.

Easa, S. (2009). Geometric Design, in W. E. Chen and J. Y. Richard Liew, Ed., *The Civil Engineering Handbook*, 2nd ed., CRC Press, Boca Raton, FL.

Jacko, R. B., and Labreche, T. M. C. (2004). Air Pollution, in W. E. Chen and J. Y. Richard Liew, Ed., *The Civil Engineering Handbook*, 2nd ed., CRC Press, Boca Raton, FL.

Reinèr, R. F. (2003). *Environmental Engineering*, Butterworth-Heinemann, Oxford, U.K.

Sykes, R. M. (2003). Water and Wastewater Planning, in W. E. Chen and J. Y. Richard Liew, Ed., *The Civil Engineering Handbook*, 2nd ed., CRC Press, Boca Raton, FL.

USACE (1984). *The Shore Protection Manual*, U.S. Army Corps of Engineers, Washington, D.C.

Vesiland, P. A., Morgan, S. M., and Heine, L. G. (2009). *Introduction to Environmental Engineering*, Cengage Learning, Stamford, CT.

Wood, W., and Meadows, G. (2003). Coastal Engineering, in W. E. Chen and J. Y. Richard Liew, Ed., *The Civil Engineering Handbook*, 2nd ed., CRC Press, Boca Raton, FL.

USEFUL RESOURCES

Airport and Seaport Systems Operations

Alderton, P. (2008). *Port Management and Operations*, Routledge, London.

Ashford, N., Coutu, P., and Beasley, J. (2012). *Airport Operations*, McGraw-Hill, New York.

Bichou, K. (2009). *Port Operations, Planning and Logistics*, Routledge, London.

de Neufville, R., Odoni, A. Belobaba, P., and Reynolds, T. (2010). *Airport Systems: Planning, Design and Management*, McGraw-Hill, New York.

Guldogan, E. U. (2011). *Port Operations and Container Terminal Management*, VDM Verlag, Saarbrucken, Germany.

Horonjeff, R., McKelvey, F., Sproule, W., and Young, S. (2010). *Planning and Design of Airports*, McGraw-Hill, New York.

Kazda, A., and Caves, R. (2013). *Airport Design and Operation*, Elsevier, Oxford, UK.

Young, S., and Wells, A. (2011). *Airport Planning and Management*, McGraw-Hill, New York.

Building Systems Operations

Clements-Croome, D. (2013). *Intelligent Buildings: Design, Management and Operation*, ICE Publishing, London.

Hensen, J. L. M., and Lamberts, R. (2011). *Building Performance Simulation for Design and Operation*, Routledge, London.

U.S. Green Building Council, (2013). *LEED Reference Guide for Green Building Operations and Maintenance*, Washington, DC.

Coastal Systems Operations

Herbich, J. (2000). *Handbook of Coastal Engineering*, McGraw-Hill, New York.

Reeve, D., Chadwick, A., and Fleming, C. (2012). *Coastal Engineering: Processes, Theory and Design Practice*, CRC Press, Boca Raton, FL.

Sato, S., and Isobe, M. (2013). *International Handbook of Coastal Engineering*, World Scientific, Singapore.

Sorensen, R. M. (2005). *Basic Coastal Engineering*, Springer, New York.

Energy Generation System Operations

Hydroelectric: Elliott, T., Chen, K., and Swanekamp, R. (1997). *Standard Handbook of Powerplant Engineering*, McGraw-Hill, New York.

Wind: Hau, E. (2005). *Wind Turbines: Fundamentals, Technologies, Application, Economics*, Springer, Berlin.

Rivkin, D., and Silk, L., (2012). *Wind Turbine Operations, Maintenance, Diagnosis, and Repair*, Jones & Bartlett Learning, Burlington, MA.

Tavner, P. (2012). *Offshore Wind Turbines: Reliability, Availability and Maintenance*, The Institution of Engineering and Technology, Stevenage, UK.

Solid Waste Management System Operations

Bagchi, A. (2004). *Design of Landfills and Integrated Solid Waste Management*, Wiley, Hoboken, NJ.

Cheremisinoff, N. P. (2003). *Handbook of Solid Waste Management and Waste Minimization Technologies*, Butterworth-Heinemann, Oxford, UK.

McDougall, F. R., White, P. R., Franke, M., and Hindle, P. (2001). *Integrated Solid Waste Management: A Life Cycle Inventory*, Wiley-Blackwell, Oxford, UK.

Singh, J. (2009). *Solid Waste Management*, AL Ramanathan Publishers.

Thomas-Hope, E. (1998). *Solid Waste Management: Critical Issues for Developing Countries*. Canoe Press, Univ. of West Indies, Barbados.

Worrell, W. A., and Vesilind, P. A. (2011). *Solid Waste Engineering*, Cengage Learning, Independence, KY.

Transportation Logistics Operations

Bookbinder, J. H., Ed., (2012). *Handbook of Global Logistics: Transportation in International Supply Chains*, Springer, New York.

Christopher, M. (2011). *Logistics and Supply Chain Management*, Financial Times Press, London.

Frazelle, E. (2001). *Supply Chain Strategy*, McGraw-Hill, New York.

Stroh, M. (2006). *A Practical Guide to Transportation and Logistics*, Logistics Network Inc., Dumont, NJ.

Urban Drainage System Operations

Ashley, R. (2014). *Urban Drainage*, ICE Publishing, FL.

Butler, D., and Davies, J. (2010). *Urban Drainage*, CRC Press, Boca Raton, FL.

Pazwash, H. (2011). *Urban Storm Water Management*, CRC Press, Boca Raton, FL.

Wastewater Systems Operations

Kerri, K. D. (2004). *Operation of Wastewater Treatment Plants*, California State Univ., Long Beach, CA.

Kreith, F., and Tchobanoglous, G. (2002). *Handbook of Solid Waste Management*, McGraw-Hill, New York.

Qasim, S. R. (1998). *Wastewater Treatment Plants: Planning, Design, and Operation*, CRC Press, Boca Raton, FL.

Spellman, F. R. (2008). *Handbook of Water and Wastewater Treatment Plant Operations*, CRC Press, Boca Raton, FL.

Water Environment Federation (2007). *Operation of Municipal Wastewater Treatment Plants*, McGraw-Hill, New York.

Public Transportation Systems Operations

Ceder, A. (2014). *Public Transit Planning and Operation: Theory, Modeling and Practice*, CRC Press, Boca Raton, FL.

Gray, G., and Hoel, L. (1991). *Public Transportation*, Prentice Hall, New York.

Schobel, A. (2006). *Optimization in Public Transportation: Stop Location, Delay Management and Tariff Zone Design in a Public Transportation Network*, Springer, New York.

Vukan, V. (2005). *Urban Transit: Operations, Planning and Economics*, Wiley, Hoboken, NJ.

Highway and Railway Systems Operations

Arnott, R., Rave, T., Schob, R. (2005). *Alleviating Urban Traffic Congestion*, The MIT Press, Cambridge, MA.

Federal Highway Administration (2006). *Coordinated Freeway and Arterial Operations Handbook*, Washington, DC.

Federal Highway Administration (2006). *Freeway Management and Operations Handbook*, Washington, DC.

Mayinger, F. (2001). *Mobility and Traffic in the 21st Century*, Springer, Berlin.

McDonald, J. F., d'Ouville, E. L., and Liu, L. N. (1999). *Economics of Urban Highway Congestion and Pricing*, Kluwer, Boston.

Ponuswamy, R. (2012). *Railway Transportation: Engineering, Operation and Management*, Alpha Science Intl. Ltd., Oxford, UK.

Profillidis, V. A. (2004). *Railway Management and Engineering*, Ashgate, Farnham, UK.

Staley, S., and Moore, A. (2008). *Mobility First: A New Vision for Transportation in a Globally Competitive 21st Century*, Rowman & Littlefield, Lanham, MD.

Bridge Systems Operations

AASHTO. (2010). *Highway Safety Manual*, American Association of State Highway and Transportation Officials, Washington, DC.

Biondini, F., and Frangopol, D., Eds., (2012). *Bridge Maintenance, Safety, Management, Resilience and Sustainability*, Procs., 6th International IABMAS Conference, Stresa, Lake Maggiore, Italy.

Ryall, M. J. (2009). *Bridge Management*, CRC Press, Boca Raton, FL.

Yanev, B. (2007). *Bridge Management*, Wiley, Hoboken, NJ.

CHAPTER 24

SYSTEM MONITORING

24.0 INTRODUCTION

Civil engineering systems are constructed using materials and components that either gradually degrade upon aging or repeated loading or fail suddenly due to an internal weakness. Also, the physical integrity (stability, component connections, orientations, and other features as designed) may degrade gradually due to internal or external factors including foundation settlements, earthquakes and so on. The failure of a system or loss of performance due to gradual or sudden threats places the system operator in a bad light of poor stewardship and accountability. So as not to be caught by surprise due to preventable failures, and also to make reliable plans for system maintenance, rehabilitation, and reconstruction, system owners are constantly seeking reliable information on the physical condition and structural integrity of all the components of their system. It is not only the physical condition that is monitored; the owner also seeks to observe the system's operational performance and the extent of system use and other characteristics of the system to ascertain that it is adequately performing the function it was meant to provide. The system use characteristics, depending on the system type in question, may include volume, flow, loading intensity, loading configuration, and user characteristics. Further, there is a need to ascertain that the environment does not pose undue risk to the system's physical structure and operations and also to ensure that the system has minimal adverse impacts on the environment. To address these knowledge gaps, it is necessary to regularly inspect the system's physical condition and monitor the system's use/loading, operational performance, and interactions with its environment.

We begin the chapter with an identification of the basic aspects and different dimensions or purposes of system monitoring. We then discuss the typical architecture of a system monitoring program; and within the system architecture, the sensing and detection mechanisms receive additional focus. Then, recognizing the impracticality of collecting monitoring data for the entire population, we present how random but representative sampling could enable cost-effective monitoring of the system. Different types of civil engineering systems have different components that deserve scrutiny during inspections and monitoring; and we isolate port facilities for a case study discussion of the specific points of monitoring in any type of civil system. Next, we show how system owners can develop a long-term plan for monitoring their system, taking into account the costs and benefits (effectiveness) of the different monitoring techniques. Lastly, the chapter discusses the monitoring of a system's users and the associated ethical conflicts that arise in this activity.

24.1 BASIC ELEMENTS OF SYSTEM MONITORING

System monitoring may be defined as the close examination, on a periodic or continuous basis, of the physical condition or operational performance, the system's impacts on the environment, and the impacts of the environment on the system. The physical condition is synonymous with the infrastructure health, while the operational performance is related to the extent to which the system's

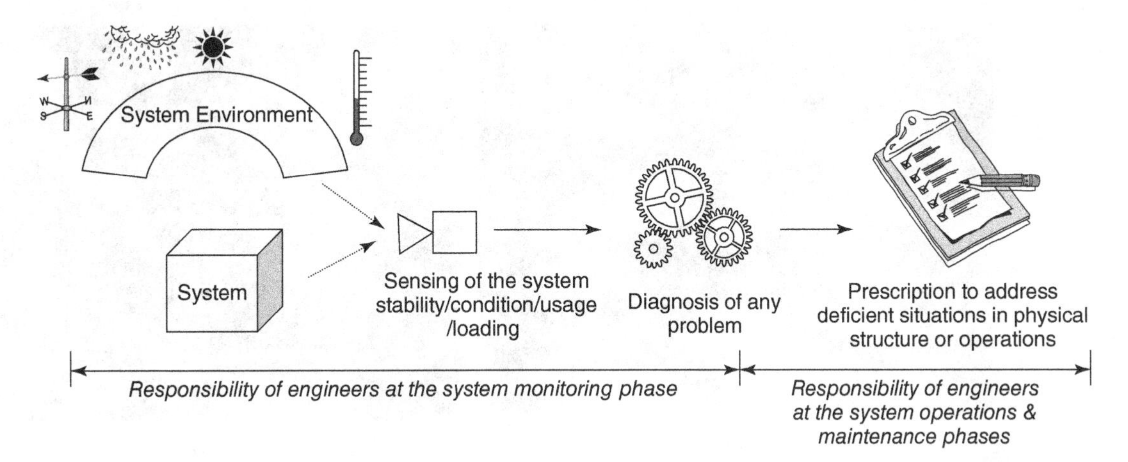

Figure 24.1 Basic elements of system monitoring.

goals are being realized, the number and characteristics of the user entities, and the patterns of use. The system monitoring phase is carried out in parallel with the system operations phase; for certain types of civil engineering systems, the same personnel are responsible for the tasks at these phases.

From the civil system monitoring practices in ancient times to the state-of-the-art monitoring practices of today, the elements of the basic architecture of monitoring have largely remained the same (Figure 24.1): image capture (a visual characterization of the subject), transmission of the monitoring data to the image processor, processing of the image, and results of the image processing (development of a statement indicating the results of the monitoring). In recent years, the field of civil system monitoring has seen dramatic improvements that include the use of embedded and wireless sensing technologies, real-time collection and processing of field data, artificial intelligence algorithms for image detection, and cost-effective analysis and storage of input and output data. As such, a typical detailed architecture for system monitoring at the current time may include the following elements: (a) a network of sensors deployed on the field (i.e., at the system location) (we will discuss sensing further in Section 24.3.1); (b) a high-performance communications system that transmits data in real time from the sensors to a central office; (c) a database with capabilities of enforcing data integrity (through error analysis and identification), and data visualization, management, and storage; (d) image interpretation using artificial intelligence techniques; (e) data analysis tools and techniques and interpretation, including probabilistic modeling and performance reliability and risk analysis; and (f) decision analysis based on the interpreted data. The last element, strictly speaking, is a part of the system maintenance phase and not the monitoring phase. In Section 24.2.3, we will discuss the components for stability monitoring of structural systems, and in Section 24.3, we will present the architecture for general monitoring.

24.2 PURPOSES OF SYSTEM MONITORING

As we mentioned in the introductory section, a civil engineering system may be monitored for at least one of four reasons. In the ensuing sections, we expand on this discussion.

24.2.1 Monitoring the Usage of the System

Monitoring the characteristics of the use of a civil engineering system includes counting the number of users (Figure 24.2*a*), the distribution of system use across time (hourly variations over a

(a)

(b)

Figure 24.2 Monitoring the Use of Systems: (*a*) Monitoring the frequency of use: Turnstiles at transit terminals help transit agencies to keep accurate counts on the extent of use of their systems. (*b*) A magnetic flow meter at the Tetley's brewery in Leeds, West Yorkshire, as seen from Crown Point Road. The meter is manufactured by 'Endress + Hauser' and is tagged 'F1' ('F' denoting flow). The meter used HART Protocol. Taken on the afternoon of Monday the 10th of May 2010. [Courtesy: (*a*). Arnold Reinhold/Wikimedia Commons and (*b*). Mtaylor848/Wikimedia Commons].

24-hour period, daily variations in a week, or variations by month or year), and the demand for the system vis-à-vis the level of use. For certain types of systems, other system use characteristics that are monitored include the physical characteristics of the user such as the user size, weight or loading (Figure 24.2*b*), the dimensions, and the user response or perceptions of the system's operational performance. The system use information is useful for *ex poste* validation of the system use predictions that were made at the planning and design phases, and thus can be used to improve the processes and assumptions made at the planning and design phases for future similar systems. Also, monitoring of the system's use provides quantitative data that can be used for management functions at the maintenance and operations phases, such as the establishment of appropriate user fees for the system, the prediction of system condition and service life on the basis of accumulated usage, and the estimation of maintenance/operations cost as a function of usage characteristics. In Section 24.4, we will discuss some of the sensors used in monitoring system usage frequency and intensity.

24.2.2 Inspection (Monitoring the Condition of the System)

Civil engineers routinely monitor the physical condition of their systems (Figure 24.3) in order to determine whether some urgent action is needed to repair the system before it suffers a catastrophic failure that may cause injury, loss of life, or property damage to the system user or the community. Inspection may be carried out manually or using sensor-equipped mobile or static devices. The physical defects that are typically uncovered during inspection are different for the different system types (see Table 25.6 in Chapter 25). The monitoring process must not only include the surface condition of infrastructure systems but also the underlying structural integrity. Often, the latter is difficult and expensive to measure. For example, for highway pavement systems, it is easier (quicker and less costly) to measure surface roughness using laser equipment compared to the measurement of structural strength using deflectometers or dynamic cone penetrometers; for bridges, it is easier to assess the bridge condition rating using visual assessments compared to load rating measurements. There is a saying that "skin condition may be adequate reflection of deeper underlying distress"

Figure 24.3 Monitoring (inspecting) the condition of various civil engineering systems: (*a*) Equipment may be necessary to help reach otherwise inaccessible areas of the system (Delaware Department of Transportation), (*b*) inspection of communications tower (Creative commons), (*c*) minihelicopters help system inspectors reach inaccessible areas of the system (Draganfly), (*d*) Slime Slime Robot pipeline inspection equipment (Courtesy of Emeritus Professor Hirose and Fukushima Laboratory of Tokyo Institute of Technology), (*e*) a technician performs MPI on a pipeline to check for stress corrosion cracking (Creative commons), (*f*) monitoring highway traffic using video, microwave, or infrared sensors (Creative commons), (*g*) undersea robot launched to monitor any threats to marine infrastructure (Creative commons), (*h*) river and canal gauging stations monitor hydrometric data (Creative commons), and (*i*) inspection and repair of damage to undersea structures (Creative commons).

and thus, a poor surface condition of infrastructure may be an indication of compromised structural integrity beneath. However, this may not always be the case. Ellingwood (2005) cautions that a facility should see continued use only when there is adequate quantitative evidence that there is no serious reduction in structural strength or stiffness due to aging. Such evidence can be obtained only through a carefully designed and implemented inspection schedule.

24.2.3 Monitoring the Stability of the System

Civil engineering structures may be rendered unstable due to changes in geotechnical conditions, design flaws, earthquakes, or other external or internal factors. Excessive movement of a structure's edges or vertices are often indicative of structural instability and can be monitored by establishing target points on the structure and using surveying techniques to ascertain the extent to which the target points move over a specified time period. For several decades, manual surveying techniques were used to detect any change in structural stability, an activity that was enhanced about two decades ago with the advent of the total station surveying equipment. The use of surveying techniques have been used to predict the failure of slopes (Han et al., 2001), monitoring of dams (Park ct al., 2001), and displacement of structures (Kang et al., 1995; Stewart and Tsakiri, 2001). In advocating for the use of photogrammtery in civil systems monitoring, some engineers have argued that the traditional surveying methods are generally unable to track the displacements of structures in real time due to variations in field conditions and due to their cumbersome processes of photographing, drawing, and analysis; Figure 24.4 presents a typical architecture for monitoring the stability of civil structures (Han et al., 2012). With the rapid advancements in image acquisition and interpretation, unprecedented opportunities exist for using digital photogrammetry to measure automatically in real time, and the displacement of target points in three dimensions can be monitored more quickly and effectively.

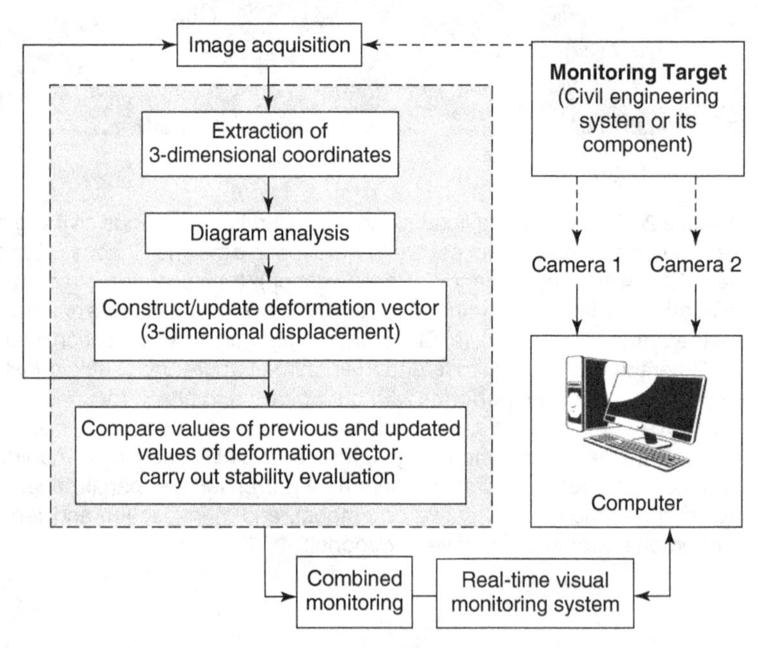

Figure 24.4 Components of stability monitoring (Brooklyn Bridge image from Wikipedia).

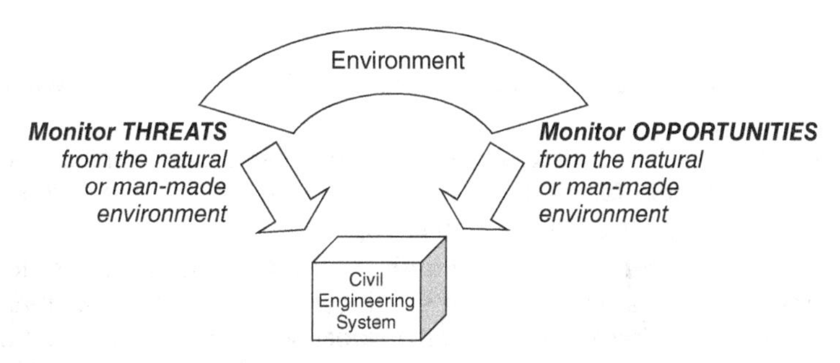

Figure 24.5 Environmental impacts on a system.

24.2.4 Monitoring the Environment's Impacts on the System

Civil engineers monitor not only the condition and performance of their systems but also any threats posed by the environment to their system. These include interactions with proximal natural and built-up environments. As we will learn in Chapter 28, the threats from the natural environment may be long term in nature (climate, soil acidity, and groundwater levels) or short term (extreme weather, hurricanes, earthquakes, and landslides). Climate-related threats include freezing (which cause the system's materials to become brittle and prone to cracking) and freeze–thaw transitions (which cause the expansion and contraction and subsequent degradation of the system materials). Also, extreme heat can cause the buckling of system components, deformation, and, in severe cases, melting of certain materials. The sun's rays, over time, cause oxidation that makes certain materials change their chemical composition and thus lose their physical strength or ductility. Climatic conditions are monitored regularly using the equipment at weather stations. Climate and weather affect not only the physical condition but also the operational performance of civil engineering systems. The system's operations can be impaired by ice, fog, and mist and hurricanes, tsunamis, earthquakes, and landslides. These environmental hazards are monitored using equipment that include meteorological station guages equipment and seismographs.

Threats from the built-up environment include the geotechnical pressures posed by newly constructed or failed/demolished neighboring structures (which can be monitored using tell-tale devices) and mobile threats such as collisions with land, sea, or airborne vehicles (which can be monitored using radar). Threats from humans (vandalism, crime, and terrorism) can be monitored by video surveillance of the system users, even though that may lead to invasion of privacy and may raise ethical questions (see Section 24.8).

As Figure 24.5 indicates, the impacts of the environment on civil engineering systems is not always unfavorable: the environment can be tapped for possible resources that can help in operating or maintaining the system. For example, a civil engineering system located in an area of high winds or high temperatures may install wind power generators or solar cells to produce energy to power the system (e.g., solar-powered street lights for urban read systems).

MONITORING THE ENVIRONMENT'S IMPACTS ON THE CIVIL ENGINEERING SYSTEM

Climate in Florida and the Orlando International Airport (OIA)

The OIA solution for monitoring the environment's impacts on its engineering system includes:

- A **total lightning sensor** for reliable detection of lightning that occurs within clouds or from clouds to the ground. Detecting in-cloud lightning helps predict the onset of severe weather events, including cloud-to-ground lightning strikes, heavy precipitation, tornadoes, and wind shear.

- An **onsite weather station** equipped with sensors that monitor real-time local conditions and make reliable forecasts of various weather parameters, such as wind direction and speed, humidity, temperature, and precipitation.
- An **Internet-based system for weather visualization and alerts** for monitoring changing weather conditions, lightning, and storm cells. This system uses real-time information from the airport's weather station as well as local weather data from global networks of weather stations.
- **Lightning alerting devices** that inform staff and officials located indoors when lightning is detected. Within the airport, these devices activate when lightning occurs within a predetermined distance and thus provide alerts or advance warning of imminent severe weather.

Source: www.earthnetworks.com

24.2.5 Monitoring the Impacts of Civil Systems on the Environment

Many owners of civil engineering systems in the private sector seem to fall into the trap of monitoring only the threats posed by the environment on their systems. As system managers, this indeed is their duty. However, consistent with the ethical principles (see Chapter 29), the first and foremost duty of civil engineers is to society. As such, engineers are increasingly concerned about the reverse direction (Figure 24.6): assessing the impacts of their system on the environment. Often, these impacts are mostly negative and include threats such as air, soil, and water pollution, noise, vibration, ecological damage, and sociocultural disruptions. Some of these threats are monitored by environmental engineers (using a variety of field and laboratory equipment) and social scientists. Specific examples of the adverse impacts that are monitored by engineers include those at highway systems (noise, air pollution, ecological disturbance), wind turbines (bird populations), dams (displacement of human settlements and habitat destruction), and nonwater pipelines (possible leakage leading to pollution of ground and surface soils and water). Figure 24.7 illustrates a station equipped with sensors that monitor the concentration of air pollutants caused by roadway systems and other sources.

Also, there may be reluctance on the part of private sector civil system owners to invest in monitoring the impacts of their system on the environment, for reasons that include lack of

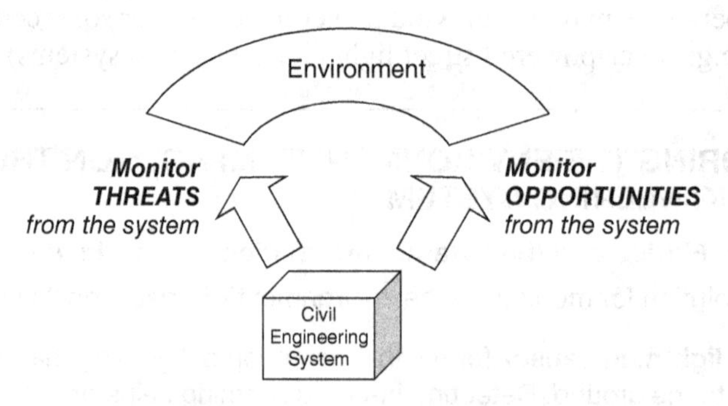

Figure 24.6 Civil system impacts on the environment.

Figure 24.7 Station monitoring the air quality impacts of road traffic and other activities (GeorgHH/Wikimedia Commons).

funds, absence of appropriate environmental quality legislation, inadequate agency goals and mission statements, or simply lack of ethics. Solutions often include the enactment of legislation or provision of funds to ensure that this is done. The impacts of civil engineering systems on their environment are not always negative; for example, if planned and designed properly, the system may enhance the aesthetic landscape and quality of life of the area. Even as the system is operating, it must be monitored for possible opportunities that could enhance its environment.

24.3 TYPICAL ARCHITECTURE OF A SYSTEM MONITORING PROGRAM

In any program that effectively monitors a civil engineering system, the condition, usage, operational characteristics, or environmental interactions for the system at any time are measured continuously or at regular intervals. The collected data are relayed to a central office where manual methods or artificial-intelligence-based algorithms are used to process the incoming data, identify the presence of any problems or defects, quantify the extent of these problems, and compare the measured defects with predefined triggers to ascertain the need for any remedial action. Thus, the key elements of system monitoring programs typically include the following: (a) sensors for the monitoring, (b) communication technologies for efficient information transfer, (c) information technologies for data mining and decision support, and (d) damage detection.

24.3.1 Sensors for Information Capture

In the simplest sense, a sensor is a sense organ. Our natural human sensors include our eyes (for seeing), ears (for hearing), nose (for smelling), and skin (for feeling by touching). For millennia, human inspectors of civil engineering systems have carried out inspections using their eyes to assess the condition or usage of civil structures. In the current era, technological devices are being used to replace or supplement human inspections. The nature of sensors used in a specific application depends on whether they are being used to monitor a system's operations, condition, or environment interactions. Generally, these devices have sensors that are sensitive to basic physical conditions (light, temperature, radiation level, etc.), or chemical or biological conditions. Their sensing capability is based on the premise that a change in system orientation, condition, strength, operations, or other attribute is evidenced by a change in the physical, chemical, or biological condition. It is critical that a sensor provide as few false positives or false negatives as possible; to have this property, a sensor should be sensitive to the property that is being measured only and insensitive to any other property likely to be encountered as it carries out its sensing task, and it must not influence the property that is being measured. Depending on the sensor type, the data collected from field sensors may be video signals, ASCII text, or other format. In recent years, the use of wireless communication has reduced the need for wiring between the various architecture elements of the monitoring system, thereby reducing its cost and enhancing its functionality, and has made it easier to monitor large and extended civil systems.

Sensor localization (i.e., specifying the location of each sensor node) is an important consideration in designing a large-scale sensor network. A number of researchers, therefore, have presented hierarchical localization methods, some of which typically consist of parent and child nodes equipped with GPS receivers and acoustic ranging devices, respectively. For the child nodes, the relative positions between can be designed using a distributed algorithm that specifies their positions while minimizing the global error accumulation simultaneously. Saeki et al. (2008) argued that the robustness and accuracy of GPS positioning is highly enhanced by using its "almost static" feature and thus is appropriate for infrastructure monitoring.

In Section 24.4, we provide further discussion on sensing mechanisms for the various purposes of system monitoring.

24.3.2 Information Technologies for Data Mining and Decision Support

These technologies include data acquisition protocols and data analysis tools. Typically, a scalable database structure is used to manage large amounts of incoming sensor data from the monitoring station. Tools for data analysis, including statistics, neural networks, and genetic algorithms, and tools for automated data mining and modal parameter extraction, can be used to interpret the data received from the sensors. Engineers worldwide continue to develop and implement flexible and scalable software architectures for monitoring their civil engineering systems. Most of these architectures, including a recent one developed by Elgamal et al. (2009), incorporate state-of-the art information systems that are not only capable of networking and integrating online real-time heterogeneous sensor data but also have powerful database and archiving systems and utilize artificial-intelligence-based image detection. Also, new-age information systems are being developed to analyze and interpret the so-called "big data" (massive amounts of information) from the sensors and to facilitate numerical simulation (particularly where the system structure or its operations are complex) and to carry out visualization of the input data or the outputs of the analysis.

24.3.3 Detection of Anomalous Situations

The engineer, by analyzing the sensor data, can ascertain the presence and extent of any anomalies in a system's physical condition, operational performance, relevant environment conditions, or in any system conditions that could threaten the environment. In order to do this, the data that streams in from the field are analyzed using the appropriate data analysis tools and techniques including probabilistic risk analysis.

There are several real-life examples of system architectures deployed to monitor civil engineering systems. The New Carquinez Suspension Bridge in Vallejo, California, uses wireless sensors to collect over 60 channels of data; physical structure attributes (acceleration of the bridge deck and towers, deck displacements), and environmental conditions (wind speed and direction, and temperature) and so on. The database is analyzed autonomously to extract data on the bridge attributes and to calculate bridge mode shapes using stochastic subspace identification techniques (Kurata et al., 2011). In a similar effort in the state of Connecticut, the state highway agency and the University of Connecticut developed a bridge monitoring program to ascertain whether a bridge component is behaving as designed and thus to assess the need for intervention (maintenance, rehabilitation, or replacement). The program uses different monitoring approaches for bridge structures in the state and provides valuable data on the integrity bridge components, including connections, diaphragms, and structural members (DeWolf et al., 1998).

24.4 SENSING AND DETECTION MECHANISMS

As we learned in Section 24.2, engineers who inspect and monitor the system are concerned about several dimensions of the monitoring phase: the impact of the system on the environment, the impact of the environment on the system, system condition and internal flaws, and system usage and performance. As we shall now discuss, there is a large variety of sensing and detection mechanisms that could be used to monitor system condition or stability, system loading intensity, or system usage frequency.

24.4.1 Mechanisms to Monitor System Condition

Sensing and detection techniques that identify material damage or defects can be categorized in many ways. These include whether they are (i) visual or nonvisual, (ii) destructive or nondestructive, (iii) intrusive or nonintrusive, (iv) contact or noncontact, (v) physical, biological, or chemical, and (vi) the nature of the target material (e.g., metal or nonmetal). In this section, we will discuss some of these mechanisms.

(a) Dye Penetrant Inspection (DPI). Dye penetrant inspection is a relatively inexpensive technique used to enhance visual detection of defects that break the surface of system components constructed of nonporous materials such as plastics, metals, and ceramics. The dye is applied and penetrates the material through the broken surface; the dye on the surface is cleaned off, leaving the marked surface defects (Figure 24.8). DPI enhances the visual detection of hairline and fatigue cracks and other surface defects.

(b) Radiographic Testing (RT). Radiographic testing identifies nonvisible flaws in a material on the basis of the ability of high-energy photons or neutrons to penetrate the material. To emit photons, the inspector uses X-ray equipment that provides short wavelength electromagnetic radiation; the performance of photon-based and neutron-based equipment differs because neutrons easily

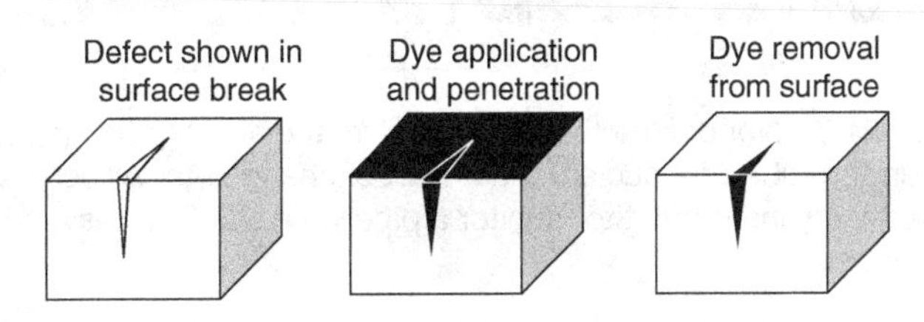

Figure 24.8 Civil system impacts on the environment.

travel through certain materials such as steel, but not so much for other nonferrous materials, such as plastics and fluids. Flaws are detected by measuring the amount of radiation emerging from the opposite side of the material; the variations in radiation intensity are a reflection of the existing thickness or composition of the material (which are reduced or modified, respectively, due to deterioration).

(c) Eddy Current Testing (EDT). Eddy current testing detects flaws in conductive materials using the principle of electromagnetic induction. Defective areas in the target material will have different magnetic permeability or electrical conductivity compared to nondefective areas of that material. The equipment may be a simple coil held on a piece of metal or special probes (Figure 24.9*a*). In the EDT test, the inspector places the equipment (a circular coil carrying an electric current) close to the target area. The alternating current in the EDT equipment generates a magnetic field, which interacts with the test specimen, producing an eddy current. The defective areas are identified by changes in the phase and amplitude of the current.

(d) Thermographic Inspection (TI). Thermographic inspection identifies the surface defects of materials by analyzing their thermal patterns at the material surface. It may use an intrusive contact technique such as applying a thin heat-sensitive layer to the material surface and measuring its temperature or infrared detection that is nonintrusive and noncontact. An energy source is used to identify a contrast in the thermal behavior of defective and nondefective areas. The energy may be delivered to the surface of the material and propagated through the material until it encounters an internal defect in the material or to stimulate the defects only.

(e) Magnetic Particle Inspection (MPI). Magnetic particle inspection (see Figure 24.3*e*) detects defects on the surface and in the small-depth subsurfaces of ferroelectric materials such as steel.

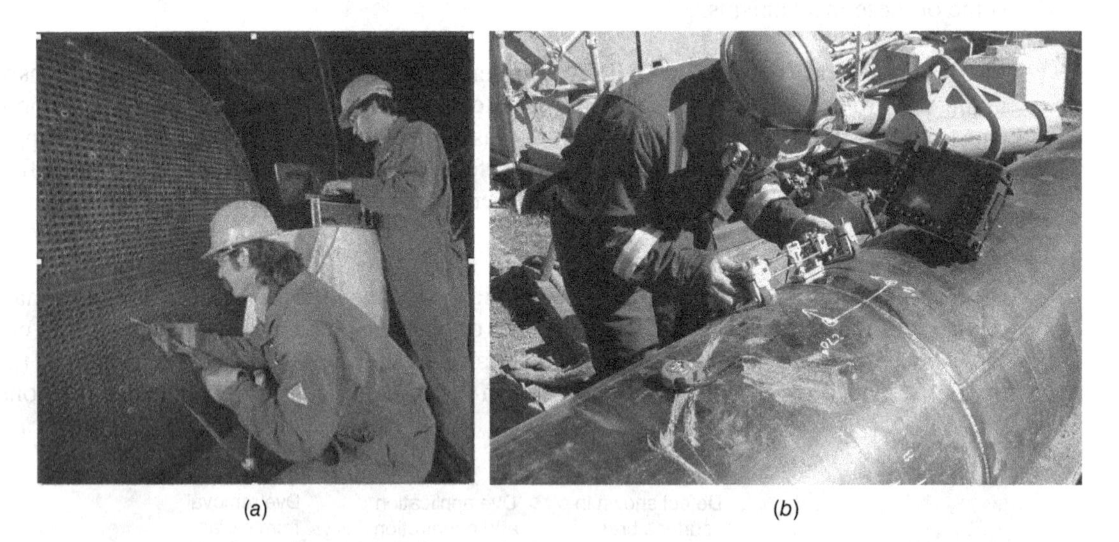

(*a*) (*b*)

Figure 24.9 Nondestructive testing of material integrity: (*a*) technicians performing EDT on the tube of a nuclear plant heat exchanger. (*b*) A technician uses an ultrasonic phased array instrument to monitor a pipeline weld (Courtesy of David Mack/Wikimedia Commons).

The inspector applies a magnetic field to the area under inspection. The magnetic flux leaks at the locations of surface or subsurface defects because defects are often discontinuities that have air gaps, and air does not accommodate as significant a magnetic field density as does metal.

(f) Ultrasonic Testing (UT). Ultrasonic testing involves the emission of short ultrasonic pulse waves into a material to detect internal flaws or to ascertain any changes in material thickness (Figure 24.9*b*); UT can be used to measure pipe corrosion. It is commonly applied to steel and other metals but may also be used on nonmetallic materials such as concrete and wood.

(g) Spectroscopy Testing. When visible light is made incident to a prism, it is dispersed according to its wavelength; a denser prism will exhibit a different pattern of dispersion. Spectroscopy tests operate on a similar premise: Defective material will show a different pattern of interaction (measured via their wavelength or frequency) with radiated energy compared with the nondefective material. One of the several types of spectroscopy, Raman spectroscopy, examines the inelastic scattering of monochromatic light (typically emitted from a laser) and observes the resulting response in the material that is manifest via upward or downward shifts in the energy of the laser photons.

24.4.2 Mechanisms to Monitor System Stability and Structural Integrity

As we learned in Section 24.2.3, the structural stability of certain types of civil engineering systems can be monitored by measuring the physical displacements of the system components and therefore can be tracked using manual surveying, photogrammetry, or other sensing techniques that measure smaller scales of displacement. For example, to measure highway and airport pavement strength using layer deflections, inspectors use the falling-weight or rolling-weight deflectometer. The displacement of structural members can be tracked using mechanical means (e.g., strain gauges) or other nontraditional means (e.g., lasers and infrared sensors). Prominent structures worldwide that are monitored closely for structural stability include the Huey Long Bridge in the United States (which has hundreds of strain gauges that monitor axial and bending stresses), the Fatih Sultan Mehmet Bridge in Turkey and the Tsing Ma Bridge in Hong Kong (Figure 24.10) (which use a

Figure 24.10 Hong Kong's Tsing Ma Bridge, one of the most instrumented civil engineering systems in the world, uses a variety of sensing mechanisms for monitoring (Minghong/Wikimedia Commons).

variety of sensor mechanisms), France's Millau Viaduc (which uses fiber-optic techniques), and Greece's Rio-Antirrio Bridge (which has over 100 sensors that monitor structural and operational performance in real time).

24.4.3 Mechanisms to Monitor System Usage Intensity (Loading) and Frequency

In Section 24.2.1, we discussed the importance of system usage monitoring. For certain types of civil engineering systems, only the usage intensity (loading) needs to be monitored; for others, only the frequency is needed; and still others, both frequency and loading. The mechanism used to monitor the system usage depends on the type of system. For example, for fluid supply and distribution systems, the equipment used includes mechanical flow meters (turbine flow, piston, gear, variable area, Woltmann, single or multiple jet, Pelton wheel, and paddle wheel), pressure-based meters (cone meters, Dall tube, orifice plate, Venturi meter, multihole pressure probe, and Pitot tube), and optical flow meters. For highway systems, loadings are typically measured using load cells, strain gauges, and fiber-optics. The mechanism used to monitor system usage frequency depends on the type of system. For example, to count their users, transit and stadia systems use mechanical means (turnstiles) or nonmechanical techniques (including microwave and infrared), and highway systems use mechanical techniques (such as manual counters) or nonmechanical techniques including microwave radar and infrared.

24.4.4 Numerical Examples

To establish programs to monitor their systems, engineers typically encounter a wide range of decision contexts including which sensor types (or which combination of sensor types) to purchase, which locations to install them, whether to carry out around-the-clock or periodic surveillance, the optimal time to install monitoring equipment, and so on. In this section, we discuss two numerical examples that illustrate two of such decision contexts.

Example 24.1

The owner of an urban drainage system seeks to install 200 pieces of equipment for channel siltation monitoring at various locations throughout a city. Equipment type A has a service life of 6 years, purchase cost of $15,000, annual operating and maintenance (O&M) cost of $1500, and a salvage cost of $1,000 each. Equipment type B has a service life of 4 years, purchase cost of $12,000, O&M cost of $1,500, and a salvage cost of $700 each. The discount rate follows a normal distribution with mean 5% and variance 2%. Determine, using an appropriate criterion for economic evaluation, identify which alternative is more economically efficient.

Solution

From the concepts of economic analysis we learned in Chapter 11, we select the equivalent uniform annual cost as the criterion for the economic efficiency evaluation. This is because the alternatives have different service lives. It may be assumed that within their respective life cycles, there is negligible difference in the effectiveness of benefits (for example, the reliability of the data they generate). Bringing all costs to present worth and annualizing this cost, we a EUAC expression to which we vary the interest rate using Monte Carlo simulation. The outcome of the simulation is provided as Figure 24.11. From the simulation, the mean EUAC can be determined as $4,309 and $4,722 for Equipment types A and B, respectively. Besides, it can be seen from Figure 24.11 that A is stochastically dominant (superior at any region of the output curve) to B. Thus, A is more economically efficient and should be selected.

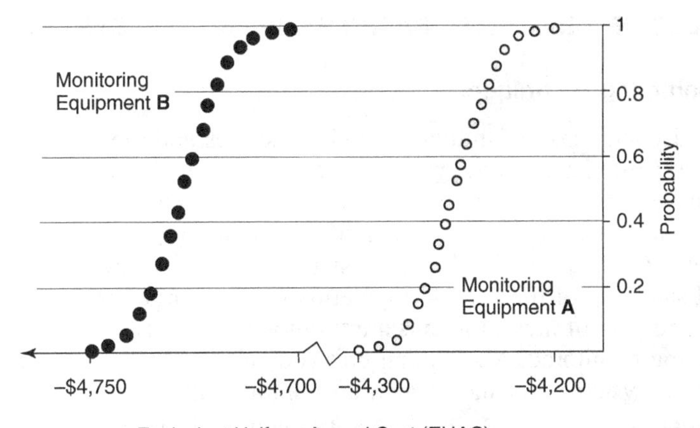

Figure 24.11 Extended-NPV and EUAC of System Monitoring Alternatives.

Example 24.2

An architectural engineer is evaluating three different sensing technologies to monitor the effects of natural climate on the operational performance of a large building system. Sensor type A has a customer rating of performance of 9 on a scale of 1 (poor) to 10 (excellent), life-cycle cost of $20,000, and 98% reliability. Sensor type B has a customer rating of performance of 7, life-cycle cost of $15,000, and 92% reliability. Sensor type C has a customer rating of performance of 5, life-cycle cost of $10,000, and 85% reliability. Customer ratings consist only of vendor customer service, maintainability, and ease of installation. Determine the optimal choice of sensor type. Assume that the relative importance values (weights) of the selection criteria are as follows: reliability = 0.5, life cycle cost = 0.2, and customer rating = 0.3. The scaling functions are as follows:

$$\text{Customer rating } (x): \text{ Utility} = 1.29x^2 - 1.97x + 0.27$$
$$\text{Life cycle cost in \$10,000s } (x): \text{ Utility} = -4.6x^2 + 100$$
$$\text{Reliability, \%, } (x): \text{ Utility} = e^{0.046x}$$

Solution

Table 24.1 presents the various aspects of the evaluation problem for the for sensor selection on the basis of the multiple criteria. A weighted-sum method (see Section 12.4.1a of Chapter 12) is used. It can be seen that Sensor type A has the largest weighted utility and thus should be selected.

Table 24.1 Multiple-Criteria Evaluation for Sensor Selection Using Weighted-Sum Method

	Raw Values			Utilities (max 100)			Weighted Utilities			
Sensor Type	Customer Rating (1–10 index)	Life-Cycle Cost (10^3)	Reliability (%)	Customer Rating	Life-Cycle Cost	Reliability	Customer Rating	Life-Cycle Cost	Reliability	Total
A	9	25	97	87.03	71.25	86.66	26.11	14.25	43.33	83.69
B	8	15	95	67.07	89.65	79.04	17.93	17.93	39.52	77.57
C	5	10	85	22.67	95.40	49.90	19.08	19.09	24.95	50.83

Sample calculation: For Sensor Type A, for customer rating; utility = 1.29 $(9)^2$ − 1.97(9) + 0.27 = 87.03; weighted utility = 87.03(0.3) = 26.11.

24.5 COST-EFFECTIVE RELIABILITY-BASED SAMPLING FOR SYSTEM MONITORING

24.5.1 Sampling Techniques

The typically large size of individual civil structures and the typically vast expanse of civil system networks often make it impractical to carry out continuous and 100% coverage monitoring of individual systems or an entire network of systems. To reach a reasonable balance between sparse coverage (where the costs are lower but little reliability is provided) and extensive coverage (which gives very good reliability but has excessive costs), inspectors of civil systems typically carry out statistical sampling of the overall population (as we learned in Chapter 6) in such a manner that a minimal number of measurements are taken without sacrificing confidence in the assessment of what is being monitored. A sampling unit is defined differently for an individual system and for a network of systems: For the former, a sampling unit is a section or component of the system; for the latter, a sampling unit is simply a system that exists in the network. In Section 6.1.2 of Chapter 6, we discussed the common sampling methods, namely, random sampling that may be simple, systematic, clustered, or stratified.

24.5.2 Sample Size Requirements for Monitoring

The objective of sampling is to achieve a healthy balance between data reliability and the cost of data acquisition. For any civil system, there is no universal standard for the sample size because it will depend on the required precision of the data and the inherent variability of the property being measured, and these attributes differ for different system owners and systems.

An important concept in the statistics of reliability is the confidence statement (assuming a symmetric distribution of the system attribute being measured). The reader may refer to Section 6.3.2 of Chapter 6, which dealt with interval estimation. The confidence statement is: "There is a probability p that a certain measured quantity will fall between a certain range." The probability is given by $1-a$ where $1-a$ is the degree of confidence, and the range is a function of the inherent variability in the data and the degree of significance. Mathematically, this is written as:

$$p(\mu_{\overline{X}} - Z_{\alpha/2} \times \sigma_{\overline{X}} < \overline{X} < \mu_{\overline{X}} + Z_{\alpha/2} \times \sigma_{\overline{X}}) = 1 - \alpha$$

For a given degree of confidence, the confidence interval associated with the estimate of a population parameter is a measure of the error or precision of that estimate.

$$\text{Specifically, Error} = {}^1\!/_2 \times \text{Confidence Interval}$$

Figure 24.12 presents the precision consequences of different sample sizes.

The range of acceptable $\overline{X}$ values is given as $\mu_{\overline{X}} \pm Z_{\alpha/2} \times \sigma_{\overline{X}}$.

The error is given by $Z_{\alpha/2} \times \sigma_{\overline{X}}$.

For the purposes of system monitoring, reliability analysis can be defined as: "Determination of the precision, minimum sample size or the level of significance associated with the estimation of a population parameter on the basis of the sampling data." Reliability analysis therefore consists of precision analysis, adequacy analysis, and level of significance analysis (Figure 24.13).

Adequacy analysis is the determination of the minimum sample size needed to ensure a certain precision or degree of confidence in the estimated value of the population parameter, given the standard deviation. The minimum sample size needed to ensure a certain precision of the estimated parameter is

$$n = \left(\frac{Z_{\alpha/2} * \sigma}{\text{Error}} \right)^2$$

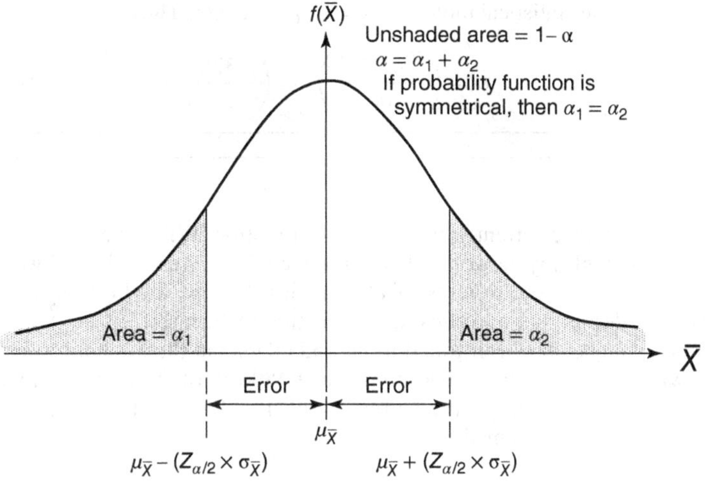

Figure 24.12 Precision consequences of different sample sizes.

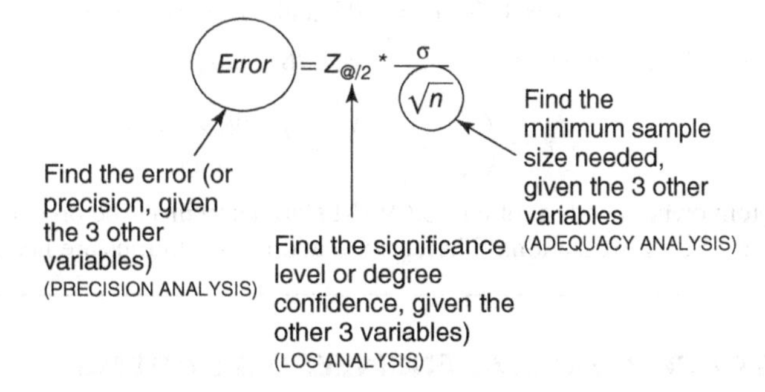

Figure 24.13 Analysis of precision, sample size adequacy, and level of significance.

where $Z_{\alpha/2}$ is the Z value corresponding to an area of $\alpha/2$ to the left; Error is the deviation of the estimated parameter for its true value; and σ is the standard deviation of the population.

Example 24.3

For a normally distributed population, how many samples should we take in order to ensure that there is a 98% chance of obtaining a sample mean falling between 105 and 115? The population mean and variance are 110 and 86, respectively.

Solution

The minimum sample size is given by

$$n = \left(\frac{Z_{\alpha/2} \times \sigma}{\text{error}_{\max}}\right)^2$$

Now $\sigma = 3.5\ \text{N/mm}^2, 1 - \alpha = 0.98$, thus, $\alpha = 0.2$, and $\alpha/2 = 0.01$; $\sigma = (86)^{0.5} = 9.87$; Error $= 0.5(115 - 105) = 5$.

From the statistical tables, $Z_{\alpha/2} = Z_{0.01} = 2.325$. Thus,

$$n = \left(\frac{Z_{\alpha/2} \times \sigma}{\text{error}_{\max}}\right)^2 = \left(\frac{2.325 \times 9.87}{5}\right)^2 = 18.58$$

Example 24.4

A certain province currently has 35 weigh-in-motion (WIM) stations located at randomly selected points on the state highway network. (This equipment is very useful in highway management because it measures and records the weights of all vehicles that use the highways, and such weights are used for pavement design, tax purposes, determination of license fees, etc.) The statewide standard deviation of truck weights has been found to be 15,000 lb, and the system owner requires that estimates of the statewide mean weights should not exceed 4800 lb. of the true mean weight at a 5% level of significance. Determine whether the current number of WIM stations is too few, just enough, or too many. State any assumptions made.

Solution

The required sample size is

$$n_{\min} = \left(\frac{Z_{\alpha/2}\sigma}{\text{error}_{\max}}\right)^2$$

$$\sigma = 1500, \alpha = 0.05, \text{and } \alpha/2 = 0.025; \text{Error} = 4800$$

From the statistical tables, $Z_{\alpha/2} = Z_{0.025} = 1.96$. Thus,

$$n = \left(\frac{Z_{\alpha/2} \times \sigma}{\text{error}_{\max}}\right)^2 = \left(\frac{1.96 \times 1500}{4800}\right)^2 = 37.51$$

The system owner needs a total of 38 WIM stations to meet the precision and confidence requirements. There are 35 existing stations. Therefore, 3 additional stations are needed.

24.6 MONITORING OF SPECIFIC CIVIL ENGINEERING SYSTEMS

The extent of progress of system monitoring is different for the different types of civil systems. For buildings, bridges, and dams, advanced sensors and instrumentation equipment and architecture have been developed; with capabilities including real-time data processing and visualization, these advancements have yielded benefits in a wide range of agency functions for these system types. Researchers including Dickenson (2007) have noted that for some other types of civil engineering systems, the development or deployment of monitoring programs has lagged behind due to factors including funding, difficulties in installing sensors at the time of the facility's construction, sensor longevity and maintenance concerns where the sensor is exposed to the rain and sun, location and access challenges, and lack of resources for efficiently acquiring, transmitting, and storing data. The next section discusses the monitoring of a specific type of civil engineering system.

24.6.1 Port Facilities

Port infrastructure may include wharfs, navigation channels, bridges, piers, other conveyance systems including cranes, and near-shore structures (such as levees, breakwaters, and jetties). At port facilities, sensors are installed to acquire information on port operations, the marine environment, physical structures, vessel impact and mooring loads, near-shore current movements, seismic responses of waterfront structures, and deformations of the ground and wharf foundations. Table 24.2 [modified from Dickenson (2007)] presents some monitoring applications for port systems.

Table 24.2 Port System Monitoring

(a) Monitoring of Environmental Phenomena Affecting Port System Operations

Measurement Objective	Cause or Phenomenon	Monitoring Mechanism/ Equipment	References
Hydraulic loads	Tides, waves, currents	Current meter, tide gauge, wave gauge	Dickenson (2007)
Corrosion of structural elements	Splash zone dynamics	Equipment that measure resistivity of metallic components, element thickness, images or element corrosion	Dickenson (2007)
Meteorological phenomena	Temperature Wind Precipitation	Thermometer, anemometer, rain gauge	Dickenson (2007)
Scour and sedimentation in navigation channels and adjacent to waterfront structures	Prop wash Natural currents	SONAR, LiDAR, multibeam, bathymetric surveying	NCHRP (1997a); NCHRP (1997b).

(b) Monitoring of Port System Performance and Structural Integrity

Measurement Objective	Cause or Phenomenon	Monitoring Mechanism/ Equipment	References
Pile integrity	Down drag Service loads Lateral soil movement	Strain gauges Inclinometers, tell-tales, sonic logging	Dickenson (2007)
Wharf, pier integrity	Mooring tension loads Vessel impact force/velocity Age-related phenomena (fatigue, creep, corrosion)	Load cells, laser, low-strain vibration monitoring, resistivity techniques	Dickenson (2007)
Bridges, cranes, and conveyance systems	Dead and live loads Fatigue	Low-strain dynamic vibration monitoring	Celebi, 2006; Masri et al., 2004
Near-shore structures (levees, breakwaters, jetties)	Foundation degradation, scour Wave action induced internal erosion Raveling of armor layers	Visual inspection, piezometers LIDAR, multibeam	Reynolds, 2002a; Reynolds, 2002b.
Changes in anchor loads or pile prestressing	Creep	Strain gauges, tell-tales	Dickenson (2007)

(c) Monitoring of Port Foundations and Structures during and after Extreme Events

Measurement Objective	Cause or Phenomenon	Monitoring Mechanism/ Equipment	References
Seismic microzonation and seismic behavior of foundation soils Behavior of structures in response to natural hazards Wharf, pier integrity Bridges	Earthquake Hurricane or flood Wave action Malicious event	Integrated network of detectors that monitor soil and structure interactions Sensors that measure elastic response, large-strain, or permanent deformations. Sensors that measure dynamic low-strain vibrations Piezometers located in foundation soils and rubble mound structures.	Donahue et al., 2004; Elgamal et al., 1996; Fletcher at al., 2001; Iai, 1998; Kokusho and Matsumoto, 1997; Tsuchida, 1990; Celebi, 2006; Masri et al., 2004.

CORROSION MONITORING OF STEEL AND REINFORCED CONCRETE SYSTEMS

Civil engineers place great premium on the monitoring of the materials used to construct their systems, particularly those in unusual or harsh environments. For steel structural systems and steel reinforcement in concrete structures, corrosion is one of the most common causes of premature end of life and thus is often a key target of monitoring efforts. The vulnerability to corrosion is particularly severe in environments where steel components or steel-reinforced concrete components are exposed to chlorides. The sources of chlorides include the concrete ingredients and concrete-penetrated chemicals from salt-laden environments (in the air at coastal areas or in surface runoff at cold regions where deicing chemicals are typically applied). Several owners of large scale civil engineering systems such as oil rigs have established corrosion monitoring programs for their systems.

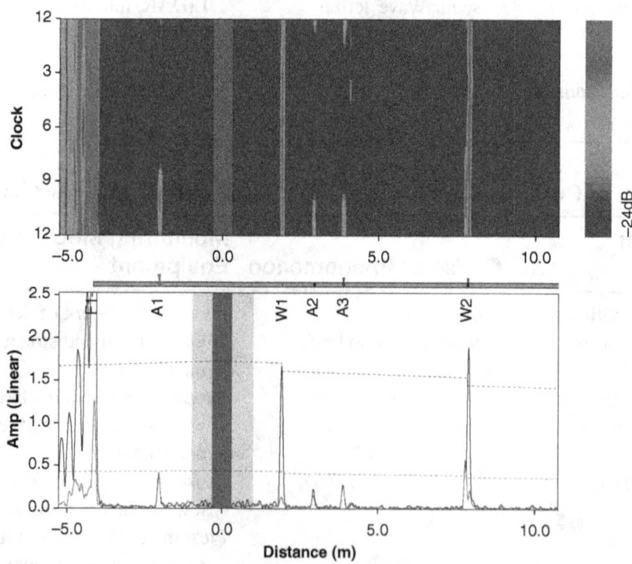

Guided wave ultrasonic testing (GWUT) is one of the several nondestructive techniques for inspecting elongated and spread-out structures such as metallic pipelines, rail guideways, rods, and metal plate structure. The portable electric unit is made to drive a signal (see top inset figure), and mechanical stress waves are generated through a transducer array mounted around the pipe surface that propagates along the structure. Because the stress is guided by the structure's boundaries, the waves travel a long distance with very little loss in energy. Along the pipe, where there is a change in local stiffness or in the pipe cross section, an echo is generated and recorded (see bottom inset figure), whereby defects due to corrosion can be identified.

24.7 PLANNING FOR SYSTEM MONITORING/INSPECTION (M/I)

24.7.1 General Considerations and Steps for M/I Planning

In a manner similar to maintenance planning (Chapter 25), the monitoring and inspection of a system can be carefully planned in a way that produces the monitoring output in a cost-effective manner. Thus, for a new system, an M/I schedule can be established for the entire life cycle; for an existing system, a new schedule can be established (or an existing schedule updated) for the rest of its life. Figure 24.14 presents the steps for developing a schedule for system monitoring.

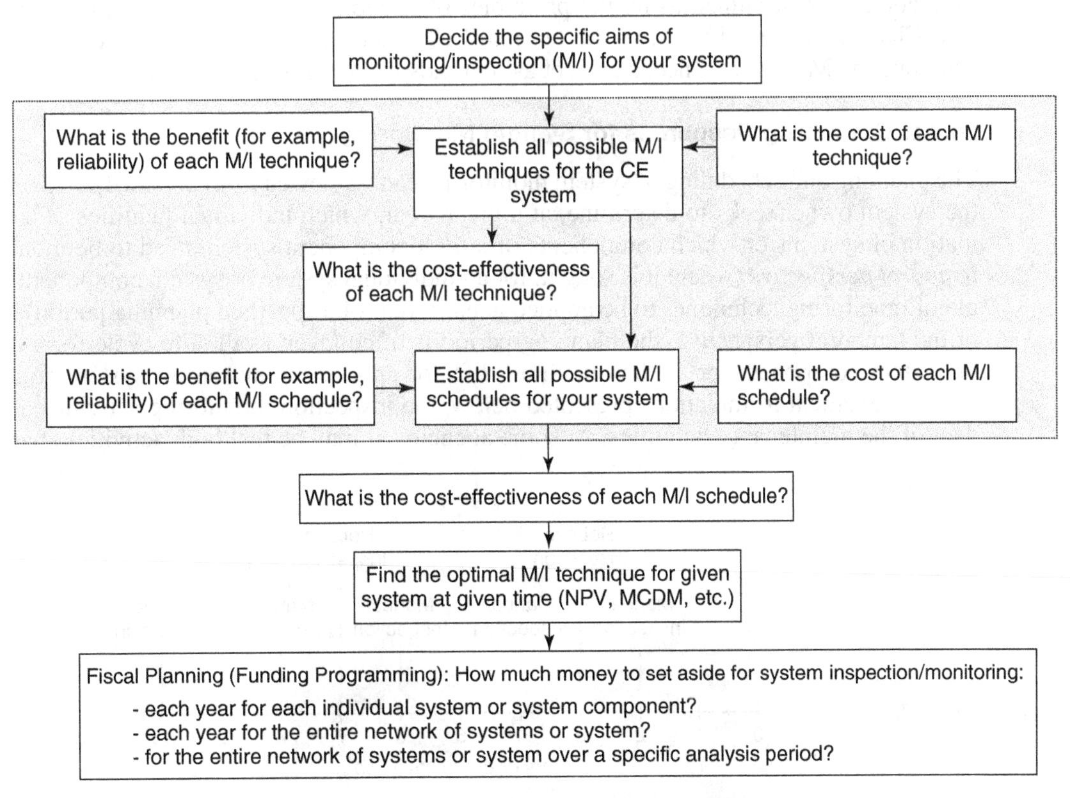

Figure 24.14 Steps for developing a schedule for system monitoring.

The first step is to establish the goals of the M/I program for the civil engineering system. In the introductory section of this chapter, we discussed M/I rationale, including tracking the system's physical condition, operational performance, or stability, the system's threat to the environment, and environmental threats to the system. The second step is to establish, based on the established goal of the M/I effort, the overall M/I architecture that includes all candidate M/I techniques for the monitoring effort; again, these should be based on the type of civil engineering system in question, the nature of the environment, and the purpose of the monitoring program. This includes the sensor type, communication and information technology (IT) features, and data acquisition and analysis tools, including AI tools for any image detection or data mining tasks, and the overall architecture of the monitoring activity.

An **M/I schedule** or program may be defined as a series of M/I techniques spaced out over a period of time (in this case, the system life or rest of life). A number of alternative M/I schedules must be established. These may be time based (years between M/I activities) or load based (user frequency or load intensity between M/I activities), as illustrated in Figure 24.15.

The benefits and costs of each type of M/I activity should be established. What constitutes a benefit or a cost may need to be defined by the system owner. Generally, benefits may be measured in terms of the system accuracy such as the number of false positives or false negatives, and the costs can be defined by the costs incurred by the agency in deploying that specific M/I technique. M/I schedules comprise M/I techniques; as such, if the benefits and costs of M/I techniques are known, the benefits and costs of M/I schedules can easily be determined either additively or after duly accounting for overlaps. Thus, the cost-effectiveness of each M/I schedule can be established. For multiple alternative M/I schedules, the optimal M/I technique for a given system at a given time could be determined using the principles of economic analysis or multiple-criteria analysis (see Chapters 11 and 12). After the optimal M/I schedule has been established, the agency can calculate the M/I funding needs and set aside funds for M/I for the system.

24.7.2 Optimizing Resources for System Monitoring

The planning and scheduling of system monitoring can be viewed from a *spatial perspective* (where the system owner seeks to determine, at a given year, which individual facilities of a larger population of systems or which components of a multicomponent system need to be monitored) or a *temporal perspective* (where it is sought, for an individual system or system component, the schedule of monitoring techniques to be applied at each year of a specified planning period). In the case of the temporal perspective, the planning period is often taken as the life cycle for a new system or system component, or as the remaining life for an existing system or system component. In the mathematical formulations presented below, the inspection/monitoring activities are independent of the maintenance activities. At some agencies, it may be that both activities share the same

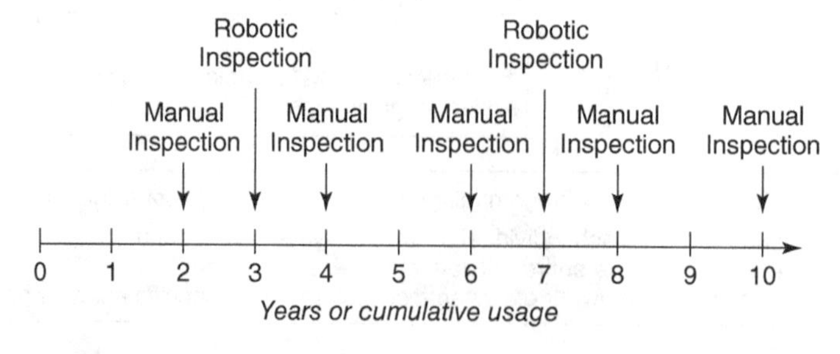

Figure 24.15 Simplified sample schedule illustration for system monitoring.

budget and planning horizons and thus could be planned together instead of separately, as discussed in Madanat and Ben-Akiva (1993).

(a) Optimizing Resources for Monitoring an Individual System over Multiple Years. In problems of this nature where the temporal perspective is concerned, the issue is: At a specific facility, which M/I technique, if any, should be deployed at each of several years within the horizon period? The horizon period could be the entire life (for a new system) or the remaining life (for an existing system. Also, note that for some years, there may be no deployment. This is a knapsack problem where the decision maker seeks the best possible monitoring schedule or set of candidate M/I techniques and timings. The word "facility" simply refers to a system or a system component. It is sought to maximize the benefit or reward over a given horizon period (typically, facility life cycle), subject to one or more constraints. The benefit could be expressed, for example, as the reliability of the monitoring exercise or the expected percentage of true outcomes (i.e., 100% less the percentage of false positives or false negatives). In a simple scenario where there is no limitation on the annual spending amount but instead a budget constraint for entire horizon period, the optimal funding allocations for the facility may be determined by solving the integer programming formulation shown as Equations (24.1)–(24.5):

$$\text{Maximize} \quad U = \sum_{i}^{I} \sum_{t}^{T} (x_{it} e_{it}) \tag{24.1}$$

$$\text{Subject to} \quad \sum_{j=1}^{J} \sum_{t=1}^{T} (x_{jt} c_i) \leq B \tag{24.2}$$

$$\sum_{j=1}^{J} \sum_{t=1}^{T} x_{jt} e_{jt} \geq E \quad \text{or} \quad \frac{1}{T} \sum_{j=1}^{J} \sum_{t=1}^{T} x_{jt} e_{jt} \geq P \tag{24.3}$$

$$x_{jt} = 0 \ \text{if} \ t < t_{cj} \tag{24.4}$$

and

$$x_{jt} = 0, 1 \tag{24.5}$$

where c_j is the cost of M/I technique j if deployed at the facility; B is the total monitoring budget for the facility over the horizon period; J is the number of candidate M/I techniques; t is the analysis year $= 1, 2, \ldots, T$; t_{cj} is the year when the facility is expected to have a need for deploying M/I technique j (also referred to as the "critical year"); e_{jt} is the benefit of the M/I technique j deployed at in year t; $x_{jt} = 1$ if M/I technique j is deployed at year t, $= 0$ otherwise; T is the length of the analysis period; E and P are the benefit thresholds over the horizon period and for each year, respectively.

Equation (24.1) is the objective function that maximizes the overall benefit of the monitoring schedule. The constraints are Equations (24.2)–(24.5). Equations (24.2) and (24.3) represent the knapsack problem's *size* constraints. Equation (24.2) specifies that the total monitoring costs over the horizon period should not exceed the budget. Equation (24.3) specifies that the total or average benefits over the horizon period should exceed. Equation (24.4) recognizes that in certain cases, there could be a "critical year" before which no deployment is necessary, and thus requires that the M/I technique j should not be deployed before its critical year, t_{cj}.

(b) Optimizing Resources for System Monitoring Solely from the Spatial Perspective. Here, the questions are: At a given year, which facility (in a given larger population of facilities) deserve deployment of some M/I technique; what percentage of facilities to monitor; and what is the appropriate or optimal annual expenditure for monitoring each facility? The word "facility" simply refers to a system (where the problem context is a larger population of systems) or a system component (where the problem context is a system that consists of multiple components).

Which Facilities Deserve Deployment of Some M/I Technique at a Specific Year. This is also a knapsack problem where the decision maker seeks the best possible set of candidate actions that need to be undertaken to maximize the overall spatial reward in a given year) and subject to a budget constraint. In the simplest scenario, there is a given budget specified for the year in question. The optimal funding allocation for monitoring the entire network of systems in the year may be determined by solving the integer programming formulation presented as Equations (24.6)–(24.9):

$$\text{Maximize} \quad U = \sum_{i}^{I} \sum_{j}^{J} (x_{ij} e_{ij}) \tag{24.6}$$

Subject to

$$\sum_{i=1}^{I} \sum_{j=1}^{J} (x_{ij} c_{ij}) \leq B \tag{24.7}$$

$$\sum_{i=1}^{I} \sum_{j=1}^{J} x_{ij} e_{ij} \geq E \quad \text{or} \quad \frac{1}{I} \sum_{i=1}^{I} \sum_{j=1}^{J} x_{ij} e_{ij} \geq P \tag{24.8}$$

$$x_{ij} = 0, 1 \tag{24.9}$$

where the symbols have their usual meaning.

Equation (24.6) is the objective function. Equations (24.7)–(24.9) are the constraints. Equation (24.7), the budget constraint, states that the total monitoring expenditure should not exceed the available budget over the analysis period. This is a *size* constraint of the knapsack problem. Equation (24.8) requires that the total or average benefits of monitoring should exceed the specified thresholds. Equation (24.9) states that there are two choices: to deploy or not to deploy.

How Much to Spend for Each Facility for Monitoring at a Specific Year. The decision that needs to be made here relates to the amount to spend on each facility of the network in given year, to maximize the network-wide reward (utility) subject to a budget constraint. Again, let us remind ourselves that the word "facility" refers to a system (where the problem context is a larger population of systems) or a system component (where the problem context is a system that consists of multiple components). The optimal funding allocation for the system monitoring may be determined by solving the following linear programming formulation:

$$\text{Maximize} \quad U = \sum_{i}^{I} \sum_{j}^{J} (e_{ij}) \tag{24.10}$$

Subject to

$$\sum_{i=1}^{I} \sum_{j=1}^{J} x_{ij} \leq B \tag{24.11}$$

$$x_{it} \geq 0 \tag{24.12}$$

where x_{ij} is the amount spent on M/T technique j for monitoring facility i. Other symbols are as defined in previous sections.

(c) Optimizing Resources for System Monitoring from both Temporal and Spatial Perspectives. Here, the question is: Which M/I technique should be deployed, at which facility, and when (in which of several programming periods)? (In this text, the programming period is assumed to be one year; at some agencies, it may be monthly, quarterly, every two years, every five years, etc.). This is a knapsack problem where the decision maker seeks the best possible *monitoring schedule* (i.e., set of candidate M/I techniques to be deployed at each facility and at each year), in order to maximize the networkwide reward) subject to a budget constraint. This can be for the problem context where the system owners seek to determine which facilities to monitor and in which year; for this problem context, there could be no limitations on the yearly amount spent, limitations on spending for entire analysis period, or limitations on yearly amounts. Also, it may be specified whether or not unspent funds in a given year can be carried over to the following year and whether there are any constraints on the monitoring performance. Problems that could be posed also include how much to spend on each system (facility) for the entire network of systems in each or all years combined of the analysis period, or what percentage of facilities to monitor in each or all years combined of the analysis period.

24.8 MONITORING THE SYSTEM USERS

24.8.1 The Ethics of System Surveillance

As we discussed in the introduction to this chapter, the system agency often needs to monitor the usage of a system for a variety of purposes that include quick response to safety hazards, real-time allocation of resources to control congestion, and prompt identification of natural or man-made threats to user security. For certain types of systems, the system operation involves direct use by humans (e.g., buildings, stadia, transit facilities, and highways) in contrast to indirect use by humans, such as water treatment plants, hydroelectric plants, and canals. As we discussed in Section 24.2, the purpose of monitoring includes the detection of user presence, counting of the users, establishing the characteristics of individual users (dimensions, weights, choices, perceptions, etc.) among other reasons. Also, the purpose of the system owner may be to watch (inspect) continuously the physical system as it is being used, with little or no direct interest in the system user per se; in some cases, however, it is difficult to separate inspection of the system from the monitoring of its users. In monitoring a system, the system owner often employs a wide range of technologies and tools that may be characterized in at least one of several ways: intrusive versus unintrusive, manual versus technology based, or blind (user identities not revealed) or nonblind. The specific technology or tool used depends on the purpose of the monitoring.

The use of monitoring tools and technologies that are described as nonblind can pose a problem to the system owner, as users may feel that such monitoring invades their privacy and therefore is unethical if not illegal. However, a system owner may contend that sustained monitoring of the system users is essential to identify defects before they become hazardous to the system users, and therefore may argue that it is in the interests of the system user to sacrifice a little privacy. Scholars in surveillance ethics, including George Orwell, Jeremy Bentham, and Kevin Macnash have discussed the ethical challenges posed by surveillance.

In Jeremy Bentham's (2011) concept of the Panopticon (a prison where each inmate was being watched without being seen), the warden communicates with the prisoners via a loudspeaker. The prisoners did not know the specific times they were being watched and thus assumed that they were being watched all the time; this arrangement encouraged the prisoners to behave well and maintain self-discipline. George Orwell is arrangement involved ordinary innocent citizenry. In 1984, Orwell (2004) described a two-way television that gave the government auditory and visual access to people's places of work and residence and other public places, and citizens were informed repeatedly that "Big Brother" was watching them. Orwell's novel gave insight into the reasons for which a government may seek to carry out ubiquitous surveillance and the resulting consequences on the behavior of individuals and the society (MacNish, 2011). In his book *Discipline and Punish*, Michel Foucault (1991) reinforced the notion that ubiquitous surveillance, similar to the persistent gaze of the prison supervisor in Benthams' Panopticon, constitutes a punishment, and therefore was not a good thing if applied to innocent citizens.

In the current era, technological advances including drones have added unprecedented complexity and mobility to surveillance (Marx, 1998). Closed circuit television (CCTV) cameras that are installed at streets, transit stations, and stadiums and license plate recognition cameras installed at traffic lights or toll booths to catch red-light runners and toll violators, respectively. These are representative of situations where the system owner becomes a hidden, anonymous watcher exercising an omnipresent and ubiquitous gaze at the system users. The situation is exacerbated by the exponentially increasing capabilities of IT systems that transmit vast quantities of information through conduits that are prone to hacking or interception. Personal information about system users collected via a system's surveillance equipment may end up on the Internet (which may be vulnerable to abuse, theft, or commodification), suffer function creep [which is defined as the extension, over a period of time, of the use of a technology or its data from the original cause to a different cause (Winner, 1977)], or simply lost by the system owner. MacNish (2011) noted that while Panopticon and Big Brother are consistent with authoritarian and negative images that corroborate the notion that surveillance is always unethical, surveillance per se is an ethically neutral concept, similar to a kitchen knife (Asimov, 2005): it can be used for great good or for great evil. It may very well be possible to encounter situations where surveillance is not only justified but even embraced by most of the people who are being watched. Thus, as MacNish duly noted, a particular instance of surveillance may be justified on the basis of its intent, the means employed, and the proportionality of the surveillance. In the post–September 11 world, the use of government surveillance has become widespread as a tool for terrorism detection and deterrent; however, this has prompted outcries from civil rights activists who generally argue that if freedom and privacy are sacrificed for security, all three, in the long run, will be lost.

SURVEILLANCE [OF SYSTEM USERS]: THE CONSEQUENTIALIST VERSUS DEONTOLOGIST PERSPECTIVES

Consequentialism is a category of normative ethical theories that postulates that the consequences of a person's conduct are the ultimate basis for any judgment regarding the rightness of that conduct; therefore, a consequentialist will argue that any act that is expected to produce a good outcome or consequence is a morally right act. On the other hand, deontological ethicists support the normative ethical perspective that assesses an action's morality on the basis of whether the action adheres established laws or rules; its proponents argue that it is not the consequences of an action that make it right or wrong but the properness of the action itself.

From the viewpoint of consequentialists, therefore, nonconsensual large-scale surveillance of the users of a civil engineering system may be considered justifiable because it leads to the greater good. Thus, even though some people may abhor the practice, surveillors may justify their actions by the ultimate benefits they bring to the society (better security and more reliable usage data for system enhancement planning). The position of the consequentialists implies that it is acceptable that the rights of the few may be overridden by the interests of the many. Deontologists strongly oppose this perspective. Any deontological justification of surveillance is likely to begin with an examination of the entity to be surveilled and an inquiry into the special features of that entity that qualifies it to be monitored in such a manner. Holding the position that surveillance is inherently harmful to those being watched, the deontologist will seek a reasonable justification to exposing the intended target to such harm.

As evidenced by the fallout following the National Security Agency leaks of June 2013, surveillance is often justified using both consequentialist and deontological viewpoints. Specifically, on one hand, the surveillor is considered justified in protecting the majority; on the other hand, the surveillor is considered justified in focusing on specific wrongdoing individuals or sets of individuals who continually pose a threat to the majority. Public location, from a public security perspective, may be considered justified on the basis of its capability of targeting only specific persons that are believed to pose a threat to the society.

Similarly, the mechanism used for surveillance could influence the justification from a consequentialist or deontological perspective. A closed circuit camera setup (CCTV), which carries out surveillance in an indiscriminate manner, will seem to be more favorable to the consequentialist. In transit stations, most people who are being watched via CCTV by the system operator have done no harm (and have no intention of doing harm) to society. Nevertheless, the CCTV benefit of detecting the minority of wrongdoers may justify the surveillance of all persons in the area. In contrast, more discriminating mechanisms of surveillance, for example, tapping the phone of a transit bombing suspect, is more likely to be supported by the deontologist because people who have given the authorities reasonable suspicion to believe they have committed a crime (or are likely to do so) are considered to have rendered themselves deserving to be surveilled in such a manner.

Finally, consequentialists and deontologists differ in the nature of their opposition to surveillance. Deontologists find surveillance to be unacceptable when it violates a person's right to privacy and other rights. On the other hand, consequentialists oppose surveillance when they are convinced that the costs of surveillance far outweigh its benefits; then consequentialists will consider as acceptable those situations where a specific surveillance instance is expected to improve the well-being of society even at the expense of the privacy of a few individuals.

Adapted from Kevin MacNish, 2011, University of Leeds, United Kingdom.

SUMMARY

In this chapter, we discussed the rationale, issues, concepts, and analytical tools and equipment for monitoring the integrity of a civil engineering system as well as its relationship to its environment. We identified the basic aspects of any monitoring effort and discussed the different dimensions or purposes of system monitoring. We then discussed the typical architecture of a system monitoring program, particularly the different types of detection and sensing that could be deployed. Given the large population of components of a system or the large number of individual systems in any network, we showed how an inspector could develop a cost-effective sample size from which confident conclusions on system attributes can be made. Using a case study of port facilities, we outlined the different components that deserve special attention during the inspection and monitoring process. We then demonstrated how engineers could develop a long-term plan for system monitoring by considering the costs and benefits (effectiveness) of the different techniques of monitoring. Finally, we discussed the rather thorny issue of system user monitoring and the associated ethical conflicts.

As a closing remark, it must be mentioned that to address the increasingly formidable challenges of monitoring civil infrastructure as a first line of defense against potential catastrophic failure, a number of engineering organizations worldwide are engaged in refining the art and science of civil engineering system monitoring. These include the University of California Davis's Nano-Engineering and Smart Structures Technologies Laboratory, the University of Michigan's Laboratory for Intelligent Structural Technology, the Indian Institute of Technology's Center for Non-Destructive Evaluation, the Virginia Institute of Technology's Center for Intelligent Material Systems and Structures, Purdue University's Bowen Laboratory, and Drexel University's Institute for Sustainable Infrastructures.

EXERCISES

1. List and explain the four purposes of civil engineering systems monitoring.
2. Discuss the basic elements of a system monitoring architecture and explain how each element could be enhanced using technology.
3. Is it adequate to monitor physical systems in a manner that is only "skin-deep"? What are the pro and cons arguments that can be offered for this debate?
4. Explain how photogrammetry could be used to monitor the stability of large civil engineering structures.
5. For any civil engineering system in your neighborhood or your interest, list and explain (i) the threats that the system faces from its environment (ii) the threats that the environment faces from the system. For each direction of monitoring, which equipment could be used to assess these threats?
6. Identify and discuss any sensing or detection mechanisms that could be used to monitor system condition, system stability and structural integrity, system usage intensity (loading) and frequency and discuss the conditions under which the mechanism is effective.
7. Discuss the goals of any sampling effort for purposes of system monitoring, and explain why there is no universal standard for the sample size even for a specific type of civil engineering system.
8. An important concept in statistical reliability is the so-called confidence statement (the premise for which is a symmetric distribution of the system attribute being measured). State fully and illustrate diagrammatically the confidence statement and write its mathematical notation. Using the notation, explain how the three aspects of statistical reliability analysis could be addressed from the perspective of sampling for system monitoring.
9. For a specific civil engineering system on your campus of your interest, discuss how the system could be monitored comprehensively. Your discussion may include the attribute or phenomenon that is being monitored, its causes, and the appropriate monitoring mechanism, equipment, or sensors.
10. The trusses of railway steel bridges in a certain city are inspected every other year. It has been determined that 5% of inspections reveal an "urgent defect" (a defect that needs to be addressed as soon as possible). What is the probability that (i) for a single railway steel bridge, an inspection reveals no urgent defect in a given year, (ii) for a single railway steel bridge, an inspection reveals no urgent defect in two consecutive years, and (iii) for two railway steel bridges, inspections reveals no urgent defect in a given year?

11. As we learned in Chapter 6, in statistics, a null hypothesis is a statement that there is no presence or effect of some phenomenon under investigation. For example, in the context of systems monitoring, the hypothesis could be: "the measured defects at a certain system is not significant". All statistical hypothesis tests have a probability of making type I and type II errors. Type I error (or, false positive) is the incorrect rejection of a true null hypothesis; in other words, the conclusion that there is a presence of defects when in fact there is none. Type II error (or, false negative) is the failure to reject a false null hypothesis; in other words, the conclusion that there is an absence of defects when in fact there are. Discuss the consequences of each type of error on the overall long-term management of the system.
12. Construct an integer programming formulation for determining the optimal funding allocation for monitoring a system under each of the following decision contexts: (a) a single facility over a given horizon period (entire life or remaining life), (b) a collection of facilities at a single year; (c) a collection of facilities over multiple years. For each context, include at least two spending constraints and two performance constraints.
13. A system owner seeks to install sensing equipment at various locations of a large civil engineering system in a bid to monitor the system's impacts on the environment. Due to budgetary constraints, the equipment may be placed at only some, not all, of the 10 candidate locations. Due to the different conditions across the locations, the benefits and costs of sensor installation differ for each location (Table 24.3). The anticipated effectiveness or benefits (in terms of the expected usefulness of the data from that location in characterizing the environmental impacts) is expressed on a scale of 0 (low) to 10 (high) and the life-cycle cost (in millions of dollars) of the sensor equipment installation and maintenance. The overall budget for that year is $2M. The system owner seeks to identify the optimal set of candidate locations for the installation; the goal is to maximize the total benefits. The total life cycle cost should not exceed $3M. (a) Write the stated objective in mathematical notation. (b) Write all the constraints in mathematical notation. (c) Use an appropriate optimization package to find the optimal solution. (d) What is the total system-wide benefit and cost that correspond to the optimal solution?

Table 24.3 Sensor Installation Location Costs and Effectiveness

Candidate Location	**A**	**B**	**C**	**D**	**E**	**F**	**G**	**H**	**I**	**J**
Installation cost ($M)	0.32	0.61	0.14	0.18	0.31	0.25	0.79	0.57	0.35	0.26
Life cycle cost ($M)	0.41	0.87	0.23	0.32	0.57	0.36	0.99	0.68	0.39	0.35
Benefit score	4.93	6.97	0.98	2.97	5.84	4.66	8.41	7.25	4.91	4.45

14. In Chapter 15, we discussed the concept of real options as a tool in decision making in environments characterized by uncertainty. Given the significant uncertainty associated with system usage/loading and the internal and external factors that contribute to system defects, explain how the flexibility in real options theory could help make prudent decisions in deferring, abandoning, expanding, contracting or staging the deployment of monitoring equipment.

REFERENCES

Asimov, I. (2005). *The Return of the Black Widowers*. Carroll & Graf, New York.

Bentham, J. (2011). *The Panopticon Writings*. Verso Books, New York.

Celebi, M. (2006). Real-Time Seismic Monitoring of the New Cape Girardeau Bridge and Preliminary Analyses of Recorded Data, *Earthquake Spectra*, 22(3), 609–630.

DeWolf, J. T., Culmo, M. P., and Lauzon, R. G. (1998). Connecticut's Bridge Infrastructure Monitoring Program for Assessment. *ASCE J. Infrastruct. Syst.* 4(2), 86–90.

Dickenson, S. E. (2007). Instrumentation and Monitoring of Port Facilities: Planning, Funding, Field Applications, and Long-Term Benefits, Ports 2007: 30 Years of Sharing Ideas, 1977-2007ASCE-TCLEE Ports Lifelines Committee.

Donahue, M. J., Dickenson, S. E., Miller, T. H., and Yim, S. C. 2004. Implications of Observed Seismic Performance of a Pile-Supported Wharf for Numerical Modeling, *Earthquake Spectra* 21(3), 617–634.

Elgamal, A., Conte, J. P., Fraser, M., Masri, S., Fountain T., Gupta, A., Trivedi, M., and El Zarki, M. (2009). Health Monitoring for Civil Infrastructure, NSF Information Technology Research, Grant 0205720 Report.

Elgamal, A. W., Zeghal, M., and Parra, E. (1996). Liquefaction of Reclaimed Island in Kobe, Japan. *ASCE J. Geotechn. Eng.* 122(1), 39–49.

Ellingwood, B. R. (2005). Risk-informed Condition Assessment of Civil Infrastructure: State of Practice and Research Issues. *Struct. Infrastruc. Eng.* 1(1), 7–118.

Fletcher, J. B., Sell, R., and Dietel, C. M. (2001). Site Response in the Port of Tacoma. EOS Transactions, American Geophysical Union, 82(47), *Proc. of the Fall AGU Meeting*.

Foucault, M. (1995). *Discipline and Punish: The Birth of the Prison*. Vintage Books, New York.

Han, J. Bae, S. H., and Oh, D. Y. (2001). Application of Photogrammetry Method to Measure Ground-Surface Displacement on the Slope. *J. Korean Env. Res Reveg. Tech.* 4(3), 10–18,

Han, J., Hong, K., and Kim, S. (2012). Application of a Photogrammetric System for Monitoring Civil Engineering Structures, Special Applications of Photogrammetry. www.intechopen.com Accessed December 10, 2012.

Iai, S. (1998). Seismic Analysis and Performance of Retaining Structures. *Proc. Geotechnical Earthquake Engineering & Soil Dynamics III*, ASCE Geotech. Special Publ. #75, 1020–1044. Ports 2007: 30 Years of Sharing Ideas, 1977–2007.

Kang, J. M., Yoon, H. C., and Bae, S. H. (1995). A Study on the 3-D Deformation Analysis for Safety Diagnosis of Bridges, *J. Korean Soc. Surveying, Geodesy, Photogram., and Cartogra.* 13(2), 69–76.

Kokusho, T., and Matsumoto, M. (1997). Nonlinear Site Response during the Hyogoken-Nanbu Earthquake Recorded by Vertical Arrays in View of Seismic Zonation Methodology, in *Seismic Behavior of Ground and Geotechnical Structures*, A. A. Balkem Publishing, pp. 61–69.

Kurata, M., Kim, J., Zhang, Y., Lynch, J. P., van der Linden, G. W., Jacob, V., Thometz, E., Hipley, P., and Sheng, L. (2011). Long-term Assessment of an Autonomous Wireless Structural Health Monitoring System at the Carquinez Suspension Bridge. *Proc. SPIE 7983, Nondestructive Characterization for Composite Materials, Aerospace Engineering, Civil Infrastructure, and Homeland Security*, San Diego, CA.

Macnish, K. (2011). Surveillance Ethics, Internet Encyclopaedia of Philosophy, www.iep.utm.edu. Accessed December, 10 2012.

Madanat, S., and Ben-Akiva, M. (1993). Optimal Inspection and Repair Policies for Transportation Facilities. *Transport. Sci.* 28, 55–62.

Marx, G. T. (1998). Ethics for the New Surveillance. *Info. Soc.* 14, 171–185.

Masri, S. F., Sheng, L. H., Caffrey, J. P., Nigbor, R. L., Wahbeh, M., and Abdel-Ghaffar, A. M. (2004). Application of a Web-enabled Real-time Structural Health Monitoring System for Civil Infrastructure Systems. *J. Smart Mat. Struct.* 13, 1269–1283.

NCHRP (1997a). *Instrumentation for Measuring Scour at Bridge Piers and Abutments*. NCHRP Report 396, National Academy Press, Washington, DC.

NCHRP (1997b). *Sonar Scour Monitor—Installation, Operation, and Fabrication Manual*. NCHRP Report 397A, National Academy Press, Washington, DC.

Orwell, G. (2004). *1984* New Edition, Penguin Classics, London.

Park, W. Y., Kim, J. S., and Lee, I. S. (2001). Monitoring of Unsafe Dam Using GPS. *J. Korean Soc. Civil Eng.* 21(3-D), 383–392.

Reynolds, G. G., Bell, D. L., Holly, D. D., and Stowe, G. (2002a). Application of Multibeam Echo Sounding with Topographic Mapping for Monitoring of Rubble Mound Breakwaters, Breakwaters '99, 1st ASCE Inter. Symp. on Monitoring of Breakwaters, pp. 122–134.

Reynolds, G. G., Bell, D., Kennedy, M., Behrns, R., Tibbetts, M., and Kohn, E. (2002b). Application of Multi-beam Echo Sounding and Side Scan Sonar for Mapping of Shoreline Protection Revetments. Breakwaters '99, 1st ASCE Inter. Symp. on Monitoring of Breakwaters, pp. 13–144.

Saeki, M., Oguni, K., Inoue, J., and Hon, M. (2008). Hierarchical Localization of Sensor Network for Infrastructure Monitoring. *ASCE J. Infrastruct. Syst.* 14(1), 15–26.

Stewart, M., and Tsakiri, M. (2001). Long-term Dam Surface Monitoring Using the Global Positioning System. *Elect. J. Geotech. Eng.* 6, 1089–3032.

Tsuchida, H. (1990). Japanese Experience with Seismic Design and Response of Port and Harbor Structures. Proc. POLA Seismic Workshop on Seismic Engineering, Port of Los Angeles, San Pedro, CA, pp. 139–164.

Winner, L. (1977). *Autonomous Technology: Technics-out-of-control as a theme for Political Thought*, The MIT Press, Cambridge, MA.

USEFUL RESOURCES

Aktan, A. E., Catbas, F. N., Grimmelsman, K. A., and Tsikos, C. J. (2000). Issues in Infrastructure Health Monitoring for Management. *ASCE J. Eng. Mech.* 126(7), 711–724.

ASCE EMI (2012). Intelligent Sensing for Structural Health Monitoring, Structural Dynamics Research at the University of Michigan. http://www.asce.org/emi/Research-Tools.

Balageas, D., Fritzen, C. P., and Guemes, A. (2006). *Structural Health Monitoring Ebook*. Wiley.

Chang, F., and Strauss, S. (2000). *Structural Health Monitoring*. CRC Press, Boca Raton, FL.

Enckell, M., Glisic, B., Myrvoll, F., and Bergstrand, B. (2011). Evaluation of a Large-scale Bridge Strain, Temperature and Crack Monitoring with Distributed Fiber Optic Sensors, *J. Civil Struct. Health Monitor.* 1(1–2), 37–46.

Encyclopedia of Structural Health Monitoring. (1999). Wiley New York.

Garevski, M. Ed., (2013). *Earthquakes and Health Monitoring of Civil Structures*. Springer, Dordrecht.

Giurgiutiu, V. (2007). *Structural Health Monitoring with Piezoelectric Wafer Active Sensors*. Academic Press, Walthanm, MA.

Grimes, C. A., Dickey, E. C., and Pishko, M. V. (2006). *Encyclopedia of Sensors*. American Scientific Publishers Valencia, CA.

Iskander, M. G., Yakubov, N., and Yu, E. (2009). A Course in Instrumentation and Monitoring of Civil Infrastructure Contemporary Topics in In-Situ Testing, Analysis, and Reliability of Foundations, *International Foundation Congress and Equipment Expo 2009*, Orlando, FL, March 15–19.

Journal of Structural Health Monitoring. Sage Publications. This journal contains articles on theoretical, experimental, and analytical investigations that cover all aspects of structural health monitoring. The scope of articles includes prognostics, self-diagnostics, wave and vibration propagation techniques for assessing damage, design or sensors, and bionanotechnology applications in structural monitoring.

Kim, J., Lynch, J. P., Lee, J. J., and Lee, C. G. (2011). Truck-based Mobile Wireless Sensor Networks for Experimental Observation of Vehicle-Bridge Interaction. *Smart Materials & Structures*, IOP, 20, 065009.

Lynch, J. P., and Loh, K. J. (2006). A Summary Review of Wireless Sensors and Sensor Networks for Structural Health Monitoring. *Shock Vibration Digest*, 38(2), 91–128.

Measures, R. M. (2001). *Structural Monitoring with Fiber Optic Technology* Academic Press, Waltham, MA.

Ohnishi, Y., Nishiyama, S., Yano, T., Matsuyama, H., and Amano, K. (2006). A Study of the Application of Digital Photogrammetry to Slope Monitoring Systems. *Int. J. Rock Mech. Mining Sci.* 43(5), 756–766.

Ostachowicz, W., and Guemes, A. (Eds.) (2013). *New Trends in Structural Health Monitoring, CISM International Center for Mechanical Sciences*, Vol. 542. Springer, London.

Ou, J., and Duan, Z. (2006). *Structural Health Monitoring and Intelligent Infrastructure*. Taylor and Francis, London, UK.

Pawar, P. M., and Ganguli, R. (2011). *Structural Health Monitoring Using Genetic Fuzzy Systems*. Springer, London.

Structural Control and Health Monitoring. Wiley & Sons. This is the official journal of the International Association for Structural Control and Monitoring and of the European Association for the Control of Structures. The journal publishes articles related to the science of structural monitoring and control.

Structural Durability & Health Monitoring. Tech Science Press. This journal contains articles on structural integrity assessment, fracture and damage mechanics, fatigue, self-diagnostics, prognostics, smart materials for damage monitoring, and assessment of damage tolerance and structural durability, integration of structural health monitoring, and control.

Talebinejad, I., Fischer, C., and Farhad, A. (2012). A Hybrid Approach for Safety Assessment of the Double Span Masonry Vaults of the Brooklyn Bridge. *J. Civil Struct. Health Monitor.* 1(1–2), 37–46,

Walia, S. K., Vinayak, H. K., Kumar, A., and Parti, R. (2012). Nodal Disparity in Opposite Trusses of a Steel Bridge: A Case Study. *J. Civil Struct. Health Monitor.* 1(1–2), 37–46,

Zimmerman, A. T., Shiraishi, M., Swartz, R. A., and Lynch, J. P. (2008). Automated Modal Parameter Estimation by Parallel Processing within Wireless Monitoring Systems, *J. Civil Struct. Health Monitor.* 4(1), 102–113.

CHAPTER 25

SYSTEM PRESERVATION (MAINTENANCE AND REHABILITATION)

25.0 INTRODUCTION

Civil engineering systems in most countries are comprised of an extensive range of mostly public-owned facilities, and the upkeep of these systems often commands a dominant share of government budgets. Agencies realize that a savings as small as 1% achieved by adopting prudent preservation practices can translate into several millions or even billions of dollars. It is therefore critical that these systems are maintained strategically and cost-effectively. This challenge has long been recognized not only by public sector agencies at all levels of government, but also by international development institutions and private sector organizations including large religious institutions, the military, utility companies, and industrial corporations that own or operate significant numbers or sizes of civil engineering infrastructure.

In order to address this challenge, engineers working at the preservation phase strive to acquire and apply the relevant tools to carry out the analytical tasks at this phase. These tasks include describing the current condition levels of the system to stakeholders, developing deterioration curves for purposes of predicting the system's future condition, gauging the effectiveness of preservation activities, evaluating and selecting optimal (cost-effective) preservation treatments and life-cycle schedules, assessing the optimal amount and timing of maintenance and rehabilitation work, and determining the funding level needed for preserving the system over a specified future horizon. In this chapter, we define preservation as any rehabilitation or maintenance activity and we often use the terms rehabilitation or maintenance synonymously with preservation.

As we discussed in Chapter 2 (see Figure 2.4), the phase of system preservation is synchronous with the phases of system operations and system monitoring; that is, as the system is being used, it is also being monitored continuously or periodically for defects and is also receiving maintenance as and when needed continuously.

We will begin this chapter by presenting the continuing motivations for and the principles of system preservation, followed by a discussion of the two key management levels at which system preservation decisions are made: the facility level and the network level. Then, we will discuss the various mechanisms used for making preservation-related decisions. Next, recognizing the all-too-common problem of inconsistent use of preservation terminologies and taxonomies across and even within agencies, this chapter further standardizes the terms used in the industry and thus provides a consistent set of terms used in the rest of this chapter. The chapter next will identify each management level in greater detail to examine the specific tasks and contexts of preservation decision making that are faced by maintenance engineers and the factors that affect these decisions. In this respect, the chapter provides mathematical formulations, numerical examples, and institutional issues regarding the decisions that maintenance engineers routinely make at the facility or network levels. For readers seeking state-of-the-art field techniques for maintaining their civil systems, the chapter provides links to a number of useful references and resources.

25.0.1 Civil System Preservation—An Ongoing Problem

In many countries, the owners and operators of civil engineering infrastructure are finding that years of neglect have rendered a significant portion of their infrastructure in need of major rehabilitation or replacement. In the United States, for example, over a quarter of the country's 600,000 bridges are considered to be structurally deficient or functionally obsolete, approximately one-third of these have exceeded their design life and are in need of maintenance, rehabilitation, or replacement (FHWA, 2011). This situation continues to be exacerbated by increasing travel demand, funding limitations or uncertainty, and increasing construction costs. Infrastructure investment in developing countries fell significantly, in the 1995–2005 period, and such countries face the challenge of addressing the large infrastructure gaps that threaten growth and impair the achievement of broad development goals (World Bank, 2005). Trapped in a Catch-22 situation, many infrastructure agencies are finding that the lack of funding for infrastructure maintenance is causing accelerated deterioration of such facilities, which in turn leads to a greater need for maintenance funds and thereby establishes a vicious cycle.

Specifically, agencies are finding that deferring preventive or corrective maintenance leads to a worsening condition of the facility to a point where rehabilitation (which is more expensive than maintenance) is needed; similarly, delaying rehabilitation invariably leads to a situation where reconstruction (which is more expensive than rehabilitation) is needed earlier than scheduled. As such, a key task at the phase of system preservation is the development of a set of timely and effective actions for preventive maintenance and rehabilitation (M&R) at specified years or at specified condition levels within the facility life cycle. Other key tasks include (i) the prediction of the physical condition at any future year as a function of age, loading, and other factors, (ii) tracking the occurrence trend of distresses or defects and implementation of proactive measures before the defects lead to accelerated deterioration of the facility and possibly, injury to the facility users, (iii) establishment of thresholds or triggers for each standard M&R treatment, and (iv) the estimation of the effectiveness of standard or experimental M&R treatments, in terms of the increase in condition or longevity of the system.

To answer these and other questions pertinent to the maintenance engineer, most agencies have established maintenance management systems (MMS). The basic foundation block for building an MMS is an effective condition inspection scheme and database for the civil engineering system. At agencies where financial and institutional problems (including cultural and administrative roadblocks, lack of research, or lack of effective condition monitoring) impede the development or implementation of an MMS, any effort to address the four questions stated above, will be seriously impaired. In this chapter, we discuss the key contexts and analytical tools associated with an effective MMS.

25.0.2 Principles of System Preservation

In their bid to preserve the physical condition of civil engineering systems through maintenance and rehabilitation, engineers are guided by a set of core principles also referred to as "good practices".

First, the preservation policy must be **driven by the policies of the agency** (system owner or operator); in other words, the preservation decisions should be based on a well-defined set of policy goals and objectives that reflect the mission or vision of the agency. Often, these include goals that are related directly to the agency (enhancing the system condition or longevity of the physical system structure), the system users (reducing the frequency of system breakdown, inconvenience and safety associated with system use, and repair downtime frequency and duration), and the community (reducing the adverse impacts related to visual quality, air quality, noise, business and social activity disruptions, and inconvenience associated with the system rehabilitation or maintenance activities).

Second, the preservation policy must be **performance based**. For this to be possible, the agency's policy objectives as stated in the previous paragraph must be translatable into specific system performance measures for operational (day to day) and tactical purposes, as well as for the strategic management of the agency's preservation resources.

Third, recognizing that multiple options often exist for preserving a system on the basis of the types and timings of preservation treatments, the preservation policy must **facilitate the analysis of preservation investment options and trade-offs**. This, often, is accomplished using life-cycle costing concepts in economic analysis (discussed in Chapter 11), multiple-criteria analysis (discussed in Chapter 12), and generation of Pareto frontiers via optimization (discussed in Chapter 9). The policy must help the agency make good decisions on how to allocate preservation funds across different system types in a network or across different components of an individual system, as well as to investigate quickly and interactively the trade-offs between different preservation projects, groups of projects, and funding levels or between preservation expenditure and performance.

Fourth, the agency's preservation policy must be **based on reliable predictions of system physical condition** using performance curves and models (that describe the trend of performance deterioration over time) and preservation treatment effectiveness models (that predict the effectiveness of preservation actions in terms of increased condition or durability). In Chapter 7, we discussed a few techniques that can be used to model system condition over time. The use of unreliable models often yields predictions of system condition that deviate far from actual conditions and thus lead to mistiming of preservation treatment applications, which constitutes a wastage of scarce funds.

Fifth, the preservation policy must be consistent with **data-driven decisions**. This data includes cost and performance information. It is only through data that the agency can effectively assess the costs and effectiveness of preservation treatments and strategies, analyze trade-offs across the performance measures, and model the trend of the system deterioration. Thus, a good data collection and management process is essential for system preservation.

Finally, the preservation policy, its expected outcomes, as well as the sensitivity of the outcomes and trade-offs among different outcomes (in terms of the overall systemwide cost and performance) must be such that they can be easily predicted or monitored, and communicated to stakeholders to enhance **clear accountability and feedback**. Such feedback can be used not only to adjust the goals and objectives of the system but also to refine the agency's resource allocation. This principle is particularly important in the current era where civil engineers increasingly seek to practice performance-based management of their systems.

25.0.3 Network-Level versus Facility-Level Management of Preservation Resources

For any system type, the management of the preservation phase can occur at one of two levels: the network or system-of-systems (SOS) level and the facility level. The network-level analysis, which involves multiple systems that share a common attribute such as type, class, or geographical location, is often aimed at the establishment of priorities for various preservation projects, determination of the optimal use of limited funds, selection of optimum maintenance policies for the entire network, assessment of the network-level impacts of alternative system preservation policies, and assessment of trade-offs at the network level. The advantage of network-level analysis is that it affords the system owner or operator a birds-eye perspective of the performance of the overall network of systems under their control, particularly, the performance consequences of different levels of the preservation budget. A disadvantage is that this level of analysis often utilizes data that are only aggregate in nature and thus does not always consider all the factors associated with system preservation at the facility level, and therefore may not be reliable for making facility-level decisions.

Facility-level (also referred to as system- or project-level) management, on the other hand, generally involves the selection and evaluation of preservation techniques or policies for a specific system type (such as a specific pipe or road segment, bridge, or water tank). Facility-level models are typically comprehensive and involve detailed information, often including specific design features of the system. At the preservation phase, a common task at the facility level is to select a preservation treatment at any given point in time or long-term strategy that will provide acceptable levels of service to system users at a minimum overall cost to the system owner or operator over a given period of time.

Network-level and facility-level preservation decisions are meant to be interdependent and synergistic, through a mechanism that may be top-down or bottom-up: The manager at the network level establishes a given budget and passes it down to the facility-level manager (top-down); based on that budget, the facility-level manager identifies specific systems on the network that are most deserving of preservation and also decides on the most appropriate preservation treatment for each deserving system. Then the list of all the deserving systems and their recommended treatment(s) and the associated costs is passed up to the network-level manager. If there are any subsequent changes in budget (due to economic conditions) or system performance thresholds (due to revised agency policy), the network-level manager feeds this information to the facility-level manager or engineer who then revise their list of recommended preservation projects. Then, at the beginning of the next planning year as the facility-level manager prepares the next list of preservation projects, they duly take note of projects that were excluded by the network-level manager for implementation and often include them in that year's project list. Table 25.1 presents the contexts of decisions typically made at each level of management.

Table 25.1 Typical Decision Contexts for System Preservation

Management Level	Decision Contexts
Facility level	What is the optimal preservation category or type to apply at any given time? (Do nothing, routine maintenance, or rehabilitation?) For a given preservation category or type that has been selected, what is the optimal specific treatment to apply at any given time? For an existing system, what is the best maintenance and rehabilitation (M&R) schedule over its remaining life? For a planned new system, what is the best M&R schedule over its full life? What is the appropriate amount to spend annually to maintain the system, given its current system age or condition? At what condition threshold should each preservation category or specific treatment be applied?
Network or SOS level	(WWW) What intervention should be undertaken, which facility, and when (year)? ($) For an entire network of systems, how much should be spent for each individual system in each year? (%) What percentage of facilities should receive some preservation action in each year of a horizon period?

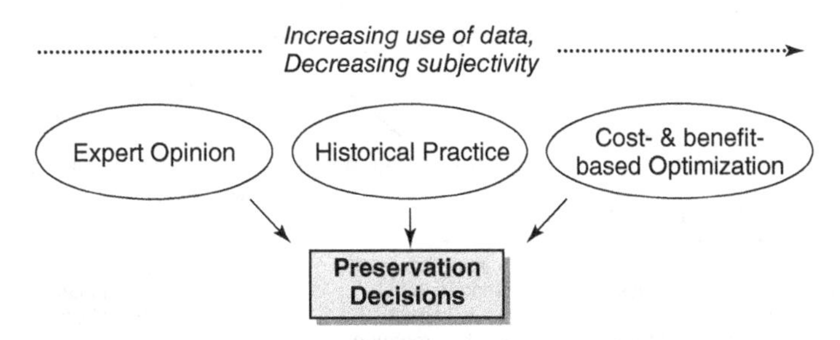

Figure 25.1 Mechanisms for preservation decision making.

25.0.4 Mechanisms for Preservation Decision Making

For the facility-level and network-level contexts of decision making presented in Table 25.1, a decision may be made using at least one of three key mechanisms: expert opinion, historical practices, or mathematical optimization (Figure 25.1). For example, the **opinion of experts** could be solicited through a questionnaire survey to determine the best treatment to apply under a given set of conditions (such as the nature of existing defects, rate of deterioration, system material and design type, and other factors). This could be further enhanced using a Delphi technique, where the results of the first survey are sent back to the experts for their consideration of the overall results and to give them an opportunity to review their initial responses; with several Delphi iterations, the experts' answers tend to converge.

The decisions can also be made using the mechanism of **historical practice**; for example, if in the past, a certain kind of treatment was applied whenever a certain defect was observed under a given set of conditions then that practice (particularly, if documented) could serve as a basis for the maintenance engineer's decisions. The problem with this decision mechanism is that historical practices are often influenced by economic conditions (i.e., funding unavailability in the past may have precluded the timely application of appropriate treatments). Thus, from a purely technical standpoint, past practices often do not adequately serve as an appropriate guide for future practice.

Ideally, any preservation decision, including those of the contexts listed in Table 25.1, should be carried out using a mechanism that is objective and should not be influenced by personal bias or historical practices. One way to do this is by using data-driven **mathematical optimization** formulations that duly consider the benefits and costs of multiple alternative decisions within that problem context. In certain cases, the recommendations from the quantitative analyses are tempered with expert opinion or data from historical practices, in order to arrive at decisions that are very practical.

In Sections 25.3 and 25.4, we shall discuss how an agency could identify the appropriate preservation treatment at a given time or preservation schedule over a long term—in including issues such as the nature of the decision problem and the basis upon which the decision is made.

25.0.5 Factors Affecting Preservation Decisions

As we learned in the preceding section, the engineer responsible for the preservation phase of a system is tasked with making systems-related decisions to enhance the effectiveness and efficiency of the system preservation efforts. As we learned in the previous section, the engineer does this using expert opinion, historical practices, or cost–benefit analysis including mathematical optimization.

Irrespective of the mechanism used or the specific problem context being addressed, the maintenance engineer's decisions are influenced by a number of factors including (Figure 25.2): the system condition (distress types, severity, and extent), age, material type, and design type; the characteristics of the preservation treatment under consideration such as the warrants, effectiveness,

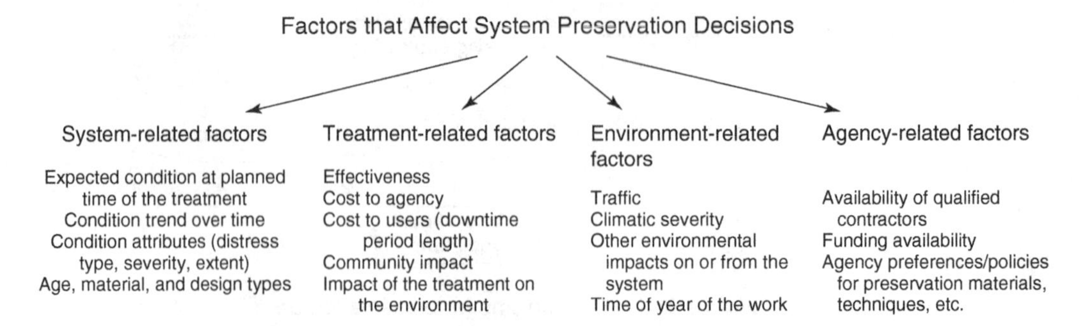

Figure 25.2 Key factors that influence system preservation decisions.

cost, downtime, community impact of the treatment, impact on the treatment on the environment, level of demand or loading, and climatic severity; and agency-related factors such as availability of qualified contractors and funding availability. These factors are explicitly used in mathematical models or expert systems that help train new maintenance engineers to make appropriate maintenance decisions.

25.0.6 Measures to Assess the Effectiveness of Preservation Treatments

For a variety of management functions including long-term planning and budgeting, infrastructure valuation, and the evaluation and comparison of materials, equipment, processes, or contractors used for the system maintenance or rehabilitation, engineers seek to measure the effectiveness (or performance) of preservation treatments applied to their systems. The effectiveness can be measured in the short term or the long term. Short-term effectiveness includes a reduction in the rate of deterioration and the immediate jump in the physical condition. In the long term, effectiveness can be measured on the basis of the effective life of the treatment, an extension in the physical life of the system due to the treatment, an increase in the physical condition of the system, a reduction in the maintenance costs subsequent to the treatment, an increase in the time taken for a specified intensity of some defect type to occur for the first time in the years following the treatment, or a decrease in the likelihood that a specified intensity of some defect type will occur in a given time period following the treatment. In Sections 25.2.9 and 25.2.10, we will discuss some of these preservation performance measures in greater detail.

25.1 BASIC TERMINOLOGIES IN SYSTEM PRESERVATION

System preservation terminology differs widely not only across different countries and agencies but even within an agency. Therefore, a clear statement of the different terms and their meanings must precede any treatise on the subject of system preservation. System preservation refers to the set of activities that ensure that a system remains in satisfactory condition. Not unexpectedly, several different definitions of the word "preservation" exist in the literature. At certain agencies, preservation means only maintenance; at others, it means only routine maintenance carried out in-house by the agency's personnel. Still, there are other agencies who use the term to include both maintenance and rehabilitation. At the extreme end are others who view the system land or right-of-way as the only permanent feature of the system and therefore define preservation to include all activities besides land acquisition, that is, all activities that ensure sustained use of the system, including reconstruction, rehabilitation, and even minor routine maintenance. In this text, we define preservation to mean rehabilitation and maintenance (Figure 25.3).

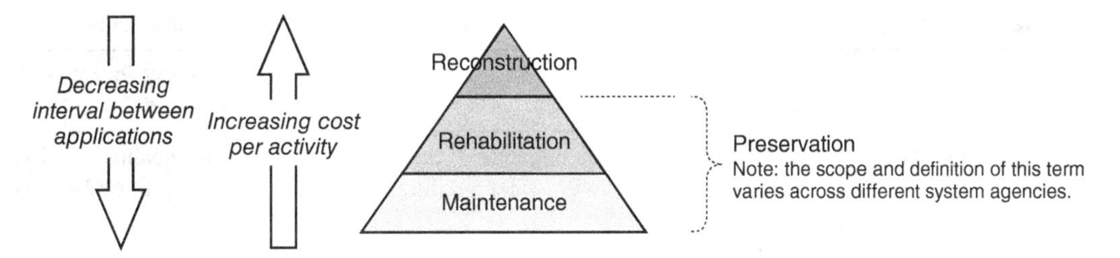

Figure 25.3 Levels of system preservation.

A system **preservation treatment** is a specific maintenance or rehabilitation activity applied at a given time. Most systems owners have established, for each system type, manuals for maintenance or rehabilitation that list a standard set of treatments to be applied to the system at a certain age or that exhibits condition (extent or severity of a specific distress). In systems preservation lingo, synonyms for treatment include: activity, action, technique, and intervention. It should be noted that certain persons use the term "strategy" to indicate treatments; while this is not inaccurate from the conceptual viewpoint, it is probably more prudent to think of a treatment as a tactical issue rather than a strategic issue.

A system **preservation schedule, strategy**, or **activity profile** is a combination of multiple preservation activities applied at various times over the system life cycle or remaining life. A system preservation schedule could be based on system age accumulated loading, system condition, and so forth. Preservation schedules typically consist of applications of a **preventive (proactive) treatment** (which are applied before the onset of significant structural deterioration (O'Brien, 1989) and typically exclude **corrective (reactive) treatments**; this is because unlike preventive treatments, corrective treatments are carried out not in anticipation of distress but to address distress that has already occurred and therefore cannot be included in a schedule that specifies when future work should be carried out.

Preventive maintenance (PM) **scheduling** or **timing** may be condition based or time based. Table 25.2 presents a few examples for bridge and pavement systems. Condition-based preventive maintenance treatments are those that are performed "as needed" and are identified through the system monitoring and inspection processes. For example, for bridge systems, such treatments often include sealing or replacement of leaking joints, installation of deck overlays; installation of cathodic protection systems, complete, spot, or zone painting/coating of steel structural elements, and installation of scour countermeasures.

Rehabilitation is any major treatment that restores the structural or functional integrity of a system, and thus improves significantly its condition, longevity, or safety. The **rehabilitation life** of a system, also referred to as the **rehabilitation interval**, is the period between (a) system (re)construction and subsequent rehabilitation, (b) two consecutive system rehabilitations, or (c) the last rehabilitation and subsequent reconstruction. For a given system type, the actual rehabilitation life could be long or short depending on agency policy, loading, climate, funding availability, and the like. Ideally, the intervals between rehabilitations should not be a function of funding availability; however, funding limitations are all too real, and cash-strapped agencies may find themselves deferring rehabilitation until a time when adequate funds are available. In any case, one of the maintenance engineer's duties is to update the agency's policy on rehabilitation intervals by using current data to establish the optimal rehabilitation intervals, for that system type, under different operating and environmental conditions.

Figure 25.4 shows the three dimensions through which system preservation could be viewed. From the perspective of **application frequency**, preservation could be routine maintenance, periodic maintenance, minor rehabilitation, or major rehabilitation. The exact time intervals for

Table 25.2 Typical Treatment Categories and Types for Two Types of Civil Engineering Systems

	Bridge Systems	**Highway Pavement Systems**
Time-based PM	Install deck overlay on concrete decks (10–25 yr) Seal concrete decks with waterproofing penetrating sealant (3–5 yr) Lubricate bearing devices (2–4 yr) Zone coat steel beam/girder ends (10–15 yr) Painting of steel bridge elements (5–10 yr)	Crack sealing (2–4 yr); Joint sealing (2–4 yr)
Condition-based PM	Sealing or replacement of leaking joints; installation of deck overlays; installation of cathodic protection systems; complete, spot, or zone painting/coating of steel structural elements; installation of scour countermeasures	Microsurfacing of rutted pavements; thin hot-mix asphalt (HMA) overlay of distressed surfaces; underdrain flushing of clogged pipes.
Corrective maintenance	Deck patching; repair of spalled areas	Shallow and deep patching
Rehabilitation	Partial or complete deck replacement; superstructure replacement; strengthening; widening without adding a travel lane.	Load transfer restoration; functional or structural resurfacing (overlay)

Sources include: FHWA (2011). Frequencies are based on FHWA's knowledge of typical state DOT practices.
Note: For bridges, functional improvements such as widening to add a travel lane or deck raising to increase vertical under clearance are classified by certain agencies as rehabilitation because they restore the functional integrity of the system.

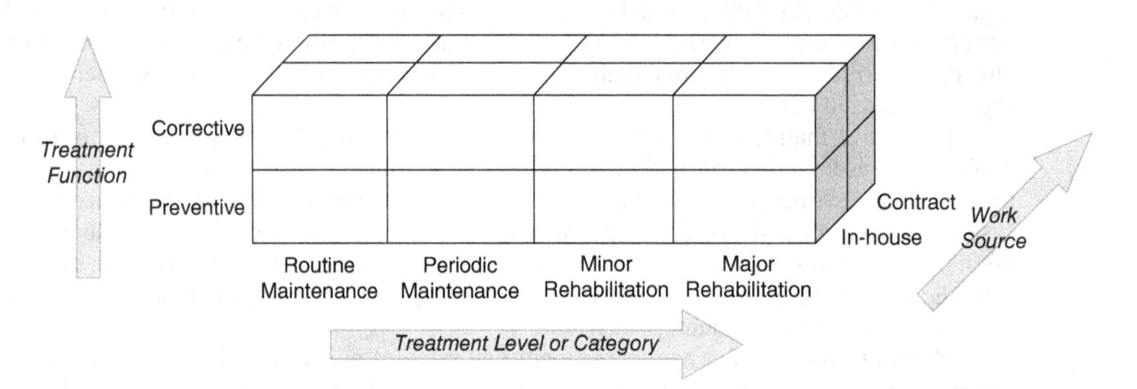

Figure 25.4 Dimensions of system preservation actions.

each of these categories vary across the system types due to the different system design lives. For highway pavement systems, for example, routine maintenance is carried out once or more annually; periodic maintenance, every 2–5 years; minor rehabilitation, every five 5–10 years; and major rehabilitation, every 10–20 years or more. The intervals of major rehabilitation could be 30 years for a dam but only 5 years for a road lighting system. From the perspective of **treatment function**, a preservation treatment could be described as preventive (applied in order to delay the onset of imminent deterioration) or corrective (to remedy an already existing defect). From the perspective of **work source**, preservation could be in-house (implemented by the system owner or operator using that agency's resources (manpower, materials, and equipment) or by contract (carried out by an agency-selected contractor).

25.2 SPECIFIC TASKS AT THE SYSTEM PRESERVATION PHASE

In Table 25.1, we learned about at least four key contexts of decision making at the facility level of management: the best M&R treatment at any given time, the best M&R schedule over the entire life cycle (for a planned or new system) or over the remaining life (for an existing system), the right amount of money to spend annually to maintain the system, the age or level of distress at which the agency should apply each standard a certain treatment. In this section, we discuss the specific tasks that are associated directly or indirectly with each context.

The tasks at the preservation phase, which are associated with systems concepts and tools that we have learned in this text so far, include establishing the goals of the system preservation efforts and measuring the extent of their attainment, developing a database for information relevant to system preservation decisions, establishing the list of possible distresses/defects and standard preservation treatment(s) for each distress/defect, and predicting system condition in terms of the physical distresses. The tasks also include establishing the threshold levels of system condition that warrant some preservation action; prediction of the costs and the short-and long-term effectiveness of preservation treatments; and selection of the best preservation treatment, schedule, or spending level over the facility life or a selected horizon period.

25.2.1 Developing the Goals of System Preservation and Measuring the Extent of Their Attainment

Establishing the goals and objectives of system preservation is the first step in developing a preservation policy/plan or reviewing an existing one. These goals and objectives are often expressed in terms of performance measures that are meant to reflect the concerns of the system owner or operator, the user, and the community. In the context of system preservation, performance measures reflect the extent to which a preservation treatment or schedule realizes its objectives. In Chapter 3, we discussed the principles of selecting an appropriate performance measure at any phase of system development. The primary goal of system preservation is to prolong system life and to ensure a certain minimum physical condition of the civil engineering system. As we learned in Section 25.0.6, for assessing the extent to which this goal is being achieved, the performance measures that can be used include: the extent to which the system life or condition increases after the treatment is applied; the extent to which the system deterioration rate is reduced; and the length of time taken after the treatment for any specified distresses to appear. At the facility level of systems preservation, performance measures are used to select the best treatment or schedule; in selecting the best action among several alternative treatments or schedules, it is sought to achieve optimal values of the established performance measures.

25.2.2 Developing a Database for Purposes of System Preservation

Key data items in any system preservation database include the physical and operational features of the system. These are the system location (such as coordinates or other referencing system), the system component dimensions, materials used to construct the different physical components, the design and construction features, and the system orientation with respect to its natural or built-up environment. Also, the current level of system usage (operations) is useful data for proper estimation of technical and economic consequences of system downtime when the preservation treatment is being applied. Data should also include the system's primary age (years since construction), secondary age (years since major rehabilitation), and history of rehabilitation and maintenance (which treatments were applied, the implementing contractor, the contract cost, and the treatment effectiveness). Relevant data on the system environment is also important: climate (average precipitation, temperature, freeze index, and freeze–thaw cycles), weather (wind speeds, storms, and so forth) geotechnical characteristics, geological conditions, and nature of the hydrogeology. Data on external threats such as earthquake intensities, ground acceleration, and flood return periods are needed for measurement of system vulnerability. For economic and financial analysis of preservation options, useful data include aggregate and disaggregate costs, the prevailing discount rate, and the inflation rate or price indices. All these data are useful for developing mathematical models that the maintenance engineer can use to carry out tasks including tracking of system condition over time, predicting the remaining life of the system physical structure, comparing the cost-effectiveness of alternative preservation treatments at a given time or alternative preservation schedules over the long term, and predicting the effectiveness of any treatment in terms of system life extension, increase in physical condition, reduced vulnerability to failure or damage, and so on.

25.2.3 Establishing the Lists of Possible Distresses/Defects

The maintenance divisions of most civil engineering system agencies have established manuals that list all the possible different physical defects that could be found on their systems (Table 25.3 and Figure 25.5 present a few of these). An example is the U.S. Federal Highway Administration's Distress Identification Manual, which provides a common, consistent, and uniform language for describing cracks, potholes, rutting, spalling, and other distresses associated with highway pavement systems. These manuals are developed using past experience (field observations, anecdotal evidence, and photographs) related to physical defects exhibited by past similar systems. For certain civil engineering systems, the maintenance manuals also provide, for each defect type, written and pictorial descriptions of the defect, the underlying causes, and treatment prescriptions. The extent or severity of each defect or distress is often expressed in terms of a quantitative *performance indicator*, for example, cracking index, corrosion index, and the percent of cracked area.

25.2.4 Identifying the Factors That Affect System Condition and Longevity

The materials used to construct civil engineering systems generally degrade over time and such deterioration is exacerbated in unfavorable environments. For example, asphalt becomes brittle due to oxidation or low temperatures and is thus rendered susceptible to cracking; concrete expands at

Table 25.3 Examples of Possible Distresses for System Types

System Type	Some Typical Distress Types
Bridges	Wearing surface friction loss, cracking, corrosion of reinforcement in deck and substructure; spalling of concrete superstructure and substructure, corrosion of elements of steel superstructure
Dams	Erosion of dam foundation, cracking of dam wall
Buildings	Settlement, cracking of columns
Water tank	Corrosion of the steel walls or of steel in reinforced concrete tanks
Levees	Erosion of levee foundation, cracking of levee wall
Water and sewer pipes	Pipe cracking, pipe corrosion, sedimentation or material deposition in inner wall
Highway and airport pavements	Portland cement pavements: surface roughness, cracking, loss of surface friction Asphalt concrete pavements: surface roughness, cracking, raveling, loss of surface friction
Windmills	Low fluid levels, bolt torque failure, blade cracking or corrosion, brake pad wear, loose cable connections, clogged filters.

Figure 25.5 Deterioration modes of some civil engineering system types [Courtesy (*a*) FHWA, (*b*) Wikimedia Commons, (*c*) Anna Frodesiak/Wikimedia Commons, and (*d*) Bidgee/Wikimedia Commons].

high temperature and cracks; reinforcement in concrete corrodes in saline environments due to coastal proximity or wintertime deicing salts; steel becomes fatigued under repetitive stress cycles over the years. The degradation of these materials translates into physical distresses and defects of the system components. Figure 25.6 presents the key factors that affect system condition and longevity. Others include man-made factors that may be intended (e.g., vandalism) or unintended (e.g., inadvertent damage to the system by its users). These factors can be used as explanatory variables in models that estimate or predict system condition or longevity, as we will discuss in Section 25.2.5.

25.2.5 Predicting System Condition and Longevity

The deterioration of any physical system begins immediately after construction and proceeds rapidly or slowly depending on a number of factors (as we discussed in Section 25.2.4) and the complex interactions between them. Thus, effective management of the physical system requires a good understanding of the causes and patterns of deterioration.

Condition and longevity prediction models play a vital role in civil system management. First, they help the maintenance engineer to track the system condition over time. Second, by tracking the different deterioration patterns for alternatives that differ by material type, design type, or contractor, these models can help compare the relative efficacies of these alternatives in terms of the system physical condition or longevity. Third, the tracking of deterioration helps ascertain the future year at which some action is needed, and thus the agency's tasks of planning, budgeting, and needs

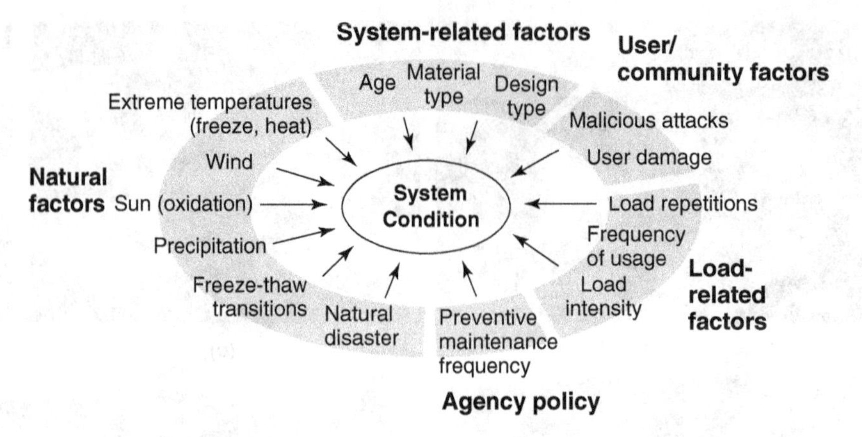

Figure 25.6 Factors that affect system condition and longevity.

assessment are facilitated and made more reliable. Fourth, these models help identify the factors that significantly affect the system condition, and thus help shape policy in mitigating the harmful factors and promoting the favorable factors.

Deterioration is measured in terms of condition indices or ratings that capture the extent and severity of various distress and defect types (Table 25.3). Generally, lower condition ratings are associated with higher levels of age, usage/loading, and climatic severity. The deterioration of most civil engineering systems generally follows one of at least three patterns illustrated in the condition-versus-time plots in Figure 25.7: convex (gentle initially but rapidly increases with age), concave (rapidly increasing rate initially but slowing down as it gets older), or S-shaped (a rate of deterioration that is gentle initially, increases rapidly in the middle years, and then slows down to a gentle rate as it nears the end of life).

There are several techniques a maintenance engineer could use to model the deterioration of a specific system or its components. The final choice of modeling technique depends on (i) the temporal nature of the condition data (i.e., whether the data is of a time series, cross-sectional, or panel nature (see Section 7.2.2 of Chapter 7), (ii) the nature of the response variable, that is, discrete versus continuous (see Section 7.1.1 of Chapter 7), and (iii) whether we seek a deterministic or probabilistic model (see Step 6 of Section 7.1.1 of Chapter 7). There are at least five common categories of techniques for modeling deterioration as a function not only of time but also of factors related to the system, its natural environment, load, and the policies of the system owner. These techniques are: regression modeling, which yields deterministic, continuous linear, or nonlinear functions; discrete outcome modeling, which yields discrete and parametric probabilistic relationships using ordinal data on system condition states; neural networks that provide deterministic,

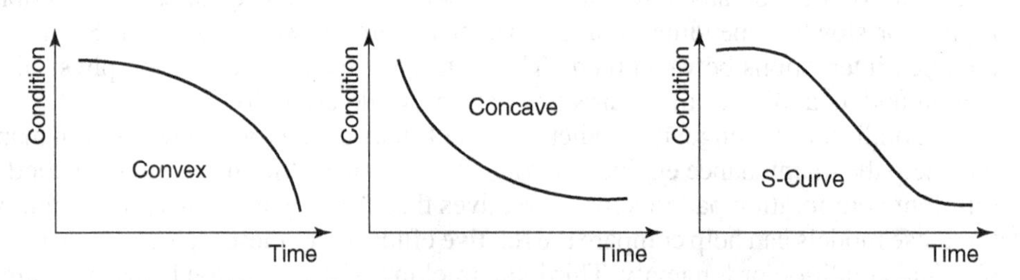

Figure 25.7 Some general deterioration–age patterns.

continuous relationships that mimic the dependencies between variables but is often considered a black box; Markov chains, which yield discrete and nonparametric probabilistic functions that predict the likelihood that the system is in a certain condition state at a given time; and duration modeling, which are probabilistic, continuous relationships known as survival or hazard curves. In this section we provide only a brief discussion of these model types. Detailed discussion of these models are outside the scope of this text but can be found in texts including Faber et al. (2011) (regression), Gurney (1997) (neural networks), Train (2009) (discrete outcomes), Stroock (2005) (Markov chains), and Rabe-Hesketh and Skrondal (2012) (duration models).

(a) Regression Modeling. Regression models continue to be popular due to the ease with which they can be developed (using widely available statistical software packages), interpreted, and applied. They are suitable where the response variable being estimated or predicted is continuous in nature. In Chapter 6, we discussed the steps for regression model building.

(b) Artificial Neural Networks. Inspired by biological neural processes such as those found in a brain, artificial neural networks, combine memory and processing and carry out predictive modeling tasks using signal flows through the connections in the network. This way, a node-link-based approach is used to model the complex relationships between variables such as the system condition on one hand and multiple, often interacting factors related to environment, user loading and frequency, agency policy, and the system features on the other hand.

(c) Discrete Response Modeling. For certain systems or system components, the physical condition is measured and/or reported using descriptions or a scale that is discrete rather than continuous. For example, ordinal response modeling could be used to estimate the probability that a given bridge system will be at a certain level within the 0–9 rating scale (0-worst to 9-best) at some future year. For describing/predicting the future condition of these systems, it is considered more appropriate to use a discrete-outcome model. The subtypes of this model include the **logit** and **probit** functional forms, largely depending on whether they are ordinal [e.g., culvert condition ratings of 0 (failed) to 9 (excellent)] or categorical (e.g., water reservoir condition characterization of structurally deficient, functionally inadequate, satisfactory). Also, these models could be **nested** (e.g., model the likelihood that a system falls in a certain deterioration type category (outer nest), and within each category, model the probability that it suffers a certain intensity of deterioration (inner nest); and whether the nature of the data inherently introduces some biases that may need to be addressed using specific modeling techniques, for example, **mixed, fixed, or random effects** models to account for temporal or spatial heterogeneity where data is longitudinal (see Section 7.2.2 of Chapter 7).

(d) Markov Chains. A Markov chain is a stochastic and sequential method of tracking the condition of a system. It involves the use of a finite number of potential condition states each of which reflect a certain level of physical deterioration of the system and hence is inherently discrete. Markov chains predict the probability that the system will be in a certain condition state after a period of discrete time blocks, often years. They are described as memoryless because they assume that the probability that the system transitions from one state to another is based only on the state at a given year and not on the states of past years; an assumption that is convenient from the computational perspective but may not always adequately represent real-world conditions. Often represented as a node-to-node graph (Figure 25.8) where the nodes are the condition states and the potential paths are the transitions, Markov chains can be analyzed using matrix multiplication.

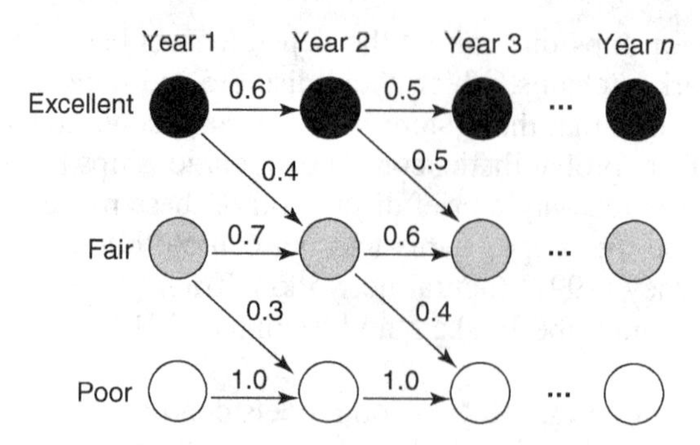

Figure 25.8 Markov chain illustration.

For the example in Figure 25.8, the transition matrix for the year 1 to year 2 transition period is

$$P = \begin{pmatrix} 0.6 & 0.4 & 0 \\ 0.7 & 0.3 & 0 \\ 0 & 0 & 1 \end{pmatrix}$$

The transition matrix presents the probabilities that the system will transition from one condition state to another after a given time period. The period, or time interval (shown as years in Figure 25.8), may be months or several years. Typically, this is taken as the interval between the system inspections. A key assumption is that the system physical condition does not transition to a state of higher physical condition. The transition probabilities can be derived using any of several techniques including expert opinion, observed frequencies of transitions. Recognizing that the system's deterioration rates are typically not the same in the early, middle, and advanced years, it is often useful to establish separate transition matrices for different age groups of the system.

Duration Modeling. In the context of system deterioration modeling for purposes of system preservation, the time between the system construction and the appearance of the first serious distress can be considered to yield useful information for preservation planning and budgeting purposes. Thus, duration, which may be referred to as **survival** or **lack of hazard**, is a probabilistic approach for predicting the likelihood that the system will not develop a certain distress (or that it will "survive") given that it has not manifested that distress at the time of the data collection [Equation (25.1) and Figure 25.9]. This is written as

$$S(t) = p(T > t)$$

where t is time or accumulated loadings, stress cycles, or climate severity exposure; T is a random variable representing the time of manifestation of the distress; and p is the probability element of the duration function. Table 25.4 presents the different ways by which duration models may be represented.

25.2.6 Establishing Condition Thresholds For Treatment Application

For each common distress type, the system maintenance engineer often seeks to establish the level of distress that should warrant the application of an appropriate preservation treatment; for example, at which level of corrosion should sandblasting and/or coating be applied? At many agencies, these thresholds are established by persons with several years of experience in such preservation work; at other agencies, this is done on the basis of past practices, that is, the thresholds that were used in the past are what the agency currently uses and will be used in the future. However, it is considered

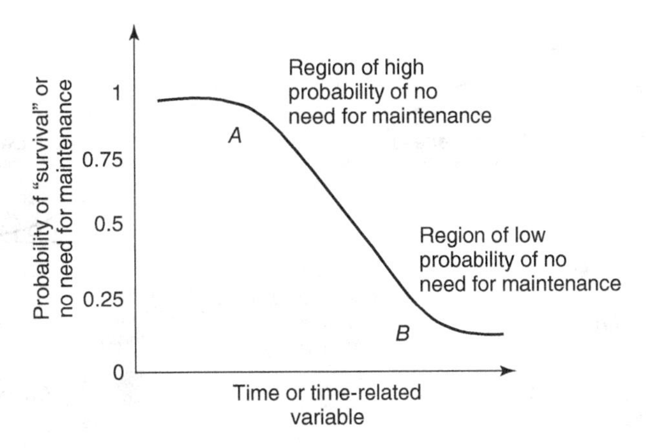

Figure 25.9 Duration model illustration.

Table 25.4 Representations of Duration Models

Representation	Relationship	Interpretation
Density function	$f(t) = \frac{dF(t)}{dt}$	Area under the curve represents the probability of "survival" that the system will not exhibit some specified intensity of a given defect or distress type with a given duration range
Cumulative function	$F(t) = \sum_{t_i<t} f(t_i)$	Probability at any point in time, that the system will not exhibit some specified intensity of a given defect or distress type
Survival function	$S(t) = 1 - F(t)$	Probability that the system will not exhibit some specified intensity of a given defect or distress type beyond any point in time
Hazard function	$h(t) = \frac{f(t)}{S(t)}$	"Failure" rate, related inversely to survival

Source: Ford et al. (2011).

most desirable to establish such optimal thresholds using multiple criteria optimization that duly considers the quantified consequences of every candidate threshold (Khurshid 2010). To do this, the maintenance engineer should first consider a range of candidate threshold levels of the system condition, for the preservation treatment. For each candidate threshold, the engineer should determine the benefits (e.g., the increased condition or life of the system); the costs to the agency, user, and/or community; and the cost-effectiveness associated with that threshold. Then plots such as that shown in Figure 25.10 can be plotted. These plots often reveal that levels of very poor or very good condition are not good candidates for treatment thresholds and that there is a certain optimal level of system condition for applying the preservation treatment.

Even after such optimal thresholds are established for each treatment, it may be the case that the agency is unable to carry out the treatment at these optimal thresholds due to reasons including political influences or lack of funding where the preservation treatment is typically applied before or after it is actually due. Therefore, it is useful for the engineer to ascertain the impacts of mistimed treatments: this can be done by setting the threshold as the baseline condition level and quantifying the consequences of the premature or deferred application of the treatment in terms of the resulting cost, effectiveness, and cost-effectiveness.

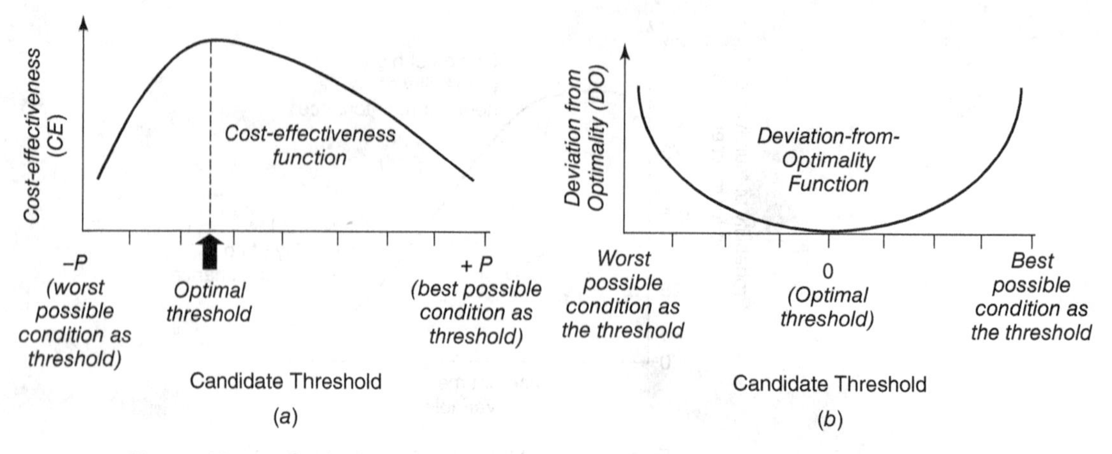

Figure 25.10 Optimal threshold determination using cost-effectiveness function.

Example 25.1

After determining the optimal threshold for applying a functional hot mix asphalt HMA treatment to highway pavements of different classes, Khurshid (2010) presented the plot of the consequences of departing from the optimal threshold (Figure 25.11). The consequences were expressed as a decrease in cost-effectiveness relative to the cost-effectiveness of the optimal threshold. From the plot, determine the consequences of applying the treatment to an interstate highway pavement (a) 2 years earlier than the optimal year and (b) 3 years after the optimal year.

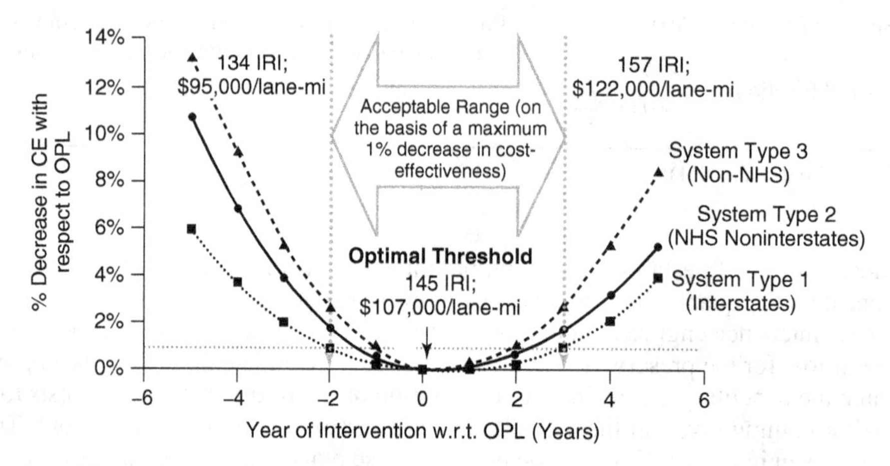

Figure 25.11 Consequences of departing from the optimal threshold.

Solution

The graph shows that if the systems agency carries out the project prematurely by 2 years, the result will be a superior system condition [international roughness index (IRI) value of 134 in./mile], but the treatment cost will be higher ($95,000 per lane-mile); also, if the agency delays the project by 3 years, the result will be a lower system condition (IRI of 157 in./mile) and a higher cost of treatment ($122,000 per lane-mile). Clearly, it is neither cost-effective to carry out the project too soon or too late.

Example 25.2

Figure 25.12 (Markow et al., 1993) presents the discounted costs versus the time at which each of four treatments was applied to address a bridge deck defect. Determine, giving reasons, the optimal threshold for each of the treatments.

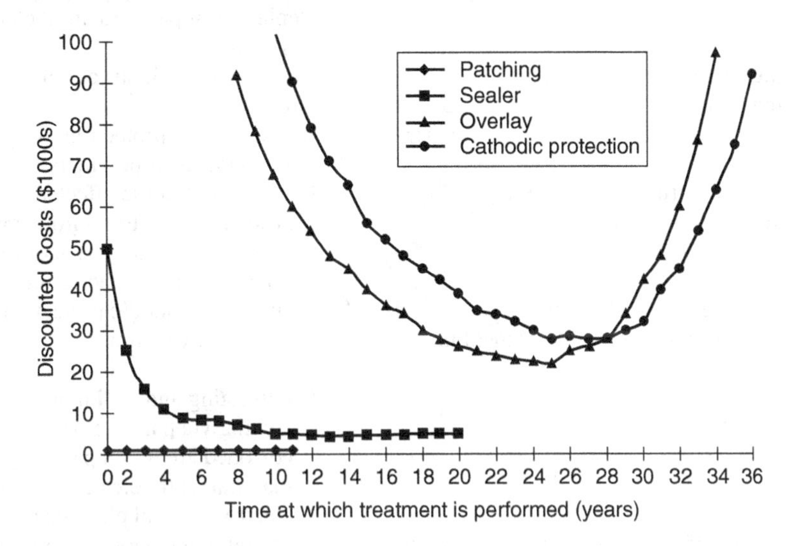

Figure 25.12 Plot of the discounted costs versus the treatment times.

Solution

From the plot, the optimal treatment thresholds are as follows: (a) cathodic protection, 28 months; (b) overlay, 25 months (for cathodic protection and overlay, applications earlier or later than these times will yield higher discounted costs); and (c) sealant application, 10 months (because applications later than this time will yield little or no reduction in discounted costs).

25.2.7 Establishing Appropriate Preservation Treatment(s) for Each Distress/Defect

As discussed in Section 25.2.3, it is important for the system owner or operator to establish and document appropriate preservation treatment(s) for each distress/defect Port Technology Group, 2011. This promotes consistent maintenance practices across the agency. Table 25.5 and Figure 25.13 presents examples of possible distresses and some common or innovative treatments that could address these defects. The owners of different types of civil engineering systems have established manuals that list the preservation treatments for addressing the different types of defects that are typically encountered. The useful resources section of this chapter lists some of these manuals.

25.2.8 Establishing the Costs of Preservation Treatments

For purposes of treatment comparisons, cost-effectiveness analysis, and budgeting, it is good practice for the maintenance engineer to estimate the cost of individual preservation treatments. As we learned in Chapter 10, such costs, in their simplest form, may be in the form of aggregate data (cost per unit dimension of a facility or per output of a preservation action) such as the cost per foot of repair of some linear system, cost per square foot of steel tank rehabilitation, cost per lane-mile of pavement resurfacing, the cost per linear foot of pipeline repair, or cost per square foot of bridge deck patching.

Table 25.5 Examples of Possible Distresses and Treatments

Distress	Recommended Treatment(s)
Removed sealant in concrete slab joints	Joint sealing
Cracking (concrete surfaces)	Fill/seal cracks if underlying damage is nonstructural Replace or repair structural element if damage is structural
Cracking (steel elements/surfaces)	Replace or repair structural element
Blocking of drainage paths for storm water flow	Dredging
Exposure of concrete surface to the weather	Install overlay protection using waterproofing membranes or polymers
Destruction of structural member(s)	Replacement of the affected member(s)
Spalling of concrete cover	If problem is due to reinforcement corrosion, then corrosion needs to be addressed followed by replacement of concrete element
Corrosion of exposed steel members (Figure 25.5) or steel reinforcement in concrete	Cathodic and anodic protection; electrochemical chloride extraction
Corrosion of steel plates	Sandblasting and/or lining the plate with inert material; coating via painting or other material, removal of corroded areas and patching by welding of new material; cathodic or anodic protection; electrochemical chloride extraction
Vulnerability of structure or natural slopes to erosive or scouring action of the wind, rain, or surface water	Stone-pitching, geotextile installation
Partial or uniform settlement of structure	Underpinning and foundation reinforcement
Exposure of sensitive elements of the structure to the sun and its oxidizing effects	Painting; provision of shade
Exposure of sensitive elements of steel structures to moisture and other elements	Coating application; in situ plating

On the other hand, the maintenance cost data may be disaggregate (data items that involve detailed and elaborate bid prices for each specific base "pay items"); in this case, pay items, priced per length, area, or volume or weight of finished product, are reported separately or combined for the factors of production (materials, labor and supervision, and equipment use). As we learned in Chapter 10, these types of cost data are obtained from specialized departments of the system owner and tend to be sensitive and confidential because they serve as a basis for developing the engineers estimates for system preservation contracts.

At any level of aggregation, cost data can be used to predict the expected future cost of a preservation treatment using the average cost or statistical regression model developed from the data (see Chapter 7). In estimating the costs of future maintenance projects on the basis of past data, the engineer must make due adjustments for the effects of inflation, economies of scale and condition, and location.

Maintenance costs are incurred not only by the system owner but also by the user (through delay and inconvenience due to system downtime) and the community (due to the extra spending incurred to prevent or mitigate the adverse community impacts arising from the maintenance work) as discussed in Chapter 10. Knowledge of all these costs are important because they can help the maintenance engineer to identify the most cost-effective treatment. For example, alternative preservation treatments or schedules may have different impacts in terms of these costs and the

Figure 25.13 Preservation treatments for civil engineering systems: (*a*) Lakes Entrance dredger *April Hamer*; (*b*) Dam maintenance and inspection; (*c*) Concrete repairs; (*d*) Deck repairs for the Riegelsville Toll-Supported Bridge Rehabilitation Project; (*e*) Jet grouting; (*f*) Wind turbine maintenance; (*g*) Navy Diver 1st Class Josh Moore welds a repair patch on the submerged bow of amphibious transport dock; (*h*) Maintenance of Swinging Bridge Dam, Forestburgh, NY; (*i*) Concrete pavement. Photos courtesy of (*a*) Dashers, Wikimedia; (*b*) Michael and Zelna Suttie; (*c*) Minnesota Department of Transportation; (*d*) Delaware River Joint Toll Bridge Commission; (*e*) Olnnu, Wikimedia; (*f*) ILA-boy, GNU General Public License; (*g*) U.S. Navy; (*h*) Christopher Ponnwitz; (*i*) U.S. Department of Transportation.

best treatment or schedule is that which generally yields the least overall cost assuming they yield the same level of repair effectiveness. With the requisite data, the modeling techniques learned in Chapter 5 can be used to establish maintenance cost models and maintenance duration models such as that shown in Example 25.3.

Example 25.3 Delay "Cost" Analysis

The downtime (days) for structural repair on fairly corroded bridges at busy freeways is a random variable X, with the following probability function:

$$f(x) = \begin{cases} \dfrac{32}{(x+4)^3} & x > 0 \\ 0 & \text{elsewhere} \end{cases}$$

It was observed last week that a vital bridge on Interstate 467 (a major freeway bypassing a city) is fairly corroded and needs structural repairs. Find the probability that such repairs will take (a) between 2 and 5 days to complete.

Solution

This is a continuous probability function. As such, the area under the curve is the probability. Thus, in order to answer this question, the function must be integrated within the given limits. The probability that such repairs will take between 2 and 5 days to complete is the area under the curve between 2 and 5 in Figure 25.14. The curve in the figure is $f(x)$.

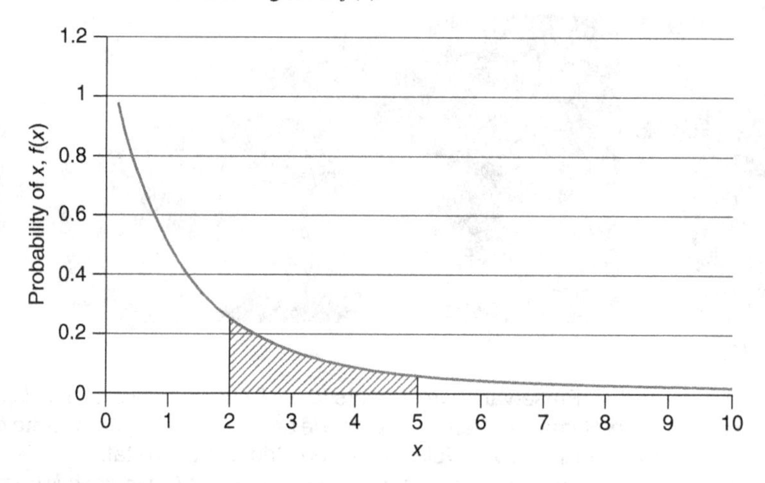

Figure 25.14 Figure for Example 25.3.

Thus, the probability that the repair will take between 2 and 5 days to complete can be calculated as

$$p = \int_2^5 f(x)\,dx = \left(1 - \frac{16}{(x+4)^2}\right)\Bigg|_2^5 = \left(1 - \frac{16}{(5+4)^2}\right) - \left(1 - \frac{16}{(2+4)^2}\right) = 0.247$$

25.2.9 Establishing the Short-Term Effectiveness of Preservation Treatments

The effectiveness of system preservation may be viewed in the short or long term. Short-term effectiveness assessment is typically used for an individual treatment or a set of treatments applied at a given point in time. The condition versus time plots shown in Figure 25.15 [adapted from Mamlouk

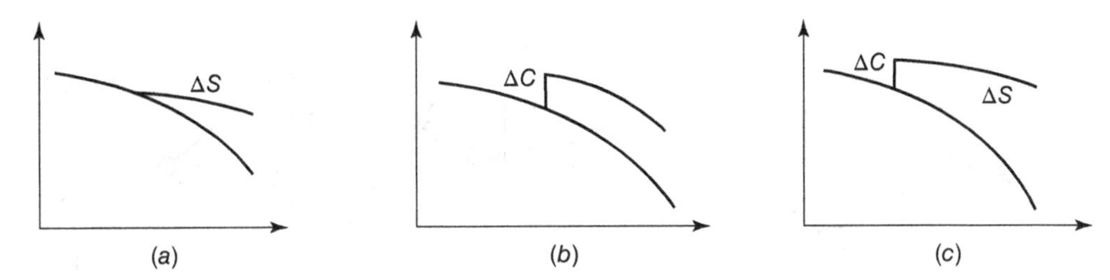

Figure 25.15 Patterns of short-term condition changes due to a preservation treatment: (*a*) no jump, reduced rate of deterioration; (*b*) jump, same rate of deterioration; and (*c*) jump, reduced rate of deterioration.

and Zaniewski (1998)] present the possible postmaintenance deterioration patterns: instantaneous change in condition, change in slope, or both.

Short-term analysis is useful in system preservation planning because it helps the maintenance engineer make quick comparisons of the benefits of alternative preservation treatments across different attributes that include treatment type, material used, procedure, or even work source (in-house versus contract).

For the maintenance engineer who seeks to assess the extent to which a preservation treatment effectively addresses system deterioration, the three basic sequential questions are: (a) How should effectiveness be measured and which condition indicator should be used for this measurement? (b) On what grounds can the preservation treatment be deemed effective? (c) If the treatment is found to be effective, can such effectiveness be modeled as a function of the attributes of the system, treatment, and the environment? The steps for assessing and analyzing the short-term effectiveness of system preservation treatments are explained in detail in Labi and Sinha (2003) and are summarized below.

Step 1. How Should Effectiveness Be Measured? This step involves the selection of an appropriate measure of short-term effectiveness (MOE), such as the increase in system physical condition or decrease in deterioration rate. The MOE values are expressed in terms of the condition indicators which are different for each system type and component. For example, the condition indicator for measuring the effectiveness of bridge deck patching treatment could be the percentage of the patched area that needs repatching a few years after the treatment.

MOE I—Performance Jump (PJ). This may simply be considered as the instantaneous elevation in the system condition due to a preservation treatment (see ΔC and ΔC_4 in Figures 25.15 and 25.16, respectively). This is computed using the values of deterioration taken just before and just after the treatment (Lytton, 1987; Markow, 1991). Unfortunately, most agencies typically do not measure the condition of their systems just before and just after preservation; therefore, it is often difficult to obtain data for PJ computation. As such, it is often necessary to extrapolate the performance curve from both directions to the point of preservation in order to estimate the jump in performance due to the treatment (i.e., *FD* in Figure 25.16).

MOE II—Deterioration Rate Reduction (DRR). This concept involves the "slowing down" of a system's physical deterioration in response to the preservation treatment. Therefore in the context of DRR, the effect of preservation is to change the steep slope associated with a rapidly deteriorating system to a slope that is relatively gentle. DRR can be calculated as the difference or the ratio of the slope of the deterioration curve before preservation and after preservation. The DRR

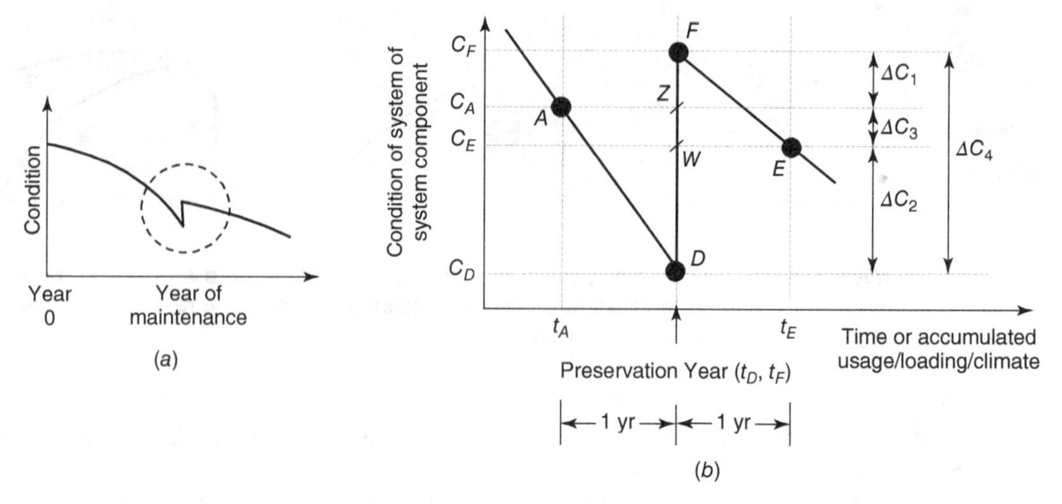

Figure 25.16 Illustrations of condition measurements and measures of effectiveness in the short term: (*a*) kink (circled) in long-term performance trend due to rehabilitation or maintenance and (*b*) closeup details of kink.

due to a specific preservation treatment (or specific combinations thereof) is best determined when the system received no other treatment in the time vicinity of the preservation application, so that the occluding effect of such "extraneous" treatments in DRR measurement, is avoided. A minimum of three data points in time (corresponding to two monitoring periods), is needed for DRR computation.

Step 2. Which Condition Indicator Should Be Used for the MOE? At this step, the engineer selects an appropriate condition indicator. This is typically the extent of spread and/or degree of severity of at least one specific distress type. The condition indicator may be aggregate (representing several distress types, their extents, and their severities) or disaggregate (representing only one distress type, such as corrosion, cracking, or loss of texture). The MOE (selected in Step 1) is expressed in terms of the condition indicator. For example, the performance jump (an MOE) can be expressed in terms of the reduction on *percent of corroded area* (a condition indicator).

Step 3. On What Grounds Can the Preservation Treatment Be Deemed Effective? This step assesses whether the preservation treatment was truly effective, on the basis of the computed values of the MOE in terms of the condition indicator. We may recall the statistical concepts we learned in Section 6.5 of Chapter 6. The reported values of the performance indicators (and consequently, the MOE values) are the average values taken across a typically large number of instances for a given system or different systems that received the treatment in question. Therefore, the distribution of the MOE values can be considered as a statistical sampling distribution of means. If this assumption holds true, then the hypothesis for the preservation treatment effectiveness, in terms of the selected MOE, can be formulated as follows:

$$H_0\colon \mu_{\text{MOE}} \leq 0 \text{ (treatment was not effective)}$$

$$H_1\colon \mu_{\text{MOE}} > 0 \text{ (treatment was effective)}$$

This is a one-sided hypothesis test with the "rejection region" in the upper tail. Assuming a normal distribution of the means of the entire population, the critical value of the test statistic is given by Z_C, which can be determined from statistical charts (see Appendix 2) as a function of the degree of significance, α. When $\alpha = 5\%, Z_C = 1.96$; when $\alpha = 10\%, Z_C = 1.64$; and when $\alpha = 20\%, Z_C = 1.28$. Also, the calculated value of the test statistic is given by

$$Z^* = \frac{\mu_{\text{MOE}} - 0}{\sigma/\sqrt{n}}$$

where σ is the standard deviation, and n is the sample size, μ_{MOE} is the mean value of the measure of effectiveness in terms of the performance indicator.

Decision: If the calculated values of the test statistic exceeds the critical value of the test statistic (which, e.g., is 1.645 if the confidence level is 95%), then the former falls in the hypothesis rejection region, thus suggesting that the preservation treatment yielded MOEs that were significantly greater than zero and therefore was effective at that confidence level. On the other hand, if the test statistic's calculated value does not exceed its critical value, then the former does not fall in the hypothesis rejection region, thus there is no evidence to suggest that the preservation treatments yielded MOEs that were significantly greater than zero; in that case, it cannot be concluded that the preservation treatment was effective at that confidence level.

Example 25.4

A new treatment type is developed that aims to reduce the deterioration rate of an undersea component of a marine structure. Tests are conducted to investigate the effect of this new treatment type. A sample of 30 newly treated components is tested, and their DRRs are calculated and their mean is determined to be 1.72. Assume that the distribution of DRR is normal with the standard deviation of 5.36. Determine whether the new treatment type is effective at a 5% significance level.

Solution

$$\text{Null hypothesis}: H_0: \mu_{\text{DRR}} \leq 0; \text{alternate hypothesis}: H_1: \mu_{\text{DRR}} > 0.$$

$$\text{The form of } H_1 \text{ implies use of an upper} - \text{tailed test with rejection } Z \geq Z_{0.05} = 1.645.$$

The test statistic value is

$$Z = \frac{\overline{\mu}_{DRR} - 0}{\sigma/\sqrt{n}} = \frac{1.72 - 0}{5.36/\sqrt{30}} = 1.758 > 1.645$$

Thus, Z falls in the rejection region, that is, the new treatment type yields a DRR that is significantly greater than zero. Therefore, there is no evidence that the new treatment is not effective at a significance level of 0.05.

Step 4: If the Treatment Is Found to Be Effective, How Could Its Effectiveness Be Modeled as a Function of the Attributes of the Treatment, the System, and the Environment? If maintenance effectiveness is thus confirmed, and MOE values are established for each system under consideration, the third stage is to develop a model to estimate this effectiveness as a function of the system attributes (e.g., system age, material type), the treatment characteristics (e.g., treatment material type, contractor expertise), and the environment (e.g., the weather at the time of treatment). In such models, the MOE is the dependent variable. Thus, Steps 2 and 3 involve the use of data from several individual systems that received the preservation treatment under investigation.

25.2.10 Establishing the Long-Term Effectiveness of Preservation Treatments

As discussed in the previous sections, there are a number of reasons why an agency will need to know how effective a specific treatment is in the short term. However, analysis of long-term effectiveness is considered more useful. Long-term assessment is typically carried out for major treatments (e.g., rehabilitation or major maintenance) or preservation schedules (multiple treatment types and timings applied over the system entire life or remaining life). System agencies develop their long range preservation plans, programs and budgets on the basis of long-term effectiveness rather than short-term effectiveness.

The maintenance engineer can assess the effectiveness of treatments in the long term using any one of several MOE's that we learned in Section 25.0.6. In this section, we focus on three of these MOEs: treatment service life; increase in the average condition of the system in the posttreatment period, relative to the condition before the treatment; and the increase in area bounded by the system's deterioration curve due to the preservation treatment.

Figure 25.17 illustrates the long-term measures of effectiveness for a typical system that receives a preservation treatment. Let x be the time-related variable against which deterioration

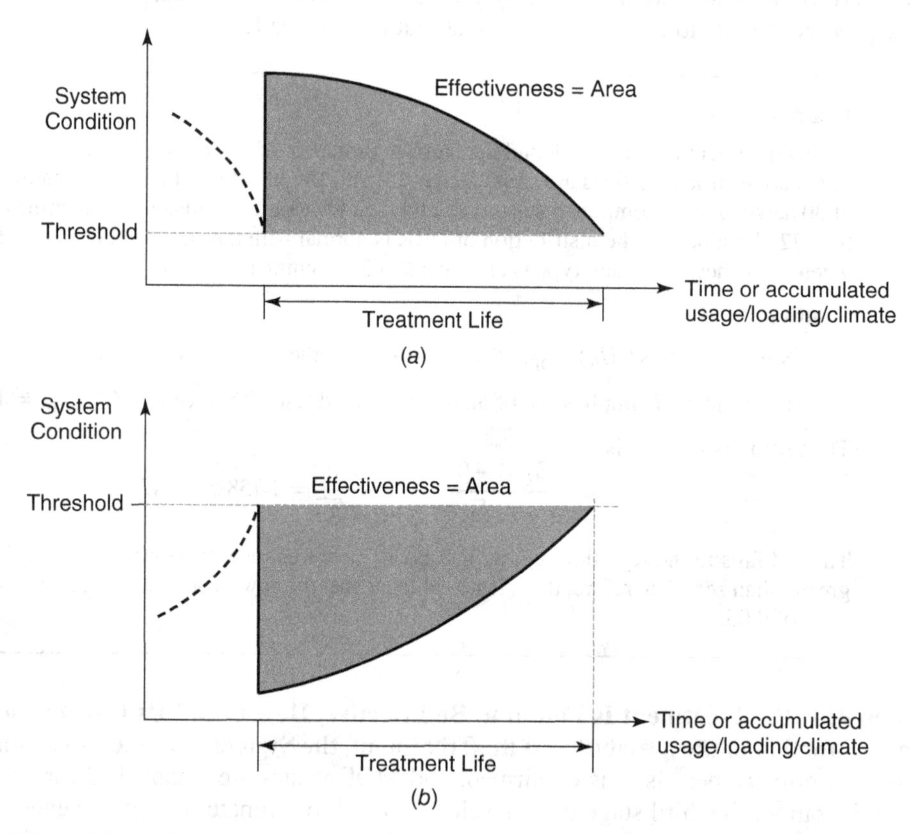

Figure 25.17 Measures of long-term effectiveness: (*a*) nonincreasing condition indicators (e.g., condition rating, health index) and (*b*) nondecreasing Indicators (e.g., corrosion, cracking, spalling).

is being monitored, such as time (years), accumulated loading (in terms of cumulative number of users), or accumulated climate effects (in terms of freeze index, freeze–thaw cycles, temperature, precipitation, etc.). The deterioration of the system after the treatment application is denoted by $f(x)$. Most agencies have developed $f(x)$ equations for each type of system or system component in their jurisdiction. As illustrated in the figure, given the deterioration curve $f(x)$ and the established threshold condition, it is a straightforward task to determine the treatment life, that is, the time taken for the system condition to reach the threshold condition.

(a) Life of the Preservation Treatment. In terms of time (years or accumulated loading or climate effects), the life of a preservation treatment (or the extension in the system life or component life due to the preservation treatment) can be determined using one of two methods: (i) estimating the time elapsed between the treatment application and the next treatment of a similar or higher level. For example, in the context of pavement systems, higher level treatments may be thick (structural) overlays and lower level treatments may be thin overlays, and (ii) estimating the time that passes before the treated facility reverts to an established condition threshold (that is, the condition before the last treatment or a prespecified condition trigger.

For the latter method, a threshold needs to be specified. Many system owners (agencies) have established condition thresholds for the various treatments for their system preservation. Where no specified thresholds exist, the average pretreatment system condition can be taken as the threshold for purposes of the effectiveness analysis; it is important to realize that the threshold, if determined in such manner, is only a reflection of the past state of practice in the agency and therefore can (and does) vary across different facilities that received the treatment at different years, and also may be subject to bias and funding limitations.

In any one of these two methods for determining the treatment life, the first step is to establish the treatment performance curve (i.e., the rate of system deterioration after the preservation treatment is applied). Assume that the performance model has the form

$$y = f(x, t)$$

where x is a vector of time-related and other variables such as annual usage or climatic severity, and t is the time in years. If y is a function of t, then making t the subject of the resulting equation yields

$$t = f^*(y, x)$$

The system condition reverts to the threshold, y_c, when the system reaches the treatment life, t_c. Therefore, solution of the above equation when $y = y_c$ yields the service life of the treatment, t_c. This can be done for each condition indicator for the system under investigation, thus yielding a number of values for the service life. The least of these service lives is taken as the actual service life of the preservation treatment.

$$t_{c,\text{ACTUAL}} = \min(t_{C1}, t_{C2}, \ldots, t_{Cn})$$

This is illustrated in Figure 25.18 for a nonincreasing condition performance indicator. Also, from the equation above, the service life can be expressed as a function of the explanatory variables represented by the vector x, such as annual usage or loading, climatic severity of the region where the system is located, and system material type.

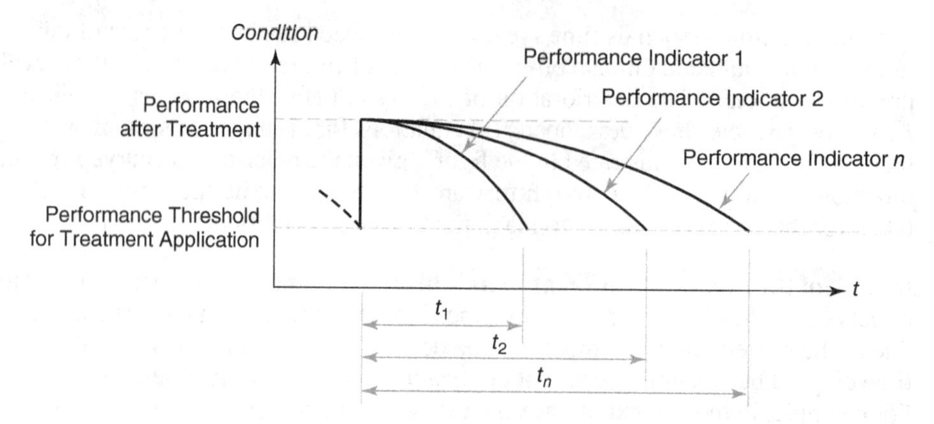

Figure 25.18 Service lives of a preservation treatment using different performance indicators.

Example 25.5

The condition of a pipe section can be measured in terms of its cracking on a scale of 0 (poor condition) to 10 (excellent condition) or corrosion, also on a similar scale. The time trend of these performance indicators is provided in Table 25.7. The threshold conditions for rehabilitation are 2 and 3 for cracking and corrosion, respectively. The pipe rehabilitation in 1998 elevated both the cracking index and the corrosion index to 10. The deterioration functions for the cracking and corrosion are

$$\text{Cracking index} = 10 - \frac{20}{2 + 80t^{-2.1}}$$

$$\text{Corrosion index} = 10 - \frac{25}{2.5 + 90t^{-2.2}}$$

Determine the expected service life of the rehabilitation treatment.

Solution

Based on the deterioration functions, the values for the cracking index and corrosion index after the rehabilitation treatment are presented in Table 25.6.

Table 25.6 Solution to Example 25.5

Year	1999	2000	2001	2002	2003	2004	2005	2006	2007	2008	2009	2010
Cracking Index	10	9	8	7	6	5	4	3	3	2	2	2
Corrosion Index	10	9	8	6	5	4	3	3	2	2	2	1

The threshold for the cracking index is 2 and for the corrosion index is 3; thus, the expected service life of the treatment is the lesser of 9 years (for cracking) and 8 years (for corrosion). Thus the service life is 8 years.

(b) Increase in Average System Condition over the Life of the Preservation Treatment. For a given indicator of system condition, y, the average value of the system condition indicator, over the preservation treatment life, can be determined using the following expression:

$$y_{\text{AVG}} = 1/t_C\,(y_0 + y_1 + \cdots + y_c)$$

If the annual field measurements of condition indicator are available for the given time interval, the y values can be determined using one of two approaches: (i) calculating the average of the y values for each year, across all the systems that received that treatment, or (ii) developing a condition model to represent the condition trend for all the systems that received the treatment, and then from the model, determining the ordinate (system condition) at each year of the treatment life.

Let y_0 and y_c represent the system condition just after the preservation treatment and at the time when the system condition reaches the threshold, respectively, y_i represents the system condition at any intervening year, i, and t_C is the service life. The increase in average system condition due to the preservation treatment can then be found by computing the percentage change in average condition relative to the condition before treatment.

$$\text{Treatment effectiveness} = 100 \times \left(\frac{y_{\text{AVG}} - y_{\text{INI}}}{y_{\text{INI}}} \right) \tag{25.1}$$

where y_{INI} is the initial condition of the system (i.e., the pretreatment condition).

An example computation is provided in Example 25.6.

(c) Area Bounded by Deterioration Curve due to Treatment. The area bounded by the deterioration curve and the threshold line embodies the effectiveness concepts of both the average system condition and the service life and can therefore be considered the best way to measure the long-term effectiveness of preservation treatments. A simple approach for determining the area under the deterioration curve is to develop a deterioration curve for the treated systems and then calculate the area bounded by the curve from time of treatment to the time of reaching a specified threshold using coordinate geometry or calculus.

As seen in Figure 25.14, for nonincreasing condition indicators such as structural condition rating or sign retroreflectivity, the treatment effectiveness is the area bounded by the curve and the horizontal line projected from the threshold condition level (i.e., the area *under* the curve); for nondecreasing indicators such as corrosion and spalling, treatment effectiveness is the area bounded by the curve and the horizontal line projected from the threshold condition level (i.e., the area *over* the curve).

The area-bounded-by-the-curve concept has seen some application in past research and practice where it has often been used to represent the effectiveness of highway preservation in terms of the reduction is user costs (Geoffroy, 1996). The concept can be applied to most other civil systems where reduction in user cost or inconvenience can represent a benefit of the system preservation.

From the figure, the treatment effectiveness in terms of the area bounded by the curve can be expressed mathematically as follows:

(i) For nondecreasing condition indicators (where effectiveness is represented by the area *over* the curve):

$$(PJ \times t_c) - \int_0^{t_c} f(t)\, dt \tag{25.2}$$

(ii) For nonincreasing performance indicators (where effectiveness is represented by the area *under* the curve):

$$\int_0^{t_c} f(t)\, dt \tag{25.3}$$

where PJ is the performance jump.

Example 25.6

The posttreatment model of the HMA structural overlay for a flexible pavement is (Irfan, 2010):

$$\mathrm{IRI} = e^{3.858+0.019\mathrm{AATT}^{*}t+0.151\mathrm{ANDX}^{*}t}$$

where IRI is the pavement international roughness index; AATA is the annual average truck traffic (in millions); ANDX is the average annual freeze index (in thousands); and t is the time (years) after the HMA functional overlay.

The maximum threshold of the IRI for the flexible pavement is 170 in./mile. Assume the AATA is 3.65 million annually and the ANDX is 500. Calculate the benefit (area bounded by the curve) of the HMA functional overlay for this flexible pavement (Figure 25.19).

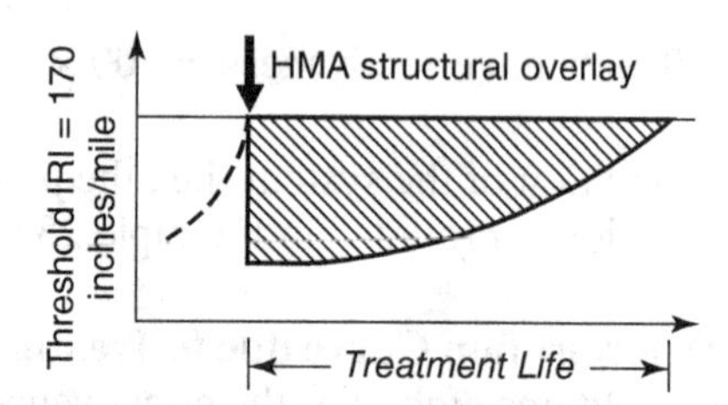

Figure 25.19 Figure for Example 25.6.

Solution

The treatment life can be estimated by solving the following equation for t_C.

$$e^{3.858+0.019\mathrm{AATT}^{*}t_c+0.151\mathrm{ANDX}^{*}t_c} = 170$$

Solving for t_C yields 8.8 years. The performance jump is calculated as

$$170 - e^{3.858+0.019\mathrm{AATT}(0)+0.151\mathrm{ANDX}(0)} = 122.63 \text{ IRI units}$$

Then the area bounded by the curve shown in the figure is

$$122.63 \times 8.8 - \int_0^{8.8} e^{3.858+0.019^{*}3.65^{*}t+0.151^{*}0.5^{*}t}dt = 1079.14 - 842.94 = 236.2 \text{ IRI-years}$$

25.2.11 Identifying the Factors That Affect Treatment Effectiveness

The success of a preservation treatment depends on a variety of factors that can be categorized as follows: system-related factors, treatment-related factors, environment-related factors, and contractor-related factors. The system-related factors include the system condition just before the treatment and the age of the system. The treatment-related factors include the type and intensity of the treatment. The environment-related factors include the nature of the weather at the time of treatment, for example, excessively hot or cold temperatures or wind, rain, or snow conditions that adversely affect the quality of workmanship or the materials being used for the preservation work. Contractor-related factors include the quality (class, experience, or grade) of the contractor.

The impact of different levels of preservation received by a civil engineering system over its life cycle on its longevity can be modeled in one of at least two ways: (i) using preservation occurrence frequency, or intensity as independent variables or (ii) developing separate posttreatment performance models for the different preservation treatments, intensities, and/or frequencies as shown in Figure 25.20.

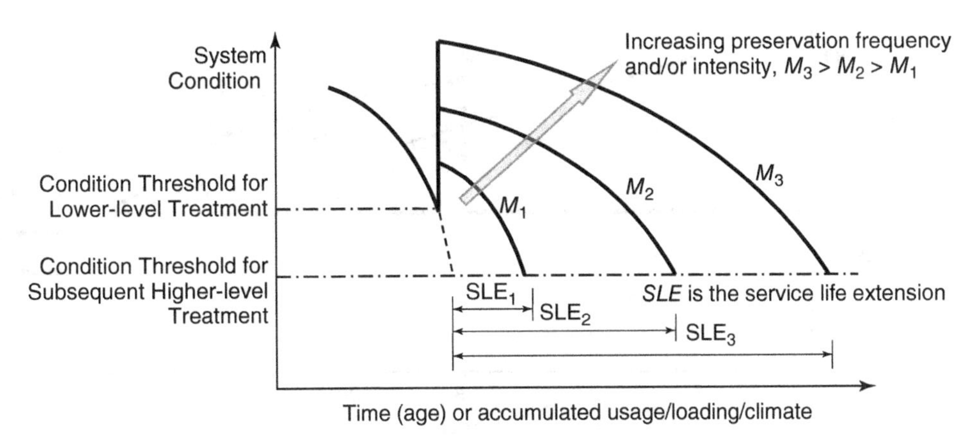

Figure 25.20 Effect of higher preservation intensity on effectiveness.

Example 25.7

A maintenance engineer is considering three options to preserve the existing Eagle Creek bridge in Westminster County. Table 25.7 shows the National Bridge Inventory (NBI) NBI ratings that are expected to correspond to each preservation strategy. Use a threshold of 40 units. The implementation costs of a minor and major rehabilitation are $160,000 and $270,000, respectively, in 2009 dollars. Assume that the annual average cost of routine maintenance is $3000 and that this is the same across all the alternative strategies.

(a) Plot on the same graph the performance trends and jumps for the three alternative preservation strategies. Use a different legend or color for each strategy.

(b) Using any one of the effectiveness criteria (service life, average condition over service life, or area bounded by the performance curve), use your graph to estimate the effectiveness of each strategy.

(c) Estimate the present worth of costs for each strategy. Use an interest rate of 5%.

(d) On the basis of cost-effectiveness (ratio of effectiveness to cost), identify the optimal preservation strategy.

Table 25.7 Details of Preservation Strategies

	Year 0 (Now)	Year 5	Year 10	Year 15	Year 20	Year 25	Year 30
Strategy 1	80	74	62	55	41	—	—
Strategy 2	80	Before minor rehab—74 After minor rehab—80	72	63	Before minor rehab—57 After minor rehab—63	52	—
Strategy 3	80	74	62	Before major rehab—55 After major rehab—76	70	62	42

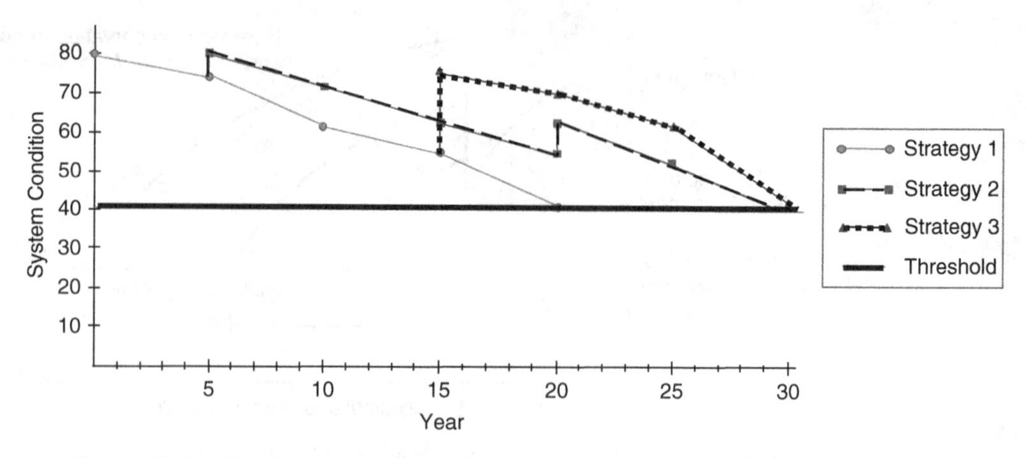

Figure 25.21 Predictions of performance under the different preservation strategies.

Solution

(a) Figure 25.21 presents the performance trends and jumps over time, for each preservation strategy.

(b) Assume the threshold condition (NBI rating) = 40.

Approach 1 (using asset condition increase as a measure of effectiveness)

Assuming the condition (NBI rating) prior to the application of strategy, $(\text{PERF}_{\text{INI}}) = 40$:

$$\text{Alternative 1}: \text{PERF}_{\text{AVG}} = \left(\frac{1}{5}\right) \times (80 + 74 + 63 + 55 + 40) = \left(\frac{313}{5}\right) = 62.4.$$

$$\text{Effectiveness} = 100\ [(\text{PERF}_{\text{AVG}} - \text{PERF}_{\text{INI}})/\text{PERF}_{\text{INI}}] = 100\ [(62.4 - 40)/40] = 56.0\%.$$

$$\text{Alternative 2}: \text{PERF}_{\text{AVG}} = \left(\frac{1}{7}\right) \times (80 + 77^{*} + 74 + 63 + 60^{*} + 52 + 40) = \left(\frac{446}{7}\right) = 63.7.$$

$$\text{Effectiveness} = 100\ [(\text{PERF}_{\text{AVG}} - \text{PERF}_{\text{INI}})/\text{PERF}_{\text{INI}}] = 100\ [(63.7 - 40)/40] = 59.25\%.$$

$$\text{Alternative 3}: \text{PERF}_{\text{AVG}} = \left(\frac{1}{7}\right) \times (80 + 74 + 62 + 65.5^{*} + 70 + 62 + 40) = (453.5/7) = 64.8.$$

$$\text{Effectiveness} = 100\ [(\text{PERF}_{\text{AVG}} - \text{PERF}_{\text{INI}})/\text{PERF}_{\text{INI}}] = 100\ [(64.8 - 40)/40] = 61.96\%.$$

*[At performance jump (year of minor/ major rehabilitation), taking average of the NBI ratings before and after the treatment]

Approach 2 (using "asset life due to the treatment strategy" as a measure of effectiveness):

Alternative 1: 20 years; Alternative 2: 29.5 years (by extrapolating the performance trend curve); Alternative 3: 30 years.

Approach 3 (using "area under curve" as a measure of effectiveness):

Using the area bounded by the performance curve effectiveness criterion, the effectiveness of each preservation method is given in Table 25.8. Assume threshold NBI rating = 40.

(c) Estimate the present worth of costs for each strategy. Use an interest rate of 5%.

Note: n = analysis period

Present Worth Cost (PWC) of Strategy 1:

Given: Annual average maintenance cost = \$3000, Interest rate = 5%

$$p_1 = 3000 \times \frac{(1 + 0.05)^{20} - 1}{0.05(1 + 0.05)^{20}} = \$37,386.63$$

Table 25.8 Area Bounded by Performance Curve, for Each Preservation Strategy

Strategy	Effectiveness (Area under Curve) for Each 5-year Duration						Total Area (Effectiveness) for each Strategy
	0–5	5–10	10–15	15–20	20–25	25–30	
Strategy 1	185	140	92.5	40	0	0	457.5
Strategy 2	185	180	137.5	100	87.5	29.4	719.4
Strategy 3	185	140	92.5	165	130	60	772.5

Present Worth Cost for Strategy 2:

PWC for annual average maintenance cost for strategy 3 = $45,774.52 (for $n = 30$).

Given: Minor rehabilitation cost = $160,000 (in 2009 dollars):

$$\text{Converting future minor rehabilitation in year 5 to PWC} = \frac{F}{(1+i)^n} = \frac{160,000}{(1+0.05)^5} = \$125,364.2$$

$$\text{Converting future minor rehabilitation in year 20 to PWC} = \frac{F}{(1+i)^n} = \frac{160000}{(1+0.05)^{20}} = \$60,302.32$$

$$\text{Total PWC for strategy 2} = \$45,774.52 + \$125,364.2 + \$60,302.32 = \$231,441.04$$

Present Worth Cost for Strategy 3:

PWC for annual average maintenance cost for Strategy 3 = $46,117.35 (for $n = 30$)

Given: Major rehabilitation = $270,000 (in 2009 dollars):

$$\text{Converting future major rehabilitation in year 15 to PWC} = \frac{F}{(1+i)^n} = \frac{270,000}{(1+0.05)^{15}} = \$129,875.62$$

$$\text{Total PWC for Strategy 3} = \$46,117.35 + \$129,875.62 = \$175,993.97$$

(d) On the basis of cost-effectiveness (ratio of effectiveness to cost), identify the optimal preservation strategy.

Using the area under the curve as the measure of effectiveness, we obtain the values of cost-effectiveness shown in Table 25.9.

Table 25.9 Details of Preservation Strategies

Strategy	Effectiveness (e)	Cost (c)	e/c
Strategy 1	457.5	$37,386.63	0.012237
Strategy 2	719.4	$231,441.04	0.003106
Strategy 3	772.5	$175,993.97	0.004389

Based on the above calculations, it can be inferred that strategy 1 is the most cost-effective. Note: For readers who used service life or increased condition as the measure of strategy effectiveness, different cost-effectiveness values will be obtained, and a different preservation strategy may be identified as being the optimal.

25.3 IDENTIFYING THE APPROPRIATE PRESERVATION TREATMENT AT A GIVEN TIME

As we learned in Table 25.1 in Section 25.0.3, one of the common decision contexts for system preservation is the choice of the best treatment to apply at any given time (Hicks et al, 1997).

We also learned from Section 25.0.4 that for making such decisions, the engineer uses a variety of mechanisms including expert opinion, historical practices, and data-driven optimization. Then in Sections 25.0.5 and 25.0.6 respectively, we discussed the factors that influence the decisions, and the measures to assess the impacts of preservation treatments.

In this section, we present a simplified use of optimization involving life-cycle cost analysis (LCCA) or multiple criteria decision making (MCDM). As we learned in Chapters 11 and 12, we use LCCA for making decisions involving monetary or monetized factors and MCDM for decisions involving monetary and/or nonmonetary factors. In this section, we will first discuss the nature of the problem and the basis upon which the decision is made. Then we will discuss the decision structures that could be established by the system owner for purposes of consistency in preservation decision making.

25.3.1 Problem Nature and Basis of the Decision

Here, the system maintenance engineer seeks to answer the question: "For a specific individual system, what is the best preservation action to carry out at a given time?" In making such decision on the basis of a single factor or performance measure such as life-cycle cost or service life extension, the answer is relatively straightforward. For several performance measures, however, the decision problem is relatively complicated and can be structured as shown in Figure 25.22. This can be solved using life-cycle cost analysis (see Chapter 11) where the engineer selects the treatment that yields the least life-cycle cost or the highest life-cycle return in monetary or multiple criteria terms (see Chapter 12).

Life-Cycle Cost Analysis (LCCA). Here, the engineer calculates the life-cycle cost (LCC_i) of each treatment alternative or candidate i as follows:

$$\text{LCC}_i = w_{\text{agency}}\text{LCC}_i^{\text{agency}} + w_{\text{user}}\text{LCC}_i^{\text{user}} + w_{\text{community}}\ \text{LCC}_i^{\text{community}}$$

For treatment i, $\text{LCC}_i^{\text{agency}}$ $\text{LCC}_i^{\text{user}}$ and $\text{LCC}_i^{\text{community}}$ are the life-cycle costs borne agency, users and community, respectively; w_{agency}, w_{user} and $w_{\text{community}}$ are the weights of agency, user, and community costs, respectively. Then the engineer identifies the treatment with the minimum LCC as the optimal treatment: Min $\{\text{LCC}_1, \text{LCC}_2, \ldots, \text{LCC}_I\}$, where I is the number of alternative treatments.

Multicriteria Decision Making (MCDM). Here, we calculate the impact of each treatment alternative or candidate i in terms of performance measure $j, j = 1, 2, \ldots J$. The J performance measures may have different units, thus there may be a need to transform all their values onto the same scale

Calculated Costs and Benefits (Agency, User, Community)

Alternative Treatments for Preserving the Asset	PM 1	PM 2	. . .	PM K	Total Combined Impact of Alt j
Treatment Alt 1					I_1
Treatment Alt 2					I_2
Treatment Alt 3					I_3
⋮					⋮
Treatment Alt *j*					I_J

Figure 25.22 Basic matrix for selecting optimal preservation treatment.

such that they have the same unit, S, or that they are all dimensionless. Also, the engineer typically attaches different levels of importance to the performance measures, so there is a need to establish their relative weights, w_j. Then, the overall or amalgamated impact of treatment i is as follows:

$$U_i = U_i^{\mathrm{PM1}} + U_i^{\mathrm{PM2}} + \cdots + U_i^{\mathrm{PMJ}}$$

$$U_i = w_{\mathrm{PM1}} S_i^{\mathrm{PM1}} + w_{\mathrm{PM2}} S_i^{\mathrm{PM2}} + \cdots + w_{\mathrm{PMJ}} S_i^{\mathrm{PMJ}}$$

Then the engineer identifies the treatment with the maximum U as the optimal treatment: Min $\{U_1, U_2, \ldots, U_I\}$, where I is the number of alternative treatments.

Example 25.8

There are two alternative treatments for an aging structural system: Treatment A and treatment B. The system owner is evaluating these treatments on the basis of the increase in condition (% increase in load-bearing capacity), extension in the life of the structure (years), and cost of the treatment (Table 25.10). Use the MCDM method to choose the optimal treatment. The scaling function for the performance measures are as follows: Condition: $U_{\mathrm{Condition}} = (300 - C)/300$; Condition: $U_{\mathrm{Extension}} = E/20$; Cost: $U_{\mathrm{Cost}} = (25 - M)/25$. Assume that the three performance measures have equal weights, and use the weighted-sum method for the amalgamation.

Table 25.10 Data for Example 25.8

	Increase in condition, C	Extension in facility life (yr), E	Cost ($M), M
Treatment A	75	10	10
Treatment B	85	13	14

Solution

Based on the scaling functions, the scaled values of the performance measures are determined (Table 25.11).

Table 25.11 Solution to Example 25.8

	Scaled values of Performance Using the Utility Functions		
	Increase in condition, U_C	Extension in facility life (yr), U_E	Cost ($M), U_M
Treatment A	0.717	0.6	0.6
Treatment B	0.75	0.75	0.44

The amalgamated value for treatment A is $U_A = 0.717 + 0.6 + 0.6 = 1.917$ and the amalgamated value for treatment B is $U_B = 0.75 + 0.75 + 0.44 = 1.940$. Thus, treatment B is the superior treatment.

25.3.2 Decision Structures

For a given physical defect of the system, and under different conditions related to the system, environment, and other factors (Figure 25.5), the maintenance engineer identifies the best preservation

IRI	Rutting	Level of Cracking		
		Light	Moderate	Severe
Excellent IRI < 80	Light	Do Nothing	Seal Cracks	N/A
	Moderate	N/A	N/A	N/A
	Severe	N/A	N/A	N/A
Good 80 < IRI < 114	Light	Do Nothing	Seal Cracks	Mill and Fill 1"
	Moderate	Thin HMA Overlay 1.5"	Mill and Fill 1.5"	Mill and Fill 2"
	Severe	Thin HMA Overlay 2"	N/A	N/A
Fair 115 < IRI < 149	Light	N/A	N/A	Mill and Fill 2.5"
	Moderate	HMA Overlay 2.5"	HMA Overlay 3"	Mill 1.5" and HMA Overlay 4"
	Severe	HMA Overlay 3.5"	Mill 1.5" and HMA Overlay	Pavement Replacement
Poor IRI > 150	Any	Pavement Replacement	Pavement Replacement	Pavement Replacement

Figure 25.23 Illustration of warrants for treatment selection.

treatment using expert opinion, historical practice, or mathematical optimization as we have seen in the preceding sections. Often, the pathway to reach the best decision for each of several combinations of existing conditions is documented in a structured, schematic manner. This documentation, which is in the form of text, tables, or figures, are called decision policy documents, **decision trees**, or **decision matrices**, and are collectively referred to as support systems for preservation treatment selection. An example is provided in Figure 25.23. Such decision tools are typically characterized by a set of sequential logical rules and criteria; in the past, these have been based largely on either the opinions of experienced maintenance engineers or past practices. However, as we have stressed in earlier sections, these pathways can be established using a data-driven optimization mechanism that examines the alternatives on the basis of life-cycle costing or multiple-criteria analysis.

25.4 DEVELOPMENT OF SYSTEM PRESERVATION SCHEDULES

25.4.1 Mechanisms for Preservation Schedule Decision Making

Similar to the case for treatment selection at a given time (Section 25.3), the system owner can identify the best schedule or long-term strategy for preserving the system (over its entire life or remaining life) using any one of three mechanisms: expert opinion, historical practices, or mathematical optimization. Each of these mechanisms, to some extent, involves data-driven or experiential analysis of the costs and benefits, implicitly or explicitly, associated with each candidate schedule.

25.4.2 Nature and Basis of the Decision

As we learned in Section 25.4.1, the system's maintenance engineer often seeks to answer the question: "For a specific individual system, what is the best schedule (set of multiple preservation treatments to be applied at various points of the horizon period). The horizon period is often the entire life of a new or planned system or the remaining life of an existing system. For example, for

	Agency, user, and community impact of each preservation schedule in terms of the performance measures						
Performance measure i / Schedule alternative j	Initial or life-cycle cost (PM_1)	Increased condition (PM_2)	Increased durability (PM_3)	Reduced workzone frequency or duration (PM_4)	...	Performance measure I (PM_I)	Overall Impact of Alternative j $O_{j,I}$
Alternative schedule 1							$O_{1,I}$
Alternative schedule 2							$O_{2,I}$
...							
Alternative schedule J							$O_{J,I}$

Figure 25.24 Basic matrix for selecting optimal preservation schedule.

an existing system in its 10th year of a 25-year service life, should we carry out only routine maintenance annually for the rest of its life? Or should we carry out routine maintenance every 3 years and rehabilitation every 6 years? The number of options is endless. Similar to the treatment selection context discussed in Section 25.3, the schedule selection context could be associated with a single performance measure such as life-cycle cost or service life extension, or several performance measures that may be monetary or nonmonetary.

The decision problem could therefore be structured as shown in Figure 25.24 and solved using LCCA or multiple-criteria techniques.

Specifically, each candidate schedule consists of one or more preservation treatments, and the total cost of these constituent treatments can be calculated for each schedule. Also, each treatment in the candidate schedule is associated with a jump in condition (which could also be translated as a reduction in the rate of deterioration) and it is therefore possible to determine the overall benefit of each preservation schedule as an increase in the area under the performance curve relative to the do-nothing scenario, extension of service life or a reduction in the condition-related operating costs relative to a base scenario such as the do-nothing strategy. For each schedule, the overall cost-effectiveness could be estimated over the system entire life (for a new or planned system) or remaining life (for an existing system), and the optimal preservation schedule could then be identified.

25.4.3 Decision Structures

On the basis of the conditions related to the system, environment, and other factors, the maintenance engineer identifies the best preservation schedule using expert opinion, historical practice, or mathematical optimization. The decision structures are similar to those described for the treatment selection context. The timings of the preservation treatments are based either on predefined intervals of time or accumulated usage/loading, or on the system condition threshold levels as shown in Figure 25.25.

25.4.4 Selecting the Best Preservation Schedule [Which Interventions to Undertake and at Which Year(s)?]

For maintaining and rehabilitating their systems, maintenance engineers adopt strategies that range from parsimonious (due to lack of funding) to excessive. As discussed in earlier sections, both extremes are not cost effective. Maintenance engineers seek to establish a preservation schedule which simply is a design of which treatment types to carry out and at which year (for several of the years in the schedule, there could be no treatment or just routine maintenance). There are at least two ways to do this: establish several alternative pre-designed schedules and evaluate each schedule on the merit of its overall impacts to the agency, user and community, as directly indicated

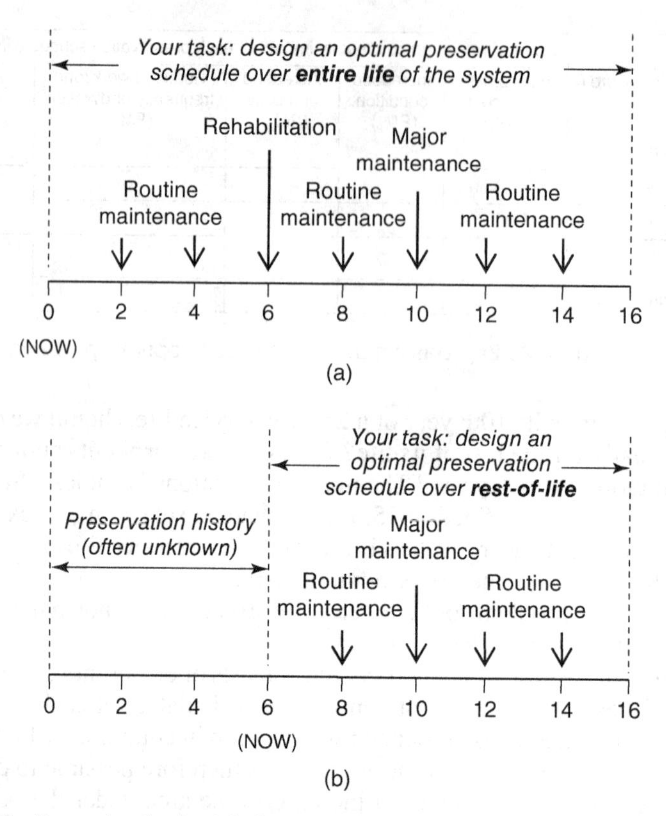

Figure 25.25 Example of schedule for system preservation: (*a*) over entire life of a new or planned system and (*b*) over remaining life of an existing system.

in Figure 25.24; here the decision variable is simply whether to choose a given schedule. The second method is more disaggregate in nature: each treatment type and its year of application are the decision variables, and the optimal solution is a specification of the best set of treatment types and timings for all years of the analysis period. The second method is explained in greater detail using the mathematical formulation below.

The second method is a knapsack problem where the decision maker seeks the best possible schedule (set of candidate treatments to be applied and at which years). It is sought to maximize reward over a given horizon period (full life or remaining life of the system or facility) in terms of single or multiple performance measures, subject to one or more constraints. Assume no budgetary constraints and only one nonbudget constraint.

The system owner seeks to develop a schedule that minimizes the agency costs to attain given performance targets at each year. The formulation is as follows:

$$\text{Minimize} \quad U = \sum_{i}^{m} \sum_{t}^{p} (x_{it} C_{it})$$

$$\text{Subject to} \quad H_t \geq (\leq) H_{\min} (H_{\max}) \qquad \text{for all } t$$

where $H_{\min}$ is the Performance target (floor) specified by the decision maker for the nonbudgetary constraint; C_{it} is the cost of treatment i in year t. It must be noted that the formulation must also include the postperformance function after each treatment; often, these functions differ across the different treatments.

It must be noted that there other variations to the problem context that lead to slightly different mathematical formulations for the knapsack problem.

Example 25.9

For a flexible pavement, the postperformance models for HMA structural and functional overlays are (Irfan, 2010)

$$\text{IRI} = e^{3.858+0.019\text{AATT}^{*}t+0.151\text{ANDX}^{*}t}$$

$$\text{IRI} = e^{4.007+0.020\text{AATT}^{*}t+0.162\text{ANDX}^{*}t}$$

where IRI is the pavement international roughness index, AATA is the annual average truck traffic (in millions), ANDX is the average annual freeze index (in thousands), and t is the time in terms of years after the HMA functional overlay. The maximum threshold of the IRI for the flexible pavement is 170 inches/mile. Assume the AATA is 3.65 million annually and the ANDX is 500. Currently, the IRI of the pavement is 170 inches/mile, which means that the pavement is in immediate need of treatment. Use a 30-year analysis period; also, assume that the only candidate treatments are HMA structural overlay and HMA functional overlay. Establish the optimal treatment schedule for this pavement. The costs of the HMA functional and structural overlays are \$10M and \$12M, respectively, in year 2015 dollars. The base year is 2015 and the discount rate is 5%.

Solution

The formulation for the problem is

$$\text{Minimize} \quad U = \sum_{i}^{2} \sum_{t}^{30} (x_{it} C_{it})$$

$$\text{Subject to} \quad \text{IRI}_t \leq 170 \qquad \text{for all } t$$

where $x_{it} = 1$ means treatment i is applied in year t; $C_{\text{it}} = 12/(1+5\%)^t$ for HMA structural overlay; $C_{\text{it}} = 10/(1+5\%)^t$ for HMA functional overlay. Thus, the optimal treatment schedule for the pavement over the 30-year period is: 1st year: HMA structural overlay; 9th year: HMA structural overlay; 18th year: HMA structural overlay; and 24th year: HMA functional overlay.

(a) For a given facility, how much to spend in each of several years or programming periods?

The optimal funding allocation may be obtained by solving the following linear programming equation:

$$\text{Maximize} \quad U = \sum_{t}^{p} (E_t - x_t)$$

$$\text{Subject to} \quad \sum_{t=1}^{p} x_t \leq B \qquad x_t \geq 0$$

The rest is left as an exercise for the reader.

The owner or operator of a civil engineering system may recognize the immense potential benefits of an effective decision-support process for rehabilitating and maintaining the system but may be unable to implement the process for a number of reasons. In the following sections, we examine some of these possible threats and barriers.

25.5 IMPEDIMENTS TO THE PRESCRIPTION OF PRESERVATION TREATMENTS OR AND SCHEDULES

25.5.1 Lack of Established Trigger Values

A trigger value is the threshold or minimum condition level at which some specific preservation treatment must be applied. These are often specific to each defect type or may be in terms of an index representing multiple defect types. Also, they are often published in system preservation manuals. The lack of established trigger values can impede the selection of the appropriate preservation treatment at a given time or the selection of the appropriate condition-based preservation schedule over the system's remaining life. For instance, at what level of corrosion should sand blasting and antirust coating be applied? At what level of patch deterioration should a bridge deck be replaced? At which level of retroreflectivity should a road sign be replaced? The difficulty in establishing objective answers to these questions probably explains why many system owners continue to use age-based timing, rather than condition-based timing, for scheduling their preservation treatments.

25.5.2 Lack of Current Data on System Condition

Successful application of established preservation schedules hinges on the availability of up-to-date data on the system physical condition. If such data are available, then the appropriate preservation treatment can be recommended for application as soon as the existing system condition reaches the threshold. On the other hand, if such data are unavailable or unreliable, the preservation treatments will likely be applied long before they are needed (leading to a waste of funds) or long after they are needed (leading to poor system condition and subsequently, high user costs of delay and inconvenience).

25.5.3 Inability to Match (or Undocumented Matching of) Defects/Distresses and Treatments

For certain systems, the maintenance engineer may seek to use disaggregate measures for designing condition-based preservation schedules. This can only be done if the deterioration curves for each distress type have been established and at least one preservation treatment has been identified to address each distress type when it reaches its threshold. At many agencies, such matching information exists in facility maintenance manuals. If such information does not exist, then it is difficult to establish a detailed and specific long-term preservation schedule.

25.5.4 Lack of Posttreatment Deterioration Functions

In several cases, agencies have only a general deterioration curve for their systems. It must be realized, however, that the pattern of deterioration of a system or component is different after it receives a treatment. Also, the post-treatment deterioration patterns are different for each treatment type. Without treatment-specific deterioration models and seperate models for preapplication and postapplication for each treatment type, it is difficult to identify which treatments are optimal in the long term or to establish life-cycle schedules for rehabilitation and maintenance of the engineering system.

25.6 CONCLUDING REMARKS

25.6.1 Communicating the Benefits of Maintenance

The rehabilitation and maintenance of civil engineering infrastructure continue to play a vital role in the life-cycle management of these facilities. Unfortunately, in many jurisdictions, maintenance is not given the recognition it deserves, for a variety of reasons including lack of funding, lack of political will, and lack of political glamour associated with maintenance projects (compared to new construction). As such, engineers working at this phase of system development need to communicate regularly to key stakeholders about the benefits of maintenance. To do this, engineers must be equipped with the analytical tools needed to develop cost-effective maintenance strategies and treatment timings and to present the results of scenario analysis (specifically, the consequences of delayed maintenance) in an effective manner to the stakeholders. The stakeholders, include the general public, legislators and other government nongovernmental development agencies that release funds for the preservation of civil engineering systems. The stakeholders are generally more willing to invest in preservation if the maintenance engineer can quantify and demonstrate clearly the benefits of preservation, specifically showing that appropriate levels of maintenance can lead to increased system longevity and much lower overall costs over the system life.

25.6.2 Importance of Feedback to Other Phases

Feedback is essential in the maintenance management of civil engineering systems. The engineer must provide feedback to the system designer regarding the extent to which different designs yield reduced frequency and/or intensity of subsequent maintenance and rehabilitation over the system life. For example, the use of stainless steel in the reinforcement of bridge decks or is other structural steel elements may lead to increased life and lower frequency of rehabilitation and replacement of the system and its components.

25.6.3 Need for Appropriate Prescriptions Based on Good Data and Analytical Tools

In order to keep their practice updated and relevant, maintenance engineers must document common distress types that occur on their systems, establish candidate treatments for each distress type, standardize their maintenance treatment terminology, and maintain a database that contains data on the system components: age, material type, loading, climatic exposure, and history of maintenance. Also, they must be equipped with requisite analytical tools that help them to identify the best treatment for a distressed system component at a given time under a given set of circumstances related to the physical condition of the system or its components and the environment, assess the benefits and costs of treatments, and develop life-cycle schedules for preserving their assets. Three categories of models that are most critical for system preservation management are *deterioration models* (often referred to as condition curves) that predict condition as a function of loading, climate, and other factors; *cost models* (that predict the cost of a treatment as a function of the factors of the treatment and the system); and decision matrices or *decision models* (optimization models or discrete statistical (empirical) models that identify the best treatment under a given set of existing conditions). Often, a robust maintenance research program established by the system owner is needed to address these issues adequately.

SUMMARY

The duties carried out by maintenance managers of the different civil engineering system types are quite similar in context even though the system types are very different in form and function. This chapter presented the good practices in civil systems maintenance management and discussed two levels of management at which maintenance decisions could be made. The decision making mechanisms and basic maintenance terminologies are then discussed. We also learned about the duties of the maintenance engineers, that involves data management, distress/defect identification, factors that affect system condition and longevity, system condition and longevity modeling, treatment thresholds for system condition, and preservation treatment(s) matchups for each distress/defect. The chapter also discussed how the impacts of alternative treatments or long-term schedules could be evaluated in terms of their costs and effectiveness, and therefore showed how to evaluate options for preserving a given system over its full life or remaining life. The chapter discussed the impediments to effective prescriptions of preservation treatment or schedule, and stressed the importance of communicating the benefits of maintenance, feeding back maintenance lessons to engineers at the preceding phases, and making appropriate prescriptions on the basis of quality data and the requisite analytical tools.

EXERCISES

1. Discuss why civil system maintenance is a critical issue in your country, state, or province.
2. Efforts by a systems owner or operator to maintain the system must be guided by a set of good practices or principles. List and discuss any three of these principles. Your discussion should include a statement of the problems that could arise if that principle is not met.
3. Discuss the differences between maintenance decision making at a system-of-systems level and at a facility level. How do these two levels interact with each other?
4. What are the three key mechanisms used for making maintenance decisions? Prepare a table showing the pros and cons of each mechanism.
5. Identify a civil engineering system type on your campus. For this system, list at least one treatment type in the following categories: time-based preventive maintenance, condition-based preventive maintenance, corrective maintenance, and rehabilitation.
6. Refer to Figure 25.4 that shows the dimensions of systems preservation actions. For each of the three dimensions, explain the difference between the dimension levels.
7. The owner of a major steel bridge in Madrid has recently awarded a contract for rehabilitation of the bridge. List the factors that could affect the success of this project. These should include the factors related to the bridge, the treatment, the environment, the contractor, and the bridge owner.
8. Explain why the establishment of a database is critical in the maintenance management of civil engineering systems.
9. What are some of the common distresses that could occur on any three of the following civil systems: bridges, dams, buildings, water tanks, levees, sewer pipes, highway pavements, and windmills? For each distress, identify a possible treatment. Tabulate your answers.
10. Identify the three categories of models that are most critical for system preservation management. Discuss the importance of each category.
11. The condition of a certain engineering system ranges from 0 (failed) to 10 (excellent or as-new). The owner seeks the condition level at which rehabilitation of the system must be carried out. Explain why both liberal and conservative thresholds are not cost-effective and describe how you would establish the optimal threshold for that system.
12. The operator of a large steel water reservoir serving a large suburb in Kano has just completed a major maintenance of that system. Describe how you would assess the effectiveness of the treatment in (a) the short term and (b) the long term.
13. Carry out a literature review to identify the key challenges and issues that are currently associated with the rehabilitation of aging infrastructure of any one of the following types of civil infrastructure: (a) water systems, (b) highway systems, (c) sewer systems, or (d) marine systems.
14. Out of every 100 bridge struts, it is observed that 2 suffer from premature corrosion, over a certain length of time. What is the probability that exactly 20 out of 200 struts of a given bridge would suffer premature corrosion within the monitoring period? Use (a) the binomial distribution and then (b) the standard normal approximation to the binomial to solve this problem. (Use standard tables.)

15. Table 25.12 presents the performance [present serviceability rating (PSR), which ranges from 0 (poor) to 5 (excellent)] for two different treatments (thin hot-mix asphalt and microsurfacing) each of which was applied to two different pavement sections. Assume that the two pavement sections are similar in all other respects (underlying material type, traffic, climatic conditions, etc.).

Table 25.12 Data for Exercise 15

Year	2001	2002	2003	2004	2005	2006	2007	2008	2009	2010	2009
Thin HMA Overlay	4	3.8	3.5	3	4.45	4.25	4	3.65	3	2.5	2
Microsurfacing	4	3.8	3.5	3	3.95	3.85	3.7	3.5	3.3	3	2.5

a. Plot a time series trend of the posttreatment performance of each treatment on the same graph.

b. Develop a simple posttreatment performance model for each treatment.

c. Using the posttreatment performance models, determine, for each treatment, the values of the following measures of long-term effectiveness: (i) service life, (ii) increase in average condition, and (iii) area bounded by under the curve. The threshold condition for the treatments is 3.0 PSR. Tabulate your results.

d. Which treatment, in your opinion, is more effective? Explain.

e. The performance models you developed use only the treatment age as the explanatory variable. Which other explanatory variables could have enhanced the models?

16. In a bid to arrest accelerated deterioration, the owner of a large network of steel railway bridges considering the adoption of Ultrex, a patented coating material, for spray application on all 3520 steel bridges in the state. You have been consulted to design a statistical experiment to ascertain the effectiveness of this treatment.

a. How would you design the sampling process for the experiment?

b. What are some of the precautions you would take to ensure that you have a good sample and also to minimize possible bias in the study conclusions?

17. As the chief engineer in charge of a proposed major civil engineering structure in your country's capital, you have a fiduciary responsibility to optimize the use of taxpayer dollars in maintaining the system. You have been tasked with developing a maintenance program for the new system from now (year 0) until the end of its service life in 75 years. For each year of this horizon period, you seek to determine whether or not to undertake some repair activity. There are three categories of repair that you could undertake: major maintenance, minor maintenance, or simple routine maintenance. Every year, you need to apply only one of these repair categories. Each repair category has an associated benefit and cost. You seek to maximize the ratio of total benefits to total costs over the life cycle. The overall maintenance budget over the life cycle is $\$C$. Other constraints are that the total benefit must be at least a certain threshold, R units. (a) What kind of knapsack problem is exemplified by this problem? Give reason for your answer. (b) Using suitable decision variables, write a simple but complete mathematical formulation for this optimization problem.

REFERENCES

Faber, M., Koehler, J., and Nishijma, K. (2011). *Applications of Statistics and Probability in Civil Engineering*, CRC Press, Boca Raton, FL.

FHWA (2011). *Bridge Preservation Guide*. Publ. Nr. HIF-11042, Federal Highway Administration, Washington, DC.

Ford, K. M., Arman, M. H., Labi, S., Sinha, K. C., Thompson, P., Shirole, A., and Li, Z. (2011). *Estimating Life Expectancies* of Highway Assets, Vol. 2., NCHRP Report 713, Transp. Res. Board, Washington, DC.

Geoffroy, D. N. (1996). *Synthesis of Highway Practice 223: Cost-effective Preventive Pavement Maintenance*. Transportation Research Board, Washington, DC.

Gurney, K. (1997). *An Introduction to Neural Networks*. CRC Press, Boca Raton, FL.

Hicks, R. G., Dunn, K., and Moulthrop, J. S. (1997). Framework for Selecting Effective Preventive Maintenance Treatments for Flexible Pavements. *Transp. Res. Rec.* 1597.

Irfan, M. (2010). A Framework for Developing Optimal Pavement Life-Cycle Activity Profiles. Ph.D. Dissertation, Purdue University, W. Lafayette, IN.

Khurshid, M. B. (2010). A Framework for Establishing Optimal Performance Thresholds for Highway Asset Interventions. Ph.D. Dissertation, Purdue University, W. Lafayette, IN.

Labi, S., and Sinha, K. C. (2003). Measures of Short-Term Effectiveness of Highway Pavement Maintenance. *ASCE J. Transp. Eng.* 129(6).

Lytton, R. L. (1987). Concepts of Pavement Performance Prediction and Modeling. Proceedings: Volume 2, Second North American Conference on Managing Pavements, Toronto, Canada.

Mamlouk, M. S., and Zaniewski, J. P. (199E). *Pavement Preventive Maintenance, Flexible Pavement Rehabilitation and Maintenance*, American Society for Testing and Materials West Conshohocken, PA.

Markow, M. J. (1991). Life-Cycle Costs Evaluations of Effects of Pavement Maintenance. *Transport. Res. Rec.* 1276.

Markow, M. J., Madanat, S. M., and Gurenrich, D. I. (1993). Optimal Rehabilitation Times for Concrete Decks, 72nd Annual Meetting of the Transp. Res. Board, Washington, DC.

O'Brien, L. G. (1989). *NCHRP Synthesis of Highway Practice 153: Evolution and Benefits of Preventive Maintenance Strategies*. Transportation Research Board, Washington, DC.

Port Technology Group (2011). Guidelines on Strategic Maintenance for Port Structures, ASEAN-Japan Partnership, Port and Airport Research Institute (PARI), Japan.

Rabe-Hesketh, S., and Skrondal, A. (2012). *Multilevel and Longitudinal Modeling Using Stata*, Vol. II, Stata Press, College Station, TX.

Stroock, D. W. (2005). *An Introduction to Markov Processes*. Springer, Berlin.

Train, K. (2009). *Discrete Choice Methods with Simulation*, 2nd ed., Cambridge University Press. Cambridge, UK.

World Bank (2005). Infrastructure and the World Bank: A Progress Report, Infrastructure Vice-Presidency, DC2005-0015 World Bank.

USEFUL RESOURCES

Agocs, Z., Brodniansky, J., Vican, J., and Ziolko, J. (2005). *Assessment and Refarbishment of Steel Structures*, CRC Press, Boca Raton, FL.

Allen, R. T., Edwards, S. C., and Shaw, D. N. (1993). *Repair of Concrete Structures, CRC Press*, Boca Raton, FL.

Bijen, J. (2003). *Durability of Engineering Structures: Design, Repair and Maintenance*, CRC Press, Boca Raton, FL.

Bridge Preservation Guide. (2011). FHWA-HIF-11042, August, Federal Highway Administration, Washington DC.

Delatte, N. (2009). *Failure, Distress, and Repair of Concrete Structures*, CRC Press, Boca Raton, FL.

Dhir, R. K., Jones, M. R., and Zheng, L. (2005). *Repair and Renovation of Concrete Structures*, Thomans Telford, London, UK.

El-Reedy M. (2007). *Steel-Reinforced Concrete Structures: Assessment, Repair and Maintenance of Corrosion*. CRC Press, Boca Raton, FL.

Guide for the Evaluation and Repair of Unbonded Post-Tensioned Concrete Structures, (2012). International Concrete Repair Institute, Post-Tensioning Institute, Farmington Hills, MI.

Guide for Maintenance of Concrete Bridge Members. (2006). ACI 345.1R-06, American Concrete Institute, Farmington Hills, MI.

Haider, S. W., Chatti, K., Buch, N. J., Lyles, R. W., Pulipaka, A. S., and Gilliland, D. (2007). Effect of Design and Site Factors on the Long-term Performance of Flexible Pavements, *ASCE J. of Pref. of Constr. Fac.*, 21(4), 378–388.

Hollaway, L., and Teng, J. G. (2008). *Strengthening and Rehabilitation of Civil Infrastructures using Fiber-reinforced Polymer (FRP) Composites*, Woodhead Publishing, Cambridge, UK.

Khan, M. (2010). *Bridge and Highway Structure Rehabilation and Repair*, McGraw-Hill, New York.

Maintenance Manual for Roadways and Bridges. (2007). American Association of State Highway and Transportation Officials. Washington, DC.

Maslehuddin, M. (2008). *Repair Manual Research Institute*, King Fahd University of Petroleum and Minerals, Saudi Arabia.

Mays, G. C. (1990). *Durability of Concrete Struectures: Investigation, Repair, and Protection*, CRC Press, Boca Raton, FL.

Morton, J. (2012). Wind Turbine Maintenance Strategy, Buildings 02/24/2012 Issue.

Perkins, P. (1997). *Repair, Protection, and Waterproofing of Concrete Structures*, CRC Press, Boca Raton, FL.

Shaw, J. D. N., Allen, R. T. L., and Edwards, S. C. (1994). *Repair of Concrete Structures*, Chapman and Hall, London, UK.

Stormwater Facility Maintenance Manual, (2009). Clark County Public Works Department. Vancouver, WA.

USACE. (2001). *Inspection, Evaluation, and Repair of Hydraulic Steel Structures*, U.S. Army Corps of Engineers, Washington, DC.

Woodson, R. D. (2009). *Concrete Structures: Protection, Repair, and Rehabilitation*, Butterworth-Heinemann, Oxford, UK.

CHAPTER 26

SYSTEM END OF LIFE

26.0 INTRODUCTION

Engineering systems are designed to last as long as possible; but the reality is that they do not last forever, and a specification of their life span is a part of the design process. Similar to all system performance outcomes, the system life is not deterministic. Some systems have actual lives that are shorter than was intended by their designer; while others last for a few more years or even decades beyond their design lives.

In this chapter, we will discuss the different definitions of system life; specifically, the different criteria by which the life of a system could be said to have ended. We will discuss why, in the context of these different definitions, the physical or functional lives in particular are most applicable for measuring system end of life.

Recognizing that, in certain cases, the system is destroyed through a natural or man-made disaster before it reaches the end of its intended or design life, we will make a clear distinction between end-of-life agents that are intended by the systems owner and those that are unintended. For those unintended by the system owner, distinctions are made between natural and man-made agents as well as between sudden and gradual agents. We will discuss the analytical tasks that are related to system end of life. These include the prediction of the system longevity and quantifying the effects of changes in longevity factors on system longevity. We then discuss the options for the system owner when the system reaches the end of its life. Lastly, the chapter presents some estimates of the typical life spans of a few civil engineering systems.

26.1 SYSTEM LIFE DEFINITIONS

26.1.1 Different Perspectives of System Life

At what point in system life is the system owner able to assert that "full life" has been reached? This question can be answered on the basis of any one of several considerations, for example, when the system is physically destroyed, or when the system starts to experience extreme conditions of deterioration, serviceability deficiencies, or functional obsolescence. For example, the structural or functional condition may be so poor that it is not cost-effective to repair it, such as an aging, severely corroded water tank. Also, a system may be in sound structural condition but could be said to have reached its end of life when its capacity is inadequate to meet the needs of current or future users even with the aid of capacity-enhancing or load-reducing technologies and policies (e.g., an inadequate number of lanes on a highway facing growing demand or an inadequate levee height in an era of greater flooding intensities). Also, changes in economic and social needs may lead to premature end of system life: Even where the system may be structurally and functionally adequate, the system may be no longer needed at its current location due to changes in developmental patterns or priorities, land-use allocations, or regulations. Furthermore, the system owner may declare that a system needs to be replaced in order to eliminate some structural vulnerability inherent in the current design that renders it vulnerable to damage from extreme natural or man-made disasters (Thompson et al., 2012).

Clearly, it is meaningless to talk about system end of life without specifying what constitutes end of life. Lemer (1996) offered a useful definition of a system life as the time between construction and replacement due to substandard performance, technological obsolescence, regulatory changes, or changes in consumer behavior and values. Thus, prior to defining the point at which end of life is reached, the civil engineer must identify the primary reason for which the system is being replaced or retired. As we have seen from the previous paragraph, these reasons may include the need to accommodate the changing nature, patterns, or levels of user demand; mitigate user or community safety or security problems associated with the system operations; and avoid excessive maintenance or operating costs associated with its current design features, which may be outdated or evolutions have occurred in development patterns that eliminate the need for the system. Therefore, it is useful to recognize that the longevity of a civil engineering system can be viewed from a number of perspectives (Kirk and Dell'isola, 1995; Lemer, 1996; Thompson et al., 2012). We discuss these below.

The **functional life** is the period during which a system exhibits no functional deficiencies that could impair the system's operational functions, such as serious deterioration, congestion arising from excess demand over capacity, or changes in design standards or institutional requirements associated with system use. Examples of systems that have reached the end of their functional life include a water tower that cannot meet the needs of a growing population, a levee built to withstand flooding levels that were lower than are currently experienced, and foundations that experience excessive (but nonthreatening) settlement.

Often, the functional life can be extended by actions that address the deficiencies. There are also a few ancient civil engineering systems that still exist today and continue to perform their originally intended function even if only modestly, such as parts of the Apian Highway in Rome (see Figure 1.6 in Chapter 1).

The **physical life** is the period when a system's physical structure is intact whether or not the system performs its intended function. Examples include a no-longer-used water reservoir or a historic highway bridge that currently carries only pedestrians. Certain engineers consider the life of a civil system as its physical life; that position certainly deserves merit, however, a counterargument could be that a physical structure that is still standing but serves no functional purpose cannot be said to be "living" in the true sense of the word. There are remains of ancient structures that still stand today but serve no operational function besides being tourist attractions—for these, the functional life has ended even though the physical life has not.

The **service life** of a system is closely related to the **functional life**; and in many cases, these terms are used synonymously. There is a dichotomy to the interpretation of service life. On one hand, as long as the system is providing service to its users, irrespective of the functional deficiencies it may be experiencing, its service life has not expired. On the other hand, however, service can be defined in terms of a specific performance measure or criterion; and when the system's performance falls below the established threshold, the service life is considered to have ended even though the system is still operating. For example, in highway pavement systems, there can be different service lives on the basis of different end-of-life criteria, such as surface roughness, texture, and cracking.

The **economic life** of a system is the period when the system owner considers it economically feasible to keep operating the system instead of retiring or replacing it. Thus, the economic life of a system ends when the incremental benefits of keeping the system in operation exceed the incremental costs of doing so.

Service life is always less than or equal to physical life. Functional life is always less than or equal to service life. Economic life is usually less than or equal to service life but may be greater if the system is removed or replaced prematurely for reasons other than economic.

Other definitions of system life include the **technological life** (the number of years until technological advances render the system obsolete); the **actual life** (the time taken to actually replace the system irrespective of its originally intended physical or functional life); the **design life** (a target life based on technical and economic considerations, which is developed during the system design phase on the basis of the strength and durability of the construction materials and the expected rates of material degradation from user loading/usage and environmental effects.

"Actual" and "estimated" are adjectives that could be applied to any of the asset life criteria (physical, service, functional, or economic), for example, actual physical life, estimated (expected) physical life, actual economic life, expected functional life, and so on.

26.1.2 End of Life Defined on the Basis of Physical or Functional Life

Of the different definitions of system life presented in Section 26.1.1, it is the end of the physical or functional life that often signals the system end of life, which then triggers a new system development cycle beginning with needs assessment. As we have learned in the previous section, when the system is physically destroyed through a disaster event or through accumulated deterioration that cannot be repaired cost-effectively, the system becomes ripe for replacement; at that point, the system owner reassesses the need for a new system and proceeds to the phases of planning, design, and so forth. Similarly, replacement becomes the viable option when the system is structurally deficient or functionally obsolete for various reasons, such as excess demand over capacity, violation of design standards or regulations that were established after the system was initially constructed, or changing nature or pattern of system users or usage. The system owner, at this time, reassesses the need for a new system, addressing first the issues related to the deficiency, followed by the subsequent phases necessary. In some cases, the functional life expires but the system continues to be operated until the obstacle preventing the replacement (e.g., funding limitations, opposition from community or environmental groups, institutional and legal barriers and so forth.) is removed or until the system reaches the end of its physical life.

From the perspective of the other definitions of life, the end of a system's life may occur during the system operations phase. For example, when the system is still at the operations phase, the economic life expires when it is no longer economically viable to keep the system in service; and the technological life expires when it is technologically obsolete, as we learned in the previous section.

26.1.3 Relationships between the Physical and Functional Life

Figure 26.1 (adapted from Thompson et al., 2012) illustrates some of the different relationships that could exist between the system's physical and functional lives. Point C in the figure is the year of system construction, PF is the expected or actual year of the system's physical failure, and FF is the year the system is expected to or actually does reach the end of its functional life. In Figure 26.1*a*, the system first reaches the end of its functional life but is replaced in year N, after the functional end of life but before it reaches the end of its physical life. The replacement of most civil engineering systems seems to fall into this category. In certain cases, the system owner replaces the system before the end of its functional or physical lives (Figure 26.1*b*) or at the exact year it is expected to reach the end of its functional life (Figure 26.1*c*). In situations where a system has reached the end of its physical or functional life but the system owner is unable to replace it due to limitations related

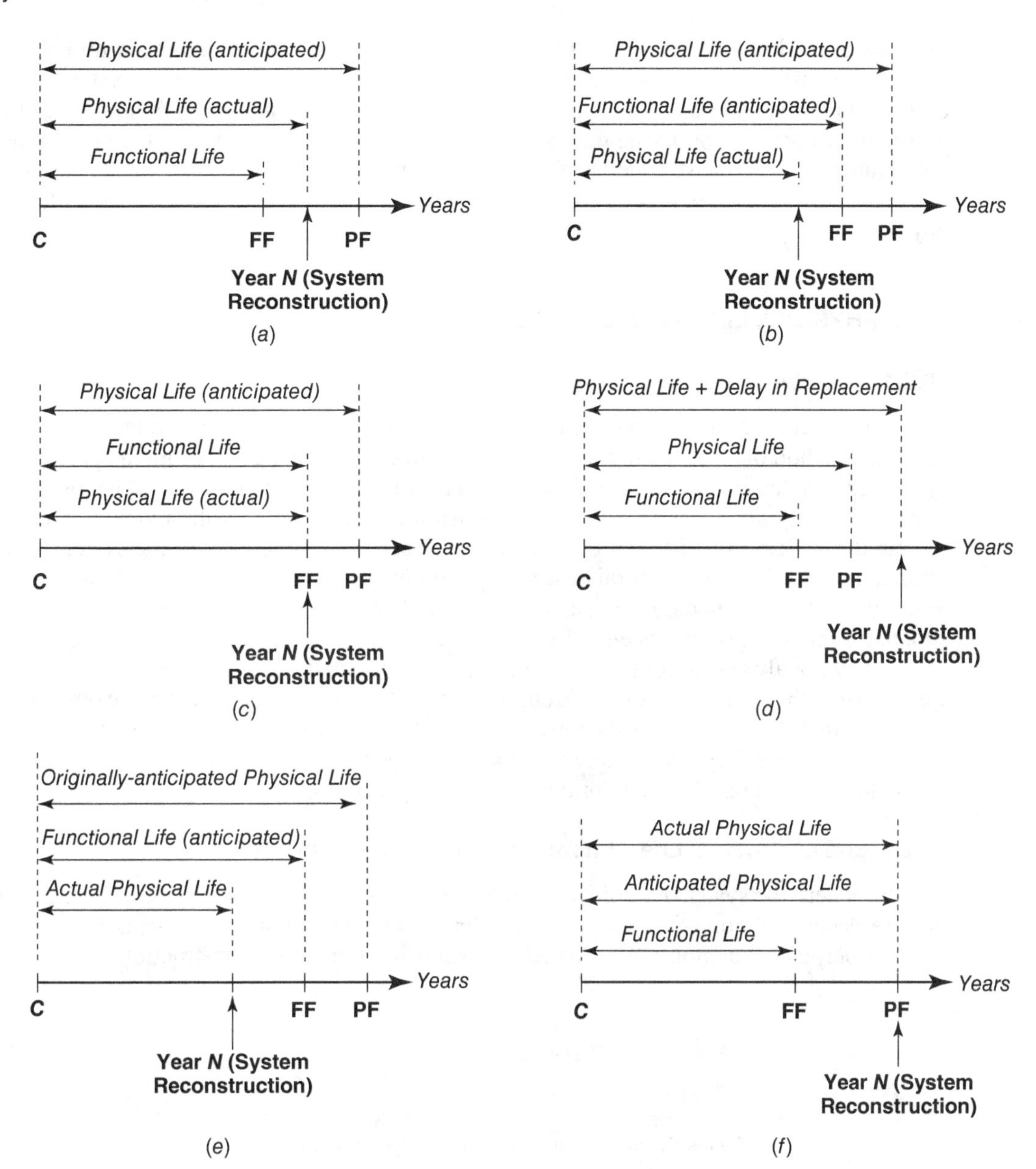

Figure 26.1 Relationships between the physical and functional lives of a system: (*a*) System reconstructed after the end of the functional life but before the end of the physical life, (*b*) system reconstructed before the end of both the functional and physical lives, (*c*) system reconstructed at the end of the functional life but before the end of the physical life, (*d*) system reconstructed after the end of both the functional and physical lives, (*e*) system reconstructed at the end of the actual physical life but before the functional life or the originally anticipated physical life, and (*f*) system reconstructed at the end of the physical life, several years after the end of the functional life.

to funding, institutional or legal issues, or other constraints, the system ends up being replaced only after considerable delay (Figure 26.1*d*). Also, there are cases where a system is physically destroyed (at year *N*) before its expected years of physical or functional failure; often this is due to design or construction flaws, natural disasters, or man-made attacks, (if the system did not fail prematurely, it would have ended its functional life at year FF and physical life at year PF). Finally, in Figure 26.1*f*, the system is replaced at the exact year that it is expected to reach the end of its physical life.

26.2 END OF PHYSICAL LIFE—THE CAUSES

26.2.1 Prelude

There are several agents that could cause a civil engineering system to reach the end of its physical life. One method of categorizing these agents is shown in Figure 26.2. From the perspective of the system owner, the system's physical life could end with or without the intent of the owner. Intended termination may arise from a desire to demolish the system (often with a view to reconstruct it), putting the system to a different use, or salvaging certain components of the system. Unintended termination may be due to natural disaster, operational error, or malicious attack by humans. If any of these agents result in damage that can be repaired and thus do not end the system's life, then the situation is not categorized as end of life. As such, in a strict sense, gradual destruction by natural agents may not always be considered as the agent of system destruction because these agents cause wear and tear that can be addressed through rehabilitation or maintenance. On the other hand, it may be argued that they can be consistently considered as end-of-life agents because the accumulated wear and tear that they cause, notwithstanding temporary fixes through maintenance, eventually facilitate the reaching of the end of the system's physical life.

26.2.2 End of Physical Life—Intended by the System Owner

In cases where the system end of life is intended by the system owner, the end-of-life mechanism may be deconstruction of the system (in order to salvage certain components for reconstruction or other purposes); demolition followed by reconstruction (or no construction if the system is no

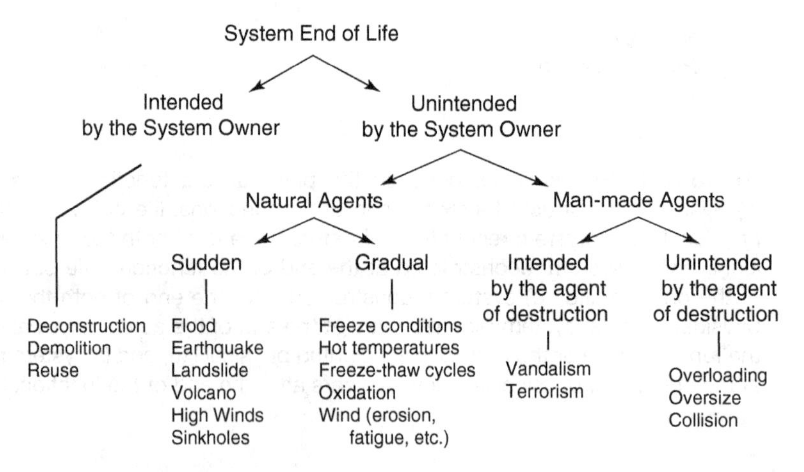

Figure 26.2 End of system physical life—mechanisms and agents.

longer needed at that location); or decommissioning and reuse at the same or different location, for example, turning an urban highway bridge into a recreational area or a pedestrian bridge that may also serve as an historic bridge.

26.2.3 End of Physical Life—Unintended by the System Owner

Unintended termination may be due to natural or man-made forces. Natural forces may be sudden (system destruction due to earthquakes, floods, and volcanoes) or gradual (oxidation, corrosion, or weathering of the system constituent material due to temperature extremes and temperature changes). Man-made forces may be deliberate (as in acts of vandalism or terrorist attacks) or not intended by the destruction agent (such as overloads or collisions). In Section 27.2, we will present a discussion of the various types of threats that could cause the end of the physical life of a civil engineering system. Therefore, in this chapter, we provide only a cursory description of these threats. Figure 26.3 shows examples of intended and unintended ends of life.

(a) Sudden Natural Agents. These are agents that may be building up over time but manifest themselves within a relatively short period of time, often with very few warning signs, thus precluding adequate preparation to protect the system. Let us first discuss **flooding**. Civil engineering systems that are located near water bodies (e.g., rivers, lakes, or the sea) are particularly vulnerable to inundation when the level of the water rises for reasons such as earthquakes, hurricanes, dam breaches, and the like. Often, owners of civil infrastructure in such areas take precautionary measures to protect their systems in the event of a flood. A cost–benefit ratio (see economic analysis in Chapter 8) can be carried out to determine the economic efficiency of different flood protection options. **Earthquakes** are another common threat to civil engineering systems. The rupture of geological faults, volcanic activity, or nuclear tests cause ground movement due to the transfer of seismic energy from deep harder soils to superficial softer soils. When these ground movements are severe, they cause significant ground acceleration and rupture, both of which constitute serious threats to the stability of civil infrastructure. The vulnerability of a structure to earthquake-induced failure can be assessed in terms of the expected seismic acceleration and soil profile which influence the degree of seismic wave amplification through the soil. Of the natural agents, earthquakes have a dubious honor of being the "mother" of most threats due to their incubatory nature: when they occur, they tend to trigger other natural agents of destruction (floods, landslides, and volcanoes). In mountainous regions of the world, **landslides** continue to pose a threat to civil engineering infrastructure. Geological phenomena that include slope failure and rock falls, particularly at hilly slopes, occur due to gravity but are triggered or facilitated by slope instability and earthquakes. Landslides can destroy systems located at or near slopes of mountains. Further, in most areas, **erosion**, **scour**, and **sedimentation** continue to damage civil infrastructure. The forces of water and wind erode natural or man-made ground and cause deposition of eroded material at areas where they are not desired. Like scouring of civil engineering foundations, erosion and silting could be considered sudden in some cases but gradual in others. Civil engineering structures that are exposed to surface waters are particularly vulnerable to erosion and scour.

(b) Gradual Natural Agents. It is easy to overlook or take for granted the deleterious actions of certain natural agents that are gradual but that surreptitiously cause civil infrastructure damage and, ultimately, destruction. These natural agents include extreme temperatures (which alter, often permanently, the physical and chemical composition of the structure material), freeze–thaw transitions (which cause destructive forces within the structure of constituent materials due to alternating cycles of expansion and contraction), oxidation (which permanently impairs the desired mechanical properties of certain materials), and wind. Wind speed, wind impulses, and variations

Figure 26.3 Illustrations of system end-of-life mechanisms: (*a*) intended—demolition of grain elevator, Minneapolis, MN, (*b*) unintended, internal flaw, I-35W Mississippi River Bridge, Minneapolis, MN, (*c*) unintended, natural—destroyed buildings after the 1906 San Francisco earthquake, (*d*) unintended, man-made—I-5 bridge destroyed by collision with oversize truck, near Seattle, WA, (*e*) intended—Tower building demolition using mechanical equipment, and (*f*) intended—spared from demolition, Singapore's historic Cavenagh bridge is being reused as a pedestrian bridge [(a) Wikimedia Commons, (*b*) Kevin Rofidal, United States Coast Guard, (*c*) Chadwick, H. D. (U.S. Gov. War Department. Office of the Chief Signal Officer), (*d*) Wikimedia Commons, (*e*) Michal Maňas/Wikimedia Commons, and (*f*) Sengkang/Wikimedia Commons]

in wind profile can cause fatigue failure in structures, including small vertically cantilevered facilities. Also, the stability of the foundations of existing structures can be threatened gradually by cyclical variations or long-term increases in subsoil pore water pressures due to rises in sea level or local groundwater level. System wear, even if gradual, needs attention because it begets further wear, a process that feeds on itself and is the reason behind the sharp increase in deterioration rates in the mid-to-late years of typical system deterioration curves.

(c) Man-made Agents. Man-made system destruction could be intended by the agent of the destruction (vandalism or terrorism) or unintended by the agent (overloading, oversize, or collision). Also, for existing structures located in the vicinity of a large structure that is being demolished, the stability of their foundations can be threatened by changes in subsoil pore water pressures due to the resulting sudden removal of surcharge arising from the removal of the building.

26.3 APPLICATIONS OF LIFE ESTIMATES IN SYSTEMS MANAGEMENT

For a variety of reasons related to system management functions, every system owner seeks knowledge about how long their system is expected to last, as shown in Figure 26.4 (adapted from Ford et al., 2012). One of these functions is to plan ahead for the appropriate end-of-life option. As we discussed in an earlier section, one option is to demolish a structurally or functionally inadequate system and construct a new one in its place. If the system owner has knowledge of when the functional or physical life ends, the system owner is in a better position to plan for the system replacement by setting aside funds for that activity. If the owner possesses multiple systems at different locations, a year-by-year replacement schedule is prepared to indicate which systems will need replacement at which year; also for each year, the total costs of the replacement of all deserving systems are summed up to yield the replacement funding needs for that year. This way, the system owner can assess the physical and fiscal needs for an entire system of systems.

System longevity models can also be used to estimate the expected life of a system corresponding to different maintenance treatments or strategies (long-term schedules) and thereby help to establish optimal replacement intervals, specify the timing, frequency, and scope of maintenance, and to compare alternative designs or materials. Estimates of system life are also useful in the assessments of economic feasibility of proposed systems and for determining the remaining life

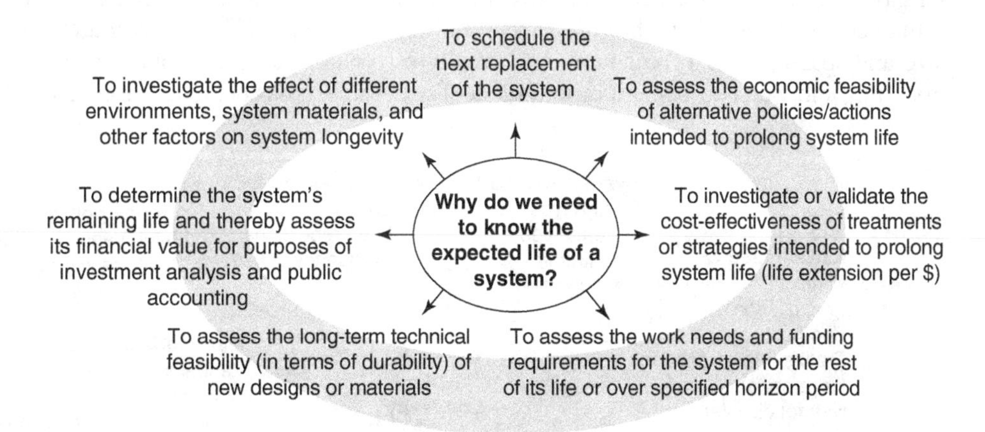

Figure 26.4 Uses of system life estimates in systems management.

of an existing system (and, therefore, its asset value, in dollars, for purposes of investment analysis and public accounting). Further, for purposes of comparative analysis, it is beneficial to know the differences in the lives of systems that are similar in all respects but their locations or environments, materials, design, or some other attribute.

26.4 PREDICTING THE LONGEVITY OF CIVIL ENGINEERING SYSTEMS

System life prediction models estimate (i) the number of years or accumulated usage for the system to reach the physical or functional end of life or (ii) the probability of that end of life occurring at any given year, as a function of the factors shown in Figure 26.5. As we learned in Section 26.3, knowledge of the relationship between system longevity and the longevity factors (i.e., attributes of the system, environment, usage, and maintenance efforts) as discussed previously in Section 26.3 can help the system owner ascertain how much extra life could be earned by changes in material type, design, maintenance amounts, usage levels, or operating polices. In this section, we shall discuss the factors that generally affect system longevity, techniques for estimating system longevity, including issues of data truncation, and a few basic computations involving system longevity.

26.4.1 Identifying the Factors That Affect System Longevity

One of the analytical tasks that face the owner of a civil engineering system is to identify the factors that influence the system life and the strengths of such influences. This question could be addressed using statistical models (see Chapter 7) that predict the system life as a function of its attributes (Figure 26.5). Often, the influential factors are those related to the system (dominant material type, design, configuration, orientation); its natural environment (freeze index, freeze–thaw cycles, wind strength, soil acidity); user characteristics (annual demand, average loading, user characteristics); agency policy (timeliness, frequency, intensity of maintenance); and man-made factors (vandalism/terrorism, collisions, errors in design or construction). The relative strengths of the influence of these factors depend on the type of system in question.

We will now discuss a few examples of the factors that specifically affect the longevity of three specific types of systems: bridges, culverts, and pavements. The longevity of bridge structures can be affected by the construction technique and geometry (number of spans, skewness, and span length). For all reinforced concrete structures, one of the main causes of end of life, corrosion, is influenced by the ambient chloride concentration, diffusion coefficient, average depth of bar cover, size and spacing of reinforcement, concrete type, type of curing, amount of air entrainment, carbonation, and water-to-cement ratio. For steel structures, it has been determined that the influential

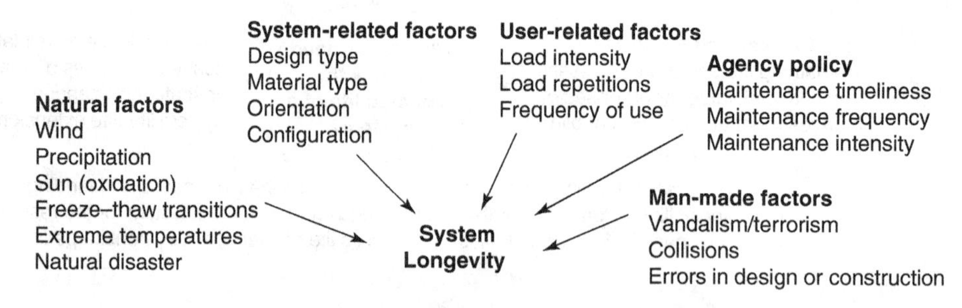

Figure 26.5 Factors that affect system longevity.

factors of longevity include loading, fatigue durability, and temperature. Research has shown that the significant factors of culvert longevity include culvert material type, backfill material type, protection coatings, pH values of the flowing water and backfill soil, pipe flow conditions, chloride content, and electrical resistivity of the backfill soil. Airport and highway pavement longevity factors include the quality and thickness of the pavement base, the surface material type (flexible, rigid, and composite) and thickness, construction quality, traffic loading, subgrade moisture conditions, and availability of subdrains. For asphaltic pavements, the longevity factors include the temperature gradient in the asphalt, the mixture properties, the aggregate quality and characteristics, the level of bonding, the layer properties, and the degree of compaction.

26.4.2 Methods, Approaches, and Techniques for Estimating the Life Expectancy of Civil Systems

Empirical (statistical evidence-based) and mechanistic (physical process-based) methods dominate the literature of life expectancy estimation. A brief review of these methods is provided below.

(a) Mechanistic Methods for Estimating Life Expectancy. Mechanistic methods generally involve the use of basic theory and data from laboratory or field observations to quantify a physical or chemical process or property that is related to the facility deterioration, such as corrosion. As we discuss below, different methods have been used for systems constructed of different dominant materials mostly, concrete and steel.

Approaches for Concrete Structures. For concrete structures, mechanistic approaches and applications have been used extensively to predict the life of civil engineering structures or their components. These are mainly physical–mathematical models that predict the longevity of a reinforced concrete structure on the basis of factors related to the concrete, the reinforcement, and the natural or man-made environment. These factors include the load-carrying capacity, deformation rate, and permeability; the breaking time of the bond between the concrete and its reinforcing steel; the temperature of the surrounding medium; the seasonal effects; and the construction quality. The key determinant of reinforced concrete life is corrosion of its reinforcement (Liang et al., 2002). Steel reinforcement corrosion is initiated when its passive layer is broken down by chloride ions, and carbonation occurs due to carbon dioxide reactions with the cement phase of the concrete (Xi et al., 2004): As the reinforcement corrodes, its volume expands, and this expansive force causes the concrete to crack and, eventually, spall. This process is exacerbated in marine environments or cold climates, where the chlorides from salt-laden air or from deicing chemicals are prevalent. The chloride travels through cracks in the concrete layer to reach the steel reinforcement, ultimately disrupting the passive surface film and establishing the conditions for corrosion, as illustrated in Figure 26.6. The corrosion occurs in three stages (Liang et al., 2002): **initiation** time—the time taken for chloride ions, transported by the alkaline hydrated cement matrix, to penetrate the concrete surface to reach the passive film surrounding the steel reinforcement; **depassivation** time—the time taken for the chloride ions to attack and destroy the passive film, leading to pitting corrosion; and **propagation** or **corrosion** time—the time taken for the corroded reinforcement to expand and cause cracking and spalling. Fick's law (Daigle et al., 2008) or Weyers technique (Sohanghpurwala, 2006) can be used to predict the duration of the corrosion stage.

Approaches for Steel Structures. In most studies and in practice, fatigue has been the principal criteria upon which the life of a steel structure is assessed. The fatigue life may be assessed using data from in-service structures or a laboratory. In the latter, the experimental setup could

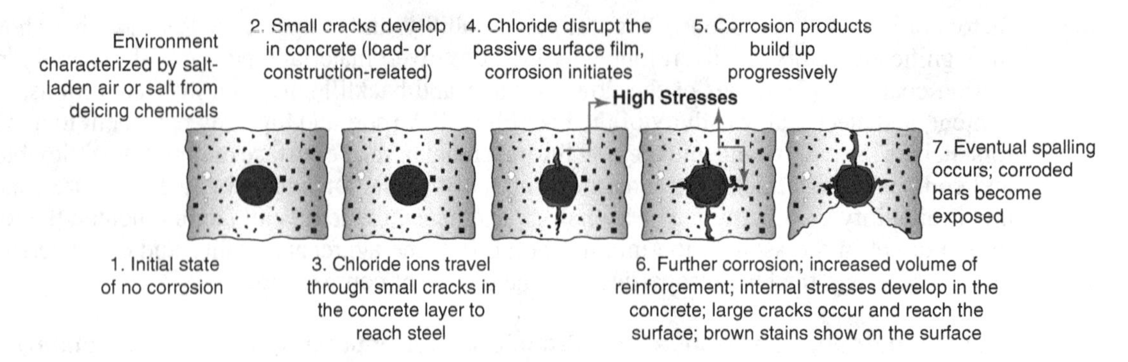

Figure 26.6 Sequence of corrosion (FDI, 2014).

be a virtual structure studied using computer simulation of the stresses and strains or an actual structure that is full scale or a smaller replica of the structure. Concepts that have been used by researchers to study fatigue life include fatigue damage theory, vibration theory, fracture mechanics, the Palmgren–Miner linear damage, Miner's hypothesis, and finite elements. Miner's hypothesis for assessing fatigue life is governed by the following equation (Tanquist, 2002):

$$\sum_{i=1}^{k} \frac{n_i}{N_i} = C$$

where N is the maximum allowable load cycles over cycle i, n is the accumulation of loads over cycle i, and C is the fraction life, at the time corresponding to n.

Approaches for General Structures. System reliability models can be used for structures with any type of dominant material. These models are inherently mechanistic-empirical. Statistical models of load and resistance are developed using data from tests. These data include material properties and structural strength properties, including material strengths in bending and shear, or live loading. An index is developed on the basis of these factors, and the life of the structure is measured as the time taken for the index to reach a certain target level.

(b) Empirical Methods for Estimating Life Expectancy. For determining the longevity of civil engineering systems using empirical methods, the two broad approaches are the condition-based approach and the age-based approach. For each of these approaches, at least one of the following specific techniques may be used: statistical regression, Markov chains, duration models, and machine learning. Some of these model techniques are discussed in detail in Chapters 6 and 25; therefore, we limit our discussion here to the longevity estimation process.

The Condition-Based Approach. The condition-based approach for life estimation is most appropriate for systems whose physical condition or functional performance is monitored and recorded on a regular basis. Using such data, it is often possible to develop a deterioration curve, and the life of the system is then taken as the number of years between the construction time and the time the condition or performance reaches (or is expected to reach) the established threshold level (Figure 26.7). Where multiple measures of condition or performance are used, multiple

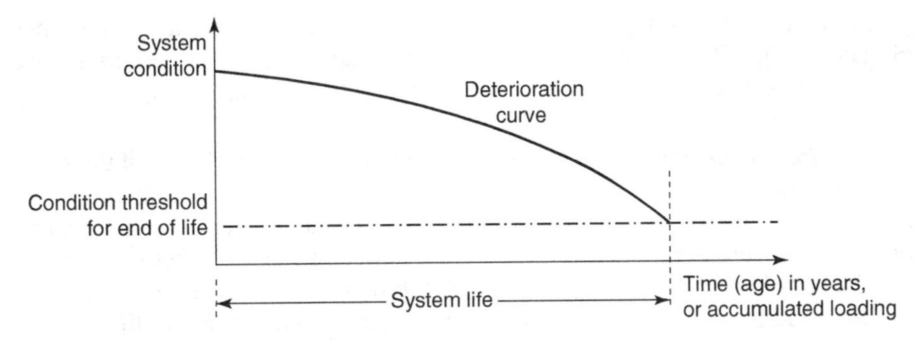

Figure 26.7 Determination of system life on the basis of system condition.

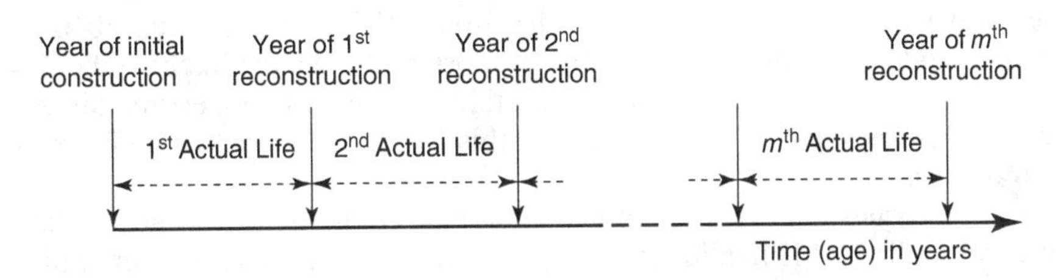

Figure 26.8 Determination of system life on the basis of system age.

deterioration curves are developed, and multiple lives can be determined; the estimated life is often taken as the minimum of these life values.

The Age-Based Approach. In the age-based approach, system condition data are not needed. Instead, historical dates for the years of the system construction and subsequent reconstruction are recorded—the difference between these two time points is the life duration of the system (Figure 26.8). For a large collection of similar systems, this data on system life durations could be collected to build a longevity model that predicts the response (system life) as a function of the attributes of the system and its environment. Also, such data could be used to build a survivor model (often referred in some literature as a reliability model) to predict the probability that the system expires given that it has not expired at a given time.

A drawback of the age-based approach is that the data may be inherently biased: some of the past observed lives for a system may be due not to the system reaching the end of its physical or functional life but for some other reason, such as a disaster event or change in socioeconomic environment that renders the system no longer needed. Thus, the recorded lives may not always be the actual lives from a physical or functional standpoint. Also, like all functions, the reliability of the models developed from recorded life data is subject to accuracy in the age data; for example, where archivists record a major rehabilitation as a reconstruction, incorrect observations of system life will be present in the data set. Generally, where a system owner lacks complete archival information that indicates the years of construction and reconstruction, it may not be possible to use the age-based approach. Finally, the application of age-based models for longevity prediction of future systems is hinged on the assumption that the future will mimic the past. However, this assumption may be unduly restrictive considering the fact that compared to their predecessors, future systems may have greater or less service life because they are likely to be built using improved materials, construction processes, and contracting approaches, or may operate in environments modified by the effects of climate change. We now discuss a number of modeling issues related to the age-based approach, namely, hazard duration modeling, censoring, and truncation.

Hazard-Based Duration Modeling of Age-Based Data. This is one of the several techniques that could be used to analyze age-based data on the longevity of individual systems. In Chapter 25, duration models were discussed in the context of system deterioration modeling for the purposes of system preservation (duration was therein defined as the time between the system construction or rehabilitation and the time of the next rehabilitation). In the current chapter, duration is defined as the period between the system construction and the system end of life.

The duration model for system longevity is a probabilistic method that predicts the conditional probability that the system's life will come to an end (or that it will not "survive") at some time t, given that its life has not ended until that time t (Washington et al., 2010). Thus, hazard-based duration models are built to account for the possibility that the likelihood of a system reaching the end of its life can change as time passes; this likelihood may become higher, lower, or constant over time. In most cases of civil systems, this likelihood increases with time due to accumulated wear and tear.

So, the question becomes: What is the pattern of change of the increasing likelihood? Is it linear or nonlinear? If nonlinear, is it quadratic, cubic, polynomial, or sigmoidal to name a few curve types? Also, is it concave or convex? The possibilities are numerous. Fortunately, for most civil systems, the patterns of such hazard follow only a few sets of well-defined trajectories.

The concept of hazard and survival are similar to those of risk and reliability as we discussed in Chapter 13. The development of a hazard function starts with the cumulative distribution function (see Chapter 6):

$$F(t) = P(T < t) \tag{26.1}$$

where F is the cumulative distribution function, T is a random variable, P is probability, and t is some specified amount of time. In the context of system longevity, Equation 26.1 represents the probability that the system life comes to an end before some transpired time t.

The density function associated with this probability distribution function, which is determined as the first derivative of the cumulative distribution with respect to time, is given by

$$f(t) = \frac{dF(t)}{dt} \tag{26.2}$$

and the hazard function is

$$h(\mathrm{t}) = \frac{f(t)}{S(t)} = \frac{f(t)}{1 - F(t)} \tag{26.3}$$

where $h(t)$ is the conditional probability that the system life will end between time t and time $t + dt$, given that such an event has not occurred up to the time t. By the definition of the word "probability," it seems clear that the function $h(t)$ represents the rate at which systems are experiencing (or the percentage of systems that experience) their end of life after they have reached a certain age t, given that they have not experienced end of life up to the time t.

The cumulative rate, or the cumulative hazard at which the systems are reaching their end of life up to or before t, is represented by $H(t)$. $H(t)$, is often referred in the literature as the "integrated" hazard function because it is obtained by integrating the hazard function, between time = 0 to time = t.

$$H(t) = \int_0^t h(t)\, dt \tag{26.4}$$

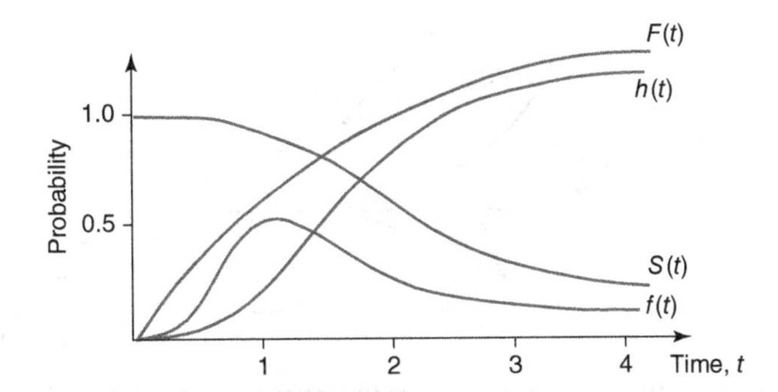

Figure 26.9 Illustration of density, cumulative distribution, hazard, survivor, and integrated hazard functions.

A concept closely related to the hazard function is the survivor function:

$$S(t) = P(T \geq t) = 1 - H(t) \tag{26.5}$$

The relationships between the density, cumulative distribution, hazard, survivor, and integrated hazard functions are such that if any one of these is known, the others can be easily obtained. The relationships, illustrated in Figure 26.9, are (Washington et al., 2010):

$$S(t) = 1 - F(t) = 1 - \int_0^t f(t)\,dt = \exp[-H(t)] \tag{26.6}$$

$$f(t) = \frac{d}{dt}F(t) = h(t)\exp[-H(t)] = -\frac{d}{dt}S(t) \tag{26.7}$$

$$H(t) = \int_0^t h(t)\,dt = -LN[S(t)] \tag{26.8}$$

$$h(t) = \frac{f(t)}{S(t)} = \frac{f(t)}{1 - F(t)} = \frac{d}{dt}[H(t)] \tag{26.9}$$

where: $f(t)$ is the density function, $F(t)$ is the cumulative distribution function, $h(t)$ is the hazard function, $H(t)$ is the integrated hazard function, $S(t)$ is the survivor function.

The gradient of the hazard function (the first derivative with respect to time) can be used to explain or describe why some system types have a propensity toward certain end-of-life trajectories. The slope is a statement of the *duration dependence*, that is, the relationship between the probability that the system experiences its end of life and the extent of life it has lived to date. Consider, for example, a system that has gone some time without experiencing its end of life, and has a certain probability of experiencing end of life.

Figure 26.10 presents four possible hazard functions for this situation. $h_a(t)$ has a negative slope, for all t, and is thus monotonically decreasing in duration; this means that the longer the system goes without experiencing end of life, the less likely that it will experience it soon. $hb(t)$ is nonmonotonic and has positive or negative slope depending on the length of time t that it has gone without an end of life. The next duration function is negative for all t and is monotonically increasing in duration. This means that the longer the system goes without experiencing end of life, the more likely that it will experience end of life soon. The fourth hazard function, $hd(t)$ has a zero slope, thus the end-of-life probability is independent of duration; hence no duration dependence exists for this category of systems (Hensher and Mannering, 2004).

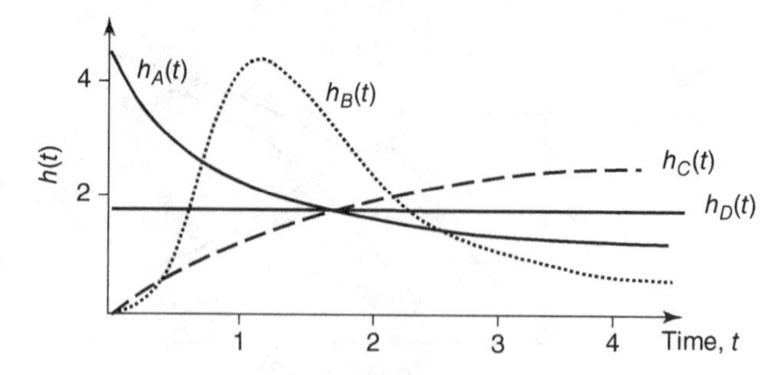

Figure 26.10 Examples of hazard functions (Adapted from Washington et al., 2010).

Table 26.1 presents the different representations of duration models, and Figure 26.11 illustrates a simple survivor model. Table 26.2 presents the survivor, hazard, and integrated hazard functions for two functional forms of the longevity (survival time) probability distributions: exponential and Weibull.

Censoring of Age-based Data. This is a common missing-data problem in duration and survival analysis. Ideally, both the construction year and the end-of-life (and hence, possibly reconstruction) year of a system are known, in which case the longevity of the system is known. At this point, there

Table 26.1 Relationships between Density, Cumulative Distribution, Hazard, and Survivor Functions

Representation	Relationship	Interpretation
Density function	$f(t) = \dfrac{dF(t)}{dt}$	Area under the curve represents the probability that the life of the civil engineering system will not end with a given duration range
Cumulative function	$F(t) = \int_0^t f(t)\,dt$	Failure probability associated with any time, t
Survival function	$S(t) = 1 - F(t)$	Probability that the life of the civil engineering system will not end beyond any point in time
Hazard function	$h(t) = \dfrac{f(t)}{S(t)}$	Failure rate, related inversely to survival

Source: Ford et al. (2012).

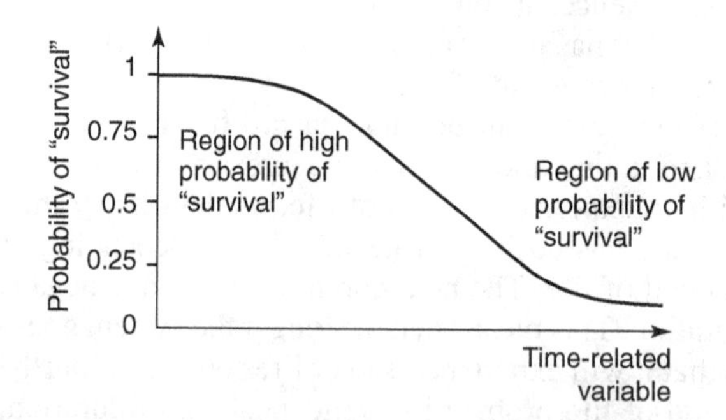

Figure 26.11 Illustration of survivor function.

Table 26.2 Survivor, Hazard, and Cumulative Hazard Functions for Exponentially and Weibull-Distributed Observations of System Longevity

	Nature of Probability Distribution of the System Longevities	
	Exponential	**Weibull**
	$f(t) = \frac{1}{\theta} e^{\left(-\frac{t}{\theta}\right)}$	$f(t) = \frac{\lambda t^{\lambda-1}}{\theta^{\lambda}} \exp\left[-\left(\frac{t}{\theta}\right)^{\lambda}\right]$
Survivor function	$S(t) = 1 - F(t) = e^{\left(-\frac{t}{\theta}\right)}$	$S(t) = \int_t^{\infty} \left(\frac{\lambda t^{\lambda-1}}{\theta^{\lambda}} \exp\left[-\left(\frac{t}{\theta}\right)^{\lambda}\right]\right) dt = \exp\left[-\left(\frac{t}{\theta}\right)^{\lambda}\right]$
Hazard function	$h(t) = \frac{f(t)}{S(t)} = \frac{\frac{1}{\theta} e^{-\frac{t}{\theta}}}{e^{-\frac{t}{\theta}}} = \frac{1}{\theta}$	$h(t) = \frac{\frac{\lambda t^{\lambda-1}}{\theta^{\lambda}} \exp\left[-\left(\frac{t}{\theta}\right)^{\lambda}\right]}{\exp\left[-\left(\frac{t}{\theta}\right)^{\lambda}\right]} = \left(\frac{\lambda}{\theta^{\lambda}}\right) t^{\lambda-1}$
Cumulative hazard function	$H(t) = \log_e(S(t)) = \log e \left(e^{\left(-\frac{t}{\theta}\right)}\right) = \frac{t}{\theta}$	$h(t) = \left(\frac{1}{\theta^{\lambda}}\right) t^{\lambda}$
Comments	When the survival times (longevities) are exponentially distributed, the hazard function is constant. This is a consequence of the memoryless property of the exponential distribution (that is, the distribution of the system's remaining survival time given that it has survived till time t does not depend on t.	When the survival times (longevities) are Weibull distributed, the hazard function depends on t. Thus, depending on whether λ is greater than or less than 1, the hazard can increase or decrease with increasing t. This is considered more realistic than the assumption of a constant hazard function (as in the exponential case). Also, note that the exponential distribution is a special case of the Weibull with $\lambda = 1$.

are two points to note. First, the end of a system's life does not necessarily mean a reconstruction will take place, so it may be problematic to use these two terms synonymously: reconstruction may take place much later than its end of life due to funding limitations or much earlier than the expected end of life due to other rationale for replacement. Second, the longevity data for a system often refers to the number of years that the system was physically standing; if the life refers to a functional life, then the life data refer to the number of years that the system was functionally adequate.

Ideally, the longevity data should be available for all systems within an agency's jurisdiction. However, this often is not the case, and the practical reality is that, for any system, the availability of life data can be characterized by one of the scenarios presented in Figure 26.12 (adapted from Washington et al., 2010). In this figure, the black dot represents the system end of life and not the system reconstruction. However, from a practical standpoint, the available longevity data mostly provide the year of reconstruction and not the year of end of life. As discussed earlier, these years could differ, particularly where the system is reconstructed for reasons other than the end of its life. Therefore, some caution is necessary in preparing the data sets for duration modeling.

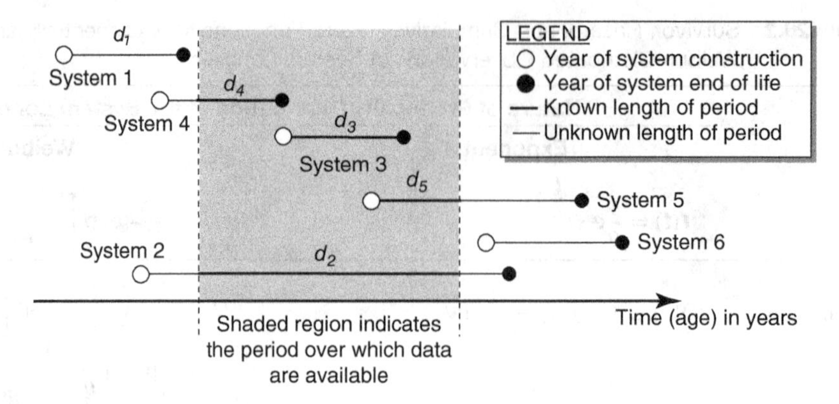

Figure 26.12 Censoring types in duration modeling.

For systems 1 and 6 in Figure 26.12, the construction dates and end-of-life dates are unknown, thus there are no life data on these systems. For system 2, the life data reflect only a fraction of the real life because the construction date and end-of-life date are not known. System 3 presents the ideal scenario, where both the construction and end-of-life dates are known. For systems 4 and 5, the construction dates and end-of-life dates are unknown. Thus, for systems 2, 4, and 5, the known life data reflect only a life that is a partial of the true life, a phenomenon that is referred to as censored data.

As seen in the figure, life data could be left censored, right censored, both left and right censored, not captured, or noncensored over the period of observation. Right censoring refers to data where the construction year is known and the end-of-life year is unknown (see d_5 in Figure 26.12); where a system's recorded longevity is known to be less than a certain specified duration, the lifetime is said to be left censored (see t_4 in the Figure 26.12); in d_2, neither the construction nor reconstruction years are known; in d_1, both years are not captured; and the systems with noncensored data are those with known years of construction and end of life (i.e., d_3). Censored data can be problematic in that they can lead to biased estimates of system longevity unless they are addressed using appropriate statistical techniques.

Truncation Issues in System Life Estimation Models. In certain cases, there are observations (systems) that are still in existence. So, while the date of construction is known at the time of analysis, the future date of end of life is not known. In this case, the observations (i.e., the life that each system has lived to date) in any longevity database are not their full lives but their truncated lives. It should be noted that truncation is different from censoring; Censoring is a characteristic of the way the data are collected as Figure 26.12 indicates, while truncation is a feature of the population under investigation (Greene, 2005). Thus, for a censored observation, the measurement exists but not known (i.e., the full life has occurred but only a part of it was measured and reported in the data), whereas for a truncated observation, the observation or measurement does not yet exist (i.e., the full life of the system, in this context).

Overcoming the Problem of Truncation. Where there are data on the trends of system condition over time, the issue of truncation may be addressed using deterioration models. In cases of truncation, the measurement (in this example, the expected full life of the system) can be extrapolated using the existing past data and the deterioration curve to yield an "extrapolated" or "simulated" full life (Figure 26.13). In that case, the extrapolated full life can be used as a normal continuous

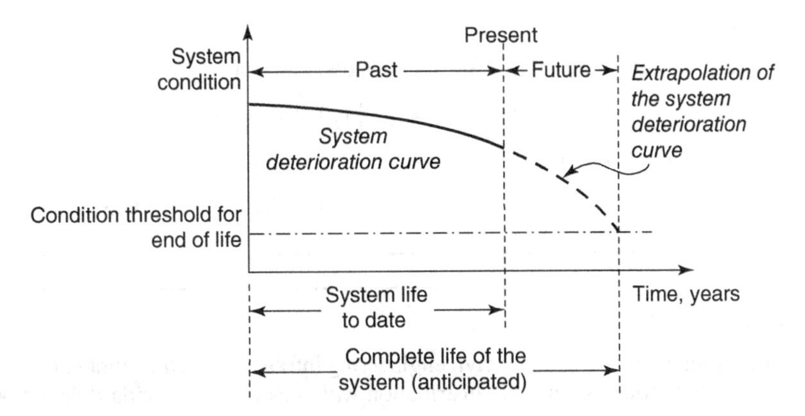

Figure 26.13 Curve extrapolation to generate simulated system life measurements, thus eliminating the need for limited response modeling.

variable without experiencing any of the statistical estimation problems associated with truncation or censoring.

26.4.3 Some General Basic Computations Involving System Life Expectancy

---

Example 26.1

The physical life of a certain type of concrete levee is described by a random variable X (years) that follows the probability density function as shown:

$$f(x) = \begin{cases} x/500 & 0 < x < 20 \text{ years} \\ 1/15 - x/750 & 20 < x < 50 \text{ years} \\ 0 & \text{otherwise} \end{cases}$$

What is the expected life of this type of levee?

Solution

The expected value is given by

$$\mu = E(X) = \int_{-\infty}^{\infty} [xf(x)]\, dx = \int_{-\infty}^{0} xf_1(x)\, dx + \int_{0}^{20} xf_2(x)\, dx + \int_{20}^{50} xf_3(x)\, dx + \int_{50}^{\infty} xf_4(x)\, dx$$

$$= 0 + \int_{0}^{20} x\frac{x}{500}dx + \int_{20}^{50} x\left(\frac{1}{15} - \frac{x}{750}\right) dx + 0 = 23 \text{ years}$$

Therefore, the expected physical life of the levee is 23.33 years.

Example 26.2

As we learned in Chapter 5, the distribution of the number of Bernoulli trials, say x, needed to elicit only one "success," is termed a geometric distribution. Ironically, in the context of this chapter, a "success" is the end of life (physical destruction) of the system. The probability that a certain engineering system fails due to a severe earth tremor is 0.40. What is the probability that six serious earth tremors will occur before it fails. Assume that each trial (earth tremor occurrence) is a Bernoulli process. Also assume that the only possible cause of failure of that system is an earthquake.

Solution

We seek the probability that x trials are needed to ensure *1* success:

$$p(X = 6) = \binom{6-1}{4-1}(0.40)^4(0.60)^{6-4} = 0.092$$

Therefore, $x = 6, k = 1$:

$$p(X = 6) = \binom{6-1}{1-1}(0.40)^1(0.60)^{6-1} = 0.031$$

Example 26.3

In a certain region with aging civil engineering infrastructure, the number of sewer pipe sections that fail every month follows a normal distribution with mean 10 and standard deviation 3. Find the probability that, at most, eight pipe sections fail in any randomly selected month:

$$p\left(\frac{x-\mu}{\sigma} \leq \frac{8-\mu}{\sigma}\right) = p\left(z < \frac{8-10}{3}\right) = p(z < -0.667)$$

Solution

We seek $p(X \leq 8)$. Standardizing this problem yields (Figure 26.14):

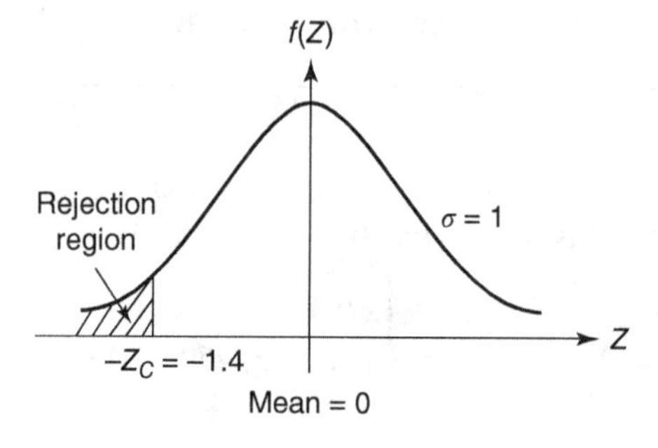

Figure 26.14

Method 1: (Using Formula)

$$p(Z < -0.667) = \int_{-\infty}^{-0.667} f(z)\,dz = \int_{-\infty}^{-0.667} \frac{e^{-0.5*z^2}}{\sqrt{2\pi}}\,dz = 0.251$$

Method 2: (Using Statistical Tables)

$$p(Z \leq -0.667) = F(-0.667) = 0.251$$

Example 26.4

The expected physical life of building cooling systems at a certain large university campus is found to be normally distributed with a mean of 34 months and a variance of 50 months. For a random sample of cooling systems from 36 buildings on this campus, determine:

(a) The level of significance if the mean of that sample is expected to fall between 32 and 36 months?

(b) The proportion of samples for which the mean of that sample can be expected to fall within the interval stated in (a)?

Solution

(a)

$$\text{Error} = Z_{\alpha/2}\left(\frac{\sigma}{\sqrt{n}}\right)$$

$$\Rightarrow \qquad \alpha = 2Z^{-1}\left(\frac{\sqrt{n}\,\text{Error}}{\sigma}\right)$$

where n is 36; error is ½(confidence interval) = ½(36 − 32) = 2; standard deviation, $\sigma = (50)^{0.5} = 7.07$.

Thus, the level of significance, α, is given by

$$\Rightarrow \quad \alpha = 2 \times Z^{-1}\left(\frac{\sqrt{36} \times 2}{7.07}\right) = 2 \times Z^{-1}(1.7) = 2 \times 0.0446 = 0.0892 = 8.92\%.$$

[Note that from Appendix 2.1 $Z^{-1}(1.7) = 0.0446$].

(b) Proportion of samples whose means fall between 32 and 36 units = probability that the mean of a randomly selected sample falls between 32 and 36 units = degree of confidence = 1 − a = 1 − 8.92% = 91.08%.

Example 26.5

As part of an inventory audit, a railway company carried out a statistical analysis of their track sections and found that the mean and variance of the remaining life of the sections are 25 and 4 years, respectively. Ten track sections in a certain jurisdiction of the inventory were subsequently chosen at random for further scrutiny. Find the 90% confidence interval for the remaining life of the track sections in that jurisdiction. Assume that the population (i.e., remaining lives of all track sections) is normally distributed.

Solution

$$\text{Error} = Z_{\alpha/2} \times \frac{\sigma}{\sqrt{n}}$$

$1 - \alpha = 0.90$ hence $\alpha = 0.1, \alpha/2 = 0.05$, and $Z_{\alpha/2} = 1.645$. Thus,

$$\text{Error} = 1.645 \times \frac{4}{\sqrt{10}} = 2.07$$

The 90% confidence interval is 25 ± 2.07 days, that is, 22.93–27.07 days.

Example 26.6

From past records, it is estimated that the physical life of a sign gantry is normally distributed with a mean of 800 days and a standard deviation of 40 days. For a random sample of 16 such systems, find the probability that the average life will be less than 775 days.

Solution

The sampling distribution of X will be approximately normal, with the mean of several sample means, $\mu_X = 800$, and the variance of several sample means $\sigma_X = 40/(16^{0.5}) = 10$. Let X be the mean of the engineer's random sample taken from the entire production population, which has a known mean μ and known variance σ^2; n is considered small (as it is less than 30), but we can still apply the central limit theorem because we are told that the population has a normal distribution.

$$p(X < 775.5) = p\left(Z < \frac{775.5 - \mu}{\sigma/\sqrt{n}}\right) = p\left(Z < \frac{775.5 - 800}{40/\sqrt{16}}\right) = p(Z < -2.45) = 0.62\%$$

Therefore, there is a 0.62% chance that the average life of a randomly selected system will be less than 775 days.

Example 26.7

A flood-prone city is considering the construction of new levees along the banks of the river. Two design are being considered. Design A has an expected life of 20 years, an initial cost of $1M per mile, annual maintenance cost of $8000 per mile, and a $20,000 per mile salvage value. Design B has an expected life of 16 years, an initial cost of $750,000 per mile, annual maintenance cost of $12,000 per mile, and salvage value of $10,000 per mile.

(a) Find the equivalent annual cost of each alternative and decide which option is more desirable. Assume a 6% interest rate.

(b) If the system longevities are not deterministic but normally distributed with means and standard deviations (20, 5) and (18, 6), use Monte Carlo simulation to make a statement about the superior alternative.

Solution

$$\begin{aligned}
\text{EUAC}_\text{A} \text{ (in thousands)} &= -1{,}000\ [(A/P, 6\%, 20)] - 8 + 20[(A/F, 6\%, 20)] \\
&= -1{,}000(0.0872) - 8 + 20(0.0272) \\
&= -\$94.656.
\end{aligned}$$

$$\begin{aligned}
\text{EUAC}_\text{B} \text{ (in thousands)} &= -750\ [(A/P, 6\%, 16)] - 12 + 10[(A/F, 6\%, 16)] \\
&= -750(0.099) - 12 + 10(0.039) \\
&= -\$85.860.
\end{aligned}$$

Thus, Design B is the more desirable alternative.

Example 26.8

A small city plans to build its own sewage treatment plant at a cost of $15M. The estimated service life follows a uniform distribution ranging from 50 to 70 years. The system will have an annual cost of operations of $5M and routine maintenance cost of $0.5M. During each replacement cycle, it will require rehabilitation costing $1M at the year and $2M in the 20th and 40th years, respectively. At the end of the replacement cycle, the reservoir will be reconstructed and the entire cycle is assumed to repeat perpetually. What is the present worth of all costs to perpetuity? Assume that the interest rate is normally distributed with mean of 5% and standard deviation 2%. Assume P' (the starting nonrecurring cost representing land acquisition and engineering services and environmental impact assessment studies is $5M).

Solution

As we learned in Chapter 11 (see Equation 11.12), the present worth of all amounts to perpetuity is given as shown in Figure 26.15:

$$PW_{R,\infty} = P' + \frac{R}{(1+i)^N} + \frac{R}{(1+i)^{2N}} + \frac{R}{(1+i)^{3N}} + \cdots = P' + \frac{R}{(1+i)^N - 1}$$

$$R = \text{compounded life-cycle cost} = 600{,}000[\text{SPCAF}(5\%, 60)] + 200{,}000[\text{SPCAF}(5\%, 40)]$$
$$+ 200{,}000[\text{SPCAF}(5\%, 20)] + 5000[\text{USCAF}(5\%, 60)] = \$14{,}914{,}087$$

$$\text{PWR}_\infty = \frac{R}{(1+i)^N - 1} = \frac{14{,}914{,}087}{(1+0.05)^{60} - 1} = \$843{,}596$$

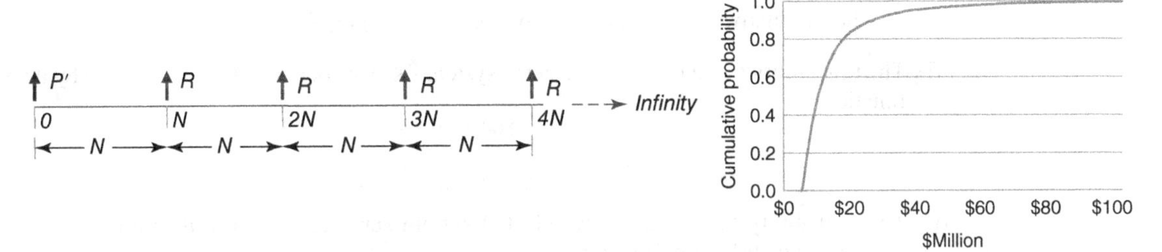

Figure 26.15 System replacement in perpetuity.

Example 26.9

A certain type of overhead water retaining structure has a mean failure time of 36 years. Assuming that the tanks have a constant probability of failure, what is the longevity reliability of these systems over a period of (a) 36 years and (b) 12 years?

Solution

$$\lambda = 1/\text{MTTF} = \frac{1}{36} = 0.028$$

Thus, R (36 years) = $R(36) = e^{-0.028(36)} = 0.368$; and R (12 years) = $R(24) = e^{-0.028(12)} = 0.717$.

Therefore, approximately 37% of the water tanks are expected to last to the mean time to failure (MTTF) (i.e., 36 years); also, 72% will last 12 years or more.

Example 26.10

The owner of a pipeline network seeks to estimate the percentage of pipeline sections that are expected to survive beyond a given number of years, so that appropriate work plans can be drawn up and budgets solicited for those that are not expected to survive as such. The owner has compiled historical records of the lives of pipe sections in the last few years. The lives were measured as the time interval between the year of pipe installation and the year of replacement. A random sample of 29 equal-length pipe sections was obtained from this population, and their individual survival durations (t) are presented below.

22.5, 24.8, 26.1, 29.9, 30.0, 35.2, 33.7, 36.3, 35.0, 39.5, 40.4, 40.9, 41.8, 42.2, 44.5, 45.4,

45.3, 45.2, 46.6, 49.6, 50.4, 52.1, 51.7, 55.5, 53.7, 56.8, 60.5, 61.2, 65.1

Assuming that the probability density function for the pipe section longevities follows an exponential distribution, (a) determine the parameters of the density function, (b) develop the following models that further describe the longevity of these pipe sections: survivor, hazard, and cumulative or integrated function, (c) use the models to find the percentage of pipe sections that are expected to survive (i) beyond 50 years, (ii) not more than 55 years.

Solution

(a) The exponential distribution has the following functional form: $f(t) = ae^{-bt}$.

Applying the given data to a standard curve fitting software, the probability density function can be determined as follows:

$f(t) = 0.148e^{-0.148t} = (1/6.74)e^{(-t/6.74)}$; Where t is the time in years.

The survivor function is: $S(t) = e^{-t/6.74}$.

The hazard function is: $h(t) = \frac{1}{6.74} = 0.148$ (recall that for exponentially-distributed density functions, the hazard function is a constant).

The cumulative or integrated function is then, $H(t) = \frac{t}{6.74}$

(i) The probability that a randomly selected system survives beyond 50 years is given by the survivor function.

$$S(t) = e^{-\frac{t}{6.74}}$$

$$S(50) = e^{-\frac{50}{6.74}} = 0.06\%$$

(ii) The probability that a randomly selected system survives not more than 55 years is given by cumulative distribution function.

$$F(t) = \int_0^t f(t)dt = \int_0^t 0.148e^{-0.148t}dt = (-1)[e^{-0.148t} - e^{-0.148\times0}] = (1 - e^{-0.148t})$$

$$F(5) = (1 - e^{0.148\times55}) = 0.999714 \text{ or } 99.97\%.$$

26.5 QUANTIFYING SYSTEM VULNERABILITY TO DESTRUCTION

Civil engineering systems are vulnerable to man-made and natural threats. As we learned in Section 26.2, natural threats include floods, earthquakes, landslides, and hurricanes; man-made threats include overloading due to excessive traffic or occupancy and accidental or malicious collisions between mobile man-made objects and the civil engineering structure. Agencies are now paying more attention to infrastructure security by continuously monitoring any imminent threats and also to the implementation of investments geared towards reducing system vulnerability. Spurred by highly publicized disaster events, such as, the 2005 hurricanes Katrina and Rita, and the 2007 I-35W Mississippi River bridge collapse, the 2012 Hurricane Sandy, and the 2013 Interstate 5 bridge collapse in Seattle, it has become more and more important to monitor the vulnerability of civil engineering systems to internal or external threats and to assess the effectiveness of interventions aimed towards reducing the vulnerability of these systems to these threats. The looming specter of global warming (IPCC, 1994) and the subsequent but gradual change in sea and groundwater levels, wind speeds, and other environmental changes are expected to have deleterious and widespread impacts on civil engineering structures (ASCE, 2007; Lenkei, 2007). Thus these constitute serious natural threats that must be addressed by civil engineers.

At the current time, however, there are few or no consistent and objective methodologies that civil engineers can use to assess, and hence constantly monitor, the vulnerability of their systems to natural and man-made threats. Thus, owners of civil engineering systems continue to seek methodologies, irrespective of the civil system type and the nature of the threat, to quantify the vulnerability of their systems and to analyze the effectiveness of investments intended to address directly or indirectly such vulnerability (Labi et al., 2011).

In Chapter 27, we will present a procedure to assess the vulnerability of a civil engineering system to externally caused destruction or damage. This rating procedure is based on the likelihood and consequence of a threat. The threat likelihood, assessed on the basis of threat type, is a function of the system environment and not of the system itself. The threat consequence, or the impact of the system destruction, is based on the possible scope of the destruction and the extent to which the system is exposed to the threat—both of these are specific to the system in question and not to the threat type. The exposure to the threat is a measure of the consequence that a system destruction will have on its users—this is related to the occupancy or usage volumes of the system and the importance of the system in the region's socioeconomic development. The system resilience is a measure of the capacity of the system to withstand the threat or to quickly recover from the effects of the threat.

26.6 END-OF-LIFE OPTIONS

When the system reaches its end of life and thus is intentionally terminated or retired by the owner, or after the system is destroyed through natural or man-made causes, a number of end-of-life options exist for the system owner. For a given system or decision context, the principles of sustainability, which we will discuss in Chapter 28, can provide useful guidelines for the system owner in deciding the most economically efficient, safe, and sustainable option for system end of life.

26.6.1 Deconstruction

Deconstruction is when the system is demolished without constructing another in its place. The system owner chooses this option when economic and social forces or changes in development patterns render it no longer needed at that location. For example, the relocation of the main employer in a town may cause mass emigration, leading to the nonuse of certain facilities; where such facilities pose a safety hazard or aesthetic blight, they may be simply torn down and not replaced. Also, from a social standpoint, an event of great tragedy at a system location (often, buildings) may cause relocation of the services offered by the system and the deliberate destruction of the structure.

26.6.2 Demolition with Reconstruction

In the last section, we discussed a situation where a civil engineering system may be no longer needed at a particular location. For a great majority of civil engineering systems, however, the practical reality is that society will need them perpetually at their existing locations, at least until there is a drastically transformational change in the way we live. For example, the adoption of flying automobiles will render most highways obsolete; even in that case; it may be argued that highways will be retained to serve as landing pads for the airborne personal vehicles in cases of emergency.

Complementing this perpetual need is the fact that engineering systems do not last forever. Thus, there is this twin situation of perpetual need by society coupled with limited life of systems. In view of this, the most appropriate and common option for system owners is to demolish and reconstruct systems that reach their end of life.

As we discussed earlier in this chapter, in certain cases, the system owner may decide to demolish and reconstruct an existing system not because it has reached the end of its design or physical life but for a number of other reasons (Thompson et al., 2012): It has reached the end of its economic life (it is no longer cost-effective to carry out repair/rehabilitation of deteriorated areas, or there are excessive maintenance costs due to the existing, often decades-old design); it has reached the end of its functional life (safety problems exist due to growing inconsistency between the system capabilities and user characteristics); or there is a need to eliminate potential vulnerability to disaster due to a recently realized design flaw.

26.6.3 Decommission and Reuse

When an engineering system is put out of service because it is no longer needed or because another (of greater capacity, strength, etc.) is being constructed nearby, the system owner may choose not to demolish it rather than put it to a different use. It is commonplace to see old schools and old factories reused as shops or offices or highway bridges reused as pedestrian bridges or recreational centers. New projects that reuse existing infrastructure can potentially enhance aesthetics, inspire citizens, and provide value in terms of open space, mobility, and development opportunities (Hellendrung, 2012). In certain cases, the decommissioned system (or parts of it) is used to enhance the quality of environment. For example, in New York City, subway transit vehicles that have reached the end of their life are stripped of any toxic components and dumped at specific locations in deep areas of the ocean to serve as artificial barrier reefs. This is important because even though barrier reefs play a vital role in the undersea ecosystem (they serve as a habitat for sea creatures), natural coral reefs are eroding very quickly as a result of global warming and human activity, including coastal construction, sewage, and fishing. Therefore, artificial barrier reef initiatives, such as that spearheaded by the Delaware Department of Natural Resources and Environmental Control, has led to the deposition of several out-of-commission transit vehicles. This initiative has created a new reef 15 miles off the state's coastline, transforming "a barren stretch of ocean floor into a bountiful oasis, carpeted in sea grasses, walled thick with blue mussels and sponges, and teeming with black sea bass and blackfish" (Chino, 2008). In Chapter 28, we will discuss the issues of sustainability in greater detail.

26.7 FEEDBACK TO PRECEDING PHASES

The end of a system's life, intended or unintended, offers significant opportunities to civil engineers involved at the various phases to learn from past mistakes and become more aware of the significance of threats to system longevity as well as opportunities to enhance longevity. Table 26.3 presents examples of the possible feedback that may be provided to engineers involved at the various phases that precede the end-of-life phase.

Table 26.3 Examples of Feedback from the End-of-Life Phase

Maintenance	Indicate which specific maintenance practices, materials, and timings helped reduce or enhance system longevity.
Inspection/ Monitoring	Make recommendations regarding the frequency, intensity, and techniques for inspection and monitoring to reliably and adequately capture true loading/usage extents and system physical conditions in order to avoid premature system failure.
Operations	Identify which specific operational policies might have led to early end of life of the system as well as those policies that had the opposite effect.
Construction	Provide advice on which construction practices enhance or reduce system longevity.
Design	Identify which design aspects (e.g., material types, dimensions, and component sizes and orientations, and maintainability) that generally seem to favor system longevity and those that do not.
Planning	Make recommendations regarding the system location, system orientation relative natural or man-made features, or other planning outputs that seems to have significant impact of system longevity.
Need Assessment	Indicate whether early failure of the system was linked to underestimation of system usage levels of loading; thus make recommendations for changes in need assessment techniques.

26.8 TYPICAL LIFE EXPECTANCIES OF COMMON CIVIL ENGINEERING SYSTEMS

The life expectancy of civil engineering structures has been found to vary by a number of factors. Some of these factors are related to the physical characteristics of the system (e.g., size, design type, material used in construction, quality of construction), operational characteristics (e.g., loading intensity and frequency, and level of abuse or misuse), maintenance policy (e.g., frequencies, intensities, and condition thresholds for maintenance or rehabilitation activities), and the operating environment (if above surface, weather and climate; if buried, soil or groundwater chemical properties, and so forth). Table 26.4 presents some typical service lives of common civil engineering systems. For each of these systems, the variability in life expectancy is due to a variety of factors including those mentioned above.

Table 26.4 Typical Service Lives of Common Civil Engineering Systems

System Type	Details	Life Estimate	Source(s)
Bridges	Reinforced concrete decks	24–48 years[a]	Estes and Frangopol (2001)
	Steel decks	37 years	Thompson et al. (2010)
	Steel rails	37 years[a]	Estes and Frangopol (2001)
	Reinforced concrete substructures	23–42 years[a]	Estes and Frangopol (2001)
	Entire structure	50–100 years	MIIC (2005); Hallberg (2005); van Noortwijk & Klatter (2004); Gion et al., 1993
Culverts	Box or pipe, General (including precast concrete)	50–70 years	Wyant (2002); Markow (2007)
	Pipe, corrugated metal	30–35 years	Meegoda et al. (2008); Markow (2007)
	Pipe, Asphalt coated corrugated metal	50 years	Markow (2007)
	Pipe, Small diameter plastic	50 years	Markow (2007)
	Pipe, High-density polyethylene	50–90 years	Perrin Jr. and Jhaveri (2004); Markow (2007)
	Box, concrete	50 years	Markow (2007)
	Box, Timber	30 years	Markow (2007)
Traffic signs		10–17 years	Immaneni et al. (2009)
Wind turbines		10–25 years	General sources
Dams		50–over 100 years[b]	General sources
Highway pavements	Concrete	30–50 years	General sources
	Asphalt	20–40 years	
Pipelines		10–30 years[c]	General sources
Buildings		20–over 100 years[d]	General sources
Parking garages		30–50 years	General sources
Levee		20–50 years	General sources
Structural steel frame tower		50–80 years	General sources

[a]For threshold NBI condition rating of 4.
[b]Generally, compacted earth dams have the shortest life spans; rockfill dams have the longest lifespans.
[c]Also depends on nature of material being transported.
[d]Also depends on building purpose (offices, residential, and so on).

26.9 CONCLUDING REMARKS

In recent years, agencies are paying increased attention to investments that reduce system vulnerability to unintended end of life. Chapter 27 presents a procedure for system agencies to quantify the level of system vulnerability to destruction at any time. The procedure quantitatively assesses the likelihood that a threat will occur, the resilience of the system to the threat, and the consequence in terms of damage and harm to the environment and population in the event that the system is destroyed by the threat. An overall hazard score is derived using this framework; Thus, a greater threat likelihood, lower system resilience, and greater exposure or consequence translate into a higher level of overall hazard.

The tool of simulation, which we learned in Chapter 8, is very useful in studies related to system end of life. Simulation can be used to analyze the likelihood of system end of life at any time or the number of years of loading cycles it will take to reach system end of life. In cases where the end of life is not intended by the system owner (e.g., system destruction due to earthquakes, floods, internal design errors, or man-made threats), simulation tools can be used to mimic end-of-life events, including "attack" processes by the end-of-life agent, the mode of system failure, the effects on neighboring structures and facilities, evacuation of the affected population, and so forth. In cases of unintended end of life, such simulation studies also can help identify areas of improvement to avoid or mitigate the disaster or to make adequate preparations to recover if it occurs. Where the end of life of a system is intended (e.g., due to system structural or functional obsolescence or the need to be demolished), simulation tools that help visualize the process and the impacts of the demolition options can help in choosing the option that minimizes the impacts on neighboring structures and the environment.

In a bid to increase the longevity of their systems, civil engineers of today are actively searching for new materials, construction techniques, construction approaches, monitoring/inspection strategies, and operational policies that are consistent with longer system lives. Promising materials include stainless steel and fiber-reinforced polymer (FRP) composites.

The concepts of system resilience and sustainability, which are discussed in Chapters 27 and 28, are vital to the discussion of system end of life for at least two reasons. First, good practices of system resilience and sustainability could increase system longevity because a system that is designed and operated and maintained in such a manner that minimizes adverse impacts to and from the environment and the sociocultural system is likely to better withstand the forces of nature and socioeconomic abandonment. Second, at the end of their lives, the components of a system can be recycled or reused; such practices enhance sustainable development.

SUMMARY

The design of any civil engineering system is accompanied by a specification of the expected life. The actual life of the system may be lower or higher than the design life due to factors including climate, loading, or sudden destruction due to disaster. When systems reach the end of their physical lives, the system owner faces end-of-life choices that include keeping the system as a historic site, deconstructing it, demolishing and replacing it with a new, larger, and/or stronger facility, decommissioning it and using it for another purpose, tearing it down and having the parts recycled or reused, or some other purpose. In certain cases, before a system reaches the end of its design life, it is destroyed through a natural or man-made disaster. In this chapter, a clear distinction was made between end-of-life agents that are intended by the system owner and those that are unintended. For the agents unintended by the system owner, distinctions were made again between those that are natural or man-made and those that are sudden or gradual. Also, in the chapter, we discussed the different definitions of system life, specifically, the different criteria by which the system life could be said to have ended, such as the physical life, functional life, service life, economic life, actual life, and design life. We then presented some computations involving system life, and listed some examples of longevity values and influencing factors for a few select types of civil engineering systems.

EXERCISES

1. A structural engineering system consists of four subcomponents. The stability of each subcomponent is statistically independent of the others. In any given hurricane event, the probability that any subcomponent is unstable is 0.006. The entire system is considered unstable if any one subcomponent is unstable. Find the probability that the entire structural system is stable (a) in any given hurricane event and (b) in five hurricane events.
2. The probability that a system recovers from a rare cold spell is 0.35. If 80 systems are known to be affected by cold spell is a certain year, what is the probability that less than 40 recover?
3. Historical records of sewer pipe failures in an old city show that during the period 2005 to 2013, there were 30 failures that were catastrophic (exploded due to the build-up of methane or other gases). Assuming that the pipe failures follow a Poisson process, find the probability that (a) 4 catastrophic failures occur within the next 6 months, (b) the interval from now till the next catastrophic pipe failure will be at most 2 months (c) The probability that no catastrophic pipe failure will occur in the next 12 months.
4. The water distribution pipe network of a certain district, consist of pipe sections made of precast concrete (70%) steel (25%), and polyvinyl chloride (PVC) (5%). Assume that the probability that a precast pipe reaches the end of its life this year is 0.16, while such probability for steel and PVC pipes are 0.01 and 0.005, respectively. What is the probability that a pipe section selected at random from this population reaches the end of its life this year?
5. Consider the problem in the previous question. (a) One day, a certain pipe section in the district experienced end of life evidenced by multiple bursts. What is the probability that it is a precast concrete pipe? (b) Given that a randomly selected pipe section is made of precast concrete, what is the probability that it will reach the end of its life this year?
6. The physical life of a certain type of concrete levee is described by a random variable X (years) that follows the probability density function as shown:

$$f(x) = \begin{cases} x/300 & 0 < x < 15 \text{ years} \\ 0.08- & 15 < x < 40 \text{ years} \\ 0 & \text{otherwise} \end{cases}$$

   What is the expected life of this type of levee?
7. The probability that a certain levee reaches the end of its physical life due to a category 5 storm (on the Saffir–Simpson hurricane scale) is 0.12. What is the probability that it fails after five category 5 storms. Assume that each trial (category 5 storm occurrence) is a Bernoulli process and thus the trials are independent of each other. From a practical standpoint, the independence-of-trials assumption might be considered unduly restrictive. Explain.
8. As part of an inventory audit, a railway company carried out a statistical analysis of their track segments and found that the mean and variance of the remaining life of their track sections are 30 and 6 years, respectively. Fifteen track sections were subsequently chosen at random for further scrutiny. Find the 90% confidence interval for the remaining life of the track sections. Assume that the population (i.e., remaining lives of all track sections) is normally distributed.
9. From past records, it is estimated that the physical life of a road sign gantry is normally distributed with a mean of 750 days and a standard deviation of 100 days. For a random sample of 16 such systems, find the probability that that the average life will be less than 765 days.

10. The owner of a power supply network has compiled historical records of the lives (time interval between construction and replacement) of steel pylons over the last few years. A random sample of 29 equal-length pipe sections was obtained from this population, and their individual survival durations (t) are presented below.

$$80.35, 79.14, 76.22, 73.37, 71.71, 67.69, 69.01, 64.13, 67.61, 62.62, 60.87, 62.69,$$
$$63.09, 54.60, 57.95, 56.00, 55.64, 53.49, 51.39, 50.75, 45.35, 49.17, 47.14, 43.96,$$
$$38.91, 38.22, 79.54, 78.59, 71.48, 70.22, 70.90, 68.29, 66.07, 62.22, 64.21, 63.45,$$
$$62.30, 61.60, 59.18, 55.62, 55.11, 58.03, 57.24, 50.67, 53.29, 46.88, 44.53, 42.41,$$
$$39.72, 37.76.$$

(a) Develop the following models to describe the longevity of these pipe sections: probability density function of their longevities, survivor function, hazard function, and the cumulative or integrated function. (b) Use the models to find the probability that a randomly selected pylon survives (i) beyond the specified pylon service life of 60 years, (ii) not more than 30 years.

REFERENCES

ASCE (2007). Impact of Global Climate Change, *ASCE Policy Statement 360*, Energy, Environment, and Water Policy Committee of ASCE, Reston, VA.

Beniston, M. (2004). *Climatic Change and its Impacts*. Kluwer, Dordrecht, The Netherlands.

Chino, M. (2008). Sunken Subway Cars to Replace Eroding Barrier Reefs, Transportation. inhabitat.com/sunken-subway-cars-form-new-reef/; accessed December 29, 2012.

Daigle, L., Cusson, D., and Lounis, Z. (2008). Extending the Service Life of High Performance Concrete Bridge Decks with Internal Curing, Procs., 8th International Conference on Durability of Concrete Structures, ISE-Shima, Japan.

Estes, A. C., and Frangopol, D. M. (2001). Bridge Lifetime System Reliability Under Multiple Limit States. *ASCE Jrnl. of Bridge Eng.*, 6 (6), 523–528.

FDI (2014). GFRB versus Black Steel, www.frbdistributors.com. Accessed January 30, 2014.

Ford, K. M., Arman, M. H. R., Labi, S., Sinha, K. C. Thompson, P. D., Shirole, A., and Li, Z. (2012). Estimating the Life Expectancy of Highway Assets. *NCHRP Rept. Volume* 2, Transp. Res. Board, Washington, DC.

Gion, L. C., Gough, J., Sinha, K. C., and Woods, R. E. (1993). *Implementation of the Indiana Bridge Management System*, Joint Transp. Res. Program, West Lafayette, IN.

Greene, W. (2005). Censored and Truncated Distributions, New York University, NY.

Hallberg, D. (2005). Development and Adaptation of a Life-cycle Management System for Constructed Work. *Licentiate Thesis*, KTH, Civil and Arch. Eng., Gavle, Sweden.

Hellendrung, J. (2012). Reusing Infrastructure. American Society of Landscape Architects (ASLA) Annual Meeting, September 28–October 1, 2012 Phoenix, AZ.

Hensher, D., and Mannering, F. (2004). Hazard-based duration models and their applications, *Transport Reviews*, 14, 63–82, Taylor & Francis, London.

Immaneni, V. P., Hummer, J. E., Harris, E. A., Yeom, C., and Rasdorf, W. J. (2009). Synthesis of Sign Deterioration Rates Across the United States. *ASCE Jrnl. of Transp. Eng.*, 135 (3), 94–103.

IPCC (1994). *Climate Change*. Cambridge University Press, Cambridge.

Kirk, S. J., and Dell'isola A. (1995). *Life Cycle Costing for Design Professionals*. McGraw-Hill, New York.

Lemer, A. C. (1996). Infrastructure Obsolescence and Design Service Life. *ASCE J. Infrastruct. Syst.* 2(4), 153–161.

Lenkei, P. (2007). Climate Change and Structural Engineering, *Period. Polytech.* 47–50.

Liang M. T., Lin, L. H., and Liang, C. H. (2002). Service Life Prediction of Existing Reinforced Concrete Bridges Exposed to Chloride Environment, *ASCE Jrnl. of Infrastruct. Sys.*, 8(3), 76–85.

Markow, M. J. (2007). Managing Selected Transportation Assets: Signals, Lighting, Signs, Pavement Markings, Culverts, and Sidewalks, *NCHRP Synthesis 371*, Transp. Res. Board, Washington, D.C.

Meegoda, J. N., Juliano, T. M., and Wadhawan, S. (2008). Estimation of the Remaining Service Life of Culverts. Procs., 87th Annual Meeting of the Transp. Res. Board, Washington, D.C.

MIIC (2005). *Infrastructure Status Report: Massachusetts Bridges*, The Engineering Center, Massachusetts Infrastructure Investment Coalition, Boston, MA.

Perrin Jr., J., and Jhaveri, C. S. (2004). The Economic Costs of Culvert Failures. Procs., 83rd Annual Meeting of the Transp. Res. Board. Washington, D.C.

Sohanghpurwala, A. A. (2006). Service Life of Corrosion Damaged Reinforced Concrete Bridge Supestructure Elements, *NCHRP Report 558*, Transp. Res. Board, Washington, DC.

Tanquist, B. A. (2002). A Quick Reliability Method for Mechanistic-Empirical Asphalt Pavement Design, Procs., 81st Annual Meeting of the Transp. Res. Board, Washington, DC.

Thompson P. D., Ford, K. M., Arman, M. H. R., Labi, S., Sinha, K. C., and Shirole, A. M. (2012). Estimating the Life Expectancies of Highway Assets. *NCHRP Report 713*, Vol. 1, Trans. Res. Board, Washington, DC.

Thompson, P. D., & Sobanjo, J. O. (2010). Estimation of Enhanced Pontis Deterioration Models in Florida. Procs., 5th Intern. Conf. on Bridge Maint., Safety, & Management, Philadelphia, PA.

van Noortwijk, J. M., and Frangopol, D. M. (2004). Deterioration and Maintenance Models for Insuring Safety of Civil Infrastructures at the Lowest Cost. (D. M. Frangopol, E. Bruhwiler, M. H. Faber, & B. Adey, Eds.) *Life-cycle Performance of Deteriorating Structures: Assessment, Design, and Management*, 384–391.

Washington, S., Mannering, F., and Karlaftis, M. (2010). Statistical and Econometric Methods for Transportation Data Analysis, CRC Press, Boca Raton, FL.

Wyant, D. C. (2002). *Assessment and Rehabilitation of Existing Culverts*, NCHRP Synthesis 303, Transp. Res. Board, Washington, D.C.

Xi, Y., Abu-Hejleh, N., Asiz, A., and Suwito, A. (2004). Performance Evaluation of Various Corrosion Protection Systems for Bridges in Colorado, Colorado Dept. of Transp., Denver, Co.

USEFUL RESOURCES

Al-Suleiman, T. I., & Shiyab, A. M. (2003). Prediction of Pavement Remaining Service Life Using Roughness Data-Case Study in Dubai. *Intern. Jrnl of Pavement Eng.*, 4(2), 121–129.

Biondini, F., Bontempi, F., Frangopol, D. M., and Malerba, P. G. (2006). Probabilistic Service Life Assessment and Maintenance Planning of Concrete Structures. *ASCE Jrnl of Struct. Eng.*, 132(5), 810–825.

Biondini, F. and Frangopol, D. (2008). *Proceedings of the International Symposium on Life-Cycle Civil Engineering, IALCCE 2008*. CRC Press, Boca Raton, FL.

Breysse, D., Domec, V., Yotte, S., and Roche, C. D. (2005). Better Assessment of Bituminous Materials Lifetime Accounting for the Influence of Rest Periods. *Road Materials and Pavement Design*, 6(2), 175–195.

Chang, J., and Garvin, M. J. (2006). A Conceptual Model of Bridge Service Life, *Bridge Structures: Assessment, Design, & Constr.*, 2 (2), 107–116.

Chen, G., Wu, J., Yu, J., Dharani, L. R., and Barker, M. (2001). Fatigue Assessment of Traffic Signal Mast Arms Based on Field Test Data Under Natural Wind Gusts. *Transp. Res. Record* 1770, 188–194.

Dossey, T., Easley, S., and McCullough, B. F. (1996). Methodology for Estimating Remaining Life of Continuously Reinforced Concrete Pavements. *Transp. Res. Record* 1525, 83–90.

Gabriel, L. H., & Moran, E. T. (1998). Service Life of Drainage Pipe, *NCHRP Synthesis 254*, Transp. Res. Board. Washington, D.C.

Georges, P., Lamy, A.-G., Nicolas, E., Quibel, G., & Roncalli, T. (2001). Multivariate Survival Modelling: A Unified Approach with Copulas. Working Paper, Credit Lyonnais, Groupe de Recherche Operationnelle, Lyon, FR.

Jiang, Y., & Sinha, K. C. (1989). Bridge Service Life Prediction Model Using Markov Chain. *Transp. Res. Record* 1223, 24–30.

Karbhari, V. M. and Lee, L. S. (Eds.) (2011). Service Life Estimation and Extension of Civil Engineering Structures, Woodhead Publishing Series in Civil and Structural Engineering.

Peil, U. and Mehdianpour M. (1999). Life Cycle Prediction via Monitoring. Proceedings of 2nd International Workshop on Structural Health Monitoring, Stanford University, Stanford, CA.

Vieira, P., and Horvath, A. (2008). Assessing the End-of-Life Impacts of Buildings. *Environm. Sci. Tech.* 42(13), 4663–4669.

PART V

OTHER TOPICS RELATED TO CIVIL SYSTEMS DEVELOPMENT

CHAPTER 27

THREATS, EXPOSURE, AND SYSTEM RESILIENCE

27.0 INTRODUCTION

27.0.1 Prelude

Civil engineering in the current era is particularly challenging. Today's civil engineers have inherited systems that were built decades ago to meet the demands of past generations. Not only are many of these systems either functionally obsolete or approaching the end of their design lives, but also current and future generations present the problems of increasing populations and unprecedented demand for the services provided by civil engineering systems. Furthermore, civil engineers face a new and different set of performance metrics such as reliable security in the face of natural or man-made disaster. As such, in most countries, many civil engineering systems are in need of expansion, rehabilitation, replacement, or retrofitting to meet the demands associated with capacity, strength, and failure resistance in the event of disasters.

The issue of inadequate, aging, or vulnerable civil infrastructure has deservedly gained international attention due to a series of well-publicized physical or functional system failures in recent years in many countries. Examples in the United States include the New Orleans levee breaches during Hurricane Katrina, the Minnesota Interstate 35 bridge collapse, transit tunnel shutdowns in New York City in the aftermath of tropical storm Sandy, China's bullet train crash in 2011, and sewer blowups in several old cities. Other less publicized failures of public or private civil systems due to design or construction error, fire, or natural threats include collapse of Delhi Metro Rail Corporation's pier 67 cap in India, sinking of offshore oil rigs (Kielland in 1980, Ocean Ranger in 1982, Piper Alpha in 1988, and Aban Pearl in 2009), the 2010 Boston water emergency, and the 1995 Kharkiv drinking water disaster. The increasing specter of tort liability and inadequate or uncertain funding for sustained preservation and renewal of these systems are also everyday realities faced by the owners or operators of civil engineering systems.

In this chapter, we will first define the basic terms of threat likelihood, system resilience, and community exposure. The chapter will then present how threats could be categorized on the basis of whether they are internal or external, sudden or gradual, natural or man-made and discusses the various types of threats. The chapter also discusses the exposure and resilience of civil engineering systems to threats, particularly, how civil engineers could increase the resilience their systems. The concept of resilience in the design of future systems is discussed. Finally, the chapter combines the trio of threat likelihood, public exposure, and system resilience to show how they could be used to develop a generalized procedure for quantifying an overall hazard or risk.

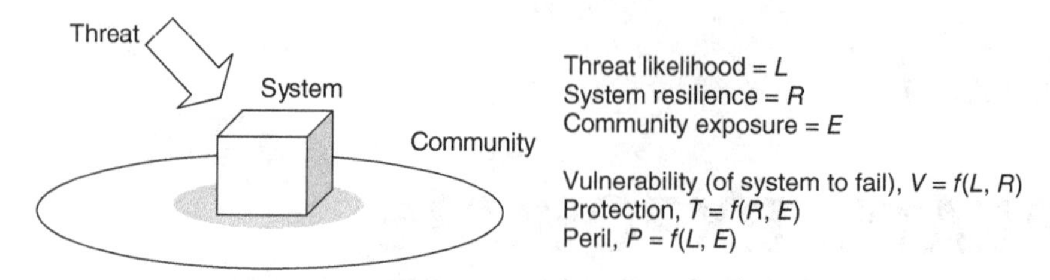

Figure 27.1 Depiction of threat–system–community.

27.0.2 Threat Likelihood, System Resilience, and Community Exposure/Consequences

There are a number of preliminary but important definitions that must be presented before going further in this chapter. There are primary terms and secondary terms. Primary terms (Figure 27.1) are those directly related or derived from only *one* entity: the threat, the system, or the community; the first is **threat likelihood**; this is the probability that the external or internal threat will occur; **community exposure** is the extent of destruction of the surrounding environment if the threat were to occur in terms of the loss of human life or injury and damage to natural assets or built-up facilities in the proximity of the system; **system resilience** is the ability of the system to withstand or recover from the threat. The secondary terms are **consequence** (the effect on the community and the environment if the system is damaged due to the threat, and is a direct function of the exposure); system **vulnerability**, which refers to the susceptibility of a system to failure/destruction, and is a the combination of threat likelihood and system resilience; and **hazard**, the overall danger posed by the threat and is a resultant effect of the threat likelihood, exposure, and system resilience.

The overall hazard, H, posed by a situation where a civil engineering system located in a certain environment is threatened by a certain type of threat (Figure 27.1) can be represented as:

$$H = f(L, C, R)$$

where: L is threat likelihood, or the probability that the threat will occur. This is a function of the threat type and the system location and orientation. This could be derived either from past data (frequency of occurrence) or from theoretical reconstruction/simulation. Examples include earthquake intensities and flood return periods. When the threat occurs, the system is said to suffer a **disruption**.

C is the consequence of the threat if it were to occur, and is a function of the environment (ecological resources, population, man-made facilities); this represents the adverse impacts caused by civil engineering system damage due to threat occurrence (injury, loss of life or property, social and business disruption, ecological damage, etc.). The consequence can be predicted using data from past observations (occurrences of similar threat types to similar system types in similar environments) or using data from computer-aided simulation of such events.

R is the resilience of the system, which is a function of the system age, condition, material type, design type, and other system attributes. There are several different types of resilience (e.g., physical or structural resilience, operational resilience, etc.).

As we noted in the preceding paragraph, the threat likelihood, system resilience, and consequence of the system failure on the community may be referred to as **primary terms**. From these, **secondary terms** that can be derived include: *vulnerability* of the system to failure which is a function of the threat likelihood and system resilience (a high threat likelihood and low system resilience translates into a high vulnerability); *protection*, which can be described loosely as a function of the system resilience and exposure or consequence (a high system resilience and low exposure or consequence translates into a high level of protection); and *peril*, which also can be described loosely as a function of the threat likelihood and the exposure or consequence (a high threat likelihood and high exposure or consequence translates into a high level of peril). Also, consequence is generally considered a function of exposure; exposure, in turn, is a function of the usage (demand, occupancy, or traffic volume) and the regional sensitivity (national or cultural importance, iconic character, environmental quality, and so on) of the system. These definitions are presented here only to provide a general picture of how these concepts may be related to each other–it is important to note that the meanings of these terms may differ across different system types, and owners or operators.

In Section 27.1, we will identify the categories of threats to civil engineering systems, and in Section 27.2, we will discuss the various threat types under each threat category.

27.1 CATEGORIES OF THREATS TO CIVIL ENGINEERING SYSTEMS

As we discussed in Chapter 19, unintended destruction or damage to a civil engineering system may occur due to a variety of threats that can be categorized broadly as internal or external (Figure 27.2). External threats may be sudden or gradual. Sudden natural external threats include floods, earthquakes, landslides, volcanoes, high winds, coastal waves, and icy conditions in a winter storm; sudden man-made threats include terrorism and vandalism (which are intentional) and overloading, oversize, and collision (which are generally not intentional). Gradual external threats are often climate related and include frigid or torrid conditions, freeze–thaw cycles, oxidation, and long-term wind action that causes erosion and fatigue; other gradual external threats are the acidity of the soil in which the civil structure is founded, weak soils that cause gradual settlement of structures, air-laden salt in marine environments, and deleterious chemicals in compounds used for deicing of

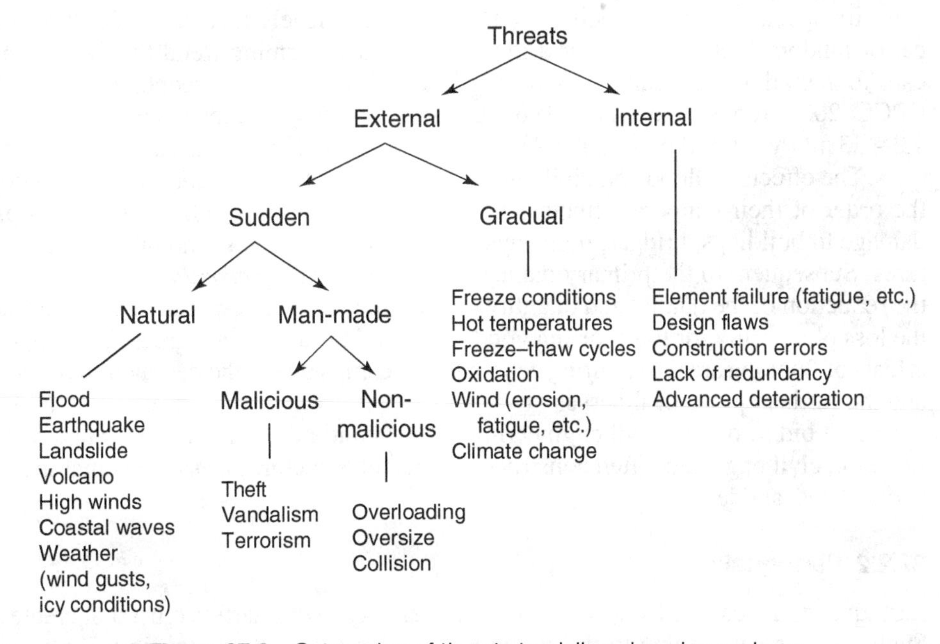

Figure 27.2 Categories of threats to civil engineering systems.

facilities in cold regions. Internal threats are those arising from within the system itself and often impair the structural integrity of the system; these may be due to design or construction flaws, unrepaired failure of some critical element due to fatigue, shear or bending, lack of redundancy in design, and corrosion, oxidation, and general weakening of the physical elements of a structure due to advanced aging of the construction materials.

27.2 THREAT TYPES

In this section, we discuss the various threat types for each threat category identified in the previous Section.

27.2.1 Flooding

Since the dawn of time, increases in the levels of lakes, rivers, and oceans have threatened the stability or operational performance of civil infrastructure located near these water bodies. These infrastructure include not only river and sea ports but also buildings, highways, and hydraulic structures such as weirs, levees, and dams. Flooding occurs when rivers or lakes become swollen with excess runoff from rain or when rainwater accumulates on soil that is already saturated.

Flooding types are often categorized by the type of water body associated with the flood. **Area flooding** occurs in low-lying, flat landscapes where the ground is saturated with moisture and there is no escape for the water or when the rate of rainfall accumulation far exceeds the rate of runoff. **Coastal flooding** is caused by severe sea storms or other hazards, including hurricanes. **Tsunami flooding** is caused by cyclones and hurricanes and earthquakes where the epicenter is offshore. **Riverine flooding** occurs when the runoff from rainfall or snow melt exceeds the capacity of the river channel. Excessive rainfall may be caused by hurricanes and monsoons; and channel capacity could be impaired by drainage obstructions including debris accumulation and silting. Flash flooding in a river may also occur due to the sudden release of water from an upstream dam.

Irrespective of the flooding type, rising water levels that cause flooding may be a cyclical or random event or a part of a long-term global warming trend that is causing the polar ice caps to melt, thus increasing the seawater levels. The Intergovernmental Panel on Climate Change (IPCC, 2007) reported that since 1961, the global average sea level has risen at an average rate of 1.3–2.3 mm/year; and between 1993 and 2003, the rate has increased to 2.4–3.8 mm/yr.

The effects of floods on civil engineering systems and the community can be categorized in the order of their temporal effects. The short-term or *primary* effects of floods are the physical damage to buildings, bridges, highways, streets, sewer systems, and other civil engineering structures. Subsequent to the primary damage, medium-term or *secondary* effects may occur, namely, the reduction of the quality and quantity of drinking water, the spread of water-borne diseases, and the loss of mobility for transporting goods and services to affected areas. The long-term or *tertiary* effects of flooding are the resulting economic hardships due to the destruction of physical structures and the further spread of illnesses.

In a bid to protect civil engineering systems and other man-made or natural structures from flooding, civil engineers often construct specialized structures or systems that include sea defense walls, dams, and levees.

27.2.2 Earthquakes

Earthquakes are caused by sudden releases of energy in the Earth's crust that create seismic waves. Such releases can be due to the rupture of geological faults, volcanic activity, or nuclear tests.

When earthquakes occur, the surface of the Earth shakes and often the ground is displaced. The focus or *hypocenter* of an earthquake is the point of the initial rupture of the earthquake; and the *epicenter*, which may be on land or in the sea, is the point at the ground level that is directly above the hypocenter. When an earthquake occurs in the sea, displacement of the seabed can cause waves of increasing amplitude as it moves away from the epicenter until it reaches land as a high wave or tsunami. Earthquakes can also lead to landslides and, in certain cases, volcanic activity. Therefore, earthquakes represent a unique type of threat because they often catalyze events of other threat types.

Earthquakes are prevalent worldwide, particularly in Asia, southern Europe, and the western United States. Figure 27.3 is a map of the earthquake epicenters in the period 1963–1998. Of the several hundreds of thousands of earthquakes that occur worldwide each year, approximately 25% are felt by humans, and a fraction of these are severe enough to pose a threat to civil engineering systems. The failure of these systems during seismic events is often due to ground movements; local amplification of movement due to transfer of seismic energy from deep harder soils to superficial softer soils; or rupture, which is a visible break and displacement of the Earth's surface along the fault line. The vulnerability of a structure to earthquake damage or failure can be assessed as the product of the structural and geotechnical features and likelihood of seismic events where the structure is located. For a bridge, the relevant structural features include (i) connections, bearings, and seat widths (which is influenced by the bearing types, support lengths, and support skew); (ii) piers, due to the pier design and shear failure and flexural failure; (iii) abutments; and (iv) soil type and properties related to liquefaction (NYSDOT, 2002). The seismic likelihood or rating is a function of the design seismic acceleration coefficient and the soil profile type (which is a predictor of soil amplification due to an earthquake).

27.2.3 Landslides

Landslides (Figure 27.4), which are a specific kind of geological phenomena, include slope failure and shallow debris flows that occur due to gravity. Landslides, which are due to slope instability, can seriously impair the structural integrity or operational functions of civil engineering structures and have been known to create safety hazards at highways and bridges in mountainous areas. Researchers believe that long-term changes in global climatic conditions are generally likely to lead to increased geotechnical activity such as rockfalls and landslides in mountainous areas, particularly when these areas experience increased rates of groundwater seepage through rock joints and increased groundwater pressure (Beniston, 2004). Landslides can be triggered or accelerated

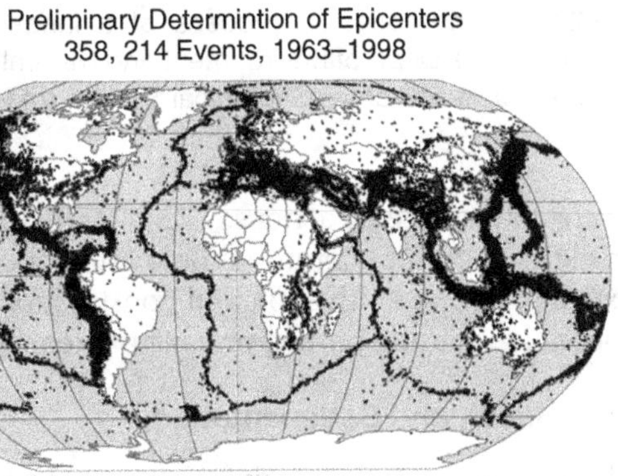

Preliminary Determintion of Epicenters
358, 214 Events, 1963–1998

Figure 27.3 Earthquake epicenters, 1963–1998 (Courtesy of NASA).

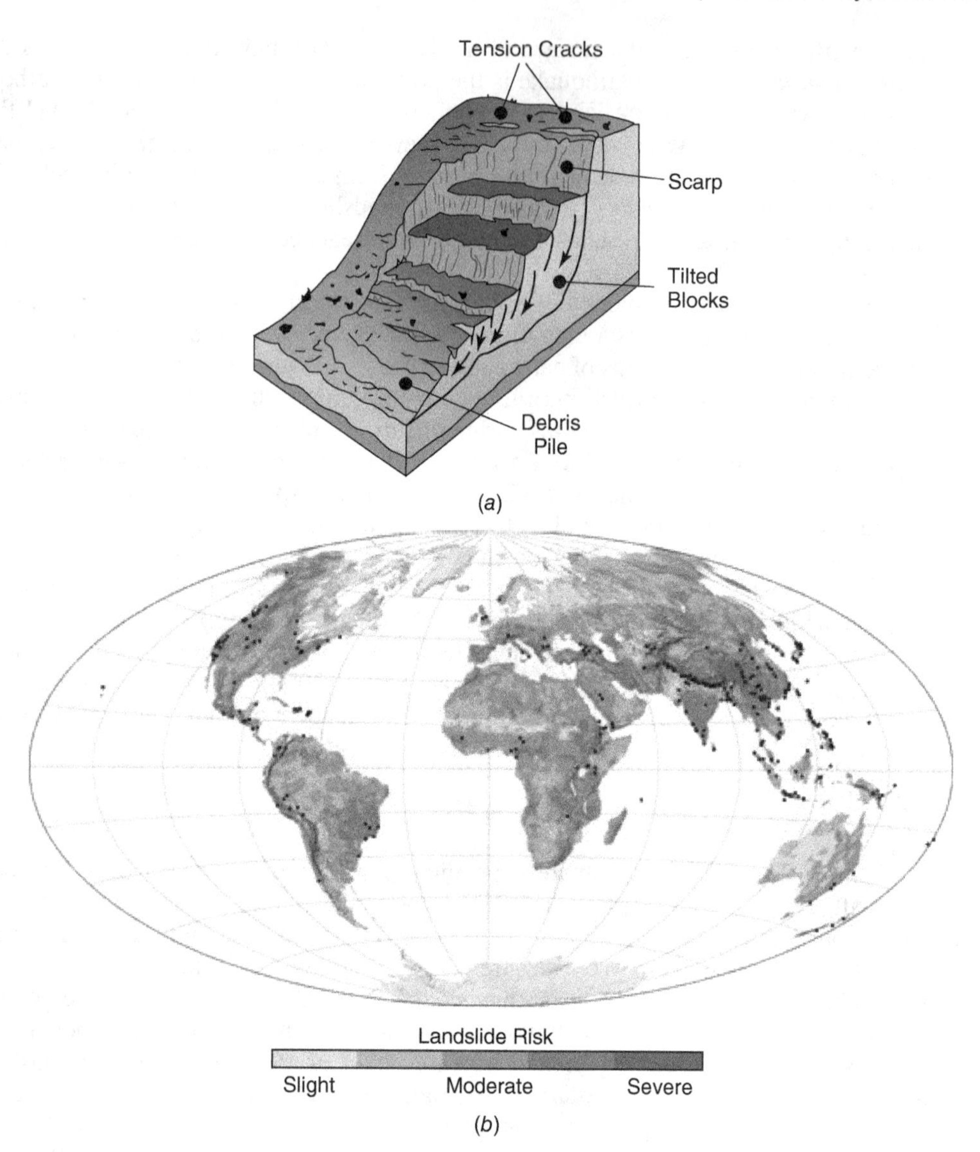

Figure 27.4 Landslides: (*a*) process of a landslide and (*b*) global landslide risks [(*a*) North Carolina Dept. of Environment and Natural Resources; Division of Energy, Mineral, and Land Resources—Geological Survey; (*b*) Robert Simmon/NASA)].

by earthquakes or anthropogenic actions. The latter include vibrations from construction or mining machinery or highway traffic, deforestation, earthworks that alter the slope or impose new surcharges on existing slopes, cultivation and construction activities that destabilize fragile slopes, and blasting operations during mining or construction (Sassa and Canuti, 2008).

27.2.4 Erosion, Scour, and Sedimentation

The structural foundations of civil engineering systems located in the proximity of water bodies or at areas of strong wind are vulnerable to the erosive forces of water and wind (Figure 27.5). Also

Figure 27.5 Threat of erosion continues to impede the management of paved and unpaved highway systems (Paul Anderson/Wikimedia Commons).

affected are civil systems downstream including any hydraulic structures where the sedimentation (deposition of the eroded material) impairs hydraulic and other functional efficiencies. Environmental features that affect the level of such threat include the type and nature of the water body (river, lake, or sea) and the erodibility of the residual soils. In the case of rivers, important factors include water volume, velocity, river slope, and the shape and nature of the river bed (Labi et al., 2011). To help predict the level of erosion/sedimentation threat to a civil engineering system, engineers typically measure its proximity to river confluence, ascertain the extent to which the system is affected by backwater, the historic maximum flood depth, and the historic scour depth, and establish the extent of availability of overflow/relief hydraulic structures. In the case of scour threat, the frequency of floating debris and ice are factors that can provide indication of the threat level and extent. For bridges, culverts, and other hydraulic structural systems, the threat posed to their foundations can be measured on the basis of the existence of scour countermeasures, whether the abutment is located at a river bend, the angle of inclination, and the embankment encroachment. For piers, the threat is influenced by the skew angle, the pier/pile bottom below the streambed, and the pier width, among other considerations. These elements of erosion and scour vulnerability were established by Shirolé and Holt (1991) and Shirolé and Loftus (1992) for bridge structures, but their conclusions could be easily adapted for other civil structures located near water bodies. In the current era where the specter of climate change looms, the increased frequency and strength of tropical storms are expected to lead to more severe soil erosion around structures and deposition of eroded material at other structures downstream. A measure that quantifies the erosion and scour threat could be derived on the basis of these considerations.

27.2.5 Sinkholes

A sinkhole is a depression on the Earth surface. Sinkholes are caused by erosion or gradual chemical dissolution of partially soluble bedrock (such as dolomite, limestone, and gypsum) by percolating water. At some locations, they are also caused by cave roof collapse or lowering of the water table. The groundwater dissolves the carbonate cement that binds the particles of sandstone, carries away the released particles, and leaves a void over a period of time. Often circular in shape, they can be as small as a faction of a meter to over 500 meters in diameter. Sinkholes develop gradually but their effect may be felt suddenly. Also, sinkholes may be due to human activity, for example, when large-scale and shallow underground mining leaves large voids under the ground.

27.2.6 Threats Related to Climate and Weather

Weather can be defined as the short-term (minutes, days, or weeks) variations of the atmospheric conditions, including temperature, humidity, precipitation, cloudiness, visibility, and wind. Weather impairs the operational performance of a civil engineering system (e.g., ice-covered runways), and in extreme cases such as high winds, can cause physical damage to the structure through sudden collapse or fatigue. On the other hand, climate is the average of conditions experienced over an extended time period (typically, 30 years) and compared to weather is a more formidable threat to the long-term durability of civil system physical structures. The threats related to climate include freeze, freeze–thaw, oxidation, and warm temperatures. Freeze conditions in wintertime cause materials to become brittle and prone to cracking, and cause ice lenses in soils which leads to soil heaving upon thawing during the spring season. Hot temperatures in torrid regions cause materials to buckle, to enter plastic phases of deformation, or in extreme cases, to melt. Also, gradual oxidation of materials such as asphalt causes them to lose their ductility and become brittle and prone to cracking. Freeze–thaw transitions, often experienced in the early and late phases of the winter season, cause expansion and contraction of materials that can lead to failure. In the very long term, climate change poses a serious threat to civil engineering systems because systems that are designed for past climatic conditions could become inadequate when these conditions change. In a separate section of this chapter, we will address the threat of climate change.

27.2.7 Internal Fatigue or Design Flaws

Fatigue may be described as the progressive and localized structural damage that occurs when a material is subjected to repeated loading and unloading until the structural element fails with little or no warning. In a subsequent section of this chapter, we will discuss the issues associated with fatigue failure, particularly with respect to concrete and steel structures. The New York State Department of Transportation (NYSDOT, 1997, 1999) developed a methodology for assessing the fatigue failure of concrete and steel bridges; with some modification, this methodology could be extended to other types of civil structures. Also, problems in design can threaten the physical or operational integrity of civil engineering systems. For example, designs that do not accommodate adequate redundancy can lead to sudden life-threatening failures of the system. For example, during the 9–11 attacks on New York City, World Trade Center Building 5 (WTC5) collapsed due to fire. It has been argued that the use of different structural detailing at the time of the building design could have significantly improved the fire resistance of the structure (LaMalva, 2002): Specifically, slotted holes could have been provided in the girder webs or wider spacings provided between the girder stub ends and the beginning of the simply supported center spans as these would have permitted greater freedom for girder rotation thus avoiding the prying action that led to the tearing of the girder webs; also, maintaining the shear connection near the column face would have helped lower the temperature of this connection.

27.2.8 Poor Condition due to Aging

In an article in *Time*, Lowe (2007) stated that urban planning experts consider America's older cities as modern-day Pompeiis—vulnerable to the "volcanoes" of infrastructure failures. Figure 27.6 shows the site of gas line explosion damage. In many countries, a significant fraction of the civil engineering systems were designed and constructed to not only standards that have become obsolete when weighed against design standards of today but also to serve populations that were far lower than their current levels. To exacerbate such structural and functional obsolescence, inadequate renewal investments and maintenance neglect have rendered many civil engineering systems unprepared to handle such imminent stresses.

27.2.9 Terrorism

Terrorism is a violent act that is intended to create fear and is perpetrated for the purpose of gaining publicity for a group, cause, or individual for political, nationalistic, religious, or ideological reasons. A key aspect of terrorism is that it uses violence indiscriminately against noncombatants, and those who perpetrate this crime base their actions on the assumption that they can leverage human fear to help achieve their objectives. Large civil engineering systems offer an attractive target for terrorists for a number of reasons. First, many of these systems often serve as symbols of national pride or icons of society or culture; for example, Brasilia's Digital TV Tower, Beijing's National Grand Theater, Egypt's Suez Canal, Switzerland's Salginatobel Bridge, Paris's Eiffel Tower, Sydney's Opera House, and London's Tower Bridge. Second, large or expansive civil engineering systems often serve a large number of users at any given time, such as large skyscrapers, mass transit systems, urban freeway bridges, and water supply reservoirs.

Figure 27.6 Site of gas line explosion damage, San Bruno, California.

Unlike most other types of threats to civil engineering systems, it is difficult to quantify the likelihood of intended man-made threats. In some countries, the departments of national security or homeland security have attempted to assign terrorism threat likelihood ratings to each major civil engineering system on the basis of criteria that include the system size, the exposure (usage level or population density in area where system is located), and the iconic value of the system in terms of national pride. Other factors that could influence the malicious man-made threat likelihood of civil engineering systems include usage-related factors (average or maximum daily levels of use), the facility's contribution to the overall network's operations, its international proximity (whether the system is located near a neighboring country), and the role of the system in national defense. Rummel et al. (2002) developed a similar set of criteria to develop a criticality index that can be used to assess the likelihood that structures experience man-made threatening events.

27.2.10 Overloading

Most civil engineering systems are designed to handle a certain capacity. Some types of civil engineering systems suffer little or no physical damage when demand exceeds capacity (such as a water supply system) or exhibit resilience to overloading; however, other system types fail or suffer serious damage when they are overloaded. For example, overloaded transit systems cause reductions in operational levels of service as well as gradual physical damage to the transit infrastructure. Bridges rarely fail suddenly upon the first instance of overloading; rather, it is often a series of overloading events that gradually culminate in some mode of failure, such as shear, bending, or fatigue in the bridge elements. These failure modes, if not detected through regular inspections, ultimately lead to some damage or in extreme cases, destruction of the structure through catastrophic failure and loss of life. As we have stated previously, civil engineers design for a certain load level, often with an appropriate factor of safety. However, it is not impractical to assume that a surge in demand due to special events or unforeseen circumstance may cause the system to experience overloading situations. In such cases, as part of the design process, it is useful for the civil engineer to carry out computer or laboratory simulation to characterize the behavior of the system under overloading situations in order to ascertain the "brittleness" of the system and make any needed recommendations to avoid or accommodate overload situations during the system's operations phase.

27.2.11 Oversize

Certain types of civil engineering systems are designed to accommodate individual users of certain dimensions. "Users" could refer to persons, cars, trucks, aircraft, ships, and the like. When the height, width, or length of an individual user exceeds the design dimensions, the system may suffer damage and/or the user may experience some adverse effects, including inconvenience, discomfort, safety hazard, injury, or even death. In the case of highway systems, for example, the adverse effects may include damage to the user's vehicle.

27.2.12 Collision

Civil engineering systems are always prone to accidental damage from its users or any activity in close proximity of the system. The level of danger in this case is influenced by the type of occupancy (e.g., traffic), the nature of the occupancy (e.g., the number of large trucks and their average speeds), the height of the structure, or the location (e.g., in a navigation channel). For bridge structural systems, for example, the NYSDOT (1996) developed a collision vulnerability rating that is based on superstructure vulnerability to a truck-under-bridge collision, pier vulnerability to truck-under-bridge collision, superstructure vulnerability to water vessel collision, pier vulnerability to water vessel collision, superstructure vulnerability to train-under-bridge collision, and pier vulnerability to train-under-bridge collision.

27.2.13 Vandalism, Theft, and Other Threats

Vandalism can be described as the intended or malicious destruction or damage of property without any political motives. For a civil engineering system, this includes damage or removal of parts of the system including lighting, electrical, and mechanical components and spray painting and graffiti defacement. Theft [often referred to as acquisitive vandalism (Cohen 1973)] is the unauthorized removal of parts of civil engineering systems, often with the intention to use or sell for profit. While such acts rarely cause complete destruction of a system or jeopardize its structural strength, it can cause aesthetic or operational problems. For example, theft of metal barriers that constitute highway guardrails can lead to impaired function of these systems and could lead to greater severity of crashes that occur at such locations. In some urban areas, vandalism and theft pose serious threats to the operations of civil engineering systems, and laws are specifically enacted to combat such social menaces. Fire, intended (arson) or unintended, can cause serious damage or complete destruction of civil engineering systems, particularly when the system is not designed adequately to accommodate high temperatures. Similar to climatic high temperatures, fire can cause materials such as steel and reinforced concrete to buckle, enter plastic phases of deformation, and even melt in extreme cases. Figure 27.7 illustrates a number of threat types to civil engineering systems.

27.3 THE THREAT OF CLIMATE CHANGE

To underscore the importance of the climate change threat, we devote this separate section for its discussion. As the Earth enters a phase of obvious climate change, it is important for future civil engineers to continually conduct performance reviews of existing civil engineering systems, revise their design processes as needed, assess the vulnerability of structures to this threat, and develop and implement requisite proactive or palliative policies and actions.

27.3.1 Evidence of Climate Change

Climate change is the variation in the statistical distribution of weather patterns over a very long period of time; this phenomenon is manifest as a distributional shift of the frequency and intensity of weather events (IPCC, 2007). The causes of climate change include plate tectonics, orbital variations, solar output, oceanic activity, and anthropogenic factors.

Global warming, a key element of climate change, is not only well documented but also accepted by most scientists as an evidence-based matter of fact. In June 2006, a U.S. National Research Council panel provided documented evidence that "the Earth is the hottest it has been in at least 400 years, and possibly even the last 2,000 years." Other studies have shown that the last hundred years have seen an increase in the average global surface temperature by approximately 0.5–1.0°F—the largest centurial increase in the Earth's surface temperature in the past 1000 years; scientists predict that the temperature increase over this century will be even greater than the last. Although the temperature increases seem to be small, they trigger large changes in climate and weather patterns. For example, there was only a 5°C difference between global temperatures at the Ice Age and the ice-free ages (Stanford Solar Centre, 2009). Figure 27.8 presents the global temperature changes from 1880–2000.

(a) (b) (c) (d) (e) (f)

Figure 27.7 Illustration—types of threats to civil engineering systems: (*a*) flooded underground subway tunnel after Superstorm Sandy, 2012, New York, NY; (*b*) flooding of highway system due to Hurricance Katrina, 2005, New Orleans, LA; (*c*) landslide damage of Freeway Nr. 3, 2010, Keelung City. Taiwan; (*d*) wind damage to coastal structure; (*e*) Guatemala City sinkhole; (*f*) Side view of support-column failure and collapsed upper deck. [Photos courtesy of (*a*) Metropolitan Transportation Authority of the State of New York/Wikimedia Commons; (*b*) U.S. Coast Guard, Petty Officer 2nd Class Kyle Niemi; (*c*) National Science Council, Taiwan; (*d*) Tim Burkitt, FEMA photo; (*e*) Paulo Raquec, Wikimedia; (*f*) H. G. Wilshire, U.S. Department of the Interior.]

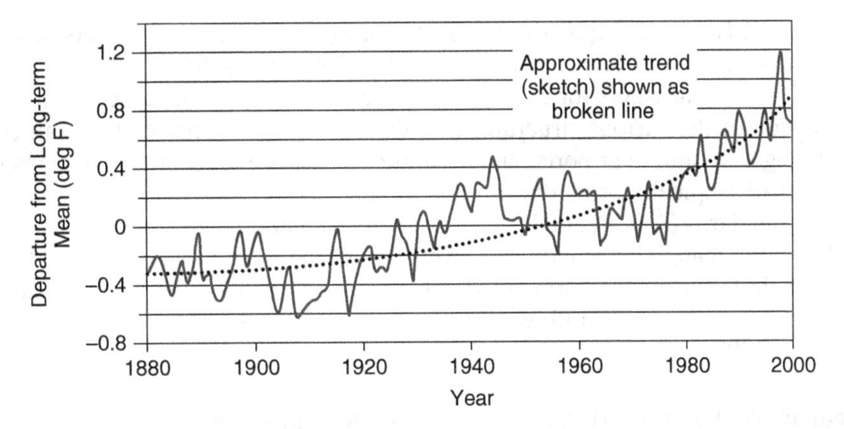

Figure 27.8 Global temperature changes, 1880–2000.

27.3.2 Primary Impacts of Climate Change on Civil Engineering Systems

The increase in global temperature is expected to lead to accelerated melting of polar ice caps and, ultimately, increased levels of sea and land groundwater and altered frequencies and intensities of extreme weather. Other primary impacts are expected to include an increased frequency and severity of droughts and freeze–thaw cycles, the warming of ocean surfaces (thus, more intense typhoons and hurricanes), larger and more abrupt floods, changing levels of groundwater, and changes in wind speed and profiles. These primary effects will translate into secondary effects on existing civil engineering systems; and future engineers, planners, and designers, by necessity, will need to contend with the impacts of climate change.

One of the most consequential primary impacts of climate change, from the perspective of civil engineering systems, is the rise in sea levels. IPCC (2007) reported that there has been a 1.3–2.3 mm/yr increase in global average sea level since 1961 and increased to 2.4–3.8 mm/yr between 1993 and 2003. It is not certain whether the increased rate of sea level rise in the 1993–2003 period can be attributed to natural variations in sea level over that time period or whether it was indicative of an underlying trend in the long term. Projecting to the end of the 21st century, IPCC (2007) projected that the average sea level could rise by 18–59 cm for the time period 2090–2099, relative to the average sea levels over the period from 1980 to 1999. Nicholls and Cazenave (2010), hypothesizing the collapse of the West Antarctic Ice Sheet and exploring the global impacts of consequent rise in sea level, determined that the effect of climate change is nonlinear; in other words, the impact of a 5-m sea level rise could be much worse than five times the impact of a 1-m-rise in sea level.

27.3.3 Secondary Impacts of Climate Change on Civil Engineering Systems

Studies have been carried out on the climate change impacts on health care, energy, agriculture, and other sectors. However, relatively little work has been done on the climate change impacts on civil engineering systems. It can be argued that the sensitivity of civil engineering systems to climate change is relatively small compared to that of other sectors because civil systems have greater capabilities for mitigation and adaption. However, another school of thought seems to hold the vulnerability of civil engineering systems to sudden changes in climate or weather, is significant (IPCC, 1994). We now discuss some of the preparations being made in each area of civil engineering to counter the secondary impacts of climate change on their systems.

Structural Engineering Systems. In preparing to adapt to the effects of climate change, structural engineers will need to review design codes for planned structures and to adopt adaptation and mitigation measures for existing structures. Increases in ocean levels, for instance, will require adaptation of the various structures that form or will form parts of overall coastal infrastructure, including seaports, river ports, and harbors. Also, it is expected that changes in the frequency and strength of tropical storms due to climate change will require amendments to civil engineering design standards (ASCE, 2007). Further, more frequent freeze–thaw cycles will influence concrete technology; changing groundwater levels will cause foundation problems; heavy abrupt snowfalls and precipitation in the form of ice clumps could damage structures; and higher wind speeds and wind impulses and different wind profiles will affect the structural design of tall buildings (Lenkei, 2007).

Geotechnical Engineering Systems. Variations in subsoil pore water pressures due to rises in sea level could threaten the stability of existing structures. Geotechnical engineers will need to retrofit the foundations of existing structures. For future structures, special foundation designs might be needed to reduce the vulnerability of civil engineering systems to such threats. Extreme precipitation events could lead to greater rates of erosion, discharge, and sedimentation; and there could be increased likelihoods of rock falls and landslides in mountainous regions caused by increased rates of groundwater seepage through rock joints and increased groundwater pressure (Beniston, 2004). Also, trigger mechanisms for landslides are associated with pressure-release joints following glaciations; and slope instabilities that could jeopardize the safety of system users at mountainous areas may need to be addressed. ASCE's Policy Statement 360, "Impact of Global Climate Change," identified changes in the permafrost conditions due to climate change, which could require retrofitting of existing foundations and alterations to foundation designs as well.

Transportation Engineering Systems. The relationship between transportation and climate change has often been viewed through a one-way lens: how transportation affects climate change. The lessons learned from that perspective are useful for developing techniques that reduce the magnitude of the climate change problem. However, the other direction of the relationship (i.e., the impact of climate change on transportation) has generated relatively little concern and deserves at least as much attention. In preparing for the effects of climate change, engineers have begun to review the designs of guideways in all modes of transportation to accommodate higher groundwater levels and increased frequency and intensity of flood events, as well as to develop operational policies for rapid evacuation of inhabitants in the event of disasters caused or facilitated by climate change. Of the man-made systems, transportation systems, particularly those located in permafrost areas and vulnerable coastal zones, are particularly sensitive to extreme events engendered by climate change (IPCC, 1994).

Environmental Engineering and Public Health Systems. Future engineers will need to tackle public-health-related problems caused or catalyzed by climate change phenomena that include the generation of stronger heat waves of greater frequency. Prolonged heat can increase smog and allergen dispersal and thus foster respiratory ailments. Also, as human infections are linked to the global climatic environment, disruptions in this environment potentially could intensify certain infectious diseases. The extent of the impact of global warming will be influenced by the interaction between human host populations and causative infectious agents. Thus, any human migration that is caused directly or indirectly by climate change will possibly lead to a shift in disease patterns to shift; and diseases may be easily transmitted where such migration are accompanied by scarcity or contamination of potable water sources (FPE, 2009).

Hydraulic Engineering Systems. The rise in sea levels and greater surface runoff volumes will necessitate performance reviews of existing hydraulic systems and new designs. For example, researchers have determined that climate change will impact the seasonal distributions of snow storage and runoff from catchment areas for hydroelectric systems. Further, the sensitivity of hydrology to climate change must be taken into account during the planning and design of future hydropower infrastructure. Further, there will be increased need for coastal engineers to reevaluate existing sea defense structures and also to plan, design, and build new ones to protect vulnerable islands and coastal areas from submerging. As noted by ASCE (2007), for hydraulic systems at other locations, the impacts of extreme climates, including droughts, floods, and other significant changes in hydrologic patterns may generate a need for changes or capacity expansions to flood management systems.

Hydrologic Engineering Systems and Water Resources. Gertner (2008) introduced the concept of "water footprint" and described the consequence of climate change on the basis of relationships between energy, carbon greenhouse gas emissions, and water, arguing that changes in the volumes and patterns of water availability would severely impact most sectors of human activity. Mathematical modeling by climate change researchers suggests that freshwater resources in several regions of the world are likely to be impacted, and current arid areas could experience further decreases in runoff. Climate change, in one or more of its several manifestations, is likely to disturb an already complex water management system. Existing drainage systems, water control structures, and conveyance and distribution systems typically are designed on the basis of flood recurrence intervals (also called return periods) and annual exceedance probabilities. These data are derived from past observations and associated levels of tolerable risk and economic consequences (IPCC, 1994). Given the significant anticipated turnover in water management infrastructure, with considerable maintenance and rehabilitation occurring every few decades, it is expected that the managers of these civil engineering systems will need to change the operating capacities of such structures to conform with evolving changes in climate and their concomitant effects. The only exceptions are where (i) any such infrastructure system has a design life of several decades, say, 50–100 years, (ii) the effects of the climate change occur sooner than expected, and (iii) the country does not have the financial wherewithal to carry out investments that increase the capacity of the water structures and systems. As the world confronts the specter of climate change in the new millennium, therefore, water engineers will be seeking strategies and techniques to confront the potential adverse impacts of climate variability on the quantity and quality of water. ASCE (2007) states that hydrologic pattern variations due to climate change are expected to generate a need for multipurpose water resource projects related to the hydroelectric industry and water supply utilities, such as increases in reservoir storage capacities.

General Civil Engineering Systems. The secondary impacts of climate change closely follow the primary impacts as discussed in a preceding section: Warming of ocean surfaces result in hurricanes or typhoons; thawing of ice reserves result in rising sea levels; changing precipitation with larger and more abrupt floods result in higher water levels in lakes and rivers; changing of groundwater level causes geotechnical problems. Also, heavy and abrupt snowfalls, higher wind speed and impulses and changes in wind profiles can cause accelerated deterioration of structural surfaces and damage to the water tightness of different elements of civil system structures. The potential devastation that could be caused to civil engineering systems by these impacts will necessitate a quick and coordinated engineering response; however, adaptation and mitigation measures might be challenging due to the dynamic character of these impacts (Lenkei, 2007).

Deglaciation, in certain locations and circumstances, is expected to result in the "accumulation of water behind unstable moraines of isolated blocks of ice that have broken off from the leading edges of retreating glaciers" (Beniston, 2004). Rapidly evolving conditions (e.g., extreme precipitation and sudden outbursts or overspills) could result in intense debris flows and flooding that could damage civil engineering systems, particularly in snow-covered mountainous areas. IPCC (1994) contends that due to climate change, certain coastal populations will experience increased threat of flooding and land loss through erosion. For example, it is estimated that a 50-cm increase in sea level will put 92 million people worldwide at the risk of flooding, and certain coastal nations and islands will face greater threat or even complete overtopping if their existing sea defense walls are inadequate. The troubling thought of the potential of rising sea levels and the concomitant effects are exemplified by the realization that downtown Boston, for example, is only a few feet above sea level and many parts of New York City are already susceptible to serious flooding.

Silver linings in this cloud include that fact that climate change occurs gradually and that the life cycle of most civil engineering systems is much shorter than the climate change cycles. Thus, adaptation of many existing civil engineering systems could occur through management and normal replacement cycles of these infrastructure (IPCC, 1994). The success of such adaptation hinges on (i) the timely and adequate delivery of information about the potential impacts to systems decision makers and (ii) the capacity of system owners and operators to respond in a timely and effective manner. Future civil engineers will therefore need to properly incorporate potential climate and sea-level changes in planning and design in order to reduce the risk to civil engineering systems and their users. However, in cases where a civil engineering system has a long life cycle (such as 100 years for channels, water supply systems, drainage systems, dams, etc.), the sudden effects of climate change can be devastating and difficult to recover from. The uncertainty in the occurrence of a climate change-induced event makes the investment decision process difficult (IPCC, 1994). The Japan Society of Civil Engineers (JSCE) stated that the effects of climate change will occur in four ways: the safety and performance of civil engineering structures and systems, the efficiency of executing civil engineering projects, the operations of civil engineering infrastructure, and the impacts on safety, security, and environmental conservation (JSCE, 2009).

27.3.4 Climate Change Impacts on Civil Engineering Systems—Studies on Mitigation and Adaptation

It has been argued that climate change can be decelerated and its adverse consequences on natural and man-made systems (including civil engineering systems) can be mitigated through environmentally friendly practices such as fewer emissions and carbon footprint reduction. *Mitigation* must also be accompanied by *adaptation*, the reduction of the adverse effects of climate change by designing civil engineering systems that will duly accommodate the rising sea levels, the increased frequency and intensity of extreme weather events, and other primary impacts of climate change. While most engineers agree that both approaches are needed, there is a split as to which approach is more cost-effective or worthy of attention and investment.

In the mitigation arena, there has been considerable research in civil engineering systems, particularly in the transportation branch. Also, there are several ways in which architectural engineers could develop energy-efficient building systems and architecture to address global warming (Ali, 2008). Further, it has been argued that greenhouse gas (GHG) emissions are partly responsible for global warming in the past century, and therefore efforts have been made to study the implications of proposed GHG mitigation policies and technologies and the degree to which they are likely to enhance our future. Other research scientists have examined strategies to reduce the rate of global warming through the use of a number of non-fossil-fuel energy options.

With regard to adaptation of climate change, future civil engineers will need to carry out continual performance review of existing civil structures, revise their design processes, assess the vulnerability of their structures to this threat, and develop and implement requisite proactive or palliative policies and actions. In the area of hydraulic engineering, JSCE (2009) recommended the pursuit of innovative techniques to evaluate flood risk, transform water resource policies, and provide support for climate change adaptive measures. In many coastal cities, such as the New York metropolitan region, civil engineering systems will inevitably be exposed to more severe coastal flooding as rising sea levels and storms become more frequent and severe with global warming Hill (2008); and thus researchers have argued more for the prevention of flooding than for flood recovery planning and have suggested ways by which the inner city could be protected, such as the installation of storm surge barriers at choke points in the waterways at strategic locations. Also, in the area of coastal engineering, JSCE (2009) advocated for mitigation and adaptation measures that take due cognizance of the anticipated timeline of the effects of climate change. In the area of environmental engineering, JSCE called for the improvement of water systems, the enhancement of evaporative functions in urban areas, and the promotion of countermeasures against pathogens and tropical diseases.

Policies geared toward both mitigation and adaptation of climate change on the basis of low-probability/high-consequence scenarios have been justified using real options (see Chapter 15), dynamic programming, and other analytical tools to assess the implications of changes in climate policy (Guillerminet et al., 2008). Also, on the premise that climate change is catalyzed by a "dangerous anthropogenic interference with the climate system," McInerney and Kelleruse (2008) used an integrated climate change assessment model to establish strategies to optimally reduce the economic risks associated with climate change impacts; they contend that risk reduction is indeed feasible but would need policies that quickly lead to reductions in CO_2 emissions. In a similar study, Felgenhauer and De Bruin (2009) identified optimal paths for mitigating and adapting to climate change under scenarios of uncertainty and certainty. Also, researchers have highlighted the distinction between natural and man-made risks in arriving at a total assessment of risk—this distinction is important for policy making and concludes that adaptation does not reduce the inherent vulnerability of the territories concerned, but enhances the resilience of man-made systems to climate change (Briguglio, 2010). In Switzerland, Hill et al. (2010) used adaptive planning to study the reduction of the vulnerability of natural and man-made systems to climate change in the Alps. In Japan, JSCE (2009) provided a wide variety of recommendations from the civil engineering perspective, to combat the problem of global warming through mitigation and adaptation. The Japan study considered making global warming countermeasures an area of major social and economic policy and including low-carbon and energy-efficiency considerations in civil engineering systems design and construction.

Prominent civil engineers have advocated the need for urgency in planning toward adaptation measures. ASCE, in 2008, published "Engineering Strategies for Global Climate Change," which identified strategies for addressing the impacts of climate change on civil engineering structures. Lenkei (2007) stated that in order to prepare for the effects of climate change, some structures will need to be strengthened by changing their structural behavior through investments that increase their static indeterminacy or make them more robust. Liao (2008) stated that engineering infrastructure solutions may take years or decades of lead time to plan, permit, design, and construct, and that by the time the engineering infrastructure can be brought to bear for its intended purpose of climate change adaptation, the future will already have become the present. In the words of Liao, "those who can envision the future—including the engineering profession that has an obligation to lead—are destined to create it."

27.4 EXPOSURE TO THREATS

For a given combination of threat type and system resilience, the exposure represents the extent of the adverse consequences to the users and the community in the event that the threat were to occur and as a result, the civil engineering system is damaged and/or put out of service. The exposure is a function of the environment, specifically, the ecological resources of flora, fauna (including endangered or threatened species) and their habitat, the human population, and the man-made facilities in the vicinity of the system. In many cases, only the population and the number of system users are used in predicting the consequence of a threat. For a given threat and system combination, the consequence can be predicted using data from past observations (past experiences of occurrence of similar threat types to similar types of system in similar environments) or using data from computer-aided simulation of such events in similar settings.

Clearly, therefore, the greater is the population density and number of users, the greater the exposure. In system hazard computations, it is common to encounter situations where it is assumed that the threat type, system resilience, and community exposure are all independent of each other. In reality, however, it may very well be the case that community exposure is related to the other two parameters. It may be possible to have two different types of systems in the same vicinity but have very different levels of exposure in case of a threat. For example, in the event of an earthquake, the exposure of residents in a specific urban location with an urban freeway and a water tank would be different in the event of a freeway failure compared to a failure of the water tank.

As we discussed earlier, exposure pertains to system users and the community, at a minimum. With regard to user exposure, the expression for the exposure level depends on the system type: for highway pavement and bridge systems, the average annual daily traffic (AADT); for airport terminal systems, the average number of passengers per day; for airport runways, the average landings and takeoffs (LTOs) per day; for levees and other flood control systems, the population of residents living in the area protected by the system; for buildings, the average daily occupancy-hours in terms of the number of users per day and the average length of time spent by each user; and for stadiums and tourist structures, the number of users during the times when the facility is open for public use.

The patronage of civil engineering systems often varies with time; as such, the exposure is not constant and depends on the time of day, day of week, and season of the year. For example, building systems typically have high levels of exposure between the hours of 9AM to 5PM; stadia have the greatest exposure when an event is in progress; highway systems have their highest levels of exposure during the morning and afternoon peak hours; and any failure of water systems affects the greatest number of users between 6AM and 8AM when water use is typically highest.

For high-vulnerability situations (i.e., systems with low resilience facing threats of high frequency and intensity), it is imperative for system managers to reduce exposure as much as practicable, particularly where funding or other limitations preclude enhancement of the system's resilience. Exposure reduction strategies and techniques include the education of system users to use any alternative or parallel systems in the vicinity or establishing policies that encourage spreading out the system use so that the average exposure per unit time is minimized.

27.5 RESILIENCE OF CIVIL ENGINEERING SYSTEMS

27.5.1 Resilience and Sustainability

In the next chapter, we will address the concept of sustainability in relation to the development of civil engineering systems. We will discuss the effect of a civil engineering system on its environment and will emphasize the need to develop systems that maximize any positive effects and minimize any adverse effects on the environment which is broadly defined here to include ecology, economy, and society. In the context of this text, therefore, sustainability focuses on the effect of a system on its environment. System resilience, on the other hand, takes the reverse direction: the effect of the environment, including threats, on a system. Thus, in this text, we allude frequently to

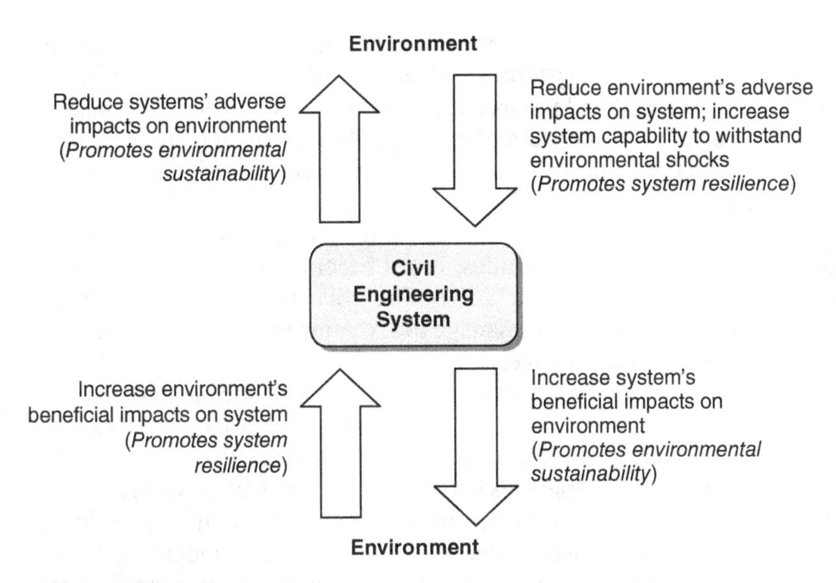

Figure 27.9 Civil engineering systems, sustainability, and resilience.

a *sustainable ecology, economy, or society* but a *resilient civil engineering system.* However, there are other contexts that look at the issue in the reverse direction: a resilient environment, economy, or society, (outside the scope of this text) and a sustainable civil engineering system.

In developing their systems, civil engineers seek to make it resilient by minimizing any adverse effects that its environment (ecology, economics, and social capital) would have on the system (Figure 27.9).

Sustainability and resilience are strongly related: good sustainability translates into good system resilience. In other words, if you ignore the negative impacts that you inflict on your environment, you are not likely to last very long yourself. This is consistent with the design philosophies of the ancient Greeks and engineers in the Taoist era in ancient China (see Chapter 1). The link between sustainability and resilience has been emphasized by a number of researchers. Recognizing that resilience is the ability to resist disorder (Gunderson and Protchard, 2002), it was suggested by Fiskel (2003) that the very essence of (or motivation for) sustainability is resilience. Brand (2009) stated that ecosystem resilience is a key requirement in the journey toward the goal of sustainable development, and Perman et al. (2003) used resilience to describe at least one concept of sustainability, indicating that a sustainable state is one that satisfies the minimum conditions for ecosystem, societal, and economic resilience over a period of time. Berkes and Folke (1998) presented a set of good practices for establishing sustainability and resilience that utilizes adaptive management strategies, institutional learning and self-organization, and local knowledge-based practices for management.

Appreciation of the synergy between sustainability and resilience has been felt as far as the corporate level where studies have been carried out to link the survival of companies to how well they adapt to and nurture their business environments. deGeus (1997) reported that in the 1990s, Royal Dutch Shell Corporation studied corporations in a bid to examine the reasons behind corporate longevity and determined that most companies die prematurely as the mean life expectancy of large corporations all over the world did not exceed 50 years. The study also found that the four key factors of corporate longevity are "sensitivity and adaptability to the business environment, cohesion and sense of identity, tolerance of diversity (decentralization), and conservative use of capital." It was noted that the list excludes profitability, which was considered to be an outcome and not a predictor of corporate longevity; indeed, a large number of companies have shown remarkable profitability only to disappear after a short period of time. The Shell study demonstrated that for a corporation to exhibit long-term survival, it must be managed as a living organism that needs to nurture its environment and not as a machine merely engineered to make profits (Fiksel, 2003).

In recent years, the concept of resilience has attracted increasing interest in civil engineering and several other disciplines and has seen applications in a wide range of sectors including physical and technical systems, and organizational, social, and economic systems. The current era, which is characterized by increasing globalization and concomitant complexity of investment ecosystems, transnational risks, and uncertainties, it has become increasingly difficult to maintain high levels of security (Brunner and Giroux, 2009). The concept of resilience can serve as a pedestal for the development of a framework to manage today's uncertain environment and to facilitate emergency preparedness and disaster recovery.

As we conclude this section (27.5.1), We need to note again that this chapter is not about economic resilience, ecological resilience, or social resilience (which, also, are very valid concerns and are the subjects of other texts), but it is rather about the resilience of civil engineering systems. Ecological resilience, for example, has been defined as the capacity of an ecosystem to recover quickly from natural or man-made perturbations including floods, fires, introduction or population surges of particular species of flora or fauna, and human activities including deforestation or construction/operations of civil engineering systems. In this regard, the sustainable development of a civil engineering system is development that reduces the adverse impact on the ecology and thus reduces the need for ecological resilience.

In the next section, we will discuss the resilience of a civil engineering system.

27.5.2 Definition of System Resilience

In the traditional civil engineering areas of strength of materials and mechanics of materials, resilience has often been defined in texts as the ability of a material to absorb energy when it is deformed by an external load within its elastic limits and its ability to release that energy after removal of the load. This is the definition that readily comes to mind for most civil engineers. As civil engineers increasingly consider the impacts of their systems on the environment and, conversely, the impacts of the environment on their systems, their mental interpretation of the term resilience extends to address not only the resilience of a civil engineering system to threats from the natural or man-made environment but also the resilience of ecosystems to the adverse effects of the civil engineering system. System resilience could be defined to consider one direction of resilience: *the ability of a system to continue to provide acceptable performance in the face of threats and challenges from the natural or built-up environment*. Gunderson and Pritchard (2002) defined engineering resilience as "the speed of return of the engineering system to the steady state following a perturbation, which implies a focus on the efficiency of the function."

These perturbations and threats can range from sudden events such as simple operational mishaps to devastating natural disasters to targeted attacks, as well as prolonged occurrences such as climate change. As such, the issue of civil engineering systems resilience includes a wide range of disciplines and subject areas.

A number of engineering organizations are actively addressing the issue of engineering system resilience. For example, the Resilient Systems Working Group (RSWG) of the International Council on Systems Engineering (INCOSE) was established to apply the principles of systems engineering to enhance systems resilience so that the recovery from disasters can be enhanced. The RSWG developed a working definition for the resilience (of a system) as follows: "the capability of a system with specific characteristics before, during, and after a disruption to absorb the disruption, recover to an acceptable level of performance, and sustain that level for an acceptable period of time" (INCOSE, 2012). Disruption is the point in time when a specific threat occurs.

The following terms also were clarified by RSWG: "Capability" is often the preferred terminology over the word "capacity" because the latter has a specific connotation from the viewpoint of design principles; "system" refers to man-made processes and products that contain hardware, software, concepts, humans, and processes; "characteristics" refer to the static features such as redundancy or the dynamic features such as corrective action; "before, during, and after" allows all three phases of disruption to be considered, specifically, the anticipatory and corrective action to be considered before the disruption occurrence, how the system will survive the effects of the disruption, and how the system will recover from the disruption (INCOSE, 2012).

In some literature, system resilience has been linked to reliability and risk management. Foster (1993) defines resilience as "a system's ability to accommodate variable and unexpected conditions without catastrophic failure," or "the capacity to absorb shocks gracefully." The Stockholm Resilience Center defines resilience as the "long-term capacity of a system to deal with change and continue to develop." C. S. "Buzz" Holling, the father of resilience theory, has studied ecology as a system, blending ecological science with systems theory, with the aid of policy analysis and simulation modeling, to develop theories that have been applied to ecosystem system resilience analysis.

Others have tried to establish a link between resilience and uncertainty and have argued that resilience reflects the uncertainty associated with future threats. In other words, it has been postulated that system agencies are not in a position to make perfect predictions of the future combination of conditions and that "if the future were predictable, resilience would lose its importance" and the system owner or operator would then simply need to plan for a single, known set of conditions with absolute confidence. The future, however, cannot be predicted with certainty; as such, there is a need for agencies to prepare for a diverse range of future possibilities, even those that are not likely but which could lead to severe adverse consequences if they occur (VTPI, 2010).

The concept of resilience could be explained in terms of real options (Chapter 15). In the area of transportation systems, communities find that it is prudent to support transit and transportation services that they currently do not use so that these options will be available if they are needed at a future time (ECONorthwest and PBQD, 2002); structural designers add redundancy in designs—there may be no need for the redundant member, but it is comforting to know that, in the event of failure of the main member, another member is there to play its role; for instance, car drivers value having a spare tire even when it might never be used. In general, researchers have maintained that a system is most resilient when its critical components have redundancy, diversity, autonomy, efficiency, and strength, thus making the system robust enough to accommodate a wide range of operating conditions or needs of the system owner or user and ensures that the system will continue to function even when a component is "out of service" due to technical adversities, including natural or man-made disruptions or the sudden lack of a particular resource input. Factors that enhance system resilience can be categorized as those that influence physical resilience and those that influence operational resilience.

The Resilience Engineering Network (REN), an open organization of persons with a shared interest in the development and application of resilience in engineering, holds the position that resilience engineering is an innovative approach for incorporating safety in system operations. REN believes that a dichotomy exists between conventional risk management and resilience engineering; the former is based on hindsight and emphasizes the documentation of errors and the computation of failure probabilities; the latter examines approaches by which organizations can develop flexible but robust processes, track risks and revise risk models accordingly, and proactively allocate resources to counter imminent threats or disruptions or to mitigate evolving pressures. As such, in resilience engineering, failures do not represent breakdowns or malfunctions of normal system processes;

instead, they represent the converse of the adaptations needed to address real-world complexities (REN, 2012).

27.5.3 Stages of Civil Engineering System Disruption

The best way to enhance the resilience of a civil engineering system to threat occurrence (disruption) is to view the disruption as a multistage process (Westrum, 2006). In this section, we will discuss two models that address the key elements of each stage of the process; and in the next section, we will present some good practices that can enhance physical or operational resilience at each stage.

(a) The Four-Stage Model. Figure 27.10, adapted from and Brunner and Giroux (2009), presents a four-stage model for the disruption of civil engineering systems and resilience applications at each stage.

Preparedness (before the Disruption). At this stage, the engineer makes efforts to prevent the disruption by eliminating the threat; where elimination is impractical, the engineer makes efforts to enhance the system resilience by increasing the system's physical strength or capacity in anticipation of the disruption. Also at this stage, the system owner prepares the community and system users and takes the necessary proactive measures to minimize exposure and thus reduce the consequences (casualties, inconvenience, and so on) due to the disruption.

Response (during the Disruption). During the disruption, the purpose of enhancing resilience is to increase survival and to reduce casualties or inconvenience during the disruption. In certain cases, the system owner or operator is required to keep the system functioning during the disruption and subsequent recovery; therefore, both the physical and operational resilience are sought for the system.

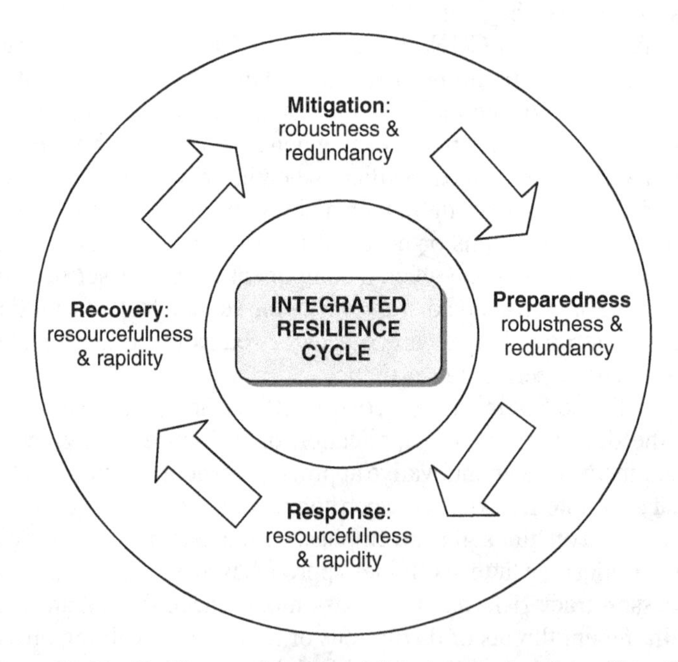

Figure 27.10 Resilience attributes at the stages of system disruption.

Recovery (after the Disruption). The biggest challenge that system owners face is the restoration of their systems after disruptions. Depending on the type of system and the nature of the disruption, the recovery may take weeks, months, or even years. Also, a concern is that for a long period of time, the restored physical or operational state may be inferior to the state before the disruption. In many cases, a system owner is able to "exploit" the damage occurrence as an opportunity to lobby for increased funding to increase the system's physical condition or capacity to a higher level. A good way to imagine the term "recovery" would be the capability of a ball to return to its original shape after a force is removed (Jackson, 2009): upon deformation, a rubber ball recovers much better compared to balls made of other materials.

Mitigation (after the Disruption). Mitigation includes learning the lessons from previous similar disruptions and making plans for long-term prevention.

(b) The Sheard and Mostashari Model. Figure 27.11, adapted from Sheard and Mostashari (2008), presents a five-stage model for the disruption of civil engineering systems and the resilience activities at each stage. Also, the Resilience Engineering Network (2007), Hale and Heijer (2006), Westrum (2006), Hollnagel et al. (2006), and Gunderson (2000) provide some resilience-enhancing initiatives and good practices at each stage of the disruption process. These are discussed below.

Long-Term Prevention. System agencies promote resilience by include the developing analytical or simulation models that help predict the occurrence and intensity of the disruption. With such knowledge, the agency is placed in a better position to anticipate or plan for disruptions, prepare the community in order to reduce the exposure, and equip itself with the necessary resources to prevent or reduce the loss of control during the disruption.

Short-Term Avoidance. Where the disruption is predicted and where there is adequate time, the system agency can manage, to some extent, the looming threat. This includes efforts to keep information systems, social network resources, and other communication tools up to date.

Immediate-Term Coping: Survive. At this stage, the system agency addresses problems associated with the disruption. Often, these are sudden problems that pertain to the operational integrity of the system; in other cases, they pertain to physical damage that must be repaired in order for

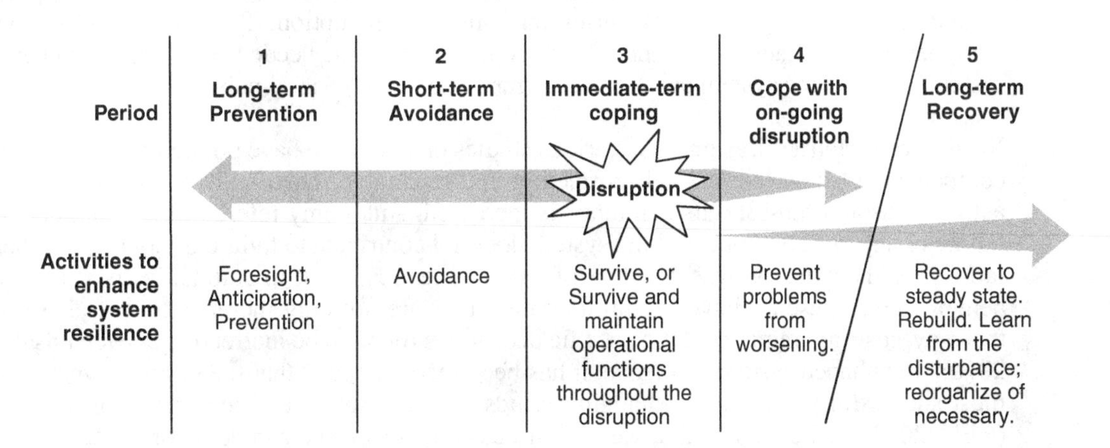

Figure 27.11 Resilience attributes within a five-stage process of system disruption.

operational performance to be restored. It is often the case that a system must be resilient enough to operate continuously throughout the disruption albeit with a lower level of functionality. A resilient system, therefore, is one that responds or reacts quickly and efficiently to disruptions, is able to recover from loss of control, and is robust enough to resist adverse situations or stimuli that could impair its functionality.

Cope with Ongoing Trouble. Not only must the system be resilient enough to survive the disruption, but it must also prevent a bad situation from becoming worse. To do this, the resilient system must be capable of enduring adverse situations during the disruption, and the system owner will need to carry out continuous monitoring of the situation to identify and address other threats or to exploit opportunities that emerge during the disruption.

Long-Term Recovery. To enhance system resilience, the system owner must carry out activities that help the community as well as the system to recover from the disruption. The system owner must recognize and learn the appropriate lessons related to the reliability of prediction of the threat occurrence and magnitude, the reduction of community exposure, and thc adequacy of the system resilience. This knowledge will assist the system owner in becoming better prepared for future similar threats.

27.5.4 How to Increase the Resilience of Civil Engineering Systems

At any of the stages of resilience discussed above, both operational and physical resilience could be enhanced using a similar set of principles. We herein discuss a number of these principles.

Threat Identification and Measurement. The first step in increasing the resilience of civil engineering systems is to identify the possible threats and to establish the appropriate criteria for measuring resilience.

Information Systems. The system owner must establish a database and management system that monitors the condition of the system as well as the system's usage, the climatic conditions, and other factors that affect suddenly or gradually, the system's physical condition.

Effective Communication. Civil engineering systems can be made more resilient when they have an effective communication system that collects, manages, and disseminates critical information related to the system operations during and after the disruption. Typically, a strong social network helps to facilitate, during and after disruption events, the needed communication between the system operator, the system users, the community, and the general public.

Resilience-Oriented Design. As various studies on resilience have pointed out, when the critical components of a system are self-correcting and repairable, have built-in redundancies, and are autonomous and fail-safe, its resilience is increased. Autonomy refers to the situation where the failure of any one component of the system does not contribute to failure of other components; and fail-safe is the situation where failure of a component causes it to automatically shift to its most benign form. These attributes assure that decisions are not only incremental but also reversible; that way, costs are minimized if a specific decision is found to be ineffective or even unsafe (VTPI, 2010). To enhance system resilience, it has been recommended that the system owner or operator must successfully predict the changing trends of risk prior to the occurrence of any disruption (REN, 2012). In Section 27.5.6, we will discuss how resilience can be incorporated at the design phase of system development.

Resilience-Oriented Operations. The resilience of civil engineering systems can be enhanced by providing the system operator the capability, even under extreme conditions, to correct operational problems and to undertake remedial measures.

Avoid Resilience Drift. In practice, system resilience is often high in the aftermath of a disruption; however, with time, there is a tendency for a gradual drift away from policies and processes that ensure system resilience and safety (Leveson et al., 2006). It is important for the system owner or operator to guard against this tendency.

The above examples are general principles for increasing the resilience of civil engineering systems. The specific strategies will differ across system types in the different civil engineering disciplines. In the field of transportation engineering, for example, a number of strategies that can increase the resilience were presented by ECONorthwest and PBQD (2002) and Husdal (2004): Enhance the diversity of transportation systems by ensuring that adequate opportunities exist for people to use other forms of transportation: cycling, walking, carsharing, ridesharing, and traveling by transit; increase the redundancy and connectivity of the area's transportation network; promote standards for system design and construction to withstand extreme natural or operating conditions; improve monitoring systems in order to quickly identify problems before they occur, including physical damage, demand surges, unsafe operating conditions, and other risks; improve communication with transportation system users, including those with special needs, even during the time of the disruption occurrence; give higher priority to transportation system resources (signal green times and road space, e.g.) to higher-value transportation activities such as evacuation or response to medical emergencies or fire.

27.5.5 Resistant versus Resilient Systems

Fiksel (2003) explained the difference between a resistant system and a resilient system. Figure 27.12 [adapted from Fiksel (2003)], is a simplified illustration of the different levels of resilience exhibited by three systems, 1, 2, and 3. For each of these systems, there is a stable state that represents the lowest possible potential energy; at this state, order is maintained. Each system is subject to disruptions that leave it in any one of several possible states, albeit very temporarily.

System 1 is designed to resist disruptions from its stable state and operates within a very narrow band of possible states. Systems having such resilience recover rapidly from small disruptions but are unlikely to survive large perturbations. Characterized as low resilience or highly resistant, these systems are typical of civil engineering systems that are designed within specific confines and engineered to be highly controlled.

System 2, compared to system 1, is considered to be a resilient system because it can survive across a broader range of possible states. These systems recover eventually from significant disruptions, gradually returning to their stable state. The ability of such systems to survive large

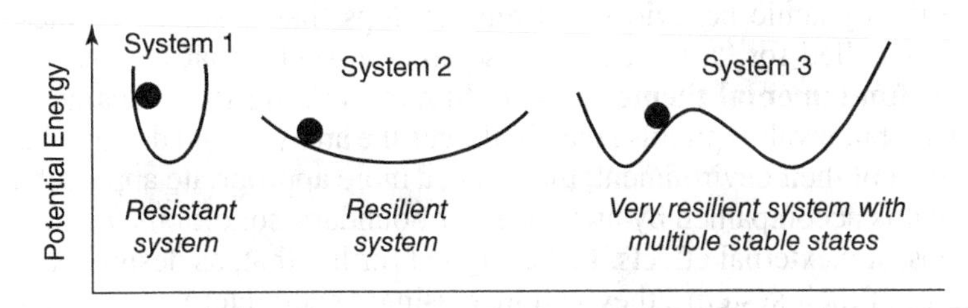

Figure 27.12 Resilience attributes at the stages of system disruption.

disruptions is often attributed to their inherent ability to evolve and adapt to new conditions. This is typical of ecosystems and social systems.

System 3 is even more resilient than system 2 because it can tolerate larger disruptions. When sufficiently disrupted, these systems may evolve into a different state of stability, a shift that is often accompanied by a fundamental change in its physical structure or operational function.

According to Fiksel (2003), the design of engineered systems has typically followed a pattern of *hierarchical decomposition*; that is, first, the overall system architecture and function are established followed by the design of the constituent subsystems and components. Systems that are hierarchically organized, including nuclear plants and aircraft systems, generally have operating parameters that are rather rigid and are resistant to stress only within relatively narrow boundaries. They are thus generally vulnerable to small, often unexpected disruptions, as reflected by system 1 in Figure 27.12. The "brittleness" in these designs is evidenced in the disruption that the system suffers when it undergoes a small disruption. On the other hand, distributed systems that are composed of independent yet interactive components have greater resilience and thus generally exhibit a functionality that is equivalent or even superior to the former. Examples include a collection of water supply towers (one in each town) distributed over a region instead of a single large water tower that supplies multiple towns in the region, as reflected by system 2 in Figure 27.12.

In connecting resilience to sustainability, Fiksel (2003) noted that if requirements that address inherent resilience are considered during system design, the end product generally tends to be more sustainable. The resilience-enhancing attributes of diversity and adaptability may not directly lead to enhanced system performance or profitability, but they are likely to contribute to the system's durability and its capability to adapt and survive in temporary or permanent conditions of disruption.

As we saw in Chapter 1, different design philosophies translated into the longevity of the systems developed by different civilizations. The ancient Romans and Persians developed systems that were built mainly for purposes of military strategy and commerce and tended to disrupt the natural patterns of nature. On the other hand, the civil engineering systems built by the ancient Greeks tended to shy away from violent interference with natural land forms and obstacles. Also, in ancient China, the Confucius philosophy of design was to confine and repress nature and thus advocated "masculine" activities such as dike construction, similar to the Roman and Persian approaches. The Taoist philosophy, on the other hand, was consistent with greater freedom for natural courses and thus advocated the use of "feminine" activities such as dredged concavities, similar to the Greek approaches (Needham, 1986). Needless to say, structures built using the resilient Greek and Taoist philosophies generally outperformed those built under the resistant Roman, Persian, and Confucian philosophies.

27.5.6 Resilience in Design: What the Future Holds

Recognizing that it is difficult to model complex, nonlinear systems by connecting a fragmented number of linear models, Fiksel (2003) advocated for new metaphors and language to adequately describe the dynamic behaviors and relationships that characterize these kinds of systems, and specifically called for "a new, multi-disciplinary toolkit that begins with connectivity and integration as **fundamental themes rather than afterthoughts**." Separateness, Fiskel contends, is a convenient (but invalid) premise that facilitates the analysis and design of entities as though they are independent of their environment; therefore, a more appropriate approach for system design should be one that is accompanied by awareness of boundary conditions, related systems, potential feedback loops, and external effects. Fiskel argued further that, as design teams continue to expand the boundaries of their systems, they will increasingly encounter the challenge of addressing the technical issues that emerge; for example, the considerations at the start of design will include system behaviors and not only the system performance outcomes; predictive modeling will be abandoned in favor of exploratory scenario analysis; and design strategies will not be based on control but on intervention. Also, design teams will seek to ensure system robustness through resilience rather than resistance; and in analyzing the risks associated with system operations, they will consider the concepts of diversity and capability to evolve to adapt to disruptions.

Researchers contend that owners of civil engineering systems will need to give due consideration to the resilience of their systems if they seek to adopt sustainable practices in the long term. However, it will not be sufficient for an agency to redesign only the systems within its control; it is believed that such outreach will result in incremental changes that may not lead to significant benefits for the system owner or for the community but also do no harm. The notion is that system owners seeking to enhance long-term resilience of their systems must instead reach beyond their own borders to acquire deeper understanding of the overall system of systems of which they are participants. If the "playing field" is widened to such an extent, the system owners will find that for the system to survive, the importance of strategic adaptation exceeds that of strategic planning, and the system owner will find value in embracing uncertainty and adapting to it rather than trying to eliminate it (Fiksel, 2003).

In the different branches of civil engineering, there have been efforts to incorporate resilience into various phases of system development including planning, design, and operations. These include the development of a framework for improving resilience during bridge design (Chavel and Yadlosky, 2011), examination of ways to increase structural resilience to earthquake threats (Takewaki et al., 2013), development of strategies to make cities more resilient (Molin Valdes et al., 2013), enhancing blast-resilient design of concrete frame buildings (Khan, 2011), and enhancing the resilience of transportation system operations (ECONorthwest and PBQD (2002).

27.5.7 Resilience and Reliability

In Chapter 13, we discussed the reliability of civil engineering systems and recognized that the concept of reliability is very much related to system resilience. As we learned in that chapter, the reliability of a civil engineering system may be described as the capability of the system (or its component) to perform its required functions or to achieve its established performance objectives under a given set of conditions and at a given point in time. Thus, reliability typically refers to situations involving normal operations of the system. However, when the system's operation (or its very existence) is threatened by a natural or man-made disruption, the concept of system resilience becomes even more important. A resilient system is one that can continue to maintain its reliability after a disruption. As we learned in Chapter 13, a system that is designed to be resilient will absorb the shocks associated with the disruption, and such disruption may be related to a large spectrum of possibilities including natural or man-made disasters, major equipment failures, and loading upsurges.

Generally, a system that is diverse, autonomous, and had redundant components has a high level of reliability, and therefore, resilience. Other characteristics of a reliable system are the efficiency of its processes and having critical components with high levels of structural and operational integrity, and in some cases, the capability for self-healing. Autonomy means that "the failure of one component or subsystem does not cause other components to fail". Thus, a reliable system is also a resilient one because such a system is able to continue to perform its function in the event that a component fails or a particular material, equipment, or human resource becomes unavailable (VTPI, 2010). At the phases of systems planning and design, engineers typically seek to enhance resilience by adopting designs that render their systems less vulnerable to damage from sudden or sustained threats. In Chapter 13, we also learned that the resilience of a civil engineering system can be enhanced by increasing its adaptive capacity either by incorporating greater redundancy in the system design to ensure continuous functioning or by increasing the ability and speed by which the civil engineering system recovers, evolves, or adapts to new often challenging situations. The concept of resilience is particularly important when seen in the context of a system of systems. Civil engineering infrastructure and systems are become increasingly networked. Besides the numerous advantages of this integration, the adverse consequence is that they are becoming more interdependent because previously independent systems are now vulnerable to disruptions at other systems; thus, there is a need for SOSs to be equipped with the capability to recover from the shocks induced by natural or anthropogenic stimuli.

27.5.8 Concluding Remarks on System Resilience

The concept of system resilience, particularly in the context of social, ecological, and management systems, has been studied in great detail by two prominent ecologists, C.S. "Buzz" Holling and Lance Gunderson, along with an international group of researchers. These scientists established a general theory of adaptive cycles by positing that the patterns exhibited by all systems are similar: gradual buildup of resources, increasing linkages, decreasing resilience, system life is punctuated by "alternating periods of crisis, transformation, and renewal." If these patterns can be understood, it will be possible for humans to devise intervention strategies that exploit these system dynamics rather than merely seeking to resist change (Fiskel, 2003). In *Collaborative Resilience: Moving through Crisis to Opportunity* [edited by Bruce Goldstein (2011)], a number of prominent scientists describe collaborative efforts in the wake of system disruptions, including natural disasters or intended or unintended man-made disasters (e.g., economic collapse, technological failure, and acts of violence), and they examine how certain communities have managed to survive, and even thrive, in the aftermath of such disruptions.

27.6 COMBINING THREAT LIKELIHOOD, PUBLIC EXPOSURE, AND SYSTEM RESILIENCE FOR DECISION MAKING

The duties of the civil engineer includes measurement of the extent of the danger or "hazard" posed by external or internal threats to the system, monitoring the hazard over time, assessing the effectiveness of actions intended to reduce the hazard, communicating this information to the general public and legislators, and making demands for appropriate resources in order to increase the resilience of the system to the threat, reduce the exposure to the threat, or facilitate recovery in the event the threat is realized. In this section, we present a generalized procedure for quantifying overall hazard to destruction or damage of civil engineering systems. This framework can be modified to suit a specific civil system or situation, and can be used by engineers in the above-mentioned duties.

27.6.1 Generalized Procedure for Quantifying Overall Hazard to Destruction or Damage

As we learned in the introduction to this chapter, the overall hazard, H, posed by a situation where a civil engineering system located in a certain environment faces a certain type of threat (Figure 27.1) can be represented as follows:

$$H = f(L, C, R)$$

where L is the likelihood of the threat, C is the consequence of the threat if it were to occur, and R is the resilience of the system.

This section of the chapter presents a hazard rating procedure synthesized from various procedures from the literature and based on the likelihood and consequence of a threat. The threat likelihood, assessed on the basis of the threat type, is a function of the system environment rather than a function of the system itself. The threat consequence, or impact of failure, is based on the possible failure scope and the extent to which the facility is exposed to the threat—both of these are specific to the facility in question and not to the threat type. The exposure to the threat is a measure of the effect that a failure of the facility will have on its users, which is related to the occupancy or traffic volume of the facility and the importance of the facility in the areas of socioeconomic development or national defense (Labi et al., 2011).

The essential elements of risk assessment are consistent with systems hazard analysis (Ezell et al., 2000): What could go wrong, what is the likelihood that it will go wrong, and what are the consequences if it goes wrong? With regard to what could go wrong, a number of studies have identified or quantified the types of threats to civil engineering systems. These threats, which are related to environmental factors or to system characteristics, include hydraulic factors, overloads, steel or concrete structural details, collisions, earthquakes, and condition-related reductions in load capacity (Shirole and Loftus, 1992; Kuprenas et al., 1998; Stein et al., 1999; Small, 1999; Monti and Nistico, 2002). Some of these studies developed methods to assess civil structure hazard and to select those in need of improvements to guard against imminent threats. Hazard assessments have been carried out in other disciplines besides civil engineering as well, and they offer valuable lessons. Luers et al. (2003) developed a framework to assess the hazards of agricultural systems in Mexico's Yaqui Valley. Phillips and Swiler (1998) presented a flexible graph-based approach to security network hazard analysis and used probability theory and various graph algorithms to identify attack paths that have the maximum probability of being realized. Eakin and Luers (2006) investigated the hazards of social-environmental systems, and Moy et al. (1986) investigated the reliability in water supply reservoir operations by exploring system hazard and system resilience. Ezell (2007) presented a model for quantifying the hazards of critical infrastructure, using hazard density functions derived from value functions and weights.

Figure 27.13, synthesized from the hazard and vulnerability assessment procedures of NYS-DOT and other literature, presents a general procedure for assessing the overall hazard due to different threat types including erosion, scour, fatigue/fracture, earthquake, and collision. This assigns a hazard "class" and is specific to the threat type under consideration. The procedure is discussed in subsequent sections.

Step 1. Computation of Threat Likelihood Level. For a given facility, the threat likelihood level is generally independent of the system (facility) type and depends on the location of the system. For

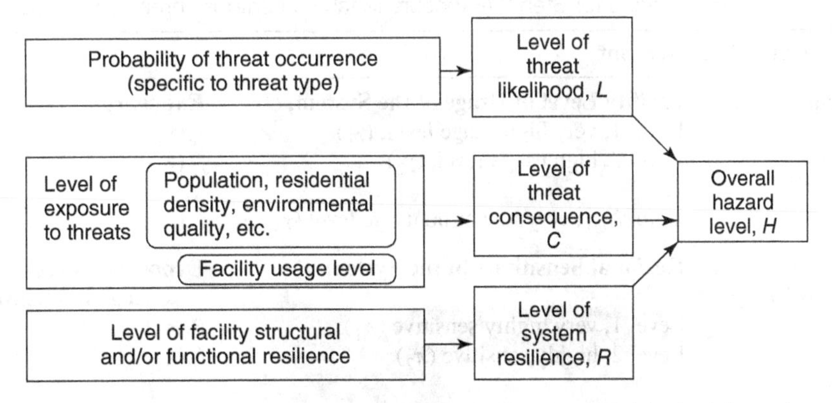

Figure 27.13 Procedure for rating the hazard associated with a civil engineering system.

instance, some systems are located in earthquake-prone areas, in floodplains, or on mountainsides where systems are vulnerable to landslide threats. Thus, generally, the threat likelihood level, L, can be expressed as a function of the past/likely occurrence locations of the threat as follows:

$$L = f_L(I) \tag{27.1}$$

where I is the past/likely occurrence locations and other threat attributes, and $f_L(I)$ can be derived from past frequencies of occurrence of that threat type, simulation, expert opinion (using direct assignment of scores by experts), or utility functions. A more objective measurement of $f_L(I)$ could be the use of geographical maps that indicate the variation of the threat level at each location.

Step 2. Computation of Consequence Level. The consequence is related not to the threat but to the environment in which the civil engineering facility is located. The consequence is a function of the regional sensitivity and the occupancy, as shown in Table 27.1.

(a) Computation of the Exposure Level. For a facility, the exposure level, is defined as a function of P, the regional sensitivity (population, residential density, environmental quality), and O, the usage level as follows:

$$\mathrm{E} = f_E(P, O) \tag{27.2}$$

where P and O can be determined from $f_P(\pi_i)$ and $f_O(\gamma_i)$, respectively. Examples of the function f_{E} include:

$$\mathrm{E} = \Phi_1 O^\alpha + \Phi_2 P^\beta \tag{27.3}$$

$$\mathrm{E} = \chi_1 O^\alpha \chi_2 P^\beta \tag{27.4}$$

NYSDOT, in its hazard rating procedure (NYSDOT, 1996–2002), uses the simple weighted linear additive form by setting $\Phi_1 = \Phi_2 = 1$, and $\alpha = \beta = 1$, as follows: Exposure level = usage score + regional sensitivity score.

(b) Computation of the Consequence Level. For any given facility, the consequence level, C, is defined as a function of the exposure level calculated in Equation (27.2) as follows:

$$C = f_C(\mathrm{E}) \tag{27.5}$$

Table 27.1 Parameters for Step 2 (Exposure Measurement for Threat Consequence Assessment)

Exposure Measurement		
Usage Level	**Facility Level of Usage of the System, O**	**Exposure Corresponding to Usage Level**
	Level 1, very high usage level, (γ_1)	$f_{\mathrm{E}}(\gamma_1)$
	Level 2, high usage level, (γ_2)	$f_{\mathrm{E}}(\gamma_2)$
	...	...
	Level N_{OT}, very low or no usage level $(\gamma_{N,O})$	$f_{\mathrm{E}}(\gamma_{N,O})$
Regional sensitivity	**Regional Sensitivity of the System, P**	**Exposure Corresponding to Regional Sensitivity Level**
	Level 1, very highly sensitive (π_1)	$f_P(\pi_1)$
	Level 2, highly sensitive (π_2)	$f_P(\pi_2)$
	...	...
	Level N_C, very low sensitivity $(\pi_{N,\mathrm{FC}})$	$f_P(\pi_{N,P})$

Source: Adapted from (NYSDOT, 1992–1997)

Examples of Equation (27.5) include:

$$C = \Omega_1 E^{\omega} \tag{27.6}$$

$$C = \Psi_1 E^{\omega} \tag{27.7}$$

Step 3. Computation of Facility Resilience Level. The level of resilience represents the ability of the facility to withstand external or internal shocks, to not fail, and to continue performing its function after the threat has occurred. This data could be obtained from field measurements or laboratory simulations.

$$R = f_R(\text{structural integrity, operational reliability, functional adequacy, etc.})$$

From the perspective of structural resilience, a weak bridge has low resilience while a strong bridge has high resilience. From the perspective of operational/functional resilience, a highway with a number of lanes that have very low demand has low resilience while a highway with lanes with excess demand has high resilience in the event it is needed to assist in evacuation from a disaster area. In certain texts, system resilience is expressed in terms of a failure score (higher failure score means lower resilience).

Step 4. Computation of Overall Hazard of the Facility. For each threat type, a hazard rating level, H, is defined as a function of the likelihood level (L), consequence level (C), and resilience level (R) as follows:

$$H = f_{\text{HR}}(L, C, R) \tag{27.8}$$

Examples of this function could include:

$$H = \Lambda_1 L^{\eta} + \Lambda_2 C^{\tau} + \Lambda_3 R^{\kappa} \tag{27.9}$$

$$H = \Theta_1 L^{\eta} * \Theta_2 C^{\tau} * \Theta_3 R^{\kappa} \tag{27.10}$$

A simple weighted linear additive form can be obtained by setting $\Lambda_i = \Theta_i = 1$, and $\eta = \tau = \kappa = 1$; Hazard rating = likelihood score + consequence score + resilience score. In general, the hazard rating can be calculated using the appropriate functional form for Equation (27.8), scaled on a 0–100 scale, and then interpreted in step 5.

Step 5. Interpretation of the Overall Hazard Rating. Figure 27.14 and Table 27.2 provide a possible interpretation of the overall hazard rating calculated using the above procedure. In the figure, the boundaries between the hazard descriptions are based on expert judgments and are for illustration purposes only. Researchers continue to establish these boundaries in a more objective manner.

27.6.2 Quantifying the Vulnerability of a Threat–Resilience Situation

The definition of vulnerability varies from agency to agency. However, for the purposes of this text, we define vulnerability as a function of the threat intensity/likelihood and system resilience. That

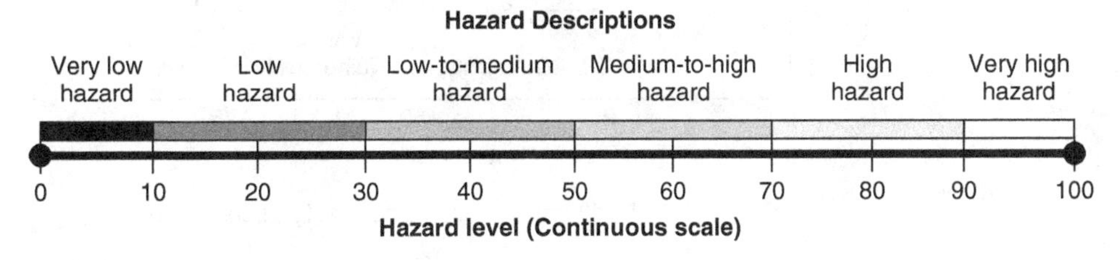

Figure 27.14 Scale for hazard interpretation.

Table 27.2 Interpretation of 0–100 Hazard Rating

Hazard Rating	Interpretation
0–9.9	Indicates little or no hazard.
10–29.9	Indicates low hazard. Often reflects the hazard of a system that is built to the current design standards.
30–49.9	Indicates low-to-medium hazard. Unexpected failure can be avoided during the remaining service life of the facility by performing standard scheduled inspections with due attention to factors that influence the system hazard.
50–69.9	Indicates medium-to-high hazard. Systems in this range can be monitored at a frequency slightly exceeding standard frequency. The risk of failure can be tolerated until a hazard-reducing retrofitting project is carried out.
70–89.9	Indicates high hazard. The agency should be ready to undertake actions to reduce the hazard of the system.
90–100	Indicates very high hazard. Immediate action should be undertaken to reduce the hazard of the system.

Source: Adapted from NYSDOT (1996–2002) and O'Connor (2000).

is, the vulnerability of a threat/system situation can be represented as follows: $V = f(L, R)$, where L is the likelihood and intensity of the threat, and R is the resilience of the system. It can be seen that this definition of vulnerability excludes the consequence, which is a function of the exposure. The vulnerability and the consequence produce the overall hazard.

Thus, the term "vulnerability" as defined in this chapter pertains to both the threat and the system. True vulnerability lies in the situation that involves a consideration of both the threat and the system together. Thus, a highly vulnerable situation is one involving high threat likelihood/intensity and low system resilience (e.g., a bridge in very poor condition in an earthquake-prone area). On the other hand, a low-vulnerability situation is one involving low threat likelihood/intensity and high system resilience (e.g., a new, well-designed bridge in an area with no risk of earthquake, flooding, or other threat types). A vulnerability matrix or nomograph can be used to ascertain the vulnerability of a situation on the basis of the threat type likelihood and intensity and the resilience of the system. An example is presented as Figure 27.15 where the threat likelihood/ intensity and system resilience are expressed on a 1–100 scale. Similar nomographs or matrices could be established for each threat type.

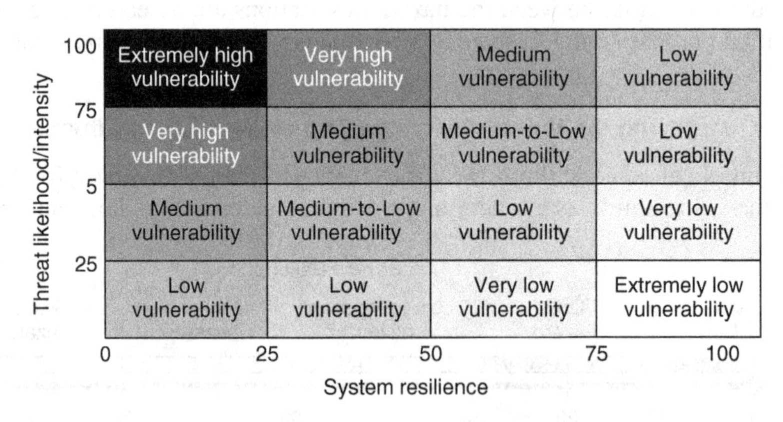

Figure 27.15 Illustration of vulnerability assessment.

SUMMARY

Civil engineering systems are vulnerable to natural and man-made threats that arise from the environment in which they are located. Natural threats include floods, landslides, earthquakes, and hurricanes; man-made threats include overloading due to excessive traffic or occupancy and accidental or malicious collisions between man-made objects (such as land, sea, or airborne vehicles) and the civil engineering structures. Increased attention has been paid by agencies. In recent years, agencies are increasingly paying attention to continuous monitoring of imminent threats and also to evaluation and implementation of investments that reduce system hazards by reducing threat likelihood, increasing the resilience of their systems, or reducing the exposure. As we discussed in this chapter, highly publicized disaster events in recent times have underscored the vital importance of monitoring the hazards to civil engineering systems to external or internal threats and to assess the effectiveness of hazard reducing investments. The looming specter of climate change and the anticipated subsequent rise in sea and groundwater levels, wind speeds, and other environmental changes are expected to lead ultimately to potentially widespread and deleterious impacts on civil structures and thus constitute a serious natural threat with which engineers will need to contend. This chapter presented a methodology, to quantify hazard for any civil engineering system type and the nature of the threat.

EXERCISES

1. Discuss the relationships between system resilience and environmental sustainability. Why are these subjects currently of great interest and attention worldwide?
2. Define the following basic or "primary" terms associated with system hazard: threat likelihood, community exposure, and system resilience. Identify a prominent civil engineering system in your neighborhood or on campus and identify the threats and their likelihoods, the community exposure in case of a threat occurrence, and the resilience of the system to the threats.
3. Define the "secondary" terms protection, peril, vulnerability, and overall hazard and provide a practical discussion of these terms in the context of the example you provided in the previous question.
4. Draw a table that shows the various ways by which threats to civil engineering systems could be categorized. For each category, list the possible threat types and provide supporting references from the literature to support your list.
5. List any three civil engineering systems each from different branches of civil engineering. For each system, identify the possible internal and external threats. For external threats, identify those that are sudden or gradual; for sudden threats, identify those that are natural or man-made; and for man-made threats, identify those that are malicious (intended) or non-malicious (unintended).
6. What can civil engineers do to reduce the dangers posed to their systems by any two of the following threat types: flooding, earthquakes, landslides, erosion, scour, and sedimentation, internal fatigue or design flaws, poor condition due to aging, terrorism, overloading and oversize, collision, threats related to climate and weather, vandalism, and theft?
7. Discuss the primary impacts and the secondary impacts of climate change on civil engineering systems, from the perspective of any one branch of the discipline.
8. Discuss the various ways in which civil engineers are seeking to mitigate the effects of climate change on their systems.
9. Comment on the following: (a) exposure to threats faced by civil engineering systems, (b) relationship between environmental resilience and system sustainability, and (c) relationship between resilience and reliability.

10. Discuss, using diagrams, any one of the two common models used to describe the stages of civil engineering system disruption. On the basis of the model, explain how civil engineers could enhance the resilience of their systems to external and internal threats.
11. Explain the difference between resistant and resilient systems, and identify, with reasons, which is more desirable.
12. Discuss the role of resilience in design of future civil engineering systems.
13. Section 27.6.1 of this chapter presents a generalized procedure for quantifying the overall hazard to system destruction or damage. Show how the elements of this framework could be modified for specific application to (a) building systems, (b) highway bridge systems, (c) water supply systems, and (d) levee systems.
14. A number of civil engineers reviewed and assigned resilience scores to 36 civil systems (Table 27.3). Res_ rating is the rating assigned to each system on a 0 (port) to 100 (excellent) scale; Mat_Type is the dominant material type used for the system construction (0 for steel, 1 for concrete), Des_Type is the type of design used for the system (0 for traditional design, 1 for newer designs), Age is the number of years since the construction, and Mtce is the amount of average annual maintenance received by the system since construction \$ per ft^2. Develop a statistical function that (i) determines, from the data, which system factors appear to be significant at 5% significance level and (ii) predicts the resilience score as a function of the system characteristics.

Table 27.3 Data for Exercise 27.14

ID	1	2	3	4	5	6	7	8	9	10	11	12	13
Mtce	2.0	2.7	3.5	3.0	2.5	3.0	2.5	3.8	4.5	3.0	2.5	4.0	2.8
Age	6	6	6	6	2	6	2	5	8	8	5	7	3
Des_Type	0	0	0	0	0	0	0	0	0	1	0	1	0
Mat_Type	1	1	1	0	0	1	1	0	0	1	0	0	0
Res_ rating	10	13	67	10	23	15	16	17	31	13	22	14	20

ID	14	15	16	17	18	19	20	21	22	23	24	25	26
Mtce	2.1	2.4	2.6	4.0	2.6	3.0	2.5	4.0	2.8	2.1	2.4	2.6	4.0
Age	6	5	5	6	6	8	5	7	3	6	5	5	6
Des_Type	0	0	0	0	0	1	0	1	0	0	0	0	0
Mat_Type	0	0	1	1	0	1	0	0	0	0	0	1	1
Res_ rating	11	11	90	5	28	13	22	14	20	11	11	90	5

ID	27	28	29	30	31	32	33	34	35	36	37	38	39
Mtce	2.6	2.5	3.0	2.4	1.3	2.1	3.5	2.3	3.0	2.5	2.5	3.0	1.8
Age	6	6	6	7	7	7	8	6	6	7	9	8	6
Des_Type	0	0	0	1	0	1	1	0	1	0	1	0	0
Mat_Type	0	0	0	0	1	0	1	0	0	0	0	1	0
Res_ rating	28	17	15	12	6	8	21	15	39	8	11	50	9

15. A certain civil engineering facility operates in a very harsh environment characterized by seismic events, high winds, and extreme temperatures, and thus was designed and constructed to withstand these multiple threats. From laboratory simulations, the system is expected to have a mean failure time of 15 years. Assuming that the system has a constant probability of failure, what is the longevity reliability of these systems over a period of (a) 10 years (b) 15 years (c) 20 years?

16. Figure 27.16 represents a system with components with both parallel and series arrangement. Each component has a reliability function governed by the expression: $R = e^{-\lambda t}$. Derive the final expression for the reliability of this system. If a number of internal and external threats lead to the following increases in failure rates of each component: *A*, 25%; *B*, 10%; *C*, 50%; *D*, 10%; *E*, 20%; and *F*, 15%, determine the overall percentage reduction in reliability due to the threats.

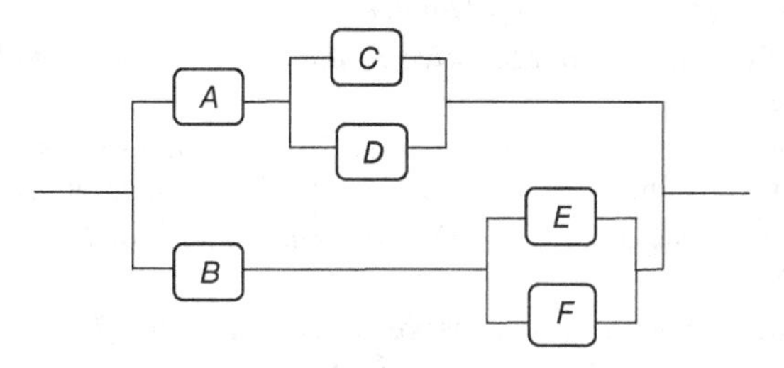

Figure 27.16 Figure for Exercise 27.16.

REFERENCES

Ali, M. (2008). Energy Efficient Architecture and Building Systems to Address Global Warming, Engineering Strategies for Global Climate Change. *ASCE Leadership Manage. Eng.* 8(3), 113–123.

ASCE (2007). Impact of Global Climate Change. ASCE Policy Statement 360, Energy, Environment, and Water Policy Committee of ASCE. Reston, VA.

Beniston, M. (2004). *Climatic Change and Its Impacts*. Kluwer, Dordrecht.

Berkes, F., and Folke, C. (1998). *Linking Social and Ecological Systems: Management Practices and Social Mechanisms for Building Resilience*, Colding, J. (Ed.). Cambridge University Press, Cambridge, UK, 1(33), 429–433.

Brand, F. (2009). Critical Natural Capital Revisited: Ecological Resilience and Sustainable Development, *Ecolog. Econ.* 68(3), 605–612.

Briguglio, L. P. (2010). Defining and Assessing the Risk of Being Harmed by Climate Change. *Int. J. Climate Change Strat. Manage.* 2(1), 23–34.

Brunner, E., and Giroux, J. (2009). Factsheet: Examining Resilience–A Concept to Improve Societal Security and Technical Safety, Center for Security Studies, ETH, Zurich, Switzerland.

Chavel, B. W., and Yadlosky, J. M. (2011). Framework for Improving Resilience of Bridge Design. Rep. Nr. FHWA-IF-11-016, Fed. Hwy. Admin., U.S. Dept. of Transp., Washington, DC.

Cohen, S. (1973). Property Destruction: Motives and Meanings, in Ward, C., Ed. *Vandalism*. Architectural Press, London.

deGeus, A. (1997). *The Living Company* Harvard Business School Press, Cambridge, MA.

Eakin, H., and Luers, A. L. (2006). Assessing the Hazard of Social-Environmental Systems. *Ann. Rev. Environ. Resourc.* 31, 365–394.

ECONorthwest and PBQD (2002). Estimating the Benefits and Costs of Public Transit Projects. TCRP Report 78, TRB, Washington, DC.

Ezell, B. C. (2007). Infrastructure Hazard Assessment Model. *Risk Anal.* 27(3), 571–583.

Ezell, B. C., Farr, J. V., and Wiese, I. (2000). Infrastructure Risk Analysis Model. *ASCE J. Infrastruct. Syst.* 6(3), 114–117.

Felgenhauer, T., and De Bruin, K. C. (2009). The Optimal Paths of Climate Change Mitigation and Adaptation under Certainty and Uncertainty. *Int. J. Global Warming* 1, 1/2/3.

Fiksel, J. (2003). Designing Resilient, Sustainable Systems. *Environ. Sci. Tech.* 37, 5330–5339.

Foster, H. (1993). Resilience Theory and System Evaluation, in Verification and Validation of Complex Systems: Human Factor Issues, in Wise, J. A., Hopkin, V. D., and Stager, P., Eds. *NATO Advanced Science Institutes, Series F: Computer and Systems Sciences*, Vol. 110. Springer, New York, pp. 35–60.

FPE (2009). Global Warming and Its Impact on Public Health. Forum of Pakistani Engineers, Islamabad, Pakistan. http://fope.net/?p=36. Accessed April 8, 2010.

Gertner, J. (2008). The Future Is Drying up: The Other Water Problem, Engineering Strategies for Global Climate Change. *ASCE Leadership Manage. Eng.* 8(3).

Goldstein, B. E., Ed. (2011). *Collaborative Resilience, Moving through Crisis to Opportunity*, MIT Press, Cambridge, MA.

Guillerminet, M.-L., Tol, R. S. J., and Vafeidis, A. T. (2008). Decision Making under Catastrophic Risk and Learning: The Case of the Possible Collapse of the West Antarctic Ice Sheet. *Climatic Change* 91, 71–191.

Gunderson, L., and Protchard, L., Jr. (2002). *Resilience and the Behavior of Large-Scale Systems*: Island Press, Washington, DC.

Hill, D. (2008). Must New York City Have Its Own Katrina? Engineering Strategies for Global Climate Change. *ASCE Leadership Manage. Eng.* 8(3).

Hill, M., Wallner, A., and Furtado, J. (2010). Reducing Hazard to Climate Change in the Swiss Alps: a Study of Adaptive Planning. *Climate Policy* 10(1), 70–86(17).

Husdal, J. (2004). Why Reliability and Hazard Should Be an Issue in Road Development Projects, *Samferdsel: J. Norwegian Inst. Transp. Econ.* www.toi.no/samferdsel; at www.husdal.com.

INCOSE (2012). Resilient Systems Working Group. www.incose.org.

IPCC (1994). *Climate Change*. Cambridge University Press, Cambridge.

IPCC (2007). *Climate Change 2007: The Physical Science Basis. Contribution of Working Group I to the Fourth Assessment Report of the Intergovernmental Panel on Climate Change*, Solomon, S. et al., Eds. Cambridge University Press, Cambridge, UK.

Jackson, S. (2009). *Architecting Resilient Systems: Accident Avoidance and Survival and Recovery from Disruptions*. Wiley, Hoboken, NJ.

JSCE (2009). *Civil Engineers Confront Global Warming, Mitigation, and Adaptation Plans in Japan to Decrease the Risks of Global Warming*. Japan Society of Civil Engineers, Tokyo, Japan.

Khan, Z. A. (2011). Basics of Blast Resilient Design of Concrete Frame Buildings. engineeringcases knovelblogs.com.

Kuprenas, J. A., Madjidi, F., Vidaurrazaga, A., and Lim, C. L. (1998). Seismic Retrofit Program for Los Angeles Bridges. *ASCE J Infrastruct. Syst.* 4(4), 185–191.

Labi, S., Bai, Q., Ahmed, A., and Anastasopoulos, P. (2011). Quantifying System Hazard as a Performance Measure for Systems Investment Evaluation and Decision-making. 1st International Conference on Hazard and Risk Analysis and Management (ICVRAM), College Park, MD.

LaMalva, K. J. (2002). Complete Report on Failure Analysis of World Trade Center, Vol. 5, Simpson Gumpertz & Heger, Civil Engineering portal, www.engineeringcivil.com.

Lenkei, P. (2007). Climate Change and Structural Engineering. *Period. Polytech.* 47–50.

Leveson, N., Dulac, N., Zipkin, D., Cutcher-Gershenfeld, J., Carroll, J., and Barrett, B. (2006). Engineering Resilience into Safety-Critical Systems, in *Resilience Engineering: Concepts and Precepts*, Hollnagel, E., Woods, D. D., Leveson, N. G. Eds. Ashgate Publishers, Aldershot, UK.

Liao, S. C. (2008). Envisioning and Creating the Future in Response to Global Climate Change, Engineering Strategies for Global Climate Change. *ASCE Leadership Manage. Eng.* 8(3).

Lowe, B. (2007). When Cities Break Down. *Time magazine*, www.time.com.

Luers, A. L., Lobell, D. B., Sklar, L. S., Addams, C. L., and Matson, P. A. (2003). A Method for Quantifying Hazard, Applied to the Agricultural System of the Yaqui Valley, Mexico, *Global Environ. Change* 13(4), 255–267.

McInerney, D., and Keller, K. (2008). Economically Optimal Risk Reduction Strategies in the Face of Uncertain Climate Thresholds. *Climatic Change* 91, 29–41.

Molin Valdes, H., Amaratunga, D., and Haigh, R. (2013). Making Cities Resilient: From Awareness to Implementation. *Int. J. Disaster Resil. Built Environ.* 4(1).

Monti, G., and Nistico, N. (2002). Simple Probability-based Assessment of Bridges under Scenario Earthquakes. *ASCE J. Bridge Eng.* 7(2), 104–114.

Moy, W.-S., Cohon, J. L., and ReVelle, C. S. (1986). A Programming Model for Analysis of the Reliability, Resilience, and Hazard of a Water Supply Reservoir. *Water Resourc. Res.* 22(4), 489–498.

Needham, J. (1986). Science and Civilization in China, Part 3, Vol. 4., Caves Books, Taipei.

Nicholls, R. J. and Cozenave, A. (2010). Sea-level Rise and Its Impact on Coastal Zones, *Science* (18), 5985, 1517–1520.

NYSDOT (1996–2002). *Hazard Manuals*. Dept. of Transp., Albany, NY.

O'Connor, J. S. (2000). Bridge Safety Assurance Measures Taken in New York State. *Transport. Res. Record* 1696, 187–192.

Perman, R., Ma, Y., McGilvray, J., and Common, M. (2003). *Natural Resource and Environmental Economics*. Longman. London.

Phillips, C., and Swiler, L. P. (1998). *A Graph-based System for Network-Hazard Analysis*. Procs., New Security Paradigms Workshop, 22–25.

REN (2012). *Resilience Engineering*. The Resilience Engineering Network, www.resilience-engineering.org.

Rosen, M. A. (2009). Combating Global Warming via Non-Fossil Fuel Energy Options. *Int. J. Global Warming* 1, 1/2/3.

Rummel, T., Hyzak, M., and Ralls, M. L. (2002). Transportation Security Activities in Texas. Procs., 19th International Bridge Conference, Pittsburgh, PA.

Sassa, K., and Canuti, P. (2008). *Landslides–Disaster Risk Reduction*. Springer, Berlin.

Sheard, S. A., and Mostashari, A. (2008). A Framework for System Resilience Discussions. Procs., 18th Annual INCOSE International Symposium, Utrecht, The Netherlands.

Shirolé, A., and Holt, R. C. (1991). Planning for a Comprehensive Bridge Safety Assurance Program. *Transport. Res. Rec.* 1290, 39–50.

Shirolé, A. M., and Loftus, M. J. (1992). Assessment of Bridge Hazard to Hydraulic Failures. *Transport. Res. Rec.* 1347, 18–24.

Small, E. P. (1999). Examination of Alternative Strategies for Integration of Seismic Risk Considerations in Bridge Management Systems. 8th International Bridge Management Conference, Denver, CO.

Stanford Solar Center (2009). Global Warming, http://solar-center.stanford.edu/sun-on-earth/glob-warm.html. Accessed April 1, 2010.

Stein, S. M., Young, G. K., Trent, R. E., and Pearson, D. R. (1999). Prioritizing Scour Vulnerable Bridges Using Risk. *ASCE J. Infrastruct. Syst.* 5(3), 95–101.

Takewaki, I., Moustafa, A., and Fujita, K. (2013). *Improving the Earthquake Resilience of Buildings*. Springer Series in Reliability Engineering, New York, NY.

VTPI (2010). *Evaluating Resilience*, Victoria Transportation Policy Institute, Victoria, BC.

Westrum, R. (2006). *A Typology of Resilience Situation, in Resilience Engineering: Concepts and Precepts*, Eds.: Hollnagel, E., Woods, D., Leveson, N. G., Ashgate, Aldershot, UK.

USEFUL RESOURCES

Chang, S., and Nojima, N. (2001). Measuring Post-Disaster Transportation System Performance: The 1995 Kobe Earthquake in Comparative Perspective. *Transport. Res. A* 35(6), 475–494.

Foster, H. (1995). Disaster Mitigation: The Role of Resilience, in Etkin, D., ed. *Proceedings of a Tri-lateral Workshop on Natural Hazards*. Merrickville, Ontario, Canada, Feb. 11–14, pp. 93–108.

Hollnagel, E., Paries, J., Woods, D., and Wreathall, J. (2011). *Resilience Engineering in Practice: A Guidebook*, Ashgate, Aldershot, UK.

TRB (2008). Potential Impacts of Climate Change on U.S. Transportation, Transportation Research Board Special Report 290, TRB, Washington, DC.

USDOT (2005). Effects of Catastrophic Events on Transportation System Management and Operations: New York City—September 11. U.S. Dept. of Transportation.

CHAPTER 28

SUSTAINABILITY

28.0 INTRODUCTION

Sustainability, which takes its roots in the ancient Latin words *tenere* (to hold) and *sus* (up), generally means the capacity of a given entity or situation to endure adverse changes over time; such capacity is enabled by actions or conditions that nourish, preserve, or renew the entity or situation. For example, one could talk of a sustainable energy situation, where a country relies mostly on energy sources that are renewable; a sustainable ecological system such as a healthy wetland; or a sustainable development of civil engineering system that is built using renewable materials, operated with renewable energy, or can be completely salvaged at the end of its service life. The term sustainability has even pervaded everyday language; one may argue, for example, that living far from campus, working a full-time job, and having several pets at the same time is not a sustainable lifestyle for a college student.

The basis of the concept of sustainability is the notion that communities are constituted of economic, social, and environmental resources and entities that constantly interact with each other and that these interactions must be maintained in a state of harmonious balance, otherwise the future survival of the community will be in jeopardy. The community may refer to a region, nation, state or province, city/town/village, or even a local area within a city, town, or village. Thus, a sustainable community is one that is expected to endure hiccups or even disaster events on its economy, social or environment, and yet continue to provide, for all its members, a way of life that is decent, respect their human rights, and provide them a sense of dignity and safety. Complete sustainability is an ideal situation; however efforts towards this ideal can be extremely beneficial. By serving as a yardstick upon which civil systems managers weigh proposed or past actions, plans, expenditures, and decisions, sustainability can play a vital role in decision-making processes for civil systems and consequently, influence greatly the longevity of these systems.

In this chapter, we begin with formal definitions of sustainability and sustainable development. Then we discuss sustainability in ancient times and motivations for the drive toward sustainability in the current era. This is followed by a listing of the principles of sustainability, a timeline of the evolution of global policy on sustainability, and identification of the elements (or pillars) of sustainability. We also discuss the indicators and measurement of overall sustainability, and the different perspectives from which sustainability could be viewed. We then see how systems concepts could be applied in sustainability modeling, and the different sustainability considerations at each phase of system development. We will review how past researchers have sought to model the impact of human actions on sustainability, and we will also examine how to incorporate sustainability into project- and network-level evaluation of civil engineering systems. We will identify a number of specific sustainability considerations in civil engineering disciplines, and will conclude the chapter with some interesting issues related to sustainability.

28.0.1 Formal Definitions

Formal definitions of sustainability, from the civil engineering systems context, include:

> A set of environmental, economic and social conditions in which all of society has the capacity and opportunity to maintain and improve its quality of life indefinitely without degrading the quantity, quality or availability of natural, economic, and social resources. (ASCE, 2012)
>
> Use of the biosphere by present generations while maintaining its potential yield (benefit) for future generations; and/or non-declining trends of economic growth and development that might be impaired by natural resource depletion and environmental degradation. [The Organization for Economic Co-operation and Development (OECD, 2012)]
>
> The ability of a society, ecosystem, or any such ongoing system to continue functioning into the indefinite future without being forced into decline through exhaustion or overloading of key resources on which the system depends. (Gillman, 1992)
>
> The property of being sustainable, the condition where human activity may be continued indefinitely without damaging the environment and where the needs of all peoples are met equally. (Dauncey, 2012).

Sustainability is a key issue in the current century, and developers of civil systems particularly are in a unique position to gear current and future development to foster global sustainability (Bell, 2011). This can be facilitated by helping to equip society with a healthy new mindset about sustainable practices and also to mitigate the adversities caused partly or wholly by unsustainable engineering practices of the past.

As we learned in Chapter 3, one of the key goals of civil systems development is the sustainable development of the communities in which we live and the maintenance of a high quality of life. However, at different parts of the world, societies are at different stages of their evolution. As such, at a given time, a specific element of sustainable development may be more appropriate at certain societies than at others. Thus, as Amekudzi et al., (2011) point out, the broader context of community or society must be emphasized in evaluating infrastructure investments that seek to promote sustainability or sustainable development.

28.0.2 Sustainable Development

Similar to sustainability, the literature on sustainable development is replete with a large number of definitions. These include: "meeting human needs for natural resources, industrial products, energy, food, transportation, shelter, and effective waste management while conserving and protecting environmental quality and the natural resource base essential for future development" (ASCE, 2010a); "development that meets the needs of the present without compromising the ability of future generations to meet their own needs" ... "[situation] where the exploitation of resources, the direction of investments, the orientation of technological development, and institutional change are all in harmony and enhance both current and future potential to meet human needs and aspirations" (Brundtland, 1987).

The ASCE and Bruntland definitions imply the need for better stewardship of the environment but recognize that economic growth is necessary to meet the needs of society. There are others who refuse to accept the term sustainable development, arguing that it is an oxymoron because in their opinion, development is inherently inimical to sustainability, citing human proclivity to material wealth accumulation, greed, or prioritization of wealth ahead of environmental stewardship or social justice.

The need for sustainable practices in engineering design, construction, and operations has been duly recognized by professional engineering organizations and embraced by international development-oriented organizations and civil engineering professional societies worldwide. The

World Bank considers sustainable development as being fundamental to the bank's mission to reduce poverty, and has established a Sustainable Development Network that works with civil society, policymakers, and the private sector to encourage practices that foster sustainability (World Bank, 2013). In the United States, the ASCE in 1996 modified its code of ethics to emphasize the importance of sustainability in the development of civil engineering systems; as we will learn in Chapter 29, the Fundamental Canon 1 of ASCE's Code of Ethics, states that "engineers shall hold paramount the safety, health and welfare of the public and shall strive to comply with the principles of **sustainable development** in the performance of their professional duties." The ASCE has established a Committee on Sustainability and a Sustainability Action Plan and has published a number of policy statements and other documents (ASCE 2009, 2010a) that support sustainable practices. These include Policy 360 (Impact of Global Climate Change), Policy 418 (The Role of the Civil Engineer in Sustainable Development), Policy 488 (Greenhouse Gases), and Policy 517 (Millennium Development Goals).

ASCE's Policy Statement 418 asserts that "sustainable development helps convert natural resources into products and services that are more profitable, productive, and useful, while maintaining or enhancing the quantity, quality, availability, and productivity of the remaining natural resource base and the ecological systems on which the products and services depend." The statement adds that the civil engineering profession recognizes the "reality of limited natural resources, the desire for sustainable practices (including life-cycle analysis and sustainable design techniques), and the need for social equity in the consumption of resources." To achieve these objectives, ASCE supports the following implementation strategies (ASCE, 2010a):

- Advance wider understanding of technical, environmental, economic, political, and social processes and issues and their relationship with sustainable development.
- Promote the knowledge, skills, and information on subjects necessary for realizing a sustainable future; including natural systems, ecologies, system flows, and the effects of all phases of the system life cycle on the ecosystem.
- Encourage economic analysis approaches that duly recognize that natural resources and the environment are capital assets.
- Encourage the establishment and consideration of goals that are multidisciplinary yet integrated, whole-system oriented, and multiobjective goals at any phase of system development including planning and design, construction, maintenance, operations, and decommissioning.
- Promote the notion that the reduction of a system's vulnerability to natural, accidental, and malicious hazards is also an aspect of sustainable development.
- Advocate for the application of performance-based guidelines and standards to serve as a base for voluntary actions, policies, and for regulations related to sustainable development, for planned or existing infrastructure.

In 2006, the presidents of the American Society of Civil Engineers, the Canadian Society for Civil Engineering, and the Institution of Civil Engineers in the United Kingdom signed a Sustainability Protocol, which states that:

> ASCE, CSCE and ICE believe that the current approach to development is unsustainable and that "We are consuming the Earth's natural resources beyond its ability to regenerate them. We are living beyond our means. This, along with security and stability, is the most critical issue facing our profession and the societies we serve. In addition to the environmental impacts of our actions, the needs of societies around the world are not being met. Our goal as civil engineers is the creation of sustainable communities in harmony with their natural environment. In doing so, we will be addressing some of the most profound problems facing humanity, for example, climate change and global poverty". (ASCE et al., 2012).

Also, in China, one of the missions of the Chinese Academy of Engineering is to facilitate sustainable economic and social development. Similar specific efforts toward sustainability awareness and incorporation in all phases of systems development have been made by professional societies in other countries worldwide. Also, agencies and organizations have been spurred to undertake initiatives that bring further awareness of the need to address sustainable development in the current era, to catalyze the initiation of explicit and appropriate actions to address the issue. These initiatives include the United Nations eight millennium development goals (the seventh of which is to ensure environmental sustainability); the U.S. National Research Council's blueprint for achieving sustainability (NRC, 1999); the Federal Highway Administration's Sustainability Guidebook (Amekudzi et al., 2011); and the ASCE's Vision of Civil Engineering in 2025, which expands the responsibilities of civil engineers to include stewardship of the natural environment and its resources (ASCE 2007).

The practice of sustainability in the design, manufacture, and distribution of products has been adopted in other sectors including the business sector in a bid to demonstrate their commitment to responsible practices that do not jeopardize the environment and society. This support comes after the business sector appears to have accepted the notions that (a) profits alone do not guarantee continuity of their existence and that sustainability can and does promote a healthy and long-term viability of their businesses (Dandy et al., 2008); and (b) environmental and social pressures accentuated by lack of sustainability can actually impede economic development, as recognized by the World Business Council on Sustainable Development (Schmidheiny, 1992). In Section 28.1, we discuss the forces spearheading the drive toward sustainability in the current era.

In recent years, repetitive and often inappropriate use of the word sustainability in some cases has led to terminological fatigue and thus a need for rebranding may exist. Occasional rebranding is not necessarily a bad idea because society needs to be reminded every now and then of the importance of sustainable development. Such reminders may lose their effectiveness if tired clichés are used for a long time.

LAKE VIEW TERRACE LIBRARY—A MODEL FOR SUSTAINABLE DESIGN, CONSTRUCTION, OPERATIONS, AND MAINTENANCE

The Lake View Terrace Library in California is known for combining the virtues of environmental sustainability, utility, and aesthetics. A platinum LEED-rated structure, this building system features a photovoltaic array that provides shade for the entrance and provides 15% of its energy needs.

To minimize artificial lighting needs, the reading room was designed to lie along an east–west axis, thereby fully exploiting available daylight. The arch structures facilitate natural ventilation thus keeping the building cool without resorting to air conditioning, even in the absence of strong natural wind. The landscaping was designed in such a way that surface runoff is reduced by 25%.

To save water in landscaping irrigation processes, the plants are mostly drought tolerant, and when the rain falls the plants are not watered. Plumbing features include aerated faucets that conserve water by mixing air with the water to maintain a strong water flow, thereby reducing water consumption. The building was constructed mostly using recycled materials and the concrete used had a significant fraction of fly ash.

Source: McGrath (2012).

Source: harleyellisdevereaux.com.

28.0.3 Sustainability in Ancient Times

Even though sustainability concerns have become a major issue in recent times, the underlying tenets of sustainability have been a concern for ages. Sustainability is not a new concept but is merely a rebranding of existing concepts that have been addressed well, if not explicitly, in past literature (Goodman and Hastak, 2007) and indeed as far back as ancient times (Dandy et al., 2008).

In their transformation from nomadic lifestyles to permanent settlements, with the accompanying need to develop structures and systems to facilitate or enhance farming, warfare, commerce, and their quality of life, humans have always interacted with their natural environment in many ways that have been conflicting (Khisty et al., 2012). Historical evidence shows that the ancient civilizations that had relatively longer lives were those that managed to live harmoniously with their natural environments (Chambers et al., 2000).

As we discussed in Chapter 1, in the design and construction of civil engineering systems across the civilizations and over the millennia, two opposing philosophies existed: those that were utilitarian in nature and those that were devotional (Straub, 1964). It is important for present-day civil engineers to learn lessons from these philosophies. The ancient Romans and Persians developed systems that were consistent with the utilitarian philosophy as they were mainly built for purposes of military strategy and commerce. On the other hand, the civil systems built by the ancient Greeks were primarily of devotional value first; and military, commerce, and other values came second. This devotional value philosophy was probably due to the ancient Greek's animistic conception of nature, ascribing a living soul to mountains, rivers, and valleys, making them shy away from violent interference with natural land forms and obstacles. A parallel to this dichotomy can be found in the two rival moralities that guided the design and construction of hydraulic and other civil engineering systems in ancient China (Needham, 1971). The first was the Confucius philosophy (confining and repressing nature, and thus advocating "masculine" activities such as dike construction), which was similar to the Roman and Persian approaches. The second was the Taoist philosophy (greater freedom for natural courses and thus advocating the use of "feminine" activities such as dredged concavities), which was similar to the Greek approach. The relatively long lives of Greek and Taoist structures compared to Roman and Confucius structures could generally be attributed to their more sustainable design philosophies.

Also in ancient times, there were other more explicit attempts to reduce environmental degradation due to man-made activities. At a certain period in ancient Rome, Emperor Julius Caesar banned all wheeled vehicles from the city between sunrise and two hours before sunset, in a bid to enhance pedestrian traffic flow and convenience. In 500 BC, the ancient Greek philosopher Plato bemoaned the consequences of unrestrained grazing and logging in the Attica region (Dandy et al., 2008). Chambers et al. (2000), Dandy et al. (2008), and Khisty et al. (2012) recognized that the history of human existence has always been characterized by cultures, laws, and lifestyles that often recognized and promoted (at least implicitly) sustainable practices by the incumbent civilizations and that any internally driven demise of ancient civilizations was largely influenced by unsustainable practices or situations.

28.1 MOTIVATIONS FOR THE DRIVE TOWARD SUSTAINABILITY IN THE CURRENT ERA

As we just discussed in an earlier section, sustainability is not a novel concept in the evaluation of system outcomes but rather a new cliché to describe what engineers have always considered at least implicitly in their systems development over the past millennia. Notwithstanding its seemingly superficial role as a new word for packaging old concepts, sustainability has a deeper purpose: It unifies the disparate outcome types (i.e., economic, environmental, and social impacts) because the effect of the sum of the outcomes is more realistic and often presents a graver situation compared to the sum of the effect of the outcomes. More importantly, the need for better stewardship of the Earth brings a greater sense of urgency to the need to minimize the adverse impacts of anthropogenic activities. This is true at the current time where such urgency is reaching unparalleled heights and where unsustainable practices potentially jeopardize the survival and well-being of future generations. A Native American proverb makes this point succinctly: "We do not inherit the Earth from our ancestors; we borrow it from our children." We now will discuss a few of the motivations for enhancing sustainability.

28.1.1 Unparalleled Scale of Human Activity

In the last decade, prominent researchers and scientists have stated that the scale and rate of the biosphere's changes and types and combinations of these changes differed from the changes at other times in planetary history, and that the outcomes of these changes are generally unfavorable terms of their impacts on the biosphere (IPCC, 2000).

28.1.2 Unprecedented Demand for Civil Systems

The demand placed on civil engineering systems is greater than ever before: Khisty et al. (2012) stated that: "in 2000, the Earth's population had doubled within a 50-year period (the last time it had doubled, that change took place over a 200-year period); in 2000, the number of old people exceeded the number of young people for the first time in history; from 2007 onward, urban dwellers outnumbered rural dwellers for the first time." Also, increased living standards in populous developing nations have increased consumption, resulting in further strain on the Earth's resources. In many countries, this is manifest by the increasing scarcity of potable water, food shortages, and urban congestion and the resulting air pollution, inability to manage solid waste, poverty, hunger, and disease. ASCE's *Vision of Civil Engineering in 2025*, and the United States National Academy of Engineering's Grand Challenges Summit in 2009 recognized that the challenges faced by engineers are greater than ever and that they need to work toward improved quality of human life in a cost-effective and sustainable manner.

28.1.3 The Reality That the Resilience of Natural Resources Is Not Infinite

The increasing prominence of efforts toward sustainability is rooted in the realization that the biosphere does not have limitless capacity to absorb the abuses heaped upon it by human actions. The Earth is limited in its capacity to yield products for human consumption and to absorb or sequestrate the wastes generated by human activity. The notion that that there is always a way to absorb externalities is flawed, and it is now understood that ecosystems are not homeostatic: After an ecological stress is removed, the ecosystem may not bounce back to its original condition, for example, polluted rivers do not necessarily return to their pristine state after pollution ceases (Adams, 2006). Because the biosphere is not infinite, its survival is critical to the sustainability of the human race and its endeavors.

28.1.4 Relationship between Sustainability and Human Well-being

In the traditional model of human development, wellbeing is interpreted solely from an economic perspective, particularly in terms of the accessibility of goods and services to humans. Adams (2006) argues that this means-based perspective of wellbeing is fallacious. Citing Amartya Sen's concept of "development as freedom" (Sen, 2000), Adams contends that not only the means but also the ends must be considered to define a society's wellbeing. As such, there is a need to establish a new understanding of human endeavor and achievement on the basis of sustainability; and a relevant metric of sustainability is not necessarily the production of material goods but enhancement of human wellbeing per unit of extraction from or imposition upon nature. Thus, it is necessary to abandon the perspective that material consumption and political security are separate from and more important than, quality of life. Also, citing David Orr's statement that "no human being has the right to diminish the life and well-being of another and no generation has the right to inflict harm on generations to come", Adams avers that the security between people is influenced largely by issues of equity within and between generations, and that security and wellbeing are both rooted in issues of justice on a global scale. As we saw in the various definitions of sustainability in the initial sections of this chapter, the importance of future generations and their wellbeing, a value that is cherished worldwide, are fundamental to the concept of sustainable development.

28.2 PRINCIPLES OF SUSTAINABILITY

The term "principle of sustainability" can have several different meanings. In this chapter, we refer to it as the characteristics that must be possessed by a sustainable action or situation. Using this term in that context, we list a number of principles presented by Monday (2002), McCuen et al. (2011), Natural Step (2013), and APWA (2013):

Effect on Natural Environments

It must be associated with an enhancement of the quality and quantity of natural resources including flora, fauna, and their habitats. A sustainable community makes active efforts to coexist with its natural environment by avoiding needless degradation of the soil, lakes, air, oceans, and other natural systems and by outlawing detrimental practices with those that facilitate reclamation, restoration, rehabilitation, or self-renewal of ecosystems. Thus, it is important to reduce the accumulation not only of certain materials extracted from the Earth's crust such as fossil fuels and heavy metals but also of chemical by-products of the society such as dioxins and polychlorobiphenyls (PCBs).

Use of Natural Resources

It must maximize the efficient use of natural resources through actions including avoidance of use, reduction of the amount used, reuse, and recycling. This is a particularly important principle from the civil engineering perspective because a large amount and variety of natural materials are used in civil systems construction and operations (geological resources, fossil fuels, metals, and timber, to name a few).

Effect of Anthropogenic Systems

It must promote (or minimize any damage to) the built-up physical environment, the social and cultural structures, and the economy. It must enhance local economic vitality.

Institutional Effects

It must not be associated with or promote adverse institutional effects including financing challenges, tort liability, and corruption.

Intergenerational Equity and Quality of Life

It must enhance (or at least maintain) the quality of life at the current time and for future generations. In a sustainable community, the opportunities and resources are available to all, irrespective of individual's attributes including age, cultural background, race, or ethnicity, gender, or religion. Also, a sustainable community does not engage in wanton depletion of its natural resources even for short-term financial gain, nor does it pass along to its progeny, environmental hazards, simmering social conflicts, or excessive debts.

Flexibility

It must provide flexibility for possible changes in stakeholder requirements or demands in the future. Flexibility could also mean that a sustainable community must be able to recover quickly from disaster events to its economy, natural or man-made capital, or environment or social fabric.

Value Based

It must address human values that include fairness (impartial to all parties) and duty (sense of responsibility), provide knowledge-based solutions, promote efficient production with minimal waste of resources, seek the general well-being of the public and the environment, and foster accountability for actions or inactions.

Inclusiveness

Actions are more sustainable when they are based on decisions developed through consensus building and participatory processes. When the decision making engages everyone who has a stake in the outcome of the action being contemplated, the concerns and issues are better identified, generation of comprehensive and innovative solutions is promoted, a sense of community is fostered, and a sense of ownership on the part of the community is encouraged.

THE TRAGEDY OF COMMONS

The tragedy of the commons refers to the eventual depletion of a shared resource by individual entities, acting rationally and independently to protect their individual self-interests, even though they are very well aware that depletion of the common resource is inimical to their long-term survival or interests. A typical example used to illustrate this concept consists of herders that share a common piece of land on which they are each entitled to let their cows graze. For each herder, it is in his interest to put an additional cow to feed on the land, even if the pasture quality is degraded due to the overgrazing. The herder receives the benefits of adding one more cow; however the pain of the degraded pasture is brone by all the herders. If all herders make the economic decision, to add one more cow each (which is rational from their individual points of view), the pasture will be depleted or even destroyed, to the detriment of all. This illustration was first presented by William F. Lloyd in Europe in 1833. The concept was recognized as far back as the time of the Greek civilization as reflected in the following quotes translated by Jowett (1885):

> "[People] devote a very small fraction of time to the consideration of any public object, most of it to the prosecution of their own objects. Meanwhile each fancies that no harm will come to his neglect, that it is the business of somebody else to look after this or that for him; and so, by the same notion being entertained by all separately, the common cause imperceptibly decays" [Thucydides (460–395 BC)] and
>
> That which is common has the least care bestowed upon it. Aristotle (384–322 BC)

The tragedy of commons has been used, often implicitly, by the two sides of the political spectrum to support their policy positions. One side argues that resources must be held in private hands because people take better care of property when it is theirs, compared to public property; and for that reason, public ownership of resources is not sustainable. The other side contends that individuals tend to seek their own selfish interests and thus if resources are left in private hands, particularly where unregulated, such individuals may prosper in the short run but the society and even the entire human race will be worse off in the end and thus is not sustainable. For these and other reasons, however, the tragedy of commons, particularly as articulated by Hardin (1968) in his herdsmen illustration, has seen a hailstorm of criticism on various aspects, including the definition of the word "commons" and the appropriateness of the assumption of inherent selfishness of the individual.

In the current era, the tragedy of commons may be used to explain the behavior of individual persons or organizations in applications including overpopulation, dumping of industrial waste and pollutants, use of cheap fossil fuels for production and subsequent greenhouse gas emissions, overfishing and over logging, consumption of the Earth's resources at the expense of future generations, and littering of public places. Also, the concept may be used to explain the current decimation of industrial towns in the United States due to outsourcing, the lack of long-term sustainability of using nonrenewable natural resources, the effectiveness of traffic control measures at congested urban freeways, and the long-term demise of purely communistic and purely capitalistic societies.

Sources: Hardin (1968), Dandy et al. 2008, and Rankin et al., (2007).

28.3 EVOLUTION OF GLOBAL POLICY ON SUSTAINABILITY

Figure 28.1 presents the watermarks in the timeline of global policy on sustainability. The issue of environmental limits to human activities on Earth started to attract international attention in the late 1960s and early 1970s by a number of events and publications. These included the Club of Rome's computer model that indicated the existence of limits to development (Meadows et al., 1972) and the 1969 International Union for Conservation of Nature and Natural Resources (IUCN) mandate. The IUCN mandate highlighted "the perpetuation and enhancement of the living world—man's natural environment—and the natural resources on which all living things depend," which referred to management of air, water, soils, minerals, and living species including man, so as to achieve the highest sustainable quality of life. This term was a key theme of the United Nations Conference on the Human Environment in Stockholm in 1972 and reaffirmed that it was possible to achieve economic growth and industrialization without environmental damage. The subsequent years saw an increasing engagement with sustainability during planning, by national governments, business

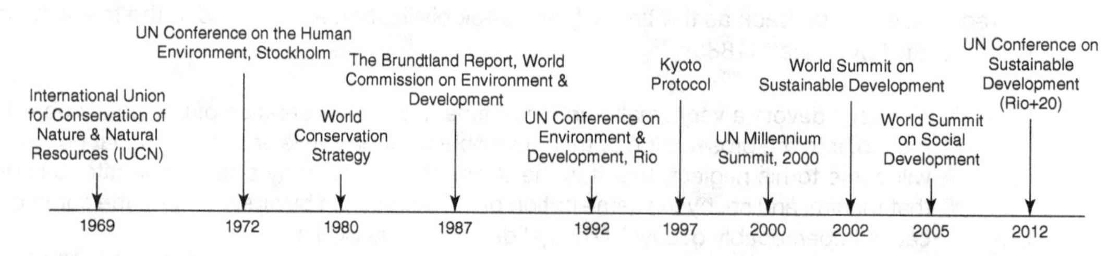

Figure 28.1 Watermarks in the timeline of global policy on sustainability.

leaders and nongovernmental organizations of all kinds. In the eighties and nineties, five major events watermarked the increasing role of sustainability in global and regional development, at least from a policy standpoint (Adams, 2006): the World Conservation Strategy in 1980, Law of the Sea in 1982, Montreal Protocol on substances that deplete the ozone later in 1987, the Brundtland Report in 1987, and the United Nations Conference on Environment and Development in Rio in 1992.

The World Conservation Strategy in 1980 presented what is probably the first explicit statement on environmental sustainability, and emphasized the need to "maintain essential ecological processes and life support systems, to preserve genetic diversity, and to ensure the sustainable utilization of species and ecosystems". The Rio conference, which followed in 1992, emphasized global environmental change, biodiversity, resource depletion, and climate change. Since the Rio conference, there has been a large increase in the number of environmental legislation at the international, regional, national, and local levels. International agreements such as the Kyoto Protocol helped to bring greater awareness of environmental change and served as an engine for further changes in global policy. At the United Nations Millennium Summit in 2000, sustainability was stated as one of the eight millennium development goals. Also, at the World Summit on Sustainable Development in 2002, the demands on the biosphere due to human development was featured prominently in the international dialogue. The United Nations Conference on Sustainable Development, Rio+, which took place in 2012 in Rio de Janeiro, Brazil, focused on green economy in the context of sustainable development and poverty eradication, and an institutional framework for sustainable development.

Worldwide, most development agencies, governments, and even corporations are increasingly showing that they recognize and appreciate the need to address the adverse social and environmental impacts associated with human development. Consistent with this development, an increasing number of private and public sector organizations are taking steps to incorporate "green" actions and materials in their business processes. They state that doing this is a key part of their social and environmental responsibilities. However, as Adams (2006) noted, in many cases, sustainability is still a "boutique concern within wider relationship management" rather than a driver of structural change in the nature or scale of their core business.

The importance of explicitly and effectively incorporating sustainability considerations in civil systems development cannot be overemphasized as a growing body of incontrovertible evidence continues to emerge on the global footprint of human development. Researchers have described as unacceptable and unprecedented the consumption of living resources either as raw material or as sinks for waste materials, (Wilson, 1992; Wackernagel and Rees, 1996; Vitousek et al., 1997). Certain scholars posit the view that the adverse effects of development can be tolerated as long as remediation actions can be undertaken to restore the damaged resources. Other researchers such as Adams (2006) take a more cautious approach and admit that developments in ecological restoration present novel and inspiring opportunities to enhance or reinstate ecosystem biodiversity; however, they admonish that humans have a limited capability to repair damaged ecosystems and that it is not possible to restore critical natural capital within realistic timeframes.

28.4 ELEMENTS OF SUSTAINABILITY

The elements of sustainability refer to the items to be considered in assessing the sustainability of an action or situation. Different organizations, professional bodies, and system agencies have established different elements of sustainability depending on their focus or mission. For example, McCuen et al. (2011) identified four elements as: energy resource, soil, environmental, and ecological. In most texts and reports, however, the elements are generally identified as **economic**,

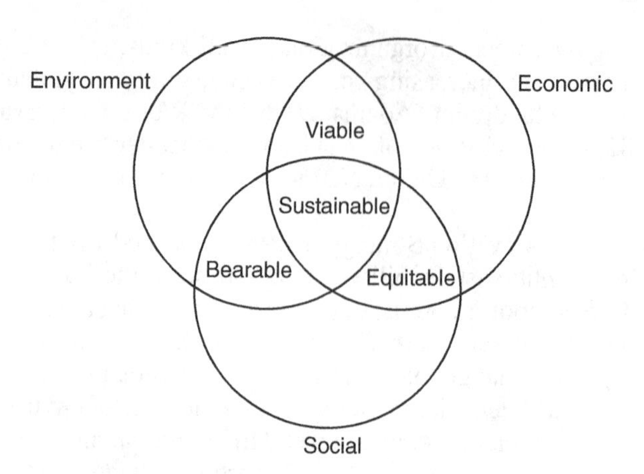

Figure 28.2 The basic pillars of sustainability.

environmental, and **social**. These are referred throughout this text as the *sustainability pillars* (Figure 28.2) or, as often referred in the literature, the *triple bottom lines* (TPLs). An action or situation is described as *viable* when it is sound from the environmental and economic perspectives; *bearable* when it is sound from the environmental and social perspectives; *equitable* when it is sound from the economic and social perspectives; and *sustainable* when it is sound from all three perspectives.

The three elements or pillars in Figure 28.2 are meant to be treated neither as being equivalent nor as distinct and the reason is twofold. First, society drives the economy, and the economy often shapes the nature of society. On the other hand, the natural environment is different because it is not created by society and analysis of trade-offs between these pillars cannot really be carried out. Also, the environment serves as a substrate for both society and the economy, as the resources available in the ultimate environment (the Earth and the solar system) effectively serve as a limiting barrier to human economic and social activity. As Adams (2006) noted, such limits are often specific, for example, the biosphere has a limited capacity (in space and time) to absorb pollutants and provide resources for social and economic development. Such consumption varies from place to place, and at areas such as warm shallow coastal waters adjacent to industrialized regions, the consumption is close to the capacity. Second, the actual relative size of each pillar in Figure 28.2 is a reflection of the influence of each circle in development decisions or outcomes and thus varies across geographical areas (countries, regions, or cities); and for each geographical area, the actual relative sizes may vary across time. Increasing consideration of the environment would mean that the actual size of that circle would be greater relative to the other two, thus providing greater outcomes in terms of the development's viability and bearability.

As we discussed in earlier sections, the elements of sustainability have been addressed in numerous studies and texts that have evaluated the feasibility of proposed civil engineering projects or systems or the overall performance of existing projects even though those studies did not explicitly use the term "sustainability." As we saw in Figure 28.2, the categories of sustainability include environmental (air quality, greenhouse gas emissions, water quality, noise, hydrology, ecology, and aesthetics), economic (employment, number of business establishments, gross domestic product), and social. It has been argued that a comprehensive view of sustainability must consider other elements such as economic efficiency and financial (initial costs, life-cycle cost/benefits, net present value, or benefit–cost ratio), legal (tort liability exposure), technical (facility condition, longevity, safety, and ability to perform its intended function), the system's resilience to natural or man-made disaster, and culture (way of life). In the next few sections, we discuss in greater detail an expanded view of the economic, environmental, and social elements of sustainability.

28.4.1 Economic Element of Sustainability

The economic dimensions of sustainability can be broadly defined to include all economics-related considerations including economic development impacts, economic efficiency, and financial feasibility. Will the new or existing civil engineering system promote increased job opportunities? Will the system's owner see positive returns for every dollar of investment? Can the system pay for itself or is there a way to finance not only its construction but also its long-term operations and maintenance? Such concerns continue to resound at the World Bank, the United Nations, the Organization for Economic Cooperation and Development (OECD), and other multilateral lending agencies. Goodman and Hastak (2007) noted that when a country or system owner fails to achieve the intended level of economic performance, that situation represents a misuse of scarce investment funds, jeopardizes the monetary position of the country or the system owner, and betrays the trust invested by the system stakeholders which may include the system users or the country's citizens in general.

With the increased legislation-driven and policy-driven emphasis on the economic development benefits of civil engineering infrastructure, the economic impacts are increasingly being used in civil engineering systems evaluation. The impacts of infrastructure in a regional economy can be measured by examining their specific roles at each stage of any economic production process. Also, impacts may be viewed from the perspective of their coverage: relating to overall area economy or relating to specific aspects of economic development. Impact types relating to the overall area economy include regional output, gross regional product, wages, number of business establishments, and employment. Impact types relating to specific aspects of economic development include capital investment, productivity, and property value appreciation. For each of these impact types, the impact mechanism may be induced, direct, or a multiplier; the spatial scope could be local, regional, or national; and the temporal scope could be short term, medium term, or long term. Indicators of sustainability from the economic viewpoint include the resource flows (i.e., the total material flows associated with economic processes) established by the World Resources Institute.

28.4.2 Environment Element of Sustainability

Of the three sustainability pillars, the environment is often considered the most fragile and, therefore, the most critical. Aspects of the environment that could jeopardize the sustainable development of a civil engineering system at any phase include air quality, water quality, noise, aesthetics, ecology, and hydrology.

(a) Basic Aspects of the Environmental Element

Air Quality

Infrastructure-related legislation over the past few decades has consistently emphasized the need to include air quality as one of the criteria in the evaluation of infrastructure operations. Thus, owners of civil engineering systems typically take great pains to ensure that at all phases, particularly during operations, air quality is not compromised.

Water Quality

During the construction and operations of infrastructure systems, the quality of surface water can be quickly degraded, and sustainable actions at this phase are those that prevent or reduce such adversity.

Noise

The noise associated with the operations of certain civil infrastructure systems has often been linked to health problems and often merits consideration in any sustainability assessment.

Hydrology

The construction and, to a lesser extent, the operations of infrastructure systems has been linked to disruptions in surface and subsurface hydrological patterns. For example, the construction of highway pavements, buildings, parking lots, and airport runways leads to reduced permeable land cover, reduced percolation rates of surface water, and consequently reduced recharge of underground aquifers. Increased surface run-off arising from the construction of such facilities leads to greater volumes of surface flow which fosters increased soil erosion and flooding. Also, the construction of bridges and culverts is often associated with forced channelization of surface water along unnatural water courses and may lead to unexpected surges in surface water flow at certain areas and flow dry-up at other areas downstream of the structure.

Ecology

The construction and operation of infrastructure facilities is often directly associated with destruction of flora and fauna and their habitat. Thus, studies of ecological impacts now include not only assessment of the threats posed by natural features to engineering structures but also the other direction of impact, namely, the effect of the operation of engineering infrastructure on the ecology.

Aesthetics

Civil infrastructure systems typically have a profound visual impact on the surrounding natural or built-up environment. Such impacts may occur as an enhanced or diminished blend with the surrounding environment or obscuring an aesthetically pleasing natural or man-made feature. Borne out of a strong federal emphasis on context-sensitive design (CSD), it is considered sustainable practice to include aesthetics in design, so that there can be a good blend of the civil engineering system with its natural environment and also to provide its patrons an aesthetically pleasing and satisfying experience while using the system.

(b) Aggregate Indicators of Environmental Sustainability

Ecological Footprint

Wackernagel and Rees (1996) developed the ecological footprint as a measure of human demand on the Earth's ecosystems, specifically, the amount of biologically productive land and sea areas needed to supply the resources needed for human consumption. For 2007, it was estimated that humanity's total ecological footprint was 1.5 "planet Earth" units; in other words, in that year, humanity used its renewable ecological resources 1.5 times as quickly as the planet could regenerate them (Global Footprint Network, 2010). Figure 28.3 presents the human welfare and ecological footprint across the various continents and select countries in 2005.

Carbon Footprint

The term "carbon footprint" is defined as the amount of carbon or greenhouse gases emitted by an activity or organization (Global Footprint Network, 2010). The carbon component of the ecological footprint is measured using the amount of productive land and sea area required

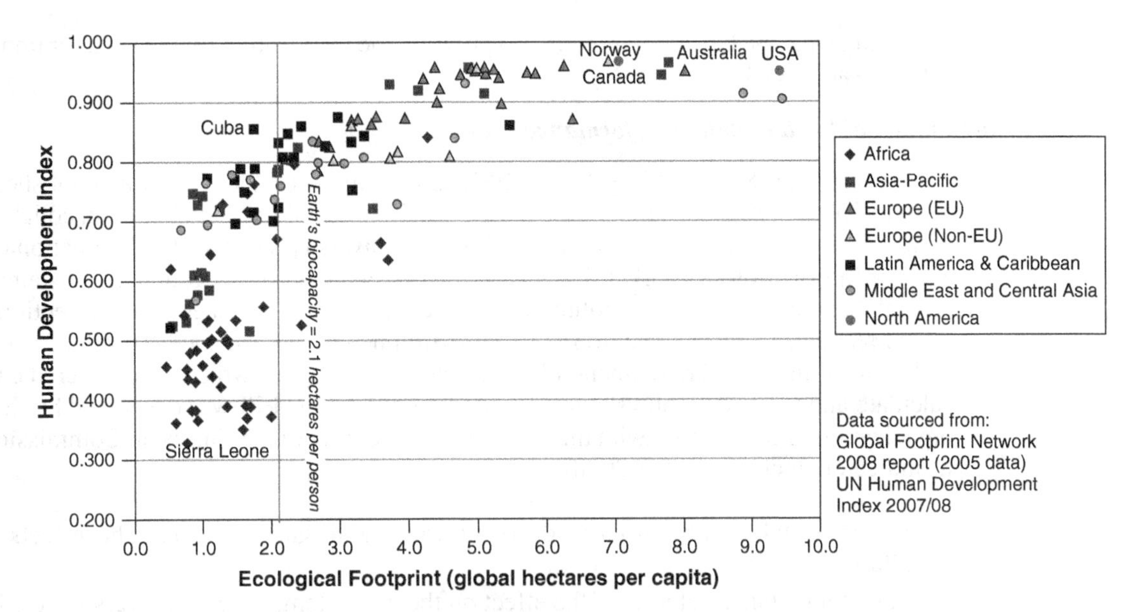

Figure 28.3 Human welfare and ecological footprint across the continents in 2005 (Travelplanner, GNU Free Documentation License).

to sequester carbon dioxide emissions. The ecological footprint reflects a comparison and interactions between carbon emissions and other elements of human demand, including the pressure on food sources, the amount of living resources needed for producing goods, and the quantity of land taken out of production due to paving for parking lots, highways, buildings, and other infrastructure. With an 11-fold increase since 1961 and constituting over 50% of humanity's ecological footprint (EF), humanity's carbon footprint is EF's most rapidly growing component (Figure 28.4). Research has shown that the demand of humanity is rapidly

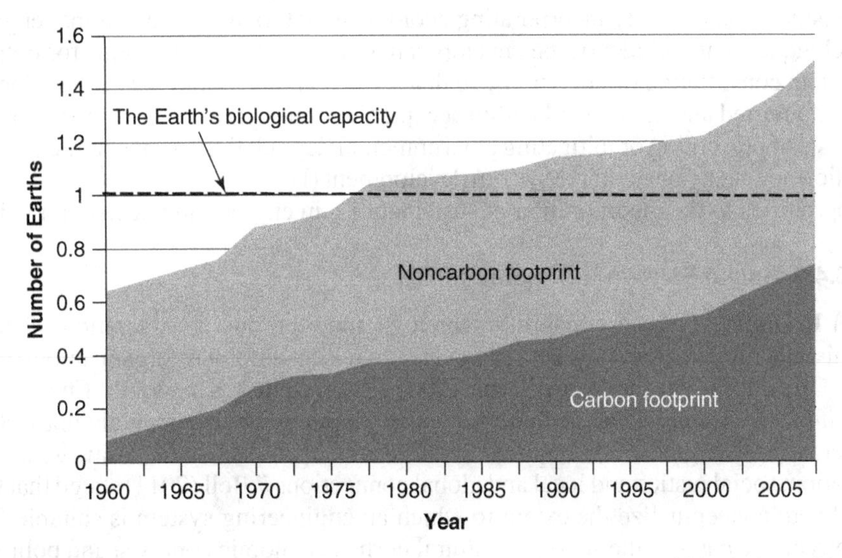

Figure 28.4 Carbon component of the ecological footprint, 1961–2005 (adapted from Global Footprint Network, 2010).

exceeding what the Earth can provide and provides the justification for the need for immediate action to address the situation.

Environmental Sustainability/Performance Index

The Environmental Sustainability Index (ESI) is a composite index that was published from 1999 to 2005 by Yale University's Center for Environmental Law and Policy in collaboration with the World Economic Forum and Columbia University's Center for International Earth Science Information Network. ESI tracked about 20 elements of environmental sustainability: natural resource endowments, pollution levels, environmental management, protection of the global resources, and capacity to improve environmental performance. From 2006 forward, ESI evolved into the Environmental Performance Index (EPI), which is considered easier to calculate and use and employs outcome-oriented indicators. EPI was developed by the same organization in conjunction with the Joint Research Center of the European Commission, and its indicators include (CLEP, 2008):

- Environmental health (disease, drinking water and sanitation, and the effects of air pollution)
- Ecosystem vitality in terms of the effect on the ecosystem, water, biodiversity, and habitat
- Productive natural resources including fisheries forestry, and agriculture
- Climate change (including emissions per capita and industrial carbon intensity)

The Environmental Pressure Index

The Environmental Pressure Index (EPI) is a composite index aggregating over 40 different indicators of environmental stressors. The European Commission has sponsored or supported work by the ecologic organizations to develop this index and to identify robust data sources for its constituent indicators.

Bell et al. (2011) indicated that the practice of engineering could be infused with sustainability considerations by duly incorporating ecological factors into conventional engineering models and techniques that include: (i) the development of technical specifications for pollution control during system construction or operations, so that the environmental impacts of engineering systems could be monitored and maintained within acceptable limits, (ii) explicit inclusion of environmental costs (costs of preventing or mitigating environmental degradation) as one of the cost criteria for resource efficiency at any phase of the system development (Hawken et al., 2000), and (iii) systemic (holistic) approaches to the incorporation of sustainability in engineering decision making.

28.4.3 Social Element of Sustainability

(a) Prelude. True sustainability cannot exclude the due consideration of the social dimension. This has been realized by key engineering and development organizations worldwide including ASCE (2009) and the World Bank (2003). The United Kingdom's Engineering Council states, in its 2009 *Guidance on Sustainability*, that "a purely environmental approach is insufficient, and increasingly engineers are required to take a wider perspective including goals such as poverty alleviation, social justice and local and global connections." Bell (2011) stated that the social dimension helps to conceptualize the extent to which an engineering system is suitable for different populations on the basis of the society's cultural norms, economic contexts, and political realities. Adding that this element of sustainability influences the technical success or failure of civil systems, Bell cited examples that include the current status of sustainability (whether or not it is adopted at a specific location or situation), the need for the phases of system planning and design to be consistent with the outcome of the needs assessment phase, and monitoring of system demand so that system operations and future designs can adequately address future patterns of consumption.

As one would expect, assessing the social and cultural sustainability of an existing or planned civil engineering system is considered an inexact science because social environments tend to be very dynamic and adaptable and also differ from place to place. In other words, these impacts depend on the manner of the social change interpretation, the level of anticipation, and the extent of the resilience of the affected humans. IOCGP (2003) defines social impacts as "the consequences to human populations of any public or private actions that alter the ways in which people live, work, play, relate to one another, organize to meet their needs, and generally cope as members of society"; FHWA (1982) describes it as the change (often destruction or disruption) of man-made resources, social values, community cohesion, and availability of public facilities and services; displacement of people, businesses, and farms; and the disruption of desirable community and regional growth. In the United States, the impetuses for including social considerations in system sustainability analysis include the 1970 Federal Highway Act, the 1973 National Environmental Policy Act (NEPA), and a number of executive orders from President Clinton during his term of office. In the developing world, multilateral lending agencies such as the World Bank require borrower countries to undertake social impact assessments to ensure that funded projects yield significant positive or minimal adverse impacts on the lives of people in those countries in terms of sociocultural, institutional, historical, and political considerations (World Bank, 2003). Since 1968 when then World Bank president Robert McNamara stressed the issue of poverty alleviation, social impact analysis has gained a prominent role in the agenda of international lending agencies and development organizations including the United Nations and the Inter-American, African, and Asian Development Banks.

The analysis of distributive effects is an important aspect of social impact assessments. These effects can refer to the variations in impact severity of the project as one is further removed from the project area, but they are more often taken to mean the variations in the impact severity across community groups, population groups, or/and ethnic groups in the overall area where the project is located. Distributive effects can also include how such distance-based and community group-based variations change over time. The analysis of distributive social effects is particularly critical when the project (i) requires an unusually large space in an urban area; (ii) would involve the displacement of a large number of households, businesses, community amenities, historic districts, and landmarks; (iii) conflicts with local land-use plans; and (iv) would unduly and unfairly reduce the welfare of vulnerable segments of the population (Sinha and Labi, 2007).

According to the World Bank (2003), social impact assessments should be a continuous process occurring throughout the cycle of systems development, including the phases of planning and implementation (construction and operations). The World Bank identified five dimensions of inquiry, or "entry points," for social impact assessments: social diversity and gender, institutions, rules and behavior, stakeholder participation, and social risk; and the bank stated that the relative scope of each dimension depends on the circumstances and context of a particular project. In countries including the United States, a number of studies have been carried out to examine the relationships between the spatial distribution of the social disbenefits of civil engineering systems and the sociodemographic attributes of affected communities, particularly those that are disadvantaged, marginalized, or disenfranchised.

(b) Target Facilities and Groups. In assessing the social impact of an existing or proposed civil engineering system, it is important to first identify the facilities and population segments that will likely be affected (Sinha and Labi, 2007). The facilities typically considered include schools, religious institutions, playgrounds, parks, recreational areas, hospitals, clinics, other medical facilities, residential and social facilities for the elderly, social service agencies, and libraries. Generally, all persons within the impact area are considered in the analysis. Where environmental justice is an issue, focus should be placed on certain specific population segments such as elderly persons, disabled persons, nondrivers and transit-dependent persons, minority groups, and low-income or poverty-afflicted individuals and households. It is useful to note that the definition of poverty extends beyond income inadequacy to include deprivation of basic capabilities (Sen, 2000). Target groups include those that are vulnerable to conflict, violence, or economic shocks.

(c) Social Indicators of Sustainability Performance. Indicators of sustainability performance in terms of social attributes can be either beneficial or adverse to varying degrees depending on the perspectives of the affected population (Sinha and Labi, 2007). These indicators may differ in scale, severity, or intensity depending on the community resources available and the nature of the community. For a given indicator, the direction (beneficial or adverse) and the intensity of the impacts may vary among different communities and population groups depending on their resilience, diversity, amount of social capital, among other resource types. For example, the construction or operations of a civil engineering system may produce generally positive social effects for certain groups or communities but may have adverse impacts for others. In many cases, low-income and other groups with relatively little effective electoral representation find themselves disproportionately affected by strategic national and regional plans and decisions that culminate in outcomes such as relocations to make way for the civil engineering system. Compared to the economic and environmental pillars, social indicators are less developed because questions still exist about the meaning and measurement of social sustainability. However, transparency, trust, corruption, and conflict could be used as indicators for assessing civil engineering system sustainability from a sociocultural standpoint (World Bank, 2003).

"Community cohesion" describes the social network and actions that provide satisfaction, security, camaraderie, and identity to members of a community or neighborhood. For many people, community cohesion is vital to the success of family life and contributes to feelings of satisfaction and fulfillment in community life (Forkenbrock and Weisbrod, 2001). It is tempting to derive some mathematical index or rating to describe the level of the indicators of social sustainability; however, such efforts must be accompanied by a great deal of circumspection. Caltrans (1997) reports that for a quarter of a century, several transportation agencies countrywide have used a "stability index" to measure levels of community cohesion. Such indices are based on the length of time residents have lived in a community (i.e., the longer the length of time in the community, the greater the stability index). The stability index computation may be biased because it may not capture renters who despite being low income and minorities and thus, frequently, movers, nevertheless, are a part of a cohesive community.

At the state of Indiana, in the final environmental impact statement for the Interstate 69 Evansville-to-Indianapolis highway (Cambridge Systematics and BLA, 2003), the criteria for assessing the cultural impacts of each alignment included the possibility and the extent of encroachment of archeological sites, historic school buildings, Amish communities, Mennonite communities, and establishments registered (or eligible to register) with the National Register of Historic Places.

In the developing world, the social impacts of civil engineering systems can be expressed in terms of the change in the social and economic assets and capabilities of people, particularly the low income and vulnerable, and the extent to which the system helps to reduce social tensions, conflict, and political unrest. That is not to say that civil engineering systems can prevent armed conflicts. However, systems that provide or improve infrastructure, including construction of roads, harnessing of energy sources, supply of potable water, and construction of sanitation facilities, could help reduce poverty and inequality, and help foster cross-ethnic interactions that are among the root causes of ethnic tensions and unrest (Sinha and Labi, 2007).

The pursuit of the social element of sustainable development is worthy; however, because perceptional boundaries continue to separate engineering and the society, this sustainability element has not been typically considered adequately in civil systems planning, thus limiting opportunities for addressing the full range of challenges associated with the sustainable development of engineering systems. Bell et al. (2011) stated that ecological modernization (the dominant policy response to environmental problems) tends to reinforce the false borders between technology and society and suggested that engineering can be viewed as a hybrid sociotechnical profession that could help remove such boundaries.

Example 28.1

Table 28.1 presents the trend of social, economic, and environmental measures for a system over 11 years. The social sustainability is measured in terms of an index ranging from 0 (very unfavorable social impacts) to 10 (very favorable). The economic sustainability is measured in terms of the costs of repairing a unit dimension in the system (higher repair costs are indicative of poor sustainability). The environmental sustainability is measured in terms of the area of habitat that is adversely affected by the system operations. Assuming that the system is considered to be unsustainable as soon as the first index reaches its threshold, determine the sustainable life of the system. Also, assume that the three indices are independent of each other. The thresholds are as follows: social, 4 units; economic (costs), $\$14/ft^2$; environmental, 5 hectares.

Table 28.1 Data for Example 28.1

Year	2003	2004	2005	2006	2007	2008	2009	2010	2011	2012	2013
Social Impact	9.05	9.12	9.12	9.05	8.91	8.71	8.43	8.08	7.67	7.18	6.62
Economic	3.38	3.08	3.05	3.29	3.79	4.55	5.57	6.86	8.41	10.23	12.30
Environmental	0.23	0.80	1.33	1.83	2.30	2.72	3.12	3.47	3.79	4.08	4.32

Solution

The sustainable life can be estimated as the time taken for the system to reach the first threshold. This can be found by extrapolating the sustainability curves until they reach the threshold, and reading from the plots (Figure 28.5) or by substituting the threshold values of indices and solving for the independent variable (years). It can be seen that the sustainable life from social perspective = 13.5 yr, from the economic perspective = 11.6 yr, and from the environmental perspective = 13 yr. The least of these (11.5 yr) is taken as the sustainable life.

28.5 COMBINED INDICATORS OF SUSTAINABILITY

A combined indicator of sustainability is one that represents the overall sustainability in terms of the constituent elements (economic, environmental, etc.). Table 28.2 presents a number of composite indicators of sustainability. In Section 28.9, we present how some of these combined indicators can be calculated using data on the environment, economy, and society.

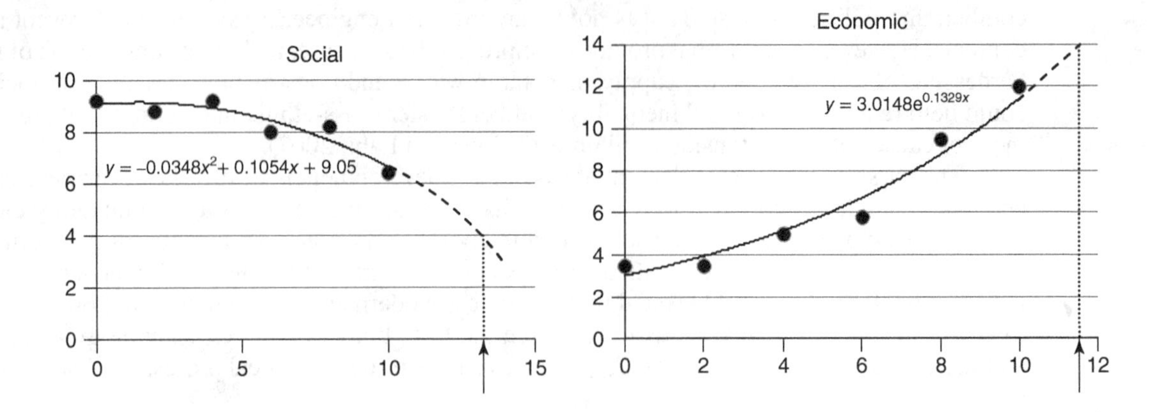

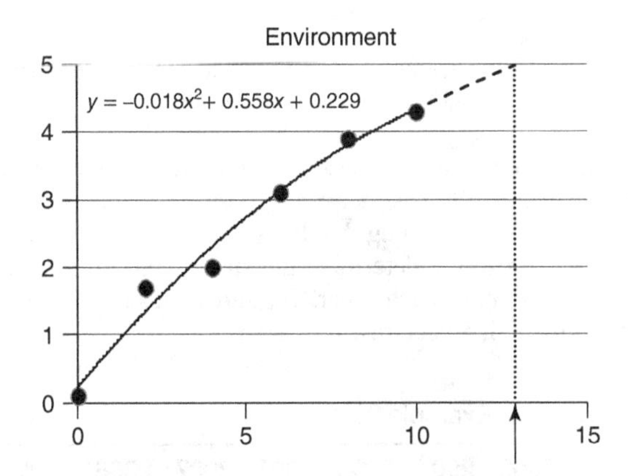

Figure 28.5 Figure for Example 28.1.

Table 28.2 Some Composite Indicators of Sustainability

Sustainability Indicator	Description (see Section 28.6 for details)
GAS-EEA (Green Accounts Systems of Environmental and Economic Accounts)	Addresses environmental and economic criteria.
Adjusted Net Savings (The World Bank)	Measures the change in wealth and accounts for resource depletion and environmental damage.
Genuine Progress Indicator	An adjusted measure of gross domestic product that reflects welfare losses on the basis of environmental and social performance.
Sustainability Footprint	Considers the ecology, economic development, and human quality of life.
The IPAT Indicators	Measures the human impact on the environment due to the population, level of affluence, production, pollution, and technology.
Triaxial Representation of Technological Sustainability (TRTS) Indicator	Addresses changes in sustainability due to changes in stakeholder satisfaction (a measure of social quality of life), and the impact on the demand for natural resources and on the ecosystem.
Quality of Life/Natural Capital Indicator	Measures overall sustainability in terms of the quality of life for humans and the ecological quality.
The Footprint Indicators	Measures sustainability as the overall human consumption of specific resources in terms of aggregate units.

Sources: World Bank (2003), Goodman and Hastak (2007), and Amekudzi et al. (2009).

Example 28.2

Table 28.3 shows the past (years 2006–2013) and expected (years 2014–2019) sustainability indices (SI) pertaining to an area that is expected to be impacted by a proposed civil system to be constructed in that area in 2013. The indices represent the average estimates and predictions from a number of ecological experts. Determine the long-term damage to sustainability that is expected to be caused by the project and its operations.

Table 28.3 Data for Example 28.2

Year	2006	2007	2008	2009	2010	2011	2012	2013	2014	2015	2016	2017	2018	2019
SI	12.75	12.61	12.22	12.07	11.99	11.69	11.69	11.39	6.27	6.72	6.16	5.34	5.77	5.49

Solution

Year 0 represents 2006. The changes in overall habitat units over the years are shown in Figure 28.6.

$$\text{Damage to sustainability} = \int_7^{13} (-0.187t + 12.704)\ dt - \int_7^{13} (-0.334t^2 + 0.5341t + 4.1519)\ dt$$

$$\left(\frac{-0.187t^2}{2} + 12.704t\right)_7^{13} - \left(\frac{-0.334t^3}{3} + \frac{0.5341t^2}{2} + 4.1519t\right)_7^{13}$$

$$= 65.004 - 36.316 = 28.69 \text{ SI-years}$$

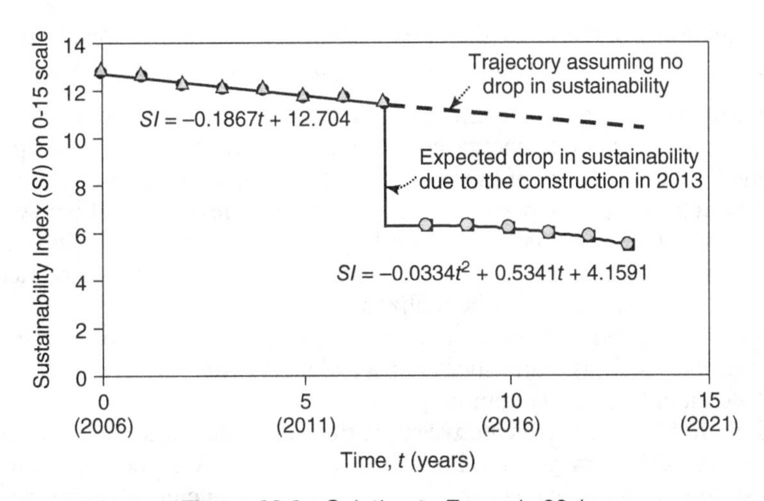

Figure 28.6 Solution to Example 28.1.

Example 28.3

Table 28.4 presents the trend of overall sustainability index for two identical civil engineering systems over a number of years over which the systems operated to date (from 2009 to 2013). The threshold level of sustainability is 0.1 unit. Action A focuses on the increased use of renewable energy for system operations, and action B focuses on increased recycling and reuse of materials for system maintenance. Assume that the two actions are mutually exclusive. (a) Determine which action is more effective in terms of the sustainability-related life of the system. (b) If action A costs \$0.5M initially and \$0.1M annually until the end of its sustainability-related life, and action B costs \$0.2M initially and \$0.15M annually until the end of its sustainability-related life, determine which action is more cost-effective in terms of sustainability. Assume a 7% discount rate.

Table 28.4 Yearly Sustainability Indices in Example 28.3

	2002	2003	2004	2005	2006	2007	2008	2009	2010	2011	2012	2013
Action A	0.82	0.67	0.59	0.52	0.41	0.39	0.38	0.32	0.28	0.28	0.22	0.20
Action B	0.60	0.60	0.60	0.59	0.59	0.58	0.57	0.55	0.55	0.52	0.48	0.47

Solution

(a) It can be seen from Figure 28.6 that Action A takes 10 years to reach the sustainability threshold while Action B takes 9 years. Thus A has a greater sustainability related life.

(b) The equivalent uniform annual cost (EUAC) of A = 0.1712 \$M; the EUAC of B = 0.2307 \$M

The cost-effectiveness of A = 10/0.1712 = 54.8 years/\$M, and the cost-effectiveness of B = 10/0.2307 = 39.01 years/\$M. Thus A is more cost-effective.

28.6 DIMENSIONS OF SUSTAINABILITY

The sustainability of a civil engineering system or of actions related to their development may be viewed from a number of perspectives, a few of which are discussed below.

Temporal Certain elements of sustainability are more applicable to the short term than to the long term. For example, increased noise from an airport construction project can be short term, while the land-use impacts of the noise from the airport's operations, such as changes in property values, are felt over a long period of time. Also, the construction of a new hydroelectric plant may result in immediate impacts (e.g., reliability of power supply, flooding of nearby villages) while longer term impacts such as increased business productivity due to reduction of electrical power outages, will take some time to be realized.

Spatial Spatial scopes, which could be categorized as global, regional, and local, include specific scopes such as statewide, countywide, citywide, areawide, corridorwide, or within the immediate vicinity of the system in question.

Stakeholder The perspectives of various affected entities and stakeholders, as to what constitutes sustainability, may differ. For example, with regard to the economic element of sustainability, the system owner may be primarily interested in financial solvency through user fees while the system users may view a system with high user fees as unsustainable.

28.7 SYSTEMS CONCEPTS IN SUSTAINABILITY MODELING

Sustainability considerations are very consistent with the concept of systems thinking. As we discussed in Chapter 2, a system can be defined as a construct of different elements that, when combined, produce results not achievable by the elements acting individually. Similarly, planet Earth or any part thereof could be considered as a system, not as a multicomponent entity but as a complex interrelated and interdependent dynamic whole (Khisty et al., 2012). In this section, we will discuss two key concepts in systems analysis where sustainability considerations are applicable.

28.7.1 Closed Loop as a Virtue Indicative of Sustainability

In nonsustainable situations (which, unfortunately, characterize most past human endeavors), the development of a product (a good, service, or in our context, a civil engineering system) has been open looped in nature. In other words, (i) the construction of the product is often carried out using materials that were not salvaged or recycled from other earlier similar products; (ii) the use of the product (i.e., the operation of the system) involves the acquisition of external often nonrenewable resources such as energy and water; and (iii) at the end of the life of the product, the product is not reused for another purpose or is not decomposed into its components for reuse.

Incorporating the principles of sustainability in the production process would mean that the end product of the consumption (otherwise to be discarded as waste) serves as the raw material for the production processes of construction (Figure 28.7). For operating the product, energy could be from a renewable source such as the sun, wind, sea waves, and deep ground (geothermal) (Figure 28.8). Such a closed-loop nature of the production process is characteristic of a self-sustaining system; and can be attained through gradual evolution over time (Roberts, 1990). As a complex system, the Earth contains not only a nearly infinite number of possible connections between its constituent parts but also several feedback loops where small changes could result in significant differences in the outcome (Khisty et al., 2012).

The Resource and Material Flows (RMF) model (Figure 28.9) is similar to the Roberts model. The RMF model considers the various phases of system development and the material and energy flows within and across the phases. With regard to energy flows, at the operations phase, energy can be generated to help in system operations (Figure 28.6), maintenance, or monitoring (e.g., solar-powered equipment that captures and transmits images showing the condition of system elements). Also, at each phase, energy used for carrying out work at that phase may be obtained from renewable or non-renewable sources. The percentage split shown below, for purposes of illustration, is 1980–2020; in a perfectly sustainable situation, this split might be 100–0). The energy emitted by the system at that phase may be recycled or simply made to dissipate; again, in a perfectly sustainable situation, this split might be 100–0. With regard to material flows, at the end-of-life phase, the system may be decommissioned and reused for another purpose, its constituent materials may be recycled to build a new system to replace the erstwhile system, transported elsewhere to help maintain another system, or disposed of in a landfill or other location.

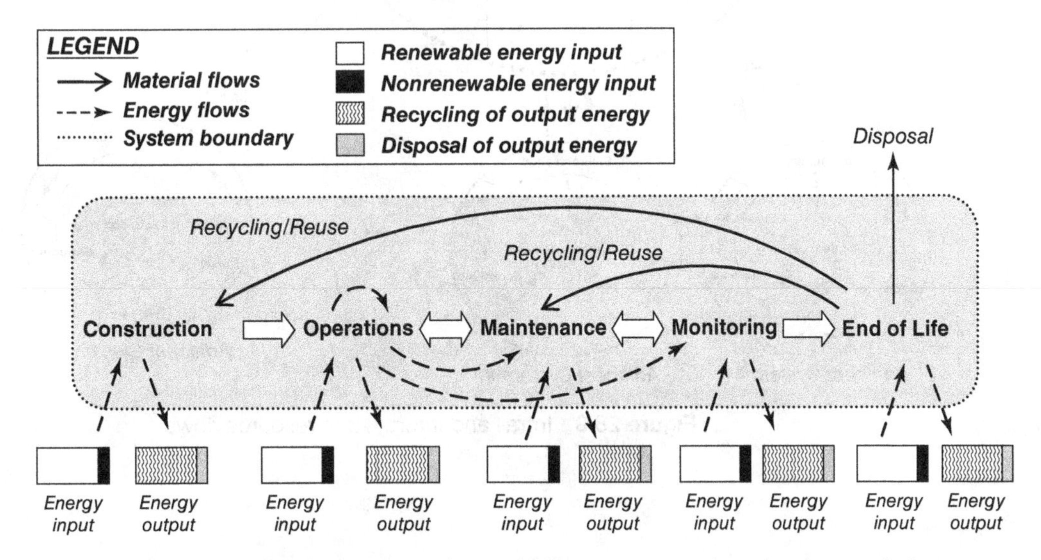

Figure 28.7 Model of resource and material flows across system development phases.

Figure 28.8 Power generated from system users may be used for lighting and monitoring purposes.

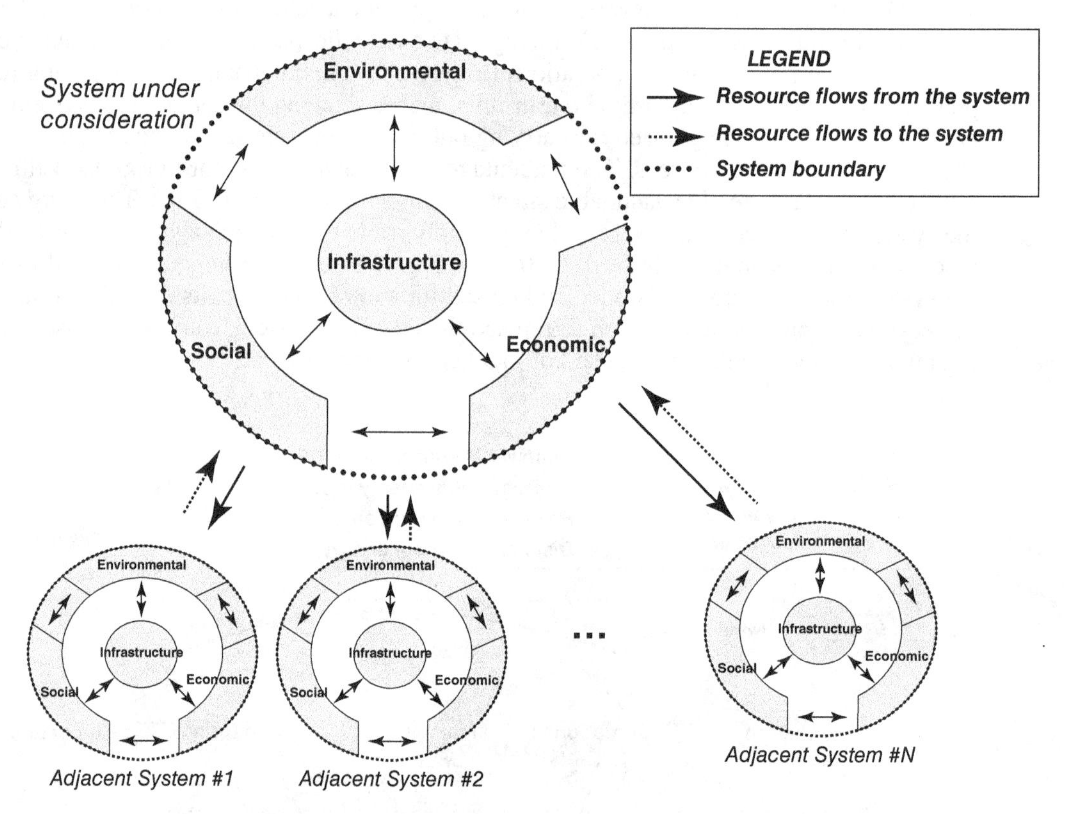

Figure 28.9 Intra- and intersystem resource flows.

28.7.2 Resource Flows within a System-of-Systems Characterization of Sustainability Relationships

Figure 28.9 presents the intra- and intersystem resource flows in a typical system that is indicative of sustainability. In the figure, infrastructure may be defined to include buildings, the water supply system, the sewer system, waste disposal system, the transportation system, and other components of a larger system of systems (SOS). The resource flows include the flow of energy, water, and finance. Resource flows occur not only from the infrastructure and the environment, economy, and society but also vice versa. Also, there are resource flows across resources and entities related to the environment, economy, and society.

28.8 SUSTAINABILITY CONSIDERATIONS AT EACH PHASE OF SYSTEM DEVELOPMENT

Figure 28.10 presents various sustainability considerations at each phase of system development. At each phase, the context of considering sustainability is twofold:

- *Impact of the phase on EEST (Environment, Economy, Social, Technical)*: Ensure that the adverse impacts of the system at any phase in terms of EEST, is minimized, and the positive EEST impacts are maximized. This is the more traditional context of sustainability consideration at each phase of system development.
- *Impact of EEST on the system at each phase*: Ensure that the successful completion of each phase is not jeopardized by economic, environment, social or technical conditions or stimuli.

28.8.1 Phase of Problem Identification and Needs Assessment

At this phase, the existence of a problem is established and the need for a solution (e.g., providing a new system or enhancing an existing system) is examined. If sustainability were defined to include

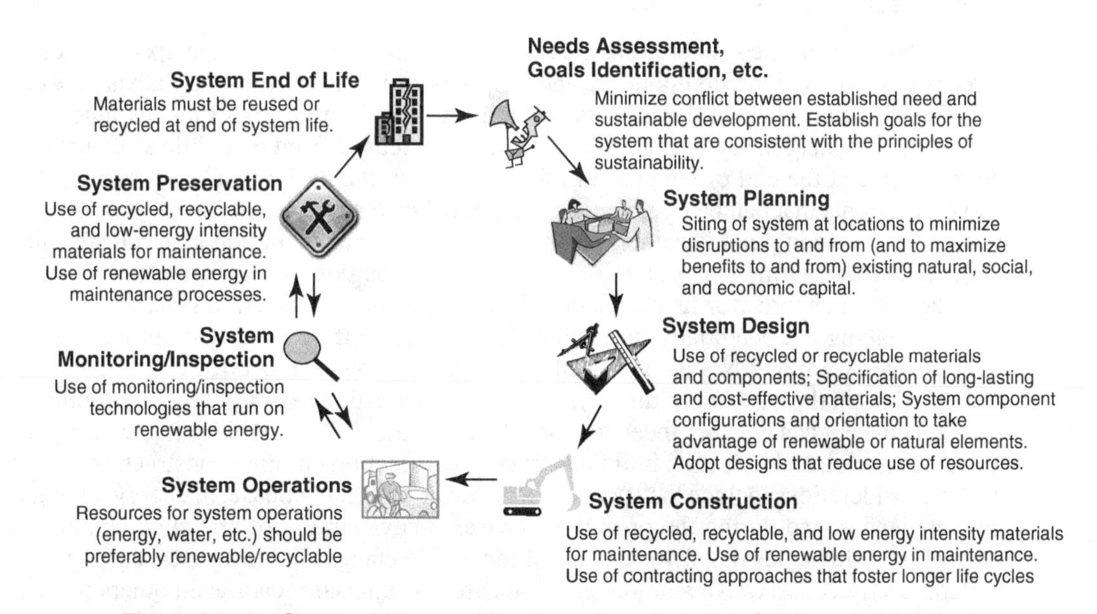

Figure 28.10 Sustainability considerations at each phase of system development.

the technical adversities caused by lack of the proposed system, then any existing lack of sustainability in that context could be cited as one of the problems and thus a rationale for proceeding with the system development. Note that by "technical adversities" we refer to the purpose for which the system is being proposed, such as mobility (for transportation systems), load-bearing support (for structural systems), and water flow control (for hydraulic systems). Most definitions of sustainability, however, exclude the technical aspects of performance, and thus preclude, or at least inhibit, the consideration of the triple bottom line as a deficiency that could generate the need for a new or enhanced system or as one of the decision criteria when evaluating infrastructure investments.

In Chapter 19, we discussed quantitative and qualitative techniques for assessing the need for a civil engineering system. The lesson we can learn from the current chapter is that such traditional needs assessments can be infused with sustainability considerations. In recent decades, the "predict-and-provide" model of delivering engineering services has faced significant challenges. Infrastructure managers increasingly seek to pursue a twin-track approach: expanding supply and reducing per capita demand. Resources cannot be indefinitely expanded, and therefore reducing per capita demand lessens the pressure on existing supplies due to population growth. Reduced per capita demand can help expand the number of people supplied by current resources and postpone, reduce, or cancel the need for investment in new supplies (Bell, 2011).

28.8.2 Planning Phase

Solutions at the planning phase resound throughout the system rest of life: planning decisions that do not adequately consider the environmental, economic, and social consequences of the system at its subsequent phases of construction, operations, and maintenance can lead to serious problems at those phases. For this reason, failure to adequately consider the sustainability elements at this phase is potentially more damaging in the long term compared to any such failings at other phases. In location selection, for example, environmental considerations should include the impact of the proposed system on the natural or man-made ecology; for example, the effects of the system construction or operations on the quality of habitats, air, water, soil, and economic and social capital of an area.

28.8.3 Design Phase

At the design phase, considerations related to sustainability include the specification of system dimensions, orientations, components, and materials. Orientations must take advantage of the intensity and direction of natural elements including sun and wind. Materials, as much as possible, should be durable, locally available, and recycled or reused from erstwhile structures and should be recyclable at the end of the system life. In addition, the design could be made to feature system components that can be operated using minimal or renewable natural resources of energy and materials. A case in point is cement-based concrete, the most common material specified by civil systems designers; this material exhibits excellent properties as a construction material and has abundant raw materials for its production; however, cement production generates large quantities of CO_2, making it a potentially unsustainable material. In their 2012 book titled *Sustainable Use of Concrete*, Sakai and Noguchi present guidelines on how to move toward sustainable concrete construction. In the case of building systems, the Leadership in Energy and Environmental Design (LEED) concept can help engineers to consider elements of sustainability in the design of residential, commercial, and industrial buildings (thus, also affecting building construction and operation). These considerations include the specification, during the design phase, of energy-saving lightbulbs and appliances, and during the operations phase, the availability of recycling containers, and the promotion of lifestyles that favor the use of renewable energy. In most cases, current design codes include overly conservative features in certain areas and unsafe features in others (Lee, 2010) and thus could be rewritten to enhance sustainability.

SUSTAINABILITY AT THE DESIGN AND CONSTRUCTION PHASES: FOCUS ON MATERIALS

Governments and the private sector have recognized that significant progress can be made in the drive toward sustainability by specifying and using materials whose performance is consistent with the principles of sustainability. In the United Kingdom, the Government Construction Client's Panel encourages the delivery of civil engineering infrastructure that "enhances the quality of life and customer satisfaction, offers the potential to address user changes in the future, provides and supports desirable natural and social environments, and maximizes the efficient use of resources". Also in the UK, the Green Guide to Specification provides guidance on the impact of different construction assemblies in terms of resource use, and, toxicity and embodies energy, durability, and other environmental criteria. In Australia, efforts include the Greenhouse Challenge Program (1995–2009), which encouraged abatement, improved greenhouse gas management, and improved emissions measurement and monitoring. In the United States, the concrete industry promotes conservative uses of energy and efficient production of concrete. In the European Union, the Cleaner Technology Solutions in the Life Cycle of Concrete Products (TESCOP) project was commissioned to help develop concrete products that are not only cost-effective but also cleaner in terms of environmental impacts.

Sources: UK GCPP (2000); DSEWPC (2009); Haugaard and Glavind (1998); Sirivivatnanon et al. (2003).

28.8.4 Construction Phase

At the construction phase of civil engineering systems development, the engineer makes decisions whose outcomes could enhance or degrade the environment, social capital, or the economy. Certain construction practices tend to be more sustainable than others; and certain contracting approaches, such as performance-based contracts, may be more sustainable, from a financial and economic standpoint, than others. Also, opportunities exist for materials reduction during construction; for example, concrete waste typically constitutes over 50% of all construction waste and could be reduced with careful planning and efficient materials delivery and utilization processes.

28.8.5 Operations Phase

The operations phase, by virtue of the fact that it is the longest phase, offers great opportunity to infuse sustainability concepts in the development of civil engineering systems, particularly with regard to the energy used to power the system. This energy could be tapped from renewable sources including wind, solar, geothermal, and motion of the system users. Other reuse and waste reduction opportunities exist in the use of water in system operations.

28.8.6 Monitoring Phase

At this phase, the engineer typically seeks effective and cost-efficient techniques for inspecting and monitoring the system. Also, the monitoring tasks must not pose a risk to the safety of the personnel carrying out that task. For example, it is not sustainable to have a traffic counting or bridge inspection program that exposes the workers to possible injury from passing traffic. The energy used in the monitoring could be tapped from renewable sources including solar and wind.

28.8.7 Maintenance Phase

At the maintenance phase, the engineer could specify the use of materials that are not only long-lasting to enhance economic sustainability but also would have little or no adverse impact of the surface waters, soils, and air. These include fuels, consummable parts, herbicides and pesticides, and deicing chemicals. Also, materials for maintaining the physical structure of civil engineering systems should preferably be those that can be reused or recycled.

28.8.8 End-of-Life Phase

The end-of-life phase of a civil engineering system often presents a unique opportunity for the engineer to apply an appropriate end-of-life treatment that minimizes material disposal in landfills and maximizes reuse and recycling (see Figure 28.11).

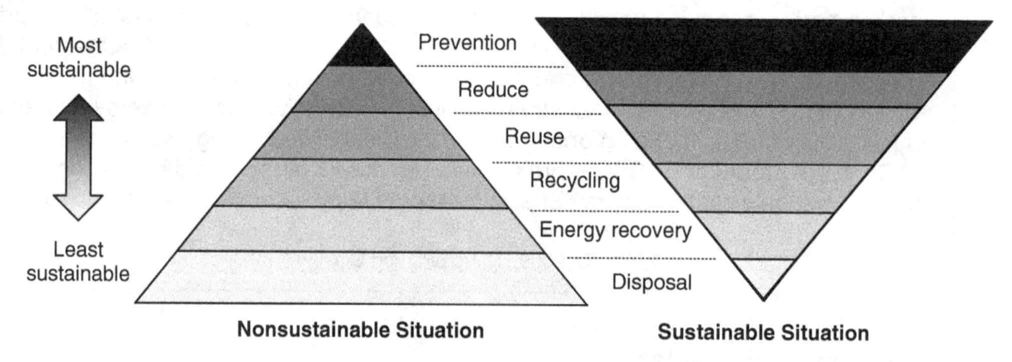

Figure 28.11 Pyramids of end-of-life choices (adapted from HS, 2012).

28.9 MODELING THE IMPACT OF HUMAN ACTIONS ON SUSTAINABILITY

The term *sustainability model* may mean one of at least two things. It may mean the use of sustainability as a factor to produce some outcome, or it may mean the estimation or prediction of sustainability in response to some factors that influence sustainability. In this section, we prefer to define a sustainability model in terms of the latter, specifically, to explain the impact of human actions on sustainability. If these actions are alternatives being considered at the phases of planning, design, construction, operations, or maintenance of a civil engineering system, then the desirability of each alternative action could be assessed in terms of sustainability, and the best option would be identified as that alternative with the greatest positive effect on sustainability or the least adverse effect on sustainability (this is elaborated further in Section 28.7).

In Section 28.4, we discussed a number of constituent elements of sustainability, and in Section 28.5 we identified a number of overall sustainability indicators that combine at least two of these elements. In this section, we discuss how these combined indicators are calculated as a function of the constituent elements, and consequently how a change in these elements (due, e.g., to civil system operations), could be estimated using the model.

If, for a given problem, sustainability is a weighted combination of different sustainability indicators, then the overall sustainability impact of each alternative is the weighted combination of these indicators, and the best alternative is duly identified. This approach is the same as the overall utility or disutility method used in multiple-criteria decision making. On the other hand, if sustainability is defined as a distinct set of sustainability indicators that must be maximized or minimized as appropriate, then the problem is a multiobjective problem, and the best alternative is that which provides the optimal solution to the set of multiple objective functions (see Chapter 12).

Models that estimate the impact of human actions on sustainability can therefore be categorized as **single objective** (where the outcome of each action is a single sustainability "measure" that is a combination of multiple indicators of sustainability) versus **multiple objective** (where the outcome of each action is a set of multiple sustainability indicators). Other ways of categorizing sustainability models include the type of indicator being used to describe/predict sustainability: whether the model is **aggregate** (models that use aggregate explanatory variables) versus **disaggregate** (models that use disaggregate explanatory variables); and **single pillar** (where only one of the sustainability elements or pillars is being predicted) versus **multiple pillar** (where two or more of the sustainability pillars are being predicted). We now present a number of models described by Khisty et al. (2012). These models can be used to estimate the impact of human actions on sustainability.

28.9.1 The IPAT Model

IPAT represents the lettering of a highly aggregate formula that describes the impact of human activity on the environment (Ehrlich and Holdren, 1971): $I = (P)(A)(T)$, where I is the human impact on the environment, P is population, A is level of affluence, and T is technology.

The impact of increased *population* on the environment includes increased land use (which leads to loss in the quantity and quality of habitat for other species of flora and fauna), increased use of resource (which leads to reduction in land cover, reduced percolation of rainwater, and increased runoff, and erosion), and increased pollution (which leads to climate change in the long term, illnesses in people and other species, and reduction of the quality of the ecosystems). *Affluence* can be represented by the average consumption of each person in the population. As people become more affluent, they consume more resources, and this can have an adverse effect on the total environmental quality. The gross domestic product (GDP) per capita (which, strictly speaking, measures production) is typically used as a surrogate for measuring consumption, with the assumption that consumption increases when production increases. *Technology* represents the resource intensity of the production of affluence; it measures the extent to which the production life cycle of a good or service (fabrication, transportation, usage, and disposal) affects the environment. The impact of technology can be reduced by innovations that enhance production. As the model components suggest, technological and behavioral improvements are vital in the drive toward sustainability (reduction of the human impacts); such changes can be fostered by government policy, regulations, and enforcement (Khisty et al., 2012) as well as sustainability-driven planning, design, and operations of engineering systems.

There are a few limitations with the IPAT model. First, the factors may not be independent of each other; there may exist some interactions and nonlinear relationships between them. Second, the measurability of variables can be an issue. Unlike the case for population, it is relatively difficult to establish proxies of affluence and technology. Variants of the IPAT model are presented in Table 28.5.

28.9.2 Triaxial Model

The triaxial representation of technological sustainability (TRTS) model (Pearce and Vanegas, 2002) is a three-dimensional representation of the sustainability of a situation or change thereof due to an action. This is estimated on the basis of the change in stakeholder satisfaction (a measure of social quality of life), and the impact on the demand for natural resources, and on the ecosystem (Figure 28.12).

28.9.3 Quality of Life/Natural Capital (QOL/NC) Model

The QOL/NC model estimates the resulting level of sustainability that is attained in response to actions explicitly associated with sustainable development. The model assumes that the primary motive of sustainability is to achieve a satisfying quality of life for all without exceeding the bounds of nature (Chambers et al., 2000). The model can be represented by four quadrants of a two-imensional chart (Figure 28.13). The first quadrant represents an ideal situation where quality of life is achieved and natural capital is protected, which is seen by some as idealistic but can serve as a target to which societies should strive. In the second quadrant, nature is protected, but the quality of

Table 28.5 Variants of the IPAT Model

Model	Description	Source
Pollution $= \text{population}\left(\frac{\text{goods}}{\text{population}}\right)\left(\frac{\text{pollutants}}{\text{goods}}\right)$	Environmental impact is the amount of emitted pollution which is a function of: population, amount of goods consumed per capita and amount of pollutant generated per unit of good produced.	Commoner (1972)
$I = aP^bA^cT^de$	Introduces stochasticity in the IPAT equation. *a*, *b*, *c*, and *d* are parameters, and *e* is an error term.	Dietz and Rosa (1994)
$I = PBAT$	*B* represents the behavioral choices of humans.	Schulze (2002)

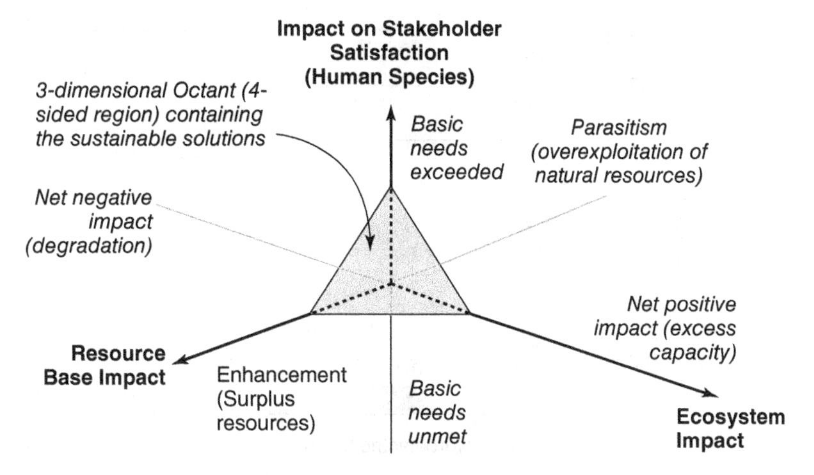

Figure 28.12 Triaxial model in the three dimensions.

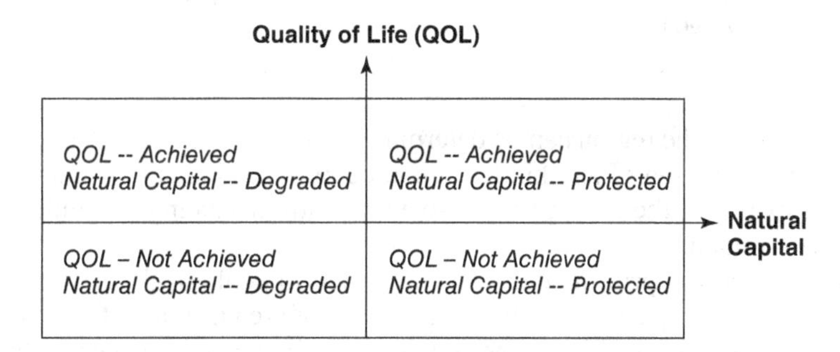

Figure 28.13 Four quadrants of the QOL/NC model.

life is poor (e.g., a resource-rich community that has been denied access to its resources). The third quadrant represents the worst situation: The environment is degraded and the quality of life is low. This is the situation found in certain villages in some developing nations where natural resources such as oil are extracted by the national government or companies without the due payment of royalties and without repairing the damage done by the extraction process. The fourth quadrant is indicative of a situation where the quality of life is high but little attention is paid to the protection of natural capital; certain developed countries that have weak regulation on environmental pollution fall into this category.

28.9.4 Footprint Models

(a) Ecological Footprint Model. As we discussed in Section 28.1.2, the ecological footprint (EF) is a measure of human demand on the Earth's ecosystems. An EF model, therefore, is one that describes or predicts the impact, in terms of EF, of a certain set of conditions. The EF can be determined for each productive entity including humans and industries. For example, the EF of an automobile manufacturing company is determined on the basis of all the materials including wastes (for example, wood, leather, steel, and plastics) that are consumed in the production of each automobile. Each of these resources is then translated into an equivalent number of global hectares. Knowing the number of automobiles produced, the total EF of that manufacturer can be determined. Also, knowing the total number of automobile manufacturers in a region, the total EF of automobile

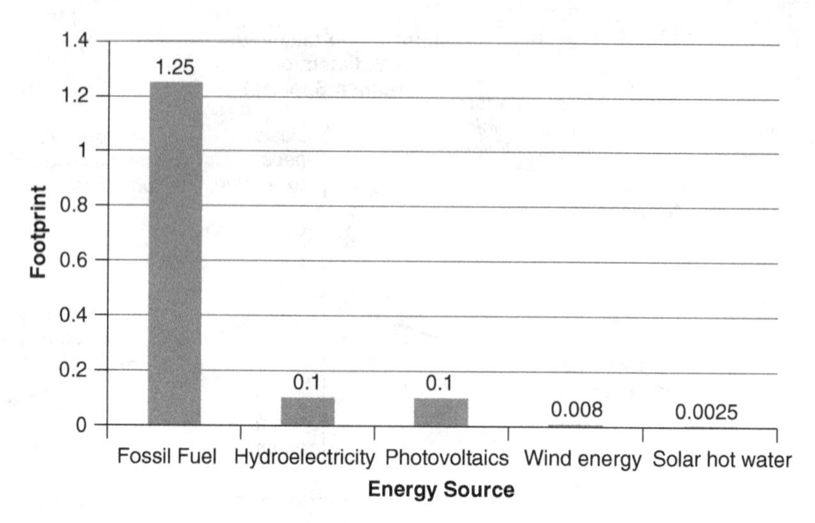

Figure 28.14 Ecological footprints of various energy sources (adapted from Wackernagel and Rees, 1996).

manufacturing in the region can be determined. The same procedure can be used for measuring the resource consumption of any kind of resource by other industries, humans, and other consumers. Thus, for example, the average EF associated with producing one unit of any product or service can be determined.

Ecological footprint models can be applied in ecological assessments by comparing the EF value with the biocapacity, i.e., the total amount of area available for production including wastes. An ecologically sustainable region is where the biocapacity exceeds the total EF. Figure 28.14 presents the EFs of various energy sources. It can be seen that to produce 100 gigajoules of fossil fuel per year, 1.25 hectares will be needed while only 0.0008 hectares will be needed in the case of wind energy.

(b) Sustainability Footprint (SF) Model. The SF model is a composite model comprised of the TRTS, EF, and QOL/NC model outcomes over a time interval, perspective of sustainability (Amekudzi et al., 2009). Assuming independence among the three outcomes, a civil system's overall sustainability impact or sustainability performance, in terms of the three measures of sustainability, can be measured over a given period of time. SF is then defined as the rate of change of some measure of civil engineering system performance related to the quality of life of the system users as a function of the life-cycle environmental and economic costs associated with the attainment of that performance.

The SF of an entity (such as a municipality), in the time period between time $t = i$ and $t = i + n$ (where n is some finite amount of time), can be expressed as follows:

$$\mathrm{SF}_{j_{t=i}}^{t=i+n} = \left[\frac{d}{dt}\left(\frac{z_j}{x_j y_j}\right), \frac{dZ_j}{dt}, \frac{dY_j}{dt}, \frac{dX_j}{dt}, Z_j, Y_j, X_j \right]$$

where Z_j is the system performance in terms of some measure of the quality of life; X_j is the resources or inputs used including associated wastes; and Y_j is the economic net benefits.

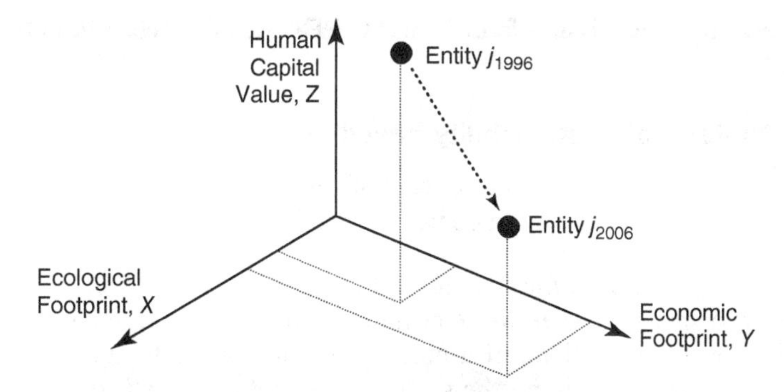

Figure 28.15 Sustainability footprint model.

Figure 28.15 (Amekudzi et al., 2009) illustrates the sustainability footprint model. As the figure shows, from the year 1996 and 2006, the entity degraded in its human capital value, increased in economic footprint, and decreased its ecological footprint. Thus, the overall sustainability and its rate of change over the time period can be assessed.

28.9.5 Capital Effects Model

Often referred to as the true sustainability index (TSI) model, this model helps assess the extent to which an action enhances any of the three basic elements or pillars of sustainability (economic, environmental, and social) at various spatial levels, namely, global, regional, or local (CSO, 2009). The index estimates the sustainability of a situation or action as a function of the impacts on resources that are termed "capitals" (i.e., natural capital and anthropogenic capital, which comprises the human, social, and constructed capitals) as shown in Table 28.6. The impacts of an action or situation on the carrying capacity of the vital capitals could be normative or actual. The TSI consists of 15 sustainability indicators that link the vital capitals to the TBL capitals. The environmental pillar, which is based on the assumption that human quality of life is tied to ecological quality, consists of indicators that include the quality of the air and water, climate, and solid waste recycling. The economic pillar comprises indicators such as livable wages, and economic institutions and infrastructure, emphasizes the impacts on the overall economy, and is linked to the economic health of individuals. The social pillar focuses on the impacts on the society, and measures all three types of anthropogenic capital and includes indicators such as human health and social institutions (McElroy, 2009).

Table 28.6 Matrix for the TSI Index Model

Impact Categories / Sustainability Pillars	Natural Capital	Man-made Capital: Human Capital	Man-made Capital: Social Capital	Man-made Capital: Constructed Capital
Environmental	X			
Social		X	X	X
Economic		X	X	X

28.10 INCORPORATING SUSTAINABILITY INTO PROJECT—AND NETWORK-LEVEL EVALUATION

28.10.1 Contexts of Sustainability-based Evaluation

For civil engineering systems, the concept of sustainability can be useful in evaluation and decision making in several contexts. Such contexts include (Sinha and Labi, 2007):

Evaluation of Proposed Investments—For the purposes of decision support, a civil engineering system owner may seek to determine the impacts, in terms of overall sustainability or one of its constituent pillars, of a number of alternatives for planning, designing, constructing, operating, or maintaining the system. The methods used for determining the impacts range from questionnaire surveys to comprehensive analytical or simulation models. The output of such studies is typically a prediction of the expected outcomes relative to base case scenarios.

Fulfillment of Regulatory or Policy Mandate—Regulations from the federal, state, or local government or the policy of a project sponsor (such as a multilateral lending agency) may require that the system development be preceded by an assessment of its impacts on sustainability from the economic, environmental, and social perspectives, at a minimum.

Postimplementation Evaluation—It is useful to assess the actual sustainability impacts that are measured after project implementation and to evaluate such findings vis-à-vis the levels predicted at the preimplementation phase, as well as the base year levels. Unfortunately, few agencies typically invest time and resources in such efforts.

Public Education—In cases of controversial projects or for the purposes of public relations, a civil engineering system owner may need to carry out evaluation in terms of sustainability impacts, with the objective of increasing general public awareness.

28.10.2 Incorporating Sustainability in Project-Level Decision Making

At any phase, facility-specific (i.e., project-level) decision-making contexts typically include the identification of the best action or sequence of actions to undertake to address a given technical situation for the system, deciding on the optimal level of some action that needs to be undertaken, and identifying the trigger level of some structural, functional condition or performance at which some action is warranted. The best action is that which maximizes some overall utility or minimizes some overall disutility. This utility or disutility could be expressed as a single index of sustainability or a general index that includes sustainability indicators among others.

28.10.3 Incorporating Sustainability in Network-Level Decision Making

At the network level, decision-making contexts at any phase typically include the identification of the best set of individual systems for which some action must be undertaken in order to maximize some overall networkwide or global utility. Similar to the project-level context, this overall utility or disutility of the network could be expressed as a single index of sustainability or a general index that includes at least one indicator of sustainability.

28.11 SPECIFIC SUSTAINABILITY CONSIDERATIONS IN CIVIL ENGINEERING DISCIPLINES

The greatest opportunities to incorporate sustainability in civil engineering occur at the phases of needs assessment, planning, design, and operations. The need for or the planned size of a new system should be determined not only by traditional prediction of the demand but also by determining how (i) to enhance capacity without physical increases in supply and (ii) reduce the demand through a battery of initiatives embodied by the demand management pyramid shown as Figure 28.9: eliminating the need, reducing the need, recycling, and reusing the system or an existing resource associated with the system.

For any civil engineering system, the instruments for implementing any aspect of the supply or demand management include policy, legal restrictions, and education campaigns and subsidies for system users to adopt operational policies or to use appliances that conserve resources. Also, capacity could be increased without additional supply but by improving the efficiency of the existing engineering system (e.g., efficient use of existing highway lanes using intelligent transportation systems). Demand management strategies are most useful during periods of congestion (when demand for the system's service exceeds capacity of the system), but as Bell (2011) indicates, these strategies are also critical for stable reductions in system demand in the long term. With regard to water and wastewater systems, demand management could include elimination of water needs (e.g., waterless sanitation), reducing water use (e.g., using appliances that use less water such as effective low-flush toilets and low-flow showers), reuse or recycling of gray water, and substitute sources (e.g., collection and use of rainwater).

With regard to building systems, the efficiency of water and energy could be promoted by implementing building codes or standards for new construction and national policy to replace inefficient fixtures such as lightbulbs, improve the efficiency of appliances, provide subsidies or promote educational campaigns for people and companies to make active efforts to conserve water and energy or to adopt renewable sources, and require manufacturers to provide product labels that give efficiency information, disaggregate metering, and billing of resource use.

For structural engineers, ASCE (2010b) has provided sustainability guidelines that offer advice on material selection and other aspects of the design process in order to enhance sustainability.

In the field of transportation engineering, significant work has been done to identify and implement sustainable practices and to establish the metrics for measuring sustainability, from all phases of transportation systems development (particularly, the construction, operations, and end-of-life phases) and also for both guideways and vehicles of various modes of transportation. The European Union Council of Ministers of Transport defines a sustainable transportation system as "one that facilitates basic access and development needs to be met in a manner consistent with safety, human, and ecosystem health and promotes equity within and between successive generations; is affordable, operates fairly and efficiently, offers a choice of transport mode, and supports a competitive economy, as well as balanced regional development; and limits emissions and waste within the Earth's ability to absorb them, uses renewable resources at or below their rates of generation, and uses nonrenewable resources at or below the rates of development of renewable substitutes, while minimizing the impact on the use of land and the generation of noise". This position is consistent with that of transportation agencies worldwide including those of the United States, Brazil, Canada, Colombia, New Zealand, and Australia (Jeon and Amekudzi, 2005).

In materials engineering systems, sustainability is increasingly being considered for purposes of design and maintenance, with an emphasis on (i) the use of materials whose production require less energy or other natural resources, and less environmental degradation, and (ii) reduction in demand for new materials due to reuse or recycling of existing materials. For example, the demand for cement can be reduced by replacing one-fifth of it with fly ash, thus reducing CO_2 emissions.

In each specific branch of civil engineering, sustainable practices can also be categorized not only by the phase of the development (see Section 28.4) but also on the basis of whether the sustainability is related to a physical entity (e.g., material) or nonphysical entity (e.g., policy).

28.12 SUSTAINABILITY—SOME INTERESTING FINALES OF THE DISCUSSION

28.12.1 Sustainable Development: Sincere Progress or Deceptive Illusion

Adams (2006) argues that sustainability and sustainable development have become effectively ethical concepts that reflect desirable economic and social outcomes from development activities. The author states that the term sustainable has therefore been "applied loosely to policies to express this aspiration or to imply that the policy choice is 'greener' than it might otherwise be, such as the idea of a sustainable road building program". Lamenting that sustainable development has become mere rhetoric and is ignored in practical decision making, Adams states that efforts at sustainable development often end up being development as usual, with professed adherents often admitting failure and exhibiting a "brief wistful and embarrassed genuflection towards the desirability of sustainability." In the absence of explicit indicators and targets for sustainability progress, and sincere political and institutional will, it is doubtful that sustainability considerations will play a meaningful role in decision making for civil systems. Generally, there seems to be optimism that ultimately humankind will be willing and able to create a sustainable society (Dresner, 2008).

28.12.2 Current Status of the Triple Bottom Line

As we have learned from this chapter, sustainable development of civil engineering systems requires due consideration of factors that are associated with the economic, environmental (including ecological), and social dimensions. At the current time, significant progress has been made in the economic and environmental dimensions as system developers have considered economic factors and increasingly consider the environmental performance of industries, technologies, and infrastructure such as energy and resource efficiency benefits, waste reduction, and pollution mitigation. However, relatively little has been done by civil engineering system developers regarding the incorporation of the social dimension of sustainability (Bell, 2011). A general consensus seems to be that future frontiers in sustainable engineering studies will go beyond economic efficiency and environmental impacts reduction toward a more explicit recognition and incorporation of the social aspects of development.

28.12.3 Relationships between Sustainability, Ethics, Law, and Morality

In Chapter 29, we will discuss the issues of ethics and legal aspects associated with the development of civil engineering systems. Also in Chapter 3 (where we discussed system goals) and the current chapter, we discuss some of the outcomes of civil engineering systems, particularly in terms of their so-called "technical" impacts, environmental and social impacts, among others. The concepts learned from these two sets of chapters are not disconnected. In order to comply with legislative mandates and to ensure consistency with their professional obligations to the society, engineers have a duty to ensure that their work at any development phase does not jeopardize the well being of current inhabitants and more importantly future generations. This is possible if engineers explicitly consider performance measures that are related to sustainability, where relevant, as they go about their various tasks at each phase of system development.

28.12.4 Sustainability Conflicts and Trade-offs between Sustainability Indicators

In an ideal situation, an engineering action has favorable impacts in terms of all indicators of sustainability. However, the practical reality is that the attainment of high levels of one indicator may be at the expense of another due to either the nature of the indicator or the diversion of the budget to certain work types or locations. This gives rise to the concept of trade-off between sustainability-related indicators (Bai, 2012). In a more general context, a "trade-off" or barter refers to the sacrifice of a physical entity of quality in return for gaining another. For example, for a given system of project, what is the state of practice trade-off (how much of sustainability indicator A is currently being earned at the expense of sustainability indicator B?) or the trade-off assuming optimal allocation of resources?

Also, there could be a trade-off between the uncertainty (or variability) associated with a sustainability indicator and the expected level of that indicator. All else being the same, the engineer prefers actions that yield the highest level of sustainability and the smallest uncertainty of achieving that level; in other words, we want to (i) achieve superior level of sustainability performance and (ii) be reasonably certain that we will achieve it. However, in some cases, a project may have expected high sustainability that has high uncertainty (with indicator levels ranging from, say, 40 to 100 with an average of 60); a rival project may have relatively low sustainability performance (which is bad) with low uncertainty (which is good), with performance ranging from, say, 55 to 60 with an average of 58. As such, it is possible to investigate the trade-offs between the mean level and the variability (uncertainty) in the manner in which they relate to project sustainability.

28.12.5 Short-Term versus Long-Term Trade-off

Related to the sustainability conflicts and trade-offs discussed in Section 28.12.4, is the issue of long-term versus short-term trade-offs. Often, in seeking increased environmental or social sustainability in the long term, there is a reduction in financial sustainability in the short term but an increase in financial sustainability in the long term. Thus, short-sighted or cash-strapped system owners may therefore be inclined not to pursue environmental or social sustainability. For example, managers of public or private civil facilities may perceive sustainable considerations to be a mandate that cost money (which they might not have), thus they may view such sustainability as unattainable or even unfavorable and may be inclined not to pursue it. However, they need to be made to understand that environmental or social sustainable choices that may increase initial costs often actually reduce the rest-of-life costs, and thus, overall life-cycle costs. For example, compact fluorescent bulbs are generally more costly upfront but use less electricity and have longer service lives and therefore can significantly reduce energy costs in the long term.

Thus, it is clear that the incorporation of sustainability in decision making at the civil systems phases of planning, design, construction, operations, and maintenance will translate into larger initial costs of providing the system or reconstructing it when it is due for replacement. In an era of tight funding, higher capital outlays may be inimical to the likelihood of the civil system development. As such, civil engineers have a duty to educate the general public, legislators, and other owners of stakeholders of civil systems that the incorporation of environmental or social sustainability in civil system development can actually lead to reduced overall costs over the life cycle; that way, the stakeholders will be more inclined to view sustainability initiatives in a more favorable light.

28.12.6 Weak Sustainability versus Strong Sustainability

With increases in global population and demand, a rise in both the total and the per capita stock of overall capital is necessary in order to maintain or even enhance quality of human life. For this to be possible, the depletion in the natural capital due to rising population must be replaced by equivalent man-made capital. In this respect, Khisty et al. (2012) discussed the dichotomy between weak sustainability and strong sustainability. The difference in the two concepts lies in the extent to which depletion in the stock of natural resources could be substituted by equivalent man-made capital (Solow, 1993). In a weak sustainable situation, natural capital may be decreasing but as long as it is being replaced by man-made capital (thus, overall stock does not decrease) society is considered sustainable. This is a compromise that maintains the dominant models of development by paying greater attention to environmental concerns (Bell, 2011). Unlike a weakly sustainable situation, a strongly sustainable situation is where stocks of natural capital are held constant independently of (and are complementary to) man-made capital, but man-made capital cannot replace lost natural capital (Wackernagel and Rees, 1996). This is a radically different characterization of the relationships between the natural world on one side and economics, development, and humans on the other side (Neumayer, 2010).

28.12.7 Local versus Global

In any discussion of sustainability, a distinction needs to be established between the local level of the efforts and global levels of the outcomes, respectively. Although sustainability discussions typically address global impacts, actions geared toward sustainability are fundamentally local in nature. The local character of efforts is further underscored by the fact that every community has unique cultural, economic, social, and environmental concerns (Monday, 2002). Further, each community has its unique levels and changing patterns of the quantity, quality, and importance of these concerns. Thus, efforts to address sustainability are mostly discussed in terms of local actions and decisions while the outcomes are mostly discussed in global terms. However, if the outcomes are sought not in terms of overall sustainability but of its components individually (economics, social, environmental, etc.), then the analysis of outcomes can be done at the local (project) level, using methodologies provided in the literature including Forkenbrock and Weisbrod (2001), Ortolano (1997), and Sinha and Labi (2007).

28.12.8 Engineers, Society, and Sustainability

To date, the focus of engineering contributions on sustainability has been to reduce the environmental impacts of development and to increase the efficiency of resource use. Bell (2011) points out the engineering profession continues to have a pivotal role in potentially achieving sustainable development through the changes that this discipline could engender in the society. The new winds of thought include a growing notion of **ecological modernization**, i.e., achieving sustainability by undertaking fundamental reforms of modern society accompanied by the development of environmental technologies. Environmental philosophers, however, have been skeptical about this notion, indicating that the very nature of modern society is associated with serious ills including the wanton destruction of nature for economic gains and the persistence of social inequality. Thus the patterns of domination and the separation of nature and culture, it seems, are central to the current crises of ecology and human development. It has been agreed that the engineering profession has a clear role to play in ecological modernization; however, its role in more radical visions of sustainability remains uncertain. A concept known as **actor-network theory**, presented by Bell (2011), assesses sociotechnical systems that avoid the separation of culture and nature, and presents a perspective of how engineering could help shape not only society but also the relationship between society and the environment. The actor network theory describes the state of world in terms of the relationships between the actors, human and nonhuman, and demonstrates that social relationships are influenced by technology and nonhuman factors.

Example 28.4 A Pearl of Sustainability: The Le Paik Nam June Media Bridge in Soeul, Korea

A 1080-meters long "sustainable bridge," the Paik Nam June Media Bridge, is currently being planned to cross the Han River in the center of Seoul, Korea, in several innovative ways both aesthetically and ecologically. Connecting a planned public cultural space and The National Assembly Building, this bridge is meant to be an extension of the city into the river. Inspired by the water strider, the bridge's overall shape is organic, with a sleek streamlined outline, and will be covered with panels to generate energy on its own. The bridge will accommodate cars, bicycles, and pedestrians; water taxis, yachts, and cruise ships will dock at the base of the bridge. The sustainable green space over the bridge is a circulated vertical and horizontal garden, and will utilize local resources—river and rain water, and natural light and ventilation. The bridge will have an organic skin that will serve as a canvas for media and video artists from all over the world, and will have an IT-equipped public museum and library.

Source: www.archdaily.com. (2012).

SUMMARY

Managers of civil engineering systems increasingly seek to consider sustainability in decision making. As Dandy et al. (2004) pointed out, consideration of sustainability does not replace assessments, over a system's life cycle, of the environmental impacts, economic efficiency or economic development impacts, and social impacts. Instead, traditional decision-making frameworks may be reinforced with an assessment of sustainability (Pope et al., 2004). Even though the underlying tenets of sustainability have always been of concern for ages, overall sustainability has become a major issue in recent times, particularly due to the unparalleled scale of human activity, the unprecedented demand for civil engineering systems, the reality that the resilience of natural resources is not infinite, and the demonstrated relationship between sustainability and human well-being. A number of international initiatives have been undertaken in the past 50 years to bring global attention to the urgency for sustainable practices in all sectors including civil engineering. This chapter also discussed the principles, elements and dimensions of sustainability, and identified a number of composite indicators of sustainability that combines economic, environmental, social, and other considerations.

The chapter also discussed the incorporation of systems concepts in sustainability modeling; first, as a closed loop as a virtue that is indicative of sustainability, and second as an accounting of resource flows within a system-of-systems characterization of sustainability relationships. Then, for each phase of civil engineering system development, the chapter discussed the impact of the triple bottom line (TBL) on the successful implementation of that phase and the TBL impact of each phase. With respect to the latter, the chapter further focused on specific models that model the impact of the phase and of human actions in general, on sustainability at a more aggregate scale. The chapter discusses how sustainability considerations could be incorporated into evaluation of civil systems and identifies specific sustainability considerations in the civil engineering branches. The chapter concluded by discussing a number of pertinent issues related to sustainability and sustainable development; including ethics, law, and morality; sustainability conflicts and trade-offs between sustainability indicators; and the relationship between engineers, society, and sustainability.

EXERCISES

1. Identify and discuss 10 ways to enhance general sustainability on your campus. Areas of discussion could include sports, the dining halls, classrooms, and the residence halls.
2. Consider a civil engineering system on campus. Discuss the impacts of the system on elements of the triple bottom line, namely: environment, economy, and the society (EST). Also, discuss how certain aspects of EST could jeopardize or foster the successful operations of the system.
3. Present a cogent argument, with illustrations in any branch of civil engineering, why evaluation and decisions based on life-cycle cost and benefit analysis compared to the analysis of initial costs only is more consistent with sustainability.
4. What is LEED certification? In striving toward this type of certification, how could a building's development be made more sustainable?
5. The values of society are not in consistent with the achievement of sustainability. Discuss.
6. Explain how the Tragedy of Commons: (a) is an important consideration in assessing the long-term sustainability of civil engineering systems. (b) has been used by each side of the political spectrum to advance their agenda.
7. As the engineer in charge of designing a new hydroelectric plant, discuss how you would, through your design, enhance sustainability of this system. Discuss your answer in the context of experiences of the Aswan Dam in Egypt and the Three Gorges Dam in China. In your answer, refer to *Dams and Development*, published in 2000 by the World Commission on Dams.
8. Discuss the Russian doll (or concentric circles) model of sustainability and contrast it with the overlapping Venn diagram (or, intersecting circles) model (Figure 28.16). Which model, in your opinion, is more effective in describing the impacts of civil engineering systems in terms of the sustainability elements? Also, argue why a civil engineering system's technical impacts or functional performance could be considered a fourth dimension as espoused by Jeon and Amekudzi (2005).

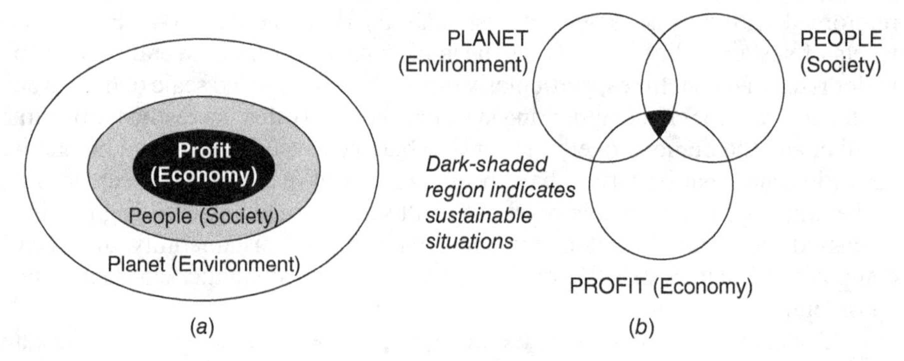

Figure 28.16 Two representations of the triple bottom line (adapted from Khisty et al., 2012): (*a*) Russian doll model and (*b*) overlaping Venn diagram model.

9. At the planning phase, two different route locations are being considered for a new highway to link two fast growing cities. The alternatives differ in terms of the topology, number of rivers to be crossed, traversing of sensitive environments (such as natural ecologies), number of residents to be displaced and relocated, and alignment distance. Discuss the sustainability-related issues that must be considered in order to throw more light on the evaluation of these alternatives.
10. Planners of a proposed nuclear plant who are concerned about long-term sustainability of the plant seek to establish the relative weights of the sustainability evaluation criteria needed for selecting the best location for the plant. The planners believe that the new plant will have profound impacts in terms of spurring economic development, increasing employment, and degradation of the area's sensitive ecology. As such, they intend to use the following evaluation criteria: economy, society, and the environment. The comparison matrix based on the decision-makers' judgments is shown in Table 28.7. Investigate the consistency of the planners judgments with regard to these criteria.

Table 28.7 Comparison Matrix of Sustainability Criteria

	Environment	Society	Economy
Environment	1	5	3
Society	—	1	1/4
Economy	—	—	1

11. For the problem in Exercise 28.9, the level of each sustainability criterion corresponding to the alternative locations of the nuclear plant is given in Table 28.8. (a) Select five of your colleagues and carry out a scaling survey using any method (see Section 12.3 of Chapter 12) and use the results of your survey to develop utility functions for each criterion. (b) Using the relative weights that were obtained for the three criteria in Exercise 28.10 and the utility functions developed in this question, identify the best location for the nuclear plant. Use the additive function for amalgamation.

Table 28.8 Expected Outcomes in Terms of Sustainability Criteria

Location	Area Likely to Be Affected (hectares)	Employment (thousands)	Economic Development ($millions of business sales)
P	3.4	2.6	2.1
Q	1.2	1.1	3.9
R	0.9	0.7	2.1
S	3.1	1.0	3.3

REFERENCES

Adams, W. M. (2006). The Future of Sustainability: Re-thinking Environment and Development in the Twenty-first Century. Report of the IUCN Renowned Thinkers Meeting, 29–31 Jan 2006. Retrieved Feb. 16, 2009.

Amekudzi, A., Khisty, C. J., and Khayesi, M. (2009). Using the Sustainability Footprint Model to Assess Development Impacts of Transportation Systems. *Transp. Res. A: Policy Practice*, Vol. 43, Issue 3, 339–348.

Amekudzi, A., Meyer, M., Ross, C., and Barrella, E. (2011). *Transportation Planning and Sustainability Guidebook*. Fed. Hwy Admin., Washington, DC.

APWA (2013). *The Principles of Sustainability*. APWA Center for Sustainability, Washington, DC.

ASCE (2007). *The Vision for Civil Engineering in 2025*, ASCE Press, Reston, VA.

ASCE (2009). Building a Sustainable Future. Ariaratnam, S. T., and Rojas, E. M. (Eds), ASCE Press, Reston, VA.

ASCE (2010a). Policy Statement 418—The Role of the Civil Engineer in Sustainable Development, ASCE.

ASCE (2010b). *Sustainability Guidelines for the Structural Engineer*. ASCE Press, Reston, VA.

ASCE (2012). ASCE and Sustainability. http://www.asce.org/sustainability/. Accessed Oct. 12, 2010.

ASCE, ICE, CSCE (2012). *Protocol: Civil Engineering for a Sustainable Future*, American Society of Civil Engineers, Institute of Civil Engineers, Canadian Society of Civil Engineers, Reston, VA.

Bai, Q. (2012). Tradeoff Analysis in Transportation Asset Management. Ph.D. Dissertation, Purdue University, W. Lafayette, IN.

Bell, S. (2011). *Engineers, Society, and Sustainability, Synthesis Lectures on Engineers, Technology, and Society*. Baillie, C. (Ed.). Morgan & Claypool, San Rafael, CA.

Bell, S., Chilvers, A., and Hillier, J. (2011). The Socio-Technology of Engineering Sustainability. *Procs., ICE—Engineering Sustainability*, 164(3), 177–184.

Brundtland, H. (1987). *Our Common Future*, Oxford University Press, for the United Nations World Commission on Environment and Development, New York.

Caltrans (1997). *Community Impact Assessment, Caltrans Environmental Handbook*, Vol. 4. Caltrans, Sacramento, CA.

CamSys, BLA. (2003). I-69 Evansville to Indianapolis—Final Environmental Impact Statement. Cambridge Systematics Inc. and Bernardin, Lochmueller & Associates Inc., prepared for the Indiana Dept. of Transp. http://www.deis.i69indyevn.org/FEIS/. Accessed Jan. 14, 2006.

CSO (Center for Sustainable Organizations) (2009). The True Sustainability Index. www.sustainable organizations.org. Accessed Nov. 25, 2012.

Chambers, N., Simmons, C., and Wackernagel, M. (2000). *Sharing Nature's Interest, Ecological Footprints as an Indicator of Sustainability*. Earthscan, London/Sterling, VA.

CLEP (2008). Center for Environmental Law & Policy. Yale Univ., Environmental Performance Index, in collaboration with the Center for International Earth Science Information Network (CIESIN), Columbia Univ., the World Economic Forum, Geneva, and Joint Research Centre of the European Commission, Ispra, Italy.

Commoner, B. (1972). The Environment Cost of Economic Growth, Population, Resources and the Environment. Ridker, R. G. (Ed). U.S. Govt. Printing Office, Washington, DC.

Dandy, G., Walker, D., Daniell, T., and Warner, R. (2008). *Planning and Design of Engineering Systems*, 2nd ed. CRC Press, Boca Raton, FL.

Dauncey, G. (2012). Towards Sustainability. Available at www.towards-sustainability.co.uk/ infodir/susquote .html. Accessed Nov. 28, 2012.

Dietz, T., and Rosa, E. (1994). Rethinking the Environmental Impacts of Population, Affluence and Technology. *Human Ecol. Rev.*, 1, 271–300.

Dresner, S. (2008). *The Principles of Sustainability*, 2nd Ed. Routledge. London, UK.

DSEWPC (2009). *Greenhouse Challenge Plus*. Department of Sustainability, Environment, Water, Population and Communities, Canberra, Australia.

Ehrlich, P. R., and Holdren, J. P. (1971). Impact of Population Growth. Science 171, 1212–1217.

FHWA (1982). *Social Impact Assessment: A Sourcebook for Highway Planners, Vol. III: Inventory of Highway Related Social Impacts*. Fed. Hwy. Admn., U.S. Dept. of Transp., Washington, DC.

Forkenbrock, D. J., and Weisbrod, G. E. (2001). *Guidebook for Assessing the Social and Economic Effects of Transportation Projects*. NCHRP Report 456, National Academy Press, Washington, DC.

Gillman, R. (1992). Sustainability, UIA/AIA Call for Sustainable Community Solutions. www.context.org/ ICLIB. Accessed Nov. 29, 2012.

Global Footprint Network (2010). *Living Planet Report 2010*. WWF, Zoological Society of London.

Goodman, A., and Hastak, M. (2007). *Infrastructure Planning Handbook*. MH/ASCE Press, Reston, VA.

Hardin, G. (1968). The Tragedy of the Commons. Science 162(3859), 1243–1248.

Haugaard, M., and Glavind, M. (1998). Cleaner Technology Solutions in the Life Cycle of Concrete Products (TESCOP). Procs., Conf. on Euro Environment, Denmark.

Hawken, P., Lovins, A. B., and Lovins, L. H. (2000). *Natural Capitalism*. Earthscan, London.

HS (2012). The 4 R's—Reduce, Reuse, Recycle, and Recover Heidi's Sharebook. http://www.care2.com/c2c/share/detail/246651.

IOCGP (2003). Principles and Guidelines for Social Impact Assessment in the USA. *Impact Assessment and Project Appraisal* 21(3), 231–250. The Interorganizational Committee on Principles and Guidelines for Social Impact Assessment. Beech Tree Publishing, London, UK.

IPCC (2000). A Report of Working Group I of the Intergovernmental Panel on Climate Change—Summary for Policymakers. IPCC. Geneva, Switzerland.

Jeon, C. M., and Amekudzi, A. (2005). Addressing Sustainability in Transportation Systems: Definitions, Indicators, and Metrics. *ASCE J. Infr. Syst.*, 31–50.

Jowett, B. (1885). *Translator, Politics, by Aristotle (384 BC–322 BC)*. Clarendon Press, Oxford, UK.

Khisty, C. J., Mohammadi, J., and Amekudzi, A. A. (2012). *Systems Engineering*, 2nd ed. J. Ross Publishing, Plantation, FL.

Lee, G. C. (2010). *Sustainable Development in Bridge Engineering: Development of Multi-Hazard Design*. Distinguished Speaker Series, State University of New York, Buffalo, NY.

McCuen, R. H., Ezzell, E. Z., and Wong, M. K. (2011). *Fundamentals of Civil Engineering—An Introduction to the ASCE Body of Knowledge*. CRC Press, Boca Raton, FL.

McElroy, M. W. (2009). *True Sustainability Index*, Center for Sustainable Orgainzations, Thetford Center, VT.

McGrath, J. (2012). Sustainable Buildings. http://dsc.discovery.com/tv-shows/curiosity/topics/10-sustainable-buildings.htm. Accessed Dec. 15, 2012.

Meadows, D., Randers, J., and Behrens, W. W. (1972). *The Limits to Growth*. Universe Books, New York.

Monday, J. L. (2002). Building Back Better—Creating a Sustainable Community after Disaster. Natural Hazards Informer, Nr. 3.

Natural Step (2013). Principles of Sustainability. http://www.naturalstep.org/en/usa/principles-sustainability.

Needham, J. (1971). *Science and Civilization in China: Vol. 4, Part 3*. Caves Books, Taipei.

Neumayer, E. (2010). *Weak versus Strong Sustainability*, 3rd ed. Edward Elgar, Cheltenham, UK.

NRC (1999). *Our Common Journey, Board on Sustainable Development*. National Research Council Washington, DC.

OECD (2012). Glossary of Statistical Terms. The Organization for Economic Co-operation and Development, http://stats.oecd.org/glossary/detail.asp?ID=2625.

Ortolano, L. (1997). *Environmental Regulation and Impact Assessment*. Wiley, New York.

Pearce, A., and Vanegas, J. A. (2002). Defining sustainability for Built Environment Systems. *Int. J. Environ. Tech. Manage.* (2)1, 94–113.

Pope, J., Annandale, D., and Morrison-Saunders, A. (2004). Conceptualizing Sustainability Assessment. *Environ. Impact Assess. Rev.* 24, 595–616.

Rankin, D. J., Bargum, K., and Kokko, H. (2007). Trends Ecol. *Evol.* 22, 643–665.

Roberts, D. V. (1990). Sustainable Development, A Challenge for the Engineering Profession. Paper presented at the Annual Conference of the International Federation of Consulting Engineers, Oslo, Norway.

Schmidheiny, S. (1992). *Changing Course: A Global Business Perspective on Development and the Environment*. MIT Press, Cambridge, MA.

Schulze, P. C. (2002). I=PBAT. *Ecol. Econ.* 40(2), 149–150.

Sen, A. (2000). *Development as Freedom*. Anchor Publishing Harpswell, ME.

Sinha, K. C., and Labi, S. (2007). *Transportation Decision Making—Principles of Project Evaluation and Programming*. Wiley, Hoboken, NJ.

Sirivivatnanon, V., Tam, C. T., and Ho, D. W. S. (2003). *Special Concrete for Sustainable Development, Civil Engineering Handbook*. Chen, W. F., and Liew, J. Y. (Eds.). Boca Raton, FL.

Solow, R. (1993). An Almost Impractical Step Towards Sustainability, *Resources Policy* 16, 162–172.

Straup, H. (1964). A History of Civil Engineering–An Outline from Ancient to Modern Times, MIT Press, Cambridge, MA.

UK GCPP (2000). Achieving Sustainability in Construction Procurement. Government Construction Clients Panel United Kingdom.

Vitousek, P. M., Mooney, H. A., Lubchenco, J., and Melillo, J. M. (1997). Human Domination of Earth's Ecosystems. Science 277, 494–499.

Wackernagel, M., and Rees, W. (1996). *Our Ecological Footprint: Reducing Human Impact on the Earth*. New Society Publishers, Gabriola Island, BC.

Wilson, E. (1992). *The Diversity of Life*. Harvard University Press, Cambridge, MA.

World Bank (2003). *Social Analysis Sourcebook: Incorporating Social Dimensions into Bank-supported Projects*. Social Dev. Dept., Washington, DC.

World Bank (2013). http://go.worldbank.org/57GVYJEEN0. Washington, DC.

USEFUL RESOURCES

Agyeman, J. (2005). *Sustainable Communities and the Challenge of Environmental Justice*. New York Univ. Press, New York. NY.

Amekudzi, A. A. (2011). Placing Carbon Management in the Context of Sustainable Development Priorities: A Global Perspective. *Carbon Manage.* 2(4), 413–423.

Baker, S. (2006). *Sustainable Development*. Routledge, London, UK.

Bond, A., Morrison-Saunders, A., Pope, J. (2012). *Sustainability Assessment: The State of the Art, Impact Assessment and Project Apprasial*, Taylor & Francis, London, UK.

Engineering Council UK (2009). Guidance on Sustainability for the Engineering Profession. http://www.engc.org.uk/about-us/sustainability. Accessed Nov. 30, 2012.

ICE (2010). Canadian Society for Civil Engineering, Institution of Civil Engineers, American Society of Civil Engineers. Protocol for Engineering: A Sustainable Future for the Planet. www.ice.org.uk/Information-resources/Document-Library. Accessed Nov. 30, 2012.

Jeon, C. M., and Amekudzi, A. (2005). Addressing Sustainability in Transportation Systems: Definitions, Indicators and Metrics. *ASCE J. Infrastr. Syst.* 11(1), 31–50.

Jeon, C. M., and Amekudzi, A. A., and Guensler, R. (2010). Evaluating Transportation System Sustainability: Atlanta Metropolitan Region. *Int. J. Sust. Transp.* 4(4), 227–247.

NRC (2009). *Sustainable Critical Infrastructure Systems, A Framework for Meeting 21st Century Imperatives*. National Academies Press, Washington, DC.

Rogers, P. Jalal, K. F., and Boyd, J. A. (2008). *An Introduction to Sustainable Development*, 2nd ed., Earthscan, London.

Sinha, K. C., Varma, A. Souba, J., and Faiz, A. (1989). Environmental and Ecological Considerations in Land Transport—A Resource Guide. World Bank Tech. Paper INU41. World Bank, Washington, D.C.

CHAPTER 29

ETHICS AND LEGAL ISSUES IN CIVIL SYSTEMS DEVELOPMENT

29.0 INTRODUCTION

In the process of carrying out the various tasks at each phase of systems development, civil engineers often encounter values-related dilemmas. Often, the properness of our decisions or actions are clear-cut from the ethical, moral, or legal perspectives; but in other situations, engineers face uncertainty as to whether a specific action constitutes a breach of the law or is a violation of ethics or morality. When an engineer's action is questionable from any of these perspectives, the planning, construction, or operation of the civil engineering system in question may be jeopardized, suspended, or even terminated early. Engineers therefore need to be knowledgeable and cognizant of the legal, ethical, and moral issues associated with their work. This chapter discusses these issues.

In this chapter, we will discuss human values, morality, ethics, and law. The section dealing with ethics describes the three main branches of ethics, discusses evolutionary ethics, and identifies civil engineering organizations that champion ethical behavior. A few ethical tenets and principles are discussed, and some potential ethical issues at the various phases of systems development are identified. In the section dealing with legal issues, the origins of laws are discussed as well as the domain and scope of U.S. law. In addition, a number of potential legal issues that may be encountered at each phase of system development are identified, and a few laws in the United States that directly affect the design, operations, and other development phases of civil engineering systems are discussed. We also examine contract law and tort law and discuss the standard mechanisms for resolving contract disputes.

29.0.1 Values and Value Systems

Values can be defined as the set of preferences regarding what is appropriate and what is not. The types include ethical values, moral values, religious values, political values, cultural values, social values, and aesthetic values. Thus, a specific value could be described as a worthwhile or desirable principle, standard, or character trait with regard to some attribute, including freedom, patriotism, equality, empathy, and love (McCuen et al., 2011). Physical assets (e.g., cars, houses, and money) may have value, but not in the frame of reference we discuss here. A value system, which typically pertains to a community, is a set of consistent values that may be drawn from one or more of the above attributes. Personal values are similar in that they constitute the sum of multiple attributes; however, unlike value systems, personal values may be less stable because they pertain to the individual person and thus may change relatively frequently with time or circumstances. Even for the same individual, what may be a value in their professional lives may not be a value in their personal lives or vice versa. Values are not necessarily related to what is legal in a society. So it is possible to have two law-abiding people with very different sets of values.

For an individual or community, values constitute an internal gauge for what is proper or improper, beneficial or harmful, useful or useless, beautiful or repulsive, important or trivial, desirable or unwanted, and so on. Over several millennia, personal values, often shaped by religion, have led to the development of customs, traditions, and laws within communities; and with the formation of professions and other organizations of common interest, values also led to the development of rules of behavior for members of that organization.

As we shall soon see in Section 29.2, values that ultimately appear in the code of ethics of most engineering professional societies include loyalty, altruism, stewardship, honesty, discipline, and devotion. For example, engineers consider as a value, loyalty to society, their clients, and their employers; ensuring that the outcome of their engineering work does not degrade the welfare of others, particularly those who have no political voice; taking better care of current-day resources to avoid shortchanging future generation; truthfulness in making any public statements or testimony about their work; and taking pains to ensure that their work is done diligently according to well laid out plans. With regard to loyalty to the society, client, and employer, there is a precedence order because in certain cases, loyalty to one party may preclude loyalty to another, and there is a preference order: first, the society; second, the employer; and third, the client.

29.0.2 Morality, Law, and Ethics

As we discussed in the last chapter, the values of an individual or community often ultimately evolve into tradition, morality, ethics, and law. In this section, we focus on the close association between these concepts. Morality is what distinguishes between actions that are considered right by society and those that are wrong. Morality is heavily influenced by religion and culture, particularly where explicit moral codes are established to guide human behavior. The most famous example of a moral code is the Golden Rule: "Treat others how you wish to be treated." Ethics is a branch of philosophy that addresses what can be considered right or wrong behavior. Also referred to as moral philosophy, ethics can be categorized as applied ethics, metaethics, and normative ethics. Law is a collection of rules to guide the behavior of individuals and organizations and thus protect the individual or natural resources from the intended or unintended actions of others. The importance of law in society is underscored by the fact that in most countries, the law-making body (or legislature) constitutes one of the three arms of government (the other two are the executive and the judiciary). As civil engineers plan, design, and operate civil systems, they constantly encounter situations related to morality, ethics, and law.

Figure 29.1 conceptualizes the overlaps between morality, ethics, and law. It is possible to encounter a situation that involves one, two or all three of these concepts. Unlike the relationship between law and ethics, there seems to be very little dichotomy between morality and ethics because ethical codes are often based on behavior that is considered moral. All three concepts often have an implicit or explicit code or set or rules to guide behavior. Society establishes laws to govern individual behavior and violations may lead to fines or imprisonment; professional bodies establish a code of ethical conduct, which may lead to sanctions or withdrawal of professional licenses when violated; and religious organizations have codes of model conduct (e.g., the Ten Commandments in Abrahamic religions) whose violation could lead to divine retribution. An action may be ethical from a professional standpoint but immoral in certain religions; for example, certain religions prohibit the charging of interest on borrowed money, so while interest charging is immoral for such people, doing so may be both ethical and legal in secular societies where such religious people

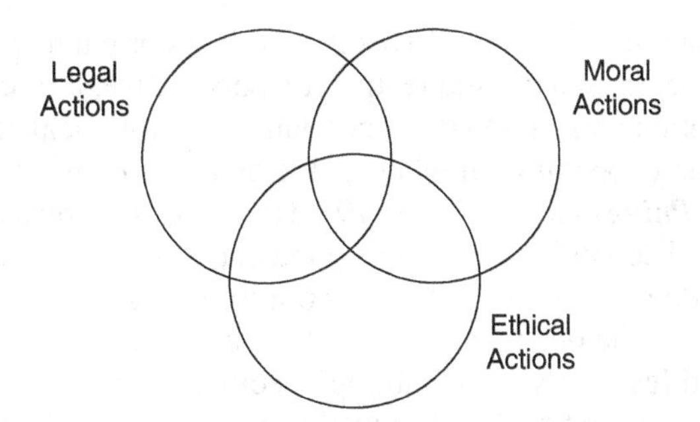

Figure 29.1 Overlaps between legal, ethical, and moral actions.

work and live. Similarly, certain actions may be unethical but not illegal, such as when an engineer awards a public contract to a family member without prior, full disclosure.

29.1 ETHICS IN CIVIL SYSTEMS DEVELOPMENT

29.1.1 Branches of Ethics

The word "ethics" takes its roots from the ancient Latin word *ethica* (moral philosophy) and the ancient Greek word *ethos* (custom or habit). Ethics, which is thus closely related to morality and can be defined as a collection of moral principles or rules of conduct that guides the behavior and attitudes of a specific group of people, such as engineering ethics, medical ethics, or religious ethics. Ethics is also described as a branch of philosophy that addresses the values associated with the behavior of humans. It deals with the wrongness or rightness of actions, the badness or goodness of motives, and the consequences of actions. Ethics is often categorized as metaethics, normative ethics, and applied ethics.

Metaethics examines the source and meaning of society's ethical principles and asks the following questions. Do they emanate from religious influences? Are they the outcome of the evolution of society's values? Are they due to the culture of society? Are they social constructs? Are they expressions of individual emotions? Students of metaethics examine so-called universal truths, the role of religion such as the will of God, and the role of logical reasoning in ethical judgments. Thus, metaethicists seek to understand how society establishes what is considered right and what are wrong. In this respect, two schools of thought exist: cognitivists, who claim that right and wrong are based on matters of fact, and noncognitivists, who assert that when one judges an issue as right or wrong, such judgment is neither true nor false as one may only be expressing one's emotional feelings about the issue (van Roojen, 2011). The study of metaethics gained prominence after G.E. Moore released *Principia Ethica* in 1903.

Normative ethics or moral theory is a more practical undertaking as it involves seeking how to develop moral standards to govern right and wrong behavior; for example, issuing statements of the good habits that need to be acquired, our individual responsibilities, and how our behavior affects others. The Golden Rule is a classic example of the principles of normative ethics: If an engineer does not want his house to be designed poorly, then it is wrong for him to design someone else's structure poorly. So, how is an action considered to be right or wrong? This depends on whether one is examining the issue from a *deontological* or a *consequentialist* viewpoint. The consequentialist would argue that the ultimate consequence of an action is what should serve as a yardstick for adjudging its rightness or wrongness; the deontologist would argue that the inherent nature of the action itself is what drives its goodness or badness, no matter how good the outcome may be. Therefore, from a consequentialist's perspective, a morally right action or inaction is one that yields a favorable result. On the other hand, deontological ethics determines the morality of an action or imission depending on whether the action is consistent with established rules (*Stanford Encyclopedia of Philosophy*) (Flew, 1979). The theory of ethics put forth by philosopher Immanuel Kant is considered deontological as he argued that morally behavior is one where people act as they are expected to do so by duty, and that morality stems not from the consequences of actions but from the motives of the one that carries out the action (Orend, 2000; Kelly, 2006).

Applied ethics involves examining specific, often controversial, issues, including *biomedical ethics* (surrogate motherhood, genetic manipulation of fetuses, the status of unused frozen embryos, patient rights, abortion, confidentiality of patient records, medical experimentation on human subjects, involuntary commitment, the rights of the mentally disabled, physician-assisted suicide, and euthanasia); *business ethics* (social responsibilities of capitalist practices, deceptive advertising, job discrimination, insider trading, affirmative action, employee rights, and whistle blowing); *environmental ethics* (animal rights, environmental resource management, and sustainability); and *social ethics* (capital punishment, gun control, recreational use of drugs, affirmative action, racism, the trade-offs between infrastructure development and environmental quality, homosexuality, and capital punishment and bribery) (NSPE, 2012; Fieser, 2009).

Of the three categories of ethics, applied ethics is that which is perhaps the most directly related to the practice of civil engineering, and thus we will further discuss it here.

According to Fieser (2009), the most common principles in applied ethical discussions include "personal benefit (acknowledge the extent to which an action produces beneficial consequences for the individual in question), social benefit (acknowledge the extent to which an action produces beneficial consequences for society), benevolence (help those in need), paternalism (assist others in pursuing their best interests when they cannot do so themselves), harm (do not harm others), honesty (do not deceive others), lawfulness (do not violate the law), autonomy (acknowledge a person's freedom over his/her actions or physical body), justice (acknowledge a person's right to due process, fair compensation for harm done, and fair distribution of benefits), and rights (acknowledge a person's rights to life, information, privacy, free expression, and safety)." These principles are indicative of a range of traditional normative principles and derive mainly from perspectives that are consequentialist and duty-based. The notions of personal and social benefit are inherently consequentialist because they addresss the consequences of an action in terms of its effect on the individual or society, while the principles of lawfulness, benevolence, honesty harm and paternalism, have a basis in our duties to others. The principles of justice, autonomy and rights are rooted in moral rightness. In the early 1970s, John Rawls published *A Theory of Justice*, a treatise that pursued moral arguments and eschewed metaethics and thus ushered in renewed interest in normative ethics. Sinha and Labi (2007), in citing the work of Khisty (1996), Bass (1998), Sen (2000), Alsnih and Stopher (2003), ITS (2003), and The World Bank (2003), suggested that Rawls' publication has profoundly influenced the consideration of environmental justice in the evaluation of civil engineering systems.

Applied ethicists using the concepts of metaethics and normative ethics, attempt to resolve controversial issues such as these but Fieser (2009) cautions that "the lines of distinction between metaethics, normative ethics, and applied ethics are often blurry". Fieser cited the example of abortion, a topic in applied ethics because it involves a specific type of "controversial" behavior; however, it is also influenced by more general normative principles; for example, the right of self-rule and the right to life serve as litmus tests for adjudging whether abortion is moral. The issue is also influenced by metaethical issues including the source of our rights and which people are entitled to certain rights in some societies.

THE "ACTS AND OMISSIONS" DOCTRINE

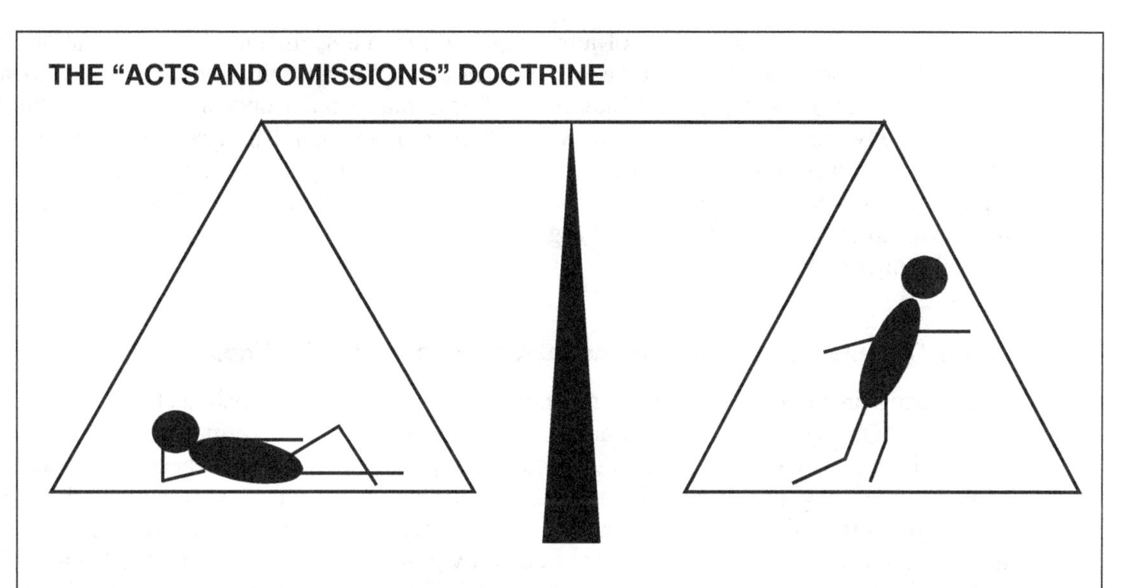

Consequentialists contend that the rightness of an action lies in the goodness of its outcome. Therefore, consequentialist theories generally hold that there is no difference between a deliberate action that yields a certain outcome and a deliberate decision not to act that yields the same outcome. For example, designing a structure poorly, leading to its ultimate failure, is no different from not taking action while having knowledge that the structure is being designed poorly by someone else. In certain professions, however, there may be a moral distinction between an action and a deliberate lack of action that lead to the same outcome; for example, in medicine, assisted suicide (actively killing patients with their consent by administering harmful drugs to them) and not acting while the patient slowly starves to death voluntarily. A true consequentialist would see no moral difference between these two situations.

Source: www.onlineown.com.

29.1.2 Evolutionary Ethics

Evolutionary ethics, an extreme consequentialist concept, is one of the most controversial areas in philosophy. In simple terms, its proponents believe that any action that enhances long-term survival of the human species can be considered ethical, even if the action leads to diminishing some individuals or species whose characteristics are perceived to be inimical to the survival of the human race as a whole. Critics contend that it is, first of all, difficult to identify which characteristics are unfavorable to long-term human survival; and second, if such positions are held by a powerful group of people on the basis of their race, religion, socioeconomic status, or genetic health, their beliefs could be used as a reason to persecute another group of people perceived to be inferior on the basis of these characteristics. On a more global scale, such perceptions have often led to maltreatment, mass slaughter, or even genocide of certain races, such as the Holocaust during World War II. On a relatively local scale, such perceptions may lead to eugenic slaughter (eradication of deformed or mentally ill people and improvement of the human race by fostering the multiplication of healthy individuals) and opposition to environmental justice concerns regarding civil engineering systems planning, construction, and operations. For example, it may be the case that officials or citizens in certain cities actively discourage the construction of sidewalks or the provision of convenient public transit for fear of attracting low-income segments of the population that may happen to be dominated by certain age groups or races. Such trepidation may be due to stated or unstated concerns about a greater likelihood of criminal or other undesired behavior, but may be unconsciously related to a larger underlying desire to deter such demographics or species in favor of others perceived to be more "desirable" or less "threatening." The planning and design of civil engineering systems to accommodate not only the mainstream population but also to serve explicitly the disabled, elderly, and indigent segments of the population can be considered a repudiation of evolutionary ethics.

29.1.3 Organizations That Promote Ethical Behavior in Civil Engineering

Engineering professional organizations in most countries have established ethical codes by which their members regulate their work habits and relationships. These organizations also urge engineers to hold themselves to the highest standards of professional and ethical conduct and recognize explicitly the obligation of individual engineers to uphold the integrity, dignity, and honor of the engineering profession by honest and impartial service to their employers, clients, and the public. In the United States, ethical behavior is steered by guidelines established by the National Council of Examiners for Engineering and Surveying (NCEES), the American Society of Civil Engineers (ASCE), and the National Society of Professional Engineers (NSPE). In Asia, the ethical codes of the Chinese Academy of Engineering, the Engineering Academy of Japan, and the National Academy of Engineering of Korea explicitly recognize the vital role of engineering in enhancing the quality of human life and environmental sustainability, and cherish the Asian cultural heritage of harmonious existence with other persons and the natural environment. In New Zealand, the Institute of Professional Engineers' code of ethics is based on fundamental ethical values that include protection of human life and community well-being, sustainability, professionalism, integrity, and competence. Similar values are espoused in Australia, where members of the Institution of Engineers are obliged to apply and uphold the cardinal principles of their code of ethics, which are encapsulated within and established by the tenets of the code of ethics. In Canada, the Council of Professional Engineers (CCPE) issues national guidelines for its constituent associations that express the guiding principles that while supporting the autonomy of the organization to administer its code within its jurisdiction. Similar codes have been established by engineering professional organizations in other countries, including Brazil, Russia, China, India, South Africa, and Ghana, to regulate the behavior of engineers. The World Federation of Engineering Organizations (WFEO) has established a code of ethics where professional engineers are expected to conduct themselves in an honorable and ethical manner and to uphold the values of integrity and honesty and to hold sacrosanct, all human life, the public welfare, and the natural environment.

In the United States, the NCEES rules of professional conduct govern the obligations of registered civil engineers to society, their employers, and their clients and to other registered engineers; and ASCE's professional and ethical conduct guidelines and NSPE's code of ethics for engineers comprise the fundamental canons and rules of practice. Members of the engineering profession have a duty to uphold high standards of integrity and honesty. As such, in providing their services, engineers must ensure honesty, fairness, impartiality, and equity, and always seek the protection of the public safety, health, and welfare. Engineers must therefore be guided by a standard of professional behavior that adheres to the highest levels of ethical behavior. Ethics are based on core societal values, including integrity, honesty, fidelity, charity, responsibility, and self-discipline (Drnevich, 2011).

The preamble of NCEES's model rules of professional conduct provides an overarching statement of the obligations of each registered engineer seeking to perform engineering and land surveying services. These rules, which are binding on every registrant, were developed to "safeguard life, health, and property; to promote the public welfare; and to maintain a high standard of integrity and practice". Further, each registrant is charged with the responsibility of adhering to the highest ethical and moral standards of conduct in all aspects of the practice of professional engineering and land surveying. The preamble also states that the practice of professional engineering and land surveying is a privilege, rather than a right. Other aspects of the NCEES preamble and the specific obligation of registered engineers are discussed in various subsections in Section 29.2.3.

The preamble of ASCE's professional and ethical conduct guidelines and NSPE's Code of Ethics for Engineers state that "members of the profession recognize that their work has a direct and vital impact on the quality of life for all people". As such, the preamble states that any service rendered by consulting engineers must be guided by honesty, impartiality, fairness, and equity and must be dedicated to the protection of public health, safety, and welfare. Further, in their professional practice, engineers must abide by a standard of professional behavior that requires adherence to the highest principles of ethical conduct on behalf of the public, their clients and employees, and the profession. ASCE's five fundamental canons and five rules of practice are discussed in various subsections in Section 29.2.3. Figure 29.2 presents the categories of NCEES model rules of professional conduct, and Figure 29.3 presents the categories of ASCE guidelines for professional and ethical conduct. Also, ASCE's fundamental canons and rules of practice are presented in Table 29.1.

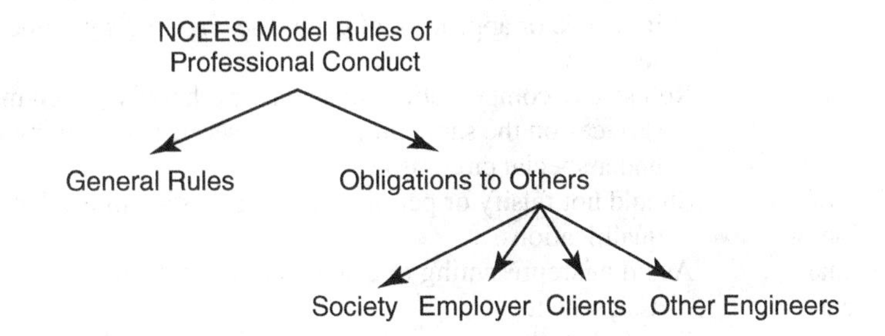

Figure 29.2 Categories of NCEES model rules of professional conduct.

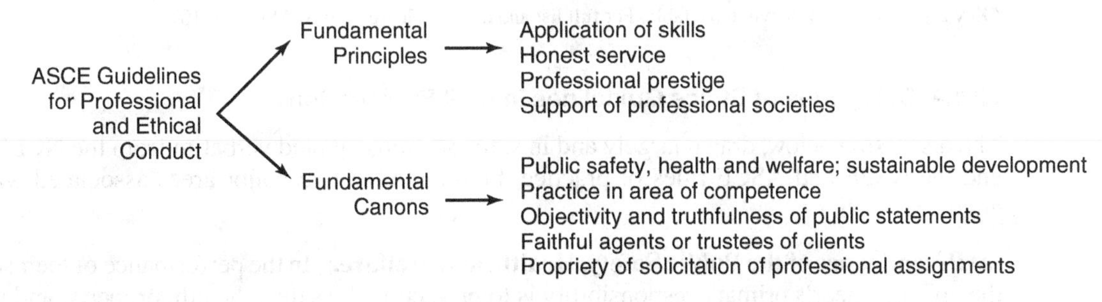

Figure 29.3 Categories of ASCE guidelines for professional and ethical conduct.

Table 29.1 ASCE Fundamental Canons and Rules of Practice (ASCE, 2010)

Fundamental Canon	Rules of Practice[a]
1. Safety, health, and welfare of the public	Primary obligation is to protect the public safety, health, property, and welfare. Strive to comply with the principles of sustainable development. Approve only work that is consistent with standards and is safe for public health, property, and welfare. Shall not reveal facts, data, or information obtained in a professional capacity without the prior consent of the client, except in special circumstances. Prohibit use of name in business ventures that may be engaging in fraudulent or dishonest business or professional practices. Report behavior that is inconsistent with the ASCE Guidelines for Professional and Ethical Conduct.
2. Practice in area of competence	Undertake assignments only when qualified by education or experience in the specific technical fields involved. Avoid affixing signature to documents dealing with subject matter in which engineer lacks competence or were not prepared under engineer's direction and control.
3. Objectivity and truthfulness of public statements	Objectiveness and truthfulness in professional reports, statements, or testimony. Express publicly a professional opinion on technical subjects only when that opinion is founded upon adequate knowledge of the facts and competence in the subject matter. Issue no statements, criticisms, or arguments on technical matters that are inspired or paid for by interested parties, unless with appropriate disclaimer.
4. Faithful agents or trustees of clients	Disclose all known/potential conflicts of interest to clients by promptly informing them of any business association, interest, or other circumstances that could influence or appear to influence the engineer's judgment or the quality of services. Refuse any compensation, financial or otherwise, from more than one party for services on the same project or for services pertaining to the same project, unless under special circumstances.
5. Propriety of solicitation of professional assignments	Should not falsify or permit misrepresentation of academic or professional qualifications. Avoid misrepresenting or exaggerating the degree of responsibility in prior assignments. Should not offer, give, solicit, or receive, either directly or indirectly, any political contribution in an amount intended to influence the award of a contract by a public authority.

[a]Key aspects only are shown in the table. For full list and details, please refer to ASCE (2010).

29.1.4 Categories of Ethics Guidelines in Civil Engineering

The discussion below, drawn largely and in some sections, quoted verbatim from the NCEES rules and the NSPE and ASCE rules of practice, highlights the five major areas associated with civil engineering ethics.

(a) Preeminence of the Public Safety, Health, and Welfare. In the performance of their services, the civil engineer's primary responsibility is to protect public safety, health, property, and welfare. As such, at any phase of development in their work, if their professional judgment is overruled under circumstances where the public safety, health, property, or welfare are endangered, the civil engineer is expected to notify the client or appropriate authority. Also, engineers are expected to approve only engineering work that, to the best of their knowledge and belief, does not endanger public health, property, and welfare and conforms to accepted engineering codes of practice and standards.

This requirement is related to the civil engineer's obligation to society (see I-a to I-h of NCEES rules), the three rules of practice of the second fundamental canon of ASCE, the rules of practice of the first fundamental canon of NSPE, and the third cardinal principle and first tenet of the Australian Institute of Engineers Code of Ethics 2000 version.

(b) Performing Services Only in the Areas of Their Competence. Civil engineers are expected to "exercise their privilege of practicing by performing services only when qualified by their education or experience, according to current standards of technical competence, and only in the areas of their competence". For example, they are expected not to affix their signatures or seals to engineering document that deals with subject matter in which they lack competence or that were not prepared under their direction or control. Similarly, they may express publicly a professional opinion on technical subjects only when that opinion is founded upon adequate knowledge of the facts and competence in the subject matter. The civil engineer may accept an assignment outside of their field of competence, but only "when their services are restricted to those specific project phases in which they are qualified and when they are satisfied that all other design segments of the phases of such project will be performed, signed, and sealed by registered or otherwise qualified associates, consultants, or employees; only in that case, may they sign the engineering documents for the overall project."

This requirement is related to the first element of the NCEES rules of the registrants' obligation to employers and clients and also their obligation to society [I(a) and I(d) of the NCEES rules], the fifth rule of practice of the first fundamental canon of the ASCE rules, the rules of practice of the second fundamental canon of NSPE, and one of the tenets of the Australian Institute of Engineers Code of Ethics.

(c) Objectivity, Truthfulness, and Nonfraudulent Actions. Engineers are expected to prohibit the use of their names or firms or the association in "business ventures with any person or firm they have reason to believe is engaging in fraudulent or dishonest business or professional practices". In their relationships with the public, engineers are expected to be objective and truthful in their professional opinions regarding civil engineering systems that are used by the public or are related to public welfare. In professional releases (statements reports, or testimony), engineers are expected to be objective and truthful and to include all relevant information in such releases. Engineers also are expected not to affix their signatures or seals to any plan or document not prepared under their direction and control. Engineers may express publicly a professional opinion on technical subjects only when that opinion is based on their sufficient knowledge of the facts and competence in the subject matter. Similarly, engineers are expected not to issue any "statements, criticisms, or arguments on technical matters that are inspired or paid for by interested parties, unless they have prefaced their comments by explicitly identifying the interested parties on whose behalf they are speaking and by revealing the existence of any interest they may have in the matter". Engineers are also expected to avoid injuring, directly or indirectly, maliciously or falsely, the professional reputation, practice, prospects, or employment of other engineers, nor indiscriminately criticizing other engineers' work. Engineers are expected to not solicit or accept a professional contract from a government body on which a principal or officer of their firm/organization serves as a member. Conversely, engineers serving as advisors, members, or employees of a government body or department, who are the principals or employees of a private concern, are not expected to participate in "decisions with respect to professional services offered or provided by said concern to the government body which they serve". Engineers having knowledge of possible violations of any of the rules of professional conduct are expected to provide information to the governing board and assistance when necessary to their final determination of such a violation.

These rules of professional conduct are related to engineers' obligations to society (I-b and I-d to I-f), to their employers and clients (II-b to II-h), and to other registrants (III-a to III-c of NCEES rules), the three rules of practice of the third fundamental canon of ASCE, the third and fifth fundamental canons of NSPE, and the second, fourth, sixth, and seventh tenets of the Australian Institute of Engineers Code of Ethics 2000 edition.

(d) Faithful Agents or Trustees of the Client. The engineer–client relationship is expected to be sacrosanct, with the only exception being where the welfare of the public is at stake. As such, engineers are expected to disclose all potential or known conflicts of interest to their clients by informing them promptly of any business association, interest, or other circumstances that could influence or appear to influence the client's judgment of the quality of their services. Also, engineers are expected not to accept any kind of compensation, including financial, from more than one party for services rendered on or pertaining to the same project, unless the circumstances are fully disclosed to, and agreed to, by all interested parties. It is also unethical for engineers to "solicit or accept financial or other valuable consideration, directly or indirectly, from persons other than their employer in connection with work for employers and clients". Further, as members of a governmental body or department, public sector engineers are expected to avoid participating in decisions with respect to professional services solicited or provided by them or their organizations in private engineering practices; similarly, engineers are expected to avoid soliciting or accepting professional contracts from a governmental body on which a principal or officer of their organization serves as a member. Engineers are expected to "avoid revealing data or information obtained in a professional capacity without the prior consent of the client except as authorized or required by law or the guidelines of professional conduct". The responsibility to society, however, overrides that to the client; as such, engineers are expected to breach the client confidentiality requirement if they feel that the safety of the public is in jeopardy.

These rules of professional conduct are related to the obligation of engineers to their employers and clients (II-a to II-h) of the NCEES rules, the five rules of practice of the fourth fundamental canon of ASCE, and the fifth tenet of the Australian Institute of Engineers Code of Ethics 2000 edition.

(e) Solicitation of Professional Assignments. In their proposals for services or in company advertising at websites or in brochures, engineers are expected not to falsify or misrepresent or exaggerate (i) the professional or academic qualifications of themselves or their associates, and (ii) their degree of responsibility in the (or complexity of the subject matter of) prior assignments. It is unethical for engineers to offer, give, solicit, or receive, either directly or indirectly, any political contribution in an amount intended to influence a contract award by a public authority, or that may be "reasonably construed by the public as having the effect or intent to influence the award". Furthermore, engineers must "refrain from offering gifts or other valuable consideration in order to secure work and must not or pay a commission, percentage, or brokerage fee in order to secure work, except to bona fide employees or bona fide established commercial marketing agencies retained by them."

This requirement is related to their obligation to the employers and clients (I-d, II-e, II-f, II-h, and III-a and III-b) of the NCEES rules, the two rules of practice of the fifth fundamental canon of the ASCE, and the rules of practice of the third and fifth fundamental canons of NSPE.

(f) Overall Ethical Responsibilities. Most engineering professional societies worldwide recognize that the practice of professional engineering is *not a right but a privilege*. Also, these societies invariably charge engineers with the responsibility of "adhering to the highest standards of ethical and moral conduct in all aspects of their professional practice", and in certain cases, their personal lives. As such, this privilege (in the form of a license to practice) could be withdrawn in the event of behavior deemed unethical or immoral by the professional society. In some countries such as Australia, there is explicit recognition in the code of ethics that engineers need to respect the dignity of the individual and to act only on the basis of a "well-informed conscience".

29.1.5 Resolving Ethical Situations

A critical aspect of resolving an ethical problem in the workplace is to recognize the dichotomy between *internal appeal* (actions taken within the organization to resolve the issue) and *external appeal* (actions taken with the participation of entities outside the organization). McCuen et al. (2011) thus presented a procedure for resolving ethical conflicts (Figure 29.4).

Also, Drnevich (2011) proposed the following steps for resolving ethical problems:

1. Realize that there is a problem.
2. Define the problem (who—stakeholders, what, when, where, why, evidence).
3. Define the options available—possible solutions.
4. Weigh the consequences (pros versus cons).
5. Compare to others—code of ethics.
6. Compare to the law.
7. Does it feel right? (The Golden Rule).
8. Ask someone else.
9. Choose what to do.
10. Act on it.
11. Learn from experience.

29.1.6 Civil Engineering Ethics Case Study

The Board of Ethical Review (BER) of the National Society of Professional Engineers (NSPE) makes available to students and the general public brief cases of ethics-related situations that are occur frequently in the practice of engineering. An example, reproduced with permission and adapted from the NSPE Case 09–12 (NSPE, 2012), is provided below.

Facts: Panos Properties sought to purchase a multistory apartment complex. However, the home inspection company, Iniesta Inc., felt that the building was not structurally sound. Therefore,

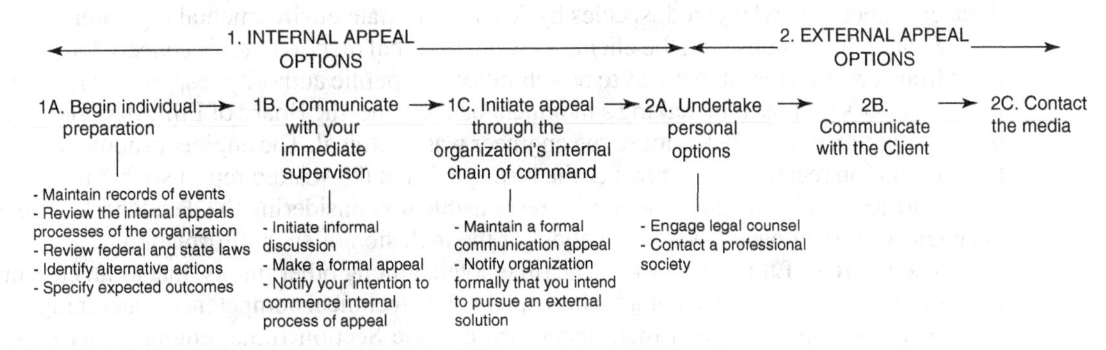

Figure 29.4 Steps to resolve ethical conflicts.

upon the recommendation of Ineista, Panos Properties awarded a contract to Erin, a qualified professional civil and structural engineer, to carry out inspection of the structural integrity of the complex including the foundation. After a thorough inspection in the presence of the home inspector, realtor, and prospective owner, Erin concluded that while the apartment complex could benefit from some minor structural maintenance, it was structurally sound and in no imminent danger of collapse. However, Erin found that there were excessive levels of moisture in the basement of the building and observed extensive instances of mildew and mold. Erin presented solutions for addressing the moisture problem; these solutions, in Erin's opinion, would also protect further the building's structural integrity. The discussion, however, drifted from the building's structural condition and began to focus on the health issues related to the apartments.

Question: Given these circumstances, what are Erin's obligations?

References: Section I.1 of the NSPE Code of Ethics states that engineers, in the fulfillment of their professional duties, shall hold paramount the safety, health, and welfare of the public; Section II.2 states that engineers shall perform services only in the areas of their competence; Section II.3.a states that engineers shall be objective and truthful in professional reports, statements, or testimony. They shall include all relevant and pertinent information in such reports, statements, or testimony, which should bear the date indicating when it was current; and Section III.2.d states that engineers are encouraged to adhere to the principles of sustainable development in order to protect the environment for future generations.

Discussion: In the current era, public and private clients and the general public increasingly rely on professional engineers to address environmental and ecological issues associated with their work. As clients endeavor to address the issues related to public health and safety, more and more engineers possess the requisite professional qualifications, education, and experience to provide the needed services in this regard. Not long ago, the NSPE Code of Ethics for Engineers was revised to include a new section (III.2.d, which is cited in the preceding paragraph) to encourage engineers to abide by the tenets of sustainable development so that environmental quality can be preserved for future generations. This purpose of the revision was to promote, in the course of making decisions at any stage of civil engineering systems development, the due consideration of environmental issues. This provision, however, was intended to be a general statement that should be interpreted in the context of the entire NSPE Code of Ethics.

A recent example of an NSPE Board of Ethical Review's interpretation of this provision is BER Case 07–6. In that case, the examining engineer Alex, a principal in an environmental engineering company, was engaged by a developer client to study the feasibility of an area adjacent to a protected wetland for development as a residential complex. During the study, Chen, one of the company's biologists reported to Alex that in his (Chen's) opinion, the condominium project could affect a particular bird species that inhabits the adjacent wetlands area. The bird is considered a threatened, but not endangered, species by federal and state environmental regulators.

In later discussions with the client, Alex made verbal mention of this concern but excluded the issue from a written report that was to be submitted to a public authority responsible for considering the developer's proposal. In making a judgment on the issue, the Board of Ethical Review at the time held the opinion that the behavior of the engineer was unethical. The engineer should have included the information regarding the threat to the bird species in the written report so that it could be duly taken into account by the public authority responsible for considering the developer's proposal, and the engineer should have advised the client of the inclusion of such information.

The Board of Ethical Review at the time noted, among other things, that as an environmental engineer aided by the services of a biologist, Alex had technical competence concerning the matter in question. The Board added that, under NSPE Code Section II.3.a, engineers have an obligation to be objective and truthful in professional reports, statements, or testimony, and are expected to include all relevant and pertinent information in such reports. The Board held that there was no excuse for excluding the stated information because it is reasonable to assume that the public authority approving the development would be interested in such information. There does not appear to be any indication that the client would treat the information as confidential so Alex therefore had an ethical obligation to include the information in the written report and advise the client of its inclusion.

The facts and circumstances of BER Case 07–6 somewhat differ from those of the case at hand; however, some similarities exist. The first similarity is the importance of having competent individuals as part of the team to determine the most appropriate course of action. In Case 07–6, the engineer Alex was assisted by a biologist. In the present case, a home inspector, who clearly understood that the issue before him (the structural soundness of the house) was beyond his area of competence, had taken the right step of bringing an engineer, Erin, into the decision-making process. A second similarity is the importance of providing objective and truthful information to assist a client or the public on matters of concern.

Regarding the issue of working in the area of one's professional competence, the Board of Ethical Review held the view that Erin had an obligation to recommend that the client seek the services of a competent professional to address the potential health issues associated with the basement conditions, such as an environmental scientist, biologist, industrial hygienist, or physician, similar to the decision by the home inspector to bring in a licensed professional engineer to assist in examining the building structure.

Conclusion: Erin has an obligation based on the facts to recommend that the client seek the services of competent and experienced experts to address the potential health issues raised by the inspection of the basement.

Board of Ethical Review: Curtis A. Beck, P.E., F.NSPE; Mark H. Dubbin, P.E., NSPE; Robert C. Gibson, P.E., F.NSPE; Monte L. Phillips, Ph.D., P.E., F.NSPE; Samuel G. Sudler III, P.E., NSPE; Mumtaz A. Usmen, Ph.D., P.E., F.NSPE; Michael L. Shirley, P.E., F.NSPE, Chair.

29.1.7 Potential Ethical Issues at the Various Phases of Systems Development

As they perform their professional duties throughout the phases of civil engineering systems development, engineers encounter situations where they are explicitly expected to protect, above all, public safety, health, and welfare and to promote sustainable development, provide services in their areas of competence only, and issue public statements in a truthful and objective manner only. Also, there are countless situations where engineers are expected to act in serve as trustees or faithful agents for each client or employer and to avoid conflict of interest, build their professional reputation solely on the merit of their services, and to avoid unfair competition with others. This includes behaving only in a manner that upholds and enhances the integrity, honor, and dignity of the profession and devoid of bribery, fraud, and corruption. Interestingly, it is also considered unethical for engineers to fail to provide opportunities for further professional development of engineers working under their supervision. In this section, we will discuss certain situations specific to the development phases where some of these ethical breaches may occur.

(a) Needs Assessment Phase. This is a delicate phase that is particularly prone to ethical problems because the engineer interacts with a large variety of stakeholders including corporations, the general public, and the news media. Thus, the engineer may be influenced financially by stakeholders who stand to incur significant loss or benefit monetarily if the project proceeds. In this case, the engineer may face the unethical situation of understating or overstating the need for the system (or expansion or enhancement of an existing system), such as embellishing demand forecasts to attract support or funding for the project. Also, in making statements to the news media, there is the danger that the client's confidential information may be revealed or that the opposite (and worse) situation occurs where information that the public ought to know, is withheld.

(b) Planning Phase. At this phase, engineers may face the ethical issues associated with the confirmation that a specific proposed system will satisfy the identified need or with choosing a system locating purposely to satisfy the will of certain social, political, or business interests. Also, unethical engineers may neglect to reveal or deliberately withhold information from the public about the adverse impacts of a system's construction or operations on the environment. Further, at the planning phase, the engineer has unparalleled opportunity to serve constructively in civic affairs, advance public safety, health, and well-being, enhance the quality of life of the general public, and protect the environment through the practice of sustainable development. For example, an ethical engineer will refuse to put forth or accept system plans that violate environmental standards.

(c) Design Phase. At the design phase, ethical lapses occur when an engineer fails to check thoroughly or approves/seals design documents that were not prepared or reviewed by the engineer or that are clearly not consistent with public safety, health and welfare improvements or levels that conform to accepted standards. In cases where the engineer's professional judgment is overruled in situations where the safety, health, and welfare of the public are left endangered or the principles of sustainable development are ignored, it is unethical to neglect informing the employer or client of the situation and the possible consequences. Also at the design phase, ethical questions may arise from design work that is based on contingency or commission, as such work may undermine the professional standing of engineers or even create a conflict of interest. In certain cases, design engineers may find themselves coerced to conceal design calculations that reveal a problem that may endanger the system users or the general public. Also, it is unethical to submit the same design to different clients and charge them separately for that work or to write specifications for civil engineering system components when the specified products are knowingly supplied by a close relative of the engineer.

(d) Construction Phase. At the construction phase, engineers may be faced with ethical issues when their superiors ignore their warnings about poor or unsafe working conditions, finished work that is of substandard quality or that falls short of the specifications, or when they are asked to make decisions that place cost, time, or work quality over the safety of construction personnel. It is unethical to engage in, promote, or turn a blind eye to lack of transparency in the procurement and execution of projects. Such transparency includes full disclosure of the names, designations, contact information, roles, and fees paid to all entities associated with the project.

(e) Operations Phase. At the operations phase where the welfare or safety of system users and the general public is at the greatest risk, engineers face a special responsibility to be vigilant and to respond quickly to potentially unsafe situations. If the system monitoring and inspection is carried out effectively, engineers will be in a better position to avoid ethical lapses associated with system operations. Thus, it is unethical for the engineer to continue operating a system that is unsafe even though management may deem it more cost-effective to wait for a few more years before replacing or repairing the system.

(f) Monitoring/Inspection Phase. To engineers who monitor the usage of civil engineering systems either to acquire knowledge on system demand/usage or to enhance system security during operations, the ethics of surveillance is relevant the ethical issues revolve around the moral aspects of the mode of surveillance (Macnish, 2011). Is it a value-neutral activity that may be used for good or ill? What are the benefits and harms of surveillance? Who is entitled to carry out surveillance, when, and under what circumstances? While constituting an invasion of privacy, surveillance may be justifiable on the basis of the consequentialist appeal to the greater good, namely, the security and safety of society is best served by full or partial monitoring. As Macnish pointed out, this justification implies that the rights of a few may be overridden by the collective interests of the many and thus is likely to be rejected by proponents of deontological ethics.

(g) Maintenance Phase. At the maintenance phase, the engineer may encounter a situation where a specific maintenance material or process may serve its intended function well and cost-effectively but has undue adverse impacts on the environment. Also, the selection of contractors for awarding maintenance contracts can serve as a major source of ethical violations for engineers.

(h) End-of-Life Phase. At this phase, the ethical issues that arise include those related to disposal of the system components. Improper disposal of system parts that leads to the release of toxic solids, liquids, or gases causing air, land, or water pollution is clearly unethical. It is ethical practice to consider recycling, reuse, or other end life strategies for individual system components.

(i) All Phases. As we have seen above, certain ethical issues arise predominantly at some specific phases of systems development. However, there are other ethical issues that could arise at any phase. Examples include "whistleblowing" or the release or knowledge of information that may be damaging to society, the use of other engineers' intellectual property (including data, proposals, theories, and analytical frameworks) without their permission, and failure to disclose a potential conflict of interest. Other ethical situations can arise when an engineer leaves employment in the public sector to work in the same area in the private sector because the engineer may have access to government information that may give an unfair advantage to his new employer. Also, when consulting engineers agree to work for a commission, ethical questions may arise about the engineer's professional judgment under the pressure of such a fee arrangement. Employment is also another area where ethical conflicts often arise; for example, an engineering firm withdraws an employment offer letter; a prospective employee notifies a company that he is no longer interested in joining them after he has previously accepted an employment offer. Also, it is unethical for engineers to promote themselves by advertising alleged negative attributes of their competitors, which often arises when a group of engineers leave their employer to establish their own firm. Another commonly encountered situation is the giving or acceptance of gifts to or from persons in positions that could influence the professional work of the engineer. In the area of academia, ethical issues may arise in joint authorship of technical papers and reports, intentional omission of research data to skew the results toward a particular direction, or failing to credit or acknowledge another engineer for their contribution to the work.

29.1.8 Evolving Trends in Civil Engineering Ethics

Initiatives to enforce ethical practice in civil engineering were spurred by major engineering failures in the early 20th century (Petroski, 1985). In some cases, ethical lapses may have contributed to technical flaws and ultimately, failure or public danger, such as the 1981 Kansas City Hyatt Regency walkway collapse, the 1980 Love Canal environmental disaster, the 1978 Citigroup Center retrofitting, the 1919 Boston Molasses disaster, the 1907 Quebec Bridge collapse, and the 1876 Ashtabula River railroad disaster. Ethics cases do not always have easy solutions; the 500 advisory opinions published by the NSPE's Board of Ethical Review are helpful in resolving some of these dilemmas. In recent years in many countries, ethics codes are being rewritten to include risk minimization, sustainable development, offshoring of engineering work, and environmental protection.

29.2 LEGAL ISSUES IN CIVIL SYSTEMS DEVELOPMENT

29.2.1 Origin and Evolution of Laws

Throughout the evolution of human society, laws (often a set of rules) have been used to govern the behavior of humans. These laws always have been intended to protect the overall well-being of society and thus largely reflect the values of society. The first known set of laws was the Cuneiform Law (2350–1400 BC), which refers to the legal codes inscribed using cuneiform symbols. These were used throughout the ancient Middle East by the Sumerians, Assyrians, Babylonians, Elamites, Hittites, Hurrians, and Kassites (BOE, 2012). King Hammurabi's 1700 BC Akkadian law code in Babylon, Mesopotamia, is the most well-known of these laws. Subsequent collections of laws spanning the time period between 2300 BC and 5th century BC include the Code of Urukagina; the Code of Ur-Nammu, king of Ur; the Laws of Eshnunna; the Codes of Lipit-Ishtar of Isin; the Code of Hammurabi; the Code of the Nesilim; the Hittite laws; the Assyrian laws (Code of the Assura); the Hebraic law (Hebrew Bible or Old Testament); Gentoo Code; the Maxims of Ptahhotep and Sharia Law; the Draconian constitution; the Gortyn code; the Twelve Tables of Roman Law; Traditional Chinese law (the laws, regulations, and rules used in China since the 11th century BC until the first decade of the 20th century when the last imperial dynasty fell); and various laws in ancient or fairly recent civilizations in regions in South America and Africa including the Inca, Maya, and Ashanti.

In many of the legal codes mentioned above, punishments ranged from fines, exile, loss of property, flogging, mutilation, cutting off a part of the body, and death. Also, most of these codes (particularly, the Code of Hammurabi) were based on or strongly related to *lex taliois* (an eye for an eye), in other words, retaliation authorized by law. The victim (an individual or the society at large), was made to receive compensation from the culprit in a bid to punish the culprit and also to compensate the victim. Also, most of these codes recognized the importance of intent; and offenders often received less punishment if they could prove that their misdeed was unintentional. This reflects a parallel with today's laws relating to tort liability arising from injury and inconvenience caused to users of civil engineering systems, namely, the system operators often are in a better legal position if they can show that the offending defects on their system were not intentional (which is easy to prove) or were not ignored (which is more difficult to prove).

However, in ancient times, absence of malice was not always a saving grace, as culprits of events caused by neglect or carelessness were severely punished; for example, for a builder whose bad construction practices led to the death of the house owner's son, his son was put to death as punishment and the builder had to rebuild the house, repair the defect, or repair the damages due to defective building; if a physician's work led to loss the patient's of life or limb, the physician's hands were cut off as a punishment; and the builder of a defective boat had to repair the defect and provide a year's warranty. Under Akkadian law, if a domestic animal killed a person on the street, the owner was responsible for damages only if the owner was aware that the animal was dangerous to humans. Again, this is somewhat similar to the current situation when the users of a defective civil engineering system incur injury or death because of some defect in the system: The legal precedent is that the system owners can be held liable if they were aware of the defect and had no plan in place to fix it.

In certain societies such as Mesopotamia, there was a clear distinction between what was illegal and what was immoral or unethical. For example, the use of false weights in commercial weighing, making untrue statements, and other bad but not illegal behavior could not be brought into court; however, such acts were frowned upon mostly through oral code and it was stipulated that the culprit would face the wrath of God (moral) and not the wrath of the king (legal) (BOE, 2012). In other societies where theocracy was practiced or where the law was closely related to the existing religious practices, there was very little distinction between legal and moral behavior.

29.2.2 Sources of Laws

In most societies, laws have evolved from generally accepted customs handed down over successive generations. The evolution of laws, and thus, the sources that generate laws, differ from country to country. In this section, we use the experience of the United States as an illustration.

U.S law was derived from English law, and the sources of law are constitutional law, administrative law, statutory law, and common law.

Constitutional law interprets and implements the U.S. Constitution. This set of laws defines the scope and application of the terms of the constitution, establishes the relationship between federal and state governments and the rights of individuals and specifies what the government can or cannot do. Constitutional law is often complex and broad, and in many cases, is even ambiguous.

Administrative law. Administrative agencies regulate the interactions between human activities (social, economic, political, etc.). Administrative law covers the legal aspects of the activities of government administrative agencies. Thus, administrative law covers law enforcement, international trade, manufacturing, the environment, taxation, immigration, public structures, engineering systems, and so forth. In the U.S. legal system, many administrative agencies, such as the Environmental Protection Agency (EPA), are under the executive branch of government and have been the source of several laws.

Statutory law is new law established by legislature (or sometimes by the executive branch). Often established in response to an observed need to clarify govement functions in a specific context, these laws (also called statutes) serve to improve civil order, to codify existing law, or to obtain special treatment for an individual or company. These include municipal law, which is established by towns and cities, and often involve ordinances covering traffic laws, zoning, and building codes.

Common law, also referred to as decisional law (derived from judicial decisions) or precedential law (based on past precedent), is a collection of documented judicial opinions at countries or jurisdictions that have common-law legal systems. Thus, common law is law that is published and thereby becomes precedent (i.e., the basis for future decisions). Common law is nonstatutory because it is not enacted by the legislature. Under common law in most countries, a formal contract has three parts: an offer, acceptance of the offer, and consideration (something of value, typically, money) that serves to bind the contract. *Promissory estoppel* is a doctrine in common law that is used by courts to enforce promises that have been made and thus were trusted subsequently. Ishibashi and Singh (2011) compared breach-of-contract versus promissory estoppel in a bid to address the question of whether they can coexist and still justly serve their purposes in the civil construction industry.

29.2.3 Domain and Scope of U.S. Law

Private versus Public Law. Private law is law that is associated with the relationships between entities or persons, without government intervention. Public law on the other band, governs the relationship between individuals or entities and the government; Its subdivisions are constitutional law, administrative law, and criminal law and include consumer protection laws, contract law, and tort law (which are all encountered in civil engineering systems development).

Criminal Law versus Civil Law. Criminal law involves the imposition of sanctions by the government for crimes committed by individuals or entities. The sanctions are imposed punitively, so that society can not only achieve justice but also maintain social order. In criminal law, the objective is to ascertain whether the defendant is guilty or not guilty. In certain countries including the United States, the defendant is assumed innocent until proven guilty beyond a reasonable doubt through a unanimous verdict of a jury of the defendant's peers, upon which an appropriate penalty is imposed, such as imprisonment or death in extreme cases in certain societies.

Civil law on the other hand, addresses disputes between two parties that are not of significant public concern. In civil cases, the question is not about guilt or innocence but whether any one of the parties acted improperly or failed to act as expected, resulting in damage to another party. In civil trials, there is a lower standard of responsibility: The requirement is to reach a verdict only on the basis of a preponderance of evidence; and the jury's verdict is not required to be unanimous. In civil cases, the winner is typically awarded monetary damages or custody of the disputed property at the expense of the loser.

29.2.4 Legal Issues Encountered at the Various Phases of Systems Development

Needs Assessment and Goals Identification Phase. At the needs assessment phase, engineers may grapple with legal issues such as determination of whether social justice is one of the goals of the proposed new civil engineering system or proposed extension to an existing system; and engineers may seek to ensure that there is reasonable equity, that is, fairness to all population segments. Often, this is an ethical or even moral issue; however, as evidenced by President Clinton's Executive Order 12898 in 1994, the need to ensure equity may be established as a statutory law. Engineers, at the needs assessment phase, therefore must include the needs of the indigent, elderly, and disabled in their needs assessment activities. This accommodation spills over into other phases, and design engineers ensure that the system is designed to facilitate access by disabled and elderly persons and in the system operations phase where engineers ensure that the elderly and disabled can use the system without undue physical strain on their person.

System Planning Phase. At the planning phase of a civil engineering system, an important issue is whether the system abides by all federal and state environmental laws. Section 29.2.5 discusses some past and current laws in the United States that affect the planning of civil engineering systems.

System Design Phase. At the design phase, legal questions that may arise include whether the civil engineering system was designed according to the appropriate design code and whether all ADA requirements are met. Often design-related legal issues arise at or after the construction phase when design flaws are revealed; and when that happens, the system designers may find themselves in a legal dispute. The adverse consequences of design flaws can include injury, loss of life or property damage; the loss of productivity; or the cost of redesigning and retrofitting a structure. It has been found that jurisdictions vary in their treatment of claims against system designers for economic loss arising out of work on construction projects. In the case of *Terracon Consultants Western, Inc. v. Mandalay Resort Group*, the Nevada Supreme Court held that the design professional was not liable for purely economic losses on the commercial property development project (Caplicki, 2010).

System Construction Phase. At the system construction phase, legal issues that arise include a contract been signed between the system owner and the contractor for a specific project. Other legal issues may arise due to variations, change orders, and unexpected site conditions. To accommodate the inevitable delays and their inherent changes to the construction schedule without disrupting work, construction engineers use variations and change orders to adjust the scope of work stated in the initial contract. A variation may be defined as a deviation from a previously agreed and well-defined scope of work, and a change order is a formal document that is used to modify the agreed contractual agreement and becomes part of the projects documents (Fisk 1997; O'Brien, 1998). Change orders may be established for changes in the construction schedule, resource allocation, project scope, and compensation. As McCormick and Singh (2010) pointed out, the change order is recognized as "a legal extension to the terms of the contract and is used when any situation arises in the project that requires alteration to the contract terms."

One of the most common sources of legal disputes at the construction phase is the variability of site conditions or "differing site conditions" (DSC), that is, where the conditions encountered on site is often different (in many cases more challenging) than that envisaged at the design phase. DSC disputes can be minimized by ensuring, at the design phase, very through geotechnical studies and site investigations that would reveal the existence of any poor pockets of material, sensitive conditions (e.g., ancient burial ground or habitats of endangered species), and locations of buried utilities facilities (e.g., water pipes, gas mains, and electricity cables). Often, disclaimers are inserted in contracts as an attempt to absolve the client of DSC responsibility. A disclaimer is a contract clause that attempts to shift the adverse consequences of certain situations or events to another contractual party and is typically applied with regard to unforeseen subsurface conditions. Disclaimers typically assign the sole responsibility of ascertaining site conditions to the contractor and further state that the owner does not guarantee that the site information provided is accurate and that the owner is not liable for any claim related to adverse site conditions. Thomas (2012) pointed out that, despite the lack of legal success with disclaimers, owners continue to put disclaimers in contracts in the hope of protecting their interests (the owner may prevail at times, particularly "when the claim amount is small or the contractor is not inclined to drag the owner into court. But, if the dispute does go to court, owners often attempt to rely on the language of the disclaimer"). In DSC disputes, disagreements and misunderstandings often arise not only over the meaning of the DSC term but also over the use of disclaimers. Citing the 2007 case of *Condon-Johnson & Associates, Inc. v. Sacramento Municipal Utility District (SMUD)* that was decided by the Third Appellate District of the California Court of Appeals, Thomas stated that construction professionals who administer the contract or attorneys that give legal advice, due to inexperience, may not fully be aware that law relative to indications and disclaimers is generally well settled.

System Operations Phase. During the operations of civil engineering systems, users are invariably subject to risks of personal injury, fatality, and property damage. If it is established that such risks were caused by inaction, carelessness, or negligent actions by the system owner or operator, the owner or operator could be held liable to tort and may be asked to offer compensation for damages. While the risks associated with system operations are generally unavoidable, prudent risk management strategies by the system owner or operator can reduce the frequency or intensity of incidents and consequently, tort liability cases and payment amounts. Also, during operations, civil engineering systems may cause undue harm to the community or the environment (air, soil, or noise pollution), thus violating any laws established purposely to protect these resources.

System Monitoring Phase. In monitoring the usage of their systems, a system owner may record the system operations on video, and in doing so, may violate laws associated with surveillance and privacy. In many countries, such monitoring may be considered as spying on the citizenry and thus may be at least unethical if not illegal.

System Preservation Phase. At the maintenance phase, legal questions may arise from the possible toxicity of materials that are intended for use for protecting the system structure from further wear (e.g., lead-based paints for buildings and bridges), enhancing system operations (e.g., glycol or saline deicers), and herbicides for vegetation control. Also, engineers working in this phase are responsible for quickly remedying defects before system users or the general public are harmed as a result of the defect. In cases where a material is found to be deleterious to public health, the vendor of the material, rather than the engineer who specified it, is often held liable, on the basis of the law of public nuisance, for the costs of abatement or environmental remediation; but in certain cases, the courts have ruled otherwise (White, 2010).

System End-of-Life Phase. Where the end of life of a system is consistent with foreseen failure or deliberate demolition, the legal question that arises is whether the system termination adversely affects the geotechnical stability of neighboring structures. In certain cases, tell-tale markers are installed across existing cracks in neighboring structures to monitor whether the cracks have widened due to the structure demolition. Also, where the system fails unexpectedly, the legal question that may arise is whether poor design, faulty construction, defective monitoring, or incompetent or inappropriate operations (such as overloads) was responsible for the system failure.

All Phases. Expert witnessing is another growing application of the legal issues in civil engineering systems. At any phase, a civil engineering systems engineer may be solicited to give testimony in a court of law. Termed "court-appointed expert," such an engineer is often a recognized professional who, on the basis of their academic training and/or expertise in that branch of civil engineering and at that phase of the civil system development, have acquired specialized knowledge that make them uniquely qualified to render well-grounded opinion to a court of justice on specific points in dispute that are the subject of the court's ruling (Chasco and Meneses, 2010).

29.2.5 Some Past/Current U.S. Laws That Affect Civil Systems Development

Table 29.2 presents a number of laws in the United States that affect civil systems development. Air quality-related legislation that affect civil systems development include the Air Pollution Control Act of 1955; the Clean Air Act (CAA) of 1963, which established emissions standards; the Air Quality Control Act of 1967, which published air quality criteria; the Intermodal Surface Transportation Efficiency Act of 1991 (ISTEA), which mandated transportation plans conform to air quality enhancement initiatives; and the Kyoto Protocol of 1997, which was ratified in 2005 by 141 countries, to reduce emissions. Equity-related legislation that affects civil engineering systems development include Title VI of the Civil Rights Act of 1964; the National Environmental Policy Act (NEPA) of 1973, which included the effects of the civil system on the social environment; the National Historic Preservation Act of 1966, which protected historic resources from demolition due to human development including physical expansion of civil engineering systems; the American Indian Religious Freedom Act of 1978, which protected Native American burial sites from physical expansion of civil engineering systems; and Executive Order 12898 in 1994 by President Clinton, which ensured that low-income and minority populations do not unduly suffer adverse the environmental effects of the construction/operations of human development.

29.2.6 Discussion of Some Selected Areas of Law in CE Systems Development

(a) Contract Law. A contract is a legal voluntary agreement where one party agrees to perform a service for another for a stated payment. The conditions for legal enforceability of contracts are: (a) the offer must be made voluntarily by one party; (b) the offer must be accepted voluntarily by the other party; (c) there must be some "consideration," that is, promise of some form of payment in return for the service; (d) the intended service must be legal; and (e) there must be an intention to create a legal relationship between the two parties. For example, if Apex Contractors enter into a contract in which they agree to widen a bridge for Topeka County for $1.5 million, both the money and service (in this case, the bridge widening) are the considerations. Contract law is generally classified under civil law.

Table 29.2 Selected Laws in the United States That Affect Civil Systems Development

Year	Legislation	Description
1899	Rivers & Harbors Appropriation Act	Regulated actions affecting navigation in U.S. waters, including wetlands.
1918	Migratory Bird Treaty Act	Provided for protection of most common wild birds.
1964	Wilderness Act	Established criteria and restrictions on activities that can be undertaken on a designated land and water in the National Wildlife Refuge System for conservation purposes.
1966	Department of Transportation Act	Required conservation of the countryside, publicly owned parks and recreation lands, wildlife and waterfowl refuges, and significant historic sites.
1969	National Environmental Policy Act	Encouraged prevention of damage to the environment and biosphere and encouraged the understanding of ecological systems and vital natural resources.
1971	Wild Horses and Burros Protection Act	Encouraged the protection of wild and free-roaming horses and burros from capture, branding, harassment, or death.
1972	Marine Mammal Protection Act	Encouraged conservation of marine mammals such as the sea otter, walrus, polar bear, dugong, manatee, cetacean, and pinniped.
1972	Federal Insecticide, Fungicide, and Rodenticide Act	Established control of pesticides application to protect habitat and wildlife.
1972	Marine Protection Research and Sanctuaries Act	Established limits on ocean dumping of any material that could adversely affect human health and welfare, amenities, the marine environment, ecological systems, or economic potential.
1973	Endangered Species Act	Encouraged conservation of threatened/endangered fauna and flora and their habitats.
1982	Coastal Barrier Resources Act (Great Lakes Coastal Barrier Act of 1988)	Established protection for undeveloped coastal barriers and related areas by prohibiting direct or indirect federal funding of projects in such areas that might support development and minimizing damage to fish, wildlife, and other natural resources in these areas.
1990	Coastal Zone Management Act Reauthorization Amendments	Established controls of nonpoint source pollution for activities located in coastal zones to protect estuarine and marine habitats and species.
1990	Coastal Wetlands Planning, Protection and Restoration Act	Provided supports and funds for coastal wetlands restoration and conservation projects, especially in Louisiana.
1991	Intermodal Surface Transportation Efficiency Act	Provided for environmental conservation through highway funds to enhance the environment, such as wetland banking and mitigation of damage to wildlife habitat.
1998	Transportation Equity Act for the 21st Century (TEA-21)	Authorized funding to conserve the environment, including water quality improvement and wetlands restoration.
2002	Homeland Security Act	Established the Department of Homeland Security; reinforced the notion that certain parts of the national infrastructure as critical to the national and economic security, and required steps to be taken to protect it.
2005	Transportation Equity Act	Provided funding to improve and maintain surface transportation infrastructure; promoted environmental stewardship.

(*continued*)

Table 29.2 *(Continued)*

Year	Legislation	Description
2006	Federal Funding Accountability and Transparency Act of 2006	Required the full disclosure to the public of all entities or organizations receiving federal funds
2007	Water Resources Development Act	Reauthorized the Water Resources Development Act (WRDA), and authorized projects and studies for clean water, flood control, navigation, and environment improvement.
2009	The American Recovery and Reinvestment Act or "The Stimulus"	Funded investments in infrastructure, education, health, and renewable energy.
2009	Omnibus Public Land Management Act	Created a National Landscape Conservation System, made new additions to the National Conservation Areas, the National Wild and Scenic Rivers System, and the National Park System; created programs for oceanic observation, research, and exploration.
2012	Moving Ahead for Progress in the 21st Century Act (MAP-21)	Provided transportation funding without increasing transportation user fees.

Source: Adapted from Canter (1995), and other sources.

We all enter contracts every day. Sometimes these are written (e.g., a letter of acceptance into a company, leasing an apartment or rental car, buying a ticket for a movie, or paying tuition and attending classes), or are oral (e.g., ordering pizza or buying soda at a shop). The general format of any formal contract is as follows:

1. Title and date
2. Preamble (introduction to the agreement)
3. Names and contact information of all parties to the contract
4. Supporting documents (engineering drawings, general and specific technical specifications, conditions of contract)
5. Contract dates (start of the work and of the expected duration)
6. Terms of payment
7. Damages to be assessed in the case of nonperformance (liquidated damages)
8. Process for dispute resolution
9. Other general provisions of the contract
10. Concluding remarks to the contract
11. Signatures of parties to the contract, witnesses, and notary public or attorneys if applicable

Contracts are often signed at each phase of civil engineering systems development. At the needs assessment phase, for example, the system owner may sign a contract with a consultant to carry out an assessment of the demand for the system, review of the existing capacity, and/or assessment of the needs at the current or future time. At the planning phase, the owner may sign a contract with a consultant to carry out feasibility studies and to develop location planning and preliminary engineering for the system. At the design phase, the owner may engage the services of a design firm to carry out the system design. At the construction phase, the owner signs a contract with a contractor to build or implement the system; also at that phase, the owner may sign a contract with a construction inspection company to supervise/inspect the contractor's work. At the monitoring phase, the owner contracts out the inspection or assessment of the system loading or usage as well as the monitoring of the system to identify any structural, functional, or operational flaws. At the operations phase, the owner may sign a contract with a contractor to operate the system. For example, certain cities in the United States contract out all traffic operations to consultants. At the preservation phase, the owner may sign a contract with a contractor to carry out maintenance or rehabilitation of the system. At the end-of-life phase, the owner may sign a contract with a consultant to investigate why the system failed and/or to render professional advice on the best way to demolish or reuse an aging or obsolete system.

A contract is **discharged** when it has been carried out to completion, to the satisfaction of all parties to the contract or when all parties agree that *force majeure* has occurred (i.e., it has become impossible to perform due to circumstances beyond the control of the contracting parties). A contract is said to have been **breached** when one party fails to perform their part of the contract; in that case, the **injured party** can resort to any one of several ways of contract dispute resolution.

Mechanisms for Resolving Contract Disputes

The contract-specified mechanisms (in order of their increasing legal responsibility, cost, and time) are: negotiation, stand-in neutral, mediation, arbitration, dispute resolution boards, litigation.

Negotiation: This is an informal discussion that has little or no cost and is often brief, quick, and efficient. Failure to agree may lead to arbitration or litigation.

Stand-in Neutral: A third party that has relevant experience and is paid by both parties, a stand-in neutral provides expert advice that is nonbinding (any one or both parties have the option of refusing to accept such advice).

Mediation: An officially trained, recognized mediator chosen by both parties is brought in to help resolve the conflict. The outcome is voluntary (nonbinding), and the disputing parties are expected to reach an agreement on the basis of the mediator's recommendation. The mediator holds relatively informal meetings in order to counsel the parties, clarify discrepancies, and gather facts. However, the mediator has no authority to enforce their verdict. This mechanism is relatively inexpensive and the proceedings are typically confidential.

Arbitration: Slower than negotiation or mediation but faster than litigation, arbitration yields an outcome that is legally binding and enforceable in certain cases. In most cases, no appeal is possible and no explanation of the outcome is required. Also, the proceedings are not confidential. Arbitration involves five steps: agree to arbitrate, select arbiter, prepare for the hearing, conduction of the hearing, and award or outcome (within 30 days of close of the hearing).

Dispute Resolution Board: A type of arbitration tool that is more commonly used in disputes involving subsurface work, this board typically consists of the following three members who are experienced construction professionals: the owner's appointee, the contractor's appointee, and a third appointee mutually agreed by both parties. The board meets regularly until an award is made.

Litigation. Typically used as a last resort, litigation is public and thus lacks confidentiality. The litigation process follows established case law and thus can be expensive and lengthy (often taking as much as 5 years or more to reach trial).

(b) Tort Law. As we have mentioned in an earlies section, the users of civil engineering systems are invariably subject to risks of personal injury, death, and property damage arising from the construction, maintenance, and operation of such facilities. Civil engineering tort liability can be defined as the compensation for damages caused by inaction, carelessness, or negligence of the civil infrastructure agency. While such risks are generally unavoidable, managing them could substantially reduce their frequency or intensity.

In the 1970s and 1980s, the legal codes of most states were amended to abandon the doctrine of sovereign immunity, thereby removing state immunity from liability in state court proceedings for damages resulting from exercise of its proprietary or governmental functions such as design, construction, or maintenance of public infrastructure. Since then, the number of infrastructure-related tort claims has increased steadily (Bair et al., 1980; Smith et al., 2000). The loss of sovereign immunity by public entities is only one way in which tort liability has evolved over the past decades; legislatures and courts of law, it seems, have long been part of a general trend to ensure that injured persons are duly compensated. Thus, a general pattern over the years has been the replacement of absolute bars to recovery (such as **contributory negligence** and assumption of risk) by the doctrine of **comparative negligence** (Smith et al., 2000, Giraud et al., 2004).

The issue of civil infrastructure maintenance and rehabilitation (and the inadequacy of funding for these interventions) continues to be a growing problem in most countries worldwide. At the current time, civil infrastructure in the United States is generally considered as being in a poor state of repair. In its 2013 Report Card for America's infrastructure, ASCE, on a scale of A (exceptional) to F (failing), assigned grades of D to Dams, D^- to levees, D to wastewater facilities, C to ports C^+ to rail, D to aviation, C^+ to bridges, D– to inland waterways, D to roads, and D+ to transit (ASCE, 2013). Many facilities have exceeded or are approaching their design lives, and their poor physical conditions render them vulnerable to natural or man-made disasters, crashes, and incidents. Ideally, agencies desire to address all problems on their systems but are constrained by inadequate funding. ASCE states that by 2020, the total investment need for all publicly owned civil engineering infrastructure is \$3.6 trillion (ASCE, 2013), far in excess of operating budgets at the current time or in the foreseeable future. Recognizing that the poor condition and operational level of service may translate into increased tort, there is a need for agencies to manage the risks associated with the use of civil infrastructure so that the greatest reduction in overall risk can be obtained with the least amount of dollars spent on infrastructure preservation. To counter the growing problem of tort liability associated with civil infrastructure, many jurisdictions have initiated steps to develop programs explicitly to manage such risks (Gittings, 1987; Hoel et al., 1991; Datta et al., 1991; Demetsky and Yu, 1993; Giraud et al., 2004).

A continuing study of tort liability costs and risk management initiatives is useful for a number of reasons. First, increases in the costs associated with tort claims against infrastructure agencies have resulted, directly or indirectly, in a reduction of available funds for vital agency functions of construction, congestion mitigation, safety enhancement, maintenance, and the like. (Smith et al., 2000). Such adverse impact on an agency's budget is considered critical in the current era, which is characterized by increasing user demand of civil systems, increasing user expectation, and increasing geriatric patronage of the systems due to incipient retirement of the baby boomer generation. The situation is further exacerbated by the aging of civil infrastructure and, most importantly, severe funding limitations for maintenance and renewal. Second, increasing tort liability claims may very well reflect (or lead to) unfavorable user perceptions of the infrastructure system and translate into public relations problems at a time of increasing calls for greater accountability of taxpayer money and stewardship of public infrastructure. Therefore, many systems owners are caught in a Catch-22 situation: The increased exposure to tort risk, which is due to inadequate funding, may in turn lead to increased tort suits and further reduction in funding for preservation and reconstruction.

At the design phase, defects in design may lead to undue delay, discomfort, or even death of the users or the general public, and thus may lead to tort. At the system construction or maintenance phase, tort may arise due to personal injuries at workzones. At the operations phase, tort cases may arise due to the negligence or incompetence of the system operator. At the end-of-life phase, individuals may sue the agency as a result of fatality, injury, or property damage in the process of the demolition, whether such demotion was intended (because the system has exceeded its service life) or unintended (because it was destroyed by a natural or man-made disaster).

Thomas (2003) contends that an infrastructure design is considered as being generally immune as a protected exercise of discretion. According to the researcher, the primary defense to a state's tort liability for negligent design and maintenance is based on the premise that certain government actions are "discretionary" in nature and therefore are immune from tort. In this regard, some public agencies that own civil engineering systems may claim immunity for decisions involving project plans and designs, even if they either contain a defective feature or omit a required feature. Thomas also stated that some courts recognize that design generally involves the consideration of broad policy factors protected by the discretionary function exemption but provide exceptions to immunity (e.g., where the plan or design was approved without due deliberation or study or where it was unreasonable or arbitrary). At a later time when the plan is found defective or inadequate (and subsequently, poses a danger to the public) due to changed physical conditions, for instance, then the public system owner may be responsible for remedying the unsafe condition or to giving adequate notice to the system users. In some jurisdictions, design immunity statutes have been enacted by legislatures, but such statutes may not absolutely protect the state from any tort action associated with its duty of designing public property.

For a public system owner to have design immunity, it is often required that the owner must establish a "causal relationship between the plan or design and the accident, discretionary approval of the plan or design prior to construction, and the existence of substantial evidence supporting the reasonableness of the adoption of the plan or design". Thomas (2003) stated that design decisions based on budgetary or other economic constraints are generally seen as discretionary in nature; In a past case, an agency in New York argued that its failure to replace a barrier was due to funding priorities. However, the agency presented no evidence on planning, ordering of priorities, or limitations on available funding and was therefore held liable for injuries caused by the defective barrier.

Many owners of civil engineering systems are beginning to take explicit measures to reduce the incidence of tort suits. Figure 29.5 presents a number of strategies to reduce or manage the tort exposure of civil engineering systems owners and operators. These include **preemptive risk management**, where actions of a legal, administrative, engineering, and enforcement nature are taken to reduce the incidence of tort liability incidents, and **palliative risk management**, which involves actions taken to lessen the impacts of tort liability incidents after they have occurred.

Preemptive risk management strategies are comprised of administrative procedures and legal actions, information and training, enforcement, and engineering. Preemptive legal action involves aggressively maintaining appropriate laws that reduce liability exposure. This includes laws concerning immunity. It is expected that enhanced communication between the agency and legislative authorities would help address this issue. *Information and training* involves education of the general public by raising its awareness of accident-prone situations in the area near or within the system. Furthermore, there is a need to educate (or continue to educate) the system owner's personnel about the tort liability issues involved in their day-to-day operations. Such training of personnel is an important part of successful risk management in many states. *Enforcement* refers to the assurance, through a variety of instruments, that system users behave so as not to constitute a hazard to other users. *Engineering* refers to the elements of a system and the construction, operations, and

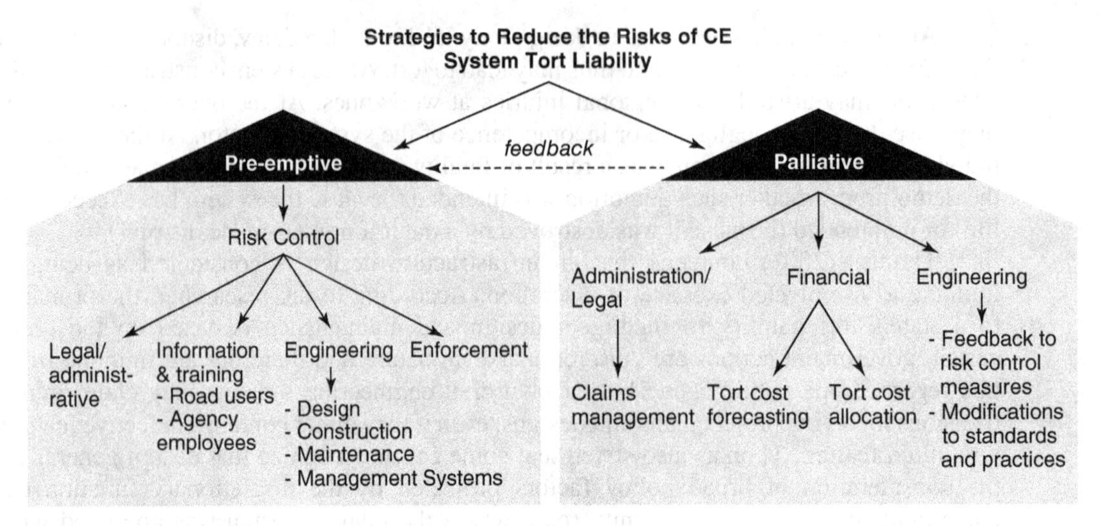

Figure 29.5 Strategies to reduce or manage the tort exposure of civil engineers.

maintenance involving system components. Reduction of tort exposure is an important element of preemptive risk management. Design and maintenance decisions based on budgetary or other economic constraints are generally seen as discretionary in nature, and therefore public system owners are generally not liable to tort in cases related to such areas. However, an owner that argues that its failure to remedy a defective design was due to funding priorities, could be held liable if it presents no evidence in planning, ordering of priorities, or limitations on available funding. In this regard, the current development of safety and maintenance management systems by public system owners will serve to provide such evidence in planning and programming of investments and will subsequently reduce the exposure of these owners to tort suits. Furthermore, it should be noted that in the current situation, risk management is implicitly involved in various aspects of facility design and operations at most agencies. For instance, the use of safety factors in civil engineering design is consistent with risk management practice.

Palliative risk management can occur in three different areas: legal/administrative, financial, and engineering. The main purpose of palliative risk management is to provide feedback to ensure a better preemptive risk management program. The legal aspect of palliative risk management involves a rigorous claims follow-up process that is termed "claims management." This process includes collecting all data related to tort claims and maintaining an accurate and up-to-date database containing as much relevant information as possible.

The need for feedback between preemptive and palliative risk management cannot be overemphasized. Such a feedback constitutes the most vital aspect of the operational framework as it allows an evaluation of preemptive strategies. Tort claim frequencies and amounts could be analyzed jointly by various divisions of the public agency. Periodic reporting of tort liability cases and outcomes should be encouraged.

The implementation of a risk management program for tort liability can help in the following ways: (a) Coordinating and tracking of all system-use-related claims and litigation against the agency (b) Processing of all system-use-related claims and managing a tort liability loss mitigation program, and in directing departmental resources to minimize the adverse effects of litigation (c) Promoting a cost-effective risk management effort statewide, development of control mechanisms through training and counseling, and fostering awareness by all employees of the risk potential associated with their actions (d) Improving the safety of system users by identifying incident types and locations associated with high tort costs and/or frequencies (e) Reducing an agency's exposure and loss due to tort liability

As evidenced in research findings, the number of tort claims and settlement amounts continue to grow. As we mentioned earlier, when system owners pay more to settle tort liability claims, less funding is left to operate and maintain their civil engineering systems; Furthermore, the ever-increasing payments to investigate, judge, and settle tort liability cases may be reflective of unfavorable user perceptions of civil engineering system and consequently, indicate public relations issues perhaps at play.

It is expected that in the 21st century, technology and innovations will increasingly play a role in causing more tort claims but also providing avenues for avoiding or resolving tort-related cases (Smith et al., 2000). Tort issues may be expected to arise when new technology is implemented and used, and others may arise when the innovation or technology fails to function properly, resulting in a loss to the user.

(c) Intellectual Property Law. Intellectual property (IP) refers to innovations for which exclusive rights to the creator are recognized by law (Rockman, 2004). Such innovations include discoveries and inventions, designs or processes, artistic and literary works, and even words, phrases, and symbols. Creators of intellectual property (IP) are granted certain *sui generis* (exclusive rights) to benefits that arise from their products. Intellectual property is typically enforced using trademarks, copyrights, patents, and trade secrets. Covered under property law, IPs can be leased or sold. These rights are based on the premise that inventors and creators have little incentive to invent unless they are legally entitled to the full social value of their inventions (Lemley, 2004). Intellectual property laws establish an incentive for inventors and authors to create and disclose their work through exchanging limited exclusive rights for disclosure of their work, thereby mutually benefiting society and patent/copyright holders.

A **copyright** is a legal instrument that provides inventors or creators of artistic, literary, musical, or other creative products the sole right to publish or sell it. **Utility patents** are a form of IP that cover the functional features of a design, while so-called **design patents** protect the aesthetic aspects of a design (such as the orientation, arrangement, configuration, or surface decoration, or shape. A **trademark** is used to protect the names or symbols (logo) of a company or its products. Designed to protect against cheap imitations, trademark protection is achieved by registration with a country's patent and trademark office or by its sustained use in the marketplace such that it achieves market recognition.

29.2.7 Evolution of Legal Aspects of Civil Systems Development

The legal side of civil engineering systems development is expected to undergo significant changes in the future due to changes in both the law and in the civil engineering profession. As Sweet and Schneier (2011) pointed out: "Law, though it seeks stability, is not static. Day in day out, new cases are ruled upon in the courts and legislatures enact new statutes in every session." The authors state that significant changes in the law continue to take place in areas including the application and enforceability of limitation of liability clauses, contractual indemnity and hold harmless clauses, the intellectual property rights of system planners and designers, and the further erosion of the economic loss rule, thereby increasing the liability of system designers to lawsuits from third parties with whom they are not in privity of contract. With regard to the evolving nature of civil engineering, we have learned from Chapter 1 how the various fields of the profession are undergoing changes that will influence the way civil engineers plan, design, operate, and maintain their systems. These changes, which are due to the environment or technological advancements, will also lead to changes in the nature, frequency, and intensity of legal interactions in the civil engineering discipline.

SUMMARY

The chapter first defined values and value systems as precursors of morality, law, and ethics. Then the relationships between these three concepts were examined. In the section dealing with ethics, the chapter discussed the branches of ethics and evolutionary ethics. The organizations that promote ethical behavior in civil engineering were identified, and the key categories of ethics guidelines in civil engineering were presented. A step-by-step guideline was provided for helping engineers resolve ethical situations, and an ethics case study in civil engineering was discussed. The chapter also identified potential ethical issues at the various phases of systems development and provided a brief statement about evolving trends in civil engineering ethics. In the section dealing with legal issues, the chapter discussed the origins of laws and the domain and scope of U.S. law. The chapter then described a number of potential legal issues that may be encountered at the various phases of systems development and lists some past/current U.S. laws that affect civil engineering systems development. Two areas of law that pertain to civil engineering were discussed, namely, contract law and tort law. In the discussion of contract law, the chapter identified a number of standard mechanisms for resolving contract disputes. Finally, some of the future trends of legal aspect of civil engineering systems development were discussed.

EXERCISES

1. In Figure 29.6, give, where possible or practicable, an example of situations faced by the civil engineer in each of the situations marked A–G.

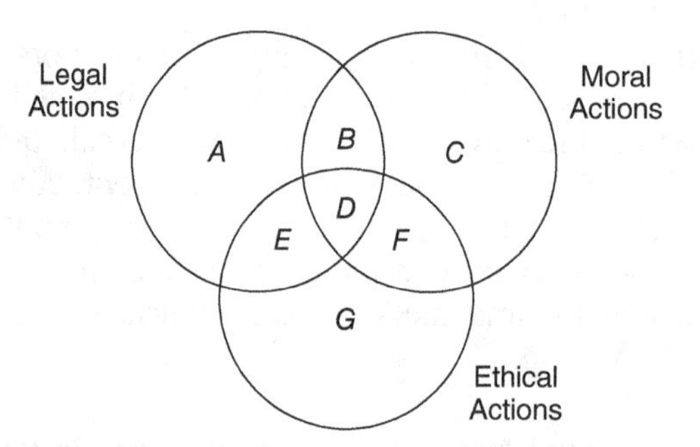

Figure 29.6 Figure for Exercise 29.1.

2. Tom Greene, an engineer working for a solid waste contractor has found that workers of his company are dumping solid wastes into a river in order to save the costs of hauling the trash to the landfill located 5 km away. He has brought the issue repeatedly to the attention of his boss but sees no change after several months. Should he just ignore the situation? Should he report the issue to the local newspaper? What should Tom do?
3. Identify elements of the ASCE, NCEES, or NSPE code of ethics which are relevant to each of the following situations: (a) An engineer responsible for selecting a supplier is taken out to lunch by one of the potential bidders. (b) Engineer Y returns from vacation and is asked hurriedly to stamp and sign a set of drawings that were prepared during her vacation. (c) Engineer Z submits a copy of the same set of drawings from a previous job to a current client. (d) To quell public outcry, Engineer W makes a statement about the extent and causes of cost overruns on his public project even though he knows that this version is not an exact representation of the truth.

4. You are the engineer in charge of operating a very busy urban interstate freeway system. Discuss your ethical responsibilities with regard to the environment.
5. What ethical responsibilities does an engineer have to her employer?
6. As a college student, what are some possible academic behaviors that could be considered unethical?
7. Give three examples where an engineer could be faced with an ethical dilemma between protecting her boss and protecting the welfare of the community.
8. Discuss the main sources of law and how they relate to the practice of civil engineering.
9. Discuss, with real or hypothetical illustrations, the legal issues encountered at each phase of the system development process.
10. Identify and discuss any two national laws that are related directly to the development of civil engineering systems.
11. How can civil engineering system developers protect themselves against tort?

REFERENCES

Alsnih, R., and Stopher, P. R. (2003). Environmental Justice Applications in Transport, in *Handbook of Transportation and the Environment*, Hensher, D. A., and Button, K. J., Eds. Elsevier, Amersterdam.

American Society of Civil Engineers (ASCE) (2010). *Code of Ethics*. ASCE Press, Reston, VA. http://www.asce.org/Leadership-and-Management/Ethics/Code-of-Ethics/.

ASCE (2013). *2013 Report Card for America's Infrastructure*, ASCE Press, Reston, VA.

Bair, B. O., Fornini, W. J., and Grubba, J. L. (1980). Highway Risk Management: A Case Study. *Transp. Res. Rec.* 742.

Bass, R. (1998). Evaluating Environmental Justice under the National Environmental Policy Act. *Environ. Impact Assess. Rev.* 18(1), 83–92.

BOE (2012). Cuneiform Law. Britannica Online Encyclopaedia, www.britannica.com. Accessed Dec. 12, 2012.

Canter, L. W. (1985). *Environmental Impact Assessment*, McGraw-Hill, New York.

Caplicki, E. (2010). Nevada Bars Negligence Claims against Design Professionals for Economic Loss. *J. Legal Affairs Dispute Resol. Eng. Construct.* 2(2), 126–127.

Chasco, F., and Meneses, A. (2010). Engineering in the Service of Law. *J. Legal Affairs Dispute Resolut. Eng. Construct.* 2(2), 100–105.

Datta, T. K., Krycinski, T. R., and Taylor, W. C. (1991). Highway Risk Management System: A Michigan Program. *ITE J.* 61(10).

Demetsky, M. J., and Yu, K. (1993). An Assessment of Risk Management Procedures and Objectives in State Departments of Transp. *Transp. Res. Rec.*1401.

Drnevich, V. (2011). *Ethics in Engineering: Ethical Problems and Solving Then*, Purdue University, West Lafayette, IN.

Ethics Center for Engineering and Research. http://www.onlineethics.org/2948.aspx.

Fieser, J. (2009). Ethics, Internet Encyclopedia of Philosophy. http://www.iep.utm.edu/ethics.

Fisk, E. R. (1997). *Construction Project Administration*, 5th ed. Prentice Hall, Upper Saddle River, NJ.

Flew, A. (1979). Consequentialism, in *A Dictionary of Philosophy*, 2nd ed. St. Martins Press, New York.

Giraud, T., Choichivien, Labi, V. T., Sinha, S., and K. C. (2004). Development of a Risk Management System for Indiana. Tech. Rep. FHWA/IN/JTRP-2003/20, Joint Transportation Research Program, Purdue University, West Lafayette, IN.

Gittings, G. L. (1987). Tort Liability and Risk Management. *J. Transp. Eng.* 113(1), Ethics-virtue, Stanford Encyclopedia of Philosophy, http://plato.stanford.edu/entries/ethics-virtue/.

Hoel, L. A., Demetsky, M. J., and Reagan, D. (1991). *Identification and Evaluation of Risk Elements for Highway Systems in Tort Liability*, University of Virginia, Charlottesville, VA.

Institution of Engineers, Australia. (2010). *Our Code of Ethics*. Engineera Australia, Barton, Australia.

Ishibashi, M., and Singh, A. (2011). Evolution of Common Law: Promissory Estoppel. *J. Legal Affairs Dispute Resolut. Eng. Construct.* 3(4), 170–177.

ITS (2003). *Environmental Justice and Transportation: A Citizen's Handbook*. UCB-ITS-M-2003-1. Prep. by S., Cairns, J. Greig, and M. Wachs. USC Berkeley Inst. of Transp. Studies.

Kelly, E. (2006). *The Basics of Western Philosophy*. Greenwood Press, Westport, CT.

Khisty, C. J. (1996). Operationalizing Concepts of Equity for Public Project Investments. *Transp. Res. Rec.* 1559, 94–99.

Lemley, M. A. (2004). Property, Intellectual Property, and Free Riding. *Tex. L. Rev.*, 83, 1031.

Macnish, K. (2011). Surveillance Ethics, Internet Encyclopedia of Philosophy. http://www.iep.utm.edu/ surv-eth/.

McCormick, M., and Singh, A. (2010). Appellate Court Reverses Judgment Based on Clear Intent of Change Orders and Claims. *J. Legal Affairs Dispute Resolut. Eng. Construct.*, 2(2), 128–131.

McCuen, R. H., Ezzell, E. Z., and Wong, M. K. (2011). *Fundamentals of Civil Engineering*, CRC Press Boca Raton, FL

National Institute for Engineering Ethics (NIEE). http://www.niee.org/murdoughCenter/.

NSPE. (2012). Ethics, Board of Ethical Review (BER) of the National Society of Professional Engineers (NSPE). http://www.nspe.org/Ethics/BoardofEthicalReview/index.html.

O'Brien, J. J. (1998). *Construction Change Orders*. McGraw-Hill, New York.

Orend, B. (2000). *War and International Justice: A Kantian Perspective*. Wilfrid Laurier University Press, West Waterloo, Ontario.

Petroski, H. (1985). *To Engineer Is Human: The Role of Failure in Successful Design*. St Martins Press, New York.

Rockman, H. B. (2004). *Intellectual Property Law for Engineers and Scientists*. Wiley, Hoboken, NJ.

Sen, A. (2000). *Development as Freedom*. Anchor Publishing, Harpswell, ME.

Sinha, K. C., and Labi, S. (2007). *Transportation Decision Making*, Wiley, Hoboken, NJ.

Smith J. L., Durant, L. A., Hill, N. M., and Lewis II, C. R. (2000). *Transportation Tort Law. A Look Forward*. Cmte. on Tort Liability and Risk Management, Transp. Res. Board, Washington, DC.

Sweet, J., and Schneier, M. M. (2011). *Legal Aspects of Architecture, Engineering, and the Construction Process*, 8th ed. Cengage Learning, Stamford, CT.

Thomas, H. (2012). Differing Site Conditions Indications and Disclaimers. *ASCE J. Legal Affairs Dispute Resolut. Eng. Constructi.* 4(3), 86–89.

Thomas, L. W. (2003). *Tort Liability of Highway Agencies, Selected Studies in Transportation Law*, Vol. 4, NCHRP. Transp. Res. Board, Washington, DC.

van Roojen, M. (2011). Moral Cognitivism vs. Non-Cognitivism, The Stanford Encyclopedia of Philosophy, Zalta, E. N., Ed. http://plato.stanford.edu/archives/spr2011/entries/moral-cognitivism/.

White, N. (2010). Lead Paint Is Not a Public Nuisance. *J. Legal Affairs Dispute Resolut. Eng. Construct.* 2(2), 132–133.

World Bank (2003). *Social Analysis Sourcebook: Incorporating Social Dimensions into Bank-supported Projects*. Social Dev. Dept., Washington, DC.

USEFUL RESOURCES

Bockrath, J., and Plotnick, F. (2010). *Contracts and the Legal Environment for Engineers and Architects*, McGraw-Hill, New York.

Ethics Center for Engineering and Research. http://www.onlineethics.org/2948.aspx.

Harris, C. E., Pritchard, M. S., Rabins, M. J., James, R., and Englehardt, E. (2013). *Engineering Ethics, Concepts and Cases*, Cengage Learning/Wadsworth, Boston.

Humphreys, K. K. (1999). *What Every Engineer should know about Ethics*, CRC Press, Boca, Raton FL.

Institution of Civil Engineers (2004). Royal Charter, By-laws, Regulations and Rules. http://www.ice-london.org.uk/london/documents/charter_and_bylaws_2005.pdf.

Kelley, G. (2012). *Construction Law: An Introduction for Engineers, Architects, and Contractors*, RS Means/Wiley, Hoboker, NJ.

Martin, M. W., and Schinzinger, R., (2004). *Ethics in Engineering*, McGraw-Hill, New York.

National Institute for Engineering Ethics (NIEE). http://www.niee.org/murdoughCenter/.

Pfatteichers, S. (2010). *Lessons Amid the Rubble: An Introduction to Post-Disaster Engineering and Ethics*, John Hopkins University Press, Baltimore, MD.

Preston, A. (2006). Analytic Philosophy, Internet Encyclopedia of Philosophy. http://www.iep.utm.edu/analytic/.

Sachs, J. (2002). *Aristotle, Nicomachean Ethics, Focus Philosophical Library*. Pullins Press, Brooklyn, NY.

Sweet, J., and Schneier, M. (2012). *Legal Aspects of Engineering and the Construction Process*, Cengage Learning/Wadsworth, Boston.

CHAPTER 30

EPILOGUE

As we come to the end of this journey through the challenging but fascinating field of civil engineering systems, a recap of the essential watermarks of the journey and how they hopefully fulfilled our initial expectations would be helpful. These watermarks included a discussion of the historical evolution of the civil engineering discipline; identification of the phases of civil engineering system development, the tasks at each phase, and the analytical tools civil engineers need in order to address these tasks effectively; and last, but not least, a discussion of the vulnerability of civil engineering systems to external threats, the concepts of resilience and sustainability, and the role of law and ethics in the development of these systems. In navigating through the chapters, we have examined not only the issues and analytical tools associated with problem solving at each phase of civil systems development but also explicitly identified the contexts in which they are applicable.

We began with a discussion of how civil engineering systems have evolved over the millennia and the external forces that catalyzed such transformations. Since the dawn of time, civilizations have continuously designed and constructed civil engineering structures and facilities to serve the basic needs of their societies, including water supply and sewerage for households, irrigation for farmlands, shelter from natural elements and protection from man-made threats, and provisions for trade and transportation, industry, defense, worship, and recreation. These basic needs have largely remained the same across the spectrum of history. In other words, no problem is really new; it is only the tools, materials, and energy sources we use to fulfill our needs that have become more "sophisticated." Philosophers and sages of the ancient times had repeatedly admonished that every problem encountered at the current time or to be encountered in the future has already been faced before; the wisdom encapsulated in this perspective is the strongest motivation for our looking back to history. Thus, for the benefit of future practice, we identified in the first chapter the lessons to be learned from history. These include the extreme ethics-related practices where builders faced harsh penalties in the event of fatal failures of their structures, the successes of the feminine philosophies in Taoist and ancient Greek design and construction, and the unsustainable masculine design philosophies of engineers in the Confucius and the ancient Roman eras. These lessons encourage civil engineers to take full responsibility for their work and also to plan, design, build, and operate systems that are more "in-sync" with their environment—weaving the civil infrastructure as part of a nature–infrastructure symbiosis geared toward enhancing the natural environment rather than exploiting it to the detriment of both these natural resources and the common future of the human race.

In this text, we devoted specific chapters to each phase. The rationale is simple. Many, if not most, engineering students after graduation will find themselves working in only one or very few of the eight phases of needs assessment, planning, design, construction, operations, monitoring, maintenance, and end-of-life. For an engineer working in a specific phase, an explicit awareness of not only the analytical tasks and tools at each phase but also of the sequential location of that phase relative to other phases and feedback responsibilities to previous phases can be beneficial to the practice.

The toolbox that civil engineers use in their current-day tasks needs to be reinforced with new tools or sharpened existing tools, particularly to address the myriad daunting challenges faced by societies in the current era. These challenges include accountability for the typically mammoth-sized expenditures in providing, operating, and maintaining civil infrastructure systems and their far-reaching impacts on socioeconomic development; the potential adverse impacts of the construction and the operations of such systems in terms of environmental degradation, community disruption, and social inequities; the growing number and loudness of stakeholder voices; the increasing tightness of government budgets; the increasing loads and demands fueled by population growth; the advanced ages and inadequate condition of many existing systems; the uncertainties in the global economic environment; and the emphasis on the security and safety of system users. In light of these challenges, engineers working at every phase of system development face unprecedented scrutiny and are required to render exemplary fiduciary stewardship of these systems. With a well-stocked toolbox that contains appropriate and effective tools to address the analytical tasks at each phase, civil engineers can be better equipped to plan, design, construct, operate, and maintain their systems in a more defensible, objective, and comprehensive manner.

Therefore, we examined in this text a significant number of analytical tools, each of which addresses at least one analytical task at any phase of system development. The task of describing or predicting a system's structure, how it works, or how it interacts with its natural or man-made environment can be accomplished using the tools explained in the chapters on modeling, simulation, system dynamics, network analysis, and queue analysis. The task of identifying the optimal or the most cost-effective action among multiple alternative actions at any phase is addressed most directly by the chapters on economic analysis, real options, decision analysis, optimization, simulation, and multiple criteria analysis. These tools are useful, for example, to engineers seeking the best system location at the planning phase, the best material type combinations and component configurations at the design phase, the best contracting approaches and labor/equipment resource combinations at the construction and operations phases, the best techniques or schedules for carrying out maintenance and monitoring of the system, or the best technique for ending the physical life of a system.

Regarding the task of incorporating uncertainty at any phase of civil engineering systems development, we discussed measuring, characterizing, and incorporating uncertainty in decision making. The probability chapter described how to measure uncertainty, the modeling and simulation chapters quantified how uncertainties in input factors translate ultimately into uncertainties in the outputs, and the risk and reliability chapter showed how to quantify the possibility that one or more threatening events occur. The chapters on real options and decision analysis showed us how to incorporate uncertainty by addressing the likelihoods of possible alternative pathways and outcomes under different potential conditions, given a set of initial or phased decision points. The task of incorporating the multiplicity of performance criteria (the chapter on multiple-criteria analysis) were addressed, as well as the influential criteria and their causal and looping interrelationships (systems dynamics chapter). This issue is important because the planning, design, and operations of civil engineering systems are not only associated with multiple performance considerations emanating from the concerns of the system owner, the user, or the community, but also the multiple factors that pertain to the system users, the system environment, or the system itself, which therefore influence some outcome of the system. Also, recognizing that engineers are encouraged to consider not only the technical aspects and consequences of their system but also its economic and financial impacts, we learned some tools for the task of costing and its associated concepts (the chapters on cost estimation and economic analysis).

Our discussion of the individual phases of civil system development not only described the general key issues at each phase but also the possible task contexts and the analytical tools that could help address the tasks at that phase. The needs assessment phase, where the engineer carries out the task of predicting the level of need for a system, determines whether to proceed with the remaining phases of development for the system. In the needs assessment chapter, we identified how tools such as statistical modeling, microeconomics, probability, and real options analysis could be applied at that phase. The chapter explaining the planning phase emphasized the realization that a well-planned system provides benefits not only to its users and the community in delivering the specific service for which it was intended but also to governments with an opportunity to meet broader national or regional objectives such as job creation, energy consumption reduction, and sustainable development. The planning chapter therefore presented the evolving and emerging contexts and the principles of systems planning and also demonstrated some analytical computations at the various steps of the system planning process. The design phase chapter followed, with a classification of design situations in civil engineering and the steps of the engineering design process and presented a number of design computation illustrations that incorporate tools to address uncertainty in the design parameters. Due to the wide range of civil engineering branches and the large number of design contexts, this text presents not the specific design techniques but the analytical tools, such as Monte Carlo simulation and optimization, that could enhance the design procedures irrespective of the branch of civil engineering in question. The chapter on the construction phase identified the various stages of this phase and discussed traditional and innovative contracting approaches. This chapter also suggested how the tools for bid evaluation, contractor selection, and other construction phase tasks could be applied. These tools include modeling to predict cost overruns, time delays, and safety incidents; critical path analysis; and statistics-based acceptance testing. The operations phase chapter provided a few specific examples of the nature of operations-related tasks and discussed some issues associated with engineering systems operations from both the traditional viewpoint and from the viewpoint of emerging challenges in the new millennium. The system monitoring chapter discussed the rationale, issues, concepts, analytical tools, and equipment available for monitoring the integrity of a civil engineering system as well as tracking any changing patterns and levels of its usage and its relationship with its environment. This chapter also demonstrated how engineers could develop a long-term plan for system monitoring by considering the costs and effectiveness of different monitoring techniques and discussed the ethical issues that arise in system user monitoring. The chapter on the maintenance phase presented the tasks of developing models to predict system condition and longevity and to evaluate the impacts of alternative treatments or long-term maintenance schedules on the basis of multiple criteria. The end-of-life phase is fertile ground for the application of uncertainty concepts; therefore, the chapter on that phase identified the various categories of system end-of-life agents, presented contexts for the application of modeling and other analytical tools to predict system longevity, and presented numerical examples on modeling uncertainty in system life durability.

This text also discussed a number of issues that are strongly related to the development of civil engineering systems. These include the system resilience to different kinds of sudden or gradual threats. Incipient global warming and the subsequent changes in sea and groundwater levels, wind speeds, and other environmental changes are expected to have potentially widespread and deleterious consequences on the integrity of civil structures and facilities, and thus constitute a serious natural threat that engineers must be equipped to address. The chapter on system resilience laid the groundwork for an appreciation and development of methodologies to assess and monitor the hazards to civil engineering systems posed by natural and man-made threats and to analyze the effectiveness of investments that address the hazards. Closely related to the issue of system resilience is system sustainability. The importance of incorporating sustainability concerns at every phase of system development is underscored by the unparalleled scale of human activity, the unprecedented demand for civil engineering systems, the reality that the resilience of natural resources is not infinite, and the demonstrated relationship between sustainability and human well-being. The chapter on sustainability identified the contexts for incorporating analytical tools in the prediction of sustainability at a broad level as both a closed loop and a balance sheet of resource flows within a system of systems. Then, for each phase of civil engineering system development, the chapter discussed the impact of the triple bottom line (environment, economy, and social) on the successful implementation of that phase; and the impact of each phase on each aspect of the triple bottom line.

In the last chapter, we acknowledged that technical skills alone are not sufficient to address the complexity of engineering problems and that civil engineers therefore need to possess a working knowledge in other areas including law and ethics. The chapter on ethics and legal issues first defined values and value systems as precursors of morality, law, and ethics; identified potential ethical issues that could arise at each phase of systems development; and presented guidelines for resolving ethics-related situations. In the legal issues section of this chapter, the origin, domain, and scope of laws were discussed as well as some of the laws that affect civil engineering systems development and potential legal issues that may be encountered at the various phases. In certain problem contexts, engineers may need to enhance their skills in other nonengineering subjects, including economics and finance (which we addressed in Chapter 11), public policy analysis, psychology, architecture, graphic arts, and ergonomics, or collaborate with experts in those fields to address engineering problems. Also, engineers will need to enhance their personal skills relating to creativity and communication and professional skills in organizational leadership, management, and administration if they are to carry out effectively their stewardship role for civil infrastructure systems.

The 29 chapters of this text were prepared for the benefit of civil engineers who will continue to develop structural, hydraulic, geotechnical, construction, environmental, transportation, architectural, and other civil systems that address the infrastructure needs of society. For these individuals, the future is bright and challenging. True, the landscape of history is littered by several, often costly or even fatal, encounters as engineers have labored to build, operate, and maintain systems that have served humankind. It is true that a number of formidable past and present challenges have plagued us and continue to stymie our efforts. It is also true that civil engineers are engaged in a profession in which achievement is taken for granted but a single failure can devastate careers. Nevertheless, we should see each imminent threat as a future opportunity to develop newer and better tools to carry out the tasks at each phase so that the overall goals of effectiveness, efficiency, and equity can be achieved without sacrificing our values, our environment, and our dignity.

We do not need a crystal ball to realize that the challenges that have faced us in the past will continue to exist in the future: (i) the need, at each phase of system development, to minimize future environmental degradation, community disruption, and social inequities; (ii) the need to ensure the security and safety of future system users; (iii) the need to optimize the use of taxpayers' dollars; and (iv) the need to carry out engineering practice in ways consistent with ethical and legal frameworks. The magnitude of these challenges will continue to be exacerbated by the decreasing renewable resources, the increasing tightness of government budgets; the increasing loads and demand due to increasing populations worldwide; the aging of existing civil engineering systems; the certainty of global economic uncertainties; and the looming specter of climate change. Engineers will be willing and, hopefully, able to address future engineering problems with the available resources; nevertheless, there will also be problems of a social and political nature that influence engineering systems development but are outside the direct control of engineers. In helping to resolve such situations, engineers, in their primary role of providing infrastructure that is primarily aimed at meeting social and economic needs, can also serve as instruments of social change: So, through our designs, we can become a voice for the voiceless and the politically disenfranchised.

Yet still, we must avoid nourishing the notion of a parental government or civil systems provider: System users must be encouraged to take personal responsibility at all times and not unduly depend on government resources. Also, safe and responsible use of civil systems must be promoted and reckless use that invites injury and subsequent litigation, discouraged. On a wider note, it is reasonable to be optimistic about the ability of engineers to address the future challenges alluded to earlier in this paragraph. New frontiers of human endeavor in other fields provide fertile and boundless opportunities for enhancing the toolbox of engineers engaged in any task at any phase for any type of civil engineering system. These opportunities include new physical and analytical tools and equipment, new materials, energy sources, and processes, and emerging technologies associated with the Internet, space exploration, and nanotechnology.

In facing the challenges of the current era and foreseeable future, civil engineers must not be intimidated by the sizes and complexities of these challenges. Current and future civil engineers can continue to draw inspiration from the audacity of engineers who developed inspiring civil engineering materials, processes, and systems worldwide across the ages from the dawn of time and medieval era to present day; from the Sumerian ziggurats in ancient Mesopotamia and the Indus Valley irrigation systems in ancient Harrapan to the Falkirk Wheel in Scotland. Future engineers must learn to adapt quickly to the vicissitudes of the times. Social and economic changes constantly create new demands not only on engineers but also on the educational systems that produce them; it is therefore imperative that engineers cultivate the skills to make informed choices, basing their decisions and actions not only on the analysis of present situations but also on the vision of a better tomorrow. In doing so, engineers must continue to incorporate the virtues of charity and humaneness that inherently characterize their profession: Unlike many other professions, the nature of civil engineering work inherently involves the stewardship of systems that have a direct impact on the quality of life of humankind in general; thus, even the most specific instances of civil engineering practice is not geared toward a particular privileged individual or narrow set of individuals only but to the entire society. Overall, the formula for successful stewardship of civil engineering infrastructure of the future includes an appreciation of the historical development of civil engineering systems, awareness of the need to address the global issues that affect (and are affected by) civil engineering systems, recognition of the phasal nature of civil systems development and the analytical tasks at each phase, and tireless acquisition of the best analytical tools to address these tasks. All these elements, reinforced by a generous amount of enthusiasm and healthy optimism, can help propel us further in our quest to develop effective and efficient civil engineering systems for today, tomorrow, and beyond.

APPENDIX 1

COMMON PROBABILITY DISTRIBUTIONS

Name	Equation	Shape of the Density Function	Comments
Normal (Gaussian)	$f(x; \mu, \sigma) = \frac{1}{\sigma\sqrt{2\pi}} e^{-(x-\mu)^2/2\sigma^2}$		Also called the Bell curve, this is the most common distribution used in in engineering due to invocation of the central limit theorem
Standard Normal	$f(z) = \frac{1}{\sigma\sqrt{2\pi}} e^{-z^2}$		Similar to the normal.
Beta	$f(x; \alpha, \beta) = \frac{1}{B(\alpha, \beta)} x^{\alpha-1}(1-x)^{\beta-1}$		Often applied to model the behavior of random variables limited to intervals of finite length in a wide variety of disciplines e.g., variability of geotechnical properties; task durations in construction project management;
Gamma	$f(x; k, \theta) = \frac{x^{k-1}e^{-x/\theta}}{\theta^k \Gamma(k)}$ for $x > 0$ and $k, \theta > 0$ $\Gamma(k)$ is the gamma function at k. Shape parameter k; scale parameter θ.		Frequently used to model the waiting time for an event to occur, for example, the number of years before a system ceases to function.
Binomial	$f(k; n, p) = p(X = k) = \binom{n}{k} p^k (1-p)^{n-k}$ where $k = 0, 1, 2, \ldots, n$		Often used to model the number of successes in a sample of size n taken from a population with replacement.
Continuous Uniform	$f(x) = \frac{1}{n}$		Also, known as the rectangular distribution, this describes a situation where every interval of the same length over the domain has the same probability of occurrence.

Name	Equation	Shape of the Density Function	Comments
Discrete Uniform	$f(x) = \begin{cases} \dfrac{1}{B-A} & \text{for } A \le x \le B \\ 0, & \text{otherwise} \end{cases}$	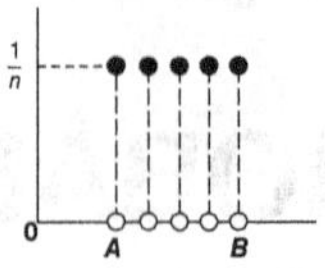	Describes a situation where every individual element in the domain has the same probability of occurrence.
Chi-square	$f(x;k) = \begin{cases} \dfrac{x^{(k/2)-1}e^{-x/2}}{2^{k/2}\Gamma(k/2)}, & \text{for } x \ge 0 \\ 0, & \text{otherwise} \end{cases}$	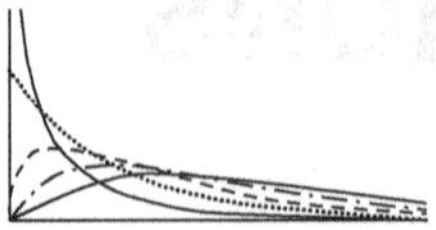	Is a probability distribution commonly applied in areas of inferential statistics including hypothesis testing and confidence intervals estimation.
Weibull (2-parameter)	$f(x;\lambda,k) = \begin{cases} \dfrac{k}{\lambda}\left(\dfrac{x}{\lambda}\right)^{k-1} e^{-(x/\lambda)^k}, & \text{for } x < 0 \\ 0, & \text{for } x \ge 0 \end{cases}$ where k and λ are the shape and scale parameters, respectively, of the distribution $k > 0$ and $\lambda > 0$	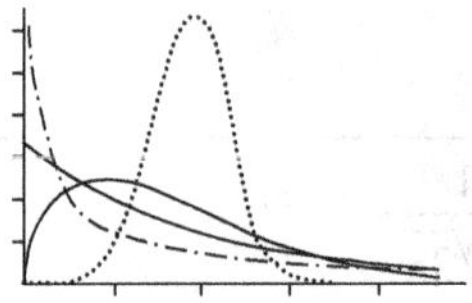	Has wide applications in hydrology, survival studies, reliability and failure analysis, insurance analysis, and extreme value analysis.
Gumbel Type 1	$f(x\|a,b) = ab.e^{-(be^{-ax}+ax)}$	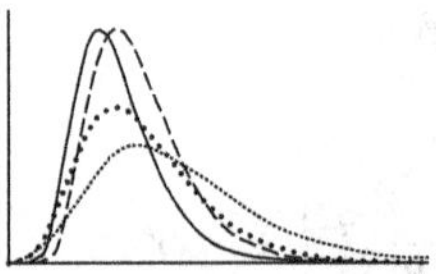	Commonly used in the analysis of extreme values and survival or duration modeling.

APPENDIX 2

STANDARD NORMAL CURVE AND STUDENT *t* DISTRIBUTION

Table A2.1 Areas under the Standard Normal Curve

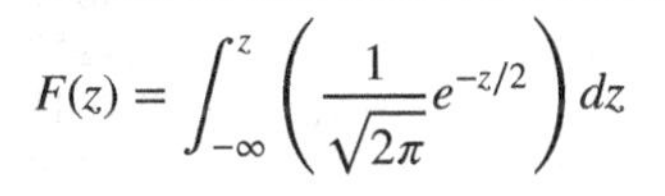

$$F(z) = \int_{-\infty}^{z} \left(\frac{1}{\sqrt{2\pi}} e^{-z/2} \right) dz$$

Examples:

$P(z < -2.14) = F(-2.14) = 0.0162$

$P(z < 1.32) = F(1.32) = 0.9066$

$P(z > 1.32) = 1 - F(1.32) = 1 - 0.9066 = 0.0034$

$Z_{0.005} = 2.575$

$Z_{0.025} = 1.96$

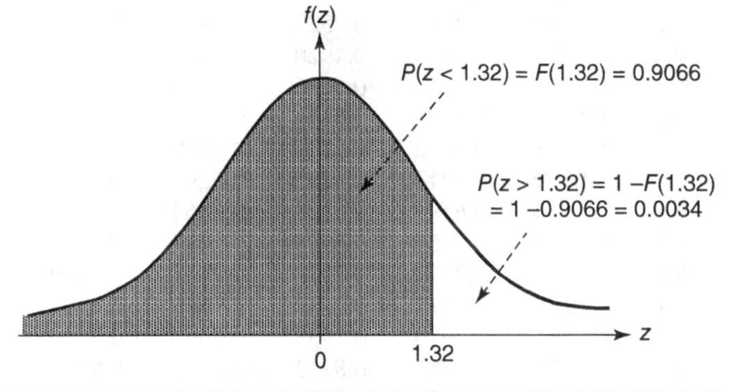

z	0.00	0.01	0.02	0.03	0.04	0.05	0.06	0.07	0.08	0.09
−3.4	0.0003	0.0003	0.0003	0.0003	0.0003	0.0003	0.0003	0.0003	0.0003	0.0002
−3.3	0.0005	0.0005	0.0005	0.0004	0.0004	0.0004	0.0004	0.0004	0.0004	0.0003
−3.2	0.0007	0.0007	0.0006	0.0006	0.0006	0.0006	0.0006	0.0005	0.0005	0.0005
−3.1	0.0010	0.0009	0.0009	0.0009	0.0008	0.0008	0.0008	0.0008	0.0007	0.0007
−3.0	0.0013	0.0013	0.0013	0.0012	0.0012	0.0011	0.0011	0.0011	0.0010	0.0010
−2.9	0.0019	0.0018	0.0017	0.0017	0.0016	0.0016	0.0015	0.0015	0.0014	0.0014
−2.8	0.0026	0.0025	0.0024	0.0023	0.0023	0.0022	0.0021	0.0021	0.0020	0.0019
−2.7	0.0035	0.0034	0.0033	0.0032	0.0031	0.0030	0.0029	0.0028	0.0027	0.0026
−2.6	0.0047	0.0045	0.0044	0.0043	0.0041	0.0040	0.0039	0.0038	0.0037	0.0036
−2.5	0.0062	0.0060	0.0059	0.0057	0.0055	0.0054	0.0052	0.0051	0.0049	0.0048
−2.4	0.0082	0.0080	0.0078	0.0075	0.0073	0.0071	0.0069	0.0068	0.0066	0.0064
−2.3	0.0107	0.0104	0.0102	0.0099	0.0096	0.0094	0.0091	0.0089	0.0087	0.0084
−2.2	0.0139	0.0136	0.0132	0.0129	0.0125	0.0122	0.0119	0.0116	0.0113	0.0110
−2.1	0.0179	0.0174	0.0170	0.0166	0.0162	0.0158	0.0154	0.0150	0.0146	0.0143
−2.0	0.0228	0.0222	0.0217	0.0212	0.0207	0.0202	0.0197	0.0192	0.0188	0.0183
−1.9	0.0287	0.0281	0.0274	0.0268	0.0262	0.0256	0.0250	0.0244	0.0239	0.0233
−1.8	0.0359	0.0352	0.0344	0.0336	0.0329	0.0322	0.0314	0.0307	0.0301	0.0294
−1.7	0.0446	0.0436	0.0427	0.0418	0.0409	0.0401	0.0392	0.0384	0.0375	0.0367
−1.6	0.0548	0.0537	0.0526	0.0516	0.0505	0.0495	0.0485	0.0475	0.0465	0.0455
−1.5	0.0668	0.0655	0.0643	0.0630	0.0618	0.0606	0.0594	0.0582	0.0571	0.0559
−1.4	0.0808	0.0793	0.0778	0.0764	0.0749	0.0735	0.0722	0.0708	0.0694	0.0681

(continued)

Table A2.1 (*continued*)

z	0.00	0.01	0.02	0.03	0.04	0.05	0.06	0.07	0.08	0.09
−1.3	0.0968	0.0951	0.0934	0.0918	0.0901	0.0885	0.0869	0.0853	0.0838	0.0823
−1.2	0.1151	0.1131	0.1112	0.1093	0.1075	0.1056	0.1038	0.1020	0.1003	0.0985
−1.1	0.1357	0.1335	0.1314	0.1292	0.1271	0.1251	0.1230	0.1210	0.1190	0.1170
−1.0	0.1587	0.1562	0.1539	0.1515	0.1492	0.1469	0.1446	0.1423	0.1401	0.1379
−0.9	0.1841	0.1814	0.1788	0.1762	0.1736	0.1711	0.1685	0.1660	0.1635	0.1611
−0.8	0.2119	0.2090	0.2061	0.2033	0.2005	0.1977	0.1949	0.1922	0.1894	0.1867
−0.7	0.2420	0.2389	0.2358	0.2327	0.2296	0.2266	0.2236	0.2206	0.2177	0.2148
−0.6	0.2743	0.2709	0.2676	0.2643	0.2611	0.2578	0.2546	0.2514	0.2483	0.2451
−0.5	0.3085	0.3050	0.3015	0.2981	0.2946	0.2912	0.2877	0.2843	0.2810	0.2776
−0.4	0.3446	0.3409	0.3372	0.3336	0.3300	0.3264	0.3228	0.3192	0.3156	0.3121
−0.3	0.3821	0.3783	0.3745	0.3707	0.3669	0.3632	0.3594	0.3557	0.3520	0.3483
−0.2	0.4207	0.4168	0.4129	0.4090	0.4052	0.4013	0.3974	0.3936	0.3897	0.3859
−0.1	0.4602	0.4562	0.4522	0.4483	0.4443	0.4404	0.4364	0.4325	0.4286	0.4247
0	0.5000	0.4960	0.4920	0.4880	0.4840	0.4801	0.4761	0.4721	0.4681	0.4641
0	0.5000	0.5040	0.5080	0.5120	0.5160	0.5199	0.5239	0.5279	0.5319	0.5359
0.1	0.5398	0.5438	0.5478	0.5517	0.5557	0.5596	0.5636	0.5675	0.5714	0.5753
0.2	0.5793	0.5832	0.5871	0.5910	0.5948	0.5987	0.6026	0.6064	0.6103	0.6141
0.3	0.6179	0.6217	0.6255	0.6293	0.6331	0.6368	0.6406	0.6443	0.6480	0.6517
0.4	0.6554	0.6591	0.6628	0.6664	0.6700	0.6736	0.6772	0.6808	0.6844	0.6879
0.5	0.6915	0.6950	0.6985	0.7019	0.7054	0.7088	0.7123	0.7157	0.7190	0.7224
0.6	0.7257	0.7291	0.7324	0.7357	0.7389	0.7422	0.7454	0.7486	0.7517	0.7549
0.7	0.7580	0.7611	0.7642	0.7673	0.7704	0.7734	0.7764	0.7794	0.7823	0.7852
0.8	0.7881	0.7910	0.7939	0.7967	0.7995	0.8023	0.8051	0.8078	0.8106	0.8133
0.9	0.8159	0.8186	0.8212	0.8238	0.8264	0.8289	0.8315	0.8340	0.8365	0.8389
1.0	0.8413	0.8438	0.8461	0.8485	0.8508	0.8531	0.8554	0.8577	0.8599	0.8621
1.1	0.8643	0.8665	0.8686	0.8708	0.8729	0.8749	0.8770	0.8790	0.8810	0.8830
1.2	0.8849	0.8869	0.8888	0.8907	0.8925	0.8944	0.8962	0.8980	0.8997	0.9015
1.3	0.9032	0.9049	0.9066	0.9082	0.9099	0.9115	0.9131	0.9147	0.9162	0.9177
1.4	0.9192	0.9207	0.9222	0.9236	0.9251	0.9265	0.9278	0.9292	0.9306	0.9319
1.5	0.9332	0.9345	0.9357	0.9370	0.9382	0.9394	0.9406	0.9418	0.9429	0.9441
1.6	0.9452	0.9463	0.9474	0.9484	0.9495	0.9505	0.9515	0.9225	0.9335	0.9545
1.7	0.9554	0.9564	0.9573	0.9582	0.9591	0.9599	0.9608	0.9616	0.9625	0.9633
1.8	0.9641	0.9649	0.9656	0.9664	0.9671	0.9678	0.9686	0.9693	0.9699	0.9706
1.9	0.9713	0.9719	0.9726	0.9732	0.9738	0.9744	0.9750	0.9756	0.9761	0.9767
2.0	0.9772	0.9778	0.9783	0.9788	0.9793	0.9798	0.9803	0.9808	0.9812	0.9817
2.1	0.9821	0.9826	0.9830	0.9834	0.9838	0.9842	0.9846	0.9850	0.9854	0.9857
2.2	0.9861	0.9864	0.9868	0.9871	0.9875	0.9878	0.9881	0.9884	0.9887	0.9890
2.3	0.9893	0.9896	0.9898	0.9901	0.9904	0.9906	0.9909	0.9911	0.9913	0.9916
2.4	0.9918	0.9920	0.9922	0.9925	0.9927	0.9929	0.9931	0.9932	0.9934	0.9936
2.5	0.9938	0.9940	0.9941	0.9943	0.9945	0.9946	0.9948	0.9949	0.9951	0.9952
2.6	0.9953	0.9950	0.9956	0.9957	0.9959	0.9960	0.9961	0.9962	0.9963	0.9964
2.7	0.9965	0.9966	0.9967	0.9968	0.9969	0.9970	0.9971	0.9972	0.9973	0.9974
2.8	0.9974	0.9975	0.9976	0.9977	0.9977	0.9978	0.9979	0.9979	0.9980	0.9981
2.9	0.9981	0.9982	0.9982	0.9983	0.9984	0.9984	0.9985	0.9985	0.9986	0.9986
3.0	0.9987	0.9987	0.9987	0.9988	0.9988	0.9989	0.9989	0.9989	0.9990	0.9990
3.1	0.9990	0.9991	0.9991	0.9991	0.9992	0.9992	0.9992	0.9992	0.9993	0.9993
3.2	0.9993	0.9993	0.9994	0.9994	0.9994	0.9994	0.9994	0.9995	0.9995	0.9995
3.3	0.9995	0.9995	0.9995	0.9996	0.9996	0.9996	0.9996	0.9996	0.9996	0.9997
3.4	0.9997	0.9997	0.9997	0.9997	0.9997	0.9997	0.9997	0.9997	0.9997	0.9998

Table A2.2 Boundary Values of the Student *t* Distribution

Assuming the number of degrees of freedom, $v = 17$, $P(t > 1.333)$ at $= 17$, $\alpha = 0.10$ or $P(t > 1.333) = 10\%$ $t_{0.005} = 2.898$; $t_{0.025} = 2.110$.

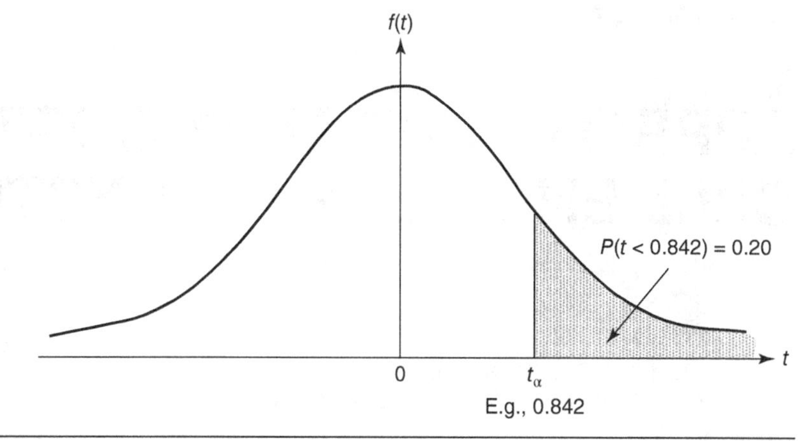

	Level of Significance, α									
v	0.40	0.20	0.10	0.05	0.025	0.02	0.015	0.0075	0.005	0.0025
1	0.325	1.376	3.078	6.314	12.706	15.895	21.205	42.433	63.657	127.321
2	0.289	1.061	1.886	2.920	4.303	4.849	5.643	8.073	9.925	14.089
3	0.277	0.978	1.638	2.353	3.182	3.482	3.896	5.047	5.841	7.453
4	0.271	0.941	1.533	2.132	2.776	2.999	3.298	4.088	4.604	5.598
5	0.267	0.920	1.476	2.015	2.571	2.757	3.003	3.634	4.032	4.773
6	0.265	0.906	1.440	1.943	2.447	2.612	2.829	3.372	3.707	4.317
7	0.263	0.896	1.415	1.895	2.365	2.517	2.715	3.203	3.499	4.029
8	0.262	0.889	1.397	1.860	2.306	2.449	2.634	3.085	3.355	3.833
9	0.261	0.883	1.383	1.833	2.262	2.398	2.574	2.998	3.250	3.690
10	0.260	0.879	1.372	1.812	2.228	2.359	2.527	2.932	3.169	3.581
11	0.260	0.876	1.363	1.796	2.201	2.328	2.491	2.879	3.106	3.497
12	0.259	0.873	1.356	1.782	2.179	2.303	2.461	2.836	3.055	3.428
13	0.259	0.870	1.350	1.771	2.160	2.282	2.436	2.801	3.012	3.372
14	0.258	0.868	1.345	1.761	2.145	2.264	2.415	2.771	2.977	3.326
15	0.258	0.866	1.341	1.753	2.131	2.249	2.397	2.746	2.947	3.286
16	0.258	0.865	1.337	1.746	2.120	2.235	2.382	2.724	2.921	3.252
17	0.257	0.863	1.333	1.740	2.110	2.224	2.368	2.706	2.898	3.222
18	0.257	0.862	1.330	1.734	2.101	2.214	2.356	2.689	2.878	3.197
19	0.257	0.861	1.328	1.729	2.093	2.205	2.346	2.674	2.861	3.174
20	0.257	0.860	1.325	1.725	2.086	2.197	2.336	2.661	2.845	3.153
21	0.257	0.859	1.323	1.721	2.080	2.189	2.328	2.649	2.831	3.135
22	0.256	0.858	1.321	1.717	2.074	2.183	2.320	2.639	2.819	3.119
23	0.256	0.858	1.319	1.714	2.069	2.177	2.313	2.629	2.807	3.104
24	0.256	0.857	1.318	1.711	2.064	2.172	2.307	2.620	2.797	3.091
25	0.256	0.856	1.316	1.708	2.060	2.167	2.301	2.612	2.787	3.078
26	0.256	0.856	1.315	1.706	2.056	2.162	2.296	2.605	2.779	3.067
27	0.256	0.855	1.314	1.703	2.052	2.158	2.291	2.598	2.771	3.057
28	0.256	0.855	1.313	1.701	2.048	2.154	2.286	2.592	2.763	3.047
29	0.256	0.854	1.311	1.699	2.045	2.150	2.282	2.586	2.756	3.038
30	0.256	0.854	1.310	1.697	2.042	2.147	2.278	2.581	2.750	3.030
60	0.254	0.848	1.296	1.671	2.000	2.099	2.223	2.504	2.660	2.915
120	0.254	0.845	1.289	1.658	1.980	2.076	2.196	2.468	2.617	2.860
∞	0.253	0.842	1.282	1.645	1.960	2.054	2.170	2.432	2.576	2.807

APPENDIX 3

APPROXIMATE UNIT COSTS OF SOME CIVIL ENGINEERING SYSTEMS

PLANNING AND DESIGN
Consultant fee = 10%–15% of project cost

CONSTRUCTION
Airport runway, \$50–60/ft^2; \$7–15/pax
Airport terminal, \$80–120/ft^2
Airport hangar, \$110–120/ft^2
Apartment complex, \$180–220/ft^2
Bridge, concrete, \$50–70 per ft^2 of deck area
Bridge, steel, \$45–65/ft^2 per ft^2 of deck area
Canal (ship bearing), \$M300–400/mile
Department store, \$120–130/ft^2
Factory, \$100–110/ft^2
Rail transit, high-speed intercity, \$M90–110/mile
Rail, transit, heavy metro, \$M280–320/mile
Rail, light, \$M 40–60/mile
Rail, mono, \$M 200–250/mile
Hospital, \$2800–320/ft^2
Hotel, \$150–200/ft^2
Office complex, \$140–180/ft^2
Pavement, asphalt, \$M 1.5–2.2/lane-mile
Pavement, concrete, \$M 2–2.5/lane-mile
Parking garage, \$50–70/ft^2
Underground parking garage, \$60–80/ft^2
Warehouse, \$90–110/ft^2
Wind turbine, \$M 1.3–1.7/MW;

OPERATIONS (Average annual costs)
Rail transit, heavy (rapid), \$0.30–0.40/pax-mile
Rail transit, light, \$0.40–0.50/pax-mile
Bus transit, \$0.40–0.55/pax-mile
Wind turbine, \$0.03–0.07/kw-hr
Canal (ship bearing), \$M 8–15/mi (O&M)

MONITORING/INSPECTION (Average annual costs)
Highway pavements \$0.008–0.02/ft^2
Airport runway, \$0.01–0.02/ft^2
Wind turbine, \$0.005–0.008/ft^2 of turbine blade

REHABILITATION AND MAINTENANCE (Average annual costs)
Wind turbine, \$0.005–0.02/kw-hr
Pavement, asphalt, \$300–700/lane-mile
Pavement, concrete, \$250–500/lane-mile

END OF LIFE (Demolition cost)
Concrete, \$3–8/ft^2

1. Amounts shown are rough averages. Actual unit costs will depend on factors including system location, material, size, operating environment, and other factors. Marginal costs are generally lower than the average costs.
2. References include: Booz-Allen & Hamilton Inc. (1991). Light Rail Transit Capital Cost Study. Prepared for Urban Mass Transportation Administration, Office of Technical Assistance and Safety; CIT. (2004). High-speed rail: international comparisons, Research Report, Commission for Integrated Transport, London, UK; Schneck, D. C., Laver R.S., Threadgill, G., Mothersole, J. (1995). The Transit Capital Cost Index Study, Report prepared for the Federal Transit Administration by Booz-Allen and Hamilton Inc. and DRI/McGraw-Hill; Wind Industry (2012). How much do wind turbines cost? www.windustry.org
3. Resources for cost estimation for specific projects are provided at the end of Chapter 10. Others include Peterson's *Construction Estimating Using Excel*; Ostwald and McLaren's *Cost Analysis and Estimating for Engineering and Management*; Ostwald's *Construction Cost Analysis and Estimating*; Holm, Schaufelberger, Griffin, and Cole's *Construction Cost Estimating: Process and Practices*; and Bartholomew's *Construction Estimating and Bidding for Heavy Construction.*

APPENDIX 4

COMPOUND INTEREST FACTORS FOR 5% INTEREST RATE

	Single Payment		Uniform Payment Series				Arithmetic Gradient		
	Compound Amount Factor	Present Worth Factor	Sinking Fund Factor	Capital Recovery Factor	Capital Amount Factor	Present Worth Factor	Gradient Uniform Series	Gradient Present Worth	
	Find F Given F/P	Find P Given F P/F	Find A Given F A/F	Find A Given P A/P	Find F Given A F/A	Find P Given A P/A	Find A Given G A/G	Find P Given G P/G	
N	**SPCAF**	**SPPWF**	**USSFDF**	**USCRF**	**USCAF**	**USPWF**	**GSUSF**	**GSPWF**	**N**
1	1.050	0.9524	1.0000	1.0500	1.000	0.952	0.000	0.000	**1**
2	1.103	0.9070	0.4878	0.5378	2.050	1.859	0.488	0.907	**2**
3	1.158	0.8638	0.3172	0.3672	3.153	2.723	0.967	2.635	**3**
4	1.216	0.8227	0.2320	0.2820	4.310	3.546	1.439	5.103	**4**
5	1.276	0.7835	0.1810	0.2310	5.526	4.329	1.903	8.237	**5**
6	1.340	0.7462	0.1470	0.1970	6.802	5.076	2.358	11.968	**6**
7	1.407	0.7107	0.1228	0.1728	8.142	5.786	2.805	16.232	**7**
8	1.477	0.6768	0.1047	0.1547	9.549	6.463	3.245	20.970	**8**
9	1.551	0.6446	0.0907	0.1407	11.027	7.108	3.676	26.127	**9**
10	1.629	0.6139	0.0795	0.1295	12.578	7.722	4.099	31.652	**10**
11	1.710	0.5847	0.0704	0.1204	14.207	8.306	4.514	37.499	**11**
12	1.796	0.5568	0.0628	0.1128	15.917	8.863	4.922	43.624	**12**
13	1.886	0.5303	0.0565	0.1065	17.713	9.394	5.322	49.988	**13**
14	1.980	0.5051	0.0510	0.1010	19.599	9.899	5.713	56.554	**14**
15	2.079	0.4810	0.0463	0.0963	21.579	10.380	6.097	63.288	**15**
16	2.183	0.4581	0.0423	0.0923	23.657	10.838	6.474	70.160	**16**
17	2.292	0.4363	0.0387	0.0887	25.840	11.274	6.842	77.140	**17**
18	2.407	0.4155	0.0355	0.0855	28.132	11.690	7.203	84.204	**18**
19	2.527	0.3957	0.0327	0.0827	30.539	12.085	7.557	91.328	**19**
20	2.653	0.3769	0.0302	0.0802	33.066	12.462	7.903	98.488	**20**

	Single Payment		Uniform Payment Series				Arithmetic Gradient		
	Compound Amount Factor	Present Worth Factor	Sinking Fund Factor	Capital Recovery Factor	Capital Amount Factor	Present Worth Factor	Gradient Uniform Series	Gradient Present Worth	
	Find F Given P F/P	Find P Given F P/F	Find A Given F A/F	Find A Given P A/P	Find F Given A F/A	Find P Given A P/A	Find A Given G A/G	Find P Given G P/G	
N	**SPCAF**	**SPPWF**	**USSFDF**	**USCRF**	**USCAF**	**USPWF**	**GSUSF**	**GSPWF**	**N**
21	2.786	0.3589	0.0280	0.0780	35.719	12.821	8.242	105.667	**21**
22	2.925	0.3418	0.0260	0.0760	38.505	13.163	8.573	112.846	**22**
23	3.072	0.3256	0.0241	0.0741	41.430	13.489	8.897	120.009	**23**
24	3.225	0.3101	0.0225	0.0725	44.502	13.799	9.214	127.140	**24**
25	3.386	0.2953	0.0210	0.0710	47.727	14.094	9.524	134.228	**25**
26	3.556	0.2812	0.0196	0.0696	51.113	14.375	9.827	141.259	**26**
27	3.733	0.2678	0.0183	0.0683	54.669	14.643	10.122	148.223	**27**
28	3.920	0.2551	0.0171	0.0671	58.403	14.898	10.411	155.110	**28**
29	4.116	0.2429	0.0160	0.0660	62.323	15.141	10.694	161.913	**29**
30	4.322	0.2314	0.0151	0.0651	66.439	15.372	10.969	168.623	**30**
31	4.538	0.2204	0.0141	0.0641	70.761	15.593	11.238	175.233	**31**
32	4.765	0.2099	0.0133	0.0633	75.299	15.803	11.501	181.739	**32**
33	5.003	0.1999	0.0125	0.0625	80.064	16.003	11.757	188.135	**33**
34	5.253	0.1904	0.0118	0.0618	85.067	16.193	12.006	194.417	**34**
35	5.516	0.1813	0.0111	0.0611	90.320	16.374	12.250	200.581	**35**
40	7.040	0.1420	0.0083	0.0583	120.800	17.159	13.377	229.545	**40**
45	8.985	0.1113	0.0063	0.0563	159.700	17.774	14.364	255.315	**45**
50	11.467	0.0872	0.0048	0.0548	209.348	18.256	15.223	277.915	**50**
55	14.636	0.0683	0.0037	0.0537	272.713	18.633	15.966	297.510	**55**
60	18.679	0.0535	0.0028	0.0528	353.584	18.929	16.606	314.343	**60**
65	23.840	0.0419	0.0022	0.0522	456.798	19.161	17.154	328.691	**65**
70	30.426	0.0329	0.0017	0.0517	588.529	19.343	17.621	340.841	**70**
75	38.833	0.0258	0.0013	0.0513	756.654	19.485	18.018	351.072	**75**
80	49.561	0.0202	0.0010	0.0510	971.229	19.596	18.353	359.646	**80**
85	63.254	0.0158	0.0008	0.0508	1,245.087	19.684	18.635	366.801	**85**
90	80.730	0.0124	0.0006	0.0506	1,594.607	19.752	18.871	372.749	**90**
95	103.035	0.0097	0.0005	0.0505	2,040.694	19.806	19.069	377.677	**95**
100	131.501	0.0076	0.0004	0.0504	2,610.025	19.848	19.234	381.749	**100**

INDEX

D

F

G

Q

T

Printed in the USA/Agawam, MA
January 3, 2023